$154.95 per copy (in United States).
Price subject to change without prior notice.

0099

RSMeans
Interior Cost Data
26th Annual Edition

NEW FOR 2009!

Line items found in this cost data that have the icon shown above are green.

The identified items fall into a broad definition of what is considered green.

Please see page ix for more information.

R.S. Means Company, Inc.
Construction Publishers & Consultants
63 Smiths Lane
Kingston, MA 02364-3008
(781) 422-5000

Copyright©2008 by R.S. Means Company, Inc.
All rights reserved.

Printed in the United States of America.
ISSN 8755-7541
ISBN 978-0-87629-178-8

The authors, editors, and engineers of RSMeans, apply diligence and judgment in locating and using reliable sources for the information published. **However, RSMeans makes no express or implied warranty or guarantee in connection with the content of the information contained herein, including the accuracy, correctness, value, sufficiency, or completeness of the data, methods, and other information contained herein. RSMeans makes no express or implied warranty of merchantability or fitness for a particular purpose.** RSMeans shall have no liability to any customer or third party for any loss, expense, or damage, including consequential, incidental, special, or punitive damage, including lost profits or lost revenue, caused directly or indirectly by any error or omission, or arising out of, or in connection with, the information contained herein.

No part of this publication may be reproduced, stored in a retrieval system, or transmitted in any form or by any means without prior written permission of R.S. Means Company, Inc. The cost data contained in this publication is valuable and proprietary information of RSMeans and others and there is no transfer to you of any ownership rights in the cost data or any license granted to you to create derivative works based on or which utilize the cost data.

Senior Editor
Barbara Balboni

Contributing Editors
Christopher Babbitt
Ted Baker
Robert A. Bastoni
John H. Chiang, PE
Gary W. Christensen
David G. Drain, PE
Cheryl Elsmore
Robert J. Kuchta
Melville J. Mossman, PE
Robert C. McNichols
Jeannene D. Murphy
Stephen C. Plotner
Eugene R. Spencer
Marshall J. Stetson
Phillip R. Waier, PE

Editorial Advisory Board

James E. Armstrong, CPE, CEM
Senior Energy Consultant
KEMA Services, Inc.

William R. Barry, CCC
Cost Consultant

Robert F. Cox, PhD
Department Head and Professor
of Building Construction Management
Purdue University

Senior Vice President & General Manager
John Ware

Vice President of Operations
Dave Walsh

Vice President of Sales & Marketing
Sev Ritchie

Marketing Director
John M. Shea

Director of Product Development
Thomas J. Dion

Engineering Manager
Bob Mewis, CCC

Roy F. Gilley, AIA
Principal
Gilley Design Associates

Kenneth K. Humphreys, PhD, PE, CCE

Patricia L. Jackson, PE
Jackson A&E Associates, Inc.

Martin F. Joyce
Executive Vice President
Bond Brothers, Inc.

Production Manager
Michael Kokernak

Production Coordinator
Jill Goodman

Technical Support
Jonathan Forgit
Mary Lou Geary
Roger Hancock
Gary L. Hoitt
Genevieve Medeiros
Debbie Panarelli
Paula Reale-Camelio
Kathryn S. Rodriguez
Sheryl A. Rose

Book & Cover Design
Norman R. Forgit

This book is printed on recycled paper (10% PCW cover, 20% PCW text), using soy-based printing ink. This book is recyclable.

Foreword

Our Mission

Since 1942, RSMeans has been actively engaged in construction cost publishing and consulting throughout North America.

Today, over 60 years after RSMeans began, our primary objective remains the same: to provide you, the construction and facilities professional, with the most current and comprehensive construction cost data possible.

Whether you are a contractor, an owner, an architect, an engineer, a facilities manager, or anyone else who needs a reliable construction cost estimate, you'll find this publication to be a highly useful and necessary tool.

With the constant flow of new construction methods and materials today, it's difficult to find the time to look at and evaluate all the different construction cost possibilities. In addition, because labor and material costs keep changing, last year's cost information is not a reliable basis for today's estimate or budget.

That's why so many construction professionals turn to RSMeans. We keep track of the costs for you, along with a wide range of other key information, from city cost indexes . . . to productivity rates . . . to crew composition . . . to contractor's overhead and profit rates.

RSMeans performs these functions by collecting data from all facets of the industry and organizing it in a format that is instantly accessible to you. From the preliminary budget to the detailed unit price estimate, you'll find the data in this book useful for all phases of construction cost determination.

The Staff, the Organization, and Our Services

When you purchase one of RSMeans' publications, you are, in effect, hiring the services of a full-time staff of construction and engineering professionals.

Our thoroughly experienced and highly qualified staff works daily at collecting, analyzing, and disseminating comprehensive cost information for your needs. These staff members have years of practical construction experience and engineering training prior to joining the firm. As a result, you can count on them not only for the cost figures, but also for additional background reference information that will help you create a realistic estimate.

The RSMeans organization is always prepared to help you solve construction problems through its four major divisions: Construction and Cost Data Publishing, Electronic Products and Services, Consulting and Business Solutions, and Professional Development Services.

Besides a full array of construction cost estimating books, RSMeans also publishes a number of other reference works for the construction industry. Subjects include construction estimating and project and business management; special topics such as HVAC, roofing, plumbing, and hazardous waste remediation; and a library of facility management references.

In addition, you can access all of our construction cost data electronically using *Means CostWorks®* CD or on the Web using Means CostWorks.com.

What's more, you can increase your knowledge and improve your construction estimating and management performance with an RSMeans Construction Seminar or In-House Training Program. These two-day seminar programs offer unparalleled opportunities for everyone in your organization to get updated on a wide variety of construction-related issues.

RSMeans is also a worldwide provider of construction cost management and analysis services for commercial and government owners.

In short, RSMeans can provide you with the tools and expertise for constructing accurate and dependable construction estimates and budgets in a variety of ways.

Robert Snow Means Established a Tradition of Quality That Continues Today

Robert Snow Means spent years building RSMeans, making certain he always delivered a quality product.

Today, at RSMeans, we do more than talk about the quality of our data and the usefulness of our books. We stand behind all of our data, from historical cost indexes to construction materials and techniques to current costs.

If you have any questions about our products or services, please call us toll-free at 1-800-334-3509. Our customer service representatives will be happy to assist you. You can also visit our Web site at www.rsmeans.com.

Table of Contents

Foreword	ii
How the Book Is Built: An Overview	v
Absolute Essentials for a Quick Start	vi
Estimating with RSMeans Unit Price Cost Data	vii
How To Use the Book: The Details	ix
Unit Price Section	1
Assemblies Section	429
Reference Section	547
Construction Equipment Rental Costs	549
Crew Listings	561
Historical Cost Indexes	591
City Cost Indexes	592
Location Factors	612
Reference Tables	618
Change Orders	648
Square Foot Costs	653
Abbreviations	663
Index	667
Other RSMeans Products and Services	694
Labor Trade Rates including Overhead & Profit	Inside Back Cover

Related RSMeans Products and Services

The engineers at RSMeans suggest the following products and services as companion information resources to *RSMeans Interior Cost Data:*

Construction Cost Data Books
Building Construction Cost Data 2009
Repair & Remodeling Cost Data 2009
Assemblies Cost Data 2009
Square Foot Costs 2009

Reference Books
Repair & Remodeling Estimating Methods, 4th Ed.
Facilities Planning & Relocation
Historic Preservation: Project Planning & Estimating
Residential & Light Commercial Construction Standards, 3rd Ed.
Project Scheduling and Management for Construction, 3rd Ed.
ADA Compliance Pricing Guide, 2nd Ed.
Estimating Building Costs
Estimating Handbook, 2nd Ed.

Seminars and In-House Training
Repair & Remodeling Estimating
Scheduling & Project Management
Means Data for Job Order Contracting (JOC)
Means CostWorks® Training
Plan Reading & Material Takeoff

RSMeans on the Internet
Visit RSMeans at **www.rsmeans.com.** The site contains useful interactive cost and reference material. Request or download **FREE** estimating software demos. Visit our bookstore for convenient ordering and to learn more about new publications and companion products.

RSMeans Electronic Data
Get the information found in RSMeans cost books electronically on *Means CostWorks®* CD or on the Web at MeansCostworks.com

RSMeans Business Solutions
Engineers and Analysts offer research studies, benchmark analysis, predictive cost modeling, analytics, job order contracting, and real property management consultation, as well as custom-designed, Web-based dashboards and calculators that apply RCD/RSMeans extensive databases. Clients include federal government agencies, architects, construction management firms, and institutional organizations such as school systems, health care facilities, associations, and corporations.

RSMeans for Job Order Contracting (JOC)
Best practice JOC tools for cost estimating and project management to help streamline delivery processes for renovation projects. Renovation is a $147 billion market in the U.S., and includes projects in school districts, municipalities, health care facilities, colleges and universities, and corporations.

- RSMeans Engineers consult in contracting methods and conduct JOC Facility Audits
- JOCWorks™ Software (Basic, Advanced, PRO levels)
- RSMeans Job Order Contracting Cost Data for the entire U.S.

Construction Costs for Software Applications
Over 25 unit price and assemblies cost databases are available through a number of leading estimating and facilities management software providers (listed below). For more information see the "Other RSMeans Products" pages at the back of this publication.

MeansData™ is also available to federal, state, and local government agencies as multi-year, multi-seat licenses.

- 4Clicks-Solutions, LLC
- ArenaSoft Estimating
- Beck Technology
- BSD – Building Systems Design, Inc.
- CMS – Construction Management Software
- Corecon Technologies, Inc.
- CorVet Systems
- Estimating Systems, Inc.
- Maximus Asset Solutions
- Sage Timberline Office
- US Cost, Inc.
- VFA – Vanderweil Facility Advisers
- WinEstimator, Inc.

How the Book Is Built: An Overview

The Construction Specifications Institute (CSI) and Construction Specifications Canada (CSC) have produced the 2004 edition of MasterFormat, a system of titles and numbers used extensively to organize construction information.

All unit price data in the RSMeans cost data books is now arranged in the 50-division MasterFormat 2004 system.

A Powerful Construction Tool

You have in your hands one of the most powerful construction tools available today. A successful project is built on the foundation of an accurate and dependable estimate. This book will enable you to construct just such an estimate.

For the casual user the book is designed to be:

- quickly and easily understood so you can get right to your estimate.
- filled with valuable information so you can understand the necessary factors that go into the cost estimate.

For the regular user, the book is designed to be:

- a handy desk reference that can be quickly referred to for key costs.
- a comprehensive, fully reliable source of current construction costs and productivity rates so you'll be prepared to estimate any project.
- a source book for preliminary project cost, product selections, and alternate materials and methods.

To meet all of these requirements we have organized the book into the following clearly defined sections.

New: Quick Start

See our new "Quick Start" instructions on the following page.

Estimating with RSMeans Unit Price Cost Data

Please see these steps to complete an estimate using RSMeans unit price cost data.

How To Use the Book: The Details

This section contains an in-depth explanation of how the book is arranged . . . and how you can use it to determine a reliable construction cost estimate. It includes information about how we develop our cost figures and how to completely prepare your estimate.

Unit Price Section

All unit price cost data has been divided into the 50 divisions according to the MasterFormat system of classification and numbering. For a listing of these divisions and an outline of their subdivisions, see the Unit Price Section Table of Contents.

Estimating tips are included at the beginning of each division.

Assemblies Section

The cost data in this section has been organized in an "Assemblies" format. These assemblies are the functional elements of a building and are arranged according to the 7 divisions of the UNIFORMAT II classification system. For a complete explanation of a typical "Assemblies" page, see "How To Use the Assemblies Cost Tables."

Reference Section

This section includes information on Equipment Rental Costs, Crew Listings, Historical Cost Indexes, City Cost Indexes, Location Factors, Reference Tables, Change Orders, Square Foot Costs, and a listing of Abbreviations.

Equipment Rental Costs: This section contains the average costs to rent and operate hundreds of pieces of construction equipment.

Crew Listings: This section lists all the crews referenced in the book. For the purposes of this book, a crew is composed of more than one trade classification and/or the addition of power equipment to any trade classification. Power equipment is included in the cost of the crew. Costs are shown both with the bare labor rates and with the installing contractor's overhead and profit added. For each, the total crew cost per eight-hour day and the composite cost per labor-hour are listed.

Historical Cost Indexes: These indexes provide you with data to adjust construction costs over time.

City Cost Indexes: All costs in this book are U.S. national averages. Costs vary because of the regional economy. You can adjust costs by CSI Division to over 316 locations throughout the U.S. and Canada by using the data in this section.

Location Factors: You can adjust total project costs to over 900 locations throughout the U.S. and Canada by using the data in this section.

Reference Tables: At the beginning of selected major classifications in the Unit Price and Assemblies sections are "reference numbers" shown in a shaded box. These numbers refer you to related information in the Reference Section. In this section, you'll find reference tables, explanations, estimating information that support how we develop the unit price data, technical data, and estimating procedures.

Change Orders: This section includes information on the factors that influence the pricing of change orders.

Square Foot Costs: This section contains costs for 59 different building types that allow you to make a rough estimate for the overall cost of a project or its major components.

Abbreviations: A listing of abbreviations used throughout this book, along with the terms they represent, is included in this section.

Index

A comprehensive listing of all terms and subjects in this book will help you quickly find what you need when you are not sure where it falls in MasterFormat.

The Scope of This Book

This book is designed to be as comprehensive and as easy to use as possible. To that end we have made certain assumptions and limited its scope in two key ways:

1. We have established material prices based on a national average.
2. We have computed labor costs based on a 30-city national average of union wage rates.

For a more detailed explanation of how the cost data is developed, see "How To Use the Book: The Details."

Project Size

This book is aimed primarily at interior construction projects costing $50,000 and greater.

With reasonable exercise of judgment the figures can be used for any interior building construction.

Absolute Essentials for a Quick Start

If you feel you are ready to use this book and don't think you will need the detailed instructions that begin on the following page, this Absolute Essentials for a Quick Start page is for you. These steps will allow you to get started estimating in a matter of minutes.

1 Scope
Think through the project that you will be estimating and identify the many individual work tasks that will need to be covered in your estimate.

2 Quantify
Determine the number of units that will be required for each work task that you identified.

3 Pricing
Locate individual Unit Price line items that match the work tasks you identified. The Unit Price Section Table of Contents that begins on page 1 and the Index in the back of the book will help you find these line items.

4 Multiply
Multiply the Total Incl O&P cost for a Unit Price line item in the book by your quantity for that item. The price you calculate will be an estimate for a completed item of work performed by a subcontractor. Keep adding line items in this manner to build your estimate.

5 Project Overhead
Include project overhead items in your estimate. These items are needed to make the job run and are typically, but not always, provided by the General Contractor. They can be found in Division 1. An alternate method of estimating project overhead costs is to apply a percentage of the total project cost.

Include rented tools not included in crews, waste, rubbish handling and cleanup.

6 Estimate Summary
Include General Contractor's markup on subcontractors, General Contractor's office overhead and profit, and sales tax on materials and equipment.

Adjust your estimate to the project's location by using the City Cost Indexes or Location Factors found in the Reference Section.

Editors' Note: We urge you to spend time reading and understanding the supporting material in the front of this book. An accurate estimate requires experience, knowledge and careful calculation. The more you know about how we at RSMeans developed the data, the more accurate your estimate will be. In addition, it is important to take into consideration the reference material in the back of the book such as Equipment Listings, Crew Listings, City Cost Indexes, Location Factors and Reference Numbers.

Estimating with RSMeans Unit Price Cost Data

Following these steps will allow you to complete an accurate estimate using RSMeans Unit Price cost data.

1 Scope out the project

- Identify the individual work tasks that will need to be covered in your estimate.
- The Unit Price data inside this book has been divided into 44 Divisions according to the CSI MasterFormat2004 – their titles are listed on the back cover of your book.
- Think through the project that you will be estimating and identify those CSI Divisions that you will need to use in your estimate.
- The Unit Price Section Table of Contents that begins on page 1 may also be helpful when scoping out your project.
- Experienced estimators find it helpful to begin an estimate with Division 2, estimating Division 1 after the full scope of the project is known.

2 Quantify

- Determine the number of units that will be required for each work task that you previously identified.
- Experienced estimators will include an allowance for waste in their quantities (waste is not included in RSMeans Unit Price line items unless so stated).

3 Price the quantities

- Use the Unit Price Table of Contents, and the Index, to locate the first individual Unit Price line item for your estimate.
- Reference Numbers indicated within a Unit Price section refer to additional information that you may find useful.
- The crew will tell you who is performing the work for that task. Crew codes are expanded in the Crew Listings in the Reference Section to include all trades and equipment that comprise the crew.
- The Daily Output is the amount of work the crew is expected to do in one day.
- The Labor-Hours value is the amount of time it will take for the average crew member to install one unit of measure.
- The abbreviated Unit designation indicates the unit of measure upon which the crew, productivity and prices are based.
- Bare Costs are shown for materials, labor, and equipment needed to complete the Unit Price line item. Bare costs do not include waste, project overhead, payroll insurance, payroll taxes, main office overhead, or profit.
- The Total Incl O&P cost is the billing rate or invoice amount of the installing contractor or subcontractor who performs the work for the Unit Price line item.

4 Multiply

- Multiply the total number of units needed for your project by the Total Incl O&P cost for the Unit Price line item.
- Be careful that the final unit of measure for your quantity of units matches the unit of measure in the Unit column in the book.
- The price you calculate will be an estimate for a completed item of work.
- Keep scoping individual tasks, determining the number of units required for those tasks, matching them up with individual Unit Price line items in the book, and multiplying quantities by Total Incl O&P costs. In this manner keep building your estimate.
- An estimate completed to this point in this manner will be priced as if a subcontractor, or set of subcontractors, will perform the work. The estimate does not yet include Project Overhead or Estimate Summary components such as General Contractor markups on subcontracted and self-performed work, General Contractor office overhead and profit, contingency, and location factor.

5 Project Overhead

- Include project overhead items from Division 1 – General Requirements.
- These are items that will be needed to make the job run. These are typically, but not always, provided by the General Contractor. They include, but are not limited to, such items as field personnel, insurance, performance bond, permits, testing, temporary utilities, field office and storage facilities, temporary scaffolding and platforms, equipment mobilization and demobilization, temporary roads and sidewalks, winter protection, temporary barricades and fencing, temporary security, temporary signs, field engineering and layout, final cleaning and commissioning.
- These items should be scoped, quantified, matched to individual Unit Price line items in Division 1, priced and added to your estimate.
- An alternate method of estimating project overhead costs is to apply a percentage of the total project cost, usually 5% to 15% with an average of 10% (see General Conditions, page ix).
- Include other project related expenses in your estimate such as:
  - Rented equipment not itemized in the Crew Listings.
  - Rubbish handling throughout the project (see 02 41 19.23).

6 Estimate Summary

- Includes sales tax on materials and equipment.
  Note: Sales tax must be added for materials in subcontracted work.
- Include the General Contractor's markup on self-performed work, usually 5% to 15% with an average of 10%.
- Include the General Contractor's markup on subcontracted work, usually 5% to 15% with an average of 10%.
- Include General Contractor's main office overhead and profit.
- RSMeans gives general guidelines on the General Contractor's main office overhead (see section 01 31 13.50 and Reference Number R013113-50).
- RSMeans gives no guidance on the General Contractor's profit.
- Markups will depend on the size of the General Contractor's operations, his projected annual revenue, the level of risk he is taking on, and on the level of competitiveness in the local area and for this project in particular.
- Include a contingency, usually 3% to 5%.
- Adjust your estimate to the project's location by using the City Cost Indexes or the Location Factors in the Reference Section.
- Look at the rules on the pages for How to Use the City Cost Indexes to see how to apply the Indexes for your location.
- When the proper Index or Factor has been identified for the project's location, convert it to a multiplier by dividing it by 100, then multiply that multiplier by your estimate total cost. Your original estimate total cost will now be adjusted up or down from the national average to a total that is appropriate for your location.

Editors' Notes:

1) We urge you to spend time reading and understanding the supporting material in the front of this book. An accurate estimate requires experience, knowledge, and careful calculation. The more you know about how we at RSMeans developed the data, the more accurate your estimate will be. In addition, it is important to take into consideration the reference material in the back of the book such as Equipment Listings, Crew Listings, City Cost Indexes, Location Factors, and Reference Numbers.

2) Contractors who are bidding or are involved in JOC, DOC, SABER, or IDIQ type contracts are cautioned that workers' compensation Insurance, federal and state payroll taxes, waste, project supervision, project overhead, main office overhead, and profit are not included in bare costs. Your coefficient or multiplier must cover these costs.

How to Use the Book: The Details

What's Behind the Numbers? The Development of Cost Data

The staff at RSMeans continuously monitors developments in the construction industry in order to ensure reliable, thorough, and up-to-date cost information.

While *overall* construction costs may vary relative to general economic conditions, price fluctuations within the industry are dependent upon many factors. Individual price variations may, in fact, be opposite to overall economic trends. Therefore, costs are continually monitored and complete updates are published yearly. Also, new items are frequently added in response to changes in materials and methods.

Costs—$ (U.S.)

All costs represent U.S. national averages and are given in U.S. dollars. The RSMeans City Cost Indexes can be used to adjust costs to a particular location. The City Cost Indexes for Canada can be used to adjust U.S. national averages to local costs in Canadian dollars. No exchange rate is necessary.

[G] The processes or products identified by the green symbol in this publication have been determined to be environmentally responsible and/or resource-efficient solely by the RSMeans engineering staff. The inclusion of the green symbol does not represent compliance with any specific industry association or standard.

Material Costs

The RSMeans staff contacts manufacturers, dealers, distributors, and contractors all across the U.S. and Canada to determine national average material costs. If you have access to current material costs for your specific location, you may wish to make adjustments to reflect differences from the national average. Included within material costs are fasteners for a normal installation. RSMeans engineers use manufacturers' recommendations, written specifications and/or standard construction practice for size and spacing of fasteners. Adjustments to material costs may be required for your specific application or location. Material costs do not include sales tax.

Labor Costs

Labor costs are based on the average of wage rates from 30 major U.S. cities. Rates are determined from labor union agreements or prevailing wages for construction trades for the current year. Rates, along with overhead and profit markups, are listed on the inside back cover of this book.

- If wage rates in your area vary from those used in this book, or if rate increases are expected within a given year, labor costs should be adjusted accordingly.

Labor costs reflect productivity based on actual working conditions. In addition to actual installation, these figures include time spent during a normal workday on tasks, such as material receiving and handling, mobilization at site, site movement, breaks, and cleanup.

Productivity data is developed over an extended period so as not to be influenced by abnormal variations and reflects a typical average.

Equipment Costs

Equipment costs include not only rental, but also operating costs for equipment under normal use. The operating costs include parts and labor for routine servicing such as repair and replacement of pumps, filters, and worn lines. Normal operating expendables, such as fuel, lubricants, tires, and electricity (where applicable), are also included. Extraordinary operating expendables with highly variable wear patterns, such as diamond bits and blades, are excluded. These costs are included under materials. Equipment rental rates are obtained from industry sources throughout North America—contractors, suppliers, dealers, manufacturers, and distributors.

Equipment costs do not include operators' wages; nor do they include the cost to move equipment to a job site (mobilization) or from a job site (demobilization).

Equipment Cost/Day—The cost of power equipment required for each crew is included in the Crew Listings in the Reference Section (small tools that are considered as essential everyday tools are not listed out separately). The Crew Listings itemize specialized tools and heavy equipment along with labor trades. The daily cost of itemized equipment included in a crew is based on dividing the weekly bare rental rate by 5 (number of working days per week) and then adding the hourly operating cost times 8 (the number of hours per day). This Equipment Cost/Day is shown in the last column of the Equipment Rental Cost pages in the Reference Section.

Mobilization/Demobilization—The cost to move construction equipment from an equipment yard or rental company to the job site and back again is not included in equipment costs. Mobilization (to the site) and demobilization (from the site) costs can be found in the Unit Price section. If a piece of equipment is already at the job site, it is not appropriate to utilize mobil./demob. costs again in an estimate.

General Conditions

Cost data in this book is presented in two ways: Bare Costs and Total Cost including O&P (Overhead and Profit). General Conditions, when applicable, should also be added to the Total Cost including O&P. The costs for General Conditions are listed in Division 1 of the Unit Price Section and the Reference Section of this book. General Conditions for the *Installing Contractor* may range from 0% to 10% of the Total Cost including O&P. For the *General* or *Prime Contractor*, costs for General Conditions may range from 5% to 15% of the Total Cost including O&P, with a figure of 10% as the most typical allowance.

Overhead and Profit

Total Cost including O&P for the *Installing Contractor* is shown in the last column on both the Unit Price and the Assemblies pages of this book. This figure is the sum of the bare material cost plus 10% for profit, the bare labor cost plus total overhead and profit, and the bare equipment cost plus 10% for profit. Details for the calculation of Overhead and Profit on labor are shown on the inside back cover and in the

Reference Section of this book. (See the "How To Use the Unit Price Pages" for an example of this calculation.)

Factors Affecting Costs

Costs can vary depending upon a number of variables. Here's how we have handled the main factors affecting costs.

Quality—The prices for materials and the workmanship upon which productivity is based represent sound construction work. They are also in line with U.S. government specifications.

Overtime—We have made no allowance for overtime. If you anticipate premium time or work beyond normal working hours, be sure to make an appropriate adjustment to your labor costs.

Productivity—The productivity, daily output, and labor-hour figures for each line item are based on working an eight-hour day in daylight hours in moderate temperatures. For work that extends beyond normal work hours or is performed under adverse conditions, productivity may decrease. (See the section in "How To Use the Unit Price Pages" for more on productivity.)

Size of Project—The size, scope of work, and type of construction project will have a significant impact on cost. Economies of scale can reduce costs for large projects. Unit costs can often run higher for small projects. Costs in this book are intended for the size and type of project as previously described in "How the Book Is Built: An Overview." Costs for projects of a significantly different size or type should be adjusted accordingly.

Location—Material prices in this book are for metropolitan areas. However, in dense urban areas, traffic and site storage limitations may increase costs. Beyond a 20-mile radius of large cities, extra trucking or transportation charges may also increase the material costs slightly. On the other hand, lower wage rates may be in effect. Be sure to consider both of these factors when preparing an estimate, particularly if the job site is located in a central city or remote rural location.

In addition, highly specialized subcontract items may require travel and per-diem expenses for mechanics.

Other Factors—
- season of year
- contractor management
- weather conditions
- local union restrictions
- building code requirements
- availability of:
  - adequate energy
  - skilled labor
  - building materials
- owner's special requirements/restrictions
- safety requirements
- environmental considerations

Unpredictable Factors—General business conditions influence "in-place" costs of all items. Substitute materials and construction methods may have to be employed. These may affect the installed cost and/or life cycle costs. Such factors may be difficult to evaluate and cannot necessarily be predicted on the basis of the job's location in a particular section of the country. Thus, where these factors apply, you may find significant but unavoidable cost variations for which you will have to apply a measure of judgment to your estimate.

Rounding of Costs

In general, all unit prices in excess of $5.00 have been rounded to make them easier to use and still maintain adequate precision of the results. The rounding rules we have chosen are in the following table.

Prices from ...	Rounded to the nearest ...
$.01 to $5.00	$.01
$5.01 to $20.00	$.05
$20.01 to $100.00	$.50
$100.01 to $300.00	$1.00
$300.01 to $1,000.00	$5.00
$1,000.01 to $10,000.00	$25.00
$10,000.01 to $50,000.00	$100.00
$50,000.01 and above	$500.00

Final Checklist

Estimating can be a straightforward process provided you remember the basics. Here's a checklist of some of the steps you should remember to complete before finalizing your estimate.

Did you remember to ...

- factor in the City Cost Index for your locale?
- take into consideration which items have been marked up and by how much?
- mark up the entire estimate sufficiently for your purposes?
- read the background information on techniques and technical matters that could impact your project time span and cost?
- include all components of your project in the final estimate?
- double check your figures for accuracy?
- call RSMeans if you have any questions about your estimate or the data you've found in our publications?

Remember, RSMeans stands behind its publications. If you have any questions about your estimate ... about the costs you've used from our books ... or even about the technical aspects of the job that may affect your estimate, feel free to call your RSMeans representative at 1-800-334-3509.

Unit Price Section

Table of Contents

Div. No.		Page
	General Requirements	
01 11	Summary of Work	8
01 21	Allowances	8
01 31	Project Management and Coordination	10
01 32	Construction Progress Documentation	12
01 41	Regulatory Requirements	12
01 51	Temporary Utilities	12
01 52	Construction Facilities	13
01 54	Construction Aids	13
01 55	Vehicular Access and Parking	15
01 56	Temporary Barriers and Enclosures	16
01 74	Cleaning and Waste Management	17
	Existing Conditions	
02 41	Demolition	20
02 82	Asbestos Remediation	22
02 83	Lead Remediation	25
02 85	Mold Remediation	27
	Concrete	
03 01	Maintenance of Concrete	30
03 05	Common Work Results for Concrete	30
03 15	Concrete Accessories	31
03 30	Cast-In-Place Concrete	31
03 31	Structural Concrete	32
03 35	Concrete Finishing	33
03 39	Concrete Curing	34
03 48	Precast Concrete Specialties	34
03 54	Cast Underlayment	35
03 81	Concrete Cutting	35
03 82	Concrete Boring	35
	Masonry	
04 01	Maintenance of Masonry	40
04 05	Common Work Results for Masonry	41
04 21	Clay Unit Masonry	44
04 22	Concrete Unit Masonry	46
04 23	Glass Unit Masonry	48
04 24	Adobe Unit Masonry	49
04 25	Unit Masonry Panels	49
04 27	Multiple-Wythe Unit Masonry	50
04 41	Dry-Placed Stone	50
04 43	Stone Masonry	50
04 57	Masonry Fireplaces	54
04 71	Manufactured Brick Masonry	54
04 73	Manufactured Stone Masonry	54
	Metals	
05 01	Maintenance of Metals	57
05 05	Common Work Results for Metals	57
05 12	Structural Steel Framing	60
05 15	Wire Rope Assemblies	61
05 21	Steel Joist Framing	64
05 31	Steel Decking	66
05 41	Structural Metal Stud Framing	66
05 42	Cold-Formed Metal Joist Framing	70
05 44	Cold-Formed Metal Trusses	75
05 51	Metal Stairs	76
05 52	Metal Railings	78
05 55	Metal Stair Treads and Nosings	79
05 58	Formed Metal Fabrications	79
05 59	Metal Specialties	80
05 71	Decorative Metal Stairs	81
05 73	Decorative Metal Railings	81
05 75	Decorative Formed Metal	82
	Wood, Plastics & Composites	
06 05	Common Work Results for Wood, Plastics and Composites	84
06 11	Wood Framing	91
06 15	Wood Decking	99
06 16	Sheathing	99
06 17	Shop-Fabricated Structural Wood	101
06 22	Millwork	101
06 25	Prefinished Paneling	103
06 26	Board Paneling	105
06 43	Wood Stairs and Railings	105
06 44	Ornamental Woodwork	107
06 48	Wood Frames	109
06 52	Plastic Structural Assemblies	110
06 63	Plastic Railings	110
	Thermal & Moisture Protection	
07 05	Common Work Results for Thermal and Moisture Protection	112
07 11	Dampproofing	112
07 16	Cementitious and Reactive Waterproofing	112
07 21	Thermal Insulation	112
07 22	Roof and Deck Insulation	115
07 26	Vapor Retarders	117
07 31	Shingles and Shakes	117
07 32	Roof Tiles	118
07 44	Faced Panels	119
07 46	Siding	119
07 61	Sheet Metal Roofing	120
07 65	Flexible Flashing	120
07 72	Roof Accessories	121
07 81	Applied Fireproofing	122
07 84	Firestopping	122
07 92	Joint Sealants	124
	Openings	
08 05	Common Work Results for Openings	128
08 11	Metal Doors and Frames	129
08 12	Metal Frames	129
08 13	Metal Doors	131
08 14	Wood Doors	133
08 16	Composite Doors	139
08 17	Integrated Door Opening Assemblies	139
08 31	Access Doors and Panels	140
08 32	Sliding Glass Doors	141
08 33	Coiling Doors and Grilles	141
08 34	Special Function Doors	142
08 36	Panel Doors	144
08 38	Traffic Doors	145
08 41	Entrances and Storefronts	146
08 42	Entrances	147
08 43	Storefronts	148
08 44	Curtain Wall and Glazed Assemblies	148
08 45	Translucent Wall and Roof Assemblies	149
08 51	Metal Windows	149
08 52	Wood Windows	151
08 53	Plastic Windows	155
08 56	Special Function Windows	155
08 62	Unit Skylights	155
08 63	Metal-Framed Skylights	156
08 71	Door Hardware	156
08 74	Access Control Hardware	165
08 75	Window Hardware	165
08 79	Hardware Accessories	165
08 81	Glass Glazing	166
08 83	Mirrors	168
08 84	Plastic Glazing	169
08 87	Glazing Surface Films	169
08 88	Special Function Glazing	170
08 95	Vents	170
	Finishes	
09 01	Maintenance of Finishes	172
09 05	Common Work Results for Finishes	172
09 21	Plaster and Gypsum Board Assemblies	174
09 22	Supports for Plaster and Gypsum Board	175
09 23	Gypsum Plastering	179
09 24	Portland Cement Plastering	180
09 26	Veneer Plastering	181
09 28	Backing Boards and Underlayments	181
09 29	Gypsum Board	182
09 30	Tiling	185
09 51	Acoustical Ceilings	187
09 53	Acoustical Ceiling Suspension Assemblies	189
09 63	Masonry Flooring	189
09 64	Wood Flooring	190
09 65	Resilient Flooring	192
09 66	Terrazzo Flooring	194
09 67	Fluid-Applied Flooring	195
09 68	Carpeting	196
09 69	Access Flooring	198
09 72	Wall Coverings	198
09 77	Special Wall Surfacing	199
09 81	Acoustic Insulation	200
09 84	Acoustic Room Components	201
09 91	Painting	201
09 93	Staining and Transparent Finishing	216
09 96	High-Performance Coatings	216
09 97	Special Coatings	217
	Specialties	
10 05	Common Work Results for Specialties	220
10 11	Visual Display Surfaces	220
10 13	Directories	223
10 14	Signage	223
10 17	Telephone Specialties	224
10 21	Compartments and Cubicles	224
10 22	Partitions	228
10 26	Wall and Door Protection	232
10 28	Toilet, Bath, and Laundry Accessories	233
10 31	Manufactured Fireplaces	234
10 32	Fireplace Specialties	235
10 35	Stoves	236
10 44	Fire Protection Specialties	236
10 51	Lockers	237
10 55	Postal Specialties	238
10 56	Storage Assemblies	239
10 57	Wardrobe and Closet Specialties	240
10 73	Protective Covers	240
	Equipment	
11 05	Common Work Results for Equipment	242
11 13	Loading Dock Equipment	244
11 14	Pedestrian Control Equipment	244
11 16	Vault Equipment	244
11 17	Teller and Service Equipment	245
11 19	Detention Equipment	245
11 21	Mercantile and Service Equipment	245
11 23	Commercial Laundry and Dry Cleaning Equipment	246
11 24	Maintenance Equipment	247
11 26	Unit Kitchens	247
11 27	Photographic Processing Equipment	248
11 31	Residential Appliances	248
11 33	Retractable Stairs	250
11 41	Food Storage Equipment	250
11 42	Food Preparation Equipment	252
11 43	Food Delivery Carts and Conveyors	253
11 44	Food Cooking Equipment	253
11 46	Food Dispensing Equipment	254
11 47	Ice Machines	255
11 48	Cleaning and Disposal Equipment	255
11 52	Audio-Visual Equipment	256
11 53	Laboratory Equipment	257
11 57	Vocational Shop Equipment	258

Table of Contents (cont.)

Div. No.		Page
11 61	Theater and Stage Equipment	258
11 66	Athletic Equipment	259
11 71	Medical Sterilizing Equipment	261
11 72	Examination and Treatment Equipment	261
11 73	Patient Care Equipment	261
11 74	Dental Equipment	262
11 76	Operating Room Equipment	262
11 77	Radiology Equipment	263
11 78	Mortuary Equipment	263
11 82	Solid Waste Handling Equipment	263
11 91	Religious Equipment	264
Furnishings		
12 05	Common Work Results for Furnishings	268
12 12	Wall Decorations	268
12 21	Window Blinds	268
12 22	Curtains and Drapes	269
12 23	Interior Shutters	272
12 24	Window Shades	273
12 32	Manufactured Wood Casework	273
12 35	Specialty Casework	278
12 36	Countertops	279
12 43	Portable Lamps	282
12 45	Bedroom Furnishings	283
12 46	Furnishing Accessories	283
12 48	Rugs and Mats	284
12 51	Office Furniture	284
12 52	Seating	287
12 54	Hospitality Furniture	289
12 55	Detention Furniture	296
12 56	Institutional Furniture	296
12 59	Systems Furniture	299
12 61	Fixed Audience Seating	301
12 63	Stadium and Arena Seating	302
12 67	Pews and Benches	302
12 92	Interior Planters and Artificial Plants	302
12 93	Site Furnishings	306
Special Construction		
13 11	Swimming Pools	310
13 17	Tubs and Pools	310
13 21	Controlled Environment Rooms	311
13 24	Special Activity Rooms	313
13 28	Athletic and Recreational Special Construction	313
13 34	Fabricated Engineered Structures	314
13 42	Building Modules	314
13 48	Sound, Vibration, and Seismic Control	314
13 49	Radiation Protection	315
Conveying Equipment		
14 11	Manual Dumbwaiters	320
14 12	Electric Dumbwaiters	320
14 21	Electric Traction Elevators	320
14 24	Hydraulic Elevators	321
14 27	Custom Elevator Cabs	322
14 28	Elevator Equipment and Controls	323
14 31	Escalators	324
14 32	Moving Walks	325

Div. No.		Page
14 42	Wheelchair Lifts	325
14 51	Correspondence and Parcel Lifts	325
14 91	Facility Chutes	326
14 92	Pneumatic Tube Systems	326
Fire Suppression		
21 05	Common Work Results for Fire Suppression	328
21 11	Facility Fire-Suppression Water-Service Piping	328
21 12	Fire-Suppression Standpipes	329
21 13	Fire-Suppression Sprinkler Systems	332
21 21	Carbon-Dioxide Fire-Extinguishing Systems	334
21 22	Clean-Agent Fire-Extinguishing Systems	334
21 31	Centrifugal Fire Pumps	334
Plumbing		
22 05	Common Work Results for Plumbing	338
22 07	Plumbing Insulation	338
22 13	Facility Sanitary Sewerage	341
22 33	Electric Domestic Water Heaters	341
22 34	Fuel-Fired Domestic Water Heaters	342
22 35	Domestic Water Heat Exchangers	343
22 41	Residential Plumbing Fixtures	343
22 42	Commercial Plumbing Fixtures	348
22 43	Healthcare Plumbing Fixtures	351
22 45	Emergency Plumbing Fixtures	352
22 46	Security Plumbing Fixtures	352
22 47	Drinking Fountains and Water Coolers	353
22 51	Swimming Pool Plumbing Systems	355
22 52	Fountain Plumbing Systems	355
Heating Ventilation Air Conditioning		
23 05	Common Work Results for HVAC	358
23 07	HVAC Insulation	359
23 09	Instrumentation and Control for HVAC	360
23 21	Hydronic Piping and Pumps	361
23 31	HVAC Ducts and Casings	362
23 33	Air Duct Accessories	363
23 34	HVAC Fans	363
23 37	Air Outlets and Inlets	367
23 38	Ventilation Hoods	369
23 41	Particulate Air Filtration	369
23 42	Gas-Phase Air Filtration	370
23 43	Electronic Air Cleaners	370
23 51	Breechings, Chimneys, and Stacks	370
23 55	Fuel-Fired Heaters	371
23 56	Solar Energy Heating Equipment	372
23 62	Packaged Compressor and Condenser Units	373
23 74	Packaged Outdoor HVAC Equipment	373
23 76	Evaporative Air-Cooling Equipment	374
23 81	Decentralized Unitary HVAC Equipment	374

Div. No.		Page
23 82	Convection Heating and Cooling Units	376
23 83	Radiant Heating Units	377
23 84	Humidity Control Equipment	378
Electrical		
26 05	Common Work Results for Electrical	380
26 24	Switchboards and Panelboards	391
26 27	Low-Voltage Distribution Equipment	391
26 28	Low-Voltage Circuit Protective Devices	394
26 33	Battery Equipment	394
26 51	Interior Lighting	395
26 52	Emergency Lighting	400
26 53	Exit Signs	401
26 54	Classified Location Lighting	401
26 55	Special Purpose Lighting	402
26 56	Exterior Lighting	402
26 61	Lighting Systems and Accessories	403
Communications		
27 41	Audio-Video Systems	406
27 51	Distributed Audio-Video Communications Systems	406
27 52	Healthcare Communications and Monitoring Systems	407
27 53	Distributed Systems	407
Electronic Safety & Security		
28 16	Intrusion Detection	410
28 23	Video Surveillance	410
28 31	Fire Detection and Alarm	411
Earthwork		
31 46	Needle Beams	414
Exterior Improvements		
32 06	Schedules for Exterior Improvements	416
32 11	Base Courses	416
32 14	Unit Paving	417
32 18	Athletic and Recreational Surfacing	418
32 31	Fences and Gates	418
32 32	Retaining Walls	420
32 34	Fabricated Bridges	420
32 91	Planting Preparation	420
32 93	Plants	421
32 94	Planting Accessories	424
32 96	Transplanting	425
Material Processing & Handling Equipment		
41 21	Conveyors	428
41 22	Cranes and Hoists	428

How to Use the Unit Price Pages

The following is a detailed explanation of a sample entry in the Unit Price Section. Next to each bold number below is the described item with the appropriate component of the sample entry following in parentheses. Some prices are listed as bare costs; others as costs that include overhead and profit of the installing contractor. In most cases, if the work is to be subcontracted, the general contractor will need to add an additional markup (RSMeans suggests using 10%) to the figures in the column "Total Incl. O&P."

1 Division Number/Title
(03 30/Cast-In-Place Concrete)

Use the Unit Price Section Table of Contents to locate specific items. The sections are classified according to the CSI MasterFormat 2004 system.

2 Line Numbers
(03 30 53.40 3920)

Each unit price line item has been assigned a unique 12-digit code based on the CSI MasterFormat classification.

3 Description
(Concrete In Place, etc.)

Each line item is described in detail. Sub-items and additional sizes are indented beneath the appropriate line items. The first line or two after the main item (in boldface) may contain descriptive information that pertains to all line items beneath this boldface listing.

Items which include the symbol **CN** are updated in The *RSMeans Quarterly Update Service* online. To obtain access to this service contact RSMeans customer service at 1-800-334-3509.

4 Reference Number Information

You'll see reference numbers shown in shaded boxes at the beginning of some sections. These refer to related items in the Reference Section, visually identified by a vertical gray bar on the page edges.

The relation may be an estimating procedure that should be read before estimating, or technical information.

The "R" designates the Reference Section. The numbers refer to the MasterFormat 2004 classification system.

It is strongly recommended that you review all reference numbers that appear within the section in which you are working.

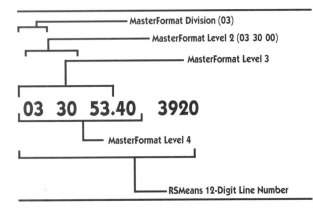

03 30 Cast-In-Place Concrete
03 30 53 – Miscellaneous Cast-In-Place Concrete

03 30 53.40 Concrete In Place	Crew	Daily Output	Labor-Hours	Unit	Material	2009 Bare Costs Labor	Equipment	Total	Total Incl O&P
0010 **CONCRETE IN PLACE**									
0020 Including forms (4 uses), reinforcing steel, concrete, placement,									
0050 and finishing unless otherwise indicated R033053-50									
0300 Beams, 5 kip per L.F., 10' span	C-14A	15.62	12.804	C.Y.	400	515	49	964	1,300
0350 25' span	"	18.55	10.782		430	430	41	901	1,200
3850 Over 5 C.Y.	↓	75	1.493		198	57	.34	255.34	305
3900 Footings, strip, 18" x 9", unreinforced	C-14L	40	2.400		120	89.50	.65	210.15	271
3920 18" x 9", reinforced	C-14C	35	3.200		157	122	.74	279.74	365
3925 20" x 10", unreinforced	C-14L	45	2.133		117	79.50	.58	197.08	252
3930 20" x 10", reinforced	C-14C	40	2.800		147	107	.64	254.64	330
3935 24" x 12", unreinforced	C-14L	55	1.745		116	65	.47	181.47	229
3940 24" x 12", reinforced	C-14C	48	2.333		147	89.50	.54	237.04	300

5 Crew (C-14C)

The "Crew" column designates the typical trade or crew used to install the item. If an installation can be accomplished by one trade and requires no power equipment, that trade and the number of workers are listed (for example, "2 Carpenters"). If an installation requires a composite crew, a crew code designation is listed (for example, "C-14C"). You'll find full details on all composite crews in the Crew Listings.

- For a complete list of all trades utilized in this book and their abbreviations, see the inside back cover.

Crews

Crew No.	Bare Costs		Incl. Subs O & P		Cost Per Labor-Hour	
Crew C-14C	Hr.	Daily	Hr.	Daily	Bare Costs	Incl. O&P
1 Carpenter Foreman (out)	$41.95	$335.60	$65.05	$520.40	$38.25	$59.61
6 Carpenters	39.95	1917.60	61.95	2973.60		
2 Rodmen (reinf.)	44.55	712.80	72.85	1165.60		
4 Laborers	31.60	1011.20	49.00	1568.00		
1 Cement Finisher	38.30	306.40	56.05	448.40		
1 Gas Engine Vibrator		26.00		28.60	.23	.26
112 L.H., Daily Totals		$4309.60		$6704.60	$38.48	$59.86

6 Productivity: Daily Output (35.0)/ Labor-Hours (3.20)

The "Daily Output" represents the typical number of units the designated crew will install in a normal 8-hour day. To find out the number of days the given crew would require to complete the installation, divide your quantity by the daily output. For example:

Quantity	÷	Daily Output	=	Duration
100 C.Y.	÷	35.0/ Crew Day	=	2.86 Crew Days

The "Labor-Hours" figure represents the number of labor-hours required to install one unit of work. To find out the number of labor-hours required for your particular task, multiply the quantity of the item times the number of labor-hours shown. For example:

Quantity	x	Productivity Rate	=	Duration
100 C.Y.	x	3.20 Labor-Hours/ C.Y.	=	320 Labor-Hours

7 Unit (C.Y.)

The abbreviated designation indicates the unit of measure upon which the price, production, and crew are based (C.Y. = Cubic Yard). For a complete listing of abbreviations, refer to the Abbreviations Listing in the Reference Section of this book.

8 Bare Costs:
Mat. (Bare Material Cost) (157)

The unit material cost is the "bare" material cost with no overhead or profit included. Costs shown reflect national average material prices for January of the current year and include delivery to the job site. No sales taxes are included.

Labor (122)

The unit labor cost is derived by multiplying bare labor-hour costs for Crew C-14C by labor-hour units. The bare labor-hour cost is found in the Crew Section under C-14C. (If a trade is listed, the hourly labor cost—the wage rate—is found on the inside back cover.)

Labor-Hour Cost Crew C-14C	x	Labor-Hour Units	=	Labor
$38.25	x	3.20	=	$122.00

Equip. (Equipment) (.74)

Costs for equipment included in each crew are listed in the Crew Listings in the Reference Section. Costs for small tools that are considered as essential everyday tools are covered in the "Overhead" column on the "Installing Contractor's Overhead and Profit" table that lists labor trades, base rates and markups. The unit equipment cost displayed here is derived by multiplying the bare equipment hourly cost in the Crew Listing by the unit Labor-Hours value.

Equipment Cost Crew C-14C	x	Labor-Hour Units	=	Equip.
$.23	x	3.20	=	$.74

Total (279.74)

The total of the bare costs is the arithmetic total of the three previous columns: mat., labor, and equip.

Material	+	Labor	+	Equip.	=	Total
$157	+	$122	+	$.74	=	$279.74

9 Total Costs Including O &P

This figure is the sum of the bare material cost plus 10% for profit; the bare labor cost plus total overhead and profit (per the inside back cover or, if a crew is listed, from the crew listings); and the bare equipment cost plus 10% for profit.

Material is Bare Material Cost + 10% = 157 + 15.70	=	$172.70
Labor for Crew C-14C = Labor-Hour Cost (59.61) x Labor-Hour Units (3.20)	=	$190.75
Equip. is Bare Equip. Cost + 10% = .74 + .07	=	$ .81
Total (Rounded)	=	$ 365

Estimating Tips

01 20 00 Price and Payment Procedures

- Allowances that should be added to estimates to cover contingencies and job conditions that are not included in the national average material and labor costs are shown in section 01 21.
- When estimating historic preservation projects (depending on the condition of the existing structure and the owner's requirements), a 15%-20% contingency or allowance is recommended, regardless of the stage of the drawings.

01 30 00 Administrative Requirements

- Before determining a final cost estimate, it is a good practice to review all the items listed in Subdivisions 01 31 and 01 32 to make final adjustments for items that may need customizing to specific job conditions.
- Requirements for initial and periodic submittals can represent a significant cost to the General Requirements of a job. Thoroughly check the submittal specifications when estimating a project to determine any costs that should be included.

01 40 00 Quality Requirements

- All projects will require some degree of Quality Control. This cost is not included in the unit cost of construction listed in each division. Depending upon the terms of the contract, the various costs of inspection and testing can be the responsibility of either the owner or the contractor. Be sure to include the required costs in your estimate.

01 50 00 Temporary Facilities and Controls

- Barricades, access roads, safety nets, scaffolding, security, and many more requirements for the execution of a safe project are elements of direct cost. These costs can easily be overlooked when preparing an estimate. When looking through the major classifications of this subdivision, determine which items apply to each division in your estimate.
- Construction Equipment Rental Costs can be found in the Reference Section in section 01 54 33. Operators' wages are not included in equipment rental costs.
- Equipment mobilization and demobilization costs are not included in equipment rental costs and must be considered separately in section 01 54 36.50.
- The cost of small tools provided by the installing contractor for his workers is covered in the "Overhead" column on the "Installing Contractor's Overhead and Profit" table that lists labor trades, base rates and markups and, therefore, is included in the "Total Incl. O&P" cost of any Unit Price line item. For those users who are constrained by contract terms to use only bare costs, there are two line items in section 01 54 39.70 for small tools as a percentage of bare labor cost. If some of those users are further constrained by contract terms to refrain from using line items from Division 1, you are advised to cover the cost of small tools within your coefficient or multiplier.

01 70 00 Execution and Closeout Requirements

- When preparing an estimate, thoroughly read the specifications to determine the requirements for Contract Closeout. Final cleaning, record documentation, operation and maintenance data, warranties and bonds, and spare parts and maintenance materials can all be elements of cost for the completion of a contract. Do not overlook these in your estimate.

Reference Numbers

Reference numbers are shown in shaded boxes at the beginning of some major classifications. These numbers refer to related items in the Reference Section. The reference information may be an estimating procedure, an alternate pricing method, or technical information.

Note: Not all subdivisions listed here necessarily appear in this publication.

01 11 Summary of Work

01 11 31 – Professional Consultants

01 11 31.10 Architectural Fees	Crew	Daily Output	Labor-Hours	Unit	Material	2009 Bare Costs Labor	Equipment	Total	Total Incl O&P
0010 **ARCHITECTURAL FEES**									
0020 For new construction									
0060 Minimum				Project					4.90%
0090 Maximum									16%
0100 For alteration work, to $500,000, add to new construction fee									50%
0150 Over $500,000, add to new construction fee									25%

01 11 31.20 Construction Management Fees									
0010 **CONSTRUCTION MANAGEMENT FEES**									
0020 $1,000,000 job, minimum				Project					4.50%
0050 Maximum				"					7.50%

01 11 31.30 Engineering Fees									
0010 **ENGINEERING FEES**									
0020 Educational planning consultant, minimum				Project					.50%
0100 Maximum				"					2.50%
0200 Electrical, minimum				Contrct					4.10%
0300 Maximum									10.10%
0600 Food service & kitchen equipment, minimum									8%
0700 Maximum									12%
1000 Mechanical (plumbing & HVAC), minimum									4.10%
1100 Maximum									10.10%

01 11 31.50 Models									
0010 **MODELS**									
0020 Cardboard & paper, 1 building, minimum				Ea.	700			700	770
0050 Maximum					1,600			1,600	1,750
0100 2 buildings, minimum					935			935	1,025
0150 Maximum					2,125			2,125	2,325
0200 Plexiglass and metal, basic layout				SF Flr.	.06			.06	.07
0210 Including equipment and personnel				"	.31			.31	.34

01 11 31.75 Renderings									
0010 **RENDERINGS** Color, matted, 20" x 30", eye level,									
0020 1 building, minimum				Ea.	1,950			1,950	2,150
0050 Average					2,775			2,775	3,075
0100 Maximum					4,450			4,450	4,900
1000 5 buildings, minimum					3,900			3,900	4,300
1100 Maximum					7,800			7,800	8,575

01 21 Allowances

01 21 16 – Contingency Allowances

01 21 16.50 Contingencies	Crew	Daily Output	Labor-Hours	Unit	Material	2009 Bare Costs Labor	Equipment	Total	Total Incl O&P
0010 **CONTINGENCIES**, Add to estimate									
0020 Conceptual stage				Project					20%
0050 Schematic stage									15%
0100 Preliminary working drawing stage (Design Dev.)									10%
0150 Final working drawing stage									3%

01 21 53 – Factors Allowance

01 21 53.50 Factors

0010 **FACTORS** Cost adjustments R012153-10									
0100 Add to construction costs for particular job requirements									
0500 Cut & patch to match existing construction, add, minimum				Costs	2%	3%			

01 21 Allowances

01 21 53 – Factors Allowance

01 21 53.50 Factors

		Crew	Daily Output	Labor-Hours	Unit	Material	2009 Bare Costs Labor	Equipment	Total	Total Incl O&P
0550	Maximum				Costs	5%	9%			
0800	Dust protection, add, minimum					1%	2%			
0850	Maximum					4%	11%			
1100	Equipment usage curtailment, add, minimum					1%	1%			
1150	Maximum					3%	10%			
1400	Material handling & storage limitation, add, minimum					1%	1%			
1450	Maximum					6%	7%			
1700	Protection of existing work, add, minimum					2%	2%			
1750	Maximum					5%	7%			
2000	Shift work requirements, add, minimum						5%			
2050	Maximum						30%			
2300	Temporary shoring and bracing, add, minimum					2%	5%			
2350	Maximum					5%	12%			
2400	Work inside prisons and high security areas, add, minimum						30%			
2450	Maximum						50%			

01 21 55 – Job Conditions Allowance

01 21 55.50 Job Conditions

		Crew	Daily Output	Labor-Hours	Unit	Material	2009 Bare Costs Labor	Equipment	Total	Total Incl O&P
0010	**JOB CONDITIONS** Modifications to applicable									
0020	cost summaries									
0100	Economic conditions, favorable, deduct				Project				2%	2%
0200	Unfavorable, add								5%	5%
0300	Hoisting conditions, favorable, deduct								2%	2%
0400	Unfavorable, add								5%	5%
0500	General Contractor management, experienced, deduct								2%	2%
0600	Inexperienced, add								10%	10%
0700	Labor availability, surplus, deduct								1%	1%
0800	Shortage, add								10%	10%
0900	Material storage area, available, deduct								1%	1%
1000	Not available, add								2%	2%
1100	Subcontractor availability, surplus, deduct								5%	5%
1200	Shortage, add								12%	12%
1300	Work space, available, deduct								2%	2%
1400	Not available, add								5%	5%

01 21 57 – Overtime Allowance

01 21 57.50 Overtime

0010	**OVERTIME** for early completion of projects or where	R012909-90								
0020	labor shortages exist, add to usual labor, up to				Costs		100%			

01 21 61 – Cost Indexes

01 21 61.10 Construction Cost Index

0010	**CONSTRUCTION COST INDEX** (Reference) over 930 zip code locations in									
0020	the U.S. and Canada, total bldg. cost, min. (Guymon, OK)				%					69.40%
0050	Average									100%
0100	Maximum (New York, NY)									130.70%

01 21 61.30 Labor Index

0010	**LABOR INDEX** (Reference) For over 930 zip code locations in									
0020	the U.S. and Canada, minimum (Del Rio, TX)				%		31.90%			
0050	Average						100%			
0100	Maximum (New York, NY)						164.60%			

01 21 61.50 Material Index

0011	**MATERIAL INDEX** For over 930 zip code locations in	R013113-50								

01 21 Allowances

01 21 61 – Cost Indexes

01 21 61.50 Material Index		Crew	Daily Output	Labor-Hours	Unit	Material	2009 Bare Costs Labor	Equipment	Total	Total Incl O&P
0020	the U.S. and Canada, minimum (Elizabethtown, KY)				%	90.10%				
0040	Average					100%				
0060	Maximum (Ketchikan, AK)					139.60%				

01 21 63 – Taxes

01 21 63.10 Taxes

0010	**TAXES**	R012909-80								
0020	Sales tax, State, average					%	4.91%			
0050	Maximum	R012909-85					7.25%			
0200	Social Security, on first $102,000 of wages							7.65%		
0300	Unemployment, combined Federal and State, minimum							.80%		
0350	Average							6.20%		
0400	Maximum							11.76%		

01 31 Project Management and Coordination

01 31 13 – Project Coordination

01 31 13.20 Field Personnel

0010	**FIELD PERSONNEL**									
0020	Clerk, average				Week		380		380	590
0100	Field engineer, minimum						895		895	1,375
0120	Average						1,165		1,165	1,800
0140	Maximum						1,350		1,350	2,100
0160	General purpose laborer, average						1,250		1,250	1,925
0180	Project manager, minimum						1,650		1,650	2,550
0200	Average						1,925		1,925	2,975
0220	Maximum						2,175		2,175	3,375
0240	Superintendent, minimum						1,600		1,600	2,475
0260	Average						1,775		1,775	2,750
0280	Maximum						2,025		2,025	3,125
0290	Timekeeper, average						1,040		1,040	1,600

01 31 13.30 Insurance

0010	**INSURANCE**	R013113-40								
0020	Builders risk, standard, minimum				Job					.24%
0050	Maximum	R013113-60								.64%
0200	All-risk type, minimum									.25%
0250	Maximum									.62%
0600	Public liability, average									2.02%
0800	Workers' compensation & employer's liability, average									
0850	by trade, carpentry, general				Payroll		17.80%			
0900	Clerical						.58%			
0950	Concrete						14.58%			
1000	Electrical						6.46%			
1050	Excavation						10.01%			
1100	Glazing						13.89%			
1150	Insulation						14.44%			
1200	Lathing						10.63%			
1250	Masonry						14.37%			
1300	Painting & decorating						12.49%			
1350	Pile driving						20.09%			
1400	Plastering						13.60%			
1450	Plumbing						7.84%			

01 31 Project Management and Coordination

01 31 13 – Project Coordination

01 31 13.30 Insurance		Crew	Daily Output	Labor-Hours	Unit	Material	2009 Bare Costs Labor	Equipment	Total	Total Incl O&P
1500	Roofing				Payroll		31.18%			
1550	Sheet metal work (HVAC)						11.57%			
1600	Steel erection, structural						37.94%			
1650	Tile work, interior ceramic						9.13%			
1700	Waterproofing, brush or hand caulking						6.69%			
1800	Wrecking						34.13%			
2000	Range of 35 trades in 50 states, excl. wrecking, min.						2.09%			
2100	Average						15.50%			
2200	Maximum						168.17%			

01 31 13.40 Main Office Expense										
0010	**MAIN OFFICE EXPENSE** Average for General Contractors									
0020	As a percentage of their annual volume									
0125	Annual volume under 1 million dollars				% Vol.				17.50%	
0145	Up to 2.5 million dollars								8%	
0150	Up to 4.0 million dollars								6.80%	
0200	Up to 7.0 million dollars								5.60%	
0250	Up to 10 million dollars								5.10%	
0300	Over 10 million dollars								3.90%	

01 31 13.50 General Contractor's Mark-Up										
0010	**GENERAL CONTRACTOR'S MARK-UP** on Change Orders									
0200	Extra work, by subcontractors, add				%					10%
0250	By General Contractor, add									15%
0400	Omitted work, by subcontractors, deduct all but									5%
0450	By General Contractor, deduct all but									7.50%
0600	Overtime work, by subcontractors, add									15%
0650	By General Contractor, add									10%

01 31 13.60 Installing Contractor's Main Office Overhead										
0010	**INSTALLING CONTRACTOR'S MAIN OFFICE OVERHEAD** R013113-50									
0020	As percent of direct costs, minimum				%				5%	
0050	Average								13%	
0100	Maximum								30%	

01 31 13.80 Overhead and Profit										
0010	**OVERHEAD & PROFIT** Allowance to add to items in this									
0020	book that do not include Subs O&P, average				%					25%
0100	Allowance to add to items in this book that									
0110	do include Subs O&P, minimum				%					5%
0150	Average									10%
0200	Maximum									15%
0300	Typical, by size of project, under $100,000									30%
0350	$500,000 project									25%
0400	$2,000,000 project									20%
0450	Over $10,000,000 project									15%

01 31 13.90 Performance Bond										
0010	**PERFORMANCE BOND**									
0020	For buildings, minimum				Job					.60%
0100	Maximum				"					2.50%

01 32 Construction Progress Documentation

01 32 13 – Scheduling of work

01 32 13.50 Scheduling

		Crew	Daily Output	Labor-Hours	Unit	Material	2009 Bare Costs Labor	Equipment	Total	Total Incl O&P
0010	**SCHEDULING**									
0020	Critical path, as % of architectural fee, minimum				%					.50%
0100	Maximum				"					1%
0300	Computer-update, micro, no plots, minimum				Ea.				455	500
0400	Including plots, maximum				"				1,455	1,600
0600	Rule of thumb, CPM scheduling, small job ($10 Million)				Job					.05%
0650	Large job ($50 Million +)									.03%
0700	Including cost control, small job									.08%
0750	Large job									.04%

01 32 33 – Photographic Documentation

01 32 33.50 Photographs

		Crew	Daily Output	Labor-Hours	Unit	Material	Labor	Equipment	Total	Total Incl O&P
0010	**PHOTOGRAPHS**									
0020	8" x 10", 4 shots, 2 prints ea., std. mounting				Set	475			475	520
0100	Hinged linen mounts					530			530	580
0200	8" x 10", 4 shots, 2 prints each, in color					415			415	460
0300	For I.D. slugs, add to all above					5.30			5.30	5.85
1500	Time lapse equipment, camera and projector, buy					3,775			3,775	4,175
1550	Rent per month					565			565	620
1700	Cameraman and film, including processing, B.&W.				Day	1,375			1,375	1,525
1720	Color				"	1,375			1,375	1,525

01 41 Regulatory Requirements

01 41 26 – Permits

01 41 26.50 Permits

		Crew	Daily Output	Labor-Hours	Unit	Material	Labor	Equipment	Total	Total Incl O&P
0010	**PERMITS**									
0020	Rule of thumb, most cities, minimum				Job					.50%
0100	Maximum				"					2%

01 51 Temporary Utilities

01 51 13 – Temporary Electricity

01 51 13.80 Temporary Utilities

		Crew	Daily Output	Labor-Hours	Unit	Material	Labor	Equipment	Total	Total Incl O&P
0010	**TEMPORARY UTILITIES**									
0100	Heat, incl. fuel and operation, per week, 12 hrs. per day	1 Skwk	100	.080	CSF Flr	27	3.27		30.27	34.50
0200	24 hrs. per day	"	60	.133		52	5.45		57.45	66
0350	Lighting, incl. service lamps, wiring & outlets, minimum	1 Elec	34	.235		2.63	11.05		13.68	19.35
0360	Maximum	"	17	.471		5.70	22		27.70	39.50
0400	Power for temp lighting only, per month, min/month 6.6 KWH								.75	.83
0450	Maximum/month 23.6 KWH								2.85	3.14
0600	Power for job duration incl. elevator, etc., minimum								47	51.70
0650	Maximum								110	121

01 52 Construction Facilities

01 52 13 – Field Offices and Sheds

01 52 13.20 Office and Storage Space

		Crew	Daily Output	Labor-Hours	Unit	Material	2009 Bare Costs Labor	Equipment	Total	Total Incl O&P
0010	**OFFICE AND STORAGE SPACE**									
0020	Trailer, furnished, no hookups, 20' x 8', buy	2 Skwk	1	16	Ea.	8,200	655		8,855	10,000
0250	Rent per month					163			163	179
1200	Storage boxes, 20' x 8', buy	2 Skwk	1.80	8.889		4,675	365		5,040	5,700
1250	Rent per month					72			72	79
1300	40' x 8', buy	2 Skwk	1.40	11.429		6,400	465		6,865	7,775
1350	Rent per month					99			99	109

01 52 13.40 Field Office Expense

		Crew	Daily Output	Labor-Hours	Unit	Material	Labor	Equipment	Total	Total Incl O&P
0010	**FIELD OFFICE EXPENSE**									
0100	Office equipment rental average				Month	155			155	171
0120	Office supplies, average				"	85			85	93.50
0125	Office trailer rental, see Div. 01 52 13.20									
0140	Telephone bill; avg. bill/month incl. long dist.				Month	80			80	88
0160	Lights & HVAC				"	150			150	165

01 54 Construction Aids

01 54 09 – Protection Equipment

01 54 09.50 Personnel Protective Equipment

		Crew	Daily Output	Labor-Hours	Unit	Material	Labor	Equipment	Total	Total Incl O&P
0010	**PERSONNEL PROTECTIVE EQUIPMENT**									
0015	Hazardous waste protection									
0020	Respirator mask only, full face, silicone				Ea.	223			223	245
0030	Half face, silicone					33			33	36.50
0040	Respirator cartridges, 2 req'd/mask, dust or asbestos					5.30			5.30	5.85
0050	Chemical vapor					4.69			4.69	5.15
0060	Combination vapor and dust					9.70			9.70	10.65
0100	Emergency escape breathing apparatus, 5 min					465			465	510
0110	10 min					500			500	550
0150	Self contained breathing apparatus with full face piece, 30 min					1,750			1,750	1,925
0160	60 min					2,925			2,925	3,225
0200	Encapsulating suits, limited use, level A					905			905	995
0210	Level B					270			270	297
0300	Over boots, latex				Pr.	6.35			6.35	7
0310	PVC					21.50			21.50	24
0320	Neoprene					41.50			41.50	46
0400	Gloves, nitrile/PVC					21			21	23.50
0410	Neoprene coated					24			24	26.50

01 54 09.60 Safety Nets

		Crew	Daily Output	Labor-Hours	Unit	Material	Labor	Equipment	Total	Total Incl O&P
0010	**SAFETY NETS**									
0020	No supports, stock sizes, nylon, 4" mesh				S.F.	1.10			1.10	1.21
0100	Polypropylene, 6" mesh					1.59			1.59	1.75
0200	Small mesh debris nets, 1/4" & 3/4" mesh, stock sizes					.74			.74	.81

01 54 23 – Temporary Scaffolding and Platforms

01 54 23.70 Scaffolding

		Crew	Daily Output	Labor-Hours	Unit	Material	Labor	Equipment	Total	Total Incl O&P
0010	**SCAFFOLDING** R015423-10									
0015	Steel tube, regular, no plank, labor only to erect & dismantle									
0460	Building interior, wall face area, up to 16' high	3 Carp	12	2	C.S.F.		80		80	124
0560	16' to 40' high		10	2.400	"		96		96	149
0800	Building interior floor area, up to 30' high		150	.160	C.C.F.		6.40		6.40	9.90
0900	Over 30' high	4 Carp	160	.200	"		8		8	12.40
0906	Complete system for face of walls, no plank, material only rent/mo				C.S.F.	35.50			35.50	39

01 54 Construction Aids

01 54 23 – Temporary Scaffolding and Platforms

01 54 23.70 Scaffolding		Crew	Daily Output	Labor-Hours	Unit	Material	2009 Bare Costs Labor	Equipment	Total	Total Incl O&P
0908	Interior spaces, no plank, material only rent/mo				C.C.F.	3.40			3.40	3.74
0910	Steel tubular, heavy duty shoring, buy									
0920	Frames 5' high 2' wide				Ea.	93			93	102
0925	5' high 4' wide					105			105	116
0930	6' high 2' wide					106			106	117
0935	6' high 4' wide					124			124	136
0940	Accessories									
0945	Cross braces				Ea.	18			18	19.80
0950	U-head, 8" x 8"					21.50			21.50	24
0955	J-head, 4" x 8"					15.80			15.80	17.40
0960	Base plate, 8" x 8"					17.60			17.60	19.35
0965	Leveling jack					38			38	41.50
1000	Steel tubular, regular, buy									
1100	Frames 3' high 5' wide				Ea.	71			71	78
1150	5' high 5' wide					83			83	91.50
1200	6'-4" high 5' wide					104			104	114
1350	7'-6" high 6' wide					179			179	197
1500	Accessories cross braces					20.50			20.50	22.50
1550	Guardrail post					18.60			18.60	20.50
1600	Guardrail 7' section					9.90			9.90	10.90
1650	Screw jacks & plates					27			27	29.50
1700	Sidearm brackets					43			43	47.50
1750	8" casters					35.50			35.50	39
1800	Plank 2" x 10" x 16'-0"					51			51	56
1900	Stairway section					325			325	360
1910	Stairway starter bar					36.50			36.50	40
1920	Stairway inside handrail					66			66	72.50
1930	Stairway outside handrail					98			98	108
1940	Walk-thru frame guardrail					47			47	51.50
2000	Steel tubular, regular, rent/mo.									
2100	Frames 3' high 5' wide				Ea.	5			5	5.50
2150	5' high 5' wide					5			5	5.50
2200	6'-4" high 5' wide					5.05			5.05	5.55
2250	7'-6" high 6' wide					7			7	7.70
2500	Accessories, cross braces					1			1	1.10
2550	Guardrail post					1			1	1.10
2600	Guardrail 7' section					1			1	1.10
2650	Screw jacks & plates					2			2	2.20
2700	Sidearm brackets					2			2	2.20
2750	8" casters					8			8	8.80
2800	Outrigger for rolling tower					3			3	3.30
2850	Plank 2" x 10" x 16'-0"					6			6	6.60
2900	Stairway section					40			40	44
2940	Walk-thru frame guardrail					2.50			2.50	2.75
3000	Steel tubular, heavy duty shoring, rent/mo.									
3250	5' high 2' & 4' wide				Ea.	5			5	5.50
3300	6' high 2' & 4' wide					5			5	5.50
3500	Accessories, cross braces					1			1	1.10
3600	U - head, 8" x 8"					1			1	1.10
3650	J - head, 4" x 8"					1			1	1.10
3700	Base plate, 8" x 8"					1			1	1.10
3750	Leveling jack					2			2	2.20
5700	Planks, 2x10x16'-0", labor only to erect & remove to 50' H	3 Carp	72	.333			13.30		13.30	20.50

01 54 Construction Aids

01 54 23 – Temporary Scaffolding and Platforms

01 54 23.70 Scaffolding		Crew	Daily Output	Labor-Hours	Unit	Material	2009 Bare Costs Labor	Equipment	Total	Total Incl O&P
5800	Over 50' high	4 Carp	80	.400	Ea.		16		16	25

01 54 23.75 Scaffolding Specialties

		Crew	Daily Output	Labor-Hours	Unit	Material	Labor	Equipment	Total	Total Incl O&P
0010	**SCAFFOLDING SPECIALTIES**									
1900	Catwalks, 20" wide, no guardrails, 7' span, buy				Ea.	145			145	160
2000	10' span, buy					203			203	223
3800	Rolling ladders with handrails, 30" wide, buy, 2 step					214			214	235
4000	7 step					655			655	720
4050	10 step					910			910	1,000
4100	Rolling towers, buy, 5' wide, 7' long, 10' high					1,350			1,350	1,500
4200	For 5' high added sections, to buy, add					207			207	228
4300	Complete incl. wheels, railings, outriggers,									
4350	21' high, to buy				Ea.	2,250			2,250	2,475
4400	Rent/month = 5% of purchase cost				"	7.50			7.50	8.25

01 54 36 – Equipment Mobilization

01 54 36.50 Mobilization or Demob.

		Crew	Daily Output	Labor-Hours	Unit	Material	Labor	Equipment	Total	Total Incl O&P
0010	**MOBILIZATION OR DEMOB.** (One or the other, unless noted)									
0015	Up to 25 mi haul dist (50 mi RT for mob/demob crew)									
0020	Dozer, loader, backhoe, excav., grader, paver, roller, 70 to 150 H.P.	B-34N	4	2	Ea.		64	117	181	227
0900	Shovel or dragline, 3/4 C.Y.	B-34K	3.60	2.222			71	218	289	350
1100	Small equipment, placed in rear of, or towed by pickup truck	A-3A	8	1			31	16.05	47.05	65
1150	Equip up to 70 HP, on flatbed trailer behind pickup truck	A-3D	4	2			62	57.50	119.50	159
2000	Mob & demob truck-mounted crane up to 75 ton, driver only	1 Eqhv	3.60	2.222			94.50		94.50	142
2200	Crawler-mounted, up to 75 ton	A-3F	2	8			298	405	703	895
2500	For each additional 5 miles haul distance, add						10%	10%		
3000	For large pieces of equipment, allow for assembly/knockdown									

01 54 39 – Construction Equipment

01 54 39.70 Small Tools

		Crew	Daily Output	Labor-Hours	Unit	Material	Labor	Equipment	Total	Total Incl O&P
0010	**SMALL TOOLS** R013113-50									
0020	As % of contractor's bare labor cost for project, minimum				Total		.50%			
0100	Maximum				"		2%			

01 55 Vehicular Access and Parking

01 55 23 – Temporary Roads

01 55 23.50 Roads and Sidewalks

		Crew	Daily Output	Labor-Hours	Unit	Material	Labor	Equipment	Total	Total Incl O&P
0010	**ROADS AND SIDEWALKS** Temporary									
1000	Ramp, 3/4" plywood on 2" x 6" joists, 16" O.C.	2 Carp	300	.053	S.F.	1.35	2.13		3.48	4.79
1100	On 2" x 10" joists, 16" O.C.	"	275	.058	"	1.82	2.32		4.14	5.60

01 56 Temporary Barriers and Enclosures

01 56 13 – Temporary Air Barriers

01 56 13.60 Tarpaulins

		Crew	Daily Output	Labor-Hours	Unit	Material	2009 Bare Costs Labor	Equipment	Total	Total Incl O&P
0010	**TARPAULINS**									
0020	Cotton duck, 10 oz. to 13.13 oz. per S.Y., minimum				S.F.	.54			.54	.59
0050	Maximum					.53			.53	.58
0100	Polyvinyl coated nylon, 14 oz. to 18 oz., minimum					.48			.48	.53
0150	Maximum					.68			.68	.75
0200	Reinforced polyethylene 3 mils thick, white					.15			.15	.17
0300	4 mils thick, white, clear or black					.20			.20	.22
0400	5.5 mils thick, clear					.21			.21	.23
0500	White, fire retardant					.35			.35	.39
0600	7.5 mils, oil resistant, fire retardant					.40			.40	.44
0700	8.5 mils, black					.53			.53	.58
0710	Woven polyethylene, 6 mils thick					.35			.35	.39
0730	Polyester reinforced w/ integral fastening system 11 mils thick					1.07			1.07	1.18
0740	Mylar polyester, non-reinforced, 7 mils thick				↓	1.17			1.17	1.29

01 56 13.90 Winter Protection

		Crew	Daily Output	Labor-Hours	Unit	Material	Labor	Equipment	Total	Total Incl O&P
0010	**WINTER PROTECTION**									
0100	Framing to close openings	2 Clab	750	.021	S.F.	.39	.67		1.06	1.48
0200	Tarpaulins hung over scaffolding, 8 uses, not incl. scaffolding		1500	.011		.25	.34		.59	.80
0250	Tarpaulin polyester reinf. w/ integral fastening system 11 mils thick		1600	.010		.80	.32		1.12	1.37
0300	Prefab fiberglass panels, steel frame, 8 uses	↓	1200	.013	↓	.85	.42		1.27	1.59

01 56 23 – Temporary Barricades

01 56 23.10 Barricades

		Crew	Daily Output	Labor-Hours	Unit	Material	Labor	Equipment	Total	Total Incl O&P
0010	**BARRICADES**									
0020	5' high, 3 rail @ 2" x 8", fixed	2 Carp	20	.800	L.F.	4.99	32		36.99	55
0150	Movable	"	30	.533	"	4.12	21.50		25.62	37.50
0300	Stock units, 6' high, 8' wide, plain, buy				Ea.	435			435	480
0350	With reflective tape, buy				"	525			525	580
0400	Break-a-way 3" PVC pipe barricade									
0410	with 3 ea. 1' x 4' reflectorized panels, buy				Ea.	305			305	335
0500	Plywood with steel legs, 32" wide					72			72	79
1300	Barricade tape, polyethylene, 7 mil, 3" wide x 500' long roll				↓	25			25	27.50

01 56 29 – Temporary Protective Walkways

01 56 29.50 Protection

		Crew	Daily Output	Labor-Hours	Unit	Material	Labor	Equipment	Total	Total Incl O&P
0010	**PROTECTION**									
0020	Stair tread, 2" x 12" planks, 1 use	1 Carp	75	.107	Tread	4.16	4.26		8.42	11.15
0100	Exterior plywood, 1/2" thick, 1 use		65	.123		1.39	4.92		6.31	9.10
0200	3/4" thick, 1 use	↓	60	.133	↓	2.49	5.35		7.84	11

01 56 32 – Temporary Security

01 56 32.50 Watchman

		Crew	Daily Output	Labor-Hours	Unit	Material	Labor	Equipment	Total	Total Incl O&P
0010	**WATCHMAN**									
0020	Service, monthly basis, uniformed person, minimum				Hr.				25	27.50
0100	Maximum				"				45.45	50

01 74 Cleaning and Waste Management

01 74 13 – Progress Cleaning

01 74 13.20 Cleaning Up	Crew	Daily Output	Labor-Hours	Unit	Material	2009 Bare Costs Labor	Equipment	Total	Total Incl O&P
0010 **CLEANING UP**									
0020 After job completion, allow, minimum				Job					.30%
0040 Maximum				"					1%
0050 Cleanup of floor area, continuous, per day, during const.	A-5	24	.750	M.S.F.	1.70	23.50	2.03	27.23	40.50
0100 Final by GC at end of job	"	11.50	1.565	"	2.71	49.50	4.23	56.44	84
0200 Rubbish removal, see Div. 02 41 19.23									

Division Notes

	CREW	DAILY OUTPUT	LABOR-HOURS	UNIT	2009 BARE COSTS				TOTAL INCL O&P
					MAT.	LABOR	EQUIP.	TOTAL	

Estimating Tips

02 30 00 Subsurface Investigation

In preparing estimates on structures involving earthwork or foundations, all information concerning soil characteristics should be obtained. Look particularly for hazardous waste, evidence of prior dumping of debris, and previous stream beds.

02 41 00 Demolition and Structure Moving

The costs shown for selective demolition do not include rubbish handling or disposal. These items should be estimated separately using RSMeans data or other sources.

- Historic preservation often requires that the contractor remove materials from the existing structure, rehab them, and replace them. The estimator must be aware of any related measures and precautions that must be taken when doing selective demolition and cutting and patching. Requirements may include special handling and storage, as well as security.
- In addition to Subdivision 02 41 00, you can find selective demolition items in each division. Example: Roofing demolition is in Division 7.

02 80 00 Hazardous Material Disposal/Remediation

This subdivision includes information on hazardous waste handling, asbestos remediation, lead remediation, and mold remediation. See reference R028213-20 and R028319-60 for further guidance in using these unit price lines.

Reference Numbers

Reference numbers are shown in shaded boxes at the beginning of some major classifications. These numbers refer to related items in the Reference Section. The reference information may be an estimating procedure, an alternate pricing method, or technical information.

Note: Not all subdivisions listed here necessarily appear in this publication.

No part of this publication may be reproduced, stored in a retrieval system, or transmitted in any form or by any means without prior written permission of R.S. Means Company, Inc.

02 41 Demolition

02 41 16 – Structure Demolition

02 41 16.13 Building Demolition

		Crew	Daily Output	Labor-Hours	Unit	Material	2009 Bare Costs Labor	2009 Bare Costs Equipment	Total	Total Incl O&P
0010	**BUILDING DEMOLITION** Large urban projects, incl. 20 mi. haul R024119-10									
0011	No foundation or dump fees, C.F. is vol. of building standing									
0020	Steel	B-8	21500	.003	C.F.		.10	.13	.23	.31
0050	Concrete		15300	.004			.15	.19	.34	.42
0080	Masonry		20100	.003			.11	.14	.25	.33
0100	Mixture of types, average		20100	.003			.11	.14	.25	.33
5000	For buildings with no interior walls, deduct				Ea.				50%	

02 41 19 – Selective Structure Demolition

02 41 19.13 Selective Building Demolition

0010	**SELECTIVE BUILDING DEMOLITION**									
0020	Costs related to selective demolition of specific building components									
0025	are included under Common Work Results (XX 05 00)									
0030	in the component's appropriate division.									

02 41 19.16 Selective Demolition, Cutout

		Crew	Daily Output	Labor-Hours	Unit	Material	Labor	Equipment	Total	Incl O&P
0010	**SELECTIVE DEMOLITION, CUTOUT** R024119-10									
0020	Concrete, elev. slab, light reinforcement, under 6 C.F.	B-9	65	.615	C.F.		19.70	2.95	22.65	33.50
0050	Light reinforcing, over 6 C.F.		75	.533	"		17.05	2.55	19.60	29.50
0200	Slab on grade to 6" thick, not reinforced, under 8 S.F.		85	.471	S.F.		15.05	2.25	17.30	26
0250	8 – 16 S.F.		175	.229	"		7.30	1.09	8.39	12.55
0600	Walls, not reinforced, under 6 C.F.		60	.667	C.F.		21.50	3.19	24.69	36.50
0650	6 – 12 C.F.		80	.500			16	2.40	18.40	27.50
1000	Concrete, elevated slab, bar reinforced, under 6 C.F.		45	.889			28.50	4.26	32.76	48.50
1050	Bar reinforced, over 6 C.F.		50	.800			25.50	3.83	29.33	43.50
1200	Slab on grade to 6" thick, bar reinforced, under 8 S.F.		75	.533	S.F.		17.05	2.55	19.60	29.50
1250	8 – 16 S.F.		150	.267	"		8.55	1.28	9.83	14.65
1400	Walls, bar reinforced, under 6 C.F.		50	.800	C.F.		25.50	3.83	29.33	43.50
1450	6 – 12 C.F.		70	.571	"		18.30	2.74	21.04	31.50
2000	Brick, to 4 S.F. opening, not including toothing									
2040	4" thick	B-9	30	1.333	Ea.		42.50	6.40	48.90	73
2060	8" thick		18	2.222			71	10.65	81.65	122
2080	12" thick		10	4			128	19.15	147.15	219
2400	Concrete block, to 4 S.F. opening, 2" thick		35	1.143			36.50	5.45	41.95	62.50
2420	4" thick		30	1.333			42.50	6.40	48.90	73
2440	8" thick		27	1.481			47.50	7.10	54.60	81.50
2460	12" thick		24	1.667			53.50	8	61.50	91.50
2600	Gypsum block, to 4 S.F. opening, 2" thick		80	.500			16	2.40	18.40	27.50
2620	4" thick		70	.571			18.30	2.74	21.04	31.50
2640	8" thick		55	.727			23.50	3.48	26.98	40
2800	Terra cotta, to 4 S.F. opening, 4" thick		70	.571			18.30	2.74	21.04	31.50
2840	8" thick		65	.615			19.70	2.95	22.65	33.50
2880	12" thick		50	.800			25.50	3.83	29.33	43.50
3000	Toothing masonry cutouts, brick, soft old mortar	1 Brhe	40	.200	V.L.F.		6.45		6.45	9.75
3100	Hard mortar		30	.267			8.55		8.55	13
3200	Block, soft old mortar		70	.114			3.67		3.67	5.55
3400	Hard mortar		50	.160			5.15		5.15	7.80
6000	Walls, interior, not including re-framing,									
6010	openings to 5 S.F.									
6100	Drywall to 5/8" thick	1 Clab	24	.333	Ea.		10.55		10.55	16.35
6200	Paneling to 3/4" thick		20	.400			12.65		12.65	19.60
6300	Plaster, on gypsum lath		20	.400			12.65		12.65	19.60
6340	On wire lath		14	.571			18.05		18.05	28
7000	Wood frame, not including re-framing, openings to 5 S.F.									

02 41 Demolition

02 41 19 – Selective Structure Demolition

02 41 19.16 Selective Demolition, Cutout

		Crew	Daily Output	Labor-Hours	Unit	Material	2009 Bare Costs Labor	Equipment	Total	Total Incl O&P
7200	Floors, sheathing and flooring to 2" thick	1 Clab	5	1.600	Ea.		50.50		50.50	78.50
7310	Roofs, sheathing to 1" thick, not including roofing		6	1.333			42		42	65.50
7410	Walls, sheathing to 1" thick, not including siding		7	1.143			36		36	56

02 41 19.19 Selective Demolition, Dump Charges

			Crew	Daily Output	Labor-Hours	Unit	Material	Labor	Equipment	Total	Total Incl O&P
0010	**SELECTIVE DEMOLITION, DUMP CHARGES**	R024119-10									
0020	Dump charges, typical urban city, tipping fees only										
0100	Building construction materials					Ton	100			100	110
0200	Trees, brush, lumber						75			75	82.50
0300	Rubbish only						90			90	99
0500	Reclamation station, usual charge						100			100	110

02 41 19.21 Selective Demolition, Gutting

			Crew	Daily Output	Labor-Hours	Unit	Material	Labor	Equipment	Total	Total Incl O&P
0010	**SELECTIVE DEMOLITION, GUTTING**	R024119-10									
0020	Building interior, including disposal, dumpster fees not included										
0500	Residential building										
0560	Minimum		B-16	400	.080	SF Flr.		2.58	1.33	3.91	5.45
0580	Maximum		"	360	.089	"		2.86	1.48	4.34	6.05
0900	Commercial building										
1000	Minimum		B-16	350	.091	SF Flr.		2.94	1.52	4.46	6.25
1020	Maximum		"	250	.128	"		4.12	2.13	6.25	8.75

02 41 19.23 Selective Demolition, Rubbish Handling

			Crew	Daily Output	Labor-Hours	Unit	Material	Labor	Equipment	Total	Total Incl O&P
0010	**SELECTIVE DEMOLITION, RUBBISH HANDLING**	R024119-10									
0020	The following are to be added to the demolition prices										
0400	Chute, circular, prefabricated steel, 18" diameter		B-1	40	.600	L.F.	47	19.35		66.35	82
0440	30" diameter		"	30	.800	"	48.50	26		74.50	93.50
0725	Dumpster, weekly rental, 1 dump/week, 20 C.Y. capacity (8 Tons)	R024119-20				Week	775			775	852.50
0800	30 C.Y. capacity (10 Tons)						1,000			1,000	1,100
0840	40 C.Y. capacity (13 Tons)						1,300			1,300	1,430
1000	Dust partition, 6 mil polyethylene, 1" x 3" frame		2 Carp	2000	.008	S.F.	.44	.32		.76	.99
1080	2" x 4" frame		"	2000	.008	"	.27	.32		.59	.80
2000	Load, haul, and dump, 50' haul		2 Clab	24	.667	C.Y.		21		21	32.50
2040	100' haul			16.50	.970			30.50		30.50	47.50
2080	Over 100' haul, add per 100 L.F.			35.50	.451			14.25		14.25	22
2120	In elevators, per 10 floors, add			140	.114			3.61		3.61	5.60
3000	Loading & trucking, including 2 mile haul, chute loaded		B-16	45	.711			23	11.85	34.85	48.50
3040	Hand loading truck, 50' haul		"	48	.667			21.50	11.10	32.60	45
3080	Machine loading truck		B-17	120	.267			8.95	5.15	14.10	19.40
5000	Haul, per mile, up to 8 C.Y. truck		B-34B	1165	.007			.22	.46	.68	.84
5100	Over 8 C.Y. truck		"	1550	.005			.16	.34	.50	.63

02 41 19.25 Selective Demolition, Saw Cutting

			Crew	Daily Output	Labor-Hours	Unit	Material	Labor	Equipment	Total	Total Incl O&P
0010	**SELECTIVE DEMOLITION, SAW CUTTING**	R024119-10									
0015	Asphalt, up to 3" deep		B-89	1050	.015	L.F.	.40	.53	.38	1.31	1.67
0020	Each additional inch of depth		"	1800	.009		.08	.31	.22	.61	.81
1200	Masonry walls, hydraulic saw, brick, per inch of depth		B-89B	300	.053		.42	1.87	2.28	4.57	5.80
1220	Block walls, solid, per inch of depth		"	250	.064		.42	2.24	2.74	5.40	6.85
2000	Brick or masonry w/hand held saw, per inch of depth		A-1	125	.064		.34	2.02	.51	2.87	4.07
5000	Wood sheathing to 1" thick, on walls		1 Carp	200	.040			1.60		1.60	2.48
5020	On roof		"	250	.032			1.28		1.28	1.98

02 41 19.27 Selective Demolition, Torch Cutting

			Crew	Daily Output	Labor-Hours	Unit	Material	Labor	Equipment	Total	Total Incl O&P
0010	**SELECTIVE DEMOLITION, TORCH CUTTING**	R024119-10									
0020	Steel, 1" thick plate		1 Clab	360	.022	L.F.	.22	.70		.92	1.33
0040	1" diameter bar		"	210	.038	Ea.		1.20		1.20	1.87

02 41 Demolition

02 41 19 – Selective Structure Demolition

02 41 19.27 Selective Demolition, Torch Cutting

		Crew	Daily Output	Labor-Hours	Unit	Material	2009 Bare Costs Labor	Equipment	Total	Total Incl O&P
1000	Oxygen lance cutting, reinforced concrete walls									
1040	12" to 16" thick walls	1 Clab	10	.800	L.F.		25.50		25.50	39
1080	24" thick walls	"	6	1.333	"		42		42	65.50

02 82 Asbestos Remediation

02 82 13 – Asbestos Abatement

02 82 13.41 Asbestos Abatement Equip.

		Crew	Daily Output	Labor-Hours	Unit	Material	2009 Bare Costs Labor	Equipment	Total	Total Incl O&P
0010	**ASBESTOS ABATEMENT EQUIP.** R028213-20									
0011	Equipment and supplies, buy									
0200	Air filtration device, 2000 C.F.M.				Ea.	795			795	875
0250	Large volume air sampling pump, minimum					305			305	335
0260	Maximum					455			455	500
0300	Airless sprayer unit, 2 gun					3,125			3,125	3,425
0350	Light stand, 500 watt					72.50			72.50	80
0400	Personal respirators									
0410	Negative pressure, 1/2 face, dual operation, min.				Ea.	25.50			25.50	28
0420	Maximum					31			31	34.50
0450	P.A.P.R., full face, minimum					124			124	137
0460	Maximum					150			150	165
0470	Supplied air, full face, incl. air line, minimum					157			157	172
0480	Maximum					239			239	263
0500	Personnel sampling pump, minimum					201			201	221
1500	Power panel, 20 unit, incl. G.F.I.					1,100			1,100	1,200
1600	Shower unit, including pump and filters					1,075			1,075	1,200
1700	Supplied air system (type C)					9,800			9,800	10,800
1750	Vacuum cleaner, HEPA, 16 gal., stainless steel, wet/dry					675			675	745
1760	55 gallon					1,700			1,700	1,875
1800	Vacuum loader, 9-18 ton/hr					90,000			90,000	99,000
1900	Water atomizer unit, including 55 gal. drum					230			230	253
2000	Worker protection, whole body, foot, head cover & gloves, plastic					6.80			6.80	7.45
2500	Respirator, single use					10.90			10.90	12
2550	Cartridge for respirator					12.30			12.30	13.55
2570	Glove bag, 7 mil, 50" x 64"					9.20			9.20	10.10
2580	10 mil, 44" x 60"					74.50			74.50	82
3000	HEPA vacuum for work area, minimum					1,325			1,325	1,475
3050	Maximum					3,100			3,100	3,425
6000	Disposable polyethylene bags, 6 mil, 3 C.F.					1.24			1.24	1.36
6300	Disposable fiber drums, 3 C.F.					12			12	13.20
6400	Pressure sensitive caution labels, 3" x 5"					1.19			1.19	1.31
6450	11" x 17"					7.05			7.05	7.75
6500	Negative air machine, 1800 C.F.M.					770			770	845

02 82 13.42 Preparation of Asbestos Containment Area

		Crew	Daily Output	Labor-Hours	Unit	Material	2009 Bare Costs Labor	Equipment	Total	Total Incl O&P
0010	**PREPARATION OF ASBESTOS CONTAINMENT AREA**									
0100	Pre-cleaning, HEPA vacuum and wet wipe, flat surfaces	A-9	12000	.005	S.F.	.02	.24		.26	.39
0200	Protect carpeted area, 2 layers 6 mil poly on 3/4" plywood	"	1000	.064		1.50	2.83		4.33	6.10
0300	Separation barrier, 2" x 4" @ 16", 1/2" plywood ea. side, 8' high	2 Carp	400	.040		1.25	1.60		2.85	3.86
0310	12' high		320	.050		1.40	2		3.40	4.64
0320	16' high		200	.080		1.50	3.20		4.70	6.60
0400	Personnel decontam. chamber, 2" x 4" @ 16", 3/4" ply ea. side		280	.057		2.50	2.28		4.78	6.30
0450	Waste decontam. chamber, 2" x 4" studs @ 16", 3/4" ply ea. side		360	.044		3	1.78		4.78	6.05
0500	Cover surfaces with polyethylene sheeting									

02 82 Asbestos Remediation

02 82 13 – Asbestos Abatement

02 82 13.42 Preparation of Asbestos Containment Area

		Crew	Daily Output	Labor-Hours	Unit	Material	2009 Bare Costs Labor	2009 Bare Costs Equipment	Total	Total Incl O&P
0501	Including glue and tape									
0550	Floors, each layer, 6 mil	A-9	8000	.008	S.F.	.06	.35		.41	.61
0551	4 mil		9000	.007		.04	.31		.35	.53
0560	Walls, each layer, 6 mil		6000	.011		.06	.47		.53	.80
0561	4 mil		7000	.009		.04	.40		.44	.67
0570	For heights above 12', add						20%			
0575	For heights above 20', add						30%			
0580	For fire retardant poly, add					100%				
0590	For large open areas, deduct					10%	20%			
0600	Seal floor penetrations with foam firestop to 36 Sq. In.	2 Carp	200	.080	Ea.	1,275	3.20		1,278.20	1,400
0610	36 Sq. In. to 72 Sq. In.		125	.128		2,550	5.10		2,555.10	2,825
0615	72 Sq. In. to 144 Sq. In.		80	.200		5,125	8		5,133	5,625
0620	Wall penetrations, to 36 square inches		180	.089		1,275	3.55		1,278.55	1,400
0630	36 Sq. In. to 72 Sq. In.		100	.160		2,550	6.40		2,556.40	2,825
0640	72 Sq. In. to 144 Sq. In.		60	.267		5,125	10.65		5,135.65	5,650
0800	Caulk seams with latex	1 Carp	230	.035	L.F.	.15	1.39		1.54	2.32
0900	Set up neg. air machine, 1-2k C.F.M./25 M.C.F. volume	1 Asbe	4.30	1.860	Ea.		82		82	129

02 82 13.43 Bulk Asbestos Removal

		Crew	Daily Output	Labor-Hours	Unit	Material	2009 Bare Costs Labor	2009 Bare Costs Equipment	Total	Total Incl O&P
0010	**BULK ASBESTOS REMOVAL**									
0020	Includes disposable tools and 2 suits and 1 respirator filter/day/worker									
0100	Beams, W 10 x 19	A-9	235	.272	L.F.	.88	12.05		12.93	19.80
0110	W 12 x 22		210	.305		.99	13.45		14.44	22
0120	W 14 x 26		180	.356		1.15	15.70		16.85	26
0130	W 16 x 31		160	.400		1.29	17.65		18.94	29
0140	W 18 x 40		140	.457		1.48	20		21.48	33
0150	W 24 x 55		110	.582		1.88	25.50		27.38	42.50
0160	W 30 x 108		85	.753		2.43	33.50		35.93	54.50
0170	W 36 x 150		72	.889		2.87	39.50		42.37	64.50
0200	Boiler insulation		480	.133	S.F.	.50	5.90		6.40	9.80
0210	With metal lath, add				%				50%	
0300	Boiler breeching or flue insulation	A-9	520	.123	S.F.	.40	5.45		5.85	8.95
0310	For active boiler, add				%				100%	
0400	Duct or AHU insulation	A-10B	440	.073	S.F.	.24	3.22		3.46	5.30
0500	Duct vibration isolation joints, up to 24 Sq. In. duct	A-9	56	1.143	Ea.	3.69	50.50		54.19	83
0520	25 Sq. In. to 48 Sq. In. duct		48	1.333		4.31	59		63.31	97
0530	49 Sq. In. to 76 Sq. In. duct		40	1.600		5.15	70.50		75.65	117
0600	Pipe insulation, air cell type, up to 4" diameter pipe		900	.071	L.F.	.23	3.14		3.37	5.15
0610	4" to 8" diameter pipe		800	.080		.26	3.53		3.79	5.85
0620	10" to 12" diameter pipe		700	.091		.30	4.04		4.34	6.70
0630	14" to 16" diameter pipe		550	.116		.38	5.15		5.53	8.45
0650	Over 16" diameter pipe		650	.098	S.F.	.32	4.35		4.67	7.15
0700	With glove bag up to 3" diameter pipe		200	.320	L.F.	2.15	14.15		16.30	24.50
1000	Pipe fitting insulation up to 4" diameter pipe		320	.200	Ea.	.65	8.85		9.50	14.55
1100	6" to 8" diameter pipe		304	.211		.68	9.30		9.98	15.30
1110	10" to 12" diameter pipe		192	.333		1.08	14.70		15.78	24
1120	14" to 16" diameter pipe		128	.500		1.62	22		23.62	36.50
1130	Over 16" diameter pipe		176	.364	S.F.	1.18	16.05		17.23	26.50
1200	With glove bag, up to 8" diameter pipe		75	.853	L.F.	6.55	37.50		44.05	66
2000	Scrape foam fireproofing from flat surface		2400	.027	S.F.	.09	1.18		1.27	1.94
2100	Irregular surfaces		1200	.053		.17	2.36		2.53	3.88
3000	Remove cementitious material from flat surface		1800	.036		.11	1.57		1.68	2.59
3100	Irregular surface		1000	.064		.15	2.83		2.98	4.59

02 82 Asbestos Remediation

02 82 13 – Asbestos Abatement

02 82 13.43 Bulk Asbestos Removal

		Crew	Daily Output	Labor-Hours	Unit	Material	2009 Bare Costs Labor	Equipment	Total	Total Incl O&P
4000	Scrape acoustical coating/fireproofing, from ceiling	A-9	3200	.020	S.F.	.06	.88		.94	1.45
5000	Remove VAT from floor by hand	↓	2400	.027		.09	1.18		1.27	1.94
5100	By machine	A-11	4800	.013		.04	.59	.01	.64	.98
5150	For 2 layers, add				%				50%	
6000	Remove contaminated soil from crawl space by hand	A-9	400	.160	C.F.	.52	7.05		7.57	11.60
6100	With large production vacuum loader	A-12	700	.091	"	.30	4.04	1.08	5.42	7.85
7000	Radiator backing, not including radiator removal	A-9	1200	.053	S.F.	.17	2.36		2.53	3.88
8000	Cement-asbestos transite board	2 Asbe	1000	.016		.14	.71		.85	1.27
8100	Transite shingle siding	A-10B	750	.043		.22	1.89		2.11	3.20
8200	Shingle roofing	"	2000	.016		.08	.71		.79	1.20
8250	Built-up, no gravel, non-friable	B-2	1400	.029		.08	.91		.99	1.51
8300	Asbestos millboard	2 Asbe	1000	.016	↓	.08	.71		.79	1.20
9000	For type C (supplied air) respirator equipment, add				%					10%

02 82 13.44 Demolition In Asbestos Contaminated Area

		Crew	Daily Output	Labor-Hours	Unit	Material	Labor	Equipment	Total	Total Incl O&P
0010	**DEMOLITION IN ASBESTOS CONTAMINATED AREA**									
0200	Ceiling, including suspension system, plaster and lath	A-9	2100	.030	S.F.	.10	1.35		1.45	2.22
0210	Finished plaster, leaving wire lath		585	.109		.35	4.83		5.18	7.95
0220	Suspended acoustical tile		3500	.018		.06	.81		.87	1.34
0230	Concealed tile grid system		3000	.021		.07	.94		1.01	1.56
0240	Metal pan grid system		1500	.043		.14	1.88		2.02	3.10
0250	Gypsum board		2500	.026		.08	1.13		1.21	1.86
0260	Lighting fixtures up to 2' x 4'	↓	72	.889	Ea.	2.87	39.50		42.37	64.50
0400	Partitions, non load bearing									
0410	Plaster, lath, and studs	A-9	690	.093	S.F.	.87	4.10		4.97	7.35
0450	Gypsum board and studs	"	1390	.046	"	.15	2.03		2.18	3.35
9000	For type C (supplied air) respirator equipment, add				%					10%

02 82 13.45 OSHA Testing

		Crew	Daily Output	Labor-Hours	Unit	Material	Labor	Equipment	Total	Total Incl O&P
0010	**OSHA TESTING**									
0100	Certified technician, minimum				Day				240	264
0110	Maximum				"				320	352
0200	Personal sampling, PCM analysis, NIOSH 7400, minimum	1 Asbe	8	1	Ea.	2.75	44		46.75	72
0210	Maximum	"	4	2	"	3	88		91	141
0300	Industrial hygienist, minimum				Day				320	352
0310	Maximum				"				480	528
1000	Cleaned area samples	1 Asbe	8	1	Ea.	2.63	44		46.63	72
1100	PCM air sample analysis, NIOSH 7400, minimum		8	1		30.50	44		74.50	103
1110	Maximum	↓	4	2		3.20	88		91.20	142
1200	TEM air sample analysis, NIOSH 7402, minimum								100	125
1210	Maximum				↓				400	500

02 82 13.46 Decontamination of Asbestos Containment Area

		Crew	Daily Output	Labor-Hours	Unit	Material	Labor	Equipment	Total	Total Incl O&P
0010	**DECONTAMINATION OF ASBESTOS CONTAINMENT AREA**									
0100	Spray exposed substrate with surfactant (bridging)									
0200	Flat surfaces	A-9	6000	.011	S.F.	.36	.47		.83	1.14
0250	Irregular surfaces		4000	.016	"	.31	.71		1.02	1.45
0300	Pipes, beams, and columns		2000	.032	L.F.	.56	1.41		1.97	2.83
1000	Spray encapsulate polyethylene sheeting		8000	.008	S.F.	.31	.35		.66	.89
1100	Roll down polyethylene sheeting		8000	.008	"		.35		.35	.55
1500	Bag polyethylene sheeting		400	.160	Ea.	.77	7.05		7.82	11.90
2000	Fine clean exposed substrate, with nylon brush		2400	.027	S.F.		1.18		1.18	1.85
2500	Wet wipe substrate		4800	.013			.59		.59	.92
2600	Vacuum surfaces, fine brush	↓	6400	.010	↓		.44		.44	.69
3000	Structural demolition									

02 82 Asbestos Remediation

02 82 13 – Asbestos Abatement

02 82 13.46 Decontamination of Asbestos Containment Area

		Crew	Daily Output	Labor-Hours	Unit	Material	2009 Bare Costs Labor	2009 Bare Costs Equipment	Total	Total Incl O&P
3100	Wood stud walls	A-9	2800	.023	S.F.		1.01		1.01	1.58
3500	Window manifolds, not incl. window replacement		4200	.015			.67		.67	1.05
3600	Plywood carpet protection		2000	.032			1.41		1.41	2.21
4000	Remove custom decontamination facility	A-10A	8	3	Ea.	15	133		148	225
4100	Remove portable decontamination facility	3 Asbe	12	2	"	12.75	88		100.75	152
5000	HEPA vacuum, shampoo carpeting	A-9	4800	.013	S.F.	.05	.59		.64	.98
9000	Final cleaning of protected surfaces	A-10A	8000	.003	"		.13		.13	.21

02 82 13.47 Asbestos Waste Pkg., Handling, and Disp.

		Crew	Daily Output	Labor-Hours	Unit	Material	Labor	Equipment	Total	Total Incl O&P
0010	**ASBESTOS WASTE PACKAGING, HANDLING, AND DISPOSAL**									
0100	Collect and bag bulk material, 3 C.F. bags, by hand	A-9	400	.160	Ea.	1.24	7.05		8.29	12.40
0200	Large production vacuum loader	A-12	880	.073		.80	3.21	.86	4.87	6.90
1000	Double bag and decontaminate	A-9	960	.067		1.24	2.94		4.18	5.95
2000	Containerize bagged material in drums, per 3 C.F. drum	"	800	.080		12	3.53		15.53	18.75
3000	Cart bags 50' to dumpster	2 Asbe	400	.040			1.76		1.76	2.76
5000	Disposal charges, not including haul, minimum				C.Y.				50	55
5020	Maximum				"				170	187
5100	Remove refrigerant from system	1 Plum	40	.200	Lb.		9.75		9.75	14.65
9000	For type C (supplied air) respirator equipment, add				%					10%

02 82 13.48 Asbestos Encapsulation With Sealants

		Crew	Daily Output	Labor-Hours	Unit	Material	Labor	Equipment	Total	Total Incl O&P
0010	**ASBESTOS ENCAPSULATION WITH SEALANTS**									
0100	Ceilings and walls, minimum	A-9	21000	.003	S.F.	.27	.13		.40	.51
0110	Maximum		10600	.006		.42	.27		.69	.88
0200	Columns and beams, minimum		13300	.005		.27	.21		.48	.63
0210	Maximum		5325	.012		.47	.53		1	1.35
0300	Pipes to 12" diameter including minor repairs, minimum		800	.080	L.F.	.37	3.53		3.90	5.95
0310	Maximum		400	.160	"	1.04	7.05		8.09	12.20

02 83 Lead Remediation

02 83 19 – Lead-Based Paint Remediation

02 83 19.23 Encapsulation of Lead-Based Paint

		Crew	Daily Output	Labor-Hours	Unit	Material	Labor	Equipment	Total	Total Incl O&P
0010	**ENCAPSULATION OF LEAD-BASED PAINT**									
0020	Interior, brushwork, trim, under 6"	1 Pord	240	.033	L.F.	2.46	1.17		3.63	4.47
0030	6" to 12" wide		180	.044		3.28	1.56		4.84	5.95
0040	Balustrades		300	.027		1.98	.94		2.92	3.59
0050	Pipe to 4" diameter		500	.016		1.19	.56		1.75	2.15
0060	To 8" diameter		375	.021		1.57	.75		2.32	2.86
0070	To 12" diameter		250	.032		2.36	1.13		3.49	4.29
0080	To 16" diameter		170	.047		3.47	1.66		5.13	6.30
0090	Cabinets, ornate design		200	.040	S.F.	2.97	1.41		4.38	5.40
0100	Simple design		250	.032	"	2.36	1.13		3.49	4.29
0110	Doors, 3' x 7', both sides, incl. frame & trim									
0120	Flush	1 Pord	6	1.333	Ea.	30.50	47		77.50	104
0130	French, 10-15 lite		3	2.667		6.05	94		100.05	148
0140	Panel		4	2		36.50	70.50		107	146
0150	Louvered		2.75	2.909		33.50	102		135.50	190
0160	Windows, per interior side, per 15 S.F.									
0170	1 to 6 lite	1 Pord	14	.571	Ea.	21	20		41	53
0180	7 to 10 lite		7.50	1.067		23	37.50		60.50	82
0190	12 lite		5.75	1.391		31	49		80	108
0200	Radiators		8	1		73.50	35		108.50	134

02 83 Lead Remediation

02 83 19 – Lead-Based Paint Remediation

02 83 19.23 Encapsulation of Lead-Based Paint

		Crew	Daily Output	Labor-Hours	Unit	Material	2009 Bare Costs Labor	Equipment	Total	Total Incl O&P
0210	Grilles, vents	1 Pord	275	.029	S.F.	2.15	1.02		3.17	3.90
0220	Walls, roller, drywall or plaster		1000	.008		.59	.28		.87	1.07
0230	With spunbonded reinforcing fabric		720	.011		.68	.39		1.07	1.34
0240	Wood		800	.010		.74	.35		1.09	1.34
0250	Ceilings, roller, drywall or plaster		900	.009		.68	.31		.99	1.22
0260	Wood		700	.011		.84	.40		1.24	1.52
0270	Exterior, brushwork, gutters and downspouts		300	.027	L.F.	1.98	.94		2.92	3.59
0280	Columns		400	.020	S.F.	1.47	.70		2.17	2.68
0290	Spray, siding		600	.013	"	.99	.47		1.46	1.79
0300	Miscellaneous									
0310	Electrical conduit, brushwork, to 2" diameter	1 Pord	500	.016	L.F.	1.19	.56		1.75	2.15
0320	Brick, block or concrete, spray		500	.016	S.F.	1.19	.56		1.75	2.15
0330	Steel, flat surfaces and tanks to 12"		500	.016		1.19	.56		1.75	2.15
0340	Beams, brushwork		400	.020		1.47	.70		2.17	2.68
0350	Trusses		400	.020		1.47	.70		2.17	2.68

02 83 19.26 Removal of Lead-Based Paint

		Crew	Daily Output	Labor-Hours	Unit	Material	2009 Bare Costs Labor	Equipment	Total	Total Incl O&P
0010	**REMOVAL OF LEAD-BASED PAINT** R028319-60									
0011	By chemicals, per application									
0050	Baseboard, to 6" wide	1 Pord	64	.125	L.F.	.66	4.40		5.06	7.35
0070	To 12" wide		32	.250	"	1.32	8.80		10.12	14.65
0200	Balustrades, one side		28	.286	S.F.	1.32	10.05		11.37	16.50
1400	Cabinets, simple design		32	.250		1.32	8.80		10.12	14.65
1420	Ornate design		25	.320		1.32	11.25		12.57	18.35
1600	Cornice, simple design		60	.133		1.32	4.69		6.01	8.50
1620	Ornate design		20	.400		5.05	14.10		19.15	26.50
2800	Doors, one side, flush		84	.095		1	3.35		4.35	6.10
2820	Two panel		80	.100		1.32	3.52		4.84	6.75
2840	Four panel		45	.178		1.32	6.25		7.57	10.85
2880	For trim, one side, add		64	.125	L.F.	.66	4.40		5.06	7.35
3000	Fence, picket, one side		30	.267	S.F.	1.32	9.40		10.72	15.50
3200	Grilles, one side, simple design		30	.267		1.32	9.40		10.72	15.50
3220	Ornate design		25	.320		1.32	11.25		12.57	18.35
4400	Pipes, to 4" diameter		90	.089	L.F.	1	3.13		4.13	5.80
4420	To 8" diameter		50	.160		2	5.65		7.65	10.65
4440	To 12" diameter		36	.222		3	7.80		10.80	15
4460	To 16" diameter		20	.400		4	14.10		18.10	25.50
4500	For hangers, add		40	.200	Ea.	2.54	7.05		9.59	13.35
4800	Siding		90	.089	S.F.	1.17	3.13		4.30	6
5000	Trusses, open		55	.145	SF Face	1.85	5.10		6.95	9.70
6200	Windows, one side only, double hung, 1/1 light, 24" x 48" high		4	2	Ea.	25.50	70.50		96	134
6220	30" x 60" high		3	2.667		34	94		128	179
6240	36" x 72" high		2.50	3.200		41	113		154	214
6280	40" x 80" high		2	4		51	141		192	267
6400	Colonial window, 6/6 light, 24" x 48" high		2	4		51	141		192	268
6420	30" x 60" high		1.50	5.333		68	188		256	355
6440	36" x 72" high		1	8		102	282		384	530
6480	40" x 80" high		1	8		102	282		384	530
6600	8/8 light, 24" x 48" high		2	4		51	141		192	268
6620	40" x 80" high		1	8		102	282		384	530
6800	12/12 light, 24" x 48" high		1	8		102	282		384	530
6820	40" x 80" high		.75	10.667		136	375		511	715
6840	Window frame & trim items, included in pricing above									

02 85 Mold Remediation

02 85 16 – Mold Remediation Preparation and Containment

02 85 16.50 Preparation of Mold Containment Area

		Crew	Daily Output	Labor-Hours	Unit	Material	2009 Bare Costs Labor	Equipment	Total	Total Incl O&P
0010	**PREPARATION OF MOLD CONTAINMENT AREA**									
0100	Pre-cleaning, HEPA vacuum and wet wipe, flat surfaces	A-9	12000	.005	S.F.	.02	.24		.26	.39
0300	Separation barrier, 2" x 4" @ 16", 1/2" plywood ea. side, 8' high	2 Carp	400	.040		1.25	1.60		2.85	3.86
0310	12' high		320	.050		1.40	2		3.40	4.64
0320	16' high		200	.080		1.50	3.20		4.70	6.60
0400	Personnel decontam. chamber, 2" x 4" @ 16", 3/4" ply ea. side		280	.057		2.50	2.28		4.78	6.30
0450	Waste decontam. chamber, 2" x 4" studs @ 16", 3/4" ply each side		360	.044		3	1.78		4.78	6.05
0500	Cover surfaces with polyethylene sheeting									
0501	Including glue and tape									
0550	Floors, each layer, 6 mil	A-9	8000	.008	S.F.	.09	.35		.44	.65
0551	4 mil		9000	.007		.05	.31		.36	.55
0560	Walls, each layer, 6 mil		6000	.011		.09	.47		.56	.84
0561	4 mil		7000	.009		.07	.40		.47	.71
0570	For heights above 12', add						20%			
0575	For heights above 20', add						30%			
0580	For fire retardant poly, add					100%				
0590	For large open areas, deduct					10%	20%			
0600	Seal floor penetrations with foam firestop to 36 sq in	2 Carp	200	.080	Ea.	1,275	3.20		1,278.20	1,400
0610	36 sq in to 72 sq in		125	.128		2,550	5.10		2,555.10	2,825
0615	72 sq in to 144 sq in		80	.200		5,125	8		5,133	5,625
0620	Wall penetrations, to 36 square inches		180	.089		1,275	3.55		1,278.55	1,400
0630	36 sq. in. to 72 sq. in.		100	.160		2,550	6.40		2,556.40	2,825
0640	72 Sq. in. to 144 sq. in.		60	.267		5,125	10.65		5,135.65	5,650
0800	Caulk seams with latex caulk	1 Carp	230	.035	L.F.	.15	1.39		1.54	2.32
0900	Set up neg. air machine, 1-2k C.F.M. /25 M.C.F. volume	1 Asbe	4.30	1.860	Ea.		82		82	129

02 85 33 – Removal and Disposal of Materials with Mold

02 85 33.50 Demolition in Mold Contaminated Area

		Crew	Daily Output	Labor-Hours	Unit	Material	2009 Bare Costs Labor	Equipment	Total	Total Incl O&P
0010	**DEMOLITION IN MOLD CONTAMINATED AREA**									
0200	Ceiling, including suspension system, plaster and lath	A-9	2100	.030	S.F.	.10	1.35		1.45	2.22
0210	Finished plaster, leaving wire lath		585	.109		.35	4.83		5.18	7.95
0220	Suspended acoustical tile		3500	.018		.06	.81		.87	1.34
0230	Concealed tile grid system		3000	.021		.07	.94		1.01	1.56
0240	Metal pan grid system		1500	.043		.14	1.88		2.02	3.10
0250	Gypsum board		2500	.026		.08	1.13		1.21	1.86
0255	Plywood		2500	.026		.08	1.13		1.21	1.86
0260	Lighting fixtures up to 2' x 4'		72	.889	Ea.	2.87	39.50		42.37	64.50
0400	Partitions, non load bearing									
0410	Plaster, lath, and studs	A-9	690	.093	S.F.	.87	4.10		4.97	7.35
0450	Gypsum board and studs		1390	.046		.15	2.03		2.18	3.35
0465	Carpet & pad		1390	.046		.15	2.03		2.18	3.35
0600	Pipe insulation, air cell type, up to 4" diameter pipe		900	.071	L.F.	.23	3.14		3.37	5.15
0610	4" to 8" diameter pipe		800	.080		.26	3.53		3.79	5.85
0620	10" to 12" diameter pipe		700	.091		.30	4.04		4.34	6.70
0630	14" to 16" diameter pipe		550	.116		.38	5.15		5.53	8.45
0650	Over 16" diameter pipe		650	.098	S.F.	.32	4.35		4.67	7.15
9000	For type C (supplied air) respirator equipment, add				%					10%

Division Notes

		CREW	DAILY OUTPUT	LABOR-HOURS	UNIT	2009 BARE COSTS				TOTAL INCL O&P
						MAT.	LABOR	EQUIP.	TOTAL	

Estimating Tips

General

- Carefully check all the plans and specifications. Concrete often appears on drawings other than structural drawings, including mechanical and electrical drawings for equipment pads. The cost of cutting and patching is often difficult to estimate. See Subdivision 03 81 for Concrete Cutting, Subdivision 02 41 19.16 for Cutout Demolition, Subdivision 03 05 05.10 for Concrete Demolition, and Subdivision 02 41 19.23 for Rubbish Handling (handling, loading and hauling of debris).
- Always obtain concrete prices from suppliers near the job site. A volume discount can often be negotiated, depending upon competition in the area. Remember to add for waste, particularly for slabs and footings on grade.

03 10 00 Concrete Forming and Accessories

- A primary cost for concrete construction is forming. Most jobs today are constructed with prefabricated forms. The selection of the forms best suited for the job and the total square feet of forms required for efficient concrete forming and placing are key elements in estimating concrete construction. Enough forms must be available for erection to make efficient use of the concrete placing equipment and crew.
- Concrete accessories for forming and placing depend upon the systems used. Study the plans and specifications to ensure that all special accessory requirements have been included in the cost estimate, such as anchor bolts, inserts, and hangers.
- Included within costs for forms-in-place are all necessary bracing and shoring.

03 20 00 Concrete Reinforcing

- Ascertain that the reinforcing steel supplier has included all accessories, cutting, bending, and an allowance for lapping, splicing, and waste. A good rule of thumb is 10% for lapping, splicing, and waste. Also, 10% waste should be allowed for welded wire fabric.
- The unit price items in the subdivision for Reinforcing In Place include the labor to install accessories such as beam and slab bolsters, high chairs, and bar ties and tie wire. The material cost for these accessories is not included; they may be obtained from the Accessories Division.

03 30 00 Cast-In-Place Concrete

- When estimating structural concrete, pay particular attention to requirements for concrete additives, curing methods, and surface treatments. Special consideration for climate, hot or cold, must be included in your estimate. Be sure to include requirements for concrete placing equipment, and concrete finishing.
- For accurate concrete estimating, the estimator must consider each of the following major components individually: forms, reinforcing steel, ready-mix concrete, placement of the concrete, and finishing of the top surface. For faster estimating, Subdivision 03 30 53.40 for Concrete-In-Place can be used; here, various items of concrete work are presented that include the costs of all five major components (unless specifically stated otherwise).

03 40 00 Precast Concrete
03 50 00 Cast Decks and Underlayment

- The cost of hauling precast concrete structural members is often an important factor. For this reason, it is important to get a quote from the nearest supplier. It may become economically feasible to set up precasting beds on the site if the hauling costs are prohibitive.

Reference Numbers

Reference numbers are shown in shaded boxes at the beginning of some major classifications. These numbers refer to related items in the Reference Section. The reference information may be an estimating procedure, an alternate pricing method, or technical information.

Note: Not all subdivisions listed here necessarily appear in this publication.

No part of this publication may be reproduced, stored in a retrieval system, or transmitted in any form or by any means without prior written permission of R.S. Means Company, Inc.

03 01 Maintenance of Concrete

03 01 30 – Maintenance of Cast-In-Place Concrete

03 01 30.62 Concrete Patching		Crew	Daily Output	Labor-Hours	Unit	Material	2009 Bare Costs Labor	Equipment	Total	Total Incl O&P
0010	**CONCRETE PATCHING**									
0100	Floors, 1/4" thick, small areas, regular grout	1 Cefi	170	.047	S.F.	1.07	1.80		2.87	3.81
0150	Epoxy grout	"	100	.080	"	6.85	3.06		9.91	12
2000	Walls, including chipping, cleaning and epoxy grout									
2100	1/4" deep	1 Cefi	65	.123	S.F.	7.50	4.71		12.21	15.15
2150	1/2" deep	↓	50	.160	↓	15.05	6.15		21.20	25.50
2200	3/4" deep	↓	40	.200	↓	22.50	7.65		30.15	36

03 05 Common Work Results for Concrete

03 05 05 – Selective Concrete Demolition

03 05 05.10 Selective Demolition, Concrete

		Crew	Daily Output	Labor-Hours	Unit	Material	2009 Bare Costs Labor	Equipment	Total	Total Incl O&P
0010	**SELECTIVE DEMOLITION, CONCRETE** R024119-10									
0012	Excludes saw cutting, torch cutting, loading or hauling									
0050	Break up into small pieces, minimum reinforcing	B-9	24	1.667	C.Y.		53.50	8	61.50	91.50
0060	Average reinforcing		16	2.500			80	12	92	137
0070	Maximum reinforcing	↓	8	5	↓		160	24	184	275
0150	Remove whole pieces, up to 2 tons per piece	E-18	36	1.111	Ea.		49.50	30	79.50	119
0160	2 – 5 tons per piece		30	1.333			59	36	95	142
0170	5 – 10 tons per piece		24	1.667			74	45	119	178
0180	10 – 15 tons per piece	↓	18	2.222			98.50	60	158.50	237
0250	Precast unit embedded in masonry, up to 1 CF	D-1	16	1			36.50		36.50	55
0260	1 – 2 CF		12	1.333			48.50		48.50	73.50
0270	2 – 5 CF		10	1.600			58		58	88
0280	5 – 10 CF	↓	8	2	↓		72.50		72.50	110

03 05 13 – Basic Concrete Materials

03 05 13.20 Concrete Admixtures and Surface Treatments

		Crew	Daily Output	Labor-Hours	Unit	Material	2009 Bare Costs Labor	Equipment	Total	Total Incl O&P
0010	**CONCRETE ADMIXTURES AND SURFACE TREATMENTS**									
0040	Abrasives, aluminum oxide, over 20 tons				Lb.	1.63			1.63	1.79
0050	1 to 20 tons					1.68			1.68	1.85
0070	Under 1 ton					1.75			1.75	1.93
0100	Silicon carbide, black, over 20 tons					2.23			2.23	2.45
0110	1 to 20 tons					2.34			2.34	2.58
0120	Under 1 ton					2.44			2.44	2.68
0500	Carbon black, liquid, 2 to 8 lbs. per bag of cement					5.90			5.90	6.45
0600	Colors, integral, 2 to 10 lb. per bag of cement, minimum					2.41			2.41	2.65
0610	Average					3.40			3.40	3.74
0620	Maximum				↓	4.80			4.80	5.30
0920	Dustproofing compound, 250 SF/gal, 5 gallon pail				Gal.	6.80			6.80	7.45
1010	Epoxy based, 125 SF/Gal, 5 gallon pail				"	45.50			45.50	50
1100	Hardeners, metallic, 55 lb. bags, natural (grey)				Lb.	.93			.93	1.02
1200	Colors					1.32			1.32	1.45
1300	Non-metallic, 55 lb. bags, natural grey					.37			.37	.41
1320	Colors				↓	.66			.66	.72
1600	Sealer, hardener and dustproofer, epoxy-based, 125 SF/gal, 5 gallon, min				Gal.	45.50			45.50	50
1620	5 gallon pail, maximum					50			50	55
1630	Sealer, solvent-based, 250 SF/gal, 55 gallon drum					15.25			15.25	16.75
1640	5 gallon pail					17.40			17.40	19.15
1650	Sealer, water based, 350 S.F., 55 gallon drum					18.30			18.30	20
1660	5 gallon pail				↓	21			21	23
2000	Waterproofing, integral 1 lb. per bag of cement				Lb.	1.38			1.38	1.52

03 05 Common Work Results for Concrete

03 05 13 – Basic Concrete Materials

03 05 13.20 Concrete Admixtures and Surface Treatments

		Crew	Daily Output	Labor-Hours	Unit	Material	2009 Bare Costs Labor	2009 Bare Costs Equipment	Total	Total Incl O&P
2100	Powdered metallic, 40 lbs. per 100 S.F., minimum				Lb.	1.63			1.63	1.79
2120	Maximum					2.28			2.28	2.51

03 15 Concrete Accessories

03 15 05 – Concrete Forming Accessories

03 15 05.85 Stair Tread Inserts

			Crew	Daily Output	Labor-Hours	Unit	Material	Labor	Equipment	Total	Total Incl O&P
0010	**STAIR TREAD INSERTS**										
0015	Cast iron, abrasive, 3" wide	G	1 Carp	90	.089	L.F.	10.65	3.55		14.20	17.20
0020	4" wide	G		80	.100		14.15	4		18.15	22
0040	6" wide	G		75	.107		21.50	4.26		25.76	30
0050	9" wide	G		70	.114		32	4.57		36.57	42
0100	12" wide	G		65	.123		42.50	4.92		47.42	54.50
0300	Cast aluminum, compared to cast iron, deduct						10%				
0500	Extruded aluminum safety tread, 3" wide	G	1 Carp	75	.107		12.10	4.26		16.36	19.90
0550	4" wide	G		75	.107		16.15	4.26		20.41	24.50
0600	6" wide	G		75	.107		24	4.26		28.26	33
0650	9" wide to resurface stairs	G		70	.114		36.50	4.57		41.07	47
1700	Cement fill for pan-type metal treads, plain		1 Cefi	115	.070	S.F.	1.87	2.66		4.53	5.95
1750	Non-slip		"	100	.080	"	2.06	3.06		5.12	6.75

03 30 Cast-In-Place Concrete

03 30 53 – Miscellaneous Cast-In-Place Concrete

03 30 53.40 Concrete In Place

		Crew	Daily Output	Labor-Hours	Unit	Material	Labor	Equipment	Total	Total Incl O&P
0010	**CONCRETE IN PLACE**									
0020	Including forms (4 uses), reinforcing steel, concrete, placement,									
0050	and finishing unless otherwise indicated									
0700	Columns, square, 12" x 12", minimum reinforcing	C-14A	11.96	16.722	C.Y.	400	670	64	1,134	1,550
0800	16" x 16", minimum reinforcing		16.22	12.330		315	495	47	857	1,175
0900	24" x 24", minimum reinforcing		23.66	8.453		275	340	32.50	647.50	865
3540	Equipment pad, 3' x 3' x 6" thick	C-14H	45	1.067	Ea.	58	42	.58	100.58	130
3550	4' x 4' x 6" thick		30	1.600		84	63	.86	147.86	191
3560	5' x 5' x 8" thick		18	2.667		140	105	1.44	246.44	320
3570	6' x 6' x 8" thick		14	3.429		188	135	1.85	324.85	420
3580	8' x 8' x 10" thick		8	6		395	236	3.24	634.24	805
3590	10' x 10' x 12" thick		5	9.600		655	380	5.20	1,040.20	1,300
3800	Footings, spread under 1 C.Y.	C-14C	28	4	C.Y.	209	153	.92	362.92	470
3825	1 C.Y to 5 C.Y.		43	2.605		221	99.50	.60	321.10	400
3850	Over 5 C.Y.		75	1.493		198	57	.34	255.34	305
3900	Footings, strip, 18" x 9", unreinforced	C-14L	40	2.400		120	89.50	.65	210.15	271
3920	18" x 9", reinforced	C-14C	35	3.200		157	122	.74	279.74	365
3925	20" x 10", unreinforced	C-14L	45	2.133		117	79.50	.58	197.08	252
3930	20" x 10", reinforced	C-14C	40	2.800		147	107	.64	254.64	330
3935	24" x 12", unreinforced	C-14L	55	1.745		116	65	.47	181.47	229
3940	24" x 12", reinforced	C-14C	48	2.333		147	89.50	.54	237.04	300
3945	36" x 12", unreinforced	C-14L	70	1.371		113	51	.37	164.37	203
3950	36" x 12", reinforced	C-14C	60	1.867		141	71.50	.43	212.93	266
4000	Foundation mat, under 10 C.Y.		38.67	2.896		255	111	.67	366.67	455
4050	Over 20 C.Y.		56.40	1.986		218	76	.46	294.46	360
4200	Wall, free-standing, 8" thick, 8' high	C-14D	45.83	4.364		195	173	16.65	384.65	500

03 30 Cast-In-Place Concrete

03 30 53 – Miscellaneous Cast-In-Place Concrete

03 30 53.40 Concrete In Place

		Crew	Daily Output	Labor-Hours	Unit	Material	2009 Bare Costs Labor	Equipment	Total	Total Incl O&P
4260	12" thick, 8' high	C-14D	64.32	3.109	C.Y.	175	124	11.90	310.90	400
4300	15" thick, 8' high	↓	80.02	2.499	↓	166	99.50	9.55	275.05	345
4520	Handicap access ramp, railing both sides, 3' wide	C-14H	14.58	3.292	L.F.	300	130	1.78	431.78	535
4525	5' wide		12.22	3.928		310	155	2.12	467.12	580
4530	With 6" curb and rails both sides, 3' wide		8.55	5.614		310	221	3.03	534.03	690
4535	5' wide	↓	7.31	6.566	↓	315	259	3.55	577.55	750
4650	Slab on grade, not including finish, 4" thick	C-14E	60.75	1.449	C.Y.	127	57	.43	184.43	230
4700	6" thick	"	92	.957	"	121	37.50	.29	158.79	193
4751	Slab on grade, incl. troweled finish, not incl. forms									
4760	or reinforcing, over 10,000 S.F., 4" thick	C-14F	3425	.021	S.F.	1.35	.76	.01	2.12	2.63
4820	6" thick		3350	.021		1.97	.78	.01	2.76	3.34
4840	8" thick		3184	.023		2.70	.82	.01	3.53	4.19
4900	12" thick		2734	.026		4.04	.96	.01	5.01	5.90
4950	15" thick	↓	2505	.029	↓	5.10	1.04	.01	6.15	7.15
6800	Stairs, not including safety treads, free standing, 3'-6" wide	C-14H	83	.578	LF Nose	6.20	23	.31	29.51	42.50
6850	Cast on ground		125	.384	"	5	15.10	.21	20.31	29
7000	Stair landings, free standing		200	.240	S.F.	5.25	9.45	.13	14.83	20.50
7050	Cast on ground	↓	475	.101	"	4	3.98	.05	8.03	10.65

03 31 Structural Concrete

03 31 05 – Normal Weight Structural Concrete

03 31 05.70 Placing Concrete

		Crew	Daily Output	Labor-Hours	Unit	Material	2009 Bare Costs Labor	Equipment	Total	Total Incl O&P
0010	**PLACING CONCRETE**									
0020	Includes labor and equipment to place and vibrate									
0050	Beams, elevated, small beams, pumped	C-20	60	1.067	C.Y.		36	13.15	49.15	70
0400	Columns, square or round, 12" thick, pumped		60	1.067			36	13.15	49.15	70
0600	18" thick, pumped		90	.711			24	8.80	32.80	46.50
0800	24" thick, pumped	↓	92	.696			23.50	8.60	32.10	45.50
4900	Walls, 8" thick, direct chute	C-6	90	.533			17.65	.58	18.23	27.50
4950	Pumped	C-20	100	.640			21.50	7.90	29.40	41.50
5050	12" thick, direct chute	C-6	100	.480			15.85	.52	16.37	25
5100	Pumped	C-20	110	.582			19.75	7.20	26.95	38
5300	15" thick, direct chute	C-6	105	.457			15.10	.49	15.59	23.50
5350	Pumped	C-20	120	.533	↓		18.10	6.60	24.70	35
5600	Wheeled concrete dumping, add to placing costs above									
5610	Walking cart, 50' haul, add	C-18	32	.281	C.Y.		8.95	1.72	10.67	15.80
5620	150' haul, add		24	.375			11.95	2.30	14.25	21
5700	250' haul, add	↓	18	.500			15.90	3.07	18.97	28
5800	Riding cart, 50' haul, add	C-19	80	.113			3.58	1.08	4.66	6.75
5810	150' haul, add		60	.150			4.77	1.44	6.21	9
5900	250' haul, add	↓	45	.200	↓		6.35	1.92	8.27	11.95

03 35 Concrete Finishing

03 35 29 – Tooled Concrete Finishing

03 35 29.30 Finishing Floors

		Crew	Daily Output	Labor-Hours	Unit	Material	2009 Bare Costs Labor	2009 Bare Costs Equipment	Total	Total Incl O&P
0010	**FINISHING FLOORS**									
0020	Manual screed finish	C-10	4800	.005	S.F.		.18		.18	.27
0100	Manual screed and bull float		4000	.006			.22		.22	.32
0125	Manual screed, bull float, manual float		2000	.012			.43		.43	.64
0150	Manual screed, bull float, manual float & broom finish		1850	.013			.47		.47	.70
0200	Manual screed, bull float, manual float, manual steel trowel		1265	.019			.68		.68	1.02
0250	Manual screed, bull float, machine float & trowel (walk-behind)	C-10C	1715	.014			.50	.02	.52	.78
0300	Power screed, bull float, machine float & trowel (walk-behind)	C-10D	2400	.010			.36	.05	.41	.59
0350	Power screed, bull float, machine float & trowel (ride-on)	C-10E	4000	.006			.22	.06	.28	.39
0400	Integral topping and finish, using 1:1:2 mix, 3/16" thick	C-10B	1000	.040		.10	1.37	.24	1.71	2.44
0450	1/2" thick		950	.042		.26	1.44	.25	1.95	2.74
0500	3/4" thick		850	.047		.39	1.61	.28	2.28	3.18
0600	1" thick		750	.053		.52	1.83	.32	2.67	3.68
0800	Granolithic topping, laid after, 1:1:1-1/2 mix, 1/2" thick		590	.068		.29	2.32	.41	3.02	4.28
0820	3/4" thick		580	.069		.43	2.36	.41	3.20	4.50
0850	1" thick		575	.070		.57	2.38	.42	3.37	4.70
0950	2" thick		500	.080		1.15	2.74	.48	4.37	5.95
1200	Heavy duty, 1:1:2, 3/4" thick, preshrunk, gray, 20 MSF		320	.125		.73	4.29	.75	5.77	8.15
1300	100 MSF		380	.105		.39	3.61	.63	4.63	6.55
1600	Exposed local aggregate finish, minimum	1 Cefi	625	.013		.22	.49		.71	.97
1650	Maximum		465	.017		.66	.66		1.32	1.69
1800	Floor abrasives, .25 psf, aluminum oxide		850	.009		.44	.36		.80	1.01
1850	Silicon carbide		850	.009		.61	.36		.97	1.20
2000	Floor hardeners, metallic, light service, .50 psf, add		850	.009		.51	.36		.87	1.09
2050	Medium service, .75 psf		750	.011		.76	.41		1.17	1.43
2100	Heavy service, 1.0 psf		650	.012		1.01	.47		1.48	1.80
2150	Extra heavy, 1.5 psf		575	.014		1.52	.53		2.05	2.45
2300	Non-metallic, light service, .50 psf		850	.009		.23	.36		.59	.78
2350	Medium service, .75 psf		750	.011		.34	.41		.75	.98
2400	Heavy service, 1.00 psf		650	.012		.45	.47		.92	1.19
2450	Extra heavy, 1.50 psf		575	.014		.68	.53		1.21	1.53
2800	Trap rock wearing surface for monolithic floors									
2810	2.0 psf	C-10B	1250	.032	S.F.	.03	1.10	.19	1.32	1.90
3000	Floor coloring, dusted on, minimum (0.6 psf), add to above	1 Cefi	1300	.006		.43	.24		.67	.81
3050	Maximum (1.0 psf), add to above	"	625	.013		.71	.49		1.20	1.50
3100	Colored powder only				Lb.	.71			.71	.78
3600	1/2" topping using 0.6 psf powdered color	C-10B	590	.068	S.F.	4.86	2.32	.41	7.59	9.30
3650	1/2" topping using 1.0 psf powdered color	"	590	.068		5.15	2.32	.41	7.88	9.60
3800	Dustproofing, solvent-based, 1 coat	1 Cefi	1900	.004		.17	.16		.33	.43
3850	2 coats		1300	.006		.61	.24		.85	1.02
4000	Epoxy-based, 1 coat		1500	.005		.15	.20		.35	.46
4050	2 coats		1500	.005		.29	.20		.49	.62
4400	Stair finish, float		275	.029			1.11		1.11	1.63
4500	Steel trowel finish		200	.040			1.53		1.53	2.24
4600	Silicon carbide finish, .25 psf		150	.053		.44	2.04		2.48	3.47

03 35 29.35 Control Joints, Saw Cut

		Crew	Daily Output	Labor-Hours	Unit	Material	Labor	Equipment	Total	Total Incl O&P
0010	**CONTROL JOINTS, SAW CUT**									
0100	Sawcut in green concrete									
0120	1" depth	C-27	2000	.008	L.F.	.07	.31	.07	.45	.60
0140	1-1/2" depth		1800	.009		.10	.34	.08	.52	.70
0160	2" depth		1600	.010		.13	.38	.09	.60	.81
0200	Clean out control joint of debris	C-28	6000	.001			.05		.05	.07

03 35 Concrete Finishing

03 35 29 – Tooled Concrete Finishing

03 35 29.35 Control Joints, Saw Cut

		Crew	Daily Output	Labor-Hours	Unit	Material	2009 Bare Costs Labor	Equipment	Total	Total Incl O&P
0300	Joint sealant									
0320	Backer rod, polyethylene, 1/4" diameter	1 Cefi	460	.017	L.F.	.04	.67		.71	1.02
0340	Sealant, polyurethane									
0360	1/4" x 1/4" (308 LF/Gal)	1 Cefi	270	.030	L.F.	.18	1.13		1.31	1.85
0380	1/4" x 1/2" (154 LF/Gal)	"	255	.031	"	.35	1.20		1.55	2.15

03 35 29.60 Finishing Walls

		Crew	Daily Output	Labor-Hours	Unit	Material	Labor	Equipment	Total	Total Incl O&P
0010	**FINISHING WALLS**									
0020	Break ties and patch voids	1 Cefi	540	.015	S.F.	.03	.57		.60	.86
0050	Burlap rub with grout		450	.018		.03	.68		.71	1.03
0100	Carborundum rub, dry		270	.030			1.13		1.13	1.66
0150	Wet rub		175	.046			1.75		1.75	2.56
0300	Bush hammer, green concrete	B-39	1000	.048			1.59	.19	1.78	2.66
0350	Cured concrete	"	650	.074			2.45	.29	2.74	4.10
0600	Float finish, 1/16" thick	1 Cefi	300	.027		.27	1.02		1.29	1.78
0700	Sandblast, light penetration	E-11	1100	.029		.70	1.04	.18	1.92	2.70
0750	Heavy penetration	"	375	.085		1.41	3.04	.52	4.97	7.20
0850	Grind form fins flush	1 Clab	700	.011	L.F.		.36		.36	.56

03 39 Concrete Curing

03 39 13 – Water Concrete Curing

03 39 13.50 Water Curing

		Crew	Daily Output	Labor-Hours	Unit	Material	Labor	Equipment	Total	Total Incl O&P
0010	**WATER CURING**									
0015	With burlap, 4 uses assumed, 7.5 oz.	2 Clab	55	.291	C.S.F.	8.65	9.20		17.85	24
0100	10 oz.	"	55	.291	"	15.55	9.20		24.75	31.50
0400	Curing blankets, 1" to 2" thick, buy, minimum				S.F.	.32			.32	.35
0450	Maximum				"	.48			.48	.53

03 39 23 – Membrane Concrete Curing

03 39 23.13 Chemical Compound Membrane Concrete Curing

		Crew	Daily Output	Labor-Hours	Unit	Material	Labor	Equipment	Total	Total Incl O&P
0010	**CHEMICAL COMPOUND MEMBRANE CONCRETE CURING**									
0300	Sprayed membrane curing compound	2 Clab	95	.168	C.S.F.	5.25	5.30		10.55	14
0700	Curing compound, solvent based, 400 S.F./gal, 55 gallon lots				Gal.	15.25			15.25	16.75
0720	5 gallon lots					17.40			17.40	19.15
0800	Curing compound, water based, 250 S.F./gal, 55 gallon lots					18.30			18.30	20
0820	5 gallon lots					21			21	23

03 48 Precast Concrete Specialties

03 48 43 – Precast Concrete Trim

03 48 43.40 Precast Lintels

		Crew	Daily Output	Labor-Hours	Unit	Material	Labor	Equipment	Total	Total Incl O&P
0010	**PRECAST LINTELS**									
0800	Precast concrete, 4" wide, 8" high, to 5' long	D-10	28	1.143	Ea.	25.50	45	21.50	92	120
0850	5'-12' long		24	1.333		82.50	52.50	25	160	198
1000	6" wide, 8" high, to 5' long		26	1.231		39.50	48.50	23	111	143
1050	5'-12' long		22	1.455		99.50	57.50	27.50	184.50	227
1200	8" wide, 8" high, to 5' long		24	1.333		47.50	52.50	25	125	160
1250	5'-12' long		20	1.600		135	63	30	228	278
1400	10" wide, 8" high U-Shape, to 14' long		18	1.778		209	70	33.50	312.50	375
1450	12" wide, 8" high U-Shape, to 19' long		16	2		275	79	37.50	391.50	465

03 48 Precast Concrete Specialties

03 48 43 – Precast Concrete Trim

03 48 43.90 Precast Window Sills

		Crew	Daily Output	Labor-Hours	Unit	Material	2009 Bare Costs Labor	Equipment	Total	Total Incl O&P
0010	**PRECAST WINDOW SILLS**									
0600	Precast concrete, 4" tapers to 3", 9" wide	D-1	70	.229	L.F.	11.75	8.30		20.05	25.50
0650	11" wide		60	.267		15.60	9.70		25.30	32
0700	13" wide, 3 1/2" tapers to 2 1/2", 12" wall	↓	50	.320	↓	15	11.65		26.65	34

03 54 Cast Underlayment

03 54 16 – Hydraulic Cement Underlayment

03 54 16.50 Cement Underlayment

		Crew	Daily Output	Labor-Hours	Unit	Material	2009 Bare Costs Labor	Equipment	Total	Total Incl O&P
0010	**CEMENT UNDERLAYMENT**									
2510	Underlayment, P.C based self-leveling, 4100 psi, pumped, 1/4"	C-8	20000	.003	S.F.	1.44	.10	.04	1.58	1.77
2520	1/2"		19000	.003		2.87	.10	.04	3.01	3.36
2530	3/4"		18000	.003		4.31	.11	.04	4.46	4.96
2540	1"		17000	.003		5.75	.12	.04	5.91	6.55
2550	1-1/2"	↓	15000	.004		8.60	.13	.05	8.78	9.75
2560	Hand mix, 1/2"	C-18	4000	.002		2.87	.07	.01	2.95	3.29
2610	Topping, P.C. based self-level/dry 6100 psi, pumped, 1/4"	C-8	20000	.003		2.21	.10	.04	2.35	2.63
2620	1/2"		19000	.003		4.43	.10	.04	4.57	5.05
2630	3/4"		18000	.003		6.65	.11	.04	6.80	7.50
2660	1"		17000	.003		8.85	.12	.04	9.01	10
2670	1-1/2"	↓	15000	.004		13.30	.13	.05	13.48	14.85
2680	Hand mix, 1/2"	C-18	4000	.002	↓	4.43	.07	.01	4.51	5

03 81 Concrete Cutting

03 81 13 – Flat Concrete Sawing

03 81 13.50 Concrete Floor/Slab Cutting

		Crew	Daily Output	Labor-Hours	Unit	Material	2009 Bare Costs Labor	Equipment	Total	Total Incl O&P
0010	**CONCRETE FLOOR/SLAB CUTTING**									
0400	Concrete slabs, mesh reinforcing, up to 3" deep	B-89	980	.016	L.F.	.49	.57	.41	1.47	1.86
0420	Each additional inch of depth	"	1600	.010	"	.16	.35	.25	.76	.99

03 81 16 – Track Mounted Concrete Wall Sawing

03 81 16.50 Concrete Wall Cutting

		Crew	Daily Output	Labor-Hours	Unit	Material	2009 Bare Costs Labor	Equipment	Total	Total Incl O&P
0010	**CONCRETE WALL CUTTING**									
0800	Concrete walls, hydraulic saw, plain, per inch of depth	B-89B	250	.064	L.F.	.44	2.24	2.74	5.42	6.90
0820	Rod reinforcing, per inch of depth	"	150	.107	"	.62	3.73	4.56	8.91	11.35

03 82 Concrete Boring

03 82 13 – Concrete Core Drilling

03 82 13.10 Core Drilling

		Crew	Daily Output	Labor-Hours	Unit	Material	2009 Bare Costs Labor	Equipment	Total	Total Incl O&P
0010	**CORE DRILLING**									
0020	Reinf. conc slab, up to 6" thick, incl. bit, layout & set up									
0100	1" diameter core	B-89A	17	.941	Ea.	3.45	34	6.50	43.95	64
0150	Each added inch thick in same hole, add		1440	.011		.63	.40	.08	1.11	1.39
0300	3" diameter core		16	1		7.40	36	6.90	50.30	72
0350	Each added inch thick in same hole, add		720	.022		1.38	.81	.15	2.34	2.94
0500	4" diameter core		15	1.067		7.40	38.50	7.35	53.25	76.50
0550	Each added inch thick in same hole, add		480	.033		1.81	1.21	.23	3.25	4.11
0700	6" diameter core	↓	14	1.143	↓	12.20	41.50	7.85	61.55	86

03 82 Concrete Boring

03 82 13 – Concrete Core Drilling

03 82 13.10 Core Drilling

		Crew	Daily Output	Labor-Hours	Unit	Material	2009 Bare Costs Labor	Equipment	Total	Total Incl O&P
0750	Each added inch thick in same hole, add	B-89A	360	.044	Ea.	2.17	1.61	.31	4.09	5.20
0900	8" diameter core		13	1.231		16.70	44.50	8.50	69.70	96.50
0950	Each added inch thick in same hole, add		288	.056		2.88	2.01	.38	5.27	6.70
1100	10" diameter core		12	1.333		22	48.50	9.20	79.70	109
1150	Each added inch thick in same hole, add		240	.067		3.62	2.42	.46	6.50	8.25
1300	12" diameter core		11	1.455		27	52.50	10	89.50	123
1350	Each added inch thick in same hole, add		206	.078		4.35	2.81	.54	7.70	9.75
1500	14" diameter core		10	1.600		33	58	11	102	138
1550	Each added inch thick in same hole, add		180	.089		5.85	3.22	.61	9.68	12.10
1700	18" diameter core		9	1.778		42.50	64.50	12.25	119.25	160
1750	Each added inch thick in same hole, add		144	.111		7.60	4.03	.77	12.40	15.45
1760	For horizontal holes, add to above						20%	20%		
1770	Prestressed hollow core plank, 8" thick									
1780	1" diameter core	B-89A	17.50	.914	Ea.	2.30	33	6.30	41.60	61
1790	Each added inch thick in same hole, add		3840	.004		.46	.15	.03	.64	.77
1800	3" diameter core		17	.941		4.81	34	6.50	45.31	65.50
1810	Each added inch thick in same hole, add		1920	.008		.83	.30	.06	1.19	1.44
1820	4" diameter core		16.50	.970		6.60	35	6.70	48.30	69
1830	Each added inch thick in same hole, add		1280	.013		1.21	.45	.09	1.75	2.12
1840	6" diameter core		15.50	1.032		8.10	37.50	7.10	52.70	74.50
1850	Each added inch thick in same hole, add		960	.017		1.38	.60	.11	2.09	2.59
1860	8" diameter core		15	1.067		11.50	38.50	7.35	57.35	81
1870	Each added inch thick in same hole, add		768	.021		1.93	.75	.14	2.82	3.45
1880	10" diameter core		14	1.143		15.25	41.50	7.85	64.60	89.50
1890	Each added inch thick in same hole, add		640	.025		2.01	.91	.17	3.09	3.80
1900	12" diameter core		13.50	1.185		17.85	43	8.15	69	95
1910	Each added inch thick in same hole, add		548	.029		3.22	1.06	.20	4.48	5.40
1999	Drilling, core, minimum labor/equipment charge		4	4	Job		145	27.50	172.50	256
3010	Bits for core drill, diamond, premium, 1" diameter				Ea.	117			117	129
3020	3" diameter					288			288	315
3040	4" diameter					320			320	350
3050	6" diameter					510			510	565
3080	8" diameter					705			705	775
3120	12" diameter					1,050			1,050	1,150
3180	18" diameter					1,900			1,900	2,075
3240	24" diameter					2,600			2,600	2,875

03 82 16 – Concrete Drilling

03 82 16.10 Concrete Impact Drilling

		Crew	Daily Output	Labor-Hours	Unit	Material	2009 Bare Costs Labor	Equipment	Total	Total Incl O&P
0010	**CONCRETE IMPACT DRILLING**									
0050	Up to 4" deep in conc/brick floor/wall, incl. bit & layout, no anchor									
0100	Holes, 1/4" diameter	1 Carp	75	.107	Ea.	.06	4.26		4.32	6.65
0150	For each additional inch of depth, add		430	.019		.01	.74		.75	1.17
0200	3/8" diameter		63	.127		.05	5.05		5.10	7.90
0250	For each additional inch of depth, add		340	.024		.01	.94		.95	1.47
0300	1/2" diameter		50	.160		.06	6.40		6.46	9.95
0350	For each additional inch of depth, add		250	.032		.01	1.28		1.29	2
0400	5/8" diameter		48	.167		.08	6.65		6.73	10.45
0450	For each additional inch of depth, add		240	.033		.02	1.33		1.35	2.08
0500	3/4" diameter		45	.178		.10	7.10		7.20	11.10
0550	For each additional inch of depth, add		220	.036		.03	1.45		1.48	2.28
0600	7/8" diameter		43	.186		.12	7.45		7.57	11.70
0650	For each additional inch of depth, add		210	.038		.03	1.52		1.55	2.39

03 82 Concrete Boring

03 82 16 – Concrete Drilling

03 82 16.10 Concrete Impact Drilling	Crew	Daily Output	Labor-Hours	Unit	Material	2009 Bare Costs Labor	Equipment	Total	Total Incl O&P
0700 1" diameter	1 Carp	40	.200	Ea.	.14	8		8.14	12.55
0750 For each additional inch of depth, add		190	.042		.04	1.68		1.72	2.65
0800 1-1/4" diameter		38	.211		.20	8.40		8.60	13.25
0850 For each additional inch of depth, add		180	.044		.05	1.78		1.83	2.81
0900 1-1/2" diameter		35	.229		.29	9.15		9.44	14.45
0950 For each additional inch of depth, add		165	.048		.07	1.94		2.01	3.08
1000 For ceiling installations, add						40%			

Division Notes

	CREW	DAILY OUTPUT	LABOR-HOURS	UNIT	2009 BARE COSTS				TOTAL INCL O&P
					MAT.	LABOR	EQUIP.	TOTAL	

Estimating Tips

04 05 00 Common Work Results for Masonry

- The terms *mortar* and *grout* are often used interchangeably, and incorrectly. Mortar is used to bed masonry units, seal the entry of air and moisture, provide architectural appearance, and allow for size variations in the units. Grout is used primarily in reinforced masonry construction and is used to bond the masonry to the reinforcing steel. Common mortar types are M(2500 psi), S(1800 psi), N(750 psi), and O(350 psi), and conform to ASTM C270. Grout is either fine or coarse and conforms to ASTM C476, and in-place strengths generally exceed 2500 psi. Mortar and grout are different components of masonry construction and are placed by entirely different methods. An estimator should be aware of their unique uses and costs.
- Waste, specifically the loss/droppings of mortar and the breakage of brick and block, is included in all masonry assemblies in this division. A factor of 25% is added for mortar and 3% for brick and concrete masonry units.
- Scaffolding or staging is not included in any of the Division 4 costs. Refer to Subdivision 01 54 23 for scaffolding and staging costs.

04 20 00 Unit Masonry

- The most common types of unit masonry are brick and concrete masonry. The major classifications of brick are building brick (ASTM C62), facing brick (ASTM C216), glazed brick, fire brick, and pavers. Many varieties of texture and appearance can exist within these classifications, and the estimator would be wise to check local custom and availability within the project area. For repair and remodeling jobs, matching the existing brick may be the most important criteria.
- Brick and concrete block are priced by the piece and then converted into a price per square foot of wall. Openings less than two square feet are generally ignored by the estimator because any savings in units used is offset by the cutting and trimming required.
- It is often difficult and expensive to find and purchase small lots of historic brick. Costs can vary widely. Many design issues affect costs, selection of mortar mix, and repairs or replacement of masonry materials. Cleaning techniques must be reflected in the estimate.
- All masonry walls, whether interior or exterior, require bracing. The cost of bracing walls during construction should be included by the estimator, and this bracing must remain in place until permanent bracing is complete. Permanent bracing of masonry walls is accomplished by masonry itself, in the form of pilasters or abutting wall corners, or by anchoring the walls to the structural frame. Accessories in the form of anchors, anchor slots, and ties are used, but their supply and installation can be by different trades. For instance, anchor slots on spandrel beams and columns are supplied and welded in place by the steel fabricator, but the ties from the slots into the masonry are installed by the bricklayer. Regardless of the installation method, the estimator must be certain that these accessories are accounted for in pricing.

Reference Numbers

Reference numbers are shown in shaded boxes at the beginning of some major classifications. These numbers refer to related items in the Reference Section. The reference information may be an estimating procedure, an alternate pricing method, or technical information.

Note: Not all subdivisions listed here necessarily appear in this publication.

No part of this publication may be reproduced, stored in a retrieval system, or transmitted in any form or by any means without prior written permission of R.S. Means Company, Inc.

04 01 Maintenance of Masonry

04 01 20 – Maintenance of Unit Masonry

04 01 20.10 Patching Masonry		Crew	Daily Output	Labor-Hours	Unit	Material	2009 Bare Costs Labor	2009 Bare Costs Equipment	Total	Total Incl O&P
0010	**PATCHING MASONRY**									
0500	Concrete patching, includes chipping, cleaning and epoxy									
0520	Minimum	1 Cefi	65	.123	S.F.	5.45	4.71		10.16	12.90
0540	Average		50	.160		8.20	6.15		14.35	17.95
0580	Maximum	↓	40	.200	↓	10.95	7.65		18.60	23.50

04 01 20.20 Pointing Masonry		Crew	Daily Output	Labor-Hours	Unit	Material	Labor	Equipment	Total	Total Incl O&P
0010	**POINTING MASONRY**									
0300	Cut and repoint brick, hard mortar, running bond	1 Bric	80	.100	S.F.	.53	4.05		4.58	6.75
0320	Common bond		77	.104		.53	4.21		4.74	7
0360	Flemish bond		70	.114		.56	4.63		5.19	7.60
0400	English bond		65	.123		.56	4.98		5.54	8.15
0600	Soft old mortar, running bond		100	.080		.53	3.24		3.77	5.50
0620	Common bond		96	.083		.53	3.37		3.90	5.70
0640	Flemish bond		90	.089		.56	3.60		4.16	6.05
0680	English bond		82	.098	↓	.56	3.95		4.51	6.60
0700	Stonework, hard mortar		140	.057	L.F.	.70	2.31		3.01	4.28
0720	Soft old mortar		160	.050	"	.70	2.03		2.73	3.84
1000	Repoint, mask and grout method, running bond		95	.084	S.F.	.70	3.41		4.11	5.90
1020	Common bond		90	.089		.70	3.60		4.30	6.20
1040	Flemish bond		86	.093		.74	3.77		4.51	6.50
1060	English bond		77	.104		.74	4.21		4.95	7.20
2000	Scrub coat, sand grout on walls, minimum		120	.067		2.84	2.70		5.54	7.20
2020	Maximum	↓	98	.082	↓	3.94	3.31		7.25	9.35

04 01 20.40 Sawing Masonry		Crew	Daily Output	Labor-Hours	Unit	Material	Labor	Equipment	Total	Total Incl O&P
0010	**SAWING MASONRY**									
0050	Brick or block by hand, per inch depth	A-1	125	.064	L.F.		2.02	.51	2.53	3.70

04 01 20.50 Toothing Masonry		Crew	Daily Output	Labor-Hours	Unit	Material	Labor	Equipment	Total	Total Incl O&P
0010	**TOOTHING MASONRY**									
0500	Brickwork, soft old mortar	1 Clab	40	.200	V.L.F.		6.30		6.30	9.80
0520	Hard mortar		30	.267			8.45		8.45	13.05
0700	Blockwork, soft old mortar		70	.114			3.61		3.61	5.60
0720	Hard mortar	↓	50	.160	↓		5.05		5.05	7.85

04 01 30 – Unit Masonry Cleaning

04 01 30.20 Cleaning Masonry		Crew	Daily Output	Labor-Hours	Unit	Material	Labor	Equipment	Total	Total Incl O&P
0010	**CLEANING MASONRY**									
0200	Chemical cleaning, new construction, brush and wash, minimum	D-1	1000	.016	S.F.	.05	.58		.63	.93
0220	Average		800	.020		.07	.73		.80	1.18
0240	Maximum		600	.027		.09	.97		1.06	1.57
0260	Light restoration, minimum		800	.020		.07	.73		.80	1.17
0270	Average		400	.040		.10	1.45		1.55	2.31
0280	Maximum		330	.048		.13	1.76		1.89	2.82
0300	Heavy restoration, minimum		600	.027		.09	.97		1.06	1.57
0310	Average		400	.040		.14	1.45		1.59	2.36
0320	Maximum	↓	250	.064		.19	2.33		2.52	3.74
0400	High pressure water only, minimum	B-9	2000	.020			.64	.10	.74	1.10
0420	Average		1500	.027			.85	.13	.98	1.46
0440	Maximum		1000	.040			1.28	.19	1.47	2.19
0800	High pressure water and chemical, minimum		1800	.022		.08	.71	.11	.90	1.31
0820	Average		1200	.033		.12	1.07	.16	1.35	1.96
0840	Maximum		800	.050		.16	1.60	.24	2	2.92
1200	Sandblast, wet system, minimum	↓	1750	.023	↓	.47	.73	.11	1.31	1.77

04 01 Maintenance of Masonry

04 01 30 – Unit Masonry Cleaning

04 01 30.20 Cleaning Masonry

		Crew	Daily Output	Labor-Hours	Unit	Material	2009 Bare Costs Labor	Equipment	Total	Total Incl O&P
1220	Average	B-9	1100	.036	S.F.	.70	1.16	.17	2.03	2.76
1240	Maximum		700	.057		.94	1.83	.27	3.04	4.17
1400	Dry system, minimum		2500	.016		.47	.51	.08	1.06	1.39
1420	Average		1750	.023		.70	.73	.11	1.54	2.02
1440	Maximum	↓	1000	.040		.94	1.28	.19	2.41	3.22
1800	For walnut shells, add					.50			.50	.55
1820	For corn chips, add					.50			.50	.55
2000	Steam cleaning, minimum	B-9	3000	.013			.43	.06	.49	.73
2020	Average		2500	.016			.51	.08	.59	.87
2040	Maximum	↓	1500	.027			.85	.13	.98	1.46
4000	Add for masking doors and windows	1 Clab	800	.010	↓	.06	.32		.38	.56
4200	Add for pedestrian protection				Job					10%

04 05 Common Work Results for Masonry

04 05 05 – Selective Masonry Demolition

04 05 05.10 Selective Demolition

			Crew	Daily Output	Labor-Hours	Unit	Material	Labor	Equipment	Total	Total Incl O&P
0010	**SELECTIVE DEMOLITION**	R024119-10									
1000	Chimney, 16" x 16", soft old mortar		1 Clab	55	.145	C.F.		4.60		4.60	7.15
1020	Hard mortar			40	.200			6.30		6.30	9.80
1030	16" x 20", soft old mortar			55	.145			4.60		4.60	7.15
1040	Hard mortar			40	.200			6.30		6.30	9.80
1050	16" x 24", soft old mortar			55	.145			4.60		4.60	7.15
1060	Hard mortar			40	.200			6.30		6.30	9.80
1080	20" x 20", soft old mortar			55	.145			4.60		4.60	7.15
1100	Hard mortar			40	.200			6.30		6.30	9.80
1110	20" x 24", soft old mortar			55	.145			4.60		4.60	7.15
1120	Hard mortar			40	.200			6.30		6.30	9.80
1140	20" x 32", soft old mortar			55	.145			4.60		4.60	7.15
1160	Hard mortar			40	.200			6.30		6.30	9.80
1200	48" x 48", soft old mortar			55	.145			4.60		4.60	7.15
1220	Hard mortar		↓	40	.200	↓		6.30		6.30	9.80
1250	Metal, high temp steel jacket, 24" diameter		E-2	130	.431	V.L.F.		18.75	13.40	32.15	47
1260	60" diameter		"	60	.933			40.50	29	69.50	102
1280	Flue lining, up to 12" x 12"		1 Clab	200	.040			1.26		1.26	1.96
1282	Up to 24" x 24"			150	.053			1.69		1.69	2.61
2000	Columns, 8" x 8", soft old mortar			48	.167			5.25		5.25	8.15
2020	Hard mortar			40	.200			6.30		6.30	9.80
2060	16" x 16", soft old mortar			16	.500			15.80		15.80	24.50
2100	Hard mortar			14	.571			18.05		18.05	28
2140	24" x 24", soft old mortar			8	1			31.50		31.50	49
2160	Hard mortar			6	1.333			42		42	65.50
2200	36" x 36", soft old mortar			4	2			63		63	98
2220	Hard mortar			3	2.667	↓		84.50		84.50	131
2230	Alternate pricing method, soft old mortar			30	.267	C.F.		8.45		8.45	13.05
2240	Hard mortar		↓	23	.348	"		11		11	17.05
3000	Copings, precast or masonry, to 8" wide										
3020	Soft old mortar		1 Clab	180	.044	L.F.		1.40		1.40	2.18
3040	Hard mortar		"	160	.050	"		1.58		1.58	2.45
3100	To 12" wide										
3120	Soft old mortar		1 Clab	160	.050	L.F.		1.58		1.58	2.45
3140	Hard mortar		"	140	.057	"		1.81		1.81	2.80

04 05 Common Work Results for Masonry

04 05 05 – Selective Masonry Demolition

04 05 05.10 Selective Demolition

		Crew	Daily Output	Labor-Hours	Unit	Material	2009 Bare Costs Labor	Equipment	Total	Total Incl O&P
4000	Fireplace, brick, 30" x 24" opening									
4020	Soft old mortar	1 Clab	2	4	Ea.		126		126	196
4040	Hard mortar		1.25	6.400			202		202	315
4100	Stone, soft old mortar		1.50	5.333			169		169	261
4120	Hard mortar		1	8	↓		253		253	390
5000	Veneers, brick, soft old mortar		140	.057	S.F.		1.81		1.81	2.80
5020	Hard mortar		125	.064			2.02		2.02	3.14
5100	Granite and marble, 2" thick		180	.044			1.40		1.40	2.18
5120	4" thick		170	.047			1.49		1.49	2.31
5140	Stone, 4" thick		180	.044			1.40		1.40	2.18
5160	8" thick		175	.046	↓		1.44		1.44	2.24
5400	Alternate pricing method, stone, 4" thick		60	.133	C.F.		4.21		4.21	6.55
5420	8" thick	↓	85	.094	"		2.97		2.97	4.61

04 05 13 – Masonry Mortaring

04 05 13.10 Cement

		Crew	Daily Output	Labor-Hours	Unit	Material	Labor	Equipment	Total	Total Incl O&P
0010	**CEMENT**									
0100	Masonry, 70 lb. bag, T.L. lots				Bag	9.05			9.05	9.95
0150	L.T.L. lots					9.60			9.60	10.55
0200	White, 70 lb. bag, T.L. lots					15.05			15.05	16.55
0250	L.T.L. lots				↓	16.05			16.05	17.65

04 05 13.20 Lime

0010	**LIME**									
0020	Masons, hydrated, 50 lb. bag, T.L. lots				Bag	8.95			8.95	9.85
0050	L.T.L. lots					9.30			9.30	10.25
0200	Finish, double hydrated, 50 lb. bag, T.L. lots					9.90			9.90	10.90
0250	L.T.L. lots				↓	11			11	12.10

04 05 13.30 Mortar

0010	**MORTAR** R040513-10									
0020	With masonry cement									
0100	Type M, 1:1:6 mix	1 Brhe	143	.056	C.F.	4.75	1.80		6.55	7.95
0200	Type N, 1:3 mix	"	143	.056	"	4.29	1.80		6.09	7.45
2000	With portland cement and lime									
2100	Type M, 1:1/4:3 mix	1 Brhe	143	.056	C.F.	8.45	1.80		10.25	12.05
2200	Type N, 1:1:6 mix, 750 psi		143	.056		6.80	1.80		8.60	10.20
2300	Type O, 1:2:9 mix (Pointing Mortar)	↓	143	.056		7.50	1.80		9.30	11
2650	Pre-mixed, type S or N					4.94			4.94	5.45
2700	Mortar for glass block	1 Brhe	143	.056	↓	10.20	1.80		12	13.95
2900	Mortar for Fire Brick, 80 lb. bag, T.L. Lots				Bag	25.50			25.50	28

04 05 13.91 Masonry Restoration Mortaring

0010	**MASONRY RESTORATION MORTARING**									
0020	Masonry restoration mix				Lb.	1.97			1.97	2.17
0050	White				"	2.27			2.27	2.50

04 05 13.93 Mortar Pigments

0010	**MORTAR PIGMENTS**, 50 lb. bags (2 bags per M bricks)									
0020	Color admixture, range 2 to 10 lb. per bag of cement, minimum				Lb.	4.50			4.50	4.95
0050	Average					5.40			5.40	5.95
0100	Maximum				↓	10.45			10.45	11.50

04 05 13.95 Sand

0010	**SAND**, screened and washed at pit									
0020	For mortar, per ton				Ton	17.80			17.80	19.55
0050	With 10 mile haul				↓	26			26	28.50

04 05 Common Work Results for Masonry

04 05 13 – Masonry Mortaring

04 05 13.95 Sand

		Crew	Daily Output	Labor-Hours	Unit	Material	2009 Bare Costs Labor	Equipment	Total	Total Incl O&P
0100	With 30 mile haul				Ton	36			36	40
0200	Screened and washed, at the pit				C.Y.	24.50			24.50	27
0250	With 10 mile haul					36			36	40
0300	With 30 mile haul					50			50	55.50

04 05 19 – Masonry Anchorage and Reinforcing

04 05 19.05 Anchor Bolts

		Crew	Daily Output	Labor-Hours	Unit	Material	Labor	Equipment	Total	Total Incl O&P
0010	**ANCHOR BOLTS**									
0020	Hooked, with nut and washer, 1/2" diam., 8" long	1 Bric	200	.040	Ea.	.83	1.62		2.45	3.37
0030	12" long		190	.042		1.26	1.71		2.97	3.98
0040	5/8" diameter, 8" long		180	.044		1.13	1.80		2.93	3.97
0050	12" long		170	.047		1.24	1.91		3.15	4.26
0060	3/4" diameter, 8" long		160	.050		1.86	2.03		3.89	5.10
0070	12" long		150	.053		2.33	2.16		4.49	5.85

04 05 19.16 Masonry Anchors

		Crew	Daily Output	Labor-Hours	Unit	Material	Labor	Equipment	Total	Total Incl O&P
0010	**MASONRY ANCHORS**									
0020	For brick veneer, galv., corrugated, 7/8" x 7", 22 Ga.	1 Bric	10.50	.762	C	9.80	31		40.80	58
0100	24 Ga.		10.50	.762		9.15	31		40.15	57
0150	16 Ga.		10.50	.762		24	31		55	73.50
0200	Buck anchors, galv., corrugated, 16 gauge, 2" bend, 8" x 2"		10.50	.762		134	31		165	194
0250	8" x 3"		10.50	.762		138	31		169	199
0670	3/16" diameter		10.50	.762		21	31		52	70
0680	1/4" diameter		10.50	.762		39.50	31		70.50	90.50
0850	8" long, 3/16" diameter		10.50	.762		24	31		55	73
0855	1/4" diameter		10.50	.762		46.50	31		77.50	98
1000	Rectangular type, galvanized, 1/4" diameter, 2" x 6"		10.50	.762		67	31		98	121
1050	4" x 6"		10.50	.762		83.50	31		114.50	139
1100	3/16" diameter, 2" x 6"		10.50	.762		37.50	31		68.50	88.50
1150	4" x 6"		10.50	.762		39	31		70	90
1500	Rigid partition anchors, plain, 8" long, 1" x 1/8"		10.50	.762		103	31		134	160
1550	1" x 1/4"		10.50	.762		170	31		201	234
1580	1-1/2" x 1/8"		10.50	.762		138	31		169	198
1600	1-1/2" x 1/4"		10.50	.762		280	31		311	355
1650	2" x 1/8"		10.50	.762		178	31		209	243
1700	2" x 1/4"		10.50	.762		325	31		356	405

04 05 23 – Masonry Accessories

04 05 23.95 Wall Plugs

		Crew	Daily Output	Labor-Hours	Unit	Material	Labor	Equipment	Total	Total Incl O&P
0010	**WALL PLUGS** (for nailing to brickwork)									
0020	26 ga., galvanized, plain	1 Bric	10.50	.762	C	31	31		62	81
0050	Wood filled	"	10.50	.762	"	104	31		135	161

04 21 Clay Unit Masonry

04 21 13 – Brick Masonry

04 21 13.13 Brick Veneer Masonry

		Crew	Daily Output	Labor-Hours	Unit	Material	2009 Bare Costs Labor	2009 Bare Costs Equipment	Total	Total Incl O&P
0010	**BRICK VENEER MASONRY**, T.L. lots, excl. scaff., grout & reinforcing									
0015	Material costs incl. 3% brick and 25% mortar waste									
2000	Standard, sel. common, 4" x 2-2/3" x 8", (6.75/S.F.)	D-8	230	.174	S.F.	3.60	6.45		10.05	13.75
2020	Standard, red, 4" x 2-2/3" x 8", running bond (6.75/SF)		220	.182		5.65	6.75		12.40	16.50
2050	Full header every 6th course (7.88/S.F.) R040513-10		185	.216		6.60	8.05		14.65	19.45
2100	English, full header every 2nd course (10.13/S.F.)		140	.286		8.50	10.60		19.10	25.50
2150	Flemish, alternate header every course (9.00/S.F.)		150	.267		7.55	9.90		17.45	23.50
2200	Flemish, alt. header every 6th course (7.13/S.F.)		205	.195		6	7.25		13.25	17.60
2250	Full headers throughout (13.50/S.F.) R040519-50		105	.381		11.30	14.15		25.45	34
2300	Rowlock course (13.50/S.F.)		100	.400		11.30	14.85		26.15	35
2400	Soldier course (6.75/S.F.)		200	.200		5.65	7.45		13.10	17.50
2450	Sailor course (4.50/S.F.)		290	.138		3.80	5.15		8.95	12
2600	Buff or gray face, running bond, (6.75/S.F.)		220	.182		6	6.75		12.75	16.85
2700	Glazed face brick, running bond		210	.190		11.15	7.10		18.25	23
2750	Full header every 6th course (7.88/S.F.)		170	.235		13	8.75		21.75	27.50
3000	Jumbo, 6" x 4" x 12" running bond (3.00/S.F.)		435	.092		4.86	3.42		8.28	10.55
3050	Norman, 4" x 2-2/3" x 12" running bond, (4.5/S.F.)		320	.125		5.65	4.65		10.30	13.30
3100	Norwegian, 4" x 3-1/5" x 12" (3.75/S.F.)		375	.107		4.14	3.96		8.10	10.55
3150	Economy, 4" x 4" x 8" (4.50/S.F.)		310	.129		4.18	4.79		8.97	11.85
3200	Engineer, 4" x 3-1/5" x 8" (5.63/S.F.)		260	.154		3.46	5.70		9.16	12.45
3250	Roman, 4" x 2" x 12" (6.00/S.F.)		250	.160		5.75	5.95		11.70	15.30
3300	SCR, 6" x 2-2/3" x 12" (4.50/S.F.)		310	.129		5.25	4.79		10.04	13
3350	Utility, 4" x 4" x 12" (3.00/S.F.)		450	.089		4.40	3.30		7.70	9.85
3400	For cavity wall construction, add					15%				
3450	For stacked bond, add					10%				
3500	For interior veneer construction, add					15%				
3550	For curved walls, add					30%				

04 21 13.35 Common Building Brick

0010	**COMMON BUILDING BRICK**, C62, TL lots, material only R042110-10									
0020	Standard, minimum				M	340			340	375
0050	Average (select)				"	430			430	475

04 21 13.40 Structural Brick

0010	**STRUCTURAL BRICK** C652, Grade SW, incl. mortar, scaffolding not incl.									
0120	Bond beam	D-8	225	.178	S.F.	9.65	6.60		16.25	20.50
0140	V cut bond beam		225	.178		10.45	6.60		17.05	21.50
0160	Stretcher quoin, 5-5/8" x 2-3/4" x 9-5/8"		245	.163		10.10	6.05		16.15	20.50
0180	Corner quoin		245	.163		11.30	6.05		17.35	21.50
0200	Corner, 45 deg, 4-5/8" x 2-3/4" x 10-7/16"		235	.170		11.70	6.35		18.05	22.50

04 21 13.45 Face Brick

0010	**FACE BRICK** Material Only, C216, TL lots									
0300	Standard modular, 4" x 2-2/3" x 8", minimum				M	730			730	805
0350	Maximum					835			835	920
0450	Economy, 4" x 4" x 8", minimum					815			815	895
0500	Maximum					1,050			1,050	1,150
0510	Economy, 4" x 4" x 12", minimum					1,200			1,200	1,325
0520	Maximum					1,550			1,550	1,725
0550	Jumbo, 6" x 4" x 12", minimum					1,475			1,475	1,625
0600	Maximum					1,875			1,875	2,075
0610	Jumbo, 8" x 4" x 12", minimum					1,475			1,475	1,625
0620	Maximum					1,875			1,875	2,075
0650	Norwegian, 4" x 3-1/5" x 12", minimum					980			980	1,075
0700	Maximum					1,275			1,275	1,400

04 21 Clay Unit Masonry

04 21 13 – Brick Masonry

04 21 13.45 Face Brick

		Crew	Daily Output	Labor-Hours	Unit	Material	2009 Bare Costs Labor	Equipment	Total	Total Incl O&P
0710	Norwegian, 6" x 3-1/5" x 12", minimum				M	1,425			1,425	1,575
0720	Maximum					1,875			1,875	2,050
0850	Standard glazed, plain colors, 4" x 2-2/3" x 8", minimum					1,525			1,525	1,675
0900	Maximum					1,975			1,975	2,175
1000	Deep trim shades, 4" x 2-2/3" x 8", minimum					1,600			1,600	1,750
1050	Maximum					1,900			1,900	2,100
1080	Jumbo utility, 4" x 4" x 12"					1,325			1,325	1,450
1120	4" x 8" x 8"					1,550			1,550	1,700
1140	4" x 8" x 16"					3,100			3,100	3,400
1260	Engineer, 4" x 3-1/5" x 8", minimum					510			510	560
1270	Maximum					625			625	690
1350	King, 4" x 2-3/4" x 10", minimum					525			525	575
1360	Maximum					560			560	620
1400	Norman, 4" x 2-3/4" x 12"					525			525	575
1450	Roman, 4" x 2" x 12"					525			525	575
1500	SCR, 6" x 2-2/3" x 12"					525			525	575
1550	Double, 4" x 5-1/3" x 8"					525			525	575
1600	Triple, 4" x 5-1/3" x 12"					525			525	575
1770	Standard modular, double glazed, 4" x 2-2/3" x 8"					1,775			1,775	1,950
1850	Jumbo, colored glazed ceramic, 6" x 4" x 12"					2,400			2,400	2,650
2050	Jumbo utility, glazed, 4" x 4" x 12"					4,050			4,050	4,450
2100	4" x 8" x 8"					4,725			4,725	5,200
2150	4" x 16" x 8"					6,200			6,200	6,825
2170	For less than truck load lots, add					15			15	16.50
2180	For buff or gray brick, add					16			16	17.60

04 21 26 – Glazed Structural Clay Tile Masonry

04 21 26.10 Structural Facing Tile

		Crew	Daily Output	Labor-Hours	Unit	Material	Labor	Equipment	Total	Total Incl O&P
0010	**STRUCTURAL FACING TILE**, std. colors, excl. scaffolding, grout, reinforcing									
0020	6T series, 5-1/3" x 12", 2.3 pieces per S.F., glazed 1 side, 2" thick	D-8	225	.178	S.F.	8.45	6.60		15.05	19.30
0100	4" thick		220	.182		11.05	6.75		17.80	22.50
0150	Glazed 2 sides		195	.205		14.15	7.60		21.75	27
0250	6" thick		210	.190		16.60	7.10		23.70	29
0300	Glazed 2 sides		185	.216		19.45	8.05		27.50	33.50
0400	8" thick		180	.222		22	8.25		30.25	36.50
0500	Special shapes, group 1		400	.100	Ea.	6.85	3.72		10.57	13.20
0550	Group 2		375	.107		10.95	3.96		14.91	18.05
0600	Group 3		350	.114		14.45	4.25		18.70	22.50
0650	Group 4		325	.123		30	4.57		34.57	40
0700	Group 5		300	.133		35.50	4.95		40.45	47
0750	Group 6		275	.145		49	5.40		54.40	62
1000	Fire rated, 4" thick, 1 hr. rating		210	.190	S.F.	15.95	7.10		23.05	28.50
1300	Acoustic, 4" thick		210	.190	"	34	7.10		41.10	48.50
1400	For designer colors, add					25%				
2000	8W series, 8" x 16", 1.125 pieces per S.F.									
2050	2" thick, glazed 1 side	D-8	360	.111	S.F.	9.35	4.13		13.48	16.50
2100	4" thick, glazed 1 side		345	.116		12	4.31		16.31	19.80
2150	Glazed 2 sides		325	.123		13.45	4.57		18.02	22
2200	6" thick, glazed 1 side		330	.121		21	4.50		25.50	30.50
2250	8" thick, glazed 1 side		310	.129		21.50	4.79		26.29	31
2500	Special shapes, group 1		300	.133	Ea.	12.95	4.95		17.90	22
2550	Group 2		280	.143		17.40	5.30		22.70	27
2600	Group 3		260	.154		18.55	5.70		24.25	29

04 21 Clay Unit Masonry

04 21 26 – Glazed Structural Clay Tile Masonry

04 21 26.10 Structural Facing Tile

		Crew	Daily Output	Labor-Hours	Unit	Material	2009 Bare Costs Labor	Equipment	Total	Total Incl O&P
2650	Group 4	D-8	250	.160	Ea.	37	5.95		42.95	49.50
2700	Group 5		240	.167		36	6.20		42.20	49
2750	Group 6		230	.174		77.50	6.45		83.95	95.50
3000	4" thick, glazed 1 side		345	.116	S.F.	11.10	4.31		15.41	18.75
3100	Acoustic, 4" thick		345	.116	"	13.95	4.31		18.26	22
3120	4W series, 8" x 8", 2.25 pieces per S.F.									
3125	2" thick, glazed 1 side	D-8	360	.111	S.F.	9.75	4.13		13.88	16.95
3130	4" thick, glazed 1 side		345	.116		10.60	4.31		14.91	18.20
3135	Glazed 2 sides		325	.123		14.45	4.57		19.02	23
3140	6" thick, glazed 1 side		330	.121		14.90	4.50		19.40	23.50
3150	8" thick, glazed 1 side		310	.129		21.50	4.79		26.29	31.50
3155	Special shapes, group I		300	.133	Ea.	6.95	4.95		11.90	15.15
3160	Group II		280	.143	"	7.90	5.30		13.20	16.75
3200	For designer colors, add					25%				
3300	For epoxy mortar joints, add				S.F.	1.58			1.58	1.74

04 21 29 – Terra Cotta Masonry

04 21 29.20 Terra Cotta Tile

		Crew	Daily Output	Labor-Hours	Unit	Material	2009 Bare Costs Labor	Equipment	Total	Total Incl O&P
0010	**TERRA COTTA TILE**, on walls, dry set, 1/2" thick									
0100	Square, hexagonal or lattice shapes, unglazed	1 Tilf	135	.059	S.F.	6.35	2.26		8.61	10.25
0300	Glazed, plain colors		130	.062		8.75	2.34		11.09	13.05
0400	Intense colors		125	.064		10.20	2.44		12.64	14.75

04 22 Concrete Unit Masonry

04 22 10 – Concrete Masonry Units

04 22 10.14 Concrete Block, Back-Up

		Crew	Daily Output	Labor-Hours	Unit	Material	2009 Bare Costs Labor	Equipment	Total	Total Incl O&P
0010	**CONCRETE BLOCK, BACK-UP**, C90, 2000 psi									
0020	Normal weight, 8" x 16" units, tooled joint 1 side									
0050	Not-reinforced, 2000 psi, 2" thick	D-8	475	.084	S.F.	1.36	3.13		4.49	6.25
0200	4" thick		460	.087		1.52	3.23		4.75	6.55
0300	6" thick		440	.091		2.12	3.38		5.50	7.45
0350	8" thick		400	.100		2.23	3.72		5.95	8.10
0400	10" thick		330	.121		3.03	4.50		7.53	10.20
0450	12" thick	D-9	310	.155		3.44	5.65		9.09	12.35

04 22 10.18 Concrete Block, Column

		Crew	Daily Output	Labor-Hours	Unit	Material	2009 Bare Costs Labor	Equipment	Total	Total Incl O&P
0010	**CONCRETE BLOCK, COLUMN** or pilaster									
0050	Including vertical reinforcing (4-#4 bars) and grout									
0160	1 piece unit, 16" x 16"	D-1	26	.615	V.L.F.	18.65	22.50		41.15	54.50
0170	2 piece units, 16" x 20"		24	.667		24.50	24		48.50	63.50
0180	20" x 20"		22	.727		35.50	26.50		62	79
0190	22" x 24"		18	.889		49.50	32.50		82	104
0200	20" x 32"		14	1.143		53.50	41.50		95	122

04 22 10.32 Concrete Block, Lintels

		Crew	Daily Output	Labor-Hours	Unit	Material	2009 Bare Costs Labor	Equipment	Total	Total Incl O&P
0010	**CONCRETE BLOCK, LINTELS**, C90, normal weight									
0100	Including grout and horizontal reinforcing									
0200	8" x 8" x 8", 1 #4 bar	D-4	300	.107	L.F.	4.86	3.84	.42	9.12	11.60
0250	2 #4 bars		295	.108		5.20	3.90	.43	9.53	12.10
0400	8" x 16" x 8", 1 #4 bar		275	.116		3.36	4.18	.46	8	10.55
0450	2 #4 bars		270	.119		3.71	4.26	.47	8.44	11.05
1000	12" x 8" x 8", 1 #4 bar		275	.116		6.75	4.18	.46	11.39	14.30
1100	2 #4 bars		270	.119		7.10	4.26	.47	11.83	14.80

04 22 Concrete Unit Masonry

04 22 10 – Concrete Masonry Units

04 22 10.32 Concrete Block, Lintels		Crew	Daily Output	Labor-Hours	Unit	Material	2009 Bare Costs Labor	Equipment	Total	Total Incl O&P
1150	2 #5 bars	D-4	270	.119	L.F.	7.50	4.26	.47	12.23	15.20
1200	2 #6 bars		265	.121		7.95	4.34	.48	12.77	15.85
1500	12" x 16" x 8", 1 #4 bar		250	.128		10.65	4.60	.51	15.76	19.20
1600	2 #3 bars		245	.131		10.70	4.70	.52	15.92	19.40
1650	2 #4 bars		245	.131		11	4.70	.52	16.22	19.75
1700	2 #5 bars	↓	240	.133	↓	11.40	4.79	.53	16.72	20.50

04 22 10.34 Concrete Block, Partitions			Crew	Daily Output	Labor-Hours	Unit	Material	Labor	Equipment	Total	Total Incl O&P
0010	**CONCRETE BLOCK, PARTITIONS**, excludes scaffolding	R040513-10									
0100	Acoustical slotted block										
0200	4" thick, type A-1		D-8	315	.127	S.F.	3.73	4.72		8.45	11.25
0250	8" thick, type Q			275	.145		8.30	5.40		13.70	17.35
0400	8" thick, type RSC/RF	R040519-50		275	.145		6.50	5.40		11.90	15.35
0500	NRC .60 type R, 8" thick			265	.151		7.95	5.60		13.55	17.25
0600	NRC .65 type RR, 8" thick	R042110-50		265	.151		7.30	5.60		12.90	16.55
0700	NRC .65 type 4R-RF, 8" thick		↓	265	.151	↓	8.20	5.60		13.80	17.55
1000	Lightweight block, tooled joints, 2 sides, hollow	R042210-20									
1100	Not reinforced, 8" x 16" x 4" thick		D-8	440	.091	S.F.	1.72	3.38		5.10	7
1150	6" thick			410	.098		2.33	3.63		5.96	8.05
1200	8" thick			385	.104		2.86	3.86		6.72	9
1250	10" thick		↓	370	.108		3.73	4.02		7.75	10.20
1300	12" thick		D-9	350	.137		4.26	4.98		9.24	12.25
1500	Reinforced alternate courses, 4" thick		D-8	435	.092		1.80	3.42		5.22	7.20
1600	6" thick			405	.099		2.42	3.67		6.09	8.20
1650	8" thick		↓	380	.105		2.95	3.91		6.86	9.20
2000	Not reinforced, 8" x 24" x 4" thick, hollow		D-9	460	.104		1.25	3.79		5.04	7.15
2100	6" thick			440	.109		1.68	3.96		5.64	7.85
2150	8" thick			415	.116		2.10	4.20		6.30	8.65
2200	10" thick			385	.125		2.72	4.53		7.25	9.85
2250	12" thick			365	.132		3.11	4.78		7.89	10.65
2400	Reinforced alternate courses, 4" thick			455	.105		1.34	3.83		5.17	7.25
2500	6" thick			435	.110		1.80	4.01		5.81	8.10
2550	8" thick			410	.117		2.19	4.25		6.44	8.85
2600	10" thick			380	.126		2.82	4.59		7.41	10.05
2650	12" thick		↓	360	.133		3.22	4.84		8.06	10.90
2800	Solid, not reinforced, 8" x 16" x 2" thick		D-8	440	.091		1.42	3.38		4.80	6.65
2900	4" thick			420	.095		2.30	3.54		5.84	7.90
2950	6" thick			390	.103		2.66	3.81		6.47	8.70
3000	8" thick			365	.110		3.57	4.07		7.64	10.10
3050	10" thick		↓	350	.114		4.20	4.25		8.45	11.05
3100	12" thick		D-9	330	.145		6.20	5.30		11.50	14.85
3300	Solid, reinforced alternate courses, 4" thick		D-8	415	.096		2.39	3.58		5.97	8.10
3400	6" thick			385	.104		2.88	3.86		6.74	9
3450	8" thick			360	.111		3.67	4.13		7.80	10.30
3500	10" thick		↓	345	.116		4.31	4.31		8.62	11.30
3550	12" thick		D-9	325	.148	↓	6.30	5.35		11.65	15.10
4000	Regular block, tooled joints, 2 sides, hollow										
4100	Not reinforced, 8" x 16" x 4" thick		D-8	430	.093	S.F.	1.43	3.46		4.89	6.85
4150	6" thick			400	.100		2.03	3.72		5.75	7.90
4200	8" thick			375	.107		2.14	3.96		6.10	8.35
4250	10" thick		↓	360	.111		2.95	4.13		7.08	9.50
4300	12" thick		D-9	340	.141		3.35	5.15		8.50	11.50
4500	Reinforced alternate courses, 8" x 16" x 4" thick		D-8	425	.094	↓	1.56	3.50		5.06	7

04 22 Concrete Unit Masonry

04 22 10 – Concrete Masonry Units

04 22 10.34 Concrete Block, Partitions

		Crew	Daily Output	Labor-Hours	Unit	Material	2009 Bare Costs Labor	Equipment	Total	Total Incl O&P
4550	6" thick	D-8	395	.101	S.F.	2.16	3.76		5.92	8.10
4600	8" thick		370	.108		2.28	4.02		6.30	8.60
4650	10" thick		355	.113		3.37	4.19		7.56	10.05
4700	12" thick	D-9	335	.143		3.50	5.20		8.70	11.75
4900	Solid, not reinforced, 2" thick	D-8	435	.092		1.31	3.42		4.73	6.65
5000	3" thick		430	.093		1.56	3.46		5.02	6.95
5050	4" thick		415	.096		1.80	3.58		5.38	7.45
5100	6" thick		385	.104		2.02	3.86		5.88	8.05
5150	8" thick		360	.111		3.19	4.13		7.32	9.75
5200	12" thick	D-9	325	.148		4.86	5.35		10.21	13.50
5500	Solid, reinforced alternate courses, 4" thick	D-8	420	.095		1.88	3.54		5.42	7.40
5550	6" thick		380	.105		2.12	3.91		6.03	8.30
5600	8" thick		355	.113		3.29	4.19		7.48	9.95
5650	12" thick	D-9	320	.150		4.40	5.45		9.85	13.10

04 22 10.42 Concrete Block, Screen Block

		Crew	Daily Output	Labor-Hours	Unit	Material	Labor	Equipment	Total	Total Incl O&P
0010	**CONCRETE BLOCK, SCREEN BLOCK**									
0200	8" x 16", 4" thick	D-8	330	.121	S.F.	1.96	4.50		6.46	9
0300	8" thick		270	.148		2.98	5.50		8.48	11.65
0350	12" x 12", 4" thick		290	.138		2.21	5.15		7.36	10.25
0500	8" thick		250	.160		3.04	5.95		8.99	12.35

04 22 10.44 Glazed Concrete Block

		Crew	Daily Output	Labor-Hours	Unit	Material	Labor	Equipment	Total	Total Incl O&P
0010	**GLAZED CONCRETE BLOCK** C744									
0100	Single face, 8" x 16" units, 2" thick	D-8	360	.111	S.F.	8.25	4.13		12.38	15.35
0200	4" thick		345	.116		8.50	4.31		12.81	15.90
0250	6" thick		330	.121		9.15	4.50		13.65	16.90
0300	8" thick		310	.129		9.20	4.79		13.99	17.40
0350	10" thick		295	.136		11.10	5.05		16.15	19.85
0400	12" thick	D-9	280	.171		11.75	6.25		18	22.50
0700	Double face, 8" x 16" units, 4" thick	D-8	340	.118		12.20	4.37		16.57	20
0750	6" thick		320	.125		14.50	4.65		19.15	23
0800	8" thick		300	.133		15.15	4.95		20.10	24
1000	Jambs, bullnose or square, single face, 8" x 16", 2" thick		315	.127	Ea.	17.50	4.72		22.22	26.50
1050	4" thick		285	.140	"	20.50	5.20		25.70	30.50
1200	Caps, bullnose or square, 8" x 16", 2" thick		420	.095	L.F.	14.15	3.54		17.69	21
1250	4" thick		380	.105		15.95	3.91		19.86	23.50
1500	Cove base, 8" x 16", 2" thick		315	.127		11	4.72		15.72	19.25
1550	4" thick		285	.140		7.15	5.20		12.35	15.75
1600	6" thick		265	.151		7.70	5.60		13.30	16.95
1650	8" thick		245	.163		8.10	6.05		14.15	18.10

04 23 Glass Unit Masonry

04 23 13 – Vertical Glass Unit Masonry

04 23 13.10 Glass Block

		Crew	Daily Output	Labor-Hours	Unit	Material	Labor	Equipment	Total	Total Incl O&P
0010	**GLASS BLOCK** R040513-10									
0100	Plain, 4" thick, under 1,000 S.F., 6" x 6"	D-8	115	.348	S.F.	20.50	12.95		33.45	42
0150	8" x 8"		160	.250		14.05	9.30		23.35	29.50
0160	end block		160	.250		39.50	9.30		48.80	57
0170	90 deg corner		160	.250		39	9.30		48.30	57
0180	45 deg corner		160	.250		19.15	9.30		28.45	35
0200	12" x 12"		175	.229		16	8.50		24.50	30.50

04 23 Glass Unit Masonry

04 23 13 – Vertical Glass Unit Masonry

04 23 13.10 Glass Block

		Crew	Daily Output	Labor-Hours	Unit	Material	2009 Bare Costs Labor	Equipment	Total	Total Incl O&P
0210	4" x 8"	D-8	160	.250	S.F.	10.25	9.30		19.55	25.50
0220	6" x 8"		160	.250		11	9.30		20.30	26
0300	1,000 to 5,000 S.F., 6" x 6"		135	.296		20	11		31	38.50
0350	8" x 8"		190	.211		13.75	7.80		21.55	27
0400	12" x 12"		215	.186		15.70	6.90		22.60	28
0410	4" x 8"		215	.186		4.43	6.90		11.33	15.35
0420	6" x 8"		215	.186		4.79	6.90		11.69	15.75
0500	Over 5,000 S.F., 6" x 6"		145	.276		19.50	10.25		29.75	37
0550	8" x 8"		215	.186		13.35	6.90		20.25	25
0600	12" x 12"		240	.167		15.20	6.20		21.40	26
0610	4" x 8"		240	.167		4.33	6.20		10.53	14.15
0620	6" x 8"		240	.167		4.74	6.20		10.94	14.60
0700	For solar reflective blocks, add					100%				
1000	Thinline, plain, 3-1/8" thick, under 1,000 S.F., 6" x 6"	D-8	115	.348	S.F.	15.05	12.95		28	36
1050	8" x 8"		160	.250		9.35	9.30		18.65	24.50
1200	Over 5,000 S.F., 6" x 6"		145	.276		16.30	10.25		26.55	33.50
1400	For cleaning block after installation (both sides), add		1000	.040		.11	1.49		1.60	2.37
4000	Accessories									
4100	Anchors, 20 ga. galv., 1-3/4" wide x 24" long				Ea.	2.95			2.95	3.25
4200	Emulsion asphalt				Gal.	8.10			8.10	8.90
4300	Expansion joint, fiberglass				L.F.	.67			.67	.74
4400	Steel mesh, double galvanized				"	.94			.94	1.03

04 24 Adobe Unit Masonry

04 24 16 – Manufactured Adobe Unit Masonry

04 24 16.06 Adobe Brick

			Crew	Daily Output	Labor-Hours	Unit	Material	Labor	Equipment	Total	Total Incl O&P
0010	**ADOBE BRICK**, Semi-stabilized, with cement mortar										
0060	Brick, 10" x 4" x 14", 2.6/S.F.	G	D-8	560	.071	S.F.	3.54	2.65		6.19	7.95
0080	12" x 4" x 16", 2.3/S.F.	G		580	.069		4.50	2.56		7.06	8.85
0100	10" x 4" x 16", 2.3/S.F.	G		590	.068		4.27	2.52		6.79	8.50
0120	8" x 4" x 16", 2.3/S.F.	G		560	.071		3.34	2.65		5.99	7.70
0140	4" x 4" x 16", 2.3/S.F.	G		540	.074		2.55	2.75		5.30	7
0160	6" x 4" x 16", 2.3/S.F.	G		540	.074		2.58	2.75		5.33	7
0180	4" x 4" x 12", 3.0/S.F.	G		520	.077		2.29	2.86		5.15	6.85
0200	8" x 4" x 12", 3.0/S.F.	G		520	.077		2.52	2.86		5.38	7.10

04 25 Unit Masonry Panels

04 25 20 – Pre-Fabricated Masonry Panels

04 25 20.10 Brick and Epoxy Mortar Panels

		Crew	Daily Output	Labor-Hours	Unit	Material	Labor	Equipment	Total	Total Incl O&P
0010	**BRICK AND EPOXY MORTAR PANELS**									
0020	Prefabricated brick & epoxy mortar, 4" thick, minimum	C-11	775	.093	S.F.	9.25	4.07	2.44	15.76	19.90
0100	Maximum	"	500	.144		10.75	6.30	3.78	20.83	27
0200	For 2" concrete back-up, add					50%				
0300	For 1" urethane & 3" concrete back-up, add					70%				

04 27 Multiple-Wythe Unit Masonry

04 27 10 – Multiple-Wythe Masonry

04 27 10.40 Steps		Crew	Daily Output	Labor-Hours	Unit	Material	2009 Bare Costs Labor	2009 Bare Costs Equipment	Total	Total Incl O&P
0010	**STEPS**									
0012	Entry steps, select common brick	D-1	.30	53.333	M	430	1,950		2,380	3,425

04 41 Dry-Placed Stone

04 41 10 – Dry Placed Stone

04 41 10.10 Rough Stone Wall

			Crew	Daily Output	Labor-Hours	Unit	Material	Labor	Equipment	Total	Total Incl O&P
0011	**ROUGH STONE WALL**, Dry										
0100	Random fieldstone, under 18" thick	G	D-12	60	.533	C.F.	11.15	19.65		30.80	42.50
0150	Over 18" thick	G	"	63	.508	"	13.40	18.70		32.10	43.50

04 43 Stone Masonry

04 43 10 – Masonry with Natural and Processed Stone

04 43 10.05 Ashlar Veneer

			Crew	Daily Output	Labor-Hours	Unit	Material	Labor	Equipment	Total	Total Incl O&P
0011	**ASHLAR VENEER** 4" + or – thk, random or random rectangular	R040513-10									
0150	Sawn face, split joints, low priced stone		D-8	140	.286	S.F.	9.40	10.60		20	26.50
0200	Medium priced stone			130	.308		11.25	11.45		22.70	29.50
0300	High priced stone			120	.333		15.05	12.40		27.45	35.50
0600	Seam face, split joints, medium price stone			125	.320		7.55	11.90		19.45	26.50
0700	High price stone			120	.333		15	12.40		27.40	35.50
1000	Split or rock face, split joints, medium price stone			125	.320		10.80	11.90		22.70	30
1100	High price stone			120	.333		15	12.40		27.40	35.50

04 43 10.10 Bluestone

			Crew	Daily Output	Labor-Hours	Unit	Material	Labor	Equipment	Total	Total Incl O&P
0010	**BLUESTONE**, cut to size	R040513-10									
0100	Paving, natural cleft, to 4', 1" thick		D-8	150	.267	S.F.	9.85	9.90		19.75	26
0150	1-1/2" thick			145	.276		9.85	10.25		20.10	26.50
0200	Smooth finish, 1" thick			150	.267		9.85	9.90		19.75	26
0250	1-1/2" thick			145	.276		9.85	10.25		20.10	26.50
0300	Thermal finish, 1" thick			150	.267		9.85	9.90		19.75	26
0350	1-1/2" thick			145	.276		9.85	10.25		20.10	26.50
0500	Sills, natural cleft, 10" wide to 6' long, 1-1/2" thick		D-11	70	.343	L.F.	13.35	13.15		26.50	34.50
0550	2" thick			63	.381		14.60	14.60		29.20	38
0600	Smooth finish, 1-1/2" thick			70	.343		25	13.15		38.15	47.50
0650	2" thick			63	.381		30	14.60		44.60	55
0800	Thermal finish, 1-1/2" thick			70	.343		30	13.15		43.15	53
0850	2" thick			63	.381		35	14.60		49.60	60.50
1000	Stair treads, natural cleft, 12" wide, 6' long, 1-1/2" thick		D-10	115	.278		30	10.95	5.25	46.20	55.50
1050	2" thick			105	.305		30	12	5.75	47.75	57.50
1100	Smooth finish, 1-1/2" thick			115	.278		15	10.95	5.25	31.20	39
1150	2" thick			105	.305		15	12	5.75	32.75	41
1300	Thermal finish			115	.278		15	10.95	5.25	31.20	39
1350	2" thick			105	.305		15	12	5.75	32.75	41
2000	Coping, finished top & 2 sides, 12" to 6'										
2100	Natural cleft, 1-1/2" thick		D-10	115	.278	L.F.	40	10.95	5.25	56.20	66.50
2150	2" thick			105	.305		40	12	5.75	57.75	68.50
2200	Smooth finish, 1-1/2" thick			115	.278		40	10.95	5.25	56.20	66.50
2250	2" thick			105	.305		40	12	5.75	57.75	68.50
2300	Thermal finish, 1-1/2" thick			115	.278		45	10.95	5.25	61.20	72
2350	2" thick			105	.305		45	12	5.75	62.75	74

04 43 Stone Masonry

04 43 10 – Masonry with Natural and Processed Stone

04 43 10.45 Granite

		Crew	Daily Output	Labor-Hours	Unit	Material	2009 Bare Costs Labor	2009 Bare Costs Equipment	Total	Total Incl O&P
0010	**GRANITE**, cut to size									
0050	Veneer, polished face, 3/4" to 1-1/2" thick									
0150	Low price, gray, light gray, etc.	D-10	130	.246	S.F.	25.50	9.70	4.63	39.83	48
0180	Medium price, pink, brown, etc.		130	.246		28	9.70	4.63	42.33	50.50
0220	High price, red, black, etc.		130	.246		40	9.70	4.63	54.33	64
0300	1-1/2" to 2-1/2" thick, veneer									
0350	Low price, gray, light gray, etc.	D-10	130	.246	S.F.	27.50	9.70	4.63	41.83	50
0500	Medium price, pink, brown, etc.		130	.246		32	9.70	4.63	46.33	55
0550	High price, red, black, etc.		130	.246		49.50	9.70	4.63	63.83	74.50
0700	2-1/2" to 4" thick, veneer									
0750	Low price, gray, light gray, etc.	D-10	110	.291	S.F.	36.50	11.45	5.45	53.40	63.50
0850	Medium price, pink, brown, etc.		110	.291		42	11.45	5.45	58.90	69.50
0950	High price, red, black, etc.		110	.291		60	11.45	5.45	76.90	89.50
1000	For bush hammered finish, deduct					5%				
1050	Coarse rubbed finish, deduct					10%				
1100	Honed finish, deduct					5%				
1150	Thermal finish, deduct					18%				
1800	Carving or bas-relief, from templates or plaster molds									
1850	Minimum	D-10	80	.400	C.F.	168	15.75	7.50	191.25	217
1900	Maximum	"	80	.400	"	490	15.75	7.50	513.25	570
2000	Intricate or hand finished pieces									
2010	Mouldings, radius cuts, bullnose edges, etc.									
2050	Add, minimum					30%				
2100	Add, maximum					300%				
2500	Steps, copings, etc., finished on more than one surface									
2550	Minimum	D-10	50	.640	C.F.	91	25	12.05	128.05	151
2600	Maximum	"	50	.640	"	146	25	12.05	183.05	211
2700	Pavers, Belgian block, 8"-13" long, 4"-6" wide, 4"-6" deep	D-11	120	.200	S.F.	25.50	7.70		33.20	39.50
2800	Pavers, 4" x 4" x 4" blocks, split face and joints									
2850	Minimum	D-11	80	.300	S.F.	12.35	11.50		23.85	31
2900	Maximum	"	80	.300	"	27.50	11.50		39	48
3000	Pavers, 4" x 4" x 4", thermal face, sawn joints									
3050	Minimum	D-11	65	.369	S.F.	24.50	14.15		38.65	48
3100	Maximum	"	65	.369	"	32	14.15		46.15	57

04 43 10.50 Lightweight Natural Stone

			Crew	Daily Output	Labor-Hours	Unit	Material	Labor	Equipment	Total	Total Incl O&P
0011	**LIGHTWEIGHT NATURAL STONE** Lava type										
0100	Veneer, rubble face, sawed back, irregular shapes	G	D-10	130	.246	S.F.	6.30	9.70	4.63	20.63	26.50
0200	Sawed face and back, irregular shapes	G	"	130	.246	"	6.30	9.70	4.63	20.63	26.50

04 43 10.55 Limestone

		Crew	Daily Output	Labor-Hours	Unit	Material	Labor	Equipment	Total	Total Incl O&P
0010	**LIMESTONE**, cut to size									
0020	Veneer facing panels									
0750	5" thick, 5' x 14' panels	D-10	275	.116	S.F.	52.50	4.59	2.19	59.28	67
1000	Sugarcube finish, 2" Thick, 3' x 5' panels		275	.116		35	4.59	2.19	41.78	48
1050	3" Thick, 4' x 9' panels		275	.116		38	4.59	2.19	44.78	51.50
1200	4" Thick, 5' x 11' panels		275	.116		41	4.59	2.19	47.78	54.50
1400	Sugarcube, textured finish, 4-1/2" thick, 5' x 12'		275	.116		45	4.59	2.19	51.78	59
1450	5" thick, 5' x 14' panels		275	.116		53.50	4.59	2.19	60.28	68
2000	Coping, sugarcube finish, top & 2 sides		30	1.067	C.F.	58.50	42	20	120.50	150
2100	Sills, lintels, jambs, trim, stops, sugarcube finish, average		20	1.600		58.50	63	30	151.50	193
2150	Detailed		20	1.600		58.50	63	30	151.50	193
2300	Steps, extra hard, 14" wide, 6" rise		50	.640	L.F.	35	25	12.05	72.05	90
3000	Quoins, plain finish, 6"x12"x12"	D-12	25	1.280	Ea.	58.50	47		105.50	136

04 43 Stone Masonry

04 43 10 – Masonry with Natural and Processed Stone

04 43 10.55 Limestone		Crew	Daily Output	Labor-Hours	Unit	Material	2009 Bare Costs Labor	Equipment	Total	Total Incl O&P
3050	6"x16"x24"	D-12	25	1.280	Ea.	78	47		125	158
04 43 10.60 Marble										
0011	**MARBLE**, ashlar, split face, 4" + or - thick, random									
0040	Lengths 1' to 4' & heights 2" to 7-1/2", average	D-8	175	.229	S.F.	16.90	8.50		25.40	31.50
0100	Base, polished, 3/4" or 7/8" thick, polished, 6" high	D-10	65	.492	L.F.	15.35	19.40	9.25	44	56.50
0300	Carvings or bas relief, from templates, average		80	.400	S.F.	137	15.75	7.50	160.25	183
0350	Maximum	↓	80	.400	"	320	15.75	7.50	343.25	380
0600	Columns, cornices, mouldings, etc.									
0650	Hand or special machine cut, average	D-10	35	.914	C.F.	137	36	17.20	190.20	224
0700	Maximum	"	35	.914	"	294	36	17.20	347.20	400
1000	Facing, polished finish, cut to size, 3/4" to 7/8" thick									
1050	Average	D-10	130	.246	S.F.	22.50	9.70	4.63	36.83	45
1100	Maximum		130	.246		52.50	9.70	4.63	66.83	77.50
1300	1-1/4" thick, average		125	.256		32.50	10.10	4.82	47.42	56.50
1350	Maximum		125	.256		65.50	10.10	4.82	80.42	92.50
1500	2" thick, average		120	.267		38	10.50	5	53.50	63
1550	Maximum	↓	120	.267	↓	65.50	10.50	5	81	93.50
1700	Rubbed finish, cut to size, 4" thick									
1740	Average	D-10	100	.320	S.F.	38	12.60	6	56.60	67
1780	Maximum	"	100	.320	"	73	12.60	6	91.60	106
2500	Flooring, polished tiles, 12" x 12" x 3/8" thick									
2510	Thin set, average	D-11	90	.267	S.F.	11.35	10.25		21.60	28
2600	Maximum		90	.267		174	10.25		184.25	208
2700	Mortar bed, average		65	.369		10.15	14.15		24.30	32.50
2740	Maximum	↓	65	.369		174	14.15		188.15	214
2780	Travertine, 3/8" thick, average	D-10	130	.246		14.40	9.70	4.63	28.73	35.50
2790	Maximum	"	130	.246	↓	35.50	9.70	4.63	49.83	59
2900	Shower or toilet partitions, 7/8" thick partitions									
3050	3/4" or 1-1/4" thick stiles, polished 2 sides, average	D-11	75	.320	S.F.	44	12.30		56.30	66.50
3201	Soffits, add to above prices				"	20%	100%			
3210	Stairs, risers, 7/8" thick x 6" high	D-10	115	.278	L.F.	14.80	10.95	5.25	31	38.50
3360	Treads, 12" wide x 1-1/4" thick	"	115	.278	"	21.50	10.95	5.25	37.70	46
3500	Thresholds, 3' long, 7/8" thick, 4" to 5" wide, plain	D-12	24	1.333	Ea.	16.30	49		65.30	92.50
3550	Beveled	↓	24	1.333	"	18.95	49		67.95	95.50
3700	Window stools, polished, 7/8" thick, 5" wide	↓	85	.376	L.F.	15.20	13.85		29.05	38
04 43 10.75 Sandstone or Brownstone										
0011	**SANDSTONE OR BROWNSTONE**									
0100	Sawed face veneer, 2-1/2" thick, to 2' x 4' panels	D-10	130	.246	S.F.	17.25	9.70	4.63	31.58	39
0150	4' thick, to 3'-6" x 8'panels		100	.320		17.25	12.60	6	35.85	44.50
0300	Split face, random sizes	↓	100	.320	↓	12.40	12.60	6	31	39.50
04 43 10.80 Slate										
0010	**SLATE**									
0040	Pennsylvania - blue gray to black									
0050	Vermont - unfading green, mottled green & purple, gray & purple									
0100	Virginia, blue black									
1000	Interior flooring, natural cleft, 1/2" thick									
1100	6" x 6" Pennsylvania	D-12	100	.320	S.F.	4	11.80		15.80	22.50
1150	Vermont		100	.320		11	11.80		22.80	30
1200	Virginia		100	.320		11.15	11.80		22.95	30
1300	24" x 24" Pennsylvania		120	.267		7.75	9.80		17.55	23.50
1350	Vermont		120	.267		25	9.80		34.80	42.50
1400	Virginia	↓	120	.267	↓	14.75	9.80		24.55	31

04 43 Stone Masonry

04 43 10 – Masonry with Natural and Processed Stone

04 43 10.80 Slate

		Crew	Daily Output	Labor-Hours	Unit	Material	2009 Bare Costs Labor	Equipment	Total	Total Incl O&P
1500	18" x 24" Pennsylvania	D-12	120	.267	S.F.	7.75	9.80		17.55	23.50
1550	Vermont		120	.267		16	9.80		25.80	32.50
1600	Virginia	↓	120	.267	↓	14.95	9.80		24.75	31.50
2000	Facing panels, 1-1/4" thick, to 4' x 4' panels									
2100	Natural cleft finish, Pennsylvania	D-10	180	.178	S.F.	34	7	3.34	44.34	51.50
2110	Vermont		180	.178		27	7	3.34	37.34	44.50
2120	Virginia	↓	180	.178		33	7	3.34	43.34	51
2150	Sand rubbed finish, surface, add					10.15			10.15	11.20
2200	Honed finish, add					7.35			7.35	8.10
2500	Ribbon, natural cleft finish, 1" thick, to 9 S.F.	D-10	80	.400		13.10	15.75	7.50	36.35	46.50
2550	Sand rubbed finish		80	.400		17.75	15.75	7.50	41	52
2600	Honed finish		80	.400		16.50	15.75	7.50	39.75	50.50
2700	1-1/2" thick		78	.410		17.05	16.20	7.70	40.95	52
2750	Sand rubbed finish		78	.410		22.50	16.20	7.70	46.40	58
2800	Honed finish		78	.410		21.50	16.20	7.70	45.40	56.50
2850	2" thick		76	.421		20.50	16.60	7.90	45	56
2900	Sand rubbed finish		76	.421		28.50	16.60	7.90	53	64.50
2950	Honed finish	↓	76	.421	↓	26	16.60	7.90	50.50	62
3500	Stair treads, sand finish, 1" thick x 12" wide									
3550	Under 3 L.F.	D-10	85	.376	L.F.	22	14.85	7.10	43.95	54.50
3600	3 L.F. to 6 L.F.	"	120	.267	"	24	10.50	5	39.50	47.50
3700	Ribbon, sand finish, 1" thick x 12" wide									
3750	To 6 L.F.	D-10	120	.267	L.F.	20	10.50	5	35.50	43.50
4000	Stools or sills, sand finish, 1" thick, 6" wide	D-12	160	.200		11.50	7.35		18.85	24
4100	Honed finish		160	.200		11	7.35		18.35	23.50
4200	10" wide		90	.356		17.75	13.10		30.85	39.50
4250	Honed finish		90	.356		16.50	13.10		29.60	38
4400	2" thick, 6" wide		140	.229		18.50	8.40		26.90	33.50
4450	Honed finish		140	.229		18.50	8.40		26.90	33.50
4600	10" wide		90	.356		29	13.10		42.10	52
4650	Honed finish	↓	90	.356		27.50	13.10		40.60	50.50
4800	For lengths over 3', add				↓	25%				

04 43 10.85 Window Sill

		Crew	Daily Output	Labor-Hours	Unit	Material	2009 Bare Costs Labor	Equipment	Total	Total Incl O&P
0010	**WINDOW SILL**									
0020	Bluestone, thermal top, 10" wide, 1-1/2" thick	D-1	85	.188	S.F.	16.05	6.85		22.90	28
0050	2" thick		75	.213	"	18.75	7.75		26.50	32.50
0100	Cut stone, 5" x 8" plain		48	.333	L.F.	11.90	12.10		24	31.50
0200	Face brick on edge, brick, 8" wide		80	.200		2.53	7.25		9.78	13.80
0400	Marble, 9" wide, 1" thick		85	.188		8.50	6.85		15.35	19.70
0900	Slate, colored, unfading, honed, 12" wide, 1" thick		85	.188		17.10	6.85		23.95	29
0950	2" thick	↓	70	.229	↓	24	8.30		32.30	38.50

04 57 Masonry Fireplaces

04 57 10 – Brick or Stone Fireplaces

04 57 10.10 Fireplace		Crew	Daily Output	Labor-Hours	Unit	Material	2009 Bare Costs Labor	Equipment	Total	Total Incl O&P
0010	**FIREPLACE**									
0100	Brick fireplace, not incl. foundations or chimneys									
0110	30" x 29" opening, incl. chamber, plain brickwork	D-1	.40	40	Ea.	535	1,450		1,985	2,800
0200	Fireplace box only (110 brick)	"	2	8	"	176	291		467	635
0300	For elaborate brickwork and details, add					35%	35%			
0400	For hearth, brick & stone, add	D-1	2	8	Ea.	197	291		488	655
0410	For steel angle, damper, cleanouts, add		4	4		138	145		283	370
0600	Plain brickwork, incl. metal circulator		.50	32		1,025	1,175		2,200	2,900
0800	Face brick only, standard size, 8" x 2-2/3" x 4"		.30	53.333	M	535	1,950		2,485	3,550

04 71 Manufactured Brick Masonry

04 71 10 – Simulated or Manufactured Brick

04 71 10.10 Simulated Brick

		Crew	Daily Output	Labor-Hours	Unit	Material	2009 Bare Costs Labor	Equipment	Total	Total Incl O&P
0010	**SIMULATED BRICK**									
0020	Aluminum, baked on colors	1 Carp	200	.040	S.F.	3.53	1.60		5.13	6.35
0050	Fiberglass panels		200	.040		3.87	1.60		5.47	6.75
0100	Urethane pieces cemented in mastic		150	.053		6.70	2.13		8.83	10.65
0150	Vinyl siding panels		200	.040		2.67	1.60		4.27	5.40
0160	Cement base, brick, incl. mastic	D-1	100	.160		4.22	5.80		10.02	13.45
0170	Corner		50	.320	V.L.F.	10.55	11.65		22.20	29.50
0180	Stone face, incl. mastic		100	.160	S.F.	10.20	5.80		16	20
0190	Corner		50	.320	V.L.F.	11.25	11.65		22.90	30

04 73 Manufactured Stone Masonry

04 73 20 – Simulated or Manufactured Stone

04 73 20.10 Simulated Stone

		Crew	Daily Output	Labor-Hours	Unit	Material	2009 Bare Costs Labor	Equipment	Total	Total Incl O&P
0010	**SIMULATED STONE**									
0100	Insulated fiberglass panels, 5/8" ply backer	L-4	200	.120	S.F.	11.40	4.48		15.88	19.45

Estimating Tips

05 05 00 Common Work Results for Metals

- Nuts, bolts, washers, connection angles, and plates can add a significant amount to both the tonnage of a structural steel job and the estimated cost. As a rule of thumb, add 10% to the total weight to account for these accessories.
- Type 2 steel construction, commonly referred to as "simple construction," consists generally of field-bolted connections with lateral bracing supplied by other elements of the building, such as masonry walls or x-bracing. The estimator should be aware, however, that shop connections may be accomplished by welding or bolting. The method may be particular to the fabrication shop and may have an impact on the estimated cost.

05 12 23 Structural Steel

- Steel items can be obtained from two sources: a fabrication shop or a metals service center. Fabrication shops can fabricate items under more controlled conditions than can crews in the field. They are also more efficient and can produce items more economically. Metal service centers serve as a source of long mill shapes to both fabrication shops and contractors.
- Most line items in this structural steel subdivision, and most items in 05 50 00 Metal Fabrications, are indicated as being shop fabricated. The bare material cost for these shop fabricated items is the "Invoice Cost" from the shop and includes the mill base price of steel plus mill extras, transportation to the shop, shop drawings and detailing where warranted, shop fabrication and handling, sandblasting and a shop coat of primer paint, all necessary structural bolts, and delivery to the job site. The bare labor cost and bare equipment cost for these shop fabricated items is for field installation or erection.
- Line items in Subdivision 05 12 23.40 Lightweight Framing, and other items scattered in Division 5, are indicated as being field fabricated. The bare material cost for these field fabricated items is the "Invoice Cost" from the metals service center and includes the mill base price of steel plus mill extras, transportation to the metals service center, material handling, and delivery of long lengths of mill shapes to the job site. Material costs for structural bolts and welding rods should be added to the estimate. The bare labor cost and bare equipment cost for these items is for both field fabrication and field installation or erection, and include time for cutting, welding and drilling in the fabricated metal items. Drilling into concrete and fasteners to fasten field fabricated items to other work are not included and should be added to the estimate.

05 20 00 Steel Joist Framing

- In any given project the total weight of open web steel joists is determined by the loads to be supported and the design. However, economies can be realized in minimizing the amount of labor used to place the joists. This is done by maximizing the joist spacing, and therefore minimizing the number of joists required to be installed on the job. Certain spacings and locations may be required by the design, but in other cases maximizing the spacing and keeping it as uniform as possible will keep the costs down.

05 30 00 Steel Decking

- The takeoff and estimating of metal deck involves more than simply the area of the floor or roof and the type of deck specified or shown on the drawings. Many different sizes and types of openings may exist. Small openings for individual pipes or conduits may be drilled after the floor/roof is installed, but larger openings may require special deck lengths as well as reinforcing or structural support. The estimator should determine who will be supplying this reinforcing. Additionally, some deck terminations are part of the deck package, such as screed angles and pour stops, and others will be part of the steel contract, such as angles attached to structural members and cast-in-place angles and plates. The estimator must ensure that all pieces are accounted for in the complete estimate.

05 50 00 Metal Fabrications

- The most economical steel stairs are those that use common materials, standard details, and most importantly, a uniform and relatively simple method of field assembly. Commonly available A36 channels and plates are very good choices for the main stringers of the stairs, as are angles and tees for the carrier members. Risers and treads are usually made by specialty shops, and it is most economical to use a typical detail in as many places as possible. The stairs should be pre-assembled and shipped directly to the site. The field connections should be simple and straightforward to be accomplished efficiently, and with minimum equipment and labor.

Reference Numbers

Reference numbers are shown in shaded boxes at the beginning of some major classifications. These numbers refer to related items in the Reference Section. The reference information may be an estimating procedure, an alternate pricing method, or technical information.
Note: Not all subdivisions listed here necessarily appear in this publication.

There's always a solution in steel.

In today's fast-paced, competitive construction industry, success is often a matter of access to the right information at the right time. The AISC Steel Solutions Center makes it easy for you to find, compare, select, and specify the right system for your project. From typical framing studies to total structural systems, **including project costs and schedules**, we can provide you with up-to-date information for your project—**for free!**

All you have to do is call 866.ASK.AISC or e-mail solutions@aisc.org.

Trademarks licensed from AISC

Your Connection to Ideas + Answers

312.670.2400 ▪ solutions@aisc.org

05 01 Maintenance of Metals

05 01 10 – Maintenance of Structural Metal Framing

05 01 10.51 Cleaning of Structural Metal Framing

		Daily Output	Labor-Hours	Unit	Material	2009 Bare Costs Labor	2009 Bare Costs Equipment	Total	Total Incl O&P
0010	**CLEANING OF STRUCTURAL METAL FRAMING**								
6125	Steel surface treatments, PDCA guidelines								
6170	Wire brush, hand (SSPC-SP2)	1 Psst / 400	.020	S.F.	.02	.72		.74	1.33
6180	Power tool (SSPC-SP3)	" / 700	.011		.06	.41		.47	.81
6215	Pressure washing, 2800-6000 S.F./day	1 Pord / 10000	.001			.03		.03	.04
6220	Steam cleaning, 2800-4000 S.F./day	2000	.004			.14		.14	.21
6225	Water blasting	2500	.003			.11		.11	.17
6230	Brush-off blast (SSPC-SP7)	E-11 / 1750	.018		.23	.65	.11	.99	1.47
6235	Com'l blast (SSPC-SP6), loose scale, fine pwder rust, 2.0 #/S.F. sand	1200	.027		.47	.95	.16	1.58	2.29
6240	Tight mill scale, little/no rust, 3.0 #/S.F. sand	1000	.032		.70	1.14	.20	2.04	2.90
6245	Exist coat blistered/pitted, 4.0 #/S.F. sand	875	.037		.94	1.30	.22	2.46	3.46
6250	Exist coat badly pitted/nodules, 6.7 #/S.F. sand	825	.039		1.57	1.38	.24	3.19	4.30
6255	Near white blast (SSPC-SP10), loose scale, fine rust, 5.6 #/S.F. sand	450	.071		1.31	2.54	.44	4.29	6.15
6260	Tight mill scale, little/no rust, 6.9 #/S.F. sand	325	.098		1.62	3.51	.60	5.73	8.30
6265	Exist coat blistered/pitted, 9.0 #/S.F. sand	225	.142		2.11	5.05	.87	8.03	11.80
6270	Exist coat badly pitted/nodules, 11.3 #/S.F. sand	150	.213		2.65	7.60	1.31	11.56	17.05

05 05 Common Work Results for Metals

05 05 05 – Selective Metals Demolition

05 05 05.10 Selective Demolition, Metals

		Crew	Daily Output	Labor-Hours	Unit	Material	Labor	Equipment	Total	Total Incl O&P
0010	**SELECTIVE DEMOLITION, METALS** R024119-10									
0015	Excludes shores, bracing, cutting, loading, hauling, dumping									
0020	Remove nuts only up to 3/4" diameter	1 Sswk	480	.017	Ea.		.75		.75	1.33
0030	7/8" to 1-1/4" diameter		240	.033			1.49		1.49	2.65
0040	1-3/8" to 2" diameter		160	.050			2.24		2.24	3.98
0060	Unbolt and remove structural bolts up to 3/4" diameter		240	.033			1.49		1.49	2.65
0070	7/8" to 2" diameter		160	.050			2.24		2.24	3.98
0140	Light weight framing members, remove whole or cut up, up to 20 lb		240	.033			1.49		1.49	2.65
0150	21 – 40 lb	2 Sswk	210	.076			3.41		3.41	6.05
0160	41 – 80 lb	3 Sswk	180	.133			5.95		5.95	10.60
0170	81 – 120 lb	4 Sswk	150	.213			9.55		9.55	17
0230	Structural members, remove whole or cut up, up to 500 lb	E-19	48	.500			21.50	22.50	44	62
0240	1/4 – 2 tons	E-18	36	1.111			49.50	30	79.50	119
0250	2 – 5 tons	E-24	30	1.067			47	26.50	73.50	110
0260	5 – 10 tons	E-20	24	2.667			117	55.50	172.50	261
0270	10 – 15 tons	E-2	18	3.111			135	96.50	231.50	340
0340	Fabricated item, remove whole or cut up, up to 20 lb	1 Sswk	96	.083			3.72		3.72	6.65
0350	21 – 40 lb	2 Sswk	84	.190			8.50		8.50	15.15
0360	41 – 80 lb	3 Sswk	72	.333			14.90		14.90	26.50
0370	81 – 120 lb	4 Sswk	60	.533			24		24	42.50
0380	121 – 500 lb	E-19	48	.500			21.50	22.50	44	62
0390	501 – 1000 lb	"	36	.667			29	30	59	82

05 05 21 – Fastening Methods for Metal

05 05 21.15 Drilling Steel

		Crew	Daily Output	Labor-Hours	Unit	Material	Labor	Equipment	Total	Total Incl O&P
0010	**DRILLING STEEL**									
1910	Drilling & layout for steel, up to 1/4" deep, no anchor									
1920	Holes, 1/4" diameter	1 Sswk	112	.071	Ea.	.08	3.19		3.27	5.80
1925	For each additional 1/4" depth, add		336	.024		.08	1.06		1.14	1.99
1930	3/8" diameter		104	.077		.09	3.44		3.53	6.25
1935	For each additional 1/4" depth, add		312	.026		.09	1.15		1.24	2.14

05 05 Common Work Results for Metals

05 05 21 – Fastening Methods for Metal

05 05 21.15 Drilling Steel

		Crew	Daily Output	Labor-Hours	Unit	Material	2009 Bare Costs Labor	Equipment	Total	Total Incl O&P
1940	1/2" diameter	1 Sswk	96	.083	Ea.	.10	3.72		3.82	6.75
1945	For each additional 1/4" depth, add		288	.028		.10	1.24		1.34	2.32
1950	5/8" diameter		88	.091		.13	4.06		4.19	7.40
1955	For each additional 1/4" depth, add		264	.030		.13	1.35		1.48	2.56
1960	3/4" diameter		80	.100		.15	4.47		4.62	8.10
1965	For each additional 1/4" depth, add		240	.033		.15	1.49		1.64	2.82
1970	7/8" diameter		72	.111		.18	4.97		5.15	9.05
1975	For each additional 1/4" depth, add		216	.037		.18	1.66		1.84	3.15
1980	1" diameter		64	.125		.20	5.60		5.80	10.15
1985	For each additional 1/4" depth, add	↓	192	.042	↓	.20	1.86		2.06	3.54
1990	For drilling up, add						40%			

05 05 21.90 Welding Steel

			Crew	Daily Output	Labor-Hours	Unit	Material	Labor	Equipment	Total	Total Incl O&P
0010	**WELDING STEEL**, Structural	R050521-20									
0020	Field welding, 1/8" E6011, cost per welder, no oper. engr		E-14	8	1	Hr.	4.73	46.50	16.80	68.03	107
0200	With 1/2 operating engineer		E-13	8	1.500		4.73	66	16.75	87.48	137
0300	With 1 operating engineer		E-12	8	2		4.73	86	16.80	107.53	166
0500	With no operating engineer, 2# weld rod per ton		E-14	8	1	Ton	4.73	46.50	16.80	68.03	107
0600	8# E6011 per ton		"	2	4		18.90	187	67	272.90	430
0800	With one operating engineer per welder, 2# E6011 per ton		E-12	8	2		4.73	86	16.80	107.53	166
0900	8# E6011 per ton		"	2	8	↓	18.90	345	67	430.90	665
1200	Continuous fillet, stick welding, incl. equipment										
1300	Single pass, 1/8" thick, 0.1#/L.F.		E-14	150	.053	L.F.	.24	2.49	.89	3.62	5.70
1400	3/16" thick, 0.2#/L.F.			75	.107		.47	4.98	1.79	7.24	11.35
1800	3 passes, 3/8" thick, 0.5#/L.F.			30	.267		1.18	12.45	4.47	18.10	28
2010	4 passes, 1/2" thick, 0.7#/L.F.		↓	22	.364		1.66	17	6.10	24.76	39
2600	For all position welding, add, minimum							20%			
2700	Maximum							300%			
2900	For semi-automatic welding, deduct, minimum							5%			
3000	Maximum					↓		15%			
4000	Cleaning and welding plates, bars, or rods										
4010	to existing beams, columns, or trusses		E-14	12	.667	L.F.	1.18	31	11.20	43.38	69

05 05 23 – Metal Fastenings

05 05 23.20 Expansion Anchors

			Crew	Daily Output	Labor-Hours	Unit	Material	Labor	Equipment	Total	Total Incl O&P
0010	**EXPANSION ANCHORS**										
2100	Hollow wall anchors for gypsum wall board, plaster or tile										
2300	1/8" diameter, short	G	1 Carp	160	.050	Ea.	.18	2		2.18	3.30
2400	Long	G		150	.053		.21	2.13		2.34	3.53
2500	3/16" diameter, short	G		150	.053		.36	2.13		2.49	3.70
2600	Long	G		140	.057		.47	2.28		2.75	4.06
2700	1/4" diameter, short	G		140	.057		.45	2.28		2.73	4.04
2800	Long	G		130	.062		.49	2.46		2.95	4.35
3000	Toggle bolts, bright steel, 1/8" diameter, 2" long	G		85	.094		.17	3.76		3.93	6.05
3100	4" long	G		80	.100		.20	4		4.20	6.40
3400	1/4" diameter, 3" long	G		75	.107		.30	4.26		4.56	6.95
3500	6" long	G		70	.114		.42	4.57		4.99	7.55
3600	3/8" diameter, 3" long	G		70	.114		.68	4.57		5.25	7.85
3700	6" long	G	↓	60	.133		1.13	5.35		6.48	9.50

05 05 23.30 Lag Screws

			Crew	Daily Output	Labor-Hours	Unit	Material	Labor	Equipment	Total	Total Incl O&P
0010	**LAG SCREWS**										
0020	Steel, 1/4" diameter, 2" long	G	1 Carp	200	.040	Ea.	.08	1.60		1.68	2.57
0100	3/8" diameter, 3" long	G		150	.053		.26	2.13		2.39	3.59
0200	1/2" diameter, 3" long	G	↓	130	.062	↓	.45	2.46		2.91	4.31

05 05 Common Work Results for Metals

05 05 23 – Metal Fastenings

05 05 23.30 Lag Screws		Crew	Daily Output	Labor-Hours	Unit	Material	2009 Bare Costs Labor	Equipment	Total	Total Incl O&P
0300	5/8" diameter, 3" long	G 1 Carp	120	.067	Ea.	1.12	2.66		3.78	5.35

05 05 23.35 Machine Screws

0010	**MACHINE SCREWS**									
0020	Steel, round head, #8 x 1" long	G 1 Carp	4.80	1.667	C	2.91	66.50		69.41	106
0110	#8 x 2" long	G	2.40	3.333		4.46	133		137.46	212
0200	#10 x 1" long	G	4	2		3.44	80		83.44	128
0300	#10 x 2" long	G	2	4		5.85	160		165.85	254

05 05 23.40 Machinery Anchors

0010	**MACHINERY ANCHORS**, heavy duty, incl. sleeve, floating base nut,									
0020	lower stud & coupling nut, fiber plug, connecting stud, washer & nut.									
0030	For flush mounted embedment in poured concrete heavy equip. pads.									
0200	Stud & bolt, 1/2" diameter	G E-16	40	.400	Ea.	67	18.30	3.36	88.66	110
0300	5/8" diameter	G	35	.457		74.50	21	3.84	99.34	123
0500	3/4" diameter	G	30	.533		86	24.50	4.47	114.97	143
0600	7/8" diameter	G	25	.640		93.50	29.50	5.35	128.35	161
0800	1" diameter	G	20	.800		98.50	36.50	6.70	141.70	180
0900	1-1/4" diameter	G	15	1.067		131	49	8.95	188.95	241

05 05 23.50 Powder Actuated Tools and Fasteners

0010	**POWDER ACTUATED TOOLS & FASTENERS**									
0020	Stud driver, .22 caliber, buy, minimum				Ea.	224			224	246
0100	Maximum				"	360			360	395
0300	Powder charges for above, low velocity				C	8.65			8.65	9.50
0400	Standard velocity					12.30			12.30	13.55
0600	Drive pins & studs, 1/4" & 3/8" diam., to 3" long, minimum	G 1 Carp	4.80	1.667		2.88	66.50		69.38	106
0700	Maximum	G "	4	2		11.25	80		91.25	136
0800	Pneumatic stud driver for 1/8" diameter studs				Ea.	1,550			1,550	1,725

05 05 23.55 Rivets

0010	**RIVETS**									
0100	Aluminum rivet & mandrel, 1/2" grip length x 1/8" diameter	G 1 Carp	4.80	1.667	C	6.15	66.50		72.65	110

05 05 23.90 Welding Rod

0010	**WELDING ROD**									
0020	Steel, type 6011, 1/8" diam., less than 500#				Lb.	2.36			2.36	2.60
0100	500# to 2,000#					2.13			2.13	2.34
0300	5/32" diameter, less than 500#					2.28			2.28	2.50
0310	500# to 2,000#					2.05			2.05	2.26
0400	3/16" diam., less than 500#					2.26			2.26	2.49
0500	500# to 2,000#					2.04			2.04	2.24
0620	Steel, type 6010, 1/8" diam., less than 500#					2.05			2.05	2.26
0630	500# to 2,000#					1.85			1.85	2.04
0640	2,000# to 5,000#					1.74			1.74	1.91
0650	Steel, type 7018 Low Hydrogen, 1/8" diam., less than 500#					2.39			2.39	2.63
0660	500# to 2,000#					2.15			2.15	2.37
0700	Steel, type 7024 Jet Weld, 1/8" diam., less than 500#					2.26			2.26	2.49
0710	500# to 2,000#					2.04			2.04	2.24
1550	Aluminum, type 4043 TIG, 1/8" diam., less than 10#					4.15			4.15	4.57
1560	10# to 60#					3.74			3.74	4.11
1570	Over 60#					3.52			3.52	3.87
1600	Aluminum, type 5356 TIG, 1/8" diam., less than 10#					4.33			4.33	4.76
1610	10# to 60#					3.90			3.90	4.29
1620	Over 60#					3.67			3.67	4.03
1900	Cast iron, type 8 Nickel, 1/8" diam., less than 500#					29.50			29.50	32.50

05 05 Common Work Results for Metals

05 05 23 – Metal Fastenings

05 05 23.90 Welding Rod		Crew	Daily Output	Labor-Hours	Unit	Material	2009 Bare Costs Labor	Equipment	Total	Total Incl O&P
1910	500# to 1,000#				Lb.	26.50			26.50	29
1920	Over 1,000#					25			25	27.50
2000	Stainless steel, type 316/316L, 1/8" diam., less than 500#					7.85			7.85	8.65
2100	500# to 1000#					7.10			7.10	7.80

05 12 Structural Steel Framing

05 12 23 – Structural Steel for Buildings

05 12 23.10 Ceiling Supports

			Crew	Daily Output	Labor-Hours	Unit	Material	Labor	Equipment	Total	Total Incl O&P
0010	**CEILING SUPPORTS**										
1000	Entrance door/folding partition supports, shop fabricated	G	E-4	60	.533	L.F.	30	24	2.23	56.23	78.50
1100	Linear accelerator door supports	G		14	2.286		137	103	9.60	249.60	345
1200	Lintels or shelf angles, hung, exterior hot dipped galv.	G		267	.120		20.50	5.40	.50	26.40	32.50
1250	Two coats primer paint instead of galv.	G		267	.120		17.75	5.40	.50	23.65	29.50
1400	Monitor support, ceiling hung, expansion bolted	G		4	8	Ea.	475	360	33.50	868.50	1,200
1450	Hung from pre-set inserts	G		6	5.333		510	241	22.50	773.50	1,025
1600	Motor supports for overhead doors	G		4	8		242	360	33.50	635.50	950
1700	Partition support for heavy folding partitions, without pocket	G		24	1.333	L.F.	68.50	60.50	5.60	134.60	188
1750	Supports at pocket only	G		12	2.667		137	121	11.15	269.15	375
2000	Rolling grilles & fire door supports	G		34	.941		58.50	42.50	3.94	104.94	145
2100	Spider-leg light supports, expansion bolted to ceiling slab	G		8	4	Ea.	195	181	16.75	392.75	555
2150	Hung from pre-set inserts	G		12	2.667	"	210	121	11.15	342.15	460
2400	Toilet partition support	G		36	.889	L.F.	68.50	40	3.72	112.22	151
2500	X-ray travel gantry support	G		12	2.667	"	234	121	11.15	366.15	485

05 12 23.17 Columns, Structural

			Crew	Daily Output	Labor-Hours	Unit	Material	Labor	Equipment	Total	Total Incl O&P
0010	**COLUMNS, STRUCTURAL**										
0015	Made from recycled materials										
0020	Shop fab'd for 100-ton, 1-2 story project, bolted connections										
4500	Structural tubing, sq, 4" x 4" x 1/4" x 12'-0"	G	E-2	58	.966	Ea.	248	42	30	320	375
4550	6" x 6" x 1/4" x 12'-0"	G		54	1.037		405	45	32	482	560
4600	8" x 8" x 3/8" x 14'-0"	G		50	1.120		880	49	35	964	1,075
4650	10" x 10" x 1/2" x 16'-0"	G		48	1.167		1,625	51	36.50	1,712.50	1,925
5100	Structural tubing, rect, 5" to 6" wide, light section	G		8000	.007	Lb.	1.50	.30	.22	2.02	2.41
5300	7" to 10" wide, light section	G		15000	.004	"	1.50	.16	.12	1.78	2.06
5500	Structural tubing, rect, 5" x 3" x 1/4" x 12'-0"	G		58	.966	Ea.	240	42	30	312	370
5550	6" x 4" x 5/16" x 12'-0"	G		54	1.037		375	45	32	452	530
5600	8" x 4" x 3/8" x 12'-0"	G		54	1.037		550	45	32	627	715
5650	10" x 6" x 3/8" x 14'-0"	G		50	1.120		880	49	35	964	1,075
5700	12" x 8" x 1/2" x 16'-0"	G		48	1.167		1,625	51	36.50	1,712.50	1,900
8090	For projects 75 to 99 tons, add					All	10%				
8092	50 to 74 tons, add						20%				
8094	25 to 49 tons, add						30%	10%			
8096	10 to 24 tons, add						50%	25%			
8098	2 to 9 tons, add						75%	50%			
8099	Less than 2 tons, add						100%	100%			

05 12 23.40 Lightweight Framing

			Crew	Daily Output	Labor-Hours	Unit	Material	Labor	Equipment	Total	Total Incl O&P
0010	**LIGHTWEIGHT FRAMING**										
0015	Made from recycled materials										
0200	For load-bearing steel studs see Div. 05 41 13.30										
0400	Angle framing, field fabricated, 4" and larger	G	E-3	440	.055	Lb.	.87	2.47	.30	3.64	5.70
0450	Less than 4" angles	G		265	.091		.90	4.11	.51	5.52	8.85

05 12 Structural Steel Framing

05 12 23 – Structural Steel for Buildings

05 12 23.40 Lightweight Framing

			Crew	Daily Output	Labor-Hours	Unit	Material	2009 Bare Costs Labor	Equipment	Total	Total Incl O&P
0600	Channel framing, field fabricated, 8" and larger	G	E-3	500	.048	Lb.	.90	2.18	.27	3.35	5.15
0650	Less than 8" channels	G	↓	335	.072		.90	3.25	.40	4.55	7.25
1000	Continuous slotted channel framing system, shop fab, minimum	G	2 Sswk	2400	.007		4.65	.30		4.95	5.65
1200	Maximum	G	"	1600	.010		5.25	.45		5.70	6.60
1300	Cross bracing, rods, shop fabricated, 3/4" diameter	G	E-3	700	.034		1.80	1.56	.19	3.55	4.96
1310	7/8" diameter	G		850	.028		1.80	1.28	.16	3.24	4.43
1320	1" diameter	G		1000	.024		1.80	1.09	.13	3.02	4.07
1330	Angle, 5" x 5" x 3/8"	G		2800	.009		1.80	.39	.05	2.24	2.72
1350	Hanging lintels, shop fabricated, average	G	↓	850	.028		1.80	1.28	.16	3.24	4.43
1380	Roof frames, shop fabricated, 3'-0" square, 5' span	G	E-2	4200	.013		1.80	.58	.41	2.79	3.43
1400	Tie rod, not upset, 1-1/2" to 4" diameter, with turnbuckle	G	2 Sswk	800	.020		1.95	.89		2.84	3.74
1420	No turnbuckle	G		700	.023		1.88	1.02		2.90	3.88
1500	Upset, 1-3/4" to 4" diameter, with turnbuckle	G		800	.020		1.95	.89		2.84	3.74
1520	No turnbuckle	G	↓	700	.023		1.88	1.02		2.90	3.88

05 12 23.45 Lintels

			Crew	Daily Output	Labor-Hours	Unit	Material	Labor	Equipment	Total	Total Incl O&P
0010	**LINTELS**										
0015	Made from recycled materials										
0020	Plain steel angles, shop fabricated, under 500 lb.	G	1 Bric	550	.015	Lb.	1.16	.59		1.75	2.16
0100	500 to 1000 lb.	G	"	640	.013		1.13	.51		1.64	2.01
0500	For built-up angles and plates, add to above	G				↓	.38			.38	.41
0700	For engineering, add to above						.15			.15	.17
2000	Steel angles, 3-1/2" x 3", 1/4" thick, 2'-6" long	G	1 Bric	47	.170	Ea.	16.20	6.90		23.10	28.50
2100	4'-6" long	G		26	.308		29	12.45		41.45	51
2600	4" x 3-1/2", 1/4" thick, 5'-0" long	G		21	.381		37	15.45		52.45	64.50
2700	9'-0" long	G	↓	12	.667		67	27		94	115

05 15 Wire Rope Assemblies

05 15 16 – Steel Wire Rope Assemblies

05 15 16.05 Accessories for Steel Wire Rope

			Crew	Daily Output	Labor-Hours	Unit	Material	Labor	Equipment	Total	Total Incl O&P
0010	**ACCESSORIES FOR STEEL WIRE ROPE**										
0015	Made from recycled materials										
1500	Thimbles, heavy duty, 1/4"	G	E-17	160	.100	Ea.	.63	4.57		5.20	8.85
1510	1/2"	G		160	.100		2.77	4.57		7.34	11.20
1520	3/4"	G		105	.152		6.30	6.95		13.25	19.30
1530	1"	G		52	.308		12.60	14.05		26.65	39
1540	1-1/4"	G		38	.421		19.35	19.25		38.60	56
1550	1-1/2"	G		13	1.231		54.50	56.50		111	160
1560	1-3/4"	G		8	2		112	91.50		203.50	287
1570	2"	G		6	2.667		164	122		286	395
1580	2-1/4"	G		4	4		221	183		404	570
1600	Clips, 1/4" diameter	G		160	.100		2.84	4.57		7.41	11.25
1610	3/8" diameter	G		160	.100		3.11	4.57		7.68	11.55
1620	1/2" diameter	G		160	.100		5	4.57		9.57	13.65
1630	3/4" diameter	G		102	.157		8.10	7.15		15.25	21.50
1640	1" diameter	G		64	.250		13.50	11.45		24.95	35.50
1650	1-1/4" diameter	G		35	.457		22	21		43	61.50
1670	1-1/2" diameter	G		26	.615		30	28		58	83
1680	1-3/4" diameter	G		16	1		69.50	45.50		115	158
1690	2" diameter	G		12	1.333		77.50	61		138.50	194
1700	2-1/4" diameter	G		10	1.600		114	73		187	255
1800	Sockets, open swage, 1/4" diameter	G	↓	160	.100		40.50	4.57		45.07	53

05 15 Wire Rope Assemblies

05 15 16 – Steel Wire Rope Assemblies

05 15 16.05 Accessories for Steel Wire Rope		Crew	Daily Output	Labor-Hours	Unit	Material	2009 Bare Costs Labor	Equipment	Total	Total Incl O&P
1810	1/2" diameter	E-17	77	.208	Ea.	58.50	9.50		68	81.50
1820	3/4" diameter		19	.842		91.50	38.50		130	169
1830	1" diameter		9	1.778		163	81		244	325
1840	1-1/4" diameter		5	3.200		227	146		373	510
1850	1-1/2" diameter		3	5.333		495	244		739	980
1860	1-3/4" diameter		3	5.333		880	244		1,124	1,400
1870	2" diameter		1.50	10.667		1,325	485		1,810	2,350
1900	Closed swage, 1/4" diameter		160	.100		23.50	4.57		28.07	34
1910	1/2" diameter		104	.154		41	7.05		48.05	57.50
1920	3/4" diameter		32	.500		61.50	23		84.50	108
1930	1" diameter		15	1.067		108	49		157	206
1940	1-1/4" diameter		7	2.286		162	104		266	365
1950	1-1/2" diameter		4	4		293	183		476	650
1960	1-3/4" diameter		3	5.333		435	244		679	910
1970	2" diameter		2	8		840	365		1,205	1,575
2000	Open spelter, galv., 1/4" diameter		160	.100		49.50	4.57		54.07	62
2010	1/2" diameter		70	.229		51.50	10.45		61.95	75
2020	3/4" diameter		26	.615		77	28		105	135
2030	1" diameter		10	1.600		214	73		287	365
2040	1-1/4" diameter		5	3.200		305	146		451	595
2050	1-1/2" diameter		4	4		650	183		833	1,050
2060	1-3/4" diameter		2	8		1,125	365		1,490	1,900
2070	2" diameter		1.20	13.333		1,300	610		1,910	2,500
2080	2-1/2" diameter		1	16		2,400	730		3,130	3,950
2100	Closed spelter, galv., 1/4" diameter		160	.100		40	4.57		44.57	52
2110	1/2" diameter		88	.182		42.50	8.30		50.80	62
2120	3/4" diameter		30	.533		64.50	24.50		89	115
2130	1" diameter		13	1.231		138	56.50		194.50	251
2140	1-1/4" diameter		7	2.286		220	104		324	430
2150	1-1/2" diameter		6	2.667		475	122		597	735
2160	1-3/4" diameter		2.80	5.714		630	261		891	1,150
2170	2" diameter		2	8		780	365		1,145	1,500
2200	Jaw & jaw turnbuckles, 1/4" x 4"		160	.100		22.50	4.57		27.07	33
2250	1/2" x 6"		96	.167		28.50	7.60		36.10	45
2260	1/2" x 9"		77	.208		38	9.50		47.50	59
2270	1/2" x 12"		66	.242		42.50	11.10		53.60	67
2300	3/4" x 6"		38	.421		56	19.25		75.25	96
2310	3/4" x 9"		30	.533		61.50	24.50		86	112
2320	3/4" x 12"		28	.571		79.50	26		105.50	134
2330	3/4" x 18"		23	.696		95	32		127	161
2350	1" x 6"		17	.941		108	43		151	196
2360	1" x 12"		13	1.231		119	56.50		175.50	231
2370	1" x 18"		10	1.600		178	73		251	325
2380	1" x 24"		9	1.778		196	81		277	360
2400	1-1/4" x 12"		7	2.286		199	104		303	405
2410	1-1/4" x 18"		6.50	2.462		247	112		359	470
2420	1-1/4" x 24"		5.60	2.857		330	131		461	600
2450	1-1/2" x 12"		5.20	3.077		335	141		476	620
2460	1-1/2" x 18"		4	4		355	183		538	715
2470	1-1/2" x 24"		3.20	5		480	229		709	930
2500	1-3/4" x 18"		3.20	5		725	229		954	1,200
2510	1-3/4" x 24"		2.80	5.714		825	261		1,086	1,375
2550	2" x 24"		1.60	10		1,125	455		1,580	2,050

05 15 Wire Rope Assemblies

05 15 16 – Steel Wire Rope Assemblies

05 15 16.50 Steel Wire Rope

			Crew	Daily Output	Labor-Hours	Unit	Material	2009 Bare Costs Labor	Equipment	Total	Total Incl O&P
0010	**STEEL WIRE ROPE**										
0015	Made from recycled materials										
0020	6 x 19, bright, fiber core, 5000' rolls, 1/2" diameter	G				L.F.	1.68			1.68	1.84
0050	Steel core	G					2.21			2.21	2.43
0100	Fiber core, 1" diameter	G					5.65			5.65	6.20
0150	Steel core	G					6.45			6.45	7.10
0300	6 x 19, galvanized, fiber core, 1/2" diameter	G					2.47			2.47	2.72
0350	Steel core	G					2.82			2.82	3.10
0400	Fiber core, 1" diameter	G					7.25			7.25	7.95
0450	Steel core	G					7.60			7.60	8.35
0500	6 x 7, bright, IPS, fiber core, <500 L.F. w/acc., 1/4" diameter	G	E-17	6400	.003		1.25	.11		1.36	1.57
0510	1/2" diameter	G		2100	.008		3.04	.35		3.39	3.96
0520	3/4" diameter	G		960	.017		5.50	.76		6.26	7.40
0550	6 x 19, bright, IPS, IWRC, <500 L.F. w/acc., 1/4" diameter	G		5760	.003		1.85	.13		1.98	2.27
0560	1/2" diameter	G		1730	.009		3	.42		3.42	4.05
0570	3/4" diameter	G		770	.021		5.20	.95		6.15	7.40
0580	1" diameter	G		420	.038		8.80	1.74		10.54	12.80
0590	1-1/4" diameter	G		290	.055		14.65	2.52		17.17	20.50
0600	1-1/2" diameter	G		192	.083		18	3.81		21.81	26.50
0610	1-3/4" diameter	G	E-18	240	.167		28.50	7.40	4.50	40.40	49.50
0620	2" diameter	G		160	.250		37	11.10	6.75	54.85	67
0630	2-1/4" diameter	G		160	.250		49	11.10	6.75	66.85	80.50
0650	6 x 37, bright, IPS, IWRC, <500 L.F. w/acc., 1/4" diameter	G	E-17	6400	.003		2.35	.11		2.46	2.78
0660	1/2" diameter	G		1730	.009		3.98	.42		4.40	5.15
0670	3/4" diameter	G		770	.021		6.45	.95		7.40	8.75
0680	1" diameter	G		430	.037		10.20	1.70		11.90	14.25
0690	1-1/4" diameter	G		290	.055		15.40	2.52		17.92	21.50
0700	1-1/2" diameter	G		190	.084		22	3.85		25.85	31
0710	1-3/4" diameter	G	E-18	260	.154		35	6.85	4.15	46	55
0720	2" diameter	G		200	.200		45.50	8.90	5.40	59.80	71.50
0730	2-1/4" diameter	G		160	.250		60	11.10	6.75	77.85	92.50
0800	6 x 19 & 6 x 37, swaged, 1/2" diameter	G	E-17	1220	.013		7.15	.60		7.75	8.90
0810	9/16" diameter	G		1120	.014		8.30	.65		8.95	10.25
0820	5/8" diameter	G		930	.017		9.85	.79		10.64	12.20
0830	3/4" diameter	G		640	.025		12.55	1.14		13.69	15.85
0840	7/8" diameter	G		480	.033		15.80	1.52		17.32	20
0850	1" diameter	G		350	.046		19.30	2.09		21.39	24.50
0860	1-1/8" diameter	G		288	.056		23.50	2.54		26.04	30.50
0870	1-1/4" diameter	G		230	.070		28.50	3.18		31.68	37
0880	1-3/8" diameter	G		192	.083		33	3.81		36.81	43.50
0890	1-1/2" diameter	G	E-18	300	.133		40.50	5.90	3.60	50	58.50

05 15 16.60 Galvanized Steel Wire Rope and Accessories

			Crew	Daily Output	Labor-Hours	Unit	Material	Labor	Equipment	Total	Total Incl O&P
0010	**GALVANIZED STEEL WIRE ROPE & ACCESSORIES**										
0015	Made from recycled materials										
3000	Aircraft cable, galvanized, 7 x 7 x 1/8"	G	E-17	5000	.003	L.F.	.23	.15		.38	.51
3100	Clamps, 1/8"	G	"	125	.128	Ea.	2.03	5.85		7.88	12.65

05 15 16.70 Temporary Cable Safety Railing

			Crew	Daily Output	Labor-Hours	Unit	Material	Labor	Equipment	Total	Total Incl O&P
0010	**TEMPORARY CABLE SAFETY RAILING**, Each 100' strand incl.										
0020	2 eyebolts, 1 turnbuckle, 100' cable, 2 thimbles, 6 clips										
0025	Made from recycled materials										
0100	One strand using 1/4" cable & accessories	G	2 Sswk	4	4	C.L.F.	194	179		373	535
0200	1/2" cable & accessories	G	"	2	8	"	390	360		750	1,075

05 21 Steel Joist Framing

05 21 13 – Deep Longspan Steel Joist Framing

05 21 13.50 Deep Longspan Joists

			Crew	Daily Output	Labor-Hours	Unit	Material	2009 Bare Costs Labor	Equipment	Total	Total Incl O&P
0010	**DEEP LONGSPAN JOISTS**										
3010	DLH series, 40-ton job lots, bolted cross bridging, shop primer										
3015	Made from recycled materials										
3020	Spans to 144' (shipped in 2 pieces), minimum	G	E-7	16	5	Ton	2,225	221	126	2,572	2,975
3040	Average	G	↓	13	6.154		2,400	271	155	2,826	3,300
3100	Maximum	G	↓	11	7.273	↓	2,900	320	183	3,403	3,950
3500	For less than 40-ton job lots										
3502	For 30 to 39 tons, add						10%				
3504	20 to 29 tons, add						20%				
3506	10 to 19 tons, add						30%				
3507	5 to 9 tons, add						50%	25%			
3508	1 to 4 tons, add						75%	50%			
3509	Less than 1 ton, add						100%	100%			
4020	Spans to 200' (shipped in 3 pieces), minimum	G	E-7	16	5	Ton	2,200	221	126	2,547	2,950
4040	Average	G	↓	13	6.154		2,475	271	155	2,901	3,375
4060	Maximum	G	↓	11	7.273	↓	2,950	320	183	3,453	3,975
6100	For less than 40-ton job lots										
6102	For 30 to 39 tons, add						10%				
6104	20 to 29 tons, add						20%				
6106	10 to 19 tons, add						30%				
6107	5 to 9 tons, add						50%	25%			
6108	1 to 4 tons, add						75%	50%			
6109	Less than 1 ton, add						100%	100%			

05 21 16 – Longspan Steel Joist Framing

05 21 16.50 Longspan Joists

			Crew	Daily Output	Labor-Hours	Unit	Material	2009 Bare Costs Labor	Equipment	Total	Total Incl O&P
0010	**LONGSPAN JOISTS**										
2000	LH series, 40-ton job lots, bolted cross bridging, shop primer										
2015	Made from recycled materials										
2020	Spans to 96', minimum	G	E-7	16	5	Ton	2,050	221	126	2,397	2,775
2040	Average	G	↓	13	6.154		2,250	271	155	2,676	3,125
2080	Maximum	G	↓	11	7.273	↓	2,675	320	183	3,178	3,675
2600	For less than 40-ton job lots										
2602	For 30 to 39 tons, add						10%				
2604	20 to 29 tons, add						20%				
2606	10 to 19 tons, add						30%				
2607	5 to 9 tons, add						50%	25%			
2608	1 to 4 tons, add						75%	50%			
2609	Less than 1 ton, add						100%	100%			
6000	For welded cross bridging, add							30%			

05 21 19 – Open Web Steel Joist Framing

05 21 19.10 Open Web Joists

			Crew	Daily Output	Labor-Hours	Unit	Material	2009 Bare Costs Labor	Equipment	Total	Total Incl O&P
0010	**OPEN WEB JOISTS**										
0015	Made from recycled materials										
0020	K series, 40-ton lots, horiz. bridging, spans to 30', shop primer, minimum	G	E-7	15	5.333	Ton	1,825	235	134	2,194	2,550
0050	Average	G	↓	12	6.667		2,050	294	167	2,511	2,950
0080	Maximum	G	↓	9	8.889		2,450	390	223	3,063	3,625
0410	Span 30' to 50', minimum	G		17	4.706		1,775	208	118	2,101	2,450
0440	Average	G		17	4.706		2,000	208	118	2,326	2,700
0460	Maximum	G	↓	10	8	↓	2,125	355	201	2,681	3,150
0800	For less than 40-ton job lots										
0802	For 30 to 39 tons, add						10%				
0804	20 to 29 tons, add						20%				

05 21 Steel Joist Framing

05 21 19 – Open Web Steel Joist Framing

05 21 19.10 Open Web Joists		Crew	Daily Output	Labor-Hours	Unit	Material	2009 Bare Costs Labor	Equipment	Total	Total Incl O&P
0806	10 to 19 tons, add					30%				
0807	5 to 9 tons, add					50%	25%			
0808	1 to 4 tons, add					75%	50%			
0809	Less than 1 ton, add					100%	100%			
1010	CS series, 40-ton job lots, horizontal bridging, shop primer									
1020	Spans to 30', minimum	G	E-7	15	5.333	Ton	1,875	235	134	2,244	2,625
1040	Average	G		12	6.667		2,100	294	167	2,561	3,000
1060	Maximum	G		9	8.889		2,475	390	223	3,088	3,650
1500	For less than 40-ton job lots									
1502	For 30 to 39 tons, add					10%				
1504	20 to 29 tons, add					20%				
1506	10 to 19 tons, add					30%				
1507	5 to 9 tons, add					50%	25%			
1508	1 to 4 tons, add					75%	50%			
1509	Less than 1 ton, add					100%	100%			
6200	For shop prime paint other than mfrs. standard, add					20%				
6300	For bottom chord extensions, add per chord	G				Ea.	44.50			44.50	48.50
6400	Individual steel bearing plate, 6" x 6" x 1/4" with J-hook	G	1 Bric	160	.050	"	9	2.03		11.03	12.95

05 21 23 – Steel Joist Girder Framing

05 21 23.50 Joist Girders

		Crew	Daily Output	Labor-Hours	Unit	Material	Labor	Equipment	Total	Total Incl O&P
0010	**JOIST GIRDERS**									
0015	Made from recycled materials									
7000	Joist girders, 40-ton job lots, shop primer, minimum	G	E-5	15	5.333	Ton	1,850	235	125	2,210	2,600
7020	Average	G		13	6.154		2,050	271	144	2,465	2,875
7040	Maximum	G		11	7.273		2,150	320	170	2,640	3,100
7100	For less than 40-ton job lots									
7102	For 30 to 39 tons, add					10%				
7104	20 to 29 tons, add					20%				
7106	10 to 19 tons, add					30%				
7107	5 to 9 tons, add					50%	25%			
7108	1 to 4 tons, add					75%	50%			
7109	Less than 1 ton, add					100%	100%			
8000	Trusses, 40-ton job lots, shop fabricated WT chords, shop primer, average	G	E-5	11	7.273	Ton	6,675	320	170	7,165	8,100
8100	For less than 40-ton job lots									
8102	For 30 to 39 tons, add					10%				
8104	20 to 29 tons, add					20%				
8106	10 to 19 tons, add					30%				
8107	5 to 9 tons, add					50%	25%			
8108	1 to 4 tons, add					75%	50%			
8109	Less than 1 ton, add					100%	100%			

05 31 Steel Decking

05 31 33 – Steel Form Decking

05 31 33.50 Form Decking

		Crew	Daily Output	Labor-Hours	Unit	Material	2009 Bare Costs Labor	Equipment	Total	Total Incl O&P
0010	**FORM DECKING**									
0015	Made from recycled materials									
6100	Slab form, steel, 28 gauge, 9/16" deep, uncoated	G	E-4	4000	.008	S.F.	1.72	.36	.03	2.11	2.57
6200	Galvanized	G		4000	.008		1.52	.36	.03	1.91	2.35
6220	24 gauge, 1" deep, uncoated	G		3900	.008		1.87	.37	.03	2.27	2.76
6240	Galvanized	G		3900	.008		2.20	.37	.03	2.60	3.12
6300	24 gauge, 1-5/16" deep, uncoated	G		3800	.008		1.99	.38	.04	2.41	2.91
6400	Galvanized	G		3800	.008		2.34	.38	.04	2.76	3.29
6500	22 gauge, 1-5/16" deep, uncoated	G		3700	.009		2.50	.39	.04	2.93	3.49
6600	Galvanized	G		3700	.009		2.55	.39	.04	2.98	3.55
6700	22 gauge, 2" deep uncoated	G		3600	.009		3.28	.40	.04	3.72	4.37
6800	Galvanized	G		3600	.009		3.22	.40	.04	3.66	4.30
7000	Sheet metal edge closure form, 12" wide with 2 bends, galv										
7100	18 gauge	G	E-14	360	.022	L.F.	5.30	1.04	.37	6.71	8.10
7200	16 gauge	G	"	360	.022	"	7.20	1.04	.37	8.61	10.15

05 41 Structural Metal Stud Framing

05 41 13 – Load-Bearing Metal Stud Framing

05 41 13.05 Bracing

		Crew	Daily Output	Labor-Hours	Unit	Material	2009 Bare Costs Labor	Equipment	Total	Total Incl O&P
0010	**BRACING**, shear wall X-bracing, per 10' x 10' bay, one face										
0015	Made of recycled materials										
0120	Metal strap, 20 ga x 4" wide	G	2 Carp	18	.889	Ea.	27	35.50		62.50	84.50
0130	6" wide	G		18	.889		42	35.50		77.50	101
0160	18 ga x 4" wide	G		16	1		39	40		79	105
0170	6" wide	G		16	1		57	40		97	125
0410	Continuous strap bracing, per horizontal row on both faces										
0420	Metal strap, 20 ga x 2" wide, studs 12" O.C.	G	1 Carp	7	1.143	C.L.F.	68	45.50		113.50	146
0430	16" O.C.	G		8	1		68	40		108	137
0440	24" O.C.	G		10	.800		68	32		100	125
0450	18 ga x 2" wide, studs 12" O.C.	G		6	1.333		94.50	53.50		148	187
0460	16" O.C.	G		7	1.143		94.50	45.50		140	175
0470	24" O.C.	G		8	1		94.50	40		134.50	166

05 41 13.10 Bridging

		Crew	Daily Output	Labor-Hours	Unit	Material	2009 Bare Costs Labor	Equipment	Total	Total Incl O&P
0010	**BRIDGING**, solid between studs w/ 1-1/4" leg track, per stud bay										
0015	Made from recycled materials										
0200	Studs 12" O.C., 18 ga x 2-1/2" wide	G	1 Carp	125	.064	Ea.	1.16	2.56		3.72	5.25
0210	3-5/8" wide	G		120	.067		1.40	2.66		4.06	5.65
0220	4" wide	G		120	.067		1.48	2.66		4.14	5.75
0230	6" wide	G		115	.070		1.93	2.78		4.71	6.45
0240	8" wide	G		110	.073		2.45	2.91		5.36	7.20
0300	16 ga x 2-1/2" wide	G		115	.070		1.44	2.78		4.22	5.90
0310	3-5/8" wide	G		110	.073		1.76	2.91		4.67	6.45
0320	4" wide	G		110	.073		1.88	2.91		4.79	6.60
0330	6" wide	G		105	.076		2.42	3.04		5.46	7.40
0340	8" wide	G		100	.080		3.08	3.20		6.28	8.35
1200	Studs 16" O.C., 18 ga x 2-1/2" wide	G		125	.064		1.49	2.56		4.05	5.60
1210	3-5/8" wide	G		120	.067		1.80	2.66		4.46	6.10
1220	4" wide	G		120	.067		1.90	2.66		4.56	6.20
1230	6" wide	G		115	.070		2.48	2.78		5.26	7.05
1240	8" wide	G		110	.073		3.14	2.91		6.05	7.95
1300	16 ga x 2-1/2" wide	G		115	.070		1.85	2.78		4.63	6.35

05 41 Structural Metal Stud Framing

05 41 13 – Load-Bearing Metal Stud Framing

05 41 13.10 Bridging

			Crew	Daily Output	Labor-Hours	Unit	Material	2009 Bare Costs Labor	Equipment	Total	Total Incl O&P
1310	3-5/8" wide	G	1 Carp	110	.073	Ea.	2.26	2.91		5.17	7
1320	4" wide	G		110	.073		2.41	2.91		5.32	7.15
1330	6" wide	G		105	.076		3.10	3.04		6.14	8.15
1340	8" wide	G		100	.080		3.94	3.20		7.14	9.30
2200	Studs 24" O.C., 18 ga x 2-1/2" wide	G		125	.064		2.15	2.56		4.71	6.30
2210	3-5/8" wide	G		120	.067		2.60	2.66		5.26	7
2220	4" wide	G		120	.067		2.75	2.66		5.41	7.15
2230	6" wide	G		115	.070		3.58	2.78		6.36	8.25
2240	8" wide	G		110	.073		4.54	2.91		7.45	9.50
2300	16 ga x 2-1/2" wide	G		115	.070		2.67	2.78		5.45	7.25
2310	3-5/8" wide	G		110	.073		3.27	2.91		6.18	8.10
2320	4" wide	G		110	.073		3.49	2.91		6.40	8.35
2330	6" wide	G		105	.076		4.49	3.04		7.53	9.65
2340	8" wide	G		100	.080		5.70	3.20		8.90	11.25
3000	Continuous bridging, per row										
3100	16 ga x 1-1/2" channel thru studs 12" O.C.	G	1 Carp	6	1.333	C.L.F.	65	53.50		118.50	154
3110	16" O.C.	G		7	1.143		65	45.50		110.50	143
3120	24" O.C.	G		8.80	.909		65	36.50		101.50	128
4100	2" x 2" angle x 18 ga, studs 12" O.C.	G		7	1.143		94.50	45.50		140	175
4110	16" O.C.	G		9	.889		94.50	35.50		130	159
4120	24" O.C.	G		12	.667		94.50	26.50		121	146
4200	16 ga, studs 12" O.C.	G		5	1.600		120	64		184	231
4210	16" O.C.	G		7	1.143		120	45.50		165.50	203
4220	24" O.C.	G		10	.800		120	32		152	182

05 41 13.25 Framing, Boxed Headers/Beams

			Crew	Daily Output	Labor-Hours	Unit	Material	Labor	Equipment	Total	Total Incl O&P
0010	**FRAMING, BOXED HEADERS/BEAMS**										
0015	Made from recycled materials										
0200	Double, 18 ga x 6" deep	G	2 Carp	220	.073	L.F.	6.70	2.91		9.61	11.90
0210	8" deep	G		210	.076		7.45	3.04		10.49	12.90
0220	10" deep	G		200	.080		9	3.20		12.20	14.90
0230	12" deep	G		190	.084		9.90	3.36		13.26	16.10
0300	16 ga x 8" deep	G		180	.089		8.55	3.55		12.10	14.90
0310	10" deep	G		170	.094		10.30	3.76		14.06	17.15
0320	12" deep	G		160	.100		11.20	4		15.20	18.50
0400	14 ga x 10" deep	G		140	.114		11.95	4.57		16.52	20.50
0410	12" deep	G		130	.123		13.10	4.92		18.02	22
1210	Triple, 18 ga x 8" deep	G		170	.094		10.80	3.76		14.56	17.75
1220	10" deep	G		165	.097		13	3.87		16.87	20.50
1230	12" deep	G		160	.100		14.35	4		18.35	22
1300	16 ga x 8" deep	G		145	.110		12.45	4.41		16.86	20.50
1310	10" deep	G		140	.114		14.85	4.57		19.42	23.50
1320	12" deep	G		135	.119		16.20	4.73		20.93	25
1400	14 ga x 10" deep	G		115	.139		16.30	5.55		21.85	26.50
1410	12" deep	G		110	.145		18.10	5.80		23.90	29

05 41 13.30 Framing, Stud Walls

			Crew	Daily Output	Labor-Hours	Unit	Material	Labor	Equipment	Total	Total Incl O&P
0010	**FRAMING, STUD WALLS** w/ top & bottom track, no openings,										
0020	Headers, beams, bridging or bracing										
0025	Made from recycled materials										
4100	8' high walls, 18 ga x 2-1/2" wide, studs 12" O.C.	G	2 Carp	54	.296	L.F.	11.25	11.85		23.10	30.50
4110	16" O.C.	G		77	.208		9	8.30		17.30	23
4120	24" O.C.	G		107	.150		6.75	5.95		12.70	16.65
4130	3-5/8" wide, studs 12" O.C.	G		53	.302		13.40	12.05		25.45	33.50

05 41 Structural Metal Stud Framing

05 41 13 – Load-Bearing Metal Stud Framing

05 41 13.30 Framing, Stud Walls		Crew	Daily Output	Labor-Hours	Unit	Material	2009 Bare Costs Labor	Equipment	Total	Total Incl O&P
4140	16" O.C.	2 Carp	76	.211	L.F.	10.70	8.40		19.10	25
4150	24" O.C.	G	105	.152		8.05	6.10		14.15	18.30
4160	4" wide, studs 12" O.C.	G	52	.308		14	12.30		26.30	34.50
4170	16" O.C.	G	74	.216		11.20	8.65		19.85	26
4180	24" O.C.	G	103	.155		8.40	6.20		14.60	18.85
4190	6" wide, studs 12" O.C.	G	51	.314		17.85	12.55		30.40	39
4200	16" O.C.	G	73	.219		14.30	8.75		23.05	29.50
4210	24" O.C.	G	101	.158		10.75	6.35		17.10	21.50
4220	8" wide, studs 12" O.C.	G	50	.320		22	12.80		34.80	44
4230	16" O.C.	G	72	.222		17.50	8.90		26.40	33
4240	24" O.C.	G	100	.160		13.20	6.40		19.60	24.50
4300	16 ga x 2-1/2" wide, studs 12" O.C.	G	47	.340		13.25	13.60		26.85	35.50
4310	16" O.C.	G	68	.235		10.50	9.40		19.90	26
4320	24" O.C.	G	94	.170		7.75	6.80		14.55	19.05
4330	3-5/8" wide, studs 12" O.C.	G	46	.348		15.85	13.90		29.75	39
4340	16" O.C.	G	66	.242		12.55	9.70		22.25	29
4350	24" O.C.	G	92	.174		9.25	6.95		16.20	21
4360	4" wide, studs 12" O.C.	G	45	.356		16.75	14.20		30.95	40.50
4370	16" O.C.	G	65	.246		13.25	9.85		23.10	30
4380	24" O.C.	G	90	.178		9.80	7.10		16.90	22
4390	6" wide, studs 12" O.C.	G	44	.364		21	14.55		35.55	45.50
4400	16" O.C.	G	64	.250		16.70	10		26.70	34
4410	24" O.C.	G	88	.182		12.35	7.25		19.60	25
4420	8" wide, studs 12" O.C.	G	43	.372		26	14.85		40.85	51.50
4430	16" O.C.	G	63	.254		20.50	10.15		30.65	38.50
4440	24" O.C.	G	86	.186		15.25	7.45		22.70	28.50
5100	10' high walls, 18 ga x 2-1/2" wide, studs 12" O.C.	G	54	.296		13.50	11.85		25.35	33
5110	16" O.C.	G	77	.208		10.70	8.30		19	24.50
5120	24" O.C.	G	107	.150		7.85	5.95		13.80	17.90
5130	3-5/8" wide, studs 12" O.C.	G	53	.302		16.05	12.05		28.10	36.50
5140	16" O.C.	G	76	.211		12.70	8.40		21.10	27
5150	24" O.C.	G	105	.152		9.35	6.10		15.45	19.75
5160	4" wide, studs 12" O.C.	G	52	.308		16.80	12.30		29.10	37.50
5170	16" O.C.	G	74	.216		13.30	8.65		21.95	28
5180	24" O.C.	G	103	.155		9.80	6.20		16	20.50
5190	6" wide, studs 12" O.C.	G	51	.314		21.50	12.55		34.05	43
5200	16" O.C.	G	73	.219		16.95	8.75		25.70	32.50
5210	24" O.C.	G	101	.158		12.55	6.35		18.90	23.50
5220	8" wide, studs 12" O.C.	G	50	.320		26	12.80		38.80	48.50
5230	16" O.C.	G	72	.222		20.50	8.90		29.40	37
5240	24" O.C.	G	100	.160		15.35	6.40		21.75	27
5300	16 ga x 2-1/2" wide, studs 12" O.C.	G	47	.340		16	13.60		29.60	38.50
5310	16" O.C.	G	68	.235		12.55	9.40		21.95	28.50
5320	24" O.C.	G	94	.170		9.10	6.80		15.90	20.50
5330	3-5/8" wide, studs 12" O.C.	G	46	.348		19.15	13.90		33.05	42.50
5340	16" O.C.	G	66	.242		15.05	9.70		24.75	31.50
5350	24" O.C.	G	92	.174		10.90	6.95		17.85	23
5360	4" wide, studs 12" O.C.	G	45	.356		20	14.20		34.20	44
5370	16" O.C.	G	65	.246		15.85	9.85		25.70	32.50
5380	24" O.C.	G	90	.178		11.50	7.10		18.60	23.50
5390	6" wide, studs 12" O.C.	G	44	.364		25.50	14.55		40.05	50.50
5400	16" O.C.	G	64	.250		19.95	10		29.95	37.50
5410	24" O.C.	G	88	.182		14.55	7.25		21.80	27.50

05 41 Structural Metal Stud Framing

05 41 13 – Load-Bearing Metal Stud Framing

05 41 13.30 Framing, Stud Walls		Crew	Daily Output	Labor-Hours	Unit	Material	2009 Bare Costs Labor	Equipment	Total	Total Incl O&P
5420	8" wide, studs 12" O.C. G	2 Carp	43	.372	L.F.	31	14.85		45.85	57.50
5430	16" O.C. G		63	.254		24.50	10.15		34.65	43
5440	24" O.C. G		86	.186		17.90	7.45		25.35	31.50
6190	12' high walls, 18 ga x 6" wide, studs 12" O.C. G		41	.390		25	15.60		40.60	51.50
6200	16" O.C. G		58	.276		19.60	11		30.60	38.50
6210	24" O.C. G		81	.198		14.30	7.90		22.20	28
6220	8" wide, studs 12" O.C. G		40	.400		30.50	16		46.50	58.50
6230	16" O.C. G		57	.281		24	11.20		35.20	44
6240	24" O.C. G		80	.200		17.50	8		25.50	31.50
6390	16 ga x 6" wide, studs 12" O.C. G		35	.457		29.50	18.25		47.75	61
6400	16" O.C. G		51	.314		23	12.55		35.55	45
6410	24" O.C. G		70	.229		16.70	9.15		25.85	32.50
6420	8" wide, studs 12" O.C. G		34	.471		36.50	18.80		55.30	69
6430	16" O.C. G		50	.320		28.50	12.80		41.30	51.50
6440	24" O.C. G		69	.232		20.50	9.25		29.75	37
6530	14 ga x 3-5/8" wide, studs 12" O.C. G		34	.471		28	18.80		46.80	60
6540	16" O.C. G		48	.333		22	13.30		35.30	44.50
6550	24" O.C. G		65	.246		15.80	9.85		25.65	32.50
6560	4" wide, studs 12" O.C. G		33	.485		29.50	19.35		48.85	62.50
6570	16" O.C. G		47	.340		23	13.60		36.60	46.50
6580	24" O.C. G		64	.250		16.65	10		26.65	34
6730	12 ga x 3-5/8" wide, studs 12" O.C. G		31	.516		39	20.50		59.50	75
6740	16" O.C. G		43	.372		30	14.85		44.85	56
6750	24" O.C. G		59	.271		21	10.85		31.85	40.50
6760	4" wide, studs 12" O.C. G		30	.533		41.50	21.50		63	79
6770	16" O.C. G		42	.381		32	15.20		47.20	59
6780	24" O.C. G		58	.276		22.50	11		33.50	42
7390	16' high walls, 16 ga x 6" wide, studs 12" O.C. G		33	.485		38.50	19.35		57.85	72
7400	16" O.C. G		48	.333		29.50	13.30		42.80	53
7410	24" O.C. G		67	.239		21	9.55		30.55	38
7420	8" wide, studs 12" O.C. G		32	.500		47	20		67	83
7430	16" O.C. G		47	.340		36.50	13.60		50.10	61
7440	24" O.C. G		66	.242		26	9.70		35.70	43.50
7560	14 ga x 4" wide, studs 12" O.C. G		31	.516		38.50	20.50		59	74.50
7570	16" O.C. G		45	.356		29.50	14.20		43.70	54.50
7580	24" O.C. G		61	.262		21	10.50		31.50	39.50
7590	6" wide, studs 12" O.C. G		30	.533		48.50	21.50		70	86.50
7600	16" O.C. G		44	.364		37.50	14.55		52.05	63.50
7610	24" O.C. G		60	.267		26.50	10.65		37.15	45.50
7760	12 ga x 4" wide, studs 12" O.C. G		29	.552		54.50	22		76.50	94
7770	16" O.C. G		40	.400		41.50	16		57.50	71
7780	24" O.C. G		55	.291		29	11.60		40.60	50
7790	6" wide, studs 12" O.C. G		28	.571		69	23		92	112
7800	16" O.C. G		39	.410		53	16.40		69.40	83.50
7810	24" O.C. G		54	.296		37	11.85		48.85	59
8590	20' high walls, 14 ga x 6" wide, studs 12" O.C. G		29	.552		59.50	22		81.50	99.50
8600	16" O.C. G		42	.381		45.50	15.20		60.70	74
8610	24" O.C. G		57	.281		32	11.20		43.20	52.50
8620	8" wide, studs 12" O.C. G		28	.571		72.50	23		95.50	116
8630	16" O.C. G		41	.390		56	15.60		71.60	85.50
8640	24" O.C. G		56	.286		39.50	11.40		50.90	60.50
8790	12 ga x 6" wide, studs 12" O.C. G		27	.593		85	23.50		108.50	130
8800	16" O.C. G		37	.432		65	17.30		82.30	98.50

05 41 Structural Metal Stud Framing

05 41 13 – Load-Bearing Metal Stud Framing

05 41 13.30 Framing, Stud Walls		Crew	Daily Output	Labor-Hours	Unit	Material	2009 Bare Costs Labor	Equipment	Total	Total Incl O&P
8810	24" O.C.	G 2 Carp	51	.314	L.F.	45	12.55		57.55	69
8820	8" wide, studs 12" O.C.	G	26	.615		103	24.50		127.50	152
8830	16" O.C.	G	36	.444		79	17.75		96.75	115
8840	24" O.C.	G	50	.320		54.50	12.80		67.30	80

05 42 Cold-Formed Metal Joist Framing

05 42 13 – Cold-Formed Metal Floor Joist Framing

05 42 13.05 Bracing

		Crew	Daily Output	Labor-Hours	Unit	Material	Labor	Equipment	Total	Total Incl O&P
0010	**BRACING**, continuous, per row, top & bottom									
0015	Made from recycled materials									
0120	Flat strap, 20 ga x 2" wide, joists at 12" O.C.	G 1 Carp	4.67	1.713	C.L.F.	71.50	68.50		140	185
0130	16" O.C.	G	5.33	1.501		69	60		129	169
0140	24" O.C.	G	6.66	1.201		66.50	48		114.50	148
0150	18 ga x 2" wide, joists at 12" O.C.	G	4	2		93.50	80		173.50	227
0160	16" O.C.	G	4.67	1.713		92	68.50		160.50	207
0170	24" O.C.	G	5.33	1.501		90.50	60		150.50	193

05 42 13.10 Bridging

		Crew	Daily Output	Labor-Hours	Unit	Material	Labor	Equipment	Total	Total Incl O&P
0010	**BRIDGING**, solid between joists w/ 1-1/4" leg track, per joist bay									
0015	Made from recycled materials									
0230	Joists 12" O.C., 18 ga track x 6" wide	G 1 Carp	80	.100	Ea.	1.93	4		5.93	8.30
0240	8" wide	G	75	.107		2.45	4.26		6.71	9.30
0250	10" wide	G	70	.114		3.02	4.57		7.59	10.45
0260	12" wide	G	65	.123		3.50	4.92		8.42	11.45
0330	16 ga track x 6" wide	G	70	.114		2.42	4.57		6.99	9.75
0340	8" wide	G	65	.123		3.08	4.92		8	11
0350	10" wide	G	60	.133		3.78	5.35		9.13	12.40
0360	12" wide	G	55	.145		4.36	5.80		10.16	13.80
0440	14 ga track x 8" wide	G	60	.133		3.87	5.35		9.22	12.50
0450	10" wide	G	55	.145		4.76	5.80		10.56	14.25
0460	12" wide	G	50	.160		5.50	6.40		11.90	15.95
0550	12 ga track x 10" wide	G	45	.178		7	7.10		14.10	18.70
0560	12" wide	G	40	.200		7.90	8		15.90	21
1230	16" O.C., 18 ga track x 6" wide	G	80	.100		2.48	4		6.48	8.90
1240	8" wide	G	75	.107		3.14	4.26		7.40	10.05
1250	10" wide	G	70	.114		3.88	4.57		8.45	11.35
1260	12" wide	G	65	.123		4.49	4.92		9.41	12.55
1330	16 ga track x 6" wide	G	70	.114		3.10	4.57		7.67	10.50
1340	8" wide	G	65	.123		3.94	4.92		8.86	11.95
1350	10" wide	G	60	.133		4.85	5.35		10.20	13.60
1360	12" wide	G	55	.145		5.60	5.80		11.40	15.15
1440	14 ga track x 8" wide	G	60	.133		4.97	5.35		10.32	13.70
1450	10" wide	G	55	.145		6.10	5.80		11.90	15.70
1460	12" wide	G	50	.160		7.05	6.40		13.45	17.65
1550	12 ga track x 10" wide	G	45	.178		9	7.10		16.10	21
1560	12" wide	G	40	.200		10.15	8		18.15	23.50
2230	24" O.C., 18 ga track x 6" wide	G	80	.100		3.58	4		7.58	10.15
2240	8" wide	G	75	.107		4.54	4.26		8.80	11.60
2250	10" wide	G	70	.114		5.60	4.57		10.17	13.25
2260	12" wide	G	65	.123		6.50	4.92		11.42	14.75
2330	16 ga track x 6" wide	G	70	.114		4.49	4.57		9.06	12.05
2340	8" wide	G	65	.123		5.70	4.92		10.62	13.90

05 42 Cold-Formed Metal Joist Framing

05 42 13 – Cold-Formed Metal Floor Joist Framing

05 42 13.10 Bridging

			Crew	Daily Output	Labor-Hours	Unit	Material	2009 Bare Costs Labor	Equipment	Total	Total Incl O&P
2350	10" wide	G	1 Carp	60	.133	Ea.	7	5.35		12.35	15.95
2360	12" wide	G		55	.145		8.10	5.80		13.90	17.90
2440	14 ga track x 8" wide	G		60	.133		7.20	5.35		12.55	16.15
2450	10" wide	G		55	.145		8.85	5.80		14.65	18.70
2460	12" wide	G		50	.160		10.20	6.40		16.60	21
2550	12 ga track x 10" wide	G		45	.178		13	7.10		20.10	25.50
2560	12" wide	G		40	.200		14.70	8		22.70	28.50

05 42 13.25 Framing, Band Joist

0010	**FRAMING, BAND JOIST** (track) fastened to bearing wall										
0015	Made from recycled materials										
0220	18 ga track x 6" deep	G	2 Carp	1000	.016	L.F.	1.58	.64		2.22	2.72
0230	8" deep	G		920	.017		2	.69		2.69	3.27
0240	10" deep	G		860	.019		2.47	.74		3.21	3.86
0320	16 ga track x 6" deep	G		900	.018		1.97	.71		2.68	3.27
0330	8" deep	G		840	.019		2.51	.76		3.27	3.94
0340	10" deep	G		780	.021		3.09	.82		3.91	4.67
0350	12" deep	G		740	.022		3.56	.86		4.42	5.25
0430	14 ga track x 8" deep	G		750	.021		3.16	.85		4.01	4.80
0440	10" deep	G		720	.022		3.89	.89		4.78	5.65
0450	12" deep	G		700	.023		4.49	.91		5.40	6.35
0540	12 ga track x 10" deep	G		670	.024		5.70	.95		6.65	7.80
0550	12" deep	G		650	.025		6.45	.98		7.43	8.65

05 42 13.30 Framing, Boxed Headers/Beams

0010	**FRAMING, BOXED HEADERS/BEAMS**										
0015	Made from recycled materials										
0200	Double, 18 ga x 6" deep	G	2 Carp	220	.073	L.F.	6.70	2.91		9.61	11.90
0210	8" deep	G		210	.076		7.45	3.04		10.49	12.90
0220	10" deep	G		200	.080		9	3.20		12.20	14.90
0230	12" deep	G		190	.084		9.90	3.36		13.26	16.10
0300	16 ga x 8" deep	G		180	.089		8.55	3.55		12.10	14.90
0310	10" deep	G		170	.094		10.30	3.76		14.06	17.15
0320	12" deep	G		160	.100		11.20	4		15.20	18.50
0400	14 ga x 10" deep	G		140	.114		11.95	4.57		16.52	20.50
0410	12" deep	G		130	.123		13.10	4.92		18.02	22
0500	12 ga x 10" deep	G		110	.145		15.80	5.80		21.60	26.50
0510	12" deep	G		100	.160		17.50	6.40		23.90	29
1210	Triple, 18 ga x 8" deep	G		170	.094		10.80	3.76		14.56	17.75
1220	10" deep	G		165	.097		13	3.87		16.87	20.50
1230	12" deep	G		160	.100		14.35	4		18.35	22
1300	16 ga x 8" deep	G		145	.110		12.45	4.41		16.86	20.50
1310	10" deep	G		140	.114		14.85	4.57		19.42	23.50
1320	12" deep	G		135	.119		16.20	4.73		20.93	25
1400	14 ga x 10" deep	G		115	.139		17.35	5.55		22.90	27.50
1410	12" deep	G		110	.145		19.15	5.80		24.95	30
1500	12 ga x 10" deep	G		90	.178		23	7.10		30.10	36.50
1510	12" deep	G		85	.188		25.50	7.50		33	40

05 42 13.40 Framing, Joists

0010	**FRAMING, JOISTS**, no band joists (track), web stiffeners, headers,										
0020	Beams, bridging or bracing										
0025	Made from recycled materials										
0030	Joists (2" flange) and fasteners, materials only										
0220	18 ga x 6" deep	G				L.F.	2.08			2.08	2.29

05 42 Cold-Formed Metal Joist Framing

05 42 13 – Cold-Formed Metal Floor Joist Framing

05 42 13.40 Framing, Joists			Crew	Daily Output	Labor-Hours	Unit	Material	2009 Bare Costs Labor	Equipment	Total	Total Incl O&P
0230	8" deep	G				L.F.	2.47			2.47	2.71
0240	10" deep	G					2.90			2.90	3.19
0320	16 ga x 6" deep	G					2.54			2.54	2.80
0330	8" deep	G					3.05			3.05	3.35
0340	10" deep	G					3.56			3.56	3.92
0350	12" deep	G					4.03			4.03	4.44
0430	14 ga x 8" deep	G					3.84			3.84	4.23
0440	10" deep	G					4.43			4.43	4.87
0450	12" deep	G					5.05			5.05	5.55
0540	12 ga x 10" deep	G					6.45			6.45	7.10
0550	12" deep	G					7.35			7.35	8.10
1010	Installation of joists to band joists, beams & headers, labor only										
1220	18 ga x 6" deep		2 Carp	110	.145	Ea.		5.80		5.80	9
1230	8" deep			90	.178			7.10		7.10	11
1240	10" deep			80	.200			8		8	12.40
1320	16 ga x 6" deep			95	.168			6.75		6.75	10.45
1330	8" deep			70	.229			9.15		9.15	14.15
1340	10" deep			60	.267			10.65		10.65	16.50
1350	12" deep			55	.291			11.60		11.60	18
1430	14 ga x 8" deep			65	.246			9.85		9.85	15.25
1440	10" deep			45	.356			14.20		14.20	22
1450	12" deep			35	.457			18.25		18.25	28.50
1540	12 ga x 10" deep			40	.400			16		16	25
1550	12" deep			30	.533			21.50		21.50	33

05 42 13.45 Framing, Web Stiffeners			Crew	Daily Output	Labor-Hours	Unit	Material	Labor	Equipment	Total	Total Incl O&P
0010	**FRAMING, WEB STIFFENERS** at joist bearing, fabricated from										
0020	Stud piece (1-5/8" flange) to stiffen joist (2" flange)										
0025	Made from recycled materials										
2120	For 6" deep joist, with 18 ga x 2-1/2" stud	G	1 Carp	120	.067	Ea.	2.49	2.66		5.15	6.85
2130	3-5/8" stud	G		110	.073		2.75	2.91		5.66	7.55
2140	4" stud	G		105	.076		2.66	3.04		5.70	7.65
2150	6" stud	G		100	.080		2.92	3.20		6.12	8.15
2160	8" stud	G		95	.084		3	3.36		6.36	8.50
2220	8" deep joist, with 2-1/2" stud	G		120	.067		2.73	2.66		5.39	7.15
2230	3-5/8" stud	G		110	.073		2.96	2.91		5.87	7.75
2240	4" stud	G		105	.076		2.91	3.04		5.95	7.90
2250	6" stud	G		100	.080		3.20	3.20		6.40	8.50
2260	8" stud	G		95	.084		3.44	3.36		6.80	9
2320	10" deep joist, with 2-1/2" stud	G		110	.073		3.85	2.91		6.76	8.75
2330	3-5/8" stud	G		100	.080		4.23	3.20		7.43	9.60
2340	4" stud	G		95	.084		4.18	3.36		7.54	9.80
2350	6" stud	G		90	.089		4.55	3.55		8.10	10.50
2360	8" stud	G		85	.094		4.62	3.76		8.38	10.95
2420	12" deep joist, with 2-1/2" stud	G		110	.073		4.07	2.91		6.98	9
2430	3-5/8" stud	G		100	.080		4.42	3.20		7.62	9.80
2440	4" stud	G		95	.084		4.34	3.36		7.70	9.95
2450	6" stud	G		90	.089		4.78	3.55		8.33	10.75
2460	8" stud	G		85	.094		5.15	3.76		8.91	11.50
3130	For 6" deep joist, with 16 ga x 3-5/8" stud	G		100	.080		2.89	3.20		6.09	8.15
3140	4" stud	G		95	.084		2.87	3.36		6.23	8.35
3150	6" stud	G		90	.089		3.15	3.55		6.70	8.95
3160	8" stud	G		85	.094		3.31	3.76		7.07	9.50

05 42 Cold-Formed Metal Joist Framing

05 42 13 – Cold-Formed Metal Floor Joist Framing

05 42 13.45 Framing, Web Stiffeners		Crew	Daily Output	Labor-Hours	Unit	Material	2009 Bare Costs Labor	Equipment	Total	Total Incl O&P
3230	8" deep joist, with 3-5/8" stud	G 1 Carp	100	.080	Ea.	3.21	3.20		6.41	8.50
3240	4" stud	G	95	.084		3.15	3.36		6.51	8.65
3250	6" stud	G	90	.089		3.49	3.55		7.04	9.35
3260	8" stud	G	85	.094		3.73	3.76		7.49	9.95
3330	10" deep joist, with 3-5/8" stud	G	85	.094		4.38	3.76		8.14	10.65
3340	4" stud	G	80	.100		4.48	4		8.48	11.10
3350	6" stud	G	75	.107		4.86	4.26		9.12	11.95
3360	8" stud	G	70	.114		5.05	4.57		9.62	12.65
3430	12" deep joist, with 3-5/8" stud	G	85	.094		4.79	3.76		8.55	11.10
3440	4" stud	G	80	.100		4.70	4		8.70	11.35
3450	6" stud	G	75	.107		5.20	4.26		9.46	12.35
3460	8" stud	G	70	.114		5.55	4.57		10.12	13.20
4230	For 8" deep joist, with 14 ga x 3-5/8" stud	G	90	.089		4.16	3.55		7.71	10.10
4240	4" stud	G	85	.094		4.24	3.76		8	10.50
4250	6" stud	G	80	.100		4.59	4		8.59	11.25
4260	8" stud	G	75	.107		4.92	4.26		9.18	12
4330	10" deep joist, with 3-5/8" stud	G	75	.107		5.85	4.26		10.11	13.05
4340	4" stud	G	70	.114		5.80	4.57		10.37	13.45
4350	6" stud	G	65	.123		6.35	4.92		11.27	14.60
4360	8" stud	G	60	.133		6.65	5.35		12	15.55
4430	12" deep joist, with 3-5/8" stud	G	75	.107		6.20	4.26		10.46	13.45
4440	4" stud	G	70	.114		6.30	4.57		10.87	14.05
4450	6" stud	G	65	.123		6.85	4.92		11.77	15.15
4460	8" stud	G	60	.133		7.35	5.35		12.70	16.35
5330	For 10" deep joist, with 12 ga x 3-5/8" stud	G	65	.123		6.15	4.92		11.07	14.40
5340	4" stud	G	60	.133		6.35	5.35		11.70	15.20
5350	6" stud	G	55	.145		7	5.80		12.80	16.70
5360	8" stud	G	50	.160		7.70	6.40		14.10	18.35
5430	12" deep joist, with 3-5/8" stud	G	65	.123		6.85	4.92		11.77	15.10
5440	4" stud	G	60	.133		6.70	5.35		12.05	15.60
5450	6" stud	G	55	.145		7.65	5.80		13.45	17.40
5460	8" stud	G	50	.160		8.80	6.40		15.20	19.55

05 42 23 – Cold-Formed Metal Roof Joist Framing

05 42 23.05 Framing, Bracing

		Crew	Daily Output	Labor-Hours	Unit	Material	Labor	Equipment	Total	Total Incl O&P
0010	**FRAMING, BRACING**									
0015	Made from recycled materials									
0020	Continuous bracing, per row									
0100	16 ga x 1-1/2" channel thru rafters/trusses @ 16" O.C.	G 1 Carp	4.50	1.778	C.L.F.	65	71		136	182
0120	24" O.C.	G	6	1.333		65	53.50		118.50	154
0300	2" x 2" angle x 18 ga, rafters/trusses @ 16" O.C.	G	6	1.333		94.50	53.50		148	187
0320	24" O.C.	G	8	1		94.50	40		134.50	166
0400	16 ga, rafters/trusses @ 16" O.C.	G	4.50	1.778		120	71		191	242
0420	24" O.C.	G	6.50	1.231		120	49		169	209

05 42 23.10 Framing, Bridging

		Crew	Daily Output	Labor-Hours	Unit	Material	Labor	Equipment	Total	Total Incl O&P
0010	**FRAMING, BRIDGING**									
0015	Made from recycled materials									
0020	Solid, between rafters w/ 1-1/4" leg track, per rafter bay									
1200	Rafters 16" O.C., 18 ga x 4" deep	G 1 Carp	60	.133	Ea.	1.90	5.35		7.25	10.35
1210	6" deep	G	57	.140		2.48	5.60		8.08	11.40
1220	8" deep	G	55	.145		3.14	5.80		8.94	12.45
1230	10" deep	G	52	.154		3.88	6.15		10.03	13.80
1240	12" deep	G	50	.160		4.49	6.40		10.89	14.85

05 42 Cold-Formed Metal Joist Framing

05 42 23 – Cold-Formed Metal Roof Joist Framing

05 42 23.10 Framing, Bridging

			Crew	Daily Output	Labor-Hours	Unit	Material	2009 Bare Costs Labor	Equipment	Total	Total Incl O&P
2200	24" O.C., 18 ga x 4" deep	G	1 Carp	60	.133	Ea.	2.75	5.35		8.10	11.25
2210	6" deep	G		57	.140		3.58	5.60		9.18	12.65
2220	8" deep	G		55	.145		4.54	5.80		10.34	14
2230	10" deep	G		52	.154		5.60	6.15		11.75	15.70
2240	12" deep	G		50	.160		6.50	6.40		12.90	17.05

05 42 23.50 Framing, Parapets

			Crew	Daily Output	Labor-Hours	Unit	Material	Labor	Equipment	Total	Total Incl O&P
0010	**FRAMING, PARAPETS**										
0015	Made from recycled materials										
0100	3' high installed on 1st story, 18 ga x 4" wide studs, 12" O.C.	G	2 Carp	100	.160	L.F.	7	6.40		13.40	17.60
0110	16" O.C.	G		150	.107		5.95	4.26		10.21	13.15
0120	24" O.C.	G		200	.080		4.92	3.20		8.12	10.35
0200	6" wide studs, 12" O.C.	G		100	.160		9	6.40		15.40	19.80
0210	16" O.C.	G		150	.107		7.65	4.26		11.91	15
0220	24" O.C.	G		200	.080		6.35	3.20		9.55	11.90
1100	Installed on 2nd story, 18 ga x 4" wide studs, 12" O.C.	G		95	.168		7	6.75		13.75	18.15
1110	16" O.C.	G		145	.110		5.95	4.41		10.36	13.40
1120	24" O.C.	G		190	.084		4.92	3.36		8.28	10.60
1200	6" wide studs, 12" O.C.	G		95	.168		9	6.75		15.75	20.50
1210	16" O.C.	G		145	.110		7.65	4.41		12.06	15.25
1220	24" O.C.	G		190	.084		6.35	3.36		9.71	12.15
2100	Installed on gable, 18 ga x 4" wide studs, 12" O.C.	G		85	.188		7	7.50		14.50	19.35
2110	16" O.C.	G		130	.123		5.95	4.92		10.87	14.15
2120	24" O.C.	G		170	.094		4.92	3.76		8.68	11.25
2200	6" wide studs, 12" O.C.	G		85	.188		9	7.50		16.50	21.50
2210	16" O.C.	G		130	.123		7.65	4.92		12.57	16
2220	24" O.C.	G		170	.094		6.35	3.76		10.11	12.80

05 42 23.60 Framing, Roof Rafters

			Crew	Daily Output	Labor-Hours	Unit	Material	Labor	Equipment	Total	Total Incl O&P
0010	**FRAMING, ROOF RAFTERS**										
0015	Made from recycled materials										
0100	Boxed ridge beam, double, 18 ga x 6" deep	G	2 Carp	160	.100	L.F.	6.70	4		10.70	13.60
0110	8" deep	G		150	.107		7.45	4.26		11.71	14.80
0120	10" deep	G		140	.114		9	4.57		13.57	17.05
0130	12" deep	G		130	.123		9.90	4.92		14.82	18.50
0200	16 ga x 6" deep	G		150	.107		7.60	4.26		11.86	14.95
0210	8" deep	G		140	.114		8.55	4.57		13.12	16.50
0220	10" deep	G		130	.123		10.30	4.92		15.22	18.90
0230	12" deep	G		120	.133		11.20	5.35		16.55	20.50
1100	Rafters, 2" flange, material only, 18 ga x 6" deep	G					2.08			2.08	2.29
1110	8" deep	G					2.47			2.47	2.71
1120	10" deep	G					2.90			2.90	3.19
1130	12" deep	G					3.37			3.37	3.71
1200	16 ga x 6" deep	G					2.54			2.54	2.80
1210	8" deep	G					3.05			3.05	3.35
1220	10" deep	G					3.56			3.56	3.92
1230	12" deep	G					4.03			4.03	4.44
2100	Installation only, ordinary rafter to 4:12 pitch, 18 ga x 6" deep		2 Carp	35	.457	Ea.		18.25		18.25	28.50
2110	8" deep			30	.533			21.50		21.50	33
2120	10" deep			25	.640			25.50		25.50	39.50
2130	12" deep			20	.800			32		32	49.50
2200	16 ga x 6" deep			30	.533			21.50		21.50	33
2210	8" deep			25	.640			25.50		25.50	39.50
2220	10" deep			20	.800			32		32	49.50

05 42 Cold-Formed Metal Joist Framing

05 42 23 – Cold-Formed Metal Roof Joist Framing

05 42 23.60 Framing, Roof Rafters

		Crew	Daily Output	Labor-Hours	Unit	Material	2009 Bare Costs Labor	Equipment	Total	Total Incl O&P
2230	12" deep	2 Carp	15	1.067	Ea.		42.50		42.50	66
8100	Add to labor, ordinary rafters on steep roofs						25%			
8110	Dormers & complex roofs						50%			
8200	Hip & valley rafters to 4:12 pitch						25%			
8210	Steep roofs						50%			
8220	Dormers & complex roofs						75%			
8300	Hip & valley jack rafters to 4:12 pitch						50%			
8310	Steep roofs						75%			
8320	Dormers & complex roofs						100%			

05 42 23.70 Framing, Soffits and Canopies

			Crew	Daily Output	Labor-Hours	Unit	Material	Labor	Equipment	Total	Total Incl O&P
0010	**FRAMING, SOFFITS & CANOPIES**										
0015	Made from recycled materials										
0130	Continuous ledger track @ wall, studs @ 16" O.C., 18 ga x 4" wide	G	2 Carp	535	.030	L.F.	1.27	1.19		2.46	3.24
0140	6" wide	G		500	.032		1.65	1.28		2.93	3.80
0150	8" wide	G		465	.034		2.09	1.37		3.46	4.43
0160	10" wide	G		430	.037		2.59	1.49		4.08	5.15
0230	Studs @ 24" O.C., 18 ga x 4" wide	G		800	.020		1.21	.80		2.01	2.57
0240	6" wide	G		750	.021		1.58	.85		2.43	3.05
0250	8" wide	G		700	.023		2	.91		2.91	3.61
0260	10" wide	G		650	.025		2.47	.98		3.45	4.24
1000	Horizontal soffit and canopy members, material only										
1030	1-5/8" flange studs, 18 ga x 4" deep	G				L.F.	1.68			1.68	1.85
1040	6" deep	G					2.12			2.12	2.34
1050	8" deep	G					2.57			2.57	2.82
1140	2" flange joists, 18 ga x 6" deep	G					2.38			2.38	2.61
1150	8" deep	G					2.82			2.82	3.10
1160	10" deep	G					3.31			3.31	3.64
4030	Installation only, 18 ga, 1-5/8" flange x 4" deep		2 Carp	130	.123	Ea.		4.92		4.92	7.60
4040	6" deep			110	.145			5.80		5.80	9
4050	8" deep			90	.178			7.10		7.10	11
4140	2" flange, 18 ga x 6" deep			110	.145			5.80		5.80	9
4150	8" deep			90	.178			7.10		7.10	11
4160	10" deep			80	.200			8		8	12.40
6010	Clips to attach facia to rafter tails, 2" x 2" x 18 ga angle	G	1 Carp	120	.067		1.12	2.66		3.78	5.35
6020	16 ga angle	G	"	100	.080		1.42	3.20		4.62	6.50

05 44 Cold-Formed Metal Trusses

05 44 13 – Cold-Formed Metal Roof Trusses

05 44 13.60 Framing, Roof Trusses

			Crew	Daily Output	Labor-Hours	Unit	Material	Labor	Equipment	Total	Total Incl O&P
0010	**FRAMING, ROOF TRUSSES**										
0015	Made from recycled materials										
0020	Fabrication of trusses on ground, Fink (W) or King Post, to 4:12 pitch										
0120	18 ga x 4" chords, 16' span	G	2 Carp	12	1.333	Ea.	78.50	53.50		132	169
0130	20' span	G		11	1.455		98	58		156	198
0140	24' span	G		11	1.455		118	58		176	219
0150	28' span	G		10	1.600		137	64		201	250
0160	32' span	G		10	1.600		157	64		221	271
0250	6" chords, 28' span	G		9	1.778		173	71		244	300
0260	32' span	G		9	1.778		198	71		269	330
0270	36' span	G		8	2		223	80		303	370
0280	40' span	G		8	2		248	80		328	395

05 44 Cold-Formed Metal Trusses

05 44 13 – Cold-Formed Metal Roof Trusses

05 44 13.60 Framing, Roof Trusses

			Crew	Daily Output	Labor-Hours	Unit	Material	2009 Bare Costs Labor	Equipment	Total	Total Incl O&P
1120	5:12 to 8:12 pitch, 18 ga x 4" chords, 16' span	G	2 Carp	10	1.600	Ea.	89.50	64		153.50	198
1130	20' span	G		9	1.778		112	71		183	233
1140	24' span	G		9	1.778		134	71		205	258
1150	28' span	G		8	2		157	80		237	296
1160	32' span	G		8	2		179	80		259	320
1250	6" chords, 28' span	G		7	2.286		198	91.50		289.50	360
1260	32' span	G		7	2.286		227	91.50		318.50	390
1270	36' span	G		6	2.667		255	107		362	445
1280	40' span	G		6	2.667		283	107		390	475
2120	9:12 to 12:12 pitch, 18 ga x 4" chords, 16' span	G		8	2		112	80		192	247
2130	20' span	G		7	2.286		140	91.50		231.50	296
2140	24' span	G		7	2.286		168	91.50		259.50	325
2150	28' span	G		6	2.667		196	107		303	380
2160	32' span	G		6	2.667		224	107		331	410
2250	6" chords, 28' span	G		5	3.200		248	128		376	470
2260	32' span	G		5	3.200		283	128		411	510
2270	36' span	G		4	4		320	160		480	600
2280	40' span	G		4	4		355	160		515	640
5120	Erection only of roof trusses, to 4:12 pitch, 16' span		F-6	48	.833			31	16	47	65
5130	20' span			46	.870			32.50	16.70	49.20	68
5140	24' span			44	.909			34	17.45	51.45	71
5150	28' span			42	.952			35.50	18.30	53.80	74.50
5160	32' span			40	1			37	19.20	56.20	78
5170	36' span			38	1.053			39	20	59	82.50
5180	40' span			36	1.111			41.50	21.50	63	87
5220	5:12 to 8:12 pitch, 16' span			42	.952			35.50	18.30	53.80	74.50
5230	20' span			40	1			37	19.20	56.20	78
5240	24' span			38	1.053			39	20	59	82.50
5250	28' span			36	1.111			41.50	21.50	63	87
5260	32' span			34	1.176			43.50	22.50	66	92.50
5270	36' span			32	1.250			46.50	24	70.50	98
5280	40' span			30	1.333			49.50	25.50	75	104
5320	9:12 to 12:12 pitch, 16' span			36	1.111			41.50	21.50	63	87
5330	20' span			34	1.176			43.50	22.50	66	92.50
5340	24' span			32	1.250			46.50	24	70.50	98
5350	28' span			30	1.333			49.50	25.50	75	104
5360	32' span			28	1.429			53	27.50	80.50	112
5370	36' span			26	1.538			57	29.50	86.50	121
5380	40' span			24	1.667			62	32	94	131

05 51 Metal Stairs

05 51 13 – Metal Pan Stairs

05 51 13.50 Pan Stairs

			Crew	Daily Output	Labor-Hours	Unit	Material	Labor	Equipment	Total	Total Incl O&P
0010	**PAN STAIRS**, shop fabricated, steel stringers										
0015	Made from recycled materials										
0200	Cement fill metal pan, picket rail, 3'-6" wide	G	E-4	35	.914	Riser	560	41.50	3.83	605.33	700
0300	4'-0" wide	G		30	1.067		635	48	4.47	687.47	790
0350	Wall rail, both sides, 3'-6" wide	G		53	.604		430	27.50	2.53	460.03	525
1500	Landing, steel pan, conventional	G		160	.200	S.F.	75	9.05	.84	84.89	99.50
1600	Pre-erected	G		255	.125	"	131	5.65	.53	137.18	155
1700	Pre-erected, steel pan tread, 3'-6" wide, 2 line pipe rail	G	E-2	87	.644	Riser	620	28	20	668	750

05 51 Metal Stairs

05 51 13 – Metal Pan Stairs

05 51 13.50 Pan Stairs		Crew	Daily Output	Labor-Hours	Unit	Material	2009 Bare Costs Labor	Equipment	Total	Total Incl O&P
1800	With flat bar picket rail	E-2	87	.644	Riser	695	28	20	743	830

05 51 16 – Metal Floor Plate Stairs

05 51 16.50 Floor Plate Stairs

		Crew	Daily Output	Labor-Hours	Unit	Material	Labor	Equipment	Total	Incl O&P
0010	**FLOOR PLATE STAIRS**, shop fabricated, steel stringers									
0015	Made from recycled materials									
0400	Cast iron tread and pipe rail, 3'-6" wide	E-4	35	.914	Riser	600	41.50	3.83	645.33	740
0500	Checkered plate tread, industrial, 3'-6" wide		28	1.143		375	51.50	4.79	431.29	505
0550	Circular, for tanks, 3'-0" wide		33	.970		410	44	4.06	458.06	535
0600	For isolated stairs, add						100%			
0800	Custom steel stairs, 3'-6" wide, minimum	E-4	35	.914		560	41.50	3.83	605.33	700
0810	Average		30	1.067		750	48	4.47	802.47	915
0900	Maximum		20	1.600		935	72.50	6.70	1,014.20	1,150
1100	For 4' wide stairs, add					5%	5%			
1300	For 5' wide stairs, add					10%	10%			

05 51 19 – Metal Grating Stairs

05 51 19.50 Grating Stairs

		Crew	Daily Output	Labor-Hours	Unit	Material	Labor	Equipment	Total	Incl O&P
0010	**GRATING STAIRS**, shop fabricated, steel stringers, safety nosing on treads									
0015	Made from recycled materials									
0020	Grating tread and pipe railing, 3'-6" wide	E-4	35	.914	Riser	375	41.50	3.83	420.33	490
0100	4'-0" wide	"	30	1.067	"	485	48	4.47	537.47	625

05 51 23 – Metal Fire Escapes

05 51 23.25 Fire Escapes

		Crew	Daily Output	Labor-Hours	Unit	Material	Labor	Equipment	Total	Incl O&P
0010	**FIRE ESCAPES**, shop fabricated									
0200	2' wide balcony, 1" x 1/4" bars 1-1/2" O.C.	1 Sswk	5	1.600	L.F.	65.50	71.50		137	199
0400	1st story cantilevered stair, standard		.09	88.889	Ea.	2,725	3,975		6,700	10,100
0700	Platform & fixed stair, 36" x 40"		.17	47.059	Flight	1,200	2,100		3,300	5,075
0900	For 3'-6" wide escapes, add to above					100%	150%			

05 51 23.50 Fire Escape Stairs

		Crew	Daily Output	Labor-Hours	Unit	Material	Labor	Equipment	Total	Incl O&P
0010	**FIRE ESCAPE STAIRS**, shop fabricated									
0020	One story, disappearing, stainless steel	2 Sswk	20	.800	V.L.F.	223	36		259	310
0100	Portable ladder				Ea.	93			93	102
1100	Fire escape, galvanized steel, 8'-0" to 10'-4" ceiling	2 Carp	1	16		1,525	640		2,165	2,675
1110	10'-6" to 13'-6" ceiling	"	1	16		2,000	640		2,640	3,200

05 51 33 – Metal Ladders

05 51 33.13 Vertical Metal Ladders

		Crew	Daily Output	Labor-Hours	Unit	Material	Labor	Equipment	Total	Incl O&P
0010	**VERTICAL METAL LADDERS**, shop fabricated									
0015	Made from recycled materials									
0020	Steel, 20" wide, bolted to concrete, with cage	E-4	50	.640	V.L.F.	78.50	29	2.68	110.18	141
0100	Without cage		85	.376		41	17	1.58	59.58	77
0300	Aluminum, bolted to concrete, with cage		50	.640		111	29	2.68	142.68	176
0400	Without cage		85	.376		64.50	17	1.58	83.08	103

05 51 33.16 Inclined Metal Ladders

		Crew	Daily Output	Labor-Hours	Unit	Material	Labor	Equipment	Total	Incl O&P
0010	**INCLINED METAL LADDERS**, shop fabricated									
3900	Industrial ships ladder, 3' W, grating treads, 2 line pipe rail	E-4	30	1.067	Riser	203	48	4.47	255.47	315
4000	Aluminum	"	30	1.067	"	310	48	4.47	362.47	435

05 51 33.23 Alternating Tread Ladders

		Crew	Daily Output	Labor-Hours	Unit	Material	Labor	Equipment	Total	Incl O&P
0010	**ALTERNATING TREAD LADDERS**, shop fabricated									
0015	Made from recycled materials									
1350	Alternating tread stair, 56/68°, steel, standard paint color	2 Sswk	50	.320	V.L.F.	168	14.30		182.30	211
1360	Non-standard paint color		50	.320		190	14.30		204.30	235

05 51 Metal Stairs

05 51 33 – Metal Ladders

05 51 33.23 Alternating Tread Ladders		Crew	Daily Output	Labor-Hours	Unit	Material	2009 Bare Costs Labor	Equipment	Total	Total Incl O&P
1370	Galvanized steel	G	2 Sswk	50	.320	V.L.F.	196	14.30		210.30	242
1380	Stainless steel	G	↓	50	.320		282	14.30		296.30	335
1390	68°, aluminum	G		50	.320	↓	206	14.30		220.30	253

05 52 Metal Railings

05 52 13 – Pipe and Tube Railings

05 52 13.50 Railings, Pipe

			Crew	Daily Output	Labor-Hours	Unit	Material	Labor	Equipment	Total	Total Incl O&P
0010	**RAILINGS, PIPE**, shop fab'd, 3'-6" high, posts @ 5' O.C.										
0015	Made from recycled materials										
0020	Aluminum, 2 rail, satin finish, 1-1/4" diameter	G	E-4	160	.200	L.F.	34	9.05	.84	43.89	54.50
0030	Clear anodized	G		160	.200		42	9.05	.84	51.89	63.50
0040	Dark anodized	G		160	.200		47.50	9.05	.84	57.39	69.50
0080	1-1/2" diameter, satin finish	G		160	.200		41	9.05	.84	50.89	62
0090	Clear anodized	G		160	.200		45.50	9.05	.84	55.39	67
0100	Dark anodized	G		160	.200		50.50	9.05	.84	60.39	72.50
0140	Aluminum, 3 rail, 1-1/4" diam., satin finish	G		137	.234		52	10.55	.98	63.53	77.50
0150	Clear anodized	G		137	.234		65.50	10.55	.98	77.03	92
0160	Dark anodized	G		137	.234		72.50	10.55	.98	84.03	99.50
0200	1-1/2" diameter, satin finish	G		137	.234		62.50	10.55	.98	74.03	89
0210	Clear anodized	G		137	.234		71	10.55	.98	82.53	98
0220	Dark anodized	G		137	.234		78	10.55	.98	89.53	105
0500	Steel, 2 rail, on stairs, primed, 1-1/4" diameter	G		160	.200		28	9.05	.84	37.89	48
0520	1-1/2" diameter	G		160	.200		30.50	9.05	.84	40.39	51
0540	Galvanized, 1-1/4" diameter	G		160	.200		39	9.05	.84	48.89	59.50
0560	1-1/2" diameter	G		160	.200		43.50	9.05	.84	53.39	65
0580	Steel, 3 rail, primed, 1-1/4" diameter	G		137	.234		41.50	10.55	.98	53.03	66
0600	1-1/2" diameter	G		137	.234		44.50	10.55	.98	56.03	68.50
0620	Galvanized, 1-1/4" diameter	G		137	.234		58.50	10.55	.98	70.03	84.50
0640	1-1/2" diameter	G		137	.234		69	10.55	.98	80.53	96
0700	Stainless steel, 2 rail, 1-1/4" diam. #4 finish	G		137	.234		101	10.55	.98	112.53	131
0720	High polish	G		137	.234		162	10.55	.98	173.53	199
0740	Mirror polish	G		137	.234		203	10.55	.98	214.53	244
0760	Stainless steel, 3 rail, 1-1/2" diam., #4 finish	G		120	.267		152	12.05	1.12	165.17	190
0770	High polish	G		120	.267		251	12.05	1.12	264.17	299
0780	Mirror finish	G		120	.267		305	12.05	1.12	318.17	360
0900	Wall rail, alum. pipe, 1-1/4" diam., satin finish	G		213	.150		19.50	6.80	.63	26.93	34.50
0905	Clear anodized	G		213	.150		24	6.80	.63	31.43	39
0910	Dark anodized	G		213	.150		29	6.80	.63	36.43	44.50
0915	1-1/2" diameter, satin finish	G		213	.150		21.50	6.80	.63	28.93	36.50
0920	Clear anodized	G		213	.150		27	6.80	.63	34.43	43
0925	Dark anodized	G		213	.150		33.50	6.80	.63	40.93	50
0930	Steel pipe, 1-1/4" diameter, primed	G		213	.150		16.95	6.80	.63	24.38	31.50
0935	Galvanized	G		213	.150		24.50	6.80	.63	31.93	40
0940	1-1/2" diameter	G		176	.182		17.45	8.20	.76	26.41	34.50
0945	Galvanized	G		213	.150		24.50	6.80	.63	31.93	40
0955	Stainless steel pipe, 1-1/2" diam., #4 finish	G		107	.299		80.50	13.50	1.25	95.25	114
0960	High polish	G		107	.299		164	13.50	1.25	178.75	205
0965	Mirror polish	G	↓	107	.299	↓	193	13.50	1.25	207.75	237
2000	Aluminum pipe & picket railing, double top rail, pickets @ 4-1/2" OC										
2010	36" high, straight & level	G	2 Sswk	80	.200	L.F.	81	8.95		89.95	105
2020	Curved & level	G	↓	60	.267		111	11.90		122.90	143

05 52 Metal Railings

05 52 13 – Pipe and Tube Railings

05 52 13.50 Railings, Pipe		Crew	Daily Output	Labor-Hours	Unit	Material	2009 Bare Costs Labor	Equipment	Total	Total Incl O&P
2030	Straight & sloped	G 2 Sswk	40	.400	L.F.	93	17.90		110.90	134

05 52 16 – Industrial Railings

05 52 16.50 Railings, Industrial		Crew	Daily Output	Labor-Hours	Unit	Material	Labor	Equipment	Total	Total Incl O&P
0010	**RAILINGS, INDUSTRIAL**, shop fab'd, 3'-6" high, posts @ 5' O.C.									
0020	2 rail, 3'-6" high, 1-1/2" pipe	G E-4	255	.125	L.F.	31	5.65	.53	37.18	44.50
0100	2" angle rail	G "	255	.125		28	5.65	.53	34.18	41.50
0200	For 4" high kick plate, 10 gauge, add	G				6.40			6.40	7.05
0300	1/4" thick, add	G				8.25			8.25	9.10
0500	For curved rails, add					30%	30%			

05 55 Metal Stair Treads and Nosings

05 55 13 – Metal Stair Treads

05 55 13.50 Stair Treads		Crew	Daily Output	Labor-Hours	Unit	Material	Labor	Equipment	Total	Total Incl O&P
0010	**STAIR TREADS**									
0020	Aluminum grating, 3' long, 1-1/2" x 3/16" rect. bars, 6" wide	G 1 Sswk	24	.333	Ea.	65.50	14.90		80.40	98.50
0100	12" wide	G	22	.364		96.50	16.25		112.75	135
0200	1-1/2" x 3/16" I-bars, 6" wide	G	24	.333		55	14.90		69.90	87
0300	12" wide	G	22	.364		88.50	16.25		104.75	127
0400	For abrasive nosings, add	G				17.15			17.15	18.85
0500	For narrow mesh, add					60%				
0600	Stair treads, not incl. stringers, see also Div. 03 15 05.85									
0700	Cast aluminum, abrasive, 3' long x 12" wide, 5/16" thick	G 1 Sswk	15	.533	Ea.	145	24		169	203
0800	3/8" thick	G	15	.533		154	24		178	213
0900	1/2" thick	G	15	.533		169	24		193	229
1000	Cast bronze, abrasive, 3/8" thick	G	8	1		660	44.50		704.50	805
1100	1/2" thick	G	8	1		865	44.50		909.50	1,025
1200	Cast iron, abrasive, 3' long x 12" wide, 3/8" thick	G	15	.533		128	24		152	183
1300	1/2" thick	G	15	.533		147	24		171	205
1400	Fiberglass reinforced plastic with safety nosing,									
1500	1-1/2" thick, 12" wide, 24" long	G 1 Sswk	22	.364	Ea.	144	16.25		160.25	188
1600	30" long	G	22	.364		151	16.25		167.25	195
1700	36" long	G	22	.364		165	16.25		181.25	211

05 58 Formed Metal Fabrications

05 58 21 – Formed Chain

05 58 21.05 Alloy Steel Chain		Crew	Daily Output	Labor-Hours	Unit	Material	Labor	Equipment	Total	Total Incl O&P
0010	**ALLOY STEEL CHAIN**, Grade 80									
0015	Self-colored, cut lengths, w/accessories, 1/4"	G E-17	4	4	C.L.F.	455	183		638	825
0020	3/8"	G	2	8		615	365		980	1,325
0030	1/2"	G	1.20	13.333		1,100	610		1,710	2,300
0040	5/8"	G	.72	22.222		1,675	1,025		2,700	3,650
0050	3/4"	G E-18	.48	83.333		2,675	3,700	2,250	8,625	11,800
0060	7/8"	G	.40	100		4,150	4,450	2,700	11,300	15,200
0070	1"	G	.35	114		5,700	5,075	3,075	13,850	18,500
0080	1-1/4"	G	.24	166		16,900	7,400	4,500	28,800	36,400
0110	Clevis slip hook, 1/4"	G			Ea.	13.10			13.10	14.40
0120	3/8"	G				18.60			18.60	20.50
0130	1/2"	G				33			33	36
0140	5/8"	G				47			47	52

05 58 Formed Metal Fabrications

05 58 21 – Formed Chain

05 58 21.05 Alloy Steel Chain

			Crew	Daily Output	Labor-Hours	Unit	Material	2009 Bare Costs Labor	Equipment	Total	Total Incl O&P
0150	3/4"	G				Ea.	102			102	112
0160	Eye/sling hook w/ hammerlock coupling, 7/8"	G					280			280	310
0170	1"	G					475			475	520
0180	1-1/4"	G					700			700	770

05 58 23 – Formed Columns

05 58 23.10 Aluminum Columns

0010	**ALUMINUM COLUMNS**										
0015	Made from recycled materials										
0020	Aluminum, extruded, stock units, no cap or base, 6" diameter	G	E-4	240	.133	L.F.	11.65	6.05	.56	18.26	24
0100	8" diameter	G		170	.188		15	8.50	.79	24.29	32.50
0200	10" diameter	G		150	.213		20.50	9.65	.89	31.04	40.50
0300	12" diameter	G		140	.229		37	10.35	.96	48.31	60
0400	15" diameter	G		120	.267		44	12.05	1.12	57.17	71
0410	Caps and bases, plain, 6" diameter	G				Set	22			22	24
0420	8" diameter	G					29			29	31.50
0430	10" diameter	G					37			37	40.50
0440	12" diameter	G					50			50	55
0450	15" diameter	G					90			90	99

05 58 25 – Formed Lamp Posts

05 58 25.40 Lamp Posts

0010	**LAMP POSTS**										
0020	Aluminum, 7' high, stock units, post only	G	1 Carp	16	.500	Ea.	28.50	20		48.50	62.50
0100	Mild steel, plain	G	"	16	.500	"	23	20		43	56.50

05 58 27 – Formed Guards

05 58 27.90 Window Guards

0010	**WINDOW GUARDS**, shop fabricated										
0015	Expanded metal, steel angle frame, permanent	G	E-4	350	.091	S.F.	26	4.13	.38	30.51	36.50
0025	Steel bars, 1/2" x 1/2", spaced 5" O.C.	G	"	290	.110	"	18	4.99	.46	23.45	29
0030	Hinge mounted, add	G				Opng.	52			52	57.50
0040	Removable type, add	G				"	33			33	36.50
0050	For galvanized guards, add					S.F.	35%				
0070	For pivoted or projected type, add						105%	40%			
0100	Mild steel, stock units, economy	G	E-4	405	.079		7.05	3.57	.33	10.95	14.45
0200	Deluxe	G	"	405	.079		14.50	3.57	.33	18.40	22.50

05 59 Metal Specialties

05 59 63 – Detention Enclosures

05 59 63.13 Enclosures, Detention

0010	**ENCLOSURES, DETENTION**										
0500	Bar front, rolling, 7/8" bars, 4" O.C., 7' high, 5' wide, with hardware	G	E-4	2	16	Ea.	5,850	725	67	6,642	7,800

05 71 Decorative Metal Stairs

05 71 13 – Fabricated Metal Spiral Stairs

05 71 13.50 Spiral Stairs

		Crew	Daily Output	Labor-Hours	Unit	Material	2009 Bare Costs Labor	2009 Bare Costs Equipment	Total	Total Incl O&P
0010	**SPIRAL STAIRS**, shop fabricated									
1820	Custom units	G E-4	45	.711	Riser	1,275	32	2.98	1,309.98	1,475
1900	Spiral, cast iron, 4'-0" diameter, ornamental, minimum	G	45	.711		605	32	2.98	639.98	725
1920	Maximum	G	25	1.280		825	58	5.35	888.35	1,025
2000	Spiral, steel, industrial checkered plate, 4' diameter	G	45	.711		605	32	2.98	639.98	725
2200	Stock units, 6'-0" diameter	G	40	.800		735	36	3.35	774.35	880
3110	Spiral steel, stock units, primed, flat metal tread, 3'-6" dia	G 2 Carp	1.60	10	Flight	1,250	400		1,650	1,975
3120	4'-0" dia	G	1.45	11.034		1,425	440		1,865	2,250
3130	4'-6" dia	G	1.35	11.852		1,575	475		2,050	2,450
3140	5'-0" dia	G	1.25	12.800		1,950	510		2,460	2,950
3210	Galvanized, 3'-6" dia	G	1.60	10		2,100	400		2,500	2,950
3220	4'-0" dia	G	1.45	11.034		2,350	440		2,790	3,250
3230	4'-6" dia	G	1.35	11.852		2,550	475		3,025	3,525
3240	5'-0" dia	G	1.25	12.800		3,050	510		3,560	4,150
3310	Checkered plate tread, 3'-6" dia	G	1.45	11.034		1,450	440		1,890	2,275
3320	4'-0" dia	G	1.35	11.852		1,650	475		2,125	2,525
3330	4'-6" dia	G	1.25	12.800		1,800	510		2,310	2,775
3340	5'-0" dia	G	1.15	13.913		2,200	555		2,755	3,275
3410	Galvanized, 3'-6" dia	G	1.45	11.034		2,400	440		2,840	3,325
3420	4'-0" dia	G	1.35	11.852		2,650	475		3,125	3,625
3430	4'-6" dia	G	1.25	12.800		2,850	510		3,360	3,925
3440	5'-0" dia	G	1.15	13.913		3,350	555		3,905	4,550
3510	Red oak tread on flat metal, 3'-6" dia		1.35	11.852		2,375	475		2,850	3,325
3520	4'-0" dia		1.25	12.800		2,575	510		3,085	3,625
3530	4'-6" dia		1.15	13.913		2,800	555		3,355	3,925
3540	5'-0" dia		1.05	15.238		3,250	610		3,860	4,525

05 73 Decorative Metal Railings

05 73 16 – Wire Rope Decorative Metal Railings

05 73 16.10 Cable Railings

		Crew	Daily Output	Labor-Hours	Unit	Material	Labor	Equipment	Total	Total Incl O&P
0010	**CABLE RAILINGS**, with 316 stainless steel 1 x 19 cable, 3/16" diameter									
0015	Made from recycled materials									
0100	1-3/4" diameter stainless steel posts x 42" high, cables 4" OC	G 2 Sswk	25	.640	L.F.	43	28.50		71.50	98

05 73 23 – Ornamental Railings

05 73 23.50 Railings, Ornamental

		Crew	Daily Output	Labor-Hours	Unit	Material	Labor	Equipment	Total	Total Incl O&P
0010	**RAILINGS, ORNAMENTAL**, shop fab'd, 3'-6" high, posts @ 5' O.C.									
0020	Bronze or stainless, minimum	G 1 Sswk	24	.333	L.F.	36.50	14.90		51.40	67
0100	Maximum	G	9	.889		365	39.50		404.50	470
0200	Aluminum ornamental rail, minimum	G	15	.533		36.50	24		60.50	83
0300	Maximum	G	8	1		110	44.50		154.50	201
0400	Hand-forged wrought iron, minimum	G	12	.667		115	30		145	179
0500	Maximum	G	8	1		345	44.50		389.50	460
0600	Composite metal/wood/glass, minimum		6	1.333		220	59.50		279.50	350
0700	Maximum		5	1.600		440	71.50		511.50	610

05 75 Decorative Formed Metal

05 75 13 – Columns

05 75 13.20 Columns, Ornamental		Crew	Daily Output	Labor-Hours	Unit	Material	2009 Bare Costs Labor	Equipment	Total	Total Incl O&P
0010	**COLUMNS, ORNAMENTAL**, shop fabricated									
6400	Mild steel, flat, 9" wide, stock units, painted, plain	G	E-4	160	.200	V.L.F.	8.90	9.05	.84	18.79	27
6450	Fancy	G		160	.200		17	9.05	.84	26.89	35.50
6500	Corner columns, painted, plain	G		160	.200		15.50	9.05	.84	25.39	34
6550	Fancy	G		160	.200		30	9.05	.84	39.89	50

Estimating Tips

06 05 00 Common Work Results for Wood, Plastics, and Composites

- Common to any wood-framed structure are the accessory connector items such as screws, nails, adhesives, hangers, connector plates, straps, angles, and hold-downs. For typical wood-framed buildings, such as residential projects, the aggregate total for these items can be significant, especially in areas where seismic loading is a concern. For floor and wall framing, the material cost is based on 10 to 25 lbs. per MBF. Hold-downs, hangers, and other connectors should be taken off by the piece.

06 10 00 Carpentry

- Lumber is a traded commodity and therefore sensitive to supply and demand in the marketplace. Even in "budgetary" estimating of wood-framed projects, it is advisable to call local suppliers for the latest market pricing.
- Common quantity units for wood-framed projects are "thousand board feet" (MBF). A board foot is a volume of wood, 1" x 1' x 1', or 144 cubic inches. Board-foot quantities are generally calculated using nominal material dimensions—dressed sizes are ignored. Board foot per lineal foot of any stick of lumber can be calculated by dividing the nominal cross-sectional area by 12. As an example, 2,000 lineal feet of 2 x 12 equates to 4 MBF by dividing the nominal area, 2 x 12, by 12, which equals 2, and multiplying by 2,000 to give 4,000 board feet. This simple rule applies to all nominal dimensioned lumber.
- Waste is an issue of concern at the quantity takeoff for any area of construction. Framing lumber is sold in even foot lengths, i.e., 10', 12', 14', 16', and depending on spans, wall heights, and the grade of lumber, waste is inevitable. A rule of thumb for lumber waste is 5%–10% depending on material quality and the complexity of the framing.
- Wood in various forms and shapes is used in many projects, even where the main structural framing is steel, concrete, or masonry. Plywood as a back-up partition material and 2x boards used as blocking and cant strips around roof edges are two common examples. The estimator should ensure that the costs of all wood materials are included in the final estimate.

06 20 00 Finish Carpentry

- It is necessary to consider the grade of workmanship when estimating labor costs for erecting millwork and interior finish. In practice, there are three grades: premium, custom, and economy. The RSMeans daily output for base and case moldings is in the range of 200 to 250 L.F. per carpenter per day. This is appropriate for most average custom-grade projects. For premium projects, an adjustment to productivity of 25%–50% should be made, depending on the complexity of the job.

Reference Numbers

Reference numbers are shown in shaded boxes at the beginning of some major classifications. These numbers refer to related items in the Reference Section. The reference information may be an estimating procedure, an alternate pricing method, or technical information.

Note: Not all subdivisions listed here necessarily appear in this publication.

No part of this publication may be reproduced, stored in a retrieval system, or transmitted in any form or by any means without prior written permission of R.S. Means Company, Inc.

06 05 Common Work Results for Wood, Plastics and Composites

06 05 05 – Selective Wood and Plastics Demolition

06 05 05.10 Selective Demolition Wood Framing		Crew	Daily Output	Labor-Hours	Unit	Material	2009 Bare Costs Labor	Equipment	Total	Total Incl O&P
0010	**SELECTIVE DEMOLITION WOOD FRAMING** R024119-10									
0100	Timber connector, nailed, small	1 Clab	96	.083	Ea.		2.63		2.63	4.08
0110	Medium		60	.133			4.21		4.21	6.55
0120	Large		48	.167			5.25		5.25	8.15
0130	Bolted, small		48	.167			5.25		5.25	8.15
0140	Medium		32	.250			7.90		7.90	12.25
0150	Large		24	.333			10.55		10.55	16.35
2958	Beams, 2" x 6"	2 Clab	1100	.015	L.F.		.46		.46	.71
2960	2" x 8"		825	.019			.61		.61	.95
2965	2" x 10"		665	.024			.76		.76	1.18
2970	2" x 12"		550	.029			.92		.92	1.43
2972	2" x 14"		470	.034			1.08		1.08	1.67
2975	4" x 8"	B-1	413	.058			1.88		1.88	2.91
2980	4" x 10"		330	.073			2.35		2.35	3.64
2985	4" x 12"		275	.087			2.82		2.82	4.37
3000	6" x 8"		275	.087			2.82		2.82	4.37
3040	6" x 10"		220	.109			3.52		3.52	5.45
3080	6" x 12"		185	.130			4.19		4.19	6.50
3120	8" x 12"		140	.171			5.55		5.55	8.60
3160	10" x 12"		110	.218			7.05		7.05	10.90
3162	Alternate pricing method		1.10	21.818	M.B.F.		705		705	1,100
3170	Blocking, in 16" OC wall framing, 2" x 4"	1 Clab	600	.013	L.F.		.42		.42	.65
3172	2" x 6"		400	.020			.63		.63	.98
3174	In 24" OC wall framing, 2" x 4"		600	.013			.42		.42	.65
3176	2" x 6"		400	.020			.63		.63	.98
3178	Alt method, wood blocking removal from wood framing		.40	20	M.B.F.		630		630	980
3179	Wood blocking removal from steel framing		.36	22.222	"		700		700	1,100
3180	Bracing, let in, 1" x 3", studs 16" OC		1050	.008	L.F.		.24		.24	.37
3181	Studs 24" OC		1080	.007			.23		.23	.36
3182	1" x 4", studs 16" OC		1050	.008			.24		.24	.37
3183	Studs 24" OC		1080	.007			.23		.23	.36
3184	1" x 6", studs 16" OC		1050	.008			.24		.24	.37
3185	Studs 24" OC		1080	.007			.23		.23	.36
3186	2" x 3", studs 16" OC		800	.010			.32		.32	.49
3187	Studs 24" OC		830	.010			.30		.30	.47
3188	2" x 4", studs 16" OC		800	.010			.32		.32	.49
3189	Studs 24" OC		830	.010			.30		.30	.47
3190	2" x 6", studs 16" OC		800	.010			.32		.32	.49
3191	Studs 24" OC		830	.010			.30		.30	.47
3192	2" x 8", studs 16" OC		800	.010			.32		.32	.49
3193	Studs 24" OC		830	.010			.30		.30	.47
3194	"T" shaped metal bracing, studs at 16" OC		1060	.008			.24		.24	.37
3195	Studs at 24" OC		1200	.007			.21		.21	.33
3196	Metal straps, studs at 16" OC		1200	.007			.21		.21	.33
3197	Studs at 24" OC		1240	.006			.20		.20	.32
3200	Columns, round, 8' to 14' tall		40	.200	Ea.		6.30		6.30	9.80
3202	Dimensional lumber sizes	2 Clab	1.10	14.545	M.B.F.		460		460	715
3250	Blocking, between joists	1 Clab	320	.025	Ea.		.79		.79	1.23
3252	Bridging, metal strap, between joists		320	.025	Pr.		.79		.79	1.23
3254	Wood, between joists		320	.025	"		.79		.79	1.23
3260	Door buck, studs, header & access., 8' high 2" x 4" wall, 3' wide		32	.250	Ea.		7.90		7.90	12.25
3261	4' wide		32	.250			7.90		7.90	12.25
3262	5' wide		32	.250			7.90		7.90	12.25

06 05 Common Work Results for Wood, Plastics and Composites

06 05 05 – Selective Wood and Plastics Demolition

06 05 05.10 Selective Demolition Wood Framing		Crew	Daily Output	Labor-Hours	Unit	Material	2009 Bare Costs Labor	Equipment	Total	Total Incl O&P
3263	6' wide	1 Clab	32	.250	Ea.		7.90		7.90	12.25
3264	8' wide		30	.267			8.45		8.45	13.05
3265	10' wide		30	.267			8.45		8.45	13.05
3266	12' wide		30	.267			8.45		8.45	13.05
3267	2" x 6" wall, 3' wide		32	.250			7.90		7.90	12.25
3268	4' wide		32	.250			7.90		7.90	12.25
3269	5' wide		32	.250			7.90		7.90	12.25
3270	6' wide		32	.250			7.90		7.90	12.25
3271	8' wide		30	.267			8.45		8.45	13.05
3273	12' wide		30	.267			8.45		8.45	13.05
3274	Window buck, studs, header & access, 8' high 2" x 4" wall, 2' wide		24	.333			10.55		10.55	16.35
3275	3' wide		24	.333			10.55		10.55	16.35
3276	4' wide		24	.333			10.55		10.55	16.35
3277	5' wide		24	.333			10.55		10.55	16.35
3278	6' wide		24	.333			10.55		10.55	16.35
3279	7' wide		24	.333			10.55		10.55	16.35
3280	8' wide		22	.364			11.50		11.50	17.80
3281	10' wide		22	.364			11.50		11.50	17.80
3282	12' wide		22	.364			11.50		11.50	17.80
3283	2" x 6" wall, 2' wide		24	.333			10.55		10.55	16.35
3284	3' wide		24	.333			10.55		10.55	16.35
3285	4' wide		24	.333			10.55		10.55	16.35
3286	5' wide		24	.333			10.55		10.55	16.35
3287	6' wide		24	.333			10.55		10.55	16.35
3288	7' wide		24	.333			10.55		10.55	16.35
3289	8' wide		22	.364			11.50		11.50	17.80
3290	10' wide		22	.364			11.50		11.50	17.80
3291	12' wide		22	.364			11.50		11.50	17.80
3400	Fascia boards, 1" x 6"		500	.016	L.F.		.51		.51	.78
3440	1" x 8"		450	.018			.56		.56	.87
3480	1" x 10"		400	.020			.63		.63	.98
3490	2" x 6"		450	.018			.56		.56	.87
3500	2" x 8"		400	.020			.63		.63	.98
3510	2" x 10"		350	.023			.72		.72	1.12
3610	Furring, on wood walls or ceiling		4000	.002	S.F.		.06		.06	.10
3620	On masonry or concrete walls or ceiling		1200	.007	"		.21		.21	.33
3800	Headers over openings, 2 @ 2" x 6"		110	.073	L.F.		2.30		2.30	3.56
3840	2 @ 2" x 8"		100	.080			2.53		2.53	3.92
3880	2 @ 2" x 10"		90	.089			2.81		2.81	4.36
3885	Alternate pricing method		.26	30.651	M.B.F.		970		970	1,500
3920	Joists, 1" x 4"		1250	.006	L.F.		.20		.20	.31
3930	1" x 6"		1135	.007			.22		.22	.35
3940	1" x 8"		1000	.008			.25		.25	.39
3950	1" x 10"		895	.009			.28		.28	.44
3960	1" x 12"		765	.010			.33		.33	.51
4200	2" x 4"	2 Clab	1000	.016			.51		.51	.78
4230	2" x 6"		970	.016			.52		.52	.81
4240	2" x 8"		940	.017			.54		.54	.83
4250	2" x 10"		910	.018			.56		.56	.86
4280	2" x 12"		880	.018			.57		.57	.89
4281	2" x 14"		850	.019			.59		.59	.92
4282	Composite joists, 9-1/2"		960	.017			.53		.53	.82
4283	11-7/8"		930	.017			.54		.54	.84

06 05 Common Work Results for Wood, Plastics and Composites

06 05 05 – Selective Wood and Plastics Demolition

06 05 05.10 Selective Demolition Wood Framing		Crew	Daily Output	Labor-Hours	Unit	Material	2009 Bare Costs Labor	Equipment	Total	Total Incl O&P
4284	14"	2 Clab	897	.018	L.F.		.56		.56	.87
4285	16"		865	.019			.58		.58	.91
4290	Wood joists, alternate pricing method		1.50	10.667	M.B.F.		335		335	525
4500	Open web joist, 12" deep		500	.032	L.F.		1.01		1.01	1.57
4505	14" deep		475	.034			1.06		1.06	1.65
4510	16" deep		450	.036			1.12		1.12	1.74
4520	18" deep		425	.038			1.19		1.19	1.84
4550	Ledger strips, 1" x 2"	1 Clab	1200	.007			.21		.21	.33
4560	1" x 3"		1200	.007			.21		.21	.33
4570	1" x 4"		1200	.007			.21		.21	.33
4580	2" x 2"		1100	.007			.23		.23	.36
4590	2" x 4"		1000	.008			.25		.25	.39
4600	2" x 6"		1000	.008			.25		.25	.39
4601	2" x 8 or 2" x 10"		800	.010			.32		.32	.49
4602	4" x 6"		600	.013			.42		.42	.65
4604	4" x 8"		450	.018			.56		.56	.87
5400	Posts, 4" x 4"	2 Clab	800	.020			.63		.63	.98
5405	4" x 6"		550	.029			.92		.92	1.43
5410	4" x 8"		440	.036			1.15		1.15	1.78
5425	4" x 10"		390	.041			1.30		1.30	2.01
5430	4" x 12"		350	.046			1.44		1.44	2.24
5440	6" x 6"		400	.040			1.26		1.26	1.96
5445	6" x 8"		350	.046			1.44		1.44	2.24
5450	6" x 10"		320	.050			1.58		1.58	2.45
5455	6" x 12"		290	.055			1.74		1.74	2.70
5480	8" x 8"		300	.053			1.69		1.69	2.61
5500	10" x 10"		240	.067			2.11		2.11	3.27
5660	Tongue and groove floor planks		2	8	M.B.F.		253		253	390
5750	Rafters, ordinary, 16" OC, 2" x 4"		880	.018	S.F.		.57		.57	.89
5755	2" x 6"		840	.019			.60		.60	.93
5760	2" x 8"		820	.020			.62		.62	.96
5770	2" x 10"		820	.020			.62		.62	.96
5780	2" x 12"		810	.020			.62		.62	.97
5785	24" OC, 2" x 4"		1170	.014			.43		.43	.67
5786	2" x 6"		1117	.014			.45		.45	.70
5787	2" x 8"		1091	.015			.46		.46	.72
5788	2" x 10"		1091	.015			.46		.46	.72
5789	2" x 12"		1077	.015			.47		.47	.73
5795	Rafters, ordinary, 2" x 4" (alternate method)		862	.019	L.F.		.59		.59	.91
5800	2" x 6" (alternate method)		850	.019			.59		.59	.92
5840	2" x 8" (alternate method)		837	.019			.60		.60	.94
5855	2" x 10" (alternate method)		825	.019			.61		.61	.95
5865	2" x 12" (alternate method)		812	.020			.62		.62	.97
5870	Sill plate, 2" x 4"	1 Clab	1170	.007			.22		.22	.34
5871	2" x 6"		780	.010			.32		.32	.50
5872	2" x 8"		586	.014			.43		.43	.67
5873	Alternate pricing method		.78	10.256	M.B.F.		325		325	505
5885	Ridge board, 1" x 4"	2 Clab	900	.018	L.F.		.56		.56	.87
5886	1" x 6"		875	.018			.58		.58	.90
5887	1" x 8"		850	.019			.59		.59	.92
5888	1" x 10"		825	.019			.61		.61	.95
5889	1" x 12"		800	.020			.63		.63	.98
5890	2" x 4"		900	.018			.56		.56	.87

06 05 Common Work Results for Wood, Plastics and Composites

06 05 05 – Selective Wood and Plastics Demolition

06 05 05.10 Selective Demolition Wood Framing		Crew	Daily Output	Labor-Hours	Unit	Material	2009 Bare Costs Labor	Equipment	Total	Total Incl O&P
5892	2" x 6"	2 Clab	875	.018	L.F.		.58		.58	.90
5894	2" x 8"		850	.019			.59		.59	.92
5896	2" x 10"		825	.019			.61		.61	.95
5898	2" x 12"		800	.020			.63		.63	.98
6050	Rafter tie, 1" x 4"		1250	.013			.40		.40	.63
6052	1" x 6"		1135	.014			.45		.45	.69
6054	2" x 4"		1000	.016			.51		.51	.78
6056	2" x 6"		970	.016			.52		.52	.81
6070	Sleepers, on concrete, 1" x 2"	1 Clab	4700	.002			.05		.05	.08
6075	1" x 3"		4000	.002			.06		.06	.10
6080	2" x 4"		3000	.003			.08		.08	.13
6085	2" x 6"		2600	.003			.10		.10	.15
6086	Sheathing from roof, 5/16"	2 Clab	1600	.010	S.F.		.32		.32	.49
6088	3/8"		1525	.010			.33		.33	.51
6090	1/2"		1400	.011			.36		.36	.56
6092	5/8"		1300	.012			.39		.39	.60
6094	3/4"		1200	.013			.42		.42	.65
6096	Board sheathing from roof		1400	.011			.36		.36	.56
6100	Sheathing, from walls, 1/4"		1200	.013			.42		.42	.65
6110	5/16"		1175	.014			.43		.43	.67
6120	3/8"		1150	.014			.44		.44	.68
6130	1/2"		1125	.014			.45		.45	.70
6140	5/8"		1100	.015			.46		.46	.71
6150	3/4"		1075	.015			.47		.47	.73
6152	Board sheathing from walls		1500	.011			.34		.34	.52
6158	Subfloor, with boards		1050	.015			.48		.48	.75
6160	Plywood, 1/2" thick		768	.021			.66		.66	1.02
6162	5/8" thick		760	.021			.67		.67	1.03
6164	3/4" thick		750	.021			.67		.67	1.05
6165	1-1/8" thick		720	.022			.70		.70	1.09
6166	Underlayment, particle board, 3/8" thick	1 Clab	780	.010			.32		.32	.50
6168	1/2" thick		768	.010			.33		.33	.51
6170	5/8" thick		760	.011			.33		.33	.52
6172	3/4" thick		750	.011			.34		.34	.52
6200	Stairs and stringers, minimum	2 Clab	40	.400	Riser		12.65		12.65	19.60
6240	Maximum	"	26	.615	"		19.45		19.45	30
6300	Components, tread	1 Clab	110	.073	Ea.		2.30		2.30	3.56
6320	Riser		80	.100	"		3.16		3.16	4.90
6390	Stringer, 2" x 10"		260	.031	L.F.		.97		.97	1.51
6400	2" x 12"		260	.031			.97		.97	1.51
6410	3" x 10"		250	.032			1.01		1.01	1.57
6420	3" x 12"		250	.032			1.01		1.01	1.57
6590	Wood studs, 2" x 3"	2 Clab	3076	.005			.16		.16	.25
6600	2" x 4"		2000	.008			.25		.25	.39
6640	2" x 6"		1600	.010			.32		.32	.49
6720	Wall framing, including studs, plates and blocking, 2" x 4"	1 Clab	600	.013	S.F.		.42		.42	.65
6740	2" x 6"		480	.017	"		.53		.53	.82
6750	Headers, 2" x 4"		1125	.007	L.F.		.22		.22	.35
6755	2" x 6"		1125	.007			.22		.22	.35
6760	2" x 8"		1050	.008			.24		.24	.37
6765	2" x 10"		1050	.008			.24		.24	.37
6770	2" x 12"		1000	.008			.25		.25	.39
6780	4" x 10"		525	.015			.48		.48	.75

06 05 Common Work Results for Wood, Plastics and Composites

06 05 05 – Selective Wood and Plastics Demolition

06 05 05.10 Selective Demolition Wood Framing

		Crew	Daily Output	Labor-Hours	Unit	Material	2009 Bare Costs Labor	Equipment	Total	Total Incl O&P
6785	4" x 12"	1 Clab	500	.016	L.F.		.51		.51	.78
6790	6" x 8"		560	.014			.45		.45	.70
6795	6" x 10"		525	.015			.48		.48	.75
6797	6" x 12"		500	.016			.51		.51	.78
8000	Soffit, T & G wood		520	.015	S.F.		.49		.49	.75
8010	Hardboard, vinyl or aluminum		640	.013			.40		.40	.61
8030	Plywood	2 Carp	315	.051			2.03		2.03	3.15

06 05 05.20 Selective Demolition Millwork and Trim

		Crew	Daily Output	Labor-Hours	Unit	Material	Labor	Equipment	Total	Total Incl O&P
0010	**SELECTIVE DEMOLITION MILLWORK AND TRIM** R024119-10									
1000	Cabinets, wood, base cabinets, per L.F.	2 Clab	80	.200	L.F.		6.30		6.30	9.80
1020	Wall cabinets, per L.F.	"	80	.200	"		6.30		6.30	9.80
1060	Remove and reset, base cabinets	2 Carp	18	.889	Ea.		35.50		35.50	55
1070	Wall cabinets	"	20	.800	"		32		32	49.50
1100	Steel, painted, base cabinets	2 Clab	60	.267	L.F.		8.45		8.45	13.05
1120	Wall cabinets		60	.267	"		8.45		8.45	13.05
1200	Casework, large area		320	.050	S.F.		1.58		1.58	2.45
1220	Selective		200	.080	"		2.53		2.53	3.92
1500	Counter top, minimum		200	.080	L.F.		2.53		2.53	3.92
1510	Maximum		120	.133			4.21		4.21	6.55
1550	Remove and reset, minimum	2 Carp	50	.320			12.80		12.80	19.80
1560	Maximum	"	40	.400			16		16	25
2000	Paneling, 4' x 8' sheets	2 Clab	2000	.008	S.F.		.25		.25	.39
2100	Boards, 1" x 4"		700	.023			.72		.72	1.12
2120	1" x 6"		750	.021			.67		.67	1.05
2140	1" x 8"		800	.020			.63		.63	.98
3000	Trim, baseboard, to 6" wide		1200	.013	L.F.		.42		.42	.65
3040	Greater than 6" and up to 12" wide		1000	.016			.51		.51	.78
3080	Remove and reset, minimum	2 Carp	400	.040			1.60		1.60	2.48
3090	Maximum	"	300	.053			2.13		2.13	3.30
3100	Ceiling trim	2 Clab	1000	.016			.51		.51	.78
3120	Chair rail		1200	.013			.42		.42	.65
3140	Railings with balusters		240	.067			2.11		2.11	3.27
3160	Wainscoting		700	.023	S.F.		.72		.72	1.12

06 05 23 – Wood, Plastic, and Composite Fastenings

06 05 23.10 Nails

		Crew	Daily Output	Labor-Hours	Unit	Material	Labor	Equipment	Total	Total Incl O&P
0010	**NAILS**, material only, based upon 50# box purchase									
0020	Copper nails, plain				Lb.	10.20			10.20	11.20
0400	Stainless steel, plain					6.65			6.65	7.30
0500	Box, 3d to 20d, bright					.91			.91	1
0520	Galvanized					1.67			1.67	1.84
0600	Common, 3d to 60d, plain					1.11			1.11	1.22
0700	Galvanized					1.53			1.53	1.68
0800	Aluminum					4.45			4.45	4.90
1000	Annular or spiral thread, 4d to 60d, plain					1.82			1.82	2
1200	Galvanized					1.99			1.99	2.19
1400	Drywall nails, plain					.79			.79	.87
1600	Galvanized					1.59			1.59	1.75
1800	Finish nails, 4d to 10d, plain					1.11			1.11	1.22
2000	Galvanized					1.75			1.75	1.93
2100	Aluminum					4.10			4.10	4.51
2300	Flooring nails, hardened steel, 2d to 10d, plain					2.04			2.04	2.24
2400	Galvanized					3.88			3.88	4.27

06 05 Common Work Results for Wood, Plastics and Composites

06 05 23 – Wood, Plastic, and Composite Fastenings

06 05 23.10 Nails		Crew	Daily Output	Labor-Hours	Unit	Material	2009 Bare Costs Labor	Equipment	Total	Total Incl O&P
2500	Gypsum lath nails, 1-1/8", 13 ga. flathead, blued				Lb.	2.64			2.64	2.90
2600	Masonry nails, hardened steel, 3/4" to 3" long, plain					1.97			1.97	2.17
2700	Galvanized					2.78			2.78	3.06
2900	Roofing nails, threaded, galvanized					1.38			1.38	1.52
3100	Aluminum					4.80			4.80	5.30
3300	Compressed lead head, threaded, galvanized					2.50			2.50	2.75
3600	Siding nails, plain shank, galvanized					1.50			1.50	1.65
3800	Aluminum					4.11			4.11	4.52
5000	Add to prices above for cement coating					.10			.10	.11
5200	Zinc or tin plating					.13			.13	.14
5500	Vinyl coated sinkers, 8d to 16d					.60			.60	.66

06 05 23.20 Pneumatic Nails										
0010	**PNEUMATIC NAILS**									
0020	Framing, per carton of 5000, 2"				Ea.	40			40	44
0100	2-3/8"					45			45	49.50
0200	Per carton of 4000, 3"					40			40	44
0300	3-1/4"					40			40	44
0400	Per carton of 5000, 2-3/8", galv.					60			60	66
0500	Per carton of 4000, 3", galv.					65			65	71.50
0600	3-1/4", galv.					82			82	90
0700	Roofing, per carton of 7200, 1"					35.50			35.50	39
0800	1-1/4"					35			35	38.50
0900	1-1/2"					38.50			38.50	42.50
1000	1-3/4"					47.50			47.50	52.50

06 05 23.40 Sheet Metal Screws										
0010	**SHEET METAL SCREWS**									
0020	Steel, standard, #8 x 3/4", plain				C	3.30			3.30	3.63
0100	Galvanized					3.39			3.39	3.73
0300	#10 x 1", plain					4.25			4.25	4.68
0400	Galvanized					4.43			4.43	4.87
0600	With washers, #14 x 1", plain					11.50			11.50	12.65
0700	Galvanized					12.45			12.45	13.70
0900	#14 x 2", plain					15			15	16.50
1000	Galvanized					17.15			17.15	18.85
1500	Self-drilling, with washers, (pinch point) #8 x 3/4", plain					8			8	8.80
1600	Galvanized					8.70			8.70	9.55
1800	#10 x 3/4", plain					10			10	11
1900	Galvanized					10.95			10.95	12.05
3000	Stainless steel w/aluminum or neoprene washers, #14 x 1", plain					32			32	35
3100	#14 x 2", plain					41			41	45

06 05 23.50 Wood Screws										
0010	**WOOD SCREWS**									
0020	#8 x 1" long, steel				C	5.10			5.10	5.60
0100	Brass					11.50			11.50	12.65
0200	#8, 2" long, steel					5.30			5.30	5.85
0300	Brass					19.50			19.50	21.50
0400	#10, 1" long, steel					3.71			3.71	4.08
0500	Brass					15.65			15.65	17.20
0600	#10, 2" long, steel					5.95			5.95	6.55
0700	Brass					23.50			23.50	26
0800	#10, 3" long, steel					9.95			9.95	10.95
1000	#12, 2" long, steel					7.90			7.90	8.70

06 05 Common Work Results for Wood, Plastics and Composites

06 05 23 – Wood, Plastic, and Composite Fastenings

06 05 23.50 Wood Screws		Crew	Daily Output	Labor-Hours	Unit	Material	2009 Bare Costs Labor	Equipment	Total	Total Incl O&P
1100	Brass				C	31.50			31.50	34.50
1500	#12, 3" long, steel					11.45			11.45	12.60
2000	#12, 4" long, steel					25.50			25.50	28

06 05 23.60 Timber Connectors

		Crew	Daily Output	Labor-Hours	Unit	Material	Labor	Equipment	Total	Total Incl O&P
0010	**TIMBER CONNECTORS**									
0020	Add up cost of each part for total cost of connection									
0100	Connector plates, steel, with bolts, straight	2 Carp	75	.213	Ea.	25.50	8.50		34	41
0110	Tee, 7 gauge		50	.320		29.50	12.80		42.30	52
0120	T-Strap, 14 gauge, 12" x 8" x 2"		50	.320		29.50	12.80		42.30	52
0150	Anchor plates, 7 ga, 9" x 7"		75	.213		25.50	8.50		34	41
0200	Bolts, machine, sq. hd. with nut & washer, 1/2" diameter, 4" long	1 Carp	140	.057		.79	2.28		3.07	4.41
0300	7-1/2" long		130	.062		1.37	2.46		3.83	5.30
0500	3/4" diameter, 7-1/2" long		130	.062		4.12	2.46		6.58	8.35
0800	Drilling bolt holes in timber, 1/2" diameter		450	.018	Inch		.71		.71	1.10
0900	1" diameter		350	.023	"		.91		.91	1.42
1100	Framing anchor, angle, 3" x 3" x 1-1/2", 12 ga		175	.046	Ea.	2.20	1.83		4.03	5.25
1150	Framing anchors, 18 gauge, 4-1/2" x 2-3/4"		175	.046		2.20	1.83		4.03	5.25
1160	Framing anchors, 18 gauge, 4-1/2" x 3"		175	.046		2.20	1.83		4.03	5.25
1170	Clip anchors plates, 18 gauge, 12" x 1-1/8"		175	.046		2.20	1.83		4.03	5.25
1250	Holdowns, 3 gauge base, 10 gauge body		8	1		21	40		61	85
1260	Holdowns, 7 gauge 11-1/16" x 3-1/4"		8	1		21	40		61	85
1270	Holdowns, 7 gauge 14-3/8" x 3-1/8"		8	1		21	40		61	85
1275	Holdowns, 12 gauge 8" x 2-1/2"		8	1		21	40		61	85
1300	Joist and beam hangers, 18 ga. galv., for 2" x 4" joist		175	.046		.61	1.83		2.44	3.50
1400	2" x 6" to 2" x 10" joist		165	.048		1.17	1.94		3.11	4.29
1600	16 ga. galv., 3" x 6" to 3" x 10" joist		160	.050		2.49	2		4.49	5.85
1700	3" x 10" to 3" x 14" joist		160	.050		4.14	2		6.14	7.65
1800	4" x 6" to 4" x 10" joist		155	.052		2.58	2.06		4.64	6.05
2300	3/16" thick, 6" x 8" joist		145	.055		54.50	2.20		56.70	63.50
2700	1/4" thick, 6" x 14" joist		130	.062		61.50	2.46		63.96	72
2900	Plywood clips, extruded aluminum H clip, for 3/4" panels					.20			.20	.22
3200	Post framing, 16 ga. galv. for 4" x 4" base, 2 piece	1 Carp	130	.062		13.80	2.46		16.26	18.95
3300	Cap		130	.062		19.15	2.46		21.61	25
3500	Rafter anchors, 18 ga. galv., 1-1/2" wide, 5-1/4" long		145	.055		.41	2.20		2.61	3.87
3600	10-3/4" long		145	.055		1.23	2.20		3.43	4.77
3800	Shear plates, 2-5/8" diameter		120	.067		2	2.66		4.66	6.35
4100	Spike grids, 3" x 6"		120	.067		.81	2.66		3.47	5
4400	Split rings, 2-1/2" diameter		120	.067		1.65	2.66		4.31	5.95
4500	4" diameter		110	.073		2.50	2.91		5.41	7.25
4550	Tie plate, 20 gauge, 7" x 3 1/8"		110	.073		2.50	2.91		5.41	7.25
4560	Tie plate, 20 gauge, 5" x 4 1/8"		110	.073		2.50	2.91		5.41	7.25
4575	Twist straps, 18 gauge, 12" x 1 1/4"		110	.073		2.50	2.91		5.41	7.25
4580	Twist straps, 18 gauge, 16" x 1 1/4"		110	.073		2.50	2.91		5.41	7.25
4600	Strap ties, 20 ga., 2-1/16" wide, 12 13/16" long		180	.044		.80	1.78		2.58	3.63
4700	Strap ties, 16 ga., 1-3/8" wide, 12" long		180	.044		.80	1.78		2.58	3.63
4800	21-5/8" x 1-1/4"		160	.050		2.49	2		4.49	5.85
5000	Toothed rings, 2-5/8" or 4" diameter		90	.089		1.49	3.55		5.04	7.15
5200	Truss plates, nailed, 20 gauge, up to 32' span		17	.471	Truss	10.85	18.80		29.65	41
5400	Washers, 2" x 2" x 1/8"				Ea.	.34			.34	.37
5500	3" x 3" x 3/16"				"	.89			.89	.98

06 05 23.70 Rough Hardware

| 0010 | **ROUGH HARDWARE**, average percent of carpentry material | | | | | | | | | |

06 05 Common Work Results for Wood, Plastics and Composites

06 05 23 – Wood, Plastic, and Composite Fastenings

06 05 23.70 Rough Hardware		Crew	Daily Output	Labor-Hours	Unit	Material	2009 Bare Costs Labor	Equipment	Total	Total Incl O&P
0020	Minimum					.50%				
0200	Maximum					1.50%				
0210	In seismic or hurricane areas, up to					10%				
06 05 23.80 Metal Bracing										
0010	**METAL BRACING**									
0302	Let-in, "T" shaped, 22 ga. galv. steel, studs at 16" O.C.	1 Carp	580	.014	L.F.	.75	.55		1.30	1.68
0402	Studs at 24" O.C.		600	.013		.75	.53		1.28	1.66
0502	16 ga. galv. steel straps, studs at 16" O.C.		600	.013		.97	.53		1.50	1.90
0602	Studs at 24" O.C.		620	.013		.97	.52		1.49	1.87

06 11 Wood Framing

06 11 10 – Framing with Dimensional, Engineered or Composite Lumber

06 11 10.02 Blocking		Crew	Daily Output	Labor-Hours	Unit	Material	2009 Bare Costs Labor	Equipment	Total	Total Incl O&P
0010	**BLOCKING**									
1950	Miscellaneous, to wood construction									
2000	2" x 4"	1 Carp	250	.032	L.F.	.29	1.28		1.57	2.30
2005	Pneumatic nailed		305	.026		.29	1.05		1.34	1.94
2050	2" x 6"		222	.036		.52	1.44		1.96	2.80
2055	Pneumatic nailed		271	.030		.52	1.18		1.70	2.40
2100	2" x 8"		200	.040		.71	1.60		2.31	3.26
2105	Pneumatic nailed		244	.033		.71	1.31		2.02	2.81
2150	2" x 10"		178	.045		.99	1.80		2.79	3.87
2155	Pneumatic nailed		217	.037		.99	1.47		2.46	3.37
2200	2" x 12"		151	.053		1.39	2.12		3.51	4.80
2205	Pneumatic nailed		185	.043		1.39	1.73		3.12	4.20
2300	To steel construction									
2320	2" x 4"	1 Carp	208	.038	L.F.	.29	1.54		1.83	2.70
2340	2" x 6"		180	.044		.52	1.78		2.30	3.32
2360	2" x 8"		158	.051		.71	2.02		2.73	3.92
2380	2" x 10"		136	.059		.99	2.35		3.34	4.73
2400	2" x 12"		109	.073		1.39	2.93		4.32	6.05
06 11 10.04 Wood Bracing										
0010	**WOOD BRACING**									
0011	Let-in, with 1" x 6" boards, studs 16" O.C.	1 Carp	1.50	5.333	C.L.F.	64	213		277	400
0200	Studs @ 24" O.C.	"	2.30	3.478	"	64	139		203	286
06 11 10.06 Bridging										
0010	**BRIDGING**									
0011	Wood, for joists 16" O.C., 1" x 3"	1 Carp	1.30	6.154	C.Pr.	51	246		297	435
0015	Pneumatic nailed		1.70	4.706		51	188		239	350
0100	2" x 3" bridging		1.30	6.154		47.50	246		293.50	430
0105	Pneumatic nailed		1.70	4.706		47.50	188		235.50	345
0300	Steel, galvanized, 18 ga., for 2" x 10" joists at 12" O.C.		1.30	6.154		158	246		404	555
0400	24" O.C.		1.40	5.714		169	228		397	540
0600	For 2" x 14" joists at 16" O.C.		1.30	6.154		170	246		416	565
0700	24" O.C.		1.40	5.714		260	228		488	640
0900	Compression type, 16" O.C., 2" x 8" joists		2	4		164	160		324	430
1000	2" x 12" joists		2	4		164	160		324	430
06 11 10.10 Beam and Girder Framing										
0010	**BEAM AND GIRDER FRAMING**									
1000	Single, 2" x 6"	2 Carp	700	.023	L.F.	.52	.91		1.43	1.99

06 11 Wood Framing

06 11 10 – Framing with Dimensional, Engineered or Composite Lumber

06 11 10.10 Beam and Girder Framing

		Crew	Daily Output	Labor-Hours	Unit	Material	2009 Bare Costs Labor	Equipment	Total	Total Incl O&P
1005	Pneumatic nailed	2 Carp	812	.020	L.F.	.52	.79		1.31	1.79
1020	2" x 8"		650	.025		.71	.98		1.69	2.31
1025	Pneumatic nailed		754	.021		.71	.85		1.56	2.09
1040	2" x 10"		600	.027		.99	1.07		2.06	2.74
1045	Pneumatic nailed		696	.023		.99	.92		1.91	2.51
1060	2" x 12"		550	.029		1.39	1.16		2.55	3.32
1065	Pneumatic nailed		638	.025		1.39	1		2.39	3.07
1080	2" x 14"		500	.032		2.14	1.28		3.42	4.33
1085	Pneumatic nailed		580	.028		2.14	1.10		3.24	4.06
1100	3" x 8"		550	.029		2.10	1.16		3.26	4.11
1120	3" x 10"		500	.032		2.94	1.28		4.22	5.20
1140	3" x 12"		450	.036		3.54	1.42		4.96	6.10
1160	3" x 14"		400	.040		4.48	1.60		6.08	7.40
1180	4" x 8"	F-3	1000	.040		2.79	1.62	.77	5.18	6.40
1200	4" x 10"		950	.042		3.68	1.70	.81	6.19	7.55
1220	4" x 12"		900	.044		4.74	1.80	.85	7.39	8.90
1240	4" x 14"		850	.047		5.60	1.90	.90	8.40	10.05
2000	Double, 2" x 6"	2 Carp	625	.026		1.04	1.02		2.06	2.74
2005	Pneumatic nailed		725	.022		1.04	.88		1.92	2.52
2020	2" x 8"		575	.028		1.42	1.11		2.53	3.28
2025	Pneumatic nailed		667	.024		1.42	.96		2.38	3.05
2040	2" x 10"		550	.029		1.98	1.16		3.14	3.98
2045	Pneumatic nailed		638	.025		1.98	1		2.98	3.73
2060	2" x 12"		525	.030		2.77	1.22		3.99	4.94
2065	Pneumatic nailed		610	.026		2.77	1.05		3.82	4.67
2080	2" x 14"		475	.034		4.28	1.35		5.63	6.80
2085	Pneumatic nailed		551	.029		4.28	1.16		5.44	6.50
3000	Triple, 2" x 6"		550	.029		1.56	1.16		2.72	3.52
3005	Pneumatic nailed		638	.025		1.56	1		2.56	3.27
3020	2" x 8"		525	.030		2.12	1.22		3.34	4.22
3025	Pneumatic nailed		609	.026		2.12	1.05		3.17	3.96
3040	2" x 10"		500	.032		2.98	1.28		4.26	5.25
3045	Pneumatic nailed		580	.028		2.98	1.10		4.08	4.98
3060	2" x 12"		475	.034		4.16	1.35		5.51	6.65
3065	Pneumatic nailed		551	.029		4.16	1.16		5.32	6.35
3080	2" x 14"		450	.036		6.40	1.42		7.82	9.25
3085	Pneumatic nailed		522	.031		6.40	1.22		7.62	8.95

06 11 10.12 Ceiling Framing

		Crew	Daily Output	Labor-Hours	Unit	Material	Labor	Equipment	Total	Total Incl O&P
0010	**CEILING FRAMING**									
6000	Suspended, 2" x 3"	2 Carp	1000	.016	L.F.	.32	.64		.96	1.34
6050	2" x 4"		900	.018		.29	.71		1	1.42
6100	2" x 6"		800	.020		.52	.80		1.32	1.81
6150	2" x 8"		650	.025		.71	.98		1.69	2.31

06 11 10.14 Posts and Columns

		Crew	Daily Output	Labor-Hours	Unit	Material	Labor	Equipment	Total	Total Incl O&P
0010	**POSTS AND COLUMNS**									
0100	4" x 4"	2 Carp	390	.041	L.F.	1.56	1.64		3.20	4.26
0150	4" x 6"		275	.058		2.34	2.32		4.66	6.15
0200	4" x 8"		220	.073		2.79	2.91		5.70	7.60
0250	6" x 6"		215	.074		4.87	2.97		7.84	9.95
0300	6" x 8"		175	.091		7.60	3.65		11.25	14
0350	6" x 10"		150	.107		10.90	4.26		15.16	18.60

06 11 Wood Framing

06 11 10 – Framing with Dimensional, Engineered or Composite Lumber

06 11 10.18 Joist Framing

		Crew	Daily Output	Labor-Hours	Unit	Material	2009 Bare Costs Labor	2009 Bare Costs Equipment	Total	Total Incl O&P
0010	**JOIST FRAMING** R061110-30									
2000	Joists, 2" x 4"	2 Carp	1250	.013	L.F.	.29	.51		.80	1.11
2005	Pneumatic nailed		1438	.011		.29	.44		.73	1.01
2100	2" x 6"		1250	.013		.52	.51		1.03	1.36
2105	Pneumatic nailed		1438	.011		.52	.44		.96	1.26
2150	2" x 8"		1100	.015		.71	.58		1.29	1.68
2155	Pneumatic nailed		1265	.013		.71	.51		1.22	1.56
2200	2" x 10"		900	.018		.99	.71		1.70	2.19
2205	Pneumatic nailed		1035	.015		.99	.62		1.61	2.05
2250	2" x 12"		875	.018		1.39	.73		2.12	2.65
2255	Pneumatic nailed		1006	.016		1.39	.64		2.03	2.51
2300	2" x 14"		770	.021		2.14	.83		2.97	3.64
2305	Pneumatic nailed		886	.018		2.14	.72		2.86	3.47
2350	3" x 6"		925	.017		1.52	.69		2.21	2.75
2400	3" x 10"		780	.021		2.94	.82		3.76	4.50
2450	3" x 12"		600	.027		3.54	1.07		4.61	5.55
2500	4" x 6"		800	.020		2.34	.80		3.14	3.81
2550	4" x 10"		600	.027		3.68	1.07		4.75	5.70
2600	4" x 12"		450	.036		4.74	1.42		6.16	7.40
2605	Sister joist, 2" x 6"		800	.020		.52	.80		1.32	1.81
2606	Pneumatic nailed		960	.017		.52	.67		1.19	1.60
2610	2" x 8"		640	.025		.71	1		1.71	2.33
2611	Pneumatic nailed		768	.021		.71	.83		1.54	2.07
2615	2" x 10"		535	.030		.99	1.19		2.18	2.94
2616	Pneumatic nailed		642	.025		.99	1		1.99	2.63
2620	2" x 12"		455	.035		1.39	1.40		2.79	3.70
2625	Pneumatic nailed		546	.029		1.39	1.17		2.56	3.34
3000	Composite wood joist 9-1/2" deep		.90	17.778	M.L.F.	2,000	710		2,710	3,300
3010	11-1/2" deep		.88	18.182		2,175	725		2,900	3,500
3020	14" deep		.82	19.512		2,800	780		3,580	4,275
3030	16" deep		.78	20.513		3,250	820		4,070	4,850
4000	Open web joist 12" deep		.88	18.182		3,050	725		3,775	4,475
4010	14" deep		.82	19.512		3,275	780		4,055	4,800
4020	16" deep		.78	20.513		3,250	820		4,070	4,850
4030	18" deep		.74	21.622		3,500	865		4,365	5,200

06 11 10.24 Miscellaneous Framing

		Crew	Daily Output	Labor-Hours	Unit	Material	2009 Bare Costs Labor	2009 Bare Costs Equipment	Total	Total Incl O&P
0010	**MISCELLANEOUS FRAMING**									
2000	Firestops, 2" x 4"	2 Carp	780	.021	L.F.	.29	.82		1.11	1.59
2005	Pneumatic nailed		952	.017		.29	.67		.96	1.36
2100	2" x 6"		600	.027		.52	1.07		1.59	2.22
2105	Pneumatic nailed		732	.022		.52	.87		1.39	1.92
5000	Nailers, treated, wood construction, 2" x 4"		800	.020		.44	.80		1.24	1.73
5005	Pneumatic nailed		960	.017		.44	.67		1.11	1.52
5100	2" x 6"		750	.021		.69	.85		1.54	2.07
5105	Pneumatic nailed		900	.018		.69	.71		1.40	1.85
5120	2" x 8"		700	.023		1.05	.91		1.96	2.57
5125	Pneumatic nailed		840	.019		1.05	.76		1.81	2.33
5200	Steel construction, 2" x 4"		750	.021		.44	.85		1.29	1.81
5220	2" x 6"		700	.023		.69	.91		1.60	2.17
5240	2" x 8"		650	.025		1.05	.98		2.03	2.68
7000	Rough bucks, treated, for doors or windows, 2" x 6"		400	.040		.69	1.60		2.29	3.23
7005	Pneumatic nailed		480	.033		.69	1.33		2.02	2.81

06 11 Wood Framing

06 11 10 – Framing with Dimensional, Engineered or Composite Lumber

06 11 10.24 Miscellaneous Framing

		Crew	Daily Output	Labor-Hours	Unit	Material	2009 Bare Costs Labor	Equipment	Total	Total Incl O&P
7100	2" x 8"	2 Carp	380	.042	L.F.	1.05	1.68		2.73	3.76
7105	Pneumatic nailed		456	.035		1.05	1.40		2.45	3.32
8000	Stair stringers, 2" x 10"		130	.123		.99	4.92		5.91	8.70
8100	2" x 12"		130	.123		1.39	4.92		6.31	9.10
8150	3" x 10"		125	.128		2.94	5.10		8.04	11.20
8200	3" x 12"		125	.128		3.54	5.10		8.64	11.85

06 11 10.26 Partitions

		Crew	Daily Output	Labor-Hours	Unit	Material	Labor	Equipment	Total	Total Incl O&P
0010	**PARTITIONS** R061110-30									
0020	Single bottom and double top plate, no waste, std. & better lumber									
0180	2" x 4" studs, 8' high, studs 12" O.C.	2 Carp	80	.200	L.F.	3.19	8		11.19	15.90
0185	12" O.C., pneumatic nailed		96	.167		3.19	6.65		9.84	13.85
0200	16" O.C.		100	.160		2.61	6.40		9.01	12.75
0205	16" O.C., pneumatic nailed		120	.133		2.61	5.35		7.96	11.10
0300	24" O.C.		125	.128		2.03	5.10		7.13	10.20
0305	24" O.C., pneumatic nailed		150	.107		2.03	4.26		6.29	8.85
0380	10' high, studs 12" O.C.		80	.200		3.77	8		11.77	16.55
0385	12" O.C., pneumatic nailed		96	.167		3.77	6.65		10.42	14.50
0400	16" O.C.		100	.160		3.04	6.40		9.44	13.25
0405	16" O.C., pneumatic nailed		120	.133		3.04	5.35		8.39	11.60
0500	24" O.C.		125	.128		2.32	5.10		7.42	10.50
0505	24" O.C., pneumatic nailed		150	.107		2.32	4.26		6.58	9.15
0580	12' high, studs 12" O.C.		65	.246		4.35	9.85		14.20	20
0585	12" O.C., pneumatic nailed		78	.205		4.35	8.20		12.55	17.50
0600	16" O.C.		80	.200		3.48	8		11.48	16.25
0605	16" O.C., pneumatic nailed		96	.167		3.48	6.65		10.13	14.20
0700	24" O.C.		100	.160		2.61	6.40		9.01	12.75
0705	24" O.C., pneumatic nailed		120	.133		2.61	5.35		7.96	11.10
0780	2" x 6" studs, 8' high, studs 12" O.C.		70	.229		5.75	9.15		14.90	20.50
0785	12" O.C., pneumatic nailed		84	.190		5.75	7.60		13.35	18.10
0800	16" O.C.		90	.178		4.69	7.10		11.79	16.15
0805	16" O.C., pneumatic nailed		108	.148		4.69	5.90		10.59	14.35
0900	24" O.C.		115	.139		3.65	5.55		9.20	12.60
0905	24" O.C., pneumatic nailed		138	.116		3.65	4.63		8.28	11.20
0980	10' high, studs 12" O.C.		70	.229		6.75	9.15		15.90	21.50
0985	12" O.C., pneumatic nailed		84	.190		6.75	7.60		14.35	19.25
1000	16" O.C.		90	.178		5.45	7.10		12.55	17
1005	16" O.C., pneumatic nailed		108	.148		5.45	5.90		11.35	15.20
1100	24" O.C.		115	.139		4.17	5.55		9.72	13.20
1105	24" O.C., pneumatic nailed		138	.116		4.17	4.63		8.80	11.80
1180	12' high, studs 12" O.C.		55	.291		7.80	11.60		19.40	26.50
1185	12" O.C., pneumatic nailed		66	.242		7.80	9.70		17.50	23.50
1200	16" O.C.		70	.229		6.25	9.15		15.40	21
1205	16" O.C., pneumatic nailed		84	.190		6.25	7.60		13.85	18.70
1300	24" O.C.		90	.178		4.69	7.10		11.79	16.15
1305	24" O.C., pneumatic nailed		108	.148		4.69	5.90		10.59	14.35
1400	For horizontal blocking, 2" x 4", add		600	.027		.29	1.07		1.36	1.97
1500	2" x 6", add		600	.027		.52	1.07		1.59	2.22
1600	For openings, add		250	.064			2.56		2.56	3.96
1700	Headers for above openings, material only, add				M.B.F.	530			530	585

06 11 10.30 Roof Framing

		Crew	Daily Output	Labor-Hours	Unit	Material	Labor	Equipment	Total	Total Incl O&P
0010	**ROOF FRAMING**									
2000	Fascia boards, 2" x 8"	2 Carp	225	.071	L.F.	.71	2.84		3.55	5.20

06 11 Wood Framing

06 11 10 – Framing with Dimensional, Engineered or Composite Lumber

06 11 10.30 Roof Framing

		Crew	Daily Output	Labor-Hours	Unit	Material	2009 Bare Costs Labor	Equipment	Total	Total Incl O&P
2100	2" x 10"	2 Carp	180	.089	L.F.	.99	3.55		4.54	6.60
5000	Rafters, to 4 in 12 pitch, 2" x 6", ordinary		1000	.016		.52	.64		1.16	1.56
5060	2" x 8", ordinary		950	.017		.71	.67		1.38	1.82
5120	2" x 10", ordinary		630	.025		.99	1.01		2	2.66
5180	2" x 12", ordinary		575	.028		1.39	1.11		2.50	3.24
5300	Hip and valley rafters, 2" x 6", ordinary		760	.021		.52	.84		1.36	1.87
5360	2" x 8", ordinary		720	.022		.71	.89		1.60	2.16
5420	2" x 10", ordinary		570	.028		.99	1.12		2.11	2.83
5460	On dormers or complex roofs		380	.042		.99	1.68		2.67	3.70
5480	2" x 12", ordinary		525	.030		1.39	1.22		2.61	3.41
5540	Hip and valley jacks, 2" x 6", ordinary		600	.027		.52	1.07		1.59	2.22
5600	2" x 8", ordinary		490	.033		.71	1.30		2.01	2.80
5660	2" x 10", ordinary		450	.036		.99	1.42		2.41	3.29
5720	2" x 12", ordinary		375	.043		1.39	1.70		3.09	4.16
5761	For slopes steeper than 4 in 12, add						30%			
5770	For dormers or complex roofs, add						50%			
5780	Rafter tie, 1" x 4", #3	2 Carp	800	.020	L.F.	.44	.80		1.24	1.72
5800	Ridge board, #2 or better, 1" x 6"		600	.027		.64	1.07		1.71	2.35
5820	1" x 8"		550	.029		.84	1.16		2	2.72
5840	1" x 10"		500	.032		1.23	1.28		2.51	3.33
5860	2" x 6"		500	.032		.52	1.28		1.80	2.55
5880	2" x 8"		450	.036		.71	1.42		2.13	2.98
5900	2" x 10"		400	.040		.99	1.60		2.59	3.57
5920	Roof cants, split, 4" x 4"		650	.025		1.56	.98		2.54	3.25
5940	6" x 6"		600	.027		4.87	1.07		5.94	7
5960	Roof curbs, untreated, 2" x 6"		520	.031		.52	1.23		1.75	2.48
5980	2" x 12"		400	.040		1.39	1.60		2.99	4
6000	Sister rafters, 2" x 6"		800	.020		.52	.80		1.32	1.81
6020	2" x 8"		640	.025		.71	1		1.71	2.33
6040	2" x 10"		535	.030		.99	1.19		2.18	2.94
6060	2" x 12"		455	.035		1.39	1.40		2.79	3.70

06 11 10.32 Sill and Ledger Framing

		Crew	Daily Output	Labor-Hours	Unit	Material	Labor	Equipment	Total	Total Incl O&P
0010	**SILL AND LEDGER FRAMING** R061110-30									
2000	Ledgers, nailed, 2" x 4"	2 Carp	755	.021	L.F.	.29	.85		1.14	1.63
2050	2" x 6"		600	.027		.52	1.07		1.59	2.22
2100	Bolted, not including bolts, 3" x 6"		325	.049		1.52	1.97		3.49	4.73
2150	3" x 12"		233	.069		3.54	2.74		6.28	8.15
2600	Mud sills, redwood, construction grade, 2" x 4"		895	.018		2.27	.71		2.98	3.60
2620	2" x 6"		780	.021		3.40	.82		4.22	5
4000	Sills, 2" x 4"		600	.027		.29	1.07		1.36	1.97
4050	2" x 6"		550	.029		.52	1.16		1.68	2.37
4080	2" x 8"		500	.032		.71	1.28		1.99	2.76
4100	2" x 10"		450	.036		.99	1.42		2.41	3.29
4120	2" x 12"		400	.040		1.39	1.60		2.99	4
4200	Treated, 2" x 4"		550	.029		.44	1.16		1.60	2.29
4220	2" x 6"		500	.032		.69	1.28		1.97	2.73
4240	2" x 8"		450	.036		1.05	1.42		2.47	3.35
4260	2" x 10"		400	.040		1.28	1.60		2.88	3.89
4280	2" x 12"		350	.046		2.03	1.83		3.86	5.05
4400	4" x 4"		450	.036		1.49	1.42		2.91	3.84
4420	4" x 6"		350	.046		2.19	1.83		4.02	5.25
4460	4" x 8"		300	.053		3.14	2.13		5.27	6.75

06 11 Wood Framing

06 11 10 – Framing with Dimensional, Engineered or Composite Lumber

06 11 10.32 Sill and Ledger Framing		Crew	Daily Output	Labor-Hours	Unit	Material	2009 Bare Costs Labor	Equipment	Total	Total Incl O&P
4480	4" x 10"	2 Carp	260	.062	L.F.	4.10	2.46		6.56	8.30
06 11 10.34 Sleepers										
0010	**SLEEPERS**									
0100	On concrete, treated, 1" x 2"	2 Carp	2350	.007	L.F.	.38	.27		.65	.84
0150	1" x 3"		2000	.008		.52	.32		.84	1.07
0200	2" x 4"		1500	.011		.44	.43		.87	1.15
0250	2" x 6"		1300	.012		.69	.49		1.18	1.51
06 11 10.36 Soffit and Canopy Framing										
0010	**SOFFIT AND CANOPY FRAMING**									
1000	Canopy or soffit framing, 1" x 4"	2 Carp	900	.018	L.F.	.44	.71		1.15	1.58
1020	1" x 6"		850	.019		.64	.75		1.39	1.87
1040	1" x 8"		750	.021		.84	.85		1.69	2.24
1100	2" x 4"		620	.026		.29	1.03		1.32	1.92
1120	2" x 6"		560	.029		.52	1.14		1.66	2.34
1140	2" x 8"		500	.032		.71	1.28		1.99	2.76
1200	3" x 4"		500	.032		.93	1.28		2.21	3
1220	3" x 6"		400	.040		1.52	1.60		3.12	4.16
1240	3" x 10"		300	.053		2.94	2.13		5.07	6.55
06 11 10.38 Treated Lumber Framing Material										
0010	**TREATED LUMBER FRAMING MATERIAL**									
0100	2" x 4"				M.B.F.	660			660	730
0110	2" x 6"					685			685	755
0120	2" x 8"					785			785	865
0130	2" x 10"					765			765	845
0140	2" x 12"					1,025			1,025	1,125
0200	4" x 4"					1,125			1,125	1,225
0210	4" x 6"					1,100			1,100	1,200
0220	4" x 8"					1,175			1,175	1,300
06 11 10.40 Wall Framing										
0010	**WALL FRAMING** R061110-30									
2000	Headers over openings, 2" x 6"	2 Carp	360	.044	L.F.	.52	1.78		2.30	3.32
2005	2" x 6", pneumatic nailed		432	.037		.52	1.48		2	2.86
2050	2" x 8"		340	.047		.71	1.88		2.59	3.70
2055	2" x 8", pneumatic nailed		408	.039		.71	1.57		2.28	3.21
2100	2" x 10"		320	.050		.99	2		2.99	4.19
2105	2" x 10", pneumatic nailed		384	.042		.99	1.66		2.65	3.67
2150	2" x 12"		300	.053		1.39	2.13		3.52	4.82
2155	2" x 12", pneumatic nailed		360	.044		1.39	1.78		3.17	4.27
2200	4" x 12"		190	.084		4.74	3.36		8.10	10.40
2205	4" x 12", pneumatic nailed		228	.070		4.74	2.80		7.54	9.55
2250	6" x 12"		140	.114		9.15	4.57		13.72	17.20
2255	6" x 12", pneumatic nailed		168	.095		9.15	3.80		12.95	16
5000	Plates, untreated, 2" x 3"		850	.019		.32	.75		1.07	1.52
5005	2" x 3", pneumatic nailed		1020	.016		.32	.63		.95	1.32
5020	2" x 4"		800	.020		.29	.80		1.09	1.56
5025	2" x 4", pneumatic nailed		960	.017		.29	.67		.96	1.35
5040	2" x 6"		750	.021		.52	.85		1.37	1.89
5045	2" x 6", pneumatic nailed		900	.018		.52	.71		1.23	1.67
5060	Treated, 2" x 3"		850	.019		.48	.75		1.23	1.70
5065	2" x 3", treated, pneumatic nailed		1020	.016		.48	.63		1.11	1.50
5080	2" x 4"		800	.020		.44	.80		1.24	1.73
5085	2" x 4", treated, pneumatic nailed		960	.017		.44	.67		1.11	1.52

06 11 Wood Framing

06 11 10 – Framing with Dimensional, Engineered or Composite Lumber

06 11 10.40	Wall Framing	Crew	Daily Output	Labor-Hours	Unit	Material	2009 Bare Costs Labor	Equipment	Total	Total Incl O&P
5100	2" x 6"	2 Carp	750	.021	L.F.	.69	.85		1.54	2.07
5103	2" x 6", treated, pneumatic nailed		900	.018		.69	.71		1.40	1.85
5120	Studs, 8' high wall, 2" x 3"		1200	.013		.32	.53		.85	1.18
5125	2" x 3", pneumatic nailed		1440	.011		.32	.44		.76	1.04
5140	2" x 4"		1100	.015		.29	.58		.87	1.22
5146	2" x 4", pneumatic nailed		1320	.012		.29	.48		.77	1.07
5160	2" x 6"		1000	.016		.52	.64		1.16	1.56
5166	2" x 6", pneumatic nailed		1200	.013		.52	.53		1.05	1.40
5180	3" x 4"		800	.020		.93	.80		1.73	2.26
5185	3" x 4", pneumatic nailed		960	.017		.93	.67		1.60	2.05
5200	Installed on second story, 2" x 3"		1170	.014		.32	.55		.87	1.20
5205	2" x 3", pneumatic nailed		1404	.011		.32	.46		.78	1.06
5220	2" x 4"		1015	.016		.29	.63		.92	1.30
5225	2" x 4", pneumatic nailed		1218	.013		.29	.52		.81	1.13
5240	2" x 6"		890	.018		.52	.72		1.24	1.68
5245	2" x 6", pneumatic nailed		1080	.015		.52	.59		1.11	1.49
5260	3" x 4"		800	.020		.93	.80		1.73	2.26
5265	3" x 4", pneumatic nailed		960	.017		.93	.67		1.60	2.05
5280	Installed on dormer or gable, 2" x 3"		1045	.015		.32	.61		.93	1.30
5285	2" x 3", pneumatic nailed		1254	.013		.32	.51		.83	1.14
5300	2" x 4"		905	.018		.29	.71		1	1.42
5305	2" x 4", pneumatic nailed		1086	.015		.29	.59		.88	1.23
5320	2" x 6"		800	.020		.52	.80		1.32	1.81
5325	2" x 6", pneumatic nailed		960	.017		.52	.67		1.19	1.60
5340	3" x 4"		700	.023		.93	.91		1.84	2.44
5345	3" x 4", pneumatic nailed		840	.019		.93	.76		1.69	2.20
5360	6' high wall, 2" x 3"		970	.016		.32	.66		.98	1.37
5365	2" x 3", pneumatic nailed		1164	.014		.32	.55		.87	1.20
5380	2" x 4"		850	.019		.29	.75		1.04	1.49
5385	2" x 4", pneumatic nailed		1020	.016		.29	.63		.92	1.29
5400	2" x 6"		740	.022		.52	.86		1.38	1.91
5405	2" x 6", pneumatic nailed		888	.018		.52	.72		1.24	1.69
5420	3" x 4"		600	.027		.93	1.07		2	2.67
5425	3" x 4", pneumatic nailed		720	.022		.93	.89		1.82	2.40
5440	Installed on second story, 2" x 3"		950	.017		.32	.67		.99	1.39
5445	2" x 3", pneumatic nailed		1140	.014		.32	.56		.88	1.22
5460	2" x 4"		810	.020		.29	.79		1.08	1.54
5465	2" x 4", pneumatic nailed		972	.016		.29	.66		.95	1.34
5480	2" x 6"		700	.023		.52	.91		1.43	1.99
5485	2" x 6", pneumatic nailed		840	.019		.52	.76		1.28	1.75
5500	3" x 4"		550	.029		.93	1.16		2.09	2.82
5505	3" x 4", pneumatic nailed		660	.024		.93	.97		1.90	2.52
5520	Installed on dormer or gable, 2" x 3"		850	.019		.32	.75		1.07	1.52
5525	2" x 3", pneumatic nailed		1020	.016		.32	.63		.95	1.32
5540	2" x 4"		720	.022		.29	.89		1.18	1.70
5545	2" x 4", pneumatic nailed		864	.019		.29	.74		1.03	1.47
5560	2" x 6"		620	.026		.52	1.03		1.55	2.17
5565	2" x 6", pneumatic nailed		744	.022		.52	.86		1.38	1.90
5580	3" x 4"		480	.033		.93	1.33		2.26	3.08
5585	3" x 4", pneumatic nailed		576	.028		.93	1.11		2.04	2.74
5600	3' high wall, 2" x 3"		740	.022		.32	.86		1.18	1.69
5605	2" x 3", pneumatic nailed		888	.018		.32	.72		1.04	1.47
5620	2" x 4"		640	.025		.29	1		1.29	1.87

06 11 Wood Framing

06 11 10 – Framing with Dimensional, Engineered or Composite Lumber

06 11 10.40 Wall Framing

		Crew	Daily Output	Labor-Hours	Unit	Material	2009 Bare Costs Labor	2009 Bare Costs Equipment	Total	Total Incl O&P
5625	2" x 4", pneumatic nailed	2 Carp	768	.021	L.F.	.29	.83		1.12	1.61
5640	2" x 6"		550	.029		.52	1.16		1.68	2.37
5645	2" x 6", pneumatic nailed		660	.024		.52	.97		1.49	2.07
5660	3" x 4"		440	.036		.93	1.45		2.38	3.27
5665	3" x 4", pneumatic nailed		528	.030		.93	1.21		2.14	2.90
5680	Installed on second story, 2" x 3"		700	.023		.32	.91		1.23	1.77
5685	2" x 3", pneumatic nailed		840	.019		.32	.76		1.08	1.53
5700	2" x 4"		610	.026		.29	1.05		1.34	1.94
5705	2" x 4", pneumatic nailed		732	.022		.29	.87		1.16	1.67
5720	2" x 6"		520	.031		.52	1.23		1.75	2.48
5725	2" x 6", pneumatic nailed		624	.026		.52	1.02		1.54	2.16
5740	3" x 4"		430	.037		.93	1.49		2.42	3.33
5745	3" x 4", pneumatic nailed		516	.031		.93	1.24		2.17	2.94
5760	Installed on dormer or gable, 2" x 3"		625	.026		.32	1.02		1.34	1.94
5765	2" x 3", pneumatic nailed		750	.021		.32	.85		1.17	1.67
5780	2" x 4"		545	.029		.29	1.17		1.46	2.14
5785	2" x 4", pneumatic nailed		654	.024		.29	.98		1.27	1.84
5800	2" x 6"		465	.034		.52	1.37		1.89	2.70
5805	2" x 6", pneumatic nailed		558	.029		.52	1.15		1.67	2.35
5820	3" x 4"		380	.042		.93	1.68		2.61	3.63
5825	3" x 4", pneumatic nailed		456	.035		.93	1.40		2.33	3.19

06 11 10.42 Furring

		Crew	Daily Output	Labor-Hours	Unit	Material	Labor	Equipment	Total	Total Incl O&P
0010	**FURRING**									
0012	Wood strips, 1" x 2", on walls, on wood	1 Carp	550	.015	L.F.	.25	.58		.83	1.18
0015	On wood, pneumatic nailed		710	.011		.25	.45		.70	.98
0300	On masonry		495	.016		.25	.65		.90	1.28
0400	On concrete		260	.031		.25	1.23		1.48	2.19
0600	1" x 3", on walls, on wood		550	.015		.34	.58		.92	1.28
0605	On wood, pneumatic nailed		710	.011		.34	.45		.79	1.08
0700	On masonry		495	.016		.34	.65		.99	1.38
0800	On concrete		260	.031		.34	1.23		1.57	2.29
0850	On ceilings, on wood		350	.023		.34	.91		1.25	1.80
0855	On wood, pneumatic nailed		450	.018		.34	.71		1.05	1.48
0900	On masonry		320	.025		.34	1		1.34	1.93
0950	On concrete		210	.038		.34	1.52		1.86	2.74

06 11 10.44 Grounds

		Crew	Daily Output	Labor-Hours	Unit	Material	Labor	Equipment	Total	Total Incl O&P
0010	**GROUNDS**									
0020	For casework, 1" x 2" wood strips, on wood	1 Carp	330	.024	L.F.	.25	.97		1.22	1.78
0100	On masonry		285	.028		.25	1.12		1.37	2.02
0200	On concrete		250	.032		.25	1.28		1.53	2.26
0400	For plaster, 3/4" deep, on wood		450	.018		.25	.71		.96	1.38
0500	On masonry		225	.036		.25	1.42		1.67	2.48
0600	On concrete		175	.046		.25	1.83		2.08	3.11
0700	On metal lath		200	.040		.25	1.60		1.85	2.76

06 15 Wood Decking

06 15 16 – Wood Roof Decking

06 15 16.10 Solid Wood Roof Decking		Crew	Daily Output	Labor-Hours	Unit	Material	2009 Bare Costs Labor	Equipment	Total	Total Incl O&P
0010	**SOLID WOOD ROOF DECKING**									
0400	Cedar planks, 3" thick	2 Carp	320	.050	S.F.	6.20	2		8.20	9.95
0500	4" thick		250	.064		8.35	2.56		10.91	13.15
0700	Douglas fir, 3" thick		320	.050		2.33	2		4.33	5.65
0800	4" thick		250	.064		3.12	2.56		5.68	7.40
1000	Hemlock, 3" thick		320	.050		2.33	2		4.33	5.65
1100	4" thick		250	.064		3.11	2.56		5.67	7.40
1300	Western white spruce, 3" thick		320	.050		2.24	2		4.24	5.55
1400	4" thick		250	.064		2.99	2.56		5.55	7.25

06 16 Sheathing

06 16 23 – Subflooring

06 16 23.10 Subfloor

			Crew	Daily Output	Labor-Hours	Unit	Material	Labor	Equipment	Total	Total Incl O&P
0010	**SUBFLOOR**	R061636-20									
0011	Plywood, CDX, 1/2" thick		2 Carp	1500	.011	SF Flr.	.46	.43		.89	1.17
0015	Pneumatic nailed			1860	.009		.46	.34		.80	1.04
0100	5/8" thick			1350	.012		.67	.47		1.14	1.47
0105	Pneumatic nailed			1674	.010		.67	.38		1.05	1.33
0200	3/4" thick			1250	.013		.83	.51		1.34	1.70
0205	Pneumatic nailed			1550	.010		.83	.41		1.24	1.55
0300	1-1/8" thick, 2-4-1 including underlayment			1050	.015		1.63	.61		2.24	2.73
0450	1" x 8", laid regular			1000	.016		1.34	.64		1.98	2.46
0460	Laid diagonal			850	.019		1.34	.75		2.09	2.64
0500	1" x 10", laid regular			1100	.015		1.54	.58		2.12	2.60
0600	Laid diagonal			900	.018		1.54	.71		2.25	2.80
8990	Subfloor adhesive, 3/8" bead		1 Carp	2300	.003	L.F.	.09	.14		.23	.32

06 16 26 – Underlayment

06 16 26.10 Wood Product Underlayment

				Crew	Daily Output	Labor-Hours	Unit	Material	Labor	Equipment	Total	Total Incl O&P
0010	**WOOD PRODUCT UNDERLAYMENT**	R061636-20										
0030	Plywood, underlayment grade, 3/8" thick			2 Carp	1500	.011	SF Flr.	1.03	.43		1.46	1.79
0070	Pneumatic nailed				1860	.009		1.03	.34		1.37	1.66
0100	1/2" thick				1450	.011		1	.44		1.44	1.78
0105	Pneumatic nailed				1798	.009		1	.36		1.36	1.65
0200	5/8" thick				1400	.011		1.29	.46		1.75	2.13
0205	Pneumatic nailed				1736	.009		1.29	.37		1.66	1.99
0300	3/4" thick				1300	.012		1.33	.49		1.82	2.22
0305	Pneumatic nailed				1612	.010		1.33	.40		1.73	2.08
0500	Particle board, 3/8" thick		G		1500	.011		.41	.43		.84	1.11
0505	Pneumatic nailed		G		1860	.009		.41	.34		.75	.98
0600	1/2" thick		G		1450	.011		.45	.44		.89	1.18
0605	Pneumatic nailed		G		1798	.009		.45	.36		.81	1.05
0800	5/8" thick		G		1400	.011		.54	.46		1	1.30
0805	Pneumatic nailed		G		1736	.009		.54	.37		.91	1.16
0900	3/4" thick		G		1300	.012		.65	.49		1.14	1.48
0905	Pneumatic nailed		G		1612	.010		.65	.40		1.05	1.34
1100	Hardboard, underlayment grade, 4' x 4', .215" thick		G		1500	.011		.50	.43		.93	1.21

06 16 Sheathing

06 16 36 – Wood Panel Product Sheathing

06 16 36.10 Sheathing		Crew	Daily Output	Labor-Hours	Unit	Material	2009 Bare Costs Labor	Equipment	Total	Total Incl O&P
0010	**SHEATHING**, plywood on roofs R061636-20									
0012	Plywood on roofs, CDX									
0030	5/16" thick	2 Carp	1600	.010	S.F.	.53	.40		.93	1.20
0035	Pneumatic nailed		1952	.008		.53	.33		.86	1.09
0050	3/8" thick		1525	.010		.45	.42		.87	1.14
0055	Pneumatic nailed		1860	.009		.45	.34		.79	1.02
0100	1/2" thick		1400	.011		.46	.46		.92	1.22
0105	Pneumatic nailed		1708	.009		.46	.37		.83	1.09
0200	5/8" thick		1300	.012		.67	.49		1.16	1.50
0205	Pneumatic nailed		1586	.010		.67	.40		1.07	1.37
0300	3/4" thick		1200	.013		.83	.53		1.36	1.74
0305	Pneumatic nailed		1464	.011		.83	.44		1.27	1.59
0500	Plywood on walls with exterior CDX, 3/8" thick		1200	.013		.45	.53		.98	1.32
0505	Pneumatic nailed		1488	.011		.45	.43		.88	1.16
0600	1/2" thick		1125	.014		.46	.57		1.03	1.39
0605	Pneumatic nailed		1395	.011		.46	.46		.92	1.22
0700	5/8" thick		1050	.015		.67	.61		1.28	1.68
0705	Pneumatic nailed		1302	.012		.67	.49		1.16	1.50
0800	3/4" thick		975	.016		.83	.66		1.49	1.93
0805	Pneumatic nailed		1209	.013		.83	.53		1.36	1.73
1000	For shear wall construction, add						20%			
1200	For structural 1 exterior plywood, add				S.F.	10%				
1400	With boards, on roof 1" x 6" boards, laid horizontal	2 Carp	725	.022		1.40	.88		2.28	2.90
1500	Laid diagonal		650	.025		1.40	.98		2.38	3.06
1700	1" x 8" boards, laid horizontal		875	.018		1.34	.73		2.07	2.60
1800	Laid diagonal		725	.022		1.34	.88		2.22	2.84
2000	For steep roofs, add						40%			
2200	For dormers, hips and valleys, add					5%	50%			
2400	Boards on walls, 1" x 6" boards, laid regular	2 Carp	650	.025		1.40	.98		2.38	3.06
2500	Laid diagonal		585	.027		1.40	1.09		2.49	3.22
2700	1" x 8" boards, laid regular		765	.021		1.34	.84		2.18	2.77
2800	Laid diagonal		650	.025		1.34	.98		2.32	3
2850	Gypsum, weatherproof, 1/2" thick		1125	.014		.61	.57		1.18	1.55
2900	With embedded glass mats		1100	.015		.49	.58		1.07	1.44
3000	Wood fiber, regular, no vapor barrier, 1/2" thick		1200	.013		.56	.53		1.09	1.45
3100	5/8" thick		1200	.013		.71	.53		1.24	1.61
3300	No vapor barrier, in colors, 1/2" thick		1200	.013		.75	.53		1.28	1.66
3400	5/8" thick		1200	.013		.80	.53		1.33	1.71
3600	With vapor barrier one side, white, 1/2" thick		1200	.013		.55	.53		1.08	1.44
3700	Vapor barrier 2 sides, 1/2" thick		1200	.013		.77	.53		1.30	1.68
3800	Asphalt impregnated, 25/32" thick		1200	.013		.28	.53		.81	1.14
3850	Intermediate, 1/2" thick		1200	.013		.18	.53		.71	1.03

06 17 Shop-Fabricated Structural Wood

06 17 53 – Shop-Fabricated Wood Trusses

06 17 53.10 Roof Trusses

		Crew	Daily Output	Labor-Hours	Unit	Material	2009 Bare Costs Labor	2009 Bare Costs Equipment	Total	Total Incl O&P
0010	**ROOF TRUSSES**									
5000	Common wood, 2" x 4" metal plate connected, 24" O.C., 4/12 slope									
5050	20' span	F-6	62	.645	Ea.	72	24	12.40	108.40	130
5100	24' span		60	.667		84	25	12.80	121.80	145
5150	26' span		57	.702		87	26	13.50	126.50	151
5200	28' span		53	.755		77.50	28	14.50	120	144
5250	32' span		50	.800		112	29.50	15.40	156.90	185
5300	36' span		46	.870		148	32.50	16.70	197.20	231
5350	8/12 pitch, 1' overhang, 20' span		57	.702		76.50	26	13.50	116	139
5400	24' span		55	.727		97	27	14	138	164
5450	26' span		52	.769		101	28.50	14.80	144.30	171
5500	28' span		49	.816		111	30.50	15.70	157.20	187
5550	32' span		45	.889		134	33	17.10	184.10	217
5600	36' span		41	.976		167	36	18.75	221.75	260
5650	38' span		40	1		182	37	19.20	238.20	278
5700	40' span		40	1		188	37	19.20	244.20	284

06 22 Millwork

06 22 13 – Standard Pattern Wood Trim

06 22 13.15 Moldings, Base

		Crew	Daily Output	Labor-Hours	Unit	Material	Labor	Equipment	Total	Total Incl O&P
0010	**MOLDINGS, BASE**									
0500	Base, stock pine, 9/16" x 3-1/2"	1 Carp	240	.033	L.F.	2.22	1.33		3.55	4.50
0550	9/16" x 4-1/2"	"	200	.040	"	2.47	1.60		4.07	5.20

06 22 13.30 Moldings, Casings

		Crew	Daily Output	Labor-Hours	Unit	Material	Labor	Equipment	Total	Total Incl O&P
0010	**MOLDINGS, CASINGS**									
0090	Apron, stock pine, 5/8" x 2"	1 Carp	250	.032	L.F.	1.24	1.28		2.52	3.34
0110	5/8" x 3-1/2"		220	.036		1.88	1.45		3.33	4.32
0300	Band, stock pine, 11/16" x 1-1/8"		270	.030		.82	1.18		2	2.74
0350	11/16" x 1-3/4"		250	.032		1.10	1.28		2.38	3.19
0700	Casing, stock pine, 11/16" x 2-1/2"		240	.033		1.33	1.33		2.66	3.52
0750	11/16" x 3-1/2"		215	.037		1.89	1.49		3.38	4.39

06 22 13.35 Moldings, Ceilings

		Crew	Daily Output	Labor-Hours	Unit	Material	Labor	Equipment	Total	Total Incl O&P
0010	**MOLDINGS, CEILINGS**									
0600	Bed, stock pine, 9/16" x 1-3/4"	1 Carp	270	.030	L.F.	.97	1.18		2.15	2.91
0650	9/16" x 2"		240	.033		1.23	1.33		2.56	3.41
1200	Cornice molding, stock pine, 9/16" x 1-3/4"		330	.024		1.02	.97		1.99	2.62
1300	9/16" x 2-1/4"		300	.027		1.24	1.07		2.31	3.01
2400	Cove scotia, stock pine, 9/16" x 1-3/4"		270	.030		.93	1.18		2.11	2.86
2500	11/16" x 2-3/4"		255	.031		1.35	1.25		2.60	3.43
2600	Crown, stock pine, 9/16" x 3-5/8"		250	.032		1.92	1.28		3.20	4.09
2700	11/16" x 4-5/8"		220	.036		3.22	1.45		4.67	5.80

06 22 13.40 Moldings, Exterior

		Crew	Daily Output	Labor-Hours	Unit	Material	Labor	Equipment	Total	Total Incl O&P
0010	**MOLDINGS, EXTERIOR**									
1500	Cornice, boards, pine, 1" x 2"	1 Carp	330	.024	L.F.	.27	.97		1.24	1.80
1600	1" x 4"		250	.032		.73	1.28		2.01	2.78
1700	1" x 6"		250	.032		1.15	1.28		2.43	3.25
1800	1" x 8"		200	.040		1.34	1.60		2.94	3.95
1900	1" x 10"		180	.044		1.78	1.78		3.56	4.71
2000	1" x 12"		180	.044		2.31	1.78		4.09	5.30
2200	Three piece, built-up, pine, minimum		80	.100		2.15	4		6.15	8.55

06 22 Millwork

06 22 13 – Standard Pattern Wood Trim

06 22 13.40 Moldings, Exterior

		Crew	Daily Output	Labor-Hours	Unit	Material	2009 Bare Costs Labor	2009 Bare Costs Equipment	Total	Total Incl O&P
2300	Maximum	1 Carp	65	.123	L.F.	5.45	4.92		10.37	13.55
3000	Corner board, sterling pine, 1" x 4"		200	.040		.69	1.60		2.29	3.24
3100	1" x 6"		200	.040		1.05	1.60		2.65	3.64
3200	2" x 6"		165	.048		1.98	1.94		3.92	5.20
3300	2" x 8"		165	.048		2.63	1.94		4.57	5.90
3350	Fascia, sterling pine, 1" x 6"		250	.032		1.05	1.28		2.33	3.14
3370	1" x 8"		225	.036		1.43	1.42		2.85	3.77
3395	Grounds, 1" x 1" redwood		300	.027		.25	1.07		1.32	1.93
3400	Trim, back band, 11/16" x 1-1/16"		250	.032		.89	1.28		2.17	2.96
3500	Casing, 11/16" x 4-1/4"		250	.032		2.19	1.28		3.47	4.39
3600	Crown, 11/16" x 4-1/4"		250	.032		3.65	1.28		4.93	6
3700	1" x 4" and 1" x 6" railing w/ balusters 4" O.C.		22	.364		22.50	14.55		37.05	47.50
3800	Insect screen framing stock, 1-1/16" x 1-3/4"		395	.020		2.03	.81		2.84	3.48
4100	Verge board, sterling pine, 1" x 4"		200	.040		.69	1.60		2.29	3.24
4200	1" x 6"		200	.040		1.05	1.60		2.65	3.64
4300	2" x 6"		165	.048		1.98	1.94		3.92	5.20
4400	2" x 8"		165	.048		2.63	1.94		4.57	5.90
4700	For redwood trim, add					200%				

06 22 13.45 Moldings, Trim

		Crew	Daily Output	Labor-Hours	Unit	Material	2009 Bare Costs Labor	2009 Bare Costs Equipment	Total	Total Incl O&P
0010	**MOLDINGS, TRIM**									
0200	Astragal, stock pine, 11/16" x 1-3/4"	1 Carp	255	.031	L.F.	1.19	1.25		2.44	3.25
0250	1-5/16" x 2-3/16"		240	.033		2.57	1.33		3.90	4.89
0800	Chair rail, stock pine, 5/8" x 2-1/2"		270	.030		1.45	1.18		2.63	3.44
0900	5/8" x 3-1/2"		240	.033		2.27	1.33		3.60	4.56
1000	Closet pole, stock pine, 1-1/8" diameter		200	.040		1.01	1.60		2.61	3.59
1100	Fir, 1-5/8" diameter		200	.040		1.67	1.60		3.27	4.32
3300	Half round, stock pine, 1/4" x 1/2"		270	.030		.24	1.18		1.42	2.10
3350	1/2" x 1"		255	.031		.61	1.25		1.86	2.61
3400	Handrail, fir, single piece, stock, hardware not included									
3450	1-1/2" x 1-3/4"	1 Carp	80	.100	L.F.	1.85	4		5.85	8.25
3470	Pine, 1-1/2" x 1-3/4"		80	.100		1.85	4		5.85	8.25
3500	1-1/2" x 2-1/2"		76	.105		2.30	4.21		6.51	9.05
3600	Lattice, stock pine, 1/4" x 1-1/8"		270	.030		.42	1.18		1.60	2.30
3700	1/4" x 1-3/4"		250	.032		.66	1.28		1.94	2.71
3800	Miscellaneous, custom, pine, 1" x 1"		270	.030		.34	1.18		1.52	2.21
3900	1" x 3"		240	.033		.69	1.33		2.02	2.82
4100	Birch or oak, nominal 1" x 1"		240	.033		.53	1.33		1.86	2.64
4200	Nominal 1" x 3"		215	.037		1.81	1.49		3.30	4.30
4400	Walnut, nominal 1" x 1"		215	.037		.87	1.49		2.36	3.27
4500	Nominal 1" x 3"		200	.040		2.61	1.60		4.21	5.35
4700	Teak, nominal 1" x 1"		215	.037		1.23	1.49		2.72	3.66
4800	Nominal 1" x 3"		200	.040		3.52	1.60		5.12	6.35
4900	Quarter round, stock pine, 1/4" x 1/4"		275	.029		.22	1.16		1.38	2.04
4950	3/4" x 3/4"		255	.031		.59	1.25		1.84	2.59
5600	Wainscot moldings, 1-1/8" x 9/16", 2' high, minimum		76	.105	S.F.	11.05	4.21		15.26	18.65
5700	Maximum		65	.123	"	20	4.92		24.92	30

06 22 13.50 Moldings, Window and Door

		Crew	Daily Output	Labor-Hours	Unit	Material	2009 Bare Costs Labor	2009 Bare Costs Equipment	Total	Total Incl O&P
0010	**MOLDINGS, WINDOW AND DOOR**									
2800	Door moldings, stock, decorative, 1-1/8" wide, plain	1 Carp	17	.471	Set	45	18.80		63.80	78.50
2900	Detailed		17	.471	"	87.50	18.80		106.30	126
2960	Clear pine door jamb, no stops, 11/16" x 4-9/16"		240	.033	L.F.	5.05	1.33		6.38	7.60
3150	Door trim set, 1 head and 2 sides, pine, 2-1/2 wide		5.90	1.356	Opng.	22.50	54		76.50	109

06 22 Millwork

06 22 13 – Standard Pattern Wood Trim

06 22 13.50 Moldings, Window and Door		Crew	Daily Output	Labor-Hours	Unit	Material	2009 Bare Costs Labor	Equipment	Total	Total Incl O&P
3170	3-1/2" wide	1 Carp	5.30	1.509	Opng.	32	60.50		92.50	129
3250	Glass beads, stock pine, 3/8" x 1/2"		275	.029	L.F.	.31	1.16		1.47	2.14
3270	3/8" x 7/8"		270	.030		.41	1.18		1.59	2.29
4850	Parting bead, stock pine, 3/8" x 3/4"		275	.029		.36	1.16		1.52	2.20
4870	1/2" x 3/4"		255	.031		.45	1.25		1.70	2.44
5000	Stool caps, stock pine, 11/16" x 3-1/2"		200	.040		2.27	1.60		3.87	4.98
5100	1-1/16" x 3-1/4"		150	.053		3.59	2.13		5.72	7.25
5300	Threshold, oak, 3' long, inside, 5/8" x 3-5/8"		32	.250	Ea.	9	10		19	25.50
5400	Outside, 1-1/2" x 7-5/8"		16	.500	"	37	20		57	71.50
5900	Window trim sets, including casings, header, stops,									
5910	stool and apron, 2-1/2" wide, minimum	1 Carp	13	.615	Opng.	32.50	24.50		57	73.50
5950	Average		10	.800		38	32		70	91.50
6000	Maximum		6	1.333		62.50	53.50		116	151

06 22 13.60 Moldings, Soffits

		Crew	Daily Output	Labor-Hours	Unit	Material	Labor	Equipment	Total	Total Incl O&P
0010	**MOLDINGS, SOFFITS**									
0200	Soffits, pine, 1" x 4"	2 Carp	420	.038	L.F.	.44	1.52		1.96	2.84
0210	1" x 6"		420	.038		.64	1.52		2.16	3.06
0220	1" x 8"		420	.038		.84	1.52		2.36	3.28
0230	1" x 10"		400	.040		1.23	1.60		2.83	3.83
0240	1" x 12"		400	.040		1.60	1.60		3.20	4.24
0250	STK cedar, 1" x 4"		420	.038		.65	1.52		2.17	3.08
0260	1" x 6"		420	.038		1.09	1.52		2.61	3.56
0270	1" x 8"		420	.038		1.52	1.52		3.04	4.03
0280	1" x 10"		400	.040		2.09	1.60		3.69	4.78
0290	1" x 12"		400	.040		2.70	1.60		4.30	5.45
1000	Exterior AC plywood, 1/4" thick		420	.038	S.F.	.85	1.52		2.37	3.30
1050	3/8" thick		420	.038		1.03	1.52		2.55	3.49
1100	1/2" thick		420	.038		1	1.52		2.52	3.46
1150	Polyvinyl chloride, white, solid	1 Carp	230	.035		1.09	1.39		2.48	3.35
1160	Perforated	"	230	.035		1.09	1.39		2.48	3.35

06 25 Prefinished Paneling

06 25 13 – Prefinished Hardboard Paneling

06 25 13.10 Paneling, Hardboard

			Crew	Daily Output	Labor-Hours	Unit	Material	Labor	Equipment	Total	Total Incl O&P
0010	**PANELING, HARDBOARD**										
0050	Not incl. furring or trim, hardboard, tempered, 1/8" thick	G	2 Carp	500	.032	S.F.	.38	1.28		1.66	2.40
0100	1/4" thick	G		500	.032		.53	1.28		1.81	2.56
0300	Tempered pegboard, 1/8" thick	G		500	.032		.47	1.28		1.75	2.50
0400	1/4" thick	G		500	.032		.65	1.28		1.93	2.70
0600	Untempered hardboard, natural finish, 1/8" thick	G		500	.032		.39	1.28		1.67	2.41
0700	1/4" thick	G		500	.032		.42	1.28		1.70	2.44
0900	Untempered pegboard, 1/8" thick	G		500	.032		.38	1.28		1.66	2.40
1000	1/4" thick	G		500	.032		.42	1.28		1.70	2.44
1200	Plastic faced hardboard, 1/8" thick	G		500	.032		.64	1.28		1.92	2.68
1300	1/4" thick	G		500	.032		.85	1.28		2.13	2.92
1500	Plastic faced pegboard, 1/8" thick	G		500	.032		.61	1.28		1.89	2.65
1600	1/4" thick	G		500	.032		.75	1.28		2.03	2.81
1800	Wood grained, plain or grooved, 1/4" thick, minimum	G		500	.032		.57	1.28		1.85	2.61
1900	Maximum	G		425	.038		1.19	1.50		2.69	3.64
2100	Moldings for hardboard, wood or aluminum, minimum			500	.032	L.F.	.38	1.28		1.66	2.40
2200	Maximum			425	.038	"	1.08	1.50		2.58	3.52

06 25 Prefinished Paneling

06 25 16 – Prefinished Plywood Paneling

06 25 16.10 Paneling, Plywood

		Crew	Daily Output	Labor-Hours	Unit	Material	2009 Bare Costs Labor	2009 Bare Costs Equipment	Total	Total Incl O&P
0010	**PANELING, PLYWOOD** R061636-20									
2400	Plywood, prefinished, 1/4" thick, 4' x 8' sheets									
2410	with vertical grooves. Birch faced, minimum	2 Carp	500	.032	S.F.	.89	1.28		2.17	2.96
2420	Average		420	.038		1.35	1.52		2.87	3.85
2430	Maximum		350	.046		1.97	1.83		3.80	5
2600	Mahogany, African		400	.040		2.51	1.60		4.11	5.25
2700	Philippine (Lauan)		500	.032		1.08	1.28		2.36	3.17
2900	Oak or Cherry, minimum		500	.032		2.10	1.28		3.38	4.29
3000	Maximum		400	.040		3.22	1.60		4.82	6
3200	Rosewood		320	.050		4.58	2		6.58	8.15
3400	Teak		400	.040		3.22	1.60		4.82	6
3600	Chestnut		375	.043		4.77	1.70		6.47	7.90
3800	Pecan		400	.040		2.06	1.60		3.66	4.75
3900	Walnut, minimum		500	.032		2.75	1.28		4.03	5
3950	Maximum		400	.040		5.20	1.60		6.80	8.25
4000	Plywood, prefinished, 3/4" thick, stock grades, minimum		320	.050		1.24	2		3.24	4.46
4100	Maximum		224	.071		5.40	2.85		8.25	10.35
4300	Architectural grade, minimum		224	.071		3.98	2.85		6.83	8.80
4400	Maximum		160	.100		6.10	4		10.10	12.90
4600	Plywood, "A" face, birch, V.C., 1/2" thick, natural		450	.036		1.89	1.42		3.31	4.28
4700	Select		450	.036		2.06	1.42		3.48	4.47
4900	Veneer core, 3/4" thick, natural		320	.050		1.99	2		3.99	5.30
5000	Select		320	.050		2.23	2		4.23	5.55
5200	Lumber core, 3/4" thick, natural		320	.050		2.99	2		4.99	6.40
5500	Plywood, knotty pine, 1/4" thick, A2 grade		450	.036		1.63	1.42		3.05	3.99
5600	A3 grade		450	.036		2.06	1.42		3.48	4.47
5800	3/4" thick, veneer core, A2 grade		320	.050		2.11	2		4.11	5.40
5900	A3 grade		320	.050		2.38	2		4.38	5.70
6100	Aromatic cedar, 1/4" thick, plywood		400	.040		2.08	1.60		3.68	4.77
6200	1/4" thick, particle board		400	.040		1.01	1.60		2.61	3.59

06 25 26 – Panel System

06 25 26.10 Panel Systems

		Crew	Daily Output	Labor-Hours	Unit	Material	2009 Bare Costs Labor	2009 Bare Costs Equipment	Total	Total Incl O&P
0010	**PANEL SYSTEMS**									
0100	Raised panel, eng. wood core w/ wood veneer, std., paint grade	2 Carp	300	.053	S.F.	11.90	2.13		14.03	16.40
0110	Oak veneer		300	.053		19.70	2.13		21.83	25
0120	Maple veneer		300	.053		25	2.13		27.13	31
0130	Cherry veneer		300	.053		32	2.13		34.13	38.50
0300	Class I fire rated, paint grade		300	.053		14.25	2.13		16.38	18.95
0310	Oak veneer		300	.053		23.50	2.13		25.63	29.50
0320	Maple veneer		300	.053		30	2.13		32.13	36.50
0330	Cherry veneer		300	.053		38.50	2.13		40.63	45.50
0510	Beadboard, 5/8" MDF, standard, primed		300	.053		8.40	2.13		10.53	12.55
0520	Oak veneer, unfinished		300	.053		14.60	2.13		16.73	19.35
0530	Maple veneer, unfinished		300	.053		17.25	2.13		19.38	22.50
0610	Rustic paneling, 5/8" MDF, standard, maple veneer, unfinished		300	.053		22	2.13		24.13	28

06 26 Board Paneling

06 26 13 – Profile Board Paneling

06 26 13.10 Paneling, Boards		Crew	Daily Output	Labor-Hours	Unit	Material	2009 Bare Costs Labor	Equipment	Total	Total Incl O&P
0010	**PANELING, BOARDS**									
6400	Wood board paneling, 3/4" thick, knotty pine	2 Carp	300	.053	S.F.	1.44	2.13		3.57	4.88
6500	Rough sawn cedar		300	.053		1.85	2.13		3.98	5.35
6700	Redwood, clear, 1" x 4" boards		300	.053		4.30	2.13		6.43	8.05
6900	Aromatic cedar, closet lining, boards		275	.058		3.32	2.32		5.64	7.25

06 43 Wood Stairs and Railings

06 43 13 – Wood Stairs

06 43 13.20 Prefabricated Wood Stairs

		Crew	Daily Output	Labor-Hours	Unit	Material	Labor	Equipment	Total	Total Incl O&P
0010	**PREFABRICATED WOOD STAIRS**									
0100	Box stairs, prefabricated, 3'-0" wide									
0110	Oak treads, up to 14 risers	2 Carp	39	.410	Riser	85	16.40		101.40	119
0600	With pine treads for carpet, up to 14 risers	"	39	.410	"	55	16.40		71.40	86
1100	For 4' wide stairs, add				Flight	25%				
1550	Stairs, prefabricated stair handrail with balusters	1 Carp	30	.267	L.F.	75	10.65		85.65	98.50
1700	Basement stairs, prefabricated, pine treads									
1710	Pine risers, 3' wide, up to 14 risers	2 Carp	52	.308	Riser	55	12.30		67.30	79.50
4000	Residential, wood, oak treads, prefabricated		1.50	10.667	Flight	1,100	425		1,525	1,850
4200	Built in place		.44	36.364	"	1,575	1,450		3,025	3,975
4400	Spiral, oak, 4'-6" diameter, unfinished, prefabricated,									
4500	incl. railing, 9' high	2 Carp	1.50	10.667	Flight	3,625	425		4,050	4,625

06 43 13.40 Wood Stair Parts

		Crew	Daily Output	Labor-Hours	Unit	Material	Labor	Equipment	Total	Total Incl O&P
0010	**WOOD STAIR PARTS**									
0020	Pin top balusters, 1-1/4", oak, 34"	1 Carp	96	.083	Ea.	9.15	3.33		12.48	15.20
0030	36"		96	.083		9.60	3.33		12.93	15.70
0040	42"		96	.083		11.65	3.33		14.98	17.95
0050	Poplar, 34"		96	.083		6.10	3.33		9.43	11.90
0060	36"		96	.083		6.50	3.33		9.83	12.30
0070	42"		96	.083		14.20	3.33		17.53	21
0080	Maple, 31"		96	.083		11.65	3.33		14.98	17.95
0090	34"		96	.083		12.95	3.33		16.28	19.40
0100	36"		96	.083		13.50	3.33		16.83	20
0110	39"		96	.083		15.45	3.33		18.78	22
0120	41"		96	.083		17.35	3.33		20.68	24.50
0130	Primed, 31"		96	.083		4.03	3.33		7.36	9.60
0140	34"		96	.083		4.33	3.33		7.66	9.90
0150	36"		96	.083		4.51	3.33		7.84	10.10
0160	39"		96	.083		5.05	3.33		8.38	10.70
0170	41"		96	.083		5.55	3.33		8.88	11.30
0180	Box top balusters, 1-1/4", oak, 34"		60	.133		9.40	5.35		14.75	18.60
0190	36"		60	.133		10.25	5.35		15.60	19.50
0200	42"		60	.133		11.45	5.35		16.80	21
0210	Poplar, 34"		60	.133		6	5.35		11.35	14.85
0220	36"		60	.133		6.75	5.35		12.10	15.65
0230	42"		60	.133		7.50	5.35		12.85	16.50
0240	Maple, 31"		60	.133		13.45	5.35		18.80	23
0250	34"		60	.133		14.15	5.35		19.50	24
0260	36"		60	.133		16.35	5.35		21.70	26.50
0270	39"		60	.133		16.60	5.35		21.95	26.50
0280	41"		60	.133		17.15	5.35		22.50	27
0290	Primed, 31"		60	.133		4.46	5.35		9.81	13.15

06 43 Wood Stairs and Railings

06 43 13 – Wood Stairs

06 43 13.40 Wood Stair Parts		Crew	Daily Output	Labor-Hours	Unit	Material	2009 Bare Costs Labor	Equipment	Total	Total Incl O&P
0300	34"	1 Carp	60	.133	Ea.	4.79	5.35		10.14	13.50
0310	36"		60	.133		4.97	5.35		10.32	13.70
0320	39"		60	.133		5.60	5.35		10.95	14.45
0330	41"		60	.133	▼	6.25	5.35		11.60	15.15
0340	Square balusters, cut from lineal stock, pine, 1-1/16" x 1-1/16"		180	.044	L.F.	1.32	1.78		3.10	4.20
0350	1-5/16" x 1-5/16"		180	.044		1.85	1.78		3.63	4.79
0360	1-5/8" x 1-5/8"		180	.044	▼	3.03	1.78		4.81	6.10
0370	Turned newel, oak, 3-1/2" square, 50" high		8	1	Ea.	119	40		159	192
0380	65" high		8	1		154	40		194	232
0390	Poplar, 3-1/2" square, 50" high		8	1		99	40		139	171
0400	65" high		8	1		128	40		168	202
0410	Maple, 3-1/2" square, 50" high		8	1		138	40		178	213
0420	65" high		8	1		160	40		200	237
0430	Square newel, oak, 3-1/2" square, 50" high		8	1		226	40		266	310
0440	65" high		8	1		242	40		282	330
0450	Poplar, 3-1/2" square, 50" high		8	1		215	40		255	298
0460	65" high		8	1		220	40		260	305
0470	Maple, 3" square, 50" high		8	1		243	40		283	330
0480	65" high		8	1	▼	265	40		305	355
0490	Railings, oak, minimum		96	.083	L.F.	7.50	3.33		10.83	13.40
0500	Average		96	.083		9.20	3.33		12.53	15.25
0510	Maximum		96	.083		9.35	3.33		12.68	15.45
0520	Maple, minimum		96	.083		8.80	3.33		12.13	14.85
0530	Average		96	.083		11	3.33		14.33	17.25
0540	Maximum		96	.083		13.20	3.33		16.53	19.70
0550	Oak, for bending rail, minimum		48	.167		21	6.65		27.65	33.50
0560	Average		48	.167		30	6.65		36.65	43.50
0570	Maximum		48	.167		32	6.65		38.65	46
0580	Maple, for bending rail, minimum		48	.167		32	6.65		38.65	46
0590	Average		48	.167		35	6.65		41.65	49
0600	Maximum		48	.167	▼	37.50	6.65		44.15	51.50
0610	Risers, oak, 3/4" x 8", 36" long		80	.100	Ea.	22	4		26	30
0620	42" long		80	.100		25.50	4		29.50	34
0630	48" long		80	.100		29.50	4		33.50	38
0640	54" long		80	.100		33	4		37	42
0650	60" long		80	.100		36.50	4		40.50	46.50
0660	72" long		80	.100		44	4		48	54.50
0670	Poplar, 3/4" x 8", 36" long		80	.100		14.35	4		18.35	22
0680	42" long		80	.100		16.75	4		20.75	24.50
0690	48" long		80	.100		19.15	4		23.15	27
0700	54" long		80	.100		21.50	4		25.50	29.50
0710	60" long		80	.100		24	4		28	32.50
0720	72" long		80	.100		28.50	4		32.50	37.50
0730	Pine, 1" x 8", 36" long		80	.100		2.51	4		6.51	8.95
0740	42" long		80	.100		2.93	4		6.93	9.40
0750	48" long		80	.100		3.34	4		7.34	9.90
0760	54" long		80	.100		3.76	4		7.76	10.35
0770	60" long		80	.100		4.18	4		8.18	10.80
0780	72" long		80	.100		5	4		9	11.70
0790	Treads, oak, no returns, 1-1/32" x 11-1/2" x 36" long		32	.250		43.50	10		53.50	63.50
0800	42" long		32	.250		50.50	10		60.50	71.50
0810	48" long		32	.250		58	10		68	79
0820	54" long		32	.250	▼	65	10		75	87

06 43 Wood Stairs and Railings

06 43 13 – Wood Stairs

06 43 13.40 Wood Stair Parts		Crew	Daily Output	Labor-Hours	Unit	Material	2009 Bare Costs Labor	Equipment	Total	Total Incl O&P
0830	60" long	1 Carp	32	.250	Ea.	72.50	10		82.50	95
0840	72" long		32	.250		87	10		97	111
0850	Mitred return one end, 1-1/32" x 11-1/2" x 36" long		24	.333		65	13.30		78.30	92
0860	42" long		24	.333		76	13.30		89.30	104
0870	48" long		24	.333		86.50	13.30		99.80	116
0880	54" long		24	.333		97.50	13.30		110.80	128
0890	60" long		24	.333		108	13.30		121.30	140
0900	72" long		24	.333		130	13.30		143.30	164
0910	Mitred return two ends, 1-1/32" x 11-1/2" x 36" long		12	.667		81	26.50		107.50	131
0920	42" long		12	.667		95	26.50		121.50	146
0930	48" long		12	.667		108	26.50		134.50	161
0940	54" long		12	.667		122	26.50		148.50	176
0950	60" long		12	.667		135	26.50		161.50	191
0960	72" long		12	.667		162	26.50		188.50	221
0970	Starting step, oak, 48", bullnose		8	1		163	40		203	242
0980	Double end bullnose		8	1		193	40		233	274
0990	Half circle		8	1		216	40		256	299
1010	Double end half circle		8	1		286	40		326	375
1030	Skirt board, pine, 1" x 10"		55	.145	L.F.	1.23	5.80		7.03	10.35
1040	1" x 12"		52	.154	"	1.60	6.15		7.75	11.30
1050	Oak landing tread, 1-1/16" thick		54	.148	S.F.	8.25	5.90		14.15	18.30
1060	Oak cove molding		96	.083	L.F.	2.75	3.33		6.08	8.20
1070	Oak stringer molding		96	.083		2.75	3.33		6.08	8.20
1080	Oak shoe molding		96	.083		2.75	3.33		6.08	8.20
1090	Rail bolt, 5/16" x 3-1/2"		48	.167	Ea.	1.65	6.65		8.30	12.15
1100	5/16" x 4-1/2"		48	.167		2.20	6.65		8.85	12.75
1120	Newel post anchor		16	.500		27.50	20		47.50	61.50
1130	Tapered plug, 1/2"		240	.033		.28	1.33		1.61	2.37
1140	1"		240	.033		.39	1.33		1.72	2.49
1150	Rail bolt kit		24	.333		11	13.30		24.30	32.50

06 43 16 – Wood Railings

06 43 16.10 Wood Handrails and Railings		Crew	Daily Output	Labor-Hours	Unit	Material	2009 Bare Costs Labor	Equipment	Total	Total Incl O&P
0010	**WOOD HANDRAILS AND RAILINGS**									
0020	Custom design, architectural grade, hardwood, minimum	1 Carp	38	.211	L.F.	9.50	8.40		17.90	23.50
0100	Maximum		30	.267		51.50	10.65		62.15	73
0300	Stock interior railing with spindles 4" O.C., 4' long		40	.200		31.50	8		39.50	47.50
0400	8' long		48	.167		31.50	6.65		38.15	45.50

06 44 Ornamental Woodwork

06 44 19 – Wood Grilles

06 44 19.10 Grilles		Crew	Daily Output	Labor-Hours	Unit	Material	2009 Bare Costs Labor	Equipment	Total	Total Incl O&P
0010	**GRILLES** and panels, hardwood, sanded									
0020	2' x 4' to 4' x 8', custom designs, unfinished, minimum	1 Carp	38	.211	S.F.	14.65	8.40		23.05	29
0050	Average		30	.267		31.50	10.65		42.15	51.50
0100	Maximum		19	.421		49	16.80		65.80	79.50
0300	As above, but prefinished, minimum		38	.211		14.65	8.40		23.05	29
0400	Maximum		19	.421		55	16.80		71.80	86.50

06 44 Ornamental Woodwork

06 44 33 – Wood Mantels

06 44 33.10 Fireplace Mantels		Crew	Daily Output	Labor-Hours	Unit	Material	2009 Bare Costs Labor	2009 Bare Costs Equipment	Total	Total Incl O&P
0010	**FIREPLACE MANTELS**									
0015	6" molding, 6' x 3'-6" opening, minimum	1 Carp	5	1.600	Opng.	228	64		292	350
0100	Maximum		5	1.600		410	64		474	555
0300	Prefabricated pine, colonial type, stock, deluxe		2	4		1,250	160		1,410	1,650
0400	Economy		3	2.667		530	107		637	750

06 44 33.20 Fireplace Mantel Beam

		Crew	Daily Output	Labor-Hours	Unit	Material	Labor	Equipment	Total	Total Incl O&P
0010	**FIREPLACE MANTEL BEAM**									
0020	Rough texture wood, 4" x 8"	1 Carp	36	.222	L.F.	6	8.90		14.90	20.50
0100	4" x 10"		35	.229	"	7	9.15		16.15	22
0300	Laminated hardwood, 2-1/4" x 10-1/2" wide, 6' long		5	1.600	Ea.	110	64		174	220
0400	8' long		5	1.600	"	150	64		214	264
0600	Brackets for above, rough sawn		12	.667	Pr.	10	26.50		36.50	52.50
0700	Laminated		12	.667	"	15	26.50		41.50	58

06 44 39 – Wood Posts and Columns

06 44 39.10 Decorative Beams

		Crew	Daily Output	Labor-Hours	Unit	Material	Labor	Equipment	Total	Total Incl O&P
0010	**DECORATIVE BEAMS**									
0020	Rough sawn cedar, non-load bearing, 4" x 4"	2 Carp	180	.089	L.F.	1.47	3.55		5.02	7.10
0100	4" x 6"		170	.094		2.21	3.76		5.97	8.30
0200	4" x 8"		160	.100		2.95	4		6.95	9.45
0300	4" x 10"		150	.107		3.68	4.26		7.94	10.65
0400	4" x 12"		140	.114		4.42	4.57		8.99	11.95
0500	8" x 8"		130	.123		7.50	4.92		12.42	15.85

06 44 39.20 Columns

		Crew	Daily Output	Labor-Hours	Unit	Material	Labor	Equipment	Total	Total Incl O&P
0010	**COLUMNS**									
0050	Aluminum, round colonial, 6" diameter	2 Carp	80	.200	V.L.F.	16	8		24	30
0100	8" diameter		62.25	.257		17	10.25		27.25	34.50
0200	10" diameter		55	.291		19.10	11.60		30.70	39
0250	Fir, stock units, hollow round, 6" diameter		80	.200		17.20	8		25.20	31.50
0300	8" diameter		80	.200		20.50	8		28.50	35
0350	10" diameter		70	.229		25.50	9.15		34.65	42.50
0400	Solid turned, to 8' high, 3-1/2" diameter		80	.200		8.40	8		16.40	21.50
0500	4-1/2" diameter		75	.213		12.80	8.50		21.30	27.50
0600	5-1/2" diameter		70	.229		16.40	9.15		25.55	32
0800	Square columns, built-up, 5" x 5"		65	.246		10.20	9.85		20.05	26.50
0900	Solid, 3-1/2" x 3-1/2"		130	.123		6.70	4.92		11.62	14.95
1600	Hemlock, tapered, T & G, 12" diam, 10' high		100	.160		31.50	6.40		37.90	44.50
1700	16' high		65	.246		46.50	9.85		56.35	66.50
1900	10' high, 14" diameter		100	.160		78.50	6.40		84.90	96.50
2000	18' high		65	.246		63.50	9.85		73.35	85.50
2200	18" diameter, 12' high		65	.246		114	9.85		123.85	141
2300	20' high		50	.320		80	12.80		92.80	108
2500	20" diameter, 14' high		40	.400		129	16		145	167
2600	20' high		35	.457		120	18.25		138.25	161
2800	For flat pilasters, deduct					33%				
3000	For splitting into halves, add				Ea.	116			116	128
4000	Rough sawn cedar posts, 4" x 4"	2 Carp	250	.064	V.L.F.	6.30	2.56		8.86	10.90
4100	4" x 6"		235	.068		15	2.72		17.72	20.50
4200	6" x 6"		220	.073		26.50	2.91		29.41	33.50
4300	8" x 8"		200	.080		60	3.20		63.20	71

06 48 Wood Frames

06 48 13 – Exterior Wood Door Frames

06 48 13.10 Exterior Wood Door Frames and Accessories

		Crew	Daily Output	Labor-Hours	Unit	Material	2009 Bare Costs Labor	Equipment	Total	Total Incl O&P
0010	**EXTERIOR WOOD DOOR FRAMES AND ACCESSORIES**									
0400	Exterior frame, incl. ext. trim, pine, 5/4 x 4-9/16" deep	2 Carp	375	.043	L.F.	5.55	1.70		7.25	8.75
0420	5-3/16" deep		375	.043		11.30	1.70		13	15.10
0440	6-9/16" deep		375	.043		8.50	1.70		10.20	12
0600	Oak, 5/4 x 4-9/16" deep		350	.046		12	1.83		13.83	16.05
0620	5-3/16" deep		350	.046		13.50	1.83		15.33	17.70
0640	6-9/16" deep		350	.046		15.30	1.83		17.13	19.70
0800	Walnut, 5/4 x 4-9/16" deep		350	.046		14	1.83		15.83	18.25
0820	5-3/16" deep		350	.046		18	1.83		19.83	22.50
0840	6-9/16" deep		350	.046		20	1.83		21.83	25
1000	Sills, 8/4 x 8" deep, oak, no horns		100	.160		19	6.40		25.40	31
1020	2" horns		100	.160		19	6.40		25.40	31
1040	3" horns		100	.160		19.95	6.40		26.35	32
1100	8/4 x 10" deep, oak, no horns		90	.178		21.50	7.10		28.60	35
1120	2" horns		90	.178		19.80	7.10		26.90	33
1140	3" horns		90	.178		20.50	7.10		27.60	34
2000	Exterior, colonial, frame & trim, 3' opng., in-swing, minimum		22	.727	Ea.	315	29		344	395
2010	Average		21	.762		470	30.50		500.50	560
2020	Maximum		20	.800		1,075	32		1,107	1,225
2100	5'-4" opening, in-swing, minimum		17	.941		360	37.50		397.50	455
2120	Maximum		15	1.067		1,075	42.50		1,117.50	1,250
2140	Out-swing, minimum		17	.941		370	37.50		407.50	465
2160	Maximum		15	1.067		1,100	42.50		1,142.50	1,300
2400	6'-0" opening, in-swing, minimum		16	1		345	40		385	440
2420	Maximum		10	1.600		1,100	64		1,164	1,325
2460	Out-swing, minimum		16	1		375	40		415	475
2480	Maximum		10	1.600		1,300	64		1,364	1,525
2600	For two sidelights, add, minimum		30	.533	Opng.	355	21.50		376.50	425
2620	Maximum		20	.800	"	1,125	32		1,157	1,300
2700	Custom birch frame, 3'-0" opening		16	1	Ea.	205	40		245	288
2750	6'-0" opening		16	1		305	40		345	395
2900	Exterior, modern, plain trim, 3' opng., in-swing, minimum		26	.615		35.50	24.50		60	77
2920	Average		24	.667		42	26.50		68.50	87.50
2940	Maximum		22	.727		50.50	29		79.50	101

06 48 16 – Interior Wood Door Frames

06 48 16.10 Interior Wood Door Jamb and Frames

		Crew	Daily Output	Labor-Hours	Unit	Material	Labor	Equipment	Total	Total Incl O&P
0010	**INTERIOR WOOD DOOR JAMB AND FRAMES**									
3000	Interior frame, pine, 11/16" x 3-5/8" deep	2 Carp	375	.043	L.F.	5.50	1.70		7.20	8.70
3020	4-9/16" deep		375	.043		6.45	1.70		8.15	9.75
3200	Oak, 11/16" x 3-5/8" deep		350	.046		4.35	1.83		6.18	7.60
3220	4-9/16" deep		350	.046		4.50	1.83		6.33	7.80
3240	5-3/16" deep		350	.046		4.58	1.83		6.41	7.90
3400	Walnut, 11/16" x 3-5/8" deep		350	.046		7.30	1.83		9.13	10.90
3420	4-9/16" deep		350	.046		7.50	1.83		9.33	11.10
3440	5-3/16" deep		350	.046		7.90	1.83		9.73	11.50
3600	Pocket door frame		16	1	Ea.	67	40		107	136
3800	Threshold, oak, 5/8" x 3-5/8" deep		200	.080	L.F.	3.83	3.20		7.03	9.15
3820	4-5/8" deep		190	.084		4.58	3.36		7.94	10.25
3840	5-5/8" deep		180	.089		7.15	3.55		10.70	13.35

06 52 Plastic Structural Assemblies

06 52 10 – Fiberglass Structural Assemblies

06 52 10.40 Fiberglass Floor Grating		Crew	Daily Output	Labor-Hours	Unit	Material	2009 Bare Costs Labor	Equipment	Total	Total Incl O&P
0010	**FIBERGLASS FLOOR GRATING**									
0100	Reinforced polyester, fire retardant, 1" x 4" grid, 1" thick	E-4	510	.063	S.F.	14.25	2.84	.26	17.35	21
0200	1-1/2" x 6" mesh, 1-1/2" thick	↓	500	.064	↓	16.65	2.89	.27	19.81	24
0300	With grit surface, 1-1/2" x 6" grid, 1-1/2" thick	↓	500	.064	↓	17	2.89	.27	20.16	24

06 63 Plastic Railings

06 63 10 – Plastic (PVC) Railings

06 63 10.10 Plastic Railings		Crew	Daily Output	Labor-Hours	Unit	Material	2009 Bare Costs Labor	Equipment	Total	Total Incl O&P
0010	**PLASTIC RAILINGS**									
0100	Horizontal PVC handrail with balusters, 3-1/2" wide, 36" high	1 Carp	96	.083	L.F.	23.50	3.33		26.83	30.50
0150	42" high		96	.083		26.50	3.33		29.83	34
0200	Angled PVC handrail with balusters, 3-1/2" wide, 36" high		72	.111		27	4.44		31.44	36.50
0250	42" high		72	.111		30	4.44		34.44	40
0300	Post sleeve for 4 x 4 post		96	.083	↓	12	3.33		15.33	18.35
0400	Post cap for 4 x 4 post, flat profile		48	.167	Ea.	11.35	6.65		18	23
0450	Newel post style profile		48	.167		21	6.65		27.65	33.50
0500	Raised corbeled profile		48	.167		22	6.65		28.65	35
0550	Post base trim for 4 x 4 post	↓	96	.083	↓	12.80	3.33		16.13	19.20

Division 7 - Thermal and Moisture Protection

Estimating Tips

07 10 00 Dampproofing and Waterproofing

- Be sure of the job specifications before pricing this subdivision. The difference in cost between waterproofing and dampproofing can be great. Waterproofing will hold back standing water. Dampproofing prevents the transmission of water vapor. Also included in this section are vapor retarding membranes.

07 20 00 Thermal Protection

- Insulation and fireproofing products are measured by area, thickness, volume or R-value. Specifications may give only what the specific R-value should be in a certain situation. The estimator may need to choose the type of insulation to meet that R-value.

07 30 00 Steep Slope Roofing
07 40 00 Roofing and Siding Panels

- Many roofing and siding products are bought and sold by the square. One square is equal to an area that measures 100 square feet. This simple change in unit of measure could create a large error if the estimator is not observant. Accessories necessary for a complete installation must be figured into any calculations for both material and labor.

07 50 00 Membrane Roofing
07 60 00 Flashing and Sheet Metal
07 70 00 Roofing and Wall Specialties and Accessories

- The items in these subdivisions compose a roofing system. No one component completes the installation, and all must be estimated. Built-up or single-ply membrane roofing systems are made up of many products and installation trades. Wood blocking at roof perimeters or penetrations, parapet coverings, reglets, roof drains, gutters, downspouts, sheet metal flashing, skylights, smoke vents, and roof hatches all need to be considered along with the roofing material. Several different installation trades will need to work together on the roofing system. Inherent difficulties in the scheduling and coordination of various trades must be accounted for when estimating labor costs.

07 90 00 Joint Protection

- To complete the weather-tight shell, the sealants and caulkings must be estimated. Where different materials meet—at expansion joints, at flashing penetrations, and at hundreds of other locations throughout a construction project—they provide another line of defense against water penetration. Often, an entire system is based on the proper location and placement of caulking or sealants. The detailed drawings that are included as part of a set of architectural plans show typical locations for these materials. When caulking or sealants are shown at typical locations, this means the estimator must include them for all the locations where this detail is applicable. Be careful to keep different types of sealants separate, and remember to consider backer rods and primers if necessary.

Reference Numbers

Reference numbers are shown in shaded boxes at the beginning of some major classifications. These numbers refer to related items in the Reference Section. The reference information may be an estimating procedure, an alternate pricing method, or technical information.

Note: Not all subdivisions listed here necessarily appear in this publication.

07 05 Common Work Results for Thermal and Moisture Protection

07 05 05 – Selective Demolition

07 05 05.10 Selective Demo., Thermal and Moist. Protection

		Crew	Daily Output	Labor-Hours	Unit	Material	2009 Bare Costs Labor	Equipment	Total	Total Incl O&P
0010	**SELECTIVE DEMOLITION, THERMAL AND MOISTURE PROTECTION**									
3170	Asphalt shingles, 1 layer	B-2	3500	.011	S.F.		.37		.37	.57
3620	5 ply		1600	.025			.80		.80	1.24
3720	5 ply, with gravel R024119-10		890	.045			1.44		1.44	2.23
4670	Wood shingles		2200	.018			.58		.58	.90
4970	Siding, horizontal wood clapboards	1 Clab	380	.021			.67		.67	1.03
5225	Horizontal strips		444	.018			.57		.57	.88
5320	Vertical strips		400	.020			.63		.63	.98
5520	Wood shingles		350	.023			.72		.72	1.12
5670	Textured plywood		725	.011			.35		.35	.54
5870	Wood, boards, vertical		400	.020			.63		.63	.98

07 11 Dampproofing

07 11 16 – Cementitious Dampproofing

07 11 16.20 Cementitious Parging

		Crew	Daily Output	Labor-Hours	Unit	Material	Labor	Equipment	Total	Total Incl O&P
0010	**CEMENTITIOUS PARGING**									
0020	Portland cement, 2 coats, 1/2" thick	D-1	250	.064	S.F.	.28	2.33		2.61	3.84
0100	Waterproofed Portland cement, 1/2" thick, 2 coats	"	250	.064	"	2.62	2.33		4.95	6.40

07 16 Cementitious and Reactive Waterproofing

07 16 16 – Crystalline Waterproofing

07 16 16.20 Cementitious Waterproofing

		Crew	Daily Output	Labor-Hours	Unit	Material	Labor	Equipment	Total	Total Incl O&P
0010	**CEMENTITIOUS WATERPROOFING**									
0020	1/8" application, sprayed on	G-2A	1000	.024	S.F.	1.62	.73	.61	2.96	3.65
0030	2 coat, cementitious/metallic slurry, troweled, 1/4" thick	1 Cefi	2.48	3.226	C.S.F.	24	124		148	208
0040	3 coat, 3/8" thick		1.84	4.348		42.50	167		209.50	291
0050	4 coat, 1/2" thick		1.20	6.667		56.50	255		311.50	435

07 21 Thermal Insulation

07 21 13 – Board Insulation

07 21 13.10 Rigid Insulation

			Crew	Daily Output	Labor-Hours	Unit	Material	Labor	Equipment	Total	Total Incl O&P
0010	**RIGID INSULATION**, for walls										
0040	Fiberglass, 1.5#/CF, unfaced, 1" thick, R4.1	G	1 Carp	1000	.008	S.F.	.44	.32		.76	.98
0060	1-1/2" thick, R6.2	G		1000	.008		.64	.32		.96	1.20
0080	2" thick, R8.3	G		1000	.008		.69	.32		1.01	1.26
0120	3" thick, R12.4	G		800	.010		.80	.40		1.20	1.50
0370	3#/CF, unfaced, 1" thick, R4.3	G		1000	.008		.49	.32		.81	1.04
0390	1-1/2" thick, R6.5	G		1000	.008		.95	.32		1.27	1.55
0400	2" thick, R8.7	G		890	.009		1.15	.36		1.51	1.83
0420	2-1/2" thick, R10.9	G		800	.010		1.01	.40		1.41	1.73
0440	3" thick, R13	G		800	.010		1.01	.40		1.41	1.73
0520	Foil faced, 1" thick, R4.3	G		1000	.008		.92	.32		1.24	1.51
0540	1-1/2" thick, R6.5	G		1000	.008		1.36	.32		1.68	2
0560	2" thick, R8.7	G		890	.009		1.71	.36		2.07	2.44
0580	2-1/2" thick, R10.9	G		800	.010		2.02	.40		2.42	2.84
0600	3" thick, R13	G		800	.010		2.19	.40		2.59	3.03
0670	6#/CF, unfaced, 1" thick, R4.3	G		1000	.008		.98	.32		1.30	1.58

07 21 Thermal Insulation

07 21 13 – Board Insulation

07 21 13.10 Rigid Insulation

		Crew	Daily Output	Labor-Hours	Unit	Material	2009 Bare Costs Labor	Equipment	Total	Total Incl O&P
0690	1-1/2" thick, R6.5 G	1 Carp	890	.009	S.F.	1.50	.36		1.86	2.21
0700	2" thick, R8.7 G		800	.010		2.12	.40		2.52	2.95
0721	2-1/2" thick, R10.9 G		800	.010		2.32	.40		2.72	3.17
0741	3" thick, R13 G		730	.011		2.78	.44		3.22	3.74
0821	Foil faced, 1" thick, R4.3 G		1000	.008		1.38	.32		1.70	2.02
0840	1-1/2" thick, R6.5 G		890	.009		1.98	.36		2.34	2.74
0850	2" thick, R8.7 G		800	.010		2.59	.40		2.99	3.47
0880	2-1/2" thick, R10.9 G		800	.010		3.11	.40		3.51	4.04
0900	3" thick, R13 G		730	.011		3.72	.44		4.16	4.77
1500	Foamglass, 1-1/2" thick, R4.5 G		800	.010		1.37	.40		1.77	2.13
1550	3" thick, R9 G		730	.011		3.29	.44		3.73	4.30
1700	Perlite, 1" thick, R2.77 G		800	.010		.30	.40		.70	.95
1750	2" thick, R5.55 G		730	.011		.60	.44		1.04	1.34
1900	Extruded polystyrene, 25 PSI compressive strength, 1" thick, R5 G		800	.010		.51	.40		.91	1.18
1940	2" thick R10 G		730	.011		1.04	.44		1.48	1.82
1960	3" thick, R15 G		730	.011		1.40	.44		1.84	2.22
2100	Expanded polystyrene, 1" thick, R3.85 G		800	.010		.24	.40		.64	.88
2120	2" thick, R7.69 G		730	.011		.61	.44		1.05	1.35
2140	3" thick, R11.49 G		730	.011		.78	.44		1.22	1.54

07 21 16 – Blanket Insulation

07 21 16.20 Blanket Insulation for Walls

		Crew	Daily Output	Labor-Hours	Unit	Material	2009 Bare Costs Labor	Equipment	Total	Total Incl O&P
0010	**BLANKET INSULATION FOR WALLS**									
0020	Kraft faced fiberglass, 3-1/2" thick, R11, 15" wide G	1 Carp	1350	.006	S.F.	.34	.24		.58	.74
0030	23" wide G		1600	.005		.34	.20		.54	.68
0060	R13, 11" wide G		1150	.007		.34	.28		.62	.80
0080	15" wide G		1350	.006		.34	.24		.58	.74
0100	23" wide G		1600	.005		.34	.20		.54	.68
0110	R-15, 11" wide G		1150	.007		.41	.28		.69	.88
0120	15" wide G		1350	.006		.41	.24		.65	.82
0130	23" wide G		1600	.005		.41	.20		.61	.76
0140	6" thick, R19, 11" wide G		1150	.007		.44	.28		.72	.91
0160	15" wide G		1350	.006		.44	.24		.68	.85
0180	23" wide G		1600	.005		.44	.20		.64	.79
0182	R21, 11" wide G		1150	.007		.50	.28		.78	.98
0184	15" wide G		1350	.006		.50	.24		.74	.92
0186	23" wide G		1600	.005		.50	.20		.70	.86
0188	9" thick, R-30, 11" wide G		985	.008		.66	.32		.98	1.23
0200	15" wide G		1150	.007		.66	.28		.94	1.16
0220	23" wide G		1350	.006		.66	.24		.90	1.10
0230	12" thick, R38, 11" wide G		985	.008		.95	.32		1.27	1.55
0240	15" wide G		1150	.007		.95	.28		1.23	1.48
0260	23" wide G		1350	.006		.95	.24		1.19	1.42
0410	Foil faced fiberglass, 3-1/2" thick, R13, 11" wide G		1150	.007		.57	.28		.85	1.06
0420	15" wide G		1350	.006		.57	.24		.81	1
0440	23" wide G		1600	.005		.57	.20		.77	.94
0442	R15, 11" wide G		1150	.007		.56	.28		.84	1.05
0444	15" wide G		1350	.006		.56	.24		.80	.99
0446	23" wide G		1600	.005		.56	.20		.76	.93
0448	6" thick, R19, 11" wide G		1150	.007		.77	.28		1.05	1.28
0460	15" wide G		1350	.006		.77	.24		1.01	1.22
0480	23" wide G		1600	.005		.77	.20		.97	1.16
0482	R-21, 11" wide G		1150	.007		.82	.28		1.10	1.33

07 21 Thermal Insulation

07 21 16 – Blanket Insulation

07 21 16.20 Blanket Insulation for Walls

		Crew	Daily Output	Labor-Hours	Unit	Material	2009 Bare Costs Labor	Equipment	Total	Total Incl O&P
0484	15" wide	1 Carp [G]	1350	.006	S.F.	.82	.24		1.06	1.27
0486	23" wide	[G]	1600	.005		.82	.20		1.02	1.21
0488	9" thick, R-30, 11" wide	[G]	985	.008		.89	.32		1.21	1.48
0500	9" thick, R30, 15" wide	[G]	1150	.007		.89	.28		1.17	1.41
0550	23" wide	[G]	1350	.006		.89	.24		1.13	1.35
0560	12" thick, R-38, 11" wide	[G]	985	.008		.90	.32		1.22	1.49
0570	15" wide	[G]	1150	.007		.90	.28		1.18	1.42
0580	23" wide	[G]	1350	.006		.90	.24		1.14	1.36
0620	Unfaced fiberglass, 3-1/2" thick, R-13, 11" wide	[G]	1150	.007		.29	.28		.57	.75
0820	15" wide	[G]	1350	.006		.29	.24		.53	.69
0830	23" wide	[G]	1600	.005		.29	.20		.49	.63
0832	R15, 11" wide	[G]	1150	.007		.37	.28		.65	.84
0836	23" wide	[G]	1600	.005		.37	.20		.57	.72
0838	6" thick, R19, 11" wide	[G]	1150	.007		.42	.28		.70	.89
0860	15" wide	[G]	1150	.007		.42	.28		.70	.89
0880	23" wide	[G]	1350	.006		.42	.24		.66	.83
0882	R-21, 11" wide	[G]	1150	.007		.46	.28		.74	.94
0886	15" wide	[G]	1350	.006		.46	.24		.70	.88
0888	23" wide	[G]	1600	.005		.46	.20		.66	.82
0890	9" thick, R30, 11" wide	[G]	985	.008		.75	.32		1.07	1.33
0900	15" wide	[G]	1150	.007		.75	.28		1.03	1.26
0920	23" wide	[G]	1350	.006		.75	.24		.99	1.20
0930	12" thick, R38, 11" wide	[G]	985	.008		1.28	.32		1.60	1.91
0940	15" wide	[G]	1000	.008		1.28	.32		1.60	1.91
0960	23" wide	[G]	1150	.007		1.28	.28		1.56	1.84
1300	Mineral fiber batts, kraft faced									
1320	3-1/2" thick, R12	1 Carp [G]	1600	.005	S.F.	.38	.20		.58	.73
1340	6" thick, R19	[G]	1600	.005		.48	.20		.68	.84
1380	10" thick, R30	[G]	1350	.006		.73	.24		.97	1.17
1850	Friction fit wire insulation supports, 16" O.C.		960	.008	Ea.	.08	.33		.41	.61

07 21 23 – Loose-Fill Insulation

07 21 23.10 Poured Loose-Fill Insulation

		Crew	Daily Output	Labor-Hours	Unit	Material	Labor	Equipment	Total	Total Incl O&P
0010	**POURED LOOSE-FILL INSULATION**									
0020	Cellulose fiber, R3.8 per inch	1 Carp [G]	200	.040	C.F.	.60	1.60		2.20	3.14
0040	Ceramic type (perlite), R3.2 per inch	[G]	200	.040		1.75	1.60		3.35	4.41
0080	Fiberglass wool, R4 per inch	[G]	200	.040		.46	1.60		2.06	2.99
0100	Mineral wool, R3 per inch	[G]	200	.040		.39	1.60		1.99	2.91
0300	Polystyrene, R4 per inch	[G]	200	.040		3.09	1.60		4.69	5.90
0400	Vermiculite or perlite, R2.7 per inch	[G]	200	.040		1.75	1.60		3.35	4.41
0700	Wood fiber, R3.85 per inch	[G]	200	.040		.70	1.60		2.30	3.25

07 21 23.20 Masonry Loose-Fill Insulation

		Crew	Daily Output	Labor-Hours	Unit	Material	Labor	Equipment	Total	Total Incl O&P
0010	**MASONRY LOOSE-FILL INSULATION**, vermiculite or perlite									
0100	In cores of concrete block, 4" thick wall, .115 CF/SF	D-1 [G]	4800	.003	S.F.	.20	.12		.32	.40
0200	6" thick wall, .175 CF/SF	[G]	3000	.005		.31	.19		.50	.63
0300	8" thick wall, .258 CF/SF	[G]	2400	.007		.45	.24		.69	.87
0400	10" thick wall, .340 CF/SF	[G]	1850	.009		.60	.31		.91	1.13
0500	12" thick wall, .422 CF/SF	[G]	1200	.013		.74	.48		1.22	1.54
0550	For sand fill, deduct from above	[G]				70%				
0600	Poured cavity wall, vermiculite or perlite, water repellant	D-1 [G]	250	.064	C.F.	1.75	2.33		4.08	5.45
0700	Foamed in place, urethane in 2-5/8" cavity	G-2A [G]	1035	.023	S.F.	.42	.71	.58	1.71	2.25
0800	For each 1" added thickness, add	" [G]	2372	.010	"	.12	.31	.26	.69	.91

07 21 Thermal Insulation

07 21 26 – Blown Insulation

07 21 26.10 Blown Insulation

		Crew	Daily Output	Labor-Hours	Unit	Material	2009 Bare Costs Labor	Equipment	Total	Total Incl O&P
0010	**BLOWN INSULATION** Ceilings, with open access									
0020	Cellulose, 3-1/2" thick, R13 [G]	G-4	5000	.005	S.F.	.19	.15	.07	.41	.52
0030	5-3/16" thick, R19 [G]		3800	.006		.28	.20	.09	.57	.73
0050	6-1/2" thick, R22 [G]		3000	.008		.36	.26	.11	.73	.92
0100	8-11/16" thick, R30 [G]		2600	.009		.48	.30	.13	.91	1.13
0120	10-7/8" thick, R38 [G]		1800	.013		.62	.43	.19	1.24	1.55
1000	Fiberglass, 5.5" thick, R11 [G]		3800	.006		.19	.20	.09	.48	.63
1050	6" thick, R12 [G]		3000	.008		.23	.26	.11	.60	.77
1100	8.8" thick, R19 [G]		2200	.011		.33	.35	.15	.83	1.08
1200	10" thick, R22 [G]		1800	.013		.38	.43	.19	1	1.29
1300	11.5" thick, R26 [G]		1500	.016		.46	.52	.22	1.20	1.55
1350	13" thick, R30 [G]		1400	.017		.50	.55	.24	1.29	1.67
1450	16" thick, R38 [G]		1145	.021		.61	.68	.29	1.58	2.04
1500	20" thick, R49 [G]		920	.026		.77	.84	.36	1.97	2.55
2000	Mineral wool, 4" thick, R12 [G]		3500	.007		.23	.22	.10	.55	.69
2050	6" thick, R17 [G]		2500	.010		.26	.31	.13	.70	.92
2100	9" thick, R23 [G]		1750	.014		.34	.44	.19	.97	1.27
2500	Wall installation, incl. drilling & patching from outside, two 1"									
2510	diam. holes @ 16" O.C., top & mid-point of wall, add to above									
2700	For masonry [G]	G-4	415	.058	S.F.	.06	1.87	.80	2.73	3.84
2800	For wood siding [G]		840	.029		.06	.92	.40	1.38	1.94
2900	For stucco/plaster [G]		665	.036		.06	1.16	.50	1.72	2.43

07 21 27 – Reflective Insulation

07 21 27.10 Reflective Insulation Options

		Crew	Daily Output	Labor-Hours	Unit	Material	2009 Bare Costs Labor	Equipment	Total	Total Incl O&P
0010	**REFLECTIVE INSULATION OPTIONS**									
0020	Aluminum foil on reinforced scrim [G]	1 Carp	19	.421	C.S.F.	14.20	16.80		31	41.50
0100	Reinforced with woven polyolefin [G]		19	.421		17.20	16.80		34	45
0500	With single bubble air space, R8.8 [G]		15	.533		25.50	21.50		47	61
0600	With double bubble air space, R9.8 [G]		15	.533		26.50	21.50		48	62

07 21 29 – Sprayed Insulation

07 21 29.10 Sprayed-On Insulation

		Crew	Daily Output	Labor-Hours	Unit	Material	2009 Bare Costs Labor	Equipment	Total	Total Incl O&P
0010	**SPRAYED-ON INSULATION**									
0020	Fibrous/cementitious, finished wall, 1" thick, R3.7 [G]	G-2	2050	.012	S.F.	.26	.39	.06	.71	.95
0100	Attic, 5.2" thick, R19 [G]	"	1550	.015	"	.42	.52	.08	1.02	1.34

07 22 Roof and Deck Insulation

07 22 16 – Roof Board Insulation

07 22 16.10 Roof Deck Insulation

		Crew	Daily Output	Labor-Hours	Unit	Material	2009 Bare Costs Labor	Equipment	Total	Total Incl O&P
0010	**ROOF DECK INSULATION**									
0020	Fiberboard low density, 1/2" thick R1.39 [G]	1 Rofc	1000	.008	S.F.	.24	.27		.51	.72
0030	1" thick R2.78 [G]		800	.010		.42	.34		.76	1.04
0080	1 1/2" thick R4.17 [G]		800	.010		.64	.34		.98	1.28
0100	2" thick R5.56 [G]		800	.010		.85	.34		1.19	1.52
0110	Fiberboard high density, 1/2" thick R1.3 [G]		1000	.008		.22	.27		.49	.70
0120	1" thick R2.5 [G]		800	.010		.44	.34		.78	1.06
0130	1-1/2" thick R3.8 [G]		800	.010		.66	.34		1	1.31
0200	Fiberglass, 3/4" thick R2.78 [G]		1000	.008		.56	.27		.83	1.08
0400	15/16" thick R3.70 [G]		1000	.008		.74	.27		1.01	1.27
0460	1-1/16" thick R4.17 [G]		1000	.008		.93	.27		1.20	1.48

07 22 Roof and Deck Insulation

07 22 16 – Roof Board Insulation

07 22 16.10 Roof Deck Insulation

			Crew	Daily Output	Labor-Hours	Unit	Material	2009 Bare Costs Labor	2009 Bare Costs Equipment	Total	Total Incl O&P
0600	1-5/16" thick R5.26	G	1 Rofc	1000	.008	S.F.	1.26	.27		1.53	1.85
0650	2-1/16" thick R8.33	G		800	.010		1.35	.34		1.69	2.07
0700	2-7/16" thick R10	G		800	.010		1.55	.34		1.89	2.29
1500	Foamglass, 1-1/2" thick R4.5	G		800	.010		1.35	.34		1.69	2.07
1530	3" thick R9	G		700	.011	↓	2.75	.39		3.14	3.69
1600	Tapered for drainage	G		600	.013	B.F.	1.16	.46		1.62	2.05
1650	Perlite, 1/2" thick R1.32	G		1050	.008	S.F.	.34	.26		.60	.81
1655	3/4" thick R2.08	G		800	.010		.37	.34		.71	.99
1660	1" thick R2.78	G		800	.010		.46	.34		.80	1.09
1670	1-1/2" thick R4.17	G		800	.010		.48	.34		.82	1.11
1680	2" thick R5.56	G		700	.011		.80	.39		1.19	1.54
1685	2-1/2" thick R6.67	G		700	.011	↓	.95	.39		1.34	1.71
1690	Tapered for drainage	G		800	.010	B.F.	.73	.34		1.07	1.38
1700	Polyisocyanurate, 2#/CF density, 3/4" thick, R5.1	G		1500	.005	S.F.	.44	.18		.62	.79
1705	1" thick R7.14	G		1400	.006		.50	.20		.70	.88
1715	1-1/2" thick R10.87	G		1250	.006		.63	.22		.85	1.06
1725	2" thick R14.29	G		1100	.007		.81	.25		1.06	1.31
1735	2-1/2" thick R16.67	G		1050	.008		1.01	.26		1.27	1.55
1745	3" thick R21.74	G		1000	.008		1.26	.27		1.53	1.85
1755	3-1/2" thick R25	G		1000	.008	↓	1.93	.27		2.20	2.58
1765	Tapered for drainage	G	↓	1400	.006	B.F.	1.93	.20		2.13	2.45
1900	Extruded Polystyrene										
1910	15 PSI compressive strength, 1" thick, R5	G	1 Rofc	1500	.005	S.F.	.49	.18		.67	.85
1920	2" thick, R10	G		1250	.006		.62	.22		.84	1.05
1940	25 PSI compressive strength, 1" thick R5	G		1500	.005		.67	.18		.85	1.05
1942	2" thick R10	G		1250	.006		1.27	.22		1.49	1.77
1944	3" thick R15	G		1000	.008		1.93	.27		2.20	2.58
1946	4" thick R20	G		1000	.008		2.72	.27		2.99	3.45
1950	40 psi compressive strength, 1" thick R5	G		1500	.005		.51	.18		.69	.87
1952	2" thick R10	G		1250	.006		.96	.22		1.18	1.43
1954	3" thick R15	G		1000	.008		1.40	.27		1.67	2
1956	4" thick R20	G		1000	.008		1.88	.27		2.15	2.53
1960	60 PSI compressive strength, 1" thick R5	G		1450	.006		.59	.19		.78	.97
1962	2" thick R10	G		1200	.007		1.05	.23		1.28	1.54
1964	3" thick R15	G		975	.008		1.55	.28		1.83	2.18
1966	4" thick R20	G		950	.008	↓	2.17	.29		2.46	2.88
1968	Tapered for drainage	G		1400	.006	B.F.	.85	.20		1.05	1.27
2010	Expanded polystyrene, 1#/CF density, 3/4" thick R2.89	G		1500	.005	S.F.	.31	.18		.49	.65
2020	1" thick R3.85	G		1500	.005		.31	.18		.49	.65
2100	2" thick R7.69	G		1250	.006		.62	.22		.84	1.05
2110	3" thick R11.49	G		1250	.006		.91	.22		1.13	1.37
2120	4" thick R15.38	G		1200	.007		.98	.23		1.21	1.46
2130	5" thick R19.23	G		1150	.007		1.05	.24		1.29	1.56
2140	6" thick R23.26	G		1150	.007	↓	1.23	.24		1.47	1.75
2150	Tapered for drainage	G		1500	.005	B.F.	.49	.18		.67	.85
2400	Composites with 2" EPS										
2410	1" fiberboard	G	1 Rofc	950	.008	S.F.	1.07	.29		1.36	1.67
2420	7/16" oriented strand board	G		800	.010		1.26	.34		1.60	1.97
2430	1/2" plywood	G		800	.010		1.37	.34		1.71	2.09
2440	1" perlite	G	↓	800	.010	↓	1.12	.34		1.46	1.81
2450	Composites with 1-1/2" polyisocyanurate										
2460	1" fiberboard	G	1 Rofc	800	.010	S.F.	1.46	.34		1.80	2.19
2470	1" perlite	G	↓	850	.009	↓	1.53	.32		1.85	2.22

07 22 Roof and Deck Insulation

07 22 16 – Roof Board Insulation

07 22 16.10 Roof Deck Insulation

		Crew	Daily Output	Labor-Hours	Unit	Material	2009 Bare Costs Labor	Equipment	Total	Total Incl O&P
2480	7/16" oriented strand board G	1 Rofc	800	.010	S.F.	1.77	.34		2.11	2.53

07 26 Vapor Retarders

07 26 10 – Vapor Retarders Made of Natural and Synthetic Materials

07 26 10.10 Vapor Retarders

		Crew	Daily Output	Labor-Hours	Unit	Material	Labor	Equipment	Total	Total Incl O&P
0010	**VAPOR RETARDERS**									
0020	Aluminum and kraft laminated, foil 1 side G	1 Carp	37	.216	Sq.	5.20	8.65		13.85	19.15
0100	Foil 2 sides G		37	.216		8.70	8.65		17.35	23
0400	Asphalt felt sheathing paper, 15#		37	.216		4.77	8.65		13.42	18.65
0450	Housewrap, exterior, spun bonded polypropylene									
0470	Small roll G	1 Carp	3800	.002	S.F.	.23	.08		.31	.38
0480	Large roll G	"	4000	.002	"	.12	.08		.20	.25
0500	Material only, 3' x 111.1' roll G				Ea.	75			75	82.50
0520	9' x 111.1' roll G				"	120			120	132
0600	Polyethylene vapor barrier, standard, .002" thick G	1 Carp	37	.216	Sq.	1.09	8.65		9.74	14.60
0700	.004" thick G		37	.216		3.54	8.65		12.19	17.30
0900	.006" thick G		37	.216		5.55	8.65		14.20	19.50
1200	.010" thick G		37	.216		6.70	8.65		15.35	21
1300	Clear reinforced, fire retardant, .008" thick G		37	.216		10.15	8.65		18.80	24.50
1350	Cross laminated type, .003" thick G		37	.216		7.10	8.65		15.75	21
1400	.004" thick G		37	.216		7.80	8.65		16.45	22
1500	Red rosin paper, 5 sq rolls, 4 lbs. per square		37	.216		2.19	8.65		10.84	15.80
1600	5 lbs. per square		37	.216		2.85	8.65		11.50	16.55
1800	Reinf. waterproof, .002" polyethylene backing, 1 side		37	.216		5.65	8.65		14.30	19.60
1900	2 sides		37	.216		7.45	8.65		16.10	21.50
2100	Asphalt felt roof deck vapor barrier, class 1 metal decks	1 Rofc	37	.216		18.65	7.40		26.05	33
2200	For all other decks	"	37	.216		13.85	7.40		21.25	28
2400	Waterproofed kraft with sisal or fiberglass fibers, minimum	1 Carp	37	.216		6.10	8.65		14.75	20
2500	Maximum	"	37	.216		15.10	8.65		23.75	30

07 31 Shingles and Shakes

07 31 13 – Asphalt Shingles

07 31 13.10 Asphalt Roof Shingles

		Crew	Daily Output	Labor-Hours	Unit	Material	Labor	Equipment	Total	Total Incl O&P
0010	**ASPHALT ROOF SHINGLES**									
0100	Standard strip shingles									
0150	Inorganic, class A, 210-235 lb/sq	1 Rofc	5.50	1.455	Sq.	50	50		100	139
0155	Pneumatic nailed		7	1.143		50	39		89	121
0200	Organic, class C, 235-240 lb/sq		5	1.600		51	55		106	149
0205	Pneumatic nailed		6.25	1.280		51	44		95	130

07 31 29 – Wood Shingles and Shakes

07 31 29.13 Wood Shingles

		Crew	Daily Output	Labor-Hours	Unit	Material	Labor	Equipment	Total	Total Incl O&P
0010	**WOOD SHINGLES** R061110-30									
0012	16" No. 1 red cedar shingles, 5" exposure, on roof	1 Carp	2.50	3.200	Sq.	330	128		458	565
0015	Pneumatic nailed		3.25	2.462		330	98.50		428.50	515
0200	7-1/2" exposure, on walls		2.05	3.902		221	156		377	485
0205	Pneumatic nailed		2.67	2.996		221	120		341	430
0300	18" No. 1 red cedar perfections, 5-1/2" exposure, on roof		2.75	2.909		330	116		446	540
0305	Pneumatic nailed		3.57	2.241		330	89.50		419.50	500
0500	7-1/2" exposure, on walls		2.25	3.556		242	142		384	485

07 31 Shingles and Shakes

07 31 29 – Wood Shingles and Shakes

07 31 29.13 Wood Shingles

		Crew	Daily Output	Labor-Hours	Unit	Material	2009 Bare Costs Labor	Equipment	Total	Total Incl O&P
0505	Pneumatic nailed	1 Carp	2.92	2.740	Sq.	242	109		351	435
0600	Resquared, and rebutted, 5-1/2" exposure, on roof		3	2.667		365	107		472	570
0605	Pneumatic nailed		3.90	2.051		365	82		447	530
0900	7-1/2" exposure, on walls		2.45	3.265		269	130		399	500
0905	Pneumatic nailed		3.18	2.516		269	101		370	450
1000	Add to above for fire retardant shingles, 16" long					63			63	69.50
1050	18" long					63			63	69.50
1060	Preformed ridge shingles	1 Carp	400	.020	L.F.	3.50	.80		4.30	5.10
2000	White cedar shingles, 16" long, extras, 5" exposure, on roof		2.40	3.333	Sq.	178	133		311	405
2005	Pneumatic nailed		3.12	2.564		178	102		280	355
2100	7-1/2" exposure, on walls		2	4		127	160		287	390
2105	Pneumatic nailed		2.60	3.077		127	123		250	330
2300	For 15# organic felt underlayment on roof, 1 layer, add		64	.125		4.77	4.99		9.76	13
2400	2 layers, add		32	.250		9.55	10		19.55	26
2600	For steep roofs (7/12 pitch or greater), add to above						50%			
2700	Panelized systems, No.1 cedar shingles on 5/16" CDX plywood									
2800	On walls, 8' strips, 7" or 14" exposure	2 Carp	700	.023	S.F.	5.55	.91		6.46	7.50
3500	On roofs, 8' strips, 7" or 14" exposure	1 Carp	3	2.667	Sq.	555	107		662	775
3505	Pneumatic nailed	"	4	2	"	555	80		635	735

07 31 29.16 Wood Shakes

		Crew	Daily Output	Labor-Hours	Unit	Material	2009 Bare Costs Labor	Equipment	Total	Total Incl O&P
0010	**WOOD SHAKES**									
1100	Hand-split red cedar shakes, 1/2" thick x 24" long, 10" exp. on roof	1 Carp	2.50	3.200	Sq.	252	128		380	475
1105	Pneumatic nailed		3.25	2.462		252	98.50		350.50	430
1110	3/4" thick x 24" long, 10" exp. on roof		2.25	3.556		252	142		394	500
1115	Pneumatic nailed		2.92	2.740		252	109		361	450
1200	1/2" thick, 18" long, 8-1/2" exp. on roof		2	4		239	160		399	510
1205	Pneumatic nailed		2.60	3.077		239	123		362	455
1210	3/4" thick x 18" long, 8 1/2" exp. on roof		1.80	4.444		239	178		417	540
1215	Pneumatic nailed		2.34	3.419		239	137		376	475
1255	10" exp. on walls		2	4		231	160		391	500
1260	10" exposure on walls, pneumatic nailed		2.60	3.077		231	123		354	445
1700	Add to above for fire retardant shakes, 24" long					63			63	69.50
1800	18" long					63			63	69.50
1810	Ridge shakes	1 Carp	350	.023	L.F.	3.50	.91		4.41	5.25

07 32 Roof Tiles

07 32 19 – Metal Roof Tiles

07 32 19.10 Metal Roof Tiles

		Crew	Daily Output	Labor-Hours	Unit	Material	2009 Bare Costs Labor	Equipment	Total	Total Incl O&P
0010	**METAL ROOF TILES**									
0020	Accessories included, .032" thick aluminum, mission tile	1 Carp	2.50	3.200	Sq.	815	128		943	1,100
0200	Spanish tiles	"	3	2.667	"	560	107		667	780

07 44 Faced Panels

07 44 73 – Metal Faced Panels

07 44 73.10 Metal Faced Panels and Accessories

		Crew	Daily Output	Labor-Hours	Unit	Material	2009 Bare Costs Labor	Equipment	Total	Total Incl O&P
0010	**METAL FACED PANELS AND ACCESSORIES**									
0400	Textured aluminum, 4' x 8' x 5/16" plywood backing, single face	2 Shee	375	.043	S.F.	3.73	2.01		5.74	7.20
0600	Double face		375	.043		4.44	2.01		6.45	8
0700	4' x 10' x 5/16" plywood backing, single face		375	.043		4.16	2.01		6.17	7.70
0900	Double face		375	.043		5.85	2.01		7.86	9.55
1000	4' x 12' x 5/16" plywood backing, single face		375	.043		5.90	2.01		7.91	9.60
1300	Smooth aluminum, 1/4" plywood panel, fluoropolymer finish, double face		375	.043		5.80	2.01		7.81	9.50
1350	Clear anodized finish, double face		375	.043		9.15	2.01		11.16	13.15
1400	Double face textured aluminum, structural panel, 1" EPS insulation		375	.043		5.05	2.01		7.06	8.65
1500	Accessories, outside corner	1 Shee	175	.046	L.F.	1.90	2.16		4.06	5.40
1600	Inside corner		175	.046		1.34	2.16		3.50	4.79
1800	Batten mounting clip		200	.040		.49	1.89		2.38	3.45
1900	Low profile batten		480	.017		.60	.79		1.39	1.87
2100	High profile batten		480	.017		1.41	.79		2.20	2.76
2200	Water table		200	.040		2.11	1.89		4	5.25
2400	Horizontal joint connector		200	.040		1.67	1.89		3.56	4.75
2500	Corner cap		200	.040		1.84	1.89		3.73	4.93
2700	H - moulding		480	.017		1.25	.79		2.04	2.59

07 46 Siding

07 46 23 – Wood Siding

07 46 23.10 Wood Board Siding

		Crew	Daily Output	Labor-Hours	Unit	Material	2009 Bare Costs Labor	Equipment	Total	Total Incl O&P
0010	**WOOD BOARD SIDING**									
3200	Wood, cedar bevel, A grade, 1/2" x 6"	1 Carp	295	.027	S.F.	4.62	1.08		5.70	6.80
3300	1/2" x 8"		330	.024		4.97	.97		5.94	6.95
3500	3/4" x 10", clear grade		375	.021		6.70	.85		7.55	8.65
3600	"B" grade		375	.021		6.25	.85		7.10	8.20
3800	Cedar, rough sawn, 1" x 4", A grade, natural		220	.036		3.77	1.45		5.22	6.40
3900	Stained		220	.036		4.17	1.45		5.62	6.85
4100	1" x 12", board & batten, #3 & Btr., natural		420	.019		3.94	.76		4.70	5.50
4200	Stained		420	.019		4.20	.76		4.96	5.80
4400	1" x 8" channel siding, #3 & Btr., natural		330	.024		2.34	.97		3.31	4.07
4500	Stained		330	.024		2.50	.97		3.47	4.25
4700	Redwood, clear, beveled, vertical grain, 1/2" x 4"		220	.036		3.10	1.45		4.55	5.65
4750	1/2" x 6"		295	.027		3.10	1.08		4.18	5.10
4800	1/2" x 8"		330	.024		3.24	.97		4.21	5.05
5000	3/4" x 10"		375	.021		3.32	.85		4.17	4.97
5200	Channel siding, 1" x 10", B grade		375	.021		3.32	.85		4.17	4.97
5250	Redwood, T&G boards, B grade, 1" x 4"		220	.036		3.10	1.45		4.55	5.65
5270	1" x 8"		330	.024		3.24	.97		4.21	5.05
5400	White pine, rough sawn, 1" x 8", natural		330	.024		2.20	.97		3.17	3.92
5500	Stained		330	.024		2.60	.97		3.57	4.36

07 46 29 – Plywood Siding

07 46 29.10 Plywood Siding Options

		Crew	Daily Output	Labor-Hours	Unit	Material	2009 Bare Costs Labor	Equipment	Total	Total Incl O&P
0010	**PLYWOOD SIDING OPTIONS**									
0900	Plywood, medium density overlaid, 3/8" thick	2 Carp	750	.021	S.F.	.96	.85		1.81	2.38
1000	1/2" thick		700	.023		1.27	.91		2.18	2.82
1100	3/4" thick		650	.025		1.31	.98		2.29	2.97
1600	Texture 1-11, cedar, 5/8" thick, natural		675	.024		2.45	.95		3.40	4.17
1700	Factory stained		675	.024		1.96	.95		2.91	3.63

07 46 Siding

07 46 29 – Plywood Siding

07 46 29.10 Plywood Siding Options

		Crew	Daily Output	Labor-Hours	Unit	Material	2009 Bare Costs Labor	Equipment	Total	Total Incl O&P
1900	Texture 1-11, fir, 5/8" thick, natural	2 Carp	675	.024	S.F.	1.13	.95		2.08	2.71
2000	Factory stained		675	.024		1.73	.95		2.68	3.37
2050	Texture 1-11, S.Y.P., 5/8" thick, natural		675	.024		1.21	.95		2.16	2.80
2100	Factory stained		675	.024		1.25	.95		2.20	2.85
2200	Rough sawn cedar, 3/8" thick, natural		675	.024		1.21	.95		2.16	2.80
2300	Factory stained		675	.024		.97	.95		1.92	2.54
2500	Rough sawn fir, 3/8" thick, natural		675	.024		.77	.95		1.72	2.32
2600	Factory stained		675	.024		.97	.95		1.92	2.54
2800	Redwood, textured siding, 5/8" thick		675	.024		1.94	.95		2.89	3.60
3000	Polyvinyl chloride coated, 3/8" thick		750	.021		1.12	.85		1.97	2.55

07 61 Sheet Metal Roofing

07 61 13 – Standing Seam Sheet Metal Roofing

07 61 13.10 Standing Seam Sheet Metal Roofing, Field Fab.

		Crew	Daily Output	Labor-Hours	Unit	Material	2009 Bare Costs Labor	Equipment	Total	Total Incl O&P
0010	**STANDING SEAM SHEET METAL ROOFING, FIELD FABRICATED**									
0400	Copper standing seam roofing, over 10 squares, 16 oz, 125 lb per sq	1 Shee	1.30	6.154	Sq.	1,075	290		1,365	1,625
0600	18 oz, 140 lb per sq		1.20	6.667		1,200	315		1,515	1,775
0700	20 oz, 150 lb per sq		1.10	7.273		1,300	345		1,645	1,975
1200	For abnormal conditions or small areas, add					25%	100%			
1300	For lead-coated copper, add					25%				

07 61 16 – Batten Seam Sheet Metal Roofing

07 61 16.10 Batten Seam Sheet Metal Roofing, Field Fabricated

		Crew	Daily Output	Labor-Hours	Unit	Material	2009 Bare Costs Labor	Equipment	Total	Total Incl O&P
0010	**BATTEN SEAM SHEET METAL ROOFING, FIELD FABRICATED**									
0012	Copper batten seam roofing, over 10 sq, 16 oz, 130 lb per sq	1 Shee	1.10	7.273	Sq.	1,075	345		1,420	1,700
0200	Copper roofing, batten seam, over 10 sq, 18 oz, 145 lb per sq		1	8		1,200	380		1,580	1,875
0300	20 oz, 160 lb per sq		1	8		1,300	380		1,680	2,025

07 61 19 – Flat Seam Sheet Metal Roofing

07 61 19.10 Flat Seam Sheet Metal Roofing, Field Fabricated

		Crew	Daily Output	Labor-Hours	Unit	Material	2009 Bare Costs Labor	Equipment	Total	Total Incl O&P
0010	**FLAT SEAM SHEET METAL ROOFING, FIELD FABRICATED**									
0900	Copper flat seam roofing, over 10 squares, 16 oz, 115 lb per sq	1 Shee	1.20	6.667	Sq.	1,075	315		1,390	1,650
1000	20 oz, 145 lb per sq	"	1.10	7.273	"	1,300	345		1,645	1,975

07 65 Flexible Flashing

07 65 10 – Sheet Metal Flashing

07 65 10.10 Sheet Metal Flashing and Counter Flashing

		Crew	Daily Output	Labor-Hours	Unit	Material	2009 Bare Costs Labor	Equipment	Total	Total Incl O&P
0010	**SHEET METAL FLASHING AND COUNTER FLASHING**									
0011	Including up to 4 bends									
0020	Aluminum, mill finish, .013" thick	1 Rofc	145	.055	S.F.	.67	1.89		2.56	3.92
0030	.016" thick		145	.055		.79	1.89		2.68	4.05
1600	Copper, 16 oz, sheets, under 1000 lbs.		115	.070		6.80	2.38		9.18	11.45
1900	20 oz sheets, under 1000 lbs.		110	.073		7.90	2.49		10.39	12.90
2200	24 oz sheets, under 1000 lbs.		105	.076		9.60	2.61		12.21	15
2500	32 oz sheets, under 1000 lbs.		100	.080		12.80	2.74		15.54	18.70
5800	Lead, 2.5 lb. per SF, up to 12" wide		135	.059		4.14	2.03		6.17	7.95
5900	Over 12" wide		135	.059		4.14	2.03		6.17	7.95
8900	Stainless steel sheets, 32 ga, .010" thick		155	.052		3.79	1.77		5.56	7.15
9000	28 ga, .015" thick		155	.052		4.70	1.77		6.47	8.15
9100	26 ga, .018" thick		155	.052		5.70	1.77		7.47	9.25

07 65 Flexible Flashing

07 65 10 – Sheet Metal Flashing

07 65 10.10 Sheet Metal Flashing and Counter Flashing	Crew	Daily Output	Labor-Hours	Unit	Material	2009 Bare Costs Labor	Equipment	Total	Total Incl O&P
9200	24 ga, .025" thick	1 Rofc	155	.052	S.F.	7.40	1.77		9.17	11.15

07 65 12 – Fabric and Mastic Flashings

07 65 12.10 Fabric and Mastic Flashing and Counter Flashing

		Crew	Daily Output	Labor-Hours	Unit	Material	Labor	Equipment	Total	Incl O&P
0010	**FABRIC AND MASTIC FLASHING AND COUNTER FLASHING**									
4900	Fabric, asphalt-saturated cotton, specification grade	1 Rofc	35	.229	S.Y.	2.37	7.85		10.22	15.80
5000	Utility grade		35	.229		1.50	7.85		9.35	14.85
5200	Open-mesh fabric, saturated, 40 oz per S.Y.		35	.229		1.67	7.85		9.52	15.05
5300	Close-mesh fabric, saturated, 17 oz per S.Y.		35	.229		1.75	7.85		9.60	15.15
5500	Fiberglass, resin-coated		35	.229		1.43	7.85		9.28	14.75

07 65 13 – Laminated Sheet Flashing

07 65 13.10 Laminated Sheet Flashing

		Crew	Daily Output	Labor-Hours	Unit	Material	Labor	Equipment	Total	Incl O&P
0010	**LAMINATED SHEET FLASHING**, Including up to 4 bends									
2800	Copper, paperbacked 1 side, 2 oz	1 Rofc	330	.024	S.F.	1.42	.83		2.25	2.96
2900	3 oz		330	.024		1.85	.83		2.68	3.44
3100	Paperbacked 2 sides, 2 oz		330	.024		1.43	.83		2.26	2.97
3150	3 oz		330	.024		1.84	.83		2.67	3.42
3400	Mastic-backed 2 sides, copper, 2 oz		330	.024		1.69	.83		2.52	3.26
3500	3 oz		330	.024		2.08	.83		2.91	3.69
3800	Fabric-backed 2 sides, copper, 2 oz		330	.024		1.79	.83		2.62	3.37
4000	3 oz		330	.024		2.31	.83		3.14	3.94
4300	Copper-clad stainless steel, .015" thick, under 500 lbs.		115	.070		5.25	2.38		7.63	9.75
6100	Lead-coated copper, fabric-backed, 2 oz		330	.024		2.42	.83		3.25	4.06

07 72 Roof Accessories

07 72 33 – Roof Hatches

07 72 33.10 Roof Hatch Options

		Crew	Daily Output	Labor-Hours	Unit	Material	Labor	Equipment	Total	Incl O&P
0010	**ROOF HATCH OPTIONS**									
0500	2'-6" x 3', aluminum curb and cover	G-3	10	3.200	Ea.	985	126		1,111	1,275
0520	Galvanized steel curb and aluminum cover		10	3.200		595	126		721	850
0540	Galvanized steel curb and cover		10	3.200		500	126		626	745
0600	2'-6" x 4'-6", aluminum curb and cover		9	3.556		850	140		990	1,150
0800	Galvanized steel curb and aluminum cover		9	3.556		600	140		740	875
0900	Galvanized steel curb and cover		9	3.556		825	140		965	1,125
1100	4' x 4' aluminum curb and cover		8	4		1,675	158		1,833	2,075
1120	Galvanized steel curb and aluminum cover		8	4		1,675	158		1,833	2,075
1140	Galvanized steel curb and cover		8	4		1,600	158		1,758	2,025
1200	2'-6" x 8'-0", aluminum curb and cover		6.60	4.848		1,675	191		1,866	2,125
1400	Galvanized steel curb and aluminum cover		6.60	4.848		1,600	191		1,791	2,050
1500	Galvanized steel curb and cover		6.60	4.848		1,200	191		1,391	1,625
1800	For plexiglass panels, 2'-6" x 3'-0", add to above					400			400	440

07 72 36 – Smoke Vents

07 72 36.10 Smoke Hatches

		Crew	Daily Output	Labor-Hours	Unit	Material	Labor	Equipment	Total	Incl O&P
0010	**SMOKE HATCHES**									
0200	For 3'-0" long, add to roof hatches from Division 07 72 33.10				Ea.	25%	5%			
0250	For 4'-0" long, add to roof hatches from Division 07 72 33.10					20%	5%			
0300	For 8'-0" long, add to roof hatches from Division 07 72 33.10					10%	5%			

07 72 36.20 Smoke Vent Options

		Crew	Daily Output	Labor-Hours	Unit	Material	Labor	Equipment	Total	Incl O&P
0010	**SMOKE VENT OPTIONS**									
0100	4' x 4' aluminum cover and frame	G-3	13	2.462	Ea.	1,900	97		1,997	2,250

07 72 Roof Accessories

07 72 36 – Smoke Vents

07 72 36.20 Smoke Vent Options

		Crew	Daily Output	Labor-Hours	Unit	Material	2009 Bare Costs Labor	Equipment	Total	Total Incl O&P
0200	Galvanized steel cover and frame	G-3	13	2.462	Ea.	1,675	97		1,772	1,975
0300	4' x 8' aluminum cover and frame		8	4		2,600	158		2,758	3,100
0400	Galvanized steel cover and frame	↓	8	4	↓	2,200	158		2,358	2,675

07 72 53 – Snow Guards

07 72 53.10 Snow Guard Options

		Crew	Daily Output	Labor-Hours	Unit	Material	2009 Bare Costs Labor	Equipment	Total	Total Incl O&P
0010	**SNOW GUARD OPTIONS**									
0100	Slate & asphalt shingle roofs	1 Rofc	160	.050	Ea.	9.40	1.71		11.11	13.25
0200	Standing seam metal roofs		48	.167		14.60	5.70		20.30	25.50
0300	Surface mount for metal roofs		48	.167	↓	7.70	5.70		13.40	18.05
0400	Double rail pipe type, including pipe	↓	130	.062	L.F.	22.50	2.11		24.61	28.50

07 81 Applied Fireproofing

07 81 16 – Cementitious Fireproofing

07 81 16.10 Sprayed Cementitious Fireproofing

		Crew	Daily Output	Labor-Hours	Unit	Material	2009 Bare Costs Labor	Equipment	Total	Total Incl O&P
0010	**SPRAYED CEMENTITIOUS FIREPROOFING**									
0050	Not incl tamping or canvas protection									
0100	1" thick, on flat plate steel	G-2	3000	.008	S.F.	.53	.27	.04	.84	1.04
0200	Flat decking		2400	.010		.53	.33	.05	.91	1.15
0400	Beams		1500	.016		.53	.53	.08	1.14	1.48
0500	Corrugated or fluted decks		1250	.019		.79	.64	.10	1.53	1.95
0700	Columns, 1-1/8" thick		1100	.022		.59	.73	.12	1.44	1.89
0800	2-3/16" thick	↓	700	.034	↓	1.13	1.14	.18	2.45	3.18
0850	For tamping, add						10%			
0900	For canvas protection, add	G-2	5000	.005	S.F.	.07	.16	.03	.26	.35
1500	Intumescent epoxy fireproofing on wire mesh, 3/16" thick									
1550	1 hour rating, exterior use	G-2	136	.176	S.F.	6.90	5.90	.93	13.73	17.55
1600	Magnesium oxychloride, 35# to 40# density, 1/4" thick		3000	.008		1.44	.27	.04	1.75	2.04
1650	1/2" thick		2000	.012		2.90	.40	.06	3.36	3.87
1700	60# to 70# density, 1/4" thick		3000	.008		1.92	.27	.04	2.23	2.57
1750	1/2" thick		2000	.012		3.85	.40	.06	4.31	4.92
2000	Vermiculite cement, troweled or sprayed, 1/4" thick		3000	.008		1.31	.27	.04	1.62	1.90
2050	1/2" thick	↓	2000	.012	↓	2.60	.40	.06	3.06	3.54

07 84 Firestopping

07 84 13 – Penetration Firestopping

07 84 13.10 Firestopping

		Crew	Daily Output	Labor-Hours	Unit	Material	2009 Bare Costs Labor	Equipment	Total	Total Incl O&P
0010	**FIRESTOPPING** R078413-30									
0100	Metallic piping, non insulated									
0110	Through walls, 2" diameter	1 Carp	16	.500	Ea.	12.40	20		32.40	44.50
0120	4" diameter		14	.571		18.90	23		41.90	56.50
0130	6" diameter		12	.667		25.50	26.50		52	69.50
0140	12" diameter		10	.800		45	32		77	99
0150	Through floors, 2" diameter		32	.250		7.50	10		17.50	24
0160	4" diameter		28	.286		10.80	11.40		22.20	29.50
0170	6" diameter		24	.333		14.20	13.30		27.50	36
0180	12" diameter	↓	20	.400	↓	24	16		40	51.50
0190	Metallic piping, insulated									
0200	Through walls, 2" diameter	1 Carp	16	.500	Ea.	17.55	20		37.55	50.50
0210	4" diameter		14	.571	↓	24	23		47	62

07 84 Firestopping

07 84 13 – Penetration Firestopping

07 84 13.10 Firestopping		Crew	Daily Output	Labor-Hours	Unit	Material	2009 Bare Costs Labor	Equipment	Total	Total Incl O&P
0220	6" diameter	1 Carp	12	.667	Ea.	30.50	26.50		57	75
0230	12" diameter		10	.800		50	32		82	105
0240	Through floors, 2" diameter		32	.250		12.70	10		22.70	29.50
0250	4" diameter		28	.286		15.95	11.40		27.35	35.50
0260	6" diameter		24	.333		19.35	13.30		32.65	42
0270	12" diameter		20	.400		24	16		40	51.50
0280	Non metallic piping, non insulated									
0290	Through walls, 2" diameter	1 Carp	12	.667	Ea.	51	26.50		77.50	97.50
0300	4" diameter		10	.800		64	32		96	120
0310	6" diameter		8	1		89	40		129	160
0330	Through floors, 2" diameter		16	.500		40	20		60	75
0340	4" diameter		6	1.333		49.50	53.50		103	137
0350	6" diameter		6	1.333		59.50	53.50		113	148
0370	Ductwork, insulated & non insulated, round									
0380	Through walls, 6" diameter	1 Carp	12	.667	Ea.	26	26.50		52.50	70
0390	12" diameter		10	.800		51.50	32		83.50	107
0400	18" diameter		8	1		84	40		124	155
0410	Through floors, 6" diameter		16	.500		14.20	20		34.20	46.50
0420	12" diameter		14	.571		26	23		49	64
0430	18" diameter		12	.667		45	26.50		71.50	91
0440	Ductwork, insulated & non insulated, rectangular									
0450	With stiffener/closure angle, through walls, 6" x 12"	1 Carp	8	1	Ea.	21.50	40		61.50	85.50
0460	12" x 24"		6	1.333		28.50	53.50		82	114
0470	24" x 48"		4	2		81.50	80		161.50	214
0480	With stiffener/closure angle, through floors, 6" x 12"		10	.800		11.65	32		43.65	62.50
0490	12" x 24"		8	1		21	40		61	85
0500	24" x 48"		6	1.333		41	53.50		94.50	128
0510	Multi trade openings									
0520	Through walls, 6" x 12"	1 Carp	2	4	Ea.	45	160		205	298
0530	12" x 24"	"	1	8		182	320		502	695
0540	24" x 48"	2 Carp	1	16		730	640		1,370	1,800
0550	48" x 96"	"	.75	21.333		2,925	850		3,775	4,550
0560	Through floors, 6" x 12"	1 Carp	2	4		45	160		205	298
0570	12" x 24"	"	1	8		182	320		502	695
0580	24" x 48"	2 Carp	.75	21.333		730	850		1,580	2,125
0590	48" x 96"	"	.50	32		2,925	1,275		4,200	5,200
0600	Structural penetrations, through walls									
0610	Steel beams, W8 x 10	1 Carp	8	1	Ea.	28.50	40		68.50	93.50
0620	W12 x 14		6	1.333		45	53.50		98.50	132
0630	W21 x 44		5	1.600		90.50	64		154.50	199
0640	W36 x 135		3	2.667		220	107		327	405
0650	Bar joists, 18" deep		6	1.333		41.50	53.50		95	128
0660	24" deep		6	1.333		51.50	53.50		105	140
0670	36" deep		5	1.600		77.50	64		141.50	185
0680	48" deep		4	2		90.50	80		170.50	224
0690	Construction joints, floor slab at exterior wall									
0700	Precast, brick, block or drywall exterior									
0710	2" wide joint	1 Carp	125	.064	L.F.	6.45	2.56		9.01	11.05
0720	4" wide joint	"	75	.107	"	12.90	4.26		17.16	21
0730	Metal panel, glass or curtain wall exterior									
0740	2" wide joint	1 Carp	40	.200	L.F.	15.30	8		23.30	29
0750	4" wide joint	"	25	.320	"	21	12.80		33.80	43
0760	Floor slab to drywall partition									

07 84 Firestopping

07 84 13 – Penetration Firestopping

07 84 13.10 Firestopping

		Crew	Daily Output	Labor-Hours	Unit	Material	2009 Bare Costs Labor	Equipment	Total	Total Incl O&P
0770	Flat joint	1 Carp	100	.080	L.F.	6.35	3.20		9.55	11.90
0780	Fluted joint		50	.160		12.90	6.40		19.30	24
0790	Etched fluted joint	↓	75	.107	↓	8.40	4.26		12.66	15.85
0800	Floor slab to concrete/masonry partition									
0810	Flat joint	1 Carp	75	.107	L.F.	14.20	4.26		18.46	22.50
0820	Fluted joint	"	50	.160	"	16.80	6.40		23.20	28.50
0830	Concrete/CMU wall joints									
0840	1" wide	1 Carp	100	.080	L.F.	7.75	3.20		10.95	13.50
0850	2" wide	↓	75	.107		14.20	4.26		18.46	22.50
0860	4" wide	↓	50	.160	↓	27	6.40		33.40	40
0870	Concrete/CMU floor joints									
0880	1" wide	1 Carp	200	.040	L.F.	3.88	1.60		5.48	6.75
0890	2" wide		150	.053		7.10	2.13		9.23	11.10
0900	4" wide	↓	100	.080	↓	13.55	3.20		16.75	19.85

07 92 Joint Sealants

07 92 10 – Caulking and Sealants

07 92 10.10 Caulking and Sealant Options

		Crew	Daily Output	Labor-Hours	Unit	Material	2009 Bare Costs Labor	Equipment	Total	Total Incl O&P
0010	**CAULKING AND SEALANT OPTIONS**									
0020	Acoustical sealant, elastomeric, cartridges				Ea.	4.60			4.60	5.05
0030	Backer rod, polyethylene, 1/4" diameter	1 Bric	4.60	1.739	C.L.F.	4.11	70.50		74.61	112
0050	1/2" diameter		4.60	1.739		7.05	70.50		77.55	115
0070	3/4" diameter		4.60	1.739		8.95	70.50		79.45	117
0090	1" diameter	↓	4.60	1.739	↓	15.95	70.50		86.45	125
0100	Acrylic latex caulk, white									
0200	11 fl. oz cartridge				Ea.	2.14			2.14	2.35
0500	1/4" x 1/2"	1 Bric	248	.032	L.F.	.17	1.31		1.48	2.17
1400	Butyl based, bulk				Gal.	26			26	28.50
1500	Cartridges				"	31.50			31.50	34.50
1700	Bulk, in place 1/4" x 1/2", 154 L.F./gal.	1 Bric	230	.035	L.F.	.17	1.41		1.58	2.33
1800	1/2" x 1/2", 77 L.F./gal.	"	180	.044	"	.34	1.80		2.14	3.10
2000	Latex acrylic based, bulk				Gal.	27			27	30
2100	Cartridges					33.50			33.50	36.50
2300	Polysulfide compounds, 1 component, bulk				↓	51			51	56
2600	1 or 2 component, in place, 1/4" x 1/4", 308 L.F./gal.	1 Bric	145	.055	L.F.	.17	2.23		2.40	3.57
2700	1/2" x 1/4", 154 L.F./gal.		135	.059		.33	2.40		2.73	4.01
2900	3/4" x 3/8", 68 L.F./gal.		130	.062		.75	2.49		3.24	4.61
3000	1" x 1/2", 38 L.F./gal.	↓	130	.062	↓	1.35	2.49		3.84	5.25
3200	Polyurethane, 1 or 2 component				Gal.	54			54	59.50
3500	Bulk, in place, 1/4" x 1/4"	1 Bric	150	.053	L.F.	.18	2.16		2.34	3.47
3600	1/2" x 1/4"		145	.055		.35	2.23		2.58	3.78
3800	3/4" x 3/8", 68 L.F./gal.		130	.062		.79	2.49		3.28	4.65
3900	1" x 1/2"	↓	110	.073	↓	1.40	2.95		4.35	6
4100	Silicone rubber, bulk				Gal.	41			41	45
4200	Cartridges				"	41			41	45
4400	Neoprene gaskets, closed cell, adhesive, 1/8" x 3/8"	1 Bric	240	.033	L.F.	.26	1.35		1.61	2.34
4500	1/4" x 3/4"		215	.037		.58	1.51		2.09	2.93
4700	1/2" x 1"		200	.040		1.30	1.62		2.92	3.89
4800	3/4" x 1-1/2"	↓	165	.048	↓	1.44	1.96		3.40	4.56
5500	Resin epoxy coating, 2 component, heavy duty				Gal.	30.50			30.50	33.50
5800	Tapes, sealant, P.V.C. foam adhesive, 1/16" x 1/4"				C.L.F.	5.65			5.65	6.20

07 92 Joint Sealants

07 92 10 - Caulking and Sealants

07 92 10.10 Caulking and Sealant Options	Crew	Daily Output	Labor-Hours	Unit	Material	2009 Bare Costs Labor	Equipment	Total	Total Incl O&P
5900 1/16" x 1/2"				C.L.F.	8.35			8.35	9.20
5950 1/16" x 1"					13.85			13.85	15.25
6000 1/8" x 1/2"				↓	9.35			9.35	10.30
6200 Urethane foam, 2 component, handy pack, 1 C.F.				Ea.	34			34	37
6300 50.0 C.F. pack				C.F.	17.20			17.20	18.95

07 92 13 - Elastomeric Joint Sealants

07 92 13.10 Masonry Joint Sealants

	Crew	Daily Output	Labor-Hours	Unit	Material	Labor	Equipment	Total	Total Incl O&P
0010 **MASONRY JOINT SEALANTS**, 1/2" x 1/2" joint									
0050 Re-caulk only, oil base	1 Bric	225	.036	L.F.	.45	1.44		1.89	2.68
0100 Acrylic latex		205	.039		.22	1.58		1.80	2.64
0200 Polyurethane		200	.040		.43	1.62		2.05	2.94
0300 Silicone		195	.041		.58	1.66		2.24	3.16
1000 Cut out and re-caulk, oil base		145	.055		.45	2.23		2.68	3.88
1050 Acrylic latex		130	.062		.22	2.49		2.71	4.02
1100 Polyurethane		125	.064		.43	2.59		3.02	4.41
1150 Silicone	↓	120	.067	↓	.58	2.70		3.28	4.74

Division Notes

		CREW	DAILY OUTPUT	LABOR-HOURS	UNIT	2009 BARE COSTS				TOTAL INCL O&P
						MAT.	LABOR	EQUIP.	TOTAL	

Estimating Tips

08 10 00 Doors and Frames

All exterior doors should be addressed for their energy conservation.

- Most metal doors and frames look alike, but there may be significant differences among them. When estimating these items, be sure to choose the line item that most closely compares to the specification or door schedule requirements regarding:
  — type of metal
  — metal gauge
  — door core material
  — fire rating
  — finish
- Wood and plastic doors vary considerably in price. The primary determinant is the veneer material. Lauan, birch, and oak are the most common veneers. Other variables include the following:
  — hollow or solid core
  — fire rating
  — flush or raised panel
  — finish

08 30 00 Specialty Doors and Frames

- There are many varieties of special doors, and they are usually priced per each. Add frames, hardware, or operators required for a complete installation.

08 40 00 Entrances, Storefronts, and Curtain Walls

- Glazed curtain walls consist of the metal tube framing and the glazing material. The cost data in this subdivision is presented for the metal tube framing alone or the composite wall. If your estimate requires a detailed takeoff of the framing, be sure to add the glazing cost.

08 50 00 Windows

- Most metal windows are delivered preglazed. However, some metal windows are priced without glass. Refer to 08 80 00 Glazing for glass pricing. The grade C indicates commercial grade windows, usually ASTM C-35.
- All wood windows are priced preglazed. The two glazing options priced are single pane float glass and insulating glass 1/2" thick. Add the cost of screens and grills if required.

08 70 00 Hardware

- Hardware costs add considerably to the cost of a door. The most efficient method to determine the hardware requirements for a project is to review the door schedule.
- Door hinges are priced by the pair, with most doors requiring 1-1/2 pairs per door. The hinge prices do not include installation labor because it is included in door installation. Hinges are classified according to the frequency of use.

08 80 00 Glazing

- Different openings require different types of glass. The three most common types are:
  — float
  — tempered
  — insulating
- Most exterior windows are glazed with insulating glass. Entrance doors and window walls, where the glass is less than 18" from the floor, are generally glazed with tempered glass. Interior windows and some residential windows are glazed with float glass.
- Coastal communities are starting to address the use of impact-resistant glass.
- The insulation or 'u' value is a strong consideration.

Reference Numbers

Reference numbers are shown in shaded boxes at the beginning of some major classifications. These numbers refer to related items in the Reference Section. The reference information may be an estimating procedure, an alternate pricing method, or technical information.

Note: Not all subdivisions listed here necessarily appear in this publication.

08 05 Common Work Results for Openings

08 05 05 – Selective Windows and Doors Demolition

08 05 05.10 Selective Demolition Doors		Crew	Daily Output	Labor-Hours	Unit	Material	2009 Bare Costs Labor	2009 Bare Costs Equipment	Total	Total Incl O&P
0010	**SELECTIVE DEMOLITION DOORS** R024119-10									
0200	Doors, exterior, 1-3/4" thick, single, 3' x 7' high	1 Clab	16	.500	Ea.		15.80		15.80	24.50
0220	Double, 6' x 7' high		12	.667			21		21	32.50
0500	Interior, 1-3/8" thick, single, 3' x 7' high		20	.400			12.65		12.65	19.60
0520	Double, 6' x 7' high		16	.500			15.80		15.80	24.50
0700	Bi-folding, 3' x 6'-8" high		20	.400			12.65		12.65	19.60
0720	6' x 6'-8" high		18	.444			14.05		14.05	22
0900	Bi-passing, 3' x 6'-8" high		16	.500			15.80		15.80	24.50
0940	6' x 6'-8" high		14	.571			18.05		18.05	28
1500	Remove and reset, minimum	1 Carp	8	1			40		40	62
1520	Maximum		6	1.333			53.50		53.50	82.50
2000	Frames, including trim, metal		8	1			40		40	62
2200	Wood	2 Carp	32	.500			20		20	31
3000	Special doors, counter doors		6	2.667			107		107	165
3100	Double acting		10	1.600			64		64	99
3200	Floor door (trap type), or access type		8	2			80		80	124
3300	Glass, sliding, including frames		12	1.333			53.50		53.50	82.50
3400	Overhead, commercial, 12' x 12' high		4	4			160		160	248
3440	up to 20' x 16' high		3	5.333			213		213	330
3445	up to 35' x 30' high		1	16			640		640	990
3500	Residential, 9' x 7' high		8	2			80		80	124
3540	16' x 7' high		7	2.286			91.50		91.50	142
3600	Remove and reset, minimum		4	4			160		160	248
3620	Maximum		2.50	6.400			256		256	395
3700	Roll-up grille		5	3.200			128		128	198
3800	Revolving door		2	8			320		320	495
3900	Storefront swing door		3	5.333			213		213	330
6600	Demo flexible transparent strip entrance	3 Shee	115	.209	SF Surf		9.85		9.85	15.15
7100	Remove double swing pneumatic doors, openers and sensors	2 Skwk	.50	32	Opng.		1,300		1,300	2,025
7110	Remove automatic operators, industrial, sliding doors, to 12' wide	"	.40	40	"		1,625		1,625	2,525

08 05 05.20 Selective Demolition of Windows

		Crew	Daily Output	Labor-Hours	Unit	Material	Labor	Equipment	Total	Total Incl O&P
0010	**SELECTIVE DEMOLITION OF WINDOWS** R024119-10									
0200	Aluminum, including trim, to 12 S.F.	1 Clab	16	.500	Ea.		15.80		15.80	24.50
0240	To 25 S.F.		11	.727			23		23	35.50
0280	To 50 S.F.		5	1.600			50.50		50.50	78.50
0320	Storm windows/screens, to 12 S.F.		27	.296			9.35		9.35	14.50
0360	To 25 S.F.		21	.381			12.05		12.05	18.65
0400	To 50 S.F.		16	.500			15.80		15.80	24.50
0600	Glass, minimum		200	.040	S.F.		1.26		1.26	1.96
0620	Maximum		150	.053	"		1.69		1.69	2.61
1000	Steel, including trim, to 12 S.F.		13	.615	Ea.		19.45		19.45	30
1020	To 25 S.F.		9	.889			28		28	43.50
1040	To 50 S.F.		4	2			63		63	98
2000	Wood, including trim, to 12 S.F.		22	.364			11.50		11.50	17.80
2020	To 25 S.F.		18	.444			14.05		14.05	22
2060	To 50 S.F.		13	.615			19.45		19.45	30
2065	To 180 S.F.		8	1			31.50		31.50	49
4300	Remove bay/bow window	2 Carp	6	2.667			107		107	165
4410	Remove skylight, plstc domes, flush/curb mtd	G-3	395	.081	S.F.		3.19		3.19	4.93
5020	Remove and reset window, minimum	1 Carp	6	1.333	Ea.		53.50		53.50	82.50
5040	Average		4	2			80		80	124
5080	Maximum		2	4			160		160	248

08 11 Metal Doors and Frames

08 11 16 – Aluminum Doors and Frames

08 11 16.10 Entrance Doors and Frames

		Crew	Daily Output	Labor-Hours	Unit	Material	2009 Bare Costs Labor	Equipment	Total	Total Incl O&P
0010	**ENTRANCE DOORS AND FRAMES** Aluminum, narrow stile									
0011	Including standard hardware, clear finish, no glass									
0020	3'-0" x 7'-0" opening	2 Sswk	2	8	Ea.	670	360		1,030	1,375
0030	3'-6" x 7'-0" opening		2	8		685	360		1,045	1,375
0100	3'-0" x 10'-0" opening, 3' high transom		1.80	8.889		1,175	395		1,570	2,000
0200	3'-6" x 10'-0" opening, 3' high transom		1.80	8.889		1,175	395		1,570	2,000
0280	5'-0" x 7'-0" opening		2	8		1,050	360		1,410	1,800
0300	6'-0" x 7'-0" opening		1.30	12.308		1,000	550		1,550	2,075
0400	6'-0" x 10'-0" opening, 3' high transom		1.10	14.545	Pr.	1,250	650		1,900	2,525
0420	7'-0" x 7'-0" opening		1	16	"	1,075	715		1,790	2,450
0520	3'-0" x 7'-0" opening, wide stile		2	8	Ea.	885	360		1,245	1,600
0540	3'-6" x 7'-0" opening		2	8		960	360		1,320	1,675
0560	5'-0" x 7'-0" opening		2	8		1,400	360		1,760	2,150
0580	6'-0" x 7'-0" opening		1.30	12.308	Pr.	1,425	550		1,975	2,550
0600	7'-0" x 7'-0" opening		1	16	"	1,525	715		2,240	2,950
1100	For full vision doors, with 1/2" glass, add				Leaf	55%				
1200	For non-standard size, add				"	67%				
1250	For installation of non-standard size, add						10%			
1300	Light bronze finish, add				Leaf	36%				
1400	Dark bronze finish, add					18%				
1500	For black finish, add					36%				
1600	Concealed panic device, add					1,050			1,050	1,175
1700	Electric striker release, add				Opng.	275			275	305
1800	Floor check, add				Leaf	810			810	890
1900	Concealed closer, add				"	540			540	595
2000	Flush 3' x 7' Insulated, 12"x 12" lite, clear finish	2 Sswk	2	8	Ea.	1,150	360		1,510	1,900

08 11 74 – Sliding Metal Grilles

08 11 74.10 Rolling Grille Supports

		Crew	Daily Output	Labor-Hours	Unit	Material	Labor	Equipment	Total	Total Incl O&P
0010	**ROLLING GRILLE SUPPORTS**									
0020	Rolling grille supports, overhead framed	E-4	36	.889	L.F.	23	40	3.72	66.72	101

08 12 Metal Frames

08 12 13 – Hollow Metal Frames

08 12 13.13 Standard Hollow Metal Frames

		Crew	Daily Output	Labor-Hours	Unit	Material	Labor	Equipment	Total	Total Incl O&P
0010	**STANDARD HOLLOW METAL FRAMES**									
0020	16 ga., up to 5-3/4" jamb depth									
0025	6'-8" high, 3'-0" wide, single	2 Carp	16	1	Ea.	123	40		163	198
0028	3'-6" wide, single		16	1		116	40		156	189
0030	4'-0" wide, single		16	1		116	40		156	189
0040	6'-0" wide, double		14	1.143		153	45.50		198.50	240
0045	8'-0" wide, double		14	1.143		159	45.50		204.50	246
0100	7'-0" high, 3'-0" wide, single		16	1		110	40		150	183
0110	3'-6" wide, single		16	1		119	40		159	193
0112	4'-0" wide, single		16	1		151	40		191	229
0140	6'-0" wide, double		14	1.143		160	45.50		205.50	247
0145	8'-0" wide, double		14	1.143		150	45.50		195.50	236
1000	16 ga., up to 4-7/8" deep, 7'-0" H, 3'-0" W, single		16	1		130	40		170	205
1140	6'-0" wide, double		14	1.143		158	45.50		203.50	244
2800	14 ga., up to 3-7/8" deep, 7'-0" high, 3'-0" wide, single		16	1		134	40		174	209
2840	6'-0" wide, double		14	1.143		161	45.50		206.50	248

08 12 Metal Frames

08 12 13 – Hollow Metal Frames

08 12 13.13 Standard Hollow Metal Frames		Crew	Daily Output	Labor-Hours	Unit	Material	2009 Bare Costs Labor	2009 Bare Costs Equipment	Total	Total Incl O&P
3000	14 ga., up to 5-3/4" deep, 6'-8" high, 3'-0" wide, single	2 Carp	16	1	Ea.	140	40		180	216
3002	3'-6" wide, single		16	1		140	40		180	216
3005	4'-0" wide, single		16	1		140	40		180	216
3600	up to 5-3/4" jamb depth, 7'-0" high, 4'-0" wide, single		15	1.067		111	42.50		153.50	188
3620	6'-0" wide, double		12	1.333		174	53.50		227.50	274
3640	8'-0" wide, double		12	1.333		164	53.50		217.50	263
3700	8'-0" high, 4'-0" wide, single		15	1.067		146	42.50		188.50	226
3740	8'-0" wide, double		12	1.333		181	53.50		234.50	282
4000	6-3/4" deep, 7'-0" high, 4'-0" wide, single		15	1.067		172	42.50		214.50	255
4020	6'-0" wide, double		12	1.333		204	53.50		257.50	305
4040	8'-0", wide double		12	1.333		222	53.50		275.50	325
4100	8'-0" high, 4'-0" wide, single		15	1.067		174	42.50		216.50	257
4140	8'-0" wide, double		12	1.333		202	53.50		255.50	305
4400	8-3/4" deep, 7'-0" high, 4'-0" wide, single		15	1.067		163	42.50		205.50	245
4440	8'-0" wide, double		12	1.333		231	53.50		284.50	340
4500	8'-0" high, 4'-0" wide, single		15	1.067		174	42.50		216.50	257
4540	8'-0" wide, double		12	1.333		203	53.50		256.50	305
4900	For welded frames, add					47.50			47.50	52.50
5400	14 ga., "B" label, up to 5-3/4" deep, 7'-0" high, 4'-0" wide, single	2 Carp	15	1.067		170	42.50		212.50	253
5440	8'-0" wide, double		12	1.333		208	53.50		261.50	310
5800	6-3/4" deep, 7'-0" high, 4'-0" wide, single		15	1.067		150	42.50		192.50	231
5840	8'-0" wide, double		12	1.333		249	53.50		302.50	355
6200	8-3/4" deep, 7'-0" high, 4'-0" wide, single		15	1.067		220	42.50		262.50	310
6240	8'-0" wide, double		12	1.333		245	53.50		298.50	355
6300	For "A" label use same price as "B" label									
6400	For baked enamel finish, add					30%	15%			
6500	For galvanizing, add					15%				
6600	For hospital stop, add				Ea.	282			282	310
7900	Transom lite frames, fixed, add	2 Carp	155	.103	S.F.	45.50	4.12		49.62	56.50
8000	Movable, add	"	130	.123	"	55	4.92		59.92	68

08 12 13.25 Channel Metal Frames

		Crew	Daily Output	Labor-Hours	Unit	Material	Labor	Equipment	Total	Total Incl O&P
0010	**CHANNEL METAL FRAMES**									
0020	Steel channels with anchors and bar stops									
0100	6" channel @ 8.2#/L.F., 3' x 7' door, weighs 150#	E-4	13	2.462	Ea.	270	111	10.30	391.30	505
0200	8" channel @ 11.5#/L.F., 6' x 8' door, weighs 275#		9	3.556		495	161	14.90	670.90	845
0300	8' x 12' door, weighs 400#		6.50	4.923		720	223	20.50	963.50	1,200
0400	10" channel @ 15.3#/L.F., 10' x 10' door, weighs 500#		6	5.333		900	241	22.50	1,163.50	1,450
0500	12' x 12' door, weighs 600#		5.50	5.818		1,075	263	24.50	1,362.50	1,700
0800	For frames without bar stops, light sections, deduct					15%				
0900	Heavy sections, deduct					10%				

08 13 Metal Doors

08 13 13 – Hollow Metal Doors

08 13 13.13 Standard Hollow Metal Doors		Crew	Daily Output	Labor-Hours	Unit	Material	2009 Bare Costs Labor	Equipment	Total	Total Incl O&P
0010	**STANDARD HOLLOW METAL DOORS** R081313-20									
0015	Flush, full panel, hollow core									
0020	1-3/8" thick, 20 ga., 2'-0"x 6'-8"	2 Carp	20	.800	Ea.	267	32		299	345
0040	2'-8" x 6'-8"		18	.889		274	35.50		309.50	355
0060	3'-0" x 6'-8"		17	.941		280	37.50		317.50	365
0100	3'-0" x 7'-0"		17	.941		295	37.50		332.50	385
0120	For vision lite, add					92.50			92.50	102
0140	For narrow lite, add					101			101	111
0320	Half glass, 20 ga., 2'-0" x 6'-8"	2 Carp	20	.800		395	32		427	485
0340	2'-8" x 6'-8"		18	.889		405	35.50		440.50	500
0360	3'-0" x 6'-8"		17	.941		410	37.50		447.50	515
0400	3'-0" x 7'-0"		17	.941		420	37.50		457.50	520
0410	1-3/8" thick, 18 ga., 2'-0"x 6'-8"		20	.800		320	32		352	400
0420	3'-0" x 6'-8"		17	.941		325	37.50		362.50	415
0425	3'-0" x 7'-0"		17	.941		330	37.50		367.50	425
0450	For vision lite, add					92.50			92.50	102
0452	For narrow lite, add					101			101	111
0460	Half glass, 18 ga., 2'-0" x 6'-8"	2 Carp	20	.800		450	32		482	545
0465	2'-8" x 6'-8"		18	.889		465	35.50		500.50	565
0470	3'-0" x 6'-8"		17	.941		455	37.50		492.50	560
0475	3'-0" x 7'-0"		17	.941		460	37.50		497.50	565
0500	Hollow core, 1-3/4" thick, full panel, 20 ga., 2'-8" x 6'-8"		18	.889		305	35.50		340.50	390
0520	3'-0" x 6'-8"		17	.941		280	37.50		317.50	365
0640	3'-0" x 7'-0"		17	.941		325	37.50		362.50	420
0680	4'-0" x 7'-0"		15	1.067		465	42.50		507.50	575
0700	4'-0" x 8'-0"		13	1.231		530	49		579	660
1000	18 ga., 2'-8" x 6'-8"		17	.941		310	37.50		347.50	400
1020	3'-0" x 6'-8"		16	1		299	40		339	390
1120	3'-0" x 7'-0"		17	.941		370	37.50		407.50	465
1180	4'-0" x 7'-0"		14	1.143		450	45.50		495.50	565
1200	4'-0" x 8'-0"		17	.941		535	37.50		572.50	645
1212	For vision lite, add					92.50			92.50	102
1214	For narrow lite, add					101			101	111
1230	Half glass, 20 ga., 2'-8" x 6'-8"	2 Carp	20	.800		430	32		462	525
1240	3'-0" x 6'-8"		18	.889		420	35.50		455.50	520
1260	3'-0" x 7'-0"		18	.889		425	35.50		460.50	520
1320	18 ga., 2'-8" x 6'-8"		18	.889		460	35.50		495.50	560
1340	3'-0" x 6'-8"		17	.941		455	37.50		492.50	560
1360	3'-0" x 7'-0"		17	.941		465	37.50		502.50	570
1380	4'-0" x 7'-0"		15	1.067		580	42.50		622.50	700
1400	4'-0" x 8'-0"		14	1.143		660	45.50		705.50	800
1720	Insulated, 1-3/4" thick, full panel, 18 ga., 3'-0" x 6'-8"		15	1.067		400	42.50		442.50	510
1740	2'-8" x 7'-0"		16	1		420	40		460	520
1760	3'-0" x 7'-0"		15	1.067		415	42.50		457.50	520
1800	4'-0" x 8'-0"		13	1.231		605	49		654	740
1805	For vision lite, add					92.50			92.50	102
1810	For narrow lite, add					101			101	111
1820	Half glass, 18 ga., 3'-0" x 6'-8"	2 Carp	16	1		535	40		575	650
1840	2'-8" x 7'-0"		17	.941		545	37.50		582.50	660
1860	3'-0" x 7'-0"		16	1		580	40		620	700
1900	4'-0" x 8'-0"		14	1.143		655	45.50		700.50	790
2000	For bottom louver, add					162			162	178
2020	For baked enamel finish, add					30%	15%			

08 13 Metal Doors

08 13 13 – Hollow Metal Doors

08 13 13.13 Standard Hollow Metal Doors		Crew	Daily Output	Labor-Hours	Unit	Material	2009 Bare Costs Labor	2009 Bare Costs Equipment	Total	Total Incl O&P
2040	For galvanizing, add					15%				
08 13 13.15 Metal Fire Doors										
0010	**METAL FIRE DOORS** R081313-20									
0015	Steel, flush, "B" label, 90 minute									
0020	Full panel, 20 ga., 2'-0" x 6'-8"	2 Carp	20	.800	Ea.	340	32		372	425
0040	2'-8" x 6'-8"		18	.889		345	35.50		380.50	435
0060	3'-0" x 6'-8"		17	.941		345	37.50		382.50	440
0080	3'-0" x 7'-0"		17	.941		360	37.50		397.50	455
0140	18 ga., 3'-0" x 6'-8"		16	1		395	40		435	495
0160	2'-8" x 7'-0"		17	.941		415	37.50		452.50	515
0180	3'-0" x 7'-0"		16	1		395	40		435	495
0200	4'-0" x 7'-0"		15	1.067		520	42.50		562.50	635
0220	For "A" label, 3 hour, 18 ga., use same price as "B" label									
0240	For vision lite, add				Ea.	112			112	124
0520	Flush, "B" label 90 min., composite, 20 ga., 2'-0" x 6'-8"	2 Carp	18	.889		430	35.50		465.50	530
0540	2'-8" x 6'-8"		17	.941		435	37.50		472.50	540
0560	3'-0" x 6'-8"		16	1		440	40		480	540
0580	3'-0" x 7'-0"		16	1		450	40		490	555
0640	Flush, "A" label 3 hour, composite, 18 ga., 3'-0" x 6'-8"		15	1.067		495	42.50		537.50	610
0660	2'-8" x 7'-0"		16	1		515	40		555	630
0680	3'-0" x 7'-0"		15	1.067		505	42.50		547.50	620
0700	4'-0" x 7'-0"		14	1.143		610	45.50		655.50	745
08 13 13.20 Residential Steel Doors										
0010	**RESIDENTIAL STEEL DOORS**									
0020	Prehung, insulated, exterior									
0030	Embossed, full panel, 2'-8" x 6'-8" [G]	2 Carp	17	.941	Ea.	277	37.50		314.50	365
0040	3'-0" x 6'-8" [G]		15	1.067		249	42.50		291.50	340
0060	3'-0" x 7'-0" [G]		15	1.067		291	42.50		333.50	385
0070	5'-4" x 6'-8", double [G]		8	2		575	80		655	760
0220	Half glass, 2'-8" x 6'-8"		17	.941		278	37.50		315.50	365
0240	3'-0" x 6'-8"		16	1		278	40		318	365
0260	3'-0" x 7'-0"		16	1		335	40		375	430
0270	5'-4" x 6'-8", double		8	2		570	80		650	755
0720	Raised plastic face, full panel, 2'-8" x 6'-8"		16	1		278	40		318	365
0740	3'-0" x 6'-8"		15	1.067		280	42.50		322.50	375
0760	3'-0" x 7'-0"		15	1.067		284	42.50		326.50	375
0780	5'-4" x 6'-8", double		8	2		525	80		605	705
0820	Half glass, 2'-8" x 6'-8"		17	.941		310	37.50		347.50	405
0840	3'-0" x 6'-8"		16	1		315	40		355	405
0860	3'-0" x 7'-0"		16	1		345	40		385	440
0880	5'-4" x 6'-8", double		8	2		670	80		750	860
1320	Flush face, full panel, 2'-8" x 6'-8"		16	1		230	40		270	315
1340	3'-0" x 6'-8"		15	1.067		230	42.50		272.50	320
1360	3'-0" x 7'-0"		15	1.067		290	42.50		332.50	385
1380	5'-4" x 6'-8", double		8	2		470	80		550	640
1420	Half glass, 2'-8" x 6'-8"		17	.941		289	37.50		326.50	380
1440	3'-0" x 6'-8"		16	1		289	40		329	380
1460	3'-0" x 7'-0"		16	1		335	40		375	430
1480	5'-4" x 6'-8", double		8	2		555	80		635	735
1500	Sidelight, full lite, 1'-0" x 6'-8" with grille					228			228	251
1510	1'-0" x 6'-8", low e					248			248	273
1520	1'-0" x 6'-8", half lite					254			254	280

08 13 Metal Doors

08 13 13 – Hollow Metal Doors

08 13 13.20 Residential Steel Doors		Crew	Daily Output	Labor-Hours	Unit	Material	2009 Bare Costs Labor	Equipment	Total	Total Incl O&P
1530	1'-0" x 6'-8", half lite, low e				Ea.	262			262	289

08 13 16 – Aluminum Doors

08 13 16.10 Commercial Aluminum Doors

		Crew	Daily Output	Labor-Hours	Unit	Material	Labor	Equipment	Total	Total Incl O&P
0010	**COMMERCIAL ALUMINUM DOORS**, no glazing									
0020	Incl. hinges, push/pull, deadlock, cyl., threshold									
1000	Narrow stile, no glazing, standard hardware, 3'-0" x 7'-0", single	2 Carp	3	5.333	Ea.	590	213		803	975
1100	Black finish		3	5.333	"	580	213		793	970
1200	Pair of 3'-0" x 7'-0"		1.70	9.412	Pr.	1,175	375		1,550	1,850
1500	3'-6" x 7'-0", single		3	5.333	Ea.	850	213		1,063	1,250
1600	Black finish		3	5.333		730	213		943	1,125
2100	Medium stile, 3'-0" x 7'-0", single		3	5.333		720	213		933	1,125
2200	Pair of 3'-0" x 7'-0"		1.70	9.412	Pr.	1,400	375		1,775	2,125
2300	3'-6" x 7'-0", single		3	5.333	Ea.	930	213		1,143	1,350

08 13 73 – Sliding Metal Doors

08 13 73.10 Steel Sliding Doors

		Crew	Daily Output	Labor-Hours	Unit	Material	Labor	Equipment	Total	Total Incl O&P
0010	**STEEL SLIDING DOORS**									
0020	Up to 50' x 18', electric, standard duty, minimum	L-5	360	.156	S.F.	26	6.95	2.19	35.14	43
0100	Maximum		340	.165		42	7.35	2.32	51.67	61.50
0500	Heavy duty, minimum		297	.189		34	8.40	2.66	45.06	55
0600	Maximum		277	.202		88	9.05	2.85	99.90	116

08 14 Wood Doors

08 14 13 – Carved Wood Doors

08 14 13.10 Types of Wood Doors, Carved

		Crew	Daily Output	Labor-Hours	Unit	Material	Labor	Equipment	Total	Total Incl O&P
0010	**TYPES OF WOOD DOORS, CARVED**									
3000	Solid wood, 1-3/4" thick stile and rail									
3020	Mahogany, 3'-0" x 7'-0", minimum	2 Carp	14	1.143	Ea.	850	45.50		895.50	1,000
3030	Maximum		10	1.600		1,350	64		1,414	1,575
3040	3'-6" x 8'-0", minimum		10	1.600		980	64		1,044	1,175
3050	Maximum		8	2		1,800	80		1,880	2,100
3100	Pine, 3'-0" x 7'-0", minimum		14	1.143		415	45.50		460.50	525
3110	Maximum		10	1.600		705	64		769	875
3120	3'-6" x 8'-0", minimum		10	1.600		750	64		814	925
3130	Maximum		8	2		1,200	80		1,280	1,450
3200	Red oak, 3'-0" x 7'-0", minimum		14	1.143		1,600	45.50		1,645.50	1,850
3210	Maximum		10	1.600		1,925	64		1,989	2,225
3220	3'-6" x 8'-0", minimum		10	1.600		1,775	64		1,839	2,050
3230	Maximum		8	2		3,150	80		3,230	3,600
4000	Hand carved door, mahogany									
4020	3'-0" x 7'-0", minimum	2 Carp	14	1.143	Ea.	1,600	45.50		1,645.50	1,825
4030	Maximum		11	1.455		3,125	58		3,183	3,550
4040	3'-6" x 8'-0", minimum		10	1.600		2,000	64		2,064	2,300
4050	Maximum		8	2		3,075	80		3,155	3,500
4200	Rose wood, 3'-0" x 7'-0", minimum		14	1.143		4,850	45.50		4,895.50	5,425
4210	Maximum		11	1.455		13,300	58		13,358	14,700
4220	3'-6" x 8'-0", minimum		10	1.600		5,475	64		5,539	6,125
4280	For 6'-8" high door, deduct from 7'-0" door					34			34	37.50
4400	For custom finish, add					370			370	405
4600	Side light, mahogany, 7'-0" x 1'-6" wide, minimum	2 Carp	18	.889		895	35.50		930.50	1,050
4610	Maximum		14	1.143		2,625	45.50		2,670.50	2,975

08 14 Wood Doors

08 14 13 – Carved Wood Doors

08 14 13.10 Types of Wood Doors, Carved		Crew	Daily Output	Labor-Hours	Unit	Material	2009 Bare Costs Labor	Equipment	Total	Total Incl O&P
4620	8'-0" x 1'-6" wide, minimum	2 Carp	14	1.143	Ea.	1,675	45.50		1,720.50	1,925
4630	Maximum		10	1.600		1,900	64		1,964	2,200
4640	Side light, oak, 7'-0" x 1'-6" wide, minimum		18	.889		1,025	35.50		1,060.50	1,175
4650	Maximum		14	1.143		1,825	45.50		1,870.50	2,100
4660	8'-0" x 1'-6" wide, minimum		14	1.143		970	45.50		1,015.50	1,150
4670	Maximum	↓	10	1.600	↓	1,825	64		1,889	2,125

08 14 16 – Flush Wood Doors

08 14 16.09 Smooth Wood Doors

		Crew	Daily Output	Labor-Hours	Unit	Material	Labor	Equipment	Total	Total Incl O&P
0010	**SMOOTH WOOD DOORS**									
0015	Flush, int., 1-3/8", 7 ply, hollow core,									
0020	Lauan face, 2'-0" x 6'-8"	2 Carp	17	.941	Ea.	32.50	37.50		70	94
0080	3'-0" x 6'-8"		17	.941		50	37.50		87.50	114
0100	4'-0" x 6'-8"		16	1		81	40		121	151
0120	Birch face, 2'-0" x 6'-8"		17	.941		44.50	37.50		82	108
0140	2'-6" x 6'-8"		17	.941		81	37.50		118.50	148
0180	3'-0" x 6'-8"		17	.941		90	37.50		127.50	158
0200	4'-0" x 6'-8"		16	1		110	40		150	183
0220	Oak face, 2'-0" x 6'-8"		17	.941		84.50	37.50		122	151
0240	2'-6" x 6'-8"		17	.941		90	37.50		127.50	158
0280	3'-0" x 6'-8"		17	.941		96	37.50		133.50	165
0300	4'-0" x 6'-8"		16	1		122	40		162	196
0320	Walnut face, 2'-0" x 6'-8"		17	.941		170	37.50		207.50	246
0340	2'-6" x 6'-8"		17	.941		174	37.50		211.50	250
0380	3'-0" x 6'-8"		17	.941		180	37.50		217.50	257
0400	4'-0" x 6'-8"	↓	16	1		204	40		244	287
0430	For 7'-0" high, add					15.20			15.20	16.70
0440	For 8'-0" high, add					28			28	31
0480	For prefinishing, clear, add					34			34	37.50
0500	For prefinishing, stain, add					45			45	49.50
1320	M.D. overlay on hardboard, 2'-0" x 6'-8"	2 Carp	17	.941		94	37.50		131.50	162
1340	2'-6" x 6'-8"		17	.941		94	37.50		131.50	162
1380	3'-0" x 6'-8"		17	.941		112	37.50		149.50	182
1400	4'-0" x 6'-8"	↓	16	1		154	40		194	231
1420	For 7'-0" high, add					8.50			8.50	9.35
1440	For 8'-0" high, add					23			23	25.50
1720	H.P. plastic laminate, 2'-0" x 6'-8"	2 Carp	16	1		224	40		264	310
1740	2'-6" x 6'-8"		16	1		224	40		264	310
1780	3'-0" x 6'-8"		15	1.067		268	42.50		310.50	360
1800	4'-0" x 6'-8"	↓	14	1.143		355	45.50		400.50	465
1820	For 7'-0" high, add					9			9	9.90
1840	For 8'-0" high, add					23			23	25.50
2020	5 ply particle core, lauan face, 2'-6" x 6'-8"	2 Carp	15	1.067		82	42.50		124.50	157
2040	3'-0" x 6'-8"		14	1.143		85.50	45.50		131	165
2080	3'-0" x 7'-0"		13	1.231		92.50	49		141.50	179
2100	4'-0" x 7'-0"		12	1.333		110	53.50		163.50	205
2120	Birch face, 2'-6" x 6'-8"		15	1.067		93.50	42.50		136	169
2140	3'-0" x 6'-8"		14	1.143		102	45.50		147.50	184
2180	3'-0" x 7'-0"		13	1.231		105	49		154	192
2200	4'-0" x 7'-0"		12	1.333		128	53.50		181.50	223
2220	Oak face, 2'-6" x 6'-8"		15	1.067		103	42.50		145.50	180
2240	3'-0" x 6'-8"		14	1.143		114	45.50		159.50	196
2280	3'-0" x 7'-0"	↓	13	1.231		117	49		166	205

08 14 Wood Doors

08 14 16 – Flush Wood Doors

08 14 16.09 Smooth Wood Doors		Crew	Daily Output	Labor-Hours	Unit	Material	2009 Bare Costs Labor	Equipment	Total	Total Incl O&P
2300	4'-0" x 7'-0"	2 Carp	12	1.333	Ea.	143	53.50		196.50	240
2320	Walnut face, 2'-0" x 6'-8"		15	1.067		114	42.50		156.50	192
2340	2'-6" x 6'-8"		14	1.143		131	45.50		176.50	215
2380	3'-0" x 6'-8"		13	1.231		147	49		196	239
2400	4'-0" x 6'-8"	↓	12	1.333		192	53.50		245.50	294
2440	For 8'-0" high, add					28			28	31
2460	For 8'-0" high walnut, add					15.35			15.35	16.90
2480	For solid wood core, add					34.50			34.50	38
2720	For prefinishing, clear, add					22.50			22.50	24.50
2740	For prefinishing, stain, add					50			50	55
3320	M.D. overlay on hardboard, 2'-6" x 6'-8"	2 Carp	14	1.143		104	45.50		149.50	185
3340	3'-0" x 6'-8"		13	1.231		109	49		158	196
3380	3'-0" x 7'-0"		12	1.333		111	53.50		164.50	205
3400	4'-0" x 7'-0"	↓	10	1.600		136	64		200	248
3440	For 8'-0" height, add					30.50			30.50	33.50
3460	For solid wood core, add					39.50			39.50	43.50
3720	H.P. plastic laminate, 2'-6" x 6'-8"	2 Carp	13	1.231		152	49		201	244
3740	3'-0" x 6'-8"		12	1.333		172	53.50		225.50	273
3780	3'-0" x 7'-0"		11	1.455		179	58		237	286
3800	4'-0" x 7'-0"	↓	8	2		217	80		297	365
3840	For 8'-0" height, add					30.50			30.50	33.50
3860	For solid wood core, add				↓	37			37	40.50

08 14 16.10 Wood Doors Decorator

		Crew	Daily Output	Labor-Hours	Unit	Material	Labor	Equipment	Total	Total Incl O&P
0010	**WOOD DOORS DECORATOR**									
0040	7 ply hollow core lauan face, 2'-6" x 6'-8"	2 Carp	17	.941	Ea.	36	37.50		73.50	98.50

08 14 16.20 Wood Fire Doors

		Crew	Daily Output	Labor-Hours	Unit	Material	Labor	Equipment	Total	Total Incl O&P
0010	**WOOD FIRE DOORS**									
0020	Particle core, 7 face plys, "B" label,									
0040	1 hour, birch face, 1-3/4" x 2'-6" x 6'-8"	2 Carp	14	1.143	Ea.	320	45.50		365.50	420
0080	3'-0" x 6'-8"		13	1.231		330	49		379	440
0090	3'-0" x 7'-0"		12	1.333		345	53.50		398.50	465
0100	4'-0" x 7'-0"		12	1.333		500	53.50		553.50	635
0140	Oak face, 2'-6" x 6'-8"		14	1.143		310	45.50		355.50	415
0180	3'-0" x 6'-8"		13	1.231		325	49		374	430
0190	3'-0" x 7'-0"		12	1.333		340	53.50		393.50	455
0200	4'-0" x 7'-0"		12	1.333		440	53.50		493.50	570
0240	Walnut face, 2'-6" x 6'-8"		14	1.143		410	45.50		455.50	520
0280	3'-0" x 6'-8"		13	1.231		420	49		469	535
0290	3'-0" x 7'-0"		12	1.333		440	53.50		493.50	565
0300	4'-0" x 7'-0"		12	1.333		590	53.50		643.50	735
0440	M.D. overlay on hardboard, 2'-6" x 6'-8"		15	1.067		274	42.50		316.50	365
0480	3'-0" x 6'-8"		14	1.143		285	45.50		330.50	385
0490	3'-0" x 7'-0"		13	1.231		300	49		349	405
0500	4'-0" x 7'-0"		12	1.333		370	53.50		423.50	490
0740	90 minutes, birch face, 1-3/4" x 2'-6" x 6'-8"		14	1.143		310	45.50		355.50	410
0780	3'-0" x 6'-8"		13	1.231		305	49		354	410
0790	3'-0" x 7'-0"		12	1.333		340	53.50		393.50	460
0800	4'-0" x 7'-0"		12	1.333		415	53.50		468.50	540
0840	Oak face, 2'-6" x 6'-8"		14	1.143		282	45.50		327.50	380
0880	3'-0" x 6'-8"		13	1.231		292	49		341	395
0890	3'-0" x 7'-0"		12	1.333		305	53.50		358.50	420
0900	4'-0" x 7'-0"	↓	12	1.333		425	53.50		478.50	550

08 14 Wood Doors

08 14 16 – Flush Wood Doors

08 14 16.20 Wood Fire Doors

		Crew	Daily Output	Labor-Hours	Unit	Material	2009 Bare Costs Labor	2009 Bare Costs Equipment	Total	Total Incl O&P
0940	Walnut face, 2'-6" x 6'-8"	2 Carp	14	1.143	Ea.	385	45.50		430.50	495
0980	3'-0" x 6'-8"		13	1.231		395	49		444	510
0990	3'-0" x 7'-0"		12	1.333		415	53.50		468.50	540
1000	4'-0" x 7'-0"		12	1.333		595	53.50		648.50	740
1140	M.D. overlay on hardboard, 2'-6" x 6'-8"		15	1.067		305	42.50		347.50	400
1180	3'-0" x 6'-8"		14	1.143		315	45.50		360.50	415
1190	3'-0" x 7'-0"		13	1.231		325	49		374	435
1200	4'-0" x 7'-0"		12	1.333		445	53.50		498.50	575
1240	For 8'-0" height, add					60			60	66
1260	For 8'-0" height walnut, add					75			75	82.50
2200	Custom architectural "B" label, flush, 1-3/4" thick, birch,									
2210	Solid core									
2220	2'-6" x 7'-0"	2 Carp	15	1.067	Ea.	289	42.50		331.50	385
2260	3'-0" x 7'-0"		14	1.143		300	45.50		345.50	400
2300	4'-0" x 7'-0"		13	1.231		395	49		444	510
2420	4'-0" x 8'-0"		11	1.455		375	58		433	505
2480	For oak veneer, add					50%				
2500	For walnut veneer, add					75%				

08 14 23 – Clad Wood Doors

08 14 23.13 Metal-Faced Wood Doors

		Crew	Daily Output	Labor-Hours	Unit	Material	Labor	Equipment	Total	Total Incl O&P
0010	**METAL-FACED WOOD DOORS**									
0020	Interior, flush type, 3' x 7'	2 Carp	4.30	3.721	Opng.	233	149		382	485

08 14 23.20 Tin Clad Wood Doors

		Crew	Daily Output	Labor-Hours	Unit	Material	Labor	Equipment	Total	Total Incl O&P
0010	**TIN CLAD WOOD DOORS**									
0020	3 ply, 6' x 7', double sliding, doors only	2 Carp	1	16	Opng.	1,825	640		2,465	3,025
1000	For electric operator, add	1 Elec	2	4	"	3,250	188		3,438	3,850

08 14 33 – Stile and Rail Wood Doors

08 14 33.10 Wood Doors Paneled

		Crew	Daily Output	Labor-Hours	Unit	Material	Labor	Equipment	Total	Total Incl O&P
0010	**WOOD DOORS PANELED**									
0020	Interior, six panel, hollow core, 1-3/8" thick									
0040	Molded hardboard, 2'-0" x 6'-8"	2 Carp	17	.941	Ea.	55	37.50		92.50	119
0060	2'-6" x 6'-8"		17	.941		59.50	37.50		97	124
0070	2'-8" x 6'-8"		17	.941		62	37.50		99.50	127
0080	3'-0" x 6'-8"		17	.941		65.50	37.50		103	131
0140	Embossed print, molded hardboard, 2'-0" x 6'-8"		17	.941		59.50	37.50		97	124
0160	2'-6" x 6'-8"		17	.941		59.50	37.50		97	124
0180	3'-0" x 6'-8"		17	.941		65.50	37.50		103	131
0540	Six panel, solid, 1-3/8" thick, pine, 2'-0" x 6'-8"		15	1.067		144	42.50		186.50	224
0560	2'-6" x 6'-8"		14	1.143		162	45.50		207.50	249
0580	3'-0" x 6'-8"		13	1.231		186	49		235	282
1020	Two panel, bored rail, solid, 1-3/8" thick, pine, 1'-6" x 6'-8"		16	1		263	40		303	350
1040	2'-0" x 6'-8"		15	1.067		345	42.50		387.50	445
1060	2'-6" x 6'-8"		14	1.143		395	45.50		440.50	505
1340	Two panel, solid, 1-3/8" thick, fir, 2'-0" x 6'-8"		15	1.067		144	42.50		186.50	224
1360	2'-6" x 6'-8"		14	1.143		162	45.50		207.50	249
1380	3'-0" x 6'-8"		13	1.231		395	49		444	510
1740	Five panel, solid, 1-3/8" thick, fir, 2'-0" x 6'-8"		15	1.067		258	42.50		300.50	350
1760	2'-6" x 6'-8"		14	1.143		415	45.50		460.50	525
1780	3'-0" x 6'-8"		13	1.231		415	49		464	530

08 14 Wood Doors

08 14 33 – Stile and Rail Wood Doors

08 14 33.20 Wood Doors Residential		Crew	Daily Output	Labor-Hours	Unit	Material	2009 Bare Costs Labor	Equipment	Total	Total Incl O&P
0010	**WOOD DOORS RESIDENTIAL**									
0200	Exterior, combination storm & screen, pine									
0220	Cross buck, 6'-9" x 2'-6" wide	2 Carp	11	1.455	Ea.	325	58		383	450
0260	2'-8" wide		10	1.600		276	64		340	405
0280	3'-0" wide		9	1.778		282	71		353	420
0300	7'-1" x 3'-0" wide		9	1.778		325	71		396	465
0400	Full lite, 6'-9" x 2'-6" wide		11	1.455		296	58		354	415
0420	2'-8" wide		10	1.600		296	64		360	425
0440	3'-0" wide		9	1.778		300	71		371	440
0500	7'-1" x 3'-0" wide		9	1.778		325	71		396	465
0700	Dutch door, pine, 1-3/4" x 6'-8" x 2'-8" wide, minimum		12	1.333		675	53.50		728.50	830
0720	Maximum		10	1.600		825	64		889	1,000
0800	3'-0" wide, minimum		12	1.333		690	53.50		743.50	840
0820	Maximum		10	1.600		900	64		964	1,100
1000	Entrance door, colonial, 1-3/4" x 6'-8" x 2'-8" wide		16	1		450	40		490	555
1020	6 panel pine, 3'-0" wide		15	1.067		430	42.50		472.50	540
1100	8 panel pine, 2'-8" wide		16	1		570	40		610	685
1120	3'-0" wide		15	1.067		555	42.50		597.50	675
1200	For tempered safety glass lites, (min of 2) add					65.50			65.50	72
1300	Flush, birch, solid core, 1-3/4" x 6'-8" x 2'-8" wide	2 Carp	16	1		102	40		142	175
1320	3'-0" wide		15	1.067		114	42.50		156.50	191
1350	7'-0" x 2'-8" wide		16	1		112	40		152	185
1360	3'-0" wide		15	1.067		119	42.50		161.50	197
1380	For tempered safety glass lites, add					97.50			97.50	107
2700	Interior, closet, bi-fold, w/hardware, no frame or trim incl.									
2720	Flush, birch, 6'-6" or 6'-8" x 2'-6" wide	2 Carp	13	1.231	Ea.	51	49		100	133
2740	3'-0" wide		13	1.231		55	49		104	137
2760	4'-0" wide		12	1.333		97.50	53.50		151	190
2780	5'-0" wide		11	1.455		100	58		158	200
2800	6'-0" wide		10	1.600		109	64		173	219
3000	Raised panel pine, 6'-6" or 6'-8" x 2'-6" wide		13	1.231		162	49		211	256
3020	3'-0" wide		13	1.231		254	49		303	355
3040	4'-0" wide		12	1.333		305	53.50		358.50	420
3060	5'-0" wide		11	1.455		365	58		423	490
3080	6'-0" wide		10	1.600		400	64		464	540
3200	Louvered, pine 6'-6" or 6'-8" x 2'-6" wide		13	1.231		106	49		155	193
3220	3'-0" wide		13	1.231		165	49		214	258
3240	4'-0" wide		12	1.333		198	53.50		251.50	300
3260	5'-0" wide		11	1.455		224	58		282	335
3280	6'-0" wide		10	1.600		248	64		312	370
4400	Bi-passing closet, incl. hardware and frame, no trim incl.									
4420	Flush, lauan, 6'-8" x 4'-0" wide	2 Carp	12	1.333	Opng.	169	53.50		222.50	268
4440	5'-0" wide		11	1.455		184	58		242	292
4460	6'-0" wide		10	1.600		198	64		262	315
4600	Flush, birch, 6'-8" x 4'-0" wide		12	1.333		213	53.50		266.50	315
4620	5'-0" wide		11	1.455		209	58		267	320
4640	6'-0" wide		10	1.600		256	64		320	380
4800	Louvered, pine, 6'-8" x 4'-0" wide		12	1.333		420	53.50		473.50	545
4820	5'-0" wide		11	1.455		390	58		448	515
4840	6'-0" wide		10	1.600		515	64		579	665
5000	Paneled, pine, 6'-8" x 4'-0" wide		12	1.333		495	53.50		548.50	630
5020	5'-0" wide		11	1.455		405	58		463	535

08 14 Wood Doors

08 14 33 – Stile and Rail Wood Doors

08 14 33.20 Wood Doors Residential

		Crew	Daily Output	Labor-Hours	Unit	Material	2009 Bare Costs Labor	Equipment	Total	Total Incl O&P
5040	6'-0" wide	2 Carp	10	1.600	Opng.	590	64		654	750
6100	Folding accordion, closet, including track and frame									
6120	Vinyl, 2 layer, stock see Div. 10 22 26.13	2 Carp	10	1.600	Ea.	51.50	64		115.50	156
6140	Woven mahogany and vinyl, stock		10	1.600		43.50	64		107.50	147
6160	Wood slats with vinyl overlay, stock		10	1.600		139	64		203	252
6180	Economy vinyl, stock		10	1.600		28.50	64		92.50	131
6200	Rigid PVC		10	1.600		32	64		96	134
6220	For custom partition, add					25%	10%			
7310	Passage doors, flush, no frame included									
7320	Hardboard, hollow core, 1-3/8" x 6'-8" x 1'-6" wide	2 Carp	18	.889	Ea.	42.50	35.50		78	102
7330	2'-0" wide		18	.889		42	35.50		77.50	101
7340	2'-6" wide		18	.889		46.50	35.50		82	106
7350	2'-8" wide		18	.889		48	35.50		83.50	108
7360	3'-0" wide		17	.941		51	37.50		88.50	115
7420	Lauan, hollow core, 1-3/8" x 6'-8" x 1'-6" wide		18	.889		30.50	35.50		66	88.50
7440	2'-0" wide		18	.889		28	35.50		63.50	85.50
7460	2'-6" wide		18	.889		31.50	35.50		67	89.50
7480	2'-8" wide		18	.889		33	35.50		68.50	91.50
7500	3'-0" wide		17	.941		34.50	37.50		72	96.50
7700	Birch, hollow core, 1-3/8" x 6'-8" x 1'-6" wide		18	.889		37	35.50		72.50	96
7720	2'-0" wide		18	.889		41	35.50		76.50	100
7740	2'-6" wide		18	.889		49	35.50		84.50	109
7760	2'-8" wide		18	.889		49.50	35.50		85	110
7780	3'-0" wide		17	.941		51	37.50		88.50	115
8000	Pine louvered, 1-3/8" x 6'-8" x 1'-6" wide		19	.842		101	33.50		134.50	163
8020	2'-0" wide		18	.889		128	35.50		163.50	195
8040	2'-6" wide		18	.889		139	35.50		174.50	208
8060	2'-8" wide		18	.889		146	35.50		181.50	216
8080	3'-0" wide		17	.941		157	37.50		194.50	231
8300	Pine paneled, 1-3/8" x 6'-8" x 1'-6" wide		19	.842		111	33.50		144.50	174
8320	2'-0" wide		18	.889		128	35.50		163.50	195
8330	2'-4" wide		18	.889		150	35.50		185.50	220
8340	2'-6" wide		18	.889		158	35.50		193.50	228
8360	2'-8" wide		18	.889		161	35.50		196.50	232
8380	3'-0" wide		17	.941		165	37.50		202.50	241
8550	For over 20 doors, deduct					15%				

08 14 40 – Interior Cafe Doors

08 14 40.10 Cafe Style Doors

		Crew	Daily Output	Labor-Hours	Unit	Material	2009 Bare Costs Labor	Equipment	Total	Total Incl O&P
0010	**CAFE STYLE DOORS**									
6520	Interior cafe doors, 2'-6" opening, stock, panel pine	2 Carp	16	1	Ea.	188	40		228	268
6540	3'-0" opening	"	16	1	"	195	40		235	277
6550	Louvered pine									
6560	2'-6" opening	2 Carp	16	1	Ea.	164	40		204	242
8000	3'-0" opening		16	1		165	40		205	244
8010	2'-6" opening, hardwood		16	1		283	40		323	370
8020	3'-0" opening		16	1		315	40		355	405

08 16 Composite Doors

08 16 13 – Fiberglass Doors

08 16 13.10 Entrance Doors, Fiberous Glass		Crew	Daily Output	Labor-Hours	Unit	Material	2009 Bare Costs Labor	Equipment	Total	Total Incl O&P
0010	**ENTRANCE DOORS, FIBEROUS GLASS**									
0020	Exterior, fiberglass, door, 2'-8" wide x 6'-8" high	2 Carp	15	1.067	Ea.	248	42.50		290.50	340
0040	3'-0" wide x 6'-8" high		15	1.067		248	42.50		290.50	340
0060	3'-0" wide x 7'-0" high		15	1.067		445	42.50		487.50	555
0080	3'-0" wide x 6'-8" high, with two lites		15	1.067		278	42.50		320.50	370
0100	3'-0" wide x 8'-0" high, with two lites		15	1.067		450	42.50		492.50	560
0110	Half glass, 3'-0" wide x 6'-8" high		15	1.067		287	42.50		329.50	380
0120	3'-0" wide x 6'-8" high, low e		15	1.067		315	42.50		357.50	415
0130	3'-0" wide x 8'-0" high		15	1.067		570	42.50		612.50	690
0140	3'-0" wide x 8'-0" high, low e		15	1.067		630	42.50		672.50	755
0150	Side lights, 1'-0" wide x 6'-8" high,					232			232	255
0160	1'-0" wide x 6'-8" high, low e					246			246	271
0180	1'-0" wide x 6'-8" high, full glass					275			275	305
0190	1'-0" wide x 6'-8" high, low e					284			284	315

08 17 Integrated Door Opening Assemblies

08 17 13 – Integrated Metal Door Opening Assemblies

08 17 13.20 Tubular Steel Swing Doors

		Crew	Daily Output	Labor-Hours	Unit	Material	Labor	Equipment	Total	Total Incl O&P
0010	**TUBULAR STEEL SWING DOORS**									
0020	Tubular steel, 7' high, single, 3'-4" opening	2 Sswk	2.50	6.400	Ea.	830	286		1,116	1,425
0100	Double, 6'-0" opening	"	2	8	Pr.	1,050	360		1,410	1,775

08 17 23 – Integrated Wood Door Opening Assemblies

08 17 23.10 Pre-Hung Doors

		Crew	Daily Output	Labor-Hours	Unit	Material	Labor	Equipment	Total	Total Incl O&P
0010	**PRE-HUNG DOORS**									
0300	Exterior, wood, comb. storm & screen, 6'-9" x 2'-6" wide	2 Carp	15	1.067	Ea.	289	42.50		331.50	385
0320	2'-8" wide		15	1.067		289	42.50		331.50	385
0340	3'-0" wide		15	1.067		296	42.50		338.50	390
0360	For 7'-0" high door, add					22.50			22.50	25
1600	Entrance door, flush, birch, solid core									
1620	4-5/8" solid jamb, 1-3/4" x 6'-8" x 2'-8" wide	2 Carp	16	1	Ea.	285	40		325	375
1640	3'-0" wide	"	16	1		350	40		390	445
1680	For 7'-0" high door, add					20.50			20.50	22.50
2000	Entrance door, colonial, 6 panel pine									
2020	4-5/8" solid jamb, 1-3/4" x 6'-8" x 2'-8" wide	2 Carp	16	1	Ea.	530	40		570	640
2040	3'-0" wide	"	16	1		530	40		570	640
2060	For 7'-0" high door, add					52.50			52.50	58
2200	For 5-5/8" solid jamb, add					40.50			40.50	44.50
2270	Pine french door, 8 panels/leaf, 6'-8" x 5'-0" wide, including frame	2 Carp	7	2.286	Pr.	1,250	91.50		1,341.50	1,525
2300	6'-0" wide	"	7	2.286	"	1,250	91.50		1,341.50	1,525
4000	Interior, passage door, 4-5/8" solid jamb									
4400	Lauan, flush, solid core, 1-3/8" x 6'-8" x 2'-6" wide	2 Carp	17	.941	Ea.	182	37.50		219.50	259
4420	2'-8" wide		17	.941		182	37.50		219.50	259
4440	3'-0" wide		16	1		195	40		235	277
4600	Hollow core, 1-3/8" x 6'-8" x 2'-6" wide		17	.941		123	37.50		160.50	194
4620	2'-8" wide		17	.941		123	37.50		160.50	194
4640	3'-0" wide		16	1		124	40		164	198
4700	For 7'-0" high door, add					24			24	26.50
5000	Birch, flush, solid core, 1-3/8" x 6'-8" x 2'-6" wide	2 Carp	17	.941		170	37.50		207.50	246
5020	2'-8" wide		17	.941		195	37.50		232.50	274
5040	3'-0" wide		16	1		205	40		245	288

08 17 Integrated Door Opening Assemblies

08 17 23 – Integrated Wood Door Opening Assemblies

08 17 23.10 Pre-Hung Doors		Crew	Daily Output	Labor-Hours	Unit	Material	2009 Bare Costs Labor	Equipment	Total	Total Incl O&P
5200	Hollow core, 1-3/8" x 6'-8" x 2'-6" wide	2 Carp	17	.941	Ea.	141	37.50		178.50	214
5220	2'-8" wide		17	.941		147	37.50		184.50	220
5240	3'-0" wide		16	1		148	40		188	224
5280	For 7'-0" high door, add					21			21	23
5500	Hardboard paneled, 1-3/8" x 6'-8" x 2'-6" wide	2 Carp	17	.941		142	37.50		179.50	215
5520	2'-8" wide		17	.941		149	37.50		186.50	222
5540	3'-0" wide		16	1		146	40		186	223
6000	Pine paneled, 1-3/8" x 6'-8" x 2'-6" wide		17	.941		247	37.50		284.50	330
6020	2'-8" wide		17	.941		267	37.50		304.50	355
6040	3'-0" wide		16	1		271	40		311	360
6500	For 5-5/8" solid jamb, add					14.60			14.60	16.05
6520	For split jamb, deduct					16.45			16.45	18.10

08 31 Access Doors and Panels

08 31 13 – Access Doors and Frames

08 31 13.10 Types of Framed Access Doors		Crew	Daily Output	Labor-Hours	Unit	Material	2009 Bare Costs Labor	Equipment	Total	Total Incl O&P
0010	**TYPES OF FRAMED ACCESS DOORS**									
1000	Fire rated door with lock									
1100	Metal, 12" x 12"	1 Carp	10	.800	Ea.	157	32		189	223
1150	18" x 18"		9	.889		205	35.50		240.50	281
1200	24" x 24"		9	.889		325	35.50		360.50	415
1250	24" x 36"		8	1		335	40		375	425
1300	24" x 48"		8	1		410	40		450	515
1350	36" x 36"		7.50	1.067		495	42.50		537.50	610
1400	48" x 48"		7.50	1.067		635	42.50		677.50	765
1600	Stainless steel, 12" x 12"		10	.800		281	32		313	360
1650	18" x 18"		9	.889		410	35.50		445.50	505
1700	24" x 24"		9	.889		500	35.50		535.50	605
1750	24" x 36"		8	1		640	40		680	765
2000	Flush door for finishing									
2100	Metal 8" x 8"	1 Carp	10	.800	Ea.	45.50	32		77.50	99.50
2150	12" x 12"	"	10	.800	"	50.50	32		82.50	105
3000	Recessed door for acoustic tile									
3100	Metal, 12" x 12"	1 Carp	4.50	1.778	Ea.	69.50	71		140.50	186
3150	12" x 24"		4.50	1.778		90.50	71		161.50	210
3200	24" x 24"		4	2		121	80		201	257
3250	24" x 36"		4	2		156	80		236	295
4000	Recessed door for drywall									
4100	Metal 12" x 12"	1 Carp	6	1.333	Ea.	77.50	53.50		131	168
4150	12" x 24"		5.50	1.455		115	58		173	216
4200	24" x 36"		5	1.600		183	64		247	300
6000	Standard door									
6100	Metal, 8" x 8"	1 Carp	10	.800	Ea.	40	32		72	93.50
6150	12" x 12"		10	.800		45.50	32		77.50	99.50
6200	18" x 18"		9	.889		62.50	35.50		98	124
6250	24" x 24"		9	.889		81.50	35.50		117	145
6300	24" x 36"		8	1		120	40		160	194
6350	36" x 36"		8	1		147	40		187	223
6500	Stainless steel, 8" x 8"		10	.800		80	32		112	138
6550	12" x 12"		10	.800		107	32		139	167
6600	18" x 18"		9	.889		197	35.50		232.50	272

08 31 Access Doors and Panels

08 31 13 – Access Doors and Frames

08 31 13.10 Types of Framed Access Doors		Crew	Daily Output	Labor-Hours	Unit	Material	2009 Bare Costs Labor	Equipment	Total	Total Incl O&P
6650	24" x 24"	1 Carp	9	.889	Ea.	258	35.50		293.50	340
7010	Aluminum cover, Ceiling hatches, 2'-6" x 2'-6", single leaf, st fr	G-3	11	2.909	↓	540	115		655	770

08 31 13.30 Commercial Floor Doors

0010	**COMMERCIAL FLOOR DOORS**									
0020	Aluminum tile, steel frame, one leaf, 2' x 2' opng.	2 Sswk	3.50	4.571	Opng.	775	204		979	1,225
0050	3'-6" x 3'-6" opening		3.50	4.571		1,100	204		1,304	1,575
0500	Double leaf, 4' x 4' opening		3	5.333		1,575	238		1,813	2,150
0550	5' x 5' opening	↓	3	5.333	↓	2,800	238		3,038	3,500

08 31 13.35 Industrial Floor Doors

0010	**INDUSTRIAL FLOOR DOORS**									
0020	Steel 300 psf L.L., single leaf, 2' x 2', 175#	2 Sswk	6	2.667	Opng.	630	119		749	905
0050	3' x 3' opening, 300#		5.50	2.909		870	130		1,000	1,200
0300	Double leaf, 4' x 4' opening, 455#		5	3.200		1,575	143		1,718	1,975
0350	5' x 5' opening, 645#		4.50	3.556		2,400	159		2,559	2,900
1000	Aluminum, 300 psf L.L., single leaf, 2' x 2', 60#		6	2.667		625	119		744	895
1050	3' x 3' opening, 100#		5.50	2.909		970	130		1,100	1,300
1500	Double leaf, 4' x 4' opening, 160#		5	3.200		1,575	143		1,718	1,975
1550	5' x 5' opening, 235#		4.50	3.556		2,400	159		2,559	2,925
2000	Aluminum, 150 psf L.L., single leaf, 2' x 2', 60#		6	2.667		570	119		689	835
2050	3' x 3' opening, 95#		5.50	2.909		850	130		980	1,175
2500	Double leaf, 4' x 4' opening, 150#		5	3.200		1,350	143		1,493	1,725
2550	5' x 5' opening, 230#	↓	4.50	3.556	↓	1,825	159		1,984	2,300

08 32 Sliding Glass Doors

08 32 19 – Sliding Wood-Framed Glass Doors

08 32 19.15 Sliding Glass Vinyl-Clad Wood Doors

0010	**SLIDING GLASS VINYL-CLAD WOOD DOORS**										
0012	Vinyl clad, 1" insul. glass, 6'-0" x 6'-10" high	G	2 Carp	4	4	Opng.	1,350	160		1,510	1,725
0030	6'-0" x 8'-0" high	G		4	4	Ea.	1,950	160		2,110	2,400
0100	8'-0" x 6'-10" high	G		4	4	Opng.	2,000	160		2,160	2,450
0500	3 leaf, 9'-0" x 6'-10" high	G		3	5.333		2,400	213		2,613	2,950
0600	12'-0" x 6'-10" high	G	↓	3	5.333	↓	3,175	213		3,388	3,825

08 33 Coiling Doors and Grilles

08 33 13 – Coiling Counter Doors

08 33 13.10 Counter Doors, Coiling Type

0010	**COUNTER DOORS, COILING TYPE**									
0020	Manual, incl. frm and hdwe, galv. stl., 4' roll-up, 6' long	2 Carp	2	8	Opng.	1,125	320		1,445	1,750
0300	Galvanized steel, UL label		1.80	8.889		1,200	355		1,555	1,875
0600	Stainless steel, 4' high roll-up, 6' long		2	8		2,050	320		2,370	2,750
0700	10' long	↓	1.80	8.889	↓	2,450	355		2,805	3,250

08 33 16 – Coiling Counter Grilles

08 33 16.10 Coiling Grilles

0010	**COILING GRILLES**									
2020	Aluminum, manual operated, mill finish	2 Sswk	82	.195	S.F.	27.50	8.70		36.20	45.50
2040	Bronze anodized		82	.195	"	43.50	8.70		52.20	63.50
2060	Steel, manual operated, 10' x 10' high		1	16	Opng.	2,450	715		3,165	3,975
2080	15' x 8' high	↓	.80	20	"	2,850	895		3,745	4,725

08 33 Coiling Doors and Grilles

08 33 16 – Coiling Counter Grilles

08 33 16.10 Coiling Grilles		Crew	Daily Output	Labor-Hours	Unit	Material	2009 Bare Costs Labor	Equipment	Total	Total Incl O&P
3000	For safety edge bottom bar, electric, add				L.F.	49			49	54
8000	For motor operation, add	2 Sswk	5	3.200	Opng.	1,250	143		1,393	1,625

08 33 23 – Overhead Coiling Doors

08 33 23.10 Rolling Service Doors

		Crew	Daily Output	Labor-Hours	Unit	Material	Labor	Equipment	Total	Total Incl O&P
0010	**ROLLING SERVICE DOORS** Steel, manual, 20 ga., incl. hardware									
0050	8' x 8' high	2 Sswk	1.60	10	Ea.	1,025	445		1,470	1,925
0100	10' x 10' high		1.40	11.429		1,750	510		2,260	2,825
0200	20' x 10' high		1	16		2,925	715		3,640	4,500
0300	12' x 12' high		1.20	13.333		1,850	595		2,445	3,075
0400	20' x 12' high		.90	17.778		1,975	795		2,770	3,600
0500	14' x 14' high		.80	20		2,825	895		3,720	4,700
0600	20' x 16' high		.60	26.667		3,225	1,200		4,425	5,675
0700	10' x 20' high		.50	32		2,300	1,425		3,725	5,075
1000	12' x 12', crank operated, crank on door side		.80	20		1,575	895		2,470	3,325
1100	Crank thru wall		.70	22.857		1,775	1,025		2,800	3,775
1300	For vision panel, add					305			305	340
1400	For 22 ga., deduct				S.F.	1.10			1.10	1.21
1600	3' x 7' pass door within rolling steel door, new construction				Ea.	1,625			1,625	1,800
1700	Existing construction	2 Sswk	2	8		1,750	360		2,110	2,550
2000	Class A fire doors, manual, 20 ga., 8' x 8' high		1.40	11.429		1,375	510		1,885	2,425
2100	10' x 10' high		1.10	14.545		1,900	650		2,550	3,225
2200	20' x 10' high		.80	20		3,900	895		4,795	5,900
2300	12' x 12' high		1	16		2,975	715		3,690	4,550
2400	20' x 12' high		.80	20		4,225	895		5,120	6,250
2500	14' x 14' high		.60	26.667		3,275	1,200		4,475	5,725
2600	20' x 16' high		.50	32		5,275	1,425		6,700	8,350
2700	10' x 20' high		.40	40		4,075	1,800		5,875	7,675
3000	For 18 ga. doors, add				S.F.	1.27			1.27	1.40
3300	For enamel finish, add				"	1.52			1.52	1.67
3600	For safety edge bottom bar, pneumatic, add				L.F.	19.80			19.80	22
3700	Electric, add					37.50			37.50	41
4000	For weatherstripping, extruded rubber, jambs, add					12.80			12.80	14.10
4100	Hood, add					9			9	9.90
4200	Sill, add					5.15			5.15	5.65
4500	Motor operators, to 14' x 14' opening	2 Sswk	5	3.200	Ea.	1,075	143		1,218	1,425
4600	Over 14' x 14', jack shaft type	"	5	3.200		1,050	143		1,193	1,400
4700	For fire door, additional fusible link, add					23.50			23.50	25.50

08 34 Special Function Doors

08 34 13 – Cold Storage Doors

08 34 13.10 Doors for Cold Area Storage

		Crew	Daily Output	Labor-Hours	Unit	Material	Labor	Equipment	Total	Total Incl O&P
0010	**DOORS FOR COLD AREA STORAGE**									
0020	Single, 20 ga. galvanized steel									
0300	Horizontal sliding, 5' x 7', manual operation, 3.5" thick	2 Carp	2	8	Ea.	3,100	320		3,420	3,900
0400	4" thick		2	8		3,800	320		4,120	4,675
0500	6" thick		2	8		3,100	320		3,420	3,925
0800	5' x 7', power operation, 2" thick		1.90	8.421		5,150	335		5,485	6,200
0900	4" thick		1.90	8.421		5,250	335		5,585	6,300
1000	6" thick		1.90	8.421		5,975	335		6,310	7,100
1300	9' x 10', manual operation, 2" insulation		1.70	9.412		4,175	375		4,550	5,150

08 34 Special Function Doors

08 34 13 – Cold Storage Doors

08 34 13.10 Doors for Cold Area Storage

		Crew	Daily Output	Labor-Hours	Unit	Material	2009 Bare Costs Labor	Equipment	Total	Total Incl O&P
1400	4" insulation	2 Carp	1.70	9.412	Ea.	4,300	375		4,675	5,300
1500	6" insulation		1.70	9.412		5,175	375		5,550	6,275
1800	Power operation, 2" insulation		1.60	10		7,175	400		7,575	8,525
1900	4" insulation		1.60	10		7,350	400		7,750	8,700
2000	6" insulation		1.70	9.412		8,325	375		8,700	9,725
2300	For stainless steel face, add					22%				
3000	Hinged, lightweight, 3' x 7'-0", galvanized 1 face, 2" thick	2 Carp	2	8	Ea.	1,350	320		1,670	1,975
3050	4" thick		1.90	8.421		1,675	335		2,010	2,350
3300	Aluminum doors, 3' x 7'-0", 4" thick		1.90	8.421		1,225	335		1,560	1,875
3350	6" thick		1.40	11.429		2,200	455		2,655	3,100
3600	Stainless steel, 3' x 7'-0", 4" thick		1.90	8.421		1,575	335		1,910	2,250
3650	6" thick		1.40	11.429		2,650	455		3,105	3,625
3900	Painted, 3' x 7'-0", 4" thick		1.90	8.421		1,125	335		1,460	1,775
3950	6" thick		1.40	11.429		2,150	455		2,605	3,075
5000	Bi-parting, electric operated									
5010	6' x 8' opening, galv. faces, 4" thick for cooler	2 Carp	.80	20	Opng.	6,750	800		7,550	8,675
5050	For freezer, 4" thick		.80	20		7,425	800		8,225	9,425
5300	For door buck framing and door protection, add		2.50	6.400		550	256		806	1,000
6000	Galvanized batten door, galvanized hinges, 4' x 7'		2	8		1,650	320		1,970	2,325
6050	6' x 8'		1.80	8.889		2,275	355		2,630	3,050
6500	Fire door, 3 hr., 6' x 8', single slide		.80	20		7,800	800		8,600	9,825
6550	Double, bi-parting		.70	22.857		12,000	915		12,915	14,600

08 34 36 – Darkroom Doors

08 34 36.10 Various Types of Darkroom Doors

		Crew	Daily Output	Labor-Hours	Unit	Material	Labor	Equipment	Total	Total Incl O&P
0010	**VARIOUS TYPES OF DARKROOM DOORS**									
0015	Revolving, standard, 2 way, 36" diameter	2 Carp	3.10	5.161	Opng.	2,075	206		2,281	2,600
0020	41" diameter		3.10	5.161		2,250	206		2,456	2,800
0050	3 way, 51" diameter		1.40	11.429		2,800	455		3,255	3,775
1000	4 way, 49" diameter		1.40	11.429		3,325	455		3,780	4,350
2000	Hinged safety, 2 way, 41" diameter		2.30	6.957		2,625	278		2,903	3,325
2500	3 way, 51" diameter		1.40	11.429		3,325	455		3,780	4,350
3000	Pop out safety, 2 way, 41" diameter		3.10	5.161		3,275	206		3,481	3,925
4000	3 way, 51" diameter		1.40	11.429		3,250	455		3,705	4,275
5000	Wheelchair-type, pop out, 51" diameter		1.40	11.429		3,300	455		3,755	4,350
5020	72" diameter		.90	17.778		6,775	710		7,485	8,550
9300	For complete darkrooms, see Div. 13 21 53.50									

08 34 59 – Vault Doors and Day Gates

08 34 59.10 Secure Storage Doors

		Crew	Daily Output	Labor-Hours	Unit	Material	Labor	Equipment	Total	Total Incl O&P
0010	**SECURE STORAGE DOORS**									
0020	Door and frame, 32" x 78", clear opening									
0100	1 hour test, 32" door, weighs 750 lbs.	2 Sswk	1.50	10.667	Opng.	3,950	475		4,425	5,175
0200	2 hour test, 32" door, weighs 950 lbs.		1.30	12.308		4,200	550		4,750	5,600
0250	40" door, weighs 1130 lbs.		1	16		4,650	715		5,365	6,400
0300	4 hour test, 32" door, weighs 1025 lbs.		1.20	13.333		4,550	595		5,145	6,075
0350	40" door, weighs 1140 lbs.		.90	17.778		5,200	795		5,995	7,125
0500	For stainless steel front, including frame, add to above					1,900			1,900	2,100
0550	Back, add					1,900			1,900	2,100
0600	For time lock, two movement, add	1 Elec	2	4	Ea.	1,450	188		1,638	1,850
0650	Three movement, add	"	2	4		1,900	188		2,088	2,375
0800	Day gate, painted, steel, 32" wide	2 Sswk	1.50	10.667		1,850	475		2,325	2,900
0850	40" wide		1.40	11.429		2,050	510		2,560	3,150
0900	Aluminum, 32" wide		1.50	10.667		2,950	475		3,425	4,100

08 34 Special Function Doors

08 34 59 – Vault Doors and Day Gates

08 34 59.10 Secure Storage Doors		Crew	Daily Output	Labor-Hours	Unit	Material	2009 Bare Costs Labor	Equipment	Total	Total Incl O&P
0950	40" wide	2 Sswk	1.40	11.429	Ea.	3,275	510		3,785	4,500
2050	Security vault door, class I, 3' wide, 3 1/2" thick	E-24	.19	166	Opng.	15,100	7,300	4,100	26,500	33,700
2100	Class II, 3' wide, 7" thick	↓	.19	166	↓	17,500	7,300	4,100	28,900	36,300
2150	Class III, 9R, 3' wide, 10" thick, minimum		.13	250		21,800	11,000	6,175	38,975	49,600

08 34 63 – Detention Doors and Frames

08 34 63.13 Steel Detention Doors and Frames

		Crew	Daily Output	Labor-Hours	Unit	Material	Labor	Equipment	Total	Total Incl O&P
0010	**STEEL DETENTION DOORS AND FRAMES**									
1000	Doors & frames, 3' x 7', complete, with hardware, single plate	E-4	4	8	Ea.	4,200	360	33.50	4,593.50	5,300
1650	Double plate	"	4	8	"	5,175	360	33.50	5,568.50	6,375

08 34 73 – Sound Control Door Assemblies

08 34 73.10 Acoustical Doors

		Crew	Daily Output	Labor-Hours	Unit	Material	Labor	Equipment	Total	Total Incl O&P
0010	**ACOUSTICAL DOORS**									
0020	Including framed seals, 3' x 7', wood, 27 STC rating	2 Carp	1.50	10.667	Ea.	400	425		825	1,100
0100	Steel, 40 STC rating		1.50	10.667		2,800	425		3,225	3,725
0200	45 STC rating		1.50	10.667		3,300	425		3,725	4,275
0300	48 STC rating		1.50	10.667		3,800	425		4,225	4,825
0400	52 STC rating	↓	1.50	10.667	↓	4,000	425		4,425	5,050

08 36 Panel Doors

08 36 13 – Sectional Doors

08 36 13.10 Overhead Commercial Doors

		Crew	Daily Output	Labor-Hours	Unit	Material	Labor	Equipment	Total	Total Incl O&P
0010	**OVERHEAD COMMERCIAL DOORS**									
1000	Stock, sectional, heavy duty, wood, 1-3/4" thick, 8' x 8' high	2 Carp	2	8	Ea.	755	320		1,075	1,325
1100	10' x 10' high		1.80	8.889		1,125	355		1,480	1,800
1200	12' x 12' high		1.50	10.667		1,625	425		2,050	2,450
1300	Chain hoist, 14' x 14' high		1.30	12.308		2,550	490		3,040	3,550
1400	12' x 16' high		1	16		2,500	640		3,140	3,750
1500	20' x 8' high		1.30	12.270		2,250	490		2,740	3,225
1600	20' x 16' high		.65	24.615		5,950	985		6,935	8,075
1800	Center mullion openings, 8' high		4	4		1,025	160		1,185	1,375
1900	20' high	↓	2	8	↓	1,825	320		2,145	2,500
2100	For medium duty custom door, deduct					5%	5%			
2150	For medium duty stock doors, deduct					10%	5%			
2300	Fiberglass and aluminum, heavy duty, sectional, 12' x 12' high	2 Carp	1.50	10.667	Ea.	2,550	425		2,975	3,450
2450	Chain hoist, 20' x 20' high		.50	32		6,450	1,275		7,725	9,075
2600	Steel, 24 ga. sectional, manual, 8' x 8' high		2	8		735	320		1,055	1,300
2650	10' x 10' high		1.80	8.889		940	355		1,295	1,575
2700	12' x 12' high		1.50	10.667		1,175	425		1,600	1,950
2800	Chain hoist, 20' x 14' high	↓	.70	22.857	↓	3,400	915		4,315	5,150
2850	For 1-1/4" rigid insulation and 26 ga. galv.									
2860	back panel, add				S.F.	3.58			3.58	3.94
2900	For electric trolley operator, 1/3 H.P., to 12' x 12', add	1 Carp	2	4	Ea.	925	160		1,085	1,275
2950	Over 12' x 12', 1/2 H.P., add	"	1	8	"	1,100	320		1,420	1,725

08 36 19 – Multi-Leaf Vertical Lift Doors

08 36 19.10 Sectional Vertical Lift Doors

		Crew	Daily Output	Labor-Hours	Unit	Material	Labor	Equipment	Total	Total Incl O&P
0010	**SECTIONAL VERTICAL LIFT DOORS**									
0020	Motorized, 14 ga. stl, incl., frm and ctrl pnl									
0050	16' x 16' high	L-10	.50	48	Ea.	21,000	2,150	1,525	24,675	28,400
0100	10' x 20' high	↓	1.30	18.462	↓	29,000	825	590	30,415	34,000

08 36 Panel Doors

08 36 19 – Multi-Leaf Vertical Lift Doors

08 36 19.10 Sectional Vertical Lift Doors

		Crew	Daily Output	Labor-Hours	Unit	Material	2009 Bare Costs Labor	Equipment	Total	Total Incl O&P
0120	15' x 20' high	L-10	1.30	18.462	Ea.	36,000	825	590	37,415	41,700
0140	20' x 20' high		1	24		41,600	1,075	770	43,445	48,400
0160	25' x 20' high		1	24		45,900	1,075	770	47,745	53,000
0170	32' x 24' high		.75	32		44,700	1,425	1,025	47,150	53,000
0180	20' x 25' high		1	24		47,600	1,075	770	49,445	55,000
0200	25' x 25' high		.70	34.286		53,500	1,525	1,100	56,125	63,000
0220	25' x 30' high		.70	34.286		62,000	1,525	1,100	64,625	72,500
0240	30' x 30' high		.70	34.286		71,500	1,525	1,100	74,125	82,500
0260	35' x 30' high		.70	34.286		78,000	1,525	1,100	80,625	89,500

08 36 23 – Telescoping Vertical Lift Doors

08 36 23.10 Telescoping Steel Doors

0010	**TELESCOPING STEEL DOORS**									
1000	Overhead, .03" thick, electric operated, 10' x 10'	E-3	.80	30	Ea.	12,500	1,350	168	14,018	16,400
2000	20' x 10'		.60	40		16,900	1,825	224	18,949	22,100
3000	20' x 16'		.40	60		18,600	2,725	335	21,660	25,700

08 38 Traffic Doors

08 38 13 – Flexible Strip Doors

08 38 13.10 Flexible Transparent Strip Doors

0010	**FLEXIBLE TRANSPARENT STRIP DOORS**									
0100	12" strip width, 2/3 overlap	3 Shee	135	.178	SF Surf	7.50	8.40		15.90	21
0200	Full overlap		115	.209		6.30	9.85		16.15	22
0220	8" strip width, 1/2 overlap		140	.171		5.75	8.10		13.85	18.80
0240	Full overlap		120	.200		7.60	9.45		17.05	23
0300	Add for suspension system, header mount				L.F.	8.90			8.90	9.80
0400	Wall mount				"	9			9	9.90

08 38 19 – Rigid Traffic Doors

08 38 19.20 Double Acting Swing Doors

0010	**DOUBLE ACTING SWING DOORS**									
1000	.063" aluminum, 7'-0" high, 4'-0" wide	2 Carp	4.20	3.810	Pr.	1,900	152		2,052	2,325
1025	6'-0" wide		4	4		2,100	160		2,260	2,550
1050	6'-8" wide		4	4		2,300	160		2,460	2,775
2000	Solid core wood, 3/4" thick, metal frame, stainless steel									
2010	base plate, 7' high opening, 4' wide	2 Carp	4	4	Pr.	2,200	160		2,360	2,675
2050	7' wide	"	3.80	4.211	"	2,500	168		2,668	3,000

08 38 19.30 Shock Absorbing Doors

0010	**SHOCK ABSORBING DOORS**									
0020	Rigid, no frame, 1-1/2" thick, 5' x 7'	2 Sswk	1.90	8.421	Opng.	1,400	375		1,775	2,225
0100	8' x 8'		1.80	8.889		2,000	395		2,395	2,900
0500	Flexible, no frame, insulated, .16" thick, economy, 5' x 7'		2	8		1,750	360		2,110	2,550
0600	Deluxe		1.90	8.421		2,625	375		3,000	3,550
1000	8' x 8' opening, economy		2	8		2,750	360		3,110	3,650
1100	Deluxe		1.90	8.421		3,500	375		3,875	4,525

08 41 Entrances and Storefronts

08 41 13 – Aluminum-Framed Entrances and Storefronts

08 41 13.20 Tube Framing		Crew	Daily Output	Labor-Hours	Unit	Material	2009 Bare Costs Labor	Equipment	Total	Total Incl O&P
0010	**TUBE FRAMING**, For window walls and store fronts, aluminum stock									
0050	Plain tube frame, mill finish, 1-3/4" x 1-3/4"	2 Glaz	103	.155	L.F.	8.70	6		14.70	18.65
0150	1-3/4" x 4"		98	.163		11.85	6.30		18.15	22.50
0200	1-3/4" x 4-1/2"		95	.168		13.80	6.50		20.30	25
0250	2" x 6"		89	.180		19.55	6.95		26.50	32
0350	4" x 4"		87	.184		19.65	7.10		26.75	32.50
0400	4-1/2" x 4-1/2"		85	.188		25	7.25		32.25	38.50
0450	Glass bead		240	.067		2.55	2.57		5.12	6.70
1000	Flush tube frame, mill finish, 1/4" glass, 1-3/4" x 4", open header		80	.200		11.45	7.70		19.15	24.50
1050	Open sill		82	.195		9	7.55		16.55	21.50
1100	Closed back header		83	.193		16.05	7.45		23.50	29
1150	Closed back sill		85	.188		15.20	7.25		22.45	27.50
1160	Tube fmg, spandrel cover both sides, alum 1" wide	1 Sswk	85	.094	S.F.	86	4.21		90.21	102
1170	Tube fmg, spandrel cover both sides, alum 2" wide	"	85	.094	"	32.50	4.21		36.71	43.50
1200	Vertical mullion, one piece	2 Glaz	75	.213	L.F.	16.90	8.25		25.15	31
1250	Two piece		73	.219		18.10	8.45		26.55	32.50
1300	90° or 180° vertical corner post		75	.213		27.50	8.25		35.75	43
1400	1-3/4" x 4-1/2", open header		80	.200		13.90	7.70		21.60	27
1450	Open sill		82	.195		11.60	7.55		19.15	24
1500	Closed back header		83	.193		16.85	7.45		24.30	30
1550	Closed back sill		85	.188		16.35	7.25		23.60	29
1600	Vertical mullion, one piece		75	.213		18.30	8.25		26.55	32.50
1650	Two piece		73	.219		19.30	8.45		27.75	34
1700	90° or 180° vertical corner post		75	.213		19.70	8.25		27.95	34
2000	Flush tube frame, mil fin.,ins. glass w/thml brk, 2" x 4-1/2", open header		75	.213		14.50	8.25		22.75	28.50
2050	Open sill		77	.208		12.25	8		20.25	25.50
2100	Closed back header		78	.205		13.75	7.90		21.65	27
2150	Closed back sill		80	.200		14.70	7.70		22.40	28
2200	Vertical mullion, one piece		70	.229		15.25	8.80		24.05	30
2250	Two piece		68	.235		16.45	9.10		25.55	32
2300	90° or 180° vertical corner post		70	.229		15.70	8.80		24.50	30.50
5000	Flush tube frame, mill fin., thermal brk., 2-1/4"x 4-1/2", open header		74	.216		15.15	8.35		23.50	29.50
5050	Open sill		75	.213		13.45	8.25		21.70	27.50
5100	Vertical mullion, one piece		69	.232		16.50	8.95		25.45	31.50
5150	Two piece		67	.239		19.15	9.20		28.35	35
5200	90° or 180° vertical corner post		69	.232		16.85	8.95		25.80	32
6980	Door stop (snap in)		380	.042		2.85	1.63		4.48	5.60
7000	For joints, 90°, clip type, add				Ea.	23			23	25
7050	Screw spline joint, add					20.50			20.50	22.50
7100	For joint other than 90°, add					43			43	47
8000	For bronze anodized aluminum, add					15%				
8020	For black finish, add					27%				
8050	For stainless steel materials, add					350%				
8100	For monumental grade, add					52%				
8150	For steel stiffener, add	2 Glaz	200	.080	L.F.	10.40	3.09		13.49	16.10
8200	For 2 to 5 stories, add per story				Story		6%			

08 41 19 – Stainless-Steel-Framed Entrances and Storefronts

08 41 19.10 Stainless-Steel and Glass Entrance Unit		Crew	Daily Output	Labor-Hours	Unit	Material	2009 Bare Costs Labor	Equipment	Total	Total Incl O&P
0010	**STAINLESS-STEEL AND GLASS ENTRANCE UNIT**, narrow stiles									
0020	3' x 7' opening, including hardware, minimum	2 Sswk	1.60	10	Opng.	6,000	445		6,445	7,400
0050	Average		1.40	11.429		6,500	510		7,010	8,050
0100	Maximum		1.20	13.333		6,900	595		7,495	8,650

08 41 Entrances and Storefronts

08 41 19 – Stainless-Steel-Framed Entrances and Storefronts

08 41 19.10 Stainless-Steel and Glass Entrance Unit	Crew	Daily Output	Labor-Hours	Unit	Material	2009 Bare Costs Labor	Equipment	Total	Total Incl O&P
1000 For solid bronze entrance units, statuary finish, add				Opng.	62%				
1100 Without statuary finish, add				↓	45%				
2000 Balanced doors, 3' x 7', economy	2 Sswk	.90	17.778	Ea.	8,100	795		8,895	10,300
2100 Premium	"	.70	22.857	"	14,000	1,025		15,025	17,200

08 41 26 – All-Glass Entrances and Storefronts

08 41 26.10 Window Walls Aluminum, Stock

	Crew	Daily Output	Labor-Hours	Unit	Material	Labor	Equipment	Total	Incl O&P
0010 **WINDOW WALLS ALUMINUM, STOCK**, including glazing									
0020 Minimum	H-2	160	.150	S.F.	39.50	5.45		44.95	52
0050 Average		140	.171		54	6.20		60.20	69
0100 Maximum	↓	110	.218	↓	152	7.90		159.90	179
0850 Cost of the above walls depends on material,									
0860 finish, repetition, and size of units.									
0870 The larger the opening, the lower the S.F. cost									
1200 Double glazed acoustical window wall for airports,									
1220 including 1" thick glass with 2" x 4-1/2" tube frame	H-2	40	.600	S.F.	99	22		121	142

08 42 Entrances

08 42 26 – All-Glass Entrances

08 42 26.10 Swinging Glass Doors

	Crew	Daily Output	Labor-Hours	Unit	Material	Labor	Equipment	Total	Incl O&P
0010 **SWINGING GLASS DOORS**									
0020 Including hardware, 1/2" thick, tempered, 3' x 7' opening	2 Glaz	2	8	Opng.	1,950	310		2,260	2,625
0100 6' x 7' opening	"	1.40	11.429	"	3,800	440		4,240	4,850

08 42 29 – Automatic Entrances

08 42 29.23 Sliding Automatic Entrances

	Crew	Daily Output	Labor-Hours	Unit	Material	Labor	Equipment	Total	Incl O&P
0010 **SLIDING AUTOMATIC ENTRANCES** 12' x 7'-6" opng., 5' x 7' door, 2 way traffic									
0020 Mat or electronic activated, panic pushout, incl. operator & hardware,									
0030 not including glass or glazing	2 Glaz	.70	22.857	Opng.	7,700	880		8,580	9,800

08 42 33 – Revolving Door Entrances

08 42 33.10 Circular Rotating Entrance Doors

	Crew	Daily Output	Labor-Hours	Unit	Material	Labor	Equipment	Total	Incl O&P
0010 **CIRCULAR ROTATING ENTRANCE DOORS**, Aluminum									
0020 6'-10" to 7' high, stock units, minimum	4 Sswk	.75	42.667	Opng.	18,800	1,900		20,700	24,000
0050 Average		.60	53.333		22,600	2,375		24,975	29,200
0100 Maximum		.45	71.111		39,100	3,175		42,275	48,700
1000 Stainless steel		.30	105		37,700	4,750		42,450	50,000
1100 Solid bronze	↓	.15	213		44,000	9,525		53,525	65,500
1500 For automatic controls, add	2 Elec	2	8	↓	14,300	375		14,675	16,300

08 42 36 – Balanced Door Entrances

08 42 36.10 Balanced Entrance Doors

	Crew	Daily Output	Labor-Hours	Unit	Material	Labor	Equipment	Total	Incl O&P
0010 **BALANCED ENTRANCE DOORS**									
0020 Hardware & frame, alum. & glass, 3' x 7', econ.	2 Sswk	.90	17.778	Ea.	5,800	795		6,595	7,800
0150 Premium	"	.70	22.857	"	7,275	1,025		8,300	9,825

08 43 Storefronts

08 43 13 – Aluminum-Framed Storefronts

08 43 13.10 Aluminum-Framed Entrance Doors

		Crew	Daily Output	Labor-Hours	Unit	Material	2009 Bare Costs Labor	Equipment	Total	Total Incl O&P
0010	**ALUMINUM-FRAMED ENTRANCE DOORS** (Frame Only)									
0020	Entrance, 3' x 7' opening, clear anodized finish	2 Sswk	7	2.286	Opng.	420	102		522	640
0040	Bronze finish		7	2.286		450	102		552	675
0500	6' x 7' opening, clear finish		6	2.667		340	119		459	585
0520	Bronze finish		6	2.667		480	119		599	740
1000	With 3' high transoms, 3' x 10' opening, clear finish		6.50	2.462		430	110		540	670
1050	Bronze finish		6.50	2.462		465	110		575	710
1100	Black finish		6.50	2.462		525	110		635	775
1500	With 3' high transoms, 6' x 10' opening, clear finish		5.50	2.909		515	130		645	795
1550	Bronze finish		5.50	2.909		555	130		685	840
1600	Black finish		5.50	2.909		635	130		765	930

08 43 13.20 Storefront Systems

		Crew	Daily Output	Labor-Hours	Unit	Material	2009 Bare Costs Labor	Equipment	Total	Total Incl O&P
0010	**STOREFRONT SYSTEMS**, aluminum frame clear 3/8" plate glass									
0020	incl. 3' x 7' door with hardware (400 sq. ft. max. wall)									
0500	Wall height to 12' high, commercial grade	2 Glaz	150	.107	S.F.	18.70	4.12		22.82	26.50
0600	Institutional grade		130	.123		23	4.75		27.75	32.50
0700	Monumental grade		115	.139		35.50	5.35		40.85	47
1000	6' x 7' door with hardware, commercial grade		135	.119		18.90	4.57		23.47	28
1100	Institutional grade		115	.139		26	5.35		31.35	36.50
1200	Monumental grade		100	.160		48	6.20		54.20	62
1500	For bronze anodized finish, add					15%				
1600	For black anodized finish, add					35%				
1700	For stainless steel framing, add to monumental					76%				

08 43 29 – Sliding Storefronts

08 43 29.10 Sliding Panels

		Crew	Daily Output	Labor-Hours	Unit	Material	2009 Bare Costs Labor	Equipment	Total	Total Incl O&P
0010	**SLIDING PANELS**									
0020	Mall fronts, aluminum & glass, 15' x 9' high	2 Glaz	1.30	12.308	Opng.	3,025	475		3,500	4,050
0100	24' x 9' high		.70	22.857		4,400	880		5,280	6,150
0200	48' x 9' high, with fixed panels		.90	17.778		8,200	685		8,885	10,000
0500	For bronze finish, add					17%				

08 44 Curtain Wall and Glazed Assemblies

08 44 13 – Glazed Aluminum Curtain Walls

08 44 13.10 Glazed Curtain Walls

		Crew	Daily Output	Labor-Hours	Unit	Material	2009 Bare Costs Labor	Equipment	Total	Total Incl O&P
0010	**GLAZED CURTAIN WALLS**, aluminum, stock, including glazing									
0020	Minimum	H-1	205	.156	S.F.	31	6.50		37.50	45.50
0050	Average, single glazed		195	.164		40.50	6.85		47.35	56
0150	Average, double glazed		180	.178		57.50	7.40		64.90	75.50
0200	Maximum		160	.200		157	8.35		165.35	186

08 45 Translucent Wall and Roof Assemblies

08 45 10 – Translucent Roof Assemblies

08 45 10.10 Skyroofs

		Crew	Daily Output	Labor-Hours	Unit	Material	2009 Bare Costs Labor	Equipment	Total	Total Incl O&P
0010	**SKYROOFS**, Translucent panels, 2-3/4" thick									
0020	Under 500 S.F.	G-3	395	.081	SF Hor.	31.50	3.19		34.69	39.50
0100	Over 5000 S.F.		465	.069		29.50	2.71		32.21	36
0300	Continuous vaulted, semi-circular, to 8' wide, double glazed	G	145	.221		49	8.70		57.70	67.50
0400	Single glazed		160	.200		33.50	7.90		41.40	48.50
0600	To 20' wide, single glazed		175	.183		37	7.20		44.20	51.50
0700	Over 20' wide, single glazed		200	.160		42.50	6.30		48.80	57
0900	Motorized opening type, single glazed, 1/3 opening		145	.221		44	8.70		52.70	61.50
1000	Full opening		130	.246		50	9.70		59.70	70
1200	Pyramid type units, self-supporting, to 30' clear opening,									
1300	square or circular, single glazed, minimum	G-3	200	.160	SF Hor.	25.50	6.30		31.80	38
1310	Average		165	.194		34.50	7.65		42.15	50
1400	Maximum		130	.246		50	9.70		59.70	70
1500	Grid type, 4' to 10' modules, single glass glazed, minimum		200	.160		33.50	6.30		39.80	47
1550	Maximum		128	.250		52.50	9.85		62.35	72.50
1600	Preformed acrylic, minimum		300	.107		40	4.20		44.20	50.50
1650	Maximum		175	.183		55	7.20		62.20	71.50
1800	Skyroofs, dome type, self-supporting, to 100' clear opening, circular									
1900	Rise to span ratio = 0.20									
1920	Minimum	G-3	197	.162	SF Hor.	15	6.40		21.40	26.50
1950	Maximum		113	.283		50	11.15		61.15	72.50
2100	Rise to span ratio = 0.33, minimum		169	.189		30	7.45		37.45	44.50
2150	Maximum		101	.317		59	12.50		71.50	84.50
2200	Rise to span ratio = 0.50, minimum		148	.216		45.50	8.50		54	63
2250	Maximum		87	.368		66	14.50		80.50	95
2400	Ridge units, continuous, to 8' wide, double	G	130	.246		115	9.70		124.70	142
2500	Single	G	200	.160		79	6.30		85.30	97
2700	Ridge and furrow units, over 4' O.C., double, minimum	G	200	.160		25	6.30		31.30	37.50
2750	Maximum	G	120	.267		46	10.50		56.50	66.50
2800	Single, minimum		214	.150		21	5.90		26.90	32
2850	Maximum		153	.209		44.50	8.25		52.75	61.50
3000	Rolling roof, translucent panels, flat roof, minimum		253	.126	S.F.	19.10	4.98		24.08	28.50
3030	Maximum	G	160	.200		38.50	7.90		46.40	54
3100	Lean-to skyroof, long span, double, minimum	G	197	.162		25.50	6.40		31.90	38
3150	Maximum	G	101	.317		51	12.50		63.50	75.50
3300	Single, minimum		321	.100		19.40	3.93		23.33	27.50
3350	Maximum		160	.200		32	7.90		39.90	47

08 51 Metal Windows

08 51 13 – Aluminum Windows

08 51 13.10 Aluminum Sash

		Crew	Daily Output	Labor-Hours	Unit	Material	2009 Bare Costs Labor	Equipment	Total	Total Incl O&P
0010	**ALUMINUM SASH**									
0020	Stock, grade C, glaze & trim not incl., casement	2 Sswk	200	.080	S.F.	36	3.58		39.58	46
0050	Double hung		200	.080		36.50	3.58		40.08	46.50
0100	Fixed casement		200	.080		15.65	3.58		19.23	23.50
0150	Picture window		200	.080		16.65	3.58		20.23	24.50
0200	Projected window		200	.080		33	3.58		36.58	42.50
0250	Single hung		200	.080		15.80	3.58		19.38	24
0300	Sliding		200	.080		20.50	3.58		24.08	29
1000	Mullions for above, tubular		240	.067	L.F.	5.45	2.98		8.43	11.25
2000	Custom aluminum sash, grade HC, glazing not included, minimum		200	.080	S.F.	38	3.58		41.58	48.50

08 51 Metal Windows

08 51 13 – Aluminum Windows

08 51 13.10 Aluminum Sash

		Crew	Daily Output	Labor-Hours	Unit	Material	2009 Bare Costs Labor	Equipment	Total	Total Incl O&P
2100	Maximum	2 Sswk	85	.188	S.F.	49.50	8.40		57.90	69.50

08 51 13.20 Aluminum Windows

		Crew	Daily Output	Labor-Hours	Unit	Material	Labor	Equipment	Total	Total Incl O&P
0010	**ALUMINUM WINDOWS**, incl. frame and glazing, commercial grade									
1000	Stock units, casement, 3'-1" x 3'-2" opening	2 Sswk	10	1.600	Ea.	355	71.50		426.50	515
1050	Add for storms					115			115	126
1600	Projected, with screen, 3'-1" x 3'-2" opening	2 Sswk	10	1.600		335	71.50		406.50	495
1700	Add for storms					112			112	123
2000	4'-5" x 5'-3" opening	2 Sswk	8	2		380	89.50		469.50	580
2100	Add for storms					120			120	132
2500	Enamel finish windows, 3'-1" x 3'-2"	2 Sswk	10	1.600		340	71.50		411.50	500
2600	4'-5" x 5'-3"		8	2		385	89.50		474.50	585
3000	Single hung, 2' x 3' opening, enameled, standard glazed		10	1.600		198	71.50		269.50	345
3100	Insulating glass		10	1.600		240	71.50		311.50	390
3300	2'-8" x 6'-8" opening, standard glazed		8	2		350	89.50		439.50	545
3400	Insulating glass		8	2		450	89.50		539.50	655
3700	3'-4" x 5'-0" opening, standard glazed		9	1.778		286	79.50		365.50	455
3800	Insulating glass		9	1.778		315	79.50		394.50	490
3890	Awning type, 3' x 3' opening standard glass		14	1.143		405	51		456	535
3900	Insulating glass		14	1.143		430	51		481	565
3910	3' x 4' opening, standard glass		10	1.600		470	71.50		541.50	645
3920	Insulating glass		10	1.600		545	71.50		616.50	720
3930	3' x 5'-4" opening, standard glass		10	1.600		570	71.50		641.50	750
3940	Insulating glass		10	1.600		670	71.50		741.50	860
3950	4' x 5'-4" opening, standard glass		9	1.778		625	79.50		704.50	825
3960	Insulating glass		9	1.778		750	79.50		829.50	965
4000	Sliding aluminum, 3' x 2' opening, standard glazed		10	1.600		209	71.50		280.50	355
4100	Insulating glass		10	1.600		224	71.50		295.50	375
4300	5' x 3' opening, standard glazed		9	1.778		320	79.50		399.50	490
4400	Insulating glass		9	1.778		370	79.50		449.50	550
4600	8' x 4' opening, standard glazed		6	2.667		335	119		454	580
4700	Insulating glass		6	2.667		545	119		664	805
5000	9' x 5' opening, standard glazed		4	4		510	179		689	880
5100	Insulating glass		4	4		820	179		999	1,225
5500	Sliding, with thermal barrier and screen, 6' x 4', 2 track		8	2		695	89.50		784.50	925
5700	4 track		8	2		880	89.50		969.50	1,125
6000	For above units with bronze finish, add					12%				
6200	For installation in concrete openings, add					6%				

08 51 23 – Steel Windows

08 51 23.10 Steel Sash

		Crew	Daily Output	Labor-Hours	Unit	Material	Labor	Equipment	Total	Total Incl O&P
0010	**STEEL SASH** Custom units, glazing and trim not included									
0100	Casement, 100% vented	2 Sswk	200	.080	S.F.	48	3.58		51.58	59
0200	50% vented		200	.080		44	3.58		47.58	55
1000	Projected, commercial, 40% vented		200	.080		48	3.58		51.58	59.50
1100	Intermediate, 50% vented		200	.080		54	3.58		57.58	65.50
1500	Industrial, horizontally pivoted		200	.080		49.50	3.58		53.08	61
1600	Fixed		200	.080		28.50	3.58		32.08	38
2000	Industrial security sash, 50% vented		200	.080		53.50	3.58		57.08	65.50
2100	Fixed		200	.080		43.50	3.58		47.08	54.50
2500	Picture window		200	.080		28	3.58		31.58	37.50
3000	Double hung		200	.080		56	3.58		59.58	68
5000	Mullions for above, open interior face		240	.067	L.F.	9.80	2.98		12.78	16.05
5100	With interior cover		240	.067	"	16.35	2.98		19.33	23.50

08 51 Metal Windows

08 51 23 – Steel Windows

08 51 23.20 Steel Windows		Crew	Daily Output	Labor-Hours	Unit	Material	2009 Bare Costs Labor	Equipment	Total	Total Incl O&P
0010	**STEEL WINDOWS** Stock, including frame, trim and insul. glass									
1000	Custom units, double hung, 2'-8" x 4'-6" opening	2 Sswk	12	1.333	Ea.	680	59.50		739.50	855
1100	2'-4" x 3'-9" opening		12	1.333		565	59.50		624.50	725
1500	Commercial projected, 3'-9" x 5'-5" opening		10	1.600		1,200	71.50		1,271.50	1,425
1600	6'-9" x 4'-1" opening		7	2.286		1,575	102		1,677	1,900
2000	Intermediate projected, 2'-9" x 4'-1" opening		12	1.333		670	59.50		729.50	840
2100	4'-1" x 5'-5" opening		10	1.600		1,350	71.50		1,421.50	1,600

08 51 66 – Metal Window Screens

08 51 66.10 Screens		Crew	Daily Output	Labor-Hours	Unit	Material	Labor	Equipment	Total	Total Incl O&P
0010	**SCREENS**									
0020	For metal sash, aluminum or bronze mesh, flat screen	2 Sswk	1200	.013	S.F.	4.10	.60		4.70	5.55
0500	Wicket screen, inside window		1000	.016		6.30	.72		7.02	8.20
0800	Security screen, aluminum frame with stainless steel cloth		1200	.013		22.50	.60		23.10	26
0900	Steel grate, painted, on steel frame		1600	.010		12.40	.45		12.85	14.45
1000	For solar louvers, add		160	.100		23.50	4.47		27.97	33.50
4000	See Div. 05 58 27.90									

08 52 Wood Windows

08 52 10 – Plain Wood Windows

08 52 10.20 Awning Window

		Crew	Daily Output	Labor-Hours	Unit	Material	Labor	Equipment	Total	Total Incl O&P
0010	**AWNING WINDOW**, Including frame, screens and grilles									
0100	Average quality, builders model, 34" x 22", double insulated glass	1 Carp	10	.800	Ea.	249	32		281	325
0200	Low E glass		10	.800		258	32		290	335
0300	40" x 28", double insulated glass		9	.889		310	35.50		345.50	395
0400	Low E Glass		9	.889		330	35.50		365.50	420
0500	48" x 36", double insulated glass		8	1		450	40		490	555
0600	Low E glass		8	1		475	40		515	585
1000	Vinyl clad, 34" x 22"		10	.800		259	32		291	335
1100	40" x 22"		10	.800		283	32		315	360
1200	36" x 28"		9	.889		300	35.50		335.50	385
1300	36" x 36"		9	.889		335	35.50		370.50	425
1400	48" x 28"		8	1		360	40		400	460
1500	60" x 36"		8	1		505	40		545	615
2200	Metal clad, 36" x 25"		9	.889		269	35.50		304.50	350
2300	40" x 30"		9	.889		335	35.50		370.50	425
2400	48" x 28"		8	1		345	40		385	440
2500	60" x 36"		8	1		365	40		405	465

08 52 10.40 Casement Window

			Crew	Daily Output	Labor-Hours	Unit	Material	Labor	Equipment	Total	Total Incl O&P
0010	**CASEMENT WINDOW**, including frame, screen and grilles										
0100	Avg. quality, bldrs. model, 2'-0" x 3'-0" H, dbl. insulated glass	G	1 Carp	10	.800	Ea.	276	32		308	355
0150	Low E glass	G		10	.800		225	32		257	298
0200	2'-0" x 4'-6" high, double insulated glass	G		9	.889		345	35.50		380.50	435
0250	Low E glass	G		9	.889		263	35.50		298.50	345
0300	2'-4" x 6'-0" high, double insulated glass	G		8	1		380	40		420	480
0350	Low E glass	G		8	1		465	40		505	575
0522	Vinyl clad, premium, double insulated glass, 2'-0" x 3'-0"	G		10	.800		266	32		298	340
0524	2'-0" x 4'-0"	G		9	.889		310	35.50		345.50	395
0525	2'-0" x 5'-0"	G		8	1		355	40		395	450
0528	2'-0" x 6'-0"	G		8	1		355	40		395	450
8100	Metal clad, deluxe, dbl. insul. glass, 2'-0" x 3'-0" high	G		10	.800		226	32		258	299

08 52 Wood Windows

08 52 10 – Plain Wood Windows

08 52 10.40 Casement Window

		Crew	Daily Output	Labor-Hours	Unit	Material	2009 Bare Costs Labor	Equipment	Total	Total Incl O&P
8120	2'-0" x 4'-0" high	1 Carp [G]	9	.889	Ea.	273	35.50		308.50	355
8140	2'-0" x 5'-0" high	[G]	8	1		310	40		350	400
8160	2'-0" x 6'-0" high	[G]	8	1		355	40		395	450
8190	For installation, add per leaf						15%			
8200	For multiple leaf units, deduct for stationary sash									
8220	2' high				Ea.	22			22	24
8240	4'-6" high					25			25	28
8260	6' high					33.50			33.50	37

08 52 10.50 Double Hung

		Crew	Daily Output	Labor-Hours	Unit	Material	Labor	Equipment	Total	Total Incl O&P
0010	**DOUBLE HUNG**, Including frame, screens and grilles									
0100	Avg. quality, bldrs. model, 2'-0" x 3'-0" high, dbl insul. glass	1 Carp [G]	10	.800	Ea.	184	32		216	252
0150	Low E glass	[G]	10	.800		208	32		240	279
0200	3'-0" x 4'-0" high, double insulated glass	[G]	9	.889		257	35.50		292.50	335
0250	Low E glass	[G]	9	.889		275	35.50		310.50	360
0300	4'-0" x 4'-6" high, double insulated glass	[G]	8	1		315	40		355	405
0350	Low E glass	[G]	8	1		340	40		380	435
1000	Vinyl clad, premium, double insulated glass, 2'-6" x 3'-0"	[G]	10	.800		225	32		257	298
1100	3'-0" x 3'-6"	[G]	10	.800		261	32		293	335
1200	3'-0" x 4'-0"	[G]	9	.889		370	35.50		405.50	460
1300	3'-0" x 4'-6"	[G]	9	.889		340	35.50		375.50	425
1400	3'-0" x 5'-0"	[G]	8	1		315	40		355	405
1500	3'-6" x 6'-0"	[G]	8	1		370	40		410	465
2000	Metal clad, deluxe, dbl. insul. glass, 2'-6" x 3'-0" high	[G]	10	.800		262	32		294	340
2100	3'-0" x 3'-6" high	[G]	10	.800		299	32		331	380
2200	3'-0" x 4'-0" high	[G]	9	.889		315	35.50		350.50	400
2300	3'-0" x 4'-6" high	[G]	9	.889		330	35.50		365.50	420
2400	3'-0" x 5'-0" high	[G]	8	1		360	40		400	455
2500	3'-6" x 6'-0" high	[G]	8	1		435	40		475	535

08 52 10.55 Picture Window

		Crew	Daily Output	Labor-Hours	Unit	Material	Labor	Equipment	Total	Total Incl O&P
0010	**PICTURE WINDOW**, Including frame and grilles									
0100	Average quality, bldrs. model, 3'-6" x 4'-0" high, dbl insulated glass	2 Carp	12	1.333	Ea.	420	53.50		473.50	550
0150	Low E glass		12	1.333		465	53.50		518.50	595
0200	4'-0" x 4'-6" high, double insulated glass		11	1.455		475	58		533	615
0250	Low E glass		11	1.455		500	58		558	640
0300	5'-0" x 4'-0" high, double insulated glass		11	1.455		555	58		613	700
0350	Low E glass		11	1.455		575	58		633	725
0400	6'-0" x 4'-6" high, double insulated glass		10	1.600		605	64		669	765
0450	Low E glass		10	1.600		610	64		674	775

08 52 10.65 Wood Sash

		Crew	Daily Output	Labor-Hours	Unit	Material	Labor	Equipment	Total	Total Incl O&P
0010	**WOOD SASH**, Including glazing but not trim									
0050	Custom, 5'-0" x 4'-0", 1" dbl. glazed, 3/16" thick lites	2 Carp	3.20	5	Ea.	167	200		367	495
0100	1/4" thick lites		5	3.200		172	128		300	385
0200	1" thick, triple glazed		5	3.200		395	128		523	630
0300	7'-0" x 4'-6" high, 1" double glazed, 3/16" thick lites		4.30	3.721		400	149		549	670
0400	1/4" thick lites		4.30	3.721		450	149		599	725
0500	1" thick, triple glazed		4.30	3.721		515	149		664	795
0600	8'-6" x 5'-0" high, 1" double glazed, 3/16" thick lites		3.50	4.571		540	183		723	875
0700	1/4" thick lites		3.50	4.571		590	183		773	935
0800	1" thick, triple glazed		3.50	4.571		595	183		778	940
0900	Window frames only, based on perimeter length				L.F.	3.68			3.68	4.05

08 52 Wood Windows

08 52 10 – Plain Wood Windows

08 52 10.70 Sliding Windows

		Crew	Daily Output	Labor-Hours	Unit	Material	2009 Bare Costs Labor	Equipment	Total	Total Incl O&P
0010	**SLIDING WINDOWS**									
0100	Average quality, bldrs. model, 3'-0" x 3'-0" high, double insulated G	1 Carp	10	.800	Ea.	244	32		276	320
0120	Low E glass G		10	.800		264	32		296	340
0200	4'-0" x 3'-6" high, double insulated G		9	.889		282	35.50		317.50	365
0220	Low E glass G		9	.889		305	35.50		340.50	390
0300	6'-0" x 5'-0" high, double insulated G		8	1		420	40		460	520
0320	Low E glass G		8	1		460	40		500	565

08 52 13 – Metal-Clad Wood Windows

08 52 13.10 Awning Windows, Metal-Clad

		Crew	Daily Output	Labor-Hours	Unit	Material	Labor	Equipment	Total	Total Incl O&P
0010	**AWNING WINDOWS, METAL-CLAD**									
2000	Metal clad, awning deluxe, double insulated glass, 34" x 22"	1 Carp	10	.800	Ea.	242	32		274	315
2100	40" x 22"	"	10	.800	"	285	32		317	365

08 52 13.35 Picture and Sliding Windows Metal-Clad

		Crew	Daily Output	Labor-Hours	Unit	Material	Labor	Equipment	Total	Total Incl O&P
0010	**PICTURE AND SLIDING WINDOWS METAL-CLAD**									
2000	Metal clad, dlx picture, dbl. insul. glass, 4'-0" x 4'-0" high	2 Carp	12	1.333	Ea.	360	53.50		413.50	485
2100	4'-0" x 6'-0" high		11	1.455		535	58		593	680
2200	5'-0" x 6'-0" high		10	1.600		590	64		654	745
2300	6'-0" x 6'-0" high		10	1.600		675	64		739	845
2400	Metal clad, dlx sliding, double insulated glass, 3'-0" x 3'-0" high G	1 Carp	10	.800		320	32		352	405
2420	4'-0" x 3'-6" high G		9	.889		395	35.50		430.50	490
2440	5'-0" x 4'-0" high G		9	.889		475	35.50		510.50	575
2460	6'-0" x 5'-0" high G		8	1		700	40		740	830

08 52 16 – Plastic-Clad Wood Windows

08 52 16.10 Bow Window

		Crew	Daily Output	Labor-Hours	Unit	Material	Labor	Equipment	Total	Total Incl O&P
0010	**BOW WINDOW** Including frames, screens, and grilles									
0020	End panels operable									
1000	Bow type, casement, wood, bldrs mdl, 8' x 5' dbl insltd glass, 4 panel	2 Carp	10	1.600	Ea.	1,500	64		1,564	1,750
1050	Low E glass		10	1.600		1,300	64		1,364	1,550
1100	10'-0" x 5'-0", double insulated glass, 6 panels		6	2.667		1,350	107		1,457	1,675
1200	Low E glass, 6 panels		6	2.667		1,450	107		1,557	1,775
1300	Vinyl clad, bldrs model, double insulated glass, 6'-0" x 4'-0", 3 panel		10	1.600		1,475	64		1,539	1,725
1340	9'-0" x 4'-0", 4 panel		8	2		1,450	80		1,530	1,725
1380	10'-0" x 6'-0", 5 panels		7	2.286		2,250	91.50		2,341.50	2,625
1420	12'-0" x 6'-0", 6 panels		6	2.667		2,925	107		3,032	3,400
1600	Metal clad, casement, bldrs mdl, 6'-0" x 4'-0", dbl insltd gls, 3 panels		10	1.600		920	64		984	1,100
1640	9'-0" x 4'-0", 4 panels		8	2		1,250	80		1,330	1,500
1680	10'-0" x 5'-0", 5 panels		7	2.286		1,725	91.50		1,816.50	2,050
1720	12'-0" x 6'-0", 6 panels		6	2.667		2,400	107		2,507	2,825
2000	Bay window, builders model, 8' x 5' dbl insul glass,		10	1.600		1,825	64		1,889	2,125
2050	Low E glass,		10	1.600		2,225	64		2,289	2,550
2100	12'-0" x 6'-0", double insulated glass, 6 panels		6	2.667		2,300	107		2,407	2,700
2200	Low E glass		6	2.667		2,175	107		2,282	2,575
2280	6'-0" x 4'-0"		11	1.455		1,100	58		1,158	1,325
2300	Vinyl clad, premium, double insulated glass, 8'-0" x 5'-0"		10	1.600		1,300	64		1,364	1,525
2340	10'-0" x 5'-0"		8	2		1,850	80		1,930	2,150
2380	10'-0" x 6'-0"		7	2.286		1,925	91.50		2,016.50	2,275
2420	12'-0" x 6'-0"		6	2.667		2,300	107		2,407	2,700
2600	Metal clad, deluxe, dbl insul. glass, 8'-0" x 5'-0" high, 4 panels		10	1.600		1,625	64		1,689	1,875
2640	10'-0" x 5'-0" high, 5 panels		8	2		1,750	80		1,830	2,050
2680	10'-0" x 6'-0" high, 5 panels		7	2.286		2,050	91.50		2,141.50	2,425
2720	12'-0" x 6'-0" high, 6 panels		6	2.667		2,850	107		2,957	3,325

08 52 Wood Windows

08 52 16 – Plastic-Clad Wood Windows

08 52 16.10 Bow Window

		Crew	Daily Output	Labor-Hours	Unit	Material	2009 Bare Costs Labor	Equipment	Total	Total Incl O&P
3000	Double hung, bldrs. model, bay, 8' x 4' high, dbl insulated glass	2 Carp	10	1.600	Ea.	1,300	64		1,364	1,525
3050	Low E glass		10	1.600		1,375	64		1,439	1,625
3100	9'-0" x 5'-0" high, double insulated glass		6	2.667		1,400	107		1,507	1,700
3200	Low E glass		6	2.667		1,475	107		1,582	1,800
3300	Vinyl clad, premium, double insulated glass, 7'-0" x 4'-6"		10	1.600		1,325	64		1,389	1,575
3340	8'-0" x 4'-6"		8	2		1,375	80		1,455	1,625
3380	8'-0" x 5'-0"		7	2.286		1,425	91.50		1,516.50	1,725
3420	9'-0" x 5'-0"		6	2.667		1,475	107		1,582	1,800
3600	Metal clad, deluxe, dbl insul. glass, 7'-0" x 4'-0" high		10	1.600		1,225	64		1,289	1,450
3640	8'-0" x 4'-0" high		8	2		1,275	80		1,355	1,525
3680	8'-0" x 5'-0" high		7	2.286		1,325	91.50		1,416.50	1,600
3720	9'-0" x 5'-0" high		6	2.667		1,400	107		1,507	1,725

08 52 16.40 Transom Windows

		Crew	Daily Output	Labor-Hours	Unit	Material	Labor	Equipment	Total	Total Incl O&P
0010	**TRANSOM WINDOWS**									
1000	Vinyl clad, premium, dbl. insul. glass, 4'-0" x 4'-0"	2 Carp	12	1.333	Ea.	485	53.50		538.50	615
1100	4'-0" x 6'-0"		11	1.455		910	58		968	1,100
1200	5'-0" x 6'-0"		10	1.600		1,000	64		1,064	1,225
1300	6'-0" x 6'-0"		10	1.600		1,025	64		1,089	1,225

08 52 16.70 Vinyl Clad, Premium, Dbl. Insulated Glass

			Crew	Daily Output	Labor-Hours	Unit	Material	Labor	Equipment	Total	Total Incl O&P
0010	**VINYL CLAD, PREMIUM, DBL. INSULATED GLASS**										
1000	Sliding, 3'-0" x 3'-0"	G	1 Carp	10	.800	Ea.	575	32		607	680
1050	4'-0" x 3'-6"	G		9	.889		670	35.50		705.50	790
1100	5'-0" x 4'-0"	G		9	.889		860	35.50		895.50	1,000
1150	6'-0" x 5'-0"	G		8	1		1,075	40		1,115	1,250

08 52 50 – Window Accessories

08 52 50.10 Window Grille or Muntin

		Crew	Daily Output	Labor-Hours	Unit	Material	Labor	Equipment	Total	Total Incl O&P
0010	**WINDOW GRILLE OR MUNTIN**, snap in type									
0020	Standard pattern interior grilles									
2000	Wood, awning window, glass size 28" x 16" high	1 Carp	30	.267	Ea.	22.50	10.65		33.15	41.50
2060	44" x 24" high		32	.250		32.50	10		42.50	51
2100	Casement, glass size, 20" x 36" high		30	.267		27.50	10.65		38.15	47
2180	20" x 56" high		32	.250		39.50	10		49.50	59
2200	Double hung, glass size, 16" x 24" high		24	.333	Set	48	13.30		61.30	73
2280	32" x 32" high		34	.235	"	128	9.40		137.40	156
2500	Picture, glass size, 48" x 48" high		30	.267	Ea.	112	10.65		122.65	140
2580	60" x 68" high		28	.286	"	160	11.40		171.40	194
2600	Sliding, glass size, 14" x 36" high		24	.333	Set	25.50	13.30		38.80	48.50
2680	36" x 36" high		22	.364	"	39	14.55		53.55	65.50

08 52 66 – Wood Window Screens

08 52 66.10 Wood Screens

		Crew	Daily Output	Labor-Hours	Unit	Material	Labor	Equipment	Total	Total Incl O&P
0010	**WOOD SCREENS**									
0020	Over 3 S.F., 3/4" frames	2 Carp	375	.043	S.F.	4.28	1.70		5.98	7.35
0100	1-1/8" frames	"	375	.043	"	7.60	1.70		9.30	11

08 53 Plastic Windows

08 53 13 – Vinyl Windows

08 53 13.30 Vinyl Double Hung Windows

		Crew	Daily Output	Labor-Hours	Unit	Material	2009 Bare Costs Labor	Equipment	Total	Total Incl O&P
0010	**VINYL DOUBLE HUNG WINDOWS**									
0300	Solid vinyl, average quality, double insulated glass, 2'-0" x 3'-0" G	1 Carp	10	.800	Ea.	279	32		311	355
0310	3'-0" x 4'-0" G		9	.889		175	35.50		210.50	248
0320	4'-0" x 4'-6" G		8	1		286	40		326	375
0330	Premium, double insulated glass, 2'-6" x 3'-0" G		10	.800		194	32		226	263
0340	3'-0" x 3'-6" G		9	.889		225	35.50		260.50	305
0350	3'-0" x 4'-0" G		9	.889		238	35.50		273.50	315
0360	3'-0" x 4'-6" G		9	.889		243	35.50		278.50	320
0370	3'-0" x 5'-0" G		8	1		261	40		301	350
0380	3'-6" x 6'-0" G		8	1		300	40		340	390

08 53 13.40 Vinyl Casement Windows

		Crew	Daily Output	Labor-Hours	Unit	Material	Labor	Equipment	Total	Total Incl O&P
0010	**VINYL CASEMENT WINDOWS**									
0340	Solid vinyl, premium, double insulated glass, 2'-0" x 3'-0" high G	1 Carp	10	.800	Ea.	252	32		284	325
0360	2'-0" x 4'-0" high G		9	.889		280	35.50		315.50	365
0380	2'-0" x 5'-0" high G		8	1		276	40		316	365

08 56 Special Function Windows

08 56 63 – Detention Windows

08 56 63.13 Visitor Cubicle Windows

		Crew	Daily Output	Labor-Hours	Unit	Material	Labor	Equipment	Total	Total Incl O&P
0010	**VISITOR CUBICLE WINDOWS**									
4000	Visitor cubicle, vision panel, no intercom	E-4	2	16	Ea.	2,925	725	67	3,717	4,600

08 62 Unit Skylights

08 62 13 – Domed Unit Skylights

08 62 13.20 Skylights

		Crew	Daily Output	Labor-Hours	Unit	Material	Labor	Equipment	Total	Total Incl O&P
0010	**SKYLIGHTS**, Plastic domes, flush or curb mounted ten or									
0100	more units									
0300	Nominal size under 10 S.F., double G	G-3	130	.246	S.F.	26.50	9.70		36.20	44
0400	Single		160	.200		23.50	7.90		31.40	37.50
0600	10 S.F. to 20 S.F., double G		315	.102		22.50	4		26.50	30.50
0700	Single		395	.081		23	3.19		26.19	30
0900	20 S.F. to 30 S.F., double		395	.081		21	3.19		24.19	28
1000	Single		465	.069		19.05	2.71		21.76	25
1200	30 S.F. to 65 S.F., double G		465	.069		21	2.71		23.71	27
1300	Single		610	.052		16.50	2.07		18.57	21.50
1500	For insulated 4" curbs, double, add					27%				
1600	Single, add					30%				
1800	For integral insulated 9" curbs, double, add					30%				
1900	Single, add					40%				
2120	Ventilating insulated plexiglass dome with									
2130	curb mounting, 36" x 36" G	G-3	12	2.667	Ea.	380	105		485	580
2150	52" x 52" G		12	2.667		570	105		675	790
2160	28" x 52" G		10	3.200		455	126		581	695
2170	36" x 52" G		10	3.200		495	126		621	740
2180	For electric opening system, add G					292			292	320
2200	Field fabricated, factory type, aluminum and wire glass G	G-3	120	.267	S.F.	15.10	10.50		25.60	33
2300	Insulated safety glass with aluminum frame G		160	.200		88	7.90		95.90	109
2400	Sandwich panels, fiberglass, for walls, 1-9/16" thick, to 250 SF G		200	.160		16	6.30		22.30	27.50
2500	250 SF and up G		265	.121		14.30	4.76		19.06	23

08 62 Unit Skylights

08 62 13 – Domed Unit Skylights

08 62 13.20 Skylights

		Crew	Daily Output	Labor-Hours	Unit	Material	2009 Bare Costs Labor	Equipment	Total	Total Incl O&P
2700	As above, but for roofs, 2-3/4" thick, to 250 SF	G-3 [G]	295	.108	S.F.	23	4.27		27.27	32
2800	250 SF and up	↓ [G]	330	.097	↓	18.90	3.82		22.72	27

08 63 Metal-Framed Skylights

08 63 23 – Ridge Metal-Framed Skylights

08 63 23.10 Prefabricated

0010	**PREFABRICATED** glass block with metal frame									
0020	Minimum	G-3 [G]	265	.121	S.F.	50	4.76		54.76	62.50
0100	Maximum	" [G]	160	.200	"	100	7.90		107.90	122

08 71 Door Hardware

08 71 13 – Automatic Door Operators

08 71 13.10 Automatic Openers Commercial

0010	**AUTOMATIC OPENERS COMMERCIAL**									
0020	Pneumatic, incl opener, motion sens, control box, tubing, compressor									
0050	For single swing door, per opening	2 Skwk	.80	20	Ea.	4,125	815		4,940	5,825
0100	Pair, per opening		.50	32	Opng.	6,875	1,300		8,175	9,600
1000	For single sliding door, per opening		.60	26.667		4,575	1,100		5,675	6,700
1300	Bi-parting pair	↓	.50	32	↓	6,900	1,300		8,200	9,625
1420	Electronic door opener incl motion sens, 12V control box, motor									
1450	For single swing door, per opening	2 Skwk	.80	20	Opng.	3,400	815		4,215	5,000
1500	Pair, per opening		.50	32		5,550	1,300		6,850	8,125
1600	For single sliding door, per opening		.60	26.667		3,675	1,100		4,775	5,725
1700	Bi-parting pair	↓	.50	32	↓	5,625	1,300		6,925	8,225
1750	Handicap actuator buttons, 2, including 12V DC wiring, add	1 Carp	1.50	5.333	Pr.	415	213		628	785

08 71 13.20 Automatic Openers Industrial

0010	**AUTOMATIC OPENERS INDUSTRIAL**									
0015	Sliding doors up to 6' wide	2 Skwk	.60	26.667	Opng.	5,400	1,100		6,500	7,600
0200	To 12' wide	"	.40	40	"	6,575	1,625		8,200	9,775
0400	Over 12' wide, add per L.F. of excess				L.F.	720			720	790
1000	Swing doors, to 5' wide	2 Skwk	.80	20	Ea.	3,150	815		3,965	4,725
1650	Bi-parting	2 Sswk	.50	32	Pr.	8,175	1,425		9,600	11,600
1860	Add for controls, wall pushbutton, 3 button	2 Skwk	4	4	Ea.	192	163		355	465
1870	Ceiling pull cord	"	4.30	3.721	"	165	152		317	415
1880	For addl elec eye for sliding door operation, one side, add					8%				
1890	Automatic opener, activating carpet, single door, one way add	2 Skwk	2.20	7.273	Ea.	720	297		1,017	1,250

08 71 20 – Hardware

08 71 20.10 Bolts, Flush

0010	**BOLTS, FLUSH**									
0020	Standard, concealed	1 Carp	7	1.143	Ea.	21	45.50		66.50	94.50
0800	Automatic fire exit	"	5	1.600		287	64		351	415
1600	For electric release, add	1 Elec	3	2.667		116	125		241	315
3000	Barrel, brass, 2" long	1 Carp	40	.200		4.90	8		12.90	17.80
3020	4" long		40	.200		3.60	8		11.60	16.35
3060	6" long	↓	40	.200	↓	10.10	8		18.10	23.50

08 71 20.15 Hardware

0009	**HARDWARE**									
0010	Average percentage for hardware, total job cost									

08 71 Door Hardware

08 71 20 – Hardware

08 71 20.15 Hardware		Crew	Daily Output	Labor-Hours	Unit	Material	2009 Bare Costs Labor	Equipment	Total	Total Incl O&P
0025	Minimum									1%
0050	Maximum									4%
0500	Total hardware for building, average distribution					85%	15%			
1000	Door hardware, apartment, interior				Door	140			140	154
1500	Hospital bedroom, minimum					310			310	340
2000	Maximum					700			700	770
2100	Pocket door				Ea.	134			134	147
2250	School, single exterior, incl. lever, not incl. panic device				Door	445			445	490
2500	Single interior, regular use, no lever included					315			315	345
2550	Including handicap lever					410			410	450
2600	Heavy use, incl. lever and closer					550			550	605
2850	Stairway, single interior					790			790	870
3100	Double exterior, with panic device				Pr.	1,075			1,075	1,200
3600	Toilet, public, single interior				Door	170			170	187
6020	Add for door holder, electro-magnetic		1 Elec	4	2	Ea.	86	94		180	235

08 71 20.30 Door Closers

		Crew	Daily Output	Labor-Hours	Unit	Material	Labor	Equipment	Total	Total Incl O&P
0010	**DOOR CLOSERS**									
0020	Adjustable backcheck, 3 way mount, all sizes, regular arm	1 Carp	6	1.333	Ea.	156	53.50		209.50	255
0040	Hold open arm		6	1.333		168	53.50		221.50	268
0100	Fusible link		6.50	1.231		130	49		179	220
0200	Non sized, regular arm		6	1.333		147	53.50		200.50	245
0240	Hold open arm		6	1.333		191	53.50		244.50	294
0400	4 way mount, non sized, regular arm		6	1.333		208	53.50		261.50	310
0440	Hold open arm		6	1.333		214	53.50		267.50	320
2000	Backcheck and adjustable power, hinge face mount									
2010	All sizes, regular arm	1 Carp	6.50	1.231	Ea.	187	49		236	283
2040	Hold open arm		6.50	1.231		235	49		284	335
2400	Top jamb mount, all sizes, regular arm		6	1.333		184	53.50		237.50	285
2440	Hold open arm		6	1.333		203	53.50		256.50	305
2800	Top face mount, all sizes, regular arm		6.50	1.231		184	49		233	279
2840	Hold open arm		6.50	1.231		203	49		252	300
4000	Backcheck, overhead concealed, all sizes, regular arm		5.50	1.455		200	58		258	310
4040	Concealed arm		5	1.600		217	64		281	340
4400	Compact overhead, concealed, all sizes, regular arm		5.50	1.455		340	58		398	460
4440	Concealed arm		5	1.600		375	64		439	515
4800	Concealed in door, all sizes, regular arm		5.50	1.455		129	58		187	232
4840	Concealed arm		5	1.600		139	64		203	252
4900	Floor concealed, all sizes, single acting		2.20	3.636		164	145		309	405
4940	Double acting		2.20	3.636		210	145		355	455
5000	For cast aluminum cylinder, deduct					17.90			17.90	19.70
5040	For delayed action, add					31			31	34
5080	For fusible link arm, add					13.20			13.20	14.50
5120	For shock absorbing arm, add					39			39	43
5160	For spring power adjustment, add					30			30	33
6000	Closer-holder, hinge face mount, all sizes, exposed arm	1 Carp	6.50	1.231		140	49		189	231
7000	Electronic closer-holder, hinge facemount, concealed arm		5	1.600		218	64		282	340
7400	With built-in detector		5	1.600		655	64		719	820

08 71 20.31 Door Closers

		Crew	Daily Output	Labor-Hours	Unit	Material	Labor	Equipment	Total	Total Incl O&P
0010	**DOOR CLOSERS**									
0015	Door closer, rack and pinion	1 Carp	6.50	1.231	Ea.	148	49		197	239

08 71 20.35 Panic Devices

0010	**PANIC DEVICES**									

08 71 Door Hardware

08 71 20 – Hardware

08 71 20.35 Panic Devices

		Crew	Daily Output	Labor-Hours	Unit	Material	2009 Bare Costs Labor	2009 Bare Costs Equipment	Total	Total Incl O&P
0015	For rim locks, single door exit only	1 Carp	6	1.333	Ea.	475	53.50		528.50	610
0020	Outside key and pull		5	1.600		520	64		584	670
0200	Bar and vertical rod, exit only		5	1.600		700	64		764	865
0210	Outside key and pull		4	2		775	80		855	980
0400	Bar and concealed rod		4	2		570	80		650	750
0600	Touch bar, exit only		6	1.333		485	53.50		538.50	620
0610	Outside key and pull		5	1.600		555	64		619	710
0700	Touch bar and vertical rod, exit only		5	1.600		630	64		694	790
0710	Outside key and pull		4	2		730	80		810	930
0800	Touch bar, low profile, exit only		6	1.333		355	53.50		408.50	475
0810	Outside key and pull		5	1.600		410	64		474	555
0900	Touch bar and vertical rod, low profile, exit only		5	1.600		515	64		579	665
0910	Outside key and pull		4	2		545	80		625	720
1000	Mortise, bar, exit only		4	2		515	80		595	690
1600	Touch bar, exit only		4	2		590	80		670	775
2000	Narrow stile, rim mounted, bar, exit only		6	1.333		620	53.50		673.50	765
2010	Outside key and pull		5	1.600		675	64		739	840
2200	Bar and vertical rod, exit only		5	1.600		635	64		699	800
2210	Outside key and pull		4	2		635	80		715	825
2400	Bar and concealed rod, exit only		3	2.667		745	107		852	980
3000	Mortise, bar, exit only		4	2		570	80		650	750
3600	Touch bar, exit only		4	2		825	80		905	1,025
6000	Trim, rim mounted, cylinder and pull		25	.320		111	12.80		123.80	142
6100	Cylinder, pull and thumb piece		20	.400		110	16		126	146
6200	Pull only		30	.267		64	10.65		74.65	86.50
6400	Mortise, cylinder and pull		25	.320		130	12.80		142.80	163
6500	Cylinder, pull and thumb piece		20	.400		125	16		141	163
6600	Pull only		30	.267		111	10.65		121.65	139

08 71 20.40 Lockset

		Crew	Daily Output	Labor-Hours	Unit	Material	Labor	Equipment	Total	Total Incl O&P
0010	**LOCKSET**, Standard duty									
0020	Non-keyed, passage	1 Carp	12	.667	Ea.	49	26.50		75.50	95
0100	Privacy		12	.667		61	26.50		87.50	109
0400	Keyed, single cylinder function		10	.800		83.50	32		115.50	142
0420	Hotel		8	1		120	40		160	194
0500	Lever handled, keyed, single cylinder function		10	.800		148	32		180	213
1000	Heavy duty with sectional trim, non-keyed, passages		12	.667		138	26.50		164.50	194
1100	Privacy		12	.667		173	26.50		199.50	232
1400	Keyed, single cylinder function		10	.800		210	32		242	281
1420	Hotel		8	1		310	40		350	400
1600	Communicating		10	.800		235	32		267	310
1690	For re-core cylinder, add					34			34	37.50
1700	Residential, interior door, minimum	1 Carp	16	.500		17	20		37	49.50
1720	Maximum		8	1		43.50	40		83.50	110
1800	Exterior, minimum		14	.571		36.50	23		59.50	75.50
1810	Average		8	1		69	40		109	138
1820	Maximum		8	1		156	40		196	234

08 71 20.41 Dead Locks

		Crew	Daily Output	Labor-Hours	Unit	Material	Labor	Equipment	Total	Total Incl O&P
0010	**DEAD LOCKS**									
0011	Mortise heavy duty outside key (security item)	1 Carp	9	.889	Ea.	150	35.50		185.50	220
0020	Double cylinder		9	.889		162	35.50		197.50	233
0100	Medium duty, outside key		10	.800		110	32		142	171
0110	Double cylinder		10	.800		142	32		174	206

08 71 Door Hardware

08 71 20 – Hardware

08 71 20.41 Dead Locks		Crew	Daily Output	Labor-Hours	Unit	Material	2009 Bare Costs Labor	Equipment	Total	Total Incl O&P
1000	Tubular, standard duty, outside key	1 Carp	10	.800	Ea.	61	32		93	117
1010	Double cylinder		10	.800		79	32		111	137
1200	Night latch, outside key	↓	10	.800	↓	78.50	32		110.50	136
08 71 20.42 Mortise Locksets										
0010	**MORTISE LOCKSETS**, Comm., wrought knobs & full escutcheon trim									
0020	Non-keyed, passage, minimum	1 Carp	9	.889	Ea.	175	35.50		210.50	248
0030	Maximum		8	1		288	40		328	375
0040	Privacy, minimum		9	.889		190	35.50		225.50	264
0050	Maximum		8	1		310	40		350	400
0100	Keyed, office/entrance/apartment, minimum		8	1		220	40		260	305
0110	Maximum		7	1.143		370	45.50		415.50	475
0120	Single cylinder, typical, minimum		8	1		189	40		229	270
0130	Maximum		7	1.143		350	45.50		395.50	455
0200	Hotel, minimum		7	1.143		220	45.50		265.50	315
0210	Maximum		6	1.333		365	53.50		418.50	485
0300	Communication, double cylinder, minimum		8	1		215	40		255	299
0310	Maximum		7	1.143		286	45.50		331.50	385
1000	Wrought knobs and sectional trim, non-keyed, passage, minimum		10	.800		119	32		151	181
1010	Maximum		9	.889		230	35.50		265.50	310
1040	Privacy, minimum		10	.800		136	32		168	200
1050	Maximum		9	.889		250	35.50		285.50	330
1100	Keyed, entrance, office/apartment, minimum		9	.889		206	35.50		241.50	282
1110	Maximum		8	1		295	40		335	385
1120	Single cylinder, typical, minimum		9	.889		200	35.50		235.50	275
1130	Maximum	↓	8	1	↓	288	40		328	375
2000	Cast knobs and full escutcheon trim									
2010	Non-keyed, passage, minimum	1 Carp	9	.889	Ea.	249	35.50		284.50	330
2020	Maximum		8	1		400	40		440	500
2040	Privacy, minimum		9	.889		305	35.50		340.50	390
2050	Maximum		8	1		440	40		480	540
2120	Keyed, single cylinder, typical, minimum		8	1		310	40		350	400
2130	Maximum		7	1.143		480	45.50		525.50	600
2200	Hotel, minimum		7	1.143		320	45.50		365.50	420
2210	Maximum		6	1.333		570	53.50		623.50	710
3000	Cast knob and sectional trim, non-keyed, passage, minimum		10	.800		196	32		228	266
3010	Maximum		10	.800		380	32		412	470
3040	Privacy, minimum		10	.800		220	32		252	292
3050	Maximum		10	.800		390	32		422	480
3100	Keyed, office/entrance/apartment, minimum		9	.889		250	35.50		285.50	330
3110	Maximum		9	.889		400	35.50		435.50	495
3120	Single cylinder, typical, minimum		9	.889		250	35.50		285.50	330
3130	Maximum	↓	9	.889		490	35.50		525.50	595
3190	For re-core cylinder, add					33			33	36.50
3800	Cipher lockset w/key pad (security item)	1 Carp	13	.615		775	24.50		799.50	890
3900	Cipher lockset, with dial for swinging doors (security item)		13	.615		1,675	24.50		1,699.50	1,900
3920	with dial for swinging doors & drill resistant plate (security item)		12	.667		2,025	26.50		2,051.50	2,275
3950	Cipher lockset with dial for safe/vault door (security item)	↓	12	.667	↓	1,300	26.50		1,326.50	1,475
3980	Keyless, pushbutton type									
4000	Residential/light commercial, deadbolt, standard	1 Carp	9	.889	Ea.	114	35.50		149.50	180
4010	Heavy duty		9	.889		134	35.50		169.50	202
4020	Industrial, heavy duty, with deadbolt		9	.889		271	35.50		306.50	355
4030	Key override	↓	9	.889		300	35.50		335.50	390

08 71 Door Hardware

08 71 20 – Hardware

08 71 20.42 Mortise Locksets		Crew	Daily Output	Labor-Hours	Unit	Material	2009 Bare Costs Labor	Equipment	Total	Total Incl O&P
4040	Lever activated handle	1 Carp	9	.889	Ea.	320	35.50		355.50	410
4050	Key override		9	.889		360	35.50		395.50	450
4060	Double sided pushbutton type		8	1		660	40		700	790
4070	Key override		8	1		655	40		695	780
08 71 20.50 Door Stops										
0010	**DOOR STOPS**									
0020	Holder & bumper, floor or wall	1 Carp	32	.250	Ea.	33.50	10		43.50	52.50
1300	Wall bumper, 4" diameter, with rubber pad, aluminum		32	.250		10.25	10		20.25	27
1600	Door bumper, floor type, aluminum		32	.250		5.85	10		15.85	22
1900	Plunger type, door mounted		32	.250		27	10		37	45
08 71 20.55 Push-Pull Plates										
0010	**PUSH-PULL PLATES**									
0100	Push plate, .050 thick, 4" x 16", aluminum	1 Carp	12	.667	Ea.	8.65	26.50		35.15	51
0500	Bronze		12	.667		20.50	26.50		47	64
1500	Pull handle and push bar, aluminum		11	.727		126	29		155	184
2000	Bronze		10	.800		163	32		195	230
3000	Push plate both sides, aluminum		14	.571		17.75	23		40.75	55
3500	Bronze		13	.615		41.50	24.50		66	83.50
4000	Door pull, designer style, cast aluminum, minimum		12	.667		67.50	26.50		94	116
5000	Maximum		8	1		365	40		405	460
6000	Cast bronze, minimum		12	.667		85.50	26.50		112	136
7000	Maximum		8	1		400	40		440	500
8000	Walnut, minimum		12	.667		65.50	26.50		92	114
9000	Maximum		8	1		365	40		405	460
08 71 20.60 Entrance Locks										
0010	**ENTRANCE LOCKS**									
0015	Cylinder, grip handle deadlocking latch	1 Carp	9	.889	Ea.	127	35.50		162.50	195
0020	Deadbolt		8	1		154	40		194	231
0100	Push and pull plate, dead bolt		8	1		146	40		186	223
0200	Push bar and pull, dead bolt, bronze		7	1.143		265	45.50		310.50	365
0240	Push bar and pull bar, dead bolt, bronze		7	1.143		415	45.50		460.50	530
0900	For handicapped lever, add					161			161	177
08 71 20.65 Thresholds										
0010	**THRESHOLDS**									
0011	Threshold 3' long saddles aluminum	1 Carp	48	.167	L.F.	4.02	6.65		10.67	14.75
0100	Aluminum, 8" wide, 1/2" thick		12	.667	Ea.	38.50	26.50		65	83.50
0500	Bronze		60	.133	L.F.	38.50	5.35		43.85	51
0600	Bronze, panic threshold, 5" wide, 1/2" thick		12	.667	Ea.	65.50	26.50		92	114
0700	Rubber, 1/2" thick, 5-1/2" wide		20	.400		37.50	16		53.50	66.50
0800	2-3/4" wide		20	.400		16	16		32	42.50
08 71 20.70 Floor Checks										
0010	**FLOOR CHECKS**									
0020	For over 3' wide doors single acting	1 Carp	2.50	3.200	Ea.	510	128		638	760
0500	Double acting	"	2.50	3.200	"	660	128		788	925
08 71 20.90 Hinges										
0010	**HINGES** R087120-10									
0012	Full mortise, avg. freq., steel base, USP, 4-1/2" x 4-1/2"				Pr.	24.50			24.50	27
0040	US26D					30.50			30.50	33.50
0080	US10A					35.50			35.50	39
0100	5" x 5", USP					41			41	45
0200	6" x 6", USP					87			87	95.50

08 71 Door Hardware

08 71 20 – Hardware

08 71 20.90 Hinges		Crew	Daily Output	Labor-Hours	Unit	Material	2009 Bare Costs Labor	Equipment	Total	Total Incl O&P
0210										
0400	Brass base, 4-1/2" x 4-1/2", US10				Pr.	50			50	55
0440	US26D					59.50			59.50	65.50
0480	US10B					66.50			66.50	73
0500	5" x 5", US10					73.50			73.50	81
0600	6" x 6", US10					125			125	137
0800	Stainless steel base, 4-1/2" x 4-1/2", US32				▼	74.50			74.50	82
0900	For non removable pin, add (security item)				Ea.	4.24			4.24	4.66
0910	For floating pin, driven tips, add					3.09			3.09	3.40
0930	For hospital type tip on pin, add					13.35			13.35	14.70
0940	For steeple type tip on pin, add				▼	11.70			11.70	12.85
0950	Full mortise, high frequency, steel base, 3-1/2" x 3-1/2", US26D				Pr.	25.50			25.50	28.50
1000	4-1/2" x 4-1/2", USP					58.50			58.50	64.50
1040	US26D					58.50			58.50	64.50
1080	US26					76.50			76.50	84
1100	5" x 5", USP					54.50			54.50	60
1200	6" x 6", USP					134			134	147
1300	8" x 8", USP					226			226	248
1400	Brass base, 3-1/2" x 3-1/2", US4					45			45	49.50
1430	4-1/2" x 4-1/2", US10					78.50			78.50	86.50
1440	US26D					96.50			96.50	106
1480	US10B					110			110	121
1500	5" x 5", US10					118			118	130
1600	6" x 6", US10					170			170	187
1700	8" x 8", US10					315			315	345
1800	Stainless steel base, 4-1/2" x 4-1/2", US32					125			125	137
1810	5" x 4-1/2", US32				▼	175			175	193
1930	For hospital type tip on pin, add				Ea.	11.65			11.65	12.80
1950	Full mortise, low frequency, steel base, 3-1/2" x 3-1/2", US26D				Pr.	10.55			10.55	11.60
2000	4-1/2" x 4-1/2", USP					12.05			12.05	13.25
2040	US26D					15.65			15.65	17.25
2080	US10A					19.65			19.65	21.50
2100	5" x 5", USP					30			30	33
2200	6" x 6", USP					59.50			59.50	65.50
2300	4-1/2" x 4-1/2", US3					17.40			17.40	19.15
2310	5" x 5", US3					43			43	47.50
2400	Brass base, 4-1/2" x 4-1/2", US10					42			42	46
2440	US26D					49			49	54
2480	US10A					45			45	49.50
2500	5" x 5", US10					63.50			63.50	70
2800	Stainless steel base, 4-1/2" x 4-1/2", US32				▼	72			72	79.50
3000	Half surface, half mortise, or full surface, average frequency									
3010	Steel base, 4-1/2" x 4-1/2", USP				Pr.	38			38	41.50
3040	US26D					60.50			60.50	66.50
3080	US10A					81			81	89
3100	5" x 5", USP					75			75	82.50
3400	Brass base, 4-1/2" x 4-1/2", US10					143			143	157
3440	US26D					154			154	169
3480	US10B					169			169	185
3500	5" x 5", US10					182			182	200
3800	Stainless steel base, 4-1/2" x 4-1/2", US32				▼	197			197	216
4000	Half surface, half mortise or full surface, high frequency									
4010	Steel base, 4-1/2" x 4-1/2", USP				Pr.	75.50			75.50	83

08 71 Door Hardware

08 71 20 – Hardware

08 71 20.90 Hinges

		Crew	Daily Output	Labor-Hours	Unit	Material	2009 Bare Costs Labor	Equipment	Total	Total Incl O&P
4040	US26D				Pr.	139			139	153
4080	US10A					168			168	185
4100	5" x 5", USP					148			148	163
4400	Brass base, 4-1/2" x 4-1/2", US10					172			172	189
4440	US26D					180			180	198
4480	US10B					205			205	226
4500	5" x 5", US10					204			204	224
4800	Stainless steel base, 4-1/2" x 4-1/2", US32				↓	226			226	249
5000	Half surface, half mortise, or full surface, low frequency									
5010	Steel base, 4-1/2" x 4-1/2", USP				Pr.	28.50			28.50	31
5040	US26D					35.50		·	35.50	39
5080	US10A				↓	57			57	62.50

08 71 20.91 Special Hinges

		Crew	Daily Output	Labor-Hours	Unit	Material	2009 Bare Costs Labor	Equipment	Total	Total Incl O&P
0010	**SPECIAL HINGES** R087120-10									
0015	Paumelle, high frequency									
0020	Steel base, 6" x 4-1/2", US10				Pr.	139			139	153
0040	US26D					173			173	190
0080	US10A					181			181	199
0100	Bronze base, 5" x 4-1/2", US10					174			174	192
0140	US26D					192			192	211
0180	US3					202			202	223
0200	Paumelle, average frequency, steel base, 4-1/2" x 3-1/2", US10					94			94	104
0240	US26D					98.50			98.50	108
0280	US10A					101			101	111
0400	Olive knuckle, low frequency, brass base, 6" x 4-1/2", US10					159			159	175
0440	US26D					179			179	197
0480	US3				↓	180			180	198
0800	Emergency door pivot, average frequency									
0810	Brass base, 4-7/8" jamb plate, USP				Pr.	64.50			64.50	71
0840	US26D					71.50			71.50	78.50
0880	For emergency door stop and hold back, add				↓	61			61	67
1000	Electric hinge with concealed conductor, average frequency									
1010	Steel base, 4-1/2" x 4-1/2", US26D				Pr.	310			310	340
1100	Bronze base, 4-1/2" x 4-1/2", US26D				"	315			315	345
1200	Electric hinge with concealed conductor, high frequency									
1210	Steel base, 4-1/2" x 4-1/2", US26D				Pr.	231			231	255
1400	Non template, full mortise, low frequency									
1410	Steel base, 4" x 4", USP				Pr.	16.45			16.45	18.10
1440	US26D					18.50			18.50	20.50
1480	US10A				↓	22			22	24.50
1500	Non template, full mortise, average frequency									
1510	Steel base, 4" x 4", USP				Pr.	34.50			34.50	38
1540	US26D					33			33	36.50
1580	US10A					42.50			42.50	47
1600	Double weight, 800 lb., steel base, removable pin, 5" x 6", USP					128			128	141
1700	Steel base-welded pin, 5" x 6", USP					158			158	174
1800	Triple weight, 2000 lb., steel base, welded pin, 5" x 6", USP					147			147	162
2000	Pivot reinf., high frequency, steel base, 7-3/4" door plate, USP					177			177	195
2040	US26D					156			156	171
2080	US10A					189			189	208
2200	Bronze base, 7-3/4" door plate, US10					222			222	245
2240	US26D				↓	232			232	255

08 71 Door Hardware

08 71 20 – Hardware

08 71 20.91 Special Hinges

		Crew	Daily Output	Labor-Hours	Unit	Material	2009 Bare Costs Labor	Equipment	Total	Total Incl O&P
2280	US10A				Pr.	266			266	293
3000	Swing clear, full mortise, full or half surface, high frequency,									
3010	Steel base, 5" high, USP				Pr.	154			154	169
3040	US26D					148			148	163
3080	US10A				↓	187			187	206
3200	Swing clear, full mortise, average frequency									
3210	Steel base, 4-1/2" high, USP				Pr.	122			122	135
3280	US10A				"	150			150	165
3400	Swing clear, half mortise, high frequency									
3410	Steel base, 4-1/2" high, USP				Pr.	154			154	170
3440	US26D					159			159	175
3480	US10A					188			188	207
4000	Wide throw, average frequency, steel base, 4-1/2" x 6", USP					93.50			93.50	103
4040	US26D					99.50			99.50	110
4080	US10A					103			103	113
4100	5" x 7", USP					114			114	126
4200	High frequency, steel base, 4-1/2" x 6", USP					143			143	157
4240	US26D					166			166	183
4280	US10A					185			185	203
4300	5" x 7", USP					146			146	160
4400	Wide throw, low frequency, steel base, 4-1/2" x 6", USP					70			70	77
4440	US26D					86			86	94.50
4480	US10A					92			92	101
4500	5" x 7", USP				↓	99.50			99.50	110
4600	Spring hinge, single acting, 6" flange, steel				Ea.	52.50			52.50	57.50
4700	Brass					92			92	101
4900	Double acting, 6" flange, steel					94			94	103
4950	Brass				↓	153			153	168
9000	Continuous hinge, steel, full mortise, heavy duty	2 Carp	64	.250	L.F.	12.40	10		22.40	29

08 71 20.95 Kick Plates

		Crew	Daily Output	Labor-Hours	Unit	Material	Labor	Equipment	Total	Total Incl O&P
0010	**KICK PLATES**									
0020	Stainless steel, 6" high, for 3' door	1 Carp	15	.533	Ea.	31	21.50		52.50	67
0500	Bronze 6" high, for 3' door	"	15	.533	"	43.50	21.50		65	81

08 71 21 – Astragals

08 71 21.10 Exterior Mouldings, Astragals

		Crew	Daily Output	Labor-Hours	Unit	Material	Labor	Equipment	Total	Total Incl O&P
0010	**EXTERIOR MOULDINGS, ASTRAGALS**									
0400	One piece, overlapping cadmium plated steel, flat, 3/16" x 2"	1 Carp	90	.089	L.F.	3.40	3.55		6.95	9.25
0600	Prime coated steel, flat, 1/8" x 3"		90	.089		4.65	3.55		8.20	10.60
0800	Stainless steel, flat, 3/32" x 1-5/8"		90	.089		17.80	3.55		21.35	25
1000	Aluminum, flat, 1/8" x 2"		90	.089		3.45	3.55		7	9.30
1200	Nail on, "T" extrusion		120	.067		.78	2.66		3.44	4.99
1300	Vinyl bulb insert		105	.076		1.25	3.04		4.29	6.10
1600	Screw on, "T" extrusion		90	.089		5.05	3.55		8.60	11.05
1700	Vinyl insert		75	.107		3.40	4.26		7.66	10.35
2000	"L" extrusion, neoprene bulbs		75	.107		1.95	4.26		6.21	8.75
2100	Neoprene sponge insert		75	.107		5.95	4.26		10.21	13.15
2200	Magnetic		75	.107		9.80	4.26		14.06	17.40
2400	Spring hinged security seal, with cam		75	.107		6.30	4.26		10.56	13.55
2600	Spring loaded locking bolt, vinyl insert		45	.178		8.60	7.10		15.70	20.50
2800	Neoprene sponge strip, "Z" shaped, aluminum		60	.133		4.10	5.35		9.45	12.75
2900	Solid neoprene strip, nail on aluminum strip	↓	90	.089	↓	3.40	3.55		6.95	9.25
3000	One piece stile protection									

08 71 Door Hardware

08 71 21 – Astragals

08 71 21.10 Exterior Mouldings, Astragals		Crew	Daily Output	Labor-Hours	Unit	Material	2009 Bare Costs Labor	Equipment	Total	Total Incl O&P
3020	Neoprene fabric loop, nail on aluminum strips	1 Carp	60	.133	L.F.	.65	5.35		6	8.95
3110	Flush mounted aluminum extrusion, 1/2" x 1-1/4"		60	.133		3.26	5.35		8.61	11.85
3140	3/4" x 1-3/8"		60	.133		4.15	5.35		9.50	12.80
3160	1-1/8" x 1-3/4"		60	.133		6.60	5.35		11.95	15.50
3300	Mortise, 9/16" x 3/4"		60	.133		3.50	5.35		8.85	12.10
3320	13/16" x 1-3/8"		60	.133		3.80	5.35		9.15	12.45
3600	Spring bronze strip, nail on type		105	.076		3	3.04		6.04	8
3620	Screw on, with retainer		75	.107		2.45	4.26		6.71	9.30
3800	Flexible stainless steel housing, pile insert, 1/2" door		105	.076		6.75	3.04		9.79	12.15
3820	3/4" door		105	.076		7.60	3.04		10.64	13.05
4000	Extruded aluminum retainer, flush mount, pile insert		105	.076		2.36	3.04		5.40	7.30
4080	Mortise, felt insert		90	.089		4.20	3.55		7.75	10.10
4160	Mortise with spring, pile insert		90	.089		3.30	3.55		6.85	9.15
4400	Rigid vinyl retainer, mortise, pile insert		105	.076		2.35	3.04		5.39	7.30
4600	Wool pile filler strip, aluminum backing		105	.076		2.36	3.04		5.40	7.30
5000	Two piece overlapping astragal, extruded aluminum retainer									
5010	Pile insert	1 Carp	60	.133	L.F.	3.06	5.35		8.41	11.60
5020	Vinyl bulb insert		60	.133		1.98	5.35		7.33	10.45
5040	Vinyl flap insert		60	.133		6.15	5.35		11.50	15.05
5060	Solid neoprene flap insert		60	.133		6.10	5.35		11.45	14.95
5080	Hypalon rubber flap insert		60	.133		6.25	5.35		11.60	15.10
5090	Snap on cover, pile insert		60	.133		7.10	5.35		12.45	16.05
5400	Magnetic aluminum, surface mounted		60	.133		24	5.35		29.35	35
5500	Interlocking aluminum, 5/8" x 1" neoprene bulb insert		45	.178		3.80	7.10		10.90	15.20
5600	Adjustable aluminum, 9/16" x 21/32", pile insert		45	.178		18.15	7.10		25.25	31
5790	For vinyl bulb, deduct					.50			.50	.55
5800	Magnetic, adjustable, 9/16" x 21/32"	1 Carp	45	.178		23.50	7.10		30.60	36.50
6000	Two piece stile protection									
6010	Cloth backed rubber loop, 1" gap, nail on aluminum strips	1 Carp	45	.178	L.F.	3.90	7.10		11	15.30
6040	Screw on aluminum strips		45	.178		6.10	7.10		13.20	17.70
6100	1-1/2" gap, screw on aluminum extrusion		45	.178		5.45	7.10		12.55	17
6240	Vinyl fabric loop, slotted aluminum extrusion, 1" gap		45	.178		1.90	7.10		9	13.10
6300	1-1/4" gap		45	.178		5.75	7.10		12.85	17.35

08 71 25 – Weatherstripping

08 71 25.10 Mechanical Seals, Weatherstripping

		Crew	Daily Output	Labor-Hours	Unit	Material	2009 Bare Costs Labor	Equipment	Total	Total Incl O&P
0010	**MECHANICAL SEALS, WEATHERSTRIPPING**									
1000	Doors, wood frame, interlocking, for 3' x 7' door, zinc	1 Carp	3	2.667	Opng.	15.50	107		122.50	182
1100	Bronze		3	2.667		24	107		131	192
1300	6' x 7' opening, zinc		2	4		16.90	160		176.90	267
1400	Bronze		2	4		32	160		192	283
1700	Wood frame, spring type, bronze									
1800	3' x 7' door	1 Carp	7.60	1.053	Opng.	21.50	42		63.50	88.50
1900	6' x 7' door	"	7	1.143	"	26.50	45.50		72	100
2200	Metal frame, spring type, bronze									
2300	3' x 7' door	1 Carp	3	2.667	Opng.	35.50	107		142.50	204
2400	6' x 7' door	"	2.50	3.200	"	46	128		174	249
2500	For stainless steel, spring type, add					133%				
2700	Metal frame, extruded sections, 3' x 7' door, aluminum	1 Carp	2	4	Opng.	45	160		205	298
2800	Bronze		2	4		114	160		274	375
3100	6' x 7' door, aluminum		1.20	6.667		57.50	266		323.50	480
3200	Bronze		1.20	6.667		134	266		400	560
3500	Threshold weatherstripping									

08 71 Door Hardware

08 71 25 – Weatherstripping

08 71 25.10 Mechanical Seals, Weatherstripping

		Crew	Daily Output	Labor-Hours	Unit	Material	2009 Bare Costs Labor	Equipment	Total	Total Incl O&P
3650	Door sweep, flush mounted, aluminum	1 Carp	25	.320	Ea.	13.30	12.80		26.10	34.50
3700	Vinyl		25	.320		15.70	12.80		28.50	37
5000	Garage door bottom weatherstrip, 12' aluminum, clear		14	.571		21.50	23		44.50	59
5010	Bronze		14	.571		81.50	23		104.50	125
5050	Bottom protection, Rubber		14	.571		22	23		45	60
5100	Threshold		14	.571		88.50	23		111.50	133

08 74 Access Control Hardware

08 74 13 – Card Key Access Control Hardware

08 74 13.50 Card Key Access

		Crew	Daily Output	Labor-Hours	Unit	Material	2009 Bare Costs Labor	Equipment	Total	Total Incl O&P
0010	**CARD KEY ACCESS**									
0020	Card type, 1 time zone, minimum				Ea.	615			615	675
0040	Maximum					1,200			1,200	1,300
0060	3 time zones, minimum					1,400			1,400	1,550
0080	Maximum					1,900			1,900	2,100
0100	System with printer, and control console, 3 zones				Total	9,150			9,150	10,100
0120	6 zones				"	12,000			12,000	13,200
0140	For each door, minimum, add				Ea.	1,400			1,400	1,550
0160	Maximum, add				"	2,000			2,000	2,200

08 75 Window Hardware

08 75 30 – Weatherstripping

08 75 30.10 Mechanical Weather Seals

		Crew	Daily Output	Labor-Hours	Unit	Material	2009 Bare Costs Labor	Equipment	Total	Total Incl O&P
0010	**MECHANICAL WEATHER SEALS**, Window, double hung, 3' X 5'									
0020	Zinc	1 Carp	7.20	1.111	Opng.	12.30	44.50		56.80	82.50
0100	Bronze		7.20	1.111		25.50	44.50		70	97
0500	As above but heavy duty, zinc		4.60	1.739		17.30	69.50		86.80	127
0600	Bronze		4.60	1.739		29	69.50		98.50	140
9000	Minimum labor/equipment charge	1 Clab	4.60	1.739	Job		55		55	85

08 79 Hardware Accessories

08 79 13 – Key Storage Equipment

08 79 13.10 Key Cabinets

		Crew	Daily Output	Labor-Hours	Unit	Material	2009 Bare Costs Labor	Equipment	Total	Total Incl O&P
0010	**KEY CABINETS**									
0020	Wall mounted, 60 key capacity	1 Carp	20	.400	Ea.	81.50	16		97.50	115
0200	Drawer type, 600 key capacity	1 Clab	15	.533		725	16.85		741.85	825
0300	2,400 key capacity		20	.400		5,500	12.65		5,512.65	6,075
0400	Tray type, 20 key capacity		50	.160		45	5.05		50.05	57.50
0500	50 key capacity		40	.200		82.50	6.30		88.80	101

08 79 20 – Door Accessories

08 79 20.10 Door Hardware Accessories

		Crew	Daily Output	Labor-Hours	Unit	Material	2009 Bare Costs Labor	Equipment	Total	Total Incl O&P
0010	**DOOR HARDWARE ACCESSORIES**									
0140	Door bolt, surface, 4"	1 Carp	32	.250	Ea.	8.65	10		18.65	25
0160	Door latch	"	12	.667	"	8.05	26.50		34.55	50.50
0200	Sliding closet door									
0220	Track and hanger, single	1 Carp	10	.800	Ea.	49.50	32		81.50	104
0240	Double		8	1		72.50	40		112.50	142

08 79 Hardware Accessories

08 79 20 – Door Accessories

08 79 20.10 Door Hardware Accessories

08 79 20.10 Door Hardware Accessories		Crew	Daily Output	Labor-Hours	Unit	Material	2009 Bare Costs Labor	Equipment	Total	Total Incl O&P
0260	Door guide, single	1 Carp	48	.167	Ea.	23.50	6.65		30.15	36.50
0280	Double		48	.167		32	6.65		38.65	46
0600	Deadbolt and lock cover plate, brass or stainless steel		30	.267		26.50	10.65		37.15	46
0620	Hole cover plate, brass or chrome		35	.229		6.95	9.15		16.10	22
2240	Mortise lockset, passage, lever handle		9	.889		194	35.50		229.50	268
4000	Security chain, standard		18	.444		7.40	17.75		25.15	35.50

08 81 Glass Glazing

08 81 10 – Float Glass

08 81 10.10 Various Types and Thickness of Float Glass

		Crew	Daily Output	Labor-Hours	Unit	Material	2009 Bare Costs Labor	Equipment	Total	Total Incl O&P
0010	**VARIOUS TYPES AND THICKNESS OF FLOAT GLASS** R088110-10									
0020	3/16" Plain	2 Glaz	130	.123	S.F.	4.48	4.75		9.23	12.15
0200	Tempered, clear		130	.123		6.05	4.75		10.80	13.85
0300	Tinted		130	.123		7.05	4.75		11.80	14.95
0600	1/4" thick, clear, plain		120	.133		5.25	5.15		10.40	13.60
0700	Tinted		120	.133		7.40	5.15		12.55	15.95
0800	Tempered, clear		120	.133		7.45	5.15		12.60	16
0900	Tinted		120	.133		9.75	5.15		14.90	18.50
1600	3/8" thick, clear, plain		75	.213		9.10	8.25		17.35	22.50
1700	Tinted		75	.213		14.80	8.25		23.05	29
1800	Tempered, clear		75	.213		15.10	8.25		23.35	29
1900	Tinted		75	.213		16.85	8.25		25.10	31
2200	1/2" thick, clear, plain		55	.291		18.15	11.25		29.40	37
2300	Tinted		55	.291		26	11.25		37.25	45.50
2400	Tempered, clear		55	.291		22	11.25		33.25	41.50
2500	Tinted		55	.291		26	11.25		37.25	45.50
2800	5/8" thick, clear, plain		45	.356		26	13.70		39.70	49.50
2900	Tempered, clear		45	.356		29.50	13.70		43.20	53.50
3200	3/4" thick, clear, plain		35	.457		33.50	17.65		51.15	63
3300	Tempered, clear		35	.457		39	17.65		56.65	69.50
3600	1" thick, clear, plain		30	.533		55.50	20.50		76	92
8900	For low emissivity coating for 3/16" & 1/4" only, add to above					16%				

08 81 13 – Decorative Glass Glazing

08 81 13.10 Beveled Glass

		Crew	Daily Output	Labor-Hours	Unit	Material	2009 Bare Costs Labor	Equipment	Total	Total Incl O&P
0010	**BEVELED GLASS**, with design patterns									
0020	Minimum	2 Glaz	150	.107	S.F.	55	4.12		59.12	66.50
0050	Average		125	.128		129	4.94		133.94	149
0100	Maximum		100	.160		229	6.20		235.20	260

08 81 13.20 Faceted Glass

		Crew	Daily Output	Labor-Hours	Unit	Material	2009 Bare Costs Labor	Equipment	Total	Total Incl O&P
0010	**FACETED GLASS**, Color tinted 3/4" thick									
0020	Minimum	2 Glaz	95	.168	S.F.	46.50	6.50		53	61
0100	Maximum	"	75	.213	"	81.50	8.25		89.75	102

08 81 13.30 Sandblasted Glass

		Crew	Daily Output	Labor-Hours	Unit	Material	2009 Bare Costs Labor	Equipment	Total	Total Incl O&P
0010	**SANDBLASTED GLASS**, float glass									
0020	1/8" thick	2 Glaz	160	.100	S.F.	10.25	3.86		14.11	17.15
0100	3/16" thick		130	.123		11.40	4.75		16.15	19.70
0500	1/4" thick		120	.133		11.80	5.15		16.95	21
0600	3/8" thick		75	.213		12.85	8.25		21.10	26.50

08 81 Glass Glazing

08 81 20 – Vision Panels

08 81 20.10 Full Vision

		Crew	Daily Output	Labor-Hours	Unit	Material	2009 Bare Costs Labor	Equipment	Total	Total Incl O&P
0010	**FULL VISION**, window system with 3/4" glass mullions									
0020	Up to 10' high	H-2	130	.185	S.F.	57.50	6.70		64.20	73
0100	10' to 20' high, minimum		110	.218		61	7.90		68.90	79
0150	Average		100	.240		65.50	8.70		74.20	86
0200	Maximum		80	.300		74	10.90		84.90	97.50

08 81 25 – Glazing Variables

08 81 25.10 Applications of Glazing

					Unit	Material	Labor	Equipment	Total	Total Incl O&P
0010	**APPLICATIONS OF GLAZING** R088110-10									
0500	For high rise glazing, exterior, add per S.F. per story				S.F.					.25
0600	For glass replacement, add				"		100%			
0700	For gasket settings, add				L.F.	5.55			5.55	6.10
0900	For sloped glazing, add				S.F.		25%			
2000	Fabrication, polished edges, 1/4" thick				Inch	.48			.48	.53
2100	1/2" thick					1.21			1.21	1.33
2500	Mitered edges, 1/4" thick					1.21			1.21	1.33
2600	1/2" thick					1.96			1.96	2.16

08 81 30 – Insulating Glass

08 81 30.10 Reduce Heat Transfer Glass

			Crew	Daily Output	Labor-Hours	Unit	Material	Labor	Equipment	Total	Total Incl O&P
0010	**REDUCE HEAT TRANSFER GLASS**, 2 lites 1/8" float, 1/2" thk under 15 S.F.										
0020	Clear R088110-10	G	2 Glaz	95	.168	S.F.	8.95	6.50		15.45	19.70
0100	Tinted	G		95	.168		12.55	6.50		19.05	23.50
0200	2 lites 3/16" float, for 5/8" thk unit, 15 to 30 S.F., clear	G		90	.178		12.95	6.85		19.80	24.50
0300	Tinted	G		90	.178		13.30	6.85		20.15	25
0400	1" thk, dbl. glazed, 1/4" float, 30-70 S.F., clear	G		75	.213		15.20	8.25		23.45	29
0500	Tinted	G		75	.213		22	8.25		30.25	37
0600	1" thick double glazed, 1/4" float, 1/4" wire	G		75	.213		21	8.25		29.25	35.50
0700	1/4" float, 1/4" tempered			75	.213		20.50	8.25		28.75	35.50
0800	1/4" wire, 1/4" tempered			75	.213		29	8.25		37.25	44.50
0900	Both lites, 1/4" wire			75	.213		35	8.25		43.25	51
2000	Both lites, light & heat reflective	G		85	.188		29.50	7.25		36.75	43.50
2500	Heat reflective, film inside, 1" thick unit, clear	G		85	.188		26	7.25		33.25	39.50
2600	Tinted	G		85	.188		28	7.25		35.25	42
3000	Film on weatherside, clear, 1/2" thick unit	G		95	.168		18.50	6.50		25	30.50
3100	5/8" thick unit	G		90	.178		18.10	6.85		24.95	30.50
3200	1" thick unit	G		85	.188		25.50	7.25		32.75	39

08 81 35 – Translucent Glass

08 81 35.10 Obscure Glass

		Crew	Daily Output	Labor-Hours	Unit	Material	Labor	Equipment	Total	Total Incl O&P
0010	**OBSCURE GLASS**, 1/8" thick									
0020	Minimum	2 Glaz	140	.114	S.F.	10.45	4.41		14.86	18.15
0100	Maximum		125	.128		12.55	4.94		17.49	21.50
0300	7/32" thick, minimum		120	.133		11.60	5.15		16.75	20.50
0400	Maximum		105	.152		14.80	5.90		20.70	25

08 81 35.20 Patterned Glass

		Crew	Daily Output	Labor-Hours	Unit	Material	Labor	Equipment	Total	Total Incl O&P
0010	**PATTERNED GLASS**, colored, 1/8" thick									
0020	Minimum	2 Glaz	140	.114	S.F.	8.60	4.41		13.01	16.10
0100	Maximum		125	.128		14.40	4.94		19.34	23.50
0300	7/32" thick, minimum		120	.133		10.70	5.15		15.85	19.55
0400	Maximum		105	.152		15.40	5.90		21.30	26

08 81 Glass Glazing

08 81 50 – Spandrel Glass

08 81 50.10 Glass for Non Vision Areas

		Crew	Daily Output	Labor-Hours	Unit	Material	2009 Bare Costs Labor	2009 Bare Costs Equipment	Total	Total Incl O&P
0010	**GLASS FOR NON VISION AREAS**, 1/4" thick standard colors									
0020	Up to 100 S.F.	2 Glaz	110	.145	S.F.	15	5.60		20.60	25
0200	1,000 to 2,000 S.F.	"	120	.133	"	13.95	5.15		19.10	23
0300	For custom colors, add				Total	10%				
0500	For 3/8" thick, add				S.F.	10.65			10.65	11.70
1000	For double coated, 1/4" thick, add					3.97			3.97	4.37
1200	For insulation on panels, add					6.55			6.55	7.20
2000	Panels, insulated, with aluminum backed fiberglass, 1" thick	2 Glaz	120	.133		15.85	5.15		21	25
2100	2" thick	"	120	.133		18.45	5.15		23.60	28.50
2500	With galvanized steel backing, add					5.50			5.50	6.05

08 81 55 – Window Glass

08 81 55.10 Sheet Glass

		Crew	Daily Output	Labor-Hours	Unit	Material	Labor	Equipment	Total	Total Incl O&P
0010	**SHEET GLASS** (window), clear float, stops, putty bed									
0015	1/8" thick, clear float	2 Glaz	480	.033	S.F.	4.27	1.29		5.56	6.65
0500	3/16" thick, clear		480	.033		5.25	1.29		6.54	7.75
0600	Tinted		480	.033		7.15	1.29		8.44	9.80
0700	Tempered		480	.033		8.65	1.29		9.94	11.45

08 81 65 – Wire Glass

08 81 65.10 Glass Reinforced With Wire

		Crew	Daily Output	Labor-Hours	Unit	Material	Labor	Equipment	Total	Total Incl O&P
0010	**GLASS REINFORCED WITH WIRE**									
0012	1/4" thick rough obscure	2 Glaz	135	.119	S.F.	13.75	4.57		18.32	22
1000	Polished wire, 1/4" thick, diamond, clear		135	.119		18.75	4.57		23.32	27.50
1500	Pinstripe, obscure		135	.119		21	4.57		25.57	30

08 83 Mirrors

08 83 13 – Mirrored Glass Glazing

08 83 13.10 Mirrors

		Crew	Daily Output	Labor-Hours	Unit	Material	Labor	Equipment	Total	Total Incl O&P
0010	**MIRRORS**, No frames, wall type, 1/4" plate glass, polished edge									
0100	Up to 5 S.F.	2 Glaz	125	.128	S.F.	9.10	4.94		14.04	17.45
0200	Over 5 S.F.		160	.100		8.80	3.86		12.66	15.55
0500	Door type, 1/4" plate glass, up to 12 S.F.		160	.100		7.85	3.86		11.71	14.50
1000	Float glass, up to 10 S.F., 1/8" thick		160	.100		5.05	3.86		8.91	11.45
1100	3/16" thick		150	.107		6.15	4.12		10.27	13
1500	12" x 12" wall tiles, square edge, clear		195	.082		1.78	3.17		4.95	6.75
1600	Veined		195	.082		4.69	3.17		7.86	9.95
2000	1/4" thick, stock sizes, one way transparent		125	.128		17.70	4.94		22.64	27
2010	Bathroom, unframed, laminated		160	.100		13.20	3.86		17.06	20.50
2500	Tempered		160	.100		16.65	3.86		20.51	24

08 83 13.15 Reflective Glass

			Crew	Daily Output	Labor-Hours	Unit	Material	Labor	Equipment	Total	Total Incl O&P
0010	**REFLECTIVE GLASS**										
0100	1/4" float with fused metallic oxide fixed	G	2 Glaz	115	.139	S.F.	15.40	5.35		20.75	25
0500	1/4" float glass with reflective applied coating	G	"	115	.139	"	12.85	5.35		18.20	22.50

08 84 Plastic Glazing

08 84 10 – Plexiglass Glazing

08 84 10.10 Plexiglass Acrylic

		Crew	Daily Output	Labor-Hours	Unit	Material	2009 Bare Costs Labor	Equipment	Total	Total Incl O&P
0010	**PLEXIGLASS ACRYLIC**, clear, masked,									
0020	1/8" thick, cut sheets	2 Glaz	170	.094	S.F.	4.69	3.63		8.32	10.65
0200	Full sheets		195	.082		2.45	3.17		5.62	7.50
0500	1/4" thick, cut sheets		165	.097		8.30	3.74		12.04	14.75
0600	Full sheets		185	.086		4.50	3.34		7.84	10
0900	3/8" thick, cut sheets		155	.103		15.15	3.98		19.13	22.50
1000	Full sheets		180	.089		8.15	3.43		11.58	14.20
1300	1/2" thick, cut sheets		135	.119		17.45	4.57		22.02	26
1400	Full sheets		150	.107		16.90	4.12		21.02	25
1700	3/4" thick, cut sheets		115	.139		62	5.35		67.35	76
1800	Full sheets		130	.123		36	4.75		40.75	46.50
2100	1" thick, cut sheets		105	.152		70	5.90		75.90	86
2200	Full sheets		125	.128		43	4.94		47.94	55
3000	Colored, 1/8" thick, cut sheets		170	.094		14.40	3.63		18.03	21.50
3200	Full sheets		195	.082		9.30	3.17		12.47	15.05
3500	1/4" thick, cut sheets		165	.097		16.10	3.74		19.84	23.50
3600	Full sheets		185	.086		11	3.34		14.34	17.15
4000	Mirrors, untinted, cut sheets, 1/8" thick		185	.086		6.75	3.34		10.09	12.50
4200	1/4" thick		180	.089		10.90	3.43		14.33	17.20

08 84 20 – Polycarbonate

08 84 20.10 Thermoplastic

		Crew	Daily Output	Labor-Hours	Unit	Material	Labor	Equipment	Total	Total Incl O&P
0010	**THERMOPLASTIC**, clear, masked, cut sheets									
0020	1/8" thick	2 Glaz	170	.094	S.F.	9.25	3.63		12.88	15.65
0500	3/16" thick		165	.097		10.55	3.74		14.29	17.25
1000	1/4" thick		155	.103		11.90	3.98		15.88	19.05
1500	3/8" thick		150	.107		21	4.12		25.12	29

08 87 Glazing Surface Films

08 87 13 – Solar Control Films

08 87 13.10 Solar Films On Glass

			Crew	Daily Output	Labor-Hours	Unit	Material	Labor	Equipment	Total	Total Incl O&P
0010	**SOLAR FILMS ON GLASS** (glass not included)										
2000	Minimum	G	2 Glaz	180	.089	S.F.	6.20	3.43		9.63	12.05
2050	Maximum	G	"	225	.071	"	14.40	2.74		17.14	19.95

08 87 16 – Safety Films

08 87 16.10 Window Protection Film

			Crew	Daily Output	Labor-Hours	Unit	Material	Labor	Equipment	Total	Total Incl O&P
0010	**WINDOW PROTECTION FILM**										
2010	Window protection film, controls blast damage	G	2 Glaz	80	.200	S.F.	6.20	7.70		13.90	18.50

08 87 53 – Security Films

08 87 53.10 Security Film

			Crew	Daily Output	Labor-Hours	Unit	Material	Labor	Equipment	Total	Total Incl O&P
0010	**SECURITY FILM**, clear, 32000 psi tensile strength, adhered to glass										
0100	.002" thick, daylight installation		H-2	950	.025	S.F.	1.45	.92		2.37	3
0150	.004" thick, daylight installation	R088110-10		800	.030		1.60	1.09		2.69	3.42
0200	.006" thick, daylight installation			700	.034		1.70	1.24		2.94	3.76
0210	Install for anchorage			600	.040		1.89	1.45		3.34	4.29
0400	.007" thick, daylight installation			600	.040		1.80	1.45		3.25	4.19
0410	Install for anchorage			500	.048		2	1.74		3.74	4.85
0500	.008" thick, daylight installation			500	.048		2.10	1.74		3.84	4.96
0510	Install for anchorage			500	.048		2.33	1.74		4.07	5.20
0600	.015" thick, daylight installation			400	.060		3.50	2.18		5.68	7.15

08 87 Glazing Surface Films

08 87 53 – Security Films

08 87 53.10 Security Film

		Crew	Daily Output	Labor-Hours	Unit	Material	2009 Bare Costs Labor	Equipment	Total	Total Incl O&P
0610	Install for anchorage	H-2	400	.060	S.F.	2.33	2.18		4.51	5.90
0900	Security film anchorage, mechanical attachment and cover plate	H-3	370	.043	L.F.	8.70	1.49		10.19	11.85
0950	Security film anchorage, wet glaze structural caulking	1 Glaz	225	.036	"	.83	1.37		2.20	2.98
1000	Adhered security film removal	1 Clab	275	.029	S.F.		.92		.92	1.43

08 88 Special Function Glazing

08 88 40 – Acoustical Glass Units

08 88 40.10 Sound Reduction Units

		Crew	Daily Output	Labor-Hours	Unit	Material	Labor	Equipment	Total	Total Incl O&P
0010	**SOUND REDUCTION UNITS**, 1 lite at 3/8", 1 lite at 3/16"									
0020	For 1" thick	2 Glaz	100	.160	S.F.	32	6.20		38.20	45
0100	For 4" thick	"	80	.200	"	55	7.70		62.70	72

08 88 56 – Ballistics-Resistant Glazing

08 88 56.10 Laminated Glass

		Crew	Daily Output	Labor-Hours	Unit	Material	Labor	Equipment	Total	Total Incl O&P
0010	**LAMINATED GLASS**									
0020	Clear float .03" vinyl 1/4"	2 Glaz	90	.178	S.F.	11.10	6.85		17.95	22.50
0100	3/8" thick		78	.205		20	7.90		27.90	34.50
0200	.06" vinyl, 1/2" thick		65	.246		23.50	9.50		33	40.50
1000	5/8" thick		90	.178		28	6.85		34.85	41
2000	Bullet-resisting, 1-3/16" thick, to 15 S.F.		16	1		69.50	38.50		108	135
2100	Over 15 S.F.		16	1		88	38.50		126.50	156
2500	2-1/4" thick, to 15 S.F.		12	1.333		91	51.50		142.50	178
2600	Over 15 S.F.		12	1.333		81	51.50		132.50	167
2700	Level 2 (.357 magnum)		12	1.333		60	51.50		111.50	144
2750	Level 3 (.44 magnum)		12	1.333		65	51.50		116.50	150
2800	Level 4 (AK-47)		12	1.333		85	51.50		136.50	172
2850	Level 5 (M-16)		12	1.333		95	51.50		146.50	183
2900	Level (7.62 Armor Piercing)		12	1.333		112	51.50		163.50	201

08 95 Vents

08 95 16 – Wall Vents

08 95 16.10 Louvers

		Crew	Daily Output	Labor-Hours	Unit	Material	Labor	Equipment	Total	Total Incl O&P
0010	**LOUVERS**									
0020	Redwood, 2'-0" diameter, full circle	1 Carp	16	.500	Ea.	159	20		179	205
0100	Half circle		16	.500		152	20		172	198
0200	Octagonal		16	.500		121	20		141	164
0300	Triangular, 5/12 pitch, 5'-0" at base		16	.500		257	20		277	315

Division 9 - Finishes

Estimating Tips

General
- Room Finish Schedule: A complete set of plans should contain a room finish schedule. If one is not available, it would be well worth the time and effort to obtain one.

09 20 00 Plaster and Gypsum Board
- Lath is estimated by the square yard plus a 5% allowance for waste. Furring, channels, and accessories are measured by the linear foot. An extra foot should be allowed for each accessory miter or stop.
- Plaster is also estimated by the square yard. Deductions for openings vary by preference, from zero deduction to 50% of all openings over 2 feet in width. The estimator should allow one extra square foot for each linear foot of horizontal interior or exterior angle located below the ceiling level. Also, double the areas of small radius work.
- Drywall accessories, studs, track, and acoustical caulking are all measured by the linear foot. Drywall taping is figured by the square foot. Gypsum wallboard is estimated by the square foot. No material deductions should be made for door or window openings under 32 S.F.

09 60 00 Flooring
- Tile and terrazzo areas are taken off on a square foot basis. Trim and base materials are measured by the linear foot. Accent tiles are listed per each. Two basic methods of installation are used. Mud set is approximately 30% more expensive than thin set. In terrazzo work, be sure to include the linear footage of embedded decorative strips, grounds, machine rubbing, and power cleanup.
- Wood flooring is available in strip, parquet, or block configuration. The latter two types are set in adhesives with quantities estimated by the square foot. The laying pattern will influence labor costs and material waste. In addition to the material and labor for laying wood floors, the estimator must make allowances for sanding and finishing these areas unless the flooring is prefinished.
- Sheet flooring is measured by the square yard. Roll widths vary, so consideration should be given to use the most economical width, as waste must be figured into the total quantity. Consider also the installation methods available, direct glue down or stretched.

09 70 00 Wall Finishes
- Wall coverings are estimated by the square foot. The area to be covered is measured, length by height of wall above baseboards, to calculate the square footage of each wall. This figure is divided by the number of square feet in the single roll which is being used. Deduct, in full, the areas of openings such as doors and windows. Where a pattern match is required allow 25%–30% waste.

09 80 00 Acoustic Treatment
- Acoustical systems fall into several categories. The takeoff of these materials should be by the square foot of area with a 5% allowance for waste. Do not forget about scaffolding, if applicable, when estimating these systems.

09 90 00 Painting and Coating
- A major portion of the work in painting involves surface preparation. Be sure to include cleaning, sanding, filling, and masking costs in the estimate.
- Protection of adjacent surfaces is not included in painting costs. When considering the method of paint application, an important factor is the amount of protection and masking required. These must be estimated separately and may be the determining factor in choosing the method of application.

Reference Numbers
Reference numbers are shown in shaded boxes at the beginning of some major classifications. These numbers refer to related items in the Reference Section. The reference information may be an estimating procedure, an alternate pricing method, or technical information.

Note: Not all subdivisions listed here necessarily appear in this publication.

No part of this publication may be reproduced, stored in a retrieval system, or transmitted in any form or by any means without prior written permission of R.S. Means Company, Inc.

09 01 Maintenance of Finishes

09 01 70 – Maintenance of Wall Finishes

09 01 70.10 Gypsum Wallboard Repairs

09 01 70.10 Gypsum Wallboard Repairs		Crew	Daily Output	Labor-Hours	Unit	Material	2009 Bare Costs Labor	2009 Bare Costs Equipment	Total	Total Incl O&P
0010	**GYPSUM WALLBOARD REPAIRS**									
0100	Fill and sand, pin / nail holes	1 Carp	960	.008	Ea.		.33		.33	.52
0110	Screw head pops		480	.017			.67		.67	1.03
0120	Dents, up to 2" square		48	.167		.01	6.65		6.66	10.35
0130	2" to 4" square		24	.333		.03	13.30		13.33	20.50
0140	Cut square, patch, sand and finish, holes, up to 2" square		12	.667		.04	26.50		26.54	41.50
0150	2" to 4" square		11	.727		.10	29		29.10	45
0160	4" to 8" square		10	.800		.25	32		32.25	50
0170	8" to 12" square		8	1		.48	40		40.48	62.50
0180	12" to 32" square		6	1.333		3.04	53.50		56.54	86
0210	16" by 48"		5	1.600		2.31	64		66.31	102
0220	32" by 48"		4	2		4.13	80		84.13	129
0230	48" square		3.50	2.286		5.85	91.50		97.35	148
0240	60" square		3.20	2.500		9.80	100		109.80	166
0500	Skim coat surface with joint compound		1600	.005	S.F.	.03	.20		.23	.34

09 05 Common Work Results for Finishes

09 05 05 – Selective Finishes Demolition

09 05 05.10 Selective Demolition, Ceilings

			Crew	Daily Output	Labor-Hours	Unit	Material	Labor	Equipment	Total	Total Incl O&P
0010	**SELECTIVE DEMOLITION, CEILINGS**	R024119-10									
0200	Ceiling, drywall, furred and nailed or screwed		2 Clab	800	.020	S.F.		.63		.63	.98
0220	On metal frame			760	.021			.67		.67	1.03
0240	On suspension system, including system			720	.022			.70		.70	1.09
1000	Plaster, lime and horse hair, on wood lath, incl. lath			700	.023			.72		.72	1.12
1020	On metal lath			570	.028			.89		.89	1.38
1100	Gypsum, on gypsum lath			720	.022			.70		.70	1.09
1120	On metal lath			500	.032			1.01		1.01	1.57
1200	Suspended ceiling, mineral fiber, 2' x 2' or 2' x 4'			1500	.011			.34		.34	.52
1250	On suspension system, incl. system			1200	.013			.42		.42	.65
1500	Tile, wood fiber, 12" x 12", glued			900	.018			.56		.56	.87
1540	Stapled			1500	.011			.34		.34	.52
1580	On suspension system, incl. system			760	.021			.67		.67	1.03
2000	Wood, tongue and groove, 1" x 4"			1000	.016			.51		.51	.78
2040	1" x 8"			1100	.015			.46		.46	.71
2400	Plywood or wood fiberboard, 4' x 8' sheets			1200	.013			.42		.42	.65

09 05 05.20 Selective Demolition, Flooring

			Crew	Daily Output	Labor-Hours	Unit	Material	Labor	Equipment	Total	Total Incl O&P
0010	**SELECTIVE DEMOLITION, FLOORING**	R024119-10									
0200	Brick with mortar		2 Clab	475	.034	S.F.		1.06		1.06	1.65
0400	Carpet, bonded, including surface scraping			2000	.008			.25		.25	.39
0440	Scrim applied			8000	.002			.06		.06	.10
0480	Tackless			9000	.002			.06		.06	.09
0550	Carpet tile, releasable adhesive			5000	.003			.10		.10	.16
0560	Permanent adhesive			1850	.009			.27		.27	.42
0600	Composition, acrylic or epoxy			400	.040			1.26		1.26	1.96
0700	Concrete, scarify skin		A-1A	225	.036			1.45	.97	2.42	3.32
0800	Resilient, sheet goods		2 Clab	1400	.011			.36		.36	.56
0820	For gym floors		"	900	.018			.56		.56	.87
0850	Vinyl or rubber cove base		1 Clab	1000	.008	L.F.		.25		.25	.39
0860	Vinyl or rubber cove base, molded corner		"	1000	.008	Ea.		.25		.25	.39
0870	For glued and caulked installation, add to labor							50%			
0900	Vinyl composition tile, 12" x 12"		2 Clab	1000	.016	S.F.		.51		.51	.78

09 05 Common Work Results for Finishes

09 05 05 – Selective Finishes Demolition

09 05 05.20 Selective Demolition, Flooring

		Crew	Daily Output	Labor-Hours	Unit	Material	2009 Bare Costs Labor	2009 Bare Costs Equipment	Total	Total Incl O&P
2000	Tile, ceramic, thin set	2 Clab	675	.024	S.F.		.75		.75	1.16
2020	Mud set		625	.026			.81		.81	1.25
2200	Marble, slate, thin set		675	.024			.75		.75	1.16
2220	Mud set		625	.026			.81		.81	1.25
2600	Terrazzo, thin set		450	.036			1.12		1.12	1.74
2620	Mud set		425	.038			1.19		1.19	1.84
2640	Terrazzo, cast in place		300	.053			1.69		1.69	2.61
3000	Wood, block, on end	1 Carp	400	.020			.80		.80	1.24
3200	Parquet		450	.018			.71		.71	1.10
3400	Strip flooring, interior, 2-1/4" x 25/32" thick		325	.025			.98		.98	1.53
3500	Exterior, porch flooring, 1" x 4"		220	.036			1.45		1.45	2.25
3800	Subfloor, tongue and groove, 1" x 6"		325	.025			.98		.98	1.53
3820	1" x 8"		430	.019			.74		.74	1.15
3840	1" x 10"		520	.015			.61		.61	.95
4000	Plywood, nailed		600	.013			.53		.53	.83
4100	Glued and nailed		400	.020			.80		.80	1.24
4200	Hardboard, 1/4" thick		760	.011			.42		.42	.65
8000	Remove flooring, bead blast, minimum	A-1A	1000	.008			.33	.22	.55	.75
8100	Maximum		400	.020			.82	.54	1.36	1.87
8150	Mastic only		1500	.005			.22	.15	.37	.50

09 05 05.30 Selective Demolition, Walls and Partitions

		Crew	Daily Output	Labor-Hours	Unit	Material	2009 Bare Costs Labor	2009 Bare Costs Equipment	Total	Total Incl O&P
0010	**SELECTIVE DEMOLITION, WALLS AND PARTITIONS** R024119-10									
0020	Walls, concrete, reinforced	B-39	120	.400	C.F.		13.25	1.60	14.85	22.50
0025	Plain	"	160	.300			9.95	1.20	11.15	16.65
0100	Brick, 4" to 12" thick	B-9	220	.182			5.80	.87	6.67	9.95
0200	Concrete block, 4" thick		1000	.040	S.F.		1.28	.19	1.47	2.19
0280	8" thick		810	.049			1.58	.24	1.82	2.71
0300	Exterior stucco 1" thick over mesh		3200	.013			.40	.06	.46	.69
1000	Drywall, nailed or screwed	1 Clab	1000	.008			.25		.25	.39
1010	2 layers		400	.020			.63		.63	.98
1020	Glued and nailed		900	.009			.28		.28	.44
1500	Fiberboard, nailed		900	.009			.28		.28	.44
1520	Glued and nailed		800	.010			.32		.32	.49
1568	Plenum barrier, sheet lead		300	.027			.84		.84	1.31
2000	Movable walls, metal, 5' high		300	.027			.84		.84	1.31
2020	8' high		400	.020			.63		.63	.98
2200	Metal or wood studs, finish 2 sides, fiberboard	B-1	520	.046			1.49		1.49	2.31
2250	Lath and plaster		260	.092			2.98		2.98	4.62
2300	Plasterboard (drywall)		520	.046			1.49		1.49	2.31
2350	Plywood		450	.053			1.72		1.72	2.67
2800	Paneling, 4' x 8' sheets	1 Clab	475	.017			.53		.53	.83
3000	Plaster, lime and horsehair, on wood lath		400	.020			.63		.63	.98
3020	On metal lath		335	.024			.75		.75	1.17
3400	Gypsum or perlite, on gypsum lath		410	.020			.62		.62	.96
3420	On metal lath		300	.027			.84		.84	1.31
3450	Plaster, interior gypsum, acoustic, or cement		60	.133	S.Y.		4.21		4.21	6.55
3500	Stucco, on masonry		145	.055			1.74		1.74	2.70
3510	Commercial 3-coat		80	.100			3.16		3.16	4.90
3520	Interior stucco		25	.320			10.10		10.10	15.70
3600	Plywood, one side	B-1	1500	.016	S.F.		.52		.52	.80
3750	Terra cotta block and plaster, to 6" thick	"	175	.137			4.43		4.43	6.85
3760	Tile, ceramic, on walls, thin set	1 Clab	300	.027			.84		.84	1.31

09 05 Common Work Results for Finishes

09 05 05 – Selective Finishes Demolition

09 05 05.30 Selective Demolition, Walls and Partitions		Crew	Daily Output	Labor-Hours	Unit	Material	2009 Bare Costs Labor	Equipment	Total	Total Incl O&P
3765	Mud set	1 Clab	250	.032	S.F.		1.01		1.01	1.57
3800	Toilet partitions, slate or marble	↓	5	1.600	Ea.		50.50		50.50	78.50
3820	Metal or plastic	↓	8	1	"		31.50		31.50	49
5000	Wallcovering, vinyl	1 Pape	700	.011	S.F.		.41		.41	.61
5010	With release agent		1500	.005			.19		.19	.28
5025	Wallpaper, 2 layers or less, by hand		250	.032			1.13		1.13	1.70
5035	3 layers or more	↓	165	.048	↓		1.72		1.72	2.57

09 21 Plaster and Gypsum Board Assemblies

09 21 13 – Plaster Assemblies

09 21 13.10 Plaster Partition Wall

		Crew	Daily Output	Labor-Hours	Unit	Material	Labor	Equipment	Total	Incl O&P
0010	**PLASTER PARTITION WALL**									
0400	Stud walls, 3.4 lb. metal lath, 3 coat gypsum plaster, 2 sides									
0600	2" x 4" wood studs, 16" O.C.	J-2	315	.152	S.F.	3.31	5.30	.40	9.01	12.05
0700	2-1/2" metal studs, 25 ga., 12" O.C.		325	.148		3.06	5.15	.39	8.60	11.50
0800	3-5/8" metal studs, 25 ga., 16" O.C.	↓	320	.150	↓	3.08	5.20	.40	8.68	11.70
0900	Gypsum lath, 2 coat vermiculite plaster, 2 sides									
1000	2" x 4" wood studs, 16" O.C.	J-2	355	.135	S.F.	3.69	4.70	.36	8.75	11.50
1200	2-1/2" metal studs, 25 ga., 12" O.C.		365	.132		3.27	4.57	.35	8.19	10.90
1300	3-5/8" metal studs, 25 ga., 16" O.C.	↓	360	.133		3.36	4.64	.35	8.35	11.05

09 21 16 – Gypsum Board Assemblies

09 21 16.23 Gypsum Board Shaft Wall Assemblies

		Crew	Daily Output	Labor-Hours	Unit	Material	Labor	Equipment	Total	Incl O&P
0010	**GYPSUM BOARD SHAFT WALL ASSEMBLIES**									
0030	1" thick coreboard wall liner on shaft side									
0040	2-hour assembly with double layer									
0060	5/8" fire rated gypsum board on room side	2 Carp	220	.073	S.F.	1.56	2.91		4.47	6.25
0100	3-hour assembly with triple layer									
0300	5/8" fire rated gypsum board on room side	2 Carp	180	.089	S.F.	1.90	3.55		5.45	7.60
0400	4-hour assembly, 1" coreboard, 5/8" fire rated gypsum board									
0600	and 3/4" galv. metal furring channels, 24" O.C., with									
0700	Double layer 5/8" fire rated gypsum board on room side	2 Carp	110	.145	S.F.	1.77	5.80		7.57	10.95
0900	For taping & finishing, add per side	1 Carp	1050	.008	"	.04	.30		.34	.52
1000	For insulation, see Div. 07 21									
5200	For work over 8' high, add	2 Carp	3060	.005	S.F.		.21		.21	.32
5300	For distribution cost over 3 stories high, add per story	"	6100	.003	"		.10		.10	.16

09 21 16.33 Partition Wall

		Crew	Daily Output	Labor-Hours	Unit	Material	Labor	Equipment	Total	Incl O&P
0010	**PARTITION WALL** Stud wall, 8' to 12' high									
0050	1/2", interior, gypsum board, std, tape & finish 2 sides									
0500	Installed on and incl., 2" x 4" wood studs, 16" O.C.	2 Carp	310	.052	S.F.	1.03	2.06		3.09	4.33
1000	Metal studs, NLB, 25 ga., 16" O.C., 3-5/8" wide		350	.046		1.16	1.83		2.99	4.10
1200	6" wide		330	.048		1.39	1.94		3.33	4.52
1400	Water resistant, on 2" x 4" wood studs, 16" O.C.		310	.052		1.37	2.06		3.43	4.71
1600	Metal studs, NLB, 25 ga., 16" O.C., 3-5/8" wide		350	.046		1.50	1.83		3.33	4.47
1800	6" wide		330	.048		1.73	1.94		3.67	4.90
2000	Fire res., 2 layers, 1-1/2 hr., on 2" x 4" wood studs, 16" O.C.		210	.076		1.69	3.04		4.73	6.60
2200	Metal studs, NLB, 25 ga., 16" O.C., 3-5/8" wide		250	.064		1.82	2.56		4.38	5.95
2400	6" wide		230	.070		2.05	2.78		4.83	6.55
2600	Fire & water res., 2 layers, 1-1/2 hr., 2" x 4" studs, 16" O.C.		210	.076		1.69	3.04		4.73	6.60
2800	Metal studs, NLB, 25 ga., 16" O.C., 3-5/8" wide		250	.064		1.82	2.56		4.38	5.95
3000	6" wide	↓	230	.070	↓	2.05	2.78		4.83	6.55

09 21 Plaster and Gypsum Board Assemblies

09 21 16 – Gypsum Board Assemblies

09 21 16.33 Partition Wall

		Crew	Daily Output	Labor-Hours	Unit	Material	2009 Bare Costs Labor	Equipment	Total	Total Incl O&P
3200	5/8", interior, gypsum board, std, tape & finish 2 sides									
3400	Installed on and including 2" x 4" wood studs, 16" O.C.	2 Carp	300	.053	S.F.	1.09	2.13		3.22	4.50
3600	24" O.C.		330	.048		1.02	1.94		2.96	4.12
3800	Metal studs, NLB, 25 ga., 16" O.C., 3-5/8" wide		340	.047		1.22	1.88		3.10	4.26
4000	6" wide		320	.050		1.45	2		3.45	4.69
4200	24" O.C., 3-5/8" wide		360	.044		1.10	1.78		2.88	3.96
4400	6" wide		340	.047		1.27	1.88		3.15	4.31
4800	Water resistant, on 2" x 4" wood studs, 16" O.C.		300	.053		1.35	2.13		3.48	4.79
5000	24" O.C.		330	.048		1.28	1.94		3.22	4.41
5200	Metal studs, NLB, 25 ga. 16" O.C., 3-5/8" wide		340	.047		1.48	1.88		3.36	4.54
5400	6" wide		320	.050		1.71	2		3.71	4.98
5600	24" O.C., 3-5/8" wide		360	.044		1.36	1.78		3.14	4.24
5800	6" wide		340	.047		1.53	1.88		3.41	4.60
6000	Fire resistant, 2 layers, 2 hr., on 2" x 4" wood studs, 16" O.C.		205	.078		1.78	3.12		4.90	6.80
6200	24" O.C.		235	.068		1.78	2.72		4.50	6.20
6400	Metal studs, NLB, 25 ga., 16" O.C., 3-5/8" wide		245	.065		2	2.61		4.61	6.25
6600	6" wide		225	.071		2.21	2.84		5.05	6.85
6800	24" O.C., 3-5/8" wide		265	.060		1.86	2.41		4.27	5.80
7000	6" wide		245	.065		2.03	2.61		4.64	6.30
7200	Fire & water resistant, 2 layers, 2 hr., 2" x 4" studs, 16" O.C.		205	.078		1.85	3.12		4.97	6.90
7400	24" O.C.		235	.068		1.78	2.72		4.50	6.20
7600	Metal studs, NLB, 25 ga., 16" O.C., 3-5/8" wide		245	.065		1.98	2.61		4.59	6.20
7800	6" wide		225	.071		2.21	2.84		5.05	6.85
8000	24" O.C., 3-5/8" wide		265	.060		1.86	2.41		4.27	5.80
8200	6" wide	↓	245	.065	↓	2.03	2.61		4.64	6.30
8600	1/2" blueboard, mesh tape both sides									
8620	Installed on and including 2" x 4" wood studs, 16" O.C.	2 Carp	300	.053	S.F.	1.09	2.13		3.22	4.50
8640	Metal studs, NLB, 25 ga., 16" O.C., 3-5/8" wide		340	.047		1.22	1.88		3.10	4.26
8660	6" wide	↓	320	.050	↓	1.45	2		3.45	4.69
9000	Exterior, 1/2" gypsum sheathing, 1/2" gypsum finished, interior,									
9100	including foil faced insulation, metal studs, 20 ga.									
9200	16" O.C., 3-5/8" wide	2 Carp	290	.055	S.F.	2.17	2.20		4.37	5.80
9400	6" wide	"	270	.059	"	2.42	2.37		4.79	6.35

09 22 Supports for Plaster and Gypsum Board

09 22 03 – Fastening Methods for Finishes

09 22 03.20 Drilling Plaster/Drywall

		Crew	Daily Output	Labor-Hours	Unit	Material	2009 Bare Costs Labor	Equipment	Total	Total Incl O&P
0010	**DRILLING PLASTER/DRYWALL**									
1100	Drilling & layout for drywall/plaster walls, up to 1" deep, no anchor									
1200	Holes, 1/4" diameter	1 Carp	150	.053	Ea.	.01	2.13		2.14	3.31
1300	3/8" diameter		140	.057		.01	2.28		2.29	3.55
1400	1/2" diameter		130	.062		.01	2.46		2.47	3.82
1500	3/4" diameter		120	.067		.01	2.66		2.67	4.14
1600	1" diameter		110	.073		.02	2.91		2.93	4.53
1700	1-1/4" diameter		100	.080		.03	3.20		3.23	4.99
1800	1-1/2" diameter	↓	90	.089	↓	.04	3.55		3.59	5.55
1900	For ceiling installations, add						40%			

09 22 13 – Metal Furring

09 22 13.13 Metal Channel Furring

0010	**METAL CHANNEL FURRING**									

09 22 Supports for Plaster and Gypsum Board

09 22 13 – Metal Furring

09 22 13.13 Metal Channel Furring

		Crew	Daily Output	Labor-Hours	Unit	Material	2009 Bare Costs Labor	Equipment	Total	Total Incl O&P
0030	Beams and columns, 7/8" channels, galvanized, 12" O.C.	1 Lath	155	.052	S.F.	.38	1.83		2.21	3.12
0050	16" O.C.		170	.047		.31	1.67		1.98	2.82
0070	24" O.C.		185	.043		.21	1.54		1.75	2.50
0100	Ceilings, on steel, 7/8" channels, galvanized, 12" O.C.		210	.038		.34	1.35		1.69	2.38
0300	16" O.C.		290	.028		.31	.98		1.29	1.79
0400	24" O.C.		420	.019		.21	.68		.89	1.23
0600	1-5/8" channels, galvanized, 12" O.C.		190	.042		.46	1.50		1.96	2.73
0700	16" O.C.		260	.031		.41	1.09		1.50	2.08
0900	24" O.C.		390	.021		.28	.73		1.01	1.38
0930	7/8" channels with sound isolation clips, 12" O.C.		120	.067		1.53	2.37		3.90	5.20
0940	16" O.C.		100	.080		2.12	2.84		4.96	6.55
0950	24" O.C.		165	.048		1.39	1.72		3.11	4.08
0960	1-5/8" channels, galvanized, 12" O.C.		110	.073		1.64	2.59		4.23	5.65
0970	16" O.C.		100	.080		2.22	2.84		5.06	6.65
0980	24" O.C.		155	.052		1.46	1.83		3.29	4.32
1000	Walls, 7/8" channels, galvanized, 12" O.C.		235	.034		.34	1.21		1.55	2.17
1200	16" O.C.		265	.030		.31	1.07		1.38	1.93
1300	24" O.C.		350	.023		.21	.81		1.02	1.43
1500	1-5/8" channels, galvanized, 12" O.C.		210	.038		.46	1.35		1.81	2.51
1600	16" O.C.		240	.033		.41	1.18		1.59	2.21
1800	24" O.C.		305	.026		.28	.93		1.21	1.68
1920	7/8" channels with sound isolation clips, 12" O.C.		125	.064		1.53	2.28		3.81	5.05
1940	16" O.C.		100	.080		2.12	2.84		4.96	6.55
1950	24" O.C.		150	.053		1.39	1.90		3.29	4.34
1960	1-5/8" channels, galvanized, 12" O.C.		115	.070		1.64	2.47		4.11	5.45
1970	16" O.C.		95	.084		2.22	2.99		5.21	6.90
1980	24" O.C.		140	.057		1.46	2.03		3.49	4.62

09 22 16 – Non-Structural Metal Framing

09 22 16.13 Metal Studs and Track

		Crew	Daily Output	Labor-Hours	Unit	Material	2009 Bare Costs Labor	Equipment	Total	Total Incl O&P
0010	**METAL STUDS AND TRACK**									
1600	Non-load bearing, galv, 8' high, 25 ga. 1-5/8" wide, 16" O.C.	1 Carp	619	.013	S.F.	.33	.52		.85	1.16
1610	24" O.C.		950	.008		.25	.34		.59	.79
1620	2-1/2" wide, 16" O.C.		613	.013		.42	.52		.94	1.27
1630	24" O.C.		938	.009		.31	.34		.65	.87
1640	3-5/8" wide, 16" O.C.		600	.013		.48	.53		1.01	1.35
1650	24" O.C.		925	.009		.36	.35		.71	.93
1660	4" wide, 16" O.C.		594	.013		.51	.54		1.05	1.39
1670	24" O.C.		925	.009		.38	.35		.73	.96
1680	6" wide, 16" O.C.		588	.014		.72	.54		1.26	1.63
1690	24" O.C.		906	.009		.54	.35		.89	1.14
1700	20 ga. studs, 1-5/8" wide, 16" O.C.		494	.016		.53	.65		1.18	1.58
1710	24" O.C.		763	.010		.40	.42		.82	1.09
1720	2-1/2" wide, 16" O.C.		488	.016		.62	.65		1.27	1.70
1730	24" O.C.		750	.011		.47	.43		.90	1.17
1740	3-5/8" wide, 16" O.C.		481	.017		.67	.66		1.33	1.77
1750	24" O.C.		738	.011		.50	.43		.93	1.22
1760	4" wide, 16" O.C.		475	.017		.80	.67		1.47	1.91
1770	24" O.C.		738	.011		.60	.43		1.03	1.33
1780	6" wide, 16" O.C.		469	.017		.94	.68		1.62	2.09
1790	24" O.C.		725	.011		.70	.44		1.14	1.45
2000	Non-load bearing, galv, 10' high, 25 ga. 1-5/8" wide, 16" O.C.		495	.016		.31	.65		.96	1.34
2100	24" O.C.		760	.011		.23	.42		.65	.90

09 22 Supports for Plaster and Gypsum Board

09 22 16 — Non-Structural Metal Framing

09 22 16.13 Metal Studs and Track		Crew	Daily Output	Labor-Hours	Unit	Material	2009 Bare Costs Labor	Equipment	Total	Total Incl O&P
2200	2-1/2" wide, 16" O.C.	1 Carp	490	.016	S.F.	.39	.65		1.04	1.44
2250	24" O.C.		750	.011		.29	.43		.72	.98
2300	3-5/8" wide, 16" O.C.		480	.017		.45	.67		1.12	1.53
2350	24" O.C.		740	.011		.33	.43		.76	1.03
2400	4" wide, 16" O.C.		475	.017		.49	.67		1.16	1.57
2450	24" O.C.		740	.011		.36	.43		.79	1.06
2500	6" wide, 16" O.C.		470	.017		.68	.68		1.36	1.80
2550	24" O.C.		725	.011		.50	.44		.94	1.23
2600	20 ga. studs, 1-5/8" wide, 16" O.C.		395	.020		.50	.81		1.31	1.80
2650	24" O.C.		610	.013		.37	.52		.89	1.22
2700	2-1/2" wide, 16" O.C.		390	.021		.59	.82		1.41	1.92
2750	24" O.C.		600	.013		.43	.53		.96	1.31
2800	3-5/8" wide, 16" OC		385	.021		.64	.83		1.47	1.99
2850	24" O.C.		590	.014		.47	.54		1.01	1.36
2900	4" wide, 16" O.C.		380	.021		.75	.84		1.59	2.13
2950	24" O.C.		590	.014		.56	.54		1.10	1.45
3000	6" wide, 16" O.C.		375	.021		.89	.85		1.74	2.30
3050	24" O.C.		580	.014		.65	.55		1.20	1.57
3060	Non-load bearing, galv, 12' high, 25 ga. 1-5/8" wide, 16" O.C.		413	.019		.30	.77		1.07	1.53
3070	24" O.C.		633	.013		.22	.51		.73	1.02
3080	2-1/2" wide, 16" O.C.		408	.020		.38	.78		1.16	1.63
3090	24" O.C.		625	.013		.27	.51		.78	1.09
3100	3-5/8" wide, 16" O.C.		400	.020		.43	.80		1.23	1.71
3110	24" O.C.		617	.013		.31	.52		.83	1.14
3120	4" wide, 16" O.C.		396	.020		.47	.81		1.28	1.76
3130	24" O.C.		617	.013		.34	.52		.86	1.17
3140	6" wide, 16" O.C.		392	.020		.65	.82		1.47	1.98
3150	24" O.C.		604	.013		.47	.53		1	1.34
3160	20 ga. studs, 1-5/8" wide, 16" O.C.		329	.024		.48	.97		1.45	2.04
3170	24" O.C.		508	.016		.35	.63		.98	1.36
3180	2-1/2" wide, 16" O.C.		325	.025		.56	.98		1.54	2.15
3190	24" O.C.		500	.016		.41	.64		1.05	1.44
3200	3-5/8" wide, 16" O.C.		321	.025		.61	1		1.61	2.21
3210	24" O.C.		492	.016		.44	.65		1.09	1.50
3220	4" wide, 16" O.C.		317	.025		.72	1.01		1.73	2.35
3230	24" O.C.		492	.016		.52	.65		1.17	1.59
3240	6" wide, 16" O.C.		313	.026		.85	1.02		1.87	2.52
3250	24" O.C.	▼	483	.017	▼	.62	.66		1.28	1.71
5000	Load bearing studs, see Div. 05 41 13.30									

09 22 26 — Suspension Systems

09 22 26.13 Ceiling Suspension Systems

		Crew	Daily Output	Labor-Hours	Unit	Material	Labor	Equipment	Total	Total Incl O&P
0010	**CEILING SUSPENSION SYSTEMS** For gypsum board or plaster									
8000	Suspended ceilings, including carriers									
8200	1-1/2" carriers, 24" O.C. with:									
8300	7/8" channels, 16" O.C.	1 Lath	215	.037	S.F.	.58	1.32		1.90	2.60
8320	24" O.C.		275	.029		.48	1.03		1.51	2.06
8400	1-5/8" channels, 16" O.C.		205	.039		.69	1.39		2.08	2.81
8420	24" O.C.	▼	265	.030	▼	.55	1.07		1.62	2.20
8600	2" carriers, 24" O.C. with:									
8700	7/8" channels, 16" O.C.	1 Lath	205	.039	S.F.	.56	1.39		1.95	2.66
8720	24" O.C.		265	.030		.46	1.07		1.53	2.09
8800	1-5/8" channels, 16" O.C.	▼	195	.041	▼	.66	1.46		2.12	2.89

09 22 Supports for Plaster and Gypsum Board

09 22 36 – Lath

09 22 36.13 Gypsum Lath

		Crew	Daily Output	Labor-Hours	Unit	Material	2009 Bare Costs Labor	2009 Bare Costs Equipment	Total	Total Incl O&P
0010	**GYPSUM LATH** R092000-50									
0020	Plain or perforated, nailed, 3/8" thick	1 Lath	85	.094	S.Y.	5.50	3.35		8.85	11
0100	1/2" thick		80	.100		5.85	3.56		9.41	11.70
0300	Clipped to steel studs, 3/8" thick		75	.107		5.50	3.79		9.29	11.65
0400	1/2" thick		70	.114		5.85	4.06		9.91	12.45
1500	For ceiling installations, add		216	.037			1.32		1.32	1.95
1600	For columns and beams, add		170	.047			1.67		1.67	2.48

09 22 36.23 Metal Lath

		Crew	Daily Output	Labor-Hours	Unit	Material	2009 Bare Costs Labor	2009 Bare Costs Equipment	Total	Total Incl O&P
0010	**METAL LATH** R092000-50									
0020	Diamond, expanded, 2.5 lb. per S.Y., painted				S.Y.	3.32			3.32	3.65
0100	Galvanized					3.47			3.47	3.82
0300	3.4 lb. per S.Y., painted					3.99			3.99	4.39
0400	Galvanized					4.09			4.09	4.50
0600	For 15# asphalt sheathing paper, add					.43			.43	.47
0900	Flat rib, 1/8" high, 2.75 lb., painted					3.62			3.62	3.98
1000	Foil backed					3.56			3.56	3.92
1200	3.4 lb. per S.Y., painted					4.61			4.61	5.05
1300	Galvanized					4.23			4.23	4.65
1500	For 15# asphalt sheathing paper, add					.43			.43	.47
1800	High rib, 3/8" high, 3.4 lb. per S.Y., painted					5.55			5.55	6.15
1900	Galvanized					5.55			5.55	6.15
2400	3/4" high, painted, .60 lb. per S.F.				S.F.	.58			.58	.64
2500	.75 lb. per S.F.				"	1.25			1.25	1.38
2800	Stucco mesh, painted, 3.6 lb.				S.Y.	4.30			4.30	4.73
3000	K-lath, perforated, absorbent paper, regular					4.43			4.43	4.87
3100	Heavy duty					5.25			5.25	5.75
3300	Waterproof, heavy duty, grade B backing					5.10			5.10	5.65
3400	Fire resistant backing					5.65			5.65	6.25
3600	2.5 lb. diamond painted, on wood framing, on walls	1 Lath	85	.094		3.32	3.35		6.67	8.60
3700	On ceilings		75	.107		3.32	3.79		7.11	9.25
3900	3.4 lb. diamond painted, on wood framing, on walls		80	.100		4.61	3.56		8.17	10.30
4000	On ceilings		70	.114		4.61	4.06		8.67	11.05
4200	3.4 lb. diamond painted, wired to steel framing		75	.107		4.61	3.79		8.40	10.65
4300	On ceilings		60	.133		4.61	4.74		9.35	12.05
4500	Columns and beams, wired to steel		40	.200		4.61	7.10		11.71	15.55
4600	Cornices, wired to steel		35	.229		4.61	8.15		12.76	17.05
4800	Screwed to steel studs, 2.5 lb.		80	.100		3.32	3.56		6.88	8.90
4900	3.4 lb.		75	.107		3.99	3.79		7.78	10
5100	Rib lath, painted, wired to steel, on walls, 2.5 lb.		75	.107		3.62	3.79		7.41	9.60
5200	3.4 lb.		70	.114		5.55	4.06		9.61	12.15
5400	4.0 lb.		65	.123		5.70	4.38		10.08	12.75
5500	For self-furring lath, add					.10			.10	.11
5700	Suspended ceiling system, incl. 3.4 lb. diamond lath, painted	1 Lath	15	.533		4.23	18.95		23.18	32.50
5800	Galvanized	"	15	.533		7.50	18.95		26.45	36.50
6000	Hollow metal stud partitions, 3.4 lb. painted lath both sides									
6010	Non-load bearing, 25 ga., w/rib lath 2-1/2" studs, 12" O.C.	1 Lath	20.30	.394	S.Y.	15.65	14		29.65	37.50
6300	16" O.C.		21.10	.379		14.70	13.50		28.20	36
6350	24" O.C.		22.70	.352		13.75	12.55		26.30	33.50
6400	3-5/8" studs, 16" O.C.		19.50	.410		15.20	14.60		29.80	38
6600	24" O.C.		20.40	.392		14.10	13.95		28.05	36
6700	4" studs, 16" O.C.		20.40	.392		15.50	13.95		29.45	37.50
6900	24" O.C.		21.60	.370		14.35	13.15		27.50	35.50

09 22 Supports for Plaster and Gypsum Board

09 22 36 – Lath

09 22 36.23 Metal Lath

		Crew	Daily Output	Labor-Hours	Unit	Material	2009 Bare Costs Labor	Equipment	Total	Total Incl O&P
7000	6" studs, 16" O.C.	1 Lath	19.50	.410	S.Y.	17.25	14.60		31.85	40.50
7100	24" O.C.		21.10	.379		15.65	13.50		29.15	37
7200	L.B. partitions, 16 ga., w/rib lath, 2-1/2" studs, 16" O.C.		20	.400		16.95	14.20		31.15	39.50
7300	3-5/8" studs, 16 ga.		19.70	.406		19.20	14.45		33.65	42.50
7500	4" studs, 16 ga.		19.50	.410		20	14.60		34.60	43.50
7600	6" studs, 16 ga.		18.70	.428		23.50	15.20		38.70	48.50

09 22 36.43 Security Mesh

		Crew	Daily Output	Labor-Hours	Unit	Material	Labor	Equipment	Total	Total Incl O&P
0010	**SECURITY MESH**, expanded metal, flat, screwed to framing									
0100	On walls, 3/4", 1.76 lb/S.F.	2 Carp	1500	.011	S.F.	1.60	.43		2.03	2.42
0110	1-1/2", 1.14 lb/S.F.		1600	.010		1.30	.40		1.70	2.05
0200	On ceilings, 3/4", 1.76 lb/S.F.		1350	.012		1.60	.47		2.07	2.49
0210	1-1/2", 1.14 lb/S.F.		1450	.011		1.30	.44		1.74	2.11

09 22 36.83 Accessories, Plaster

		Crew	Daily Output	Labor-Hours	Unit	Material	Labor	Equipment	Total	Total Incl O&P
0010	**ACCESSORIES, PLASTER**									
0020	Casing bead, expanded flange, galvanized	1 Lath	2.70	2.963	C.L.F.	46.50	105		151.50	208
0900	Channels, cold rolled, 16 ga., 3/4" deep, galvanized					34			34	37.50
1200	1-1/2" deep, 16 ga., galvanized					46			46	50.50
1620	Corner bead, expanded bullnose, 3/4" radius, #10, galvanized	1 Lath	2.60	3.077		34	109		143	200
1650	#1, galvanized		2.55	3.137		55	112		167	226
1670	Expanded wing, 2-3/4" wide, #1, galvanized		2.65	3.019		34.50	107		141.50	197
1700	Inside corner (corner rite), 3" x 3", painted		2.60	3.077		30	109		139	195
1750	Strip-ex, 4" wide, painted		2.55	3.137		30	112		142	198
1800	Expansion joint, 3/4" grounds, limited expansion, galv., 1 piece		2.70	2.963		109	105		214	276
2100	Extreme expansion, galvanized, 2 piece		2.60	3.077		173	109		282	350

09 23 Gypsum Plastering

09 23 13 – Acoustical Gypsum Plastering

09 23 13.10 Perlite or Vermiculite Plaster

		Crew	Daily Output	Labor-Hours	Unit	Material	Labor	Equipment	Total	Total Incl O&P
0010	**PERLITE OR VERMICULITE PLASTER** R092000-50									
0020	In 100 lb. bags, under 200 bags				Bag	14.40			14.40	15.85
0100	Over 200 bags				"	13.10			13.10	14.40
0300	2 coats, no lath included, on walls	J-1	92	.435	S.Y.	4.05	15.05	1.38	20.48	28.50
0400	On ceilings	"	79	.506		4.05	17.50	1.61	23.16	32.50
0600	On and incl. 3/8" gypsum lath, on metal studs	J-2	84	.571		9.60	19.85	1.51	30.96	42.50
0700	On ceilings	"	70	.686		9.60	24	1.81	35.41	48.50
0900	3 coats, no lath included, on walls	J-1	74	.541		6.30	18.70	1.71	26.71	37
1000	On ceilings	"	63	.635		6.30	22	2.01	30.31	42
1200	On and incl. painted metal lath, on metal studs	J-2	72	.667		11.85	23	1.76	36.61	50
1300	On ceilings		61	.787		11.85	27.50	2.08	41.43	56.50
1500	On and incl. suspended metal lath ceiling		37	1.297		10.55	45	3.42	58.97	83.50
1700	For irregular or curved surfaces, add to above						30%			
1800	For columns and beams, add to above						50%			
1900	For soffits, add to ceiling prices						40%			

09 23 20 – Gypsum Plaster

09 23 20.10 Gypsum Plaster On Walls and Ceilings

		Crew	Daily Output	Labor-Hours	Unit	Material	Labor	Equipment	Total	Total Incl O&P
0010	**GYPSUM PLASTER ON WALLS AND CEILINGS** R092000-50									
0020	80# bag, less than 1 ton				Bag	15.20			15.20	16.70
0100	Over 1 ton				"	14.90			14.90	16.40
0300	2 coats, no lath included, on walls	J-1	105	.381	S.Y.	3.77	13.20	1.21	18.18	25.50
0400	On ceilings	"	92	.435		3.77	15.05	1.38	20.20	28

09 23 Gypsum Plastering

09 23 20 – Gypsum Plaster

09 23 20.10 Gypsum Plaster On Walls and Ceilings

		Crew	Daily Output	Labor-Hours	Unit	Material	2009 Bare Costs Labor	Equipment	Total	Total Incl O&P
0600	On and incl. 3/8" gypsum lath on steel, on walls	J-2	97	.495	S.Y.	9.25	17.20	1.31	27.76	37.50
0700	On ceilings	"	83	.578		9.25	20	1.53	30.78	42
0900	3 coats, no lath included, on walls	J-1	87	.460		5.40	15.90	1.46	22.76	31.50
1000	On ceilings	"	78	.513		5.40	17.75	1.63	24.78	34.50
1200	On and including painted metal lath, on wood studs	J-2	86	.558		10.80	19.40	1.47	31.67	42.50
1300	On ceilings	"	76.50	.627		10.80	22	1.66	34.46	46.50
1600	For irregular or curved surfaces, add						30%			
1800	For columns & beams, add						50%			

09 23 20.20 Gauging Plaster

0010	GAUGING PLASTER R092000-50									
0020	100 lb. bags, less than 1 ton				Bag	19.30			19.30	21
0100	Over 1 ton				"	16.90			16.90	18.60

09 23 20.30 Keenes Cement

		Crew	Daily Output	Labor-Hours	Unit	Material	Labor	Equipment	Total	Total Incl O&P
0010	KEENES CEMENT R092000-50									
0020	In 100 lb. bags, less than 1 ton				Bag	27.50			27.50	30
0100	Over 1 ton				"	26			26	28.50
0300	Finish only, add to plaster prices, standard	J-1	215	.186	S.Y.	2.50	6.45	.59	9.54	13.10
0400	High quality	"	144	.278	"	2.53	9.60	.88	13.01	18.25

09 24 Portland Cement Plastering

09 24 23 – Portland Cement Stucco

09 24 23.40 Stucco

		Crew	Daily Output	Labor-Hours	Unit	Material	Labor	Equipment	Total	Total Incl O&P
0010	STUCCO R092000-50									
0015	3 coats 1" thick, float finish, with mesh, on wood frame	J-2	63	.762	S.Y.	6.70	26.50	2.01	35.21	49.50
0100	On masonry construction, no mesh incl.	J-1	67	.597		2.41	20.50	1.89	24.80	35.50
0300	For trowel finish, add	1 Plas	170	.047			1.70		1.70	2.57
0400	For 3/4" thick, on masonry, deduct	J-1	880	.045		.62	1.57	.14	2.33	3.21
0600	For coloring and special finish, add, minimum		685	.058		.40	2.02	.19	2.61	3.69
0700	Maximum		200	.200		1.39	6.90	.63	8.92	12.70
0900	For soffits, add	J-2	155	.310		2.15	10.75	.82	13.72	19.45
1000	Exterior stucco, with bonding agent, 3 coats, on walls, no mesh incl.	J-1	200	.200		3.61	6.90	.63	11.14	15.10
1200	Ceilings		180	.222		3.61	7.70	.70	12.01	16.35
1300	Beams		80	.500		3.61	17.30	1.59	22.50	31.50
1500	Columns		100	.400		3.61	13.85	1.27	18.73	26.50
1600	Mesh, painted, nailed to wood, 1.8 lb.	1 Lath	60	.133		5.75	4.74		10.49	13.35
1800	3.6 lb.		55	.145		4.30	5.15		9.45	12.40
1900	Wired to steel, painted, 1.8 lb.		53	.151		5.75	5.35		11.10	14.30
2100	3.6 lb.		50	.160		4.30	5.70		10	13.15

09 26 Veneer Plastering

09 26 13 – Gypsum Veneer Plastering

09 26 13.20 Blueboard

		Crew	Daily Output	Labor-Hours	Unit	Material	2009 Bare Costs Labor	2009 Bare Costs Equipment	Total	Total Incl O&P
0010	**BLUEBOARD** For use with thin coat									
0100	plaster application (see Div. 09 26 13.80)									
1000	3/8" thick, on walls or ceilings, standard, no finish included	2 Carp	1900	.008	S.F.	.28	.34		.62	.83
1100	With thin coat plaster finish		875	.018		.36	.73		1.09	1.53
1400	On beams, columns, or soffits, standard, no finish included		675	.024		.32	.95		1.27	1.82
1450	With thin coat plaster finish		475	.034		.40	1.35		1.75	2.53
3000	1/2" thick, on walls or ceilings, standard, no finish included		1900	.008		.33	.34		.67	.88
3100	With thin coat plaster finish		875	.018		.41	.73		1.14	1.58
3300	Fire resistant, no finish included		1900	.008		.33	.34		.67	.88
3400	With thin coat plaster finish		875	.018		.41	.73		1.14	1.58
3450	On beams, columns, or soffits, standard, no finish included		675	.024		.38	.95		1.33	1.89
3500	With thin coat plaster finish		475	.034		.46	1.35		1.81	2.60
3700	Fire resistant, no finish included		675	.024		.38	.95		1.33	1.89
3800	With thin coat plaster finish		475	.034		.46	1.35		1.81	2.60
5000	5/8" thick, on walls or ceilings, fire resistant, no finish included		1900	.008		.34	.34		.68	.89
5100	With thin coat plaster finish		875	.018		.42	.73		1.15	1.59
5500	On beams, columns, or soffits, no finish included		675	.024		.39	.95		1.34	1.90
5600	With thin coat plaster finish		475	.034		.47	1.35		1.82	2.61
6000	For high ceilings, over 8' high, add		3060	.005			.21		.21	.32
6500	For over 3 stories high, add per story		6100	.003			.10		.10	.16

09 26 13.80 Thin Coat Plaster

		Crew	Daily Output	Labor-Hours	Unit	Material	2009 Bare Costs Labor	2009 Bare Costs Equipment	Total	Total Incl O&P
0010	**THIN COAT PLASTER**									
0012	1 coat veneer, not incl. lath	J-1	3600	.011	S.F.	.08	.38	.04	.50	.71
1000	In 50 lb. bags				Bag	11.10			11.10	12.20

09 28 Backing Boards and Underlayments

09 28 13 – Cementitious Backing Boards

09 28 13.10 Cementitious Backerboard

		Crew	Daily Output	Labor-Hours	Unit	Material	2009 Bare Costs Labor	2009 Bare Costs Equipment	Total	Total Incl O&P
0010	**CEMENTITIOUS BACKERBOARD**									
0070	Cementitious backerboard, on floor, 3' x 4' x 1/2" sheets	2 Carp	525	.030	S.F.	.79	1.22		2.01	2.75
0080	3' x 5' x 1/2" sheets		525	.030		.78	1.22		2	2.74
0090	3' x 6' x 1/2" sheets		525	.030		.72	1.22		1.94	2.68
0100	3' x 4' x 5/8" sheets		525	.030		1.04	1.22		2.26	3.03
0110	3' x 5' x 5/8" sheets		525	.030		1.03	1.22		2.25	3.02
0120	3' x 6' x 5/8" sheets		525	.030		.96	1.22		2.18	2.95
0150	On wall, 3' x 4' x 1/2" sheets		350	.046		.79	1.83		2.62	3.69
0160	3' x 5' x 1/2" sheets		350	.046		.78	1.83		2.61	3.68
0170	3' x 6' x 1/2" sheets		350	.046		.72	1.83		2.55	3.62
0180	3' x 4' x 5/8" sheets		350	.046		1.04	1.83		2.87	3.97
0190	3' x 5' x 5/8" sheets		350	.046		1.03	1.83		2.86	3.96
0200	3' x 6' x 5/8" sheets		350	.046		.96	1.83		2.79	3.89
0250	On counter, 3' x 4' x 1/2" sheets		180	.089		.79	3.55		4.34	6.35
0260	3' x 5' x 1/2" sheets		180	.089		.78	3.55		4.33	6.35
0270	3' x 6' x 1/2" sheets		180	.089		.72	3.55		4.27	6.30
0300	3' x 4' x 5/8" sheets		180	.089		1.04	3.55		4.59	6.65
0310	3' x 5' x 5/8" sheets		180	.089		1.03	3.55		4.58	6.65
0320	3' x 6' x 5/8" sheets		180	.089		.96	3.55		4.51	6.55

09 29 Gypsum Board

09 29 10 – Gypsum Board Panels

09 29 10.10 Gypsum Board Ceilings		Crew	Daily Output	Labor-Hours	Unit	Material	2009 Bare Costs Labor	Equipment	Total	Total Incl O&P
0010	**GYPSUM BOARD CEILINGS**, fire rated, finished									
0100	Screwed to grid, channel or joists, 1/2" thick	2 Carp	765	.021	S.F.	.32	.84		1.16	1.65
0150	Mold resistant		765	.021		.54	.84		1.38	1.89
0200	5/8" thick		765	.021		.36	.84		1.20	1.70
0250	Mold resistant		765	.021		.66	.84		1.50	2.03
0300	Over 8' high, 1/2" thick		615	.026		.32	1.04		1.36	1.96
0350	Mold resistant		615	.026		.54	1.04		1.58	2.20
0400	5/8" thick		615	.026		.36	1.04		1.40	2.01
0450	Mold resistant		615	.026		.66	1.04		1.70	2.34
0600	Grid suspension system, direct hung									
0700	1-1/2" C.R.C., with 7/8" hi hat furring channel, 16" O.C.	2 Carp	1025	.016	S.F.	1.57	.62		2.19	2.70
0800	24" O.C.		1300	.012		1.43	.49		1.92	2.33
0900	3-5/8" C.R.C., with 7/8" hi hat furring channel, 16" O.C.		1025	.016		1.66	.62		2.28	2.80
1000	24" O.C.		1300	.012		1.44	.49		1.93	2.34
09 29 10.30 Gypsum Board										
0010	**GYPSUM BOARD** on walls & ceilings R092910-10									
0100	Nailed or screwed to studs unless otherwise noted									
0150	3/8" thick, on walls, standard, no finish included	2 Carp	2000	.008	S.F.	.31	.32		.63	.84
0200	On ceilings, standard, no finish included		1800	.009		.31	.36		.67	.89
0250	On beams, columns, or soffits, no finish included		675	.024		.31	.95		1.26	1.81
0300	1/2" thick, on walls, standard, no finish included		2000	.008		.31	.32		.63	.84
0350	Taped and finished (level 4 finish)		965	.017		.35	.66		1.01	1.42
0390	With compound skim coat (level 5 finish)		775	.021		.40	.83		1.23	1.72
0400	Fire resistant, no finish included		2000	.008		.32	.32		.64	.85
0450	Taped and finished (level 4 finish)		965	.017		.36	.66		1.02	1.43
0490	With compound skim coat (level 5 finish)		775	.021		.41	.83		1.24	1.73
0500	Water resistant, no finish included		2000	.008		.48	.32		.80	1.03
0550	Taped and finished (level 4 finish)		965	.017		.52	.66		1.18	1.60
0590	With compound skim coat (level 5 finish)		775	.021		.57	.83		1.40	1.90
0600	Prefinished, vinyl, clipped to studs		900	.018		.83	.71		1.54	2.01
0700	Mold resistant, no finish included		2000	.008		.49	.32		.81	1.04
0710	Taped and finished (level 4 finish)		965	.017		.53	.66		1.19	1.62
0720	With compound skim coat (level 5 finish)		775	.021		.58	.83		1.41	1.91
1000	On ceilings, standard, no finish included		1800	.009		.31	.36		.67	.89
1050	Taped and finished (level 4 finish)		765	.021		.35	.84		1.19	1.69
1090	With compound skim coat (level 5 finish)		610	.026		.40	1.05		1.45	2.06
1100	Fire resistant, no finish included		1800	.009		.32	.36		.68	.90
1150	Taped and finished (level 4 finish)		765	.021		.36	.84		1.20	1.70
1195	With compound skim coat (level 5 finish)		610	.026		.41	1.05		1.46	2.07
1200	Water resistant, no finish included		1800	.009		.48	.36		.84	1.08
1250	Taped and finished (level 4 finish)		765	.021		.52	.84		1.36	1.87
1290	With compound skim coat (level 5 finish)		610	.026		.57	1.05		1.62	2.24
1310	Mold resistant, no finish included		1800	.009		.49	.36		.85	1.09
1320	Taped and finished (level 4 finish)		765	.021		.53	.84		1.37	1.89
1330	With compound skim coat (level 5 finish)		610	.026		.58	1.05		1.63	2.25
1500	On beams, columns, or soffits, standard, no finish included		675	.024		.36	.95		1.31	1.86
1550	Taped and finished (level 4 finish)		475	.034		.35	1.35		1.70	2.48
1590	With compound skim coat (level 5 finish)		540	.030		.40	1.18		1.58	2.28
1600	Fire resistant, no finish included		675	.024		.32	.95		1.27	1.82
1650	Taped and finished (level 4 finish)		475	.034		.36	1.35		1.71	2.49
1690	With compound skim coat (level 5 finish)		540	.030		.41	1.18		1.59	2.29
1700	Water resistant, no finish included		675	.024		.55	.95		1.50	2.08

RSMeans Business Solutions

Leaders in Cost Engineering for the Design, Construction, and Facilities Management Industry...

Market Analysis and Third Party Validation – RSMeans conducts research studies from numerous perspectives, including escalation costs in local markets, cost estimates for legal damages, and trend research for industry associations, as well as product adoption and productivity studies for building product manufacturers. Means is often called upon to produce "third party validation" construction cost estimates for owners, architects, and boards of directors.

Predictive Cost Models for Facility Owners – From early planning, cost modeling is driven by close attention to standards, codes, methods, and business processes that impact cost and operational effectiveness. Predictive cost models offer cost control and standardization for facility owners, particularly in repetitive building environments. RSMeans custom models are based upon actual architectural plans and historical budgets, with cost data and location factors adjusted to regions anywhere in North America.

Custom Cost Databases and Web-Based Dashboards – Custom applications of the RSMeans cost databases use the building blocks of construction cost engineering, including a standard code of accounts, activity descriptions, material, labor, and equipment costs. RSMeans has in-house technical expertise for rapid deployment of cost databases, online cost calculators, and web-based analytic dashboards.

Job Order Contracting and FacilitiesWorks™ – A delivery method for construction renovation to reduce cycle times and overhead costs. RSMeans offers consulting in JOC management practices, FacilitiesWorks™ software for renovation cost estimating and project management, and on-site training programs. JOC is available to government agencies, colleges and universities, school districts, municipalities, and health care systems.

Life Cycle Cost Analysis and Sustainability – Building upon its extensive facility cost databases, RSMeans offers life cycle cost studies and sustainability models to analyze total cost of ownership (TCO) for high performance buildings. Analyses include capital costs, life cycle and operations costs, as well as studies specific to integrated building automation, energy efficiency and security systems. Cost savings realized in operations, maintenance and replacement are analyzed enterprise-wide.

CORPORATE • HEALTH CARE • EDUCATION • GOVERNMENT • MUNICIPAL • RETAIL • ASSOCIATIONS • BUILDING PRODUCT MANUFACTURERS • LEGAL

For more information contact RSMeans Business Solutions at 781-422-5101 or visit our website at www.rsmeans.com or email consulting@rsmeans.com

RSMeans

09 29 Gypsum Board

09 29 10 – Gypsum Board Panels

09 29 10.30 Gypsum Board		Crew	Daily Output	Labor-Hours	Unit	Material	2009 Bare Costs Labor	Equipment	Total	Total Incl O&P
1750	Taped and finished (level 4 finish)	2 Carp	475	.034	S.F.	.52	1.35		1.87	2.66
1790	With compound skim coat (level 5 finish)		540	.030		.57	1.18		1.75	2.46
1800	Mold resistant, no finish included		675	.024		.56	.95		1.51	2.09
1810	Taped and finished (level 4 finish)		475	.034		.53	1.35		1.88	2.68
1820	With compound skim coat (level 5 finish)		540	.030		.58	1.18		1.76	2.47
2000	5/8" thick, on walls, standard, no finish included		2000	.008		.34	.32		.66	.87
2050	Taped and finished (level 4 finish)		965	.017		.38	.66		1.04	1.45
2090	With compound skim coat (level 5 finish)		775	.021		.43	.83		1.26	1.75
2100	Fire resistant, no finish included		2000	.008		.36	.32		.68	.90
2150	Taped and finished (level 4 finish)		965	.017		.40	.66		1.06	1.47
2195	With compound skim coat (level 5 finish)		775	.021		.45	.83		1.28	1.77
2200	Water resistant, no finish included		2000	.008		.47	.32		.79	1.02
2250	Taped and finished (level 4 finish)		965	.017		.51	.66		1.17	1.59
2290	With compound skim coat (level 5 finish)		775	.021		.56	.83		1.39	1.89
2510	Mold resistant, no finish included		2000	.008		.60	.32		.92	1.16
2520	Taped and finished (level 4 finish)		965	.017		.64	.66		1.30	1.74
2530	With compound skim coat (level 5 finish)		775	.021		.69	.83		1.52	2.04
3000	On ceilings, standard, no finish included		1800	.009		.34	.36		.70	.92
3050	Taped and finished (level 4 finish)		765	.021		.38	.84		1.22	1.72
3090	With compound skim coat (level 5 finish)		615	.026		.43	1.04		1.47	2.08
3100	Fire resistant, no finish included		1800	.009		.36	.36		.72	.95
3150	Taped and finished (level 4 finish)		765	.021		.40	.84		1.24	1.74
3190	With compound skim coat (level 5 finish)		615	.026		.45	1.04		1.49	2.10
3200	Water resistant, no finish included		1800	.009		.47	.36		.83	1.07
3250	Taped and finished (level 4 finish)		765	.021		.51	.84		1.35	1.86
3290	With compound skim coat (level 5 finish)		615	.026		.56	1.04		1.60	2.22
3300	Mold resistant, no finish included		1800	.009		.60	.36		.96	1.21
3310	Taped and finished (level 4 finish)		765	.021		.64	.84		1.48	2.01
3320	With compound skim coat (level 5 finish)		615	.026		.69	1.04		1.73	2.37
3500	On beams, columns, or soffits, no finish included		675	.024		.39	.95		1.34	1.90
3550	Taped and finished (level 4 finish)		475	.034		.44	1.35		1.79	2.57
3590	With compound skim coat (level 5 finish)		380	.042		.49	1.68		2.17	3.15
3600	Fire resistant, no finish included		675	.024		.41	.95		1.36	1.93
3650	Taped and finished (level 4 finish)		475	.034		.46	1.35		1.81	2.60
3690	With compound skim coat (level 5 finish)		380	.042		.45	1.68		2.13	3.10
3700	Water resistant, no finish included		675	.024		.54	.95		1.49	2.06
3750	Taped and finished (level 4 finish)		475	.034		.59	1.35		1.94	2.74
3790	With compound skim coat (level 5 finish)		380	.042		.56	1.68		2.24	3.22
3800	Mold resistant, no finish included		675	.024		.69	.95		1.64	2.23
3810	Taped and finished (level 4 finish)		475	.034		.74	1.35		2.09	2.90
3820	With compound skim coat (level 5 finish)		380	.042		.69	1.68		2.37	3.37
4000	Fireproofing, beams or columns, 2 layers, 1/2" thick, incl finish		330	.048		.68	1.94		2.62	3.75
4010	Mold resistant		330	.048		1.02	1.94		2.96	4.12
4050	5/8" thick		300	.053		.80	2.13		2.93	4.19
4060	Mold resistant		300	.053		1.28	2.13		3.41	4.71
4100	3 layers, 1/2" thick		225	.071		1	2.84		3.84	5.50
4110	Mold resistant		225	.071		1.51	2.84		4.35	6.05
4150	5/8" thick		210	.076		1.21	3.04		4.25	6.05
4160	Mold resistant		210	.076		1.93	3.04		4.97	6.85
5050	For 1" thick coreboard on columns		480	.033		.88	1.33		2.21	3.03
5100	For foil-backed board, add					.14			.14	.15
5200	For work over 8' high, add	2 Carp	3060	.005			.21		.21	.32
5270	For textured spray, add	2 Lath	1600	.010		.08	.36		.44	.62

09 29 Gypsum Board

09 29 10 – Gypsum Board Panels

09 29 10.30 Gypsum Board

		Crew	Daily Output	Labor-Hours	Unit	Material	2009 Bare Costs Labor	Equipment	Total	Total Incl O&P
5300	For distribution cost over 3 stories high, add per story	2 Carp	6100	.003	S.F.		.10		.10	.16
5350	For finishing inner corners, add		950	.017	L.F.	.09	.67		.76	1.14
5355	For finishing outer corners, add		1250	.013		.21	.51		.72	1.02
5500	For acoustical sealant, add per bead	1 Carp	500	.016		.04	.64		.68	1.04
5550	Sealant, 1 quart tube				Ea.	6.85			6.85	7.55
5600	Sound deadening board, 1/4" gypsum	2 Carp	1800	.009	S.F.	.37	.36		.73	.96
5650	1/2" wood fiber	"	1800	.009	"	.28	.36		.64	.86

09 29 10.50 High Abuse Gypsum Board

		Crew	Daily Output	Labor-Hours	Unit	Material	Labor	Equipment	Total	Total Incl O&P
0010	**HIGH ABUSE GYPSUM BOARD**, fiber reinforced, nailed or									
0100	screwed to studs unless otherwise noted									
0110	1/2" thick, on walls, no finish included	2 Carp	1800	.009	S.F.	.87	.36		1.23	1.51
0120	Taped and finished (level 4 finish)		870	.018		.91	.73		1.64	2.14
0130	With compound skim coat (level 5 finish)		700	.023		.96	.91		1.87	2.47
0150	On ceilings, no finish included		1620	.010		.87	.39		1.26	1.57
0160	Taped and finished (level 4 finish)		690	.023		.91	.93		1.84	2.44
0170	With compound skim coat (level 5 finish)		550	.029		.96	1.16		2.12	2.85
0210	5/8" thick, on walls, no finish included		1800	.009		.94	.36		1.30	1.58
0220	Taped and finished (level 4 finish)		870	.018		.98	.73		1.71	2.22
0230	With compound skim coat (level 5 finish)		700	.023		1.03	.91		1.94	2.55
0250	On ceilings, no finish included		1620	.010		.94	.39		1.33	1.64
0260	Taped and finished (level 4 finish)		690	.023		.98	.93		1.91	2.52
0270	With compound skim coat (level 5 finish)		550	.029		1.03	1.16		2.19	2.93
0310	5/8" thick, on walls, very high impact, no finish included		1800	.009		.96	.36		1.32	1.61
0320	Taped and finished (level 4 finish)		870	.018		1	.73		1.73	2.24
0330	With compound skim coat (level 5 finish)		700	.023		1.05	.91		1.96	2.57
0350	On ceilings, no finish included		1620	.010		.96	.39		1.35	1.67
0360	Taped and finished (level 4 finish)		690	.023		1	.93		1.93	2.54
0370	With compound skim coat (level 5 finish)		550	.029		1.05	1.16		2.21	2.95
0400	High abuse, gypsum core, paper face									
0410	1/2" thick, on walls, no finish included	2 Carp	1800	.009	S.F.	.75	.36		1.11	1.38
0420	Taped and finished (level 4 finish)		870	.018		.79	.73		1.52	2.01
0430	With compound skim coat (level 5 finish)		700	.023		.84	.91		1.75	2.34
0450	On ceilings, no finish included		1620	.010		.75	.39		1.14	1.44
0460	Taped and finished (level 4 finish)		690	.023		.79	.93		1.72	2.31
0470	With compound skim coat (level 5 finish)		550	.029		.84	1.16		2	2.72
0510	5/8" thick, on walls, no finish included		1800	.009		.76	.36		1.12	1.39
0520	Taped and finished (level 4 finish)		870	.018		.80	.73		1.53	2.02
0530	With compound skim coat (level 5 finish)		700	.023		.85	.91		1.76	2.35
0550	On ceilings, no finish included		1620	.010		.76	.39		1.15	1.45
0560	Taped and finished (level 4 finish)		690	.023		.80	.93		1.73	2.32
0570	With compound skim coat (level 5 finish)		550	.029		.85	1.16		2.01	2.73
1000	For high ceilings, over 8' high, add		2750	.006			.23		.23	.36
1010	For distribution cost over 3 stories high, add per story		5500	.003			.12		.12	.18

09 29 15 – Gypsum Board Accessories

09 29 15.10 Accessories, Gypsum Board

		Crew	Daily Output	Labor-Hours	Unit	Material	Labor	Equipment	Total	Total Incl O&P
0010	**ACCESSORIES, GYPSUM BOARD**									
0020	Casing bead, galvanized steel	1 Carp	2.90	2.759	C.L.F.	22	110		132	196
0100	Vinyl		3	2.667		22.50	107		129.50	190
0300	Corner bead, galvanized steel, 1" x 1"		4	2		13.95	80		93.95	139
0400	1-1/4" x 1-1/4"		3.50	2.286		18.15	91.50		109.65	162
0600	Vinyl		4	2		17.80	80		97.80	144
0900	Furring channel, galv. steel, 7/8" deep, standard		2.60	3.077		27.50	123		150.50	222

09 29 Gypsum Board

09 29 15 – Gypsum Board Accessories

	09 29 15.10 Accessories, Gypsum Board	Crew	Daily Output	Labor-Hours	Unit	Material	2009 Bare Costs Labor	Equipment	Total	Total Incl O&P
1000	Resilient	1 Carp	2.55	3.137	C.L.F.	26.50	125		151.50	224
1100	J trim, galvanized steel, 1/2" wide		3	2.667		22.50	107		129.50	190
1120	5/8" wide		2.95	2.712		24.50	108		132.50	195
1140	L trim, galvanized		3	2.667		19.35	107		126.35	187
1150	U trim, galvanized		2.95	2.712		19.40	108		127.40	190
1160	Screws #6 x 1" A				M	7.20			7.20	7.90
1170	#6 x 1-5/8" A				"	10.55			10.55	11.60
1200	For stud partitions, see Div. 05 41 13.30 and 09 22 16.13									
1500	Z stud, galvanized steel, 1-1/2" wide	1 Carp	2.60	3.077	C.L.F.	37	123		160	232
1600	2" wide	"	2.55	3.137	"	55	125		180	255

09 30 Tiling

09 30 13 – Ceramic Tiling

09 30 13.10 Ceramic Tile

		Crew	Daily Output	Labor-Hours	Unit	Material	2009 Bare Costs Labor	Equipment	Total	Total Incl O&P
0010	**CERAMIC TILE**									
0050	Base, using 1' x 4" high pc. with 1" x 1" tiles, mud set	D-7	82	.195	L.F.	4.95	6.65		11.60	15.20
0100	Thin set	"	128	.125		4.53	4.26		8.79	11.25
0300	For 6" high base, 1" x 1" tile face, add					.72			.72	.79
0400	For 2" x 2" tile face, add to above					.38			.38	.42
0600	Cove base, 4-1/4" x 4-1/4" high, mud set	D-7	91	.176		3.93	6		9.93	13.10
0700	Thin set		128	.125		4.07	4.26		8.33	10.75
0900	6" x 4-1/4" high, mud set		100	.160		3.67	5.45		9.12	12.05
1000	Thin set		137	.117		3.47	3.98		7.45	9.65
1200	Sanitary cove base, 6" x 4-1/4" high, mud set		93	.172		4	5.85		9.85	13
1300	Thin set		124	.129		3.84	4.40		8.24	10.65
1500	6" x 6" high, mud set		84	.190		4.94	6.50		11.44	14.95
1600	Thin set		117	.137		4.74	4.66		9.40	12
1800	Bathroom accessories, average		82	.195	Ea.	10.70	6.65		17.35	21.50
1900	Bathtub, 5', rec. 4-1/4" x 4-1/4" tile wainscot, adhesive set 6' high		2.90	5.517		156	188		344	445
2100	7' high wainscot		2.50	6.400		179	218		397	515
2200	8' high wainscot		2.20	7.273		190	248		438	575
2400	Bullnose trim, 4-1/4" x 4-1/4", mud set		82	.195	L.F.	3.68	6.65		10.33	13.80
2500	Thin set		128	.125		3.30	4.26		7.56	9.85
2700	6" x 4-1/4" bullnose trim, mud set		84	.190		2.88	6.50		9.38	12.65
2800	Thin set		124	.129		2.74	4.40		7.14	9.45
3255	Floors, glazed, thin set, 6" x 6", color group 1		200	.080	S.F.	3.73	2.73		6.46	8.10
3260	8" x 8" tile		250	.064		3.73	2.18		5.91	7.30
3270	12" x 12" tile		325	.049		4.62	1.68		6.30	7.55
3280	16" x 16" tile		550	.029		6.30	.99		7.29	8.40
3285	Border, 6" x 12" tile		275	.058		11.70	1.98		13.68	15.75
3290	3" x 12" tile		200	.080		34	2.73		36.73	41.50
3300	Porcelain type, 1 color, color group 2, 1" x 1"		183	.087		4.71	2.98		7.69	9.55
3310	2" x 2" or 2" x 1", thin set		190	.084		5.40	2.87		8.27	10.15
3350	For random blend, 2 colors, add					.88			.88	.97
3360	4 colors, add					1.24			1.24	1.36
3370	For color group 3, add					.50			.50	.55
3380	For abrasive non-slip tile, add					.49			.49	.54
4300	Specialty tile, 4-1/4" x 4-1/4" x 1/2", decorator finish	D-7	183	.087		10	2.98		12.98	15.35
4500	Add for epoxy grout, 1/16" joint, 1" x 1" tile		800	.020		.62	.68		1.30	1.68
4600	2" x 2" tile		820	.020		.59	.66		1.25	1.62
4800	Pregrouted sheets, walls, 4-1/4" x 4-1/4", 6" x 4-1/4"									

09 30 Tiling

09 30 13 – Ceramic Tiling

09 30 13.10 Ceramic Tile

		Crew	Daily Output	Labor-Hours	Unit	Material	2009 Bare Costs Labor	Equipment	Total	Total Incl O&P
4810	and 8-1/2" x 4-1/4", 4 S.F. sheets, silicone grout	D-7	240	.067	S.F.	4.73	2.27		7	8.55
5100	Floors, unglazed, 2 S.F. sheets,									
5110	Urethane adhesive	D-7	180	.089	S.F.	4.71	3.03		7.74	9.65
5400	Walls, interior, thin set, 4-1/4" x 4-1/4" tile		190	.084		2.49	2.87		5.36	6.95
5500	6" x 4-1/4" tile		190	.084		2.86	2.87		5.73	7.35
5700	8-1/2" x 4-1/4" tile		190	.084		3.93	2.87		6.80	8.50
5800	6" x 6" tile		200	.080		3.34	2.73		6.07	7.65
5810	8" x 8" tile		225	.071		4.36	2.42		6.78	8.35
5820	12" x 12" tile		300	.053		3.62	1.82		5.44	6.65
5830	16" x 16" tile		500	.032		3.90	1.09		4.99	5.90
6000	Decorated wall tile, 4-1/4" x 4-1/4", minimum		270	.059		3.66	2.02		5.68	7
6100	Maximum		180	.089		42	3.03		45.03	50.50
6300	Exterior walls, frostproof, mud set, 4-1/4" x 4-1/4"		102	.157		7.30	5.35		12.65	15.85
6400	1-3/8" x 1-3/8"		93	.172		4.94	5.85		10.79	14.05
6600	Crystalline glazed, 4-1/4" x 4-1/4", mud set, plain		100	.160		4.43	5.45		9.88	12.85
6700	4-1/4" x 4-1/4", scored tile		100	.160		5.25	5.45		10.70	13.75
6900	6" x 6" plain		93	.172		5.65	5.85		11.50	14.80
7000	For epoxy grout, 1/16" joints, 4-1/4" tile, add		800	.020		.39	.68		1.07	1.43
7200	For tile set in dry mortar, add		1735	.009			.31		.31	.46
7300	For tile set in Portland cement mortar, add		290	.055		.27	1.88		2.15	3.05
9000	Regrout tile 4-1/2 x 4-1/2, or larger, wall	1 Tilf	100	.080		.14	3.05		3.19	4.61
9220	Floor	"	125	.064		.15	2.44		2.59	3.74
9300	Ceramic tiles, recycled glass, standard colors, 2" x 2" thru 6" x 6" G	D-7	190	.084		16.70	2.87		19.57	22.50
9310	6" x 6" G		200	.080		16.70	2.73		19.43	22.50
9320	8" x 8" G		225	.071		18.10	2.42		20.52	23.50
9330	12"x12" G		300	.053		18.10	1.82		19.92	22.50
9340	Earthtones, 2" x 2" to 4" x 8" G		190	.084		21	2.87		23.87	27
9350	6" x 6" G		200	.080		21	2.73		23.73	27
9360	8" x 8" G		225	.071		22.50	2.42		24.92	28
9370	12" x 12" G		300	.053		22.50	1.82		24.32	27
9380	Deep colors, 2" x 2" to 4" x 8" G		190	.084		30	2.87		32.87	37
9390	6" x 6" G		200	.080		30	2.73		32.73	37
9400	8" x 8" G		225	.071		31.50	2.42		33.92	38.50
9410	12" x 12" G		300	.053		31.50	1.82		33.32	37.50

09 30 16 – Quarry Tiling

09 30 16.10 Quarry Tile

		Crew	Daily Output	Labor-Hours	Unit	Material	2009 Bare Costs Labor	Equipment	Total	Total Incl O&P
0010	**QUARRY TILE**									
0100	Base, cove or sanitary, mud set, to 5" high, 1/2" thick	D-7	110	.145	L.F.	5.60	4.96		10.56	13.40
0300	Bullnose trim, red, mud set, 6" x 6" x 1/2" thick		120	.133		4.36	4.54		8.90	11.45
0400	4" x 4" x 1/2" thick		110	.145		4.79	4.96		9.75	12.50
0600	4" x 8" x 1/2" thick, using 8" as edge		130	.123		4.25	4.19		8.44	10.80
0700	Floors, mud set, 1,000 S.F. lots, red, 4" x 4" x 1/2" thick		120	.133	S.F.	7.50	4.54		12.04	14.90
0900	6" x 6" x 1/2" thick		140	.114		7.05	3.90		10.95	13.45
1000	4" x 8" x 1/2" thick		130	.123		7.50	4.19		11.69	14.40
1300	For waxed coating, add					.67			.67	.74
1500	For non-standard colors, add					.40			.40	.44
1600	For abrasive surface, add					.47			.47	.52
1800	Brown tile, imported, 6" x 6" x 3/4"	D-7	120	.133		8.45	4.54		12.99	15.90
1900	8" x 8" x 1"		110	.145		9.10	4.96		14.06	17.25
2100	For thin set mortar application, deduct		700	.023			.78		.78	1.14
2200	For epoxy grout & mortar, 6" x 6" x 1/2", add		350	.046		1.91	1.56		3.47	4.38
2700	Stair tread, 6" x 6" x 3/4", plain		50	.320		5.25	10.90		16.15	22

09 30 Tiling

09 30 16 – Quarry Tiling

09 30 16.10 Quarry Tile

		Crew	Daily Output	Labor-Hours	Unit	Material	2009 Bare Costs Labor	Equipment	Total	Total Incl O&P
2800	Abrasive	D-7	47	.340	S.F.	5.30	11.60		16.90	23
3000	Wainscot, 6" x 6" x 1/2", thin set, red		105	.152		4.24	5.20		9.44	12.25
3100	Non-standard colors		105	.152		4.69	5.20		9.89	12.75
3300	Window sill, 6" wide, 3/4" thick		90	.178	L.F.	5.25	6.05		11.30	14.65
3400	Corners		80	.200	Ea.	5.85	6.80		12.65	16.45

09 30 23 – Glass Mosaic Tiling

09 30 23.10 Glass Mosaics

		Crew	Daily Output	Labor-Hours	Unit	Material	Labor	Equipment	Total	Total Incl O&P
0010	**GLASS MOSAICS** 3/4" tile on 12" sheets, standard grout									
0300	Color group 1 & 2	D-7	73	.219	S.F.	17.35	7.45		24.80	30
0350	Color group 3		73	.219		18.25	7.45		25.70	31
0400	Color group 4		73	.219		25	7.45		32.45	38.50
0450	Color group 5		73	.219		28	7.45		35.45	42
0500	Color group 6		73	.219		34	7.45		41.45	48
0600	Color group 7		73	.219		39	7.45		46.45	54
0700	Color group 8, golds, silvers & specialties		64	.250		56.50	8.50		65	75

09 30 29 – Metal Tiling

09 30 29.10 Metal Tile

		Crew	Daily Output	Labor-Hours	Unit	Material	Labor	Equipment	Total	Total Incl O&P
0010	**METAL TILE** 4' x 4' sheet, 24 ga., tile pattern, nailed									
0200	Stainless steel	2 Carp	512	.031	S.F.	24.50	1.25		25.75	29
0400	Aluminized steel	"	512	.031	"	13.15	1.25		14.40	16.40

09 51 Acoustical Ceilings

09 51 23 – Acoustical Tile Ceilings

09 51 23.10 Suspended Acoustic Ceiling Tiles

		Crew	Daily Output	Labor-Hours	Unit	Material	Labor	Equipment	Total	Total Incl O&P
0010	**SUSPENDED ACOUSTIC CEILING TILES**, not including									
0100	suspension system									
0300	Fiberglass boards, film faced, 2' x 2' or 2' x 4', 5/8" thick	1 Carp	625	.013	S.F.	.73	.51		1.24	1.59
0400	3/4" thick		600	.013		1.69	.53		2.22	2.69
0500	3" thick, thermal, R11		450	.018		1.53	.71		2.24	2.78
0600	Glass cloth faced fiberglass, 3/4" thick		500	.016		3.04	.64		3.68	4.33
0700	1" thick		485	.016		2.16	.66		2.82	3.40
0820	1-1/2" thick, nubby face		475	.017		2.66	.67		3.33	3.97
1110	Mineral fiber tile, lay-in, 2' x 2' or 2' x 4', 5/8" thick, fine texture		625	.013		.57	.51		1.08	1.42
1115	Rough textured		625	.013		1.26	.51		1.77	2.18
1125	3/4" thick, fine textured		600	.013		1.56	.53		2.09	2.55
1130	Rough textured		600	.013		1.88	.53		2.41	2.90
1135	Fissured		600	.013		2.45	.53		2.98	3.53
1150	Tegular, 5/8" thick, fine textured		470	.017		1.45	.68		2.13	2.65
1155	Rough textured		470	.017		1.90	.68		2.58	3.14
1165	3/4" thick, fine textured		450	.018		2.08	.71		2.79	3.39
1170	Rough textured		450	.018		2.35	.71		3.06	3.69
1175	Fissured		450	.018		3.63	.71		4.34	5.10
1180	For aluminum face, add					6.05			6.05	6.65
1185	For plastic film face, add					.94			.94	1.03
1190	For fire rating, add					.45			.45	.50
1300	Mirror faced panels, 15/16" thick, 2' x 2'	1 Carp	500	.016		10.45	.64		11.09	12.50
1900	Eggcrate, acrylic, 1/2" x 1/2" x 1/2" cubes		500	.016		1.82	.64		2.46	2.99
2100	Polystyrene eggcrate, 3/8" x 3/8" x 1/2" cubes		510	.016		1.53	.63		2.16	2.65
2200	1/2" x 1/2" x 1/2" cubes		500	.016		2.04	.64		2.68	3.23
2400	Luminous panels, prismatic, acrylic		400	.020		2.22	.80		3.02	3.68

09 51 Acoustical Ceilings

09 51 23 – Acoustical Tile Ceilings

09 51 23.10 Suspended Acoustic Ceiling Tiles		Crew	Daily Output	Labor-Hours	Unit	Material	2009 Bare Costs Labor	Equipment	Total	Total Incl O&P
2500	Polystyrene	1 Carp	400	.020	S.F.	1.14	.80		1.94	2.49
2700	Flat white acrylic		400	.020		3.86	.80		4.66	5.50
2800	Polystyrene		400	.020		2.65	.80		3.45	4.16
3000	Drop pan, white, acrylic		400	.020		5.65	.80		6.45	7.50
3100	Polystyrene		400	.020		4.73	.80		5.53	6.45
3600	Perforated aluminum sheets, .024" thick, corrugated, painted		490	.016		2.25	.65		2.90	3.49
3700	Plain		500	.016		3.76	.64		4.40	5.15
3720	Mineral fiber, 24" x 24" or 48", reveal edge, painted, 5/8" thick		600	.013		1.38	.53		1.91	2.35
3740	3/4" thick		575	.014		2.24	.56		2.80	3.32
5020	66 – 78% recycled content, 3/4" thick [G]		600	.013		1.82	.53		2.35	2.83
5040	Mylar, 42% recycled content, 3/4" thick [G]		600	.013		4.28	.53		4.81	5.55

09 51 23.30 Suspended Ceilings, Complete		Crew	Daily Output	Labor-Hours	Unit	Material	Labor	Equipment	Total	Total Incl O&P
0010	**SUSPENDED CEILINGS, COMPLETE** Including standard									
	suspension system but not incl. 1-1/2" carrier channels									
0600	Fiberglass ceiling board, 2' x 4' x 5/8", plain faced	1 Carp	500	.016	S.F.	1.36	.64		2	2.49
0700	Offices, 2' x 4' x 3/4"		380	.021		2.32	.84		3.16	3.85
0800	Mineral fiber, on 15/16" T bar susp. 2' x 2' x 3/4" lay-in board		345	.023		2.35	.93		3.28	4.02
0810	2' x 4' x 5/8" tile		380	.021		1.20	.84		2.04	2.62
0820	Tegular, 2' x 2' x 5/8" tile on 9/16" grid		250	.032		2.40	1.28		3.68	4.61
0830	2' x 4' x 3/4" tile		275	.029		2.87	1.16		4.03	4.96
0900	Luminous panels, prismatic, acrylic		255	.031		2.85	1.25		4.10	5.10
1200	Metal pan with acoustic pad, steel		75	.107		4.16	4.26		8.42	11.20
1300	Painted aluminum		75	.107		2.88	4.26		7.14	9.75
1500	Aluminum, degreased finish		75	.107		4.80	4.26		9.06	11.90
1600	Stainless steel		75	.107		9.05	4.26		13.31	16.55
1800	Tile, Z bar suspension, 5/8" mineral fiber tile		150	.053		2.16	2.13		4.29	5.70
1900	3/4" mineral fiber tile		150	.053		2.31	2.13		4.44	5.85
2402	For strip lighting, see Div. 26 51 13.50									
2500	For rooms under 500 S.F., add				S.F.		25%			

09 51 53 – Direct-Applied Acoustical Ceilings

09 51 53.10 Ceiling Tile		Crew	Daily Output	Labor-Hours	Unit	Material	Labor	Equipment	Total	Total Incl O&P
0010	**CEILING TILE**, Stapled or cemented									
0100	12" x 12" or 12" x 24", not including furring									
0600	Mineral fiber, vinyl coated, 5/8" thick	1 Carp	300	.027	S.F.	1.85	1.07		2.92	3.69
0700	3/4" thick		300	.027		1.87	1.07		2.94	3.71
0900	Fire rated, 3/4" thick, plain faced		300	.027		1.40	1.07		2.47	3.19
1000	Plastic coated face		300	.027		1.18	1.07		2.25	2.95
1200	Aluminum faced, 5/8" thick, plain		300	.027		1.51	1.07		2.58	3.31
3700	Wall application of above, add		1000	.008			.32		.32	.50
3900	For ceiling primer, add					.13			.13	.14
4000	For ceiling cement, add					.37			.37	.41

09 53 Acoustical Ceiling Suspension Assemblies

09 53 23 – Metal Acoustical Ceiling Suspension Assemblies

09 53 23.30 Ceiling Suspension Systems

		Crew	Daily Output	Labor-Hours	Unit	Material	2009 Bare Costs Labor	Equipment	Total	Total Incl O&P
0010	**CEILING SUSPENSION SYSTEMS** for boards and tile									
0050	Class A suspension system, 15/16" T bar, 2' x 4' grid	1 Carp	800	.010	S.F.	.63	.40		1.03	1.31
0300	2' x 2' grid		650	.012		.79	.49		1.28	1.62
0310	25% recycled steel, 2' x 4' grid G		800	.010		.73	.40		1.13	1.42
0320	2' x 2' grid G		650	.012		.92	.49		1.41	1.77
0350	For 9/16" grid, add					.16			.16	.18
0360	For fire rated grid, add					.09			.09	.10
0370	For colored grid, add					.21			.21	.23
0400	Concealed Z bar suspension system, 12" module	1 Carp	520	.015		.65	.61		1.26	1.67
0600	1-1/2" carrier channels, 4' O.C., add	"	470	.017		.15	.68		.83	1.22
0700	Carrier channels for ceilings with									
0900	recessed lighting fixtures, add	1 Carp	460	.017	S.F.	.28	.69		.97	1.38
1040	Hanging wire, 12 ga., 4' long		65	.123	C.S.F.	7.65	4.92		12.57	16
1080	8' long		65	.123	"	15.25	4.92		20.17	24.50
1200	Seismic bracing, 2-1/2" compression post with four bracing wires		50	.160	Ea.	2.74	6.40		9.14	12.90
5010	Four way support for seismic compression strut # 12 wire		32	.250		40	10		50	59.50
5015	# 9 wire		24	.333		68	13.30		81.30	95.50
5020	Seismic manual compression strut for suspended ceilings		32	.250		120	10		130	148
5030	Hydraulic		32	.250		90	10		100	115
5040	Seismic wall clips for suspended ceilings, 4' O.C.		48	.167		1	6.65		7.65	11.45

09 63 Masonry Flooring

09 63 13 – Brick Flooring

09 63 13.10 Miscellaneous Brick Flooring

		Crew	Daily Output	Labor-Hours	Unit	Material	2009 Bare Costs Labor	Equipment	Total	Total Incl O&P
0010	**MISCELLANEOUS BRICK FLOORING**									
0020	Acid proof shales, red, 8" x 3-3/4" x 1-1/4" thick	D-7	.43	37.209	M	820	1,275		2,095	2,750
0200	Acid proof clay brick, 8" x 3-3/4" x 2-1/4" thick	D-1	.40	40	"	870	1,450		2,320	3,150
0250	9" x 4-1/2" x 3"	"	95	.168	S.F.	3.93	6.10		10.03	13.60
0260	Cast ceramic, pressed, 4" x 8" x 1/2", unglazed	D-7	100	.160		5.95	5.45		11.40	14.55
0270	Glazed		100	.160		7.95	5.45		13.40	16.70
0280	Hand molded flooring, 4" x 8" x 3/4", unglazed		95	.168		7.85	5.75		13.60	17.05
0290	Glazed		95	.168		9.85	5.75		15.60	19.25
0300	8" hexagonal, 3/4" thick, unglazed		85	.188		8.60	6.40		15	18.85
0310	Glazed		85	.188		15.55	6.40		21.95	26.50
0400	Heavy duty industrial, cement mortar bed, 2" thick, not incl. brick	D-1	80	.200		.79	7.25		8.04	11.85
0450	Acid proof joints, 1/4" wide	"	65	.246		1.37	8.95		10.32	15.05
0500	Pavers, 8" x 4", 1" to 1-1/4" thick, red	D-7	95	.168		3.47	5.75		9.22	12.20
0510	Ironspot	"	95	.168		4.89	5.75		10.64	13.80
0540	1-3/8" to 1-3/4" thick, red	D-1	95	.168		3.34	6.10		9.44	12.95
0560	Ironspot		95	.168		4.84	6.10		10.94	14.60
0580	2-1/4" thick, red		90	.178		3.40	6.45		9.85	13.55
0590	Ironspot		90	.178		5.30	6.45		11.75	15.60
0700	Paver, adobe brick, 6" x 12", 1/2" joint		42	.381		.90	13.85		14.75	22
0710	Mexican red, 12" x 12"	1 Tilf	48	.167		1.61	6.35		7.96	11.05
0720	Saltillo, 12" x 12"	"	48	.167		1.13	6.35		7.48	10.55
0800	For sidewalks and patios with pavers, see Div. 32 14 16.10									
0870	For epoxy joints, add	D-1	600	.027	S.F.	2.59	.97		3.56	4.32
0880	For Furan underlayment, add	"	600	.027		2.14	.97		3.11	3.82
0890	For waxed surface, steam cleaned, add	A-1H	1000	.008		.19	.25	.05	.49	.66

09 63 Masonry Flooring

09 63 40 – Stone Flooring

09 63 40.10 Marble

		Crew	Daily Output	Labor-Hours	Unit	Material	2009 Bare Costs Labor	2009 Bare Costs Equipment	Total	Total Incl O&P
0010	**MARBLE**									
0020	Thin gauge tile, 12" x 6", 3/8", white Carara	D-7	60	.267	S.F.	11.15	9.10		20.25	25.50
0100	Travertine		60	.267		12.20	9.10		21.30	26.50
0200	12" x 12" x 3/8", thin set, floors		60	.267		6.05	9.10		15.15	19.95
0300	On walls		52	.308	↓	9.35	10.50		19.85	25.50
1000	Marble threshold, 4" wide x 36" long x 5/8" thick, white	↓	60	.267	Ea.	6.05	9.10		15.15	19.95

09 63 40.20 Slate Tile

		Crew	Daily Output	Labor-Hours	Unit	Material	Labor	Equipment	Total	Total Incl O&P
0010	**SLATE TILE**									
0020	Vermont, 6" x 6" x 1/4" thick, thin set	D-7	180	.089	S.F.	6.80	3.03		9.83	11.90

09 64 Wood Flooring

09 64 16 – Wood Block Flooring

09 64 16.10 End Grain Block Flooring

		Crew	Daily Output	Labor-Hours	Unit	Material	Labor	Equipment	Total	Total Incl O&P
0010	**END GRAIN BLOCK FLOORING**									
0020	End grain flooring, coated, 2" thick	1 Carp	295	.027	S.F.	3.29	1.08		4.37	5.30
0400	Natural finish, 1" thick, fir		125	.064		3.40	2.56		5.96	7.70
0600	1-1/2" thick, pine		125	.064		3.34	2.56		5.90	7.65
0700	2" thick, pine	↓	125	.064	↓	4.64	2.56		7.20	9.05

09 64 19 – Wood Composition Flooring

09 64 19.10 Wood Composition

		Crew	Daily Output	Labor-Hours	Unit	Material	Labor	Equipment	Total	Total Incl O&P
0010	**WOOD COMPOSITION** Gym floors									
0100	2-1/4" x 6-7/8" x 3/8", on 2" grout setting bed	D-7	150	.107	S.F.	5.85	3.64		9.49	11.75
0200	Thin set, on concrete	"	250	.064	↓	5.35	2.18		7.53	9.10
0300	Sanding and finishing, add	1 Carp	200	.040		.79	1.60		2.39	3.35

09 64 23 – Wood Parquet Flooring

09 64 23.10 Wood Parquet

		Crew	Daily Output	Labor-Hours	Unit	Material	Labor	Equipment	Total	Total Incl O&P
0010	**WOOD PARQUET** flooring									
5200	Parquetry, standard, 5/16" thick, not incl. finish, oak, minimum	1 Carp	160	.050	S.F.	4.41	2		6.41	7.95
5300	Maximum		100	.080		4.95	3.20		8.15	10.40
5500	Teak, minimum		160	.050		4.79	2		6.79	8.35
5600	Maximum		100	.080		8.35	3.20		11.55	14.15
5650	13/16" thick, select grade oak, minimum		160	.050		9.20	2		11.20	13.25
5700	Maximum		100	.080		14	3.20		17.20	20.50
5800	Custom parquetry, including finish, minimum		100	.080		15.45	3.20		18.65	22
5900	Maximum		50	.160		20.50	6.40		26.90	32.50
6700	Parquetry, prefinished white oak, 5/16" thick, minimum		160	.050		3.61	2		5.61	7.05
6800	Maximum		100	.080		7.30	3.20		10.50	13
7000	Walnut or teak, parquetry, minimum		160	.050		5.35	2		7.35	9
7100	Maximum	↓	100	.080	↓	9.35	3.20		12.55	15.25
7200	Acrylic wood parquet blocks, 12" x 12" x 5/16",									
7210	Irradiated, set in epoxy	1 Carp	160	.050	S.F.	7.80	2		9.80	11.70

09 64 29 – Wood Strip and Plank Flooring

09 64 29.10 Wood

		Crew	Daily Output	Labor-Hours	Unit	Material	Labor	Equipment	Total	Total Incl O&P
0010	**WOOD**									
0020	Fir, vertical grain, 1" x 4", not incl. finish, grade B & better	1 Carp	255	.031	S.F.	2.74	1.25		3.99	4.95
0100	C grade & better		255	.031		2.58	1.25		3.83	4.78
4000	Maple, strip, 25/32" x 2-1/4", not incl. finish, select		170	.047		5.25	1.88		7.13	8.70
4100	#2 & better	↓	170	.047	↓	3.32	1.88		5.20	6.55

09 64 Wood Flooring

09 64 29 – Wood Strip and Plank Flooring

09 64 29.10 Wood		Crew	Daily Output	Labor-Hours	Unit	Material	2009 Bare Costs Labor	Equipment	Total	Total Incl O&P
4300	33/32" x 3-1/4", not incl. finish, #1 grade	1 Carp	170	.047	S.F.	4.19	1.88		6.07	7.55
4400	#2 & better	↓	170	.047	↓	3.73	1.88		5.61	7
4600	Oak, white or red, 25/32" x 2-1/4", not incl. finish									
4700	#1 common	1 Carp	170	.047	S.F.	3.18	1.88		5.06	6.40
4900	Select quartered, 2-1/4" wide		170	.047		2.94	1.88		4.82	6.15
5000	Clear		170	.047		4.01	1.88		5.89	7.35
6100	Prefinished, white oak, prime grade, 2-1/4" wide		170	.047		6.40	1.88		8.28	9.95
6200	3-1/4" wide		185	.043		8.20	1.73		9.93	11.70
6400	Ranch plank		145	.055		7.90	2.20		10.10	12.10
6500	Hardwood blocks, 9" x 9", 25/32" thick		160	.050		5.75	2		7.75	9.45
7400	Yellow pine, 3/4" x 3-1/8", T & G, C & better, not incl. finish	↓	200	.040		2.35	1.60		3.95	5.05
7500	Refinish wood floor, sand, 2 cts poly, wax, soft wood, min.	1 Clab	400	.020		.78	.63		1.41	1.84
7600	Hard wood, max		130	.062		1.16	1.94		3.10	4.30
7800	Sanding and finishing, 2 coats polyurethane	↓	295	.027	↓	.78	.86		1.64	2.19
7900	Subfloor and underlayment, see Div. 06 16									
8015	Transition molding, 2 1/4" wide, 5' long	1 Carp	19.20	.417	Ea.	15	16.65		31.65	42.50
8300	Floating floor, wood composition strip, complete	1 Clab	133	.060	S.F.	4.23	1.90		6.13	7.60
8310	Components, T & G wood composite strips					3.66			3.66	4.03
8320	Film					.15			.15	.16
8330	Foam					.30			.30	.33
8340	Adhesive					.17			.17	.18
8350	Installation kit				↓	.17			.17	.19
8360	Trim, 2" wide x 3' long				L.F.	2.49			2.49	2.74
8370	Reducer moulding				"	4.29			4.29	4.72
8600	Flooring, wood, bamboo strips, unfinished, 5/8" x 4" x 3' G	1 Carp	255	.031	S.F.	4.44	1.25		5.69	6.80
8610	5/8" x 4" x 4' G		275	.029		4.61	1.16		5.77	6.85
8620	5/8" x 4" x 6' G		295	.027		5.05	1.08		6.13	7.25
8630	Finished, 5/8" x 4" x 3' G		255	.031		4.89	1.25		6.14	7.35
8640	5/8" x 4" x 4' G		275	.029		5.10	1.16		6.26	7.45
8650	5/8" x 4" x 6' G		295	.027	↓	5.15	1.08		6.23	7.40
8660	Stair treads, unfinished, 1-1/16" x 11-1/2" x 4' G		18	.444	Ea.	42.50	17.75		60.25	74.50
8670	Finished, 1-1/16" x 11-1/2" x 4' G		18	.444		65	17.75		82.75	99
8680	Stair risers, unfinished, 5/8" x 7-1/2" x 4' G		18	.444		15.75	17.75		33.50	45
8690	Finished, 5/8" x 7-1/2" x 4' G		18	.444		30	17.75		47.75	60.50
8700	Stair nosing, unfinished, 6' long G		16	.500		35	20		55	69.50
8710	Finished, 6' long G	↓	16	.500	↓	44	20		64	79

09 64 66 – Wood Athletic Flooring

09 64 66.10 Gymnasium Flooring

		Crew	Daily Output	Labor-Hours	Unit	Material	Labor	Equipment	Total	Total Incl O&P
0010	**GYMNASIUM FLOORING**									
0600	Gym floor, in mastic, over 2 ply felt, #2 & better									
0700	25/32" thick maple	1 Carp	100	.080	S.F.	4.03	3.20		7.23	9.40
0900	33/32" thick maple		98	.082		4.65	3.26		7.91	10.15
1000	For 1/2" corkboard underlayment, add	↓	750	.011		.90	.43		1.33	1.65
1300	For #1 grade maple, add				↓	.47			.47	.52
1600	Maple flooring, over sleepers, #2 & better									
1700	25/32" thick	1 Carp	85	.094	S.F.	4.32	3.76		8.08	10.60
1900	33/32" thick	"	83	.096		5.05	3.85		8.90	11.50
2000	For #1 grade, add					.50			.50	.55
2200	For 3/4" subfloor, add	1 Carp	350	.023		1.12	.91		2.03	2.65
2300	With two 1/2" subfloors, 25/32" thick	"	69	.116	↓	5.40	4.63		10.03	13.15
2500	Maple, incl. finish, #2 & btr., 25/32" thick, on rubber									
2600	Sleepers, with two 1/2" subfloors	1 Carp	76	.105	S.F.	5.80	4.21		10.01	12.90

09 64 Wood Flooring

09 64 66 – Wood Athletic Flooring

09 64 66.10 Gymnasium Flooring

09 64 66.10 Gymnasium Flooring	Crew	Daily Output	Labor-Hours	Unit	Material	2009 Bare Costs Labor	Equipment	Total	Total Incl O&P
2800 With steel spline, double connection to channels	1 Carp	73	.110	S.F.	6.20	4.38		10.58	13.60
2900 For 33/32" maple, add					.67			.67	.74
3100 For #1 grade maple, add					.50			.50	.55
3500 For termite proofing all of the above, add					.27			.27	.30
3700 Portable hardwood, prefinished panels	1 Carp	83	.096		7.75	3.85		11.60	14.45
3720 Insulated with polystyrene, 1" thick, add		165	.048		.67	1.94		2.61	3.74
3750 Running tracks, Sitka spruce surface, 25/32" x 2-1/4"		62	.129		14.15	5.15		19.30	23.50
3770 3/4" plywood surface, finished		100	.080		3.36	3.20		6.56	8.65

09 65 Resilient Flooring

09 65 10 – Resilient Tile Underlayment

09 65 10.10 Latex Underlayment

	Crew	Daily Output	Labor-Hours	Unit	Material	Labor	Equipment	Total	Total Incl O&P
0010 **LATEX UNDERLAYMENT**									
3600 Latex underlayment, 1/8" thk., cementitious for resilient flooring	1 Tilf	160	.050	S.F.	1.72	1.91		3.63	4.68
4000 Liquid, fortified				Gal.	38.50			38.50	42

09 65 13 – Resilient Base and Accessories

09 65 13.13 Resilient Base

	Crew	Daily Output	Labor-Hours	Unit	Material	Labor	Equipment	Total	Total Incl O&P
0010 **RESILIENT BASE**									
0800 Base, cove, rubber or vinyl									
1100 Standard colors, 0.080" thick, 2-1/2" high	1 Tilf	315	.025	L.F.	.75	.97		1.72	2.25
1150 4" high		315	.025		.93	.97		1.90	2.44
1200 6" high		315	.025		1.25	.97		2.22	2.80
1450 1/8" thick, 2-1/2" high		315	.025		.75	.97		1.72	2.25
1500 4" high		315	.025		.71	.97		1.68	2.20
1550 6" high		315	.025		1.30	.97		2.27	2.85
1600 Corners, 2-1/2" high		315	.025	Ea.	1.45	.97		2.42	3.02
1630 4" high		315	.025		2.75	.97		3.72	4.45
1660 6" high		315	.025		2.85	.97		3.82	4.56

09 65 13.23 Resilient Stair Treads and Risers

	Crew	Daily Output	Labor-Hours	Unit	Material	Labor	Equipment	Total	Total Incl O&P
0010 **RESILIENT STAIR TREADS AND RISERS**									
0300 Rubber, molded tread, 12" wide, 5/16" thick, black	1 Tilf	115	.070	L.F.	11.75	2.65		14.40	16.85
0400 Colors		115	.070		12.25	2.65		14.90	17.40
0600 1/4" thick, black		115	.070		11.50	2.65		14.15	16.55
0700 Colors		115	.070		12.25	2.65		14.90	17.40
0900 Grip strip safety tread, colors, 5/16" thick		115	.070		16	2.65		18.65	21.50
1000 3/16" thick		120	.067		11.15	2.54		13.69	16
1200 Landings, smooth sheet rubber, 1/8" thick		120	.067	S.F.	5.75	2.54		8.29	10.05
1300 3/16" thick		120	.067	"	7.25	2.54		9.79	11.70
1500 Nosings, 3" wide, 3/16" thick, black		140	.057	L.F.	3.15	2.18		5.33	6.65
1600 Colors		140	.057		3.11	2.18		5.29	6.60
1800 Risers, 7" high, 1/8" thick, flat		250	.032		6.15	1.22		7.37	8.60
1900 Coved		250	.032		5.05	1.22		6.27	7.40
2100 Vinyl, molded tread, 12" wide, colors, 1/8" thick		115	.070		5.65	2.65		8.30	10.10
2200 1/4" thick		115	.070		7.50	2.65		10.15	12.15
2300 Landing material, 1/8" thick		200	.040	S.F.	5.65	1.52		7.17	8.45
2400 Riser, 7" high, 1/8" thick, coved		175	.046	L.F.	3.25	1.74		4.99	6.15
2500 Tread and riser combined, 1/8" thick		80	.100	"	8.85	3.81		12.66	15.35

09 65 Resilient Flooring

09 65 16 – Resilient Sheet Flooring

09 65 16.10 Rubber and Vinyl Sheet Flooring

		Crew	Daily Output	Labor-Hours	Unit	Material	2009 Bare Costs Labor	Equipment	Total	Total Incl O&P
0010	**RUBBER AND VINYL SHEET FLOORING**									
5500	Linoleum, sheet goods	1 Tilf	360	.022	S.F.	4.55	.85		5.40	6.25
5900	Rubber, sheet goods, 36" wide, 1/8" thick		120	.067		6.25	2.54		8.79	10.60
5950	3/16" thick		100	.080		8.50	3.05		11.55	13.80
6000	1/4" thick		90	.089		10	3.39		13.39	15.95
8000	Vinyl sheet goods, backed, .065" thick, minimum		250	.032		3.50	1.22		4.72	5.65
8050	Maximum		200	.040		3.39	1.52		4.91	5.95
8100	.080" thick, minimum		230	.035		3.50	1.33		4.83	5.80
8150	Maximum		200	.040		4.95	1.52		6.47	7.70
8200	.125" thick, minimum		230	.035		4	1.33		5.33	6.35
8250	Maximum		200	.040		5.95	1.52		7.47	8.80
8700	Adhesive cement, 1 gallon per 200 to 300 S.F.				Gal.	24.50			24.50	26.50
8800	Asphalt primer, 1 gallon per 300 S.F.					11.65			11.65	12.80
8900	Emulsion, 1 gallon per 140 S.F.					15			15	16.50

09 65 19 – Resilient Tile Flooring

09 65 19.10 Miscellaneous Resilient Tile Flooring

			Crew	Daily Output	Labor-Hours	Unit	Material	Labor	Equipment	Total	Total Incl O&P
0010	**MISCELLANEOUS RESILIENT TILE FLOORING**										
2200	Cork tile, standard finish, 1/8" thick	G	1 Tilf	315	.025	S.F.	5.75	.97		6.72	7.75
2250	3/16" thick	G		315	.025		5.75	.97		6.72	7.75
2300	5/16" thick	G		315	.025		7.15	.97		8.12	9.25
2350	1/2" thick	G		315	.025		8.45	.97		9.42	10.70
2500	Urethane finish, 1/8" thick	G		315	.025		6.85	.97		7.82	8.95
2550	3/16" thick	G		315	.025		7.20	.97		8.17	9.30
2600	5/16" thick	G		315	.025		9.20	.97		10.17	11.55
2650	1/2" thick	G		315	.025		12.40	.97		13.37	15.05
6050	Rubber tile, marbleized colors, 12" x 12", 1/8" thick			400	.020		6.20	.76		6.96	7.90
6100	3/16" thick			400	.020		9.25	.76		10.01	11.30
6300	Special tile, plain colors, 1/8" thick			400	.020		9.25	.76		10.01	11.30
6350	3/16" thick			400	.020		9.50	.76		10.26	11.55
6410	Raised, radial or square, minimum			400	.020		11.50	.76		12.26	13.75
6430	Maximum			400	.020		11.50	.76		12.26	13.75
6450	For golf course, skating rink, etc., 1/4" thick			275	.029		11.50	1.11		12.61	14.25
6700	Synthetic turf, 3/8" thick			90	.089		5.65	3.39		9.04	11.15
6750	Interlocking 2' x 2' squares, 1/2" thick, not										
6810	cemented, for playgrounds, minimum		1 Tilf	210	.038	S.F.	4.50	1.45		5.95	7.10
6850	Maximum			190	.042		11.25	1.60		12.85	14.75
7000	Vinyl composition tile, 12" x 12", 1/16" thick			500	.016		.87	.61		1.48	1.85
7050	Embossed			500	.016		2.07	.61		2.68	3.17
7100	Marbleized			500	.016		2.07	.61		2.68	3.17
7150	Solid			500	.016		2.66	.61		3.27	3.82
7200	3/32" thick, embossed			500	.016		1.30	.61		1.91	2.32
7250	Marbleized			500	.016		2.37	.61		2.98	3.50
7300	Solid			500	.016		2.20	.61		2.81	3.31
7350	1/8" thick, marbleized			500	.016		1.70	.61		2.31	2.76
7400	Solid			500	.016		2.14	.61		2.75	3.24
7450	Conductive			500	.016		5.35	.61		5.96	6.80
7500	Vinyl tile, 12" x 12", .050" thick, minimum			500	.016		2.95	.61		3.56	4.14
7550	Maximum			500	.016		5.90	.61		6.51	7.40
7600	1/8" thick, minimum			500	.016		4.20	.61		4.81	5.50
7650	Solid colors			500	.016		5.85	.61		6.46	7.35
7700	Marbleized or Travertine pattern			500	.016		5.65	.61		6.26	7.10
7750	Florentine pattern			500	.016		5.70	.61		6.31	7.20

09 65 Resilient Flooring

09 65 19 – Resilient Tile Flooring

09 65 19.10 Miscellaneous Resilient Tile Flooring	Crew	Daily Output	Labor-Hours	Unit	Material	2009 Bare Costs Labor	2009 Bare Costs Equipment	Total	Total Incl O&P
7800 Maximum	1 Tilf	500	.016	S.F.	12.35	.61		12.96	14.50

09 65 33 – Conductive Resilient Flooring

09 65 33.10 Conductive Rubber and Vinyl Flooring

0010 **CONDUCTIVE RUBBER AND VINYL FLOORING**									
1700 Conductive flooring, rubber tile, 1/8" thick	1 Tilf	315	.025	S.F.	5	.97		5.97	6.90
1800 Homogeneous vinyl tile, 1/8" thick	"	315	.025	"	6.95	.97		7.92	9.05

09 66 Terrazzo Flooring

09 66 13 – Portland Cement Terrazzo Flooring

09 66 13.10 Portland Cement Terrazzo

	Crew	Daily Output	Labor-Hours	Unit	Material	Labor	Equipment	Total	Total Incl O&P
0010 **PORTLAND CEMENT TERRAZZO**, cast-in-place									
0020 Cove base, 6" high, 16ga. zinc	1 Mstz	20	.400	L.F.	3.32	15.10		18.42	25.50
0100 Curb, 6" high and 6" wide		6	1.333		5.25	50.50		55.75	79.50
0300 Divider strip for floors, 14 ga., 1-1/4" deep, zinc		375	.021		1.21	.80		2.01	2.51
0400 Brass		375	.021		2.23	.80		3.03	3.63
0600 Heavy top strip 1/4" thick, 1-1/4" deep, zinc		300	.027		1.76	1.01		2.77	3.41
0900 Galv. bottoms, brass		300	.027		3.04	1.01		4.05	4.81
1200 For thin set floors, 16 ga., 1/2" x 1/2", zinc		350	.023		.75	.86		1.61	2.09
1300 Brass		350	.023		1.64	.86		2.50	3.06
1500 Floor, bonded to concrete, 1-3/4" thick, gray cement	J-3	75	.213	S.F.	2.87	7.30	3.38	13.55	17.55
1600 White cement, mud set		75	.213		3.27	7.30	3.38	13.95	18
1800 Not bonded, 3" total thickness, gray cement		70	.229		3.59	7.85	3.62	15.06	19.40
1900 White cement, mud set		70	.229		3.92	7.85	3.62	15.39	19.75
2100 For Venetian terrazzo, 1" topping, add					50%	50%			
2200 For heavy duty abrasive terrazzo, add					50%	50%			
2700 Monolithic terrazzo, 1/2" thick									
2710 10' panels	J-3	125	.128	S.F.	2.61	4.39	2.03	9.03	11.50
3000 Stairs, cast in place, pan filled treads		30	.533	L.F.	2.61	18.30	8.45	29.36	39
3100 Treads and risers		14	1.143	"	5.40	39	18.10	62.50	83.50
3300 For stair landings, add to floor prices						50%			
3400 Stair stringers and fascia	J-3	30	.533	S.F.	4.37	18.30	8.45	31.12	41
3600 For abrasive metal nosings on stairs, add		150	.107	L.F.	7.70	3.66	1.69	13.05	15.70
3700 For abrasive surface finish, add		600	.027	S.F.	1.03	.91	.42	2.36	2.93
3900 For raised abrasive strips, add		150	.107	L.F.	.98	3.66	1.69	6.33	8.30
4000 Wainscot, bonded, 1-1/2" thick		30	.533	S.F.	3.32	18.30	8.45	30.07	40
4200 Epoxy terrazzo, 1/4" thick		40	.400	"	4.88	13.70	6.35	24.93	32.50
4300 Stone chips, onyx gemstone, per 50 lb. bag				Bag	16.30			16.30	17.95

09 66 13.30 Terrazzo, Precast

0010 **TERRAZZO, PRECAST**									
0020 Base, 6" high, straight	1 Mstz	70	.114	L.F.	10.40	4.31		14.71	17.75
0100 Cove		60	.133		12.15	5.05		17.20	20.50
0300 8" high, straight		60	.133		10.85	5.05		15.90	19.25
0400 Cove		50	.160		15.95	6.05		22	26.50
0600 For white cement, add					.43			.43	.47
0700 For 16 ga. zinc toe strip, add					1.60			1.60	1.76
0900 Curbs, 4" x 4" high	1 Mstz	40	.200		30.50	7.55		38.05	44.50
1000 8" x 8" high	"	30	.267		34.50	10.05		44.55	52.50
1200 Floor tiles, non-slip, 1" thick, 12" x 12"	D-1	60	.267	S.F.	17.70	9.70		27.40	34
1300 1-1/4" thick, 12" x 12"		60	.267		19.60	9.70		29.30	36
1500 16" x 16"		50	.320		21.50	11.65		33.15	41

09 66 Terrazzo Flooring

09 66 13 – Portland Cement Terrazzo Flooring

09 66 13.30 Terrazzo, Precast		Crew	Daily Output	Labor-Hours	Unit	Material	2009 Bare Costs Labor	Equipment	Total	Total Incl O&P
1600	1-1/2" thick, 16" x 16"	D-1	45	.356	S.F.	19.45	12.90		32.35	41
1800	For Venetian terrazzo, add					5.85			5.85	6.40
1900	For white cement, add					.55			.55	.61
2400	Stair treads, 1-1/2" thick, non-slip, three line pattern	2 Mstz	70	.229	L.F.	39.50	8.60		48.10	56
2500	Nosing and two lines		70	.229		39.50	8.60		48.10	56
2700	2" thick treads, straight		60	.267		43	10.05		53.05	61.50
2800	Curved		50	.320		56	12.05		68.05	79
3000	Stair risers, 1" thick, to 6" high, straight sections		60	.267		9.80	10.05		19.85	25.50
3100	Cove		50	.320		13.90	12.05		25.95	33
3300	Curved, 1" thick, to 6" high, vertical		48	.333		19.05	12.55		31.60	39.50
3400	Cove		38	.421		36	15.85		51.85	62.50
3600	Stair tread and riser, single piece, straight, minimum		60	.267		50	10.05		60.05	69.50
3700	Maximum		40	.400		65	15.10		80.10	93
3900	Curved tread and riser, minimum		40	.400		70	15.10		85.10	99
4000	Maximum		32	.500		88.50	18.85		107.35	125
4200	Stair stringers, notched, 1" thick		25	.640		28.50	24		52.50	67
4300	2" thick		22	.727		34	27.50		61.50	77
4500	Stair landings, structural, non-slip, 1-1/2" thick		85	.188	S.F.	31	7.10		38.10	45
4600	3" thick		75	.213		44.50	8.05		52.55	60.50
4800	Wainscot, 12" x 12" x 1" tiles	1 Mstz	12	.667		6.50	25		31.50	44
4900	16" x 16" x 1-1/2" tiles	"	8	1		13.70	37.50		51.20	70

09 66 16 – Terrazzo Floor Tile

09 66 16.10 Tile or Terrazzo Base

		Crew	Daily Output	Labor-Hours	Unit	Material	Labor	Equipment	Total	Total Incl O&P
0010	**TILE OR TERRAZZO BASE**									
0020	Scratch coat only	1 Mstz	150	.053	S.F.	.41	2.01		2.42	3.39
0500	Scratch and brown coat only	"	75	.107	"	.78	4.02		4.80	6.75

09 66 33 – Conductive Terrazzo Flooring

09 66 33.10 Conductive Terrazzo

		Crew	Daily Output	Labor-Hours	Unit	Material	Labor	Equipment	Total	Total Incl O&P
0010	**CONDUCTIVE TERRAZZO**									
2400	Bonded conductive floor for hospitals	J-3	90	.178	S.F.	3.92	6.10	2.81	12.83	16.30
2500	Epoxy terrazzo, 1/4" thick, minimum		100	.160		4.43	5.50	2.53	12.46	15.70
2550	Average		75	.213		4.88	7.30	3.38	15.56	19.75
2600	Maximum		60	.267		5.35	9.15	4.22	18.72	24

09 67 Fluid-Applied Flooring

09 67 20 – Epoxy-Marble Chip Flooring

09 67 20.13 Elastomeric Liquid Flooring

		Crew	Daily Output	Labor-Hours	Unit	Material	Labor	Equipment	Total	Total Incl O&P
0010	**ELASTOMERIC LIQUID FLOORING**									
0020	Cementitious acrylic, 1/4" thick	C-6	520	.092	S.F.	1.45	3.05	.10	4.60	6.40
0100	3/8" thick	"	450	.107		1.85	3.53	.12	5.50	7.55
0200	Methyl methachrylate, 1/4" thick	C-8A	3000	.016		5.40	.55		5.95	6.80
0210	1/8" thick	"	3000	.016		2.97	.55		3.52	4.10
0300	Cupric oxychloride, on bond coat, minimum	C-6	480	.100		3.15	3.31	.11	6.57	8.65
0400	Maximum		420	.114		5.25	3.78	.12	9.15	11.75
2400	Mastic, hot laid, 2 coat, 1-1/2" thick, standard, minimum		690	.070		3.65	2.30	.08	6.03	7.65
2500	Maximum		520	.092		4.69	3.05	.10	7.84	9.95
2700	Acidproof, minimum		605	.079		4.69	2.62	.09	7.40	9.25
2800	Maximum		350	.137		6.50	4.53	.15	11.18	14.25
3000	Neoprene, trowelled on, 1/4" thick, minimum		545	.088		3.59	2.91	.10	6.60	8.50
3100	Maximum		430	.112		4.94	3.69	.12	8.75	11.25

09 67 Fluid-Applied Flooring

09 67 20 – Epoxy-Marble Chip Flooring

09 67 20.13 Elastomeric Liquid Flooring		Crew	Daily Output	Labor-Hours	Unit	Material	2009 Bare Costs Labor	Equipment	Total	Total Incl O&P
4300	Polyurethane, with suspended vinyl chips, minimum	C-6	1065	.045	S.F.	7	1.49	.05	8.54	10.05
4500	Maximum	↓	860	.056	↓	10.05	1.84	.06	11.95	14

09 67 20.16 Epoxy Terrazzo

0010	**EPOXY TERRAZZO**									
1800	Epoxy terrazzo, 1/4" thick, chemical resistant, minimum	J-3	200	.080	S.F.	5.25	2.74	1.27	9.26	11.20
1900	Maximum		150	.107		7.70	3.66	1.69	13.05	15.70
2100	Conductive, minimum		210	.076		6.90	2.61	1.21	10.72	12.75
2200	Maximum	↓	160	.100	↓	8.95	3.43	1.58	13.96	16.60

09 67 20.19 Polyacrylate Terrazzo

0010	**POLYACRYLATE TERRAZZO**									
3150	Polyacrylate, 1/4" thick, minimum	C-6	735	.065	S.F.	2.98	2.16	.07	5.21	6.65
3170	Maximum		480	.100		3.68	3.31	.11	7.10	9.20
3200	3/8" thick, minimum		620	.077		3.88	2.56	.08	6.52	8.30
3220	Maximum		480	.100		5.15	3.31	.11	8.57	10.80
3300	Conductive, 1/4" thick, minimum		450	.107		6.25	3.53	.12	9.90	12.40
3330	Maximum		305	.157		7.75	5.20	.17	13.12	16.70
3350	3/8" thick, minimum		365	.132		8.20	4.35	.14	12.69	15.80
3370	Maximum		255	.188		9.70	6.20	.20	16.10	20.50
3450	Granite, conductive, 1/4" thick, minimum		695	.069		7.90	2.28	.07	10.25	12.30
3470	Maximum		420	.114		10.20	3.78	.12	14.10	17.15
3500	3/8" thick, minimum		695	.069		11.55	2.28	.07	13.90	16.30
3520	Maximum	↓	380	.126	↓	13.95	4.17	.14	18.26	22

09 67 20.26 Quartz Flooring

0010	**QUARTZ FLOORING**									
0600	Epoxy, with colored quartz chips, broadcast, minimum	C-6	675	.071	S.F.	2.49	2.35	.08	4.92	6.40
0700	Maximum		490	.098		3.02	3.24	.11	6.37	8.40
0900	Trowelled, minimum		560	.086		3.21	2.83	.09	6.13	7.95
1000	Maximum	↓	480	.100	↓	4.69	3.31	.11	8.11	10.30
1200	Heavy duty epoxy topping, 1/4" thick,									
1300	500 to 1,000 S.F.	C-6	420	.114	S.F.	4.67	3.78	.12	8.57	11.10
1500	1,000 to 2,000 S.F.		450	.107		3.87	3.53	.12	7.52	9.80
1600	Over 10,000 S.F.		480	.100		3.62	3.31	.11	7.04	9.15
3600	Polyester, with colored quartz chips, 1/16" thick, minimum		1065	.045		2.67	1.49	.05	4.21	5.25
3700	Maximum		560	.086		3.63	2.83	.09	6.55	8.45
3900	1/8" thick, minimum		810	.059		3.12	1.96	.06	5.14	6.50
4000	Maximum		675	.071		4.22	2.35	.08	6.65	8.30
4200	Polyester, heavy duty, compared to epoxy, add	↓	2590	.019	↓	1.27	.61	.02	1.90	2.36

09 68 Carpeting

09 68 05 – Carpet Accessories

09 68 05.11 Flooring Transition Strip

0010	**FLOORING TRANSITION STRIP**									
0107	Clamp down brass divider, 12' strip, vinyl to carpet	1 Tilf	31.25	.256	Ea.	26.50	9.75		36.25	43.50
0117	Vinyl to hard surface	"	31.25	.256	"	26.50	9.75		36.25	43.50

09 68 10 – Carpet Pad

09 68 10.10 Commercial Grade Carpet Pad

0010	**COMMERCIAL GRADE CARPET PAD**									
9000	Sponge rubber pad, minimum	1 Tilf	150	.053	S.Y.	4.20	2.03		6.23	7.60
9100	Maximum		150	.053		10	2.03		12.03	14
9200	Felt pad, minimum	↓	150	.053	↓	4.03	2.03		6.06	7.40

09 68 Carpeting

09 68 10 – Carpet Pad

09 68 10.10 Commercial Grade Carpet Pad		Crew	Daily Output	Labor-Hours	Unit	Material	2009 Bare Costs Labor	Equipment	Total	Total Incl O&P
9300	Maximum	1 Tilf	150	.053	S.Y.	7.05	2.03		9.08	10.75
9400	Bonded urethane pad, minimum		150	.053		4.57	2.03		6.60	8.05
9500	Maximum		150	.053		7.80	2.03		9.83	11.60
9600	Prime urethane pad, minimum		150	.053		2.62	2.03		4.65	5.85
9700	Maximum		150	.053		4.84	2.03		6.87	8.30

09 68 13 – Tile Carpeting

09 68 13.10 Carpet Tile

		Crew	Daily Output	Labor-Hours	Unit	Material	Labor	Equipment	Total	Total Incl O&P
0010	**CARPET TILE**									
0100	Tufted nylon, 18" x 18", hard back, 20 oz.	1 Tilf	150	.053	S.Y.	22	2.03		24.03	27.50
0110	26 oz.		150	.053		38	2.03		40.03	44.50
0200	Cushion back, 20 oz.		150	.053		28	2.03		30.03	33.50
0210	26 oz.		150	.053		43.50	2.03		45.53	51
1100	Tufted, 24" x 24", 24 oz. nylon		80	.100		29.50	3.81		33.31	37.50
1180	35 oz.		80	.100		34	3.81		37.81	43
5060	42 oz.		80	.100		45.50	3.81		49.31	55.50

09 68 16 – Sheet Carpeting

09 68 16.10 Sheet Carpet

		Crew	Daily Output	Labor-Hours	Unit	Material	Labor	Equipment	Total	Total Incl O&P
0010	**SHEET CARPET**									
0700	Nylon, level loop, 26 oz., light to medium traffic	1 Tilf	75	.107	S.Y.	26	4.06		30.06	34.50
0720	28 oz., light to medium traffic		75	.107		27	4.06		31.06	35.50
0900	32 oz., medium traffic		75	.107		32.50	4.06		36.56	41.50
1100	40 oz., medium to heavy traffic		75	.107		48	4.06		52.06	59
2920	Nylon plush, 30 oz., medium traffic		57	.140		24	5.35		29.35	34.50
3000	36 oz., medium traffic		75	.107		32	4.06		36.06	41
3100	42 oz., medium to heavy traffic		70	.114		36	4.35		40.35	46
3200	46 oz., medium to heavy traffic		70	.114		42	4.35		46.35	52.50
3300	54 oz., heavy traffic		70	.114		46	4.35		50.35	57
3340	60 oz., heavy traffic		70	.114		64	4.35		68.35	77
3665	Olefin, 24 oz., light to medium traffic		75	.107		14.50	4.06		18.56	22
3670	26 oz., medium traffic		75	.107		15.50	4.06		19.56	23
3680	28 oz., medium to heavy traffic		75	.107		17	4.06		21.06	24.50
3700	32 oz., medium to heavy traffic		75	.107		22	4.06		26.06	30
3730	42 oz., heavy traffic		70	.114		28	4.35		32.35	37
4110	Wool, level loop, 40 oz., medium traffic		75	.107		107	4.06		111.06	124
4500	50 oz., medium to heavy traffic		75	.107		110	4.06		114.06	127
4700	Patterned, 32 oz., medium to heavy traffic		70	.114		109	4.35		113.35	126
4900	48 oz., heavy traffic		70	.114		112	4.35		116.35	129
5000	For less than full roll (approx. 1500 S.F.), add					25%				
5100	For small rooms, less than 12' wide, add						25%			
5200	For large open areas (no cuts), deduct						25%			
5600	For bound carpet baseboard, add	1 Tilf	300	.027	L.F.	2.25	1.02		3.27	3.97
5610	For stairs, not incl. price of carpet, add	"	30	.267	Riser		10.15		10.15	14.90
5620	For borders and patterns, add to labor						18%			
8950	For tackless, stretched installation, add padding to above									
9850	For "branded" fiber, add				S.Y.	25%				

09 68 20 – Athletic Carpet

09 68 20.10 Indoor Athletic Carpet

		Crew	Daily Output	Labor-Hours	Unit	Material	Labor	Equipment	Total	Total Incl O&P
0010	**INDOOR ATHLETIC CARPET**									
3700	Polyethylene, in rolls, no base incl., landscape surfaces	1 Tilf	275	.029	S.F.	3.05	1.11		4.16	4.98
3800	Nylon action surface, 1/8" thick		275	.029		3.20	1.11		4.31	5.15
3900	1/4" thick		275	.029		4.61	1.11		5.72	6.65

09 68 Carpeting

09 68 20 – Athletic Carpet

09 68 20.10 Indoor Athletic Carpet		Crew	Daily Output	Labor-Hours	Unit	Material	2009 Bare Costs Labor	Equipment	Total	Total Incl O&P
4000	3/8" thick	1 Tilf	275	.029	S.F.	5.80	1.11		6.91	7.95
4100	Golf tee surface with foam back		235	.034		5.75	1.30		7.05	8.20
4200	Practice putting, knitted nylon surface	↓	235	.034	↓	4.86	1.30		6.16	7.25
4400	Polyurethane, thermoset, prefabricated in place, indoor									
4500	3/8" thick for basketball, gyms, etc.	1 Tilf	100	.080	S.F.	4.58	3.05		7.63	9.50
4600	1/2" thick for professional sports		95	.084		6.15	3.21		9.36	11.45
4700	Outdoor, 1/4" thick, smooth, for tennis		100	.080		4.76	3.05		7.81	9.70
4800	Rough, for track, 3/8" thick		95	.084		5.50	3.21		8.71	10.75
5000	Poured in place, indoor, with finish, 1/4" thick		80	.100		3.38	3.81		7.19	9.30
5050	3/8" thick		65	.123		4.11	4.69		8.80	11.35
5100	1/2" thick		50	.160		5.50	6.10		11.60	15
5500	Polyvinyl chloride, sheet goods for gyms, 1/4" thick		80	.100		7.50	3.81		11.31	13.85
5600	3/8" thick	↓	60	.133	↓	8.50	5.10		13.60	16.80

09 69 Access Flooring

09 69 13 – Rigid-Grid Access Flooring

09 69 13.10 Access Floors		Crew	Daily Output	Labor-Hours	Unit	Material	2009 Bare Costs Labor	Equipment	Total	Total Incl O&P
0010	**ACCESS FLOORS**									
0015	Access floor package including panel, pedestal, stringers & laminate cover									
0100	Computer room, greater than 6,000 S.F.	4 Carp	750	.043	S.F.	8.65	1.70		10.35	12.15
0110	Less than 6,000 S.F.	2 Carp	375	.043		9.95	1.70		11.65	13.60
0120	Office, greater than 6,000 S.F.	4 Carp	1050	.030		4.56	1.22		5.78	6.90
0250	Panels, particle board or steel, 1250# load, no covering, under 6,000 S.F.	2 Carp	600	.027		4.24	1.07		5.31	6.30
0300	Over 6,000 S.F.		640	.025		3.72	1		4.72	5.65
0400	Aluminum, 24" panels	↓	500	.032		31.50	1.28		32.78	36.50
0600	For carpet covering, add					8.55			8.55	9.40
0700	For vinyl floor covering, add					6.70			6.70	7.35
0900	For high pressure laminate covering, add					5.45			5.45	6
0910	For snap on stringer system, add	2 Carp	1000	.016	↓	1.44	.64		2.08	2.57
0950	Office applications, steel or concrete panels,									
0960	no covering, over 6,000 S.F.	2 Carp	960	.017	S.F.	10.10	.67		10.77	12.15
1000	Machine cutouts after initial installation	1 Carp	50	.160	Ea.	5.05	6.40		11.45	15.45
1050	Pedestals, 6" to 12"	2 Carp	85	.188		7.95	7.50		15.45	20.50
1100	Air conditioning grilles, 4" x 12"	1 Carp	17	.471		66	18.80		84.80	102
1150	4" x 18"	"	14	.571	↓	90.50	23		113.50	135
1200	Approach ramps, minimum	2 Carp	60	.267	S.F.	25.50	10.65		36.15	44.50
1300	Maximum	"	40	.400	"	33.50	16		49.50	61.50
1500	Handrail, 2 rail, aluminum	1 Carp	15	.533	L.F.	99.50	21.50		121	142

09 72 Wall Coverings

09 72 23 – Wallpapering

09 72 23.10 Wallpaper		Crew	Daily Output	Labor-Hours	Unit	Material	2009 Bare Costs Labor	Equipment	Total	Total Incl O&P
0010	**WALLPAPER** including sizing; add 10-30 percent waste @ takeoff R097223-10									
0050	Aluminum foil	1 Pape	275	.029	S.F.	.97	1.03		2	2.61
0100	Copper sheets, .025" thick, vinyl backing		240	.033		5.20	1.18		6.38	7.45
0300	Phenolic backing		240	.033		6.75	1.18		7.93	9.15
0600	Cork tiles, light or dark, 12" x 12" x 3/16"		240	.033		4.25	1.18		5.43	6.45
0700	5/16" thick		235	.034		3.62	1.21		4.83	5.80
0900	1/4" basketweave	↓	240	.033	↓	5.55	1.18		6.73	7.90

09 72 Wall Coverings

09 72 23 – Wallpapering

09 72 23.10 Wallpaper

		Crew	Daily Output	Labor-Hours	Unit	Material	2009 Bare Costs Labor	Equipment	Total	Total Incl O&P
1000	1/2" natural, non-directional pattern	1 Pape	240	.033	S.F.	6.70	1.18		7.88	9.10
1100	3/4" natural, non-directional pattern		240	.033		10.90	1.18		12.08	13.75
1200	Granular surface, 12" x 36", 1/2" thick		385	.021		1.20	.74		1.94	2.42
1300	1" thick		370	.022		1.55	.77		2.32	2.86
1500	Polyurethane coated, 12" x 12" x 3/16" thick		240	.033		3.76	1.18		4.94	5.90
1600	5/16" thick		235	.034		5.35	1.21		6.56	7.70
1800	Cork wallpaper, paperbacked, natural		480	.017		1.90	.59		2.49	2.98
1900	Colors		480	.017		2.65	.59		3.24	3.81
2100	Flexible wood veneer, 1/32" thick, plain woods		100	.080		2.25	2.84		5.09	6.75
2200	Exotic woods		95	.084		3.41	2.99		6.40	8.20
2400	Gypsum-based, fabric-backed, fire resistant									
2500	for masonry walls, minimum, 21 oz./S.Y.	1 Pape	800	.010	S.F.	.80	.35		1.15	1.41
2600	Average		720	.011		1.17	.39		1.56	1.88
2700	Maximum (small quantities)		640	.013		1.30	.44		1.74	2.09
2750	Acrylic, modified, semi-rigid PVC, .028" thick	2 Carp	330	.048		1.08	1.94		3.02	4.19
2800	.040" thick	"	320	.050		1.42	2		3.42	4.66
3000	Vinyl wall covering, fabric-backed, lightweight (12-15 oz./S.Y.)	1 Pape	640	.013		.65	.44		1.09	1.38
3300	Medium weight, type 2 (20-24 oz./S.Y.)		480	.017		.77	.59		1.36	1.74
3400	Heavy weight, type 3 (28 oz./S.Y.)		435	.018		1.59	.65		2.24	2.73
3600	Adhesive, 5 gal. lots (18SY/Gal.)				Gal.	10.15			10.15	11.15
3700	Wallpaper, average workmanship, solid pattern, low cost paper	1 Pape	640	.013	S.F.	.35	.44		.79	1.05
3900	basic patterns (matching required), avg. cost paper		535	.015		.63	.53		1.16	1.48
4000	Paper at $85 per double roll, quality workmanship		435	.018		2.35	.65		3	3.57
4150	Flame treatment, minimum					.85			.85	.94
4180	Maximum					1.48			1.48	1.63
4200	Grass cloths with lining paper, minimum	1 Pape	400	.020		.74	.71		1.45	1.87
4300	Maximum	"	350	.023		2.37	.81		3.18	3.82

09 77 Special Wall Surfacing

09 77 33 – Fiberglass Reinforced Panels

09 77 33.10 Fiberglass Reinforced Plastic Panels

		Crew	Daily Output	Labor-Hours	Unit	Material	2009 Bare Costs Labor	Equipment	Total	Total Incl O&P
0010	**FIBERGLASS REINFORCED PLASTIC PANELS**, .090" thick									
0020	On walls, adhesive mounted, embossed surface	2 Carp	640	.025	S.F.	1.23	1		2.23	2.90
0030	Smooth surface		640	.025		1.51	1		2.51	3.21
0040	Fire rated, embossed surface		640	.025		2.19	1		3.19	3.96
0050	Nylon rivet mounted, on drywall, embossed surface		480	.033		1.15	1.33		2.48	3.33
0060	Smooth surface		480	.033		1.43	1.33		2.76	3.63
0070	Fire rated, embossed surface		480	.033		2.11	1.33		3.44	4.38
0080	On masonry, embossed surface		320	.050		1.15	2		3.15	4.37
0090	Smooth surface		320	.050		1.43	2		3.43	4.67
0100	Fire rated, embossed surface		320	.050		2.11	2		4.11	5.40
0110	Nylon rivet and adhesive mounted, on drywall, embossed surface		240	.067		1.30	2.66		3.96	5.55
0120	Smooth surface		240	.067		1.58	2.66		4.24	5.85
0130	Fire rated, embossed surface		240	.067		2.26	2.66		4.92	6.60
0140	On masonry, embossed surface		190	.084		1.30	3.36		4.66	6.65
0150	Smooth surface		190	.084		1.58	3.36		4.94	6.95
0160	Fire rated, embossed surface		190	.084		2.26	3.36		5.62	7.70
0170	For moldings add	1 Carp	250	.032	L.F.	.28	1.28		1.56	2.29
0180	On ceilings, for lay in grid system, embossed surface		400	.020	S.F.	1.23	.80		2.03	2.59
0190	Smooth surface		400	.020		1.51	.80		2.31	2.90
0200	Fire rated, embossed surface		400	.020		2.19	.80		2.99	3.65

09 77 Special Wall Surfacing

09 77 43 – Panel Systems

09 77 43.20 Slatwall Panels and Accessories

		Crew	Daily Output	Labor-Hours	Unit	Material	2009 Bare Costs Labor	2009 Bare Costs Equipment	Total	Total Incl O&P
0010	**SLATWALL PANELS AND ACCESSORIES**									
0100	Slatwall panel, 4' x 8' x 3/4" T, MDF, paint grade	1 Carp	500	.016	S.F.	1.60	.64		2.24	2.75
0110	Melamine finish		500	.016		2.28	.64		2.92	3.50
0120	High pressure plastic laminate finish		500	.016		3.70	.64		4.34	5.05
0130	Aluminum channel inserts, add					3.57			3.57	3.93
0200	Accessories, corner forms, 8' L				L.F.	4.89			4.89	5.40
0210	T-connector, 8' L					5.95			5.95	6.55
0220	J-mold, 8' L					1.59			1.59	1.75
0230	Edge cap, 8' L					1.16			1.16	1.28
0240	Finish end cap, 8' L					3.66			3.66	4.03
0300	Display hook, metal, 4" L				Ea.	1.09			1.09	1.20
0310	6" L					1.19			1.19	1.31
0320	8" L					1.27			1.27	1.40
0330	10" L					1.43			1.43	1.57
0340	12" L					1.58			1.58	1.74
0350	Acrylic, 4" L					.91			.91	1
0360	6" L					1.05			1.05	1.16
0370	8" L					1.10			1.10	1.21
0380	10" L					1.21			1.21	1.33
0400	Waterfall hanger, metal, 12" - 16"					6.35			6.35	6.95
0410	Acrylic					10.60			10.60	11.65
0500	Shelf bracket, metal, 8"					6.70			6.70	7.35
0510	10"					6.90			6.90	7.60
0520	12"					7.15			7.15	7.90
0530	14"					7.60			7.60	8.35
0540	16"					9.05			9.05	10
0550	Acrylic, 8"					3.41			3.41	3.75
0560	10"					3.82			3.82	4.20
0570	12"					4.23			4.23	4.65
0580	14"					4.79			4.79	5.25
0600	Shelf, acrylic, 12" x 16" x 1/4"					27			27	29.50
0610	12" x 24" x 1/4"					45.50			45.50	50

09 81 Acoustic Insulation

09 81 16 – Acoustic Blanket Insulation

09 81 16.10 Sound Attenuation Blanket

		Crew	Daily Output	Labor-Hours	Unit	Material	2009 Bare Costs Labor	2009 Bare Costs Equipment	Total	Total Incl O&P
0010	**SOUND ATTENUATION BLANKET**									
0020	Blanket, 1" thick	1 Carp	925	.009	S.F.	.25	.35		.60	.82
0500	1-1/2" thick		920	.009		.25	.35		.60	.82
1000	2" thick		915	.009		.34	.35		.69	.91
1500	3" thick		910	.009		.50	.35		.85	1.09
3000	Thermal or acoustical batt above ceiling, 2" thick		900	.009		.49	.36		.85	1.09
3100	3" thick		900	.009		.74	.36		1.10	1.36
3200	4" thick		900	.009		.92	.36		1.28	1.56
3400	Urethane plastic foam, open cell, on wall, 2" thick	2 Carp	2050	.008		3.05	.31		3.36	3.84
3500	3" thick		1550	.010		4.05	.41		4.46	5.10
3600	4" thick		1050	.015		5.70	.61		6.31	7.20
3700	On ceiling, 2" thick		1700	.009		3.04	.38		3.42	3.92
3800	3" thick		1300	.012		4.05	.49		4.54	5.20
3900	4" thick		900	.018		5.70	.71		6.41	7.35

09 81 Acoustic Insulation

09 81 16 – Acoustic Blanket Insulation

09 81 16.10 Sound Attenuation Blanket

		Crew	Daily Output	Labor-Hours	Unit	Material	2009 Bare Costs Labor	Equipment	Total	Total Incl O&P
4000	Nylon matting 0.4" thick, with carbon black spinerette									
4010	plus polyester fabric, on floor	D-7	4000	.004	S.F.	2.26	.14		2.40	2.69
4200	Fiberglass reinf. backer board underlayment, 7/16" thick, on floor	"	800	.020	"	1.90	.68		2.58	3.09

09 84 Acoustic Room Components

09 84 13 – Fixed Sound-Absorptive Panels

09 84 13.10 Fixed Panels

		Crew	Daily Output	Labor-Hours	Unit	Material	Labor	Equipment	Total	Total Incl O&P
0010	**FIXED PANELS** Perforated steel facing, painted with									
0100	Fiberglass or mineral filler, no backs, 2-1/4" thick, modular									
0200	space units, ceiling or wall hung, white or colored	1 Carp	100	.080	S.F.	10	3.20		13.20	15.95
0300	Fiberboard sound deadening panels, 1/2" thick	"	600	.013	"	.32	.53		.85	1.18
0500	Fiberglass panels, 4' x 8' x 1" thick, with									
0600	glass cloth face for walls, cemented	1 Carp	155	.052	S.F.	7	2.06		9.06	10.90
0700	1-1/2" thick, dacron covered, inner aluminum frame,									
0710	wall mounted	1 Carp	300	.027	S.F.	8.75	1.07		9.82	11.25
0900	Mineral fiberboard panels, fabric covered, 30"x 108",									
1000	3/4" thick, concealed spline, wall mounted	1 Carp	150	.053	S.F.	6.25	2.13		8.38	10.20

09 84 36 – Sound-Absorbing Ceiling Units

09 84 36.10 Barriers

		Crew	Daily Output	Labor-Hours	Unit	Material	Labor	Equipment	Total	Total Incl O&P
0010	**BARRIERS** Plenum									
0600	Aluminum foil, fiberglass reinf., parallel with joists	1 Carp	275	.029	S.F.	.85	1.16		2.01	2.74
0700	Perpendicular to joists		180	.044		.85	1.78		2.63	3.69
0900	Aluminum mesh, kraft paperbacked		275	.029		.77	1.16		1.93	2.65
0970	Fiberglass batts, kraft faced, 3-1/2" thick		1400	.006		.36	.23		.59	.75
0980	6" thick		1300	.006		.65	.25		.90	1.10
1000	Sheet lead, 1 lb., 1/64" thick, perpendicular to joists		150	.053		8.40	2.13		10.53	12.50
1100	Vinyl foam reinforced, 1/8" thick, 1.0 lb. per S.F.		150	.053		9.80	2.13		11.93	14.05

09 91 Painting

09 91 03 – Paint Restoration

09 91 03.20 Sanding

			Crew	Daily Output	Labor-Hours	Unit	Material	Labor	Equipment	Total	Total Incl O&P
0010	**SANDING** and putting interior trim, compared to	R099100-10									
0100	Painting 1 coat, on quality work						L.F.		100%			
0300	Medium work								50%			
0400	Industrial grade								25%			
0500	Surface protection, placement and removal											
0510	Basic drop cloths		1 Pord	6400	.001	S.F.		.04		.04	.07
0520	Masking with paper			800	.010		.06	.35		.41	.60
0530	Volume cover up (using plastic sheathing, or building paper)			16000	.001			.02		.02	.03

09 91 03.30 Exterior Surface Preparation

			Crew	Daily Output	Labor-Hours	Unit	Material	Labor	Equipment	Total	Total Incl O&P
0010	**EXTERIOR SURFACE PREPARATION**	R099100-10									
0015	Doors, per side, not incl. frames or trim										
0020	Scrape & sand										
0030	Wood, flush		1 Pord	616	.013	S.F.		.46		.46	.69
0040	Wood, detail			496	.016			.57		.57	.85
0050	Wood, louvered			280	.029			1.01		1.01	1.51
0060	Wood, overhead			616	.013			.46		.46	.69
0070	Wire brush										

09 91 Painting

09 91 03 – Paint Restoration

09 91 03.30 Exterior Surface Preparation		Crew	Daily Output	Labor-Hours	Unit	Material	2009 Bare Costs Labor	Equipment	Total	Total Incl O&P
0080	Metal, flush	1 Pord	640	.013	S.F.		.44		.44	.66
0090	Metal, detail		520	.015			.54		.54	.81
0100	Metal, louvered		360	.022			.78		.78	1.17
0110	Metal or fibr., overhead		640	.013			.44		.44	.66
0120	Metal, roll up		560	.014			.50		.50	.75
0130	Metal, bulkhead		640	.013			.44		.44	.66
0140	Power wash, based on 2500 lb. operating pressure									
0150	Metal, flush	A-1H	2240	.004	S.F.		.11	.02	.13	.20
0160	Metal, detail		2120	.004			.12	.02	.14	.21
0170	Metal, louvered		2000	.004			.13	.03	.16	.23
0180	Metal or fibr., overhead		2400	.003			.11	.02	.13	.18
0190	Metal, roll up		2400	.003			.11	.02	.13	.18
0200	Metal, bulkhead		2200	.004			.12	.02	.14	.21
0400	Windows, per side, not incl. trim									
0410	Scrape & sand									
0420	Wood, 1-2 lite	1 Pord	320	.025	S.F.		.88		.88	1.32
0430	Wood, 3-6 lite		280	.029			1.01		1.01	1.51
0440	Wood, 7-10 lite		240	.033			1.17		1.17	1.76
0450	Wood, 12 lite		200	.040			1.41		1.41	2.11
0460	Wood, Bay / Bow		320	.025			.88		.88	1.32
0470	Wire brush									
0480	Metal, 1-2 lite	1 Pord	480	.017	S.F.		.59		.59	.88
0490	Metal, 3-6 lite		400	.020			.70		.70	1.06
0500	Metal, Bay / Bow		480	.017			.59		.59	.88
0510	Power wash, based on 2500 lb. operating pressure									
0520	1-2 lite	A-1H	4400	.002	S.F.		.06	.01	.07	.10
0530	3-6 lite		4320	.002			.06	.01	.07	.10
0540	7-10 lite		4240	.002			.06	.01	.07	.10
0550	12 lite		4160	.002			.06	.01	.07	.10
0560	Bay / Bow		4400	.002			.06	.01	.07	.10
0600	Siding, scrape and sand, light=10-30%, med.=30-70%									
0610	Heavy=70-100% of surface to sand									
0650	Texture 1-11, light	1 Pord	480	.017	S.F.		.59		.59	.88
0660	Med.		440	.018			.64		.64	.96
0670	Heavy		360	.022			.78		.78	1.17
0680	Wood shingles, shakes, light		440	.018			.64		.64	.96
0690	Med.		360	.022			.78		.78	1.17
0700	Heavy		280	.029			1.01		1.01	1.51
0710	Clapboard, light		520	.015			.54		.54	.81
0720	Med.		480	.017			.59		.59	.88
0730	Heavy		400	.020			.70		.70	1.06
0740	Wire brush									
0750	Aluminum, light	1 Pord	600	.013	S.F.		.47		.47	.70
0760	Med.		520	.015			.54		.54	.81
0770	Heavy		440	.018			.64		.64	.96
0780	Pressure wash, based on 2500 lb. operating pressure									
0790	Stucco	A-1H	3080	.003	S.F.		.08	.02	.10	.15
0800	Aluminum or vinyl		3200	.003			.08	.02	.10	.14
0810	Siding, masonry, brick & block		2400	.003			.11	.02	.13	.18
1300	Miscellaneous, wire brush									
1310	Metal, pedestrian gate	1 Pord	100	.080	S.F.		2.82		2.82	4.22
8000	For chemical washing, see Div. 04 01 30.20									
8010	For steam cleaning, see Div. 04 01 30.20									

09 91 Painting

09 91 03 – Paint Restoration

09 91 03.30 Exterior Surface Preparation

		Crew	Daily Output	Labor-Hours	Unit	Material	2009 Bare Costs Labor	Equipment	Total	Total Incl O&P
8020	For sand blasting, see Div. 05 01 10.51 and 03 35 29.60									

09 91 03.40 Interior Surface Preparation

		Crew	Daily Output	Labor-Hours	Unit	Material	Labor	Equipment	Total	Total Incl O&P
0010	**INTERIOR SURFACE PREPARATION** R099100-10									
0020	Doors, per side, not incl. frames or trim									
0030	Scrape & sand									
0040	Wood, flush	1 Pord	616	.013	S.F.		.46		.46	.69
0050	Wood, detail		496	.016			.57		.57	.85
0060	Wood, louvered	↓	280	.029	↓		1.01		1.01	1.51
0070	Wire brush									
0080	Metal, flush	1 Pord	640	.013	S.F.		.44		.44	.66
0090	Metal, detail		520	.015			.54		.54	.81
0100	Metal, louvered	↓	360	.022			.78		.78	1.17
0110	Hand wash									
0120	Wood, flush	1 Pord	2160	.004	S.F.		.13		.13	.20
0130	Wood, detailed		2000	.004			.14		.14	.21
0140	Wood, louvered		1360	.006			.21		.21	.31
0150	Metal, flush		2160	.004			.13		.13	.20
0160	Metal, detail		2000	.004			.14		.14	.21
0170	Metal, louvered	↓	1360	.006	↓		.21		.21	.31
0400	Windows, per side, not incl. trim									
0410	Scrape & sand									
0420	Wood, 1-2 lite	1 Pord	360	.022	S.F.		.78		.78	1.17
0430	Wood, 3-6 lite		320	.025			.88		.88	1.32
0440	Wood, 7-10 lite		280	.029			1.01		1.01	1.51
0450	Wood, 12 lite		240	.033			1.17		1.17	1.76
0460	Wood, Bay / Bow	↓	360	.022	↓		.78		.78	1.17
0470	Wire brush									
0480	Metal, 1-2 lite	1 Pord	520	.015	S.F.		.54		.54	.81
0490	Metal, 3-6 lite		440	.018			.64		.64	.96
0500	Metal, Bay / Bow	↓	520	.015	↓		.54		.54	.81
0600	Walls, sanding, light=10-30%, medium - 30-70%,									
0610	heavy=70-100% of surface to sand									
0650	Walls, sand									
0660	Drywall, gypsum, plaster, light	1 Pord	3077	.003	S.F.		.09		.09	.14
0670	Drywall, gypsum, plaster, med.		2160	.004			.13		.13	.20
0680	Drywall, gypsum, plaster, heavy		923	.009			.31		.31	.46
0690	Wood, T&G, light		2400	.003			.12		.12	.18
0700	Wood, T&G, med.		1600	.005			.18		.18	.26
0710	Wood, T&G, heavy	↓	800	.010	↓		.35		.35	.53
0720	Walls, wash									
0730	Drywall, gypsum, plaster	1 Pord	3200	.003	S.F.		.09		.09	.13
0740	Wood, T&G		3200	.003			.09		.09	.13
0750	Masonry, brick & block, smooth		2800	.003			.10		.10	.15
0760	Masonry, brick & block, coarse	↓	2000	.004	↓		.14		.14	.21
8010	For steam cleaning, see Div. 04 01 30.20									
8020	For sand blasting, see Div. 03 35 29.60 and 05 01 10.51									

09 91 03.41 Scrape After Fire Damage

		Crew	Daily Output	Labor-Hours	Unit	Material	Labor	Equipment	Total	Total Incl O&P
0010	**SCRAPE AFTER FIRE DAMAGE**									
0050	Boards, 1" x 4"	1 Pord	336	.024	L.F.		.84		.84	1.26
0060	1" x 6"		260	.031			1.08		1.08	1.62
0070	1" x 8"		207	.039			1.36		1.36	2.04
0080	1" x 10"	↓	174	.046	↓		1.62		1.62	2.43

09 91 Painting

09 91 03 – Paint Restoration

09 91 03.41 Scrape After Fire Damage

		Crew	Daily Output	Labor-Hours	Unit	Material	2009 Bare Costs Labor	2009 Bare Costs Equipment	Total	Total Incl O&P
0500	Framing, 2" x 4"	1 Pord	265	.030	L.F.		1.06		1.06	1.59
0510	2" x 6"		221	.036			1.27		1.27	1.91
0520	2" x 8"		190	.042			1.48		1.48	2.22
0530	2" x 10"		165	.048			1.71		1.71	2.56
0540	2" x 12"		144	.056			1.96		1.96	2.93
1000	Heavy framing, 3" x 4"		226	.035			1.25		1.25	1.87
1010	4" x 4"		210	.038			1.34		1.34	2.01
1020	4" x 6"		191	.042			1.47		1.47	2.21
1030	4" x 8"		165	.048			1.71		1.71	2.56
1040	4" x 10"		144	.056			1.96		1.96	2.93
1060	4" x 12"		131	.061			2.15		2.15	3.22
2900	For sealing, minimum		825	.010	S.F.	.14	.34		.48	.66
2920	Maximum		460	.017	"	.28	.61		.89	1.23

09 91 13 – Exterior Painting

09 91 13.30 Fences

		Crew	Daily Output	Labor-Hours	Unit	Material	Labor	Equipment	Total	Total Incl O&P
0010	**FENCES** R099100-20									
0100	Chain link or wire metal, one side, water base									
0110	Roll & brush, first coat	1 Pord	960	.008	S.F.	.05	.29		.34	.50
0120	Second coat		1280	.006		.05	.22		.27	.39
0130	Spray, first coat		2275	.004		.05	.12		.17	.25
0140	Second coat		2600	.003		.05	.11		.16	.22
0150	Picket, water base									
0160	Roll & brush, first coat	1 Pord	865	.009	S.F.	.06	.33		.39	.55
0170	Second coat		1050	.008		.06	.27		.33	.46
0180	Spray, first coat		2275	.004		.06	.12		.18	.25
0190	Second coat		2600	.003		.06	.11		.17	.22
0200	Stockade, water base									
0210	Roll & brush, first coat	1 Pord	1040	.008	S.F.	.06	.27		.33	.47
0220	Second coat		1200	.007		.06	.23		.29	.41
0230	Spray, first coat		2275	.004		.06	.12		.18	.25
0240	Second coat		2600	.003		.06	.11		.17	.22

09 91 13.42 Miscellaneous, Exterior

		Crew	Daily Output	Labor-Hours	Unit	Material	Labor	Equipment	Total	Total Incl O&P
0010	**MISCELLANEOUS, EXTERIOR** R099100-20									
0015	For painting metals, see Div. 09 97 13.23									
0100	Railing, ext., decorative wood, incl. cap & baluster									
0110	Newels & spindles @ 12" O.C.									
0120	Brushwork, stain, sand, seal & varnish									
0130	First coat	1 Pord	90	.089	L.F.	.61	3.13		3.74	5.35
0140	Second coat	"	120	.067	"	.61	2.35		2.96	4.19
0150	Rough sawn wood, 42" high, 2" x 2" verticals, 6" O.C.									
0160	Brushwork, stain, each coat	1 Pord	90	.089	L.F.	.18	3.13		3.31	4.89
0170	Wrought iron, 1" rail, 1/2" sq. verticals									
0180	Brushwork, zinc chromate, 60" high, bars 6" O.C.									
0190	Primer	1 Pord	130	.062	L.F.	.60	2.17		2.77	3.91
0200	Finish coat		130	.062		.19	2.17		2.36	3.46
0210	Additional coat		190	.042		.22	1.48		1.70	2.46
0220	Shutters or blinds, single panel, 2' x 4', paint all sides									
0230	Brushwork, primer	1 Pord	20	.400	Ea.	.59	14.10		14.69	21.50
0240	Finish coat, exterior latex		20	.400		.44	14.10		14.54	21.50
0250	Primer & 1 coat, exterior latex		13	.615		.89	21.50		22.39	33.50
0260	Spray, primer		35	.229		.85	8.05		8.90	13
0270	Finish coat, exterior latex		35	.229		.93	8.05		8.98	13.05

09 91 Painting

09 91 13 – Exterior Painting

09 91 13.42 Miscellaneous, Exterior

		Crew	Daily Output	Labor-Hours	Unit	Material	2009 Bare Costs Labor	2009 Bare Costs Equipment	Total	Total Incl O&P
0280	Primer & 1 coat, exterior latex	1 Pord	20	.400	Ea.	.91	14.10		15.01	22
0290	For louvered shutters, add				S.F.	10%				
0300	Stair stringers, exterior, metal									
0310	Roll & brush, zinc chromate, to 14", each coat	1 Pord	320	.025	L.F.	.06	.88		.94	1.39
0320	Rough sawn wood, 4" x 12"									
0330	Roll & brush, exterior latex, each coat	1 Pord	215	.037	L.F.	.06	1.31		1.37	2.03
0340	Trellis/lattice, 2" x 2" @ 3" O.C. with 2" x 8" supports									
0350	Spray, latex, per side, each coat	1 Pord	475	.017	S.F.	.06	.59		.65	.96
0450	Decking, ext., sealer, alkyd, brushwork, sealer coat		1140	.007		.09	.25		.34	.47
0460	1st coat		1140	.007		.07	.25		.32	.45
0470	2nd coat		1300	.006		.05	.22		.27	.38
0500	Paint, alkyd, brushwork, primer coat		1140	.007		.07	.25		.32	.45
0510	1st coat		1140	.007		.07	.25		.32	.45
0520	2nd coat		1300	.006		.05	.22		.27	.38
0600	Sand paint, alkyd, brushwork, 1 coat		150	.053		.11	1.88		1.99	2.93

09 91 13.60 Siding Exterior

		Crew	Daily Output	Labor-Hours	Unit	Material	Labor	Equipment	Total	Total Incl O&P
0010	**SIDING EXTERIOR**, Alkyd (oil base) R099100-10									
0450	Steel siding, oil base, paint 1 coat, brushwork	2 Pord	2015	.008	S.F.	.06	.28		.34	.48
0500	Spray R099100-20		4550	.004		.09	.12		.21	.29
0800	Paint 2 coats, brushwork		1300	.012		.11	.43		.54	.78
1000	Spray		2750	.006		.17	.20		.37	.50
1200	Stucco, rough, oil base, paint 2 coats, brushwork		1300	.012		.11	.43		.54	.78
1400	Roller		1625	.010		.12	.35		.47	.65
1600	Spray		2925	.005		.13	.19		.32	.43
1800	Texture 1-11 or clapboard, oil base, primer coat, brushwork		1300	.012		.09	.43		.52	.75
2000	Spray		4550	.004		.09	.12		.21	.29
2100	Paint 1 coat, brushwork		1300	.012		.08	.43		.51	.74
2200	Spray		4550	.004		.08	.12		.20	.28
2400	Paint 2 coats, brushwork		810	.020		.17	.70		.87	1.22
2600	Spray		2600	.006		.19	.22		.41	.52
3000	Stain 1 coat, brushwork		1520	.011		.06	.37		.43	.63
3200	Spray		5320	.003		.07	.11		.18	.24
3400	Stain 2 coats, brushwork		950	.017		.12	.59		.71	1.03
4000	Spray		3050	.005		.14	.18		.32	.43
4200	Wood shingles, oil base primer coat, brushwork		1300	.012		.08	.43		.51	.74
4400	Spray		3900	.004		.08	.14		.22	.31
4600	Paint 1 coat, brushwork		1300	.012		.07	.43		.50	.73
4800	Spray		3900	.004		.09	.14		.23	.32
5000	Paint 2 coats, brushwork		810	.020		.14	.70		.84	1.19
5200	Spray		2275	.007		.13	.25		.38	.52
5800	Stain 1 coat, brushwork		1500	.011		.06	.38		.44	.63
6000	Spray		3900	.004		.06	.14		.20	.29
6500	Stain 2 coats, brushwork		950	.017		.12	.59		.71	1.03
7000	Spray		2660	.006		.17	.21		.38	.51
8000	For latex paint, deduct					10%				
8100	For work over 12' H, from pipe scaffolding, add						15%			
8200	For work over 12' H, from extension ladder, add						25%			
8300	For work over 12' H, from swing staging, add						35%			

09 91 13.62 Siding, Misc.

		Crew	Daily Output	Labor-Hours	Unit	Material	Labor	Equipment	Total	Total Incl O&P
0010	**SIDING, MISC.**, latex paint R099100-10									
0100	Aluminum siding									
0110	Brushwork, primer R099100-20	2 Pord	2275	.007	S.F.	.05	.25		.30	.43

09 91 Painting

09 91 13 – Exterior Painting

09 91 13.62 Siding, Misc.

		Crew	Daily Output	Labor-Hours	Unit	Material	2009 Bare Costs Labor	Equipment	Total	Total Incl O&P
0120	Finish coat, exterior latex	2 Pord	2275	.007	S.F.	.04	.25		.29	.42
0130	Primer & 1 coat exterior latex		1300	.012		.10	.43		.53	.76
0140	Primer & 2 coats exterior latex	↓	975	.016	↓	.15	.58		.73	1.03
0150	Mineral fiber shingles									
0160	Brushwork, primer	2 Pord	1495	.011	S.F.	.09	.38		.47	.66
0170	Finish coat, industrial enamel		1495	.011		.13	.38		.51	.70
0180	Primer & 1 coat enamel		810	.020		.22	.70		.92	1.28
0190	Primer & 2 coats enamel		540	.030		.35	1.04		1.39	1.95
0200	Roll, primer		1625	.010		.10	.35		.45	.63
0210	Finish coat, industrial enamel		1625	.010		.14	.35		.49	.68
0220	Primer & 1 coat enamel		975	.016		.24	.58		.82	1.14
0230	Primer & 2 coats enamel		650	.025		.38	.87		1.25	1.72
0240	Spray, primer		3900	.004		.08	.14		.22	.31
0250	Finish coat, industrial enamel		3900	.004		.12	.14		.26	.35
0260	Primer & 1 coat enamel		2275	.007		.20	.25		.45	.58
0270	Primer & 2 coats enamel		1625	.010		.31	.35		.66	.86
0280	Waterproof sealer, first coat		4485	.004		.08	.13		.21	.28
0290	Second coat	↓	5235	.003	↓	.07	.11		.18	.24
0300	Rough wood incl. shingles, shakes or rough sawn siding									
0310	Brushwork, primer	2 Pord	1280	.013	S.F.	.12	.44		.56	.79
0320	Finish coat, exterior latex		1280	.013		.07	.44		.51	.74
0330	Primer & 1 coat exterior latex		960	.017		.19	.59		.78	1.09
0340	Primer & 2 coats exterior latex		700	.023		.27	.80		1.07	1.50
0350	Roll, primer		2925	.005		.16	.19		.35	.46
0360	Finish coat, exterior latex		2925	.005		.09	.19		.28	.39
0370	Primer & 1 coat exterior latex		1790	.009		.25	.31		.56	.74
0380	Primer & 2 coats exterior latex		1300	.012		.34	.43		.77	1.02
0390	Spray, primer		3900	.004		.13	.14		.27	.37
0400	Finish coat, exterior latex		3900	.004		.07	.14		.21	.30
0410	Primer & 1 coat exterior latex		2600	.006		.20	.22		.42	.54
0420	Primer & 2 coats exterior latex		2080	.008		.27	.27		.54	.71
0430	Waterproof sealer, first coat		4485	.004		.14	.13		.27	.34
0440	Second coat	↓	4485	.004	↓	.08	.13		.21	.28
0450	Smooth wood incl. butt, T&G, beveled, drop or B&B siding									
0460	Brushwork, primer	2 Pord	2325	.007	S.F.	.09	.24		.33	.45
0470	Finish coat, exterior latex		1280	.013		.07	.44		.51	.74
0480	Primer & 1 coat exterior latex		800	.020		.16	.70		.86	1.24
0490	Primer & 2 coats exterior latex		630	.025		.23	.89		1.12	1.60
0500	Roll, primer		2275	.007		.09	.25		.34	.47
0510	Finish coat, exterior latex		2275	.007		.08	.25		.33	.46
0520	Primer & 1 coat exterior latex		1300	.012		.18	.43		.61	.84
0530	Primer & 2 coats exterior latex		975	.016		.26	.58		.84	1.15
0540	Spray, primer		4550	.004		.07	.12		.19	.27
0550	Finish coat, exterior latex		4550	.004		.07	.12		.19	.27
0560	Primer & 1 coat exterior latex		2600	.006		.14	.22		.36	.48
0570	Primer & 2 coats exterior latex		1950	.008		.21	.29		.50	.66
0580	Waterproof sealer, first coat		5230	.003		.08	.11		.19	.25
0590	Second coat	↓	5980	.003	↓	.08	.09		.17	.23
0600	For oil base paint, add					10%				

09 91 13.70 Doors and Windows, Exterior

0010	**DOORS AND WINDOWS, EXTERIOR** R099100-10									
0100	Door frames & trim, only									

09 91 Painting

09 91 13 – Exterior Painting

09 91 13.70 Doors and Windows, Exterior

		Crew	Daily Output	Labor-Hours	Unit	Material	2009 Bare Costs Labor	Equipment	Total	Total Incl O&P
0110	Brushwork, primer R099100-20	1 Pord	512	.016	L.F.	.05	.55		.60	.88
0120	Finish coat, exterior latex		512	.016		.05	.55		.60	.88
0130	Primer & 1 coat, exterior latex		300	.027		.11	.94		1.05	1.53
0140	Primer & 2 coats, exterior latex		265	.030		.16	1.06		1.22	1.77
0150	Doors, flush, both sides, incl. frame & trim									
0160	Roll & brush, primer	1 Pord	10	.800	Ea.	4.05	28		32.05	46.50
0170	Finish coat, exterior latex		10	.800		4.15	28		32.15	46.50
0180	Primer & 1 coat, exterior latex		7	1.143		8.20	40		48.20	69.50
0190	Primer & 2 coats, exterior latex		5	1.600		12.35	56.50		68.85	98
0200	Brushwork, stain, sealer & 2 coats polyurethane		4	2		21	70.50		91.50	129
0210	Doors, French, both sides, 10-15 lite, incl. frame & trim									
0220	Brushwork, primer	1 Pord	6	1.333	Ea.	2.03	47		49.03	72.50
0230	Finish coat, exterior latex		6	1.333		2.07	47		49.07	73
0240	Primer & 1 coat, exterior latex		3	2.667		4.10	94		98.10	146
0250	Primer & 2 coats, exterior latex		2	4		6.05	141		147.05	218
0260	Brushwork, stain, sealer & 2 coats polyurethane		2.50	3.200		7.70	113		120.70	178
0270	Doors, louvered, both sides, incl. frame & trim									
0280	Brushwork, primer	1 Pord	7	1.143	Ea.	4.05	40		44.05	65
0290	Finish coat, exterior latex		7	1.143		4.15	40		44.15	65
0300	Primer & 1 coat, exterior latex		4	2		8.20	70.50		78.70	115
0310	Primer & 2 coats, exterior latex		3	2.667		12.10	94		106.10	154
0320	Brushwork, stain, sealer & 2 coats polyurethane		4.50	1.778		21	62.50		83.50	117
0330	Doors, panel, both sides, incl. frame & trim									
0340	Roll & brush, primer	1 Pord	6	1.333	Ea.	4.05	47		51.05	75
0350	Finish coat, exterior latex		6	1.333		4.15	47		51.15	75
0360	Primer & 1 coat, exterior latex		3	2.667		8.20	94		102.20	150
0370	Primer & 2 coats, exterior latex		2.50	3.200		12.10	113		125.10	182
0380	Brushwork, stain, sealer & 2 coats polyurethane		3	2.667		21	94		115	164
0400	Windows, per ext. side, based on 15 SF									
0410	1 to 6 lite									
0420	Brushwork, primer	1 Pord	13	.615	Ea.	.80	21.50		22.30	33.50
0430	Finish coat, exterior latex		13	.615		.82	21.50		22.32	33.50
0440	Primer & 1 coat, exterior latex		8	1		1.62	35		36.62	55
0450	Primer & 2 coats, exterior latex		6	1.333		2.39	47		49.39	73
0460	Stain, sealer & 1 coat varnish		7	1.143		3.04	40		43.04	64
0470	7 to 10 lite									
0480	Brushwork, primer	1 Pord	11	.727	Ea.	.80	25.50		26.30	39.50
0490	Finish coat, exterior latex		11	.727		.82	25.50		26.32	39.50
0500	Primer & 1 coat, exterior latex		7	1.143		1.62	40		41.62	62.50
0510	Primer & 2 coats, exterior latex		5	1.600		2.39	56.50		58.89	87
0520	Stain, sealer & 1 coat varnish		6	1.333		3.04	47		50.04	74
0530	12 lite									
0540	Brushwork, primer	1 Pord	10	.800	Ea.	.80	28		28.80	43
0550	Finish coat, exterior latex		10	.800		.82	28		28.82	43
0560	Primer & 1 coat, exterior latex		6	1.333		1.62	47		48.62	72.50
0570	Primer & 2 coats, exterior latex		5	1.600		2.39	56.50		58.89	87
0580	Stain, sealer & 1 coat varnish		6	1.333		3.06	47		50.06	74
0590	For oil base paint, add					10%				

09 91 13.80 Trim, Exterior

0010	**TRIM, EXTERIOR** R099100-10									
0100	Door frames & trim (see Doors, interior or exterior)									
0110	Fascia, latex paint, one coat coverage R099100-20									

09 91 Painting

09 91 13 – Exterior Painting

09 91 13.80 Trim, Exterior		Crew	Daily Output	Labor-Hours	Unit	Material	2009 Bare Costs Labor	2009 Bare Costs Equipment	Total	Total Incl O&P
0120	1" x 4", brushwork	1 Pord	640	.013	L.F.	.02	.44		.46	.68
0130	Roll		1280	.006		.02	.22		.24	.35
0140	Spray		2080	.004		.01	.14		.15	.22
0150	1" x 6" to 1" x 10", brushwork		640	.013		.06	.44		.50	.72
0160	Roll		1230	.007		.06	.23		.29	.41
0170	Spray		2100	.004		.05	.13		.18	.25
0180	1" x 12", brushwork		640	.013		.06	.44		.50	.72
0190	Roll		1050	.008		.06	.27		.33	.47
0200	Spray		2200	.004		.05	.13		.18	.24
0210	Gutters & downspouts, metal, zinc chromate paint									
0220	Brushwork, gutters, 5", first coat	1 Pord	640	.013	L.F.	.07	.44		.51	.73
0230	Second coat		960	.008		.06	.29		.35	.51
0240	Third coat		1280	.006		.05	.22		.27	.39
0250	Downspouts, 4", first coat		640	.013		.07	.44		.51	.73
0260	Second coat		960	.008		.06	.29		.35	.51
0270	Third coat		1280	.006		.05	.22		.27	.39
0280	Gutters & downspouts, wood									
0290	Brushwork, gutters, 5", primer	1 Pord	640	.013	L.F.	.05	.44		.49	.72
0300	Finish coat, exterior latex		640	.013		.05	.44		.49	.71
0310	Primer & 1 coat exterior latex		400	.020		.11	.70		.81	1.18
0320	Primer & 2 coats exterior latex		325	.025		.16	.87		1.03	1.48
0330	Downspouts, 4", primer		640	.013		.05	.44		.49	.72
0340	Finish coat, exterior latex		640	.013		.05	.44		.49	.71
0350	Primer & 1 coat exterior latex		400	.020		.11	.70		.81	1.18
0360	Primer & 2 coats exterior latex		325	.025		.08	.87		.95	1.39
0370	Molding, exterior, up to 14" wide									
0380	Brushwork, primer	1 Pord	640	.013	L.F.	.06	.44		.50	.73
0390	Finish coat, exterior latex		640	.013		.06	.44		.50	.72
0400	Primer & 1 coat exterior latex		400	.020		.13	.70		.83	1.20
0410	Primer & 2 coats exterior latex		315	.025		.13	.89		1.02	1.48
0420	Stain & fill		1050	.008		.07	.27		.34	.48
0430	Shellac		1850	.004		.08	.15		.23	.32
0440	Varnish		1275	.006		.09	.22		.31	.43

09 91 13.90 Walls, Masonry (CMU), Exterior

		Crew	Daily Output	Labor-Hours	Unit	Material	Labor	Equipment	Total	Total Incl O&P
0350	**WALLS, MASONRY (CMU), EXTERIOR** R099100-10									
0360	Concrete masonry units (CMU), smooth surface									
0370	Brushwork, latex, first coat	1 Pord	640	.013	S.F.	.04	.44		.48	.71
0380	Second coat		960	.008		.04	.29		.33	.48
0390	Waterproof sealer, first coat		736	.011		.23	.38		.61	.83
0400	Second coat		1104	.007		.23	.26		.49	.64
0410	Roll, latex, paint, first coat		1465	.005		.05	.19		.24	.35
0420	Second coat		1790	.004		.04	.16		.20	.28
0430	Waterproof sealer, first coat		1680	.005		.23	.17		.40	.51
0440	Second coat		2060	.004		.23	.14		.37	.46
0450	Spray, latex, paint, first coat		1950	.004		.04	.14		.18	.27
0460	Second coat		2600	.003		.03	.11		.14	.20
0470	Waterproof sealer, first coat		2245	.004		.23	.13		.36	.45
0480	Second coat		2990	.003		.23	.09		.32	.40
0490	Concrete masonry unit (CMU), porous									
0500	Brushwork, latex, first coat	1 Pord	640	.013	S.F.	.09	.44		.53	.76
0510	Second coat		960	.008		.04	.29		.33	.49
0520	Waterproof sealer, first coat		736	.011		.23	.38		.61	.83

09 91 Painting

09 91 13 – Exterior Painting

09 91 13.90 Walls, Masonry (CMU), Exterior

	Crew	Daily Output	Labor-Hours	Unit	Material	2009 Bare Costs Labor	Equipment	Total	Total Incl O&P
0530 Second coat	1 Pord	1104	.007	S.F.	.23	.26		.49	.64
0540 Roll latex, first coat		1465	.005		.07	.19		.26	.36
0550 Second coat		1790	.004		.04	.16		.20	.29
0560 Waterproof sealer, first coat		1680	.005		.23	.17		.40	.51
0570 Second coat		2060	.004		.23	.14		.37	.46
0580 Spray latex, first coat		1950	.004		.05	.14		.19	.27
0590 Second coat		2600	.003		.03	.11		.14	.20
0600 Waterproof sealer, first coat		2245	.004		.23	.13		.36	.45
0610 Second coat		2990	.003		.23	.09		.32	.40

09 91 23 – Interior Painting

09 91 23.20 Cabinets and Casework

	Crew	Daily Output	Labor-Hours	Unit	Material	Labor	Equipment	Total	Total Incl O&P
0010 **CABINETS AND CASEWORK** R099100-10									
1000 Primer coat, oil base, brushwork	1 Pord	650	.012	S.F.	.05	.43		.48	.71
2000 Paint, oil base, brushwork, 1 coat R099100-20		650	.012		.06	.43		.49	.72
2500 2 coats		400	.020		.12	.70		.82	1.19
3000 Stain, brushwork, wipe off		650	.012		.06	.43		.49	.72
4000 Shellac, 1 coat, brushwork		650	.012		.07	.43		.50	.73
4500 Varnish, 3 coats, brushwork, sand after 1st coat		325	.025		.23	.87		1.10	1.55
5000 For latex paint, deduct					10%				

09 91 23.33 Doors and Windows, Interior Alkyd (Oil Base)

	Crew	Daily Output	Labor-Hours	Unit	Material	Labor	Equipment	Total	Total Incl O&P
0010 **DOORS AND WINDOWS, INTERIOR ALKYD (OIL BASE)** R099100-10									
0500 Flush door & frame, 3' x 7', oil, primer, brushwork	1 Pord	10	.800	Ea.	2.34	28		30.34	44.50
1000 Paint, 1 coat R099100-20		10	.800		2.25	28		30.25	44.50
1200 2 coats		6	1.333		3.70	47		50.70	74.50
1220 3 coats		5	1.600		8.70	56.50		65.20	94
1400 Stain, brushwork, wipe off		18	.444		1.29	15.65		16.94	25
1600 Shellac, 1 coat, brushwork		25	.320		1.48	11.25		12.73	18.50
1800 Varnish, 3 coats, brushwork, sand after 1st coat		9	.889		4.77	31.50		36.27	52.50
2000 Panel door & frame, 3' x 7', oil, primer, brushwork		6	1.333		1.96	47		48.96	72.50
2200 Paint, 1 coat		6	1.333		2.25	47		49.25	73
2400 2 coats		3	2.667		6.45	94		100.45	148
2420 3 coats		2	4		8.45	141		149.45	220
2600 Stain, brushwork, panel door, 3' x 7', not incl. frame		16	.500		1.29	17.60		18.89	28
2800 Shellac, 1 coat, brushwork		22	.364		1.48	12.80		14.28	21
3000 Varnish, 3 coats, brushwork, sand after 1st coat		7.50	1.067		4.77	37.50		42.27	62
4400 Windows, including frame and trim, per side									
4600 Colonial type, 6/6 lites, 2' x 3', oil, primer, brushwork	1 Pord	14	.571	Ea.	.31	20		20.31	30.50
5800 Paint, 1 coat		14	.571		.36	20		20.36	30.50
6000 2 coats		9	.889		.69	31.50		32.19	48
6010 3 coats		7	1.143		1.03	40		41.03	61.50
6200 3' x 5' opening, 6/6 lites, primer coat, brushwork		12	.667		.78	23.50		24.28	36
6400 Paint, 1 coat		12	.667		.89	23.50		24.39	36
6600 2 coats		7	1.143		1.73	40		41.73	62.50
6610 3 coats		6	1.333		2.56	47		49.56	73.50
6800 4' x 8' opening, 6/6 lites, primer coat, brushwork		8	1		1.65	35		36.65	55
7000 Paint, 1 coat		8	1		1.90	35		36.90	55
7200 2 coats		5	1.600		3.68	56.50		60.18	88.50
7210 3 coats		4	2		5.45	70.50		75.95	112
7500 Standard, 6/6 lites, 2' x 3', primer		14	.571		.31	20		20.31	30.50
7520 Paint 1 coat		14	.571		.36	20		20.36	30.50
7540 2 coats		9	.889		.69	31.50		32.19	48
7560 3 coats		7	1.143		1.03	40		41.03	61.50

09 91 Painting

09 91 23 – Interior Painting

09 91 23.33 Doors and Windows, Interior Alkyd (Oil Base)

		Crew	Daily Output	Labor-Hours	Unit	Material	2009 Bare Costs Labor	2009 Bare Costs Equipment	Total	Total Incl O&P
7580	3' x 5', 6/6 lites, primer	1 Pord	12	.667	Ea.	.78	23.50		24.28	36
7600	Paint 1 coat		12	.667		.89	23.50		24.39	36
7620	2 coats		7	1.143		1.73	40		41.73	62.50
7640	3 coats		6	1.333		2.56	47		49.56	73.50
7660	4' x 8', primer		8	1		1.65	35		36.65	55
7680	Paint 1 coat		8	1		1.90	35		36.90	55
7700	2 coats		5	1.600		3.68	56.50		60.18	88.50
7720	3 coats		4	2		5.45	70.50		75.95	112
8000	Single lite type, 2' x 3', oil base, primer coat, brushwork		33	.242		.31	8.55		8.86	13.15
8200	Paint, 1 coat		33	.242		.36	8.55		8.91	13.20
8400	2 coats		20	.400		.69	14.10		14.79	22
8410	3 coats		16	.500		1.03	17.60		18.63	27.50
8600	3' x 5' opening, primer coat, brushwork		20	.400		.78	14.10		14.88	22
8800	Paint, 1 coat		20	.400		.89	14.10		14.99	22
8900	2 coats		13	.615		1.73	21.50		23.23	34.50
9010	3 coats		10	.800		2.56	28		30.56	45
9200	4' x 8' opening, primer coat, brushwork		14	.571		1.65	20		21.65	32
9400	Paint, 1 coat		14	.571		1.90	20		21.90	32
9600	2 coats		8	1		3.68	35		38.68	57
9610	3 coats		7	1.143		5.45	40		45.45	66.50

09 91 23.35 Doors and Windows, Interior Latex

			Crew	Daily Output	Labor-Hours	Unit	Material	Labor	Equipment	Total	Total Incl O&P
0010	**DOORS & WINDOWS, INTERIOR LATEX**	R099100-10									
0100	Doors, flush, both sides, incl. frame & trim										
0110	Roll & brush, primer	R099100-20	1 Pord	10	.800	Ea.	3.64	28		31.64	46
0120	Finish coat, latex			10	.800		4.21	28		32.21	46.50
0130	Primer & 1 coat latex			7	1.143		7.85	40		47.85	69
0140	Primer & 2 coats latex			5	1.600		11.80	56.50		68.30	97.50
0160	Spray, both sides, primer			20	.400		3.83	14.10		17.93	25
0170	Finish coat, latex			20	.400		4.42	14.10		18.52	26
0180	Primer & 1 coat latex			11	.727		8.30	25.50		33.80	47.50
0190	Primer & 2 coats latex			8	1		12.50	35		47.50	67
0200	Doors, French, both sides, 10-15 lite, incl. frame & trim										
0210	Roll & brush, primer		1 Pord	6	1.333	Ea.	1.82	47		48.82	72.50
0220	Finish coat, latex			6	1.333		2.10	47		49.10	73
0230	Primer & 1 coat latex			3	2.667		3.92	94		97.92	145
0240	Primer & 2 coats latex			2	4		5.90	141		146.90	218
0260	Doors, louvered, both sides, incl. frame & trim										
0270	Roll & brush, primer		1 Pord	7	1.143	Ea.	3.64	40		43.64	64.50
0280	Finish coat, latex			7	1.143		4.21	40		44.21	65
0290	Primer & 1 coat, latex			4	2		7.65	70.50		78.15	114
0300	Primer & 2 coats, latex			3	2.667		12.05	94		106.05	154
0320	Spray, both sides, primer			20	.400		3.83	14.10		17.93	25
0330	Finish coat, latex			20	.400		4.42	14.10		18.52	26
0340	Primer & 1 coat, latex			11	.727		8.30	25.50		33.80	47.50
0350	Primer & 2 coats, latex			8	1		12.75	35		47.75	67
0360	Doors, panel, both sides, incl. frame & trim										
0370	Roll & brush, primer		1 Pord	6	1.333	Ea.	3.83	47		50.83	74.50
0380	Finish coat, latex			6	1.333		4.21	47		51.21	75
0390	Primer & 1 coat, latex			3	2.667		7.85	94		101.85	150
0400	Primer & 2 coats, latex			2.50	3.200		12.05	113		125.05	182
0420	Spray, both sides, primer			10	.800		3.83	28		31.83	46
0430	Finish coat, latex			10	.800		4.42	28		32.42	47

09 91 Painting

09 91 23 – Interior Painting

09 91 23.35 Doors and Windows, Interior Latex

		Crew	Daily Output	Labor-Hours	Unit	Material	2009 Bare Costs Labor	Equipment	Total	Total Incl O&P
0440	Primer & 1 coat, latex	1 Pord	5	1.600	Ea.	8.30	56.50		64.80	93.50
0450	Primer & 2 coats, latex	↓	4	2	↓	12.75	70.50		83.25	120
0460	Windows, per interior side, based on 15 SF									
0470	1 to 6 lite									
0480	Brushwork, primer	1 Pord	13	.615	Ea.	.72	21.50		22.22	33.50
0490	Finish coat, enamel		13	.615		.83	21.50		22.33	33.50
0500	Primer & 1 coat enamel		8	1		1.55	35		36.55	54.50
0510	Primer & 2 coats enamel	↓	6	1.333	↓	2.38	47		49.38	73
0530	7 to 10 lite									
0540	Brushwork, primer	1 Pord	11	.727	Ea.	.72	25.50		26.22	39.50
0550	Finish coat, enamel		11	.727		.83	25.50		26.33	39.50
0560	Primer & 1 coat enamel		7	1.143		1.55	40		41.55	62
0570	Primer & 2 coats enamel	↓	5	1.600	↓	2.38	56.50		58.88	87
0590	12 lite									
0600	Brushwork, primer	1 Pord	10	.800	Ea.	.72	28		28.72	43
0610	Finish coat, enamel		10	.800		.83	28		28.83	43
0620	Primer & 1 coat enamel		6	1.333		1.55	47		48.55	72
0630	Primer & 2 coats enamel	↓	5	1.600	↓	2.38	56.50		58.88	87
0650	For oil base paint, add					10%				

09 91 23.39 Doors and Windows, Interior Latex, Zero Voc

		Crew	Daily Output	Labor-Hours	Unit	Material	Labor	Equipment	Total	Total Incl O&P
0010	**DOORS & WINDOWS, INTERIOR LATEX, ZERO VOC**									
0100	Doors flush, both sides, incl. frame & trim									
0110	Roll & brush, primer [G]	1 Pord	10	.800	Ea.	4.75	28		32.75	47.50
0120	Finish coat, latex [G]		10	.800		4.82	28		32.82	47.50
0130	Primer & 1 coat latex [G]		7	1.143		9.55	40		49.55	71
0140	Primer & 2 coats latex [G]		5	1.600		14.10	56.50		70.60	100
0160	Spray, both sides, primer [G]		20	.400		5	14.10		19.10	26.50
0170	Finish coat, latex [G]		20	.400		5.05	14.10		19.15	26.50
0180	Primer & 1 coat latex [G]		11	.727		10.10	25.50		35.60	49.50
0190	Primer & 2 coats latex [G]	↓	8	1	↓	14.95	35		49.95	69.50
0200	Doors, French, both sides, 10-15 lite, incl. frame & trim									
0210	Roll & brush, primer [G]	1 Pord	6	1.333	Ea.	2.38	47		49.38	73
0220	Finish coat, latex [G]		6	1.333		2.41	47		49.41	73
0230	Primer & 1 coat latex [G]		3	2.667		4.79	94		98.79	146
0240	Primer & 2 coats latex [G]	↓	2	4	↓	7.05	141		148.05	219
0360	Doors, panel, both sides, incl. frame & trim									
0370	Roll & brush, primer [G]	1 Pord	6	1.333	Ea.	5	47		52	76
0380	Finish coat, latex [G]		6	1.333		4.82	47		51.82	76
0390	Primer & 1 coat, latex [G]		3	2.667		9.55	94		103.55	152
0400	Primer & 2 coats, latex [G]		2.50	3.200		14.40	113		127.40	185
0420	Spray, both sides, primer [G]		10	.800		5	28		33	47.50
0430	Finish coat, latex [G]		10	.800		5.05	28		33.05	47.50
0440	Primer & 1 coat, latex [G]		5	1.600		10.10	56.50		66.60	95.50
0450	Primer & 2 coats, latex [G]	↓	4	2	↓	15.25	70.50		85.75	123
0460	Windows, per interior side, based on 15 SF									
0470	1 to 6 lite									
0480	Brushwork, primer [G]	1 Pord	13	.615	Ea.	.94	21.50		22.44	33.50
0490	Finish coat, enamel [G]		13	.615		.95	21.50		22.45	33.50
0500	Primer & 1 coat enamel [G]		8	1		1.89	35		36.89	55
0510	Primer & 2 coats enamel [G]	↓	6	1.333	↓	2.84	47		49.84	73.50

09 91 Painting

09 91 23 – Interior Painting

09 91 23.40 Floors, Interior

		Crew	Daily Output	Labor-Hours	Unit	Material	2009 Bare Costs Labor	Equipment	Total	Total Incl O&P
0010	**FLOORS, INTERIOR** R099100-10									
0100	Concrete paint, latex									
0110	Brushwork									
0120	1st coat	1 Pord	975	.008	S.F.	.14	.29		.43	.59
0130	2nd coat		1150	.007		.10	.25		.35	.48
0140	3rd coat	↓	1300	.006	↓	.08	.22		.30	.40
0150	Roll									
0160	1st coat	1 Pord	2600	.003	S.F.	.19	.11		.30	.37
0170	2nd coat		3250	.002		.12	.09		.21	.26
0180	3rd coat	↓	3900	.002	↓	.09	.07		.16	.20
0190	Spray									
0200	1st coat	1 Pord	2600	.003	S.F.	.17	.11		.28	.34
0210	2nd coat		3250	.002		.09	.09		.18	.23
0220	3rd coat	↓	3900	.002	↓	.07	.07		.14	.19
0300	Acid stain and sealer									
0310	Stain, one coat	1 Pord	650	.012	S.F.	.11	.43		.54	.77
0320	Two coats		570	.014		.22	.49		.71	.98
0330	Acrylic sealer, one coat		2600	.003		.15	.11		.26	.33
0340	Two coats	↓	1400	.006	↓	.30	.20		.50	.63

09 91 23.52 Miscellaneous, Interior

		Crew	Daily Output	Labor-Hours	Unit	Material	2009 Bare Costs Labor	Equipment	Total	Total Incl O&P
0010	**MISCELLANEOUS, INTERIOR** R099100-10									
2400	Floors, conc./wood, oil base, primer/sealer coat, brushwork	2 Pord	1950	.008	S.F.	.06	.29		.35	.49
2450	Roller		5200	.003		.06	.11		.17	.22
2600	Spray		6000	.003		.06	.09		.15	.20
2650	Paint 1 coat, brushwork		1950	.008		.08	.29		.37	.52
2800	Roller		5200	.003		.09	.11		.20	.26
2850	Spray		6000	.003		.09	.09		.18	.24
3000	Stain, wood floor, brushwork, 1 coat		4550	.004		.06	.12		.18	.26
3200	Roller		5200	.003		.06	.11		.17	.23
3250	Spray		6000	.003		.06	.09		.15	.21
3400	Varnish, wood floor, brushwork		4550	.004		.08	.12		.20	.27
3450	Roller		5200	.003		.08	.11		.19	.25
3600	Spray	↓	6000	.003	↓	.08	.09		.17	.23
3650	For dust proofing or anti skid, see Div. 03 35 29.30									
3800	Grilles, per side, oil base, primer coat, brushwork	1 Pord	520	.015	S.F.	.10	.54		.64	.92
3850	Spray		1140	.007		.11	.25		.36	.49
3880	Paint 1 coat, brushwork		520	.015		.12	.54		.66	.94
3900	Spray		1140	.007		.13	.25		.38	.51
3920	Paint 2 coats, brushwork		325	.025		.23	.87		1.10	1.55
3940	Spray	↓	650	.012	↓	.26	.43		.69	.94
5000	Pipe, 1" – 4" diameter, primer or sealer coat, oil base, brushwork	2 Pord	1250	.013	L.F.	.06	.45		.51	.74
5100	Spray		2165	.007		.06	.26		.32	.45
5200	Paint 1 coat, brushwork		1250	.013		.06	.45		.51	.75
5300	Spray		2165	.007		.06	.26		.32	.45
5350	Paint 2 coats, brushwork		775	.021		.11	.73		.84	1.21
5400	Spray		1240	.013		.12	.45		.57	.82
5450	5" - 8" diameter, primer or sealer coat, brushwork		620	.026		.12	.91		1.03	1.49
5500	Spray		1085	.015		.19	.52		.71	.99
5550	Paint 1 coat, brushwork		620	.026		.17	.91		1.08	1.55
5600	Spray		1085	.015		.19	.52		.71	.99
5650	Paint 2 coats, brushwork		385	.042		.22	1.46		1.68	2.44
5700	Spray	↓	620	.026	↓	.25	.91		1.16	1.63

09 91 Painting

09 91 23 – Interior Painting

09 91 23.52 Miscellaneous, Interior		Crew	Daily Output	Labor-Hours	Unit	Material	2009 Bare Costs Labor	Equipment	Total	Total Incl O&P
5750	9" - 12" diameter, primer or sealer coat, brushwork	2 Pord	415	.039	L.F.	.18	1.36		1.54	2.22
5800	Spray		725	.022		.20	.78		.98	1.38
5850	Paint 1 coat, brushwork		415	.039		.17	1.36		1.53	2.22
6000	Spray		725	.022		.19	.78		.97	1.37
6200	Paint 2 coats, brushwork		260	.062		.34	2.17		2.51	3.62
6250	Spray		415	.039		.37	1.36		1.73	2.44
6300	13" - 16" diameter, primer or sealer coat, brushwork		310	.052		.24	1.82		2.06	2.98
6350	Spray		540	.030		.26	1.04		1.30	1.85
6400	Paint 1 coat, brushwork		310	.052		.23	1.82		2.05	2.97
6450	Spray		540	.030		.26	1.04		1.30	1.84
6500	Paint 2 coats, brushwork		195	.082		.45	2.89		3.34	4.82
6550	Spray		310	.052		.50	1.82		2.32	3.27
7000	Trim, wood, incl. puttying, under 6" wide									
7200	Primer coat, oil base, brushwork	1 Pord	650	.012	L.F.	.03	.43		.46	.68
7250	Paint, 1 coat, brushwork		650	.012		.03	.43		.46	.68
7400	2 coats		400	.020		.06	.70		.76	1.12
7450	3 coats		325	.025		.09	.87		.96	1.39
7500	Over 6" wide, primer coat, brushwork		650	.012		.05	.43		.48	.71
7550	Paint, 1 coat, brushwork		650	.012		.06	.43		.49	.72
7600	2 coats		400	.020		.12	.70		.82	1.19
7650	3 coats		325	.025		.17	.87		1.04	1.49
8000	Cornice, simple design, primer coat, oil base, brushwork		650	.012	S.F.	.05	.43		.48	.71
8250	Paint, 1 coat		650	.012		.06	.43		.49	.72
8300	2 coats		400	.020		.12	.70		.82	1.19
8350	Ornate design, primer coat		350	.023		.05	.80		.85	1.27
8400	Paint, 1 coat		350	.023		.06	.80		.86	1.28
8450	2 coats		400	.020		.12	.70		.82	1.19
8600	Balustrades, primer coat, oil base, brushwork		520	.015		.05	.54		.59	.87
8650	Paint, 1 coat		520	.015		.06	.54		.60	.88
8700	2 coats		325	.025		.12	.87		.99	1.43
8900	Trusses and wood frames, primer coat, oil base, brushwork		800	.010		.05	.35		.40	.59
8950	Spray		1200	.007		.05	.23		.28	.41
9000	Paint 1 coat, brushwork		750	.011		.06	.38		.44	.63
9200	Spray		1200	.007		.07	.23		.30	.42
9220	Paint 2 coats, brushwork		500	.016		.12	.56		.68	.97
9240	Spray		600	.013		.13	.47		.60	.84
9260	Stain, brushwork, wipe off		600	.013		.06	.47		.53	.77
9280	Varnish, 3 coats, brushwork		275	.029		.23	1.02		1.25	1.78
9350	For latex paint, deduct					10%				

09 91 23.72 Walls and Ceilings, Interior

		Crew	Daily Output	Labor-Hours	Unit	Material	2009 Bare Costs Labor	Equipment	Total	Total Incl O&P
0010	**WALLS AND CEILINGS, INTERIOR** R099100-10									
0100	Concrete, drywall or plaster, oil base, primer or sealer coat R099100-20									
0200	Smooth finish, brushwork	1 Pord	1150	.007	S.F.	.05	.25		.30	.43
0240	Roller		1350	.006		.05	.21		.26	.36
0280	Spray		2750	.003		.04	.10		.14	.19
0300	Sand finish, brushwork		975	.008		.05	.29		.34	.48
0340	Roller		1150	.007		.05	.25		.30	.43
0380	Spray		2275	.004		.04	.12		.16	.24
0400	Paint 1 coat, smooth finish, brushwork		1200	.007		.06	.23		.29	.41
0440	Roller		1300	.006		.06	.22		.28	.38
0480	Spray		2275	.004		.05	.12		.17	.25
0500	Sand finish, brushwork		1050	.008		.06	.27		.33	.46

09 91 Painting

09 91 23 – Interior Painting

09 91 23.72 Walls and Ceilings, Interior		Crew	Daily Output	Labor-Hours	Unit	Material	2009 Bare Costs Labor	Equipment	Total	Total Incl O&P
0540	Roller	1 Pord	1600	.005	S.F.	.06	.18		.24	.32
0580	Spray		2100	.004		.05	.13		.18	.26
0800	Paint 2 coats, smooth finish, brushwork		680	.012		.11	.41		.52	.75
0840	Roller		800	.010		.12	.35		.47	.66
0880	Spray		1625	.005		.10	.17		.27	.37
0900	Sand finish, brushwork		605	.013		.11	.47		.58	.82
0940	Roller		1020	.008		.12	.28		.40	.54
0980	Spray		1700	.005		.10	.17		.27	.36
1200	Paint 3 coats, smooth finish, brushwork		510	.016		.17	.55		.72	1.01
1240	Roller		650	.012		.18	.43		.61	.84
1280	Spray		1625	.005		.15	.17		.32	.43
1300	Sand finish, brushwork		454	.018		.17	.62		.79	1.11
1340	Roller		680	.012		.18	.41		.59	.81
1380	Spray		1133	.007		.15	.25		.40	.54
1600	Glaze coating, 2 coats, spray, clear		1200	.007		.49	.23		.72	.89
1640	Multicolor		1200	.007		.95	.23		1.18	1.40
1660	Painting walls, complete, including surface prep, primer &									
1670	2 coats finish, on drywall or plaster, with roller	1 Pord	325	.025	S.F.	.18	.87		1.05	1.50
1700	For latex paint, deduct					10%				
1800	For ceiling installations, add						25%			
2000	Masonry or concrete block, oil base, primer or sealer coat									
2100	Smooth finish, brushwork	1 Pord	1224	.007	S.F.	.06	.23		.29	.42
2180	Spray		2400	.003		.09	.12		.21	.28
2200	Sand finish, brushwork		1089	.007		.10	.26		.36	.50
2280	Spray		2400	.003		.09	.12		.21	.28
2400	Paint 1 coat, smooth finish, brushwork		1100	.007		.10	.26		.36	.49
2480	Spray		2400	.003		.09	.12		.21	.28
2500	Sand finish, brushwork		979	.008		.10	.29		.39	.54
2580	Spray		2400	.003		.09	.12		.21	.28
2800	Paint 2 coats, smooth finish, brushwork		756	.011		.20	.37		.57	.78
2880	Spray		1360	.006		.18	.21		.39	.51
2900	Sand finish, brushwork		672	.012		.20	.42		.62	.85
2980	Spray		1360	.006		.18	.21		.39	.51
3200	Paint 3 coats, smooth finish, brushwork		560	.014		.30	.50		.80	1.08
3280	Spray		1088	.007		.28	.26		.54	.69
3300	Sand finish, brushwork		498	.016		.30	.57		.87	1.18
3380	Spray		1088	.007		.28	.26		.54	.69
3600	Glaze coating, 3 coats, spray, clear		900	.009		.70	.31		1.01	1.24
3620	Multicolor		900	.009		1.10	.31		1.41	1.68
4000	Block filler, 1 coat, brushwork		425	.019		.13	.66		.79	1.14
4100	Silicone, water repellent, 2 coats, spray		2000	.004		.30	.14		.44	.54
4120	For latex paint, deduct					10%				
8200	For work 8 – 15' H, add						10%			
8300	For work over 15' H, add						20%			
8400	For light textured surfaces, add						10%			
8410	Heavy textured, add						25%			

09 91 23.74 Walls and Ceilings, Interior, Zero VOC Latex										
0010	**WALLS AND CEILINGS, INTERIOR, ZERO VOC LATEX**									
0100	Concrete, dry wall or plaster, latex, primer or sealer coat									
0200	Smooth finish, brushwork	G 1 Pord	1150	.007	S.F.	.06	.25		.31	.44
0240	Roller	G	1350	.006		.06	.21		.27	.37
0280	Spray	G	2750	.003		.05	.10		.15	.20

09 91 Painting

09 91 23 – Interior Painting

09 91 23.74 Walls and Ceilings, Interior, Zero VOC Latex

		Crew	Daily Output	Labor-Hours	Unit	Material	2009 Bare Costs Labor	Equipment	Total	Total Incl O&P
0300	Sand finish, brushwork [G]	1 Pord	975	.008	S.F.	.06	.29		.35	.49
0340	Roller [G]		1150	.007		.07	.25		.32	.44
0380	Spray [G]		2275	.004		.05	.12		.17	.25
0400	Paint 1 coat, smooth finish, brushwork [G]		1200	.007		.06	.23		.29	.42
0440	Roller [G]		1300	.006		.06	.22		.28	.39
0480	Spray [G]		2275	.004		.05	.12		.17	.25
0500	Sand finish, brushwork [G]		1050	.008		.06	.27		.33	.46
0540	Roller [G]		1600	.005		.06	.18		.24	.33
0580	Spray [G]		2100	.004		.05	.13		.18	.26
0800	Paint 2 coats, smooth finish, brushwork [G]		680	.012		.12	.41		.53	.75
0840	Roller [G]		800	.010		.12	.35		.47	.66
0880	Spray [G]		1625	.005		.10	.17		.27	.37
0900	Sand finish, brushwork [G]		605	.013		.11	.47		.58	.83
0940	Roller [G]		1020	.008		.12	.28		.40	.54
0980	Spray [G]		1700	.005		.10	.17		.27	.36
1200	Paint 3 coats, smooth finish, brushwork [G]		510	.016		.17	.55		.72	1.02
1240	Roller [G]		650	.012		.18	.43		.61	.85
1280	Spray [G]		1625	.005		.16	.17		.33	.43
1800	For ceiling installations, add [G]						25%			
8200	For work 8 – 15' H, add						10%			
8300	For work over 15' H, add						20%			

09 91 23.75 Dry Fall Painting

			Crew	Daily Output	Labor-Hours	Unit	Material	Labor	Equipment	Total	Total Incl O&P
0010	**DRY FALL PAINTING**	R099100-10									
0100	Walls										
0200	Wallboard and smooth plaster, one coat, brush	R099100-20	1 Pord	910	.009	S.F.	.05	.31		.36	.51
0210	Roll			1560	.005		.05	.18		.23	.32
0220	Spray			2600	.003		.05	.11		.16	.21
0230	Two coats, brush			520	.015		.10	.54		.64	.92
0240	Roll			877	.009		.10	.32		.42	.59
0250	Spray			1560	.005		.10	.18		.28	.38
0260	Concrete or textured plaster, one coat, brush			747	.011		.05	.38		.43	.62
0270	Roll			1300	.006		.05	.22		.27	.37
0280	Spray			1560	.005		.05	.18		.23	.32
0290	Two coats, brush			422	.019		.10	.67		.77	1.11
0300	Roll			747	.011		.10	.38		.48	.68
0310	Spray			1300	.006		.10	.22		.32	.43
0320	Concrete block, one coat, brush			747	.011		.05	.38		.43	.62
0330	Roll			1300	.006		.05	.22		.27	.37
0340	Spray			1560	.005		.05	.18		.23	.32
0350	Two coats, brush			422	.019		.10	.67		.77	1.11
0360	Roll			747	.011		.10	.38		.48	.68
0370	Spray			1300	.006		.10	.22		.32	.43
0380	Wood, one coat, brush			747	.011		.05	.38		.43	.62
0390	Roll			1300	.006		.05	.22		.27	.37
0400	Spray			877	.009		.05	.32		.37	.53
0410	Two coats, brush			487	.016		.10	.58		.68	.98
0420	Roll			747	.011		.10	.38		.48	.68
0430	Spray			650	.012		.10	.43		.53	.76
0440	Ceilings										
0450	Wallboard and smooth plaster, one coat, brush		1 Pord	600	.013	S.F.	.05	.47		.52	.75
0460	Roll			1040	.008		.05	.27		.32	.46
0470	Spray			1560	.005		.05	.18		.23	.32

09 91 Painting

09 91 23 – Interior Painting

09 91 23.75 Dry Fall Painting

		Crew	Daily Output	Labor-Hours	Unit	Material	2009 Bare Costs Labor	Equipment	Total	Total Incl O&P
0480	Two coats, brush	1 Pord	341	.023	S.F.	.10	.83		.93	1.35
0490	Roll		650	.012		.10	.43		.53	.76
0500	Spray		1300	.006		.10	.22		.32	.43
0510	Concrete or textured plaster, one coat, brush		487	.016		.05	.58		.63	.92
0520	Roll		877	.009		.05	.32		.37	.53
0530	Spray		1560	.005		.05	.18		.23	.32
0540	Two coats, brush		276	.029		.10	1.02		1.12	1.64
0550	Roll		520	.015		.10	.54		.64	.92
0560	Spray		1300	.006		.10	.22		.32	.43
0570	Structural steel, bar joists or metal deck, one coat, spray		1560	.005		.05	.18		.23	.32
0580	Two coats, spray		1040	.008		.10	.27		.37	.52

09 93 Staining and Transparent Finishing

09 93 23 – Interior Staining and Finishing

09 93 23.10 Varnish

		Crew	Daily Output	Labor-Hours	Unit	Material	Labor	Equipment	Total	Total Incl O&P
0010	**VARNISH**									
0012	1 coat + sealer, on wood trim, brush, no sanding included	1 Pord	400	.020	S.F.	.07	.70		.77	1.14
0100	Hardwood floors, 2 coats, no sanding included, roller	"	1890	.004	"	.15	.15		.30	.39

09 96 High-Performance Coatings

09 96 23 – Graffiti-Resistant Coatings

09 96 23.10 Graffiti Resistant Treatments

		Crew	Daily Output	Labor-Hours	Unit	Material	Labor	Equipment	Total	Total Incl O&P
0010	**GRAFFITI RESISTANT TREATMENTS**, sprayed on walls									
0100	Non-sacrificial, permanent non-stick coating, clear, on metals	1 Pord	2000	.004	S.F.	1.59	.14		1.73	1.96
0200	Concrete		2000	.004		1.81	.14		1.95	2.20
0300	Concrete block		2000	.004		2.34	.14		2.48	2.79
0400	Brick		2000	.004		2.66	.14		2.80	3.13
0500	Stone		2000	.004		2.66	.14		2.80	3.13
0600	Unpainted wood		2000	.004		3.06	.14		3.20	3.58
2000	Semi-permanent cross linking polymer primer, on metals		2000	.004		.40	.14		.54	.65
2100	Concrete		2000	.004		.48	.14		.62	.74
2200	Concrete block		2000	.004		.60	.14		.74	.87
2300	Brick		2000	.004		.48	.14		.62	.74
2400	Stone		2000	.004		.48	.14		.62	.74
2500	Unpainted wood		2000	.004		.67	.14		.81	.94
3000	Top coat, on metals		2000	.004		.43	.14		.57	.68
3100	Concrete		2000	.004		.49	.14		.63	.74
3200	Concrete block		2000	.004		.68	.14		.82	.96
3300	Brick		2000	.004		.57	.14		.71	.83
3400	Stone		2000	.004		.57	.14		.71	.83
3500	Unpainted wood		2000	.004		.68	.14		.82	.96
5000	Sacrificial, water based, on metal		2000	.004		.21	.14		.35	.44
5100	Concrete		2000	.004		.21	.14		.35	.44
5200	Concrete block		2000	.004		.21	.14		.35	.44
5300	Brick		2000	.004		.21	.14		.35	.44
5400	Stone		2000	.004		.21	.14		.35	.44
5500	Unpainted wood		2000	.004		.21	.14		.35	.44
8000	Cleaner for use after treatment									
8100	Towels or wipes, per package of 30				Ea.	.80			.80	.88

09 96 High-Performance Coatings

09 96 23 – Graffiti-Resistant Coatings

09 96 23.10 Graffiti Resistant Treatments	Crew	Daily Output	Labor-Hours	Unit	Material	2009 Bare Costs Labor	Equipment	Total	Total Incl O&P
8200 Aerosol spray, 24 oz. can				Ea.	19			19	21

09 96 56 – Epoxy Coatings

09 96 56.20 Wall Coatings

		Crew	Daily Output	Labor-Hours	Unit	Material	Labor	Equipment	Total	Total Incl O&P
0010	**WALL COATINGS**									
0100	Acrylic glazed coatings, minimum	1 Pord	525	.015	S.F.	.28	.54		.82	1.11
0200	Maximum		305	.026		.60	.92		1.52	2.04
0300	Epoxy coatings, minimum		525	.015		.37	.54		.91	1.21
0400	Maximum		170	.047		1.12	1.66		2.78	3.71
0600	Exposed aggregate, troweled on, 1/16" to 1/4", minimum		235	.034		.56	1.20		1.76	2.42
0700	Maximum (epoxy or polyacrylate)		130	.062		1.20	2.17		3.37	4.57
0900	1/2" to 5/8" aggregate, minimum		130	.062		1.11	2.17		3.28	4.47
1000	Maximum		80	.100		1.90	3.52		5.42	7.40
1200	1" aggregate size, minimum		90	.089		1.93	3.13		5.06	6.80
1300	Maximum		55	.145		2.95	5.10		8.05	10.90
1500	Exposed aggregate, sprayed on, 1/8" aggregate, minimum		295	.027		.51	.95		1.46	1.99
1600	Maximum		145	.055		.96	1.94		2.90	3.97
1800	High build epoxy, 50 mil, minimum		390	.021		.62	.72		1.34	1.76
1900	Maximum		95	.084		1.06	2.96		4.02	5.60
2100	Laminated epoxy with fiberglass, minimum		295	.027		.67	.95		1.62	2.17
2200	Maximum		145	.055		1.19	1.94		3.13	4.22
2400	Sprayed perlite or vermiculite, 1/16" thick, minimum		2935	.003		.24	.10		.34	.40
2500	Maximum		640	.013		.68	.44		1.12	1.41
2700	Vinyl plastic wall coating, minimum		735	.011		.30	.38		.68	.90
2800	Maximum		240	.033		.75	1.17		1.92	2.59
3000	Urethane on smooth surface, 2 coats, minimum		1135	.007		.24	.25		.49	.63
3100	Maximum		665	.012		.53	.42		.95	1.21
3300	3 coat, minimum		840	.010		.32	.34		.66	.85
3400	Maximum		470	.017		.72	.60		1.32	1.69
3600	Ceramic-like glazed coating, cementitious, minimum		440	.018		.44	.64		1.08	1.44
3700	Maximum		345	.023		.74	.82		1.56	2.03
3900	Resin base, minimum		640	.013		.30	.44		.74	.99
4000	Maximum	↓	330	.024	↓	.49	.85		1.34	1.82

09 97 Special Coatings

09 97 13 – Steel Coatings

09 97 13.23 Exterior Steel Coatings

		Crew	Daily Output	Labor-Hours	Unit	Material	Labor	Equipment	Total	Total Incl O&P
0010	**EXTERIOR STEEL COATINGS**									
6510	Paints & protective coatings, sprayed in field									
6520	Alkyds, primer	2 Psst	3600	.004	S.F.	.06	.16		.22	.35
6540	Gloss topcoats		3200	.005		.05	.18		.23	.38
6560	Silicone alkyd		3200	.005		.12	.18		.30	.46
6610	Epoxy, primer		3000	.005		.24	.19		.43	.61
6630	Intermediate or topcoat		2800	.006		.21	.21		.42	.61
6650	Enamel coat		2800	.006		.27	.21		.48	.66
6700	Epoxy ester, primer		2800	.006		.52	.21		.73	.94
6720	Topcoats		2800	.006		.18	.21		.39	.57
6810	Latex primer		3600	.004		.05	.16		.21	.35
6830	Topcoats		3200	.005		.06	.18		.24	.39
6910	Universal primers, one part, phenolic, modified alkyd		2000	.008		.10	.29		.39	.63
6940	Two part, epoxy spray	↓	2000	.008	↓	.28	.29		.57	.83

Division Notes

		CREW	DAILY OUTPUT	LABOR-HOURS	UNIT	2009 BARE COSTS				TOTAL INCL O&P
						MAT.	LABOR	EQUIP.	TOTAL	

Estimating Tips

General

- The items in this division are usually priced per square foot or each.
- Many items in Division 10 require some type of support system or special anchors that are not usually furnished with the item. The required anchors must be added to the estimate in the appropriate division.
- Some items in Division 10, such as lockers, may require assembly before installation. Verify the amount of assembly required. Assembly can often exceed installation time.

10 20 00 Interior Specialties

- Support angles and blocking are not included in the installation of toilet compartments, shower/dressing compartments, or cubicles. Appropriate line items from Divisions 5 or 6 may need to be added to support the installations.
- Toilet partitions are priced by the stall. A stall consists of a side wall, pilaster, and door with hardware. Toilet tissue holders and grab bars are extra.
- The required acoustical rating of a folding partition can have a significant impact on costs. Verify the sound transmission coefficient rating of the panel priced to the specification requirements.
- Grab bar installation does not include supplemental blocking or backing to support the required load. When grab bars are installed at an existing facility, provisions must be made to attach the grab bars to solid structure.

Reference Numbers

Reference numbers are shown in shaded boxes at the beginning of some major classifications. These numbers refer to related items in the Reference Section. The reference information may be an estimating procedure, an alternate pricing method, or technical information.

Note: Not all subdivisions listed here necessarily appear in this publication.

10 05 Common Work Results for Specialties

10 05 05 – Selective Specialties Demolition

10 05 05.10 Selective Demolition, Specialties

		Crew	Daily Output	Labor-Hours	Unit	Material	2009 Bare Costs Labor	2009 Bare Costs Equipment	Total	Total Incl O&P
0010	**SELECTIVE DEMOLITION, SPECIALTIES**									
1100	Boards and panels, wall mounted	2 Clab	15	1.067	Ea.		33.50		33.50	52.50
1200	Cases, for directory and/or bulletin boards, incl doors		24	.667			21		21	32.50
1850	Shower partitions, cabinet or stall, incl base and door		8	2			63		63	98
1855	Shower receptor, terrazzo or concrete	1 Clab	14	.571			18.05		18.05	28
1900	Curtain track or rod, hospital type, ceiling mounted or suspended	"	220	.036	L.F.		1.15		1.15	1.78
1910	Toilet Cubicles, remove	2 Clab	8	2	Ea.		63		63	98
1920	Toilet Cubicles, remove	"	16	1	"		31.50		31.50	49
2650	Wall guard, misc. wall or corner protection	1 Clab	320	.025	L.F.		.79		.79	1.23
2750	Access floor, metal panel system, incl pedestals, covering	2 Clab	850	.019	S.F.		.59		.59	.92
3050	Fireplace, prefab, freestanding or wall hung, incl hood and screen	1 Clab	2	4	Ea.		126		126	196
3054	Chimney top, simulated brick, 4' high	"	15	.533			16.85		16.85	26
3200	Stove, woodburning, cast iron	2 Clab	2	8			253		253	390
3440	Weathervane, residential	1 Clab	12	.667			21		21	32.50
3500	Flagpole, groundset, to 70' high, excl base/fndn	K-1	1	16			565	243	808	1,150
3555	To 30' high	"	2.50	6.400			227	97	324	455
4300	Letter, signs or plaques, exterior on wall	1 Clab	20	.400			12.65		12.65	19.60
4320	Door signs interior on door 6" x 6", selective demolition	"	20	.400			12.65		12.65	19.60
4550	Turnstiles, manual or electric	2 Clab	2	8			253		253	390
5050	Lockers	"	15	1.067	Opng.		33.50		33.50	52.50
5250	Cabinets, recessed	Q-12	12	1.333	Ea.		58		58	86.50
5260	Mail boxes, Horiz., Key Lock, front loading, Remove	1 Carp	34	.235	"		9.40		9.40	14.60
5350	Awning, fabric, including frame	2 Clab	100	.160	S.F.		5.05		5.05	7.85
6050	Partition, woven wire		1400	.011			.36		.36	.56
6100	Folding gate, security, door or window		500	.032			1.01		1.01	1.57
6580	Acoustic air wall		650	.025			.78		.78	1.21
7550	Telephone enclosure, exterior, post mounted		3	5.333	Ea.		169		169	261
8850	Scale, platform, excludes foundation or pit		.25	64	"		2,025		2,025	3,125

10 11 Visual Display Surfaces

10 11 13 – Chalkboards

10 11 13.13 Fixed Chalkboards

		Crew	Daily Output	Labor-Hours	Unit	Material	2009 Bare Costs Labor	2009 Bare Costs Equipment	Total	Total Incl O&P
0010	**FIXED CHALKBOARDS** Porcelain enamel steel									
3900	Wall hung									
4000	Aluminum frame and chalktrough									
4200	3' x 4'	2 Carp	16	1	Ea.	181	40		221	261
4300	3' x 5'		15	1.067		242	42.50		284.50	330
4500	4' x 8'		14	1.143		345	45.50		390.50	450
4600	4' x 12'		13	1.231		505	49		554	630
4700	Wood frame and chalktrough									
4800	3' x 4'	2 Carp	16	1	Ea.	169	40		209	248
5000	3' x 5'		15	1.067		240	42.50		282.50	330
5100	4' x 5'		14	1.143		248	45.50		293.50	345
5300	4' x 8'		13	1.231		340	49		389	445
5400	Liquid chalk, white porcelain enamel, wall hung									
5420	Deluxe units, aluminum trim and chalktrough									
5450	4' x 4'	2 Carp	16	1	Ea.	202	40		242	284
5500	4' x 8'		14	1.143		310	45.50		355.50	415
5550	4' x 12'		12	1.333		470	53.50		523.50	600
5700	Wood trim and chalktrough									
5900	4' x 4'	2 Carp	16	1	Ea.	460	40		500	570

10 11 Visual Display Surfaces

10 11 13 – Chalkboards

10 11 13.13 Fixed Chalkboards		Crew	Daily Output	Labor-Hours	Unit	Material	2009 Bare Costs Labor	Equipment	Total	Total Incl O&P
6000	4' x 6'	2 Carp	15	1.067	Ea.	550	42.50		592.50	670
6200	4' x 8'	↓	14	1.143		635	45.50		680.50	770
6300	Liquid chalk, felt tip markers					1.62			1.62	1.78
6500	Erasers					1.24			1.24	1.36
6600	Board cleaner, 8 oz. bottle				↓	4.28			4.28	4.71
10 11 13.23 Modular-Support-Mounted Chalkboards										
0010	**MODULAR-SUPPORT-MOUNTED CHALKBOARDS**									
0400	Sliding chalkboards									
0450	Vertical, one sliding board with back panel, wall mounted									
0500	8' x 4'	2 Carp	8	2	Ea.	1,825	80		1,905	2,150
0520	8' x 8'	↓	7.50	2.133		2,575	85		2,660	2,950
0540	8' x 12'	↓	7	2.286	↓	3,125	91.50		3,216.50	3,600
0600	Two sliding boards, with back panel									
0620	8' x 4'	2 Carp	8	2	Ea.	2,750	80		2,830	3,150
0640	8' x 8'		7.50	2.133		3,925	85		4,010	4,450
0660	8' x 12'	↓	7	2.286	↓	4,775	91.50		4,866.50	5,400
0700	Horizontal, two track									
0800	4' x 8', 2 sliding panels	2 Carp	8	2	Ea.	1,700	80		1,780	2,000
0820	4' x 12', 2 sliding panels		7.50	2.133		2,175	85		2,260	2,525
0840	4' x 16', 4 sliding panels	↓	7	2.286	↓	2,925	91.50		3,016.50	3,375
0900	Four track, four sliding panels									
0920	4' x 8'	2 Carp	8	2	Ea.	2,725	80		2,805	3,125
0940	4' x 12'		7.50	2.133		3,500	85		3,585	3,975
0960	4' x 16'	↓	7	2.286	↓	4,425	91.50		4,516.50	5,025
1200	Vertical, motor operated									
1400	One sliding panel with back panel									
1450	10' x 4'	2 Carp	4	4	Ea.	6,125	160		6,285	7,000
1500	10' x 10'		3.75	4.267		7,175	170		7,345	8,150
1550	10' x 16'	↓	3.50	4.571	↓	8,225	183		8,408	9,325
1700	Two sliding panels with back panel									
1750	10' x 4'	2 Carp	4	4	Ea.	9,825	160		9,985	11,000
1800	10' x 10'		3.75	4.267		11,200	170		11,370	12,600
1850	10' x 16'	↓	3.50	4.571		12,700	183		12,883	14,200
2000	Three sliding panels with back panel									
2100	10' x 4'	2 Carp	4	4	Ea.	12,800	160		12,960	14,300
2150	10' x 10'		3.75	4.267		14,900	170		15,070	16,700
2200	10' x 16'	↓	3.50	4.571	↓	17,000	183		17,183	19,000
2400	For projection screen, glass beaded, add				S.F.	10.20			10.20	11.20
2500	For remote control, 1 panel control, add				Ea.	480			480	525
2600	2 panel control, add				"	825			825	905
2800	For units without back panels, deduct				S.F.	6.45			6.45	7.10
2850	For liquid chalk porcelain panels, add				"	6.90			6.90	7.60
3000	Swing leaf, any comb. of chalkboard & cork, aluminum frame									
3100	Floor style, 6 panels									
3150	30" x 40" panels				Ea.	1,475			1,475	1,625
3200	48" x 40" panels				"	2,100			2,100	2,300
3300	Wall mounted, 6 panels									
3400	30" x 40" panels	2 Carp	16	1	Ea.	1,350	40		1,390	1,550
3450	48" x 40" panels	"	16	1	"	1,750	40		1,790	1,975
3600	Extra panels for swing leaf units									
3700	30" x 40" panels				Ea.	282			282	310
3750	48" x 40" panels				"	335			335	370

10 11 Visual Display Surfaces

10 11 13 – Chalkboards

10 11 13.43 Portable Chalkboards	Crew	Daily Output	Labor-Hours	Unit	Material	2009 Bare Costs Labor	Equipment	Total	Total Incl O&P
0010 **PORTABLE CHALKBOARDS**									
0100 Freestanding, reversible									
0120 Economy, wood frame, 4' x 6'									
0140 Chalkboard both sides				Ea.	565			565	620
0160 Chalkboard one side, cork other side				"	485			485	535
0200 Standard, lightweight satin finished aluminum, 4' x 6'									
0220 Chalkboard both sides				Ea.	650			650	715
0240 Chalkboard one side, cork other side				"	580			580	640
0300 Deluxe, heavy duty extruded aluminum, 4' x 6'									
0320 Chalkboard both sides				Ea.	1,050			1,050	1,150
0340 Chalkboard one side, cork other side				"	940			940	1,025

10 11 23 – Tackboards

10 11 23.10 Fixed Tackboards

	Crew	Daily Output	Labor-Hours	Unit	Material	Labor	Equipment	Total	Total Incl O&P
0010 **FIXED TACKBOARDS**									
0020 Cork sheets, unbacked, no frame, 1/4" thick	2 Carp	290	.055	S.F.	3.60	2.20		5.80	7.40
0100 1/2" thick		290	.055		6.30	2.20		8.50	10.35
0300 Fabric-face, no frame, on 7/32" cork underlay		290	.055		5.35	2.20		7.55	9.30
0400 On 1/4" cork on 1/4" hardboard		290	.055		7.50	2.20		9.70	11.65
0600 With edges wrapped		290	.055		7.90	2.20		10.10	12.10
0700 On 7/16" fire retardant core		290	.055		7.35	2.20		9.55	11.50
0900 With edges wrapped		290	.055		11.90	2.20		14.10	16.50
1000 Designer fabric only, cut to size					2.02			2.02	2.22
1200 1/4" vinyl cork, on 1/4" hardboard, no frame	2 Carp	290	.055		8	2.20		10.20	12.20
1300 On 1/4" coreboard		290	.055		7.95	2.20		10.15	12.15
2000 For map and display rail, economy, add		385	.042	L.F.	2.27	1.66		3.93	5.05
2100 Deluxe, add		350	.046	"	3.23	1.83		5.06	6.40
2120 Prefabricated, 1/4" cork, 3' x 5' with aluminum frame		16	1	Ea.	109	40		149	182
2140 Wood frame		16	1		139	40		179	215
2160 4' x 4' with aluminum frame		16	1		92.50	40		132.50	163
2180 Wood frame		16	1		120	40		160	194
2200 4' x 8' with aluminum frame		14	1.143		162	45.50		207.50	249
2210 With wood frame		14	1.143		144	45.50		189.50	229
2220 4' x 12' with aluminum frame		12	1.333		240	53.50		293.50	345
2230 Bulletin board case, single glass door, with lock									
2240 36" x 24", economy	2 Carp	12	1.333	Ea.	258	53.50		311.50	365
2250 Deluxe		12	1.333		460	53.50		513.50	595
2260 42" x 30", economy		12	1.333		290	53.50		343.50	405
2270 Deluxe		12	1.333		430	53.50		483.50	560
2300 Glass enclosed cabinets, alum., cork panel, hinged doors									
2400 3' x 3', 1 door	2 Carp	12	1.333	Ea.	535	53.50		588.50	675
2500 4' x 4', 2 door		11	1.455		895	58		953	1,075
2600 4' x 7', 3 door		10	1.600		1,500	64		1,564	1,750
2800 4' x 10', 4 door		8	2		1,950	80		2,030	2,275
2900 For lights, add per door opening	1 Elec	13	.615		175	29		204	236
3100 Horizontal sliding units, 4 doors, 4' x 8', 8' x 4'	2 Carp	9	1.778		1,700	71		1,771	1,975
3200 4' x 12'		7	2.286		2,175	91.50		2,266.50	2,550
3400 8 doors, 4' x 16'		5	3.200		2,900	128		3,028	3,400
3500 4' x 24'		4	4		3,900	160		4,060	4,550

10 11 23.20 Control Boards

	Crew	Daily Output	Labor-Hours	Unit	Material	Labor	Equipment	Total	Total Incl O&P
0010 **CONTROL BOARDS**									
0020 Magnetic, porcelain finish, 18" x 24", framed	2 Carp	8	2	Ea.	199	80		279	345

10 11 Visual Display Surfaces

10 11 23 – Tackboards

10 11 23.20 Control Boards		Crew	Daily Output	Labor-Hours	Unit	Material	2009 Bare Costs Labor	Equipment	Total	Total Incl O&P
0100	24" x 36"	2 Carp	7.50	2.133	Ea.	255	85		340	415
0200	36" x 48"		7	2.286		375	91.50		466.50	555
0300	48" x 72"		6	2.667		720	107		827	955
0400	48" x 96"		5	3.200		845	128		973	1,125

10 13 Directories

10 13 10 – Building Directories

10 13 10.10 Directory Boards

		Crew	Daily Output	Labor-Hours	Unit	Material	Labor	Equipment	Total	Total Incl O&P
0010	**DIRECTORY BOARDS**									
0050	Plastic, glass covered, 30" x 20"	2 Carp	3	5.333	Ea.	242	213		455	595
0100	36" x 48"		2	8		875	320		1,195	1,450
0300	Grooved cork, 30" x 20"		3	5.333		385	213		598	755
0400	36" x 48"		2	8		520	320		840	1,075
0600	Black felt, 30" x 20"		3	5.333		184	213		397	530
0700	36" x 48"		2	8		365	320		685	895
1800	Indoor, economy, open face, 18" x 24"		7	2.286		132	91.50		223.50	288
1900	24" x 36"		7	2.286		144	91.50		235.50	300
2000	36" x 24"		6	2.667		164	107		271	345
2100	36" x 48"		6	2.667		246	107		353	435
2400	Building directory, alum., black felt panels, 1 door, 24" x 18"		4	4		315	160		475	600
2500	36" x 24"		3.50	4.571		355	183		538	675
2600	48" x 32"		3	5.333		590	213		803	980
2700	2 door, 36" x 48"		2.50	6.400		645	256		901	1,100
2800	36" x 60"		2	8		865	320		1,185	1,450
2900	48" x 60"		1	16		940	640		1,580	2,025
3100	For bronze enamel finish, add					15%				
3200	For bronze anodized finish, add					25%				
3400	For illuminated directory, single door unit, add					175			175	193
3500	For 6" header panel, 6 letters per foot, add				L.F.	29			29	31.50

10 14 Signage

10 14 19 – Dimensional Letter Signage

10 14 19.10 Exterior Signs

		Crew	Daily Output	Labor-Hours	Unit	Material	Labor	Equipment	Total	Total Incl O&P
0010	**EXTERIOR SIGNS**									
0020	Letters, 2" high, 3/8" deep, cast bronze	1 Carp	24	.333	Ea.	22	13.30		35.30	45
0140	1/2" deep, cast aluminum		18	.444		22	17.75		39.75	52
0160	Cast bronze		32	.250		29	10		39	47.50
0300	6" high, 5/8" deep, cast aluminum		24	.333		27.50	13.30		40.80	51
0400	Cast bronze		24	.333		56.50	13.30		69.80	82.50
0600	8" high, 3/4" deep, cast aluminum		14	.571		33.50	23		56.50	72.50
0700	Cast bronze		20	.400		80.50	16		96.50	114
0900	10" high, 1" deep, cast aluminum		18	.444		49.50	17.75		67.25	82
1000	Bronze		18	.444		106	17.75		123.75	145
1200	12" high, 1-1/4" deep, cast aluminum		12	.667		52.50	26.50		79	99
1500	Cast bronze		18	.444		123	17.75		140.75	164
1600	14" high, 2-5/16" deep, cast aluminum		12	.667		75	26.50		101.50	124
1800	Fabricated stainless steel, 6" high, 2" deep		20	.400		53	16		69	83
1900	12" high, 3" deep		18	.444		118	17.75		135.75	157
2100	18" high, 3" deep		12	.667		228	26.50		254.50	292

10 14 Signage

10 14 19 – Dimensional Letter Signage

10 14 19.10 Exterior Signs

		Crew	Daily Output	Labor-Hours	Unit	Material	2009 Bare Costs Labor	Equipment	Total	Total Incl O&P
2200	24" high, 4" deep	1 Carp	10	.800	Ea.	385	32		417	470
2700	Acrylic, on high density foam, 12" high, 2" deep		20	.400		20.50	16		36.50	47.50
2800	18" high, 2" deep		18	.444		51	17.75		68.75	83.50
3900	Plaques, custom, 20" x 30", for up to 450 letters, cast aluminum	2 Carp	4	4		1,050	160		1,210	1,400
4000	Cast bronze		4	4		1,475	160		1,635	1,875
4200	30" x 36", up to 900 letters cast aluminum		3	5.333		1,875	213		2,088	2,375
4300	Cast bronze		3	5.333		3,475	213		3,688	4,150
4500	36" x 48", for up to 1300 letters, cast bronze		2	8		4,050	320		4,370	4,950

10 14 23 – Panel Signage

10 14 23.13 Engraved Interior Panel Signage

		Crew	Daily Output	Labor-Hours	Unit	Material	Labor	Equipment	Total	Total Incl O&P
0010	**ENGRAVED INTERIOR PANEL SIGNAGE**									
1010	Flexible door sign, adhesive back, w/Braille, 5/8" letters, 4" x 4"	1 Clab	32	.250	Ea.	26.50	7.90		34.40	42
1050	6" x 6"		32	.250		39.50	7.90		47.40	56
1100	8" x 2"		32	.250		27.50	7.90		35.40	42.50
1150	8" x 4"		32	.250		35	7.90		42.90	51
1200	8" x 8"		32	.250		48	7.90		55.90	65.50
1250	12" x 2"		32	.250		35	7.90		42.90	51
1300	12" x 6"		32	.250		50	7.90		57.90	67.50
1350	12" x 12"		32	.250		99	7.90		106.90	121
1500	Graphic symbols, 2" x 2"		32	.250		12	7.90		19.90	25.50
1550	6" x 6"		32	.250		30	7.90		37.90	45.50
1600	8" x 8"		32	.250		30	7.90		37.90	45.50

10 17 Telephone Specialties

10 17 16 – Telephone Enclosures

10 17 16.10 Commercial Telephone Enclosures

		Crew	Daily Output	Labor-Hours	Unit	Material	Labor	Equipment	Total	Total Incl O&P
0010	**COMMERCIAL TELEPHONE ENCLOSURES**									
0300	Shelf type, wall hung, minimum	2 Carp	5	3.200	Ea.	1,300	128		1,428	1,625
0400	Maximum		5	3.200		2,725	128		2,853	3,175
0600	Booth type, painted steel, indoor or outdoor, minimum		1.50	10.667		3,475	425		3,900	4,475
0700	Maximum (stainless steel)		1.50	10.667		11,600	425		12,025	13,400
2200	Directory shelf, wall mounted, stainless steel									
2300	3 binders	2 Carp	8	2	Ea.	1,100	80		1,180	1,350
2500	4 binders		7	2.286		1,275	91.50		1,366.50	1,550
2600	5 binders		6	2.667		1,625	107		1,732	1,975
2800	Table type, stainless steel, 4 binders		8	2		1,300	80		1,380	1,575
2900	7 binders		7	2.286		1,625	91.50		1,716.50	1,950

10 21 Compartments and Cubicles

10 21 13 – Toilet Compartments

10 21 13.13 Metal Toilet Compartments

		Crew	Daily Output	Labor-Hours	Unit	Material	Labor	Equipment	Total	Total Incl O&P
0010	**METAL TOILET COMPARTMENTS**									
0110	Cubicles, ceiling hung									
0200	Powder coated steel	2 Carp	4	4	Ea.	465	160		625	765
0500	Stainless steel	"	4	4		1,775	160		1,935	2,200
0600	For handicap units, incl. 52" grab bars, add					390			390	430
0900	Floor and ceiling anchored									
1000	Powder coated steel	2 Carp	5	3.200	Ea.	485	128		613	735
1300	Stainless steel	"	5	3.200		1,700	128		1,828	2,050

10 21 Compartments and Cubicles

10 21 13 – Toilet Compartments

10 21 13.13 Metal Toilet Compartments

		Crew	Daily Output	Labor-Hours	Unit	Material	2009 Bare Costs Labor	Equipment	Total	Total Incl O&P
1400	For handicap units, incl. 52" grab bars, add				Ea.	276			276	305
1610	Floor anchored									
1700	Powder coated steel	2 Carp	7	2.286	Ea.	545	91.50		636.50	735
2000	Stainless steel	"	7	2.286		1,525	91.50		1,616.50	1,825
2100	For handicap units, incl. 52" grab bars, add					273			273	300
2200	For juvenile units, deduct					39.50			39.50	43.50
2450	Floor anchored, headrail braced									
2500	Powder coated steel	2 Carp	6	2.667	Ea.	445	107		552	650
2804	Stainless steel	"	4.60	3.478		1,125	139		1,264	1,475
2900	For handicap units, incl. 52" grab bars, add					330			330	365
3000	Wall hung partitions, powder coated steel	2 Carp	7	2.286		595	91.50		686.50	795
3300	Stainless steel	"	7	2.286		1,575	91.50		1,666.50	1,875
3400	For handicap units, incl. 52" grab bars, add					330			330	365
4000	Screens, entrance, floor mounted, 58" high, 48" wide									
4200	Powder coated steel	2 Carp	15	1.067	Ea.	233	42.50		275.50	320
4500	Stainless steel	"	15	1.067	"	850	42.50		892.50	1,000
4650	Urinal screen, 18" wide									
4704	Powder coated steel	2 Carp	6.15	2.602	Ea.	184	104		288	365
5004	Stainless steel	"	6.15	2.602	"	580	104		684	800
5100	Floor mounted, head rail braced									
5300	Powder coated steel	2 Carp	8	2	Ea.	199	80		279	345
5600	Stainless steel	"	8	2	"	555	80		635	740
5750	Pilaster, flush									
5800	Powder coated steel	2 Carp	10	1.600	Ea.	259	64		323	385
6100	Stainless steel		10	1.600		655	64		719	820
6300	Post braced, powder coated steel		10	1.600		182	64		246	299
6600	Stainless steel		10	1.600		485	64		549	635
6700	Wall hung, bracket supported									
6800	Powder coated steel	2 Carp	10	1.600	Ea.	289	64		353	420
7100	Stainless steel		10	1.600		315	64		379	450
7400	Flange supported, powder coated steel		10	1.600		95	64		159	203
7700	Stainless steel		10	1.600		276	64		340	405
7800	Wedge type, powder coated steel		10	1.600		128	64		192	240
8100	Stainless steel		10	1.600		540	64		604	695

10 21 13.16 Plastic-Laminate-Clad Toilet Compartments

		Crew	Daily Output	Labor-Hours	Unit	Material	Labor	Equipment	Total	Total Incl O&P
0010	**PLASTIC-LAMINATE-CLAD TOILET COMPARTMENTS**									
0110	Cubicles, ceiling hung									
0300	Plastic laminate on particle board	2 Carp	4	4	Ea.	605	160		765	915
0600	For handicap units, incl. 52" grab bars, add				"	390			390	430
0900	Floor and ceiling anchored									
1100	Plastic laminate on particle board	2 Carp	5	3.200	Ea.	615	128		743	875
1400	For handicap units, incl. 52" grab bars, add				"	276			276	305
1610	Floor mounted									
1800	Plastic laminate on particle board	2 Carp	7	2.286	Ea.	545	91.50		636.50	740
2450	Floor mounted, headrail braced									
2600	Plastic laminate on particle board	2 Carp	6	2.667	Ea.	575	107		682	800
3400	For handicap units, incl. 52" grab bars, add					330			330	365
4300	Entrance screen, floor mtd., plas. lam., 58" high, 48" wide	2 Carp	15	1.067		480	42.50		522.50	595
4800	Urinal screen, 18" wide, ceiling braced, plastic laminate		8	2		180	80		260	320
5400	Floor mounted, headrail braced		8	2		365	80		445	525
5900	Pilaster, flush, plastic laminate		10	1.600		400	64		464	540
6400	Post braced, plastic laminate		10	1.600		288	64		352	415

10 21 Compartments and Cubicles

10 21 13 – Toilet Compartments

10 21 13.16 Plastic-Laminate-Clad Toilet Compartments

		Crew	Daily Output	Labor-Hours	Unit	Material	2009 Bare Costs Labor	Equipment	Total	Total Incl O&P
6700	Wall hung, bracket supported									
6900	Plastic laminate on particle board	2 Carp	10	1.600	Ea.	98	64		162	207
7450	Flange supported									
7500	Plastic laminate on particle board	2 Carp	10	1.600	Ea.	203	64		267	320

10 21 13.19 Plastic Toilet Compartments

		Crew	Daily Output	Labor-Hours	Unit	Material	Labor	Equipment	Total	Total Incl O&P
0010	**PLASTIC TOILET COMPARTMENTS**									
0110	Cubicles, ceiling hung									
0250	Phenolic	2 Carp	4	4	Ea.	965	160		1,125	1,325
0600	For handicap units, incl. 52" grab bars, add				"	390			390	430
0900	Floor and ceiling anchored									
1050	Phenolic	2 Carp	5	3.200	Ea.	1,025	128		1,153	1,325
1400	For handicap units, incl. 52" grab bars, add				"	276			276	305
1610	Floor mounted									
1750	Phenolic	2 Carp	7	2.286	Ea.	1,275	91.50		1,366.50	1,550
2100	For handicap units, incl. 52" grab bars, add					273			273	300
2200	For juvenile units, deduct					39.50			39.50	43.50
2450	Floor mounted, headrail braced									
2550	Phenolic	2 Carp	6	2.667	Ea.	815	107		922	1,050

10 21 13.40 Stone Toilet Compartments

		Crew	Daily Output	Labor-Hours	Unit	Material	Labor	Equipment	Total	Total Incl O&P
0010	**STONE TOILET COMPARTMENTS**									
0100	Cubicles, ceiling hung, marble	2 Marb	2	8	Ea.	1,675	310		1,985	2,300
0600	For handicap units, incl. 52" grab bars, add					390			390	430
0800	Floor & ceiling anchored, marble	2 Marb	2.50	6.400		1,825	247		2,072	2,375
1400	For handicap units, incl. 52" grab bars, add					276			276	305
1600	Floor mounted, marble	2 Marb	3	5.333		1,075	206		1,281	1,500
2400	Floor mounted, headrail braced, marble	"	3	5.333		1,025	206		1,231	1,450
2900	For handicap units, incl. 52" grab bars, add					330			330	365
4100	Entrance screen, floor mounted marble, 58" high, 48" wide	2 Marb	9	1.778		675	68.50		743.50	850
4600	Urinal screen, 18" wide, ceiling braced, marble	D-1	6	2.667		675	97		772	890
5100	Floor mounted, head rail braced									
5200	Marble	D-1	6	2.667	Ea.	575	97		672	775
5700	Pilaster, flush, marble		9	1.778		750	64.50		814.50	925
6200	Post braced, marble		9	1.778		745	64.50		809.50	920

10 21 16 – Shower and Dressing Compartments

10 21 16.10 Partitions, Shower

		Crew	Daily Output	Labor-Hours	Unit	Material	Labor	Equipment	Total	Total Incl O&P
0010	**PARTITIONS, SHOWER** Floor mounted, no plumbing									
0100	Cabinet, incl. base, no door, painted steel, 1" thick walls	2 Shee	5	3.200	Ea.	905	151		1,056	1,225
0300	With door, fiberglass		4.50	3.556		705	168		873	1,025
0600	Galvanized and painted steel, 1" thick walls		5	3.200		955	151		1,106	1,275
0800	Stall, 1" thick wall, no base, enameled steel		5	3.200		980	151		1,131	1,300
1200	Stainless steel		5	3.200		1,700	151		1,851	2,100
1400	For double entry type, no doors, deduct					10%				
1500	Circular fiberglass, cabinet 36" diameter,	2 Shee	4	4		725	189		914	1,100
1700	One piece, 36" diameter, less door		4	4		610	189		799	960
1800	With door		3.50	4.571		1,000	216		1,216	1,425
2000	Curved shell shower, no door needed		3	5.333		875	252		1,127	1,350
2300	For fiberglass seat, add to both above					124			124	136
2400	Glass stalls, with doors, no receptors, chrome on brass	2 Shee	3	5.333		1,475	252		1,727	2,000
2700	Anodized aluminum	"	4	4		1,025	189		1,214	1,425
2900	Marble shower stall, stock design, with shower door	2 Marb	1.20	13.333		2,200	515		2,715	3,200
3000	With curtain		1.30	12.308		1,950	475		2,425	2,875
3200	Receptors, precast terrazzo, 32" x 32"		14	1.143		242	44		286	335

10 21 Compartments and Cubicles

10 21 16 – Shower and Dressing Compartments

10 21 16.10 Partitions, Shower		Crew	Daily Output	Labor-Hours	Unit	Material	2009 Bare Costs Labor	Equipment	Total	Total Incl O&P
3300	48" x 34"	2 Marb	9.50	1.684	Ea.	330	65		395	460
3500	Plastic, simulated terrazzo receptor, 32" x 32"		14	1.143		100	44		144	177
3600	32" x 48"		12	1.333		190	51.50		241.50	287
3800	Precast concrete, colors, 32" x 32"		14	1.143		187	44		231	273
3900	48" x 48"		8	2		390	77.50		467.50	545
4100	Shower doors, economy plastic, 24" wide	1 Shee	9	.889		128	42		170	205
4200	Tempered glass door, economy		8	1		169	47		216	259
4400	Folding, tempered glass, aluminum frame		6	1.333		385	63		448	520
4500	Sliding, tempered glass, 48" opening		6	1.333		255	63		318	380
4700	Deluxe, tempered glass, chrome on brass frame, minimum		8	1		223	47		270	320
4800	Maximum		1	8		880	380		1,260	1,550
4850	On anodized aluminum frame, minimum		2	4		144	189		333	450
4900	Maximum		1	8		520	380		900	1,150
5100	Shower enclosure, tempered glass, anodized alum. frame									
5120	2 panel & door, corner unit, 32" x 32"	1 Shee	2	4	Ea.	505	189		694	850
5140	Neo-angle corner unit, 16" x 24" x 16"	"	2	4		930	189		1,119	1,325
5200	Shower surround, 3 wall, polypropylene, 32" x 32"	1 Carp	4	2		275	80		355	430
5220	PVC, 32" x 32"		4	2		340	80		420	500
5240	Fiberglass		4	2		370	80		450	535
5250	2 wall, polypropylene, 32" x 32"		4	2		282	80		362	435
5270	PVC		4	2		350	80		430	510
5290	Fiberglass		4	2		355	80		435	515
5300	Tub doors, tempered glass & frame, minimum	1 Shee	8	1		204	47		251	297
5400	Maximum		6	1.333		475	63		538	620
5600	Chrome plated, brass frame, minimum		8	1		270	47		317	370
5700	Maximum		6	1.333		660	63		723	820
5900	Tub/shower enclosure, temp. glass, alum. frame, minimum		2	4		365	189		554	690
6200	Maximum		1.50	5.333		760	252		1,012	1,225
6500	On chrome-plated brass frame, minimum		2	4		505	189		694	845
6600	Maximum		1.50	5.333		1,075	252		1,327	1,550
6800	Tub surround, 3 wall, polypropylene	1 Carp	4	2		228	80		308	375
6900	PVC		4	2		345	80		425	505
7000	Fiberglass, minimum		4	2		360	80		440	520
7100	Maximum		3	2.667		605	107		712	830

10 21 23 – Cubicles

10 21 23.16 Cubicle Track and Hardware

0010	**CUBICLE TRACK AND HARDWARE**									
0020	Curtain track, box channel, ceiling mounted	1 Carp	135	.059	L.F.	5.70	2.37		8.07	9.95
0100	Suspended	"	100	.080	"	7	3.20		10.20	12.65
0300	Curtains, nylon mesh tops, fire resistant, 11 oz. per lineal yard									
0310	Polyester oxford cloth, 9' ceiling height	1 Carp	425	.019	L.F.	11.65	.75		12.40	13.95
0500	8' ceiling height		425	.019		9.85	.75		10.60	12
0700	Designer oxford cloth		425	.019		25	.75		25.75	28.50
0800	I.V. track systems									
0820	I.V. track, oval	1 Carp	135	.059	L.F.	5.10	2.37		7.47	9.25
0830	I.V. trolley		32	.250	Ea.	35	10		45	54
0840	I.V. pendent, (tree, 5 hook)		32	.250	"	125	10		135	154

10 22 Partitions

10 22 13 – Wire Mesh Partitions

10 22 13.10 Partitions, Woven Wire

		Crew	Daily Output	Labor-Hours	Unit	Material	2009 Bare Costs Labor	Equipment	Total	Total Incl O&P
0010	**PARTITIONS, WOVEN WIRE** For tool or stockroom enclosures									
0100	Channel frame, 1-1/2" diamond mesh, 10 ga. wire, painted									
0300	Wall panels, 4'-0" wide, 7' high	2 Carp	25	.640	Ea.	132	25.50		157.50	185
0400	8' high		23	.696		135	28		163	192
0600	10' high	↓	18	.889		169	35.50		204.50	241
0700	For 5' wide panels, add					5%				
0900	Ceiling panels, 10' long, 2' wide	2 Carp	25	.640		118	25.50		143.50	170
1000	4' wide		15	1.067		171	42.50		213.50	254
1200	Panel with service window & shelf, 5' wide, 7' high		20	.800		315	32		347	400
1300	8' high		15	1.067		325	42.50		367.50	425
1500	Sliding doors, full height, 3' wide, 7' high		6	2.667		410	107		517	615
1600	10' high		5	3.200		420	128		548	665
1800	6' wide sliding door, 7' full height		5	3.200		615	128		743	880
1900	10' high		4	4		750	160		910	1,075
2100	Swinging doors, 3' wide, 7' high, no transom		6	2.667		291	107		398	485
2200	7' high, 3' transom	↓	5	3.200	↓	350	128		478	585

10 22 16 – Folding Gates

10 22 16.10 Security Gates

		Crew	Daily Output	Labor-Hours	Unit	Material	Labor	Equipment	Total	Total Incl O&P
0010	**SECURITY GATES** For roll up type, see Div. 08 33 13.10									
0300	Scissors type folding gate, ptd. steel, single, 6-1/2' high, 5-1/2' wide	2 Sswk	4	4	Opng.	180	179		359	520
0350	6-1/2' wide		4	4		187	179		366	525
0400	7-1/2' wide		4	4		185	179		364	525
0600	Double gate, 8' high, 8' wide		2.50	6.400		300	286		586	840
0650	10' wide		2.50	6.400		310	286		596	850
0700	12' wide		2	8		455	360		815	1,125
0750	14' wide		2	8		540	360		900	1,225
0900	Door gate, folding steel, 4' wide, 61" high		4	4		94.50	179		273.50	425
1000	71" high		4	4		98	179		277	430
1200	81" high		4	4		106	179		285	435
1300	Window gates, 2' to 4' wide, 31" high		4	4		55	179		234	380
1500	55" high		3.75	4.267		86	191		277	435
1600	79" high	↓	3.50	4.571	↓	99.50	204		303.50	475

10 22 19 – Demountable Partitions

10 22 19.43 Demountable Composite Partitions

		Crew	Daily Output	Labor-Hours	Unit	Material	Labor	Equipment	Total	Total Incl O&P
0010	**DEMOUNTABLE COMPOSITE PARTITIONS**, add for doors									
0100	Do not deduct door openings from total L.F.									
0900	Demountable gypsum system on 2" to 2-1/2"									
1000	steel studs, 9' high, 3" to 3-3/4" thick									
1200	Vinyl clad gypsum	2 Carp	48	.333	L.F.	56.50	13.30		69.80	83
1300	Fabric clad gypsum		44	.364		141	14.55		155.55	178
1500	Steel clad gypsum	↓	40	.400	↓	153	16		169	193
1600	1.75 system, aluminum framing, vinyl clad hardboard,									
1800	paper honeycomb core panel, 1-3/4" to 2-1/2" thick									
1900	9' high	2 Carp	48	.333	L.F.	95	13.30		108.30	126
2100	7' high		60	.267		85.50	10.65		96.15	111
2200	5' high	↓	80	.200	↓	72	8		80	92
2250	Unitized gypsum system									
2300	Unitized panel, 9' high, 2" to 2-1/2" thick									
2350	Vinyl clad gypsum	2 Carp	48	.333	L.F.	123	13.30		136.30	156
2400	Fabric clad gypsum	"	44	.364	"	202	14.55		216.55	245
2500	Unitized mineral fiber system									
2510	Unitized panel, 9' high, 2-1/4" thick, aluminum frame									

10 22 Partitions

10 22 19 – Demountable Partitions

10 22 19.43 Demountable Composite Partitions		Crew	Daily Output	Labor-Hours	Unit	Material	2009 Bare Costs Labor	Equipment	Total	Total Incl O&P
2550	Vinyl clad mineral fiber	2 Carp	48	.333	L.F.	122	13.30		135.30	155
2600	Fabric clad mineral fiber	"	44	.364	"	181	14.55		195.55	223
2800	Movable steel walls, modular system									
2900	Unitized panels, 9' high, 48" wide									
3100	Baked enamel, pre-finished	2 Carp	60	.267	L.F.	138	10.65		148.65	169
3200	Fabric clad steel		56	.286	"	199	11.40		210.40	237
5310	Trackless wall, cork finish, semi-acoustic, 1-5/8" thick, minimum		325	.049	S.F.	38.50	1.97		40.47	45.50
5320	Maximum		190	.084		36.50	3.36		39.86	45
5330	Acoustic, 2" thick, minimum		305	.052		31	2.10		33.10	37.50
5340	Maximum		225	.071		53.50	2.84		56.34	63.50
5500	For acoustical partitions, add, minimum					2.22			2.22	2.44
5550	Maximum					10.35			10.35	11.40
5700	For doors, see Div. 08 11 & 08 16									
5800	For door hardware, see Div. 08 71									
6100	In-plant modular office system, w/prehung hollow core door									
6200	3" thick polystyrene core panels									
6250	12' x 12', 2 wall	2 Clab	3.80	4.211	Ea.	3,125	133		3,258	3,650
6300	4 wall		1.90	8.421		4,400	266		4,666	5,250
6350	16' x 16', 2 wall		3.60	4.444		4,600	140		4,740	5,275
6400	4 wall		1.80	8.889		6,200	281		6,481	7,225

10 22 23 – Portable Partitions, Screens, and Panels

10 22 23.13 Wall Screens

		Crew	Daily Output	Labor-Hours	Unit	Material	2009 Bare Costs Labor	Equipment	Total	Total Incl O&P
0010	**WALL SCREENS**, divider panels, free standing, fiber core									
0020	Fabric face straight									
0100	3'-0" long, 4'-0" high	2 Carp	100	.160	L.F.	106	6.40		112.40	126
0200	5'-0" high		90	.178		111	7.10		118.10	133
0500	6'-0" high		75	.213		114	8.50		122.50	138
0900	5'-0" long, 4'-0" high		175	.091		76.50	3.65		80.15	90
1000	5'-0" high		150	.107		88	4.26		92.26	103
1500	6"-0" high		125	.128		91	5.10		96.10	108
1600	6'-0" long, 5'-0" high		162	.099		76	3.95		79.95	89.50
3200	Economical panels, fabric face, 4'-0" long, 5'-0" high		132	.121		42.50	4.84		47.34	54
3250	6'-0" high		112	.143		45	5.70		50.70	58
3300	5'-0" long, 5'-0" high		150	.107		36	4.26		40.26	46
3350	6'-0" high		125	.128		38	5.10		43.10	49.50
3450	Acoustical panels, 60 to 90 NRC, 3'-0" long, 5'-0" high		90	.178		79.50	7.10		86.60	98.50
3550	6'-0" high		75	.213		89.50	8.50		98	112
3600	5'-0" long, 5'-0" high		150	.107		63	4.26		67.26	76
3650	6'-0" high		125	.128		61	5.10		66.10	75
3700	6'-0" long, 5'-0" high		162	.099		56	3.95		59.95	67.50
3750	6'-0" high		138	.116		56	4.63		60.63	68.50
3800	Economy acoustical panels, 40 N.R.C., 4'-0" long, 5'-0" high		132	.121		42.50	4.84		47.34	54
3850	6'-0" high		112	.143		45	5.70		50.70	58
3900	5'-0" long, 6'-0" high		125	.128		38	5.10		43.10	49.50
3950	6'-0" long, 5'-0" high		162	.099		30	3.95		33.95	39
4000	Metal chalkboard, 6'-6" high, chalkboard, 1 side		125	.128		101	5.10		106.10	119
4100	Metal chalkboard, 2 sides		120	.133		115	5.35		120.35	135
4300	Tackboard, both sides		123	.130		91	5.20		96.20	108

10 22 23.23 Movable Panel Systems

0010	**MOVABLE PANEL SYSTEMS**									
0030	Fabric panel, class A fire rated									
0040	Minimum N.R.C. 0.95, S.T.C. 28, fabric edged									

10 22 Partitions

10 22 23 – Portable Partitions, Screens, and Panels

10 22 23.23 Movable Panel Systems		Crew	Daily Output	Labor-Hours	Unit	Material	2009 Bare Costs Labor	Equipment	Total	Total Incl O&P
0200	42" high, 24" wide	2 Clab	30	.533	Ea.	135	16.85		151.85	174
0220	36" wide		28	.571		186	18.05		204.05	233
0240	48" wide		26	.615		192	19.45		211.45	241
0260	60" wide		25	.640		299	20		319	360
0400	54" high, 24" wide		30	.533		141	16.85		157.85	181
0420	36" wide		28	.571		169	18.05		187.05	214
0440	48" wide		26	.615		195	19.45		214.45	245
0460	60" wide		25	.640		305	20		325	365
0600	60" high, 24" wide		29	.552		186	17.45		203.45	232
0620	36" wide		27	.593		290	18.75		308.75	350
0640	48" wide		25	.640		299	20		319	360
0660	60" wide		24	.667		310	21		331	375
0800	72" high, 24" wide		29	.552		285	17.45		302.45	340
1000	36" wide		27	.593		296	18.75		314.75	355
1200	48" wide		25	.640		277	20		297	335
1400	60" wide		24	.667		455	21		476	535
2000	Hardwood edged, 42" high, 24" wide		30	.533		310	16.85		326.85	365
2100	36" wide		28	.571		335	18.05		353.05	395
2120	48" wide		26	.615		360	19.45		379.45	425
2180	60" wide		25	.640		395	20		415	465
2200	54" high, 24" wide		30	.533		320	16.85		336.85	375
2220	36" wide		28	.571		340	18.05		358.05	405
2240	48" wide		26	.615		395	19.45		414.45	465
2300	60" wide		25	.640		420	20		440	490
2400	60" high, 24" wide		29	.552		330	17.45		347.45	385
2420	36" wide		27	.593		375	18.75		393.75	445
2440	48" wide		25	.640		475	20		495	555
2460	60" wide		24	.667		520	21		541	605
2800	72" high, 24" wide		29	.552		345	17.45		362.45	405
2820	36" wide		27	.593		395	18.75		413.75	460
2840	48" wide		25	.640		505	20		525	585
2860	60" wide		24	.667		505	21		526	590
3000	Fabric panel, straight, N.R.C. <= .50									
3110	60" high, 24" wide	2 Clab	29	.552	Ea.	159	17.45		176.45	202
3120	30" wide		28	.571		162	18.05		180.05	206
3130	36" wide		27	.593		169	18.75		187.75	215
3140	48" wide		26	.615		170	19.45		189.45	216
3150	60" wide		25	.640		180	20		200	229
3160	72" wide		24	.667		179	21		200	230
3210	72" high, 24" wide		29	.552		165	17.45		182.45	209
3220	30" wide		28	.571		169	18.05		187.05	214
3230	36" wide		27	.593		169	18.75		187.75	215
3240	48" wide		26	.615		179	19.45		198.45	227
3250	60" wide		25	.640		189	20		209	240
3260	72" wide		24	.667		199	21		220	252
3310	N.R.C. => .75, 42" high, 24" wide		30	.533		179	16.85		195.85	223
3320	30" wide		29	.552		190	17.45		207.45	236
3330	36" wide		28	.571		195	18.05		213.05	243
3340	48" wide		27	.593		215	18.75		233.75	265
3350	60" wide		26	.615		240	19.45		259.45	293
3360	72" wide		25	.640		250	20		270	305
3410	60" high, 24" wide		29	.552		209	17.45		226.45	257
3420	30" wide		28	.571		219	18.05		237.05	269

10 22 Partitions

10 22 23 – Portable Partitions, Screens, and Panels

10 22 23.23 Movable Panel Systems		Crew	Daily Output	Labor-Hours	Unit	Material	2009 Bare Costs Labor	Equipment	Total	Total Incl O&P
3430	36" wide	2 Clab	27	.593	Ea.	239	18.75		257.75	292
3440	48" wide		26	.615		259	19.45		278.45	315
3450	60" wide		25	.640		315	20		335	375
3460	72" wide		24	.667		335	21		356	405
3510	72" high, 24" wide		29	.552		239	17.45		256.45	290
3520	30" wide		28	.571		255	18.05		273.05	310
3530	36" wide		27	.593		269	18.75		287.75	325
3540	48" wide		26	.615		289	19.45		308.45	350
3550	60" wide		25	.640		305	20		325	365
3560	72" wide		24	.667		335	21		356	405
3710	Clear glass, straight, 60" high, 30" wide		28	.571		330	18.05		348.05	395
3720	36" wide		27	.593		340	18.75		358.75	405
3730	48" wide		26	.615		380	19.45		399.45	445
3740	72" high, 60" wide		25	.640		425	20		445	495
3810	Fabric and clear glass, straight, 60" high, 30" wide		28	.571		268	18.05		286.05	325
3820	36" wide		27	.593		281	18.75		299.75	340
3830	48" wide		26	.615		325	19.45		344.45	390
3840	72" high, 60" wide		25	.640		470	20		490	550
3910	Connector kit, straight					20			20	22
3920	Corner, 2-way					14.35			14.35	15.75
3930	3-way					11			11	12.10
3940	4-way					10.50			10.50	11.55
3950	T-leg					18.50			18.50	20.50

10 22 26 – Operable Partitions

10 22 26.13 Accordion Folding Partitions

		Crew	Daily Output	Labor-Hours	Unit	Material	Labor	Equipment	Total	Total Incl O&P
0010	**ACCORDION FOLDING PARTITIONS**									
0100	Vinyl covered, over 150 S.F., frame not included									
0300	Residential, 1.25 lb. per S.F., 8' maximum height	2 Carp	300	.053	S.F.	19.35	2.13		21.48	25
0400	Commercial, 1.75 lb. per S.F., 8' maximum height		225	.071		22	2.84		24.84	29
0600	2 lb. per S.F., 17' maximum height		150	.107		23	4.26		27.26	31.50
0700	Industrial, 4 lb. per S.F., 20' maximum height		75	.213		32.50	8.50		41	49
0900	Acoustical, 3 lb. per S.F., 17' maximum height		100	.160		25.50	6.40		31.90	38
1200	5 lb. per S.F., 20' maximum height		95	.168		35.50	6.75		42.25	49.50
1300	5.5 lb. per S.F., 17' maximum height		90	.178		41.50	7.10		48.60	57
1400	Fire rated, 4.5 psf, 20' maximum height		160	.100		41.50	4		45.50	52
1500	Vinyl clad wood or steel, electric operation, 5.0 psf		160	.100		48.50	4		52.50	59
1900	Wood, non-acoustic, birch or mahogany, to 10' high		300	.053		26	2.13		28.13	32

10 22 26.33 Folding Panel Partitions

		Crew	Daily Output	Labor-Hours	Unit	Material	Labor	Equipment	Total	Total Incl O&P
0010	**FOLDING PANEL PARTITIONS**, acoustic, wood									
0100	Vinyl faced, to 18' high, 6 psf, minimum	2 Carp	60	.267	S.F.	49	10.65		59.65	70.50
0150	Average		45	.356		58.50	14.20		72.70	86.50
0200	Maximum		30	.533		75.50	21.50		97	117
0400	Formica or hardwood finish, minimum		60	.267		50.50	10.65		61.15	72
0500	Maximum		30	.533		54	21.50		75.50	92.50
0600	Wood, low acoustical type, 4.5 psf, to 14' high		50	.320		37	12.80		49.80	60.50
1100	Steel, acoustical, 9 to 12 lb. per S.F., vinyl faced, minimum		60	.267		52.50	10.65		63.15	74
1200	Maximum		30	.533		64	21.50		85.50	103
1700	Aluminum framed, acoustical, to 12' high, 5.5 psf, minimum		60	.267		35	10.65		45.65	55.50
1800	Maximum		30	.533		42.50	21.50		64	80
2000	6.5 lb. per S.F., minimum		60	.267		37	10.65		47.65	57
2100	Maximum		30	.533		45.50	21.50		67	83.50

10 22 Partitions

10 22 26 – Operable Partitions

10 22 26.43 Sliding Partitions

		Crew	Daily Output	Labor-Hours	Unit	Material	2009 Bare Costs Labor	Equipment	Total	Total Incl O&P
0010	**SLIDING PARTITIONS**									
0020	Acoustic air wall, 1-5/8" thick, minimum	2 Carp	375	.043	S.F.	30	1.70		31.70	35.50
0100	Maximum		365	.044		51.50	1.75		53.25	59
0300	2-1/4" thick, minimum		360	.044		33.50	1.78		35.28	40
0400	Maximum	↓	330	.048	↓	59	1.94		60.94	68
0600	For track type, add to above				L.F.	110			110	121
0700	Overhead track type, acoustical, 3" thick, 11 psf, minimum	2 Carp	350	.046	S.F.	75.50	1.83		77.33	86
0800	Maximum	"	300	.053	"	91	2.13		93.13	103

10 26 Wall and Door Protection

10 26 13 – Corner Guards

10 26 13.10 Metal Corner Guards

		Crew	Daily Output	Labor-Hours	Unit	Material	Labor	Equipment	Total	Total Incl O&P
0010	**METAL CORNER GUARDS**									
0020	Steel angle w/anchors, 1" x 1" x 1/4", 1.5#/L.F.	2 Carp	160	.100	L.F.	5.75	4		9.75	12.55
0100	2" x 2" x 1/4" angles, 3.2#/L.F.		150	.107		10.45	4.26		14.71	18.10
0200	3" x 3" x 5/16" angles, 6.1#/L.F.		140	.114		15.55	4.57		20.12	24
0300	4" x 4" x 5/16" angles, 8.2#/L.F.	↓	120	.133		18.95	5.35		24.30	29.50
0350	For angles drilled and anchored to masonry, add					15%	120%			
0370	Drilled and anchored to concrete, add					20%	170%			
0400	For galvanized angles, add					35%				
0450	For stainless steel angles, add				↓	100%				

10 26 13.20 Corner Protection

		Crew	Daily Output	Labor-Hours	Unit	Material	Labor	Equipment	Total	Total Incl O&P
0010	**CORNER PROTECTION**									
0100	Stainless steel, 16 ga., adhesive mount, 3-1/2" leg	1 Sswk	80	.100	L.F.	18.05	4.47		22.52	28
0200	12 ga. stainless, adhesive mount	"	80	.100		21	4.47		25.47	31
0300	For screw mount, add						10%			
0500	Vinyl acrylic, adhesive mount, 3" leg	1 Carp	128	.063		7.60	2.50		10.10	12.25
0550	1-1/2" leg		160	.050		4.67	2		6.67	8.25
0600	Screw mounted, 3" leg		80	.100		7.75	4		11.75	14.75
0650	1-1/2" leg		100	.080		4.12	3.20		7.32	9.50
0700	Clear plastic, screw mounted, 2-1/2"		60	.133		3.65	5.35		9	12.25
1000	Vinyl cover, alum. retainer, surface mount, 3" x 3"		48	.167		8.95	6.65		15.60	20
1050	2" x 2"		48	.167		9.30	6.65		15.95	20.50
1100	Flush mounted, 3" x 3"		32	.250		19.25	10		29.25	36.50
1150	2" x 2"	↓	32	.250	↓	15.65	10		25.65	33

10 26 16 – Bumper Guards

10 26 16.10 Wallguard

		Crew	Daily Output	Labor-Hours	Unit	Material	Labor	Equipment	Total	Total Incl O&P
0010	**WALLGUARD**									
0400	Rub rail, vinyl, adhesive mounted	1 Carp	185	.043	L.F.	7.70	1.73		9.43	11.15
0500	Neoprene, aluminum backing, 1-1/2" x 2"		110	.073		8.65	2.91		11.56	14
1000	Trolley rail, PVC, clipped to wall, 5" high		185	.043		7.65	1.73		9.38	11.10
1050	8" high		180	.044	↓	10.45	1.78		12.23	14.20
1200	Bed bumper, vinyl acrylic, alum. retainer, 21" long		10	.800	Ea.	39.50	32		71.50	93
1300	53" long with aligner		9	.889	"	100	35.50		135.50	165
1400	Bumper, vinyl cover, alum. retain., cush. mnt., 1-1/2" x 2-3/4"		80	.100	L.F.	13.95	4		17.95	21.50
1500	2" x 4-1/4"		80	.100		20.50	4		24.50	28.50
1600	Surface mounted, 1-3/4" x 3-5/8"		80	.100		11.70	4		15.70	19.10
2000	Crash rail, vinyl cover, alum. retainer, 1" x 4"		110	.073		10.80	2.91		13.71	16.40
2100	1" x 8"		90	.089		16.90	3.55		20.45	24
2150	Vinyl inserts, aluminum plate, 1" x 2-1/2"	↓	110	.073	↓	14.60	2.91		17.51	20.50

10 26 Wall and Door Protection

10 26 16 – Bumper Guards

10 26 16.10 Wallguard		Crew	Daily Output	Labor-Hours	Unit	Material	2009 Bare Costs Labor	Equipment	Total	Total Incl O&P
2200	1" x 5"	1 Carp	90	.089	L.F.	22.50	3.55		26.05	30.50
3000	Handrail/bumper, vinyl cover, alum. retainer									
3010	Bracket mounted, flat rail, 5-1/2"	1 Carp	80	.100	L.F.	17.55	4		21.55	25.50
3100	6-1/2"	↓	80	.100		22	4		26	30
3200	Bronze bracket, 1-3/4" diam. rail	↓	80	.100	↓	15.85	4		19.85	23.50

10 28 Toilet, Bath, and Laundry Accessories

10 28 13 – Toilet Accessories

10 28 13.13 Commercial Toilet Accessories

		Crew	Daily Output	Labor-Hours	Unit	Material	Labor	Equipment	Total	Total Incl O&P
0010	**COMMERCIAL TOILET ACCESSORIES**									
0200	Curtain rod, stainless steel, 5' long, 1" diameter	1 Carp	13	.615	Ea.	41	24.50		65.50	83.50
0300	1-1/4" diameter		13	.615		36.50	24.50		61	78
0400	Diaper changing station, horizontal, wall mounted, plastic	↓	10	.800	↓	280	32		312	355
0500	Dispenser units, combined soap & towel dispensers,									
0510	mirror and shelf, flush mounted	1 Carp	10	.800	Ea.	295	32		327	375
0600	Towel dispenser and waste receptacle,									
0610	18 gallon capacity	1 Carp	10	.800	Ea.	395	32		427	485
0800	Grab bar, straight, 1-1/4" diameter, stainless steel, 18" long		24	.333		29.50	13.30		42.80	53
0900	24" long		23	.348		32	13.90		45.90	56.50
1000	30" long		22	.364		34.50	14.55		49.05	60.50
1100	36" long		20	.400		36	16		52	64.50
1105	42" long		20	.400		42	16		58	71
1200	1-1/2" diameter, 24" long		23	.348		34.50	13.90		48.40	59.50
1300	36" long		20	.400		43.50	16		59.50	73
1310	42" long		18	.444		47	17.75		64.75	79
1500	Tub bar, 1-1/4" diameter, 24" x 36"		14	.571		109	23		132	155
1600	Plus vertical arm		12	.667		122	26.50		148.50	176
1900	End tub bar, 1" diameter, 90° angle, 16" x 32"		12	.667		153	26.50		179.50	210
2010	Tub/shower/toilet, 2-wall, 36" x 24"		12	.667		140	26.50		166.50	196
2300	Hand dryer, surface mounted, electric, 115 volt, 20 amp		4	2		380	80		460	540
2400	230 volt, 10 amp		4	2		815	80		895	1,025
2600	Hat and coat strip, stainless steel, 4 hook, 36" long		24	.333		54	13.30		67.30	80
2700	6 hook, 60" long		20	.400		59	16		75	90
3000	Mirror, with stainless steel 3/4" square frame, 18" x 24"		20	.400		54.50	16		70.50	85
3100	36" x 24"		15	.533		137	21.50		158.50	184
3200	48" x 24"		10	.800		173	32		205	240
3300	72" x 24"		6	1.333		330	53.50		383.50	450
3500	With 5" stainless steel shelf, 18" x 24"		20	.400		227	16		243	274
3600	36" x 24"		15	.533		300	21.50		321.50	365
3700	48" x 24"		10	.800		335	32		367	420
3800	72" x 24"		6	1.333		375	53.50		428.50	500
4100	Mop holder strip, stainless steel, 5 holders, 48" long		20	.400		80.50	16		96.50	114
4200	Napkin/tampon dispenser, recessed		15	.533		820	21.50		841.50	940
4300	Robe hook, single, regular		36	.222		15.50	8.90		24.40	31
4400	Heavy duty, concealed mounting		36	.222		15.55	8.90		24.45	31
4600	Soap dispenser, chrome, surface mounted, liquid		20	.400		47.50	16		63.50	77
4700	Powder		20	.400		46.50	16		62.50	76.50
5000	Recessed stainless steel, liquid		10	.800		140	32		172	204
5100	Powder		10	.800		213	32		245	284
5300	Soap tank, stainless steel, 1 gallon		10	.800		190	32		222	259
5400	5 gallon	↓	5	1.600	↓	262	64		326	385

10 28 Toilet, Bath, and Laundry Accessories

10 28 13 – Toilet Accessories

10 28 13.13 Commercial Toilet Accessories

		Crew	Daily Output	Labor-Hours	Unit	Material	2009 Bare Costs Labor	Equipment	Total	Total Incl O&P
5600	Shelf, stainless steel, 5" wide, 18 ga., 24" long	1 Carp	24	.333	Ea.	63.50	13.30		76.80	90
5700	48" long		16	.500		100	20		120	141
5800	8" wide shelf, 18 ga., 24" long		22	.364		77.50	14.55		92.05	108
5900	48" long		14	.571		124	23		147	173
6000	Toilet seat cover dispenser, stainless steel, recessed		20	.400		143	16		159	182
6050	Surface mounted		15	.533		34.50	21.50		56	71
6100	Toilet tissue dispenser, surface mounted, SS, single roll		30	.267		19.45	10.65		30.10	38
6200	Double roll		24	.333		24	13.30		37.30	46.50
6400	Towel bar, stainless steel, 18" long		23	.348		39.50	13.90		53.40	65
6500	30" long		21	.381		101	15.20		116.20	135
6700	Towel dispenser, stainless steel, surface mounted		16	.500		42	20		62	77
6800	Flush mounted, recessed		10	.800		270	32		302	345
7000	Towel holder, hotel type, 2 guest size		20	.400		54.50	16		70.50	85
7200	Towel shelf, stainless steel, 24" long, 8" wide		20	.400		93	16		109	127
7400	Tumbler holder, tumbler only		30	.267		38.50	10.65		49.15	59
7500	Soap, tumbler & toothbrush		30	.267		25	10.65		35.65	44
7700	Wall urn ash receiver, surface mount, 11" long		12	.667		124	26.50		150.50	178
7800	7-1/2", long		18	.444		121	17.75		138.75	161
8000	Waste receptacles, stainless steel, with top, 13 gallon		10	.800		330	32		362	410
8100	36 gallon		8	1		515	40		555	625

10 28 16 – Bath Accessories

10 28 16.20 Medicine Cabinets

		Crew	Daily Output	Labor-Hours	Unit	Material	Labor	Equipment	Total	Total Incl O&P
0010	**MEDICINE CABINETS**									
0020	With mirror, st. st. frame, 16" x 22", unlighted	1 Carp	14	.571	Ea.	81.50	23		104.50	125
0100	Wood frame		14	.571		113	23		136	160
0300	Sliding mirror doors, 20" x 16" x 4-3/4", unlighted		7	1.143		101	45.50		146.50	182
0400	24" x 19" x 8-1/2", lighted		5	1.600		159	64		223	274
0600	Triple door, 30" x 32", unlighted, plywood body		7	1.143		239	45.50		284.50	335
0700	Steel body		7	1.143		315	45.50		360.50	415
0900	Oak door, wood body, beveled mirror, single door		7	1.143		154	45.50		199.50	240
1000	Double door		6	1.333		375	53.50		428.50	495
1200	Hotel cabinets, stainless, with lower shelf, unlighted		10	.800		202	32		234	272
1300	Lighted		5	1.600		300	64		364	430

10 28 23 – Laundry Accessories

10 28 23.13 Built-In Ironing Boards

		Crew	Daily Output	Labor-Hours	Unit	Material	Labor	Equipment	Total	Total Incl O&P
0010	**BUILT-IN IRONING BOARDS**									
0020	Including cabinet, board & light, minimum	1 Carp	2	4	Ea.	299	160		459	580
0100	Maximum, see also Div. 11 23 13.13	"	1.50	5.333	"	500	213		713	880

10 31 Manufactured Fireplaces

10 31 13 – Manufactured Fireplace Chimneys

10 31 13.10 Fireplace Chimneys

		Crew	Daily Output	Labor-Hours	Unit	Material	Labor	Equipment	Total	Total Incl O&P
0010	**FIREPLACE CHIMNEYS**									
0500	Chimney dbl. wall, all stainless, over 8'-6", 7" diam., add to fireplace	1 Carp	33	.242	V.L.F.	70	9.70		79.70	92
0600	10" diameter, add to fireplace		32	.250		93	10		103	118
0700	12" diameter, add to fireplace		31	.258		111	10.30		121.30	138
0800	14" diameter, add to fireplace		30	.267		180	10.65		190.65	215
1000	Simulated brick chimney top, 4' high, 16" x 16"		10	.800	Ea.	244	32		276	320
1100	24" x 24"		7	1.143	"	455	45.50		500.50	570

10 31 Manufactured Fireplaces

10 31 13 – Manufactured Fireplace Chimneys

10 31 13.20 Chimney Accessories

		Crew	Daily Output	Labor-Hours	Unit	Material	2009 Bare Costs Labor	2009 Bare Costs Equipment	Total	Total Incl O&P
0010	**CHIMNEY ACCESSORIES**									
0020	Chimney screens, galv., 13" x 13" flue	1 Bric	8	1	Ea.	52	40.50		92.50	119
0050	24" x 24" flue		5	1.600		121	65		186	232
0200	Stainless steel, 13" x 13" flue		8	1		320	40.50		360.50	410
0250	20" x 20" flue		5	1.600		435	65		500	580
2400	Squirrel and bird screens, galvanized, 8" x 8" flue		16	.500		45.50	20.50		66	80.50
2450	13" x 13" flue		12	.667		49	27		76	95

10 31 16 – Manufactured Fireplace Forms

10 31 16.10 Fireplace Forms

		Crew	Daily Output	Labor-Hours	Unit	Material	Labor	Equipment	Total	Total Incl O&P
0010	**FIREPLACE FORMS**									
1800	Fireplace forms, no accessories, 32" opening	1 Bric	3	2.667	Ea.	620	108		728	845
1900	36" opening		2.50	3.200		785	130		915	1,050
2000	40" opening		2	4		1,050	162		1,212	1,400
2100	78" opening		1.50	5.333		1,525	216		1,741	2,000

10 31 23 – Prefabricated Fireplaces

10 31 23.10 Fireplace, Prefabricated

		Crew	Daily Output	Labor-Hours	Unit	Material	Labor	Equipment	Total	Total Incl O&P
0010	**FIREPLACE, PREFABRICATED**, free standing or wall hung									
0100	With hood & screen, minimum	1 Carp	1.30	6.154	Ea.	1,175	246		1,421	1,650
0150	Average		1	8		1,625	320		1,945	2,300
0200	Maximum		.90	8.889		3,125	355		3,480	3,975
1500	Simulated logs, gas fired, 40,000 BTU, 2' long, minimum		7	1.143	Set	400	45.50		445.50	510
1600	Maximum		6	1.333		940	53.50		993.50	1,100
1700	Electric, 1,500 BTU, 1'-6" long, minimum		7	1.143		166	45.50		211.50	254
1800	11,500 BTU, maximum		6	1.333		370	53.50		423.50	495

10 32 Fireplace Specialties

10 32 13 – Fireplace Dampers

10 32 13.10 Dampers

		Crew	Daily Output	Labor-Hours	Unit	Material	Labor	Equipment	Total	Total Incl O&P
0010	**DAMPERS**									
0800	Damper, rotary control, steel, 30" opening	1 Bric	6	1.333	Ea.	77.50	54		131.50	167
0850	Cast iron, 30" opening		6	1.333		86	54		140	177
0880	36" opening		6	1.333		93.50	54		147.50	185
0900	48" opening		6	1.333		133	54		187	229
0920	60" opening		6	1.333		299	54		353	410
0950	72" opening		5	1.600		360	65		425	495
1000	84" opening, special order		5	1.600		770	65		835	945
1050	96" opening, special order		4	2		780	81		861	985
1200	Steel plate, poker control, 60" opening		8	1		273	40.50		313.50	360
1250	84" opening, special order		5	1.600		495	65		560	645
1400	"Universal" type, chain operated, 32" x 20" opening		8	1		212	40.50		252.50	295
1450	48" x 24" opening		5	1.600		315	65		380	450

10 32 23 – Fireplace Doors

10 32 23.10 Doors

		Crew	Daily Output	Labor-Hours	Unit	Material	Labor	Equipment	Total	Total Incl O&P
0010	**DOORS**									
0400	Cleanout doors and frames, cast iron, 8" x 8"	1 Bric	12	.667	Ea.	36.50	27		63.50	81
0450	12" x 12"		10	.800		44	32.50		76.50	97.50
0500	18" x 24"		8	1		127	40.50		167.50	202
0550	Cast iron frame, steel door, 24" x 30"		5	1.600		275	65		340	400
1600	Dutch Oven door and frame, cast iron, 12" x 15" opening		13	.615		111	25		136	160

10 32 Fireplace Specialties

10 32 23 – Fireplace Doors

10 32 23.10 Doors		Crew	Daily Output	Labor-Hours	Unit	Material	2009 Bare Costs Labor	Equipment	Total	Total Incl O&P
1650	Copper plated, 12" x 15" opening	1 Bric	13	.615	Ea.	218	25		243	277

10 35 Stoves

10 35 13 – Heating Stoves

10 35 13.10 Woodburning Stoves

		Crew	Daily Output	Labor-Hours	Unit	Material	Labor	Equipment	Total	Total Incl O&P
0010	**WOODBURNING STOVES**									
0015	Cast iron, minimum	2 Carp	1.30	12.308	Ea.	1,125	490		1,615	2,000
0020	Average	↓	1	16		1,525	640		2,165	2,675
0030	Maximum		.80	20	↓	2,750	800		3,550	4,275
0050	For gas log lighter, add					37.50			37.50	41.50

10 44 Fire Protection Specialties

10 44 13 – Fire Extinguisher Cabinets

10 44 13.53 Fire Equipment Cabinets

		Crew	Daily Output	Labor-Hours	Unit	Material	Labor	Equipment	Total	Total Incl O&P
0010	**FIRE EQUIPMENT CABINETS**, not equipped, 20 ga. steel box									
0040	recessed, D.S. glass in door, box size given									
1000	Portable extinguisher, single, 8" x 12" x 27", alum. door & frame	Q-12	8	2	Ea.	124	86.50		210.50	267
1100	Steel door and frame		8	2		86.50	86.50		173	226
1200	Stainless steel door and frame		8	2		185	86.50		271.50	335
2000	Portable extinguisher, large, 8" x 12" x 36", alum. door & frame		8	2		212	86.50		298.50	365
2100	Steel door and frame		8	2		134	86.50		220.50	278
2200	Stainless steel door and frame		8	2		222	86.50		308.50	375
2500	8" x 16" x 38", aluminum door & frame		8	2		219	86.50		305.50	370
2600	Steel door and frame	↓	8	2	↓	157	86.50		243.50	300
3000	Hose rack assy., 1-1/2" valve & 100' hose, 24" x 40" x 5-1/2"									
3100	Aluminum door and frame	Q-12	6	2.667	Ea.	305	116		421	510
3200	Steel door and frame		6	2.667		272	116		388	470
3300	Stainless steel door and frame	↓	6	2.667		530	116		646	760
4000	Hose rack assy., 2-1/2" x 1-1/2" valve, 100' hose, 24" x 40" x 8"									
4100	Aluminum door and frame	Q-12	6	2.667	Ea.	305	116		421	510
4200	Steel door and frame		6	2.667		208	116		324	400
4300	Stainless steel door and frame	↓	6	2.667		410	116		526	625
5000	Hose rack assy., 2-1/2" x 1-1/2" valve, 100' hose									
5010	and extinguisher, 30" x 40" x 8"									
5100	Aluminum door and frame	Q-12	5	3.200	Ea.	390	139		529	640
5200	Steel door and frame		5	3.200		208	139		347	435
5300	Stainless steel door and frame	↓	5	3.200	↓	710	139		849	995
7000	Hose rack assy., 1-1/2" valve & 100' hose, 2-1/2" FD valve									
7010	and extinguisher, 30" x 44" x 8"									
7100	Aluminum door and frame	Q-12	5	3.200	Ea.	410	139		549	660
7200	Steel door and frame		5	3.200		510	139		649	775
7300	Stainless steel door and frame	↓	5	3.200	↓	700	139		839	980
8000	Valve cabinet for 2-1/2" FD angle valve, 18" x 18" x 8"									
8100	Aluminum door and frame	Q-12	12	1.333	Ea.	135	58		193	236
8200	Steel door and frame		12	1.333		111	58		169	210
8300	Stainless steel door and frame	↓	12	1.333	↓	182	58		240	287

10 44 16 – Fire Extinguishers

10 44 16.13 Portable Fire Extinguishers

| 0010 | **PORTABLE FIRE EXTINGUISHERS** | | | | | | | | | |

10 44 Fire Protection Specialties

10 44 16 – Fire Extinguishers

10 44 16.13 Portable Fire Extinguishers

		Crew	Daily Output	Labor-Hours	Unit	Material	2009 Bare Costs Labor	Equipment	Total	Total Incl O&P
0120	CO_2, portable with swivel horn, 5 lb.				Ea.	159			159	175
0140	With hose and "H" horn, 10 lb.					215			215	237
0180	20 lb.					305			305	335
1000	Dry chemical, pressurized									
1040	Standard type, portable, painted, 2-1/2 lb.				Ea.	35			35	38.50
1060	5 lb.					61.50			61.50	67.50
1080	10 lb.					91			91	100
1100	20 lb.					135			135	149
1120	30 lb.					295			295	325
1300	Standard type, wheeled, 150 lb.					2,200			2,200	2,425
2000	ABC all purpose type, portable, 2-1/2 lb.					34			34	37
2060	5 lb.					49			49	54
2080	9-1/2 lb.					75			75	82.50
2100	20 lb.					108			108	119
3000	Dry chemical, outside cartridge to -65° F, painted, 9 lb.					220			220	242
3060	26 lb.					275			275	305
5000	Pressurized water, 2-1/2 gallon, stainless steel					107			107	118
5060	With anti-freeze					149			149	163
9400	Installation of extinguishers, 12 or more, on wood	1 Carp	30	.267			10.65		10.65	16.50
9420	On masonry or concrete	"	15	.533			21.50		21.50	33

10 44 16.16 Wheeled Fire Extinguisher Units

		Crew	Daily Output	Labor-Hours	Unit	Material	Labor	Equipment	Total	Total Incl O&P
0010	**WHEELED FIRE EXTINGUISHER UNITS**									
0350	CO_2, portable, with swivel horn									
0360	Wheeled type, cart mounted, 50 lb.				Ea.	1,325			1,325	1,450
0400	100 lb.				"	2,650			2,650	2,900
2200	ABC all purpose type									
2300	Wheeled, 45 lb.				Ea.	715			715	785
2360	150 lb.				"	1,750			1,750	1,925

10 51 Lockers

10 51 13 – Metal Lockers

10 51 13.10 Lockers

		Crew	Daily Output	Labor-Hours	Unit	Material	Labor	Equipment	Total	Total Incl O&P
0011	**LOCKERS** Steel, baked enamel									
0110	Single tier box locker, 12" x 15" x 72"	1 Shee	8	1	Ea.	183	47		230	274
0120	18" x 15" x 72"		8	1		205	47		252	298
0130	12" x 18" x 72"		8	1		198	47		245	291
0140	18" x 18" x 72"		8	1		212	47		259	305
0410	Double tier, 12" x 15" x 36"		21	.381		199	18		217	247
0420	18" x 15" x 36"		21	.381		251	18		269	305
0430	12" x 18" x 36"		21	.381		201	18		219	249
0440	18" x 18" x 36"		21	.381		263	18		281	315
0500	Two person, 18" x 15" x 72"		8	1		253	47		300	350
0510	18" x 18" x 72"		8	1		271	47		318	370
0520	Duplex, 15" x 15" x 72"		8	1		299	47		346	405
0530	15" x 21" x 72"		8	1		325	47		372	430
0600	5 tier box lockers, minimum		30	.267	Opng.	41.50	12.60		54.10	65
0700	Maximum		24	.333		54	15.75		69.75	83.50
0900	6 tier box lockers, minimum		36	.222		30.50	10.50		41	49.50
1000	Maximum		30	.267		44	12.60		56.60	68
1100	Wire meshed wardrobe, floor. mtd., open front varsity type		7.50	1.067	Ea.	217	50.50		267.50	315
2400	16-person locker unit with clothing rack									

10 51 Lockers

10 51 13 – Metal Lockers

10 51 13.10 Lockers

		Crew	Daily Output	Labor-Hours	Unit	Material	2009 Bare Costs Labor	Equipment	Total	Total Incl O&P
2500	72 wide x 15" deep x 72" high	1 Shee	8	1	Ea.	450	47		497	570
2550	18" deep	"	8	1	"	470	47		517	590
3000	Wall mounted lockers, 4 person, with coat bar									
3100	48" wide x 18" deep x 12" high	1 Shee	8	1	Ea.	292	47		339	395
3250	Rack w/ 24 wire mesh baskets		1.50	5.333	Set	345	252		597	765
3260	30 baskets		1.25	6.400		267	300		567	760
3270	36 baskets		.95	8.421		350	395		745	995
3280	42 baskets		.80	10		385	470		855	1,150
3300	For built-in lock with 2 keys, add				Ea.	7.10			7.10	7.80
3600	For hanger rods, add				"	1.86			1.86	2.05

10 51 26 – Plastic Lockers

10 51 26.13 Lockers

			Crew	Daily Output	Labor-Hours	Unit	Material	Labor	Equipment	Total	Total Incl O&P
0011	**LOCKERS** Plastic, 30% recycled										
0110	Single tier box locker, 12" x 12" x 72"	G	1 Shee	8	1	Ea.	400	47		447	515
0120	12" x 15" x 72"	G		8	1		420	47		467	535
0130	12" x 18" x 72"	G		8	1		430	47		477	550
0410	Double tier, 12" x 12" x 72"	G		21	.381		400	18		418	470
0420	12" x 15" x 72"	G		21	.381		400	18		418	470
0430	12" x 18" x 72"	G		21	.381		420	18		438	490

10 51 53 – Locker Room Benches

10 51 53.10 Benches

		Crew	Daily Output	Labor-Hours	Unit	Material	Labor	Equipment	Total	Total Incl O&P
0010	**BENCHES**									
2100	Locker bench, laminated maple, top only	1 Shee	100	.080	L.F.	13.90	3.78		17.68	21
2200	Pedestals, steel pipe	"	25	.320	Ea.	36	15.10		51.10	63.50

10 55 Postal Specialties

10 55 23 – Mail Boxes

10 55 23.10 Commercial Mail Boxes

		Crew	Daily Output	Labor-Hours	Unit	Material	Labor	Equipment	Total	Total Incl O&P
0010	**COMMERCIAL MAIL BOXES**									
0020	Horiz., key lock, 5"H x 6"W x 15"D, alum., rear load	1 Carp	34	.235	Ea.	35	9.40		44.40	53
0100	Front loading		34	.235		36	9.40		45.40	54
0200	Double, 5"H x 12"W x 15"D, rear loading		26	.308		71.50	12.30		83.80	98
0300	Front loading		26	.308		66	12.30		78.30	91.50
0500	Quadruple, 10"H x 12"W x 15"D, rear loading		20	.400		91	16		107	125
0600	Front loading		20	.400		108	16		124	144
0800	Vertical, front load, 15"H x 5"W x 6"D, alum., per compartment		34	.235		30	9.40		39.40	47.50
0900	Bronze, duranodic finish		34	.235		40	9.40		49.40	58.50
1000	Steel, enameled		34	.235		30	9.40		39.40	47.50
1700	Alphabetical directories, 120 names		10	.800		126	32		158	188
1800	Letter collection box		6	1.333		705	53.50		758.50	860
1830	Lobby collection boxes, aluminum	2 Shee	5	3.200		2,025	151		2,176	2,450
1840	Bronze or stainless	"	4.50	3.556		2,500	168		2,668	3,000
1900	Letter slot, residential	1 Carp	20	.400		86	16		102	120
2000	Post office type		8	1		110	40		150	183
2200	Post office counter window, with grille		2	4		550	160		710	855
2250	Key keeper, single key, aluminum		26	.308		38.50	12.30		50.80	61.50
2300	Steel, enameled		26	.308		107	12.30		119.30	137

10 55 Postal Specialties

10 55 91 – Mail Chutes

10 55 91.10 Commercial Mail Chutes	Crew	Daily Output	Labor-Hours	Unit	Material	2009 Bare Costs Labor	Equipment	Total	Total Incl O&P
0010 **COMMERCIAL MAIL CHUTES**									
0020 Aluminum & glass, 14-1/4" wide, 4-5/8" deep	2 Shee	4	4	Floor	705	189		894	1,075
0100 8-5/8" deep		3.80	4.211		855	199		1,054	1,250
0300 8-3/4" x 3-1/2", aluminum		5	3.200		650	151		801	945
0400 Bronze or stainless		4.50	3.556		1,000	168		1,168	1,350

10 56 Storage Assemblies

10 56 13 – Metal Storage Shelving

10 56 13.10 Shelving

		Crew	Daily Output	Labor-Hours	Unit	Material	Labor	Equipment	Total	Total Incl O&P
0010	**SHELVING**									
0020	Metal, industrial, cross-braced, 3' wide, 12" deep	1 Sswk	175	.046	SF Shlf	6.55	2.04		8.59	10.85
0100	24" deep		330	.024		4.70	1.08		5.78	7.10
0300	4' wide, 12" deep		185	.043		6.05	1.93		7.98	10.10
0400	24" deep		380	.021		4.39	.94		5.33	6.50
1200	Enclosed sides, cross-braced back, 3' wide, 12" deep		175	.046		10.45	2.04		12.49	15.15
1300	24" deep		290	.028		9.20	1.23		10.43	12.35
1500	Fully enclosed, sides and back, 3' wide, 12" deep		150	.053		11.45	2.38		13.83	16.85
1600	24" deep		255	.031		7.65	1.40		9.05	10.95
1800	4' wide, 12" deep		150	.053		8.80	2.38		11.18	13.90
1900	24" deep		290	.028		6.45	1.23		7.68	9.30
2200	Wide span, 1600 lb. capacity per shelf, 6' wide, 24" deep		380	.021		7	.94		7.94	9.40
2400	36" deep		440	.018		8.30	.81		9.11	10.55
2600	8' wide, 24" deep		440	.018		8.65	.81		9.46	10.95
2800	36" deep		520	.015		7	.69		7.69	8.95
3000	Residential, vinyl covered wire, wardrobe, 12" deep	1 Carp	195	.041	L.F.	5.90	1.64		7.54	9.05
3100	16" deep		195	.041		8	1.64		9.64	11.35
3200	Standard, 6" deep		195	.041		3.23	1.64		4.87	6.10
3400	12" deep		195	.041		6.45	1.64		8.09	9.65
3500	16" deep		195	.041		9.90	1.64		11.54	13.45
3600	20" deep		195	.041		13.15	1.64		14.79	17
3700	Support bracket		80	.100	Ea.	4.10	4		8.10	10.70
4000	Pallet racks, steel frame 5,000 lb. capacity, 8' long, 36" deep	2 Sswk	450	.036	SF Shlf	8.60	1.59		10.19	12.30
4200	42" deep		500	.032		7.35	1.43		8.78	10.65
4400	48" deep		520	.031		6.95	1.38		8.33	10.10

10 56 13.20 Parts Bins

		Crew	Daily Output	Labor-Hours	Unit	Material	Labor	Equipment	Total	Total Incl O&P
0010	**PARTS BINS** Metal, gray baked enamel finish									
0100	6'-3" high, 3' wide									
0300	12 bins, 18" wide x 12" high, 12" deep	2 Clab	10	1.600	Ea.	281	50.50		331.50	390
0400	24" deep		10	1.600		355	50.50		405.50	470
0600	72 bins, 6" wide x 6" high, 12" deep		8	2		410	63		473	550
0700	18" deep		8	2		645	63		708	810
1000	7'-3" high, 3' wide									
1200	14 bins, 18" wide x 12" high, 12" deep	2 Clab	10	1.600	Ea.	274	50.50		324.50	380
1300	24" deep		10	1.600		360	50.50		410.50	475
1500	84 bins, 6" wide x 6" high, 12" deep		8	2		650	63		713	815
1600	24" deep		8	2		815	63		878	995

10 57 Wardrobe and Closet Specialties

10 57 13 – Hat and Coat Racks

10 57 13.10 Coat Racks and Wardrobes

		Crew	Daily Output	Labor-Hours	Unit	Material	2009 Bare Costs Labor	Equipment	Total	Total Incl O&P
0010	**COAT RACKS AND WARDROBES**									
0020	Hat & coat rack, floor model, 6 hangers									
0050	Standing, beech wood, 21" x 21" x 72", chrome				Ea.	230			230	253
0100	18 gauge tubular steel, 21" x 21" x 69", wood walnut				"	266			266	293
0500	16 gauge steel frame, 22 gauge steel shelves									
0650	Single pedestal, 30" x 18" x 63"				Ea.	234			234	257
0800	Single face rack, 29" x 18-1/2" x 62"					221			221	243
0900	51" x 18-1/2" x 70"					365			365	400
0910	Double face rack, 39" x 26" x 70"					310			310	340
0920	63" x 26" x 70"					415			415	455
0940	For 2" ball casters, add				Set	85.50			85.50	94
1400	Utility hook strips, 3/8" x 2-1/2" x 18", 6 hooks	1 Carp	48	.167	Ea.	57.50	6.65		64.15	74
1500	34" long, 12 hooks	"	48	.167	"	52.50	6.65		59.15	68.50
1650	Wall mounted racks, 16 gauge steel frame, 22 gauge steel shelves									
1850	12" x 15" x 26", 6 hangers	1 Carp	32	.250	Ea.	153	10		163	184
2000	12" x 15" x 50", 12 hangers	"	32	.250	"	179	10		189	213
2150	Wardrobe cabinet, steel, baked enamel finish									
2300	36" x 21" x 78", incl. top shelf & hanger rod				Ea.	273			273	300

10 57 23 – Closet and Utility Shelving

10 57 23.19 Wood Closet and Utility Shelving

		Crew	Daily Output	Labor-Hours	Unit	Material	Labor	Equipment	Total	Total Incl O&P
0010	**WOOD CLOSET AND UTILITY SHELVING**									
0020	Pine, clear grade, no edge band, 1" x 8"	1 Carp	115	.070	L.F.	1.98	2.78		4.76	6.50
0100	1" x 10"		110	.073		2.22	2.91		5.13	6.95
0200	1" x 12"		105	.076		2.64	3.04		5.68	7.60
0600	Plywood, 3/4" thick with lumber edge, 12" wide		75	.107		1.67	4.26		5.93	8.45
0700	24" wide		70	.114		3	4.57		7.57	10.40
0900	Bookcase, clear grade pine, shelves 12" O.C., 8" deep, /SF shelf		70	.114	S.F.	6.45	4.57		11.02	14.20
1000	12" deep shelves		65	.123	"	8.60	4.92		13.52	17.05
1200	Adjustable closet rod and shelf, 12" wide, 3' long		20	.400	Ea.	8.50	16		24.50	34.50
1300	8' long		15	.533	"	24	21.50		45.50	59.50
1500	Prefinished shelves with supports, stock, 8" wide		75	.107	L.F.	3.80	4.26		8.06	10.80
1600	10" wide		70	.114	"	4.21	4.57		8.78	11.75

10 73 Protective Covers

10 73 16 – Canopies

10 73 16.20 Metal Canopies

		Crew	Daily Output	Labor-Hours	Unit	Material	Labor	Equipment	Total	Total Incl O&P
0010	**METAL CANOPIES**									
0020	Wall hung, .032", aluminum, prefinished, 8' x 10'	K-2	1.30	18.462	Ea.	2,150	755	187	3,092	3,850
0300	8' x 20'		1.10	21.818		4,275	890	221	5,386	6,475
0500	10' x 10'		1.30	18.462		2,400	755	187	3,342	4,125
0700	10' x 20'		1.10	21.818		4,450	890	221	5,561	6,675
1000	12' x 20'		1	24		5,475	980	243	6,698	7,975
1360	12' x 30'		.80	30		8,050	1,225	305	9,580	11,300
1700	12' x 40'		.60	40		9,800	1,625	405	11,830	14,000
1900	For free standing units, add					20%	10%			

Estimating Tips
General
- The items in this division are usually priced per square foot or each. Many of these items are purchased by the owner for installation by the contractor. Check the specifications for responsibilities and include time for receiving, storage, installation, and mechanical and electrical hookups in the appropriate divisions.
- Many items in Division 11 require some type of support system that is not usually furnished with the item. Examples of these systems include blocking for the attachment of casework and support angles for ceiling-hung projection screens. The required blocking or supports must be added to the estimate in the appropriate division.
- Some items in Division 11 may require assembly or electrical hookups. Verify the amount of assembly required or the need for a hard electrical connection and add the appropriate costs.

Reference Numbers
Reference numbers are shown in shaded boxes at the beginning of some major classifications. These numbers refer to related items in the Reference Section. The reference information may be an estimating procedure, an alternate pricing method, or technical information.

Note: Not all subdivisions listed here necessarily appear in this publication.

No part of this publication may be reproduced, stored in a retrieval system, or transmitted in any form or by any means without prior written permission of R.S. Means Company, Inc.

11 05 Common Work Results for Equipment

11 05 05 – Selective Equipment Demolition

11 05 05.10 Selective Demolition	Crew	Daily Output	Labor-Hours	Unit	Material	2009 Bare Costs Labor	Equipment	Total	Total Incl O&P
0010 **SELECTIVE DEMOLITION**									
0130 Central vacuum, motor unit, residential or commercial	1 Clab	2	4	Ea.		126		126	196
0210 Vault door and frame	2 Skwk	2	8			325		325	505
0215 Day gate, for vault	"	3	5.333			218		218	335
0380 Bank equipment, teller window, bullet resistant	1 Clab	1.20	6.667			211		211	325
0381 Counter	2 Clab	1.50	10.667	Station		335		335	525
0382 Drive-up window, including drawer and glass		1.50	10.667	"		335		335	525
0383 Thru-wall boxes and chests, selective demolition		2.50	6.400	Ea.		202		202	315
0384 Bullet resistant partitions		20	.800	L.F.		25.50		25.50	39
0385 Pneumatic tube system, 2 lane drive-up	L-3	.45	35.556	Ea.		1,550		1,550	2,375
0386 Safety deposit box	1 Clab	50	.160	Opng.		5.05		5.05	7.85
0387 Surveillance system, video, complete	2 Elec	2	8	Ea.		375		375	560
0410 Church equipment, misc moveable fixtures	2 Clab	1	16			505		505	785
0412 Steeple, to 28' high	F-3	3	13.333			540	256	796	1,100
0414 40' to 60' high	"	.80	50			2,025	960	2,985	4,175
0510 Library equipment, bookshelves, wood, to 90" high	1 Clab	20	.400	L.F.		12.65		12.65	19.60
0515 Carrels, hardwood, 36" x 24"	"	9	.889	Ea.		28		28	43.50
0630 Stage equipment, light control panel	1 Elec	1	8	"		375		375	560
0632 Border lights		40	.200	L.F.		9.40		9.40	14
0634 Spotlights		8	1	Ea.		47		47	70
0636 Telescoping platforms and risers	2 Clab	175	.091	SF Stg.		2.89		2.89	4.48
1020 Barber equipment, hydraulic chair	1 Clab	40	.200	Ea.		6.30		6.30	9.80
1030 Checkout counter, supermarket or warehouse conveyor	2 Clab	18	.889			28		28	43.50
1040 Food cases, refrigerated or frozen	Q-5	6	2.667			118		118	178
1190 Laundry equipment, commercial	L-6	3	4			193		193	288
1360 Movie equipment, lamphouse, to 4000 watt, incl rectifier	1 Elec	4	2			94		94	140
1365 Sound system, incl amplifier	"	1.25	6.400			300		300	450
1410 Air compressor, to 5 H.P.	2 Clab	2.50	6.400			202		202	315
1412 Lubrication equipment, automotive, 3 reel type, incl pump, excl piping	L-4	1	24	Set		895		895	1,375
1414 Booth, spray paint, complete, to 26' long	"	.80	30	"		1,125		1,125	1,725
1560 Parking equipment, cashier booth	B-22	2	15	Ea.		550	97	647	955
1600 Loading dock equipment, dock bumpers, rubber	1 Clab	50	.160	"		5.05		5.05	7.85
1610 Door seal for door perimeter	"	50	.160	L.F.		5.05		5.05	7.85
1620 Platform lifter, fixed, 6' x 8', 5000 lb capacity	E-16	1.50	10.667	Ea.		485	89.50	574.50	970
1630 Dock leveller	"	2	8			365	67	432	725
1640 Lights, single or double arm	1 Elec	8	1			47		47	70
1650 Shelter, fabric, truck or train	1 Clab	1.50	5.333			169		169	261
1790 Waste handling equipment, commercial compactor	L-4	2	12			450		450	695
1792 Commercial or municipal incinerator, gas	"	2	12			450		450	695
1795 Crematory, excluding building	Q-3	.25	128			5,950		5,950	8,925
1910 Detection equipment, cell bar front	E-4	4	8			360	33.50	393.50	680
1912 Cell door and frame		8	4			181	16.75	197.75	340
1914 Prefab cell, 4' to 5' wide, 7' to 8' high, 7' deep		8	4			181	16.75	197.75	340
1916 Cot, bolted, single		40	.800			36	3.35	39.35	68
1918 Visitor cubicle		4	8			360	33.50	393.50	680
2850 Hydraulic gates, canal, flap, knife, slide or sluice, to 18" diameter	L-5A	8	4			179	89.50	268.50	405
2852 19" to 36" diameter		6	5.333			238	119	357	540
2854 37" to 48" diameter		2	16			715	360	1,075	1,625
2856 49" to 60" diameter		1	32			1,425	715	2,140	3,250
2858 Over 60" diameter		.30	106			4,775	2,400	7,175	10,800
3100 Sewage pumping system, prefabricated, to 1000 GPM	C-17D	.20	420			17,400	3,250	20,650	30,400
3110 Sewage treatment, holding tank for recirc chemical water closet	1 Plum	8	1			49		49	73
3900 Wastewater treatment system, to 1500 gallons	B-21	2	14			510	64.50	574.50	855

11 05 Common Work Results for Equipment

11 05 05 – Selective Equipment Demolition

11 05 05.10 Selective Demolition		Crew	Daily Output	Labor-Hours	Unit	Material	2009 Bare Costs Labor	2009 Bare Costs Equipment	Total	Total Incl O&P
4050	Food storage equipment, walk-in refrigerator/freezer	2 Clab	64	.250	S.F.		7.90		7.90	12.25
4052	Shelving, stainless steel, 4 tier or dunnage rack	1 Clab	12	.667	Ea.		21		21	32.50
4100	Food preparation equipment, small countertop		18	.444			14.05		14.05	22
4150	Food delivery carts, heated cabinets		18	.444			14.05		14.05	22
4200	Cooking equipment, commercial range	Q-1	12	1.333			58.50		58.50	88
4250	Hood and ventilation equipment, kitchen exhaust hood, excl fire prot	1 Clab	3	2.667			84.50		84.50	131
4255	Fire protection system	Q-1	3	5.333			234		234	350
4300	Food dispensing equipment, countertop items	1 Clab	15	.533			16.85		16.85	26
4310	Serving counter	"	65	.123	L.F.		3.89		3.89	6.05
4350	Ice machine, ice cube maker, flakers and storage bins, to 2000 lb/day	Q-1	1.60	10	Ea.		440		440	660
4400	Cleaning and disposal, dishwasher, commercial, rack type	L-6	1	12			580		580	865
4410	Dishwasher hood	2 Clab	5	3.200			101		101	157
4420	Garbage disposal, commercial, to 5 H.P.	L-1	8	2			96		96	143
4540	Water heater, residential, to 80 gal/day	"	5	3.200			153		153	229
4542	Water softener, automatic	2 Plum	10	1.600			78		78	117
4544	Disappearing stairway, to 15' floor height	2 Clab	6	2.667			84.50		84.50	131
4710	Darkroom equipment, light	L-7	10	2.800			108		108	166
4712	Heavy	"	1.50	18.667			720		720	1,100
4720	Doors	2 Clab	3.50	4.571	Opng.		144		144	224
4830	Bowling alley, complete, incl pinsetter, scorer, counters, misc supplies	4 Clab	.40	80	Lane		2,525		2,525	3,925
4840	Health club equipment, circuit training apparatus	2 Clab	2	8	Set		253		253	390
4842	Squat racks	"	10	1.600	Ea.		50.50		50.50	78.50
4860	School equipment, basketball backstop	L-2	2	8			281		281	435
4862	Table and benches, folding, in wall, 14' long	L-4	4	6			224		224	345
4864	Bleachers, telescoping, to 30 tier	F-5	120	.267	Seat		10.80		10.80	16.75
4866	Boxing ring, elevated	L-4	.20	120	Ea.		4,475		4,475	6,925
4867	Boxing ring, floor level	"	2	12			450		450	695
4868	Exercise equipment	1 Clab	6	1.333			42		42	65.50
4870	Gym divider	L-4	1000	.024	S.F.		.90		.90	1.39
4875	Scoreboard	R-3	2	10	Ea.		465	64.50	529.50	760
4880	Shooting range, incl bullet traps, targets, excl structure	L-9	1	36	Point		1,300		1,300	2,100
5200	Vocational shop equipment	2 Clab	8	2	Ea.		63		63	98
6200	Fume hood, incl countertop, excl HVAC	"	6	2.667	L.F.		84.50		84.50	131
7100	Medical sterilizing, distiller, water, steam heated, 50 gal capacity	1 Plum	2.80	2.857	Ea.		139		139	209
7200	Medical equipment, surgery table, minor	1 Clab	1	8			253		253	390
7210	Surgical lights, doctors office, single or double arm	2 Elec	3	5.333			251		251	375
7300	Physical therapy, table	2 Clab	4	4			126		126	196
7310	Whirlpool bath, fixed, incl mixing valves	1 Plum	4	2			97.50		97.50	146
7400	Dental equipment, chair, electric or hydraulic	1 Clab	.75	10.667			335		335	525
7410	Central suction system	1 Plum	2	4			195		195	293
7420	Drill console with accessories	1 Clab	3.20	2.500			79		79	123
7430	X-ray unit	"	4	2			63		63	98
7440	X-ray developer	1 Plum	10	.800			39		39	58.50

11 13 Loading Dock Equipment

11 13 13 – Loading Dock Bumpers

11 13 13.10 Dock Bumpers		Crew	Daily Output	Labor-Hours	Unit	Material	2009 Bare Costs Labor	Equipment	Total	Total Incl O&P
0010	**DOCK BUMPERS** Bolts not included									
0020	2" x 6" to 4" x 8", average	1 Carp	.30	26.667	M.B.F.	1,200	1,075		2,275	2,975

11 14 Pedestrian Control Equipment

11 14 13 – Pedestrian Gates

11 14 13.13 Portable Posts and Railings

		Crew	Daily Output	Labor-Hours	Unit	Material	Labor	Equipment	Total	Total Incl O&P
0010	**PORTABLE POSTS AND RAILINGS**									
0020	Portable for pedestrian traffic control, standard, minimum				Ea.	104			104	114
0100	Maximum					178			178	196
0300	Deluxe posts, minimum					170			170	187
0400	Maximum					310			310	340
0600	Ropes for above posts, plastic covered, 1-1/2" diameter				L.F.	10.15			10.15	11.15
0700	Chain core				"	10.35			10.35	11.40
1500	Portable security or safety barrier, black with 7' yellow strap				Ea.	235			235	258
1510	12' yellow strap					275			275	300
1550	Sign holder, standard design					122			122	134

11 14 13.19 Turnstiles

		Crew	Daily Output	Labor-Hours	Unit	Material	Labor	Equipment	Total	Total Incl O&P
0010	**TURNSTILES**									
0020	One way, 4 arm, 46" diameter, economy, manual	2 Carp	5	3.200	Ea.	780	128		908	1,050
0100	Electric		1.20	13.333		1,275	535		1,810	2,225
0300	High security, galv., 5'-5" diameter, 7' high, manual		1	16		3,850	640		4,490	5,250
0350	Electric		.60	26.667		4,350	1,075		5,425	6,450
0420	Three arm, 24" opening, light duty, manual		2	8		1,225	320		1,545	1,850
0450	Heavy duty		1.50	10.667		3,025	425		3,450	4,000
0460	Manual, with registering & controls, light duty		2	8		2,650	320		2,970	3,425
0470	Heavy duty		1.50	10.667		3,200	425		3,625	4,175
0480	Electric, heavy duty		1.10	14.545		3,700	580		4,280	4,975
0500	For coin or token operating, add					595			595	650
1200	One way gate with horizontal bars, 5'-5" diameter									
1300	7' high, recreation or transit type	2 Carp	.80	20	Ea.	4,275	800		5,075	5,975
1500	For electronic counter, add				"	204			204	225

11 16 Vault Equipment

11 16 16 – Safes

11 16 16.10 Commercial Safes

		Crew	Daily Output	Labor-Hours	Unit	Material	Labor	Equipment	Total	Total Incl O&P
0010	**COMMERCIAL SAFES**									
0200	Office, 1 hr. rating, 30" x 18" x 18"				Ea.	2,175			2,175	2,400
0250	40" x 18" x 18"					4,025			4,025	4,425
0300	60" x 36" x 18", double door					7,950			7,950	8,750
0600	1 hr. rating, 27" x 19" x 16"					7,425			7,425	8,150
0700	63" x 34" x 16"					14,100			14,100	15,500
0800	Money, "B" label, 9" x 14" x 14"					525			525	580
0900	Tool resistive, 24" x 24" x 20"					3,575			3,575	3,925
1050	Tool and torch resistive, 24" x 24" x 20"					9,150			9,150	10,100
1150	Jewelers, 23" x 20" x 18"					11,700			11,700	12,800
1200	63" x 25" x 18"					22,400			22,400	24,700
1300	For handling into building, add, minimum	A-2	8.50	2.824			88.50	23	111.50	162
1400	Maximum	"	.78	30.769			965	249	1,214	1,775

11 17 Teller and Service Equipment

11 17 13 – Teller Equipment Systems

11 17 13.10 Bank Equipment

		Crew	Daily Output	Labor-Hours	Unit	Material	2009 Bare Costs Labor	2009 Bare Costs Equipment	Total	Total Incl O&P
0010	**BANK EQUIPMENT**									
0020	Alarm system, police	2 Elec	1.60	10	Ea.	4,675	470		5,145	5,850
0100	With vault alarm	"	.40	40		18,600	1,875		20,475	23,200
0400	Bullet resistant teller window, 44" x 60"	1 Glaz	.60	13.333		3,225	515		3,740	4,325
0500	48" x 60"	"	.60	13.333		4,100	515		4,615	5,300
3000	Counters for banks, frontal only	2 Carp	1	16	Station	1,750	640		2,390	2,925
3100	Complete with steel undercounter	"	.50	32	"	3,400	1,275		4,675	5,725
4600	Door and frame, bullet-resistant, with vision panel, minimum	2 Sswk	1.10	14.545	Ea.	4,550	650		5,200	6,150
4700	Maximum		1.10	14.545		6,175	650		6,825	7,925
4800	Drive-up window, drawer & mike, not incl. glass, minimum		1	16		5,575	715		6,290	7,425
4900	Maximum		.50	32		8,875	1,425		10,300	12,300
5000	Night depository, with chest, minimum		1	16		7,250	715		7,965	9,250
5100	Maximum		.50	32		10,300	1,425		11,725	13,900
5200	Package receiver, painted		3.20	5		1,300	224		1,524	1,825
5300	Stainless steel		3.20	5		2,225	224		2,449	2,825
5400	Partitions, bullet-resistant, 1-3/16" glass, 8' high	2 Carp	10	1.600	L.F.	189	64		253	305
5450	Acrylic	"	10	1.600	"	360	64		424	495
5500	Pneumatic tube systems, 2 lane drive-up, complete	L-3	.25	64	Total	24,600	2,775		27,375	31,400
5550	With T.V. viewer	"	.20	80	"	47,600	3,475		51,075	58,000
5570	Safety deposit boxes, minimum	1 Sswk	44	.182	Opng.	54.50	8.15		62.65	74.50
5580	Maximum, 10" x 15" opening	"	19	.421	"	116	18.80		134.80	161
5600	Pass thru, bullet-res. window, painted steel, 24" x 36"	2 Sswk	1.60	10	Ea.	2,250	445		2,695	3,275
5700	48" x 48"		1.20	13.333		2,550	595		3,145	3,850
5800	72" x 40"		.80	20		3,450	895		4,345	5,375
5900	For stainless steel frames, add					20%				
6100	Surveillance system, video camera, complete	2 Elec	1	16	Ea.	10,400	750		11,150	12,600
6110	For each additional camera, add				"	945			945	1,050
6120	CCTV system, see Div. 27 41 19.10									
6200	Twenty-four hour teller, single unit,									
6300	automated deposit, cash and memo	L-3	.25	64	Ea.	42,800	2,775		45,575	51,500
7000	Vault front, see Div. 08 34 59.10									

11 19 Detention Equipment

11 19 30 – Detention Cell Equipment

11 19 30.10 Cell Equipment

		Crew	Daily Output	Labor-Hours	Unit	Material	2009 Bare Costs Labor	2009 Bare Costs Equipment	Total	Total Incl O&P
0010	**CELL EQUIPMENT**									
3000	Toilet apparatus including wash basin, average	L-8	1.50	13.333	Ea.	3,350	555		3,905	4,550

11 21 Mercantile and Service Equipment

11 21 13 – Cash Registers and Checking Equipment

11 21 13.10 Checkout Counter

		Crew	Daily Output	Labor-Hours	Unit	Material	2009 Bare Costs Labor	2009 Bare Costs Equipment	Total	Total Incl O&P
0010	**CHECKOUT COUNTER**									
0020	Supermarket conveyor, single belt	2 Clab	10	1.600	Ea.	2,475	50.50		2,525.50	2,800
0100	Double belt, power take-away		9	1.778		4,175	56		4,231	4,675
0400	Double belt, power take-away, incl. side scanning		7	2.286		5,200	72		5,272	5,800
0800	Warehouse or bulk type		6	2.667		6,075	84.50		6,159.50	6,800
1000	Scanning system, 2 lanes, w/registers, scan gun & memory				System	15,200			15,200	16,700
1100	10 lanes, single processor, full scan, with scales				"	144,500			144,500	158,500
2000	Register, restaurant, minimum				Ea.	615			615	675

11 21 Mercantile and Service Equipment

11 21 13 – Cash Registers and Checking Equipment

11 21 13.10 Checkout Counter	Crew	Daily Output	Labor-Hours	Unit	Material	2009 Bare Costs Labor	Equipment	Total	Total Incl O&P
2100 Maximum				Ea.	2,700			2,700	2,975
2150 Store, minimum					615			615	675
2200 Maximum					2,700			2,700	2,975

11 21 53 – Barber and Beauty Shop Equipment

11 21 53.10 Barber Equipment

0010 **BARBER EQUIPMENT**									
0020 Chair, hydraulic, movable, minimum	1 Carp	24	.333	Ea.	405	13.30		418.30	465
0050 Maximum	"	16	.500		2,750	20		2,770	3,050
0200 Wall hung styling station with mirrors, minimum	L-2	8	2		355	70.50		425.50	500
0300 Maximum	"	4	4		1,825	141		1,966	2,250
0500 Sink, hair washing basin, rough plumbing not incl.	1 Plum	8	1		395	49		444	505
1000 Sterilizer, liquid solution for tools					128			128	140
1100 Total equipment, rule of thumb, per chair, minimum	L-8	1	20		1,525	835		2,360	2,950
1150 Maximum	"	1	20		4,075	835		4,910	5,775

11 23 Commercial Laundry and Dry Cleaning Equipment

11 23 13 – Dry Cleaning Equipment

11 23 13.13 Dry Cleaners

0010 **DRY CLEANERS** Not incl. rough-in									
2000 Dry cleaners, electric, 20 lb. capacity	L-1	.20	80	Ea.	33,300	3,825		37,125	42,300
2050 25 lb. capacity		.17	94.118		45,100	4,500		49,600	56,500
2100 30 lb. capacity		.15	106		47,600	5,100		52,700	60,000
2150 60 lb. capacity		.09	177		73,000	8,500		81,500	92,500

11 23 16 – Drying and Conditioning Equipment

11 23 16.13 Dryers

0010 **DRYERS**, Not including rough-in									
1500 Industrial, 30 lb. capacity	1 Plum	2	4	Ea.	2,975	195		3,170	3,575
1600 50 lb. capacity	"	1.70	4.706		3,200	229		3,429	3,875
4700 Lint collector, ductwork not included, 8,000 to 10,000 C.F.M.	Q-10	.30	80		8,575	3,525		12,100	14,900

11 23 19 – Finishing Equipment

11 23 19.13 Folders and Spreaders

0010 **FOLDERS AND SPREADERS**									
3500 Folders, blankets & sheets, minimum	1 Elec	.17	47.059	Ea.	30,600	2,200		32,800	37,000
3700 King size with automatic stacker		.10	80		55,500	3,750		59,250	66,500
3800 For conveyor delivery, add		.45	17.778		6,000	835		6,835	7,850
4900 Spreader feeders, 240V, 2 station	L-6	.70	17.143		51,000	825		51,825	57,000
4920 4 station	"	.35	34.286		61,500	1,650		63,150	70,000

11 23 23 – Commercial Ironing Equipment

11 23 23.13 Irons and Pressers

0010 **IRONS AND PRESSERS**									
4500 Ironers, institutional, 110", single roll	1 Elec	.20	40	Ea.	29,700	1,875		31,575	35,500
4800 Pressers, low capacity air operated	L-6	1.75	6.857		8,775	330		9,105	10,100
4820 Hand operated		1.75	6.857		8,125	330		8,455	9,450
4840 Ironer 48", 240V		3.50	3.429		103,000	165		103,165	113,500
6600 Hand operated presser		.70	17.143		5,525	825		6,350	7,300
6620 Mushroom press 115V		.70	17.143		6,925	825		7,750	8,825

11 23 Commercial Laundry and Dry Cleaning Equipment

11 23 26 – Commercial Washers and Extractors

11 23 26.13 Washers and Extractors

		Crew	Daily Output	Labor-Hours	Unit	Material	2009 Bare Costs Labor	2009 Bare Costs Equipment	Total	Total Incl O&P
0010	**WASHERS AND EXTRACTORS**, not including rough-in									
6000	Combination washer/extractor, 20 lb. capacity	L-6	1.50	8	Ea.	5,350	385		5,735	6,475
6100	30 lb. capacity		.80	15		8,500	725		9,225	10,400
6200	50 lb. capacity		.68	17.647		9,925	850		10,775	12,200
6300	75 lb. capacity		.30	40		18,600	1,925		20,525	23,400
6350	125 lb. capacity		.16	75		24,900	3,625		28,525	32,800
6380	Washer extractor/dryer, 110 lb., 240V		1	12		8,150	580		8,730	9,850
6400	Washer extractor, 135 lb, 240V		1	12		23,600	580		24,180	26,900
6450	Pass through		1	12		64,000	580		64,580	71,500
6500	200 lb. washer extractor		1	12		66,500	580		67,080	74,500
6550	Pass through		1	12		69,500	580		70,080	77,500
6600	Extractor, low capacity		1.75	6.857		6,625	330		6,955	7,775

11 23 33 – Coin-Operated Laundry Equipment

11 23 33.13 Coin Operated Washers and Dryers

		Crew	Daily Output	Labor-Hours	Unit	Material	Labor	Equipment	Total	Total Incl O&P
0010	**COIN OPERATED WASHERS AND DRYERS**									
0990	Dryer, gas fired									
1000	Commercial, 30 lb. capacity, coin operated, single	1 Plum	3	2.667	Ea.	3,100	130		3,230	3,600
1100	Double stacked	"	2	4		6,275	195		6,470	7,200
4860	Coin dry cleaner 20 lb.	L-6	1.75	6.857		25,900	330		26,230	29,000
5290	Clothes washer									
5300	Commercial, coin operated, average	1 Plum	3	2.667	Ea.	1,125	130		1,255	1,450

11 24 Maintenance Equipment

11 24 19 – Vacuum Cleaning Systems

11 24 19.10 Vacuum Cleaning

		Crew	Daily Output	Labor-Hours	Unit	Material	Labor	Equipment	Total	Total Incl O&P
0010	**VACUUM CLEANING**									
0020	Central, 3 inlet, residential	1 Skwk	.90	8.889	Total	925	365		1,290	1,575
0200	Commercial		.70	11.429		1,050	465		1,515	1,900
0400	5 inlet system, residential		.50	16		1,225	655		1,880	2,350
0600	7 inlet system, commercial		.40	20		1,400	815		2,215	2,800
0800	9 inlet system, residential		.30	26.667		3,000	1,100		4,100	4,975
4010	Rule of thumb: First 1200 S.F., installed								1,150	1,275
4020	For each additional S.F., add				S.F.					.21

11 26 Unit Kitchens

11 26 13 – Metal Unit Kitchens

11 26 13.10 Commercial Unit Kitchens

		Crew	Daily Output	Labor-Hours	Unit	Material	Labor	Equipment	Total	Total Incl O&P
0010	**COMMERCIAL UNIT KITCHENS**									
1500	Combination range, refrigerator and sink, 30" wide, minimum	L-1	2	8	Ea.	900	385		1,285	1,550
1550	Maximum		1	16		3,175	765		3,940	4,650
1570	60" wide, average		1.40	11.429		2,725	545		3,270	3,800
1590	72" wide, average		1.20	13.333		3,700	640		4,340	5,025
1600	Office model, 48" wide		2	8		2,750	385		3,135	3,600
1620	Refrigerator and sink only		2.40	6.667		2,225	320		2,545	2,900
1640	Combination range, refrigerator, sink, microwave									
1660	Oven and ice maker	L-1	.80	20	Ea.	3,950	960		4,910	5,775

11 27 Photographic Processing Equipment

11 27 13 – Darkroom Processing Equipment

11 27 13.10 Darkroom Equipment		Crew	Daily Output	Labor-Hours	Unit	Material	2009 Bare Costs Labor	2009 Bare Costs Equipment	Total	Total Incl O&P
0010	**DARKROOM EQUIPMENT**									
0020	Developing sink, 5" deep, 24" x 48"	Q-1	2	8	Ea.	5,025	350		5,375	6,050
0050	48" x 52"		1.70	9.412		5,100	415		5,515	6,225
0200	10" deep, 24" x 48"		1.70	9.412		6,475	415		6,890	7,750
0250	24" x 108"		1.50	10.667		8,525	470		8,995	10,100
0500	Dryers, dehumidified filtered air, 36" x 25" x 68" high	L-7	6	4.667		4,800	180		4,980	5,550
0550	48" x 25" x 68" high		5	5.600		9,000	216		9,216	10,200
2000	Processors, automatic, color print, minimum		4	7		13,100	270		13,370	14,800
2050	Maximum		.60	46.667		20,600	1,800		22,400	25,500
2300	Black and white print, minimum		2	14		9,950	540		10,490	11,700
2350	Maximum		.80	35		59,500	1,350		60,850	67,500
2600	Manual processor, 16" x 20" maximum print size		2	14		9,175	540		9,715	10,900
2650	20" x 24" maximum print size		1	28		8,725	1,075		9,800	11,300
3000	Viewing lights, 20" x 24"		6	4.667		380	180		560	690
3100	20" x 24" with color correction		6	4.667		535	180		715	865
3500	Washers, round, minimum sheet 11" x 14"	Q-1	2	8		3,075	350		3,425	3,900
3550	Maximum sheet 20" x 24"		1	16		3,475	700		4,175	4,875
3800	Square, minimum sheet 20" x 24"		1	16		3,125	700		3,825	4,475
3900	Maximum sheet 50" x 56"		.80	20		4,850	880		5,730	6,650
4500	Combination tank sink, tray sink, washers, with									
4510	Dry side tables, average	Q-1	.45	35.556	Ea.	9,600	1,550		11,150	13,000

11 31 Residential Appliances

11 31 13 – Residential Kitchen Appliances

11 31 13.13 Cooking Equipment

		Crew	Daily Output	Labor-Hours	Unit	Material	Labor	Equipment	Total	Total Incl O&P
0010	**COOKING EQUIPMENT**									
0020	Cooking range, 30" free standing, 1 oven, minimum	2 Clab	10	1.600	Ea.	320	50.50		370.50	430
0050	Maximum	"	4	4		1,775	126		1,901	2,150
0100	Downdraft, with grille	L-3	4	4		1,975	174		2,149	2,450
0700	Free-standing, 1 oven, 21" wide range, minimum	2 Clab	10	1.600		315	50.50		365.50	425
0900	Countertop cooktops, 4 burner, standard, minimum	1 Elec	6	1.333		244	62.50		306.50	360
0950	Maximum		3	2.667		585	125		710	830
1250	Microwave oven, minimum		4	2		104	94		198	254
1300	Maximum		2	4		450	188		638	775

11 31 13.23 Refrigeration Equipment

		Crew	Daily Output	Labor-Hours	Unit	Material	Labor	Equipment	Total	Total Incl O&P
0010	**REFRIGERATION EQUIPMENT**									
5200	Icemaker, automatic, 20 lb. per day	1 Plum	7	1.143	Ea.	850	55.50		905.50	1,025
5350	51 lb. per day	"	2	4		1,175	195		1,370	1,600
5450	Refrigerator, no frost, 6 C.F.	2 Clab	15	1.067		238	33.50		271.50	315
5500	Refrigerator, no frost, 10 C.F. to 12 C.F. minimum		10	1.600		445	50.50		495.50	570
5600	Maximum		6	2.667		520	84.50		604.50	705
5800	Maximum		5	3.200		620	101		721	840
5950	18 C.F. to 20 C.F., minimum		8	2		575	63		638	735
6000	Maximum		4	4		1,000	126		1,126	1,300
6200	Maximum		3	5.333		2,950	169		3,119	3,500
6790	Energy-star qualified, 18 CF, minimum [G]	2 Carp	4	4		525	160		685	830
6795	Maximum [G]		2	8		1,050	320		1,370	1,650
6797	21.7 CF, minimum [G]		4	4		875	160		1,035	1,225
6799	Maximum [G]		4	4		1,150	160		1,310	1,525

11 31 Residential Appliances

11 31 13 – Residential Kitchen Appliances

11 31 13.33 Kitchen Cleaning Equipment		Crew	Daily Output	Labor-Hours	Unit	Material	2009 Bare Costs Labor	2009 Bare Costs Equipment	Total	Total Incl O&P
0010	**KITCHEN CLEANING EQUIPMENT**									
2750	Dishwasher, built-in, 2 cycles, minimum	L-1	4	4	Ea.	206	192		398	515
2800	Maximum		2	8		310	385		695	910
3100	Energy-star qualified, minimum	G	4	4		315	192		507	630
3110	Maximum	G	2	8		1,000	385		1,385	1,675

11 31 13.43 Waste Disposal Equipment

		Crew	Daily Output	Labor-Hours	Unit	Material	Labor	Equipment	Total	Total Incl O&P
0010	**WASTE DISPOSAL EQUIPMENT**									
1750	Compactor, residential size, 4 to 1 compaction, minimum	1 Carp	5	1.600	Ea.	485	64		549	630
1800	Maximum	"	3	2.667		515	107		622	730
3300	Garbage disposal, sink type, minimum	L-1	10	1.600		66.50	76.50		143	188
3350	Maximum	"	10	1.600		192	76.50		268.50	325

11 31 13.53 Kitchen Ventilation Equipment

		Crew	Daily Output	Labor-Hours	Unit	Material	Labor	Equipment	Total	Total Incl O&P
0010	**KITCHEN VENTILATION EQUIPMENT**									
4150	Hood for range, 2 speed, vented, 30" wide, minimum	L-3	5	3.200	Ea.	52.50	139		191.50	271
4200	Maximum	"	3	5.333		810	232		1,042	1,250
4500	For ventless hood, 2 speed, add					16.55			16.55	18.20

11 31 23 – Residential Laundry Appliances

11 31 23.13 Washers

		Crew	Daily Output	Labor-Hours	Unit	Material	Labor	Equipment	Total	Total Incl O&P
0010	**WASHERS**									
5000	Residential, 4 cycle, average	1 Plum	3	2.667	Ea.	800	130		930	1,075
6650	Washing machine, automatic, minimum		3	2.667		315	130		445	540
6700	Maximum		1	8		1,100	390		1,490	1,775
6750	Energy star qualified, front loading, minimum	G	3	2.667		510	130		640	755
6760	Maximum	G	1	8		1,600	390		1,990	2,325
6764	Top loading, minimum	G	3	2.667		485	130		615	730
6766	Maximum	G	3	2.667		990	130		1,120	1,300

11 31 23.23 Dryers

		Crew	Daily Output	Labor-Hours	Unit	Material	Labor	Equipment	Total	Total Incl O&P
0010	**DRYERS**									
0500	Gas fired residential, 16 lb. capacity, average	1 Plum	3	2.667	Ea.	625	130		755	885
6770	Electric, front loading, energy-star qualified, minimum	G	L-2	3	5.333		255	187		442	570
6780	Maximum	G	"	2	8		1,425	281		1,706	1,975
7450	Vent kits for dryers		1 Carp	10	.800		31.50	32		63.50	84.50

11 31 33 – Miscellaneous Residential Appliances

11 31 33.13 Sump Pumps

		Crew	Daily Output	Labor-Hours	Unit	Material	Labor	Equipment	Total	Total Incl O&P
0010	**SUMP PUMPS**									
6400	Cellar drainer, pedestal, 1/3 H.P., molded PVC base	1 Plum	3	2.667	Ea.	119	130		249	325
6450	Solid brass	"	2	4	"	263	195		458	585

11 31 33.23 Water Heaters

		Crew	Daily Output	Labor-Hours	Unit	Material	Labor	Equipment	Total	Total Incl O&P
0010	**WATER HEATERS**									
6900	Electric, glass lined, 30 gallon, minimum	L-1	5	3.200	Ea.	420	153		573	695
6950	Maximum		3	5.333		585	255		840	1,025
7100	80 gallon, minimum		2	8		760	385		1,145	1,400
7150	Maximum		1	16		1,050	765		1,815	2,300
7180	Gas, glass lined, 30 gallon, minimum	2 Plum	5	3.200		565	156		721	855
7220	Maximum		3	5.333		785	260		1,045	1,250
7260	50 gallon, minimum		2.50	6.400		845	310		1,155	1,400
7300	Maximum		1.50	10.667		1,175	520		1,695	2,050
7310	Water heater, see also Div. 22 33 30.13									

11 31 Residential Appliances

11 31 33 – Miscellaneous Residential Appliances

11 31 33.43 Air Quality

		Crew	Daily Output	Labor-Hours	Unit	Material	2009 Bare Costs Labor	2009 Bare Costs Equipment	Total	Total Incl O&P
0010	**AIR QUALITY**									
2450	Dehumidifier, portable, automatic, 15 pint				Ea.	150			150	165
2550	40 pint					179			179	197
3550	Heater, electric, built-in, 1250 watt, ceiling type, minimum	1 Elec	4	2		96	94		190	245
3600	Maximum		3	2.667		156	125		281	360
3700	Wall type, minimum		4	2		143	94		237	297
3750	Maximum		3	2.667		154	125		279	355
3900	1500 watt wall type, with blower		4	2		143	94		237	297
3950	3000 watt		3	2.667		292	125		417	505
4850	Humidifier, portable, 8 gallons per day					160			160	176
5000	15 gallons per day					193			193	212

11 33 Retractable Stairs

11 33 10 – Disappearing Stairs

11 33 10.10 Disappearing Stairway

		Crew	Daily Output	Labor-Hours	Unit	Material	2009 Bare Costs Labor	2009 Bare Costs Equipment	Total	Total Incl O&P
0010	**DISAPPEARING STAIRWAY** No trim included									
0100	Custom grade, pine, 8'-6" ceiling, minimum	1 Carp	4	2	Ea.	202	80		282	345
0150	Average		3.50	2.286		215	91.50		306.50	380
0200	Maximum		3	2.667		275	107		382	470
0500	Heavy duty, pivoted, from 7'-7" to 12'-10" floor to floor		3	2.667		650	107		757	880
0600	16'-0" ceiling		2	4		1,325	160		1,485	1,700
0800	Economy folding, pine, 8'-6" ceiling		4	2		195	80		275	340
0900	9'-6" ceiling		4	2		225	80		305	370
1100	Automatic electric, aluminum, floor to floor height, 8' to 9'	2 Carp	1	16		8,250	640		8,890	10,100
1400	11' to 12'		.90	17.778		8,875	710		9,585	10,900
1700	14' to 15'		.70	22.857		9,475	915		10,390	11,800

11 41 Food Storage Equipment

11 41 13 – Refrigerated Food Storage Cases

11 41 13.10 Refrigerated Food Cases

		Crew	Daily Output	Labor-Hours	Unit	Material	2009 Bare Costs Labor	2009 Bare Costs Equipment	Total	Total Incl O&P
0010	**REFRIGERATED FOOD CASES**									
0030	Dairy, multi-deck, 12' long	Q-5	3	5.333	Ea.	9,925	237		10,162	11,300
0100	For rear sliding doors, add					1,425			1,425	1,550
0200	Delicatessen case, service deli, 12' long, single deck	Q-5	3.90	4.103		6,675	182		6,857	7,625
0300	Multi-deck, 18 S.F. shelf display		3	5.333		7,425	237		7,662	8,525
0400	Freezer, self-contained, chest-type, 30 C.F.		3.90	4.103		6,700	182		6,882	7,650
0500	Glass door, upright, 78 C.F.		3.30	4.848		9,275	215		9,490	10,500
0600	Frozen food, chest type, 12' long		3.30	4.848		6,700	215		6,915	7,700
0700	Glass door, reach-in, 5 door		3	5.333		12,900	237		13,137	14,500
0800	Island case, 12' long, single deck		3.30	4.848		7,600	215		7,815	8,675
0900	Multi-deck		3	5.333		16,100	237		16,337	18,100
1000	Meat case, 12' long, single deck		3.30	4.848		5,525	215		5,740	6,400
1050	Multi-deck		3.10	5.161		9,500	229		9,729	10,700
1100	Produce, 12' long, single deck		3.30	4.848		7,300	215		7,515	8,350
1200	Multi-deck		3.10	5.161		7,975	229		8,204	9,125

11 41 13.20 Refrigerated Food Storage Equipment

		Crew	Daily Output	Labor-Hours	Unit	Material	2009 Bare Costs Labor	2009 Bare Costs Equipment	Total	Total Incl O&P
0010	**REFRIGERATED FOOD STORAGE EQUIPMENT**									
1000	Beverage chiller, small	1 Clab	13	.615	Ea.	1,325	19.45		1,344.45	1,500

11 41 Food Storage Equipment

11 41 13 – Refrigerated Food Storage Cases

11 41 13.20 Refrigerated Food Storage Equipment

		Crew	Daily Output	Labor-Hours	Unit	Material	2009 Bare Costs Labor	Equipment	Total	Total Incl O&P
1010	Large	1 Clab	13	.615	Ea.	1,950	19.45		1,969.45	2,175
2350	Cooler, reach-in, beverage, 6' long	Q-1	6	2.667		4,300	117		4,417	4,925
2360	Cooler, bottle, 6' long, minimum		.80	20		1,525	880		2,405	3,000
2370	Maximum		.80	20		1,700	880		2,580	3,200
2380	10' long, minimum		.50	32		2,100	1,400		3,500	4,400
2390	Maximum		.50	32		2,575	1,400		3,975	4,925
4300	Freezers, reach-in, 44 C.F.		4	4		3,150	176		3,326	3,725
4500	68 C.F.		3	5.333		4,350	234		4,584	5,125
4600	Freezer, pre-fab, 8' x 8' w/refrigeration	2 Carp	.45	35.556		10,200	1,425		11,625	13,400
4620	8' x 12'		.35	45.714		12,100	1,825		13,925	16,100
4640	8' x 16'		.25	64		15,000	2,550		17,550	20,500
4660	8' x 20'		.17	94.118		18,100	3,750		21,850	25,700
4680	Reach-in, 1 compartment	Q-1	4	4		1,550	176		1,726	1,975
4700	2 compartment		3	5.333		4,100	234		4,334	4,850
8300	Refrigerators, reach-in type, 44 C.F.		5	3.200		6,775	140		6,915	7,650
8310	With glass doors, 68 C.F.		4	4		6,625	176		6,801	7,575
8320	Refrigerator, reach-in, 1 compartment	R-18	7.80	3.333		1,950	123		2,073	2,325
8330	2 compartment		6.20	4.194		2,900	154		3,054	3,400
8340	3 compartment		5.60	4.643		3,925	171		4,096	4,550
8350	Pre-fab, with refrigeration, 8' x 8'	2 Carp	.45	35.556		5,375	1,425		6,800	8,100
8360	8' x 12'		.35	45.714		7,125	1,825		8,950	10,700
8370	8' x 16'		.25	64		10,400	2,550		12,950	15,400
8380	8' x 20'		.17	94.118		13,100	3,750		16,850	20,200
8390	Pass-thru/roll-in, 1 compartment	R-18	7.80	3.333		3,550	123		3,673	4,075
8400	2 compartment		6.24	4.167		4,975	153		5,128	5,675
8410	3 compartment		5.60	4.643		5,950	171		6,121	6,800
8420	Walk-in, alum, door & floor only, no refrig, 6' x 6' x 7'-6"	2 Carp	1.40	11.429		8,150	455		8,605	9,650
8430	10' x 6' x 7'-6"		.55	29.091		11,700	1,150		12,850	14,600
8440	12' x 14' x 7'-6"		.25	64		16,300	2,550		18,850	21,900
8450	12' x 20' x 7'-6"		.17	94.118		20,000	3,750		23,750	27,800
8455	For 8'-6" high, add					12%				
8460	Refrigerated cabinets, mobile				Ea.	3,425			3,425	3,775
8470	Refrigerator/freezer, reach-in, 1 compartment	R-18	5.60	4.643		4,425	171		4,596	5,125
8480	2 compartment	"	4.80	5.417		6,250	199		6,449	7,175

11 41 13.30 Wine Cellar

		Crew	Daily Output	Labor-Hours	Unit	Material	2009 Bare Costs Labor	Equipment	Total	Total Incl O&P
0010	**WINE CELLAR**, refrigerated, Redwood interior, carpeted, walk-in type									
0020	6'-8" high, including racks									
0200	80" W x 48" D for 900 bottles	2 Carp	1.50	10.667	Ea.	3,350	425		3,775	4,325
0250	80" W x 72" D for 1300 bottles		1.33	12.030		4,425	480		4,905	5,600
0300	80" W x 94" D for 1900 bottles		1.17	13.675		5,725	545		6,270	7,150
0400	80" W x 124" D for 2500 bottles		1	16		6,925	640		7,565	8,625
0600	Portable cabinets, red oak, reach-in temp.& humidity controlled									
0650	26-5/8"W x 26-1/2"D x 68"H for 235 bottles				Ea.	2,200			2,200	2,425
0660	32"W x 21-1/2"D x 73-1/2"H for 144 bottles					3,500			3,500	3,850
0670	32"W x 29-1/2"D x 73-1/2"H for 288 bottles					2,525			2,525	2,775
0680	39-1/2"W x 29-1/2"D x 86-1/2"H for 440 bottles					3,125			3,125	3,450
0690	52-1/2"W x 29-1/2"D x 73-1/2"H for 468 bottles					3,725			3,725	4,075
0700	52-1/2"W x 29-1/2"D x 86-1/2"H for 572 bottles					3,825			3,825	4,225
0730	Portable, red oak, can be built-in with glass door									
0750	23-7/8"W x 24"D x 34-1/2"H for 50 bottles				Ea.	955			955	1,050

11 41 Food Storage Equipment

11 41 33 – Food Storage Shelving

11 41 33.20 Metal Food Storage Shelving

		Crew	Daily Output	Labor-Hours	Unit	Material	2009 Bare Costs Labor	Equipment	Total	Total Incl O&P
0010	**METAL FOOD STORAGE SHELVING**									
8600	Stainless steel shelving, louvered 4-tier, 20" x 3'	1 Clab	6	1.333	Ea.	1,175	42		1,217	1,350
8605	20" x 4'		6	1.333		1,675	42		1,717	1,900
8610	20" x 6'		6	1.333		2,425	42		2,467	2,725
8615	24" x 3'		6	1.333		1,525	42		1,567	1,750
8620	24" x 4'		6	1.333		1,775	42		1,817	2,025
8625	24" x 6'		6	1.333		2,575	42		2,617	2,925
8630	Flat 4-tier, 20" x 3'		6	1.333		980	42		1,022	1,150
8635	20" x 4'		6	1.333		1,225	42		1,267	1,425
8640	20" x 5'		6	1.333		1,350	42		1,392	1,575
8645	24" x 3'		6	1.333		1,075	42		1,117	1,250
8650	24" x 4'		6	1.333		1,875	42		1,917	2,150
8655	24" x 6'		6	1.333		2,250	42		2,292	2,550
8700	Galvanized shelving, louvered 4-tier, 20" x 3'		6	1.333		595	42		637	715
8705	20" x 4'		6	1.333		660	42		702	795
8710	20" x 6'		6	1.333		700	42		742	835
8715	24" x 3'		6	1.333		525	42		567	640
8720	24" x 4'		6	1.333		710	42		752	845
8725	24" x 6'		6	1.333		1,025	42		1,067	1,200
8730	Flat 4-tier, 20" x 3'		6	1.333		425	42		467	535
8735	20" x 4'		6	1.333		480	42		522	595
8740	20" x 6'		6	1.333		835	42		877	980
8745	24" x 3'		6	1.333		430	42		472	540
8750	24" x 4'		6	1.333		615	42		657	745
8755	24" x 6'		6	1.333		865	42		907	1,025
8760	Stainless steel dunnage rack, 24" x 3'		8	1		395	31.50		426.50	485
8765	24" x 4'		8	1		775	31.50		806.50	905
8770	Galvanized dunnage rack, 24" x 3'		8	1		162	31.50		193.50	227
8775	24" x 4'		8	1		181	31.50		212.50	248

11 42 Food Preparation Equipment

11 42 10 – Commercial Food Preparation Equipment

11 42 10.10 Choppers, Mixers and Misc. Equipment

		Crew	Daily Output	Labor-Hours	Unit	Material	2009 Bare Costs Labor	Equipment	Total	Total Incl O&P
0010	**CHOPPERS, MIXERS AND MISC. EQUIPMENT**									
1700	Choppers, 5 pounds	R-18	7	3.714	Ea.	1,625	137		1,762	1,975
1720	16 pounds		5	5.200		1,975	191		2,166	2,475
1740	35 to 40 pounds		4	6.500		2,975	239		3,214	3,625
1840	Coffee brewer, 5 burners	1 Plum	3	2.667		1,175	130		1,305	1,500
1850	Coffee urn, twin 6 gallon urns		2	4		2,525	195		2,720	3,075
1860	Single, 3 gallon		3	2.667		1,800	130		1,930	2,200
3000	Fast food equipment, total package, minimum	6 Skwk	.08	600		172,000	24,500		196,500	227,500
3100	Maximum	"	.07	685		234,500	28,000		262,500	301,500
3800	Food mixers, 20 quarts	L-7	7	4		2,700	154		2,854	3,200
3810	Food mixers, bench type									
3820	25 quarts	L-7	5.40	5.185	Ea.	2,700	200		2,900	3,275
3850	40 quarts		5.40	5.185		6,950	200		7,150	7,950
3900	60 quarts		5	5.600		8,925	216		9,141	10,200
4040	80 quarts		3.90	7.179		11,000	277		11,277	12,500
4080	130 quarts		2.20	12.727		16,600	490		17,090	19,000
4100	Floor type, 20 quarts		15	1.867		3,125	72		3,197	3,525

11 42 Food Preparation Equipment

11 42 10 – Commercial Food Preparation Equipment

11 42 10.10 Choppers, Mixers and Misc. Equipment		Crew	Daily Output	Labor-Hours	Unit	Material	2009 Bare Costs Labor	2009 Bare Costs Equipment	Total	Total Incl O&P
4120	60 quarts	L-7	14	2	Ea.	7,725	77		7,802	8,625
4140	80 quarts		12	2.333		9,575	90		9,665	10,600
4160	140 quarts	↓	8.60	3.256		18,200	126		18,326	20,200
6700	Peelers, small	R-18	8	3.250		1,450	119		1,569	1,775
6720	Large	"	6	4.333		4,375	159		4,534	5,075
6800	Pulper/extractor, close coupled, 5 HP	1 Plum	1.90	4.211		3,325	205		3,530	3,975
8580	Slicer with table	R-18	9	2.889	↓	4,675	106		4,781	5,275

11 43 Food Delivery Carts and Conveyors

11 43 13 – Food Delivery Carts

11 43 13.10 Mobile Carts, Racks and Trays		Crew	Daily Output	Labor-Hours	Unit	Material	Labor	Equipment	Total	Total Incl O&P
0010	**MOBILE CARTS, RACKS AND TRAYS**									
1650	Cabinet, heated, 1 compartment, reach-in	R-18	5.60	4.643	Ea.	3,250	171		3,421	3,825
1655	Pass-thru roll-in		5.60	4.643		5,125	171		5,296	5,875
1660	2 compartment, reach-in		4.80	5.417		6,750	199		6,949	7,725
1665	Pass-thru, roll-in	↓	4.80	5.417		7,850	199		8,049	8,925
1670	Mobile					3,000			3,000	3,300
6850	Mobile rack w/pan slide					1,450			1,450	1,600
9180	Tray and silver dispenser, mobile	1 Clab	16	.500	↓	775	15.80		790.80	875

11 44 Food Cooking Equipment

11 44 13 – Commercial Ranges

11 44 13.10 Cooking Equipment		Crew	Daily Output	Labor-Hours	Unit	Material	Labor	Equipment	Total	Total Incl O&P
0010	**COOKING EQUIPMENT**									
0020	Bake oven, gas, one section	Q-1	8	2	Ea.	4,300	88		4,388	4,875
0300	Two sections		7	2.286		10,100	100		10,200	11,400
0600	Three sections	↓	6	2.667		13,200	117		13,317	14,700
0900	Electric convection, single deck	L-7	4	7		5,400	270		5,670	6,350
1300	Broiler, without oven, standard	Q-1	8	2		3,550	88		3,638	4,025
1550	Infrared	L-7	4	7		9,250	270		9,520	10,600
4750	Fryer, with twin baskets, modular model	Q-1	7	2.286		1,450	100		1,550	1,750
5000	Floor model, on 6" legs	"	5	3.200		2,400	140		2,540	2,825
5100	Extra single basket, large					99			99	109
5300	Griddle, SS, 24" plate, w/4" legs, elec, 208V, 3 phase, 3' long	Q-1	7	2.286		1,125	100		1,225	1,375
5550	4' long	"	6	2.667		1,450	117		1,567	1,775
6200	Iced tea brewer	1 Plum	3.44	2.326		670	113		783	905
6350	Kettle, w/steam jacket, tilting, w/positive lock, SS, 20 gallons	L-7	7	4		6,675	154		6,829	7,600
6600	60 gallons	"	6	4.667		8,750	180		8,930	9,875
6680	Kettle fillers	1 Plum	5.50	1.455		231	71		302	360
6900	Range, restaurant type, 6 burners and 1 standard oven, 36" wide	Q-1	7	2.286		2,325	100		2,425	2,700
6950	Convection		7	2.286		4,700	100		4,800	5,325
7150	2 standard ovens, 24" griddle, 60" wide		6	2.667		7,450	117		7,567	8,375
7200	1 standard, 1 convection oven		6	2.667		6,700	117		6,817	7,550
7450	Heavy duty, single 34" standard oven, open top		5	3.200		4,850	140		4,990	5,525
7500	Convection oven		5	3.200		3,425	140		3,565	3,975
7700	Griddle top		6	2.667		2,625	117		2,742	3,075
7750	Convection oven	↓	6	2.667		5,950	117		6,067	6,700
8850	Steamer, electric 27 KW	L-7	7	4		7,375	154		7,529	8,375
9100	Electric, 10 KW or gas 100,000 BTU	"	5	5.600		5,350	216		5,566	6,200

11 44 Food Cooking Equipment

11 44 13 – Commercial Ranges

11 44 13.10 Cooking Equipment		Crew	Daily Output	Labor-Hours	Unit	Material	2009 Bare Costs Labor	Equipment	Total	Total Incl O&P
9150	Toaster, conveyor type, 16-22 slices per minute				Ea.	1,100			1,100	1,200
9160	Pop-up, 2 slot				↓	535			535	585
9200	For deluxe models of above equipment, add					75%				
9400	Rule of thumb: Equipment cost based									
9410	on kitchen work area									
9420	Office buildings, minimum	L-7	77	.364	S.F.	76.50	14.05		90.55	106
9450	Maximum		58	.483		129	18.60		147.60	171
9550	Public eating facilities, minimum		77	.364		100	14.05		114.05	133
9600	Maximum		46	.609		163	23.50		186.50	216
9750	Hospitals, minimum		58	.483		103	18.60		121.60	143
9800	Maximum	↓	39	.718	↓	190	27.50		217.50	252

11 46 Food Dispensing Equipment

11 46 16 – Service Line Equipment

11 46 16.10 Commercial Food Dispensing Equipment

		Crew	Daily Output	Labor-Hours	Unit	Material	Labor	Equipment	Total	Total Incl O&P
0010	**COMMERCIAL FOOD DISPENSING EQUIPMENT**									
1050	Butter pat dispenser	1 Clab	13	.615	Ea.	865	19.45		884.45	985
1100	Bread dispenser, counter top	"	13	.615		825	19.45		844.45	940
1600	Beverage dispenser, carbonated	R-18	3.20	8.125		1,875	299		2,174	2,500
1620	With ice holder, 90 pounds		2.80	9.286		3,525	340		3,865	4,400
1640	150 pounds	↓	2	13		3,325	480		3,805	4,375
1900	Cup and glass dispenser, drop in	1 Clab	4	2		1,200	63		1,263	1,425
1920	Disposable cup, drop in		16	.500		405	15.80		420.80	470
2650	Dish dispenser, drop in, 12"		11	.727		1,850	23		1,873	2,050
2660	Mobile	↓	10	.800		2,225	25.50		2,250.50	2,500
3300	Food warmer, counter, 1.2 KW					670			670	735
3550	1.6 KW					1,850			1,850	2,025
3600	Well, hot food, built-in, rectangular, 12" x 20"	R-30	10	2.600		590	98		688	795
3610	Circular, 7 qt		10	2.600		330	98		428	510
3620	Refrigerated, 2 compartments		10	2.600		2,725	98		2,823	3,125
3630	3 compartments		9	2.889		3,000	109		3,109	3,475
3640	4 compartments		8	3.250		3,700	122		3,822	4,250
4720	Frost cold plate	↓	9	2.889		15,600	109		15,709	17,300
5700	Hot chocolate dispenser	1 Plum	4	2		1,025	97.50		1,122.50	1,275
6250	Jet spray dispenser	R-18	4.50	5.778		3,000	212		3,212	3,625
6300	Juice dispenser, concentrate	"	4.50	5.778		2,100	212		2,312	2,625
6690	Milk dispenser, bulk, 2 flavor	R-30	8	3.250		1,425	122		1,547	1,750
6695	3 flavor	"	8	3.250	↓	1,950	122		2,072	2,325
8800	Serving counter, straight	1 Carp	40	.200	L.F.	745	8		753	830
8820	Curved section	"	30	.267	"	935	10.65		945.65	1,050
8825	Solid surface, see Div. 12 36 61.16									
8830	Soft serve ice cream machine, medium	R-18	11	2.364	Ea.	9,850	87		9,937	10,900
8840	Large	"	9	2.889	"	17,300	106		17,406	19,200
9130	Table, basic, minimum	1 Clab	14	.571	L.F.	269	18.05		287.05	325
9135	Maximum	"	11	.727		555	23		578	645
9140	With sink, minimum	W-41E	24	.833		360	35		395	450
9145	Maximum	"	18	1.111	↓	615	47		662	750

11 47 Ice Machines

11 47 10 – Commercial Ice Machines

11 47 10.10 Commercial Ice Equipment		Crew	Daily Output	Labor-Hours	Unit	Material	2009 Bare Costs Labor	Equipment	Total	Total Incl O&P
0010	**COMMERCIAL ICE EQUIPMENT**									
2200	Compressor/evaporator, 1 HP	Q-1	6	2.667	Ea.	4,050	117		4,167	4,625
2220	1-1/2 HP		6	2.667		5,800	117		5,917	6,550
2240	2 HP		6	2.667		6,500	117		6,617	7,325
2260	3 HP		6	2.667		7,625	117		7,742	8,575
5800	Ice cube maker, 50 pounds per day		6	2.667		1,425	117		1,542	1,750
5900	250 pounds per day		1.20	13.333		2,400	585		2,985	3,525
6050	500 pounds per day		4	4		3,000	176		3,176	3,575
6060	With bin		1.20	13.333		2,775	585		3,360	3,925
6090	1000 pounds per day, with bin		1	16		4,775	700		5,475	6,300
6100	Ice flakers, 300 pounds per day		1.60	10		3,050	440		3,490	4,000
6120	600 pounds per day		.95	16.842		3,875	740		4,615	5,350
6130	1000 pounds per day		.75	21.333		4,325	935		5,260	6,150
6140	2000 pounds per day		.65	24.615		18,200	1,075		19,275	21,700
6160	Ice storage bin, 500 pound capacity	Q-5	1	16		940	710		1,650	2,100
6180	1000 pound	"	.56	28.571		1,800	1,275		3,075	3,875

11 48 Cleaning and Disposal Equipment

11 48 13 – Commercial Dishwashers

11 48 13.10 Dishwashers		Crew	Daily Output	Labor-Hours	Unit	Material	2009 Bare Costs Labor	Equipment	Total	Total Incl O&P
0010	**DISHWASHERS**									
2700	Dishwasher, commercial, rack type									
2720	10 to 12 racks per hour	Q-1	3.20	5	Ea.	4,200	219		4,419	4,950
2750	Semi-automatic 38 to 50 racks per hour	"	1.30	12.308		8,350	540		8,890	9,975
2800	Automatic, 190 to 230 racks per hour	L-6	.35	34.286		12,400	1,650		14,050	16,200
2810	2 tanks, 234 racks per hour		1.40	8.571		21,100	415		21,515	23,800
2820	235 to 275 racks per hour		.25	48		27,400	2,300		29,700	33,700
2840	8,750 to 12,500 dishes per hour		.10	120		48,000	5,775		53,775	61,500
2850	100 meals per hour		2.40	5		6,050	241		6,291	7,000
2900	Machine rack		1.40	8.571		5,525	415		5,940	6,700
2910	Pre-rinse spray	1 Plum	5	1.600		251	78		329	395
2920	Dish table, basic	W-41E	19	1.053	L.F.	252	44.50		296.50	345
2930	With trough		13	1.538		575	65		640	735
2940	Maximum		9.50	2.105		530	89		619	715
2950	Dishwasher hood, canopy type	L-3A	10	1.200		770	52.50		822.50	925
2960	Pant leg type	"	2.50	4.800	Ea.	7,375	210		7,585	8,425
5200	Garbage disposal 1.5 HP, 100 GPH	L-1	4.80	3.333		1,825	160		1,985	2,250
5210	3 HP, 120 GPH		4.60	3.478		2,300	167		2,467	2,775
5220	5 HP, 250 GPH		4.50	3.556		3,550	170		3,720	4,150
6750	Pot sink, 3 compartment	1 Plum	7.25	1.103	L.F.	845	54		899	1,000
6760	Pot washer, small		1.60	5	Ea.	8,900	244		9,144	10,200
6770	Large		1.20	6.667		14,600	325		14,925	16,600
9170	Trash compactor, small, up to 125 lb. compacted weight	L-4	4	6		20,800	224		21,024	23,200
9171	Trash Compactor for Furniture Row		12	2		16,800	74.50		16,874.50	18,600
9175	Large, up to 175 lb. compacted weight		3	8		25,400	299		25,699	28,400
9190	Wash-down hose	1 Plum	4.50	1.778		1,025	86.50		1,111.50	1,250
9195	Hose reel	"	4.50	1.778		1,275	86.50		1,361.50	1,525

11 52 Audio-Visual Equipment

11 52 13 – Projection Screens

11 52 13.10 Projection Screens, Wall or Ceiling Hung	Crew	Daily Output	Labor-Hours	Unit	Material	2009 Bare Costs Labor	Equipment	Total	Total Incl O&P
0010	**PROJECTION SCREENS, WALL OR CEILING HUNG**, matte white									
0100	Manually operated, economy	2 Carp	500	.032	S.F.	5.85	1.28		7.13	8.45
0300	Intermediate		450	.036		6.80	1.42		8.22	9.70
0400	Deluxe		400	.040	↓	9.45	1.60		11.05	12.85
0600	Electric operated, matte white, 25 S.F., economy		5	3.200	Ea.	855	128		983	1,150
0700	Deluxe		4	4		1,775	160		1,935	2,200
0900	50 S.F., economy		3	5.333		975	213		1,188	1,400
1000	Deluxe		2	8		1,975	320		2,295	2,675
1200	Heavy duty, electric operated, 200 S.F.		1.50	10.667		3,900	425		4,325	4,950
1300	400 S.F.	↓	1	16	↓	4,800	640		5,440	6,300
1500	Rigid acrylic in wall, for rear projection, 1/4" thick	2 Glaz	30	.533	S.F.	47	20.50		67.50	83
1600	1/2" thick (maximum size 10' x 20')	"	25	.640	"	83.50	24.50		108	129

11 52 16 – Projectors

11 52 16.10 Movie Equipment

		Crew	Daily Output	Labor-Hours	Unit	Material	Labor	Equipment	Total	Total Incl O&P
0010	**MOVIE EQUIPMENT**									
0020	Changeover, minimum				Ea.	465			465	510
0100	Maximum					905			905	995
0400	Film transport, incl. platters and autowind, minimum					5,025			5,025	5,525
0500	Maximum					14,300			14,300	15,700
0800	Lamphouses, incl. rectifiers, xenon, 1,000 watt	1 Elec	2	4		6,650	188		6,838	7,575
0900	1,600 watt		2	4		7,100	188		7,288	8,075
1000	2,000 watt		1.50	5.333		7,625	251		7,876	8,750
1100	4,000 watt	↓	1.50	5.333		9,425	251		9,676	10,800
1400	Lenses, anamorphic, minimum					1,275			1,275	1,400
1500	Maximum					2,875			2,875	3,150
1800	Flat 35 mm, minimum					1,100			1,100	1,225
1900	Maximum					1,725			1,725	1,900
2200	Pedestals, for projectors					1,475			1,475	1,625
2300	Console type					10,700			10,700	11,800
2600	Projector mechanisms, incl. soundhead, 35 mm, minimum					11,100			11,100	12,200
2700	Maximum				↓	15,200			15,200	16,700
3000	Projection screens, rigid, in wall, acrylic, 1/4" thick	2 Glaz	195	.082	S.F.	42	3.17		45.17	51
3100	1/2" thick	"	130	.123	"	48.50	4.75		53.25	60.50
3300	Electric operated, heavy duty, 400 S.F.	2 Carp	1	16	Ea.	2,950	640		3,590	4,250
3320	Theater projection screens, matte white, including frames	"	200	.080	S.F.	6.65	3.20		9.85	12.25
3400	Also see Div. 11 52 13.10									
3700	Sound systems, incl. amplifier, mono, minimum	1 Elec	.90	8.889	Ea.	3,300	420		3,720	4,250
3800	Dolby/Super Sound, maximum		.40	20		18,100	940		19,040	21,300
4100	Dual system, 2 channel, front surround, minimum		.70	11.429		4,625	535		5,160	5,900
4200	Dolby/Super Sound, 4 channel, maximum	↓	.40	20		16,500	940		17,440	19,600
4500	Sound heads, 35 mm					5,300			5,300	5,825
4900	Splicer, tape system, minimum					740			740	815
5000	Tape type, maximum					1,325			1,325	1,450
5300	Speakers, recessed behind screen, minimum	1 Elec	2	4		1,050	188		1,238	1,450
5400	Maximum	"	1	8		3,100	375		3,475	3,950
5700	Seating, painted steel, upholstered, minimum	2 Carp	35	.457		132	18.25		150.25	174
5800	Maximum	"	28	.571		420	23		443	500
6100	Rewind tables, minimum					2,650			2,650	2,900
6200	Maximum				↓	4,725			4,725	5,175
7000	For automation, varying sophistication, minimum	1 Elec	1	8	System	2,375	375		2,750	3,175
7100	Maximum	2 Elec	.30	53.333	"	5,575	2,500		8,075	9,850

11 53 Laboratory Equipment

11 53 03 – Laboratory Test Equipment

11 53 03.13 Test Equipment

		Crew	Daily Output	Labor-Hours	Unit	Material	2009 Bare Costs Labor	Equipment	Total	Total Incl O&P
0010	**TEST EQUIPMENT**									
1700	Thermometer, electric, portable				Ea.	340			340	370
1800	Titration unit, four 2000 ml reservoirs				"	5,500			5,500	6,050

11 53 13 – Laboratory Fume Hoods

11 53 13.13 Recirculating Laboratory Fume Hoods

		Crew	Daily Output	Labor-Hours	Unit	Material	Labor	Equipment	Total	Total Incl O&P
0010	**RECIRCULATING LABORATORY FUME HOODS**									
0600	Fume hood, with countertop & base, not including HVAC									
0610	Simple, minimum	2 Carp	5.40	2.963	L.F.	520	118		638	755
0620	Complex, including fixtures		2.40	6.667		1,675	266		1,941	2,275
0630	Special, maximum	↓	1.70	9.412	↓	1,825	375		2,200	2,575
0670	Service fixtures, average				Ea.	273			273	300
0680	For sink assembly with hot and cold water, add	1 Plum	1.40	5.714		705	279		984	1,200
0750	Glove box, fiberglass, bacteriological					16,900			16,900	18,600
0760	Controlled atmosphere					17,500			17,500	19,200
0770	Radioisotope					16,900			16,900	18,600
0780	Carcinogenic				↓	16,900			16,900	18,600

11 53 13.23 Exhaust Hoods

		Crew	Daily Output	Labor-Hours	Unit	Material	Labor	Equipment	Total	Total Incl O&P
0010	**EXHAUST HOODS**									
0650	Ductwork, minimum	2 Shee	1	16	Hood	3,575	755		4,330	5,075
0660	Maximum	"	.50	32	"	5,500	1,500		7,000	8,375

11 53 16 – Laboratory Incubators

11 53 16.13 Incubators

		Crew	Daily Output	Labor-Hours	Unit	Material	Labor	Equipment	Total	Total Incl O&P
0010	**INCUBATORS**									
1000	Incubators, minimum				Ea.	3,250			3,250	3,575
1010	Maximum				"	19,200			19,200	21,100

11 53 19 – Laboratory Sterilizers

11 53 19.13 Sterilizers

		Crew	Daily Output	Labor-Hours	Unit	Material	Labor	Equipment	Total	Total Incl O&P
0010	**STERILIZERS**									
0700	Glassware washer, undercounter, minimum	L-1	1.80	8.889	Ea.	5,450	425		5,875	6,625
0710	Maximum	"	1	16		12,100	765		12,865	14,600
1850	Utensil washer-sanitizer	1 Plum	2	4	↓	10,400	195		10,595	11,700

11 53 23 – Laboratory Refrigerators

11 53 23.13 Refrigerators

		Crew	Daily Output	Labor-Hours	Unit	Material	Labor	Equipment	Total	Total Incl O&P
0010	**REFRIGERATORS**									
1200	Blood bank, 28.6 C.F. emergency signal				Ea.	7,350			7,350	8,075
1210	Reach-in, 16.9 C.F.				"	7,375			7,375	8,100

11 53 33 – Emergency Safety Appliances

11 53 33.13 Emergency Equipment

		Crew	Daily Output	Labor-Hours	Unit	Material	Labor	Equipment	Total	Total Incl O&P
0010	**EMERGENCY EQUIPMENT**									
1400	Safety equipment, eye wash, hand held				Ea.	405			405	445
1450	Deluge shower				"	725			725	800

11 53 43 – Service Fittings and Accessories

11 53 43.13 Fittings

		Crew	Daily Output	Labor-Hours	Unit	Material	Labor	Equipment	Total	Total Incl O&P
0010	**FITTINGS**									
1600	Sink, one piece plastic, flask wash, hose, free standing	1 Plum	1.60	5	Ea.	1,800	244		2,044	2,350
1610	Epoxy resin sink, 25" x 16" x 10"	"	2	4	"	198	195		393	510
1950	Utility table, acid resistant top with drawers	2 Carp	30	.533	L.F.	147	21.50		168.50	194
8000	Alternate pricing method: as percent of lab furniture									
8050	Installation, not incl. plumbing & duct work				% Furn.				20%	22%

11 53 Laboratory Equipment

11 53 43 – Service Fittings and Accessories

11 53 43.13 Fittings	Crew	Daily Output	Labor-Hours	Unit	Material	2009 Bare Costs Labor	Equipment	Total	Total Incl O&P
8100 Plumbing, final connections, simple system				% Furn.				9.09%	10%
8110 Moderately complex system								13.64%	15%
8120 Complex system								18.18%	20%
8150 Electrical, simple system								9.09%	10%
8160 Moderately complex system								18.18%	20%
8170 Complex system								31.80%	35%

11 57 Vocational Shop Equipment

11 57 10 – Shop Equipment

11 57 10.10 Vocational School Shop Equipment

	Crew	Daily Output	Labor-Hours	Unit	Material	Labor	Equipment	Total	Incl O&P
0010 **VOCATIONAL SCHOOL SHOP EQUIPMENT**									
0020 Benches, work, wood, average	2 Carp	5	3.200	Ea.	505	128		633	755
0100 Metal, average		5	3.200		370	128		498	605
0400 Combination belt & disc sander, 6"		4	4		1,275	160		1,435	1,650
0700 Drill press, floor mounted, 12", 1/2 H.P.		4	4		400	160		560	685
0800 Dust collector, not incl. ductwork, 6" diameter	1 Shee	1.10	7.273		3,000	345		3,345	3,825
1000 Grinders, double wheel, 1/2 H.P.	2 Carp	5	3.200		245	128		373	465
1300 Jointer, 4", 3/4 H.P.		4	4		1,250	160		1,410	1,625
1600 Kilns, 16 C.F., to 2000°		4	4		2,250	160		2,410	2,725
1900 Lathe, woodworking, 10", 1/2 H.P.		4	4		765	160		925	1,100
2200 Planer, 13" x 6"		4	4		1,250	160		1,410	1,625
2500 Potter's wheel, motorized		4	4		1,125	160		1,285	1,475
2800 Saws, band, 14", 3/4 H.P.		4	4		850	160		1,010	1,175
3100 Metal cutting band saw, 14"		4	4		2,000	160		2,160	2,450
3400 Radial arm saw, 10", 2 H.P.		4	4		975	160		1,135	1,325
3700 Scroll saw, 24"		4	4		575	160		735	885
4000 Table saw, 10", 3 H.P.		4	4		1,925	160		2,085	2,350
4300 Welder AC arc, 30 amp capacity		4	4		1,950	160		2,110	2,375

11 61 Theater and Stage Equipment

11 61 23 – Folding and Portable Stages

11 61 23.10 Portable Stages

	Crew	Daily Output	Labor-Hours	Unit	Material	Labor	Equipment	Total	Incl O&P
0010 **PORTABLE STAGES**									
1500 Flooring, portable oak parquet, 3' x 3' sections				S.F.	13.40			13.40	14.70
1600 Cart to carry 225 S.F. of flooring				Ea.	400			400	440
5000 Stages, portable with steps, folding legs, stock, 8" high				SF Stg.	25.50			25.50	28.50
5100 16" high					23.50			23.50	26
5200 32" high					35.50			35.50	39
5300 40" high					59.50			59.50	65
6000 Telescoping platforms, extruded alum., straight, minimum	4 Carp	157	.204		28	8.15		36.15	43
6100 Maximum		77	.416		37.50	16.60		54.10	67.50
6500 Pie-shaped, minimum		150	.213		60.50	8.50		69	79.50
6600 Maximum		70	.457		67.50	18.25		85.75	103
6800 For 3/4" plywood covered deck, deduct					3.85			3.85	4.24
7000 Band risers, steel frame, plywood deck, minimum	4 Carp	275	.116		30.50	4.65		35.15	40.50
7100 Maximum	"	138	.232		63.50	9.25		72.75	84
7500 Chairs for above, self-storing, minimum	2 Carp	43	.372	Ea.	99.50	14.85		114.35	132
7600 Maximum	"	40	.400	"	172	16		188	214

11 61 Theater and Stage Equipment

11 61 33 – Rigging Systems and Controls

11 61 33.10 Controls		Crew	Daily Output	Labor-Hours	Unit	Material	2009 Bare Costs Labor	Equipment	Total	Total Incl O&P
0010	**CONTROLS**									
0050	Control boards with dimmers and breakers, minimum	1 Elec	1	8	Ea.	11,300	375		11,675	13,100
0100	Average		.50	16		35,500	750		36,250	40,200
0150	Maximum		.20	40		119,000	1,875		120,875	134,000
8000	Rule of thumb: total stage equipment, minimum	4 Carp	100	.320	SF Stg.	97	12.80		109.80	127
8100	Maximum	"	25	1.280	"	545	51		596	680

11 61 43 – Stage Curtains

11 61 43.10 Curtains		Crew	Daily Output	Labor-Hours	Unit	Material	2009 Bare Costs Labor	Equipment	Total	Total Incl O&P
0010	**CURTAINS**									
0500	Curtain track, straight, light duty	2 Carp	20	.800	L.F.	25.50	32		57.50	77.50
0600	Heavy duty		18	.889		59	35.50		94.50	120
0700	Curved sections		12	1.333		164	53.50		217.50	263
1000	Curtains, velour, medium weight		600	.027	S.F.	7.90	1.07		8.97	10.35
1150	Silica based yarn, inherently fire retardant		50	.320	"	15.20	12.80		28	36.50

11 66 Athletic Equipment

11 66 13 – Exercise Equipment

11 66 13.10 Physical Training Equipment		Crew	Daily Output	Labor-Hours	Unit	Material	2009 Bare Costs Labor	Equipment	Total	Total Incl O&P
0010	**PHYSICAL TRAINING EQUIPMENT**									
0020	Abdominal rack, 2 board capacity				Ea.	470			470	515
0050	Abdominal board, upholstered					525			525	575
0200	Bicycle trainer, minimum					815			815	895
0300	Deluxe, electric					4,150			4,150	4,575
0400	Barbell set, chrome plated steel, 25 lbs.					244			244	269
0420	100 lbs.					370			370	405
0450	200 lbs.					715			715	785
0500	Weight plates, cast iron, per lb.				Lb.	5			5	5.50
0520	Storage rack, 10 station				Ea.	790			790	870
0600	Circuit training apparatus, 12 machines minimum	2 Clab	1.25	12.800	Set	27,600	405		28,005	30,900
0700	Average		1	16		33,800	505		34,305	38,000
0800	Maximum		.75	21.333		40,100	675		40,775	45,200
0820	Dumbbell set, cast iron, with rack and 5 pair					570			570	630
0900	Squat racks	2 Clab	5	3.200	Ea.	765	101		866	995
1200	Multi-station gym machine, 5 station					7,950			7,950	8,750
1250	9 station					11,200			11,200	12,300
1280	Rowing machine, hydraulic					1,475			1,475	1,625
1300	Treadmill, manual					1,450			1,450	1,575
1320	Motorized					3,500			3,500	3,850
1340	Electronic					5,450			5,450	6,000
1360	Cardio-testing					7,775			7,775	8,550
1400	Treatment/massage tables, minimum					535			535	590
1420	Deluxe, with accessories					685			685	755
4150	Exercise equipment, bicycle trainer					595			595	655
4180	Chinning bar, adjustable, wall mounted	1 Carp	5	1.600		279	64		343	405
4200	Exercise ladder, 16' x 1'-7", suspended	L-2	3	5.333		1,125	187		1,312	1,550
4210	High bar, floor plate attached	1 Carp	4	2		1,100	80		1,180	1,350
4240	Parallel bars, adjustable		4	2		2,475	80		2,555	2,825
4270	Uneven parallel bars, adjustable		4	2		2,675	80		2,755	3,050
4280	Wall mounted, adjustable	L-2	1.50	10.667	Set	835	375		1,210	1,500
4300	Rope, ceiling mounted, 18' long	1 Carp	3.66	2.186	Ea.	220	87.50		307.50	375

11 66 Athletic Equipment

11 66 13 – Exercise Equipment

11 66 13.10 Physical Training Equipment

		Crew	Daily Output	Labor-Hours	Unit	Material	2009 Bare Costs Labor	Equipment	Total	Total Incl O&P
4330	Side horse, vaulting	1 Carp	5	1.600	Ea.	1,700	64		1,764	1,975
4360	Treadmill, motorized, deluxe, training type	↓	5	1.600		3,500	64		3,564	3,950
4390	Weight lifting multi-station, minimum	2 Clab	1	16		395	505		900	1,225
4450	Maximum	"	.50	32	↓	14,700	1,000		15,700	17,800

11 66 23 – Gymnasium Equipment

11 66 23.13 Basketball Equipment

		Crew	Daily Output	Labor-Hours	Unit	Material	Labor	Equipment	Total	Total Incl O&P
0010	**BASKETBALL EQUIPMENT**									
1000	Backstops, wall mtd., 6' extended, fixed, minimum	L-2	1	16	Ea.	1,150	560		1,710	2,125
1100	Maximum		1	16		1,675	560		2,235	2,725
1200	Swing up, minimum		1	16		1,275	560		1,835	2,275
1250	Maximum		1	16		5,350	560		5,910	6,775
1300	Portable, manual, heavy duty, spring operated		1.90	8.421		11,400	296		11,696	13,000
1400	Ceiling suspended, stationary, minimum		.78	20.513		2,850	720		3,570	4,275
1450	Fold up, with accessories, maximum	↓	.40	40		6,100	1,400		7,500	8,900
1600	For electrically operated, add	1 Elec	1	8	↓	2,000	375		2,375	2,750
5800	Wall pads, 1-1/2" thick	2 Carp	640	.025	S.F.	6.80	1		7.80	9.05

11 66 23.19 Boxing Ring

		Crew	Daily Output	Labor-Hours	Unit	Material	Labor	Equipment	Total	Total Incl O&P
0010	**BOXING RING**									
4100	Elevated, 22' x 22'	L-4	.10	240	Ea.	6,700	8,950		15,650	21,300
4110	For cellular plastic foam padding, add		.10	240		850	8,950		9,800	14,800
4120	Floor level, including posts and ropes only, 20' x 20'		.80	30		2,375	1,125		3,500	4,325
4130	Canvas, 30' x 30'	↓	5	4.800	↓	1,200	179		1,379	1,575

11 66 23.47 Gym Mats

		Crew	Daily Output	Labor-Hours	Unit	Material	Labor	Equipment	Total	Total Incl O&P
0010	**GYM MATS**									
5500	2" thick, naugahyde covered				S.F.	3.59			3.59	3.95
5600	Vinyl/nylon covered					6.20			6.20	6.85
6000	Wrestling mats, 1" thick, heavy duty				↓	5.50			5.50	6.05

11 66 43 – Interior Scoreboards

11 66 43.10 Scoreboards

		Crew	Daily Output	Labor-Hours	Unit	Material	Labor	Equipment	Total	Total Incl O&P
0010	**SCOREBOARDS**									
7000	Baseball, minimum	R-3	1.30	15.385	Ea.	3,225	710	99.50	4,034.50	4,700
7200	Maximum		.05	400		15,600	18,500	2,575	36,675	47,700
7300	Football, minimum		.86	23.256		3,975	1,075	150	5,200	6,150
7400	Maximum		.20	100		12,600	4,625	645	17,870	21,500
7500	Basketball (one side), minimum		2.07	9.662		2,225	445	62.50	2,732.50	3,175
7600	Maximum		.30	66.667		5,550	3,075	430	9,055	11,200
7700	Hockey-basketball (four sides), minimum		.25	80		5,675	3,700	515	9,890	12,300
7800	Maximum	↓	.15	133		5,600	6,175	860	12,635	16,300

11 66 53 – Gymnasium Dividers

11 66 53.10 Divider Curtains

		Crew	Daily Output	Labor-Hours	Unit	Material	Labor	Equipment	Total	Total Incl O&P
0010	**DIVIDER CURTAINS**									
4500	Gym divider curtain, mesh top, vinyl bottom, manual	L-4	500	.048	S.F.	8.05	1.79		9.84	11.60
4700	Electric roll up	L-7	400	.070	"	11.75	2.70		14.45	17.10

11 71 Medical Sterilizing Equipment

11 71 10 – Medical Sterilizers & Distillers

11 71 10.10 Sterilizers and Distillers		Crew	Daily Output	Labor-Hours	Unit	Material	2009 Bare Costs Labor	2009 Bare Costs Equipment	Total	Total Incl O&P
0010	**STERILIZERS AND DISTILLERS**									
0700	Distiller, water, steam heated, 50 gal. capacity	1 Plum	1.40	5.714	Ea.	17,600	279		17,879	19,700
5600	Sterilizers, floor loading, 26" x 62" x 42", single door, steam					151,000			151,000	166,000
5650	Double door, steam					194,000			194,000	213,500
5800	General purpose, 20" x 20" x 38", single door					14,600			14,600	16,000
6000	Portable, counter top, steam, minimum					3,625			3,625	3,975
6020	Maximum					5,650			5,650	6,225
6050	Portable, counter top, gas, 17" x 15" x 32-1/2"					37,400			37,400	41,200
6150	Manual washer/sterilizer, 16" x 16" x 26"	1 Plum	2	4	↓	51,500	195		51,695	57,000
6200	Steam generators, electric 10 kW to 180 kW, freestanding									
6250	Minimum	1 Elec	3	2.667	Ea.	7,625	125		7,750	8,550
6300	Maximum	"	.70	11.429		27,400	535		27,935	31,000
8200	Bed pan washer-sanitizer	1 Plum	2	4	↓	6,800	195		6,995	7,775

11 72 Examination and Treatment Equipment

11 72 13 – Examination Equipment

11 72 13.13 Examination Equipment

		Crew	Daily Output	Labor-Hours	Unit	Material	Labor	Equipment	Total	Total Incl O&P
0010	**EXAMINATION EQUIPMENT**									
4400	Scale, physician's, with height rod				Ea.	305			305	335

11 72 53 – Treatment Equipment

11 72 53.13 Medical Treatment Equipment

		Crew	Daily Output	Labor-Hours	Unit	Material	Labor	Equipment	Total	Total Incl O&P
0010	**MEDICAL TREATMENT EQUIPMENT**									
6500	Surgery table, minor minimum	1 Sswk	.70	11.429	Ea.	13,200	510		13,710	15,400
6520	Maximum	"	.50	16		24,800	715		25,515	28,600
6700	Surgical lights, doctor's office, single arm	2 Elec	2	8		1,225	375		1,600	1,900
6750	Dual arm	"	1	16	↓	5,525	750		6,275	7,200

11 73 Patient Care Equipment

11 73 10 – Patient Treatment Equipment

11 73 10.10 Treatment Equipment

		Crew	Daily Output	Labor-Hours	Unit	Material	Labor	Equipment	Total	Total Incl O&P
0010	**TREATMENT EQUIPMENT**									
1800	Heat therapy unit, humidified, 26" x 78" x 28"				Ea.	3,225			3,225	3,550
2100	Hubbard tank with accessories, stainless steel,									
2110	125 GPM at 45 psi water pressure				Ea.	25,100			25,100	27,600
2150	For electric overhead hoist, add					2,725			2,725	3,000
2900	K-Module for heat therapy, 20 oz. capacity, 75° to 110° F					385			385	425
3600	Paraffin bath, 126° F, auto controlled					1,400			1,400	1,525
3900	Parallel bars for walking training, 12'-0"					1,150			1,150	1,275
4600	Station, dietary, medium, with ice					15,300			15,300	16,800
4700	Medicine					6,900			6,900	7,575
7000	Tables, physical therapy, walk off, electric	2 Carp	3	5.333		3,050	213		3,263	3,675
7150	Standard, vinyl top with base cabinets, minimum		3	5.333		1,250	213		1,463	1,700
7200	Maximum	↓	2	8		3,675	320		3,995	4,550
8400	Whirlpool bath, mobile, sst, 18" x 24" x 60"					4,275			4,275	4,725
8450	Fixed, incl. mixing valves	1 Plum	2	4	↓	8,775	195		8,970	9,950

11 74 Dental Equipment

11 74 10 – Dental Office Equipment

11 74 10.10 Diagnostic and Treatment Equipment

		Crew	Daily Output	Labor-Hours	Unit	Material	2009 Bare Costs Labor	Equipment	Total	Total Incl O&P
0010	**DIAGNOSTIC AND TREATMENT EQUIPMENT**									
0020	Central suction system, minimum	1 Plum	1.20	6.667	Ea.	1,425	325		1,750	2,050
0100	Maximum	"	.90	8.889		3,700	435		4,135	4,725
0300	Air compressor, minimum	1 Skwk	.80	10		2,975	410		3,385	3,900
0400	Maximum		.50	16		6,850	655		7,505	8,550
0600	Chair, electric or hydraulic, minimum		.50	16		2,675	655		3,330	3,925
0700	Maximum		.25	32		13,700	1,300		15,000	17,100
1000	Drill console with accessories, minimum		1.60	5		1,500	204		1,704	1,975
1100	Maximum		1.60	5		4,750	204		4,954	5,550
2000	Light, ceiling mounted, minimum		8	1		1,400	41		1,441	1,600
2100	Maximum		8	1		1,900	41		1,941	2,150
2200	Unit light, minimum	2 Skwk	5.33	3.002		600	123		723	850
2210	Maximum		5.33	3.002		1,625	123		1,748	2,000
2220	Track light, minimum		3.20	5		2,050	204		2,254	2,600
2230	Maximum		3.20	5		3,550	204		3,754	4,225
2300	Sterilizers, steam portable, minimum					1,225			1,225	1,350
2350	Maximum					9,900			9,900	10,900
2600	Steam, institutional					4,250			4,250	4,675
2650	Dry heat, electric, portable, 3 trays					1,150			1,150	1,275
2700	Ultra-sonic cleaner, portable, minimum					244			244	268
2750	Maximum (institutional)					1,450			1,450	1,600
3000	X-ray unit, wall, minimum	1 Skwk	4	2		2,200	81.50		2,281.50	2,525
3010	Maximum		4	2		4,900	81.50		4,981.50	5,500
3100	Panoramic unit		.60	13.333		13,900	545		14,445	16,100
3105	Deluxe, minimum	2 Skwk	1.60	10		16,000	410		16,410	18,200
3110	Maximum	"	1.60	10		51,500	410		51,910	57,000
3500	Developers, X-ray, average	1 Plum	5.33	1.501		4,725	73		4,798	5,300
3600	Maximum	"	5.33	1.501		7,700	73		7,773	8,575

11 76 Operating Room Equipment

11 76 10 – Operating Room Equipment

11 76 10.10 Surgical Equipment

		Crew	Daily Output	Labor-Hours	Unit	Material	2009 Bare Costs Labor	Equipment	Total	Total Incl O&P
0010	**SURGICAL EQUIPMENT**									
5000	Scrub, surgical, stainless steel, single station, minimum	1 Plum	3	2.667	Ea.	3,925	130		4,055	4,525
5100	Maximum					7,900			7,900	8,700
6550	Major surgery table, minimum	1 Sswk	.50	16		29,900	715		30,615	34,200
6570	Maximum	"	.50	16		99,000	715		99,715	110,000
6800	Surgical lights, major operating room, dual head, minimum	2 Elec	1	16		16,600	750		17,350	19,400
6850	Maximum	"	1	16		27,700	750		28,450	31,600

11 77 Radiology Equipment

11 77 10 – Radiology Equipment

11 77 10.10 X-Ray Equipment	Crew	Daily Output	Labor-Hours	Unit	Material	2009 Bare Costs Labor	Equipment	Total	Total Incl O&P
0010 **X-RAY EQUIPMENT**									
8700　X-ray, mobile, minimum				Ea.	12,800			12,800	14,000
8750　　Maximum					71,500			71,500	79,000
8900　Stationary, minimum					39,900			39,900	43,800
8950　　Maximum					207,000			207,000	228,000
9150　Developing processors, minimum					11,100			11,100	12,200
9200　　Maximum					29,100			29,100	32,000

11 78 Mortuary Equipment

11 78 13 – Mortuary Refrigerators

11 78 13.10 Mortuary and Autopsy Equipment

	Crew	Daily Output	Labor-Hours	Unit	Material	Labor	Equipment	Total	Total Incl O&P
0010 **MORTUARY AND AUTOPSY EQUIPMENT**									
0015　Autopsy table, standard	1 Plum	1	8	Ea.	8,450	390		8,840	9,850
0020　　Deluxe	"	.60	13.333		16,800	650		17,450	19,400
3200　Mortuary refrigerator, end operated, 2 capacity					12,400			12,400	13,600
3300　　6 capacity					22,200			22,200	24,500

11 78 16 – Crematorium Equipment

11 78 16.10 Crematory

	Crew	Daily Output	Labor-Hours	Unit	Material	Labor	Equipment	Total	Total Incl O&P
0010 **CREMATORY**									
1500　Crematory, not including building, 1 place	Q-3	.20	160	Ea.	59,500	7,425		66,925	76,500
1750　　2 place	"	.10	320	"	85,000	14,900		99,900	116,000

11 82 Solid Waste Handling Equipment

11 82 19 – Packaged Incinerators

11 82 19.10 Packaged Gas Fired Incinerators

	Crew	Daily Output	Labor-Hours	Unit	Material	Labor	Equipment	Total	Total Incl O&P
0010 **PACKAGED GAS FIRED INCINERATORS**									
4400　Incinerator, gas, not incl. chimney, elec. or pipe, 50#/hr., minimum	Q-3	.80	40	Ea.	11,200	1,850		13,050	15,100
4420　　Maximum		.70	45.714		35,000	2,125		37,125	41,700
4440　200 lb. per hr., minimum (batch type)		.60	53.333		62,000	2,475		64,475	71,500
4460　　Maximum (with feeder)		.50	64		85,000	2,975		87,975	98,000
4480　400 lb. per hr., minimum (batch type)		.30	106		68,500	4,950		73,450	83,000
4500　　Maximum (with feeder)		.25	128		90,000	5,950		95,950	108,000
4520　800 lb. per hr., with feeder, minimum		.20	160		112,000	7,425		119,425	134,000
4540　　Maximum		.17	188		150,000	8,750		158,750	178,000
4700　For heat recovery system, add, minimum		.25	128		69,000	5,950		74,950	85,000
4710　　Add, maximum		.11	290		220,500	13,500		234,000	263,000
4720　For automatic ash conveyer, add		.50	64		29,000	2,975		31,975	36,400

11 82 23 – Recycling Equipment

11 82 23.10 Recycling Bins

	Crew	Daily Output	Labor-Hours	Unit	Material	Labor	Equipment	Total	Total Incl O&P
0010 **RECYCLING BINS**									
6500　Polypropylene, 19 gallon, 24" x 15-1/2" x 16-1/2"	1 Clab	288	.028	Ea.	11.60	.88		12.48	14.15

11 82 26 – Waste Compactors and Destructors

11 82 26.10 Compactors

	Crew	Daily Output	Labor-Hours	Unit	Material	Labor	Equipment	Total	Total Incl O&P
0010 **COMPACTORS**									
0020　Compactors, 115 volt, 250#/hr., chute fed	L-4	1	24	Ea.	10,900	895		11,795	13,400
0100　　Hand fed		2.40	10		7,975	375		8,350	9,350
0300　Multi-bag, 230 volt, 600#/hr, chute fed		1	24		9,900	895		10,795	12,300
0400　　Hand fed		1	24		8,500	895		9,395	10,700

11 82 Solid Waste Handling Equipment

11 82 26 – Waste Compactors and Destructors

11 82 26.10 Compactors		Crew	Daily Output	Labor-Hours	Unit	Material	2009 Bare Costs Labor	2009 Bare Costs Equipment	Total	Total Incl O&P
0500	Containerized, hand fed, 2 to 6 C.Y. containers, 250#/hr.	L-4	1	24	Ea.	10,500	895		11,395	13,000
0550	For chute fed, add per floor		1	24		1,200	895		2,095	2,700
1000	Heavy duty industrial compactor, 0.5 C.Y. capacity		1	24		7,100	895		7,995	9,175
1050	1.0 C.Y. capacity		1	24		10,800	895		11,695	13,300
1400	For handling hazardous waste materials, 55 gallon drum packer, std.					16,800			16,800	18,500
1410	55 gallon drum packer w/HEPA filter					20,900			20,900	23,000
1420	55 gallon drum packer w/charcoal & HEPA filter					28,000			28,000	30,800
1430	All of the above made explosion proof, add					12,800			12,800	14,000
5800	Shredder, industrial, minimum					20,600			20,600	22,700
5850	Maximum					110,500			110,500	121,500
5900	Baler, industrial, minimum					8,250			8,250	9,075
5950	Maximum					482,500			482,500	531,000
6000	Transfer station compactor, with power unit									
6050	and pedestal, not including pit, 50 ton per hour				Ea.	165,000			165,000	181,500

11 91 Religious Equipment

11 91 13 – Baptistries

11 91 13.10 Baptistry

		Crew	Daily Output	Labor-Hours	Unit	Material	Labor	Equipment	Total	Total Incl O&P
0010	**BAPTISTRY**									
0150	Fiberglass, 3'-6" deep, x 13'-7" long,									
0160	steps at both ends, incl. plumbing, minimum	L-8	1	20	Ea.	3,825	835		4,660	5,475
0200	Maximum	"	.70	28.571		6,250	1,200		7,450	8,700
0250	Add for filter, heater and lights					1,375			1,375	1,500

11 91 23 – Sanctuary Equipment

11 91 23.10 Sanctuary Furnishings

		Crew	Daily Output	Labor-Hours	Unit	Material	Labor	Equipment	Total	Total Incl O&P
0010	**SANCTUARY FURNISHINGS**									
0020	Altar, wood, custom design, plain	1 Carp	1.40	5.714	Ea.	1,925	228		2,153	2,450
0050	Deluxe	"	.20	40		9,225	1,600		10,825	12,700
0070	Granite or marble, average	2 Marb	.50	32		10,700	1,225		11,925	13,600
0090	Deluxe	"	.20	80		30,400	3,100		33,500	38,100
0500	Reconciliation room, wood, prefabricated, single, plain	1 Carp	.60	13.333		2,425	535		2,960	3,500
0550	Deluxe		.40	20		6,675	800		7,475	8,600
0650	Double, plain		.40	20		4,850	800		5,650	6,600
0700	Deluxe		.20	40		15,200	1,600		16,800	19,200
1000	Lecterns, wood, plain		5	1.600		750	64		814	925
1100	Deluxe		2	4		5,050	160		5,210	5,800
2000	Pulpits, hardwood, prefabricated, plain		2	4		1,225	160		1,385	1,600
2100	Deluxe		1.60	5		8,325	200		8,525	9,450
2500	Railing, hardwood, average		25	.320	L.F.	166	12.80		178.80	203
3000	Seating, individual, oak, contour, laminated		21	.381	Person	147	15.20		162.20	185
3100	Cushion seat		21	.381		133	15.20		148.20	170
3200	Fully upholstered		21	.381		130	15.20		145.20	167
3300	Combination, self-rising		21	.381		335	15.20		350.20	390
3500	For cherry, add					30%				
5000	Wall cross, aluminum, extruded, 2" x 2" section	1 Carp	34	.235	L.F.	132	9.40		141.40	160
5150	4" x 4" section		29	.276		190	11		201	226
5300	Bronze, extruded, 1" x 2" section		31	.258		260	10.30		270.30	300
5350	2-1/2" x 2-1/2" section		34	.235		395	9.40		404.40	450
5450	Solid bar stock, 1/2" x 3" section		29	.276		515	11		526	585
5600	Fiberglass, stock		34	.235		107	9.40		116.40	132

11 91 Religious Equipment

11 91 23 – Sanctuary Equipment

11 91 23.10 Sanctuary Furnishings		Crew	Daily Output	Labor-Hours	Unit	Material	2009 Bare Costs Labor	Equipment	Total	Total Incl O&P
5700	Stainless steel, 4" deep, channel section	1 Carp	29	.276	L.F.	420	11		431	475
5800	4" deep box section	↓	29	.276	↓	570	11		581	640

Division Notes

	CREW	DAILY OUTPUT	LABOR-HOURS	UNIT	MAT.	LABOR	EQUIP.	TOTAL	TOTAL INCL O&P

Estimating Tips

General

- The items in this division are usually priced per square foot or each. Most of these items are purchased by the owner and placed by the supplier. Do not assume the items in Division 12 will be purchased and installed by the supplier. Check the specifications for responsibilities and include receiving, storage, installation, and mechanical and electrical hookups in the appropriate divisions.

- Some items in this division require some type of support system that is not usually furnished with the item. Examples of these systems include blocking for the attachment of casework and heavy drapery rods. The required blocking must be added to the estimate in the appropriate division.

Reference Numbers

Reference numbers are shown in shaded boxes at the beginning of some major classifications. These numbers refer to related items in the Reference Section. The reference information may be an estimating procedure, an alternate pricing method, or technical information.

Note: Not all subdivisions listed here necessarily appear in this publication.

No part of this publication may be reproduced, stored in a retrieval system, or transmitted in any form or by any means without prior written permission of R.S. Means Company, Inc.

12 05 Common Work Results for Furnishings

12 05 13 – Fabrics

12 05 13.10 Upholstery Materials

		Crew	Daily Output	Labor-Hours	Unit	Material	2009 Bare Costs Labor	2009 Bare Costs Equipment	Total	Total Incl O&P
0010	**UPHOLSTERY MATERIALS**									
1000	Fabrics, fabric blends, minimum				S.Y.	34			34	37.50
1020	Maximum					84			84	92.50
1200	Nylons, minimum					31.50			31.50	34.50
1220	Maximum					47			47	51.50
1400	Polyester, minimum					24.50			24.50	27
1420	Maximum					59.50			59.50	65.50
1600	Silk, minimum					76			76	83.50
1620	Maximum					131			131	144
1800	Wool, minimum					54.50			54.50	60
1820	Maximum					99			99	109
3000	Treatment, acoustic, 1/8" flame retardant					9.35			9.35	10.25
3100	Acrylic backing					4.18			4.18	4.60
3200	Flame proofing					8			8	8.80
3300	Paper backing					7.10			7.10	7.80
3400	Scotchguard or teflon					2.47			2.47	2.72
5000	Leather, minimum				S.F.	8.95			8.95	9.85
5020	Maximum				"	10.30			10.30	11.35
6000	Vinyl, minimum				S.Y.	22.50			22.50	24.50
6020	Maximum				"	36.50			36.50	40

12 12 Wall Decorations

12 12 19 – Framed Prints

12 12 19.10 Art Work

		Crew	Daily Output	Labor-Hours	Unit	Material	2009 Bare Costs Labor	2009 Bare Costs Equipment	Total	Total Incl O&P
0010	**ART WORK** framed									
1000	Photography, minimum	1 Carp	36	.222	Ea.	74.50	8.90		83.40	96
1050	Maximum		30	.267		495	10.65		505.65	560
2000	Posters, minimum		36	.222		40	8.90		48.90	58
2050	Maximum		30	.267		565	10.65		575.65	640
3000	Reproductions, minimum		36	.222		75	8.90		83.90	96.50
3050	Maximum		30	.267		695	10.65		705.65	780

12 21 Window Blinds

12 21 13 – Horizontal Louver Blinds

12 21 13.13 Metal Horizontal Louver Blinds

		Crew	Daily Output	Labor-Hours	Unit	Material	2009 Bare Costs Labor	2009 Bare Costs Equipment	Total	Total Incl O&P
0010	**METAL HORIZONTAL LOUVER BLINDS**									
0020	Horizontal, 1" aluminum slats, solid color, stock	1 Carp	590	.014	S.F.	4.75	.54		5.29	6.10
0090	Custom, minimum		590	.014		5.25	.54		5.79	6.65
0100	Maximum		440	.018		7.15	.73		7.88	9
0250	2" aluminum slats, custom, minimum		590	.014		8.10	.54		8.64	9.80
0350	Maximum		440	.018		13.15	.73		13.88	15.65
0450	Stock, minimum		590	.014		4.86	.54		5.40	6.20
0500	Maximum		440	.018		7.90	.73		8.63	9.85
0600	2" steel slats, stock, minimum		590	.014		1.84	.54		2.38	2.86
0630	Maximum		440	.018		5.15	.73		5.88	6.85
0750	Custom, minimum		590	.014		1.67	.54		2.21	2.68
0850	Maximum		400	.020		8.25	.80		9.05	10.35
1000	Alternate method of figuring:									
1300	1" aluminum slats, 48" wide, 48" high	1 Carp	30	.267	Ea.	50.50	10.65		61.15	72

12 21 Window Blinds

12 21 13 – Horizontal Louver Blinds

12 21 13.13 Metal Horizontal Louver Blinds		Crew	Daily Output	Labor-Hours	Unit	Material	2009 Bare Costs Labor	Equipment	Total	Total Incl O&P
1320	72" high	1 Carp	29	.276	Ea.	67	11		78	90.50
1340	96" high		28	.286		74	11.40		85.40	98.50
1400	72" wide, 72" high		25	.320		105	12.80		117.80	136
1420	96" high		23	.348		106	13.90		119.90	139
1480	96" wide, 96" high		20	.400		149	16		165	189
1490	For special colors, add					12%				

12 21 16 – Vertical Louver Blinds

12 21 16.13 Metal Vertical Louver Blinds

		Crew	Daily Output	Labor-Hours	Unit	Material	Labor	Equipment	Total	Total Incl O&P
0010	**METAL VERTICAL LOUVER BLINDS**									
1500	Vertical, 3" to 5" PVC or cloth strips, minimum	1 Carp	460	.017	S.F.	5.40	.69		6.09	7.05
1600	Maximum		400	.020		11.75	.80		12.55	14.15
1800	4" aluminum slats, minimum		460	.017		7.20	.69		7.89	9.05
1900	Maximum		400	.020		9.10	.80		9.90	11.25
1950	Mylar mirror-finish strips, to 8" wide, minimum		460	.017		4.62	.69		5.31	6.20
1970	Maximum		400	.020		10.50	.80		11.30	12.80
1990	Alternate method of figuring:									
2000	2" aluminum slats, 48" wide, 48" high	1 Carp	30	.267	Ea.	86	10.65		96.65	111
2050	72" high		29	.276		110	11		121	138
2100	96" high		28	.286		134	11.40		145.40	166
2200	72" wide, 72" high		25	.320		133	12.80		145.80	167
2250	96" high		23	.348		161	13.90		174.90	199
2300	96" wide, 96" high		20	.400		217	16		233	264
2500	Mirror finish, 48" wide, 48" high		30	.267		87	10.65		97.65	112
2550	72" high		29	.276		105	11		116	132
2600	96" high		28	.286		124	11.40		135.40	155
2650	72" wide, 72" high		25	.320		147	12.80		159.80	182
2700	96" high		23	.348		176	13.90		189.90	216
2750	96" wide, 96" high		20	.400		234	16		250	282
2800	Decorative printed finish, 48" wide, 48" high		30	.267		85.50	10.65		96.15	111
2850	72" high		29	.276		104	11		115	131
2900	96" high		28	.286		123	11.40		134.40	153
2950	72" wide, 72" high		25	.320		144	12.80		156.80	178
2980	96" high		23	.348		179	13.90		192.90	219
2990	96" wide, 96" high		20	.400		234	16		250	282

12 22 Curtains and Drapes

12 22 13 – Draperies

12 22 13.10 Custom Draperies

		Crew	Daily Output	Labor-Hours	Unit	Material	Labor	Equipment	Total	Total Incl O&P
0010	**CUSTOM DRAPERIES** (Material only)									
0050	Lined, minimum				S.F.	5			5	5.50
0100	Maximum					14.25			14.25	15.70
0200	Unlined, minimum					2.77			2.77	3.05
0300	Maximum					9.35			9.35	10.25
0400	Lightproof type, add, minimum					1.39			1.39	1.53
0500	Maximum					8.30			8.30	9.15
0600	Fire resistant, add, minimum					1.06			1.06	1.17
0700	Maximum					3.59			3.59	3.95
0800	Valances, pleated, 10" to 18" depth, minimum				L.F.	1.55			1.55	1.71
0900	Maximum				"	8.95			8.95	9.85
2000	Alternate method, lined overlaps and returns, average									

12 22 Curtains and Drapes

12 22 13 – Draperies

12 22 13.10 Custom Draperies		Crew	Daily Output	Labor-Hours	Unit	Material	2009 Bare Costs Labor	Equipment	Total	Total Incl O&P
2020	32" thru 48" wide, 26" to 39" long				Ea.	153			153	168
2040	40" to 63" long					155			155	170
2060	64" to 72" long					159			159	175
2080	73" to 81" long					180			180	198
2100	82" to 90" long					186			186	204
2200	91" to 99" long					199			199	219
2400	100" to 108" long					208			208	229
2600	109" to 120" long					219			219	241
2800	121" to 130" long					223			223	245
3000	48" thru 72" wide, 26" to 39" long					223			223	245
3050	40" to 63" long					238			238	262
3100	64" to 72" long					249			249	274
3150	73" to 81" long					261			261	287
3200	82" to 90" long					286			286	315
3250	91" to 99" long					296			296	325
3300	100" to 108" long					305			305	340
3350	109" to 120" long					330			330	365
3400	121" to 130" long					355			355	390
3450	64" thru 96" wide, 26" to 39" long					297			297	325
3500	40" to 63" long					320			320	350
3550	64" to 72" long					330			330	365
3600	73" to 81" long					355			355	390
3650	82" to 90" long					375			375	415
3700	91" to 99" long					400			400	440
3750	100" to 108" long					410			410	450
3800	109" to 120" long					435			435	480
3820	121" to 130" long					465			465	515
3840	80" thru 120" wide, 26" to 39" long					375			375	415
3860	40" to 63" long					400			400	440
3880	64" to 72" long					410			410	450
4000	73" to 81" long					445			445	490
4020	82" to 90" long					465			465	510
4080	91" to 99" long					490			490	540
4100	100" to 108" long					525			525	575
4150	109" to 120" long					540			540	595
4200	121" to 130" long					565			565	625
4250	96" thru 144" wide, 26" to 39" long					445			445	490
4300	40" to 63" long					475			475	525
4400	64" to 72" long					490			490	535
4500	73" to 81" long					520			520	570
4600	82" to 90" long					560			560	615
4700	91" to 99" long					590			590	650
4800	100" to 108" long					625			625	685
4900	109" to 120" long					660			660	725
5000	121" to 130" long					705			705	775
5100	112" thru 168" wide, 26" to 39" long					525			525	575
5200	40" to 63" long					560			560	615
5300	64" to 72" long					580			580	640
5400	73" to 81" long					615			615	675
5500	82" to 90" long					650			650	715
5600	91" to 99" long					690			690	760
5700	100" to 108" long					730			730	805
5800	109" to 120" long					820			820	900

12 22 Curtains and Drapes

12 22 13 – Draperies

12 22 13.10 Custom Draperies		Crew	Daily Output	Labor-Hours	Unit	Material	2009 Bare Costs Labor	Equipment	Total	Total Incl O&P
6000	121" to 130" long				Ea.	875			875	965
6100	128" thru 192" wide, 26" to 39" long					590			590	650
6200	40" to 63" long					640			640	700
6300	64" to 72" long					660			660	730
6400	73" to 81" long					705			705	775
6450	82" to 90" long					740			740	815
6500	91" to 99" long					785			785	865
6550	100" to 108" long					830			830	915
6600	109" to 120" long					880			880	965
6650	121" to 130" long					935			935	1,025
6700	144" thru 216" wide, 26" to 39" long					675			675	740
6750	40" to 63" long					740			740	815
6800	64" to 72" long					790			790	865
6850	73" to 81" long					810			810	890
6880	82" to 90" long					845			845	930
6900	91" to 99" long					890			890	980
6920	100" to 108" long					935			935	1,025
6980	109" to 120" long					980			980	1,075
7000	121" to 130" long					1,050			1,050	1,175
7100	160" thru 240" wide, 26" to 39" long					740			740	815
7150	40" to 63" long					800			800	880
7200	64" to 72" long					820			820	900
7250	73" to 81" long					880			880	965
7300	82" to 90" long					935			935	1,025
7350	91" to 99" long					990			990	1,100
7400	100" to 108" long					1,025			1,025	1,150
7500	109" to 120" long					1,100			1,100	1,200
7600	121" to 130" long					1,175			1,175	1,300
8800	Drapery installation, hardware & drapes,									
9000	Labor cost only, minimum	1 Clab	75	.107	L.F.		3.37		3.37	5.25
9100	Maximum	"	20	.400	"		12.65		12.65	19.60

12 22 16 – Drapery Track and Accessories

12 22 16.10 Drapery Hardware

		Crew	Daily Output	Labor-Hours	Unit	Material	Labor	Equipment	Total	Total Incl O&P
0010	**DRAPERY HARDWARE**									
0030	Standard traverse, per foot, minimum	1 Carp	59	.136	L.F.	3.35	5.40		8.75	12.10
0100	Maximum		51	.157	"	7.90	6.25		14.15	18.35
0200	Decorative traverse, 28"-48", minimum		22	.364	Ea.	25	14.55		39.55	50
0220	Maximum		21	.381		43	15.20		58.20	71
0300	48"-84", minimum		20	.400		20.50	16		36.50	47.50
0320	Maximum		19	.421		58.50	16.80		75.30	90
0400	66"-120", minimum		18	.444		62	17.75		79.75	95.50
0420	Maximum		17	.471		78	18.80		96.80	115
0500	84"-156", minimum		16	.500		65	20		85	103
0520	Maximum		15	.533		98.50	21.50		120	142
0600	130"-240", minimum		14	.571		31	23		54	69.50
0620	Maximum		13	.615		134	24.50		158.50	185
0700	Slide rings, each, minimum					1.41			1.41	1.55
0720	Maximum					2.32			2.32	2.55
3000	Ripplefold, snap-a-pleat system, 3' or less, minimum	1 Carp	15	.533		52	21.50		73.50	90.50
3020	Maximum	"	14	.571		70.50	23		93.50	114
3200	Each additional foot, add, minimum				L.F.	2.19			2.19	2.41
3220	Maximum				"	6.65			6.65	7.30

12 22 Curtains and Drapes

12 22 16 – Drapery Track and Accessories

12 22 16.10 Drapery Hardware

		Crew	Daily Output	Labor-Hours	Unit	Material	2009 Bare Costs Labor	2009 Bare Costs Equipment	Total	Total Incl O&P
4000	Traverse rods, adjustable, 28" to 48"	1 Carp	22	.364	Ea.	20.50	14.55		35.05	45
4020	48" to 84"		20	.400		29	16		45	57
4040	66" to 120"		18	.444		40	17.75		57.75	71.50
4060	84" to 156"		16	.500		41	20		61	76
4080	100" to 180"		14	.571		42	23		65	81.50
4090	156" to 228"		13	.615		45	24.50		69.50	88
4100	228" to 312"		13	.615		58.50	24.50		83	102
4200	Double rods, adjustable, 30" to 48"		9	.889		37	35.50		72.50	96
4220	48" to 86"		9	.889		50	35.50		85.50	110
4240	86" to 150"		8	1		54.50	40		94.50	122
4260	100" to 180"		7	1.143		57	45.50		102.50	134
4300	Tray and curtain rod, adjustable, 30" to 48"		9	.889		22.50	35.50		58	80
4320	48" to 86"		9	.889		34	35.50		69.50	92.50
4340	86" to 150"		8	1		41	40		81	107
4360	100" to 180"		7	1.143		45.50	45.50		91	121
4600	Valance, pinch pleated fabric, 12" deep, up to 54" long, minimum					36			36	39.50
4610	Maximum					90			90	99
4620	Up to 77" long, minimum					55			55	60.50
4630	Maximum					145			145	160
5000	Stationary rods, first 2'					7.45			7.45	8.20
5020	Each additional foot, add				L.F.	3.57			3.57	3.93

12 22 16.20 Blast Curtains

0010	**BLAST CURTAINS**, per LF horizontal opening width, off-white or gray fabric									
0100	Blast curtains, drapery system, complete, including hardware, minimum				L.F.				123	135
0120	Average								133	146
0140	Maximum								142	157

12 23 Interior Shutters

12 23 10 – Wood Interior Shutters

12 23 10.13 Wood Panels

		Crew	Daily Output	Labor-Hours	Unit	Material	Labor	Equipment	Total	Total Incl O&P
0010	**WOOD PANELS**									
3000	Wood folding panels with movable louvers, 7" x 20" each	1 Carp	17	.471	Pr.	46	18.80		64.80	79.50
3300	8" x 28" each		17	.471		60	18.80		78.80	95
3450	9" x 36" each		17	.471		74	18.80		92.80	111
3600	10" x 40" each		17	.471		80	18.80		98.80	117
4000	Fixed louver type, stock units, 8" x 20" each		17	.471		79.50	18.80		98.30	117
4150	10" x 28" each		17	.471		88	18.80		106.80	126
4300	12" x 36" each		17	.471		104	18.80		122.80	143
4450	18" x 40" each		17	.471		168	18.80		186.80	214
5000	Insert panel type, stock, 7" x 20" each		17	.471		19.25	18.80		38.05	50
5150	8" x 28" each		17	.471		35.50	18.80		54.30	68
5300	9" x 36" each		17	.471		45	18.80		63.80	78.50
5450	10" x 40" each		17	.471		48	18.80		66.80	82
5600	Raised panel type, stock, 10" x 24" each		17	.471		134	18.80		152.80	177
5650	12" x 26" each		17	.471		156	18.80		174.80	201
5700	14" x 30" each		17	.471		210	18.80		228.80	260
5750	16" x 36" each		17	.471		288	18.80		306.80	345
6000	For custom built pine, add					22%				
6500	For custom built hardwood blinds, add					42%				

12 24 Window Shades

12 24 13 – Roller Window Shades

12 24 13.10 Shades

		Crew	Daily Output	Labor-Hours	Unit	Material	2009 Bare Costs Labor	2009 Bare Costs Equipment	Total	Total Incl O&P
0010	**SHADES**									
0020	Basswood, roll-up, stain finish, 3/8" slats	1 Carp	300	.027	S.F.	13.65	1.07		14.72	16.70
0200	7/8" slats		300	.027		12.90	1.07		13.97	15.85
0300	Vertical side slide, stain finish, 3/8" slats		300	.027		17.70	1.07		18.77	21
0400	7/8" slats	↓	300	.027		17.70	1.07		18.77	21
0500	For fire retardant finishes, add					16%				
0600	For "B" rated finishes, add					20%				
0900	Mylar, single layer, non-heat reflective	1 Carp	685	.012		6.05	.47		6.52	7.40
1000	Double layered, heat reflective		685	.012		9.10	.47		9.57	10.70
1100	Triple layered, heat reflective	↓	685	.012	↓	10.55	.47		11.02	12.35
1200	For metal roller instead of wood, add per				Shade	4.12			4.12	4.53
1300	Vinyl coated cotton, standard	1 Carp	685	.012	S.F.	3.36	.47		3.83	4.42
1400	Lightproof decorator shades		685	.012		2.10	.47		2.57	3.03
1500	Vinyl, lightweight, 4 gauge		685	.012		.55	.47		1.02	1.33
1600	Heavyweight, 6 gauge		685	.012		1.67	.47		2.14	2.56
1700	Vinyl laminated fiberglass, 6 ga., translucent		685	.012		2.36	.47		2.83	3.32
1800	Lightproof		685	.012		3.83	.47		4.30	4.93
3000	Woven aluminum, 3/8" thick, lightproof and fireproof	↓	350	.023	↓	5.40	.91		6.31	7.30

12 32 Manufactured Wood Casework

12 32 16 – Manufactured Plastic-Laminate-Clad Casework

12 32 16.20 Plastic Laminate Casework Doors

		Crew	Daily Output	Labor-Hours	Unit	Material	Labor	Equipment	Total	Total Incl O&P
0010	**PLASTIC LAMINATE CASEWORK DOORS**									
1000	For casework frames, see Div. 12 32 23.15									
1100	For casework hardware, see Div. 12 32 23.35									
6000	Plastic laminate on particle board									
6100	12" wide, 18" high	1 Carp	25	.320	Ea.	10.50	12.80		23.30	31.50
6120	24" high		24	.333		14	13.30		27.30	36
6140	30" high		23	.348		17.50	13.90		31.40	41
6160	36" high		21	.381		21	15.20		36.20	46.50
6200	48" high		16	.500		28	20		48	62
6250	60" high		13	.615		35	24.50		59.50	76.50
6300	72" high		12	.667		42	26.50		68.50	87.50
6320	15" wide, 18" high		24.50	.327		13.10	13.05		26.15	34.50
6340	24" high		23.50	.340		17.50	13.60		31.10	40.50
6360	30" high		22.50	.356		22	14.20		36.20	46
6380	36" high		20.50	.390		26.50	15.60		42.10	53
6400	48" high		15.50	.516		35	20.50		55.50	70.50
6450	60" high		12.50	.640		44	25.50		69.50	88
6480	72" high		11.50	.696		52.50	28		80.50	101
6500	18" wide, 18" high		24	.333		15.75	13.30		29.05	38
6550	24" high		23	.348		21	13.90		34.90	44.50
6600	30" high		22	.364		26.50	14.55		41.05	51.50
6650	36" high		20	.400		31.50	16		47.50	59.50
6700	48" high		15	.533		42	21.50		63.50	79
6750	60" high		12	.667		52.50	26.50		79	99.50
6800	72" high	↓	11	.727	↓	63	29		92	115

12 32 16.25 Plastic Laminate Drawer Fronts

		Crew	Daily Output	Labor-Hours	Unit	Material	Labor	Equipment	Total	Total Incl O&P
0010	**PLASTIC LAMINATE DRAWER FRONTS**									
2800	Plastic laminate on particle board front									
3000	4" high, 12" wide	1 Carp	17	.471	Ea.	4.62	18.80		23.42	34

12 32 Manufactured Wood Casework

12 32 16 – Manufactured Plastic-Laminate-Clad Casework

12 32 16.25 Plastic Laminate Drawer Fronts		Crew	Daily Output	Labor-Hours	Unit	Material	2009 Bare Costs Labor	Equipment	Total	Total Incl O&P
3200	18" wide	1 Carp	16	.500	Ea.	6.95	20		26.95	38.50
3600	24" wide		15	.533		9.25	21.50		30.75	43
3800	6" high, 12" wide		16	.500		7	20		27	38.50
4000	18" wide		15	.533		10.50	21.50		32	44.50
4500	24" wide		14	.571		14	23		37	51
4800	9" high, 12" wide		15	.533		10.50	21.50		32	44.50
5000	18" wide		14	.571		15.75	23		38.75	53
5200	24" wide	▼	13	.615	▼	21	24.50		45.50	61

12 32 23 – Hardwood Casework

12 32 23.10 Manufactured Wood Casework, Stock Units

		Crew	Daily Output	Labor-Hours	Unit	Material	Labor	Equipment	Total	Total Incl O&P
0010	**MANUFACTURED WOOD CASEWORK, STOCK UNITS**									
0700	Kitchen base cabinets, hardwood, not incl. counter tops,									
0710	24" deep, 35" high, prefinished									
0800	One top drawer, one door below, 12" wide	2 Carp	24.80	.645	Ea.	227	26		253	290
0820	15" wide		24	.667		272	26.50		298.50	340
0840	18" wide		23.30	.687		297	27.50		324.50	370
0860	21" wide		22.70	.705		310	28		338	385
0880	24" wide		22.30	.717		360	28.50		388.50	445
1000	Four drawers, 12" wide		24.80	.645		335	26		361	405
1020	15" wide		24	.667		365	26.50		391.50	440
1040	18" wide		23.30	.687		385	27.50		412.50	465
1060	24" wide		22.30	.717		410	28.50		438.50	495
1200	Two top drawers, two doors below, 27" wide		22	.727		370	29		399	455
1220	30" wide		21.40	.748		370	30		400	455
1240	33" wide		20.90	.766		375	30.50		405.50	465
1260	36" wide		20.30	.788		390	31.50		421.50	480
1280	42" wide		19.80	.808		420	32.50		452.50	510
1300	48" wide		18.90	.847		450	34		484	550
1500	Range or sink base, two doors below, 30" wide		21.40	.748		300	30		330	375
1520	33" wide		20.90	.766		320	30.50		350.50	405
1540	36" wide		20.30	.788		340	31.50		371.50	420
1560	42" wide		19.80	.808		360	32.50		392.50	450
1580	48" wide	▼	18.90	.847		380	34		414	470
1800	For sink front units, deduct					60			60	66
2000	Corner base cabinets, 36" wide, standard	2 Carp	18	.889		630	35.50		665.50	745
2100	Lazy Susan with revolving door	"	16.50	.970	▼	540	38.50		578.50	655
4000	Kitchen wall cabinets, hardwood, 12" deep with two doors									
4050	12" high, 30" wide	2 Carp	24.80	.645	Ea.	197	26		223	257
4100	36" wide		24	.667		228	26.50		254.50	293
4400	15" high, 30" wide		24	.667		198	26.50		224.50	260
4420	33" wide		23.30	.687		223	27.50		250.50	288
4440	36" wide		22.70	.705		243	28		271	310
4450	42" wide		22.70	.705		284	28		312	355
4700	24" high, 30" wide		23.30	.687		310	27.50		337.50	385
4720	36" wide		22.70	.705		345	28		373	425
4740	42" wide		22.30	.717		335	28.50		363.50	410
5000	30" high, one door, 12" wide		22	.727		191	29		220	256
5020	15" wide		21.40	.748		210	30		240	278
5040	18" wide		20.90	.766		234	30.50		264.50	305
5060	24" wide		20.30	.788		271	31.50		302.50	345
5300	Two doors, 27" wide		19.80	.808		289	32.50		321.50	370
5320	30" wide	▼	19.30	.829		320	33		353	400

12 32 Manufactured Wood Casework

12 32 23 – Hardwood Casework

12 32 23.10 Manufactured Wood Casework, Stock Units		Crew	Daily Output	Labor-Hours	Unit	Material	2009 Bare Costs Labor	Equipment	Total	Total Incl O&P
5340	36" wide	2 Carp	18.80	.851	Ea.	360	34		394	450
5360	42" wide		18.50	.865		385	34.50		419.50	480
5380	48" wide		18.40	.870		400	34.50		434.50	495
6000	Corner wall, 30" high, 24" wide		18	.889		209	35.50		244.50	285
6050	30" wide		17.20	.930		260	37		297	345
6100	36" wide		16.50	.970		290	38.50		328.50	380
6500	Revolving Lazy Susan		15.20	1.053		390	42		432	495
7000	Broom cabinet, 84" high, 24" deep, 18" wide		10	1.600		515	64		579	665
7500	Oven cabinets, 84" high, 24" deep, 27" wide		8	2	↓	785	80		865	985
7750	Valance board trim	↓	396	.040	L.F.	12.25	1.61		13.86	16
9000	For deluxe models of all cabinets, add					40%				
9500	For custom built in place, add					25%	10%			
9550	Rule of thumb, kitchen cabinets not including									
9560	appliances & counter top, minimum	2 Carp	30	.533	L.F.	130	21.50		151.50	176
9600	Maximum	"	25	.640	"	310	25.50		335.50	385
9610	For metal cabinets, see Div. 12 35 70.13									

12 32 23.15 Manufactured Wood Casework Frames		Crew	Daily Output	Labor-Hours	Unit	Material	Labor	Equipment	Total	Total Incl O&P
0010	**MANUFACTURED WOOD CASEWORK FRAMES**									
0050	Base cabinets, counter storage, 36" high									
0100	One bay, 18" wide	1 Carp	2.70	2.963	Ea.	131	118		249	330
0200	24" wide		2.50	3.200		154	128		282	365
0300	36" wide		2.30	3.478		183	139		322	415
0400	Two bay, 36" wide		2.20	3.636		200	145		345	445
1000	48" wide		2	4		216	160		376	485
1050	72" wide		1.80	4.444		254	178		432	555
1100	Three bay, 54" wide		1.50	5.333		238	213		451	590
1200	72" wide		1.30	6.154		293	246		539	700
1800	108" wide		1.10	7.273		340	291		631	825
2000	Four bay, 72" wide		1.10	7.273		330	291		621	815
2100	96" wide		1	8		370	320		690	900
2500	144" wide		.90	8.889		445	355		800	1,050
2800	Bookcases, one bay, 7' high, 18" wide		2.40	3.333		154	133		287	375
3000	24" wide		2.30	3.478		177	139		316	410
3100	36" wide		2.20	3.636		216	145		361	465
3500	Two bay, 36" wide		1.60	5		224	200		424	555
3800	48" wide		1.50	5.333		278	213		491	635
4000	72" wide		1.40	5.714		340	228		568	730
4100	Three bay, 54" wide		1.20	6.667		370	266		636	820
4200	72" wide		1.10	7.273		410	291		701	900
4500	108" wide		1	8		525	320		845	1,075
4600	Four bay, 72" wide		.95	8.421		415	335		750	980
4700	96" wide		.90	8.889		510	355		865	1,100
5000	144" wide		.85	9.412		695	375		1,070	1,350
5100	Coat racks, one bay, 7' high, 24" wide		4.50	1.778		154	71		225	279
5200	36" wide		4.30	1.860		168	74.50		242.50	300
5300	Two bay, 48" wide		2.75	2.909		214	116		330	415
5600	72" wide		2.50	3.200		246	128		374	470
5800	Three bay, 72" wide		2.10	3.810		315	152		467	580
6000	108" wide		1.90	4.211		340	168		508	635
6100	Wall mounted cabinet, one bay, 24" high, 18" wide		3.60	2.222		85	89		174	232
6400	24" wide		3.50	2.286		92	91.50		183.50	243
6600	36" wide	↓	3.40	2.353	↓	116	94		210	274

12 32 Manufactured Wood Casework

12 32 23 – Hardwood Casework

12 32 23.15 Manufactured Wood Casework Frames		Crew	Daily Output	Labor-Hours	Unit	Material	2009 Bare Costs Labor	Equipment	Total	Total Incl O&P
6800	Two bay, 36" wide	1 Carp	2.20	3.636	Ea.	124	145		269	360
7000	48" wide		2.10	3.810		139	152		291	390
7200	72" wide		2	4		177	160		337	445
7400	Three bay, 54" wide		1.70	4.706		154	188		342	460
7600	72" wide		1.60	5		185	200		385	515
7800	108" wide		1.50	5.333		231	213		444	585
8000	Four bay, 72" wide		1.40	5.714		184	228		412	560
8100	96" wide		1.30	6.154		216	246		462	620
8200	144" wide		1.20	6.667		290	266		556	735
8400	30" high, one bay, 18" wide		3.60	2.222		92	89		181	239
8600	24" wide		3.40	2.353		107	94		201	263
8800	36" wide		3.20	2.500		122	100		222	290
9000	Two bay, 36" wide		2.15	3.721		122	149		271	365
9100	48" wide		2	4		145	160		305	410
9200	72" wide		1.85	4.324		176	173		349	460
9400	Three bay, 54" wide		1.60	5		153	200		353	480
9600	72" wide		1.50	5.333		183	213		396	530
9650	108" wide		1.40	5.714		405	228		633	800
9700	Four bay, 72" wide		1.30	6.154		191	246		437	590
9750	96" wide		1.22	6.557		229	262		491	655
9780	144" wide		1.15	6.957		310	278		588	770
9800	Wardrobe, 7' high, single, 24" wide		2.70	2.963		170	118		288	370
9850	36" wide		2.50	3.200		255	128		383	480
9880	Partition & adjustable shelves, 48" wide		1.70	4.706		216	188		404	530
9890	72" wide		1.55	5.161		345	206		551	700
9950	Partition, adjustable shelves & drawers, 48" wide		1.40	5.714		325	228		553	710
9960	72" wide		1.25	6.400		425	256		681	860

12 32 23.20 Manufactured Hardwood Casework Doors		Crew	Daily Output	Labor-Hours	Unit	Material	2009 Bare Costs Labor	Equipment	Total	Total Incl O&P
0010	**MANUFACTURED HARDWOOD CASEWORK DOORS**									
2000	Glass panel, hardwood frame									
2200	12" wide, 18" high	1 Carp	34	.235	Ea.	18.75	9.40		28.15	35
2400	24" high		33	.242		25	9.70		34.70	42.50
2600	30" high		32	.250		31.50	10		41.50	50
2800	36" high		30	.267		37.50	10.65		48.15	58
3000	48" high		23	.348		50	13.90		63.90	76.50
3200	60" high		17	.471		62.50	18.80		81.30	98
3400	72" high		15	.533		75	21.50		96.50	116
3600	15" wide, 18" high		33	.242		23.50	9.70		33.20	41
3800	24" high		32	.250		31.50	10		41.50	50
4000	30" high		30	.267		39	10.65		49.65	59.50
4250	36" high		28	.286		47	11.40		58.40	69
4300	48" high		22	.364		62.50	14.55		77.05	91.50
4350	60" high		16	.500		78	20		98	117
4400	72" high		14	.571		94	23		117	139
4450	18" wide, 18" high		32	.250		28	10		38	46.50
4500	24" high		30	.267		37.50	10.65		48.15	58
4550	30" high		29	.276		47	11		58	68.50
4600	36" high		27	.296		56.50	11.85		68.35	80.50
4700	60" high		15	.533		94	21.50		115.50	136
4750	72" high		13	.615		113	24.50		137.50	162
5000	Hardwood, raised panel									
5100	12" wide, 18" high	1 Carp	16	.500	Ea.	27	20		47	60.50

12 32 Manufactured Wood Casework

12 32 23 – Hardwood Casework

12 32 23.20 Manufactured Hardwood Casework Doors

		Crew	Daily Output	Labor-Hours	Unit	Material	2009 Bare Costs Labor	Equipment	Total	Total Incl O&P
5150	24" high	1 Carp	15.50	.516	Ea.	36	20.50		56.50	71.50
5200	30" high		15	.533		45	21.50		66.50	82.50
5250	36" high		14	.571		54	23		77	95
5300	48" high		11	.727		72	29		101	124
5320	60" high		8	1		90	40		130	161
5340	72" high		7	1.143		108	45.50		153.50	190
5360	15" wide, 18" high		15.50	.516		34	20.50		54.50	69
5380	24" high		15	.533		45	21.50		66.50	82.50
5400	30" high		14.50	.552		56.50	22		78.50	96
5420	36" high		13.50	.593		67.50	23.50		91	111
5440	48" high		10.50	.762		90	30.50		120.50	146
5460	60" high		7.50	1.067		113	42.50		155.50	190
5480	72" high		6.50	1.231		135	49		184	226
5500	18" wide, 18" high		15	.533		40.50	21.50		62	77.50
5550	24" high		14.50	.552		54	22		76	93.50
5600	30" high		14	.571		67.50	23		90.50	110
5650	36" high		13	.615		81	24.50		105.50	127
5700	48" high		10	.800		108	32		140	169
5750	60" high		7	1.143		135	45.50		180.50	220
5800	72" high		6	1.333		162	53.50		215.50	261

12 32 23.25 Manufactured Wood Casework Drawer Fronts

		Crew	Daily Output	Labor-Hours	Unit	Material	Labor	Equipment	Total	Total Incl O&P
0010	**MANUFACTURED WOOD CASEWORK DRAWER FRONTS**									
0100	Solid hardwood front									
1000	4" high, 12" wide	1 Carp	17	.471	Ea.	2.66	18.80		21.46	32
1200	18" wide		16	.500		4	20		24	35.50
1400	24" wide		15	.533		5.30	21.50		26.80	39
1600	6" high, 12" wide		16	.500		4	20		24	35.50
1800	18" wide		15	.533		6	21.50		27.50	39.50
2000	24" wide		14	.571		8	23		31	44.50
2200	9" high, 12" wide		15	.533		6	21.50		27.50	39.50
2400	18" wide		14	.571		9	23		32	45.50
2600	24" wide		13	.615		12	24.50		36.50	51

12 32 23.30 Manufactured Wood Casework Vanities

		Crew	Daily Output	Labor-Hours	Unit	Material	Labor	Equipment	Total	Total Incl O&P
0010	**MANUFACTURED WOOD CASEWORK VANITIES**									
8000	Vanity bases, 2 doors, 30" high, 21" deep, 24" wide	2 Carp	20	.800	Ea.	275	32		307	355
8050	30" wide		16	1		315	40		355	410
8100	36" wide		13.33	1.200		345	48		393	455
8150	48" wide		11.43	1.400		425	56		481	555
9000	For deluxe models of all vanities, add to above					40%				
9500	For custom built in place, add to above					25%	10%			

12 32 23.35 Manufactured Wood Casework Hardware

		Crew	Daily Output	Labor-Hours	Unit	Material	Labor	Equipment	Total	Total Incl O&P
0010	**MANUFACTURED WOOD CASEWORK HARDWARE**									
1000	Catches, minimum	1 Carp	235	.034	Ea.	1	1.36		2.36	3.21
1020	Average		119.40	.067		3.29	2.68		5.97	7.75
1040	Maximum		80	.100		6.25	4		10.25	13.05
2000	Door/drawer pulls, handles									
2200	Handles and pulls, projecting, metal, minimum	1 Carp	160	.050	Ea.	4.43	2		6.43	7.95
2220	Average		95.24	.084		6.90	3.36		10.26	12.75
2240	Maximum		68	.118		9.40	4.70		14.10	17.60
2300	Wood, minimum		160	.050		4.67	2		6.67	8.25
2320	Average		95.24	.084		6.25	3.36		9.61	12.05
2340	Maximum		68	.118		8.55	4.70		13.25	16.75

12 32 Manufactured Wood Casework

12 32 23 – Hardwood Casework

12 32 23.35 Manufactured Wood Casework Hardware		Crew	Daily Output	Labor-Hours	Unit	Material	2009 Bare Costs Labor	Equipment	Total	Total Incl O&P
2600	Flush, metal, minimum	1 Carp	160	.050	Ea.	4.67	2		6.67	8.25
2620	Average		95.24	.084		6.25	3.36		9.61	12.05
2640	Maximum		68	.118	↓	8.55	4.70		13.25	16.75
3000	Drawer tracks/glides, minimum		48	.167	Pr.	7.90	6.65		14.55	19.05
3020	Average		32	.250		13.40	10		23.40	30.50
3040	Maximum		24	.333		23	13.30		36.30	46
4000	Cabinet hinges, minimum		160	.050		2.69	2		4.69	6.05
4020	Average		95.24	.084		6	3.36		9.36	11.80
4040	Maximum		68	.118	↓	10.15	4.70		14.85	18.50
5000	Cabinet locks, minimum		47.90	.167	Ea.	5.05	6.65		11.70	15.90
5020	Average		23.95	.334		7.90	13.35		21.25	29
5040	Maximum	↓	16	.500	↓	19.15	20		39.15	52

12 35 Specialty Casework

12 35 50 – Educational/Library Casework

12 35 50.13 Educational Casework

		Crew	Daily Output	Labor-Hours	Unit	Material	Labor	Equipment	Total	Total Incl O&P
0010	**EDUCATIONAL CASEWORK**									
5000	School, 24" deep, metal, 84" high units	2 Carp	15	1.067	L.F.	365	42.50		407.50	470
5150	Counter height units		20	.800		246	32		278	320
5450	Wood, custom fabricated, 32" high counter		20	.800		205	32		237	275
5600	Add for counter top		56	.286		22	11.40		33.40	41.50
5800	84" high wall units	↓	15	1.067	↓	400	42.50		442.50	505
6000	Laminated plastic finish is same price as wood									

12 35 53 – Laboratory Casework

12 35 53.13 Metal Laboratory Casework

		Crew	Daily Output	Labor-Hours	Unit	Material	Labor	Equipment	Total	Total Incl O&P
0010	**METAL LABORATORY CASEWORK**									
0020	Cabinets, base, door units, metal	2 Carp	18	.889	L.F.	183	35.50		218.50	256
0300	Drawer units		18	.889		410	35.50		445.50	505
0700	Tall storage cabinets, open, 7' high		20	.800		390	32		422	480
0900	With glazed doors		20	.800		470	32		502	565
1300	Wall cabinets, metal, 12-1/2" deep, open		20	.800		129	32		161	192
1500	With doors	↓	20	.800	↓	268	32		300	345

12 35 59 – Display Casework

12 35 59.10 Display Cases

		Crew	Daily Output	Labor-Hours	Unit	Material	Labor	Equipment	Total	Total Incl O&P
0010	**DISPLAY CASES** Free standing, all glass									
0020	Aluminum frame, 42" high x 36" x 12" deep	2 Carp	8	2	Ea.	1,200	80		1,280	1,450
0100	70" high x 48" x 18" deep	"	6	2.667		2,400	107		2,507	2,825
0500	For wood bases, add					9%				
0600	For hardwood frames, deduct					8%				
0700	For bronze, baked enamel finish, add				↓	10%				
2000	Wall mounted, glass front, aluminum frame									
2010	Non-illuminated, one section 3' x 4' x 1'-4"	2 Carp	5	3.200	Ea.	2,125	128		2,253	2,550
2100	5' x 4' x 1'-4"		5	3.200		2,300	128		2,428	2,750
2200	6' x 4' x 1'-4"		4	4		2,725	160		2,885	3,250
2500	Two sections, 8' x 4' x 1'-4"		2	8		1,775	320		2,095	2,450
2600	10' x 4' x 1'-4"		2	8		2,075	320		2,395	2,800
3000	Three sections, 16' x 4' x 1'-4"	↓	1.50	10.667	↓	3,600	425		4,025	4,625
3500	For fluorescent lights, add				Section	299			299	330
4000	Table exhibit cases, 2' wide, 3' high, 4' long, flat top	2 Carp	5	3.200	Ea.	1,050	128		1,178	1,350
4100	3' wide, 3' high, 4' long, sloping top	"	3	5.333	"	1,250	213		1,463	1,700

12 35 Specialty Casework

12 35 70 – Healthcare Casework

12 35 70.13 Hospital Casework

		Crew	Daily Output	Labor-Hours	Unit	Material	2009 Bare Costs Labor	2009 Bare Costs Equipment	Total	Total Incl O&P
0010	**HOSPITAL CASEWORK**									
0500	Base cabinets, laminated plastic	2 Carp	10	1.600	L.F.	235	64		299	355
0700	Enameled steel		10	1.600		216	64		280	335
1000	Stainless steel		10	1.600		430	64		494	575
1200	For all drawers, add					24.50			24.50	27
1300	Cabinet base trim, 4" high, enameled steel	2 Carp	200	.080		39.50	3.20		42.70	48.50
1400	Stainless steel		200	.080		78.50	3.20		81.70	91.50
1450	Countertop, laminated plastic, no backsplash		40	.400		40.50	16		56.50	69.50
1650	With backsplash		40	.400		50.50	16		66.50	80.50
1800	For sink cutout, add		12.20	1.311	Ea.		52.50		52.50	81.50
1900	Stainless steel counter top		40	.400	L.F.	131	16		147	169
2000	For drop-in stainless 43" x 21" sink, add				Ea.	860			860	950
2500	Wall cabinets, laminated plastic	2 Carp	15	1.067	L.F.	176	42.50		218.50	259
2600	Enameled steel		15	1.067		216	42.50		258.50	305
2700	Stainless steel		15	1.067		430	42.50		472.50	540
2800	For glass doors, add					29.50			29.50	32.50

12 35 70.16 Nurse Station Casework

		Crew	Daily Output	Labor-Hours	Unit	Material	Labor	Equipment	Total	Total Incl O&P
0010	**NURSE STATION CASEWORK**									
2100	Door type, laminated plastic	2 Carp	10	1.600	L.F.	272	64		336	400
2200	Enameled steel		10	1.600		260	64		324	385
2300	Stainless steel		10	1.600		520	64		584	670
2400	For drawer type, add					221			221	243

12 35 80 – Commercial Kitchen Casework

12 35 80.13 Metal Kitchen Casework

		Crew	Daily Output	Labor-Hours	Unit	Material	Labor	Equipment	Total	Total Incl O&P
0010	**METAL KITCHEN CASEWORK**									
3500	Base cabinets, metal, minimum	2 Carp	30	.533	L.F.	63.50	21.50		85	103
3600	Maximum		25	.640		161	25.50		186.50	217
3700	Wall cabinets, metal, minimum		30	.533		63.50	21.50		85	103
3800	Maximum		25	.640		145	25.50		170.50	200

12 36 Countertops

12 36 23 – Plastic Countertops

12 36 23.13 Plastic-Laminate-Clad Countertops

		Crew	Daily Output	Labor-Hours	Unit	Material	Labor	Equipment	Total	Total Incl O&P
0010	**PLASTIC-LAMINATE-CLAD COUNTERTOPS**									
0020	Stock, 24" wide w/ backsplash, minimum	1 Carp	30	.267	L.F.	8.65	10.65		19.30	26
0100	Maximum	"	25	.320	"	18	12.80		30.80	39.50

12 36 23.30 Plastic Laminate Countertop Components

		Crew	Daily Output	Labor-Hours	Unit	Material	Labor	Equipment	Total	Total Incl O&P
0010	**PLASTIC LAMINATE COUNTERTOP COMPONENTS**									
1000	Edging, 24" wide, 1-1/2" thick									
1500	Plastic laminate, minimum	1 Carp	25	.320	L.F.	3.40	12.80		16.20	23.50
1520	Average		24	.333		6.25	13.30		19.55	27.50
1540	Maximum		22	.364		9.35	14.55		23.90	33
2500	Hardwood, minimum		20	.400		4.20	16		20.20	29.50
2520	Average		18	.444		6.75	17.75		24.50	35
2540	Maximum		16	.500		9.65	20		29.65	41.50
2600	Backsplash, add to above, minimum		36	.222		1.36	8.90		10.26	15.25
2620	Average		35	.229		2.27	9.15		11.42	16.65
2640	Maximum		34	.235		3.18	9.40		12.58	18.10
2700	For metal cove, add					.80			.80	.88

12 36 Countertops

12 36 23 – Plastic Countertops

12 36 23.30 Plastic Laminate Countertop Components		Crew	Daily Output	Labor-Hours	Unit	Material	2009 Bare Costs Labor	Equipment	Total	Total Incl O&P
2900	Postformed backsplash, add to above									
2920	Minimum	1 Carp	21	.381	L.F.	1.23	15.20		16.43	25
2940	Average		20	.400		2.44	16		18.44	27.50
2960	Maximum		19	.421	↓	3.68	16.80		20.48	30
3500	Well openings (for typewriters etc.)		2.50	3.200	Ea.		128		128	198
3900	Cutouts for sinks, lavatories	↓	12	.667	"		26.50		26.50	41.50

12 36 40 – Stone Countertops

12 36 40.10 Natural Stone Countertops

		Crew	Daily Output	Labor-Hours	Unit	Material	Labor	Equipment	Total	Total Incl O&P
0010	**NATURAL STONE COUNTERTOPS**									
2500	Marble, stock, with splash, 1/2" thick, minimum	1 Bric	17	.471	L.F.	36.50	19.05		55.55	69
2700	3/4" thick, maximum		13	.615		91	25		116	138
2720	Marble, 24" wide, no splash		10	.800		28.50	32.50		61	80.50
2740	4" backsplash		10	.800		32	32.50		64.50	84.50
2800	Granite, average, 1-1/4" thick, 24" wide, no splash	↓	13.01	.615	↓	125	25		150	175

12 36 53 – Laboratory Countertops

12 36 53.10 Laboratory Countertops and Sinks

		Crew	Daily Output	Labor-Hours	Unit	Material	Labor	Equipment	Total	Total Incl O&P
0010	**LABORATORY COUNTERTOPS AND SINKS**									
0020	Countertops, not incl. base cabinets, acidproof, minimum	2 Carp	82	.195	S.F.	31.50	7.80		39.30	46.50
0030	Maximum		70	.229		41	9.15		50.15	59.50
0040	Stainless steel	↓	82	.195	↓	95.50	7.80		103.30	117

12 36 61 – Simulated Stone Countertops

12 36 61.16 Solid Surface Countertops

		Crew	Daily Output	Labor-Hours	Unit	Material	Labor	Equipment	Total	Total Incl O&P
0010	**SOLID SURFACE COUNTERTOPS**, Acrylic polymer									
0020	Pricing for orders of 100 L.F. or greater									
0100	25" wide, solid colors	2 Carp	28	.571	L.F.	44.50	23		67.50	84.50
0200	Patterned colors		28	.571		56.50	23		79.50	97.50
0300	Premium patterned colors		28	.571		70.50	23		93.50	114
0400	With silicone attached 4" backsplash, solid colors		27	.593		49	23.50		72.50	90.50
0500	Patterned colors		27	.593		62	23.50		85.50	105
0600	Premium patterned colors		27	.593		77	23.50		100.50	122
0700	With hard seam attached 4" backsplash, solid colors		23	.696		49	28		77	97
0800	Patterned colors		23	.696		62	28		90	111
0900	Premium patterned colors	↓	23	.696	↓	77	28		105	128
1000	Pricing for order of 51 – 99 L.F.									
1100	25" wide, solid colors	2 Carp	24	.667	L.F.	51.50	26.50		78	98
1200	Patterned colors		24	.667		65	26.50		91.50	113
1300	Premium patterned colors		24	.667		81.50	26.50		108	131
1400	With silicone attached 4" backsplash, solid colors		23	.696		56.50	28		84.50	105
1500	Patterned colors		23	.696		71.50	28		99.50	122
1600	Premium patterned colors		23	.696		89	28		117	141
1700	With hard seam attached 4" backsplash, solid colors		20	.800		56.50	32		88.50	112
1800	Patterned colors		20	.800		71.50	32		103.50	128
1900	Premium patterned colors	↓	20	.800	↓	89	32		121	147
2000	Pricing for order of 1 – 50 L.F.									
2100	25" wide, solid colors	2 Carp	20	.800	L.F.	60	32		92	116
2200	Patterned colors		20	.800		76.50	32		108.50	134
2300	Premium patterned colors		20	.800		95.50	32		127.50	155
2400	With silicone attached 4" backsplash, solid colors		19	.842		66	33.50		99.50	125
2500	Patterned colors		19	.842		83.50	33.50		117	144
2600	Premium patterned colors		19	.842		104	33.50		137.50	167
2700	With hard seam attached 4" backsplash, solid colors	↓	15	1.067	↓	66	42.50		108.50	139

12 36 Countertops

12 36 61 – Simulated Stone Countertops

12 36 61.16 Solid Surface Countertops

		Crew	Daily Output	Labor-Hours	Unit	Material	2009 Bare Costs Labor	Equipment	Total	Total Incl O&P
2800	Patterned colors	2 Carp	15	1.067	L.F.	83.50	42.50		126	158
2900	Premium patterned colors	↓	15	1.067	↓	104	42.50		146.50	181
3000	Sinks, pricing for order of 100 or greater units									
3100	Single bowl, hard seamed, solid colors, 13" x 17"	1 Carp	3	2.667	Ea.	300	107		407	495
3200	10" x 15"		7	1.143		139	45.50		184.50	224
3300	Cutouts for sinks	↓	8	1	↓		40		40	62
3400	Sinks, pricing for order of 51 – 99 units									
3500	Single bowl, hard seamed, solid colors, 13" x 17"	1 Carp	2.55	3.137	Ea.	345	125		470	575
3600	10" x 15"		6	1.333		160	53.50		213.50	259
3700	Cutouts for sinks	↓	7	1.143	↓		45.50		45.50	71
3800	Sinks, pricing for order of 1 – 50 units									
3900	Single bowl, hard seamed, solid colors, 13" x 17"	1 Carp	2	4	Ea.	405	160		565	695
4000	10" x 15"		4.55	1.758		188	70		258	315
4100	Cutouts for sinks		5.25	1.524			61		61	94.50
4200	Cooktop cutouts, pricing for 100 or greater units		4	2		22	80		102	149
4300	51 – 99 units		3.40	2.353		25.50	94		119.50	174
4400	1 – 50 units	↓	3	2.667	↓	30	107		137	198

12 36 61.17 Solid Surface Vanity Tops

		Crew	Daily Output	Labor-Hours	Unit	Material	2009 Bare Costs Labor	Equipment	Total	Total Incl O&P
0010	**SOLID SURFACE VANITY TOPS**									
0015	Solid surface, center bowl, 17" x 19"	1 Carp	12	.667	Ea.	189	26.50		215.50	250
0020	19" x 25"		12	.667		229	26.50		255.50	294
0030	19" x 31"		12	.667		278	26.50		304.50	345
0040	19" x 37"		12	.667		325	26.50		351.50	395
0050	22" x 25"		10	.800		202	32		234	272
0060	22" x 31"		10	.800		236	32		268	310
0070	22" x 37"		10	.800		274	32		306	350
0080	22" x 43"		10	.800		315	32		347	395
0090	22" x 49"		10	.800		345	32		377	430
0110	22" x 55"		8	1		395	40		435	495
0120	22" x 61"		8	1		450	40		490	555
0220	Double bowl, 22" x 61"		8	1		505	40		545	620
0230	Double bowl, 22" x 73"	↓	8	1	↓	705	40		745	835
0240	For aggregate colors, add					35%				
0250	For faucets and fittings, see Div. 22 41 39.10									

12 36 61.19 Engineered Stone Countertops

		Crew	Daily Output	Labor-Hours	Unit	Material	2009 Bare Costs Labor	Equipment	Total	Total Incl O&P
0010	**ENGINEERED STONE COUNTERTOPS**									
0100	25" wide, 4" backsplash, color group A, minimum	2 Carp	15	1.067	L.F.	17	42.50		59.50	84.50
0110	Maximum		15	1.067		42	42.50		84.50	112
0120	Color group B, minimum		15	1.067		22	42.50		64.50	90
0130	Maximum		15	1.067		49	42.50		91.50	120
0140	Color group C, minimum		15	1.067		29.50	42.50		72	98.50
0150	Maximum		15	1.067		60	42.50		102.50	132
0160	Color group D, minimum		15	1.067		37	42.50		79.50	107
0170	Maximum	↓	15	1.067	↓	70	42.50		112.50	143

12 43 Portable Lamps

12 43 13 – Lamps

12 43 13.23 Miscellaneous Lamps

		Crew	Daily Output	Labor-Hours	Unit	Material	2009 Bare Costs Labor	Equipment	Total	Total Incl O&P
0010	**MISCELLANEOUS LAMPS**									
1000	Ceramic, desk, minimum				Ea.	49			49	54
1020	Maximum					130			130	143
1200	End table, minimum					45			45	49.50
1220	Maximum					108			108	119
1300	Night stand, minimum					40			40	44
1320	Maximum					84			84	92.50
1400	Wall, minimum					54			54	59.50
1420	Maximum					200			200	220
2000	Glass, desk, minimum					191			191	210
2020	Maximum					555			555	610
2100	End table, minimum					185			185	204
2120	Maximum					730			730	805
2200	Floor, minimum					465			465	510
2250	Maximum					1,100			1,100	1,200
2280	Night stand, minimum					292			292	320
2300	Maximum					555			555	610
2350	Pendant, minimum					47.50			47.50	52
2400	Maximum					905			905	995
3000	Metal, desk, minimum					43			43	47.50
3100	Maximum					117			117	129
3200	End table, minimum					70			70	77
3250	Maximum					92			92	101
3300	Wall, minimum					67			67	73.50
3350	Maximum					123			123	135
3400	Floor, minimum					90			90	99
3450	Maximum					400			400	440
3500	Night stand, minimum					40			40	44
3550	Maximum					105			105	115
3600	Pendant, minimum					35			35	38.50
3650	Maximum					340			340	375
4000	Stone, desk, minimum					80			80	88
4100	Maximum					117			117	129
4200	End table, minimum					62			62	68
4300	Maximum					92			92	101
4400	Night stand, minimum					62			62	68
4500	Maximum					70			70	77
5000	Wall, minimum					58.50			58.50	64.50
6000	Maximum					92			92	101
8000	Replacement shades, lamp, desk, minimum					14			14	15.40
8020	Maximum					50			50	55
8040	End table, minimum					15.50			15.50	17.05
8060	Maximum					41			41	45
8100	Night stand minimum					13			13	14.30
8120	Maximum					36			36	39.50
8140	Wall, minimum					9			9	9.90
8160	Maximum					14.50			14.50	15.95

12 45 Bedroom Furnishings

12 45 13 – Bed Linens

12 45 13.13 Blankets		Crew	Daily Output	Labor-Hours	Unit	Material	2009 Bare Costs Labor	Equipment	Total	Total Incl O&P
0010	**BLANKETS**									
7700	Bedspreads, unquilted, twin, minimum				Ea.	33			33	36
7800	Maximum					286			286	315
7850	Full, minimum					39			39	43
7900	Maximum					410			410	450
8000	Queen, minimum					53.50			53.50	59
8100	Maximum					435			435	475
8200	Quilted, twin, minimum					77			77	85
8300	Maximum					365			365	400
8400	Full, minimum					96			96	106
8500	Maximum					490			490	540
8600	Queen, minimum					94.50			94.50	104
8700	Maximum					525			525	575

12 46 Furnishing Accessories

12 46 13 – Ash Receptacles

12 46 13.10 Ash/Trash Receivers

		Crew	Daily Output	Labor-Hours	Unit	Material	Labor	Equipment	Total	Total Incl O&P
0010	**ASH/TRASH RECEIVERS**									
1000	Ash urn, cylindrical metal									
1020	8" diameter, 20" high	1 Clab	60	.133	Ea.	125	4.21		129.21	144
1040	8" dia., 25" high		60	.133		190	4.21		194.21	216
1060	10" diameter, 26" high		60	.133		110	4.21		114.21	128
1080	12" dia., 30" high		60	.133		160	4.21		164.21	183
2000	Combination ash/trash urn, metal									
2020	8" diameter, 20" high	1 Clab	60	.133	Ea.	80	4.21		84.21	94.50
2040	8" dia., 25" high		60	.133		220	4.21		224.21	249
2050	10" diameter, 26" high		60	.133		95	4.21		99.21	111
2060	12" dia., 30" high		60	.133		240	4.21		244.21	271

12 46 19 – Clocks

12 46 19.50 Wall Clocks

		Crew	Daily Output	Labor-Hours	Unit	Material	Labor	Equipment	Total	Total Incl O&P
0010	**WALL CLOCKS**									
0080	12" diameter, single face	1 Elec	8	1	Ea.	81	47		128	159
0100	Double face	"	6.20	1.290	"	154	60.50		214.50	261

12 46 33 – Waste Receptacles

12 46 33.13 Trash Receptacles

		Crew	Daily Output	Labor-Hours	Unit	Material	Labor	Equipment	Total	Total Incl O&P
0010	**TRASH RECEPTACLES**									
4000	Trash receptacle, metal									
4020	8" diameter, 15" high	1 Clab	60	.133	Ea.	67	4.21		71.21	80
4040	10" diameter, 18" high		60	.133		125	4.21		129.21	144
4060	16" dia., 16" high		60	.133		150	4.21		154.21	172
4100	18" dia., 32" high		60	.133		270	4.21		274.21	305
5000	Plastic, fire resistant									
5020	Rectangular 11" x 8" x 12" high	1 Clab	60	.133	Ea.	13	4.21		17.21	21
5040	16" x 8" x 14" high	"	60	.133	"	19	4.21		23.21	27.50
5500	Plastic, with lid									
5520	35 gallon	1 Clab	60	.133	Ea.	211	4.21		215.21	239
5540	45 gallon		60	.133		280	4.21		284.21	315
5550	Plastic recycling barrel, w/lid & wheels, 32 gal [G]		60	.133		60	4.21		64.21	72.50
5560	65 gal [G]		60	.133		107	4.21		111.21	125
5570	95 gal [G]		60	.133		310	4.21		314.21	345

12 46 Furnishing Accessories

12 46 36 – Desk Accessories

12 46 36.10 Commercial Desk Accessories		Crew	Daily Output	Labor-Hours	Unit	Material	2009 Bare Costs Labor	Equipment	Total	Total Incl O&P
0010	**COMMERCIAL DESK ACCESSORIES**									
0300	Bookends, minimum				Ea.	7.80			7.80	8.55
0320	Maximum					14			14	15.40
1000	Calendar with pad, minimum					8.90			8.90	9.80
1020	Maximum					35			35	38.50
1200	Carafe, tray, minimum					60.50			60.50	66.50
1220	Maximum					71			71	78
1300	Desk pad, minimum					22			22	24
1320	Maximum					75			75	82.50
1400	Double pen set with pens, minimum					50			50	55
1420	Maximum					85			85	93.50
1500	Letter tray, minimum					13			13	14.30
1520	Maximum					33			33	36.50
1600	Memo box, minimum					9			9	9.90
1620	Maximum					13			13	14.30

12 48 Rugs and Mats

12 48 13 – Entrance Floor Mats and Frames

12 48 13.13 Entrance Floor Mats

		Crew	Daily Output	Labor-Hours	Unit	Material	Labor	Equipment	Total	Total Incl O&P
0010	**ENTRANCE FLOOR MATS**									
0020	Recessed, in-laid black rubber, 3/8" thick, solid	1 Clab	155	.052	S.F.	19.70	1.63		21.33	24
0050	Perforated		155	.052		13.40	1.63		15.03	17.30
0100	1/2" thick, solid		155	.052		16.10	1.63		17.73	20
0150	Perforated		155	.052		21.50	1.63		23.13	26
0200	In colors, 3/8" thick, solid		155	.052		17.35	1.63		18.98	21.50
0250	Perforated		155	.052		17.90	1.63		19.53	22
0300	1/2" thick, solid		155	.052		22.50	1.63		24.13	27
0350	Perforated		155	.052		23	1.63		24.63	28
2000	Recycled rubber tire tile, 12" x 12" x 3/8" thick [G]		125	.064		8.40	2.02		10.42	12.40
2510	Natural cocoa fiber, 1/2" thick [G]		125	.064		7.95	2.02		9.97	11.90
2520	3/4" thick [G]		125	.064		6.25	2.02		8.27	10
2530	1" thick [G]		125	.064		6.90	2.02		8.92	10.75

12 51 Office Furniture

12 51 16 – Case Goods

12 51 16.13 Metal Case Goods

		Crew	Daily Output	Labor-Hours	Unit	Material	Labor	Equipment	Total	Total Incl O&P
0010	**METAL CASE GOODS**									
0020	Desks, 29" high, double pedestal, 30" x 60", metal, minimum				Ea.	450			450	495
0030	Maximum					1,125			1,125	1,225
0400	36" x 72", metal, minimum					645			645	710
0420	Maximum					1,400			1,400	1,550
0600	Desks, single pedestal, 30" x 60", metal, minimum					405			405	445
0620	Maximum					905			905	995
0720	Desks, secretarial, 30" x 60", metal, minimum					385			385	425
0730	Maximum					640			640	705
0740	Return, 20" x 42", minimum					263			263	290
0750	Maximum					405			405	445
0940	59" x 12" x 23" high, steel, minimum					219			219	241
0960	Maximum					280			280	305

12 51 Office Furniture

12 51 16 – Case Goods

12 51 16.13 Metal Case Goods

		Crew	Daily Output	Labor-Hours	Unit	Material	2009 Bare Costs Labor	Equipment	Total	Total Incl O&P
0970	Keyboard shelf, standard				Ea.	80.50			80.50	88.50
0980	Articulating					380			380	420
0990	Center drawer, 18" wide					105			105	115
1020	Credenza, 18" to 22" x 60" to 72" metal, minimum					380			380	420
1040	Maximum					1,050			1,050	1,150
1240	Bookcase, 36" x 12" x 29" high					120			120	131
1260	52" high					180			180	197
1320	Computer stand, mobile, 25" x 24" x 38" high					280			280	305
1370	Computer desk with hutch, 47" x 24" x 46" high					310			310	340
1400	Printer stand, 22" x 24" x 34" high					202			202	222

12 51 16.16 Wood Case Goods

		Crew	Daily Output	Labor-Hours	Unit	Material	Labor	Equipment	Total	Total Incl O&P
0010	**WOOD CASE GOODS**									
0150	Desk, 29" high, double pedestal, 30" x 60"									
0160	Wood, minimum				Ea.	395			395	435
0180	Maximum				"	2,175			2,175	2,375
0550	Desk, 29" high, double pedestal, 36" x 72"									
0560	Wood, minimum				Ea.	500			500	550
0580	Maximum				"	2,925			2,925	3,200
0630	Single pedestal, 30" x 60"									
0640	Wood, minimum				Ea.	525			525	575
0650	Maximum					680			680	750
0670	Executive return, 24" x 42", with box, file, wood, minimum					355			355	390
0680	Maximum					680			680	750
0790	Desk, 29" high, secretarial, 30" x 60"									
0800	Wood, minimum				Ea.	435			435	475
0810	Maximum					2,200			2,200	2,425
0820	Return, 20" x 42", minimum					266			266	292
0830	Maximum					900			900	990
0900	Desktop organizer, 72" x 14" x 36" high, wood, minimum					160			160	176
0920	Maximum					380			380	420
1110	Furniture, credenza, 29" high, 18" to 22" x 60" to 72"									
1120	Wood, minimum				Ea.	610			610	670
1140	Maximum					2,750			2,750	3,025
1150	Hutch/bookcase, minimum					345			345	380
1160	Maximum					1,700			1,700	1,875
1200	Bookcase, 36" x 12" x 30" high, wood					149			149	164
1220	48" high					252			252	277
1300	Computer stand, mobile, wood, 25" x 24" x 38" high					415			415	455
1360	Computer desk with hutch, wood, 47" x 24" x 46" high					420			420	465

12 51 16.26 Resinite Case Goods

		Crew	Daily Output	Labor-Hours	Unit	Material	Labor	Equipment	Total	Total Incl O&P
0010	**RESINITE CASE GOODS**									
1340	Computer stand, resinite				Ea.	360			360	395
1380	Desk with hutch				"	395			395	435

12 51 19 – Filing Cabinets

12 51 19.13 Lateral Filing Cabinets

		Crew	Daily Output	Labor-Hours	Unit	Material	Labor	Equipment	Total	Total Incl O&P
0010	**LATERAL FILING CABINETS**, metal, baked enamel finish									
1060	Lateral, 36" wide, minimum				Ea.	400			400	440
1080	Maximum					430			430	475
1160	Lateral, 36" wide, minimum					420			420	460
1180	Maximum					635			635	695
1200	Wood, 2 drawer, lateral, minimum					445			445	490
1210	Maximum					795			795	875

12 51 Office Furniture

12 51 19 – Filing Cabinets

12 51 19.16 Vertical Filing Cabinets

		Crew	Daily Output	Labor-Hours	Unit	Material	2009 Bare Costs Labor	Equipment	Total	Total Incl O&P
0010	**VERTICAL FILING CABINETS**, metal, baked enamel finish									
1000	2 drawer, vertical, minimum				Ea.	145			145	159
1020	Maximum					315			315	350
1100	4 drawer, vertical, minimum					170			170	187
1120	Maximum				↓	465			465	515

12 51 19.23 Flat Files

0010	**FLAT FILES**, metal, baked enamel finish									
2010	Steel, 5 drawer, 40" wide x 27" deep				Ea.	660			660	725
2020	46" wide x 33" deep					740			740	815
2030	8 drawer, 40" wide x 27" deep					870			870	955
2040	46" wide x 33" deep					1,250			1,250	1,400
2050	Base, 40" wide x 27" deep x 5" high					187			187	206
2060	46" wide x 33" deep				↓	201			201	221

12 51 19.26 Hanging Files

0010	**HANGING FILES**, metal, baked enamel finish									
2100	File stand				Ea.	345			345	380
2110	Clamps, package of 6				"	139			139	152

12 51 23 – Office Tables

12 51 23.13 Wood Tables

0010	**WOOD TABLES**									
5000	Sled base, laminate top, coffee, minimum				Ea.	187			187	205
5100	Maximum					455			455	500
5150	End, minimum					149			149	164
5200	Maximum					315			315	345
5250	Wood cube, coffee					530			530	585
5350	End					555			555	610
5700	Designer table, Mies Barcelona, 40" sq., glass top, s.s. legs					2,050			2,050	2,250
5750	Saarinen, 42" round, plastic top, metal pedestal base					1,175			1,175	1,300
5800	Aulenti coffee table, 45" square, marble top and leg base					7,400			7,400	8,150
5840	Bruer, Laccio, 21-1/2" x 19", plastic top, tubular legs					430			430	475
5860	F. knoll, 24" square, glass top, solid steel leg base, chrome				↓	1,400			1,400	1,550

12 51 23.23 Metal Tables

0010	**METAL TABLES**									
5400	All metal drum, 14" diameter, 14" high				Ea.	385			385	420
5420	21" high					440			440	485
5480	18" diameter, 18" high					415			415	455
5500	22" diameter, 15" high					540			540	590
5550	20" high				↓	590			590	650
5600	For 3/4" glass top, add				S.F.	52.50			52.50	58
5650	For 1" marble top, add				"	52.50			52.50	58
7500	Table, training, with modesty panel, 30" x 60"				Ea.	540			540	595
8000	Tables, modular									
8010	Rectangular, 24" x 48"				Ea.	585			585	640
8020	24" x 60"					650			650	715
8030	24" x 72"					370			370	410
8040	30" x 60"					815			815	895
8050	30" x 72"					855			855	945
8100	Corner triangle, 24" deep					259			259	285
8110	30" deep					291			291	320
8200	Half round, 30" x 60"					705			705	775
8300	Trapezoid, 30" x 60"				↓	580			580	635

12 51 Office Furniture

12 51 23 – Office Tables

12 51 23.23 Metal Tables		Crew	Daily Output	Labor-Hours	Unit	Material	2009 Bare Costs Labor	Equipment	Total	Total Incl O&P
8400	Crescent, 30" x 60"				Ea.	830			830	915

12 51 23.33 Conference Tables

		Crew	Daily Output	Labor-Hours	Unit	Material	Labor	Equipment	Total	Total Incl O&P
0010	**CONFERENCE TABLES**									
6010	Segmented, oval, 72" x 144"				Ea.	3,500			3,500	3,850
6050	Boat, 96" x 42", minimum					740			740	810
6150	Maximum					2,800			2,800	3,075
6200	120" x 48", minimum					1,000			1,000	1,100
6250	Maximum					3,400			3,400	3,750
6300	144" x 48", minimum					1,175			1,175	1,300
6350	Maximum					4,350			4,350	4,775
6400	168" x 60", minimum					1,975			1,975	2,175
6450	Maximum					6,225			6,225	6,850
6500	192" x 60", minimum					2,300			2,300	2,525
6550	Maximum					6,925			6,925	7,600
6680	240" x 60", minimum					3,350			3,350	3,675
6700	Maximum					13,200			13,200	14,500
6720	Rectangle, 96" x 42", minimum					965			965	1,075
6740	Maximum					2,800			2,800	3,075
6760	120" x 48", minimum					2,100			2,100	2,300
6780	Maximum					3,600			3,600	3,950
6800	144" x 60", minimum					3,175			3,175	3,500
6820	Maximum					5,175			5,175	5,700
6840	168" x 60", minimum					3,875			3,875	4,275
6860	Maximum					6,225			6,225	6,850
6880	192" x 60", minimum					4,525			4,525	4,975
6900	Maximum					6,925			6,925	7,600
6960	240" x 60", minimum					6,000			6,000	6,600
6980	Maximum					10,600			10,600	11,700
7010	Powered, 52" x 108"					2,625			2,625	2,900
7020	52" x 120"					2,675			2,675	2,925
7030	52" x 132"				▼	2,850			2,850	3,125

12 52 Seating

12 52 13 – Chairs

12 52 13.10 Chairs, Folding and Stack

		Crew	Daily Output	Labor-Hours	Unit	Material	Labor	Equipment	Total	Total Incl O&P
0010	**CHAIRS, FOLDING & STACK**									
2000	Folding, all steel, baked enamel finish,									
2100	Form fitting seat and backrests				Ea.	24.50			24.50	27
2300	Polypropylene seat and back w/frame					35.50			35.50	39.50
2500	Chair caddy for above				▼	254			254	279
5000	Stack chair									
5300	Hardwood frame, seat and back									
5320	Minimum				Ea.	46.50			46.50	51
5340	Maximum				"	223			223	246
5400	Upholstered seat and back									
5420	Minimum				Ea.	72			72	79.50
5440	Maximum				"	335			335	365
5600	Metal frame, upholstered seat and back									
5620	Minimum				Ea.	43			43	47.50
5640	Maximum				"	300			300	330
5700	Plastic shell, metal legs									

12 52 Seating

12 52 13 – Chairs

12 52 13.10 Chairs, Folding and Stack

		Crew	Daily Output	Labor-Hours	Unit	Material	2009 Bare Costs Labor	Equipment	Total	Total Incl O&P
5720	Minimum				Ea.	56			56	61.50
5740	Maximum					127			127	140
5800	Chair caddy for above					235			235	259
5900	Tablet arms, minimum					149			149	164
5920	Maximum					193			193	212

12 52 19 – Upholstered Seating

12 52 19.13 Upholstered Office Seating

		Crew	Daily Output	Labor-Hours	Unit	Material	2009 Bare Costs Labor	Equipment	Total	Total Incl O&P
0010	**UPHOLSTERED OFFICE SEATING**									
3800	Lounge chair, upholstered, minimum				Ea.	208			208	229
3850	Maximum					1,750			1,750	1,900
4000	Sofa, two seat, upholstered, minimum					620			620	680
4100	Maximum					2,525			2,525	2,775
4650	Three seat, minimum					900			900	990
4680	Maximum					3,550			3,550	3,900
4700	Modular seating, lounge chair unit, upholstered, minimum					1,425			1,425	1,575
4750	Maximum					2,075			2,075	2,275
4780	Corner unit, minimum					980			980	1,075
4800	Maximum					2,275			2,275	2,500

12 52 23 – Office Seating

12 52 23.13 Office Chairs

		Crew	Daily Output	Labor-Hours	Unit	Material	2009 Bare Costs Labor	Equipment	Total	Total Incl O&P
0010	**OFFICE CHAIRS**									
2000	Standard office chair, executive, minimum				Ea.	255			255	280
2150	Maximum					1,750			1,750	1,950
2200	Management, minimum					183			183	201
2250	Maximum					2,000			2,000	2,200
2280	Task, minimum					117			117	128
2290	Maximum					550			550	605
2300	Arm kit, minimum					55.50			55.50	61
2320	Maximum					83			83	91.50
2340	Ergonomic, executive, minimum					405			405	445
2380	Maximum					850			850	935
2390	Management, minimum					340			340	375
2400	Maximum					1,175			1,175	1,300
2450	Task, minimum					152			152	167
2500	Maximum					570			570	625
2550	Arm kit, minimum					36			36	40
2600	Maximum					78.50			78.50	86
3000	Side/guest chairs, upholstered, sled base, metal, min.					145			145	159
3100	Maximum					490			490	540
3200	Wood, minimum					181			181	199
3250	Maximum					935			935	1,025
3400	Traditional, wood leg, minimum					144			144	158
3500	Maximum					840			840	925
3600	Conference chair, upholstered, metal leg, minimum					99			99	109
3700	Maximum					685			685	755

12 52 23.23 Multiple Office Seating Units

		Crew	Daily Output	Labor-Hours	Unit	Material	2009 Bare Costs Labor	Equipment	Total	Total Incl O&P
0010	**MULTIPLE OFFICE SEATING UNITS**									
1000	Area seating, full upholstered, 3 seat straight unit, minimum				Ea.	1,025			1,025	1,125
1020	Maximum					1,675			1,675	1,850
1100	Four seat with corner table, minimum					1,700			1,700	1,875
1120	Maximum					2,275			2,275	2,500

12 52 Seating

12 52 23 – Office Seating

12 52 23.23 Multiple Office Seating Units		Crew	Daily Output	Labor-Hours	Unit	Material	2009 Bare Costs Labor	Equipment	Total	Total Incl O&P
1200	2 seat with in-line table, minimum				Ea.	1,025			1,025	1,125
1220	Maximum					1,175			1,175	1,300
2000	Individual seat with table, minimum					675			675	745
2020	Maximum					830			830	915

12 54 Hospitality Furniture

12 54 13 – Hotel and Motel Furniture

12 54 13.10 Hotel Furniture

		Crew	Daily Output	Labor-Hours	Unit	Material	Labor	Equipment	Total	Total Incl O&P
0010	**HOTEL FURNITURE**									
0020	Standard quality set, minimum				Room	2,325			2,325	2,550
0200	Maximum				"	8,300			8,300	9,125
0300	Bed frame				Ea.	48			48	53
0400	Bench, upholstered, 42" x 18" x 18"					185			185	203
0420	18" x 18" x 23"					160			160	176
0500	Desk section, one drawer, 34" x 20" x 30"					190			190	209
0600	Free standing, 42" x 22" x 30"					250			250	275
0700	Desk chair, upholstered, minimum					80			80	88
0720	Maximum					140			140	154
1000	Dressers, uniplex, 2 drawer					330			330	365
1100	3 drawer					360			360	395
2000	Guest tables, 30" diameter					105			105	115
2100	34" diameter					135			135	148
3000	Headboards, free standing, twin					110			110	121
3050	Full					210			210	231
3100	Queen					220			220	242
3200	Wall mounted, twin					80			80	88
3250	Full					175			175	192
3300	Queen					195			195	214
4000	Lounge chair, full upholstered, minimum					310			310	340
4050	Maximum					435			435	480
4200	Open arms, minimum					85			85	93.50
4250	Maximum					390			390	430
5000	Mattress/box springs, twin					360			360	395
5050	Full					430			430	475
5100	Queen					490			490	540
5150	Mirror, framed, 29" x 45"					155			155	170
6000	Sleep sofas, twin, minimum					580			580	640
6050	Maximum					660			660	725
6100	Full, minimum					600			600	660
6150	Maximum					1,075			1,075	1,175
6200	Queen, minimum					610			610	670
6250	Maximum					1,100			1,100	1,225
7000	Table, wood top, cocktail, 54" x 24" x 16" high					295			295	325
7020	Corner, 30" x 30" x 21" high					255			255	280
7040	End, 22" x 28" x 21" high					285			285	315

12 54 13.20 Mattress and Box Springs

		Crew	Daily Output	Labor-Hours	Unit	Material	Labor	Equipment	Total	Total Incl O&P
0010	**MATTRESS & BOX SPRINGS** per set									
1000	Hospital, 34" x 84"				Ea.	440			440	480
2000	Hotel/motel, twin, minimum					360			360	395
2020	Maximum					560			560	615
2200	Full, minimum					430			430	475

12 54 Hospitality Furniture

12 54 13 – Hotel and Motel Furniture

12 54 13.20 Mattress and Box Springs	Crew	Daily Output	Labor-Hours	Unit	Material	2009 Bare Costs Labor	2009 Bare Costs Equipment	Total	Total Incl O&P
2220	Maximum				Ea.	680			680	750
2400	Queen, minimum					490			490	540
2420	Maximum					800			800	880
3000	Stow away bed with head board, twin size					380			380	420

12 54 16 – Restaurant Furniture

12 54 16.10 Tables, Folding

		Crew	Daily Output	Labor-Hours	Unit	Material	Labor	Equipment	Total	Total Incl O&P
0010	**TABLES, FOLDING** Laminated plastic tops									
1000	Tubular steel legs with glides									
1020	18" x 60", minimum				Ea.	210			210	231
1040	Maximum					1,175			1,175	1,300
1100	18" x 72", minimum					254			254	280
1120	Maximum					1,250			1,250	1,350
1200	18" x 96", minimum					282			282	310
1220	Maximum					1,500			1,500	1,650
1400	30" x 48", minimum					183			183	201
1420	Maximum					975			975	1,075
1500	30" x 60", minimum					179			179	197
1520	Maximum					1,400			1,400	1,525
1600	30" x 72", minimum					252			252	277
1620	Maximum					1,600			1,600	1,775
1700	30" x 96", minimum					218			218	239
1720	Maximum					2,025			2,025	2,250
1800	36" x 72", minimum					218			218	240
1820	Maximum					1,800			1,800	2,000
1840	36" x 96", minimum					257			257	283
1860	Maximum					2,375			2,375	2,600
2000	Round, wood stained, plywood top, 60" diameter, minimum					365			365	400
2020	Maximum					1,275			1,275	1,400
2200	72" diameter, minimum					325			325	360
2220	Maximum					2,300			2,300	2,550
4000	Mobile storage carts									
4020	For 72" tables, flat, maximum 10 tables				Ea.	575			575	635
4040	96" tables, flat, maximum 10 tables					510			510	560
4060	Rounds, on edge, maximum 10 tables					830			830	915

12 54 16.20 Furniture, Restaurant

		Crew	Daily Output	Labor-Hours	Unit	Material	Labor	Equipment	Total	Total Incl O&P
0010	**FURNITURE, RESTAURANT**									
0020	Bars, built-in, front bar	1 Carp	5	1.600	L.F.	238	64		302	360
0200	Back bar	"	5	1.600	"	173	64		237	289
0300	Booth seating, see Div. 12 54 16.70									
2000	Chair, bentwood side chair, metal, minimum				Ea.	76			76	83.50
2020	Maximum					89			89	98
2100	Wood, minimum					159			159	175
2120	Maximum					185			185	204
2400	Bruer Cesca, cane seat & back, arms, minimum					440			440	480
2420	Maximum					515			515	565
2500	Side, minimum					440			440	480
2520	Maximum					515			515	565
2600	Upholstered seat & back, arms, minimum					124			124	136
2620	Maximum					350			350	385
2640	Side, minimum					340			340	375
2660	Maximum					400			400	440
2700	Corbusier, arm chair, cane seat and back, minimum					170			170	187

12 54 Hospitality Furniture

12 54 16 – Restaurant Furniture

12 54 16.20 Furniture, Restaurant		Crew	Daily Output	Labor-Hours	Unit	Material	2009 Bare Costs Labor	Equipment	Total	Total Incl O&P
2720	Maximum				Ea.	185			185	204
2740	Fledermaus, fabric seat, ash frame, minimum					157			157	173
2760	Maximum					193			193	212
2780	Hoffman arm, fabric seat, cane back, minimum					179			179	197
2800	Maximum					196			196	215
2820	Side, minimum					140			140	154
2840	Maximum					151			151	166
2860	Lombard, minimum					96.50			96.50	106
2880	Maximum					126			126	138
2900	Mies, side chair, leather seat and back, minimum					415			415	455
2920	Maximum					610			610	670
2940	Napoleon, upholstered seat, wood back, minimum					69			69	76
2960	Maximum					88			88	97
3000	Prague, arm, upholstered, minimum					128			128	141
3020	Maximum					276			276	305
3040	Side, upholstered, minimum					128			128	141
3060	Maximum					157			157	173
3080	Contemporary leather seat & back, chrome frame, min.					257			257	282
3100	Maximum					375			375	410
3120	Foam padded seat & back, s. steel barstock, min.					685			685	755
3140	Maximum					730			730	800
3160	Chrome tube cantilever frame, padded arms, min.					106			106	116
3180	Maximum					271			271	298
4200	"Tub" style, minimum					181			181	199
4220	Maximum					410			410	450
4280	Bent wood arms and legs with back bow, minimum					50.50			50.50	55.50
4300	Maximum					335			335	365
4320	Sled base, minimum					106			106	116
4340	Maximum					335			335	365
4360	Armchair foam padded seat & back, carved wood frame min.					305			305	335
4380	Maximum					585			585	645
4400	Ornate wood frame and legs, minimum					330			330	360
4420	Maximum					385			385	425
4440	Heavy wood frame and legs, tailored back, minimum					520			520	575
4460	Maximum					605			605	665
4480	Button tufted back, minimum					340			340	370
4500	Maximum					780			780	860
4520	"Dining" wood frame and legs, minimum					252			252	277
4550	Maximum					294			294	325
4600	"Queen Ann" wood frame and legs, minimum					345			345	380
4620	Maximum					380			380	415
4660	Upholstered arms and wood legs, minimum					480			480	530
4680	Maximum					565			565	625
4700	Wicker, foam padded seat and back, barrel design, minimum					425			425	465
4720	Maximum					480			480	525
4740	Couch design, minimum					315			315	345
4760	Maximum					385			385	425
4780	Chrome tube round cantilever base, minimum					181			181	199
4800	Maximum					400			400	440
4820	Chrome tube square cantilever base, minimum					174			174	191
4840	Maximum					390			390	425
4860	Misc. chair, upholst. seat, wood arms, bow back & legs, min.					142			142	156
4880	Maximum					197			197	217

12 54 Hospitality Furniture

12 54 16 – Restaurant Furniture

12 54 16.20 Furniture, Restaurant

		Crew	Daily Output	Labor-Hours	Unit	Material	2009 Bare Costs Labor	Equipment	Total	Total Incl O&P
4900	Open curved wood back, minimum				Ea.	320			320	350
4950	Maximum					355			355	390
5000	Upholstered seat and open back, wood frame arms, min.					293			293	320
5100	Maximum					340			340	375
5150	Wood frame sled base, minimum					164			164	181
5200	Maximum					272			272	299
5250	Wood saddle seat, "Windsor", wood frame & legs, min.					168			168	185
5300	Maximum					196			196	215
5350	Brass nail trim leather, minimum					197			197	217
5400	Maximum					225			225	247
5450	Wood frame, fabric, minimum					169			169	185
5500	Maximum					199			199	219
6000	Stools, upholst. seat & back, chrome tube cantilever frame, min.					264			264	291
6150	Swivel on chrome post mount with spread base, minimum					155			155	170
6250	"Vienna" wood back and legs, minimum					207			207	228
6350	"Napoleon" wood back and legs, minimum					71			71	78
6450	Wood swivel seat and curved spindle back, leg base					123			123	135
6600	Upholstered seat no back, bent wood legs					248			248	272
7000	With back & straight wood frame and legs					246			246	271
7200	Wood swivel seat, "Captain" wood back and legs					157			157	172
7400	Veneer seat and curved wood back with arms, leg base					149			149	164
7600	Upholstered swivel seat no back, wood legs					59			59	65
8000	With back & wood legs					430			430	475
8200	Upholstered seat & wood back, wood legs					410			410	450
8400	Upholstered swivel seat and back, wood legs					167			167	184
8600	Square tube frame and legs					42.50			42.50	47
8800	Metal frame and legs					42.50			42.50	47

12 54 16.30 Table Bases

		Crew	Daily Output	Labor-Hours	Unit	Material	Labor	Equipment	Total	Total Incl O&P
0010	**TABLE BASES**									
0040	Dining height									
1000	Metal disk design, minimum				Ea.	78			78	85.50
1200	Maximum					207			207	227
1400	Wood, minimum					330			330	365
1600	Maximum					365			365	400
2000	Fluted wheel, minimum					90.50			90.50	99.50
2200	Maximum					192			192	212
2400	Heavy cast iron, minimum					79			79	87
2600	Maximum					192			192	211
2800	Hobnail design, minimum					93			93	102
3000	Maximum					283			283	310
3200	Manhole design, minimum					92.50			92.50	102
3400	Maximum					248			248	272
3600	Ring design, minimum					129			129	142
3800	Maximum					213			213	234
4000	Trumpet design, minimum					233			233	256
4200	Maximum					256			256	281
4400	Tubular, round or rectangular shape, minimum					92			92	101
4600	Maximum					193			193	212
5000	Cocktail height bases, minimum					107			107	117
5200	Maximum					118			118	129
5400	For foot ring, add, minimum					11.20			11.20	12.30
5600	Maximum					25.50			25.50	28

12 54 Hospitality Furniture

12 54 16 – Restaurant Furniture

12 54 16.40 Table Tops

		Crew	Daily Output	Labor-Hours	Unit	Material	2009 Bare Costs Labor	Equipment	Total	Total Incl O&P
0010	**TABLE TOPS** Laminated plastic top and edge									
0040	24" wide, 24" long, minimum				Ea.	60.50			60.50	66.50
1000	Maximum					235			235	258
1200	30" long, minimum					70			70	77
1400	Maximum					345			345	380
1600	36" long, minimum					88.50			88.50	97.50
1800	Maximum					365			365	405
2000	48" long, minimum					107			107	117
2200	Maximum					435			435	475
2400	60" long, minimum					173			173	190
2600	Maximum					675			675	745
2800	72" long, minimum					211			211	232
3000	Maximum					710			710	785
3050	30" wide, 30" long, minimum					73.50			73.50	81
3150	Maximum					435			435	475
3200	36" long, minimum					92.50			92.50	102
3250	Maximum					435			435	475
3400	42" long, minimum					95.50			95.50	105
3600	Maximum					250			250	275
3800	48" long, minimum					120			120	132
4000	Maximum					475			475	525
4200	60" long, minimum					170			170	187
4400	Maximum					720			720	795
4600	72" long, minimum					208			208	229
4800	Maximum					720			720	795
5000	36" wide, 36" long, minimum					98			98	108
5200	Maximum					700			700	770
5400	48" long, minimum					142			142	157
5600	Maximum					560			560	615
5800	60" long, minimum					202			202	223
6000	Maximum					720			720	795
6200	72" long, minimum					231			231	254
6400	Maximum					850			850	935
6500	42" wide, 42" long, minimum					176			176	194
6600	Maximum					700			700	770
6700	Round, 24" diameter, minimum					66.50			66.50	73
6800	Maximum					290			290	320
6900	30" diameter, minimum					87.50			87.50	96
7000	Maximum					435			435	475
7200	36" diameter, minimum					113			113	124
7400	Maximum					495			495	545
7600	42" diameter, minimum					180			180	198
8000	Maximum					710			710	785
8200	48" diameter, minimum					202			202	223
8400	Maximum					800			800	880
8800	54" diameter, minimum					325			325	355
9000	Maximum					1,050			1,050	1,150
9200	60" diameter, minimum					345			345	380
9400	Maximum					1,300			1,300	1,425

12 54 16.50 Table Tops

		Crew	Daily Output	Labor-Hours	Unit	Material	2009 Bare Costs Labor	Equipment	Total	Total Incl O&P
0010	**TABLE TOPS** laminate top and hardwood edge									
0040	24" wide, 24" long, minimum				Ea.	192			192	211

12 54 Hospitality Furniture

12 54 16 – Restaurant Furniture

12 54 16.50 Table Tops

		Crew	Daily Output	Labor-Hours	Unit	Material	2009 Bare Costs Labor	Equipment	Total	Total Incl O&P
1000	Maximum				Ea.	610			610	675
1200	30" long, minimum					212			212	234
1400	Maximum					655			655	720
1600	36" long, minimum					199			199	219
1800	Maximum					690			690	755
2000	48" long, minimum					280			280	310
2100	Maximum					755			755	835
2200	60" long, minimum					400			400	440
2400	Maximum					830			830	910
2600	72" long, minimum					455			455	500
2800	Maximum					925			925	1,025
3000	30" wide, 30" long, minimum					232			232	255
3100	Maximum					695			695	765
3200	36" long, minimum					275			275	305
3400	Maximum					750			750	825
3600	42" long, minimum					273			273	300
3800	Maximum					925			925	1,025
4000	48" long, minimum					305			305	335
4100	Maximum					805			805	890
4200	60" long, minimum					410			410	450
4300	Maximum					865			865	950
4400	72" long, minimum					330			330	360
4600	Maximum					905			905	995
4800	36" wide, 36" long, minimum					282			282	310
5000	Maximum					765			765	845
5200	48" long, minimum					330			330	365
5400	Maximum					875			875	965
5600	60" long, minimum					470			470	515
5800	Maximum					970			970	1,075
6000	72" long, minimum					510			510	565
6200	Maximum					1,075			1,075	1,175
6400	42" wide, 42" long, minimum					340			340	375
6600	Maximum					930			930	1,025
6800	Round, 24" diameter, minimum					350			350	385
6900	Maximum					650			650	715
7000	30" diameter, minimum					385			385	425
7200	Maximum					730			730	805
7400	36" diameter, minimum					425			425	465
7600	Maximum					810			810	895
7800	42" diameter, minimum					490			490	540
7900	Maximum					995			995	1,100
8000	48" diameter, minimum					530			530	580
8150	Maximum					1,050			1,050	1,175
8200	54" diameter, minimum					790			790	870
8250	Maximum					1,400			1,400	1,525
8300	60" diameter, minimum					800			800	880
8350	Maximum				↓	1,400			1,400	1,525

12 54 16.60 Table Tops

		Crew	Daily Output	Labor-Hours	Unit	Material	Labor	Equipment	Total	Total Incl O&P
0010	**TABLE TOPS** Polyester resin top & edge									
0050	Add inlay top material to cost									
0100	24" wide, 24" long				Ea.	365			365	400
1200	30" long				↓	365			365	400

12 54 Hospitality Furniture

12 54 16 – Restaurant Furniture

12 54 16.60 Table Tops

		Crew	Daily Output	Labor-Hours	Unit	Material	2009 Bare Costs Labor	Equipment	Total	Total Incl O&P
1600	36" long				Ea.	405			405	445
2000	48" long					415			415	460
2400	60" long					495			495	545
2800	72" long					580			580	635
3200	30" wide, 30" long					410			410	450
3600	36" long					485			485	535
4000	42" long					510			510	560
4400	48" long					535			535	590
4800	60" long					610			610	670
5200	72" long					740			740	815
5600	36" wide, 36" long					485			485	535
6000	48" long					585			585	645
6400	60" long					715			715	785
6800	72" long					855			855	945
7200	42" wide, 42" long					570			570	630
7600	Round, 24" diameter					545			545	600
8000	30" diameter					565			565	620
8400	36" diameter					670			670	735
8800	42" diameter					900			900	990
9150	48" diameter					775			775	855
9250	54" diameter					1,350			1,350	1,475
9350	60" diameter					1,400			1,400	1,550

12 54 16.70 Booths

		Crew	Daily Output	Labor-Hours	Unit	Material	2009 Bare Costs Labor	Equipment	Total	Total Incl O&P
0010	**BOOTHS**									
1000	Banquet, upholstered seat and back, custom									
1500	Straight, minimum	2 Carp	40	.400	L.F.	161	16		177	202
1520	Maximum		36	.444		315	17.75		332.75	375
1600	"L" or "U" shape, minimum		35	.457		164	18.25		182.25	210
1620	Maximum		30	.533		292	21.50		313.50	355
1800	Upholstered outside finished backs for									
1810	single booths and custom banquets									
1820	Minimum	2 Carp	44	.364	L.F.	18.55	14.55		33.10	43
1840	Maximum	"	40	.400	"	55.50	16		71.50	86
3000	Fixed seating, one piece plastic chair and									
3010	plastic laminate table top									
3100	Two seat, 24" x 24" table, minimum	F-7	30	1.067	Ea.	650	38		688	775
3120	Maximum		26	1.231		925	44		969	1,100
3200	Four seat, 24" x 48" table, minimum		28	1.143		645	41		686	770
3220	Maximum		24	1.333		1,100	47.50		1,147.50	1,275
3300	Six seat, 24" x 76" table, minimum		26	1.231		1,175	44		1,219	1,350
3320	Maximum		22	1.455		1,600	52		1,652	1,850
3400	Eight seat, 24" x 102" table, minimum		20	1.600		1,550	57.50		1,607.50	1,800
3420	Maximum		18	1.778		2,025	63.50		2,088.50	2,325
4000	Free standing, wood fiber core with									
4010	plastic laminate face, single booth									
4100	24" wide	2 Carp	38	.421	Ea.	305	16.80		321.80	365
4150	48" wide		34	.471		370	18.80		388.80	440
4200	60" wide		30	.533		455	21.50		476.50	535
4300	Double booth, 24" wide		32	.500		455	20		475	530
4350	48" wide		28	.571		605	23		628	700
4400	60" wide		26	.615		765	24.50		789.50	880
4600	Upholstered seat and back									

12 54 Hospitality Furniture

12 54 16 – Restaurant Furniture

12 54 16.70 Booths		Crew	Daily Output	Labor-Hours	Unit	Material	2009 Bare Costs Labor	2009 Bare Costs Equipment	Total	Total Incl O&P
4650	Foursome, single booth, minimum	2 Carp	38	.421	Ea.	345	16.80		361.80	405
4700	Maximum		30	.533		1,325	21.50		1,346.50	1,475
4800	Double booth, minimum		32	.500		765	20		785	870
4850	Maximum		26	.615		2,025	24.50		2,049.50	2,300
5000	Mount in floor, wood fiber core with									
5010	plastic laminate face, single booth									
5050	24" wide	F-7	30	1.067	Ea.	249	38		287	335
5100	48" wide		28	1.143		315	41		356	415
5150	60" wide		26	1.231		455	44		499	570
5200	Double booth, 24" wide		26	1.231		400	44		444	510
5250	48" wide		24	1.333		510	47.50		557.50	640
5300	60" wide		22	1.455		725	52		777	880

12 55 Detention Furniture

12 55 13 – Detention Bunks

12 55 13.13 Cots

		Crew	Daily Output	Labor-Hours	Unit	Material	2009 Bare Costs Labor	2009 Bare Costs Equipment	Total	Total Incl O&P
0010	**COTS**									
2500	Bolted, single, painted steel	E-4	20	1.600	Ea.	330	72.50	6.70	409.20	500
2700	Stainless steel	"	20	1.600	"	960	72.50	6.70	1,039.20	1,175

12 56 Institutional Furniture

12 56 33 – Classroom Furniture

12 56 33.10 Furniture, School

		Crew	Daily Output	Labor-Hours	Unit	Material	2009 Bare Costs Labor	2009 Bare Costs Equipment	Total	Total Incl O&P
0010	**FURNITURE, SCHOOL**									
0500	Classroom, movable chair & desk type, minimum				Set				73.50	81
0600	Maximum				"				134	148
1000	Chair, molded plastic									
1100	Integral tablet arm, minimum				Ea.	72.50			72.50	79.50
1150	Maximum					148			148	163
2000	Desk, single pedestal, top book compartment, minimum					53.50			53.50	59
2020	Maximum					67.50			67.50	74
2100	Side book compartment, minimum					77			77	84.50
2120	Maximum					97			97	107
2200	Flip top, minimum					95.50			95.50	105
2220	Maximum					122			122	134
3000	Preschool, moulded plastic chairs, minimum					28			28	30.50
3020	Maximum					46.50			46.50	51.50
3800	Tables, plastic laminate top, 24" wide, 36" long					73			73	80
3820	48" long					92			92	101
3840	60" long					104			104	115
3900	30" wide, 48" long					97.50			97.50	107
3920	60" long					140			140	155
3940	72" long					157			157	173
4000	36" wide, 36" long					85.50			85.50	94
4020	60" long					123			123	136
4040	72" long					217			217	239
4200	Round, 36" dia.					85			85	93
4220	42" dia.					97.50			97.50	107
4230	48" dia.					122			122	135

12 56 Institutional Furniture

12 56 33 – Classroom Furniture

12 56 33.10 Furniture, School

		Crew	Daily Output	Labor-Hours	Unit	Material	2009 Bare Costs Labor	Equipment	Total	Total Incl O&P
4240	60" dia.				Ea.	169			169	186

12 56 43 – Dormitory Furniture

12 56 43.10 Dormitory Furnishings

		Crew	Daily Output	Labor-Hours	Unit	Material	Labor	Equipment	Total	Total Incl O&P
0010	**DORMITORY FURNISHINGS**									
0200	Bookcase, two shelf				Ea.	145			145	159
0300	Bunkable bed, twin, minimum					270			270	297
0320	Maximum					650			650	715
1000	Chest, four drawer, minimum					420			420	460
1020	Maximum					550			550	605
1050	Built-in, minimum	2 Carp	13	1.231	L.F.	135	49		184	226
1150	Maximum		10	1.600		250	64		314	375
1200	Desk top, built-in, laminated plastic, 24" deep, minimum		50	.320		37	12.80		49.80	60.50
1300	Maximum		40	.400		111	16		127	147
1450	30" deep, minimum		50	.320		47.50	12.80		60.30	72
1550	Maximum		40	.400		207	16		223	253
1750	Dressing unit, built-in, minimum		12	1.333		203	53.50		256.50	305
1850	Maximum		8	2		610	80		690	795
2000	Desk, single pedestal				Ea.	460			460	505
2100	Hutch/bookcase, with light					222			222	244
3000	Ladder					179			179	197
4000	Mirror, with frame					104			104	114
5000	Nightstand					220			220	243
6000	Wall unit, open shelving					530			530	580
7000	Wardrobe					550			550	605
8000	Rule of thumb: total cost for furniture, minimum				Student				2,125	2,350
8050	Maximum				"				4,100	4,450

12 56 51 – Library Furniture

12 56 51.10 Library Furnishings

		Crew	Daily Output	Labor-Hours	Unit	Material	Labor	Equipment	Total	Total Incl O&P
0010	**LIBRARY FURNISHINGS**									
0100	Attendant desk, 36" x 62" x 29" high	1 Carp	16	.500	Ea.	2,575	20		2,595	2,850
0200	Book display, "A" frame display, both sides, 42" x 42" x 60" high		16	.500		1,475	20		1,495	1,650
0220	Table with bulletin board, 42" x 24" x 49" high		16	.500		1,300	20		1,320	1,450
0300	Book trucks, descending platform,									
0320	Small, 14" x 30" x 35" high	1 Carp	16	.500	Ea.	730	20		750	830
0340	Large, 14" x 40" x 42" high		16	.500		765	20		785	870
0800	Card catalogue, 30 tray unit		16	.500		2,775	20		2,795	3,075
0840	60 tray unit		16	.500		5,075	20		5,095	5,600
0880	72 tray unit	2 Carp	16	1		6,225	40		6,265	6,875
0960	120 tray unit	"	16	1		9,475	40		9,515	10,500
1000	Carrels, single face, initial unit	1 Carp	16	.500		700	20		720	800
1050	Additional unit	"	16	.500		720	20		740	820
1500	Double face, initial unit	2 Carp	16	1		1,100	40		1,140	1,275
1550	Additional unit		16	1		935	40		975	1,075
1600	Cloverleaf		11	1.455		1,900	58		1,958	2,175
1710	Carrels, hardwood, 36" x 24", minimum	1 Carp	5	1.600		510	64		574	660
1720	Maximum		4	2		785	80		865	990
1730	Metal, minimum		5	1.600		241	64		305	365
1740	Maximum		4	2		575	80		655	760
2000	Chairs, sled base, arms, minimum		24	.333		245	13.30		258.30	290
2050	Maximum		16	.500		273	20		293	330
2100	No arms, minimum		24	.333		200	13.30		213.30	241
2150	Maximum		16	.500		223	20		243	276

12 56 Institutional Furniture

12 56 51 – Library Furniture

12 56 51.10 Library Furnishings		Crew	Daily Output	Labor-Hours	Unit	Material	2009 Bare Costs Labor	Equipment	Total	Total Incl O&P
2500	Standard leg base, arms, minimum	1 Carp	24	.333	Ea.	233	13.30		246.30	277
2520	Maximum		16	.500		375	20		395	445
2600	No arms, minimum		24	.333		163	13.30		176.30	200
2620	Maximum		16	.500		280	20		300	340
2700	Card catalog file, 60 trays, complete					5,050			5,050	5,550
2720	Alternate method: each tray					84			84	92.50
3000	Charge desk, modular unit, 35" x 27" x 39" high									
3020	Wood front and edges, plastic laminate tops									
3100	Book return	1 Carp	16	.500	Ea.	865	20		885	985
3150	Book truck port		16	.500		1,000	20		1,020	1,125
3200	Card file drawer, 5 drawers		16	.500		2,025	20		2,045	2,250
3250	10 drawers		16	.500		2,425	20		2,445	2,700
3280	15 drawers		16	.500		2,675	20		2,695	2,975
3300	Card & legal file		16	.500		2,925	20		2,945	3,250
3350	Charging machine		16	.500		2,150	20		2,170	2,400
3400	Corner		16	.500		965	20		985	1,075
3450	Cupboard		16	.500		1,375	20		1,395	1,525
3550	Gate		16	.500		510	20		530	590
3600	Knee space		16	.500		1,325	20		1,345	1,475
3650	Open storage		16	.500		1,275	20		1,295	1,425
3700	Station charge		16	.500		2,550	20		2,570	2,825
3750	Work station		16	.500		2,250	20		2,270	2,500
3800	Charging desk, built-in, with counter, plastic laminated top		7	1.143	L.F.	260	45.50		305.50	355
4000	Dictionary stand, stationary		16	.500	Ea.	835	20		855	950
4020	Revolving		16	.500		335	20		355	400
4200	Exhibit case, table style, 60" x 28" x 36"		11	.727		2,725	29		2,754	3,050
4500	Globe stand		16	.500		1,025	20		1,045	1,150
4800	Magazine rack		16	.500		1,425	20		1,445	1,600
5000	Newspaper rack		16	.500		930	20		950	1,050
6010	Bookshelf, mtl, 90" high, 10" shelf, dbl face		11.50	.696	L.F.	118	28		146	173
6020	Single face		12	.667	"	113	26.50		139.50	166
6050	For 8" shelving, subtract from above					10%				
6060	For 12" shelving, add to above					10%				
6070	For 42" high with countertop, subtract from above					20%				
6100	Mobile compacted shelving, hand crank, 9'-0" high									
6110	Double face, including track, 3' section				Ea.	1,125			1,125	1,250
6150	For electrical operation, add					25%				
6200	Magazine shelving, 82" high, 12" deep, single face	1 Carp	11.50	.696	L.F.	109	28		137	163
6210	Double face		11.50	.696	"	105	28		133	158
7000	Tables, card catalog reference, 24" x 60" x 42"		16	.500	Ea.	945	20		965	1,075
7050	24" x 60" x 72"		16	.500		1,250	20		1,270	1,425
7100	Index, single tier, 48" x 72"		16	.500		2,025	20		2,045	2,250
7150	Double tier, 48" x 72"		16	.500		2,275	20		2,295	2,525
7200	Reading table, laminated top, 60" x 36"					645			645	710
8000	Parsons table, 29" high, plastic lam. top, wood legs & edges									
8010	36" x 36"	2 Carp	16	1	Ea.	735	40		775	865
8020	36" x 60"		16	1		740	40		780	875
8030	36" x 72"		16	1		830	40		870	970
8040	36" x 84"		16	1		1,175	40		1,215	1,350
8050	42" x 90"		16	1		1,125	40		1,165	1,300
8060	48" x 72"		16	1		985	40		1,025	1,125
8070	48" x 120"		16	1		2,200	40		2,240	2,475
8100	Round, leg or pedestal base, 36" diameter		16	1		615	40		655	735

12 56 Institutional Furniture

12 56 51 – Library Furniture

12 56 51.10 Library Furnishings		Crew	Daily Output	Labor-Hours	Unit	Material	2009 Bare Costs Labor	Equipment	Total	Total Incl O&P
8110	42" diameter	2 Carp	16	1	Ea.	850	40		890	995
8120	48" diameter		16	1		935	40		975	1,075
8130	60" diameter		16	1		1,200	40		1,240	1,350
8500	Study, panel ends, plastic laminate surfaces 29" high, 36" x 60"		16	1		915	40		955	1,050
8510	36" x 72"		16	1		980	40		1,020	1,125
8515	36" x 90"		16	1		1,025	40		1,065	1,200
8525	48" x 72"		16	1		1,075	40		1,115	1,250

12 56 70 – Healthcare Furniture

12 56 70.10 Furniture, Hospital

		Crew	Daily Output	Labor-Hours	Unit	Material	Labor	Equipment	Total	Total Incl O&P
0010	**FURNITURE, HOSPITAL**									
0020	Beds, manual, minimum				Ea.	730			730	800
0100	Maximum					2,500			2,500	2,750
0300	Manual and electric beds, minimum					925			925	1,025
0400	Maximum					2,825			2,825	3,125
0600	All electric hospital beds, minimum					1,325			1,325	1,450
0700	Maximum					4,100			4,100	4,525
0900	Manual, nursing home beds, minimum					810			810	890
1000	Maximum					1,675			1,675	1,850
1020	Overbed table, laminated top, minimum					345			345	380
1040	Maximum					860			860	945
1100	Patient wall systems, not incl. plumbing, minimum				Room	1,025			1,025	1,125
1200	Maximum				"	1,900			1,900	2,100
2000	Geriatric chairs, minimum				Ea.	470			470	515
2020	Maximum				"	705			705	780

12 59 Systems Furniture

12 59 13 – Panel-Hung Component System Furniture

12 59 13.10 Furniture, Office Systems

		Crew	Daily Output	Labor-Hours	Unit	Material	Labor	Equipment	Total	Total Incl O&P
0010	**FURNITURE, OFFICE SYSTEMS**, Panel hung									
0100	Acoustic panel, 43" high, 24" wide, NRC .85				Ea.	230			230	253
0110	30" wide					257			257	282
0120	36" wide					284			284	310
0130	48" wide					340			340	375
0200	64" high, 30" wide					305			305	335
0210	36" wide					325			325	360
0220	42" wide					380			380	415
0230	48" wide					400			400	440
0240	60" wide					440			440	485
0400	Bookshelf, 36" wide					106			106	117
0410	42" wide					112			112	123
0420	48" wide					115			115	127
1000	Connectors, brackets and supports									
1200	Bracket, cantilever, 20" deep, 24" deep				Ea.	36			36	39.50
1220	Countertop, per pair					14.95			14.95	16.45
1240	Flat, 20" deep, 24" wide					18.45			18.45	20.50
1260	Worksurface kit, per pair					14.95			14.95	16.45
1300	Connector kit, ell, 43" high, 90" deep					64			64	70.50
1320	64" high					38			38	41.50
1340	Straight, 43" high					26.50			26.50	29
1360	64" high					26.50			26.50	29

12 59 Systems Furniture

12 59 13 – Panel-Hung Component System Furniture

12 59 13.10 Furniture, Office Systems		Crew	Daily Output	Labor-Hours	Unit	Material	2009 Bare Costs Labor	Equipment	Total	Total Incl O&P
1400	Support column for peninsula				Ea.	92			92	101
2000	Countertop, 15" deep, 36" wide					121			121	133
2020	48" wide					135			135	149
3000	Duplex receptacle circuit 1					13.20			13.20	14.50
3010	Circuit 2					13.20			13.20	14.50
3100	Electric power harness, 36" wide					86			86	94.50
3120	42" wide					181			181	199
3140	60" wide					90.50			90.50	99.50
4000	End cover, panel, 43" high					26.50			26.50	29
4020	64" high					26.50			26.50	29
4100	Finish, variable height, 2-way					39.50			39.50	43.50
4600	Overhead cabinet with door, 42" wide					250			250	274
4620	60" wide					380			380	415
5000	Pedestal spacer, 22" deep, 15" wide					43			43	47.50
5100	28" deep, 15" wide					52			52	57
5200	Box, box, file, 22" deep, 26" high					300			300	330
5300	28" deep					315			315	345
5400	File, file, 22" deep, 26" high					280			280	310
5600	Lateral file, 2 drawer, 30" wide					395			395	435
6000	Power, base in-feed cable					96			96	105
7400	Task light, recessed, 30" - 36" wide					117			117	129
7420	42" - 48" wide					127			127	139
7600	60" wide					137			137	151
8000	Worksurface, radius edge, 24" deep, 36" wide					126			126	138
8020	42" wide					155			155	170
8040	48" wide					165			165	182
8060	60" wide					207			207	228
8080	72" wide					234			234	257
8200	30" deep, 42" wide					198			198	218
8220	60" wide					231			231	254
8240	72" wide					264			264	290
8400	Corner, 24" deep, 36" wide					305			305	335
8420	42" wide					345			345	380
8500	Peninsula, 36" wide, 66" long					395			395	435
9000	For installation of systems furniture, add 5%									

12 59 16 – Free-Standing Component System Furniture

12 59 16.10 Furniture, Office Systems

		Crew	Daily Output	Labor-Hours	Unit	Material	Labor	Equipment	Total	Total Incl O&P
0010	**FURNITURE, OFFICE SYSTEMS**, Freestanding									
0100	Desk table, 24" deep x 48" wide				Ea.	540			540	595
0110	30" deep x 48" wide					550			550	605
0120	30" deep x 60" wide					560			560	615
0130	30" deep x 72" wide					620			620	680
0200	File, mobile, 2 drawer, 22" deep					375			375	415
0210	Suspended, 2 file					395			395	435
0220	Undercounter					430			430	470
0230	2 box, 1 file					460			460	505
0400	Keyboard platform with articulating arm					142			142	156
0500	End support leg					127			127	140
0800	Modesty panel, 48" wide					72.50			72.50	80
0810	Return					79.50			79.50	87
0820	Pencil/utility drawer					105			105	116
0830	Privacy screen, 48" wide x 17-20" high					211			211	233

12 59 Systems Furniture

12 59 16 – Free-Standing Component System Furniture

12 59 16.10 Furniture, Office Systems	Crew	Daily Output	Labor-Hours	Unit	Material	2009 Bare Costs Labor	Equipment	Total	Total Incl O&P
0840	72" wide				Ea.	335			335	370
0910	Printer stand, 30" deep x 36" wide					450			450	495
0920	Universal					460			460	505
0930	Paper basket					53			53	58
1010	Storage unit with doors, 60" wide					515			515	565
1020	72" wide					755			755	830
1210	Table, peninsula, 30" x 72"					580			580	640
1220	Return, 24" x 48"					400			400	440
1230	Round conference return, 42" diameter					395			395	435
1510	Work surface, 24" deep x 48" wide					251			251	276
1520	30" deep x 30" wide					231			231	254
1530	60" wide					310			310	345
1540	72" wide					325			325	360
1550	Corner bridge, 30" x 42"					174			174	191
1560	Peninsula, 30" x 60"					480			480	530
1570	Connector plate					13.20			13.20	14.55

12 59 23 – Desk System Furniture

12 59 23.10 Work Stations

		Crew	Daily Output	Labor-Hours	Unit	Material	Labor	Equipment	Total	Total Incl O&P
0010	**WORK STATIONS**									
1000	Secretarial work station, minimum				Ea.	2,975			2,975	3,250
1020	Maximum					10,100			10,100	11,100
1400	Management work station, minimum					3,375			3,375	3,725
1420	Maximum					10,900			10,900	12,000
1800	Executive work station, minimum					7,500			7,500	8,250
1820	Maximum					19,300			19,300	21,200
1840	For installation of systems furniture, add 15%									

12 61 Fixed Audience Seating

12 61 13 – Upholstered Audience Seating

12 61 13.13 Auditorium Chairs

		Crew	Daily Output	Labor-Hours	Unit	Material	Labor	Equipment	Total	Total Incl O&P
0010	**AUDITORIUM CHAIRS**									
2000	All veneer construction	2 Carp	22	.727	Ea.	175	29		204	238
2200	Veneer back, padded seat		22	.727		221	29		250	288
2350	Fully upholstered, spring seat		22	.727		211	29		240	277
2450	For tablet arms, add					58			58	63.50
2500	For fire retardancy, CATB-133, add					25			25	27.50

12 61 13.23 Lecture Hall Seating

		Crew	Daily Output	Labor-Hours	Unit	Material	Labor	Equipment	Total	Total Incl O&P
0010	**LECTURE HALL SEATING**									
1000	Pedestal type, minimum	2 Carp	22	.727	Ea.	165	29		194	227
1200	Maximum	"	14.50	1.103	"	555	44		599	680

12 63 Stadium and Arena Seating

12 63 13 – Stadium and Arena Bench Seating

12 63 13.13 Bleachers

		Crew	Daily Output	Labor-Hours	Unit	Material	2009 Bare Costs Labor	2009 Bare Costs Equipment	Total	Total Incl O&P
0010	**BLEACHERS**									
3000	Telescoping, manual to 15 tier, minimum	F-5	65	.492	Seat	76	19.90		95.90	115
3100	Maximum		60	.533		114	21.50		135.50	160
3300	16 to 20 tier, minimum		60	.533		183	21.50		204.50	235
3400	Maximum		55	.582		228	23.50		251.50	288
3600	21 to 30 tier, minimum		50	.640		190	26		216	249
3700	Maximum		40	.800		228	32.50		260.50	300
3900	For integral power operation, add, minimum	2 Elec	300	.053		38	2.51		40.51	45.50
4000	Maximum	"	250	.064		61	3.01		64.01	71.50
5000	Benches, folding, in wall, 14' table, 2 benches	L-4	2	12	Set	645	450		1,095	1,400

12 67 Pews and Benches

12 67 13 – Pews

12 67 13.13 Sanctuary Pews

		Crew	Daily Output	Labor-Hours	Unit	Material	Labor	Equipment	Total	Total Incl O&P
0010	**SANCTUARY PEWS**									
1500	Bench type, hardwood, minimum	1 Carp	20	.400	L.F.	74	16		90	107
1550	Maximum	"	15	.533		147	21.50		168.50	194
1570	For kneeler, add					18.15			18.15	19.95

12 92 Interior Planters and Artificial Plants

12 92 13 – Interior Artificial Plants

12 92 13.10 Plants

		Crew	Daily Output	Labor-Hours	Unit	Material	Labor	Equipment	Total	Total Incl O&P
0010	**PLANTS** Permanent only, weighted									
0020	Preserved or polyester leaf, natural wood									
0030	Or molded trunk. For pots see Div. 12 92 33.10									
0100	Plants, acuba, 5' high				Ea.	210			210	231
0200	Apidistra, 4' high					198			198	217
0300	Beech, variegated, 5' high					193			193	212
0400	Birds nest fern, 6' high					345			345	380
0500	Croton, 4' high					84.50			84.50	93
0600	Diffenbachia, 3' high					78			78	86
0700	Ficus Benjamina, 3' high					92.50			92.50	102
0720	6' high					234			234	257
0760	Nitida, 3' high					208			208	229
0780	6' high					365			365	400
0800	Helicona					193			193	212
1200	Palm, green date fan, 5' high					201			201	221
1220	7' high					254			254	279
1280	Green chamdora date, 5' high					182			182	200
1300	7' high					278			278	305
1400	Green giant, 7' high					430			430	475
1600	Rubber plant, 5' high					147			147	161
1800	Schefflera, 3' high					100			100	110
1820	4' high					124			124	136
1840	5' high					193			193	212
1860	6' high					263			263	289
1880	7' high					400			400	440
1900	Spathiphyllum, 3' high					139			139	153
4000	Trees, polyester or preserved, with									

12 92 Interior Planters and Artificial Plants

12 92 13 – Interior Artificial Plants

12 92 13.10 Plants

		Crew	Daily Output	Labor-Hours	Unit	Material	2009 Bare Costs Labor	Equipment	Total	Total Incl O&P
4020	Natural trunks									
4100	Acuba, 10' high				Ea.	1,100			1,100	1,200
4120	12' high					1,525			1,525	1,700
4140	14' high					2,150			2,150	2,375
4400	Bamboo, 10' high					480			480	525
4420	12' high					555			555	610
4800	Beech, 10' high					1,475			1,475	1,625
4820	12' high					1,700			1,700	1,875
4840	14' high					2,325			2,325	2,550
5000	Birch, 10' high					1,475			1,475	1,625
5020	12' high					1,700			1,700	1,875
5040	14' high					1,825			1,825	2,025
5060	16' high					2,325			2,325	2,550
5080	18' high					2,600			2,600	2,850
5500	Ficus Benjamina, 10' high					1,075			1,075	1,200
5520	12' high					1,400			1,400	1,525
5540	14' high					1,550			1,550	1,700
5560	16' high					2,150			2,150	2,375
5580	18' high					3,250			3,250	3,575
6000	Magnolia, 10' high					1,575			1,575	1,750
6020	12' high					2,250			2,250	2,475
6040	14' high					3,050			3,050	3,375
6500	Maple, 10' high					169			169	186
6520	12' high					2,000			2,000	2,200
6540	14' high					2,475			2,475	2,725
7000	Palms, chamadora, 10' high					310			310	340
7020	12' high					395			395	435
7040	Date, 10' high					425			425	465
7060	12' high					485			485	535

12 92 33 – Interior Planters

12 92 33.10 Planters

		Crew	Daily Output	Labor-Hours	Unit	Material	2009 Bare Costs Labor	Equipment	Total	Total Incl O&P
0010	**PLANTERS**									
1000	Fiberglass, hanging, 12" diameter, 7" high				Ea.	70			70	77
1100	15" diameter, 7" high					104			104	115
1200	36" diameter, 8" high					300			300	330
1500	Rectangular, 48" long, 16" high x 15" wide					395			395	435
1550	16" high x 24" wide					410			410	450
1600	24" high x 24" wide					610			610	670
1650	60" long, 30" high, 28" wide					535			535	585
1700	72" long, 16" high, 15" wide					875			875	965
1750	21" high, 24" wide					965			965	1,075
1800	30" high, 24" wide					1,225			1,225	1,350
2000	Round, 12" diameter, 13" high					62.50			62.50	68.50
2050	25" high					65.50			65.50	72
2150	14" diameter, 15" high					89			89	98
2200	16" diameter, 16" high					107			107	117
2250	18" diameter, 19" high					128			128	141
2300	23" high					106			106	116
2350	20" diameter, 16" high					101			101	111
2400	18" high					202			202	222
2450	21" high					108			108	118
2500	22" diameter, 10" high					210			210	231

12 92 Interior Planters and Artificial Plants

12 92 33 – Interior Planters

12 92 33.10 Planters		Crew	Daily Output	Labor-Hours	Unit	Material	2009 Bare Costs Labor	Equipment	Total	Total Incl O&P
2550	24" diameter, 16" high				Ea.	103			103	113
2600	19" high					160			160	176
2650	25" high					177			177	194
2700	36" high					234			234	257
2750	48" high					525			525	575
2800	30" diameter, 16" high					185			185	204
2850	18" high					192			192	211
2900	21" high					249			249	274
3000	24" high					330			330	360
3350	27" high					150			150	165
3400	36" diameter, 16" high					177			177	195
3450	18" high					180			180	198
3500	21" high					182			182	200
3550	24" high					192			192	212
3600	27" high					206			206	226
3650	30" high					226			226	249
3700	48" diameter, 16" high					395			395	435
3750	21" high					236			236	260
3800	24" high					258			258	284
3850	27" high					279			279	305
3900	30" high					420			420	460
3950	36" high					1,000			1,000	1,100
4000	60" diameter, 16" high					910			910	1,000
4100	21" high					1,025			1,025	1,150
4150	27" high					1,225			1,225	1,350
4200	30" high					1,475			1,475	1,600
4250	33" high					1,400			1,400	1,550
4300	36" high					1,400			1,400	1,550
4400	39" high					1,525			1,525	1,675
5000	Square, 10" side, 20" high					150			150	165
5100	14" side, 15" high					178			178	196
5200	18" side, 19" high					199			199	218
5300	20" side, 16" high					246			246	271
5320	18" high					299			299	330
5340	21" high					325			325	355
5400	24" side, 16" high					390			390	425
5420	21" high					420			420	465
5440	25" high					460			460	505
5460	30" side, 16" high					520			520	575
5480	24" high					590			590	650
5490	27" high					740			740	815
5500	Round, 36" diameter, 16" high					720			720	790
5510	18" high					735			735	810
5520	21" high					750			750	825
5530	24" high					765			765	840
5540	27" high					775			775	850
5550	30" high					890			890	980
5800	48" diameter, 16" high					905			905	995
5820	21" high					940			940	1,025
5840	24" high					1,125			1,125	1,225
5860	27" high					1,175			1,175	1,300
5880	30" high					1,300			1,300	1,425
5900	60" diameter, 16" high					1,225			1,225	1,350

12 92 Interior Planters and Artificial Plants

12 92 33 – Interior Planters

12 92 33.10 Planters		Crew	Daily Output	Labor-Hours	Unit	Material	2009 Bare Costs Labor	Equipment	Total	Total Incl O&P
5920	21" high				Ea.	1,350			1,350	1,475
5940	27" high					1,475			1,475	1,625
5960	30" high					1,675			1,675	1,850
5980	36" high					1,975			1,975	2,175
6000	Metal bowl, 32" diameter, 8" high, minimum					445			445	490
6050	Maximum					590			590	650
6100	Rectangle, 30" long x 12" wide, 6" high, minimum					273			273	300
6200	Maximum					535			535	590
6300	36" long 12" wide, 6" high, minimum					540			540	590
6400	Maximum					560			560	620
6500	Square, 15" side, minimum					540			540	590
6600	Maximum					590			590	650
6700	20" side, minimum					850			850	935
6800	Maximum					925			925	1,025
6900	Round, 6" diameter x 6" high, minimum					39.50			39.50	43.50
7000	Maximum					47			47	52
7100	8" diameter x 8" high, minimum					53			53	58
7200	Maximum					56.50			56.50	62
7300	10" diameter x 11" high, minimum					84.50			84.50	93
7400	Maximum					119			119	130
7420	12" diameter x 13" high, minimum					93.50			93.50	103
7440	Maximum					136			136	149
7500	14" diameter x 15" high, minimum					125			125	137
7550	Maximum					175			175	193
7580	16" diameter x 17" high, minimum					135			135	149
7600	Maximum					193			193	212
7620	18" diameter x 19" high, minimum					164			164	180
7640	Maximum					236			236	260
7680	22" diameter x 20" high, minimum					209			209	230
7700	Maximum					300			300	330
7750	24" diameter x 21" high, minimum					274			274	300
7800	Maximum					405			405	445
7850	31" diameter x 18" high, minimum					625			625	685
7900	Maximum					1,200			1,200	1,325
7950	38" diameter x 24" high, minimum					1,000			1,000	1,100
8000	Maximum					2,150			2,150	2,375
8050	48" diameter x 24" high, minimum					1,850			1,850	2,025
8150	Maximum					2,075			2,075	2,275
8500	Plastic laminate faced, fiberglass liner, square									
8520	14" sq., 15" high				Ea.	400			400	440
8540	24" sq., 16" high					505			505	555
8580	36" sq., 21" high					710			710	780
8600	Rectangle 36" long, 12" wide, 10" high					430			430	475
8650	48" long, 12" wide, 10" high					480			480	530
8700	48" long, 12" wide, 24" high					710			710	780
8750	Wood, fiberglass liner, square									
8780	14" square, 15" high, minimum				Ea.	288			288	315
8800	Maximum					355			355	390
8820	24" sq., 16" high, minimum					355			355	390
8840	Maximum					470			470	515
8860	36" sq., 21" high, minimum					455			455	500
8880	Maximum					685			685	750
9000	Rectangle, 36" long x 12" wide, 10" high, minimum					315			315	350

12 92 Interior Planters and Artificial Plants

12 92 33 – Interior Planters

12 92 33.10 Planters		Crew	Daily Output	Labor-Hours	Unit	Material	2009 Bare Costs Labor	Equipment	Total	Total Incl O&P
9050	Maximum				Ea.	410			410	455
9100	48" long x 12" wide, 10" high, minimum					340			340	375
9120	Maximum					430			430	475
9200	48" long x 12" wide, 24" high, minimum					410			410	455
9300	Maximum					600			600	660
9400	Plastic cylinder, molded, 10" diameter, 10" high					11.65			11.65	12.85
9500	11" diameter, 11" high					37			37	40.50
9600	13" diameter, 12" high					48.50			48.50	53.50
9700	16" diameter, 14" high					50.50			50.50	55.50

12 93 Site Furnishings

12 93 23 – Trash and Litter Receptors

12 93 23.10 Trash Receptacles

		Crew	Daily Output	Labor-Hours	Unit	Material	Labor	Equipment	Total	Total Incl O&P
0010	**TRASH RECEPTACLES**									
0020	Fiberglass, 2' square, 18" high	2 Clab	30	.533	Ea.	320	16.85		336.85	380
0100	2' square, 2'-6" high		30	.533		450	16.85		466.85	520
0300	Circular, 2' diameter, 18" high		30	.533		289	16.85		305.85	345
0400	2' diameter, 2'-6" high		30	.533		385	16.85		401.85	450
0500	Recycled plastic, var colors, round, 32 Gal, 28" x 38" H G		5	3.200		290	101		391	475
0510	32 Gal, 31" x 32" H G		5	3.200		360	101		461	550
9020	Fiberglass, circular, 15 gal. capacity		15	1.067		440	33.50		473.50	535
9030	20" dia, 32" high, 30 gal. capacity		15	1.067		485	33.50		518.50	585
9040	24" dia, 30" high, 30 gal. capacity		15	1.067		485	33.50		518.50	585
9050	Ash receptacle		15	1.067		285	33.50		318.50	370

12 93 33 – Manufactured Planters

12 93 33.10 Planters

		Crew	Daily Output	Labor-Hours	Unit	Material	Labor	Equipment	Total	Total Incl O&P
0010	**PLANTERS**									
0012	Concrete, sandblasted, precast, 48" diameter, 24" high	2 Clab	15	1.067	Ea.	635	33.50		668.50	755
0100	Fluted, precast, 7' diameter, 36" high		10	1.600		1,075	50.50		1,125.50	1,250
0300	Fiberglass, circular, 36" diameter, 24" high		15	1.067		470	33.50		503.50	575
0320	36" diameter, 27" high		12	1.333		500	42		542	615
0330	33" high		15	1.067		515	33.50		548.50	620
0335	24" diameter, 36" high		15	1.067		420	33.50		453.50	520
0340	60" diameter, 39" high		8	2		1,125	63		1,188	1,325
0400	60" diameter, 24" high		10	1.600		800	50.50		850.50	955
0600	Square, 24" side, 36" high		15	1.067		455	33.50		488.50	560
0620	24" side, 16" high		20	.800		330	25.50		355.50	405
0700	48" side, 36" high		15	1.067		810	33.50		843.50	945
0740	48" side, 39" high		15	1.067		990	33.50		1,023.50	1,150
0800	Rectangular, 48" x 12" sides, 18" high		12	1.333		385	42		427	490
0820	48" x 24" sides, 21" high		12	1.333		495	42		537	610
0900	Planter/bench, 72" square, 36" high		5	3.200		1,275	101		1,376	1,550
1000	96" square, 27" high		5	3.200		1,950	101		2,051	2,300
1200	Wood, square, 48" side, 24" high		15	1.067		1,000	33.50		1,033.50	1,150
1300	Circular, 48" diameter, 30" high		10	1.600		820	50.50		870.50	985
1500	72" diameter, 30" high		10	1.600		1,450	50.50		1,500.50	1,675
1600	Planter/bench, 72"		5	3.200		3,225	101		3,326	3,700

12 93 Site Furnishings

12 93 43 – Site Seating and Tables

12 93 43.13 Site Seating		Crew	Daily Output	Labor-Hours	Unit	Material	2009 Bare Costs Labor	Equipment	Total	Total Incl O&P
0010	**SITE SEATING**									
0012	Seating, benches, park, precast conc, w/backs, wood rails, 4' long	2 Clab	5	3.200	Ea.	430	101		531	630
0100	8' long		4	4		860	126		986	1,150
0300	Fiberglass, without back, one piece, 4' long		10	1.600		565	50.50		615.50	705
0400	8' long		7	2.286		1,175	72		1,247	1,375
0500	Steel barstock pedestals w/backs, 2" x 3" wood rails, 4' long		10	1.600		1,050	50.50		1,100.50	1,225
0510	8' long		7	2.286		1,250	72		1,322	1,475
0520	3" x 8" wood plank, 4' long		10	1.600		1,050	50.50		1,100.50	1,250
0530	8' long		7	2.286		1,100	72		1,172	1,325
0540	Backless, 4" x 4" wood plank, 4' square		10	1.600		1,025	50.50		1,075.50	1,200
0550	8' long		7	2.286		975	72		1,047	1,175
0600	Aluminum pedestals, with backs, aluminum slats, 8' long		8	2		335	63		398	470
0610	15' long		5	3.200		500	101		601	705
0620	Portable, aluminum slats, 8' long		8	2		335	63		398	470
0630	15' long		5	3.200		655	101		756	875
0800	Cast iron pedestals, back & arms, wood slats, 4' long		8	2		370	63		433	510
0820	8' long		5	3.200		1,075	101		1,176	1,350
0840	Backless, wood slats, 4' long		8	2		645	63		708	810
0860	8' long		5	3.200		815	101		916	1,050
1700	Steel frame, fir seat, 10' long	↓	10	1.600	↓	239	50.50		289.50	340

Division Notes

	CREW	DAILY OUTPUT	LABOR-HOURS	UNIT	2009 BARE COSTS				TOTAL INCL O&P
					MAT.	LABOR	EQUIP.	TOTAL	

Estimating Tips

General

- The items and systems in this division are usually estimated, purchased, supplied, and installed as a unit by one or more subcontractors. The estimator must ensure that all parties are operating from the same set of specifications and assumptions, and that all necessary items are estimated and will be provided. Many times the complex items and systems are covered, but the more common ones, such as excavation or a crane, are overlooked for the very reason that everyone assumes nobody could miss them. The estimator should be the central focus and be able to ensure that all systems are complete.
- Another area where problems can develop in this division is at the interface between systems. The estimator must ensure, for instance, that anchor bolts, nuts, and washers are estimated and included for the air-supported structures and pre-engineered buildings to be bolted to their foundations. Utility supply is a common area where essential items or pieces of equipment can be missed or overlooked due to the fact that each subcontractor may feel it is another's responsibility. The estimator should also be aware of certain items which may be supplied as part of a package but installed by others, and ensure that the installing contractor's estimate includes the cost of installation. Conversely, the estimator must also ensure that items are not costed by two different subcontractors, resulting in an inflated overall estimate.

13 30 00 Special Structures

- The foundations and floor slab, as well as rough mechanical and electrical, should be estimated, as this work is required for the assembly and erection of the structure. Generally, as noted in the book, the pre-engineered building comes as a shell and additional features, such as windows and doors, must be included by the estimator. Here again, the estimator must have a clear understanding of the scope of each portion of the work and all the necessary interfaces.

Reference Numbers

Reference numbers are shown in shaded boxes at the beginning of some major classifications. These numbers refer to related items in the Reference Section. The reference information may be an estimating procedure, an alternate pricing method, or technical information.

Note: Not all subdivisions listed here necessarily appear in this publication.

13 11 Swimming Pools

13 11 46 – Swimming Pool Accessories

13 11 46.50 Swimming Pool Equipment

		Crew	Daily Output	Labor-Hours	Unit	Material	2009 Bare Costs Labor	2009 Bare Costs Equipment	Total	Total Incl O&P
0010	**SWIMMING POOL EQUIPMENT**									
0020	Diving stand, stainless steel, 3 meter	2 Carp	.40	40	Ea.	9,375	1,600		10,975	12,800
0300	1 meter		2.70	5.926		5,475	237		5,712	6,400
0600	Diving boards, 16' long, aluminum		2.70	5.926		3,075	237		3,312	3,750
0700	Fiberglass		2.70	5.926		2,425	237		2,662	3,050
1100	Gutter system, stainless steel, with grating, stock,									
1110	contains supply and drainage system	E-1	20	1.200	L.F.	205	52	6.70	263.70	320
1120	Integral gutter and 5' high wall system, stainless steel	"	10	2.400	"	310	104	13.40	427.40	530
1200	Ladders, heavy duty, stainless steel, 2 tread	2 Carp	7	2.286	Ea.	490	91.50		581.50	680
1500	4 tread		6	2.667		655	107		762	885
1800	Lifeguard chair, stainless steel, fixed		2.70	5.926		2,275	237		2,512	2,875
2100	Lights, underwater, 12 volt, with transformer, 300 watt	1 Elec	1	8		243	375		618	830
2200	110 volt, 500 watt, standard		1	8		201	375		576	780
2400	Low water cutoff type		1	8		234	375		609	815
3000	Pool covers, reinforced vinyl	3 Clab	1800	.013	S.F.	.36	.42		.78	1.05
3050	Automatic, electric								8.77	9.65
3100	Vinyl water tube	3 Clab	3200	.008		.46	.24		.70	.88
3200	Maximum	"	3000	.008		.58	.25		.83	1.03
3300	Slides, tubular, fiberglass, aluminum handrails & ladder, 5'-0", straight	2 Carp	1.60	10	Ea.	3,625	400		4,025	4,625
3320	8'-0", curved		3	5.333		14,600	213		14,813	16,400
3400	10'-0", curved		1	16		21,600	640		22,240	24,700
3420	12'-0", straight with platform		1.20	13.333		5,100	535		5,635	6,450
4500	Hydraulic lift, movable pool bottom, single ram									
4520	Under 1,000 S.F. area	L-9	.03	1200	Ea.	195,000	43,600		238,600	284,500
4600	Four ram lift, over 1,000 S.F.	"	.02	1800		280,000	65,500		345,500	413,000
5000	Removable access ramp, stainless steel	2 Clab	2	8		8,475	253		8,728	9,725
5500	Removable stairs, stainless steel, collapsible	"	2	8		4,450	253		4,703	5,275

13 17 Tubs and Pools

13 17 13 – Redwood Hot Tub System

13 17 13.10 Redwood Hot Tub System

		Crew	Daily Output	Labor-Hours	Unit	Material	2009 Bare Costs Labor	2009 Bare Costs Equipment	Total	Total Incl O&P
0010	**REDWOOD HOT TUB SYSTEM**									
7050	4' diameter x 4' deep	Q-1	1	16	Ea.	2,025	700		2,725	3,275
7150	6' diameter x 4' deep		.80	20		3,125	880		4,005	4,775
7200	8' diameter x 4' deep		.80	20		4,350	880		5,230	6,100

13 17 33 – Whirlpool Tubs

13 17 33.10 Whirlpool Bath

		Crew	Daily Output	Labor-Hours	Unit	Material	2009 Bare Costs Labor	2009 Bare Costs Equipment	Total	Total Incl O&P
0010	**WHIRLPOOL BATH**									
6000	Whirlpool, bath with vented overflow, molded fiberglass									
6100	66" x 48" x 24"	Q-1	1	16	Ea.	3,050	700		3,750	4,425
6400	72" x 36" x 24"		1	16		3,000	700		3,700	4,350
6500	60" x 30" x 21"		1	16		2,575	700		3,275	3,875
6600	72" x 42" x 22"		1	16		4,025	700		4,725	5,500
6700	83" x 65"		.30	53.333		5,250	2,350		7,600	9,275
6711	For designer colors and trim add					25%				

13 21 Controlled Environment Rooms

13 21 13 – Clean Rooms

13 21 13.50 Clean Room Components	Crew	Daily Output	Labor-Hours	Unit	Material	2009 Bare Costs Labor	2009 Bare Costs Equipment	Total	Total Incl O&P
0010	**CLEAN ROOM COMPONENTS**								
1100	Clean room, soft wall, 12' x 12', Class 100	1 Carp	.18	44.444	Ea.	15,200	1,775		16,975	19,500
1110	Class 1,000		.18	44.444		12,300	1,775		14,075	16,300
1120	Class 10,000		.21	38.095		10,800	1,525		12,325	14,200
1130	Class 100,000	↓	.21	38.095		10,100	1,525		11,625	13,500
2800	Ceiling grid support, slotted channel struts 4'-0" O.C., ea. way				S.F.				5.91	6.50
3000	Ceiling panel, vinyl coated foil on mineral substrate									
3020	Sealed, non-perforated				S.F.				1.27	1.40
4000	Ceiling panel seal, silicone sealant, 150 L.F./gal.	1 Carp	150	.053	L.F.	.28	2.13		2.41	3.61
4100	Two sided adhesive tape	"	240	.033	"	.13	1.33		1.46	2.20
4200	Clips, one per panel				Ea.	1.09			1.09	1.20
6000	HEPA filter, 2'x4', 99.97% eff., 3" dp beveled frame (silicone seal)					355			355	390
6040	6" deep skirted frame (channel seal)					375			375	415
6100	99.99% efficient, 3" deep beveled frame (silicone seal)					370			370	405
6140	6" deep skirted frame (channel seal)					395			395	430
6200	99.999% efficient, 3" deep beveled frame (silicone seal)					385			385	425
6240	6" deep skirted frame (channel seal)				↓	410			410	450
7000	Wall panel systems, including channel strut framing									
7020	Polyester coated aluminum, particle board				S.F.				18.18	20
7100	Porcelain coated aluminum, particle board								31.82	35
7400	Wall panel support, slotted channel struts, to 12' high				↓				16.36	18

13 21 26 – Cold Storage Rooms

13 21 26.50 Refrigeration	Crew	Daily Output	Labor-Hours	Unit	Material	2009 Bare Costs Labor	2009 Bare Costs Equipment	Total	Total Incl O&P
0010	**REFRIGERATION**									
0020	Curbs, 12" high, 4" thick, concrete	2 Carp	58	.276	L.F.	4.14	11		15.14	21.50
1000	Doors, see Div. 08 34 13.10									
2400	Finishes, 2 coat portland cement plaster, 1/2" thick	1 Plas	48	.167	S.F.	1.14	6.05		7.19	10.35
2500	For galvanized reinforcing mesh, add	1 Lath	335	.024		.76	.85		1.61	2.10
2700	3/16" thick latex cement	1 Plas	88	.091		1.94	3.29		5.23	7.10
2900	For glass cloth reinforced ceilings, add	"	450	.018		.46	.64		1.10	1.48
3100	Fiberglass panels, 1/8" thick	1 Carp	149.45	.054		2.57	2.14		4.71	6.15
3200	Polystyrene, plastic finish ceiling, 1" thick		274	.029		2.37	1.17		3.54	4.42
3400	2" thick		274	.029		2.70	1.17		3.87	4.78
3500	4" thick	↓	219	.037		3	1.46		4.46	5.55
3800	Floors, concrete, 4" thick	1 Cefi	93	.086		1.13	3.29		4.42	6.05
3900	6" thick	"	85	.094	↓	1.68	3.60		5.28	7.15
4000	Insulation, 1" to 6" thick, cork				B.F.	1.10			1.10	1.21
4100	Urethane					1.08			1.08	1.19
4300	Polystyrene, regular					.73			.73	.80
4400	Bead board				↓	.55			.55	.61
4600	Installation of above, add per layer	2 Carp	657.60	.024	S.F.	.36	.97		1.33	1.91
4700	Wall and ceiling juncture		298.90	.054	L.F.	1.77	2.14		3.91	5.25
4900	Partitions, galvanized sandwich panels, 4" thick, stock		219.20	.073	S.F.	7.45	2.92		10.37	12.65
5000	Aluminum or fiberglass	↓	219.20	.073	"	8.10	2.92		11.02	13.45
5200	Prefab walk-in, 7'-6" high, aluminum, incl. door & floors,									
5210	not incl. partitions or refrigeration, 6' x 6' O.D. nominal	2 Carp	54.80	.292	SF Flr.	133	11.65		144.65	164
5500	10' x 10' O.D. nominal		82.20	.195		107	7.80		114.80	129
5700	12' x 14' O.D. nominal		109.60	.146		96	5.85		101.85	114
5800	12' x 20' O.D. nominal		109.60	.146		83.50	5.85		89.35	101
6100	For 8'-6" high, add					5%				
6300	Rule of thumb for complete units, w/o doors & refrigeration, cooler	2 Carp	146	.110		120	4.38		124.38	139
6400	Freezer	↓	109.60	.146	↓	142	5.85		147.85	166

13 21 Controlled Environment Rooms

13 21 26 – Cold Storage Rooms

13 21 26.50 Refrigeration	Crew	Daily Output	Labor-Hours	Unit	Material	2009 Bare Costs Labor	Equipment	Total	Total Incl O&P
6600 Shelving, plated or galvanized, steel wire type	2 Carp	360	.044	SF Hor.	10.55	1.78		12.33	14.35
6700 Slat shelf type	↓	375	.043		13	1.70		14.70	16.95
6900 For stainless steel shelving, add				↓	300%				
7000 Vapor barrier, on wood walls	2 Carp	1644	.010	S.F.	.15	.39		.54	.77
7200 On masonry walls	"	1315	.012	"	.40	.49		.89	1.19

13 21 48 – Sound-Conditioned Rooms

13 21 48.10 Anechoic Chambers

	Crew	Daily Output	Labor-Hours	Unit	Material	Labor	Equipment	Total	Total Incl O&P
0010 **ANECHOIC CHAMBERS** Standard units, 7' ceiling heights									
0100 Area for pricing is net inside dimensions									
0300 200 cycles per second cutoff, 25 S.F. floor area				SF Flr.	1,600			1,600	1,750
0400 50 S.F.								1,045	1,150
0600 75 S.F.								1,000	1,100
0700 100 S.F.					1,200			1,200	1,325
0900 For 150 cycles per second cutoff, add to 100 S.F. room									30%
1000 For 100 cycles per second cutoff, add to 100 S.F. room				↓					45%

13 21 48.15 Audiometric Rooms

	Crew	Daily Output	Labor-Hours	Unit	Material	Labor	Equipment	Total	Total Incl O&P
0010 **AUDIOMETRIC ROOMS**									
0020 Under 500 S.F. surface	4 Carp	98	.327	SF Surf	51.50	13.05		64.55	76.50
0100 Over 500 S.F. surface	"	120	.267	"	49	10.65		59.65	70

13 21 53 – Darkrooms

13 21 53.50 Darkrooms

	Crew	Daily Output	Labor-Hours	Unit	Material	Labor	Equipment	Total	Total Incl O&P
0010 **DARKROOMS**									
0020 Shell, complete except for door, 64 S.F., 8' high	2 Carp	128	.125	SF Flr.	53.50	4.99		58.49	66.50
0100 12' high		64	.250		68.50	10		78.50	90.50
0500 120 S.F. floor, 8' high		120	.133		37.50	5.35		42.85	49.50
0600 12' high		60	.267		51.50	10.65		62.15	73
0800 240 S.F. floor, 8' high		120	.133		27	5.35		32.35	38
0900 12' high		60	.267	↓	38	10.65		48.65	58
1200 Mini-cylindrical, revolving, unlined, 4' diameter		3.50	4.571	Ea.	2,825	183		3,008	3,375
1400 5'-6" diameter	↓	2.50	6.400		3,850	256		4,106	4,625
1600 Add for lead lining, inner cylinder, 1/32" thick					1,475			1,475	1,625
1700 1/16" thick					3,975			3,975	4,375
1800 Add for lead lining, inner and outer cylinder, 1/32" thick					2,725			2,725	3,000
1900 1/16" thick				↓	6,100			6,100	6,700
2000 For darkroom door, see Div. 08 34 36.10									

13 21 56 – Music Rooms

13 21 56.50 Music Rooms

	Crew	Daily Output	Labor-Hours	Unit	Material	Labor	Equipment	Total	Total Incl O&P
0010 **MUSIC ROOMS**									
0020 Practice room, modular, perforated steel, under 500 S.F.	2 Carp	70	.229	SF Surf	31.50	9.15		40.65	48.50
0100 Over 500 S.F.	"	80	.200	"	26.50	8		34.50	42

13 24 Special Activity Rooms

13 24 16 – Saunas

13 24 16.50 Saunas and Heaters

		Crew	Daily Output	Labor-Hours	Unit	Material	2009 Bare Costs Labor	2009 Bare Costs Equipment	Total	Total Incl O&P
0010	**SAUNAS AND HEATERS**									
0020	Prefabricated, incl. heater & controls, 7' high, 6' x 4', C/C	L-7	2.20	12.727	Ea.	4,625	490		5,115	5,850
0050	6' x 4', C/P		2	14		4,275	540		4,815	5,525
0400	6' x 5', C/C		2	14		5,175	540		5,715	6,500
0450	6' x 5', C/P		2	14		4,800	540		5,340	6,125
0600	6' x 6', C/C		1.80	15.556		5,475	600		6,075	6,950
0650	6' x 6', C/P		1.80	15.556		5,125	600		5,725	6,550
0800	6' x 9', C/C		1.60	17.500		6,800	675		7,475	8,525
0850	6' x 9', C/P		1.60	17.500		6,300	675		6,975	8,000
1000	8' x 12', C/C		1.10	25.455		10,600	980		11,580	13,200
1050	8' x 12', C/P		1.10	25.455		9,725	980		10,705	12,200
1200	8' x 8', C/C		1.40	20		8,100	770		8,870	10,100
1250	8' x 8', C/P		1.40	20		7,600	770		8,370	9,575
1400	8' x 10', C/C		1.20	23.333		8,975	900		9,875	11,300
1450	8' x 10', C/P		1.20	23.333		8,350	900		9,250	10,600
1600	10' x 12', C/C		1	28		11,200	1,075		12,275	14,000
1650	10' x 12', C/P		1	28		10,100	1,075		11,175	12,800
1700	Door only, cedar, 2'x6', with tempered insulated glass window	2 Carp	3.40	4.706		610	188		798	960
1800	Prehung, incl. jambs, pulls & hardware	"	12	1.333		615	53.50		668.50	760
2500	Heaters only (incl. above), wall mounted, to 200 C.F.					605			605	665
2750	To 300 C.F.					770			770	845
3000	Floor standing, to 720 C.F., 10,000 watts, w/controls	1 Elec	3	2.667		2,025	125		2,150	2,400
3250	To 1,000 C.F., 16,000 watts	"	3	2.667		2,525	125		2,650	2,950

13 24 26 – Steam Baths

13 24 26.50 Steam Baths and Components

		Crew	Daily Output	Labor-Hours	Unit	Material	2009 Bare Costs Labor	2009 Bare Costs Equipment	Total	Total Incl O&P
0010	**STEAM BATHS AND COMPONENTS**									
0020	Heater, timer & head, single, to 140 C.F.	1 Plum	1.20	6.667	Ea.	1,400	325		1,725	2,050
0500	To 300 C.F.		1.10	7.273		1,625	355		1,980	2,300
1000	Commercial size, with blow-down assembly, to 800 C.F.		.90	8.889		5,200	435		5,635	6,375
1500	To 2500 C.F.		.80	10		8,025	490		8,515	9,550
2000	Multiple, motels, apts., 2 baths, w/ blow-down assm., 500 C.F.	Q-1	1.30	12.308		5,900	540		6,440	7,300
2500	4 baths	"	.70	22.857		7,450	1,000		8,450	9,700

13 28 Athletic and Recreational Special Construction

13 28 33 – Athletic and Recreational Court Walls

13 28 33.50 Sport Court

		Crew	Daily Output	Labor-Hours	Unit	Material	2009 Bare Costs Labor	2009 Bare Costs Equipment	Total	Total Incl O&P
0010	**SPORT COURT**									
0020	Floors, No. 2 & better maple, 25/32" thick				SF Flr.				6.05	6.65
0100	Walls, laminated plastic bonded to galv. steel studs				SF Wall				7.68	8.45
0150	Laminated fiberglass surfacing, minimum								1.91	2.10
0180	Maximum								2.15	2.37
0300	Squash, regulation court in existing building, minimum				Court	35,000			35,000	38,500
0400	Maximum				"	39,000			39,000	42,900
0450	Rule of thumb for components:									
0470	Walls	3 Carp	.15	160	Court	10,500	6,400		16,900	21,500
0500	Floor	"	.25	96		8,300	3,825		12,125	15,100
0550	Lighting	2 Elec	.60	26.667		2,000	1,250		3,250	4,075
0600	Handball, racquetball court in existing building, minimum	C-1	.20	160		37,900	6,050		43,950	51,000
0800	Maximum	"	.10	320		41,000	12,100		53,100	64,000
0900	Rule of thumb for components: walls	3 Carp	.12	200		12,000	8,000		20,000	25,600

13 28 Athletic and Recreational Special Construction

13 28 33 – Athletic and Recreational Court Walls

13 28 33.50 Sport Court		Crew	Daily Output	Labor-Hours	Unit	Material	2009 Bare Costs Labor	Equipment	Total	Total Incl O&P
1000	Floor	3 Carp	.25	96	Court	8,300	3,825		12,125	15,100
1100	Ceiling	↓	.33	72.727		4,000	2,900		6,900	8,900
1200	Lighting	2 Elec	.60	26.667	↓	2,100	1,250		3,350	4,175

13 34 Fabricated Engineered Structures

13 34 13 – Glazed Structures

13 34 13.13 Greenhouses

		Crew	Daily Output	Labor-Hours	Unit	Material	2009 Bare Costs Labor	Equipment	Total	Total Incl O&P
0010	**GREENHOUSES**, Shell only, stock units, not incl. 2' stub walls,									
0020	foundation, floors, heat or compartments									
0300	Residential type, free standing, 8'-6" long x 7'-6" wide	2 Carp	59	.271	SF Flr.	47.50	10.85		58.35	69.50
0400	10'-6" wide		85	.188		37	7.50		44.50	52.50
0600	13'-6" wide		108	.148		35	5.90		40.90	47
0700	17'-0" wide		160	.100		35	4		39	44.50
0900	Lean-to type, 3'-10" wide		34	.471		38.50	18.80		57.30	71.50
1000	6'-10" wide	↓	58	.276		37.50	11		48.50	58
1500	Commercial, custom, truss frame, incl. equip., plumbing, elec.,									
1510	benches and controls, under 2,000 S.F., minimum				SF Flr.				27.27	30
1550	Maximum					31.50			31.50	34.50
1700	Over 5,000 S.F., minimum					21.50			21.50	23.50
1750	Maximum				↓				27.27	30
3600	For 1/4" clear plate glass, add				SF Surf	1.86			1.86	2.05
3700	For 1/4" tempered glass, add				"	4.21			4.21	4.63
3900	For cooling, add, minimum				SF Flr.	2.81			2.81	3.09
4000	Maximum					6.95			6.95	7.65
4200	For heaters, 13.6 MBH, add					5.35			5.35	5.85
4300	60 MBH, add				↓	2			2	2.20
4800	For controls, add, minimum				Total	2,400			2,400	2,650
4900	Maximum				"	14,300			14,300	15,700
5200	For vinyl shading, add				S.F.	1.29			1.29	1.42

13 42 Building Modules

13 42 63 – Detention Cell Modules

13 42 63.16 Steel Detention Cell Modules

		Crew	Daily Output	Labor-Hours	Unit	Material	2009 Bare Costs Labor	Equipment	Total	Total Incl O&P
0010	**STEEL DETENTION CELL MODULES**									
2000	Cells, prefab., 5' to 6' wide, 7' to 8' high, 7' to 8' deep,									
2010	bar front, cot, not incl. plumbing	E-4	1.50	21.333	Ea.	9,650	965	89.50	10,704.50	12,400

13 48 Sound, Vibration, and Seismic Control

13 48 13 – Manufactured Sound and Vibration Control Components

13 48 13.50 Audio Masking

		Crew	Daily Output	Labor-Hours	Unit	Material	2009 Bare Costs Labor	Equipment	Total	Total Incl O&P
0010	**AUDIO MASKING**, acoustical enclosure, 4" thick wall and ceiling									
0020	8# per S.F., up to 12' span	3 Carp	72	.333	SF Surf	32	13.30		45.30	56
0300	Better quality panels, 10.5# per S.F.		64	.375		36.50	15		51.50	63
0400	Reverb-chamber, 4" thick, parallel walls		60	.400		45.50	16		61.50	75
0600	Skewed wall, parallel roof, 4" thick panels		55	.436		52	17.45		69.45	84
0700	Skewed walls, skewed roof, 4" layers, 4" air space		48	.500		58.50	20		78.50	95.50
0900	Sound-absorbing panels, pntd mtl, 2'-6" x 8', under 1,000 S.F.	↓	215	.112	↓	12.10	4.46		16.56	20

13 48 Sound, Vibration, and Seismic Control

13 48 13 – Manufactured Sound and Vibration Control Components

13 48 13.50 Audio Masking		Crew	Daily Output	Labor-Hours	Unit	Material	2009 Bare Costs Labor	Equipment	Total	Total Incl O&P
1100	Over 1000 S.F.	3 Carp	240	.100	SF Surf	11.65	4		15.65	19
1200	Fabric faced	↓	240	.100		9.45	4		13.45	16.55
1500	Flexible transparent curtain, clear	3 Shee	215	.112		7.30	5.25		12.55	16.10
1600	50% foam		215	.112		10.20	5.25		15.45	19.30
1700	75% foam		215	.112		10.20	5.25		15.45	19.30
1800	100% foam	↓	215	.112	↓	10.20	5.25		15.45	19.30
3100	Audio masking system, including speakers, amplification									
3110	and signal generator									
3200	Ceiling mounted, 5,000 S.F.	2 Elec	2400	.007	S.F.	1.25	.31		1.56	1.85
3300	10,000 S.F.		2800	.006		1.01	.27		1.28	1.51
3400	Plenum mounted, 5,000 S.F.		3800	.004		1.06	.20		1.26	1.46
3500	10,000 S.F.	↓	4400	.004	↓	.72	.17		.89	1.04

13 49 Radiation Protection

13 49 13 – Lead Sheet

13 49 13.50 Lead Sheets

		Crew	Daily Output	Labor-Hours	Unit	Material	Labor	Equipment	Total	Total Incl O&P
0010	**LEAD SHEETS**									
0300	Lead sheets, 1/16" thick	2 Lath	135	.119	S.F.	7.15	4.21		11.36	14.15
0400	1/8" thick		120	.133		14.45	4.74		19.19	23
0500	Lead shielding, 1/4" thick		135	.119		34	4.21		38.21	44
0550	1/2" thick	↓	120	.133	↓	59.50	4.74		64.24	72.50
0950	Lead headed nails (average 1 lb. per sheet)				Lb.	7.15			7.15	7.90
1000	Butt joints in 1/8" lead or thicker, 2" batten strip x 7' long	2 Lath	240	.067	Ea.	16.95	2.37		19.32	22
1200	X-ray protection, average radiography or fluoroscopy									
1210	room, up to 300 S.F. floor, 1/16" lead, minimum	2 Lath	.25	64	Total	8,075	2,275		10,350	12,300
1500	Maximum, 7'-0" walls	"	.15	106	"	9,700	3,800		13,500	16,300
1600	Deep therapy X-ray room, 250 kV capacity,									
1800	up to 300 S.F. floor, 1/4" lead, minimum	2 Lath	.08	200	Total	22,600	7,100		29,700	35,300
1900	Maximum, 7'-0" walls	"	.06	266	"	27,800	9,475		37,275	44,600

13 49 19 – Lead-Lined Materials

13 49 19.50 Shielding Lead

		Crew	Daily Output	Labor-Hours	Unit	Material	Labor	Equipment	Total	Total Incl O&P
0010	**SHIELDING LEAD**									
0100	Laminated lead in wood doors, 1/16" thick, no hardware				S.F.	45			45	49.50
0200	Lead lined door frame, not incl. hardware,									
0210	1/16" thick lead, butt prepared for hardware	1 Lath	2.40	3.333	Ea.	575	119		694	805
0850	Window frame with 1/16" lead and voice passage, 36" x 60"	2 Glaz	2	8		1,125	310		1,435	1,700
0870	24" x 36" frame		8	2	↓	900	77		977	1,100
0900	Lead gypsum board, 5/8" thick with 1/16" lead		160	.100	S.F.	7.25	3.86		11.11	13.85
0910	1/8" lead	↓	140	.114		14.10	4.41		18.51	22
0930	1/32" lead	2 Lath	200	.080	↓	4.43	2.84		7.27	9.10

13 49 21 – Lead Glazing

13 49 21.50 Lead Glazing

		Crew	Daily Output	Labor-Hours	Unit	Material	Labor	Equipment	Total	Total Incl O&P
0010	**LEAD GLAZING**									
0600	Lead glass, 1/4" thick, 2.0 mm LE, 12" x 16"	2 Glaz	13	1.231	Ea.	265	47.50		312.50	365
0700	24" x 36"		8	2		980	77		1,057	1,200
0800	36" x 60"	↓	2	8	↓	3,850	310		4,160	4,725
2000	X-ray viewing panels, clear lead plastic									
2010	7 mm thick, 0.3 mm LE, 2.3 lbs/S.F.	H-3	139	.115	S.F.	143	3.97		146.97	164
2020	12 mm thick, 0.5 mm LE, 3.9 lbs/S.F.		82	.195		194	6.70		200.70	223
2030	18 mm thick, 0.8 mm LE, 5.9 lbs/S.F.	↓	54	.296		210	10.20		220.20	247

13 49 Radiation Protection

13 49 21 – Lead Glazing

13 49 21.50 Lead Glazing		Crew	Daily Output	Labor-Hours	Unit	Material	2009 Bare Costs Labor	2009 Bare Costs Equipment	Total	Total Incl O&P
2040	22 mm thick, 1.0 mm LE, 7.2 lbs/S.F.	H-3	44	.364	S.F.	215	12.55		227.55	255
2050	35 mm thick, 1.5 mm LE, 11.5 lbs/S.F.		28	.571		241	19.70		260.70	295
2060	46 mm thick, 2.0 mm LE, 15.0 lbs/S.F.	↓	21	.762	↓	315	26.50		341.50	390
2090	For panels 12 S.F. to 48 S.F., add crating charge				Ea.					50

13 49 23 – Modular Shielding Partitions

13 49 23.50 Modular Shielding Partitions

		Crew	Daily Output	Labor-Hours	Unit	Material	Labor	Equipment	Total	Total Incl O&P
0010	**MODULAR SHIELDING PARTITIONS**									
4000	X-ray barriers, modular, panels mounted within framework for									
4002	attaching to floor, wall or ceiling, upper portion is clear lead									
4005	plastic window panels 48"H, lower portion is opaque leaded									
4008	steel panels 36"H, structural supports not incl.									
4010	1-section barrier, 36"W x 84"H overall									
4020	0.5 mm LE panels	H-3	6.40	2.500	Ea.	4,525	86		4,611	5,100
4030	0.8 mm LE panels		6.40	2.500		4,825	86		4,911	5,425
4040	1.0 mm LE panels		5.33	3.002		4,900	103		5,003	5,550
4050	1.5 mm LE panels	↓	5.33	3.002	↓	5,225	103		5,328	5,900
4060	2-section barrier, 72"W x 84"H overall									
4070	0.5 mm LE panels	H-3	4	4	Ea.	9,250	138		9,388	10,400
4080	0.8 mm LE panels		4	4		9,775	138		9,913	11,000
4090	1.0 mm LE panels		3.56	4.494		9,975	155		10,130	11,200
5000	1.5 mm LE panels	↓	3.20	5	↓	11,300	172		11,472	12,800
5010	3-section barrier, 108"W x 84"H overall									
5020	0.5 mm LE panels	H-3	3.20	5	Ea.	13,900	172		14,072	15,600
5030	0.8 mm LE panels		3.20	5		14,700	172		14,872	16,400
5040	1.0 mm LE panels		2.67	5.993		14,900	206		15,106	16,700
5050	1.5 mm LE panels	↓	2.46	6.504		16,200	224		16,424	18,100
7000	X-ray barriers, mobile, mounted within framework w/casters on									
7005	bottom, clear lead plastic window panels on upper portion,									
7010	opaque on lower, 30"W x 75"H overall, incl. framework									
7020	24"H upper w/0.5 mm LE, 48"H lower w/0.8 mm LE	1 Carp	16	.500	Ea.	2,400	20		2,420	2,675
7030	48"W x 75"H overall, incl. framework									
7040	36"H upper w/0.5 mm LE, 36"H lower w/0.8 mm LE	1 Carp	16	.500	Ea.	4,425	20		4,445	4,875
7050	36"H upper w/1.0 mm LE, 36"H lower w/1.5 mm LE	"	16	.500	"	5,250	20		5,270	5,800
7060	72"W x 75"H overall, incl. framework									
7070	36"H upper w/0.5 mm LE, 36"H lower w/0.8 mm LE	1 Carp	16	.500	Ea.	5,225	20		5,245	5,775
7080	36"H upper w/1.0 mm LE, 36"H lower w/1.5 mm LE	"	16	.500	"	6,550	20		6,570	7,225

13 49 33 – Radio Frequency Shielding

13 49 33.50 Shielding, Radio Frequency

		Crew	Daily Output	Labor-Hours	Unit	Material	Labor	Equipment	Total	Total Incl O&P
0010	**SHIELDING, RADIO FREQUENCY**									
0020	Prefabricated or screen-type copper or steel, minimum	2 Carp	180	.089	SF Surf	31.50	3.55		35.05	40
0100	Average		155	.103		33.50	4.12		37.62	43.50
0150	Maximum	↓	145	.110	↓	35.50	4.41		39.91	46
0200	RF modular shielding panels, walls & ceilings	E-1	180	.133	S.F.	17.30	5.80	.75	23.85	29.50
0210	Floor liner, steel sheet, 14 ga.		430	.056		2.51	2.43	.31	5.25	7.20
0215	11 ga.		140	.171		3.46	7.45	.96	11.87	17.50
0220	Wall liner, steel sheet, 14 ga.		180	.133		2.61	5.80	.75	9.16	13.55
0225	11 ga.		140	.171		3.60	7.45	.96	12.01	17.65
0230	Steel plate, 1/4"		90	.267		3.61	11.60	1.49	16.70	25.50
0235	Ceiling liner, steel sheet, 14 ga.		180	.133	↓	2.51	5.80	.75	9.06	13.45
0250	Ceiling hangers		45	.533	Ea.	29	23	2.98	54.98	74.50
0275	Wall supports		45	.533	"	29	23	2.98	54.98	74.50
0300	Shielding transition, 11 ga. preformed angles	↓	1365	.018	L.F.	29	.76	.10	29.86	33

13 49 Radiation Protection

13 49 33 – Radio Frequency Shielding

13 49 33.50 Shielding, Radio Frequency		Crew	Daily Output	Labor-Hours	Unit	Material	2009 Bare Costs Labor	Equipment	Total	Total Incl O&P
0350	11 ga. galvanized with steel bar	E-1	895	.027	L.F.	2.24	1.17	.15	3.56	4.60
0400	Tubing sub frame, single door		1.44	16.667	Ea.	86.50	725	93	904.50	1,425
0450	Double door		.88	27.273		130	1,175	152	1,457	2,325
0500	Radiofreq prot, door, 3' W x 7' H, to 120 Db elec/plane wave	Q-11	9	3.556		6,150	160		6,310	7,025
0550	Double door, 6' W x 7' H		6	5.333		11,900	240		12,140	13,500
0600	Wave guide vents, 2" diameter		67	.478		48	21.50		69.50	86
0610	12" diameter		17	1.882		350	84.50		434.50	515
0620	12" x 6"		17	1.882		225	84.50		309.50	375
0630	12" x 12"		17	1.882		360	84.50		444.50	530
0640	15" x 15"		17	1.882		405	84.50		489.50	580
0650	30" x 14"		11	2.909		760	131		891	1,025
0660	48" x 15"		8	4		1,025	180		1,205	1,400
0670	60" x 14"		8	4		1,275	180		1,455	1,675
0700	RF filter, 100 Db, 2 x 30 amp		7	4.571		535	206		741	905
1000	Hi-hat for lighting		13	2.462		835	111		946	1,100

Division Notes

		CREW	DAILY OUTPUT	LABOR-HOURS	UNIT	2009 BARE COSTS				TOTAL INCL O&P
						MAT.	LABOR	EQUIP.	TOTAL	

Estimating Tips

General
- Many products in Division 14 will require some type of support or blocking for installation not included with the item itself. Examples are supports for conveyors or tube systems, attachment points for lifts, and footings for hoists or cranes. Add these supports in the appropriate division.

14 10 00 Dumbwaiters
14 20 00 Elevators
- Dumbwaiters and elevators are estimated and purchased in a method similar to buying a car. The manufacturer has a base unit with standard features. Added to this base unit price will be whatever options the owner or specifications require. Increased load capacity, additional vertical travel, additional stops, higher speed, and cab finish options are items to be considered. When developing an estimate for dumbwaiters and elevators, remember that some items needed by the installers may have to be included as part of the general contract. Examples are:
  — shaftway
  — rail support brackets
  — machine room
  — electrical supply
  — sill angles
  — electrical connections
  — pits
  — roof penthouses
  — pit ladders

  Check the job specifications and drawings before pricing.
- Installation of elevators and handicapped lifts in historic structures can require significant additional costs. The associated structural requirements may involve cutting into and repairing finishes, mouldings, flooring, etc. The estimator must account for these special conditions.

14 30 00 Escalators and Moving Walks
- Escalators and moving walks are specialty items installed by specialty contractors. There are numerous options associated with these items. For specific options, contact a manufacturer or contractor. In a method similar to estimating dumbwaiters and elevators, you should verify the extent of general contract work and add items as necessary.

14 40 00 Lifts
14 90 00 Other Conveying Equipment
- Products such as correspondence lifts, chutes, and pneumatic tube systems, as well as other items specified in this subdivision, may require trained installers. The general contractor might not have any choice as to who will perform the installation or when it will be performed. Long lead times are often required for these products, making early decisions in scheduling necessary.

Reference Numbers
Reference numbers are shown in shaded boxes at the beginning of some major classifications. These numbers refer to related items in the Reference Section. The reference information may be an estimating procedure, an alternate pricing method, or technical information.

Note: Not all subdivisions listed here necessarily appear in this publication.

No part of this publication may be reproduced, stored in a retrieval system, or transmitted in any form or by any means without prior written permission of R.S. Means Company, Inc.

14 11 Manual Dumbwaiters

14 11 10 – Hand Operated Dumbwaiters

14 11 10.20 Manual Dumbwaiters

		Crew	Daily Output	Labor-Hours	Unit	Material	2009 Bare Costs Labor	Equipment	Total	Total Incl O&P
0010	**MANUAL DUMBWAITERS**									
0020	2 stop, minimum	2 Elev	.75	21.333	Ea.	2,925	1,200		4,125	5,025
0100	Maximum	↓	.50	32	"	6,650	1,800		8,450	10,000
0300	For each additional stop, add	↓	.75	21.333	Stop	1,050	1,200		2,250	2,975

14 12 Electric Dumbwaiters

14 12 10 – Dumbwaiters

14 12 10.10 Electric Dumbwaiters

		Crew	Daily Output	Labor-Hours	Unit	Material	Labor	Equipment	Total	Total Incl O&P
0010	**ELECTRIC DUMBWAITERS**									
0020	2 stop, minimum	2 Elev	.13	123	Ea.	7,375	6,975		14,350	18,500
0100	Maximum	↓	.11	145	"	22,200	8,225		30,425	36,700
0600	For each additional stop, add	↓	.54	29.630	Stop	3,250	1,675		4,925	6,075
0750	Correspondence lift, 1 floor, 2 stop, 45 lb capacity	2 Elec	.20	80	Ea.	8,350	3,750		12,100	14,800

14 21 Electric Traction Elevators

14 21 13 – Electric Traction Freight Elevators

14 21 13.10 Electric Traction Freight Elevators and Options

		Crew	Daily Output	Labor-Hours	Unit	Material	Labor	Equipment	Total	Total Incl O&P
0010	**ELECTRIC TRACTION FREIGHT ELEVATORS AND OPTIONS** R142000-10									
0425	Electric freight, base unit, 4000 lb, 200 fpm, 4 stop, std. fin.	2 Elev	.05	320	Ea.	99,500	18,100		117,600	136,500
0450	For 5000 lb capacity, add R142000-40					5,600			5,600	6,150
0500	For 6000 lb capacity, add					17,400			17,400	19,200
0525	For 7000 lb capacity, add					21,500			21,500	23,600
0550	For 8000 lb capacity, add					29,600			29,600	32,600
0575	For 10000 lb capacity, add					32,500			32,500	35,800
0600	For 12000 lb capacity, add					39,400			39,400	43,300
0625	For 16000 lb capacity, add					50,000			50,000	55,000
0650	For 20000 lb capacity, add					59,500			59,500	65,500
0675	For increased speed, 250 fpm, add					22,500			22,500	24,700
0700	300 fpm, geared electric, add					28,400			28,400	31,300
0725	350 fpm, geared electric, add					34,300			34,300	37,700
0750	400 fpm, geared electric, add					39,900			39,900	43,800
0775	500 fpm, gearless electric, add					47,300			47,300	52,000
0800	600 fpm, gearless electric, add					57,500			57,500	63,000
0825	700 fpm, gearless electric, add					68,500			68,500	75,500
0850	800 fpm, gearless electric, add					79,000			79,000	87,000
0875	For class "B" loading, add					6,525			6,525	7,175
0900	For class "C-1" loading, add					9,075			9,075	9,975
0925	For class "C-2" loading, add					10,300			10,300	11,400
0950	For class "C-3" loading, add		↓			13,000			13,000	14,400
0975	For travel over 40 V.L.F., add	2 Elev	7.25	2.207	V.L.F.	540	125		665	780
1000	For number of stops over 4, add	"	.27	59.259	Stop	4,150	3,350		7,500	9,575

14 21 23 – Electric Traction Passenger Elevators

14 21 23.10 Electric Traction Passenger Elevators and Options

		Crew	Daily Output	Labor-Hours	Unit	Material	Labor	Equipment	Total	Total Incl O&P
0010	**ELECTRIC TRACTION PASSENGER ELEVATORS AND OPTIONS**									
1625	Electric pass., base unit, 2000 lb, 200 fpm, 4 stop, std. fin.	2 Elev	.05	320	Ea.	91,500	18,100		109,600	127,500
1650	For 2500 lb capacity, add					3,800			3,800	4,200
1675	For 3000 lb capacity, add					4,550			4,550	5,000
1700	For 3500 lb capacity, add					7,675			7,675	8,425
1725	For 4000 lb capacity, add		↓			7,100			7,100	7,800

14 21 Electric Traction Elevators

14 21 23 – Electric Traction Passenger Elevators

14 21 23.10 Electric Traction Passenger Elevators and Options	Crew	Daily Output	Labor-Hours	Unit	Material	2009 Bare Costs Labor	Equipment	Total	Total Incl O&P
1750	For 4500 lb capacity, add				Ea.	8,975			8,975	9,875
1775	For 5000 lb capacity, add					10,800			10,800	11,900
1800	For increased speed, 250 fpm, geared electric, add					7,000			7,000	7,700
1825	300 fpm, geared electric, add					10,600			10,600	11,700
1850	350 fpm, geared electric, add					12,100			12,100	13,300
1875	400 fpm, geared electric, add					15,400			15,400	17,000
1900	500 fpm, gearless electric, add					32,600			32,600	35,900
1925	600 fpm, gearless electric, add					36,100			36,100	39,700
1950	700 fpm, gearless electric, add					41,000			41,000	45,100
1975	800 fpm, gearless electric, add					46,800			46,800	51,500
2000	For travel over 40 V.L.F., add	2 Elev	7.25	2.207	V.L.F.	650	125		775	900
2025	For number of stops over 4, add		.27	59.259	Stop	7,825	3,350		11,175	13,600
2400	Electric hospital, base unit, 4000 lb, 200 fpm, 4 stop, std fin.		.05	320	Ea.	79,000	18,100		97,100	114,000
2425	For 4500 lb capacity, add					5,650			5,650	6,200
2450	For 5000 lb capacity, add					6,775			6,775	7,475
2475	For increased speed, 250 fpm, geared electric, add					7,125			7,125	7,850
2500	300 fpm, geared electric, add					10,600			10,600	11,700
2525	350 fpm, geared electric, add					12,200			12,200	13,400
2550	400 fpm, geared electric, add					15,400			15,400	17,000
2575	500 fpm, gearless electric, add					31,700			31,700	34,900
2600	600 fpm, gearless electric, add					36,100			36,100	39,700
2625	700 fpm, gearless electric, add					40,900			40,900	45,000
2650	800 fpm, gearless electric, add					46,800			46,800	51,500
2675	For travel over 40 V.L.F., add	2 Elev	7.25	2.207	V.L.F.	665	125		790	920
2700	For number of stops over 4, add	"	.27	59.259	Stop	4,150	3,350		7,500	9,575

14 21 33 – Electric Traction Residential Elevators

14 21 33.20 Residential Elevators

		Crew	Daily Output	Labor-Hours	Unit	Material	Labor	Equipment	Total	Total Incl O&P
0010	**RESIDENTIAL ELEVATORS**									
7000	Residential, cab type, 1 floor, 2 stop, minimum	2 Elev	.20	80	Ea.	11,000	4,525		15,525	18,900
7100	Maximum		.10	160		18,600	9,050		27,650	34,000
7200	2 floor, 3 stop, minimum		.12	133		16,400	7,550		23,950	29,200
7300	Maximum		.06	266		26,700	15,100		41,800	52,000

14 24 Hydraulic Elevators

14 24 13 – Hydraulic Freight Elevators

14 24 13.10 Hydraulic Freight Elevators and Options

		Crew	Daily Output	Labor-Hours	Unit	Material	Labor	Equipment	Total	Total Incl O&P
0010	**HYDRAULIC FREIGHT ELEVATORS AND OPTIONS**									
1025	Hydraulic freight, base unit, 2000 lb, 50 fpm, 2 stop, std. fin.	2 Elev	.10	160	Ea.	82,000	9,050		91,050	103,500
1050	For 2500 lb capacity, add					5,025			5,025	5,525
1075	For 3000 lb capacity, add					8,850			8,850	9,750
1100	For 3500 lb capacity, add					16,000			16,000	17,600
1125	For 4000 lb capacity, add					17,100			17,100	18,800
1150	For 4500 lb capacity, add					20,000			20,000	22,000
1175	For 5000 lb capacity, add					25,300			25,300	27,800
1200	For 6000 lb capacity, add					31,500			31,500	34,600
1225	For 7000 lb capacity, add					37,900			37,900	41,700
1250	For 8000 lb capacity, add					41,700			41,700	45,900
1275	For 10000 lb capacity, add					52,500			52,500	57,500
1300	For 12000 lb capacity, add					68,500			68,500	75,500
1325	For 16000 lb capacity, add					94,000			94,000	103,000

14 24 Hydraulic Elevators

14 24 13 – Hydraulic Freight Elevators

14 24 13.10 Hydraulic Freight Elevators and Options

		Crew	Daily Output	Labor-Hours	Unit	Material	2009 Bare Costs Labor	Equipment	Total	Total Incl O&P
1350	For 20000 lb capacity, add				Ea.	114,000			114,000	125,500
1375	For increased speed, 100 fpm, add					2,150			2,150	2,375
1400	125 fpm, add					4,650			4,650	5,100
1425	150 fpm, add					5,525			5,525	6,075
1450	175 fpm, add					9,400			9,400	10,400
1475	For class "B" loading, add					6,500			6,500	7,150
1500	For class "C-1" loading, add					9,025			9,025	9,925
1525	For class "C-2" loading, add					10,300			10,300	11,300
1550	For class "C-3" loading, add					13,000			13,000	14,300
1575	For travel over 20 V.L.F., add	2 Elev	7.25	2.207	V.L.F.	1,675	125		1,800	2,025
1600	For number of stops over 2, add	"	.27	59.259	Stop	4,575	3,350		7,925	10,000

14 24 23 – Hydraulic Passenger Elevators

14 24 23.10 Hydraulic Passenger Elevators and Options

		Crew	Daily Output	Labor-Hours	Unit	Material	2009 Bare Costs Labor	Equipment	Total	Total Incl O&P
0010	**HYDRAULIC PASSENGER ELEVATORS AND OPTIONS**									
2050	Hyd. pass., base unit, 1500 lb, 100 fpm, 2 stop, std. fin.	2 Elev	.10	160	Ea.	39,900	9,050		48,950	57,500
2075	For 2000 lb capacity, add					2,025			2,025	2,250
2100	For 2500 lb. capacity, add					3,225			3,225	3,550
2125	For 3000 lb capacity, add					4,275			4,275	4,700
2150	For 3500 lb capacity, add					6,225			6,225	6,850
2175	For 4000 lb capacity, add					7,600			7,600	8,350
2200	For 4500 lb capacity, add					9,000			9,000	9,900
2225	For 5000 lb capacity, add					11,900			11,900	13,100
2250	For increased speed, 125 fpm, add					2,975			2,975	3,275
2275	150 fpm, add					3,775			3,775	4,150
2300	175 fpm, add					5,650			5,650	6,225
2325	200 fpm, add					7,850			7,850	8,625
2350	For travel over 12 V.L.F., add	2 Elev	7.25	2.207	V.L.F.	890	125		1,015	1,175
2375	For number of stops over 2, add		.27	59.259	Stop	2,575	3,350		5,925	7,825
2725	Hydraulic hospital, base unit, 4000 lb, 100 fpm, 2 stop, std. fin.		.10	160	Ea.	62,000	9,050		71,050	81,500
2750	For 4000 lb capacity, add					7,150			7,150	7,850
2775	For 4500 lb capacity, add					6,900			6,900	7,600
2800	For 5000 lb capacity, add					9,150			9,150	10,100
2825	For increased speed, 125 fpm, add					3,450			3,450	3,775
2850	150 fpm, add					4,150			4,150	4,550
2875	175 fpm, add					6,000			6,000	6,600
2900	200 fpm, add					7,275			7,275	8,000
2925	For travel over 12 V.L.F., add	2 Elev	7.25	2.207	V.L.F.	745	125		870	1,000
2950	For number of stops over 2, add	"	.27	59.259	Stop	4,200	3,350		7,550	9,625

14 27 Custom Elevator Cabs

14 27 13 – Custom Elevator Cab Finishes

14 27 13.10 Cab Finishes

		Crew	Daily Output	Labor-Hours	Unit	Material	2009 Bare Costs Labor	Equipment	Total	Total Incl O&P
0010	**CAB FINISHES**									
3325	Passenger elevator cab finishes (based on 3500 lb cab size)									
3350	Acrylic panel ceiling				Ea.	780			780	860
3375	Aluminum eggcrate ceiling					1,225			1,225	1,350
3400	Stainless steel doors					2,975			2,975	3,250
3425	Carpet flooring					650			650	715
3450	Epoxy flooring					480			480	530
3475	Quarry tile flooring					825			825	910

14 27 Custom Elevator Cabs

14 27 13 – Custom Elevator Cab Finishes

14 27 13.10 Cab Finishes	Crew	Daily Output	Labor-Hours	Unit	Material	2009 Bare Costs Labor	Equipment	Total	Total Incl O&P
3500 Slate flooring				Ea.	1,525			1,525	1,675
3525 Textured rubber flooring					200			200	220
3550 Stainless steel walls					5,175			5,175	5,700
3575 Stainless steel returns at door					1,200			1,200	1,325
4450 Hospital elevator cab finishes (based on 3500 lb cab size)									
4475 Aluminum eggcrate ceiling				Ea.	1,200			1,200	1,325
4500 Stainless steel doors					2,675			2,675	2,950
4525 Epoxy flooring					375			375	415
4550 Quarry tile flooring					845			845	930
4575 Textured rubber flooring					720			720	790
4600 Stainless steel walls					5,175			5,175	5,700
4625 Stainless steel returns at door					1,200			1,200	1,325

14 28 Elevator Equipment and Controls

14 28 10 – Elevator Equipment and Control Options

14 28 10.10 Elevator Controls and Doors	Crew	Daily Output	Labor-Hours	Unit	Material	2009 Bare Costs Labor	Equipment	Total	Total Incl O&P
0010 **ELEVATOR CONTROLS AND DOORS**									
2975 Passenger elevator options									
3000 2 car group automatic controls	2 Elev	.66	24.242	Ea.	6,650	1,375		8,025	9,350
3025 3 car group automatic controls		.44	36.364		13,000	2,050		15,050	17,400
3050 4 car group automatic controls		.33	48.485		26,500	2,750		29,250	33,200
3075 5 car group automatic controls		.26	61.538		48,100	3,475		51,575	58,000
3100 6 car group automatic controls		.22	72.727		98,500	4,125		102,625	114,500
3125 Intercom service		3	5.333		1,375	300		1,675	1,975
3150 Duplex car selective collective		.66	24.242		12,200	1,375		13,575	15,600
3175 Center opening 1 speed doors		2	8		1,850	455		2,305	2,725
3200 Center opening 2 speed doors		2	8		2,550	455		3,005	3,475
3225 Rear opening doors (opposite front)		2	8		3,125	455		3,580	4,125
3250 Side opening 2 speed doors		2	8		4,900	455		5,355	6,050
3275 Automatic emergency power switching		.66	24.242		1,200	1,375		2,575	3,375
3300 Manual emergency power switching		8	2		500	113		613	720
3625 Hall finishes, stainless steel doors					2,000			2,000	2,200
3650 Stainless steel frames					1,450			1,450	1,600
3675 12 month maintenance contract								3,600	3,960
3700 Signal devices, hall lanterns	2 Elev	8	2		540	113		653	765
3725 Position indicators, up to 3		9.40	1.702		650	96.50		746.50	855
3750 Position indicators, per each over 3		32	.500		620	28.50		648.50	725
3775 High speed heavy duty door opener					4,200			4,200	4,625
3800 Variable voltage, O.H. gearless machine, min.	2 Elev	.16	100		70,500	5,650		76,150	86,000
3815 Maximum		.07	228		105,500	12,900		118,400	135,500
3825 Basement installed geared machine		.33	48.485		72,000	2,750		74,750	83,500
3850 Freight elevator options									
3875 Doors, bi-parting	2 Elev	.66	24.242	Ea.	9,975	1,375		11,350	13,100
3900 Power operated door and gate	"	.66	24.242		23,400	1,375		24,775	27,900
3925 Finishes, steel plate floor					2,475			2,475	2,725
3950 14 ga. 1/4" x 4' steel plate walls					1,900			1,900	2,100
3975 12 month maintenance contract								3,600	3,960
4000 Signal devices, hall lanterns	2 Elev	8	2		540	113		653	765
4025 Position indicators, up to 3		9.40	1.702		650	96.50		746.50	855
4050 Position indicators, per each over 3		32	.500		620	28.50		648.50	725
4075 Variable voltage basement installed geared machine		.66	24.242		19,600	1,375		20,975	23,700

14 28 Elevator Equipment and Controls

14 28 10 – Elevator Equipment and Control Options

14 28 10.10 Elevator Controls and Doors		Crew	Daily Output	Labor-Hours	Unit	Material	2009 Bare Costs Labor	Equipment	Total	Total Incl O&P
4100	Hospital elevator options									
4125	2 car group automatic controls	2 Elev	.66	24.242	Ea.	6,650	1,375		8,025	9,350
4150	3 car group automatic controls		.44	36.364		13,000	2,050		15,050	17,400
4175	4 car group automatic controls		.33	48.485		26,500	2,750		29,250	33,200
4200	5 car group automatic controls		.26	61.538		48,100	3,475		51,575	58,000
4225	6 car group automatic controls		.22	72.727		98,500	4,125		102,625	114,500
4250	Intercom service		3	5.333		1,375	300		1,675	1,975
4275	Duplex car selective collective		.66	24.242		12,200	1,375		13,575	15,600
4300	Center opening 1 speed doors		2	8		1,850	455		2,305	2,725
4325	Center opening 2 speed doors		2	8		2,525	455		2,980	3,475
4350	Rear opening doors (opposite front)		2	8		3,125	455		3,580	4,125
4375	Side opening 2 speed doors		2	8		3,925	455		4,380	5,000
4400	Automatic emergency power switching		.66	24.242		1,200	1,375		2,575	3,375
4425	Manual emergency power switching		8	2		500	113		613	720
4675	Hall finishes, stainless steel doors					2,000			2,000	2,200
4700	Stainless steel frames					1,450			1,450	1,600
4725	12 month maintenance contract								3,600	3,960
4750	Signal devices, hall lanterns	2 Elev	8	2		540	113		653	765
4775	Position indicators, up to 3		9.40	1.702		650	96.50		746.50	855
4800	Position indicators, per each over 3		32	.500		620	28.50		648.50	725
4825	High speed heavy duty door opener					4,200			4,200	4,625
4850	Variable voltage, O.H. gearless machine, min.	2 Elev	.16	100		70,500	5,650		76,150	86,000
4865	Maximum		.07	228		105,500	12,900		118,400	135,500
4875	Basement installed geared machine		.33	48.485		26,900	2,750		29,650	33,600
5000	Drilling for piston, casing included, 18" diameter	B-48	80	.700	V.L.F.	45	24.50	36	105.50	127

14 31 Escalators

14 31 10 – Glass and Steel Escalators

14 31 10.10 Escalators

0010	**ESCALATORS**	R143110-10									
1000	Glass, 32" wide x 10' floor to floor height		M-1	.07	457	Ea.	90,500	24,600	815	115,915	137,000
1010	48" wide x 10' floor to floor height			.07	457		98,000	24,600	815	123,415	145,500
1020	32" wide x 15' floor to floor height			.06	533		96,000	28,700	950	125,650	149,500
1030	48" wide x 15' floor to floor height			.06	533		102,000	28,700	950	131,650	156,000
1040	32" wide x 20' floor to floor height			.05	653		101,000	35,100	1,150	137,250	165,000
1050	48" wide x 20' floor to floor height			.05	653		110,000	35,100	1,150	146,250	175,000
1060	32" wide x 25' floor to floor height			.04	800		107,000	43,000	1,425	151,425	183,000
1070	48" wide x 25' floor to floor height			.04	800		124,000	43,000	1,425	168,425	202,000
1080	Enameled steel, 32" wide x 10' floor to floor height			.07	457		95,500	24,600	815	120,915	143,000
1090	48" wide x 10' floor to floor height			.07	457		104,000	24,600	815	129,415	152,000
1110	32" wide x 15' floor to floor height			.06	533		104,000	28,700	950	133,650	158,500
1120	48" wide x 15' floor to floor height			.06	533		104,000	28,700	950	133,650	158,500
1130	32" wide x 20' floor to floor height			.05	653		107,500	35,100	1,150	143,750	172,000
1140	48" wide x 20' floor to floor height			.05	653		115,000	35,100	1,150	151,250	180,500
1150	32" wide x 25' floor to floor height			.04	800		115,000	43,000	1,425	159,425	192,000
1160	48" wide x 25' floor to floor height			.04	800		115,000	43,000	1,425	159,425	192,000
1170	Stainless steel, 32" wide x 10' floor to floor height			.07	457		115,000	24,600	815	140,415	164,000
1180	48" wide x 10' floor to floor height			.07	457		115,000	24,600	815	140,415	164,000
1500	32" wide x 15' floor to floor height			.06	533		106,000	28,700	950	135,650	160,500
1700	48" wide x 15' floor to floor height			.06	533		107,500	28,700	950	137,150	162,500
2300	32" wide x 25' floor to floor height			.04	800		114,500	43,000	1,425	158,925	191,500

14 31 Escalators

14 31 10 – Glass and Steel Escalators

14 31 10.10 Escalators		Crew	Daily Output	Labor-Hours	Unit	Material	2009 Bare Costs Labor	Equipment	Total	Total Incl O&P
2500	48" wide x 25' floor to floor height	M-1	.04	800	Ea.	138,500	43,000	1,425	182,925	217,500

14 32 Moving Walks

14 32 10 – Moving Walkways

14 32 10.10 Moving Walks

			Crew	Daily Output	Labor-Hours	Unit	Material	Labor	Equipment	Total	Total Incl O&P
0010	**MOVING WALKS**	R143210-20									
0020	Walk, 27" tread width, minimum		M-1	6.50	4.923	L.F.	850	265	8.75	1,123.75	1,350
0100	300' to 500', maximum			4.43	7.223		1,175	390	12.85	1,577.85	1,900
0300	48" tread width walk, minimum			4.43	7.223		1,925	390	12.85	2,327.85	2,725
0400	100' to 350', maximum			3.82	8.377		2,250	450	14.90	2,714.90	3,150
0600	Ramp, 12° incline, 36" tread width, minimum			5.27	6.072		1,575	325	10.80	1,910.80	2,225
0700	70' to 90' maximum			3.82	8.377		2,250	450	14.90	2,714.90	3,150
0900	48" tread width, minimum			3.57	8.964		2,300	480	15.95	2,795.95	3,275
1000	40' to 70', maximum			2.91	10.997		2,900	590	19.55	3,509.55	4,100

14 42 Wheelchair Lifts

14 42 13 – Inclined Wheelchair Lifts

14 42 13.10 Inclined Wheelchair Lifts and Stairclimbers

		Crew	Daily Output	Labor-Hours	Unit	Material	Labor	Equipment	Total	Total Incl O&P
0010	**INCLINED WHEELCHAIR LIFTS AND STAIRCLIMBERS**									
7700	Stair climber (chair lift), single seat, minimum	2 Elev	1	16	Ea.	5,325	905		6,230	7,225
7800	Maximum		.20	80		7,350	4,525		11,875	14,800
8700	Stair lift, minimum		1	16		14,500	905		15,405	17,300
8900	Maximum		.20	80		22,900	4,525		27,425	32,000

14 42 16 – Vertical Wheelchair Lifts

14 42 16.10 Wheelchair Lifts

		Crew	Daily Output	Labor-Hours	Unit	Material	Labor	Equipment	Total	Total Incl O&P
0010	**WHEELCHAIR LIFTS**									
8000	Wheelchair lift, minimum	2 Elev	1	16	Ea.	7,325	905		8,230	9,400
8500	Maximum	"	.50	32	"	17,300	1,800		19,100	21,700

14 51 Correspondence and Parcel Lifts

14 51 10 – Electric Correspondence and Parcel Lifts

14 51 10.10 Correspondence Lifts

		Crew	Daily Output	Labor-Hours	Unit	Material	Labor	Equipment	Total	Total Incl O&P
0010	**CORRESPONDENCE LIFTS**									
0020	1 floor, 2 stop, 25 lb capacity, electric	2 Elev	.20	80	Ea.	5,975	4,525		10,500	13,300
0100	Hand, 5 lb capacity	"	.20	80	"	2,250	4,525		6,775	9,250

14 51 10.20 Parcel Lifts

		Crew	Daily Output	Labor-Hours	Unit	Material	Labor	Equipment	Total	Total Incl O&P
0010	**PARCEL LIFTS**									
0020	20" x 20", 100 lb capacity, electric, per floor	2 Mill	.25	64	Ea.	9,350	2,650		12,000	14,200

14 91 Facility Chutes

14 91 33 – Laundry and Linen Chutes

14 91 33.10 Chutes

		Crew	Daily Output	Labor-Hours	Unit	Material	2009 Bare Costs Labor	Equipment	Total	Total Incl O&P
0011	**CHUTES**, linen, trash or refuse									
0050	Aluminized steel, 16 ga., 18" diameter	2 Shee	3.50	4.571	Floor	1,100	216		1,316	1,550
0100	24" diameter		3.20	5		1,175	236		1,411	1,650
0200	30" diameter		3	5.333		1,325	252		1,577	1,825
0300	36" diameter		2.80	5.714		1,500	270		1,770	2,075
0400	Galvanized steel, 16 ga., 18" diameter		3.50	4.571		1,000	216		1,216	1,425
0500	24" diameter		3.20	5		1,125	236		1,361	1,625
0600	30" diameter		3	5.333		1,275	252		1,527	1,775
0700	36" diameter		2.80	5.714		1,500	270		1,770	2,075
0800	Stainless steel, 18" diameter		3.50	4.571		2,350	216		2,566	2,900
0900	24" diameter		3.20	5		2,550	236		2,786	3,175
1000	30" diameter		3	5.333		3,100	252		3,352	3,775
1005	36" diameter		2.80	5.714		3,450	270		3,720	4,225
1200	Linen chute bottom collector, aluminized steel		4	4	Ea.	1,400	189		1,589	1,850
1300	Stainless steel		4	4		1,800	189		1,989	2,275
1500	Refuse, bottom hopper, aluminized steel, 18" diameter		3	5.333		890	252		1,142	1,375
1600	24" diameter		3	5.333		1,075	252		1,327	1,550
1800	36" diameter		3	5.333		1,800	252		2,052	2,350

14 91 82 – Trash Chutes

14 91 82.10 Trash Chutes and Accessories

		Crew	Daily Output	Labor-Hours	Unit	Material	Labor	Equipment	Total	Total Incl O&P
0010	**TRASH CHUTES AND ACCESSORIES**									
2900	Package chutes, spiral type, minimum	2 Shee	4.50	3.556	Floor	2,425	168		2,593	2,925
3000	Maximum	"	1.50	10.667	"	6,325	505		6,830	7,750

14 92 Pneumatic Tube Systems

14 92 10 – Conventional, Automatic and Computer Controlled Pneumatic Tube Systems

14 92 10.10 Pneumatic Tube Systems

		Crew	Daily Output	Labor-Hours	Unit	Material	Labor	Equipment	Total	Total Incl O&P
0010	**PNEUMATIC TUBE SYSTEMS**									
0020	100' long, single tube, 2 stations, stock									
0100	3" diameter	2 Stpi	.12	133	Total	6,600	6,575		13,175	17,100
0300	4" diameter	"	.09	177	"	7,450	8,775		16,225	21,400
0400	Twin tube, two stations or more, conventional system									
0600	2-1/2" round	2 Stpi	62.50	.256	L.F.	13.30	12.65		25.95	33.50
0700	3" round		46	.348		15.25	17.15		32.40	43
0900	4" round		49.60	.323		16.90	15.90		32.80	42.50
1000	4" x 7" oval		37.60	.426		24.50	21		45.50	58.50
1050	Add for blower		2	8	System	5,200	395		5,595	6,325
1110	Plus for each round station, add		7.50	2.133	Ea.	585	105		690	805
1150	Plus for each oval station, add		7.50	2.133	"	585	105		690	805
1200	Alternate pricing method: base cost, minimum		.75	21.333	Total	6,375	1,050		7,425	8,600
1300	Maximum		.25	64	"	12,700	3,150		15,850	18,700
1500	Plus total system length, add, minimum		93.40	.171	L.F.	8.20	8.45		16.65	22
1600	Maximum		37.60	.426	"	24.50	21		45.50	58
1800	Completely automatic system, 4" round, 15 to 50 stations		.29	55.172	Station	20,100	2,725		22,825	26,200
2200	51 to 144 stations		.32	50		15,600	2,475		18,075	20,900
2400	6" round or 4" x 7" oval, 15 to 50 stations		.24	66.667		25,200	3,300		28,500	32,600
2800	51 to 144 stations		.23	69.565		21,100	3,425		24,525	28,400

Division 21 - Fire Suppression

Estimating Tips

Pipe for fire protection and all uses is located in Subdivision 22 11 13.

The labor adjustment factors listed in Subdivision 22 01 02.20 also apply to Division 21.

Many, but not all, areas require backflow protection in the fire system. It is advisable to check local building codes for specific requirements.

For your reference, the following is a list of the most applicable Fire Codes and Standards which may be purchased from the NFPA, 1 Batterymarch Park, Quincy, MA 02169-7471.

NFPA 1: Uniform Fire Code
NFPA 10: Portable Fire Extinguishers
NFPA 11: Low-, Medium-, and High-Expansion Foam
NFPA 12: Carbon Dioxide Extinguishing Systems (Also companion 12A)
NFPA 13: Installation of Sprinkler Systems (Also companion 13D, 13E, and 13R)
NFPA 14: Installation of Standpipe and Hose Systems
NFPA 15: Water Spray Fixed Systems for Fire Protection
NFPA 16: Installation of Foam-Water Sprinkler and Foam-Water Spray Systems
NFPA 17: Dry Chemical Extinguishing Systems (Also companion 17A)
NFPA 18: Wetting Agents
NFPA 20: Installation of Stationary Pumps for Fire Protection
NFPA 22: Water Tanks for Private Fire Protection
NFPA 24: Installation of Private Fire Service Mains and their Appurtenances
NFPA 25: Inspection, Testing and Maintenance of Water-Based Fire Protection

Reference Numbers

Reference numbers are shown in shaded boxes at the beginning of some major classifications. These numbers refer to related items in the Reference Section. The reference information may be an estimating procedure, an alternate pricing method, or technical information.

Note: Not all subdivisions listed here necessarily appear in this publication.

Note: **Trade Service,** *in part, has been used as a reference source for some of the material prices used in Division 21.*

21 05 Common Work Results for Fire Suppression

21 05 23 – General-Duty Valves for Water-Based Fire-Suppression Piping

21 05 23.50 General-Duty Valves for Water-Based Fire Supp.

		Crew	Daily Output	Labor-Hours	Unit	Material	2009 Bare Costs Labor	2009 Bare Costs Equipment	Total	Total Incl O&P
0010	**GENERAL-DUTY VALVES FOR WATER-BASED FIRE SUPPRESSION PIPING**									
6200	Valves and components									
6210	Alarm, includes									
6220	retard chamber, trim, gauges, alarm line strainer									
6260	3" size	Q-12	3	5.333	Ea.	1,225	231		1,456	1,700
6280	4" size	"	2	8		1,275	345		1,620	1,925
6300	6" size	Q-13	4	8		1,425	365		1,790	2,125
6320	8" size	"	3	10.667		1,725	490		2,215	2,625
6500	Check, swing, C.I. body, brass fittings, auto. ball drip									
6520	4" size	Q-12	3	5.333	Ea.	255	231		486	625
6540	6" size	Q-13	4	8		470	365		835	1,075
6580	8" size	"	3	10.667		890	490		1,380	1,725
6800	Check, wafer, butterfly type, C.I. body, bronze fittings									
6820	4" size	Q-12	4	4	Ea.	283	173		456	570

21 11 Facility Fire-Suppression Water-Service Piping

21 11 16 – Facility Fire Hydrants

21 11 16.50 Fire Hydrants for Buildings

		Crew	Daily Output	Labor-Hours	Unit	Material	Labor	Equipment	Total	Total Incl O&P
0010	**FIRE HYDRANTS FOR BUILDINGS**									
3750	Hydrants, wall, w/caps, single, flush, polished brass									
3800	2-1/2" x 2-1/2"	Q-12	5	3.200	Ea.	175	139		314	400
3840	2-1/2" x 3"		5	3.200		355	139		494	600
3860	3" x 3"		4.80	3.333		287	144		431	530
3900	For polished chrome, add					20%				
3950	Double, flush, polished brass									
4000	2-1/2" x 2-1/2" x 4"	Q-12	5	3.200	Ea.	470	139		609	730
4080	3" x 3" x 4"	"	4.90	3.265		1,000	141		1,141	1,300
4200	For polished chrome, add					10%				
4350	Double, projecting, polished brass									
4400	2-1/2" x 2-1/2" x 4"	Q-12	5	3.200	Ea.	209	139		348	440
4460	Valve control, dbl. flush/projecting hydrant, cap &									
4470	chain, extension rod & cplg., escutcheon, polished brass	Q-12	8	2	Ea.	249	86.50		335.50	405
4480	Four-way square, flush, polished brass									

21 11 19 – Fire-Department Connections

21 11 19.50 Connections for the Fire-Department

		Crew	Daily Output	Labor-Hours	Unit	Material	Labor	Equipment	Total	Total Incl O&P
0010	**CONNECTIONS FOR THE FIRE-DEPARTMENT**									
6000	Roof manifold, horiz., brass, without valves & caps									
6040	2-1/2" x 2-1/2" x 4"	Q-12	4.80	3.333	Ea.	139	144		283	370
6060	2-1/2" x 2-1/2" x 6"		4.60	3.478		146	151		297	385
6080	2-1/2" x 2-1/2" x 2-1/2" x 4"		4.60	3.478		261	151		412	515
6090	2-1/2" x 2-1/2" x 2-1/2" x 6"		4.60	3.478		230	151		381	480
7000	Sprinkler line tester, cast brass					19.80			19.80	22
7140	Standpipe connections, wall, w/plugs & chains									
7160	Single, flush, brass, 2-1/2" x 2-1/2"	Q-12	5	3.200	Ea.	123	139		262	345
7240	For polished chrome, add					15%				
7280	Double, flush, polished brass									
7300	2-1/2" x 2-1/2" x 4"	Q-12	5	3.200	Ea.	425	139		564	680
7330	2-1/2" x 2-1/2" x 6"		4.60	3.478		595	151		746	880
7340	3" x 3" x 4"		4.90	3.265		775	141		916	1,075
7370	3" x 3" x 6"		4.50	3.556		765	154		919	1,075

21 11 Facility Fire-Suppression Water-Service Piping

21 11 19 – Fire-Department Connections

21 11 19.50 Connections for the Fire-Department		Crew	Daily Output	Labor-Hours	Unit	Material	2009 Bare Costs Labor	2009 Bare Costs Equipment	Total	Total Incl O&P
7400	For polished chrome, add					15%				
7440	For sill cock combination, add				Ea.	50.50			50.50	56
7580	Double projecting, polished brass									
7600	2-1/2" x 2-1/2" x 4"	Q-12	5	3.200	Ea.	390	139		529	640
7630	2-1/2" x 2-1/2" x 6"	"	4.60	3.478	"	660	151		811	950
7680	For polished chrome, add					15%				
7900	Three way, flush, polished brass									
7920	2-1/2" (3) x 4"	Q-12	4.80	3.333	Ea.	1,175	144		1,319	1,500
7930	2-1/2" (3) x 6"	"	4.80	3.333		1,150	144		1,294	1,475
8000	For polished chrome, add					9%				
8020	Three way, projecting, polished brass									
8040	2-1/2"(3) x 4"	Q-12	4.80	3.333	Ea.	985	144		1,129	1,300
8070	2-1/2" (3) x 6"	"	4.60	3.478		990	151		1,141	1,325
8100	For polished chrome, add					12%				
8200	Four way, square, flush, polished brass,									
8240	2-1/2"(4) x 6"	Q-12	3.60	4.444	Ea.	2,475	193		2,668	3,025
8300	For polished chrome, add				"	10%				
8550	Wall, vertical, flush, cast brass									
8600	Two way, 2-1/2" x 2-1/2" x 4"	Q-12	5	3.200	Ea.	580	139		719	850
8660	Four way, 2-1/2"(4) x 6"		3.80	4.211		2,125	182		2,307	2,600
8680	Six way, 2-1/2"(6) x 6"		3.40	4.706		3,275	204		3,479	3,900
8700	For polished chrome, add					10%				

21 12 Fire-Suppression Standpipes

21 12 13 – Fire-Suppression Hoses and Nozzles

21 12 13.50 Fire Hoses and Nozzles

		Crew	Daily Output	Labor-Hours	Unit	Material	2009 Bare Costs Labor	2009 Bare Costs Equipment	Total	Total Incl O&P
0010	**FIRE HOSES AND NOZZLES**									
0200	Adapters, rough brass, straight hose threads									
0220	One piece, female to male, rocker lugs									
0240	1" x 1"				Ea.	39.50			39.50	43.50
0320	2" x 2"					40.50			40.50	44.50
0380	2-1/2" x 2-1/2"					18.40			18.40	20
0420	3" x 3"					81			81	89
0500	For polished brass, add					50%				
0520	For polished chrome, add					75%				
1100	Swivel, female to female, pin lugs									
1120	1-1/2" x 1-1/2"				Ea.	50.50			50.50	55.50
1200	2-1/2" x 2-1/2"					98.50			98.50	108
1260	For polished brass, add					50%				
1280	For polished chrome, add					75%				
1400	Couplings, sngl & dbl jacket, pin lug or rocker lug, cast brass									
1410	1-1/2"				Ea.	43.50			43.50	48
1420	2-1/2"				"	57.50			57.50	63
1500	For polished brass, add					20%				
1520	For polished chrome, add					40%				
1580	Reducing, F x M, interior installation, cast brass									
1590	2" x 1-1/2"				Ea.	54			54	59.50
1600	2-1/2" x 1-1/2"					11.75			11.75	12.95
1680	For polished brass, add					50%				
1720	For polished chrome, add					75%				
2200	Hose, less couplings									

21 12 Fire-Suppression Standpipes

21 12 13 – Fire-Suppression Hoses and Nozzles

21 12 13.50 Fire Hoses and Nozzles		Crew	Daily Output	Labor-Hours	Unit	Material	2009 Bare Costs Labor	Equipment	Total	Total Incl O&P
2260	Synthetic jacket, lined, 300 lb. test, 1-1/2" diameter	Q-12	2600	.006	L.F.	2.39	.27		2.66	3.03
2280	2-1/2" diameter		2200	.007		3.98	.32		4.30	4.85
2360	High strength, 500 lb. test, 1-1/2" diameter		2600	.006		2.47	.27		2.74	3.12
2380	2-1/2" diameter	↓	2200	.007	↓	4.29	.32		4.61	5.20
5000	Nipples, straight hose to tapered iron pipe, brass									
5060	Female to female, 1-1/2" x 1-1/2"				Ea.	17.40			17.40	19.15
5080	2" x 2"					18			18	19.80
5100	2-1/2" x 2-1/2"					28.50			28.50	31
5190	For polished chrome, add					75%				
5200	Double male or male to female, 1" x 1"					34.50			34.50	38
5220	1-1/2" x 1"					50.50			50.50	56
5230	1-1/2" x 1-1/2"					11.40			11.40	12.55
5260	2" x 1-1/2"					55			55	60.50
5270	2" x 2"					69.50			69.50	76.50
5300	2-1/2" x 2"					37.50			37.50	41.50
5310	2-1/2" x 2-1/2"					21.50			21.50	23.50
5340	For polished chrome, add				↓	75%				
5600	Nozzles, brass									
5620	Adjustable fog, 3/4" booster line				Ea.	88.50			88.50	97.50
5630	1" booster line					104			104	115
5640	1-1/2" leader line					63.50			63.50	70
5660	2-1/2" direct connection					168			168	185
5680	2-1/2" playpipe nozzle				↓	176			176	193
5780	For chrome plated, add					8%				
5850	Electrical fire, adjustable fog, no shock									
5900	1-1/2"				Ea.	340			340	375
5920	2-1/2"					455			455	505
5980	For polished chrome, add					6%				
6200	Heavy duty, comb. adj. fog and str. stream, with handle									
6210	1" booster line				Ea.	325			325	360
6240	1-1/2"					365			365	405
6260	2-1/2", for playpipe					655			655	720
6280	2-1/2" direct connection					455			455	505
6300	2-1/2" playpipe combination					635			635	700
6480	For polished chrome, add					7%				
6500	Plain fog, polished brass, 1-1/2"					86.50			86.50	95
6540	Chrome plated, 1-1/2"					83			83	91.50
6700	Plain stream, polished brass, 1-1/2" x 10"					39.50			39.50	43.50
6760	2-1/2" x 15" x 7/8" or 1-1/2"				↓	71.50			71.50	78.50
6860	For polished chrome, add					20%				
7000	Underwriters playpipe, 2-1/2" x 30" with 1-1/8" tip				Ea.	203			203	224
9340	Tools, crowbar and brackets	1 Carp	12	.667		96	26.50		122.50	148
9360	Combination hydrant wrench and spanner				↓	25.50			25.50	28.50
9380	Fire axe and brackets									
9400	6 lb.	1 Carp	12	.667	Ea.	100	26.50		126.50	152

21 12 16 – Fire-Suppression Hose Reels

21 12 16.50 Fire-Suppression Hose Reels		Crew	Daily Output	Labor-Hours	Unit	Material	Labor	Equipment	Total	Total Incl O&P
0010	**FIRE -SUPPRESSION HOSE REELS**									
2990	Hose reel, swinging, for 1-1/2" polyester neoprene lined hose									
3000	50' long	Q-12	14	1.143	Ea.	108	49.50		157.50	193
3020	100' long		14	1.143		151	49.50		200.50	241
3060	For 2-1/2" cotton rubber hose, 75' long	↓	14	1.143	↓	175	49.50		224.50	267

21 12 Fire-Suppression Standpipes

21 12 16 – Fire-Suppression Hose Reels

21 12 16.50 Fire-Suppression Hose Reels		Crew	Daily Output	Labor-Hours	Unit	Material	2009 Bare Costs Labor	2009 Bare Costs Equipment	Total	Total Incl O&P
3100	150' long	Q-12	14	1.143	Ea.	202	49.50		251.50	297

21 12 19 – Fire-Suppression Hose Racks

21 12 19.50 Fire Hose Racks

		Crew	Daily Output	Labor-Hours	Unit	Material	Labor	Equipment	Total	Total Incl O&P
0010	**FIRE HOSE RACKS**									
2600	Hose rack, swinging, for 1-1/2" diameter hose,									
2620	Enameled steel, 50' & 75' lengths of hose	Q-12	20	.800	Ea.	48.50	34.50		83	105
2640	100' and 125' lengths of hose		20	.800		49	34.50		83.50	106
2680	Chrome plated, 50' and 75' lengths of hose		20	.800		60.50	34.50		95	119
2700	100' and 125' lengths of hose		20	.800		60.50	34.50		95	119
2780	For hose rack nipple, 1-1/2" polished brass, add					21.50			21.50	24
2820	2-1/2" polished brass, add					42.50			42.50	47
2840	1-1/2" polished chrome, add					32			32	35
2860	2-1/2" polished chrome, add					45			45	49.50

21 12 23 – Fire-Suppression Hose Valves

21 12 23.70 Fire Hose Valves

		Crew	Daily Output	Labor-Hours	Unit	Material	Labor	Equipment	Total	Total Incl O&P
0010	**FIRE HOSE VALVES**									
0020	Angle, combination pressure adjust/restricting, rough brass									
0030	1-1/2"	1 Spri	12	.667	Ea.	70	32		102	125
0040	2-1/2"	"	7	1.143	"	139	55		194	236
0042	Nonpressure adjustable/restricting, rough brass									
0044	1-1/2"	1 Spri	12	.667	Ea.	43.50	32		75.50	96
0046	2-1/2"	"	7	1.143	"	74	55		129	164
0050	For polished brass, add					30%				
0060	For polished chrome, add					40%				
0080	Wheel handle, 300 lb., 1-1/2"	1 Spri	12	.667	Ea.	58	32		90	112
0090	2-1/2"	"	7	1.143	"	60	55		115	149
0100	For polished brass, add					35%				
0110	For polished chrome, add					50%				
1000	Ball drip, automatic, rough brass, 1/2"	1 Spri	20	.400	Ea.	13.25	19.25		32.50	43.50
1010	3/4"	"	20	.400		15.05	19.25		34.30	45.50
1400	Caps, polished brass with chain, 3/4"					34.50			34.50	38
1420	1"					44			44	48.50
1440	1-1/2"					11.40			11.40	12.55
1460	2-1/2"					17.65			17.65	19.40
1480	3"					27.50			27.50	30.50
1900	Escutcheon plate, for angle valves, polished brass, 1-1/2"					11.05			11.05	12.15
1920	2-1/2"					26			26	29
1940	3"					24			24	26.50
1980	For polished chrome, add					15%				
3000	Gate, hose, wheel handle, N.R.S., rough brass, 1-1/2"	1 Spri	12	.667		111	32		143	170
3040	2-1/2", 300 lb.	"	7	1.143		156	55		211	254
3080	For polished brass, add					40%				
3090	For polished chrome, add					50%				
3800	Hydrant, screw type, crank handle, brass									
3840	2-1/2" size	Q-12	11	1.455	Ea.	244	63		307	365
3880	For chrome, same price									
4200	Hydrolator, vent and draining, rough brass, 1-1/2"	1 Spri	12	.667	Ea.	59	32		91	113
4280	For polished brass, add					50%				
4290	For polished chrome, add					90%				
5000	Pressure restricting, adjustable rough brass, 1-1/2"	1 Spri	12	.667		99	32		131	157
5020	2-1/2"	"	7	1.143		139	55		194	235
5080	For polished brass, add					30%				

21 12 Fire-Suppression Standpipes

21 12 23 – Fire-Suppression Hose Valves

21 12 23.70 Fire Hose Valves		Crew	Daily Output	Labor-Hours	Unit	Material	2009 Bare Costs Labor	Equipment	Total	Total Incl O&P
5090	For polished chrome, add				Ea.	45%				
8000	Wye, leader line, ball type, swivel female x male x male									
8040	2-1/2" x 1-1/2" x 1-1/2" polished brass				Ea.	214			214	236
8060	2-1/2" x 1-1/2" x 1-1/2" polished chrome				"	208			208	228

21 13 Fire-Suppression Sprinkler Systems

21 13 13 – Wet-Pipe Sprinkler Systems

21 13 13.50 Wet-Pipe Sprinkler System Components

		Crew	Daily Output	Labor-Hours	Unit	Material	2009 Bare Costs Labor	Equipment	Total	Total Incl O&P
0010	**WET-PIPE SPRINKLER SYSTEM COMPONENTS**									
1100	Alarm, electric pressure switch (circuit closer)	1 Spri	26	.308	Ea.	241	14.80		255.80	287
1140	For explosion proof, max 20 PSI, contacts close or open		26	.308		480	14.80		494.80	550
1220	Water motor, complete with gong		4	2		305	96.50		401.50	480
1800	Firecycle system, controls, includes panel,									
1820	batteries, solenoid valves and pressure switches	Q-13	1	32	Ea.	10,800	1,475		12,275	14,100
1860	Detector	1 Spri	16	.500	"	525	24		549	610
1900	Flexible sprinkler head connectors									
1910	Braided stainless steel hose with mounting bracket									
1920	1/2" and 1" outlet size									
1930	24" length	1 Spri	32	.250	Ea.	24	12.05		36.05	44.50
1940	36" length		30	.267		28	12.85		40.85	50
1950	48" length		26	.308		31.50	14.80		46.30	56.50
1960	60" length		22	.364		35.50	17.50		53	65.50
1970	72" length		18	.444		39	21.50		60.50	75
1982	May replace hard-pipe armovers									
1984	For wet, pre-action, deluge or dry pipe systems									
2000	Release, emergency, manual, for hydraulic or pneumatic system	1 Spri	12	.667	Ea.	161	32		193	225
2060	Release, thermostatic, for hydraulic or pneumatic release line		20	.400		500	19.25		519.25	580
2200	Sprinkler cabinets, 6 head capacity		16	.500		59	24		83	101
2260	12 head capacity		16	.500		62.50	24		86.50	105
2340	Sprinkler head escutcheons, standard, brass tone, 1" size		40	.200		2.55	9.65		12.20	17.25
2360	Chrome, 1" size		40	.200		3.05	9.65		12.70	17.80
2400	Recessed type, brass tone		40	.200		2.67	9.65		12.32	17.40
2440	Chrome or white enamel		40	.200		2.68	9.65		12.33	17.40
2600	Sprinkler heads, not including supply piping									
3700	Standard spray, pendent or upright, brass, 135° to 286° F									
3720	1/2" NPT, 3/8" orifice	1 Spri	16	.500	Ea.	11.45	24		35.45	48.50
3780	3/4" NPT, 17/32" orifice	"	16	.500	"	8.85	24		32.85	46
3800	For open sprinklers, deduct					15%				
3840	For chrome, add				Ea.	2.38			2.38	2.62
3860	For wax and lead coating, add					24.50			24.50	26.50
3880	For wax coating, add					14.50			14.50	15.95
3900	For lead coating, add					15.65			15.65	17.20
3920	For 360° F, same cost									
3930	For 400° F, add				Ea.	48.50			48.50	53
4280	3/4" NPT, 17/32" orifice	1 Spri	16	.500		52.50	24		76.50	93.50
4360	For satin chrome, add					1.75			1.75	1.93
4400	For 360° F, same cost									
4410	For 400° F, add				Ea.	10.90			10.90	11.95
4500	Sidewall, horizontal, brass, 135° to 286° F									
4520	1/2" NPT, 1/2" orifice	1 Spri	16	.500	Ea.	15	24		39	52.50
4540	For 360° F, same cost									

21 13 Fire-Suppression Sprinkler Systems

21 13 13 – Wet-Pipe Sprinkler Systems

21 13 13.50 Wet-Pipe Sprinkler System Components		Crew	Daily Output	Labor-Hours	Unit	Material	2009 Bare Costs Labor	Equipment	Total	Total Incl O&P
4800	Recessed pendent, brass, 135° to 286° F									
4820	1/2" NPT, 3/8" orifice	1 Spri	10	.800	Ea.	40	38.50		78.50	102
4860	1/2" NPT, 17/32" orifice	"	10	.800		33.50	38.50		72	94.50
4900	For satin chrome, add					2.65			2.65	2.92
5000	Recessed-vertical sidewall, brass, 135° to 286° F									
5020	1/2" NPT, 3/8" orifice	1 Spri	10	.800	Ea.	24	38.50		62.50	84
5030	1/2" NPT, 7/16" orifice		10	.800		24	38.50		62.50	84
5040	1/2" NPT, 1/2" orifice		10	.800		12.50	38.50		51	72
5100	For bright nickel, same cost									
5600	Concealed, complete with cover plate									
5620	1/2" NPT, 1/2" orifice, 135° F to 212° F	1 Spri	9	.889	Ea.	34	43		77	102
5800	Window, brass, 1/2" NPT, 1/4" orifice		16	.500		25.50	24		49.50	64
5820	1/2" NPT, 3/8" orifice		16	.500		25.50	24		49.50	64
5840	1/2" NPT, 1/2" orifice		16	.500		27	24		51	65.50
5860	For polished chrome, add					2.69			2.69	2.96
5880	3/4" NPT, 5/8" orifice	1 Spri	16	.500		28	24		52	67
5890	3/4 NPT, 3/4" orifice	"	16	.500		28	24		52	67
6000	Sprinkler head guards, bright zinc, 1/2" NPT					4.52			4.52	4.97
6020	Bright zinc, 3/4" NPT					4.52			4.52	4.97
6100	Sprinkler head wrenches, standard head					26			26	28.50
6120	Recessed head					34			34	37.50

21 13 16 – Dry-Pipe Sprinkler Systems

21 13 16.50 Dry-Pipe Sprinkler System Components

		Crew	Daily Output	Labor-Hours	Unit	Material	Labor	Equipment	Total	Total Incl O&P
0010	**DRY-PIPE SPRINKLER SYSTEM COMPONENTS**									
0600	Accelerator	1 Spri	8	1	Ea.	560	48		608	690
0800	Air compressor for dry pipe system, automatic, complete R211313-20									
0820	280 gal. system capacity, 3/4 HP	1 Spri	1.30	6.154	Ea.	700	296		996	1,200
0860	520 gal. system capacity, 1 HP		1.30	6.154		735	296		1,031	1,250
0910	650 gal. system capacity, 1-1/2 HP		1.30	6.154		745	296		1,041	1,275
0920	790 gal. system capacity, 2 HP		1.30	6.154		805	296		1,101	1,325
0960	Air pressure maintenance control		24	.333		256	16.05		272.05	305
1600	Dehydrator package, incl. valves and nipples		12	.667		525	32		557	630
2600	Sprinkler heads, not including supply piping									
2640	Dry, pendent, 1/2" orifice, 3/4" or 1" NPT									
2660	1/2" to 6" length	1 Spri	14	.571	Ea.	90.50	27.50		118	141
2700	15-1/4" to 18" length		14	.571		106	27.50		133.50	159
2760	33-1/4" to 36" length		13	.615		131	29.50		160.50	189
6330	Valves and components									

21 13 26 – Deluge Fire-Suppression Sprinkler Systems

21 13 26.50 Deluge Fire-Suppression Sprinkler System Components

		Crew	Daily Output	Labor-Hours	Unit	Material	Labor	Equipment	Total	Total Incl O&P
0010	**DELUGE FIRE-SUPPRESSION SPRINKLER SYSTEM COMPONENTS**									
1400	Deluge system, monitoring panel w/deluge valve & trim	1 Spri	18	.444	Ea.	8,300	21.50		8,321.50	9,175
6200	Valves and components									
7000	Deluge, assembly, incl. trim, pressure									
7020	operated relief, emergency release, gauges									
7040	2" size	Q-12	2	8	Ea.	2,000	345		2,345	2,725
7080	4" size	"	1	16	"	2,750	695		3,445	4,075

21 13 Fire-Suppression Sprinkler Systems

21 13 39 – Foam-Water Systems

21 13 39.50 Foam-Water System Components

	Foam-Water System Components	Crew	Daily Output	Labor-Hours	Unit	Material	2009 Bare Costs Labor	Equipment	Total	Total Incl O&P
0010	**FOAM-WATER SYSTEM COMPONENTS**									
2600	Sprinkler heads, not including supply piping									
3600	Foam-water, pendent or upright, 1/2" NPT	1 Spri	12	.667	Ea.	130	32		162	190

21 21 Carbon-Dioxide Fire-Extinguishing Systems

21 21 16 – Carbon-Dioxide Fire-Extinguishing Equipment

21 21 16.50 CO2 Fire Extinguishing System

		Crew	Daily Output	Labor-Hours	Unit	Material	Labor	Equipment	Total	Total Incl O&P
0010	**CO_2 FIRE EXTINGUISHING SYSTEM**									
0042	For detectors and control stations, see Div. 28 31 23.50									
0100	Control panel, single zone with batteries (2 zones det., 1 suppr.)	1 Elec	1	8	Ea.	1,350	375		1,725	2,025
0150	Multizone (4) with batteries (8 zones det., 4 suppr.)	"	.50	16		2,575	750		3,325	3,950
1000	Dispersion nozzle, CO_2, 3" x 5"	1 Plum	18	.444		52.50	21.50		74	90.50
2000	Extinguisher, CO_2 system, high pressure, 75 lb. cylinder	Q-1	6	2.667		1,000	117		1,117	1,275
2100	100 lb. cylinder	"	5	3.200		1,025	140		1,165	1,325
3000	Electro/mechanical release	L-1	4	4		131	192		323	430
3400	Manual pull station	1 Plum	6	1.333		47.50	65		112.50	150
4000	Pneumatic damper release	"	8	1		175	49		224	266

21 22 Clean-Agent Fire-Extinguishing Systems

21 22 16 – Clean-Agent Fire-Extinguishing Equipment

21 22 16.50 FM200 Fire Extinguishing System

		Crew	Daily Output	Labor-Hours	Unit	Material	Labor	Equipment	Total	Total Incl O&P
0010	**FM200 FIRE EXTINGUISHING SYSTEM**									
1100	Dispersion nozzle FM200, 1-1/2"	1 Plum	14	.571	Ea.	52.50	28		80.50	100
2400	Extinguisher, FM 200 system, filled, with mounting bracket									
2460	26 lb. container	Q-1	8	2	Ea.	1,800	88		1,888	2,100
2500	63 lb. container		6	2.667		2,800	117		2,917	3,250
2540	196 lb. container		4	4		6,100	176		6,276	6,975
6000	Average FM200 system, minimum				C.F.	1.38			1.38	1.52
6020	Maximum				"	2.75			2.75	3.03

21 31 Centrifugal Fire Pumps

21 31 13 – Electric-Drive, Centrifugal Fire Pumps

21 31 13.50 Electric-Drive Fire Pumps

		Crew	Daily Output	Labor-Hours	Unit	Material	Labor	Equipment	Total	Total Incl O&P
0010	**ELECTRIC-DRIVE FIRE PUMPS** Including controller, fittings and relief valve									
3100	250 GPM, 55 psi, 15 HP, 3,550 RPM, 2" pump	Q-13	.70	45.714	Ea.	16,000	2,100		18,100	20,700
3200	500 GPM, 50 psi, 27 HP, 1770 RPM, 4" pump		.68	47.059		19,800	2,150		21,950	25,100
3250	500 GPM, 100 psi, 47 HP, 3550 RPM, 3" pump		.66	48.485		24,000	2,225		26,225	29,700
3350	750 GPM, 50 psi, 44 HP, 1,770 RPM, 5" pump		.64	50		28,900	2,300		31,200	35,300
3450	750 GPM, 165 psi, 120 HP, 3550 RPM, 4" pump		.56	57.143		33,400	2,625		36,025	40,600
3500	1000 GPM, 50 psi, 48 HP 1770 RPM, 5" pump		.60	53.333		28,900	2,450		31,350	35,500
3550	1000 GPM, 100 psi, 86 HP, 3550 RPM, 5" pump		.54	59.259		32,400	2,725		35,125	39,800
3650	1000 GPM, 200 psi, 245 HP, 1770 RPM, 6" pump		.36	88.889		79,500	4,075		83,575	93,500
3660	1250 GPM, 75 psi, 75 HP, 1770 RPM, 5" pump		.55	58.182		29,700	2,675		32,375	36,700
3700	1500 GPM, 50 psi, 66 HP, 1770 RPM, 6" pump		.50	64		31,200	2,925		34,125	38,700
3750	1500 GPM, 100 psi, 139 HP, 1770 RPM, 6" pump		.46	69.565		35,200	3,200		38,400	43,600
3850	1500 GPM, 200 psi, 279 HP, 1770 RPM, 6" pump		.32	100		82,000	4,575		86,575	97,500
3900	2000 GPM, 100 psi, 167 HP, 1770 RPM, 6" pump		.34	94.118		45,100	4,325		49,425	56,000

21 31 Centrifugal Fire Pumps

21 31 16 – Diesel-Drive, Centrifugal Fire Pumps

21 31 16.50 Diesel-Drive Fire Pumps		Crew	Daily Output	Labor-Hours	Unit	Material	2009 Bare Costs Labor	Equipment	Total	Total Incl O&P
0010	**DIESEL-DRIVE FIRE PUMPS** Including controller, fittings and relief valve									
0050	500 GPM, 50 psi, 27 HP, 4" pump	Q-13	.64	50	Ea.	66,000	2,300		68,300	76,000
0100	500 GPM, 100 psi, 62 HP, 4" pump		.60	53.333		70,000	2,450		72,450	80,500
0200	750 GPM, 50 psi, 44 HP, 5" pump		.60	53.333		70,500	2,450		72,950	81,000
0300	750 GPM, 165 psi, 203 HP, 5" pump		.52	61.538		76,000	2,825		78,825	87,500
0350	1000 GPM, 50 psi, 48 HP, 5" pump		.58	55.172		71,000	2,525		73,525	82,000
0400	1000 GPM, 100 psi, 89 HP, 4" pump		.56	57.143		75,500	2,625		78,125	87,000
0470	1000 GPM, 200 psi, 280 HP, 5" pump		.40	80		93,500	3,675		97,175	108,500
0480	1250 GPM, 75 psi, 75 HP, 5" pump		.54	59.259		71,500	2,725		74,225	83,000
0500	1500 GPM, 50 psi, 66 HP, 6" pump		.50	64		76,500	2,925		79,425	88,500
0550	1500 GPM, 100 psi, 140 HP, 6" pump		.46	69.565		78,000	3,200		81,200	91,000
0650	1500 GPM, 200 psi, 279 HP, 6" pump		.38	84.211		138,000	3,850		141,850	158,000
0700	2,000 GPM, 100 psi, 167 HP, 6" pump		.34	94.118		80,000	4,325		84,325	94,500

Division Notes

	CREW	DAILY OUTPUT	LABOR-HOURS	UNIT	2009 BARE COSTS MAT.	LABOR	EQUIP.	TOTAL	TOTAL INCL O&P

Estimating Tips

22 10 00 Plumbing Piping and Pumps

This subdivision is primarily basic pipe and related materials. The pipe may be used by any of the mechanical disciplines, i.e., plumbing, fire protection, heating, and air conditioning.

- The labor adjustment factors listed in Subdivision 22 01 02.20 apply throughout Divisions 21, 22, and 23. CAUTION: the correct percentage may vary for the same items. For example, the percentage add for the basic pipe installation should be based on the maximum height that the craftsman must install for that particular section. If the pipe is to be located 14' above the floor but it is suspended on threaded rod from beams, the bottom flange of which is 18' high (4' rods), then the height is actually 18' and the add is 20%. The pipe coverer, however, does not have to go above the 14', and so his or her add should be 10%.
- Most pipe is priced first as straight pipe with a joint (coupling, weld, etc.) every 10' and a hanger usually every 10'. There are exceptions with hanger spacing such as for cast iron pipe (5') and plastic pipe (3 per 10'). Following each type of pipe there are several lines listing sizes and the amount to be subtracted to delete couplings and hangers. This is for pipe that is to be buried or supported together on trapeze hangers. The reason that the couplings are deleted is that these runs are usually long, and frequently longer lengths of pipe are used. By deleting the couplings, the estimator is expected to look up and add back the correct reduced number of couplings.
- When preparing an estimate, it may be necessary to approximate the fittings. Fittings usually run between 25% and 50% of the cost of the pipe. The lower percentage is for simpler runs, and the higher number is for complex areas, such as mechanical rooms.
- For historic restoration projects, the systems must be as invisible as possible, and pathways must be sought for pipes, conduit, and ductwork. While installations in accessible spaces (such as basements and attics) are relatively straightforward to estimate, labor costs may be more difficult to determine when delivery systems must be concealed.

22 40 00 Plumbing Fixtures

- Plumbing fixture costs usually require two lines: the fixture itself and its "rough-in, supply, and waste."
- In the Assemblies Section (Plumbing D2010) for the desired fixture, the System Components Group at the center of the page shows the fixture on the first line. The rest of the list (fittings, pipe, tubing, etc.) will total up to what we refer to in the Unit Price section as "Rough-in, supply, waste, and vent." Note that for most fixtures we allow a nominal 5' of tubing to reach from the fixture to a main or riser.
- Remember that gas- and oil-fired units need venting.

Reference Numbers

Reference numbers are shown in shaded boxes at the beginning of some major classifications. These numbers refer to related items in the Reference Section. The reference information may be an estimating procedure, an alternate pricing method, or technical information.

Note: Not all subdivisions listed here necessarily appear in this publication.

Note: **Trade Service,** *in part, has been used as a reference source for some of the material prices used in Division 22.*

22 05 Common Work Results for Plumbing

22 05 05 – Selective Plumbing Demolition

22 05 05.10 Plumbing Demolition

	Crew	Daily Output	Labor-Hours	Unit	Material	2009 Bare Costs Labor	Equipment	Total	Total Incl O&P
0010 **PLUMBING DEMOLITION**									
1020 Fixtures, including 10' piping									
1100 Bath tubs, cast iron	1 Plum	4	2	Ea.		97.50		97.50	146
1200 Lavatory, wall hung		10	.800			39		39	58.50
1220 Counter top		8	1			49		49	73
1300 Sink, single compartment		8	1			49		49	73
1320 Double compartment		7	1.143			55.50		55.50	83.50
1400 Water closet, floor mounted		8	1			49		49	73
1420 Wall mounted		7	1.143			55.50		55.50	83.50
1500 Urinal, floor mounted		4	2			97.50		97.50	146
1520 Wall mounted		7	1.143			55.50		55.50	83.50
1600 Water fountains, free standing		8	1			49		49	73
1620 Wall or deck mounted		6	1.333			65		65	97.50
2000 Piping, metal, up thru 1-1/2" diameter		200	.040	L.F.		1.95		1.95	2.93
2050 2" thru 3-1/2" diameter		150	.053			2.60		2.60	3.90
2100 4" thru 6" diameter	2 Plum	100	.160			7.80		7.80	11.70
2150 8" thru 14" diameter	"	60	.267			13		13	19.50
2153 16" thru 20" diameter	Q-18	70	.343			15.80	.80	16.60	24.50
2155 24" thru 26" diameter		55	.436			20	1.01	21.01	31
2156 30" thru 36" diameter		40	.600			27.50	1.39	28.89	43
2212 Deduct for salvage, aluminum scrap				Ton				1,474	1,620
2214 Brass scrap								3,440	3,780
2216 Copper scrap								5,114	5,620
2250 Water heater, 40 gal.	1 Plum	6	1.333	Ea.		65		65	97.50

22 07 Plumbing Insulation

22 07 16 – Plumbing Equipment Insulation

22 07 16.10 Insulation for Plumbing Equipment

	Crew	Daily Output	Labor-Hours	Unit	Material	2009 Bare Costs Labor	Equipment	Total	Total Incl O&P
0010 **INSULATION FOR PLUMBING EQUIPMENT**									
2900 Domestic water heater wrap kit									
2920 1-1/2" with vinyl jacket, 20-60 gal. G	1 Plum	8	1	Ea.	16.05	49		65.05	90.50

22 07 19 – Plumbing Piping Insulation

22 07 19.10 Piping Insulation

	Crew	Daily Output	Labor-Hours	Unit	Material	2009 Bare Costs Labor	Equipment	Total	Total Incl O&P
0010 **PIPING INSULATION**									
2930 Insulated protectors, (ADA)									
2935 For exposed piping under sinks or lavatories									
2940 Vinyl coated foam, velcro tabs									
2945 P Trap, 1-1/4" or 1-1/2"	1 Plum	32	.250	Ea.	18.50	12.20		30.70	39
2960 Valve and supply cover									
2965 1/2", 3/8", and 7/16" pipe size	1 Plum	32	.250	Ea.	17.60	12.20		29.80	37.50
2970 Extension drain cover									
2975 1-1/4", or 1-1/2" pipe size	1 Plum	32	.250	Ea.	19.30	12.20		31.50	40
2980 Tailpiece offset (wheelchair)									
2985 1-1/4" pipe size	1 Plum	32	.250	Ea.	21.50	12.20		33.70	42
4000 Pipe covering (price copper tube one size less than IPS)									
4280 Cellular glass, closed cell foam, all service jacket, sealant,									
4281 working temp. (-450° F to +900° F), 0 water vapor transmission									
4284 1" wall,									
4286 1/2" iron pipe size G	Q-14	120	.133	L.F.	1.82	5.30		7.12	10.30
4300 1-1/2" wall,									

22 07 Plumbing Insulation

22 07 19 – Plumbing Piping Insulation

22 07 19.10 Piping Insulation		Crew	Daily Output	Labor-Hours	Unit	Material	2009 Bare Costs Labor	Equipment	Total	Total Incl O&P
4301	1" iron pipe size G	Q-14	105	.152	L.F.	2.93	6.05		8.98	12.70
4304	2-1/2" iron pipe size G		90	.178		4.91	7.05		11.96	16.45
4306	3" iron pipe size G		85	.188		4.99	7.45		12.44	17.20
4308	4" iron pipe size G		70	.229		6.60	9.05		15.65	21.50
4310	5" iron pipe size G		65	.246		7.60	9.75		17.35	23.50
4320	2" wall,									
4322	1" iron pipe size G	Q-14	100	.160	L.F.	4.35	6.35		10.70	14.75
4324	2-1/2" iron pipe size G		85	.188		6.45	7.45		13.90	18.80
4326	3" iron pipe size G		80	.200		6.60	7.95		14.55	19.75
4328	4" iron pipe size G		65	.246		7.40	9.75		17.15	23.50
4330	5" iron pipe size G		60	.267		8.90	10.60		19.50	26.50
4332	6" iron pipe size G		50	.320		10.55	12.70		23.25	31.50
4336	8" iron pipe size G		40	.400		13.60	15.90		29.50	40
4338	10" iron pipe size G		35	.457		14.05	18.15		32.20	44
4350	2-1/2" wall,									
4360	12" iron pipe size G	Q-14	32	.500	L.F.	22.50	19.85		42.35	55.50
4362	14" iron pipe size G	"	28	.571	"	24	22.50		46.50	62
4370	3" wall,									
4378	6" iron pipe size G	Q-14	48	.333	L.F.	14.05	13.25		27.30	36
4380	8" iron pipe size G		38	.421		16.40	16.70		33.10	44
4382	10" iron pipe size G		33	.485		21.50	19.25		40.75	54
4384	16" iron pipe size G		25	.640		28	25.50		53.50	71
4386	18" iron pipe size G		22	.727		30.50	29		59.50	79
4388	20" iron pipe size G		20	.800		36	32		68	89.50
4400	3-1/2" wall,									
4412	12" iron pipe size G	Q-14	27	.593	L.F.	31.50	23.50		55	72
4414	14" iron pipe size G	"	25	.640	"	33.50	25.50		59	77
4430	4" wall,									
4446	16" iron pipe size G	Q-14	22	.727	L.F.	34	29		63	82.50
4448	18" iron pipe size G		20	.800		44	32		76	98.50
4450	20" iron pipe size G		18	.889		45.50	35.50		81	106
4480	Fittings, average with fabric and mastic									
4484	1" wall,									
4486	1/2" iron pipe size G	1 Asbe	40	.200	Ea.	3.23	8.80		12.03	17.35
4500	1-1/2" wall,									
4502	1" iron pipe size G	1 Asbe	38	.211	Ea.	5.55	9.30		14.85	20.50
4504	2-1/2" iron pipe size G		32	.250		6.90	11.05		17.95	25
4506	3" iron pipe size G		30	.267		7.60	11.75		19.35	27
4508	4" iron pipe size G		28	.286		11.70	12.60		24.30	32.50
4510	5" iron pipe size G		24	.333		12.90	14.70		27.60	37
4520	2" wall,									
4522	1" iron pipe size G	1 Asbe	36	.222	Ea.	6.45	9.80		16.25	22.50
4524	2-1/2" iron pipe size G		30	.267		7.60	11.75		19.35	27
4526	3" iron pipe size G		28	.286		9.15	12.60		21.75	30
4528	4" iron pipe size G		24	.333		12.30	14.70		27	36.50
4530	5" iron pipe size G		22	.364		15.50	16.05		31.55	42
4532	6" iron pipe size G		20	.400		18.15	17.65		35.80	47.50
4536	8" iron pipe size G		12	.667		27.50	29.50		57	76.50
4538	10" iron pipe size G		8	1		34	44		78	107
4550	2-1/2" wall,									
4560	12" iron pipe size G	1 Asbe	6	1.333	Ea.	58.50	59		117.50	157
4562	14" iron pipe size G	"	4	2	"	68	88		156	213
4570	3" wall,									

22 07 Plumbing Insulation

22 07 19 – Plumbing Piping Insulation

22 07 19.10 Piping Insulation

		Crew	Daily Output	Labor-Hours	Unit	Material	2009 Bare Costs Labor	Equipment	Total	Total Incl O&P
4578	6" iron pipe size	G 1 Asbe	16	.500	Ea.	21	22		43	57.50
4580	8" iron pipe size	G	10	.800		31	35.50		66.50	89.50
4582	10" iron pipe size	G	6	1.333		44	59		103	141
4584	16" iron pipe size	G	2	4		105	176		281	390
4586	18" iron pipe size	G	2	4		136	176		312	425
4588	20" iron pipe size	G	2	4		160	176		336	455
4600	3-1/2" wall,									
4612	12" iron pipe size	G 1 Asbe	4	2	Ea.	76	88		164	222
4614	14" iron pipe size	G "	2	4	"	84.50	176		260.50	370
4630	4" wall,									
4646	16" iron pipe size	G 1 Asbe	2	4	Ea.	126	176		302	415
4648	18" iron pipe size	G	2	4		155	176		331	445
4650	20" iron pipe size	G	2	4		180	176		356	475
4900	Calcium silicate, with cover									
5100	1" wall, 1/2" iron pipe size	G Q-14	170	.094	L.F.	2.56	3.74		6.30	8.65
5140	1" iron pipe size	G	170	.094		2.50	3.74		6.24	8.60
5170	2" iron pipe size	G	160	.100		2.98	3.97		6.95	9.50
5190	3" iron pipe size	G	150	.107		3.43	4.23		7.66	10.40
5200	4" iron pipe size	G	140	.114		4.25	4.54		8.79	11.80
5900	1-1/2" wall, 1/2" iron pipe size	G	160	.100		2.46	3.97		6.43	8.90
5920	3/4" iron pipe size	G	160	.100		2.50	3.97		6.47	8.95
5930	1" iron pipe size	G	160	.100		2.73	3.97		6.70	9.20
5940	1-1/4" iron pipe size	G	155	.103		2.94	4.10		7.04	9.65
5950	1-1/2" iron pipe size	G	155	.103		3.16	4.10		7.26	9.90
5960	2" iron pipe size	G	150	.107		3.48	4.23		7.71	10.50
5970	2-1/2" iron pipe size	G	150	.107		3.80	4.23		8.03	10.85
5980	3" iron pipe size	G	145	.110		3.97	4.38		8.35	11.20
5990	4" iron pipe size	G	135	.119		4.60	4.71		9.31	12.40
6600	Fiberglass, with all service jacket									
6640	1/2" wall, 1/2" iron pipe size	G Q-14	250	.064	L.F.	.73	2.54		3.27	4.78
6670	1" iron pipe size	G	230	.070		.86	2.76		3.62	5.30
6700	2" iron pipe size	G	210	.076		1.12	3.02		4.14	5.95
6840	1" wall, 1/2" iron pipe size	G	240	.067		.89	2.65		3.54	5.15
6870	1" iron pipe size	G	220	.073		1.04	2.89		3.93	5.65
6900	2" iron pipe size	G	200	.080		1.31	3.18		4.49	6.40
6920	3" iron pipe size	G	180	.089		1.59	3.53		5.12	7.30
6940	4" iron pipe size	G	150	.107		2.11	4.23		6.34	8.95
7879	Rubber tubing, flexible closed cell foam									
7880	3/8" wall, 1/4" iron pipe size	G 1 Asbe	120	.067	L.F.	.55	2.94		3.49	5.20
7910	1/2" iron pipe size	G	115	.070		.70	3.07		3.77	5.60
7930	1" iron pipe size	G	110	.073		.98	3.21		4.19	6.15
8100	1/2" wall, 1/4" iron pipe size	G	90	.089		.44	3.92		4.36	6.65
8130	1/2" iron pipe size	G	89	.090		.54	3.96		4.50	6.80
8150	1" iron pipe size	G	88	.091		.67	4.01		4.68	7.05
8180	2" iron pipe size	G	86	.093		1.22	4.10		5.32	7.80
8200	3" iron pipe size	G	85	.094		1.72	4.15		5.87	8.40
8220	4" iron pipe size	G	80	.100		2.55	4.41		6.96	9.70

22 13 Facility Sanitary Sewerage

22 13 19 – Sanitary Waste Piping Specialties

22 13 19.15 Sink Waste Treatment	Crew	Daily Output	Labor-Hours	Unit	Material	2009 Bare Costs Labor	Equipment	Total	Total Incl O&P
0010 **SINK WASTE TREATMENT**, System for commercial kitchens									
0100 includes clock timer, & fittings									
0200 System less chemical, wall mounted cabinet	1 Plum	16	.500	Ea.	490	24.50		514.50	575
2000 Chemical, 1 gallon, add					57			57	62.50
2100 6 gallons, add					256			256	281
2200 15 gallons, add					695			695	765
2300 30 gallons, add					1,300			1,300	1,425
2400 55 gallons, add					2,225			2,225	2,450

22 33 Electric Domestic Water Heaters

22 33 13 – Instantaneous Electric Domestic Water Heaters

22 33 13.10 Hot Water Dispensers

	Crew	Daily Output	Labor-Hours	Unit	Material	Labor	Equipment	Total	Total Incl O&P
0010 **HOT WATER DISPENSERS**									
0160 Commercial, 100 cup, 11.3 amp	1 Plum	14	.571	Ea.	400	28		428	480
3180 Household, 60 cup	"	14	.571	"	190	28		218	250

22 33 30 – Residential, Electric Domestic Water Heaters

22 33 30.13 Residential, Small-Capacity Elec. Water Heaters

	Crew	Daily Output	Labor-Hours	Unit	Material	Labor	Equipment	Total	Total Incl O&P
0010 **RESIDENTIAL, SMALL-CAPACITY ELECTRIC DOMESTIC WATER HEATERS**									
1000 Residential, electric, glass lined tank, 5 yr, 10 gal., single element	1 Plum	2.30	3.478	Ea.	315	170		485	600
1040 20 gallon, single element		2.20	3.636		360	177		537	660
1080 40 gallon, double element		2	4		500	195		695	840
1100 52 gallon, double element		2	4		560	195		755	910

22 33 33 – Light-Commercial Electric Domestic Water Heaters

22 33 33.10 Commercial Electric Water Heaters

	Crew	Daily Output	Labor-Hours	Unit	Material	Labor	Equipment	Total	Total Incl O&P
0010 **COMMERCIAL ELECTRIC WATER HEATERS**									
4000 Commercial, 100° rise. NOTE: for each size tank, a range of									
4010 heaters between the ones shown are available									
4020 Electric									
4100 5 gal., 3 kW, 12 GPH, 208V	1 Plum	2	4	Ea.	2,675	195		2,870	3,250
4120 10 gal., 6 kW, 25 GPH, 208V		2	4		2,975	195		3,170	3,575
4140 50 gal., 9 kW, 37 GPH, 208V		1.80	4.444		4,075	217		4,292	4,800
4160 50 gal., 36 kW, 148 GPH, 208V		1.80	4.444		6,325	217		6,542	7,275
4180 80 gal., 12 kW, 49 GPH, 208V		1.50	5.333		5,025	260		5,285	5,925
4200 80 gal., 36 kW, 148 GPH, 208V		1.50	5.333		7,050	260		7,310	8,150
4220 100 gal., 36 kW, 148 GPH, 208V		1.20	6.667		7,350	325		7,675	8,575
4260 150 gal., 15 kW, 61 GPH, 480V		1	8		18,000	390		18,390	20,400
4280 150 gal., 120 kW, 490 GPH, 480V		1	8		25,300	390		25,690	28,400
4300 200 gal., 15 kW, 61 GPH, 480V	Q-1	1.70	9.412		19,700	415		20,115	22,200
4320 200 gal., 120 kW, 490 GPH, 480V		1.70	9.412		26,900	415		27,315	30,100
4380 300 gal., 30 kW, 123 GPH, 480V		1.30	12.308		21,900	540		22,440	24,900
4400 300 gal., 180 kW, 738 GPH, 480V		1.30	12.308		33,300	540		33,840	37,400
4460 400 gal., 30 kW, 123 GPH, 480V		1	16		27,100	700		27,800	30,900
4480 400 gal., 210 kW, 860 GPH, 480V		1	16		37,600	700		38,300	42,500
5400 Modulating step control, 2-5 steps	1 Elec	5.30	1.509		1,975	71		2,046	2,275
5440 6-10 steps		3.20	2.500		2,550	118		2,668	3,000
5460 11-15 steps		2.70	2.963		3,100	139		3,239	3,600
5480 16-20 steps		1.60	5		3,600	235		3,835	4,300

22 34 Fuel-Fired Domestic Water Heaters

22 34 30 – Residential Gas Domestic Water Heaters

22 34 30.13 Residential, Atmos, Gas Domestic Wtr Heaters		Crew	Daily Output	Labor-Hours	Unit	Material	2009 Bare Costs Labor	Equipment	Total	Total Incl O&P
0010	**RESIDENTIAL, ATMOSPHERIC, GAS DOMESTIC WATER HEATERS**									
2000	Gas fired, foam lined tank, 10 yr, vent not incl.,									
2060	40 gallon	1 Plum	1.90	4.211	Ea.	895	205		1,100	1,300
2080	50 gallon	"	1.80	4.444	"	935	217		1,152	1,350

22 34 36 – Commercial Gas Domestic Water Heaters

22 34 36.13 Commercial, Atmos., Gas Domestic Water Htrs.

		Crew	Daily Output	Labor-Hours	Unit	Material	Labor	Equipment	Total	Total Incl O&P
0010	**COMMERCIAL, ATMOSPHERIC, GAS DOMESTIC WATER HEATERS**									
6000	Gas fired, flush jacket, std. controls, vent not incl.									
6040	75 MBH input, 73 GPH	1 Plum	1.40	5.714	Ea.	1,725	279		2,004	2,325
6060	98 MBH input, 95 GPH		1.40	5.714		3,975	279		4,254	4,800
6080	120 MBH input, 110 GPH		1.20	6.667		4,575	325		4,900	5,525
6140	155 MBH input, 150 GPH		.80	10		5,900	490		6,390	7,200
6180	200 MBH input, 192 GPH	↓	.60	13.333		6,275	650		6,925	7,900
6220	260 MBH input, 250 GPH	Q-1	.80	20		7,125	880		8,005	9,175
6260	500 MBH input, 480 GPH	"	.70	22.857		11,800	1,000		12,800	14,500
6900	For low water cutoff, add	1 Plum	8	1		275	49		324	375
6960	For bronze body hot water circulator, add	"	4	2	↓	1,575	97.50		1,672.50	1,875

22 34 36.45 Commercial Packaged Water Heater Systems

		Crew	Daily Output	Labor-Hours	Unit	Material	Labor	Equipment	Total	Total Incl O&P
0010	**COMMERCIAL PACKAGED WATER HEATER SYSTEMS**									
1000	Car wash package, continuous duty, high recovery, gas fired									
1040	100° rise, 180 MBH input, 174 GPH	1 Plum	3	2.667	Ea.	4,575	130		4,705	5,250
1080	400 MBH input, 386 GPH	"	2.50	3.200		5,625	156		5,781	6,425
3960	For unit base, add				↓	360			360	395
5000	Coin laundry units, gas fired, 100° rise									
5020	Single heater,									
5040	280 MBH input, 270 GPH	1 Plum	1.60	5	Ea.	5,350	244		5,594	6,250
5060	400 MBH input, 386 GPH	"	1.30	6.154		7,050	300		7,350	8,200
5100	605 MBH input, 584 GPH	Q-1	1.40	11.429		8,750	500		9,250	10,400
5140	1000 MBH input, 966 GPH		1	16		10,000	700		10,700	12,100
5180	1400 MBH input, 1353 GPH	↓	.70	22.857	↓	14,400	1,000		15,400	17,400
6000	Multiple heater									
6040	560 MBH input, 540 GPH	1 Plum	1	8	Ea.	8,475	390		8,865	9,875
6060	800 MBH input, 772 GPH	Q-1	1.30	12.308	"	11,000	540		11,540	12,900
6800	Standard system sizing is based on									
6820	80% of the washers operating at one time.									

22 34 46 – Oil-Fired Domestic Water Heaters

22 34 46.10 Residential Oil-Fired Water Heaters

		Crew	Daily Output	Labor-Hours	Unit	Material	Labor	Equipment	Total	Total Incl O&P
0010	**RESIDENTIAL OIL-FIRED WATER HEATERS**									
3000	Oil fired, glass lined tank, 5 yr, vent not included, 30 gallon	1 Plum	2	4	Ea.	665	195		860	1,025
3040	50 gallon		1.80	4.444		1,175	217		1,392	1,600
3060	70 gallon	↓	1.50	5.333	↓	1,725	260		1,985	2,300

22 34 46.20 Commercial Oil-Fired Water Heaters

		Crew	Daily Output	Labor-Hours	Unit	Material	Labor	Equipment	Total	Total Incl O&P
0010	**COMMERCIAL OIL-FIRED WATER HEATERS**									
8000	Oil fired, glass lined, UL listed, std. controls, vent not incl.									
8060	140 gal., 140 MBH input, 134 GPH	Q-1	2.13	7.512	Ea.	14,200	330		14,530	16,100
8120	140 gal., 270 MBH input, 259 GPH		1.20	13.333		18,600	585		19,185	21,300
8160	140 gal., 540 MBH input, 519 GPH		.96	16.667		19,900	730		20,630	23,000
8200	221 gal., 300 MBH input, 288 GPH		.88	18.182		26,800	800		27,600	30,700
8220	221 gal., 600 MBH input, 576 GPH		.86	18.605		29,900	815		30,715	34,100
8240	221 gal., 800 MBH input, 768 GPH	↓	.82	19.512		30,100	855		30,955	34,400
8280	201 gal., 1250 MBH input, 1200 GPH	Q-2	1.22	19.672	↓	31,100	895		31,995	35,600

22 34 Fuel-Fired Domestic Water Heaters

22 34 46 – Oil-Fired Domestic Water Heaters

22 34 46.20 Commercial Oil-Fired Water Heaters	Crew	Daily Output	Labor-Hours	Unit	Material	2009 Bare Costs Labor	Equipment	Total	Total Incl O&P
8900	For low water cutoff, add	1 Plum	8	1	Ea.	275	49		324	375
8960	For bronze body hot water circulator, add	"	4	2		595	97.50		692.50	800

22 35 Domestic Water Heat Exchangers

22 35 30 – Water Heating by Steam

22 35 30.10 Water Heating Transfer Package

		Crew	Daily Output	Labor-Hours	Unit	Material	Labor	Equipment	Total	Total Incl O&P
0010	**WATER HEATING TRANSFER PACKAGE**, Complete controls,									
0020	expansion tank, converter, air separator									
1000	Hot water, 180° F enter, 200° F leaving, 15# steam									
1010	One pump system, 28 GPM	Q-6	.75	32	Ea.	15,600	1,475		17,075	19,400
1020	35 GPM		.70	34.286		17,200	1,575		18,775	21,300
1040	55 GPM		.65	36.923		19,500	1,700		21,200	24,100
1060	130 GPM		.55	43.636		24,200	2,000		26,200	29,600
1080	255 GPM		.40	60		32,900	2,775		35,675	40,400
1100	550 GPM		.30	80		41,600	3,675		45,275	51,500

22 41 Residential Plumbing Fixtures

22 41 06 – Plumbing Fixtures General

22 41 06.10 Plumbing Fixture Notes

		Crew	Daily Output	Labor-Hours	Unit	Material	Labor	Equipment	Total	Total Incl O&P
0010	**PLUMBING FIXTURE NOTES**, Incl. trim fittings unless otherwise noted R224000-40									
0080	For rough-in, supply, waste, and vent, see add for each type									
0122	For electric water coolers, see Div. 22 47 16.10									
0160	For color, unless otherwise noted, add				Ea.	20%				

22 41 13 – Residential Water Closets, Urinals, and Bidets

22 41 13.10 Bidets

		Crew	Daily Output	Labor-Hours	Unit	Material	Labor	Equipment	Total	Total Incl O&P
0010	**BIDETS**									
0180	Vitreous china, with trim on fixture	Q-1	5	3.200	Ea.	650	140		790	925
0200	With trim for wall mounting	"	5	3.200		650	140		790	925
9591	For designer colors and trim add					50%				
9600	For rough-in, supply, waste and vent, add	Q-1	1.78	8.989		258	395		653	875

22 41 13.40 Water Closets

		Crew	Daily Output	Labor-Hours	Unit	Material	Labor	Equipment	Total	Total Incl O&P
0010	**WATER CLOSETS** R224000-30									
0022	For seats, see Div. 22 41 13.44									
0032	For automatic flush, see Div. 22 42 39.10 0972									
0150	Tank type, vitreous china, incl. seat, supply pipe w/stop									
0200	Wall hung									
0400	Two piece, close coupled	Q-1	5.30	3.019	Ea.	555	132		687	810
0960	For rough-in, supply, waste, vent and carrier		2.73	5.861		630	257		887	1,075
1000	Floor mounted, one piece		5.30	3.019		530	132		662	785
1020	One piece, low profile		5.30	3.019		400	132		532	645
1100	Two piece, close coupled		5.30	3.019		184	132		316	400
1961	For designer colors and trim, add					55%				
1980	For rough-in, supply, waste and vent	Q-1	3.05	5.246	Ea.	275	230		505	645

22 41 13.44 Toilet Seats

		Crew	Daily Output	Labor-Hours	Unit	Material	Labor	Equipment	Total	Total Incl O&P
0010	**TOILET SEATS**									
0100	Molded composition, white									
0150	Industrial, w/o cover, open front, regular bowl	1 Plum	24	.333	Ea.	21	16.25		37.25	47.50
0200	With self-sustaining hinge		24	.333		22.50	16.25		38.75	49

22 41 Residential Plumbing Fixtures

22 41 13 – Residential Water Closets, Urinals, and Bidets

22 41 13.44 Toilet Seats

		Crew	Daily Output	Labor-Hours	Unit	Material	2009 Bare Costs Labor	2009 Bare Costs Equipment	Total	Total Incl O&P
0220	With self-sustaining check hinge	1 Plum	24	.333	Ea.	22.50	16.25		38.75	49
0240	Extra heavy, with check hinge	↓	24	.333	↓	21	16.25		37.25	47.50
0260	Elongated bowl, same price									
0300	Junior size, w/o cover, open front	1 Plum	24	.333	Ea.	37	16.25		53.25	65
0320	Regular primary bowl, open front		24	.333		36	16.25		52.25	64
0340	Regular baby bowl, open front, check hinge		24	.333		37	16.25		53.25	65
0380	Open back & front, w/o cover, reg. or elongated bowl	↓	24	.333	↓	29	16.25		45.25	56.50
0400	Residential									
0420	Regular bowl, w/cover, closed front	1 Plum	24	.333	Ea.	29	16.25		45.25	56
0440	Open front	"	24	.333	"	26	16.25		42.25	53.50
0460	Elongated bowl, add					25%				
0500	Self-raising hinge, w/o cover, open front									
0520	Regular bowl	1 Plum	24	.333	Ea.	101	16.25		117.25	136
0540	Elongated bowl	"	24	.333	"	21	16.25		37.25	47.50
0700	Molded wood, white, with cover									
0720	Closed front, regular bowl, square back	1 Plum	24	.333	Ea.	9.90	16.25		26.15	35.50
0800	Open front	"	24	.333	"	16.45	16.25		32.70	42.50
0850	Decorator styles									
0890	Vinyl top, patterned	1 Plum	24	.333	Ea.	20.50	16.25		36.75	47
0900	Vinyl padded, plain colors, regular bowl		24	.333		20.50	16.25		36.75	47
0930	Elongated bowl	↓	24	.333	↓	21.50	16.25		37.75	48
1000	Solid plastic, white									
1030	Industrial, w/o cover, open front, regular bowl	1 Plum	24	.333	Ea.	26	16.25		42.25	53.50
1080	Extra heavy, concealed check hinge		24	.333		17.25	16.25		33.50	43.50
1100	Self-sustaining hinge		24	.333		21	16.25		37.25	47.50
1150	Elongated bowl		24	.333		29	16.25		45.25	56.50
1170	Concealed check		24	.333		17.25	16.25		33.50	43.50
1190	Self-sustaining hinge, concealed check		24	.333		47	16.25		63.25	76.50
1220	Residential, with cover, closed front, regular bowl		24	.333		41.50	16.25		57.75	70
1240	Elongated bowl		24	.333		50	16.25		66.25	80
1260	Open front, regular bowl		24	.333		36.50	16.25		52.75	64.50
1280	Elongated bowl	↓	24	.333	↓	44.50	16.25		60.75	73.50

22 41 16 – Residential Lavatories and Sinks

22 41 16.10 Lavatories

			Crew	Daily Output	Labor-Hours	Unit	Material	Labor	Equipment	Total	Total Incl O&P
0010	**LAVATORIES**, With trim, white unless noted otherwise	R224000-30									
0500	Vanity top, porcelain enamel on cast iron										
0640	33" x 19" oval	R224000-40	Q-1	6.40	2.500	Ea.	505	110		615	720
0720	19" round		"	6.40	2.500	"	218	110		328	405
0860	For color, add						25%				
1000	Cultured marble, 19" x 17", single bowl		Q-1	6.40	2.500	Ea.	169	110		279	350
1160	37" x 22", single bowl		"	6.40	2.500	"	216	110		326	400
1580	For color, same price										
1900	Stainless steel, self-rimming, 25" x 22", single bowl, ledge		Q-1	6.40	2.500	Ea.	330	110		440	525
1960	17" x 22", single bowl			6.40	2.500		320	110		430	515
2040	18-3/4" round			6.40	2.500		555	110		665	775
2600	Steel, enameled, 20" x 17", single bowl			5.80	2.759		159	121		280	355
2720	18" round		↓	5.80	2.759	↓	137	121		258	335
2860	For color, add						10%				
2900	Vitreous china, 20" x 16", single bowl		Q-1	5.40	2.963	Ea.	260	130		390	480
3020	19" round, single bowl		"	5.40	2.963	"	174	130		304	385
3560	For color, add						50%				
3580	Rough-in, supply, waste and vent for all above lavatories		Q-1	2.30	6.957	Ea.	224	305		529	705

22 41 Residential Plumbing Fixtures

22 41 16 – Residential Lavatories and Sinks

22 41 16.10 Lavatories		Crew	Daily Output	Labor-Hours	Unit	Material	2009 Bare Costs Labor	Equipment	Total	Total Incl O&P
4000	Wall hung									
4040	Porcelain enamel on cast iron, 16" x 14", single bowl	Q-1	8	2	Ea.	410	88		498	580
4180	20" x 18", single bowl		8	2		297	88		385	455
4240	22" x 19", single bowl		8	2		545	88		633	730
4580	For color, add					30%				
6000	Vitreous china, 18" x 15", single bowl with backsplash	Q-1	7	2.286	Ea.	231	100		331	405
6120	24" x 20", single bowl		7	2.286		480	100		580	675
6210	27" x 20", wheelchair type		7	2.286		365	100		465	550
6500	For color, add					30%				
6960	Rough-in, supply, waste and vent for above lavatories	Q-1	1.66	9.639	Ea.	365	425		790	1,025

22 41 16.30 Sinks

		Crew	Daily Output	Labor-Hours	Unit	Material	Labor	Equipment	Total	Total Incl O&P
0010	**SINKS**, With faucets and drain									
2000	Kitchen, counter top style, P.E. on C.I., 24" x 21" single bowl	Q-1	5.60	2.857	Ea.	272	125		397	485
2200	32" x 21" double bowl	"	4.80	3.333	"	370	146		516	630
2311	For designer colors and trim, add					50%				
3000	Stainless steel, self rimming, 19" x 18" single bowl	Q-1	5.60	2.857	Ea.	510	125		635	750
3100	25" x 22" single bowl		5.60	2.857		570	125		695	815
3200	33" x 22" double bowl		4.80	3.333		830	146		976	1,125
3300	43" x 22" double bowl		4.80	3.333		965	146		1,111	1,275
3400	22" x 43" triple bowl		4.40	3.636		975	160		1,135	1,325
4000	Steel, enameled, with ledge, 24" x 21" single bowl		5.60	2.857		145	125		270	345
4100	32" x 21" double bowl		4.80	3.333		166	146		312	400
4960	For color sinks except stainless steel, add					10%				
4980	For rough-in, supply, waste and vent, counter top sinks	Q-1	2.14	7.477		266	330		596	785
5000	Kitchen, raised deck, P.E. on C.I.									
5100	32" x 21", dual level, double bowl	Q-1	2.60	6.154	Ea.	289	270		559	725
5200	42" x 21", double bowl & disposer well	"	2.20	7.273		805	320		1,125	1,375
5700	For color, add					20%				
5790	For rough-in, supply, waste & vent, sinks	Q-1	1.85	8.649		266	380		646	865

22 41 19 – Residential Bathtubs

22 41 19.10 Baths

		Crew	Daily Output	Labor-Hours	Unit	Material	Labor	Equipment	Total	Total Incl O&P
0010	**BATHS** R224000-30									
0100	Tubs, recessed porcelain enamel on cast iron, with trim									
0180	48" x 42"	Q-1	4	4	Ea.	1,800	176		1,976	2,250
0220	72" x 36"	"	3	5.333	"	1,850	234		2,084	2,400
0300	Mat bottom									
0340	4'-6" long	Q-1	5	3.200	Ea.	920	140		1,060	1,200
0380	5' long		4.40	3.636		810	160		970	1,125
0420	5'-6" long		4	4		1,250	176		1,426	1,650
0480	Above floor drain, 5' long		4	4		800	176		976	1,150
0750	For color, add					30%				
0760	For designer colors & trim, add					60%				
2000	Enameled formed steel, 4'-6" long	Q-1	5.80	2.759	Ea.	365	121		486	580
2300	Above floor drain, 5' long	"	5.50	2.909	"	335	128		463	560
2350	For color, add					10%				
4000	Soaking, acrylic with pop-up drain, 60" x 32" x 21" deep	Q-1	5.50	2.909	Ea.	875	128		1,003	1,150
4100	60" x 48" x 18-1/2" deep		5	3.200		810	140		950	1,100
4200	72" x 45" x 21-1/2" deep		4.80	3.333		1,025	146		1,171	1,350
4311	For designer colors & trim, add					20%				
4600	Module tub & showerwall surround, molded fiberglass									
4610	5' long x 34" wide x 76" high	Q-1	4	4	Ea.	560	176		736	880
4621	For designer colors and trim add				"	25%				

22 41 Residential Plumbing Fixtures

22 41 19 – Residential Bathtubs

22 41 19.10 Baths

		Crew	Daily Output	Labor-Hours	Unit	Material	2009 Bare Costs Labor	Equipment	Total	Total Incl O&P
4750	Handicap with 1-1/2" OD grab bar, antiskid bottom									
4760	60" x 32-3/4" x 72" high	Q-1	4	4	Ea.	670	176		846	1,000
4770	60" x 30" x 71" high with molded seat		3.50	4.571		655	201		856	1,025
9600	Rough-in, supply, waste and vent, for all above tubs, add		2.07	7.729		300	340		640	840

22 41 23 – Residential Shower Receptors and Basins

22 41 23.20 Showers

		Crew	Daily Output	Labor-Hours	Unit	Material	2009 Bare Costs Labor	Equipment	Total	Total Incl O&P
0010	**SHOWERS**									
1500	Stall, with drain only. Add for valve and door/curtain									
1510	Baked enamel, molded stone receptor, 30" square	Q-1	5.20	3.077	Ea.	730	135		865	1,000
1520	32" square		5	3.200		365	140		505	610
1540	Terrazzo receptor, 32" square		5	3.200		740	140		880	1,025
1560	36" square		4.80	3.333		1,150	146		1,296	1,500
1580	36" corner angle		4.80	3.333		1,275	146		1,421	1,625
1600	For color, add					10%				
1604	For thermostatic valve add				Ea.	345			345	380
3000	Fiberglass, one piece, with 3 walls, 32" x 32" square	Q-1	5.50	2.909		460	128		588	700
3100	36" x 36" square	"	5.50	2.909		525	128		653	765
3200	Handicap, 1-1/2" O.D. grab bars, nonskid floor									
3210	48" x 34-1/2" x 72" corner seat	Q-1	5	3.200	Ea.	615	140		755	885
3220	60" x 34-1/2" x 72" corner seat		4	4		1,275	176		1,451	1,675
3230	48" x 34-1/2" x 72" fold up seat		5	3.200		2,125	140		2,265	2,550
3250	64" x 65-3/4" x 81-1/2" fold. seat, whlchr.		3.80	4.211		2,775	185		2,960	3,325
3260	For thermostatic valve add					345			345	380
4000	Polypropylene, stall only, w/ molded-stone floor, 30" x 30"	Q-1	2	8		500	350		850	1,075
4200	Rough-in, supply, waste and vent for above showers	"	2.05	7.805		395	340		735	950

22 41 23.40 Shower System Components

		Crew	Daily Output	Labor-Hours	Unit	Material	2009 Bare Costs Labor	Equipment	Total	Total Incl O&P
0010	**SHOWER SYSTEM COMPONENTS**									
4520	Showers, fiberglass receptor only	1 Plum	8	1	Ea.	187	49		236	278
5000	Built-in, head, arm, 2.5 GPM valve		4	2		208	97.50		305.50	375
5200	Head, arm, by-pass, integral stops, handles		3.60	2.222		198	108		306	380
5500	Head, water economizer	G	24	.333		49.50	16.25		65.75	79
5800	Mixing valve, built-in		6	1.333		121	65		186	231

22 41 36 – Residential Laundry Trays

22 41 36.10 Laundry Sinks

		Crew	Daily Output	Labor-Hours	Unit	Material	2009 Bare Costs Labor	Equipment	Total	Total Incl O&P
0010	**LAUNDRY SINKS**, With trim									
0020	Porcelain enamel on cast iron, black iron frame									
0050	24" x 21", single compartment	Q-1	6	2.667	Ea.	415	117		532	630
2000	Molded stone, on wall hanger or legs									
2020	22" x 23", single compartment	Q-1	6	2.667	Ea.	143	117		260	335
2100	45" x 21", double compartment	"	5	3.200	"	213	140		353	445
3000	Plastic, on wall hanger or legs									
3020	18" x 23", single compartment	Q-1	6.50	2.462	Ea.	99.50	108		207.50	272
3200	36" x 23", double compartment		5.50	2.909		158	128		286	365
5000	Stainless steel, counter top, 22" x 17" single compartment		6	2.667		58	117		175	240
5200	33" x 22", double compartment		5	3.200		70	140		210	288
9600	Rough-in, supply, waste and vent, for all laundry sinks		2.14	7.477		266	330		596	785

22 41 39 – Residential Faucets, Supplies and Trim

22 41 39.10 Faucets and Fittings

		Crew	Daily Output	Labor-Hours	Unit	Material	2009 Bare Costs Labor	Equipment	Total	Total Incl O&P
0010	**FAUCETS AND FITTINGS**									
0150	Bath, faucets, diverter spout combination, sweat	1 Plum	8	1	Ea.	104	49		153	187
0200	For integral stops, IPS unions, add					109			109	120

22 41 Residential Plumbing Fixtures

22 41 39 – Residential Faucets, Supplies and Trim

22 41 39.10 Faucets and Fittings

		Crew	Daily Output	Labor-Hours	Unit	Material	2009 Bare Costs Labor	2009 Bare Costs Equipment	Total	Total Incl O&P
0300	Three valve combinations, spout, head, arm, flange, sweat	1 Plum	6	1.333	Ea.	52.50	65		117.50	155
0400	For integral stops, IPS unions, add				Pr.	102			102	112
0420	Bath, press-bal mix valve w/diverter, spout, shower hd, arm/flange	1 Plum	8	1	Ea.	137	49		186	224
0500	Drain, central lift, 1-1/2" IPS male		20	.400		40.50	19.50		60	74
0600	Trip lever, 1-1/2" IPS male		20	.400		41	19.50		60.50	74.50
0700	Pop up, 1-1/2" IPS male		18	.444		48.50	21.50		70	86
0800	Chain and stopper, 1-1/2" IPS male		24	.333		32.50	16.25		48.75	60
0810	Bidet									
0812	Fitting, over the rim, swivel spray/pop-up drain	1 Plum	8	1	Ea.	175	49		224	265
1000	Kitchen sink faucets, top mount, cast spout		10	.800		55	39		94	119
1100	For spray, add		24	.333		16.15	16.25		32.40	42.50
1200	Wall type, swing tube spout		10	.800		74.50	39		113.50	140
1240	For soap dish, add					3.60			3.60	3.96
1250	For basket strainer w/tail piece, add					40			40	44
1300	Single control lever handle									
1310	With pull out spray									
1320	Polished chrome	1 Plum	10	.800	Ea.	192	39		231	270
1330	Polished brass		10	.800		230	39		269	310
1340	White		10	.800		211	39		250	291
1348	With spray thru escutcheon									
1350	Polished chrome	1 Plum	10	.800	Ea.	152	39		191	226
1360	Polished brass		10	.800		183	39		222	260
1370	White		10	.800		164	39		203	239
2000	Laundry faucets, shelf type, IPS or copper unions		12	.667		45.50	32.50		78	99.50
2100	Lavatory faucet, centerset, without drain		10	.800		44.50	39		83.50	108
2120	With pop-up drain		6.66	1.201		61.50	58.50		120	156
2130	For acrylic handles, add					5.15			5.15	5.65
2150	Concealed, 12" centers	1 Plum	10	.800		101	39		140	170
2160	With pop-up drain	"	6.66	1.201		118	58.50		176.50	218
2210	Porcelain cross handles and pop-up drain									
2220	Polished chrome	1 Plum	6.66	1.201	Ea.	126	58.50		184.50	227
2230	Polished brass	"	6.66	1.201	"	189	58.50		247.50	296
2260	Single lever handle and pop-up drain									
2270	Black nickel	1 Plum	6.66	1.201	Ea.	220	58.50		278.50	330
2280	Polished brass		6.66	1.201		220	58.50		278.50	330
2290	Polished chrome		6.66	1.201		169	58.50		227.50	274
2600	Shelfback, 4" to 6" centers, 17 Ga. tailpiece		10	.800		78	39		117	145
2650	With pop-up drain		6.66	1.201		95	58.50		153.50	192
2700	Shampoo faucet with supply tube		24	.333		48.50	16.25		64.75	78
2800	Self-closing, center set		10	.800		131	39		170	203
2810	Automatic sensor and operator, with faucet head	G	6.15	1.301		355	63.50		418.50	485
4000	Shower by-pass valve with union		18	.444		68.50	21.50		90	108
4100	Shower arm with flange and head		22	.364		69	17.75		86.75	103
4140	Shower, hand held, pin mount, massage action, chrome		22	.364		69	17.75		86.75	103
4142	Polished brass		22	.364		97	17.75		114.75	134
4144	Shower, hand held, wall mtd, adj. spray, 2 wall mounts, chrome		20	.400		119	19.50		138.50	160
4146	Polished brass		20	.400		242	19.50		261.50	296
4148	Shower, hand held head, bar mounted 24", adj. spray, chrome		20	.400		158	19.50		177.50	203
4150	Polished brass		20	.400		385	19.50		404.50	450
4200	Shower thermostatic mixing valve, concealed		8	1		320	49		369	425
4204	For inlet strainer, check, and stops, add					45			45	49.50
4220	Shower pressure balancing mixing valve,									
4230	With shower head, arm, flange and diverter tub spout									

22 41 Residential Plumbing Fixtures

22 41 39 – Residential Faucets, Supplies and Trim

22 41 39.10 Faucets and Fittings		Crew	Daily Output	Labor-Hours	Unit	Material	2009 Bare Costs Labor	Equipment	Total	Total Incl O&P
4240	Chrome	1 Plum	6.14	1.303	Ea.	170	63.50		233.50	283
4250	Polished brass		6.14	1.303		238	63.50		301.50	360
4260	Satin		6.14	1.303		238	63.50		301.50	360
4270	Polished chrome/brass		6.14	1.303		196	63.50		259.50	310
5000	Sillcock, compact, brass, IPS or copper to hose		24	.333		8.05	16.25		24.30	33.50
6000	Stop and waste valves, bronze									
6100	Angle, solder end 1/2"	1 Plum	24	.333	Ea.	7.05	16.25		23.30	32.50
6110	3/4"		20	.400		7.75	19.50		27.25	38
6300	Straightway, solder end 3/8"		24	.333		6.15	16.25		22.40	31.50
6310	1/2"		24	.333		2.81	16.25		19.06	27.50
6320	3/4"		20	.400		3.48	19.50		22.98	33.50
6330	1"		19	.421		19.95	20.50		40.45	53
6400	Straightway, threaded 3/8"		24	.333		7.15	16.25		23.40	32.50
6410	1/2"		24	.333		3.82	16.25		20.07	28.50
6420	3/4"		20	.400		6.35	19.50		25.85	36.50
6430	1"		19	.421		12.60	20.50		33.10	45
7800	Water closet, wax gasket		96	.083		1.42	4.06		5.48	7.65
7820	Gasket toilet tank to bowl		32	.250		1.24	12.20		13.44	19.65
8000	Water supply stops, polished chrome plate									
8200	Angle, 3/8"	1 Plum	24	.333	Ea.	7.35	16.25		23.60	32.50
8300	1/2"		22	.364		7.35	17.75		25.10	34.50
8400	Straight, 3/8"		26	.308		7.75	15		22.75	31
8500	1/2"		24	.333		7.75	16.25		24	33
8600	Water closet, angle, w/flex riser, 3/8"		24	.333		28	16.25		44.25	55.50

22 42 Commercial Plumbing Fixtures

22 42 13 – Commercial Water Closets, Urinals, and Bidets

22 42 13.30 Urinals

		Crew	Daily Output	Labor-Hours	Unit	Material	Labor	Equipment	Total	Total Incl O&P
0010	**URINALS**									
3000	Wall hung, vitreous china, with hanger & self-closing valve									
3100	Siphon jet type	Q-1	3	5.333	Ea.	248	234		482	625
3120	Blowout type		3	5.333		385	234		619	775
3300	Rough-in, supply, waste & vent		2.83	5.654		240	248		488	635
5000	Stall type, vitreous china, includes valve		2.50	6.400		575	281		856	1,050
6980	Rough-in, supply, waste and vent		1.99	8.040		350	355		705	915

22 42 13.40 Water Closets

		Crew	Daily Output	Labor-Hours	Unit	Material	Labor	Equipment	Total	Total Incl O&P
0010	**WATER CLOSETS**									
3000	Bowl only, with flush valve, seat									
3100	Wall hung	Q-1	5.80	2.759	Ea.	284	121		405	495
3200	For rough-in, supply, waste and vent, single WC		2.56	6.250		680	274		954	1,150
3300	Floor mounted		5.80	2.759		305	121		426	515
3350	With wall outlet		5.80	2.759		420	121		541	640
3400	For rough-in, supply, waste and vent, single WC		2.84	5.634		320	247		567	725
3500	Gang side by side carrier system, rough-in, supply, waste & vent									
3510	For single hook-up	Q-1	1.97	8.122	Ea.	840	355		1,195	1,450
3520	For each additional hook-up, add	"	2.14	7.477	"	790	330		1,120	1,350
3550	Gang back to back carrier system, rough-in, supply, waste & vent									
3560	For pair hook-up	Q-1	1.76	9.091	Pr.	1,300	400		1,700	2,050
3570	For each additional pair hook-up, add	"	1.81	8.840	"	1,250	390		1,640	1,950
4000	Water conserving systems									
4900	2 quart flush, round, residential [G]	Q-1	5.40	2.963	Ea.	1,275	130		1,405	1,600

22 42 Commercial Plumbing Fixtures

22 42 13 – Commercial Water Closets, Urinals, and Bidets

22 42 13.40 Water Closets		Crew	Daily Output	Labor-Hours	Unit	Material	2009 Bare Costs Labor	Equipment	Total	Total Incl O&P
5100	2 quart flush, elongated	G Q-1	4.60	3.478	Ea.	1,275	153		1,428	1,625
5140	For rough-in, supply, waste and vent	G	1.94	8.247		300	360		660	875
5200	For remote valve, add	G	24	.667		565	29.50		594.50	665
5300	For residential air compressor	G	6	2.667		730	117		847	980
5500	For heavy duty industrial air compressor	G	1	16		975	700		1,675	2,125

22 42 16 – Commercial Lavatories and Sinks

22 42 16.10 Handwasher-Dryer Module

		Crew	Daily Output	Labor-Hours	Unit	Material	Labor	Equipment	Total	Total Incl O&P
0010	**HANDWASHER-DRYER MODULE**									
0030	Wall mounted									
0040	With electric dryer									
0050	Sensor operated	Q-1	8	2	Ea.	3,650	88		3,738	4,125
0110	Sensor operated (ADA)		8	2		3,650	88		3,738	4,150
0140	Sensor operated (ADA), surface mounted		8	2		4,400	88		4,488	4,975
0150	With paper towels									
0180	Sensor operated	Q-1	8	2	Ea.	3,550	88		3,638	4,025

22 42 16.14 Lavatories

| 0010 | **LAVATORIES**, With trim, white unless noted otherwise |
| 0020 | Commercial lavatories same as residential. See Div. 22 41 16.10 |

22 42 16.34 Laboratory Countertops and Sinks

		Crew	Daily Output	Labor-Hours	Unit	Material	Labor	Equipment	Total	Total Incl O&P
0010	**LABORATORY COUNTERTOPS AND SINKS**									
0050	Laboratory sinks, corrosion resistant									
1000	Polyethylene, single sink, bench mounted, with									
1020	plug & waste fitting with 1-1/2" straight threads									
1030	2 drainboards, backnut & strainer									
1050	18-1/2" x 15-1/2" x 12-1/2" sink, 54" x 24" O.D.	Q-1	3	5.333	Ea.	1,025	234		1,259	1,500
1100	Single drainboard, backnut & strainer									
1130	18-1/2" x 15-1/2" x 12-1/2" sink, 47" x 24" O.D.	Q-1	3	5.333	Ea.	1,225	234		1,459	1,700
1150	18-1/2" x 15-1/2" x 12-1/2" sink, 70" x 24" O.D.	"	3	5.333	"	1,100	234		1,334	1,575
1290	Flanged 1-1/4" wide, rectangular with strainer									
1300	plug & waste fitting, 1-1/2" straight threads									
1320	12" x 12" x 8" sink, 14-1/2" x 14-1/2" O.D.	Q-1	4	4	Ea.	205	176		381	490
1340	16" x 16" x 8" sink, 18-1/2" x 18-1/2" O.D.		4	4		288	176		464	580
1360	21" x 18" x 10" sink, 23-1/2" x 20-1/2" O.D.		4	4		325	176		501	620
1490	For rough-in, supply, waste & vent, add		2.02	7.921		157	350		507	695
1600	Polypropylene									
1620	Cup sink, oval, integral strainers									
1640	6" x 3" I.D., 7" x 4" O.D.	Q-1	6	2.667	Ea.	105	117		222	291
1660	9" x 3" I.D., 10" x 4-1/2" O.D.	"	6	2.667		90	117		207	275
1740	1-1/2" diam. x 11" long					35			35	38.50
1980	For rough-in, supply, waste & vent, add	Q-1	1.70	9.412		150	415		565	785

22 42 16.40 Service Sinks

		Crew	Daily Output	Labor-Hours	Unit	Material	Labor	Equipment	Total	Total Incl O&P
0010	**SERVICE SINKS**									
6650	Service, floor, corner, P.E. on C.I., 28" x 28"	Q-1	4.40	3.636	Ea.	625	160		785	930
6750	Vinyl coated rim guard, add					67.50			67.50	74
6760	Mop sink, molded stone, 24" x 36"	1 Plum	3.33	2.402		248	117		365	450
6770	Mop sink, molded stone, 24" x 36", w/rim 3 sides	"	3.33	2.402		233	117		350	435
6790	For rough-in, supply, waste & vent, floor service sinks	Q-1	1.64	9.756		705	430		1,135	1,425
7000	Service, wall, P.E. on C.I., roll rim, 22" x 18"		4	4		565	176		741	890
7100	24" x 20"		4	4		625	176		801	950
7600	For stainless steel rim guard, add					105			105	116
7800	For stainless steel rim guard, front only, add					41			41	45

22 42 Commercial Plumbing Fixtures

22 42 16 – Commercial Lavatories and Sinks

22 42 16.40 Service Sinks		Crew	Daily Output	Labor-Hours	Unit	Material	2009 Bare Costs Labor	Equipment	Total	Total Incl O&P
8600	Vitreous china, 22" x 20"	Q-1	4	4	Ea.	465	176		641	780
8960	For stainless steel rim guard, front or side, add					32.50			32.50	35.50
8980	For rough-in, supply, waste & vent, wall service sinks	Q-1	1.30	12.308		940	540		1,480	1,825

22 42 23 – Commercial Shower Receptors and Basins

22 42 23.30 Group Showers

0010	**GROUP SHOWERS**									
6000	Group, w/pressure balancing valve, rough-in and rigging not included									
6800	Column, 6 heads, no receptors, less partitions	Q-1	3	5.333	Ea.	2,925	234		3,159	3,575
6900	With stainless steel partitions		1	16		8,975	700		9,675	10,900
7600	5 heads, no receptors, less partitions		3	5.333		2,400	234		2,634	3,000
7620	4 heads (1 handicap) no receptors, less partitions		3	5.333		3,850	234		4,084	4,600
7700	With stainless steel partitions		1	16		7,375	700		8,075	9,175

22 42 33 – Wash Fountains

22 42 33.20 Commercial Wash Fountains

0010	**COMMERCIAL WASH FOUNTAINS**									
1900	Group, foot control									
2000	Precast terrazzo, circular, 36" diam., 5 or 6 persons	Q-2	3	8	Ea.	4,075	365		4,440	5,025
2100	54" diameter for 8 or 10 persons		2.50	9.600		5,050	435		5,485	6,225
3000	Stainless steel, circular, 36" diameter		3.50	6.857		3,975	310		4,285	4,850
3100	54" diameter		2.80	8.571		5,150	390		5,540	6,250
5000	Thermoplastic, pre-assembled, circular, 36" diameter		6	4		3,000	182		3,182	3,550
5100	54" diameter		4	6		3,475	273		3,748	4,225
5610	Group, infrared control, barrier free									
5614	Precast terrazzo									
5620	Semi-circular 36" diam. for 3 persons	Q-2	3	8	Ea.	5,725	365		6,090	6,825
5630	46" diam. for 4 persons		2.80	8.571		6,225	390		6,615	7,425
5640	Circular, 54" diam. for 8 persons, button control		2.50	9.600		7,475	435		7,910	8,875
5700	Rough-in, supply, waste and vent for above wash fountains	Q-1	1.82	8.791		410	385		795	1,025
6200	Duo for small washrooms, stainless steel		2	8		2,625	350		2,975	3,425
6400	Bowl with backsplash		2	8		2,175	350		2,525	2,925
6500	Rough-in, supply, waste & vent for duo fountains		2.02	7.921		227	350		577	770

22 42 39 – Commercial Faucets, Supplies, and Trim

22 42 39.10 Faucets and Fittings

0010	**FAUCETS AND FITTINGS**									
0840	Flush valves, with vacuum breaker									
0850	Water closet									
0860	Exposed, rear spud	1 Plum	8	1	Ea.	130	49		179	216
0870	Top spud		8	1		120	49		169	205
0880	Concealed, rear spud		8	1		171	49		220	262
0890	Top spud		8	1		141	49		190	228
0900	Wall hung		8	1		151	49		200	239
0920	Urinal									
0930	Exposed, stall	1 Plum	8	1	Ea.	120	49		169	205
0940	Wall, (washout)		8	1		120	49		169	205
0950	Pedestal, top spud		8	1		122	49		171	208
0960	Concealed, stall		8	1		138	49		187	224
0970	Wall (washout)		8	1		148	49		197	236
0971	Automatic flush sensor and operator for									
0972	urinals or water closets	1 Plum	8	1	Ea.	405	49		454	520
3000	Service sink faucet, cast spout, pail hook, hose end	"	14	.571	"	80	28		108	130

22 43 Healthcare Plumbing Fixtures

22 43 13 – Healthcare Water Closets

22 43 13.40 Water Closets		Crew	Daily Output	Labor-Hours	Unit	Material	2009 Bare Costs Labor	Equipment	Total	Total Incl O&P
0010	**WATER CLOSETS**									
1000	Floor mounted, 1 piece, w/ seat and flush valve, ADA compliant 18" high									
1150	With wall outlet	Q-1	5.30	3.019	Ea.	365	132		497	605
1180	For rough-in, supply, waste and vent		2.84	5.634		320	247		567	725
1200	With floor outlet		5.30	3.019		274	132		406	500
1800	For rough-in, supply, waste and vent	↓	3.05	5.246	↓	275	230		505	645
3000	Bowl only, with flush valve, seat									
3100	Wall hung	Q-1	5.80	2.759	Ea.	284	121		405	495
3150	Hospital type, slotted rim for bed pan									
3160	Elongated bowl	Q-1	5.80	2.759	Ea.	315	121		436	530
3200	For rough-in, supply, waste and vent, single WC	"	2.56	6.250	"	680	274		954	1,150
3300	Floor mounted									
3360	Hospital type, slotted rim for bed pan									
3370	Elongated bowl, top spud	Q-1	5	3.200	Ea.	305	140		445	550
3380	Elongated bowl, rear spud		5	3.200		320	140		460	565
3500	For rough-in, supply, waste and vent	↓	3.05	5.246	↓	275	230		505	645

22 43 16 – Healthcare Sinks

22 43 16.10 Sinks

0010	**SINKS**									
0020	Vitreous china									
6702	Hospital type, without trim (see Div. 22 41 39.10)									
6710	20" x 18", contoured splash shield	Q-1	8	2	Ea.	130	88		218	275
6730	28" x 20", surgeon, side decks		8	2		515	88		603	700
6740	28" x 22", surgeon scrub-up, deep bowl		8	2		685	88		773	880
6750	20" x 27", patient, wheelchair		7	2.286		255	100		355	430
6760	30" x 22", all purpose		7	2.286		565	100		665	775
6770	30" x 22", plaster work		7	2.286		565	100		665	775
6820	20" x 24" clinic service, liquid/solid waste	↓	6	2.667	↓	755	117		872	1,000

22 43 19 – Healthcare Bathtubs and Showers

22 43 19.10 Bathtubs

0010	**BATHTUBS**									
5002	Hospital type, with trim (see Div. 22 41 39.10)									
5050	Bathing pool, porcelain enamel on cast iron, grab bars									
5100	Perineal (sitz), vitreous china	Q-1	3	5.333	Ea.	915	234		1,149	1,350
5120	For pedestal, vitreous china, add		8	2		135	88		223	281
5180	Pier tub, porcelain enamel on cast iron, 66-3/4" x 30"		3	5.333		2,375	234		2,609	2,950
5200	Base, porcelain enamel on cast iron		8	2		1,100	88		1,188	1,350
5300	Whirlpool, porcelain enamel on cast iron, 72" x 36"	↓	1	16	↓	2,950	700		3,650	4,300
5311	For designer colors and trim add					15%				

22 43 19.20 Showers

0010	**SHOWERS**									
5950	Module, handicap, SS panel, fixed & hand held head, control									
5960	valves, grab bar, curtain & rod, folding seat	1 Plum	4	2	Ea.	1,850	97.50		1,947.50	2,200

22 43 39 – Healthcare Faucets, Supplies, and Trim

22 43 39.10 Faucets and Fittings

0010	**FAUCETS AND FITTINGS**									
2850	Medical, bedpan cleanser, with pedal valve,	1 Plum	12	.667	Ea.	600	32.50		632.50	710
2860	With screwdriver stop valve		12	.667		310	32.50		342.50	395
2870	With self-closing spray valve		12	.667		234	32.50		266.50	305
2900	Faucet, gooseneck spout, wrist handles, grid drain		10	.800		124	39		163	196
2940	Mixing valve, knee action, screwdriver stops	↓	4	2	↓	485	97.50		582.50	680

22 45 Emergency Plumbing Fixtures

22 45 13 – Emergency Showers

22 45 13.10 Emergency Showers

		Crew	Daily Output	Labor-Hours	Unit	Material	2009 Bare Costs Labor	2009 Bare Costs Equipment	Total	Total Incl O&P
0010	**EMERGENCY SHOWERS**, Rough-in not included									
5000	Shower, single head, drench, ball valve, pull, freestanding	Q-1	4	4	Ea.	320	176		496	620
5200	Horizontal or vertical supply		4	4		460	176		636	770
6000	Multi-nozzle, eye/face wash combination		4	4		590	176		766	915
6400	Multi-nozzle, 12 spray, shower only		4	4		1,600	176		1,776	2,025
6600	For freeze-proof, add		6	2.667		335	117		452	540

22 45 16 – Eyewash Equipment

22 45 16.10 Eyewash Safety Equipment

		Crew	Daily Output	Labor-Hours	Unit	Material	Labor	Equipment	Total	Total Incl O&P
0010	**EYEWASH SAFETY EQUIPMENT**, Rough-in not included									
1000	Eye wash fountain									
1400	Plastic bowl, pedestal mounted	Q-1	4	4	Ea.	232	176		408	520
1600	Unmounted		4	4		160	176		336	440
1800	Wall mounted		4	4		166	176		342	445
2000	Stainless steel, pedestal mounted		4	4		360	176		536	660
2200	Unmounted		4	4		226	176		402	510
2400	Wall mounted		4	4		216	176		392	500

22 45 19 – Self-Contained Eyewash Equipment

22 45 19.10 Self-Contained Eyewash Safety Equipment

		Crew	Daily Output	Labor-Hours	Unit	Material	Labor	Equipment	Total	Total Incl O&P
0010	**SELF-CONTAINED EYEWASH SAFETY EQUIPMENT**									
3000	Eye wash, portable, self-contained				Ea.	455			455	500

22 45 26 – Eye/Face Wash Equipment

22 45 26.10 Eye/Face Wash Safety Equipment

		Crew	Daily Output	Labor-Hours	Unit	Material	Labor	Equipment	Total	Total Incl O&P
0010	**EYE/FACE WASH SAFETY EQUIPMENT**, Rough-in not included									
4000	Eye and face wash, combination fountain									
4200	Stainless steel, pedestal mounted	Q-1	4	4	Ea.	490	176		666	805
4400	Unmounted		4	4		345	176		521	645
4600	Wall mounted		4	4		370	176		546	670

22 46 Security Plumbing Fixtures

22 46 13 – Security Water Closets and Urinals

22 46 13.10 Security Water Closets and Urinals

		Crew	Daily Output	Labor-Hours	Unit	Material	Labor	Equipment	Total	Total Incl O&P
0010	**SECURITY WATER CLOSETS AND URINALS**, Stainless steel									
2000	Urinal, back supply and flush									
2200	Wall hung	Q-1	4	4	Ea.	1,200	176		1,376	1,575
2240	Stall		2.50	6.400		1,350	281		1,631	1,925
2300	For urinal rough-in, supply, waste and vent		1.49	10.738		320	470		790	1,050
3000	Water closet, integral seat, back supply and flush									
3300	Wall hung, wall outlet	Q-1	5.80	2.759	Ea.	935	121		1,056	1,200
3400	Floor mount, wall outlet		5.80	2.759		1,075	121		1,196	1,350
3440	Floor mount, floor outlet		5.80	2.759		1,175	121		1,296	1,475
3480	For recessed tissue holder, add					90			90	99
3500	For water closet rough-in, supply, waste and vent	Q-1	1.19	13.445		295	590		885	1,200
5000	Water closet and lavatory units, push button filler valves,									
5010	soap & paper holders, seat									
5300	Wall hung	Q-1	5	3.200	Ea.	2,200	140		2,340	2,625
5400	Floor mount		5	3.200		2,100	140		2,240	2,500
6300	For unit rough-in, supply, waste and vent		1	16		325	700		1,025	1,400

22 46 Security Plumbing Fixtures

22 46 16 – Security Lavatories and Sinks

22 46 16.10 Security Lavatories		Crew	Daily Output	Labor-Hours	Unit	Material	2009 Bare Costs Labor	Equipment	Total	Total Incl O&P
0010	**SECURITY LAVATORIES,** Stainless steel									
1000	Lavatory, wall hung, push button filler valve									
1100	Rectangular bowl	Q-1	8	2	Ea.	990	88		1,078	1,225
1200	Oval bowl		8	2		955	88		1,043	1,175
1300	For lavatory rough-in, supply, waste and vent	↓	1.50	10.667	↓	273	470		743	1,000

22 46 63 – Security Service Sink

22 46 63.10 Security Service Sink										
0010	**SECURITY SERVICE SINK,** Stainless steel									
1700	Service sink, with soap dish									
1740	24" x 19" size	Q-1	3	5.333	Ea.	1,650	234		1,884	2,150
1790	For sink rough-in, supply, waste and vent	"	.89	17.978	"	630	790		1,420	1,875

22 46 73 – Security Shower

22 46 73.10 Security Shower										
0010	**SECURITY SHOWER,** Stainless steel									
1800	Shower cabinet, unitized									
1840	36" x 36" x 88"	Q-1	2.20	7.273	Ea.	4,400	320		4,720	5,300
1900	Shower package for built-in									
1940	Hot & cold valves, recessed soap dish	Q-1	6	2.667	Ea.	555	117		672	785

22 47 Drinking Fountains and Water Coolers

22 47 13 – Drinking Fountains

22 47 13.10 Drinking Water Fountains

			Crew	Daily Output	Labor-Hours	Unit	Material	Labor	Equipment	Total	Total Incl O&P
0010	**DRINKING WATER FOUNTAINS,** For connection to cold water supply	R224000-30									
1000	Wall mounted, non-recessed										
1200	Aluminum,	R224000-40									
1280	Dual bubbler type		1 Plum	3.20	2.500	Ea.	1,950	122		2,072	2,325
1400	Bronze, with no back	R224000-50		4	2		935	97.50		1,032.50	1,175
1600	Cast iron, enameled, low back, single bubbler			4	2		580	97.50		677.50	785
1800	Cast aluminum, enameled, for correctional institutions			4	2		760	97.50		857.50	985
2000	Fiberglass, 12" back, single bubbler unit			4	2		1,150	97.50		1,247.50	1,425
2200	Polymarble, no back, single bubbler			4	2		610	97.50		707.50	815
2400	Precast stone, no back			4	2		615	97.50		712.50	825
2700	Stainless steel, single bubbler, no back			4	2		1,200	97.50		1,297.50	1,475
2740	With back			4	2		2,000	97.50		2,097.50	2,350
2780	Dual handle & wheelchair projection type			4	2		625	97.50		722.50	830
2820	Dual level for handicapped type			3.20	2.500		1,275	122		1,397	1,575
2840	Vandal resistant type		↓	4	2	↓	560	97.50		657.50	760
3300	Vitreous china										
3340	7" back		1 Plum	4	2	Ea.	510	97.50		607.50	705
3940	For vandal-resistant bottom plate, add						59.50			59.50	65.50
3980	For rough-in, supply and waste, add		1 Plum	2.21	3.620	↓	171	176		347	455
4000	Wall mounted, semi-recessed										
4200	Poly-marble, single bubbler		1 Plum	4	2	Ea.	780	97.50		877.50	1,000
4600	Stainless steel, satin finish, single bubbler			4	2		905	97.50		1,002.50	1,150
4900	Vitreous china, single bubbler			4	2		610	97.50		707.50	815
5980	For rough-in, supply and waste, add		↓	1.83	4.372	↓	171	213		384	510
6000	Wall mounted, fully recessed										
6400	Poly-marble, single bubbler		1 Plum	4	2	Ea.	890	97.50		987.50	1,125
6800	Stainless steel, single bubbler		↓	4	2	↓	875	97.50		972.50	1,100

22 47 Drinking Fountains and Water Coolers

22 47 13 – Drinking Fountains

22 47 13.10 Drinking Water Fountains

		Crew	Daily Output	Labor-Hours	Unit	Material	2009 Bare Costs Labor	Equipment	Total	Total Incl O&P
7580	For rough-in, supply and waste, add	1 Plum	1.83	4.372	Ea.	171	213		384	510
7600	Floor mounted, pedestal type									
7700	Aluminum, architectural style, C.I. base	1 Plum	2	4	Ea.	530	195		725	875
7780	Wheelchair handicap unit		2	4		1,300	195		1,495	1,750
8000	Bronze, architectural style		2	4		1,375	195		1,570	1,800
8040	Enameled steel cylindrical column style		2	4		1,150	195		1,345	1,575
8200	Precast stone/concrete, cylindrical column		1	8		970	390		1,360	1,650
8240	Wheelchair handicap unit		1	8		1,275	390		1,665	1,975
8400	Stainless steel, architectural style		2	4		1,375	195		1,570	1,800
8600	Enameled iron, heavy duty service, 2 bubblers		2	4		1,275	195		1,470	1,700
8900	For rough-in, supply and waste, add		1.83	4.372		171	213		384	510
9100	Deck mounted									
9500	Stainless steel, circular receptor	1 Plum	4	2	Ea.	350	97.50		447.50	530
9540	14" x 9" receptor		4	2		289	97.50		386.50	465
9760	White enameled steel, 14" x 9" receptor		4	2		300	97.50		397.50	475
9980	For rough-in, supply and waste, add		1.83	4.372		171	213		384	510

22 47 16 – Pressure Water Coolers

22 47 16.10 Electric Water Coolers

		Crew	Daily Output	Labor-Hours	Unit	Material	2009 Bare Costs Labor	Equipment	Total	Total Incl O&P
0010	**ELECTRIC WATER COOLERS** R224000-30									
0100	Wall mounted, non-recessed									
0180	8.2 GPH	Q-1	4	4	Ea.	645	176		821	975
0600	For hot and cold water, add					165			165	182
0640	For stainless steel cabinet, add					84			84	92
1000	Dual height, 8.2 GPH	Q-1	3.80	4.211		930	185		1,115	1,300
1240	For stainless steel cabinet, add					131			131	144
2600	Wheelchair type, 8 GPH	Q-1	4	4		1,650	176		1,826	2,075
3000	Simulated recessed, 8 GPH	"	4	4		845	176		1,021	1,200
3200	For glass filler, add					84			84	92
3240	For stainless steel cabinet, add					58			58	64
3300	Semi-recessed, 8.1 GPH	Q-1	4	4		965	176		1,141	1,325
3340	For glass filler, add					84			84	92
3360	For stainless steel cabinet, add					58			58	64
3400	Full recessed, stainless steel, 8 GPH	Q-1	3.50	4.571		1,525	201		1,726	1,975
3460	For glass filler, add					154			154	169
3600	For mounting can only					168			168	185
4600	Floor mounted, flush-to-wall									
4680	8.2 GPH	1 Plum	3	2.667	Ea.	695	130		825	960
4960	14 GPH hot and cold water	"	3	2.667		1,000	130		1,130	1,300
4980	For stainless steel cabinet, add					128			128	141
5000	Dual height, 8.2 GPH	1 Plum	2	4		1,000	195		1,195	1,400
5120	For stainless steel cabinet, add					168			168	185
5600	Explosion Proof, 16 GPH	1 Plum	3	2.667		1,475	130		1,605	1,825
5640	For stainless steel cabinet, add					110			110	121
6000	Refrigerator Compartment Type, 4.5 GPH	1 Plum	3	2.667		1,275	130		1,405	1,600
9800	For supply, waste & vent, all coolers	"	2.21	3.620		171	176		347	455

22 51 Swimming Pool Plumbing Systems

22 51 19 – Swimming Pool Water Treatment Equipment

22 51 19.50 Swimming Pool Filtration Equipment		Crew	Daily Output	Labor-Hours	Unit	Material	2009 Bare Costs Labor	Equipment	Total	Total Incl O&P
0010	**SWIMMING POOL FILTRATION EQUIPMENT**									
0900	Filter system, sand or diatomite type, incl. pump, 6,000 gal./hr.	2 Plum	1.80	8.889	Total	1,475	435		1,910	2,250
1020	Add for chlorination system, 800 S.F. pool		3	5.333	Ea.	78.50	260		338.50	475
1040	5,000 S.F. pool		3	5.333	"	1,950	260		2,210	2,525

22 52 Fountain Plumbing Systems

22 52 16 – Fountain Pumps

22 52 16.10 Fountain Water Pumps

		Crew	Daily Output	Labor-Hours	Unit	Material	Labor	Equipment	Total	Total Incl O&P
0010	**FOUNTAIN WATER PUMPS**									
0100	Pump w/controls									
0200	Single phase, 100' cord, 1/2 H.P. pump	2 Skwk	4.40	3.636	Ea.	1,800	149		1,949	2,200
0300	3/4 H.P. pump		4.30	3.721		2,050	152		2,202	2,475
0400	1 H.P. pump		4.20	3.810		3,800	156		3,956	4,425
0500	1-1/2 H.P. pump		4.10	3.902		3,500	159		3,659	4,100
0600	2 H.P. pump		4	4		4,725	163		4,888	5,450
0700	Three phase, 200' cord, 5 H.P. pump		3.90	4.103		5,000	168		5,168	5,750
0800	7-1/2 H.P. pump		3.80	4.211		8,500	172		8,672	9,625
0900	10 H.P. pump		3.70	4.324		9,275	177		9,452	10,500
1000	15 H.P. pump		3.60	4.444		11,300	182		11,482	12,800

22 52 33 – Fountain Ancillary

22 52 33.10 Fountain Miscellaneous

		Crew	Daily Output	Labor-Hours	Unit	Material	Labor	Equipment	Total	Total Incl O&P
0010	**FOUNTAIN MISCELLANEOUS**									
1100	Nozzles, minimum	2 Skwk	8	2	Ea.	150	81.50		231.50	292
1200	Maximum		8	2		300	81.50		381.50	455
1300	Lights w/mounting kits, 200 watt		18	.889		1,000	36.50		1,036.50	1,150
1400	300 watt		18	.889		1,200	36.50		1,236.50	1,375
1500	500 watt		18	.889		1,300	36.50		1,336.50	1,475
1600	Color blender		12	1.333		530	54.50		584.50	670

Division Notes

Estimating Tips

The labor adjustment factors listed in Subdivision 22 01 02.20 also apply to Division 23.

23 10 00 Facility Fuel Systems

- The prices in this subdivision for above- and below-ground storage tanks do not include foundations or hold-down slabs. The estimator should refer to Divisions 3 and 31 for foundation system pricing. In addition to the foundations, required tank accessories, such as tank gauges, leak detection devices, and additional manholes and piping, must be added to the tank prices.

23 50 00 Central Heating Equipment

- When estimating the cost of an HVAC system, check to see who is responsible for providing and installing the temperature control system. It is possible to overlook controls, assuming that they would be included in the electrical estimate.
- When looking up a boiler, be careful on specified capacity. Some manufacturers rate their products on output while others use input.
- Include HVAC insulation for pipe, boiler, and duct (wrap and liner).
- Be careful when looking up mechanical items to get the correct pressure rating and connection type (thread, weld, flange).

23 70 00 Central HVAC Equipment

- Combination heating and cooling units are sized by the air conditioning requirements. (See Reference No. R236000-20 for preliminary sizing guide.)
- A ton of air conditioning is nominally 400 CFM.
- Rectangular duct is taken off by the linear foot for each size, but its cost is usually estimated by the pound. Remember that SMACNA standards now base duct on internal pressure.
- Prefabricated duct is estimated and purchased like pipe: straight sections and fittings.
- Note that cranes or other lifting equipment are not included on any lines in Division 23. For example, if a crane is required to lift a heavy piece of pipe into place high above a gym floor, or to put a rooftop unit on the roof of a four-story building, etc., it must be added. Due to the potential for extreme variation—from nothing additional required to a major crane or helicopter—we feel that including a nominal amount for "lifting contingency" would be useless and detract from the accuracy of the estimate. When using equipment rental cost data from RSMeans, do not forget to include the cost of the operator(s).

Reference Numbers

Reference numbers are shown in shaded boxes at the beginning of some major classifications. These numbers refer to related items in the Reference Section. The reference information may be an estimating procedure, an alternate pricing method, or technical information.

Note: Not all subdivisions listed here necessarily appear in this publication.

Note: **Trade Service,** *in part, has been used as a reference source for some of the material prices used in Division 23.*

23 05 Common Work Results for HVAC

23 05 05 – Selective HVAC Demolition

23 05 05.10 HVAC Demolition

		Crew	Daily Output	Labor-Hours	Unit	Material	2009 Bare Costs Labor	2009 Bare Costs Equipment	Total	Total Incl O&P
0010	**HVAC DEMOLITION**									
0100	Air conditioner, split unit, 3 ton	Q-5	2	8	Ea.		355		355	535
0150	Package unit, 3 ton	Q-6	3	8	"		370		370	555
0298	Boilers									
0300	Electric, up thru 148 kW	Q-19	2	12	Ea.		545		545	815
0310	150 thru 518 kW	"	1	24			1,075		1,075	1,625
0320	550 thru 2000 kW	Q-21	.40	80			3,700		3,700	5,550
0330	2070 kW and up	"	.30	106			4,950		4,950	7,400
0340	Gas and/or oil, up thru 150 MBH	Q-7	2.20	14.545			685		685	1,025
0350	160 thru 2000 MBH		.80	40			1,875		1,875	2,825
0360	2100 thru 4500 MBH		.50	64			3,000		3,000	4,525
0370	4600 thru 7000 MBH		.30	106			5,025		5,025	7,525
0380	7100 thru 12,000 MBH		.16	200			9,400		9,400	14,100
0390	12,200 thru 25,000 MBH		.12	266			12,500		12,500	18,800
1000	Ductwork, 4" high, 8" wide	1 Clab	200	.040	L.F.		1.26		1.26	1.96
1100	6" high, 8" wide		165	.048			1.53		1.53	2.38
1200	10" high, 12" wide		125	.064			2.02		2.02	3.14
3000	Mechanical equipment, light items. Unit is weight, not cooling.	Q-5	.90	17.778	Ton		790		790	1,175
3600	Heavy items	"	1.10	14.545	"		645		645	970
3700	Deduct for salvage (when applicable), minimum				Job				73	80
3710	Maximum				"				455	500

23 05 93 – Testing, Adjusting, and Balancing for HVAC

23 05 93.10 Balancing, Air

		Crew	Daily Output	Labor-Hours	Unit	Material	2009 Bare Costs Labor	2009 Bare Costs Equipment	Total	Total Incl O&P
0010	**BALANCING, AIR**, (Subcontractor's quote incl. material and labor.)									
0900	Heating and ventilating equipment									
1000	Centrifugal fans, utility sets				Ea.				317.34	348.72
1100	Heating and ventilating unit								476	523.08
1200	In-line fan								476	523.08
1300	Propeller and wall fan								89.91	98.80
1400	Roof exhaust fan								211.56	232.48
2000	Air conditioning equipment, central station								687.56	755.56
2100	Built-up low pressure unit								634.67	697.44
2200	Built-up high pressure unit								740.45	813.68
2500	Multi-zone A.C. and heating unit								476	523.08
2600	For each zone over one, add								105.78	116.24
2700	Package A.C. unit								264.45	290.60
2800	Rooftop heating and cooling unit								370.22	406.84
3000	Supply, return, exhaust, registers & diffusers, avg. height ceiling								63.47	69.74
3100	High ceiling								95.20	104.62
3200	Floor height								52.89	58.12

23 05 93.20 Balancing, Water

		Crew	Daily Output	Labor-Hours	Unit	Material	2009 Bare Costs Labor	2009 Bare Costs Equipment	Total	Total Incl O&P
0010	**BALANCING, WATER**, (Subcontractor's quote incl. material and labor.)									
0050	Air cooled condenser				Ea.				188.68	207.34
0080	Boiler								379.52	417.05
0100	Cabinet unit heater								64.69	71.09
0200	Chiller								458.22	503.54
0300	Convector								53.91	59.24
0400	Converter								269.54	296.20
0500	Cooling tower								350.40	385.06
0600	Fan coil unit, unit ventilator								97.04	106.63
0700	Fin tube and radiant panels								107.82	118.48
0800	Main and duct re-heat coils								99.73	109.59

23 05 Common Work Results for HVAC

23 05 93 – Testing, Adjusting, and Balancing for HVAC

23 05 93.20 Balancing, Water		Crew	Daily Output	Labor-Hours	Unit	Material	2009 Bare Costs Labor	Equipment	Total	Total Incl O&P
0810	Heat exchanger				Ea.				99.73	109.59
1000	Pumps								237.20	260.66
1100	Unit heater								75.47	82.94

23 07 HVAC Insulation

23 07 13 – Duct Insulation

23 07 13.10 Duct Thermal Insulation

			Crew	Daily Output	Labor-Hours	Unit	Material	Labor	Equipment	Total	Total Incl O&P
0010	**DUCT THERMAL INSULATION**										
3000	Ductwork										
3020	Blanket type, fiberglass, flexible										
3140	FSK vapor barrier wrap, .75 lb. density										
3160	1" thick	G	Q-14	350	.046	S.F.	.17	1.81		1.98	3.03
3170	1-1/2" thick	G		320	.050		.20	1.99		2.19	3.33
3180	2" thick	G		300	.053		.26	2.12		2.38	3.61
3190	3" thick	G		260	.062		.39	2.44		2.83	4.26
3200	4" thick	G		242	.066		.56	2.63		3.19	4.73
3212	Vinyl Jacket, .75 lb. density, 1-1/2" thick	G		320	.050		.20	1.99		2.19	3.33
3280	Unfaced, 1 lb. density										
3310	1" thick	G	Q-14	360	.044	S.F.	.16	1.76		1.92	2.94
3320	1-1/2" thick	G		330	.048		.19	1.92		2.11	3.23
3330	2" thick	G		310	.052		.26	2.05		2.31	3.50
3400	FSK facing, 1 lb. density										
3420	1-1/2" thick	G	Q-14	310	.052	S.F.	.17	2.05		2.22	3.40
3430	2" thick	G		300	.053		.19	2.12		2.31	3.53
3470	1-1/2" thick	G		300	.053		.25	2.12		2.37	3.60
3480	2" thick	G		290	.055		.32	2.19		2.51	3.78
3795	Finishes										
3800	Stainless steel woven mesh		Q-14	100	.160	S.F.	.84	6.35		7.19	10.85
3820	18 oz. fiberglass cloth, pasted on			170	.094		.77	3.74		4.51	6.70
3900	8 oz. canvas, pasted on			180	.089		.49	3.53		4.02	6.10
9600	Minimum labor/equipment charge		1 Stpi	4	2	Job		98.50		98.50	148

23 07 16 – HVAC Equipment Insulation

23 07 16.10 HVAC Equipment Thermal Insulation

			Crew	Daily Output	Labor-Hours	Unit	Material	Labor	Equipment	Total	Total Incl O&P
0010	**HVAC EQUIPMENT THERMAL INSULATION**										
0100	Rule of thumb, as a percentage of total mechanical costs					Job				10%	
0110	Insulation req'd is based on the surface size/area to be covered										
1000	Boiler, 1-1/2" calcium silicate only	G	Q-14	110	.145	S.F.	2.65	5.75		8.40	11.95
1020	Plus 2" fiberglass	G	"	80	.200	"	3.51	7.95		11.46	16.30
2000	Breeching, 2" calcium silicate										
2020	Rectangular	G	Q-14	42	.381	S.F.	5.20	15.10		20.30	29
2040	Round	G	"	38.70	.413	"	5.45	16.40		21.85	31.50
2300	Calcium silicate block, + 200° F to + 1200° F										
2310	On irregular surfaces, valves and fittings										
2340	1" thick	G	Q-14	30	.533	S.F.	2.35	21		23.35	35.50
2360	1-1/2" thick	G		25	.640		2.65	25.50		28.15	43
2380	2" thick	G		22	.727		3.48	29		32.48	49
2400	3" thick	G		18	.889		5.35	35.50		40.85	61.50
2410	On plane surfaces										
2420	1" thick	G	Q-14	126	.127	S.F.	2.35	5.05		7.40	10.50
2430	1-1/2" thick	G		120	.133		2.65	5.30		7.95	11.20

23 07 HVAC Insulation

23 07 16 – HVAC Equipment Insulation

23 07 16.10 HVAC Equipment Thermal Insulation

	23 07 16.10 HVAC Equipment Thermal Insulation		Crew	Daily Output	Labor-Hours	Unit	Material	2009 Bare Costs Labor	Equipment	Total	Total Incl O&P
2440	2" thick	G	Q-14	100	.160	S.F.	3.48	6.35		9.83	13.80
2450	3" thick	G	↓	70	.229	↓	5.35	9.05		14.40	20
9610	Minimum labor/equipment charge		1 Stpi	4	2	Job		98.50		98.50	148

23 09 Instrumentation and Control for HVAC

23 09 33 – Electric and Electronic Control System for HVAC

23 09 33.10 Electronic Control Systems

		Crew	Daily Output	Labor-Hours	Unit	Material	Labor	Equipment	Total	Total Incl O&P
0010	**ELECTRONIC CONTROL SYSTEMS**									
0020	For electronic costs, add to Div. 23 09 43.10				Ea.					15%

23 09 43 – Pneumatic Control System for HVAC

23 09 43.10 Pneumatic Control Systems

			Crew	Daily Output	Labor-Hours	Unit	Material	Labor	Equipment	Total	Total Incl O&P
0010	**PNEUMATIC CONTROL SYSTEMS**										
0011	Including a nominal 50 Ft. of tubing. Add control panelboard if req'd.										
0100	Heating and ventilating, split system										
0200	Mixed air control, economizer cycle, panel readout, tubing										
0220	Up to 10 tons	G	Q-19	.68	35.294	Ea.	3,575	1,600		5,175	6,325
0240	For 10 to 20 tons	G		.63	37.915		3,825	1,725		5,550	6,775
0260	For over 20 tons	G		.58	41.096		4,150	1,850		6,000	7,325
0270	Enthalpy cycle, up to 10 tons			.50	48.387		3,950	2,200		6,150	7,625
0280	For 10 to 20 tons			.46	52.174		4,250	2,350		6,600	8,200
0290	For over 20 tons		↓	.42	56.604	↓	4,625	2,575		7,200	8,900
0300	Heating coil, hot water, 3 way valve,										
0320	Freezestat, limit control on discharge, readout		Q-5	.69	23.088	Ea.	2,650	1,025		3,675	4,475
0500	Cooling coil, chilled water, room										
0520	Thermostat, 3 way valve		Q-5	2	8	Ea.	1,175	355		1,530	1,825
0600	Cooling tower, fan cycle, damper control,										
0620	Control system including water readout in/out at panel		Q-19	.67	35.821	Ea.	4,700	1,625		6,325	7,600
1000	Unit ventilator, day/night operation,										
1100	freezestat, ASHRAE, cycle 2		Q-19	.91	26.374	Ea.	2,600	1,200		3,800	4,625
2000	Compensated hot water from boiler, valve control,										
2100	readout and reset at panel, up to 60 GPM		Q-19	.55	43.956	Ea.	4,875	2,000		6,875	8,325
2120	For 120 GPM			.51	47.059		5,200	2,125		7,325	8,925
2140	For 240 GPM			.49	49.180		5,450	2,225		7,675	9,325
3000	Boiler room combustion air, damper to 5 SF, controls			1.37	17.582		2,350	795		3,145	3,775
3500	Fan coil, heating and cooling valves, 4 pipe control system			3	8		1,050	360		1,410	1,725
3600	Heat exchanger system controls		↓	.86	27.907	↓	2,275	1,275		3,550	4,400
3900	Multizone control (one per zone), includes thermostat, damper										
3910	motor and reset of discharge temperature		Q-5	.51	31.373	Ea.	2,350	1,400		3,750	4,675
4000	Pneumatic thermostat, including controlling room radiator valve		"	2.43	6.593		705	293		998	1,225
4040	Program energy saving optimizer	G	Q-19	1.21	19.786		5,875	895		6,770	7,800
4060	Pump control system		"	3	8		1,075	360		1,435	1,750
4080	Reheat coil control system, not incl coil		Q-5	2.43	6.593	↓	920	293		1,213	1,450
4500	Air supply for pneumatic control system										
4600	Tank mounted duplex compressor, starter, alternator,										
4620	piping, dryer, PRV station and filter										
4630	1/2 HP		Q-19	.68	35.139	Ea.	8,725	1,600		10,325	12,000
4660	1-1/2 HP			.58	41.739		10,700	1,900		12,600	14,500
4690	5 HP		↓	.42	57.143	↓	25,300	2,575		27,875	31,800

23 09 Instrumentation and Control for HVAC

23 09 53 – Pneumatic and Electric Control System for HVAC

23 09 53.10 Control Components		Crew	Daily Output	Labor-Hours	Unit	Material	2009 Bare Costs Labor	Equipment	Total	Total Incl O&P
0010	**CONTROL COMPONENTS**									
5000	Thermostats									
5030	Manual	1 Shee	8	1	Ea.	27	47		74	102
5040	1 set back, electric, timed G		8	1		95	47		142	177
5050	2 set back, electric, timed G		8	1		209	47		256	300
5100	Locking cover					21			21	23
5200	24 hour, automatic, clock G	1 Shee	8	1		115	47		162	200
5220	Electric, low voltage, 2 wire	1 Elec	13	.615		18	29		47	63

23 21 Hydronic Piping and Pumps

23 21 20 – Hydronic HVAC Piping Specialties

23 21 20.18 Automatic Air Vent

		Crew	Daily Output	Labor-Hours	Unit	Material	Labor	Equipment	Total	Total Incl O&P
0010	**AUTOMATIC AIR VENT**									
0020	Cast iron body, stainless steel internals, float type									
0060	1/2" NPT inlet, 300 psi	1 Stpi	12	.667	Ea.	88	33		121	146
0220	3/4" NPT inlet, 250 psi	"	10	.800		275	39.50		314.50	360
0340	1-1/2" NPT inlet, 250 psi	Q-5	12	1.333		870	59		929	1,050

23 21 20.42 Expansion Joints

		Crew	Daily Output	Labor-Hours	Unit	Material	Labor	Equipment	Total	Total Incl O&P
0010	**EXPANSION JOINTS**									
0100	Bellows type, neoprene cover, flanged spool									
0140	6" face to face, 1-1/4" diameter	1 Stpi	11	.727	Ea.	248	36		284	325
0200	3" diameter	Q-5	11.40	1.404	"	291	62.50		353.50	415

23 21 20.58 Hydronic Heating Control Valves

		Crew	Daily Output	Labor-Hours	Unit	Material	Labor	Equipment	Total	Total Incl O&P
0010	**HYDRONIC HEATING CONTROL VALVES**									
0050	Hot water, nonelectric, thermostatic									
0100	Radiator supply, 1/2" diameter	1 Stpi	24	.333	Ea.	52.50	16.45		68.95	82
0120	3/4" diameter		20	.400		54	19.75		73.75	89
0140	1" diameter		19	.421		68.50	21		89.50	107
0160	1-1/4" diameter		15	.533		98	26.50		124.50	148
0500	For low pressure steam, add					25%				

23 21 20.70 Steam Traps

		Crew	Daily Output	Labor-Hours	Unit	Material	Labor	Equipment	Total	Total Incl O&P
0010	**STEAM TRAPS**									
0030	Cast iron body, threaded									
1000	Float & thermostatic, 15 psi									
1010	3/4" pipe size	1 Stpi	16	.500	Ea.	103	24.50		127.50	150
1020	1" pipe size		15	.533		124	26.50		150.50	176
1030	1-1/4" pipe size		13	.615		150	30.50		180.50	211
1040	1-1/2" pipe size		9	.889		218	44		262	305
1060	2" pipe size		6	1.333		400	66		466	540

23 21 20.78 Strainers, Y Type, Iron Body

		Crew	Daily Output	Labor-Hours	Unit	Material	Labor	Equipment	Total	Total Incl O&P
0010	**STRAINERS, Y TYPE, IRON BODY**									
0050	Screwed, 250 lb., 1/4" pipe size	1 Stpi	20	.400	Ea.	7.75	19.75		27.50	38
0070	3/8" pipe size		20	.400		7.75	19.75		27.50	38
0100	1/2" pipe size		20	.400		7.75	19.75		27.50	38
0140	1" pipe size		16	.500		12.90	24.50		37.40	51
0160	1-1/2" pipe size		12	.667		21	33		54	72.50
0180	2" pipe size		8	1		32	49.50		81.50	109

23 21 Hydronic Piping and Pumps

23 21 20 – Hydronic HVAC Piping Specialties

23 21 20.88 Venturi Flow

		Crew	Daily Output	Labor-Hours	Unit	Material	2009 Bare Costs Labor	2009 Bare Costs Equipment	Total	Total Incl O&P
0010	**VENTURI FLOW**, Measuring device									
0050	1/2" diameter	1 Stpi	24	.333	Ea.	228	16.45		244.45	276
0120	1" diameter		19	.421		225	21		246	279
0140	1-1/4" diameter		15	.533		280	26.50		306.50	350
0160	1-1/2" diameter		13	.615		293	30.50		323.50	365
0180	2" diameter		11	.727		299	36		335	385

23 31 HVAC Ducts and Casings

23 31 13 – Metal Ducts

23 31 13.13 Rectangular Metal Ducts

		Crew	Daily Output	Labor-Hours	Unit	Material	2009 Bare Costs Labor	2009 Bare Costs Equipment	Total	Total Incl O&P
0010	**RECTANGULAR METAL DUCTS**									
0020	Fabricated rectangular, includes fittings, joints, supports,									
0030	allowance for flexible connections, no insulation									
0031	NOTE: Fabrication and installation are combined									
0040	as LABOR cost. Approx. 25% fittings assumed.									
0100	Aluminum, alloy 3003-H14, under 100 lb.	Q-10	75	.320	Lb.	3.72	14.10		17.82	25.50
0160	Over 5,000 lb.		145	.166		2.15	7.30		9.45	13.55
0500	Galvanized steel, under 200 lb.		235	.102		1.12	4.50		5.62	8.15
0520	200 to 500 lb.		245	.098		.99	4.32		5.31	7.75
0540	500 to 1,000 lb.		255	.094		.97	4.15		5.12	7.45
0560	1,000 to 2,000 lb.		265	.091		.97	3.99		4.96	7.20
0580	Over 5,000 lb.		285	.084		.93	3.71		4.64	6.70
1000	Stainless steel, type 304, under 100 lb.		165	.145		4.35	6.40		10.75	14.65
1060	Over 5,000 lb.		235	.102		3.24	4.50		7.74	10.45
1100	For medium pressure ductwork, add						15%			
1200	For high pressure ductwork, add						40%			
1210	For welded ductwork, add						85%			
1220	For 30% fittings, add						11%			
1224	For 40% fittings, add						34%			
1228	For 50% fittings, add						56%			
1232	For 60% fittings, add						79%			
1236	For 70% fittings, add						101%			
1240	For 80% fittings, add						124%			
1244	For 90% fittings, add						147%			
1248	For 100% fittings, add						169%			
1252	Note: Fittings add includes time for detailing and installation.									

23 31 16 – Nonmetal Ducts

23 31 16.13 Fibrous-Glass Ducts

		Crew	Daily Output	Labor-Hours	Unit	Material	2009 Bare Costs Labor	2009 Bare Costs Equipment	Total	Total Incl O&P
0010	**FIBROUS-GLASS DUCTS**									
3490	Rigid fiberglass duct board, foil reinf. kraft facing									
3500	Rectangular, 1" thick, alum. faced, (FRK), std. weight	Q-10	350	.069	SF Surf	.73	3.02		3.75	5.45

23 33 Air Duct Accessories

23 33 46 – Flexible Ducts

23 33 46.10 Flexible Air Ducts		Crew	Daily Output	Labor-Hours	Unit	Material	2009 Bare Costs Labor	Equipment	Total	Total Incl O&P
0010	**FLEXIBLE AIR DUCTS**									
1300	Flexible, coated fiberglass fabric on corr. resist. metal helix									
1400	pressure to 12" (WG) UL-181									
1500	Non-insulated, 3" diameter	Q-9	400	.040	L.F.	.99	1.70		2.69	3.71
1540	5" diameter		320	.050		1.21	2.12		3.33	4.60
1560	6" diameter		280	.057		1.49	2.43		3.92	5.40
1600	8" diameter		200	.080		2.04	3.40		5.44	7.50
1660	12" diameter		120	.133		3.08	5.65		8.73	12.10
1900	Insulated, 1" thick, PE jacket, 3" diameter G		380	.042		2.09	1.79		3.88	5.05
1910	4" diameter G		340	.047		2.09	2		4.09	5.40
1940	6" diameter G		260	.062		2.31	2.61		4.92	6.55
1980	8" diameter G		180	.089		2.92	3.78		6.70	9
2040	12" diameter G		100	.160		4.24	6.80		11.04	15.10

23 33 53 – Duct Liners

23 33 53.10 Duct Liner Board		Crew	Daily Output	Labor-Hours	Unit	Material	Labor	Equipment	Total	Total Incl O&P
0010	**DUCT LINER BOARD**									
3490	Board type, fiberglass liner, 3 lb. density									
3500	Fire resistant, black pigmented, 1 side									
3520	1" thick G	Q-14	150	.107	S.F.	.73	4.23		4.96	7.45
3540	1-1/2" thick G		130	.123		.92	4.89		5.81	8.65
3560	2" thick G		120	.133		1.12	5.30		6.42	9.55
3600	FSK vapor barrier									
3620	1" thick G	Q-14	150	.107	S.F.	.73	4.23		4.96	7.45
3630	1-1/2" thick G		130	.123		.92	4.89		5.81	8.65
3640	2" thick G		120	.133		1.12	5.30		6.42	9.55
3680	No finish									
3700	1" thick G	Q-14	170	.094	S.F.	.38	3.74		4.12	6.25
3710	1-1/2" thick G		140	.114		.57	4.54		5.11	7.75
3720	2" thick G		130	.123		.77	4.89		5.66	8.50
3940	Board type, non-fibrous foam									
3950	Temperature, bacteria and fungi resistant									
3960	1" thick G	Q-14	150	.107	S.F.	2.25	4.23		6.48	9.15
3970	1-1/2" thick G		130	.123		3	4.89		7.89	10.95
3980	2" thick G		120	.133		3.60	5.30		8.90	12.25

23 34 HVAC Fans

23 34 13 – Axial HVAC Fans

23 34 13.10 Axial Flow HVAC Fans		Crew	Daily Output	Labor-Hours	Unit	Material	Labor	Equipment	Total	Total Incl O&P
0010	**AXIAL FLOW HVAC FANS**									
0020	Air conditioning and process air handling									
0030	Axial flow, compact, low sound, 2.5" S.P.									
0050	3,800 CFM, 5 HP	Q-20	3.40	5.882	Ea.	4,375	255		4,630	5,200
0080	6,400 CFM, 5 HP		2.80	7.143		4,875	310		5,185	5,850
0100	10,500 CFM, 7-1/2 HP		2.40	8.333		6,075	360		6,435	7,225
0120	15,600 CFM, 10 HP		1.60	12.500		7,650	540		8,190	9,250
0140	23,000 CFM, 15 HP		.70	28.571		11,700	1,250		12,950	14,800
1500	Vaneaxial, low pressure, 2000 CFM, 1/2 HP		3.60	5.556		1,825	241		2,066	2,375
1520	4,000 CFM, 1 HP		3.20	6.250		2,100	271		2,371	2,725
1540	8,000 CFM, 2 HP		2.80	7.143		2,675	310		2,985	3,400
1560	16,000 CFM, 5 HP		2.40	8.333		3,850	360		4,210	4,775

23 34 HVAC Fans

23 34 14 – Blower HVAC Fans

23 34 14.10 Blower Type HVAC Fans

	23 34 14.10 Blower Type HVAC Fans	Crew	Daily Output	Labor-Hours	Unit	Material	2009 Bare Costs Labor	2009 Bare Costs Equipment	Total	Total Incl O&P
0010	**BLOWER TYPE HVAC FANS**									
2000	Blowers, direct drive with motor, complete									
2020	1045 CFM @ .5" S.P., 1/5 HP	Q-20	18	1.111	Ea.	208	48		256	305
2060	1640 CFM @ .5" S.P., 1/3 HP	"	18	1.111	"	232	48		280	330
2500	Ceiling fan, right angle, extra quiet, 0.10" S.P.									
2520	95 CFM	Q-20	20	1	Ea.	228	43.50		271.50	320
2540	210 CFM		19	1.053		270	45.50		315.50	365
2560	385 CFM		18	1.111		345	48		393	450
2580	885 CFM		16	1.250		675	54		729	825
2600	1,650 CFM		13	1.538		935	66.50		1,001.50	1,125
2620	2,960 CFM		11	1.818		1,250	79		1,329	1,500
2640	For wall or roof cap, add	1 Shee	16	.500		228	23.50		251.50	288
2660	For straight thru fan, add					10%				
2680	For speed control switch, add	1 Elec	16	.500		125	23.50		148.50	172
7500	Utility set, steel construction, pedestal, 1/4" S.P.									
7520	Direct drive, 150 CFM, 1/8 HP	Q-20	6.40	3.125	Ea.	735	136		871	1,000
7540	485 CFM, 1/6 HP		5.80	3.448		925	150		1,075	1,250
7560	1950 CFM, 1/2 HP		4.80	4.167		1,075	181		1,256	1,475
7580	2410 CFM, 3/4 HP		4.40	4.545		2,000	197		2,197	2,500
7600	3328 CFM, 1-1/2 HP		3	6.667		2,225	289		2,514	2,900
7680	V-belt drive, drive cover, 3 phase									
7700	800 CFM, 1/4 HP	Q-20	6	3.333	Ea.	740	145		885	1,025
7720	1,300 CFM, 1/3 HP		5	4		775	174		949	1,125
7740	2,000 CFM, 1 HP		4.60	4.348		915	189		1,104	1,300
7760	2,900 CFM, 3/4 HP		4.20	4.762		1,250	207		1,457	1,675
7780	3,600 CFM, 3/4 HP		4	5		1,525	217		1,742	2,000
7800	4,800 CFM, 1 HP		3.50	5.714		1,800	248		2,048	2,350

23 34 16 – Centrifugal HVAC Fans

23 34 16.10 Centrifugal Type HVAC Fans

		Crew	Daily Output	Labor-Hours	Unit	Material	Labor	Equipment	Total	Total Incl O&P
0010	**CENTRIFUGAL TYPE HVAC FANS**									
0200	In-line centrifugal, supply/exhaust booster									
0220	aluminum wheel/hub, disconnect switch, 1/4" S.P.									
0240	500 CFM, 10" diameter connection	Q-20	3	6.667	Ea.	1,125	289		1,414	1,675
0260	1,380 CFM, 12" diameter connection		2	10		1,175	435		1,610	1,975
0280	1,520 CFM, 16" diameter connection		2	10		1,300	435		1,735	2,100
3500	Centrifugal, airfoil, motor and drive, complete									
3520	1000 CFM, 1/2 HP	Q-20	2.50	8	Ea.	1,525	345		1,870	2,225
3540	2,000 CFM, 1 HP		2	10		1,675	435		2,110	2,525
3560	4,000 CFM, 3 HP		1.80	11.111		2,350	480		2,830	3,300
3580	8,000 CFM, 7-1/2 HP		1.40	14.286		3,625	620		4,245	4,950
3600	12,000 CFM, 10 HP		1	20		4,375	870		5,245	6,150
7000	Roof exhauster, centrifugal, aluminum housing, 12" galvanized									
7020	curb, bird screen, back draft damper, 1/4" S.P.									
7100	Direct drive, 320 CFM, 11" sq. damper	Q-20	7	2.857	Ea.	510	124		634	755
7120	600 CFM, 11" sq. damper		6	3.333		520	145		665	790
7160	1450 CFM, 13" sq. damper		4.20	4.762		665	207		872	1,050
7200	V-belt drive, 1650 CFM, 12" sq. damper		6	3.333		915	145		1,060	1,225
7220	2750 CFM, 21" sq. damper		5	4		1,075	174		1,249	1,450
7230	3500 CFM, 21" sq. damper		4.50	4.444		1,175	193		1,368	1,600
7240	4910 CFM, 23" sq. damper		4	5		1,450	217		1,667	1,925
7260	8525 CFM, 28" sq. damper		3	6.667		1,825	289		2,114	2,450
7280	13,760 CFM, 35" sq. damper		2	10		2,500	435		2,935	3,425

23 34 HVAC Fans

23 34 16 – Centrifugal HVAC Fans

23 34 16.10 Centrifugal Type HVAC Fans	Crew	Daily Output	Labor-Hours	Unit	Material	2009 Bare Costs Labor	Equipment	Total	Total Incl O&P
7300 20,558 CFM, 43" sq. damper	Q-20	1	20	Ea.	5,300	870		6,170	7,175
7320 For 2 speed winding, add					15%				
7340 For explosionproof motor, add					375			375	415
7360 For belt driven, top discharge, add				↓	15%				
8500 Wall exhausters, centrifugal, auto damper, 1/8" S.P.									
8520 Direct drive, 610 CFM, 1/20 HP	Q-20	14	1.429	Ea.	330	62		392	460
8540 796 CFM, 1/12 HP		13	1.538		675	66.50		741.50	845
8560 822 CFM, 1/6 HP		12	1.667		840	72.50		912.50	1,025
8620 1983 CFM, 1/4 HP	↓	10	2	↓	1,000	87		1,087	1,250
9500 V-belt drive, 3 phase									
9520 2,800 CFM, 1/4 HP	Q-20	9	2.222	Ea.	1,500	96.50		1,596.50	1,800
9540 3,740 CFM, 1/2 HP		8	2.500		1,550	108		1,658	1,900
9560 4400 CFM, 3/4 HP		7	2.857		1,575	124		1,699	1,950
9580 5700 CFM, 1-1/2 HP	↓	6	3.333	↓	1,650	145		1,795	2,025

23 34 23 – HVAC Power Ventilators

23 34 23.10 HVAC Power Circulators and Ventilators		Crew	Daily Output	Labor-Hours	Unit	Material	2009 Bare Costs Labor	Equipment	Total	Total Incl O&P
0010	**HVAC POWER CIRCULATORS AND VENTILATORS**									
3000	Paddle blade air circulator, 3 speed switch									
3020	42", 5,000 CFM high, 3000 CFM low	G	1 Elec	2.40	3.333	Ea.	89	157		246	330
3040	52", 6,500 CFM high, 4000 CFM low	G	"	2.20	3.636	"	96.50	171		267.50	360
3100	For antique white motor, same cost										
3200	For brass plated motor, same cost										
3300	For light adaptor kit, add	G				Ea.	29			29	31.50
6000	Propeller exhaust, wall shutter, 1/4" S.P.										
6020	Direct drive, two speed										
6100	375 CFM, 1/10 HP		Q-20	10	2	Ea.	410	87		497	585
6120	730 CFM, 1/7 HP			9	2.222		455	96.50		551.50	645
6140	1000 CFM, 1/8 HP			8	2.500		635	108		743	865
6160	1890 CFM, 1/4 HP			7	2.857		645	124		769	900
6180	3275 CFM, 1/2 HP			6	3.333		660	145		805	945
6200	4720 CFM, 1 HP		↓	5	4	↓	1,025	174		1,199	1,400
6300	V-belt drive, 3 phase										
6320	6175 CFM, 3/4 HP		Q-20	5	4	Ea.	855	174		1,029	1,200
6340	7500 CFM, 3/4 HP			5	4		900	174		1,074	1,250
6360	10,100 CFM, 1 HP			4.50	4.444		1,100	193		1,293	1,500
6380	14,300 CFM, 1-1/2 HP			4	5		1,300	217		1,517	1,750
6420	26,250 CFM, 3 HP			2.60	7.692		1,775	335		2,110	2,450
6480	51,500 CFM, 10 HP		↓	1.80	11.111	↓	2,325	480		2,805	3,300
6490	V-belt drive, 115V., residential, whole house										
6500	Ceiling-wall, 5200 CFM, 1/4 HP, 30" x 30"		1 Shee	6	1.333	Ea.	445	63		508	585
6530	13,200 CFM, 1/3 HP, 48" x 48"			4	2		540	94.50		634.50	740
6550	17,025 CFM, 1/2 HP, 54" x 54"		↓	4	2	↓	1,075	94.50		1,169.50	1,350
6560	For two speed motor, add						20%				
6570	Shutter, automatic, ceiling/wall										
6580	30" x 30"		1 Shee	8	1	Ea.	145	47		192	233
6610	48" x 48"			7	1.143		225	54		279	330
6630	Timer, shut off, to 12 Hr.		↓	20	.400	↓	48	18.90		66.90	81.50
6650	Residential, bath exhaust, grille, back draft damper										
6660	50 CFM		Q-20	24	.833	Ea.	40.50	36		76.50	99.50
6670	110 CFM			22	.909		66.50	39.50		106	134
6680	Light combination, squirrel cage, 100 watt, 70 CFM			24	.833	↓	73.50	36		109.50	136
6700	Light/heater combination, ceiling mounted										

23 34 HVAC Fans

23 34 23 – HVAC Power Ventilators

23 34 23.10 HVAC Power Circulators and Ventilators

		Crew	Daily Output	Labor-Hours	Unit	Material	2009 Bare Costs Labor	Equipment	Total	Total Incl O&P
6710	70 CFM, 1450 watt	Q-20	24	.833	Ea.	89	36		125	153
6800	Heater combination, recessed, 70 CFM		24	.833		42	36		78	102
6820	With 2 infrared bulbs		23	.870		64	37.50		101.50	128
6900	Kitchen exhaust, grille, complete, 160 CFM		22	.909		76	39.50		115.50	145
6910	180 CFM		20	1		66.50	43.50		110	140
6920	270 CFM		18	1.111		117	48		165	202
6930	350 CFM	↓	16	1.250	↓	91	54		145	183
6940	Residential roof jacks and wall caps									
6944	Wall cap with back draft damper									
6946	3" & 4" dia. round duct	1 Shee	11	.727	Ea.	16.15	34.50		50.65	71
6948	6" dia. round duct	"	11	.727	"	39	34.50		73.50	96
6958	Roof jack with bird screen and back draft damper									
6960	3" & 4" dia. round duct	1 Shee	11	.727	Ea.	15	34.50		49.50	69.50
6962	3-1/4" x 10" rectangular duct	"	10	.800	"	28.50	38		66.50	89.50
6980	Transition									
6982	3-1/4" x 10" to 6" dia. round	1 Shee	20	.400	Ea.	18.55	18.90		37.45	49.50

23 34 33 – Air Curtains

23 34 33.10 Air Barrier Curtains

		Crew	Daily Output	Labor-Hours	Unit	Material	2009 Bare Costs Labor	Equipment	Total	Total Incl O&P
0010	**AIR BARRIER CURTAINS**, Incl. motor starters, transformers,									
0050	door switches & temperature controls									
0100	Shipping and receiving doors, unheated, minimal wind stoppage									
0150	8' high, multiples of 3' wide	2 Shee	6	2.667	L.F.	410	126		536	650
0160	5' wide		10	1.600		350	75.50		425.50	500
0210	10' high, multiples of 4' wide		8	2		440	94.50		534.50	625
0250	12' high, 3'-6" wide		7	2.286		194	108		302	380
0260	12' wide		6	2.667		250	126		376	470
0350	16' high, 3'-6" wide		7	2.286		194	108		302	380
0360	12' wide	↓	6	2.667	↓	250	126		376	470
0500	Maximum wind stoppage									
0550	10' high, multiples of 4' wide	2 Shee	8	2	L.F.	320	94.50		414.50	495
0650	14' high, multiples of 4' wide		8	2		365	94.50		459.50	545
0750	20' high, multiples of 8' wide	↓	6	2.667	↓	950	126		1,076	1,250
1100	Heated, maximum wind stoppage, steam heat									
1150	10' high, multiples of 4' wide	2 Shee	6	2.667	L.F.	855	126		981	1,125
1250	14' high, multiples of 4' wide		6	2.667		365	126		491	595
1350	20' high, multiples of 8' wide	↓	4	4	↓	1,250	189		1,439	1,675
1500	Customer entrance doors, unheated, minimal wind stoppage									
1550	10' high, multiples of 3' wide	2 Shee	6	2.667	L.F.	305	126		431	530
1560	5' wide		10	1.600		350	75.50		425.50	500
1650	Maximum wind stoppage, 12' high, multiples of 4' wide	↓	8	2	↓	340	94.50		434.50	520
1700	Heated, minimal wind stoppage, electric heat									
1750	8' high, multiples of 3' wide	2 Shee	6	2.667	L.F.	380	126		506	615
1760	Multiples of 5' wide		10	1.600		310	75.50		385.50	460
1850	10' high, multiples of 3' wide		6	2.667		815	126		941	1,100
1860	Multiples of 5' wide	↓	10	1.600	↓	535	75.50		610.50	700
1950	Maximum wind stoppage, steam heat									
1960	12' high, multiples of 4' wide	2 Shee	8	2	L.F.	655	94.50		749.50	865
2000	Walk-in coolers and freezers, ambient air, minimal wind stoppage									
2050	8' high, multiples of 3' wide	2 Shee	6	2.667	L.F.	360	126		486	590
2060	Multiples of 5' wide		10	1.600		238	75.50		313.50	380
2250	Maximum wind stoppage, 12' high, multiples of 3' wide		6	2.667		222	126		348	440
2450	Conveyor openings or service windows, unheated, 5' high	↓	5	3.200		228	151		379	485

23 34 HVAC Fans

23 34 33 – Air Curtains

23 34 33.10 Air Barrier Curtains

		Crew	Daily Output	Labor-Hours	Unit	Material	2009 Bare Costs Labor	2009 Bare Costs Equipment	Total	Total Incl O&P
2460	Heated, electric, 5' high, 2'-6" wide	2 Shee	5	3.200	L.F.	146	151		297	395

23 37 Air Outlets and Inlets

23 37 13 – Diffusers, Registers, and Grilles

23 37 13.10 Diffusers

		Crew	Daily Output	Labor-Hours	Unit	Material	2009 Bare Costs Labor	2009 Bare Costs Equipment	Total	Total Incl O&P
0010	**DIFFUSERS**, Aluminum, opposed blade damper unless noted									
0100	Ceiling, linear, also for sidewall									
0500	Perforated, 24" x 24" lay-in panel size, 6" x 6"	1 Shee	16	.500	Ea.	56	23.50		79.50	98
0520	8" x 8"		15	.533		57	25		82	102
0530	9" x 9"		14	.571		61	27		88	109
0540	10" x 10"		14	.571		63	27		90	111
0560	12" x 12"		12	.667		64	31.50		95.50	119
0590	16" x 16"		11	.727		106	34.50		140.50	170
0600	18" x 18"		10	.800		112	38		150	181
0610	20" x 20"		10	.800		129	38		167	200
0620	24" x 24"		9	.889		161	42		203	242
1000	Rectangular, 1 to 4 way blow, 6" x 6"		16	.500		45	23.50		68.50	86
1010	8" x 8"		15	.533		54	25		79	98.50
1014	9" x 9"		15	.533		59.50	25		84.50	105
1016	10" x 10"		15	.533		64	25		89	110
1020	12" x 6"		15	.533		65.50	25		90.50	111
1040	12" x 9"		14	.571		71.50	27		98.50	120
1060	12" x 12"		12	.667		72	31.50		103.50	128
1070	14" x 6"		13	.615		71	29		100	123
1074	14" x 14"		12	.667		101	31.50		132.50	160
1150	18" x 18"		9	.889		126	42		168	204
1160	21" x 21"		8	1		180	47		227	271
1170	24" x 12"		10	.800		134	38		172	205
1500	Round, butterfly damper, steel, diffuser size, 6" diameter		18	.444		22	21		43	56.50
1520	8" diameter		16	.500		23.50	23.50		47	62.50
1540	10" diameter		14	.571		29	27		56	73.50
1560	12" diameter		12	.667		38.50	31.50		70	91
1580	14" diameter		10	.800		49	38		87	112
2000	T bar mounting, 24" x 24" lay-in frame, 6" x 6"		16	.500		42.50	23.50		66	83.50
2020	8" x 8"		14	.571		44.50	27		71.50	90
2040	12" x 12"		12	.667		52	31.50		83.50	106
2060	16" x 16"		11	.727		70	34.50		104.50	130
2080	18" x 18"		10	.800		78	38		116	144
6000	For steel diffusers instead of aluminum, deduct					10%				

23 37 13.30 Grilles

		Crew	Daily Output	Labor-Hours	Unit	Material	2009 Bare Costs Labor	2009 Bare Costs Equipment	Total	Total Incl O&P
0010	**GRILLES**									
0020	Aluminum									
1000	Air return, 6" x 6"	1 Shee	26	.308	Ea.	17.05	14.50		31.55	41.50
1020	10" x 6"		24	.333		17.05	15.75		32.80	43
1080	16" x 8"		22	.364		21.50	17.15		38.65	50
1100	12" x 12"		22	.364		23	17.15		40.15	52
1120	24" x 12"		18	.444		32	21		53	67.50
1220	24" x 18"		16	.500		40	23.50		63.50	80.50
1280	36" x 24"		14	.571		69	27		96	117
2000	Door grilles, 12" x 12"		22	.364		52.50	17.15		69.65	84
2020	18" x 12"		22	.364		59.50	17.15		76.65	92

23 37 Air Outlets and Inlets

23 37 13 – Diffusers, Registers, and Grilles

23 37 13.30 Grilles

		Crew	Daily Output	Labor-Hours	Unit	Material	2009 Bare Costs Labor	2009 Bare Costs Equipment	Total	Total Incl O&P
2040	24" x 12"	1 Shee	18	.444	Ea.	64.50	21		85.50	104
3950	Eggcrate, framed, 6" x 6" opening		26	.308		18.70	14.50		33.20	43
3970	12" x 12" opening		22	.364		26	17.15		43.15	55
3990	18" x 18" opening		21	.381		45.50	18		63.50	77.50
4040	24" x 24" opening		15	.533		75	25		100	122
4060	Eggcrate, lay-in, T-bar system									
4070	48" x 24" sheet	1 Shee	40	.200	Ea.	90	9.45		99.45	114
6000	For steel grilles instead of aluminum in above, deduct				"	10%				
6200	Plastic, eggcrate, lay-in, T-bar system									
6210	48" x 24" sheet	1 Shee	50	.160	Ea.	22.50	7.55		30.05	36.50

23 37 13.60 Registers

		Crew	Daily Output	Labor-Hours	Unit	Material	2009 Bare Costs Labor	2009 Bare Costs Equipment	Total	Total Incl O&P
0010	**REGISTERS**									
0980	Air supply									
1000	Ceiling/wall, O.B. damper, anodized aluminum									
1010	One or two way deflection, adj. curved face bars									
1020	8" x 4"	1 Shee	26	.308	Ea.	30.50	14.50		45	56
1120	12" x 12"		18	.444		44.50	21		65.50	81.50
1240	20" x 6"		18	.444		50	21		71	87.50
1340	24" x 8"		13	.615		67.50	29		96.50	119
1350	24" x 18"		12	.667		125	31.50		156.50	187
2700	Above registers in steel instead of aluminum, deduct					10%				
4980	Air return									
5000	Ceiling or wall, fixed 45° face blades									
5010	Adjustable O.B. damper, anodized aluminum									
5020	4" x 8"	1 Shee	26	.308	Ea.	23	14.50		37.50	48
5060	6" x 10"		19	.421		26.50	19.85		46.35	59.50
5280	24" x 24"		11	.727		89	34.50		123.50	151
5300	24" x 36"		8	1		140	47		187	227

23 37 23 – HVAC Gravity Ventilators

23 37 23.10 HVAC Gravity Air Ventilators

		Crew	Daily Output	Labor-Hours	Unit	Material	2009 Bare Costs Labor	2009 Bare Costs Equipment	Total	Total Incl O&P
0010	**HVAC GRAVITY AIR VENTILATORS**, Incl. base and damper									
1280	Rotary ventilators, wind driven, galvanized									
1340	6" neck diameter	Q-9	16	1	Ea.	39	42.50		81.50	109
1400	12" neck diameter		10	1.600		56.50	68		124.50	168
1500	24" neck diameter, 3,100 CFM		8	2		165	85		250	315
1540	36" neck diameter, 5,500 CFM		6	2.667		485	113		598	710
1600	For aluminum, add					300%				
1620	For stainless steel, add					600%				
2000	Stationary, gravity, syphon, galvanized									
2160	6" neck diameter, 66 CFM	Q-9	16	1	Ea.	30	42.50		72.50	98.50
2240	12" neck diameter, 160 CFM		10	1.600		63.50	68		131.50	175
2340	24" neck diameter, 900 CFM		8	2		173	85		258	320
2380	36" neck diameter, 2,000 CFM		6	2.667		540	113		653	770
2500	For aluminum, add					300%				
2520	For stainless steel, add					600%				
4200	Stationary mushroom, aluminum, 16" orifice diameter	Q-9	10	1.600		510	68		578	665
4220	26" orifice diameter		6.15	2.602		755	111		866	1,000
4230	30" orifice diameter		5.71	2.802		1,100	119		1,219	1,400
4240	38" orifice diameter		5	3.200		1,575	136		1,711	1,950
4250	42" orifice diameter		4.70	3.404		2,100	145		2,245	2,525
4260	50" orifice diameter		4.44	3.604		2,500	153		2,653	2,975
5000	Relief vent									

23 37 Air Outlets and Inlets

23 37 23 – HVAC Gravity Ventilators

23 37 23.10 HVAC Gravity Air Ventilators

		Crew	Daily Output	Labor-Hours	Unit	Material	2009 Bare Costs Labor	Equipment	Total	Total Incl O&P
5500	Rectangular, aluminum, galvanized curb									
5510	intake/exhaust, 0.033" SP									
5580	500 CFM, 12" x 12"	Q-9	8.60	1.860	Ea.	590	79		669	770
5600	600 CFM, 12"x16"		8	2		665	85		750	860
5640	1000 CFM, 12" x 24"		6.60	2.424		750	103		853	985
5680	3000 CFM, 20" x 42"		4	4		1,325	170		1,495	1,700
5740	10,000 CFM, 48" x 60"		1.80	8.889		2,600	380		2,980	3,425
5800	15,000 CFM, 60" x 72"		1.30	12.308		3,625	525		4,150	4,800
5820	18,000 CFM, 72" x 72"		1.20	13.333		4,125	565		4,690	5,425
5880	Size is throat area, volume is at 500 FPM									
7000	Note: sizes based on exhaust. Intake, with 0.125" SP									
7100	loss, approximately twice listed capacity.									

23 38 Ventilation Hoods

23 38 13 – Commercial-Kitchen Hoods

23 38 13.10 Hood and Ventilation Equipment

		Crew	Daily Output	Labor-Hours	Unit	Material	2009 Bare Costs Labor	Equipment	Total	Total Incl O&P
0010	**HOOD AND VENTILATION EQUIPMENT**									
2970	Exhaust hood, sst, gutter on all sides, 4' x 4' x 2'	1 Carp	1.80	4.444	Ea.	4,050	178		4,228	4,725
2980	4' x 4' x 7'	"	1.60	5	"	6,350	200		6,550	7,300
7800	Vent hood, wall canopy with fire protection	L-3A	9	1.333	L.F.	355	58.50		413.50	480
7810	Without fire protection		10	1.200		248	52.50		300.50	355
7820	Island canopy with fire protection		7	1.714		375	75		450	530
7830	Without fire protection		8	1.500		272	65.50		337.50	400
7840	Back shelf with fire protection		11	1.091		355	47.50		402.50	465
7850	Without fire protection		12	1		248	43.50		291.50	340
7860	Range hood & CO_2 system, minimum	1 Carp	2.50	3.200	Ea.	2,600	128		2,728	3,075
7870	Maximum	"	1	8		31,400	320		31,720	35,000
7950	Hood fire protection system, minimum	Q-1	3	5.333		4,100	234		4,334	4,850
8050	Maximum	"	1	16		31,600	700		32,300	35,900

23 41 Particulate Air Filtration

23 41 13 – Panel Air Filters

23 41 13.10 Panel Type Air Filters

		Crew	Daily Output	Labor-Hours	Unit	Material	2009 Bare Costs Labor	Equipment	Total	Total Incl O&P
0010	**PANEL TYPE AIR FILTERS**									
2950	Mechanical media filtration units									
3000	High efficiency type, with frame, non-supported [G]				MCFM	45			45	49.50
3100	Supported type [G]				"	60			60	66
5500	Throwaway glass or paper media type				Ea.	3.35			3.35	3.69

23 41 16 – Renewable-Media Air Filters

23 41 16.10 Disposable Media Air Filters

		Crew	Daily Output	Labor-Hours	Unit	Material	2009 Bare Costs Labor	Equipment	Total	Total Incl O&P
0010	**DISPOSABLE MEDIA AIR FILTERS**									
5000	Renewable disposable roll				MCFM	250			250	275

23 41 19 – Washable Air Filters

23 41 19.10 Permanent Air Filters

		Crew	Daily Output	Labor-Hours	Unit	Material	2009 Bare Costs Labor	Equipment	Total	Total Incl O&P
0010	**PERMANENT AIR FILTERS**									
4500	Permanent washable [G]				MCFM	20			20	22

23 41 Particulate Air Filtration

23 41 23 – Extended Surface Filters

23 41 23.10 Expanded Surface Filters

		Crew	Daily Output	Labor-Hours	Unit	Material	2009 Bare Costs Labor	Equipment	Total	Total Incl O&P
0010	**EXPANDED SURFACE FILTERS**									
4000	Medium efficiency, extended surface G				MCFM	5.50			5.50	6.05

23 42 Gas-Phase Air Filtration

23 42 13 – Activated-Carbon Air Filtration

23 42 13.10 Charcoal Type Air Filtration

		Crew	Daily Output	Labor-Hours	Unit	Material	Labor	Equipment	Total	Total Incl O&P
0010	**CHARCOAL TYPE AIR FILTRATION**									
0050	Activated charcoal type, full flow				MCFM	600			600	660
0060	Full flow, impregnated media 12" deep					225			225	248
0070	HEPA filter & frame for field erection					300			300	330
0080	HEPA filter-diffuser, ceiling install.				↓	275			275	305

23 42 16 – Chemically-Impregnated Adsorption Air Filtration

23 42 16.10 Chemical Adsorption Air Filtration

		Crew	Daily Output	Labor-Hours	Unit	Material	Labor	Equipment	Total	Total Incl O&P
0010	**CHEMICAL ADSORPTION AIR FILTRATION**									
0500	Chemical media filtration type									
1110	corrosion resistant PVC construction									
1130	500 CFM	Q-9	14	1.143	Ea.	5,200	48.50		5,248.50	5,800
1150	2000 CFM		8	2		7,400	85		7,485	8,275
1170	5000 CFM	↓	5	3.200	↓	11,200	136		11,336	12,500

23 43 Electronic Air Cleaners

23 43 13 – Washable Electronic Air Cleaners

23 43 13.10 Electronic Air Cleaners

		Crew	Daily Output	Labor-Hours	Unit	Material	Labor	Equipment	Total	Total Incl O&P
0010	**ELECTRONIC AIR CLEANERS**									
2000	Electronic air cleaner, duct mounted									
2150	400 – 1000 CFM	1 Shee	2.30	3.478	Ea.	1,000	164		1,164	1,350

23 51 Breechings, Chimneys, and Stacks

23 51 26 – All-Fuel Vent Chimneys

23 51 26.30 All-Fuel Vent Chimneys, Double Wall, St. Stl.

		Crew	Daily Output	Labor-Hours	Unit	Material	Labor	Equipment	Total	Total Incl O&P
0010	**ALL-FUEL VENT CHIMNEYS, DOUBLE WALL, STAINLESS STEEL**									
7800	All fuel, double wall, stainless steel, 6" diameter	Q-9	60	.267	V.L.F.	56.50	11.35		67.85	79.50
7802	7" diameter		56	.286		73	12.15		85.15	99
7804	8" diameter		52	.308		86.50	13.05		99.55	115
7806	10" diameter		48	.333		125	14.15		139.15	159
7808	12" diameter		44	.364		167	15.45		182.45	208
7810	14" diameter	↓	42	.381	↓	219	16.20		235.20	266
8000	All fuel, double wall, stainless steel fittings									
8010	Roof support 6" diameter	Q-9	30	.533	Ea.	103	22.50		125.50	149
8020	7" diameter		28	.571		117	24.50		141.50	166
8030	8" diameter		26	.615		130	26		156	183
8040	10" diameter		24	.667		150	28.50		178.50	209
8050	12" diameter		22	.727		181	31		212	247
8060	14" diameter		21	.762		229	32.50		261.50	300
8100	Elbow 15°, 6" diameter		30	.533		165	22.50		187.50	217
8120	7" diameter	↓	28	.571		186	24.50		210.50	243

23 51 Breechings, Chimneys, and Stacks

23 51 26 – All-Fuel Vent Chimneys

23 51 26.30 All-Fuel Vent Chimneys, Double Wall, St. Stl.		Crew	Daily Output	Labor-Hours	Unit	Material	2009 Bare Costs Labor	Equipment	Total	Total Incl O&P
8140	8" diameter	Q-9	26	.615	Ea.	212	26		238	273
8160	10" diameter		24	.667		278	28.50		306.50	350
8180	12" diameter		22	.727		335	31		366	420
8200	14" diameter		21	.762		405	32.50		437.50	495
8300	Insulated tee with insulated tee cap, 6" diameter		30	.533		156	22.50		178.50	207
8340	7" diameter		28	.571		205	24.50		229.50	263
8360	8" diameter		26	.615		231	26		257	294
8380	10" diameter		24	.667		325	28.50		353.50	405
8400	12" diameter		22	.727		455	31		486	550
8420	14" diameter		21	.762		600	32.50		632.50	710
8500	Joist shield, 6" diameter		30	.533		47.50	22.50		70	87.50
8510	7" diameter		28	.571		51.50	24.50		76	94.50
8520	8" diameter		26	.615		63.50	26		89.50	110
8530	10" diameter		24	.667		86	28.50		114.50	138
8540	12" diameter		22	.727		107	31		138	166
8550	14" diameter		21	.762		133	32.50		165.50	196
8600	Round top, 6" diameter		30	.533		53	22.50		75.50	93
8620	7" diameter		28	.571		72	24.50		96.50	117
8640	8" diameter		26	.615		97	26		123	147
8660	10" diameter		24	.667		180	28.50		208.50	242
8680	12" diameter		22	.727		249	31		280	320
8700	14" diameter		21	.762		330	32.50		362.50	415
8800	Adjustable roof flashing, 6" diameter		30	.533		63	22.50		85.50	104
8820	7" diameter		28	.571		72	24.50		96.50	117
8840	8" diameter		26	.615		78	26		104	126
8860	10" diameter		24	.667		100	28.50		128.50	154
8880	12" diameter		22	.727		130	31		161	190
8900	14" diameter		21	.762		162	32.50		194.50	228

23 55 Fuel-Fired Heaters

23 55 33 – Fuel-Fired Unit Heaters

23 55 33.13 Oil-Fired Unit Heaters

		Crew	Daily Output	Labor-Hours	Unit	Material	Labor	Equipment	Total	Total Incl O&P
0010	**OIL-FIRED UNIT HEATERS**, Cabinet, grilles, fan, ctrl, burner, no piping									
6000	Oil fired, suspension mounted, 94 MBH output	Q-5	4	4	Ea.	3,400	178		3,578	4,000

23 55 33.16 Gas-Fired Unit Heaters

		Crew	Daily Output	Labor-Hours	Unit	Material	Labor	Equipment	Total	Total Incl O&P
0010	**GAS-FIRED UNIT HEATERS**, Cabinet, grilles, fan, ctrls., burner, no piping									
0022	thermostat, no piping. For flue see Div. 23 51 23.10									
1000	Gas fired, floor mounted R235000-30									
1100	60 MBH output	Q-5	10	1.600	Ea.	805	71		876	990
1140	100 MBH output		8	2		890	89		979	1,125
1180	180 MBH output		6	2.667		1,275	118		1,393	1,575
2000	Suspension mounted, propeller fan, 20 MBH output		8.50	1.882		985	83.50		1,068.50	1,200
2040	60 MBH output		7	2.286		1,200	102		1,302	1,475
2060	80 MBH output		6	2.667		1,225	118		1,343	1,525
2100	130 MBH output		5	3.200		1,500	142		1,642	1,875
2240	320 MBH output		2	8		2,625	355		2,980	3,425
2500	For powered venter and adapter, add					355			355	390
5000	Wall furnace, 17.5 MBH output	Q-5	6	2.667		615	118		733	860
5020	24 MBH output		5	3.200		640	142		782	920
5040	35 MBH output		4	4		770	178		948	1,125

23 56 Solar Energy Heating Equipment

23 56 16 – Packaged Solar Heating Equipment

23 56 16.40 Solar Heating Systems

			Crew	Daily Output	Labor-Hours	Unit	Material	2009 Bare Costs Labor	2009 Bare Costs Equipment	Total	Total Incl O&P
0010	**SOLAR HEATING SYSTEMS**										
0020	System/Package prices, not including connecting										
0030	pipe, insulation, or special heating/plumbing fixtures										
0500	Hot water, standard package, low temperature										
0540	1 collector, circulator, fittings, 65 gal. tank	G	Q-1	.50	32	Ea.	1,125	1,400		2,525	3,325
0580	2 collectors, circulator, fittings, 120 gal. tank	G		.40	40		1,825	1,750		3,575	4,650
0620	3 collectors, circulator, fittings, 120 gal. tank	G		.34	47.059		2,500	2,075		4,575	5,850
0700	Medium temperature package										
0720	1 collector, circulator, fittings, 80 gal. tank	G	Q-1	.50	32	Ea.	1,850	1,400		3,250	4,125
0740	2 collectors, circulator, fittings, 120 gal. tank	G		.40	40		2,625	1,750		4,375	5,525
0780	3 collectors, circulator, fittings, 120 gal. tank	G		.30	53.333		3,400	2,350		5,750	7,250
0980	For each additional 120 gal. tank, add	G					1,150			1,150	1,250

23 56 19 – Solar Heating Components

23 56 19.50 Solar Heating Ancillary

			Crew	Daily Output	Labor-Hours	Unit	Material	Labor	Equipment	Total	Total Incl O&P
0010	**SOLAR HEATING ANCILLARY**										
2300	Circulators, air										
2310	Blowers										
2400	Reversible fan, 20" diameter, 2 speed	G	Q-9	18	.889	Ea.	108	38		146	177
2520	Space & DHW system, less duct work	G	"	.50	32		1,475	1,350		2,825	3,725
2870	1/12 HP, 30 GPM	G	Q-1	10	1.600		330	70		400	470
3000	Collector panels, air with aluminum absorber plate										
3010	Wall or roof mount										
3040	Flat black, plastic glazing										
3080	4' x 8'	G	Q-9	6	2.667	Ea.	635	113		748	870
3200	Flush roof mount, 10' to 16' x 22" wide	G	"	96	.167	L.F.	550	7.10		557.10	615
3300	Collector panels, liquid with copper absorber plate										
3330	Alum. frame, 4' x 8', 5/32" single glazing	G	Q-1	9.50	1.684	Ea.	665	74		739	840
3390	Alum. frame, 4' x 10', 5/32" single glazing	G		6	2.667		805	117		922	1,050
3450	Flat black, alum. frame, 3.5' x 7.5'	G		9	1.778		560	78		638	735
3500	4' x 8'	G		5.50	2.909		690	128		818	945
3520	4' x 10'	G		10	1.600		830	70		900	1,025
3540	4' x 12.5'	G		5	3.200		1,025	140		1,165	1,325
3600	Liquid, full wetted, plastic, alum. frame, 3' x 10'	G		5	3.200		234	140		374	470
3650	Collector panel mounting, flat roof or ground rack	G		7	2.286		217	100		317	390
3670	Roof clamps	G		70	.229	Set	2.35	10.05		12.40	17.65
3700	Roof strap, teflon	G	1 Plum	205	.039	L.F.	19.95	1.90		21.85	25
3900	Differential controller with two sensors										
3930	Thermostat, hard wired	G	1 Plum	8	1	Ea.	79.50	49		128.50	160
4100	Five station with digital read-out	G	"	3	2.667	"	218	130		348	435
4300	Heat exchanger										
4580	Fluid to fluid package includes two circulating pumps										
4590	expansion tank, check valve, relief valve										
4600	controller, high temperature cutoff and sensors	G	Q-1	2.50	6.400	Ea.	695	281		976	1,175
4650	Heat transfer fluid										
4700	Propylene glycol, inhibited anti-freeze	G	1 Plum	28	.286	Gal.	13.85	13.95		27.80	36.50
8250	Water storage tank with heat exchanger and electric element										
8300	80 gal. with 2" x 2 lb. density insulation	G	1 Plum	1.60	5	Ea.	1,050	244		1,294	1,525
8380	120 gal. with 2" x 2 lb. density insulation	G	"	1.40	5.714	"	1,175	279		1,454	1,700

23 62 Packaged Compressor and Condenser Units

23 62 13 – Packaged Air-Cooled Refrigerant Compressor and Condenser Units

23 62 13.10 Packaged Air-Cooled Refrig. Condensing Units		Crew	Daily Output	Labor-Hours	Unit	Material	2009 Bare Costs Labor	Equipment	Total	Total Incl O&P
0010	**PACKAGED AIR-COOLED REFRIGERANT CONDENSING UNITS**									
0020	Condensing unit									
0030	Air cooled, compressor, standard controls									
0050	1.5 ton	Q-5	2.50	6.400	Ea.	1,125	284		1,409	1,675
0100	2 ton		2.10	7.619		1,150	340		1,490	1,775
0200	2.5 ton		1.70	9.412		1,250	420		1,670	2,000
0300	3 ton		1.30	12.308		1,300	545		1,845	2,250
0350	3.5 ton		1.10	14.545		1,500	645		2,145	2,625
0400	4 ton		.90	17.778		1,700	790		2,490	3,025
0500	5 ton		.60	26.667		2,050	1,175		3,225	4,025

23 74 Packaged Outdoor HVAC Equipment

23 74 33 – Packaged, Outdoor, Heating and Cooling Makeup Air-Conditioners

23 74 33.10 Roof Top Air Conditioners

		Crew	Daily Output	Labor-Hours	Unit	Material	Labor	Equipment	Total	Total Incl O&P
0010	**ROOF TOP AIR CONDITIONERS**, Standard controls, curb, economizer.									
1000	Single zone, electric cool, gas heat									
1090	2 ton cooling, 55 MBH heating	Q-5	.93	17.204	Ea.	3,100	765		3,865	4,550
1100	3 ton cooling, 60 MBH heating		.70	22.857		3,200	1,025		4,225	5,050
1120	4 ton cooling, 95 MBH heating		.61	26.403		3,750	1,175		4,925	5,875
1140	5 ton cooling, 112 MBH heating		.56	28.521		4,150	1,275		5,425	6,475
1145	6 ton cooling, 140 MBH heating		.52	30.769		4,900	1,375		6,275	7,450
1150	7.5 ton cooling, 170 MBH heating		.50	32.258		6,275	1,425		7,700	9,050
1160	10 ton cooling, 200 MBH heating	Q-6	.67	35.982		8,300	1,650		9,950	11,600
1170	12.5 ton cooling, 230 MBH heating		.63	37.975		9,400	1,750		11,150	12,900
1180	15 ton cooling, 270 MBH heating		.57	42.032		12,500	1,925		14,425	16,700
1190	18 ton cooling, 330 MBH heating		.52	45.889		15,400	2,125		17,525	20,100
1200	20 ton cooling, 360 MBH heating	Q-7	.67	47.976		20,100	2,250		22,350	25,500
1210	25 ton cooling, 450 MBH heating		.56	57.554		23,700	2,700		26,400	30,100
1220	30 ton cooling, 540 MBH heating		.47	68.376		25,800	3,225		29,025	33,100
1240	40 ton cooling, 675 MBH heating		.35	91.168		32,900	4,275		37,175	42,600
2000	Multizone, electric cool, gas heat, economizer									
2100	15 ton cooling, 360 MBH heating	Q-7	.61	52.545	Ea.	53,500	2,475		55,975	62,500
2120	20 ton cooling, 360 MBH heating		.53	60.038		57,500	2,825		60,325	67,500
2140	25 ton cooling, 450 MBH heating		.45	71.910		69,000	3,375		72,375	81,000
2160	28 ton cooling, 450 MBH heating		.41	79.012		78,500	3,725		82,225	92,000
2180	30 ton cooling, 540 MBH heating		.37	85.562		88,500	4,025		92,525	103,500
2200	40 ton cooling, 540 MBH heating		.28	113		101,000	5,350		106,350	119,000
2210	50 ton cooling, 540 MBH heating		.23	142		125,500	6,675		132,175	148,000
2220	70 ton cooling, 1500 MBH heating		.16	198		136,000	9,350		145,350	163,500
2240	80 ton cooling, 1500 MBH heating		.14	228		155,500	10,700		166,200	187,000
2260	90 ton cooling, 1500 MBH heating		.13	256		163,000	12,000		175,000	197,500
2280	105 ton cooling, 1500 MBH heating		.11	290		179,500	13,700		193,200	218,000
2400	For hot water heat coil, deduct					5%				
2500	For steam heat coil, deduct					2%				
2600	For electric heat, deduct					3%	5%			

23 76 Evaporative Air-Cooling Equipment

23 76 16 – Indirect Evaporative Air Coolers

23 76 16.10 Evaporators		Crew	Daily Output	Labor-Hours	Unit	Material	2009 Bare Costs Labor	Equipment	Total	Total Incl O&P
0010	**EVAPORATORS**, DX coils, remote compressors not included.									
2000	Coolers, reach-in and walk-in types, above freezing temperatures									
2600	Two-way discharge, ceiling mount, 150-4100 CFM,									
2610	above 34° F applications, air defrost									
2630	900 BTUH, 7 fins per inch, 8" fan	Q-5	4	4	Ea.	271	178		449	565
2660	2500 BTUH, 7 fins per inch, 10" fan		2.60	6.154		535	273		808	995
2690	5500 BTUH, 7 fins per inch, 12" fan		1.70	9.412		1,050	420		1,470	1,775
2720	8500 BTUH, 8 fins per inch, 16" fan		1.20	13.333		1,300	590		1,890	2,350
2750	15,000 BTUH, 7 fins per inch, 18" fan		1.10	14.545		1,825	645		2,470	2,975
2770	24,000 BTUH, 7 fins per inch, two 16" fans		1	16		2,750	710		3,460	4,075
2790	30,000 BTUH, 7 fins per inch, two 18" fans	↓	.90	17.778	↓	3,400	790		4,190	4,900
2850	Two-way discharge, low profile, ceiling mount,									
2860	8 fins per inch, 200-570 CFM, air defrost									
2880	800 BTUH, one 6" fan	Q-5	4	4	Ea.	289	178		467	585
2890	1300 BTUH, two 6" fans		3.50	4.571		340	203		543	680
2900	1800 BTUH, two 6" fans		3.30	4.848		360	215		575	720
2910	2700 BTUH, three 6" fans	↓	2.70	5.926	↓	425	263		688	860

23 81 Decentralized Unitary HVAC Equipment

23 81 13 – Packaged Terminal Air-Conditioners

23 81 13.10 Packaged Cabinet Type Air-Conditioners		Crew	Daily Output	Labor-Hours	Unit	Material	2009 Bare Costs Labor	Equipment	Total	Total Incl O&P
0010	**PACKAGED CABINET TYPE AIR-CONDITIONERS**, Cabinet, wall sleeve,									
0100	louver, electric heat, thermostat, manual changeover, 208 V									
0200	6,000 BTUH cooling, 8800 BTU heat	Q-5	6	2.667	Ea.	625	118		743	870
0220	9,000 BTUH cooling, 13,900 BTU heat		5	3.200		720	142		862	1,000
0240	12,000 BTUH cooling, 13,900 BTU heat		4	4		780	178		958	1,125
0260	15,000 BTUH cooling, 13,900 BTU heat		3	5.333		865	237		1,102	1,300
0320	30,000 BTUH cooling, 10 KW heat		1.40	11.429		1,875	510		2,385	2,825
0380	48,000 BTUH cooling, 10 KW heat	↓	.90	17.778	↓	2,600	790		3,390	4,025
0500	For hot water coil, increase heat by 10%, add					5%	10%			
1000	For steam, increase heat output by 30%, add					8%	10%			

23 81 23 – Computer-Room Air-Conditioners

23 81 23.10 Computer Room Units		Crew	Daily Output	Labor-Hours	Unit	Material	2009 Bare Costs Labor	Equipment	Total	Total Incl O&P
0010	**COMPUTER ROOM UNITS**									
1000	Air cooled, includes remote condenser but not									
1020	interconnecting tubing or refrigerant									
1080	3 ton	Q-5	.50	32	Ea.	15,600	1,425		17,025	19,300
1120	5 ton		.45	35.556		16,700	1,575		18,275	20,800
1160	6 ton		.30	53.333		29,500	2,375		31,875	36,100
1200	8 ton		.27	59.259		31,400	2,625		34,025	38,500
1240	10 ton	↓	.25	64		32,800	2,850		35,650	40,400
1300	20 ton	Q-6	.26	92.308		43,200	4,250		47,450	54,000
1320	22 ton		.24	100		43,800	4,600		48,400	55,000
1360	30 ton	↓	.21	114	↓	54,000	5,275		59,275	67,500
2200	Chilled water, for connection to									
2220	existing chiller system of adequate capacity									
2260	5 ton	Q-5	.74	21.622	Ea.	12,000	960		12,960	14,700
2280	6 ton		.52	30.769		12,000	1,375		13,375	15,300
2300	8 ton		.50	32		12,100	1,425		13,525	15,400
2320	10 ton	↓	.49	32.653	↓	12,200	1,450		13,650	15,600

23 81 Decentralized Unitary HVAC Equipment

23 81 23 – Computer-Room Air-Conditioners

23 81 23.10 Computer Room Units		Crew	Daily Output	Labor-Hours	Unit	Material	2009 Bare Costs Labor	Equipment	Total	Total Incl O&P
2330	12 ton	Q-5	.49	32.990	Ea.	12,500	1,475		13,975	16,000

23 81 26 – Split-System Air-Conditioners

23 81 26.10 Split Ductless Systems

0010	**SPLIT DUCTLESS SYSTEMS**									
0100	Cooling only, single zone									
0110	Wall mount									
0120	3/4 ton cooling	Q-5	2	8	Ea.	1,175	355		1,530	1,800
0130	1 ton cooling		1.80	8.889		1,300	395		1,695	2,025
0140	1-1/2 ton cooling		1.60	10		1,800	445		2,245	2,650
0150	2 ton cooling		1.40	11.429		2,125	510		2,635	3,100
1000	Ceiling mount									
1020	2 ton cooling	Q-5	1.40	11.429	Ea.	1,450	510		1,960	2,325
1030	3 ton cooling	"	1.20	13.333	"	3,900	590		4,490	5,200
2000	T-Bar mount									
2010	2 ton cooling	Q-5	1.40	11.429	Ea.	3,400	510		3,910	4,500
2020	3 ton cooling		1.20	13.333		3,800	590		4,390	5,075
2030	3-1/2 ton cooling		1.10	14.545		4,200	645		4,845	5,600
3000	Multizone									
3010	Wall mount									
3020	2 @ 3/4 ton cooling	Q-5	1.80	8.889	Ea.	1,475	395		1,870	2,225
5000	Cooling / Heating									
5010	Wall mount									
5110	1 ton cooling	Q-5	1.70	9.412	Ea.	1,100	420		1,520	1,825
5120	1-1/2 ton cooling	"	1.50	10.667	"	1,750	475		2,225	2,625
5300	Ceiling mount									
5310	3 ton cooling	Q-5	1	16	Ea.	4,125	710		4,835	5,625
7000	Accessories for all split ductless systems									
7010	Add for ambient frost control	Q-5	8	2	Ea.	130	89		219	276
7020	Add for tube / wiring kit									
7030	15' kit	Q-5	32	.500	Ea.	38.50	22		60.50	76
7040	35' kit	"	24	.667	"	125	29.50		154.50	182

23 81 43 – Air-Source Unitary Heat Pumps

23 81 43.10 Air-Source Heat Pumps

0010	**AIR-SOURCE HEAT PUMPS**, Not including interconnecting tubing.									
1000	Air to air, split system, not including curbs, pads, fan coil and ductwork									
1012	Outside condensing unit only, for fan coil see Div. 23 82 19.10									
1020	2 ton cooling, 8.5 MBH heat @ 0° F	Q-5	2	8	Ea.	1,650	355		2,005	2,350
1060	5 ton cooling, 27 MBH heat @ 0° F		.50	32		2,675	1,425		4,100	5,075
1080	7.5 ton cooling, 33 MBH heat @ 0° F		.45	35.556		5,050	1,575		6,625	7,925
1100	10 ton cooling, 50 MBH heat @ 0° F	Q-6	.64	37.500		7,050	1,725		8,775	10,400
1120	15 ton cooling, 64 MBH heat @ 0° F		.50	48		10,200	2,200		12,400	14,500
1130	20 ton cooling, 85 MBH heat @ 0° F		.35	68.571		15,300	3,150		18,450	21,600
1140	25 ton cooling, 119 MBH heat @ 0° F		.25	96		18,400	4,425		22,825	26,900
1500	Single package, not including curbs, pads, or plenums									
1520	2 ton cooling, 6.5 MBH heat @ 0° F	Q-5	1.50	10.667	Ea.	2,575	475		3,050	3,525
1580	4 ton cooling, 13 MBH heat @ 0° F		.96	16.667		3,650	740		4,390	5,125
1640	7.5 ton cooling, 35 MBH heat @ 0° F		.40	40		6,450	1,775		8,225	9,775

23 81 46 – Water-Source Unitary Heat Pumps

23 81 46.10 Water Source Heat Pumps

0010	**WATER SOURCE HEAT PUMPS**, Not incl. connecting tubing or water source									
2000	Water source to air, single package									

23 81 Decentralized Unitary HVAC Equipment

23 81 46 – Water-Source Unitary Heat Pumps

23 81 46.10 Water Source Heat Pumps		Crew	Daily Output	Labor-Hours	Unit	Material	2009 Bare Costs Labor	Equipment	Total	Total Incl O&P
2100	1 ton cooling, 13 MBH heat @ 75° F	Q-5	2	8	Ea.	1,300	355		1,655	1,950
2140	2 ton cooling, 19 MBH heat @ 75° F		1.70	9.412		1,550	420		1,970	2,325
2220	5 ton cooling, 29 MBH heat @ 75° F		.90	17.778		2,450	790		3,240	3,875
3960	For supplementary heat coil, add					10%				
4000	For increase in capacity thru use									
4020	of solar collector, size boiler at 60%									

23 82 Convection Heating and Cooling Units

23 82 16 – Air Coils

23 82 16.20 Duct Heaters

		Crew	Daily Output	Labor-Hours	Unit	Material	Labor	Equipment	Total	Incl O&P
0010	**DUCT HEATERS**, Electric, 480 V, 3 Ph.									
0020	Finned tubular insert, 500 ° F									
0100	8" wide x 6" high, 4.0 kW	Q-20	16	1.250	Ea.	705	54		759	860
0120	12" high, 8.0 kW		15	1.333		1,175	58		1,233	1,375
0140	18" high, 12.0 kW		14	1.429		1,650	62		1,712	1,900
0160	24" high, 16.0 kW		13	1.538		2,100	66.50		2,166.50	2,425
0180	30" high, 20.0 kW		12	1.667		2,575	72.50		2,647.50	2,925
0300	12" wide x 6" high, 6.7 kW		15	1.333		750	58		808	915
0320	12" high, 13.3 kW		14	1.429		1,200	62		1,262	1,425
0340	18" high, 20.0 kW		13	1.538		1,700	66.50		1,766.50	1,975
0360	24" high, 26.7 kW		12	1.667		2,175	72.50		2,247.50	2,500
0380	30" high, 33.3 kW		11	1.818		2,650	79		2,729	3,050
0500	18" wide x 6" high, 13.3 kW		14	1.429		785	62		847	960
0520	12" high, 26.7 kW		13	1.538		1,375	66.50		1,441.50	1,625
0540	18" high, 40.0 kW		12	1.667		1,850	72.50		1,922.50	2,125
0560	24" high, 53.3 kW		11	1.818		2,450	79		2,529	2,825
0580	30" high, 66.7 kW		10	2		3,050	87		3,137	3,500
0900	30" wide x 6" high, 22.2 kW		12	1.667		920	72.50		992.50	1,100
0960	24" high, 88.9 kW		9	2.222		2,675	96.50		2,771.50	3,100
8000	To obtain BTU multiply kW by 3413									

23 82 19 – Fan Coil Units

23 82 19.10 Fan Coil Air Conditioning

		Crew	Daily Output	Labor-Hours	Unit	Material	Labor	Equipment	Total	Incl O&P
0010	**FAN COIL AIR CONDITIONING** Cabinet mounted, filters, controls									
0100	Chilled water, 1/2 ton cooling	Q-5	8	2	Ea.	675	89		764	880
0120	1 ton cooling		6	2.667		775	118		893	1,025
0140	1.5 ton cooling		5.50	2.909		875	129		1,004	1,150
0160	2.5 ton cooling		5	3.200		1,450	142		1,592	1,825
0180	3 ton cooling		4	4		1,650	178		1,828	2,100
0262	For hot water coil, add					40%	10%			
0940	Direct expansion, for use w/air cooled condensing unit, 1.5 ton cooling	Q-5	5	3.200	Ea.	540	142		682	810
1000	5 ton cooling	"	3	5.333		1,025	237		1,262	1,475
1040	10 ton cooling	Q-6	2.60	9.231		2,750	425		3,175	3,675
1060	20 ton cooling		.70	34.286		4,850	1,575		6,425	7,700
1080	30 ton cooling		.60	40		6,850	1,850		8,700	10,300
1500	For hot water coil, add					40%	10%			
1512	For condensing unit add see Div. 23 62									

23 82 29 – Radiators

23 82 29.10 Hydronic Heating

| 0010 | **HYDRONIC HEATING**, Terminal units, not incl. main supply pipe |
| 1000 | Radiation |

23 82 Convection Heating and Cooling Units

23 82 29 – Radiators

23 82 29.10 Hydronic Heating	Crew	Daily Output	Labor-Hours	Unit	Material	2009 Bare Costs Labor	Equipment	Total	Total Incl O&P
1100 Panel, baseboard, C.I., including supports, no covers	Q-5	46	.348	L.F.	32	15.45		47.45	58

23 82 36 – Finned-Tube Radiation Heaters

23 82 36.10 Finned Tube Radiation

0010 **FINNED TUBE RADIATION**, Terminal units, not incl. main supply pipe									
1150 Fin tube, wall hung, 14" slope top cover, with damper									
1200 1-1/4" copper tube, 4-1/4" alum. fin	Q-5	38	.421	L.F.	39	18.70		57.70	70.50
1250 1-1/4" steel tube, 4-1/4" steel fin	"	36	.444	"	36.50	19.75		56.25	70

23 83 Radiant Heating Units

23 83 33 – Electric Radiant Heaters

23 83 33.10 Electric Heating

0010 **ELECTRIC HEATING**, not incl. conduit or feed wiring.									
1100 Rule of thumb: Baseboard units, including control	1 Elec	4.40	1.818	kW	92.50	85.50		178	229
1300 Baseboard heaters, 2' long, 375 watt		8	1	Ea.	40	47		87	114
1400 3' long, 500 watt		8	1		46	47		93	121
1600 4' long, 750 watt		6.70	1.194		54	56		110	143
1800 5' long, 935 watt		5.70	1.404		68	66		134	173
2000 6' long, 1125 watt	↓	5	1.600	↓	74	75		149	194
5600 Unit heaters, heavy duty, with fan & mounting bracket									
5650 Single phase, 208-240-277 volt, 3 kW	1 Elec	3.20	2.500	Ea.	405	118		523	620
5750 5 kW		2.40	3.333		420	157		577	695
5800 7 kW		1.90	4.211		645	198		843	1,000
5850 10 kW		1.30	6.154		735	289		1,024	1,250
5950 15 kW		.90	8.889		1,200	420		1,620	1,925
6000 480 volt, 3 kW		3.30	2.424		400	114		514	610
6020 4 kW		3	2.667		455	125		580	685
6040 5 kW		2.60	3.077		465	145		610	725
6060 7 kW		2	4		690	188		878	1,050
6080 10 kW		1.40	5.714		755	269		1,024	1,225
6100 13 kW		1.10	7.273		1,200	340		1,540	1,800
6120 15 kW		1	8		1,200	375		1,575	1,850
6140 20 kW		.90	8.889		1,550	420		1,970	2,325
6300 3 phase, 208-240 volt, 5 kW		2.40	3.333		395	157		552	670
6320 7 kW		1.90	4.211		610	198		808	965
6340 10 kW		1.30	6.154		660	289		949	1,150
6360 15 kW		.90	8.889		1,100	420		1,520	1,850
6380 20 kW		.70	11.429		1,575	535		2,110	2,525
6400 25 kW		.50	16		1,850	750		2,600	3,150
6500 480 volt, 5 kW		2.60	3.077		550	145		695	820
6520 7 kW		2	4		695	188		883	1,050
6540 10 kW		1.40	5.714		735	269		1,004	1,200
6560 13 kW		1.10	7.273		1,200	340		1,540	1,800
6580 15 kW		1	8		1,200	375		1,575	1,850
6600 20 kW		.90	8.889		1,550	420		1,970	2,325
6620 25 kW		.60	13.333		1,850	625		2,475	2,950
6630 30 kW		.70	11.429		2,150	535		2,685	3,150
6640 40 kW		.60	13.333		2,725	625		3,350	3,925
6650 50 kW		.50	16		3,300	750		4,050	4,750
7410 Sill height convector heaters, 5" high x 2' long, 500 watt		6.70	1.194		297	56		353	410
7420 3' long, 750 watt		6.50	1.231	↓	350	58		408	470

23 83 Radiant Heating Units

23 83 33 – Electric Radiant Heaters

23 83 33.10 Electric Heating

		Crew	Daily Output	Labor-Hours	Unit	Material	2009 Bare Costs Labor	Equipment	Total	Total Incl O&P
7430	4' long, 1000 watt	1 Elec	6.20	1.290	Ea.	405	60.50		465.50	535
7440	5' long, 1250 watt		5.50	1.455		460	68.50		528.50	605
7450	6' long, 1500 watt		4.80	1.667		520	78.50		598.50	690
7460	8' long, 2000 watt		3.60	2.222		710	104		814	935
7470	10' long, 2500 watt		3	2.667		880	125		1,005	1,150
7900	Cabinet convector heaters, 240 volt									
7920	3' long, 2000 watt	1 Elec	5.30	1.509	Ea.	1,775	71		1,846	2,075
7940	3000 watt		5.30	1.509		1,850	71		1,921	2,150
7960	4000 watt		5.30	1.509		1,925	71		1,996	2,200
7980	6000 watt		4.60	1.739		2,000	81.50		2,081.50	2,300
8000	8000 watt		4.60	1.739		2,075	81.50		2,156.50	2,400
8020	4' long, 4000 watt		4.60	1.739		1,950	81.50		2,031.50	2,250
8040	6000 watt		4	2		2,025	94		2,119	2,375
8060	8000 watt		4	2		2,100	94		2,194	2,450
8080	10,000 watt		4	2		2,100	94		2,194	2,450

23 84 Humidity Control Equipment

23 84 13 – Humidifiers

23 84 13.10 Humidifier Units

		Crew	Daily Output	Labor-Hours	Unit	Material	2009 Bare Costs Labor	Equipment	Total	Total Incl O&P
0010	**HUMIDIFIER UNITS**									
0520	Steam, room or duct, filter, regulators, auto. controls, 220 V									
0540	11 lb. per hour	Q-5	6	2.667	Ea.	2,275	118		2,393	2,675
0560	22 lb. per hour		5	3.200		2,525	142		2,667	3,000
0580	33 lb. per hour		4	4		2,575	178		2,753	3,125
0600	50 lb. per hour		4	4		3,175	178		3,353	3,750
0620	100 lb. per hour		3	5.333		3,800	237		4,037	4,525
0700	With blower									
0720	11 lb. per hour	Q-5	5.50	2.909	Ea.	3,175	129		3,304	3,700
0760	33 lb. per hour		3.75	4.267		3,475	190		3,665	4,100
0800	100 lb. per hour		2.75	5.818		4,725	259		4,984	5,600

Estimating Tips

26 05 00 Common Work Results for Electrical

- Conduit should be taken off in three main categories—power distribution, branch power, and branch lighting—so the estimator can concentrate on systems and components, therefore making it easier to ensure all items have been accounted for.
- For cost modifications for elevated conduit installation, add the percentages to labor according to the height of installation, and only to the quantities exceeding the different height levels, not to the total conduit quantities.
- Remember that aluminum wiring of equal ampacity is larger in diameter than copper and may require larger conduit.
- If more than three wires at a time are being pulled, deduct percentages from the labor hours of that grouping of wires.
- When taking off grounding systems, identify separately the type and size of wire, and list each unique type of ground connection.
- The estimator should take the weights of materials into consideration when completing a takeoff. Topics to consider include: How will the materials be supported? What methods of support are available? How high will the support structure have to reach? Will the final support structure be able to withstand the total burden? Is the support material included or separate from the fixture, equipment, and material specified?
- Do not overlook the costs for equipment used in the installation. If scaffolding or highlifts are available in the field, contractors may use them in lieu of the proposed ladders and rolling staging.

26 20 00 Low-Voltage Electrical Transmission

- Supports and concrete pads may be shown on drawings for the larger equipment, or the support system may be only a piece of plywood for the back of a panelboard. In either case, it must be included in the costs.

26 40 00 Electrical and Cathodic Protection

- When taking off cathodic protections systems, identify the type and size of cable, and list each unique type of anode connection.

26 50 00 Lighting

- Fixtures should be taken off room by room, using the fixture schedule, specifications, and the ceiling plan. For large concentrations of lighting fixtures in the same area, deduct the percentages from labor hours.

Reference Numbers

Reference numbers are shown in shaded boxes at the beginning of some major classifications. These numbers refer to related items in the Reference Section. The reference information may be an estimating procedure, an alternate pricing method, or technical information.

Note: Not all subdivisions listed here necessarily appear in this publication.

Note: **Trade Service,** *in part, has been used as a reference source for some of the material prices used in Division 26.*

26 05 Common Work Results for Electrical

26 05 05 – Selective Electrical Demolition

26 05 05.10 Electrical Demolition		Crew	Daily Output	Labor-Hours	Unit	Material	2009 Bare Costs Labor	2009 Bare Costs Equipment	Total	Total Incl O&P
0010	**ELECTRICAL DEMOLITION**									
0020	Conduit to 15' high, including fittings & hangers									
0100	Rigid galvanized steel, 1/2" to 1" diameter	1 Elec	242	.033	L.F.		1.55		1.55	2.31
0120	1-1/4" to 2"	"	200	.040			1.88		1.88	2.80
0140	2-1/2" to 3-1/2"	2 Elec	302	.053			2.49		2.49	3.71
0200	Electric metallic tubing (EMT), 1/2" to 1"	1 Elec	394	.020			.95		.95	1.42
0220	1-1/4" to 1-1/2"	↓	326	.025			1.15		1.15	1.72
0240	2" to 3"		236	.034	↓		1.59		1.59	2.37
0400	Wiremold raceway, including fittings & hangers									
0420	No. 3000	1 Elec	250	.032	L.F.		1.50		1.50	2.24
0440	No. 4000	"	217	.037	"		1.73		1.73	2.58
0500	Channels, steel, including fittings & hangers									
0520	3/4" x 1-1/2"	1 Elec	308	.026	L.F.		1.22		1.22	1.82
0540	1-1/2" x 1-1/2"	"	269	.030	"		1.40		1.40	2.08
0600	Copper bus duct, indoor, 3 phase									
0610	Including hangers & supports									
0620	225 amp	2 Elec	135	.119	L.F.		5.55		5.55	8.30
0640	400 amp		106	.151			7.10		7.10	10.55
0660	600 amp		86	.186			8.75		8.75	13
0680	1000 amp		60	.267			12.55		12.55	18.65
0700	1600 amp		40	.400			18.80		18.80	28
0720	3000 amp	↓	10	1.600	↓		75		75	112
0800	Plug-in switches, 600V 3 ph, incl. disconnecting									
0820	wire, conduit terminations, 30 amp	1 Elec	15.50	.516	Ea.		24.50		24.50	36
0840	60 amp		13.90	.576			27		27	40.50
0850	100 amp		10.40	.769			36		36	54
0860	200 amp	↓	6.20	1.290			60.50		60.50	90.50
0940	1200 amp	2 Elec	2	8			375		375	560
0960	1600 amp	"	1.70	9.412	↓		440		440	660
1010	Safety switches, 250 or 600V, incl. disconnection									
1050	of wire & conduit terminations									
1100	30 amp	1 Elec	12.30	.650	Ea.		30.50		30.50	45.50
1120	60 amp		8.80	.909			42.50		42.50	63.50
1140	100 amp		7.30	1.096			51.50		51.50	76.50
1160	200 amp	↓	5	1.600	↓		75		75	112
1210	Panel boards, incl. removal of all breakers,									
1220	conduit terminations & wire connections									
1230	3 wire, 120/240V, 100A, to 20 circuits	1 Elec	2.60	3.077	Ea.		145		145	215
1240	200 amps, to 42 circuits	2 Elec	2.60	6.154			289		289	430
1260	4 wire, 120/208V, 125A, to 20 circuits	1 Elec	2.40	3.333			157		157	233
1270	200 amps, to 42 circuits	2 Elec	2.40	6.667	↓		315		315	465
1300	Transformer, dry type, 1 phase, incl. removal of									
1320	supports, wire & conduit terminations									
1340	1 kVA	1 Elec	7.70	1.039	Ea.		49		49	72.50
1360	5 kVA	"	4.70	1.702	↓		80		80	119
1420	75 kVA	2 Elec	2.50	6.400	↓		300		300	450
1440	3 phase to 600V, primary									
1460	3 kVA	1 Elec	3.85	2.078	Ea.		97.50		97.50	145
1480	15 kVA	2 Elec	4.20	3.810			179		179	266
1500	30 kVA	"	3.50	4.571			215		215	320
1530	112.5 kVA	R-3	2.90	6.897			320	44.50	364.50	525
1560	500 kVA		1.40	14.286			660	92.50	752.50	1,075
1570	750 kVA	↓	1.10	18.182	↓		840	117	957	1,375

26 05 Common Work Results for Electrical

26 05 05 – Selective Electrical Demolition

26 05 05.10 Electrical Demolition	Crew	Daily Output	Labor-Hours	Unit	Material	2009 Bare Costs Labor	2009 Bare Costs Equipment	Total	Total Incl O&P
1600 Pull boxes & cabinets, sheet metal, incl. removal									
1620 of supports and conduit terminations									
1640 6" x 6" x 4"	1 Elec	31.10	.257	Ea.		12.10		12.10	18
1660 12" x 12" x 4"		23.30	.343			16.15		16.15	24
1720 Junction boxes, 4" sq. & oct.		80	.100			4.70		4.70	7
1740 Handy box		107	.075			3.51		3.51	5.25
1760 Switch box		107	.075			3.51		3.51	5.25
1780 Receptacle & switch plates		257	.031			1.46		1.46	2.18
1800 Wire, THW-THWN-THHN, removed from									
1810 in place conduit, to 15' high									
1830 #14	1 Elec	65	.123	C.L.F.		5.80		5.80	8.60
1840 #12		55	.145			6.85		6.85	10.15
1850 #10		45.50	.176			8.25		8.25	12.30
1880 #4	2 Elec	53	.302			14.20		14.20	21
1890 #3		50	.320			15.05		15.05	22.50
1910 1/0		33.20	.482			22.50		22.50	33.50
1920 2/0		29.20	.548			26		26	38.50
1930 3/0		25	.640			30		30	45
1980 400 kcmil		17	.941			44		44	66
1990 500 kcmil		16.20	.988			46.50		46.50	69
2000 Interior fluorescent fixtures, incl. supports									
2010 & whips, to 15' high									
2100 Recessed drop-in 2' x 2', 2 lamp	2 Elec	35	.457	Ea.		21.50		21.50	32
2120 2' x 4', 2 lamp		33	.485			23		23	34
2140 2' x 4', 4 lamp		30	.533			25		25	37.50
2160 4' x 4', 4 lamp		20	.800			37.50		37.50	56
2180 Surface mount, acrylic lens & hinged frame									
2200 1' x 4', 2 lamp	2 Elec	44	.364	Ea.		17.10		17.10	25.50
2220 2' x 2', 2 lamp		44	.364			17.10		17.10	25.50
2260 2' x 4', 4 lamp		33	.485			23		23	34
2280 4' x 4', 4 lamp		23	.696			32.50		32.50	48.50
2300 Strip fixtures, surface mount									
2320 4' long, 1 lamp	2 Elec	53	.302	Ea.		14.20		14.20	21
2340 4' long, 2 lamp		50	.320			15.05		15.05	22.50
2360 8' long, 1 lamp		42	.381			17.90		17.90	26.50
2380 8' long, 2 lamp		40	.400			18.80		18.80	28
2400 Pendant mount, industrial, incl. removal									
2410 of chain or rod hangers, to 15' high									
2420 4' long, 2 lamp	2 Elec	35	.457	Ea.		21.50		21.50	32
2440 8' long, 2 lamp	"	27	.593	"		28		28	41.50
2460 Interior incandescent, surface, ceiling									
2470 or wall mount, to 12' high									
2480 Metal cylinder type, 75 Watt	2 Elec	62	.258	Ea.		12.15		12.15	18.05
2500 150 Watt	"	62	.258	"		12.15		12.15	18.05
2520 Metal halide, high bay									
2540 400 Watt	2 Elec	15	1.067	Ea.		50		50	74.50
2560 1000 Watt		12	1.333			62.50		62.50	93.50
2580 150 Watt, low bay		20	.800			37.50		37.50	56
2600 Exterior fixtures, incandescent, wall mount									

26 05 19 – Low-Voltage Electrical Power Conductors and Cables

26 05 19.13 Undercarpet Electrical Power Cables

0010 **UNDERCARPET ELECTRICAL POWER CABLES**									

26 05 Common Work Results for Electrical

26 05 19 – Low-Voltage Electrical Power Conductors and Cables

26 05 19.13 Undercarpet Electrical Power Cables

		Crew	Daily Output	Labor-Hours	Unit	Material	2009 Bare Costs Labor	Equipment	Total	Total Incl O&P
0020	Power System									
0100	Cable flat, 3 conductor, #12, w/attached bottom shield	1 Elec	982	.008	L.F.	4.94	.38		5.32	6
0200	Shield, top, steel		1768	.005	"	5.30	.21		5.51	6.10
0250	Splice, 3 conductor		48	.167	Ea.	16.10	7.85		23.95	29.50
0300	Top shield		96	.083		1.43	3.92		5.35	7.40
0350	Tap		40	.200		20.50	9.40		29.90	37
0400	Insulating patch, splice, tap, & end		48	.167		51	7.85		58.85	67.50
0450	Fold		230	.035			1.63		1.63	2.43
0500	Top shield, tap & fold		96	.083		1.43	3.92		5.35	7.40
0700	Transition, block assembly		77	.104		74.50	4.88		79.38	89
0750	Receptacle frame & base		32	.250		40.50	11.75		52.25	62.50
0800	Cover receptacle		120	.067		3.47	3.13		6.60	8.50
0850	Cover blank		160	.050		4.07	2.35		6.42	8
0860	Receptacle, direct connected, single		25	.320		87	15.05		102.05	119
0870	Dual		16	.500		143	23.50		166.50	192
0880	Combination Hi & Lo, tension		21	.381		105	17.90		122.90	143
0900	Box, floor with cover		20	.400		88.50	18.80		107.30	125
0920	Floor service w/barrier		4	2		250	94		344	415
1000	Wall, surface, with cover		20	.400		58	18.80		76.80	92
1100	Wall, flush, with cover		20	.400	▼	40.50	18.80		59.30	73
1450	Cable flat, 5 conductor #12, w/attached bottom shield		800	.010	L.F.	8.10	.47		8.57	9.60
1550	Shield, top, steel		1768	.005	"	8.05	.21		8.26	9.15
1600	Splice, 5 conductor		48	.167	Ea.	26	7.85		33.85	40
1650	Top shield		96	.083		1.43	3.92		5.35	7.40
1700	Tap		48	.167		34	7.85		41.85	49
1750	Insulating patch, splice tap, & end		83	.096		51	4.53		55.53	63
1800	Transition, block assembly		77	.104		53.50	4.88		58.38	66
1850	Box, wall, flush with cover		20	.400	▼	54	18.80		72.80	87.50
1900	Cable flat, 4 conductor, #12		933	.009	L.F.	6.55	.40		6.95	7.80
1950	3 conductor #10		982	.008		5.65	.38		6.03	6.80
1960	4 conductor #10		933	.009		7.45	.40		7.85	8.75
1970	5 conductor #10	▼	884	.009	▼	9.10	.43		9.53	10.65
2500	Telephone System									
2510	Transition fitting wall box, surface	1 Elec	24	.333	Ea.	43	15.65		58.65	71
2520	Flush		24	.333		43	15.65		58.65	71
2530	Flush, for PC board		24	.333		43	15.65		58.65	71
2540	Floor service box	▼	4	2		227	94		321	390
2550	Cover, surface					13.90			13.90	15.30
2560	Flush					13.90			13.90	15.30
2570	Flush for PC board					13.90			13.90	15.30
2700	Floor fitting w/duplex jack & cover	1 Elec	21	.381		43	17.90		60.90	73.50
2720	Low profile		53	.151		14.95	7.10		22.05	27
2740	Miniature w/duplex jack		53	.151		23	7.10		30.10	36
2760	25 pair kit		21	.381		45	17.90		62.90	76
2780	Low profile		53	.151		15.30	7.10		22.40	27.50
2800	Call director kit for 5 cable		19	.421		72	19.80		91.80	109
2820	4 pair kit		19	.421		85.50	19.80		105.30	124
2840	3 pair kit		19	.421		90.50	19.80		110.30	129
2860	Comb. 25 pair & 3 conductor power		21	.381		72	17.90		89.90	106
2880	5 conductor power		21	.381		82	17.90		99.90	117
2900	PC board, 8 per 3 pair		161	.050		62.50	2.34		64.84	72
2920	6 per 4 pair		161	.050		62.50	2.34		64.84	72
2940	3 pair adapter	▼	161	.050	▼	56.50	2.34		58.84	66

26 05 Common Work Results for Electrical

26 05 19 – Low-Voltage Electrical Power Conductors and Cables

26 05 19.13 Undercarpet Electrical Power Cables

		Crew	Daily Output	Labor-Hours	Unit	Material	2009 Bare Costs Labor	Equipment	Total	Total Incl O&P
2950	Plug	1 Elec	77	.104	Ea.	2.64	4.88		7.52	10.15
2960	Couplers		321	.025	↓	7.45	1.17		8.62	9.90
3000	Bottom shield for 25 pair cable		4420	.002	L.F.	.74	.09		.83	.94
3020	4 pair		4420	.002		.35	.09		.44	.52
3040	Top shield for 25 pair cable		4420	.002	↓	.74	.09		.83	.94
3100	Cable assembly, double-end, 50', 25 pair		11.80	.678	Ea.	219	32		251	289
3110	3 pair		23.60	.339		64.50	15.95		80.45	94.50
3120	4 pair		23.60	.339	↓	72	15.95		87.95	103
3140	Bulk 3 pair		1473	.005	L.F.	1.10	.26		1.36	1.59
3160	4 pair	↓	1473	.005	"	1.38	.26		1.64	1.90
3500	Data System									
3520	Cable 25 conductor w/connection 40', 75 ohm	1 Elec	14.50	.552	Ea.	65	26		91	110
3530	Single lead		22	.364		177	17.10		194.10	221
3540	Dual lead	↓	22	.364	↓	220	17.10		237.10	268
3560	Shields same for 25 conductor as 25 pair telephone									
3570	Single & dual, none required									
3590	BNC coax connectors, Plug	1 Elec	40	.200	Ea.	9.65	9.40		19.05	24.50
3600	TNC coax connectors, Plug	"	40	.200	"	12.40	9.40		21.80	27.50
3700	Cable-bulk									
3710	Single lead	1 Elec	1473	.005	L.F.	2.37	.26		2.63	2.99
3720	Dual lead	"	1473	.005	"	3.41	.26		3.67	4.13
3730	Hand tool crimp				Ea.	470			470	520
3740	Hand tool notch				"	16.90			16.90	18.60
3750	Boxes & floor fitting same as telephone									
3790	Data cable notching, 90°	1 Elec	97	.082	Ea.		3.88		3.88	5.75
3800	180°		60	.133			6.25		6.25	9.35
8100	Drill floor		160	.050	↓	1.87	2.35		4.22	5.55
8200	Marking floor		1600	.005	L.F.		.24		.24	.35
8300	Tape, hold down		6400	.001	"	.14	.06		.20	.24
8350	Tape primer, 500 ft. per can	↓	96	.083	Ea.	28.50	3.92		32.42	37.50
8400	Tool, splicing				"	228			228	250

26 05 19.20 Armored Cable

		Crew	Daily Output	Labor-Hours	Unit	Material	Labor	Equipment	Total	Total Incl O&P
0010	**ARMORED CABLE**									
0050	600 volt, copper (BX), #14, 2 conductor, solid	1 Elec	2.40	3.333	C.L.F.	77.50	157		234.50	320
0100	3 conductor, solid		2.20	3.636		121	171		292	385
0120	4 conductor, solid		2	4		162	188		350	460
0150	#12, 2 conductor, solid		2.30	3.478		79	163		242	330
0200	3 conductor, solid		2	4		128	188		316	420
0220	4 conductor, solid		1.80	4.444		169	209		378	495
0240	#12, 19 conductor, stranded		1.10	7.273		915	340		1,255	1,500
0250	#10, 2 conductor, solid		2	4		150	188		338	445
0300	3 conductor, solid		1.60	5		221	235		456	595
0320	4 conductor, solid		1.40	5.714		350	269		619	785
0350	#8, 3 conductor, solid		1.30	6.154		450	289		739	925
9010	600 volt, copper (MC) steel clad, #14, 2 wire		2.40	3.333		77.50	157		234.50	320
9020	3 wire		2.20	3.636		120	171		291	385
9030	4 wire		2	4		162	188		350	460
9040	#12, 2 wire		2.30	3.478		79	163		242	330
9050	3 wire		2	4		128	188		316	420
9060	4 wire		1.80	4.444		169	209		378	495
9070	#10, 2 wire		2	4		150	188		338	445
9080	3 wire	↓	1.60	5	↓	221	235		456	595

26 05 Common Work Results for Electrical

26 05 19 – Low-Voltage Electrical Power Conductors and Cables

26 05 19.20 Armored Cable		Crew	Daily Output	Labor-Hours	Unit	Material	2009 Bare Costs Labor	Equipment	Total	Total Incl O&P
9090	4 wire	1 Elec	1.40	5.714	C.L.F.	350	269		619	785
9100	#8, 2 wire, stranded		1.80	4.444		305	209		514	645
9110	3 wire, stranded		1.30	6.154		455	289		744	930

26 05 19.90 Wire

0010	**WIRE**									
0020	600 volt type THW, copper, solid, #14	1 Elec	13	.615	C.L.F.	10.30	29		39.30	54.50
0920	Type THWN-THHN, copper, solid, #14		13	.615		10.30	29		39.30	54.50
0940	#12		11	.727		15.90	34		49.90	68.50
0960	#10		10	.800		25	37.50		62.50	83.50
1000	Stranded, #14		13	.615		10.40	29		39.40	54.50
1200	#12		11	.727		16.25	34		50.25	69
1250	#10		10	.800		25	37.50		62.50	83.50
1300	#8		8	1		43.50	47		90.50	118
1350	#6		6.50	1.231		67.50	58		125.50	161

26 05 33 – Raceway and Boxes for Electrical Systems

26 05 33.05 Conduit

0010	**CONDUIT** To 15' high, includes 2 terminations, 2 elbows,									
0020	11 beam clamps, and 11 couplings per 100 L.F.									
5000	Electric metallic tubing (EMT), 1/2" diameter	1 Elec	170	.047	L.F.	.66	2.21		2.87	4.01
5020	3/4" diameter		130	.062		1.05	2.89		3.94	5.45
5040	1" diameter		115	.070		1.84	3.27		5.11	6.90
5060	1-1/4" diameter		100	.080		2.95	3.76		6.71	8.85
5080	1-1/2" diameter		90	.089		3.78	4.18		7.96	10.35
5100	2" diameter		80	.100		4.88	4.70		9.58	12.35
5120	2-1/2" diameter		60	.133		11.70	6.25		17.95	22
5140	3" diameter	2 Elec	100	.160		13.75	7.50		21.25	26.50
5160	3-1/2" diameter		90	.178		17.40	8.35		25.75	31.50
5180	4" diameter		80	.200		18.80	9.40		28.20	34.50
5700	Elbows, 1" diameter	1 Elec	40	.200	Ea.	8.20	9.40		17.60	23
5720	1-1/4" diameter		32	.250		10.15	11.75		21.90	28.50
5740	1-1/2" diameter		24	.333		11.85	15.65		27.50	36.50
5760	2" diameter		20	.400		17.40	18.80		36.20	47
5780	2-1/2" diameter		12	.667		42.50	31.50		74	93
5800	3" diameter		9	.889		63	42		105	132
5820	3-1/2" diameter		7	1.143		84.50	53.50		138	173
5840	4" diameter		6	1.333		99.50	62.50		162	203
6200	Couplings, set screw, steel, 1/2" diameter					1.70			1.70	1.87
6220	3/4" diameter					2.56			2.56	2.82
6240	1" diameter					4.28			4.28	4.71
6260	1-1/4" diameter					9			9	9.90
6280	1-1/2" diameter					12.50			12.50	13.75
6300	2" diameter					16.75			16.75	18.45
6320	2-1/2" diameter					53.50			53.50	59
6340	3" diameter					57.50			57.50	63
6360	3-1/2" diameter					63.50			63.50	70
6380	4" diameter					68.50			68.50	75.50
6500	Box connectors, set screw, steel, 1/2" diameter	1 Elec	120	.067		1.10	3.13		4.23	5.85
6520	3/4" diameter		110	.073		1.79	3.42		5.21	7.05
6540	1" diameter		90	.089		3.36	4.18		7.54	9.90
6560	1-1/4" diameter		70	.114		6.60	5.35		11.95	15.25
6580	1-1/2" diameter		60	.133		9.50	6.25		15.75	19.80
6600	2" diameter		50	.160		13.15	7.50		20.65	25.50

26 05 Common Work Results for Electrical

26 05 33 – Raceway and Boxes for Electrical Systems

26 05 33.05 Conduit

		Crew	Daily Output	Labor-Hours	Unit	Material	2009 Bare Costs Labor	Equipment	Total	Total Incl O&P
6620	2-1/2" diameter	1 Elec	36	.222	Ea.	46	10.45		56.45	66
6640	3" diameter		27	.296		55.50	13.95		69.45	81.50
6680	3-1/2" diameter		21	.381		77	17.90		94.90	111
6700	4" diameter		16	.500		86	23.50		109.50	130
6740	Insulated box connectors, set screw, steel, 1/2" diameter		120	.067		1.51	3.13		4.64	6.30
6760	3/4" diameter		110	.073		2.38	3.42		5.80	7.70
6780	1" diameter		90	.089		4.38	4.18		8.56	11
6800	1-1/4" diameter		70	.114		8	5.35		13.35	16.80
6820	1-1/2" diameter		60	.133		11.25	6.25		17.50	22
6840	2" diameter		50	.160		16.35	7.50		23.85	29
6860	2-1/2" diameter		36	.222		78	10.45		88.45	102
6880	3" diameter		27	.296		91.50	13.95		105.45	122
6900	3-1/2" diameter		21	.381		128	17.90		145.90	168
6920	4" diameter		16	.500		140	23.50		163.50	189

26 05 33.50 Outlet Boxes

		Crew	Daily Output	Labor-Hours	Unit	Material	Labor	Equipment	Total	Total Incl O&P
0010	**OUTLET BOXES**									
2000	Poke-thru fitting, fire rated, for 3-3/4" floor	1 Elec	6.80	1.176	Ea.	111	55.50		166.50	206
2040	For 7" floor		6.80	1.176		111	55.50		166.50	206
2100	Pedestal, 15 amp, duplex receptacle & blank plate		5.25	1.524		122	71.50		193.50	241
2120	Duplex receptacle and telephone plate		5.25	1.524		122	71.50		193.50	241
2140	Pedestal, 20 amp, duplex recept. & phone plate		5	1.600		123	75		198	247
2160	Telephone plate, both sides		5.25	1.524		117	71.50		188.50	236
2200	Abandonment plate		32	.250		35.50	11.75		47.25	56.50

26 05 33.60 Outlet Boxes, Plastic

		Crew	Daily Output	Labor-Hours	Unit	Material	Labor	Equipment	Total	Total Incl O&P
0010	**OUTLET BOXES, PLASTIC**									
0050	4" diameter, round with 2 mounting nails	1 Elec	25	.320	Ea.	2.96	15.05		18.01	26
0100	Bar hanger mounted		25	.320		5.35	15.05		20.40	28.50
0200	4", square with 2 mounting nails		25	.320		4.44	15.05		19.49	27.50
0300	Plaster ring		64	.125		1.84	5.90		7.74	10.75
0400	Switch box with 2 mounting nails, 1 gang		30	.267		2.01	12.55		14.56	21
0500	2 gang		25	.320		4.12	15.05		19.17	27
0600	3 gang		20	.400		6.50	18.80		25.30	35
0700	Old work box		30	.267		3.96	12.55		16.51	23

26 05 90 – Residential Applications

26 05 90.10 Residential Wiring

		Crew	Daily Output	Labor-Hours	Unit	Material	Labor	Equipment	Total	Total Incl O&P
0010	**RESIDENTIAL WIRING**									
0020	20' avg. runs and #14/2 wiring incl. unless otherwise noted									
1000	Service & panel, includes 24' SE-AL cable, service eye, meter,									
1010	Socket, panel board, main bkr., ground rod, 15 or 20 amp									
1020	1-pole circuit breakers, and misc. hardware									
1100	100 amp, with 10 branch breakers	1 Elec	1.19	6.723	Ea.	585	315		900	1,100
1110	With PVC conduit and wire		.92	8.696		660	410		1,070	1,350
1120	With RGS conduit and wire		.73	10.959		840	515		1,355	1,700
1150	150 amp, with 14 branch breakers		1.03	7.767		895	365		1,260	1,525
1170	With PVC conduit and wire		.82	9.756		1,050	460		1,510	1,825
1180	With RGS conduit and wire		.67	11.940		1,400	560		1,960	2,350
1200	200 amp, with 18 branch breakers	2 Elec	1.80	8.889		1,175	420		1,595	1,925
1220	With PVC conduit and wire		1.46	10.959		1,325	515		1,840	2,250
1230	With RGS conduit and wire		1.24	12.903		1,800	605		2,405	2,875
1800	Lightning surge suppressor for above services, add	1 Elec	32	.250		50.50	11.75		62.25	73
2000	Switch devices									
2100	Single pole, 15 amp, Ivory, with a 1-gang box, cover plate,									

26 05 Common Work Results for Electrical

26 05 90 – Residential Applications

26 05 90.10 Residential Wiring		Crew	Daily Output	Labor-Hours	Unit	Material	2009 Bare Costs Labor	Equipment	Total	Total Incl O&P
2110	Type NM (Romex) cable	1 Elec	17.10	.468	Ea.	11.70	22		33.70	45.50
2120	Type MC (BX) cable		14.30	.559		28	26.50		54.50	70
2130	EMT & wire		5.71	1.401		34.50	66		100.50	136
2150	3-way, #14/3, type NM cable		14.55	.550		17.25	26		43.25	57.50
2170	Type MC cable		12.31	.650		39	30.50		69.50	88.50
2180	EMT & wire		5	1.600		38.50	75		113.50	155
2200	4-way, #14/3, type NM cable		14.55	.550		33.50	26		59.50	75.50
2220	Type MC cable		12.31	.650		55.50	30.50		86	107
2230	EMT & wire		5	1.600		55	75		130	173
2250	S.P., 20 amp, #12/2, type NM cable		13.33	.600		19.70	28		47.70	63.50
2270	Type MC cable		11.43	.700		33	33		66	85.50
2280	EMT & wire		4.85	1.649		44.50	77.50		122	164
2290	S.P. rotary dimmer, 600W, no wiring		17	.471		19.90	22		41.90	55
2300	S.P. rotary dimmer, 600W, type NM cable		14.55	.550		26.50	26		52.50	67.50
2320	Type MC cable		12.31	.650		43	30.50		73.50	92.50
2330	EMT & wire		5	1.600		51.50	75		126.50	169
2350	3-way rotary dimmer, type NM cable		13.33	.600		22.50	28		50.50	66.50
2370	Type MC cable		11.43	.700		39	33		72	92
2380	EMT & wire		4.85	1.649		47.50	77.50		125	167
2400	Interval timer wall switch, 20 amp, 1-30 min., #12/2									
2410	Type NM cable	1 Elec	14.55	.550	Ea.	53	26		79	96.50
2420	Type MC cable		12.31	.650		60	30.50		90.50	112
2430	EMT & wire		5	1.600		77.50	75		152.50	198
2500	Decorator style									
2510	S.P., 15 amp, type NM cable	1 Elec	17.10	.468	Ea.	16.15	22		38.15	50.50
2520	Type MC cable		14.30	.559		32.50	26.50		59	75
2530	EMT & wire		5.71	1.401		39	66		105	141
2550	3-way, #14/3, type NM cable		14.55	.550		21.50	26		47.50	62.50
2570	Type MC cable		12.31	.650		43.50	30.50		74	93
2580	EMT & wire		5	1.600		43	75		118	160
2600	4-way, #14/3, type NM cable		14.55	.550		38	26		64	80.50
2620	Type MC cable		12.31	.650		60	30.50		90.50	111
2630	EMT & wire		5	1.600		59.50	75		134.50	178
2650	S.P., 20 amp, #12/2, type NM cable		13.33	.600		24	28		52	68.50
2670	Type MC cable		11.43	.700		37.50	33		70.50	90.50
2680	EMT & wire		4.85	1.649		49	77.50		126.50	169
2700	S.P., slide dimmer, type NM cable		17.10	.468		30	22		52	66
2720	Type MC cable		14.30	.559		46.50	26.50		73	90.50
2730	EMT & wire		5.71	1.401		55	66		121	159
2750	S.P., touch dimmer, type NM cable		17.10	.468		27	22		49	62.50
2770	Type MC cable		14.30	.559		43.50	26.50		70	87
2780	EMT & wire		5.71	1.401		52	66		118	156
2800	3-way touch dimmer, type NM cable		13.33	.600		47	28		75	94
2820	Type MC cable		11.43	.700		63.50	33		96.50	119
2830	EMT & wire		4.85	1.649		72	77.50		149.50	194
3000	Combination devices									
3100	S.P. switch/15 amp recpt., Ivory, 1-gang box, plate									
3110	Type NM cable	1 Elec	11.43	.700	Ea.	22.50	33		55.50	74
3120	Type MC cable		10	.800		39	37.50		76.50	99
3130	EMT & wire		4.40	1.818		47.50	85.50		133	179
3150	S.P. switch/pilot light, type NM cable		11.43	.700		23.50	33		56.50	74.50
3170	Type MC cable		10	.800		40	37.50		77.50	100
3180	EMT & wire		4.43	1.806		48	85		133	179

26 05 Common Work Results for Electrical

26 05 90 – Residential Applications

26 05 90.10 Residential Wiring		Crew	Daily Output	Labor-Hours	Unit	Material	2009 Bare Costs Labor	Equipment	Total	Total Incl O&P
3190	2-S.P. switches, 2-#14/2, no wiring	1 Elec	14	.571	Ea.	7.80	27		34.80	48.50
3200	2-S.P. switches, 2-#14/2, type NM cables		10	.800		28	37.50		65.50	87
3220	Type MC cable		8.89	.900		53.50	42.50		96	122
3230	EMT & wire		4.10	1.951		52.50	91.50		144	194
3250	3-way switch/15 amp recpt., #14/3, type NM cable		10	.800		31.50	37.50		69	91
3270	Type MC cable		8.89	.900		53.50	42.50		96	122
3280	EMT & wire		4.10	1.951		53	91.50		144.50	195
3300	2-3 way switches, 2-#14/3, type NM cables		8.89	.900		44.50	42.50		87	112
3320	Type MC cable		8	1		80	47		127	158
3330	EMT & wire		4	2		62	94		156	208
3350	S.P. switch/20 amp recpt., #12/2, type NM cable		10	.800		36	37.50		73.50	95.50
3370	Type MC cable		8.89	.900		43	42.50		85.50	111
3380	EMT & wire		4.10	1.951		60.50	91.50		152	203
3400	Decorator style									
3410	S.P. switch/15 amp recpt., type NM cable	1 Elec	11.43	.700	Ea.	27	33		60	78.50
3420	Type MC cable		10	.800		43.50	37.50		81	104
3430	EMT & wire		4.40	1.818		52	85.50		137.50	184
3450	S.P. switch/pilot light, type NM cable		11.43	.700		28	33		61	79.50
3470	Type MC cable		10	.800		44.50	37.50		82	105
3480	EMT & wire		4.40	1.818		52.50	85.50		138	185
3500	2-S.P. switches, 2-#14/2, type NM cables		10	.800		32.50	37.50		70	92
3520	Type MC cable		8.89	.900		58	42.50		100.50	127
3530	EMT & wire		4.10	1.951		57	91.50		148.50	199
3550	3-way/15 amp recpt., #14/3, type NM cable		10	.800		36	37.50		73.50	95.50
3570	Type MC cable		8.89	.900		58	42.50		100.50	127
3580	EMT & wire		4.10	1.951		57.50	91.50		149	200
3650	2-3 way switches, 2-#14/3, type NM cables		8.89	.900		48.50	42.50		91	117
3670	Type MC cable		8	1		84.50	47		131.50	163
3680	EMT & wire		4	2		66.50	94		160.50	213
3700	S.P. switch/20 amp recpt., #12/2, type NM cable		10	.800		40.50	37.50		78	101
3720	Type MC cable		8.89	.900		47.50	42.50		90	116
3730	EMT & wire		4.10	1.951		65	91.50		156.50	208
4000	Receptacle devices									
4010	Duplex outlet, 15 amp recpt., Ivory, 1-gang box, plate									
4015	Type NM cable	1 Elec	14.55	.550	Ea.	10.20	26		36.20	49.50
4020	Type MC cable		12.31	.650		26.50	30.50		57	75
4030	EMT & wire		5.33	1.501		33	70.50		103.50	142
4050	With #12/2, type NM cable		12.31	.650		13.65	30.50		44.15	60.50
4070	Type MC cable		10.67	.750		27	35		62	82
4080	EMT & wire		4.71	1.699		38.50	80		118.50	162
4100	20 amp recpt., #12/2, type NM cable		12.31	.650		22.50	30.50		53	70
4120	Type MC cable		10.67	.750		35.50	35		70.50	91.50
4130	EMT & wire		4.71	1.699		47	80		127	171
4140	For GFI see Div. 26 05 90.10 line 4300 below									
4150	Decorator style, 15 amp recpt., type NM cable	1 Elec	14.55	.550	Ea.	14.65	26		40.65	54.50
4170	Type MC cable		12.31	.650		31	30.50		61.50	79.50
4180	EMT & wire		5.33	1.501		37.50	70.50		108	146
4200	With #12/2, type NM cable		12.31	.650		18.10	30.50		48.60	65.50
4220	Type MC cable		10.67	.750		31.50	35		66.50	87
4230	EMT & wire		4.71	1.699		43	80		123	166
4250	20 amp recpt. #12/2, type NM cable		12.31	.650		26.50	30.50		57	75
4270	Type MC cable		10.67	.750		40	35		75	96.50
4280	EMT & wire		4.71	1.699		51.50	80		131.50	176

26 05 Common Work Results for Electrical

26 05 90 – Residential Applications

26 05 90.10 Residential Wiring		Crew	Daily Output	Labor-Hours	Unit	Material	2009 Bare Costs Labor	Equipment	Total	Total Incl O&P
4300	GFI, 15 amp recpt., type NM cable	1 Elec	12.31	.650	Ea.	41.50	30.50		72	91
4320	Type MC cable		10.67	.750		58	35		93	116
4330	EMT & wire		4.71	1.699		64.50	80		144.50	190
4350	GFI with #12/2, type NM cable		10.67	.750		45	35		80	102
4370	Type MC cable		9.20	.870		58.50	41		99.50	125
4380	EMT & wire		4.21	1.900		69.50	89.50		159	210
4400	20 amp recpt., #12/2 type NM cable		10.67	.750		47	35		82	104
4420	Type MC cable		9.20	.870		60	41		101	127
4430	EMT & wire		4.21	1.900		71.50	89.50		161	212
4500	Weather-proof cover for above receptacles, add	↓	32	.250	↓	5.25	11.75		17	23.50
4550	Air conditioner outlet, 20 amp-240 volt recpt.									
4560	30' of #12/2, 2 pole circuit breaker									
4570	Type NM cable	1 Elec	10	.800	Ea.	63	37.50		100.50	125
4580	Type MC cable		9	.889		79	42		121	149
4590	EMT & wire		4	2		87.50	94		181.50	236
4600	Decorator style, type NM cable		10	.800		67.50	37.50		105	130
4620	Type MC cable		9	.889		84	42		126	154
4630	EMT & wire	↓	4	2	↓	92	94		186	241
4650	Dryer outlet, 30 amp-240 volt recpt., 20' of #10/3									
4660	2 pole circuit breaker									
4670	Type NM cable	1 Elec	6.41	1.248	Ea.	75	58.50		133.50	170
4680	Type MC cable		5.71	1.401		83	66		149	190
4690	EMT & wire		3.48	2.299		89	108		197	259
4700	Range outlet, 50 amp-240 volt recpt., 30' of #8/3									
4710	Type NM cable	1 Elec	4.21	1.900	Ea.	110	89.50		199.50	253
4720	Type MC cable		4	2		195	94		289	355
4730	EMT & wire		2.96	2.703		121	127		248	320
4750	Central vacuum outlet, Type NM cable		6.40	1.250		69.50	59		128.50	164
4770	Type MC cable		5.71	1.401		94	66		160	201
4780	EMT & wire	↓	3.48	2.299	↓	96.50	108		204.50	267
4800	30 amp-110 volt locking recpt., #10/2 circ. bkr.									
4810	Type NM cable	1 Elec	6.20	1.290	Ea.	80.50	60.50		141	179
4820	Type MC cable		5.40	1.481		111	69.50		180.50	226
4830	EMT & wire	↓	3.20	2.500	↓	110	118		228	296
4900	Low voltage outlets									
4910	Telephone recpt., 20' of 4/C phone wire	1 Elec	26	.308	Ea.	10.30	14.45		24.75	33
4920	TV recpt., 20' of RG59U coax wire, F type connector	"	16	.500	"	18.35	23.50		41.85	55
4950	Door bell chime, transformer, 2 buttons, 60' of bellwire									
4970	Economy model	1 Elec	11.50	.696	Ea.	75	32.50		107.50	131
4980	Custom model		11.50	.696		116	32.50		148.50	177
4990	Luxury model, 3 buttons	↓	9.50	.842	↓	310	39.50		349.50	400
6000	Lighting outlets									
6050	Wire only (for fixture), type NM cable	1 Elec	32	.250	Ea.	8.45	11.75		20.20	27
6070	Type MC cable		24	.333		18.55	15.65		34.20	44
6080	EMT & wire		10	.800		23.50	37.50		61	81.50
6100	Box (4"), and wire (for fixture), type NM cable		25	.320		17.45	15.05		32.50	41.50
6120	Type MC cable		20	.400		27.50	18.80		46.30	58.50
6130	EMT & wire	↓	11	.727		32.50	34		66.50	86.50
6200	Fixtures (use with lines 6050 or 6100 above)									
6210	Canopy style, economy grade	1 Elec	40	.200	Ea.	29	9.40		38.40	46
6220	Custom grade		40	.200		51	9.40		60.40	70
6250	Dining room chandelier, economy grade		19	.421		82.50	19.80		102.30	121
6260	Custom grade		19	.421		243	19.80		262.80	297

26 05 Common Work Results for Electrical

26 05 90 – Residential Applications

26 05 90.10 Residential Wiring

		Crew	Daily Output	Labor-Hours	Unit	Material	2009 Bare Costs Labor	Equipment	Total	Total Incl O&P
6270	Luxury grade	1 Elec	15	.533	Ea.	540	25		565	635
6310	Kitchen fixture (fluorescent), economy grade		30	.267		61.50	12.55		74.05	86
6320	Custom grade		25	.320		168	15.05		183.05	208
6350	Outdoor, wall mounted, economy grade		30	.267		29.50	12.55		42.05	51
6360	Custom grade		30	.267		108	12.55		120.55	138
6370	Luxury grade		25	.320		244	15.05		259.05	291
6410	Outdoor PAR floodlights, 1 lamp, 150 watt		20	.400		32	18.80		50.80	63
6420	2 lamp, 150 watt each		20	.400		52	18.80		70.80	85
6425	Motion sensing, 2 lamp, 150 watt each		20	.400		70	18.80		88.80	105
6430	For infrared security sensor, add		32	.250		111	11.75		122.75	140
6450	Outdoor, quartz-halogen, 300 watt flood		20	.400		41	18.80		59.80	73
6600	Recessed downlight, round, pre-wired, 50 or 75 watt trim		30	.267		42	12.55		54.55	64.50
6610	With shower light trim		30	.267		52	12.55		64.55	75.50
6620	With wall washer trim		28	.286		62.50	13.45		75.95	89
6630	With eye-ball trim		28	.286		62.50	13.45		75.95	89
6640	For direct contact with insulation, add					2.10			2.10	2.31
6700	Porcelain lamp holder	1 Elec	40	.200		4.10	9.40		13.50	18.50
6710	With pull switch		40	.200		4.59	9.40		13.99	19.05
6750	Fluorescent strip, 1-20 watt tube, wrap around diffuser, 24"		24	.333		54	15.65		69.65	83
6760	1-40 watt tube, 48"		24	.333		66	15.65		81.65	96
6770	2-40 watt tubes, 48"		20	.400		80	18.80		98.80	116
6780	With residential ballast		20	.400		90	18.80		108.80	127
6800	Bathroom heat lamp, 1-250 watt		28	.286		46	13.45		59.45	70.50
6810	2-250 watt lamps		28	.286		71.50	13.45		84.95	98.50
6820	For timer switch, see Div. 26 05 90.10 line 2400									
6900	Outdoor post lamp, incl. post, fixture, 35' of #14/2									
6910	Type NMC cable	1 Elec	3.50	2.286	Ea.	203	107		310	385
6920	Photo-eye, add		27	.296		32.50	13.95		46.45	56.50
6950	Clock dial time switch, 24 hr., w/enclosure, type NM cable		11.43	.700		63	33		96	118
6970	Type MC cable		11	.727		79.50	34		113.50	138
6980	EMT & wire		4.85	1.649		85.50	77.50		163	209
7000	Alarm systems									
7050	Smoke detectors, box, #14/3, type NM cable	1 Elec	14.55	.550	Ea.	37.50	26		63.50	79.50
7070	Type MC cable		12.31	.650		53	30.50		83.50	104
7080	EMT & wire		5	1.600		52.50	75		127.50	170
7090	For relay output to security system, add					12.90			12.90	14.20
8000	Residential equipment									
8050	Disposal hook-up, incl. switch, outlet box, 3' of flex									
8060	20 amp-1 pole circ. bkr., and 25' of #12/2									
8070	Type NM cable	1 Elec	10	.800	Ea.	33.50	37.50		71	93
8080	Type MC cable		8	1		48.50	47		95.50	124
8090	EMT & wire		5	1.600		61	75		136	180
8100	Trash compactor or dishwasher hook-up, incl. outlet box,									
8110	3' of flex, 15 amp-1 pole circ. bkr., and 25' of #14/2									
8120	Type NM cable	1 Elec	10	.800	Ea.	24.50	37.50		62	83
8130	Type MC cable		8	1		43.50	47		90.50	118
8140	EMT & wire		5	1.600		52.50	75		127.50	170
8150	Hot water sink dispensor hook-up, use line 8100									
8200	Vent/exhaust fan hook-up, type NM cable	1 Elec	32	.250	Ea.	8.45	11.75		20.20	27
8220	Type MC cable		24	.333		18.55	15.65		34.20	44
8230	EMT & wire		10	.800		23.50	37.50		61	81.50
8250	Bathroom vent fan, 50 CFM (use with above hook-up)									
8260	Economy model	1 Elec	15	.533	Ea.	24.50	25		49.50	64.50

26 05 Common Work Results for Electrical

26 05 90 – Residential Applications

26 05 90.10 Residential Wiring		Crew	Daily Output	Labor-Hours	Unit	Material	2009 Bare Costs Labor	Equipment	Total	Total Incl O&P
8270	Low noise model	1 Elec	15	.533	Ea.	32	25		57	72.50
8280	Custom model	↓	12	.667	↓	117	31.50		148.50	176
8300	Bathroom or kitchen vent fan, 110 CFM									
8310	Economy model	1 Elec	15	.533	Ea.	66.50	25		91.50	111
8320	Low noise model	"	15	.533	"	86	25		111	132
8350	Paddle fan, variable speed (w/o lights)									
8360	Economy model (AC motor)	1 Elec	10	.800	Ea.	105	37.50		142.50	172
8370	Custom model (AC motor)		10	.800		182	37.50		219.50	256
8380	Luxury model (DC motor)		8	1		360	47		407	465
8390	Remote speed switch for above, add	↓	12	.667	↓	28.50	31.50		60	78
8500	Whole house exhaust fan, ceiling mount, 36", variable speed									
8510	Remote switch, incl. shutters, 20 amp-1 pole circ. bkr.									
8520	30' of #12/2, type NM cable	1 Elec	4	2	Ea.	985	94		1,079	1,225
8530	Type MC cable		3.50	2.286		1,000	107		1,107	1,250
8540	EMT & wire	↓	3	2.667	↓	1,025	125		1,150	1,300
8600	Whirlpool tub hook-up, incl. timer switch, outlet box									
8610	3' of flex, 20 amp-1 pole GFI circ. bkr.									
8620	30' of #12/2, type NM cable	1 Elec	5	1.600	Ea.	127	75		202	251
8630	Type MC cable		4.20	1.905		135	89.50		224.50	282
8640	EMT & wire	↓	3.40	2.353	↓	146	111		257	325
8650	Hot water heater hook-up, incl. 1-2 pole circ. bkr., box;									
8660	3' of flex, 20' of #10/2, type NM cable	1 Elec	5	1.600	Ea.	36	75		111	152
8670	Type MC cable		4.20	1.905		58	89.50		147.50	197
8680	EMT & wire	↓	3.40	2.353	↓	55.50	111		166.50	226
9000	Heating/air conditioning									
9050	Furnace/boiler hook-up, incl. firestat, local on-off switch									
9060	Emergency switch, and 40' of type NM cable	1 Elec	4	2	Ea.	57	94		151	203
9070	Type MC cable		3.50	2.286		82.50	107		189.50	251
9080	EMT & wire	↓	1.50	5.333	↓	94	251		345	480
9100	Air conditioner hook-up, incl. local 60 amp disc. switch									
9110	3' sealtite, 40 amp, 2 pole circuit breaker									
9130	40' of #8/2, type NM cable	1 Elec	3.50	2.286	Ea.	229	107		336	410
9140	Type MC cable		3	2.667		360	125		485	580
9150	EMT & wire	↓	1.30	6.154	↓	268	289		557	725
9200	Heat pump hook-up, 1-40 & 1-100 amp 2 pole circ. bkr.									
9210	Local disconnect switch, 3' sealtite									
9220	40' of #8/2 & 30' of #3/2									
9230	Type NM cable	1 Elec	1.30	6.154	Ea.	595	289		884	1,075
9240	Type MC cable		1.08	7.407		865	350		1,215	1,475
9250	EMT & wire	↓	.94	8.511	↓	675	400		1,075	1,350
9500	Thermostat hook-up, using low voltage wire									
9520	Heating only, 25' of #18-3	1 Elec	24	.333	Ea.	7.70	15.65		23.35	32
9530	Heating/cooling, 25' of #18-4	"	20	.400	"	9.15	18.80		27.95	38

26 24 Switchboards and Panelboards

26 24 16 – Panelboards

26 24 16.30 Panelboards Commercial Applications	Crew	Daily Output	Labor-Hours	Unit	Material	2009 Bare Costs Labor	Equipment	Total	Total Incl O&P
0010 **PANELBOARDS COMMERCIAL APPLICATIONS**									
0050 NQOD, w/20 amp 1 pole bolt-on circuit breakers									
0100 3 wire, 120/240 volts, 100 amp main lugs									
0150 10 circuits	1 Elec	1	8	Ea.	495	375		870	1,100
0200 14 circuits		.88	9.091		580	425		1,005	1,275
0250 18 circuits		.75	10.667		635	500		1,135	1,450
0300 20 circuits		.65	12.308		715	580		1,295	1,650
0600 4 wire, 120/208 volts, 100 amp main lugs, 12 circuits		1	8		560	375		935	1,175
0650 16 circuits		.75	10.667		645	500		1,145	1,450
0700 20 circuits		.65	12.308		750	580		1,330	1,675
0750 24 circuits		.60	13.333		815	625		1,440	1,825
0800 30 circuits		.53	15.094		940	710		1,650	2,075
1600 NQOD panel, w/20 amp, 1 pole, circuit breakers									
1650 3 wire, 120/240 volt with main circuit breaker									
1700 100 amp main, 12 circuits	1 Elec	.80	10	Ea.	685	470		1,155	1,450
1750 20 circuits	"	.60	13.333	"	875	625		1,500	1,900
2000 4 wire, 120/208 volts with main circuit breaker									
2050 100 amp main, 24 circuits	1 Elec	.47	17.021	Ea.	1,025	800		1,825	2,325
2100 30 circuits	"	.40	20	"	1,150	940		2,090	2,675

26 27 Low-Voltage Distribution Equipment

26 27 23 – Indoor Service Poles

26 27 23.40 Surface Raceway

		Crew	Daily Output	Labor-Hours	Unit	Material	Labor	Equipment	Total	Total Incl O&P
0010	**SURFACE RACEWAY**									
0090	Metal, straight section									
0100	No. 500	1 Elec	100	.080	L.F.	.98	3.76		4.74	6.70
0110	No. 700		100	.080		1.10	3.76		4.86	6.80
0400	No. 1500, small pancake		90	.089		1.99	4.18		6.17	8.40
0670	No. 2200, base & cover, blank		80	.100		3	4.70		7.70	10.30
0700	Receptacle, 18" O.C.		36	.222		7.40	10.45		17.85	23.50
0720	30" O.C.		40	.200		7	9.40		16.40	21.50
0800	No. 3000, base & cover, blank		75	.107		3.91	5		8.91	11.75
0810	Receptacle, 6" O.C.		45	.178		35	8.35		43.35	51
0820	12" O.C.		62	.129		19.80	6.05		25.85	31
0830	18" O.C.		64	.125		16	5.90		21.90	26.50
0840	24" O.C.		66	.121		12	5.70		17.70	21.50
0850	30" O.C.		68	.118		10.80	5.55		16.35	20
0860	60" O.C.		70	.114		7.90	5.35		13.25	16.70
2400	Fittings, elbows, No. 500		40	.200	Ea.	1.76	9.40		11.16	15.95
2800	Elbow cover, No. 2000		40	.200		3.35	9.40		12.75	17.70
2880	Tee, No. 500		42	.190		3.40	8.95		12.35	17.05
2900	No. 2000		27	.296		11.15	13.95		25.10	33
3000	Switch box, No. 500		16	.500		11.55	23.50		35.05	47.50
3400	Telephone outlet, No. 1500		16	.500		13.80	23.50		37.30	50
3600	Junction box, No. 1500		16	.500		9.40	23.50		32.90	45.50
3800	Plugmold wired sections, No. 2000									
4000	1 circuit, 6 outlets, 3 ft. long	1 Elec	8	1	Ea.	35.50	47		82.50	110
4100	2 circuits, 8 outlets, 6 ft. long	"	5.30	1.509	"	57	71		128	169
4300	Overhead distribution systems, 125 volt									
4800	No. 2000, entrance end fitting	1 Elec	20	.400	Ea.	4.75	18.80		23.55	33.50
5000	Blank end fitting		40	.200		2.10	9.40		11.50	16.30

26 27 Low-Voltage Distribution Equipment

26 27 23 – Indoor Service Poles

26 27 23.40 Surface Raceway		Crew	Daily Output	Labor-Hours	Unit	Material	2009 Bare Costs Labor	Equipment	Total	Total Incl O&P
5200	Supporting clip	1 Elec	40	.200	Ea.	1.20	9.40		10.60	15.30
5800	No. 3000, entrance end fitting		20	.400		8.60	18.80		27.40	37.50
6000	Blank end fitting		40	.200		2.45	9.40		11.85	16.70
6020	Internal elbow		20	.400		11.20	18.80		30	40.50
6030	External elbow		20	.400		15.90	18.80		34.70	45.50
6040	Device bracket		53	.151		3.80	7.10		10.90	14.75
6400	Hanger clamp		32	.250		5.80	11.75		17.55	24
7000	No. 4000 Base		90	.089	L.F.	4.10	4.18		8.28	10.70
7200	Divider		100	.080	"	.80	3.76		4.56	6.50
7400	Entrance end fitting		16	.500	Ea.	22	23.50		45.50	59
7600	Blank end fitting		40	.200		5.80	9.40		15.20	20.50
7610	Recp. & tele. cover		53	.151		9.80	7.10		16.90	21.50
7620	External elbow		16	.500		32	23.50		55.50	70.50
7630	Coupling		53	.151		4.90	7.10		12	15.95
7640	Divider clip & coupling		80	.100		.97	4.70		5.67	8.05
7650	Panel connector		16	.500		20.50	23.50		44	57.50
7800	Take off connector		16	.500		67.50	23.50		91	110
8000	No. 6000, take off connector		16	.500		80.50	23.50		104	124
8100	Take off fitting		16	.500		59.50	23.50		83	101
8200	Hanger clamp		32	.250		14.45	11.75		26.20	33.50
8230	Coupling					7.10			7.10	7.80
8240	One gang device plate	1 Elec	53	.151		7.70	7.10		14.80	19
8250	Two gang device plate		40	.200		9.45	9.40		18.85	24.50
8260	Blank end fitting		40	.200		8.15	9.40		17.55	23
8270	Combination elbow		14	.571		33.50	27		60.50	76.50
8300	Panel connector		16	.500		14.65	23.50		38.15	51
8500	Chan-L-Wire system installed in 1-5/8" x 1-5/8" strut. Strut									
8600	not incl., 30 amp, 4 wire, 3 phase	1 Elec	200	.040	L.F.	4.60	1.88		6.48	7.85
8700	Junction box		8	1	Ea.	31	47		78	104
8800	Insulating end cap		40	.200		8.20	9.40		17.60	23
8900	Strut splice plate		40	.200		10.85	9.40		20.25	26
9000	Tap		40	.200		21.50	9.40		30.90	37.50
9100	Fixture hanger		60	.133		9.30	6.25		15.55	19.60
9200	Pulling tool					83			83	91.50

26 27 26 – Wiring Devices

26 27 26.10 Low Voltage Switching

		Crew	Daily Output	Labor-Hours	Unit	Material	Labor	Equipment	Total	Total Incl O&P
0010	**LOW VOLTAGE SWITCHING**									
3600	Relays, 120 V or 277 V standard	1 Elec	12	.667	Ea.	31	31.50		62.50	80.50
3800	Flush switch, standard		40	.200		10.75	9.40		20.15	26
4000	Interchangeable		40	.200		14.05	9.40		23.45	29.50
4100	Surface switch, standard		40	.200		7.90	9.40		17.30	22.50
4200	Transformer 115 V to 25 V		12	.667		110	31.50		141.50	168
4400	Master control, 12 circuit, manual		4	2		112	94		206	264
4500	25 circuit, motorized		4	2		122	94		216	274
4600	Rectifier, silicon		12	.667		36.50	31.50		68	86.50
4800	Switchplates, 1 gang, 1, 2 or 3 switch, plastic		80	.100		3.59	4.70		8.29	10.95
5000	Stainless steel		80	.100		9.65	4.70		14.35	17.65
5400	2 gang, 3 switch, stainless steel		53	.151		18.60	7.10		25.70	31
5500	4 switch, plastic		53	.151		8	7.10		15.10	19.35
5600	2 gang, 4 switch, stainless steel		53	.151		19.45	7.10		26.55	32
5700	6 switch, stainless steel		53	.151		43	7.10		50.10	57.50
6100	Relay gang boxes, flush or surface, 6 gang		5.30	1.509		65.50	71		136.50	178

26 27 Low-Voltage Distribution Equipment

26 27 26 – Wiring Devices

26 27 26.10 Low Voltage Switching

		Crew	Daily Output	Labor-Hours	Unit	Material	2009 Bare Costs Labor	Equipment	Total	Total Incl O&P
6400	18 gang	1 Elec	4	2	Ea.	85.50	94		179.50	235
6500	Frame, to hold up to 6 relays		12	.667	↓	57	31.50		88.50	110
7200	Control wire, 2 conductor		6.30	1.270	C.L.F.	27.50	59.50		87	119
7400	3 conductor		5	1.600		39	75		114	155
7600	19 conductor		2.50	3.200		315	150		465	570
7800	26 conductor	↓	2	4	↓	395	188		583	715

26 27 26.20 Wiring Devices Elements

		Crew	Daily Output	Labor-Hours	Unit	Material	Labor	Equipment	Total	Total Incl O&P
0010	**WIRING DEVICES ELEMENTS**									
0200	Toggle switch, quiet type, single pole, 15 amp	1 Elec	40	.200	Ea.	5.35	9.40		14.75	19.90
0500	20 amp		27	.296		7.40	13.95		21.35	28.50
0600	3 way, 15 amp		23	.348		7.65	16.35		24	33
0800	20 amp		18	.444		8.90	21		29.90	41
0900	4 way, 15 amp		15	.533		25	25		50	65
1000	20 amp		11	.727		39.50	34		73.50	94.50
1100	Toggle switch, quiet type, double pole, 15 amp		15	.533		16.25	25		41.25	55.50
1650	Dimmer switch, 120 volt, incandescent, 600 watt, 1 pole G		16	.500		13.10	23.50		36.60	49.50
1700	600 watt, 3 way G		12	.667		9.10	31.50		40.60	56.50
1750	1000 watt, 1 pole G		16	.500		46	23.50		69.50	85.50
1800	1000 watt, 3 way G		12	.667		64	31.50		95.50	117
2000	1500 watt, 1 pole G		11	.727		87	34		121	147
2100	2000 watt, 1 pole G		8	1		130	47		177	213
2110	Fluorescent, 600 watt G		15	.533		71.50	25		96.50	117
2120	1000 watt G		15	.533		89.50	25		114.50	136
2130	1500 watt G		10	.800		168	37.50		205.50	241
2460	Receptacle, duplex, 120 volt, grounded, 15 amp		40	.200		1.35	9.40		10.75	15.50
2470	20 amp		27	.296		9.95	13.95		23.90	31.50
2480	Ground fault interrupting, 15 amp		27	.296		32.50	13.95		46.45	56.50
2490	Dryer, 30 amp		15	.533		6.05	25		31.05	44
2500	Range, 50 amp		11	.727		13.10	34		47.10	65.50
2600	Wall plates, stainless steel, 1 gang		80	.100		2.30	4.70		7	9.55
2800	2 gang		53	.151		4.20	7.10		11.30	15.15
3000	3 gang		32	.250		6.50	11.75		18.25	24.50
3100	4 gang		27	.296		10.80	13.95		24.75	32.50
3110	Brown plastic, 1 gang		80	.100		.34	4.70		5.04	7.35
3120	2 gang		53	.151		.83	7.10		7.93	11.45
3130	3 gang		32	.250		1.24	11.75		12.99	18.85
3140	4 gang		27	.296		3.27	13.95		17.22	24
3150	Brushed brass, 1 gang		80	.100		4.80	4.70		9.50	12.30
3160	Anodized aluminum, 1 gang		80	.100		2.70	4.70		7.40	9.95
3200	Lampholder, keyless		26	.308		12.30	14.45		26.75	35
3400	Pullchain with receptacle	↓	22	.364	↓	13.35	17.10		30.45	40

26 27 73 – Door Chimes

26 27 73.10 Doorbell System

		Crew	Daily Output	Labor-Hours	Unit	Material	Labor	Equipment	Total	Total Incl O&P
0010	**DOORBELL SYSTEM**, incl. transformer, button & signal									
0100	6" bell	1 Elec	4	2	Ea.	107	94		201	257
0200	Buzzer		4	2		84.50	94		178.50	233
1000	Door chimes, 2 notes, minimum		16	.500		29.50	23.50		53	67
1020	Maximum		12	.667		150	31.50		181.50	212
1100	Tube type, 3 tube system		12	.667		222	31.50		253.50	292
1180	4 tube system		10	.800		355	37.50		392.50	445
1900	For transformer & button, minimum add		5	1.600		16.15	75		91.15	130
1960	Maximum, add	↓	4.50	1.778		48.50	83.50		132	178

26 27 Low-Voltage Distribution Equipment

26 27 73 – Door Chimes

26 27 73.10 Doorbell System		Crew	Daily Output	Labor-Hours	Unit	Material	2009 Bare Costs Labor	Equipment	Total	Total Incl O&P
3000	For push button only, minimum	1 Elec	24	.333	Ea.	3.27	15.65		18.92	27
3100	Maximum	↓	20	.400	↓	25.50	18.80		44.30	56

26 28 Low-Voltage Circuit Protective Devices

26 28 16 – Enclosed Switches and Circuit Breakers

26 28 16.10 Circuit Breakers

		Crew	Daily Output	Labor-Hours	Unit	Material	Labor	Equipment	Total	Total Incl O&P
0010	**CIRCUIT BREAKERS** (in enclosure)									
0100	Enclosed (NEMA 1), 600 volt, 3 pole, 30 amp	1 Elec	3.20	2.500	Ea.	610	118		728	845
0200	60 amp		2.80	2.857		610	134		744	870
0400	100 amp		2.30	3.478		700	163		863	1,025
0600	225 amp	↓	1.50	5.333		1,625	251		1,876	2,150
0700	400 amp	2 Elec	1.60	10		2,750	470		3,220	3,725

26 28 16.20 Safety Switches

0010	**SAFETY SWITCHES**									
0100	General duty 240 volt, 3 pole NEMA 1, fusible, 30 amp	1 Elec	3.20	2.500	Ea.	113	118		231	300
0200	60 amp		2.30	3.478		192	163		355	455
0300	100 amp		1.90	4.211		330	198		528	655
0400	200 amp	↓	1.30	6.154		710	289		999	1,200
0500	400 amp	2 Elec	1.80	8.889	↓	1,775	420		2,195	2,600

26 28 16.40 Time Switches

0010	**TIME SWITCHES**									
0400	Astronomic dial with reserve power	1 Elec	3.30	2.424	Ea.	595	114		709	825
0500	7 day calendar dial		3.30	2.424		146	114		260	330
0700	Photo cell 2000 watt	↓	8	1	↓	19	47		66	91

26 33 Battery Equipment

26 33 53 – Static Uninterruptible Power Supply

26 33 53.10 Uninterruptible Power Supply/Conditioner Transformers

		Crew	Daily Output	Labor-Hours	Unit	Material	Labor	Equipment	Total	Total Incl O&P
0010	**UNINTERRUPTIBLE POWER SUPPLY/CONDITIONER TRANSFORMERS**									
0100	Volt. regulating, isolating trans., w/invert. & 10 min. battery pack									
0110	Single-phase, 120 V, 0.35 kVA	1 Elec	2.29	3.493	Ea.	1,000	164		1,164	1,350
0120	0.5 kVA		2	4		1,050	188		1,238	1,425
0130	For additional 55 min. battery, add to .35 kVA		2.29	3.493		590	164		754	895
0140	Add to 0.5 kVA		1.14	7.018		620	330		950	1,175
0150	Single-phase, 120 V, 0.75 kVA		.80	10		1,325	470		1,795	2,150
0160	1.0 kVA	↓	.80	10		1,900	470		2,370	2,800
0170	1.5 kVA	2 Elec	1.14	14.035		3,275	660		3,935	4,575
0180	2 kVA	"	.89	17.978		3,550	845		4,395	5,150
0190	3 kVA	R-3	.63	31.746		4,300	1,475	205	5,980	7,150
0200	5 kVA		.42	47.619		6,225	2,200	310	8,735	10,500
0210	7.5 kVA		.33	60.606		7,950	2,800	390	11,140	13,400
0220	10 kVA		.28	71.429		8,450	3,300	460	12,210	14,700
0230	15 kVA	↓	.22	90.909	↓	11,700	4,200	585	16,485	19,800
0500	For options & accessories add to above, minimum									10%
0520	Maximum									35%
0600	For complex & special design systems to meet specific									
0610	requirements, obtain quote from vendor									

26 51 Interior Lighting

26 51 13 – Interior Lighting Fixtures, Lamps, and Ballasts

26 51 13.10 Fixture Hangers

		Crew	Daily Output	Labor-Hours	Unit	Material	2009 Bare Costs Labor	Equipment	Total	Total Incl O&P
0010	**FIXTURE HANGERS**									
0220	Box hub cover	1 Elec	32	.250	Ea.	3.90	11.75		15.65	22
0240	Canopy		12	.667		8.80	31.50		40.30	56
0260	Connecting block		40	.200		4.10	9.40		13.50	18.50
0280	Cushion hanger		16	.500		19.50	23.50		43	56.50
0300	Box hanger, with mounting strap		8	1		7.80	47		54.80	78.50
0320	Connecting block		40	.200		1.90	9.40		11.30	16.10
0340	Flexible, 1/2" diameter, 4" long		12	.667		10.80	31.50		42.30	58.50
0360	6" long		12	.667		11.55	31.50		43.05	59
0380	8" long		12	.667		12.75	31.50		44.25	60.50
0400	10" long		12	.667		13.45	31.50		44.95	61.50
0420	12" long		12	.667		14.50	31.50		46	62.50
0440	15" long		12	.667		15.25	31.50		46.75	63.50
0460	18" long		12	.667		17.30	31.50		48.80	65.50
0480	3/4" diameter, 4" long		10	.800		12.85	37.50		50.35	70
0500	6" long		10	.800		14.30	37.50		51.80	72
0520	8" long		10	.800		15.55	37.50		53.05	73
0540	10" long		10	.800		16.40	37.50		53.90	74
0560	12" long		10	.800		17.80	37.50		55.30	75.50
0580	15" long		10	.800		19.50	37.50		57	77.50
0600	18" long		10	.800		21.50	37.50		59	80

26 51 13.40 Interior HID Fixtures

		Crew	Daily Output	Labor-Hours	Unit	Material	2009 Bare Costs Labor	Equipment	Total	Total Incl O&P
0010	**INTERIOR HID FIXTURES** Incl. lamps, and mounting hardware									
0700	High pressure sodium, recessed, round, 70 watt	1 Elec	3.50	2.286	Ea.	420	107		527	620
0720	100 watt		3.50	2.286		445	107		552	650
0740	150 watt		3.20	2.500		495	118		613	720
0760	Square, 70 watt		3.60	2.222		420	104		524	615
0780	100 watt		3.60	2.222		445	104		549	645
0820	250 watt		3	2.667		605	125		730	850
0840	1000 watt	2 Elec	4.80	3.333		1,200	157		1,357	1,550
0860	Surface, round, 70 watt	1 Elec	3	2.667		595	125		720	840
0880	100 watt		3	2.667		610	125		735	855
0900	150 watt		2.70	2.963		640	139		779	910
0920	Square, 70 watt		3	2.667		580	125		705	820
0940	100 watt		3	2.667		610	125		735	855
0980	250 watt		2.50	3.200		645	150		795	935
1040	Pendent, round, 70 watt		3	2.667		635	125		760	885
1060	100 watt		3	2.667		610	125		735	855
1080	150 watt		2.70	2.963		620	139		759	885
1100	Square, 70 watt		3	2.667		650	125		775	900
1120	100 watt		3	2.667		660	125		785	910
1140	150 watt		2.70	2.963		675	139		814	950
1160	250 watt		2.50	3.200		915	150		1,065	1,225
1180	400 watt		2.40	3.333		960	157		1,117	1,275
1220	Wall, round, 70 watt		3	2.667		545	125		670	785
1240	100 watt		3	2.667		555	125		680	795
1260	150 watt		2.70	2.963		570	139		709	835
1300	Square, 70 watt		3	2.667		580	125		705	820
1320	100 watt		3	2.667		610	125		735	855
1340	150 watt		2.70	2.963		645	139		784	915
1360	250 watt		2.50	3.200		690	150		840	985
1380	400 watt	2 Elec	4.80	3.333		850	157		1,007	1,175

26 51 Interior Lighting

26 51 13 – Interior Lighting Fixtures, Lamps, and Ballasts

26 51 13.40 Interior HID Fixtures

		Crew	Daily Output	Labor-Hours	Unit	Material	2009 Bare Costs Labor	2009 Bare Costs Equipment	Total	Total Incl O&P
1400	1000 watt	2 Elec	3.60	4.444	Ea.	1,200	209		1,409	1,625
1500	Metal halide, recessed, round, 175 watt	1 Elec	3.40	2.353		420	111		531	630
1520	250 watt	"	3.20	2.500		465	118		583	690
1540	400 watt	2 Elec	5.80	2.759		665	130		795	925
1580	Square, 175 watt	1 Elec	3.40	2.353		385	111		496	585
1640	Surface, round, 175 watt		2.90	2.759		515	130		645	765
1660	250 watt	↓	2.70	2.963		815	139		954	1,100
1680	400 watt	2 Elec	4.80	3.333		915	157		1,072	1,225
1720	Square, 175 watt	1 Elec	2.90	2.759		560	130		690	815
1800	Pendent, round, 175 watt		2.90	2.759		610	130		740	865
1820	250 watt	↓	2.70	2.963		885	139		1,024	1,175
1840	400 watt	2 Elec	4.80	3.333		980	157		1,137	1,300
1880	Square, 175 watt	1 Elec	2.90	2.759		595	130		725	850
1900	250 watt	"	2.70	2.963		600	139		739	865
1920	400 watt	2 Elec	4.80	3.333		970	157		1,127	1,300
1980	Wall, round, 175 watt	1 Elec	2.90	2.759		515	130		645	765
2000	250 watt	"	2.70	2.963		815	139		954	1,100
2020	400 watt	2 Elec	4.80	3.333		835	157		992	1,150
2060	Square, 175 watt	1 Elec	2.90	2.759		545	130		675	795
2080	250 watt	"	2.70	2.963		590	139		729	855
2100	400 watt	2 Elec	4.80	3.333		830	157		987	1,150
2800	High pressure sodium, vaporproof, recessed, 70 watt	1 Elec	3.50	2.286		560	107		667	775
2820	100 watt		3.50	2.286		570	107		677	790
2840	150 watt		3.20	2.500		585	118		703	820
2900	Surface, 70 watt		3	2.667		630	125		755	880
2920	100 watt		3	2.667		660	125		785	910
2940	150 watt		2.70	2.963		685	139		824	960
3000	Pendent, 70 watt		3	2.667		625	125		750	870
3020	100 watt		3	2.667		645	125		770	895
3040	150 watt		2.70	2.963		680	139		819	955
3100	Wall, 70 watt		3	2.667		670	125		795	925
3120	100 watt		3	2.667		700	125		825	955
3140	150 watt		2.70	2.963		730	139		869	1,000
3200	Metal halide, vaporproof, recessed, 175 watt		3.40	2.353		540	111		651	760
3220	250 watt	↓	3.20	2.500		575	118		693	810
3240	400 watt	2 Elec	5.80	2.759		720	130		850	985
3260	1000 watt	"	4.80	3.333		1,325	157		1,482	1,675
3280	Surface, 175 watt	1 Elec	2.90	2.759		525	130		655	770
3300	250 watt	"	2.70	2.963		820	139		959	1,100
3320	400 watt	2 Elec	4.80	3.333		1,025	157		1,182	1,350
3340	1000 watt	"	3.60	4.444		1,475	209		1,684	1,925
3360	Pendent, 175 watt	1 Elec	2.90	2.759		545	130		675	795
3380	250 watt	"	2.70	2.963		845	139		984	1,125
3400	400 watt	2 Elec	4.80	3.333		1,000	157		1,157	1,325
3420	1000 watt	"	3.60	4.444		1,625	209		1,834	2,100
3440	Wall, 175 watt	1 Elec	2.90	2.759		595	130		725	850
3460	250 watt	"	2.70	2.963		890	139		1,029	1,175
3480	400 watt	2 Elec	4.80	3.333		1,050	157		1,207	1,400
3500	1000 watt	"	3.60	4.444		1,700	209		1,909	2,150

26 51 13.50 Interior Lighting Fixtures

0010	**INTERIOR LIGHTING FIXTURES** Including lamps, mounting									
0030	hardware and connections									

26 51 Interior Lighting

26 51 13 – Interior Lighting Fixtures, Lamps, and Ballasts

26 51 13.50 Interior Lighting Fixtures		Crew	Daily Output	Labor-Hours	Unit	Material	2009 Bare Costs Labor	Equipment	Total	Total Incl O&P
0100	Fluorescent, C.W. lamps, troffer, recess mounted in grid, RS									
0200	Acrylic lens, 1'W x 4'L, two 40 watt	1 Elec	5.70	1.404	Ea.	53	66		119	156
0210	1'W x 4'L, three 40 watt		5.40	1.481		61.50	69.50		131	172
0300	2'W x 2'L, two U40 watt		5.70	1.404		56	66		122	160
0400	2'W x 4'L, two 40 watt		5.30	1.509		56	71		127	168
0500	2'W x 4'L, three 40 watt		5	1.600		60.50	75		135.50	179
0600	2'W x 4'L, four 40 watt		4.70	1.702		64	80		144	190
0700	4'W x 4'L, four 40 watt	2 Elec	6.40	2.500		335	118		453	545
0800	4'W x 4'L, six 40 watt		6.20	2.581		345	121		466	560
0900	4'W x 4'L, eight 40 watt		5.80	2.759		360	130		490	590
0910	Acrylic lens, 1'W x 4'L, two 32 watt G	1 Elec	5.70	1.404		66	66		132	171
0930	2'W x 2'L, two U32 watt G		5.70	1.404		80	66		146	186
0940	2'W x 4'L, two 32 watt G		5.30	1.509		72	71		143	186
0950	2'W x 4'L, three 32 watt G		5	1.600		76.50	75		151.50	196
0960	2'W x 4'L, four 32 watt G		4.70	1.702		78.50	80		158.50	206
1000	Surface mounted, RS									
1030	Acrylic lens with hinged & latched door frame									
1100	1'W x 4'L, two 40 watt	1 Elec	7	1.143	Ea.	74	53.50		127.50	162
1110	1'W x 4'L, three 40 watt		6.70	1.194		76.50	56		132.50	168
1200	2'W x 2'L, two U40 watt		7	1.143		79.50	53.50		133	167
1300	2'W x 4'L, two 40 watt		6.20	1.290		91	60.50		151.50	191
1400	2'W x 4'L, three 40 watt		5.70	1.404		92	66		158	199
1500	2'W x 4'L, four 40 watt		5.30	1.509		94	71		165	209
1600	4'W x 4'L, four 40 watt	2 Elec	7.20	2.222		465	104		569	665
1700	4'W x 4'L, six 40 watt		6.60	2.424		505	114		619	725
1800	4'W x 4'L, eight 40 watt		6.20	2.581		525	121		646	755
1900	2'W x 8'L, four 40 watt		6.40	2.500		182	118		300	375
2000	2'W x 8'L, eight 40 watt		6.20	2.581		196	121		317	395
2010	Acrylic wrap around lens									
2020	6"W x 4'L, one 40 watt	1 Elec	8	1	Ea.	74.50	47		121.50	152
2030	6"W x 8'L, two 40 watt	2 Elec	8	2		81.50	94		175.50	230
2040	11"W x 4'L, two 40 watt	1 Elec	7	1.143		49.50	53.50		103	134
2050	11"W x 8'L, four 40 watt	2 Elec	6.60	2.424		82	114		196	261
2060	16"W x 4'L, four 40 watt	1 Elec	5.30	1.509		82	71		153	197
2070	16"W x 8'L, eight 40 watt	2 Elec	6.40	2.500		179	118		297	370
2080	2'W x 2'L, two U40 watt	1 Elec	7	1.143		100	53.50		153.50	190
2100	Strip fixture									
2200	4' long, one 40 watt, RS	1 Elec	8.50	.941	Ea.	31	44		75	100
2300	4' long, two 40 watt, RS		8	1		33.50	47		80.50	107
2400	4' long, one 40 watt, SL		8	1		45	47		92	120
2500	4' long, two 40 watt, SL		7	1.143		61.50	53.50		115	148
2600	8' long, one 75 watt, SL	2 Elec	13.40	1.194		46.50	56		102.50	135
2700	8' long, two 75 watt, SL	"	12.40	1.290		56	60.50		116.50	152
2800	4' long, two 60 watt, HO	1 Elec	6.70	1.194		90	56		146	183
2900	8' long, two 110 watt, HO	2 Elec	10.60	1.509		95	71		166	211
2910	4' long, two 115 watt, VHO	1 Elec	6.50	1.231		125	58		183	224
2920	8' long, two 215 watt, VHO	2 Elec	10.40	1.538		135	72.50		207.50	257
3000	Strip, pendent mounted, industrial, white porcelain enamel									
3100	4' long, two 40 watt, RS	1 Elec	5.70	1.404	Ea.	51	66		117	154
3200	4' long, two 60 watt, HO	"	5	1.600		81	75		156	201
3300	8' long, two 75 watt, SL	2 Elec	8.80	1.818		96	85.50		181.50	233
3400	8' long, two 110 watt, HO	"	8	2		123	94		217	275
3410	Acrylic finish, 4' long, two 40 watt, RS	1 Elec	5.70	1.404		84	66		150	191

26 51 Interior Lighting

26 51 13 – Interior Lighting Fixtures, Lamps, and Ballasts

26 51 13.50 Interior Lighting Fixtures		Crew	Daily Output	Labor-Hours	Unit	Material	2009 Bare Costs Labor	Equipment	Total	Total Incl O&P
3420	4' long, two 60 watt, HO	1 Elec	5	1.600	Ea.	155	75		230	283
3430	4' long, two 115 watt, VHO	↓	4.80	1.667		200	78.50		278.50	335
3440	8' long, two 75 watt, SL	2 Elec	8.80	1.818		164	85.50		249.50	305
3450	8' long, two 110 watt, HO		8	2		185	94		279	345
3460	8' long, two 215 watt, VHO	↓	7.60	2.105		262	99		361	435
3470	Troffer, air handling, 2'W x 4'L with four 40 watt, RS	1 Elec	4	2		97	94		191	247
3480	2'W x 2'L with two U40 watt RS		5.50	1.455		84.50	68.50		153	195
3510	Troffer parabolic lay-in, 1'W x 4'L with one 32 W [G]		5.70	1.404		112	66		178	221
3520	1'W x 4'L with two 32 W [G]		5.30	1.509		115	71		186	233
3530	2'W x 4'L with three 32 W [G]	↓	5	1.600	↓	127	75		202	252
4220	Metal halide, integral ballast, ceiling, recess mounted									
4230	prismatic glass lens, floating door									
4240	2'W x 2'L, 250 watt	1 Elec	3.20	2.500	Ea.	310	118		428	515
4250	2'W x 2'L, 400 watt	2 Elec	5.80	2.759		355	130		485	585
4260	Surface mounted, 2'W x 2'L, 250 watt	1 Elec	2.70	2.963		335	139		474	575
4270	400 watt	2 Elec	4.80	3.333	↓	395	157		552	670
4280	High bay, aluminum reflector,									
4290	Single unit, 400 watt	2 Elec	4.60	3.478	Ea.	405	163		568	690
4300	Single unit, 1000 watt		4	4		580	188		768	920
4310	Twin unit, 400 watt	↓	3.20	5		805	235		1,040	1,225
4320	Low bay, aluminum reflector, 250W DX lamp	1 Elec	3.20	2.500		355	118		473	565
4330	400 watt lamp	2 Elec	5	3.200	↓	515	150		665	790
4340	High pressure sodium integral ballast ceiling, recess mounted									
4350	prismatic glass lens, floating door									
4360	2'W x 2'L, 150 watt lamp	1 Elec	3.20	2.500	Ea.	380	118		498	590
4370	2'W x 2'L, 400 watt lamp	2 Elec	5.80	2.759		450	130		580	690
4380	Surface mounted, 2'W x 2'L, 150 watt lamp	1 Elec	2.70	2.963		460	139		599	715
4390	400 watt lamp	2 Elec	4.80	3.333		515	157		672	805
4400	High bay, aluminum reflector,									
4410	Single unit, 400 watt lamp	2 Elec	4.60	3.478	Ea.	370	163		533	655
4430	Single unit, 1000 watt lamp	"	4	4		535	188		723	870
4440	Low bay, aluminum reflector, 150 watt lamp	1 Elec	3.20	2.500	↓	320	118		438	530
4450	Incandescent, high hat can, round alzak reflector, prewired									
4470	100 watt	1 Elec	8	1	Ea.	69	47		116	146
4480	150 watt		8	1		102	47		149	182
4500	300 watt		6.70	1.194		236	56		292	345
4520	Round with reflector and baffles, 150 watt		8	1		49	47		96	124
4540	Round with concentric louver, 150 watt PAR	↓	8	1	↓	77	47		124	155
4600	Square glass lens with metal trim, prewired									
4630	100 watt	1 Elec	6.70	1.194	Ea.	52.50	56		108.50	142
4700	200 watt		6.70	1.194		95	56		151	189
4800	300 watt		5.70	1.404		140	66		206	252
4810	500 watt		5	1.600		275	75		350	415
4900	Ceiling/wall, surface mounted, metal cylinder, 75 watt		10	.800		53.50	37.50		91	115
4920	150 watt		10	.800		77	37.50		114.50	141
4930	300 watt		8	1		164	47		211	250
5000	500 watt		6.70	1.194		350	56		406	470
5010	Square, 100 watt		8	1		107	47		154	188
5020	150 watt		8	1		117	47		164	199
5030	300 watt		7	1.143		325	53.50		378.50	440
5040	500 watt	↓	6	1.333		330	62.50		392.50	460
5200	Ceiling, surface mounted, opal glass drum									
5300	8", one 60 watt lamp	1 Elec	10	.800	Ea.	43.50	37.50		81	104

26 51 Interior Lighting

26 51 13 – Interior Lighting Fixtures, Lamps, and Ballasts

26 51 13.50 Interior Lighting Fixtures		Crew	Daily Output	Labor-Hours	Unit	Material	2009 Bare Costs Labor	2009 Bare Costs Equipment	Total	Total Incl O&P
5400	10", two 60 watt lamps	1 Elec	8	1	Ea.	49	47		96	124
5500	12", four 60 watt lamps		6.70	1.194		67	56		123	157
5510	Pendent, round, 100 watt		8	1		107	47		154	188
5520	150 watt		8	1		118	47		165	200
5530	300 watt		6.70	1.194		165	56		221	266
5540	500 watt		5.50	1.455		315	68.50		383.50	445
5550	Square, 100 watt		6.70	1.194		146	56		202	245
5560	150 watt		6.70	1.194		152	56		208	251
5570	300 watt		5.70	1.404		225	66		291	345
5580	500 watt		5	1.600		305	75		380	445
5600	Wall, round, 100 watt		8	1		62	47		109	138
5620	300 watt		8	1		110	47		157	191
5630	500 watt		6.70	1.194		370	56		426	490
5640	Square, 100 watt		8	1		101	47		148	181
5650	150 watt		8	1		103	47		150	183
5660	300 watt		7	1.143		160	53.50		213.50	256
5670	500 watt		6	1.333		285	62.50		347.50	410
6010	Vapor tight, incandescent, ceiling mounted, 200 watt		6.20	1.290		64.50	60.50		125	161
6020	Recessed, 200 watt		6.70	1.194		104	56		160	198
6030	Pendent, 200 watt		6.70	1.194		64.50	56		120.50	154
6040	Wall, 200 watt		8	1		73	47		120	151
6100	Fluorescent, surface mounted, 2 lamps, 4'L, RS, 40 watt		3.20	2.500		113	118		231	300
6110	Industrial, 2 lamps 4' long in tandem, 430 MA		2.20	3.636		208	171		379	485
6130	2 lamps 4' long, 800 MA		1.90	4.211		178	198		376	490
6160	Pendent, indust, 2 lamps 4'L in tandem, 430 MA		1.90	4.211		243	198		441	565
6170	2 lamps 4' long, 430 MA		2.30	3.478		163	163		326	420
6180	2 lamps 4' long, 800 MA		1.70	4.706		205	221		426	555
6850	Vandalproof, surface mounted, fluorescent, two 40 watt		3.20	2.500		239	118		357	440
6860	Incandescent, one 150 watt		8	1		73	47		120	150
6900	Mirror light, fluorescent, RS, acrylic enclosure, two 40 watt		8	1		116	47		163	198
6910	One 40 watt		8	1		94	47		141	173
6920	One 20 watt		12	.667		74	31.50		105.50	128
7000	Low bay, aluminum reflector. 70 watt, high pressure sodium		4	2		265	94		359	430
7010	250 watt		3.20	2.500		355	118		473	565
7020	400 watt	2 Elec	5	3.200		360	150		510	620
7990	Decorator									
8000	Pendent RLM in colors, shallow dome, 12" diam. 100 W	1 Elec	8	1	Ea.	79	47		126	157
8010	Regular dome, 12" diam., 100 watt		8	1		81.50	47		128.50	160
8020	16" diam., 200 watt		7	1.143		83	53.50		136.50	172
8030	18" diam., 500 watt		6	1.333		100	62.50		162.50	204
8100	Picture framing light, minimum		16	.500		76	23.50		99.50	119
8110	Maximum		16	.500		100	23.50		123.50	145
8150	Miniature low voltage, recessed, pinhole		8	1		159	47		206	245
8160	Star		8	1		131	47		178	214
8170	Adjustable cone		8	1		178	47		225	265
8180	Eyeball		8	1		145	47		192	229
8190	Cone		8	1		127	47		174	210
8200	Coilex baffle		8	1		125	47		172	207
8210	Surface mounted, adjustable cylinder		8	1		129	47		176	211
8250	Chandeliers, incandescent									
8260	24" diam. x 42" high, 6 light candle	1 Elec	6	1.333	Ea.	390	62.50		452.50	525
8270	24" diam. x 42" high, 6 light candle w/glass shade		6	1.333		360	62.50		422.50	490
8280	17" diam. x 12" high, 8 light w/glass panels		8	1		277	47		324	375

26 51 Interior Lighting

26 51 13 – Interior Lighting Fixtures, Lamps, and Ballasts

26 51 13.50 Interior Lighting Fixtures

		Crew	Daily Output	Labor-Hours	Unit	Material	2009 Bare Costs Labor	Equipment	Total	Total Incl O&P
8290	32" diam. x 48"H, 10 light bohemian lead crystal	1 Elec	4	2	Ea.	655	94		749	860
8300	27" diam. x 29"H, 10 light bohemian lead crystal		4	2		570	94		664	770
8310	21" diam. x 9" high 6 light sculptured ice crystal	↓	8	1	↓	445	47		492	560
8500	Accent lights, on floor or edge, 0.5W low volt incandescent									
8520	incl. transformer & fastenings, based on 100' lengths									
8550	Lights in clear tubing, 12" on center	1 Elec	230	.035	L.F.	8.45	1.63		10.08	11.75
8560	6" on center		160	.050		11	2.35		13.35	15.60
8570	4" on center		130	.062		16.85	2.89		19.74	23
8580	3" on center		125	.064		18.70	3.01		21.71	25
8590	2" on center		100	.080		27	3.76		30.76	35.50
8600	Carpet, lights both sides 6" OC, in alum. extrusion		270	.030		25.50	1.39		26.89	30
8610	In bronze extrusion		270	.030		29	1.39		30.39	33.50
8620	Carpet-bare floor, lights 18" OC, in alum. extrusion		270	.030		20.50	1.39		21.89	24.50
8630	In bronze extrusion		270	.030		24	1.39		25.39	28
8640	Carpet edge-wall, lights 6" OC in alum. extrusion		270	.030		25.50	1.39		26.89	30
8650	In bronze extrusion		270	.030		29	1.39		30.39	33.50
8660	Bare floor, lights 18" OC, in aluminum extrusion		300	.027		20.50	1.25		21.75	24.50
8670	In bronze extrusion		300	.027		24	1.25		25.25	28
8680	Bare floor conduit, aluminum extrusion		300	.027		6.75	1.25		8	9.25
8690	In bronze extrusion		300	.027	↓	13.45	1.25		14.70	16.65
8700	Step edge to 36", lights 6" OC, in alum. extrusion		100	.080	Ea.	68	3.76		71.76	80
8710	In bronze extrusion		100	.080		70.50	3.76		74.26	83
8720	Step edge to 54", lights 6" OC, in alum. extrusion		100	.080		102	3.76		105.76	118
8730	In bronze extrusion		100	.080		107	3.76		110.76	124
8740	Step edge to 72", lights 6" OC, in alum. extrusion		100	.080		136	3.76		139.76	155
8750	In bronze extrusion		100	.080		148	3.76		151.76	169
8760	Connector, male		32	.250		2.51	11.75		14.26	20.50
8770	Female with pigtail		32	.250		5.30	11.75		17.05	23.50
8780	Clamps		400	.020		.50	.94		1.44	1.95
8790	Transformers, 50 watt		8	1		69.50	47		116.50	146
8800	250 watt		4	2		229	94		323	390
8810	1000 watt	↓	2.70	2.963	↓	440	139		579	690

26 51 13.70 Residential Fixtures

		Crew	Daily Output	Labor-Hours	Unit	Material	Labor	Equipment	Total	Total Incl O&P
0010	**RESIDENTIAL FIXTURES**									
0400	Fluorescent, interior, surface, circline, 32 watt & 40 watt	1 Elec	20	.400	Ea.	91	18.80		109.80	128
0500	2' x 2', two U 40 watt		8	1		110	47		157	191
0700	Shallow under cabinet, two 20 watt		16	.500		47.50	23.50		71	87.50
0900	Wall mounted, 4'L, one 40 watt, with baffle		10	.800		135	37.50		172.50	205
2600	Interior pendent, globe with shade, 150 watt	↓	20	.400	↓	152	18.80		170.80	195

26 52 Emergency Lighting

26 52 13 – Emergency Lighting Equipments

26 52 13.10 Emergency Lighting and Battery Units

		Crew	Daily Output	Labor-Hours	Unit	Material	Labor	Equipment	Total	Total Incl O&P
0010	**EMERGENCY LIGHTING AND BATTERY UNITS**									
0300	Emergency light units, battery operated									
0350	Twin sealed beam light, 25 watt, 6 volt each									
0500	Lead battery operated	1 Elec	4	2	Ea.	129	94		223	282
0700	Nickel cadmium battery operated		4	2		605	94		699	805
0900	Self-contained fluorescent lamp pack	↓	10	.800	↓	146	37.50		183.50	217

26 53 Exit Signs

26 53 13 – Exit Lighting

26 53 13.10 Exit Lighting Fixtures

		Crew	Daily Output	Labor-Hours	Unit	Material	2009 Bare Costs Labor	2009 Bare Costs Equipment	Total	Total Incl O&P
0010	**EXIT LIGHTING FIXTURES**									
0080	Exit light ceiling or wall mount, incandescent, single face	1 Elec	8	1	Ea.	42.50	47		89.50	117
0100	Double face		6.70	1.194		49	56		105	138
0120	Explosion proof		3.80	2.105		480	99		579	675
0150	Fluorescent, single face		8	1		65	47		112	142
0160	Double face		6.70	1.194		73	56		129	164
0200	L.E.D. standard, single face G		8	1		65.50	47		112.50	142
0220	Double face G		6.70	1.194		73	56		129	164
0230	L.E.D. vandal-resistant, single face G		7.27	1.100		210	51.50		261.50	310
0240	L.E.D. w/battery unit, single face G		4.40	1.818		138	85.50		223.50	278
0260	Double face G		4	2		140	94		234	294
0262	L.E.D. w/battery unit, vandal-resistant, single face G		4.40	1.818		345	85.50		430.50	505
0270	Combination emergency light units and exit sign		4	2		188	94		282	345

26 54 Classified Location Lighting

26 54 13 – Classified Lighting

26 54 13.20 Explosionproof

		Crew	Daily Output	Labor-Hours	Unit	Material	2009 Bare Costs Labor	2009 Bare Costs Equipment	Total	Total Incl O&P
0010	**EXPLOSIONPROOF**, incl lamps, mounting hardware and connections									
6310	Metal halide with ballast, ceiling, surface mounted, 175 watt	1 Elec	2.90	2.759	Ea.	865	130		995	1,150
6320	250 watt	"	2.70	2.963		1,050	139		1,189	1,350
6330	400 watt	2 Elec	4.80	3.333		1,100	157		1,257	1,450
6340	Ceiling, pendent mounted, 175 watt	1 Elec	2.60	3.077		820	145		965	1,125
6350	250 watt	"	2.40	3.333		995	157		1,152	1,325
6360	400 watt	2 Elec	4.20	3.810		1,075	179		1,254	1,450
6370	Wall, surface mounted, 175 watt	1 Elec	2.90	2.759		930	130		1,060	1,225
6380	250 watt	"	2.70	2.963		1,100	139		1,239	1,425
6390	400 watt	2 Elec	4.80	3.333		1,175	157		1,332	1,525
6400	High pressure sodium, ceiling surface mounted, 70 watt	1 Elec	3	2.667		935	125		1,060	1,200
6410	100 watt		3	2.667		970	125		1,095	1,250
6420	150 watt		2.70	2.963		1,000	139		1,139	1,300
6430	Pendent mounted, 70 watt		2.70	2.963		895	139		1,034	1,175
6440	100 watt		2.70	2.963		925	139		1,064	1,225
6450	150 watt		2.40	3.333		955	157		1,112	1,275
6460	Wall mounted, 70 watt		3	2.667		1,000	125		1,125	1,275
6470	100 watt		3	2.667		1,025	125		1,150	1,300
6480	150 watt		2.70	2.963		1,075	139		1,214	1,375
6510	Incandescent, ceiling mounted, 200 watt		4	2		835	94		929	1,050
6520	Pendent mounted, 200 watt		3.50	2.286		715	107		822	950
6530	Wall mounted, 200 watt		4	2		830	94		924	1,050
6600	Fluorescent, RS, 4' long, ceiling mounted, two 40 watt		2.70	2.963		2,000	139		2,139	2,400
6610	Three 40 watt		2.20	3.636		2,900	171		3,071	3,425
6620	Four 40 watt		1.90	4.211		3,725	198		3,923	4,400
6630	Pendent mounted, two 40 watt		2.30	3.478		2,325	163		2,488	2,825
6640	Three 40 watt		1.90	4.211		3,325	198		3,523	3,950
6650	Four 40 watt		1.70	4.706		4,375	221		4,596	5,125

26 55 Special Purpose Lighting

26 55 59 – Display Lighting

26 55 59.10 Track Lighting		Crew	Daily Output	Labor-Hours	Unit	Material	2009 Bare Costs Labor	Equipment	Total	Total Incl O&P
0010	**TRACK LIGHTING**									
0080	Track, 1 circuit, 4' section	1 Elec	6.70	1.194	Ea.	49.50	56		105.50	138
0100	8' section		5.30	1.509		81	71		152	196
0200	12' section		4.40	1.818		131	85.50		216.50	271
0300	3 circuits, 4' section		6.70	1.194		68	56		124	159
0400	8' section		5.30	1.509		106	71		177	223
0500	12' section		4.40	1.818		176	85.50		261.50	320
1000	Feed kit, surface mounting		16	.500		12.20	23.50		35.70	48.50
1100	End cover		24	.333		4.75	15.65		20.40	29
1200	Feed kit, stem mounting, 1 circuit		16	.500		36.50	23.50		60	75.50
1300	3 circuit		16	.500		36.50	23.50		60	75.50
2000	Electrical joiner, for continuous runs, 1 circuit		32	.250		17.60	11.75		29.35	37
2100	3 circuit		32	.250		42.50	11.75		54.25	64
2200	Fixtures, spotlight, 75W PAR halogen		16	.500		105	23.50		128.50	151
2210	50W MR16 halogen		16	.500		128	23.50		151.50	176
3000	Wall washer, 250 watt tungsten halogen		16	.500		122	23.50		145.50	169
3100	Low voltage, 25/50 watt, 1 circuit		16	.500		123	23.50		146.50	170
3120	3 circuit		16	.500		127	23.50		150.50	175

26 55 61 – Theatrical Lighting

26 55 61.10 Lights		Crew	Daily Output	Labor-Hours	Unit	Material	Labor	Equipment	Total	Total Incl O&P
0010	**LIGHTS**									
2000	Lights, border, quartz, reflector, vented,									
2100	colored or white	1 Elec	20	.400	L.F.	173	18.80		191.80	218
2500	Spotlight, follow spot, with transformer, 2,100 watt	"	4	2	Ea.	3,075	94		3,169	3,525
2600	For no transformer, deduct					910			910	1,000
3000	Stationary spot, fresnel quartz, 6" lens	1 Elec	4	2		131	94		225	284
3100	8" lens		4	2		231	94		325	395
3500	Ellipsoidal quartz, 1,000W, 6" lens		4	2		335	94		429	510
3600	12" lens		4	2		600	94		694	800
4000	Strobe light, 1 to 15 flashes per second, quartz		3	2.667		750	125		875	1,025
4500	Color wheel, portable, five hole, motorized		4	2		194	94		288	355

26 56 Exterior Lighting

26 56 26 – Landscape Lighting

26 56 26.20 Landscape Fixtures		Crew	Daily Output	Labor-Hours	Unit	Material	Labor	Equipment	Total	Total Incl O&P
0010	**LANDSCAPE FIXTURES**									
7380	Landscape recessed uplight, incl. housing, ballast, transformer									
7390	& reflector, not incl. conduit, wire, trench									
7420	Incandescent, 250 watt	1 Elec	5	1.600	Ea.	550	75		625	715
7440	Quartz, 250 watt		5	1.600		525	75		600	685
7460	500 watt		4	2		535	94		629	730

26 56 33 – Walkway Lighting

26 56 33.10 Walkway Luminaire		Crew	Daily Output	Labor-Hours	Unit	Material	Labor	Equipment	Total	Total Incl O&P
0010	**WALKWAY LUMINAIRE**									
6500	Bollard light, lamp & ballast, 42" high with polycarbonate lens									
6800	Metal halide, 175 watt	1 Elec	3	2.667	Ea.	750	125		875	1,000
6900	High pressure sodium, 70 watt		3	2.667		770	125		895	1,025
7000	100 watt		3	2.667		770	125		895	1,025
7100	150 watt		3	2.667		750	125		875	1,000
7200	Incandescent, 150 watt		3	2.667		540	125		665	780

26 56 Exterior Lighting

26 56 33 – Walkway Lighting

26 56 33.10 Walkway Luminaire		Crew	Daily Output	Labor-Hours	Unit	Material	2009 Bare Costs Labor	Equipment	Total	Total Incl O&P
8230	Lantern, high pressure sodium, 70 watt	1 Elec	3	2.667	Ea.	490	125		615	725
8240	100 watt		3	2.667		530	125		655	765
8250	150 watt		3	2.667		500	125		625	730
8260	250 watt		2.70	2.963		695	139		834	965
8270	Incandescent, 300 watt		3.50	2.286		365	107		472	560

26 61 Lighting Systems and Accessories

26 61 13 – Lighting Accessories

26 61 13.10 Energy Saving Lighting Devices

			Crew	Daily Output	Labor-Hours	Unit	Material	Labor	Equipment	Total	Total Incl O&P
0010	**ENERGY SAVING LIGHTING DEVICES**										
0100	Occupancy sensors infrared, ceiling mounted	G	1 Elec	7	1.143	Ea.	104	53.50		157.50	194
0150	Automatic wall switches	G		24	.333		64	15.65		79.65	94
0200	Remote power pack	G		10	.800		33.50	37.50		71	93

26 61 23 – Lamps Applications

26 61 23.10 Lamps

			Crew	Daily Output	Labor-Hours	Unit	Material	Labor	Equipment	Total	Total Incl O&P
0010	**LAMPS**										
0080	Fluorescent, rapid start, cool white, 2' long, 20 watt		1 Elec	1	8	C	335	375		710	925
0100	4' long, 40 watt			.90	8.889		365	420		785	1,025
0120	3' long, 30 watt			.90	8.889		415	420		835	1,075
0125	3' long, 25 watt energy saver	G		.90	8.889		780	420		1,200	1,475
0150	U-40 watt			.80	10		1,100	470		1,570	1,925
0155	U-34 watt energy saver	G		.80	10		1,100	470		1,570	1,925
0170	4' long, 34 watt energy saver	G		.90	8.889		345	420		765	1,000
0176	2' long, T8, 17 W energy saver	G		1	8		490	375		865	1,100
0178	3' long, T8, 25 W energy saver	G		.90	8.889		490	420		910	1,150
0180	4' long, T8, 32 watt energy saver	G		.90	8.889		315	420		735	965
0200	Slimline, 4' long, 40 watt			.90	8.889		825	420		1,245	1,525
0210	4' long, 30 watt energy saver	G		.90	8.889		825	420		1,245	1,525
0300	8' long, 75 watt			.80	10		950	470		1,420	1,750
0400	High output, 4' long, 60 watt			.90	8.889		945	420		1,365	1,650
0410	8' long, 95 watt energy saver	G		.80	10		960	470		1,430	1,750
0500	8' long, 110 watt			.80	10		960	470		1,430	1,750
0512	2' long, T5, 14 watt energy saver	G		1	8		1,400	375		1,775	2,100
0514	3' long, T5, 21 watt energy saver	G		.90	8.889		1,400	420		1,820	2,175
0516	4' long, T5, 28 watt energy saver	G		.90	8.889		1,225	420		1,645	1,950
0517	4' long, T5, 54 watt energy saver	G		.90	8.889		1,500	420		1,920	2,275
0520	Very high output, 4' long, 110 watt			.90	8.889		2,450	420		2,870	3,325
0525	8' long, 195 watt energy saver	G		.70	11.429		2,575	535		3,110	3,625
0550	8' long, 215 watt			.70	11.429		2,525	535		3,060	3,575
0560	Twin tube compact lamp	G		.90	8.889		575	420		995	1,250
0570	Double twin tube compact lamp	G		.80	10		1,550	470		2,020	2,425
0600	Mercury vapor, mogul base, deluxe white, 100 watt			.30	26.667		3,100	1,250		4,350	5,300
0650	175 watt			.30	26.667		2,250	1,250		3,500	4,325
0700	250 watt			.30	26.667		4,400	1,250		5,650	6,725
0800	400 watt			.30	26.667		3,550	1,250		4,800	5,775
0900	1000 watt			.20	40		8,275	1,875		10,150	11,900
1000	Metal halide, mogul base, 175 watt			.30	26.667		3,900	1,250		5,150	6,150
1100	250 watt			.30	26.667		4,400	1,250		5,650	6,725
1200	400 watt			.30	26.667		4,175	1,250		5,425	6,475
1300	1000 watt			.20	40		10,100	1,875		11,975	13,900

26 61 Lighting Systems and Accessories

26 61 23 – Lamps Applications

26 61 23.10 Lamps		Crew	Daily Output	Labor-Hours	Unit	Material	2009 Bare Costs Labor	Equipment	Total	Total Incl O&P
1320	1000 watt, 125,000 initial lumens	1 Elec	.20	40	C	19,200	1,875		21,075	24,000
1330	1500 watt		.20	40		18,000	1,875		19,875	22,700
1350	High pressure sodium, 70 watt		.30	26.667		4,525	1,250		5,775	6,850
1360	100 watt		.30	26.667		4,750	1,250		6,000	7,100
1370	150 watt		.30	26.667		4,850	1,250		6,100	7,225
1380	250 watt		.30	26.667		5,175	1,250		6,425	7,550
1400	400 watt		.30	26.667		5,300	1,250		6,550	7,700
1450	1000 watt		.20	40		14,600	1,875		16,475	18,900
1500	Low pressure sodium, 35 watt		.30	26.667		9,650	1,250		10,900	12,500
1550	55 watt		.30	26.667		9,725	1,250		10,975	12,600
1600	90 watt		.30	26.667		11,300	1,250		12,550	14,300
1650	135 watt		.20	40		15,700	1,875		17,575	20,100
1700	180 watt		.20	40		17,100	1,875		18,975	21,600
1750	Quartz line, clear, 500 watt		1.10	7.273		1,000	340		1,340	1,600
1760	1500 watt		.20	40		4,250	1,875		6,125	7,475
1762	Spot, MR 16, 50 watt		1.30	6.154		900	289		1,189	1,425
1800	Incandescent, interior, A21, 100 watt		1.60	5		180	235		415	550
1900	A21, 150 watt		1.60	5		176	235		411	545
2000	A23, 200 watt		1.60	5		246	235		481	620
2200	PS 35, 300 watt		1.60	5		680	235		915	1,100
2210	PS 35, 500 watt		1.60	5		1,150	235		1,385	1,625
2230	PS 52, 1000 watt		1.30	6.154		2,250	289		2,539	2,900
2240	PS 52, 1500 watt		1.30	6.154		6,050	289		6,339	7,075
2300	R30, 75 watt		1.30	6.154		595	289		884	1,075
2400	R40, 100 watt		1.30	6.154		605	289		894	1,100
2500	Exterior, PAR 38, 75 watt		1.30	6.154		1,350	289		1,639	1,900
2600	PAR 38, 150 watt		1.30	6.154		1,550	289		1,839	2,125
2700	PAR 46, 200 watt		1.10	7.273		2,950	340		3,290	3,750
2800	PAR 56, 300 watt		1.10	7.273		3,850	340		4,190	4,725
3000	Guards, fluorescent lamp, 4' long		1	8		1,025	375		1,400	1,675
3200	8' long		.90	8.889		2,050	420		2,470	2,875

Estimating Tips

27 20 00 Data Communications
27 30 00 Voice Communications
27 40 00 Audio-Video Communications

When estimating material costs for special systems, it is always prudent to obtain manufacturers' quotations for equipment prices and special installation requirements which will affect the total costs.

Reference Numbers

Reference numbers are shown in shaded boxes at the beginning of some major classifications. These numbers refer to related items in the Reference Section. The reference information may be an estimating procedure, an alternate pricing method, or technical information.

Note: Not all subdivisions listed here necessarily appear in this publication.

Note: **Trade Service,** *in part, has been used as a reference source for some of the material prices used in Division 27.*

27 41 Audio-Video Systems

27 41 19 – Portable Audio-Video Equipment

27 41 19.10 T.V. Systems

		Crew	Daily Output	Labor-Hours	Unit	Material	2009 Bare Costs Labor	2009 Bare Costs Equipment	Total	Total Incl O&P
0010	**T.V. SYSTEMS**, not including rough-in wires, cables & conduits									
0100	Master TV antenna system									
0200	VHF reception & distribution, 12 outlets	1 Elec	6	1.333	Outlet	203	62.50		265.50	315
0400	30 outlets		10	.800		134	37.50		171.50	203
0600	100 outlets		13	.615		137	29		166	194
0800	VHF & UHF reception & distribution, 12 outlets		6	1.333		202	62.50		264.50	315
1000	30 outlets		10	.800		134	37.50		171.50	203
1200	100 outlets		13	.615		136	29		165	193
1400	School and deluxe systems, 12 outlets		2.40	3.333		266	157		423	525
1600	30 outlets		4	2		233	94		327	395
1800	80 outlets		5.30	1.509		224	71		295	350

27 51 Distributed Audio-Video Communications Systems

27 51 16 – Public Address and Mass Notification Systems

27 51 16.10 Public Address System

		Crew	Daily Output	Labor-Hours	Unit	Material	2009 Bare Costs Labor	2009 Bare Costs Equipment	Total	Total Incl O&P
0010	**PUBLIC ADDRESS SYSTEM**									
0100	Conventional, office	1 Elec	5.33	1.501	Speaker	127	70.50		197.50	244
0200	Industrial	"	2.70	2.963	"	244	139		383	475

27 51 19 – Sound Masking Systems

27 51 19.10 Sound System

		Crew	Daily Output	Labor-Hours	Unit	Material	2009 Bare Costs Labor	2009 Bare Costs Equipment	Total	Total Incl O&P
0010	**SOUND SYSTEM**, not including rough-in wires, cables & conduits									
0100	Components, projector outlet	1 Elec	8	1	Ea.	58.50	47		105.50	135
0200	Microphone		4	2		65	94		159	212
0400	Speakers, ceiling or wall		8	1		110	47		157	191
0600	Trumpets		4	2		205	94		299	365
0800	Privacy switch		8	1		82	47		129	160
1000	Monitor panel		4	2		365	94		459	540
1200	Antenna, AM/FM		4	2		204	94		298	365
1400	Volume control		8	1		82	47		129	160
1600	Amplifier, 250 watts		1	8		1,625	375		2,000	2,350
1800	Cabinets		1	8		795	375		1,170	1,425
2000	Intercom, 25 station capacity, master station	2 Elec	2	8		1,925	375		2,300	2,650
2020	11 station capacity	"	4	4		900	188		1,088	1,275
2200	Remote station	1 Elec	8	1		153	47		200	239
2400	Intercom outlets		8	1		90	47		137	169
2600	Handset		4	2		298	94		392	470
2800	Emergency call system, 12 zones, annunciator		1.30	6.154		900	289		1,189	1,425
3000	Bell		5.30	1.509		93	71		164	208
3200	Light or relay		8	1		46.50	47		93.50	121
3400	Transformer		4	2		204	94		298	365
3600	House telephone, talking station		1.60	5		440	235		675	830
3800	Press to talk, release to listen		5.30	1.509		102	71		173	218
4000	System-on button					61			61	67
4200	Door release	1 Elec	4	2		109	94		203	260
4400	Combination speaker and microphone		8	1		186	47		233	274
4600	Termination box		3.20	2.500		58.50	118		176.50	240
4800	Amplifier or power supply		5.30	1.509		670	71		741	845
5000	Vestibule door unit		16	.500	Name	124	23.50		147.50	171
5200	Strip cabinet		27	.296	Ea.	233	13.95		246.95	278
5400	Directory		16	.500	"	110	23.50		133.50	156

27 52 Healthcare Communications and Monitoring Systems

27 52 23 – Nurse Call/Code Blue Systems

27 52 23.10 Nurse Call Systems

		Crew	Daily Output	Labor-Hours	Unit	Material	2009 Bare Costs Labor	Equipment	Total	Total Incl O&P
0010	**NURSE CALL SYSTEMS**									
0100	Single bedside call station	1 Elec	8	1	Ea.	220	47		267	310
0200	Ceiling speaker station		8	1		66	47		113	143
0400	Emergency call station		8	1		111	47		158	192
0600	Pillow speaker		8	1		205	47		252	296
0800	Double bedside call station		4	2		221	94		315	385
1000	Duty station		4	2		170	94		264	325
1200	Standard call button		8	1		90	47		137	169
1400	Lights, corridor, dome or zone indicator		8	1		61	47		108	137
1600	Master control station for 20 stations	2 Elec	.65	24.615	Total	3,900	1,150		5,050	6,025

27 53 Distributed Systems

27 53 13 – Clock Systems

27 53 13.50 Clock Equipments

		Crew	Daily Output	Labor-Hours	Unit	Material	2009 Bare Costs Labor	Equipment	Total	Total Incl O&P
0010	**CLOCK EQUIPMENTS**, not including wires & conduits									
0100	Time system components, master controller	1 Elec	.33	24.242	Ea.	1,750	1,150		2,900	3,625
0200	Program bell		8	1		57.50	47		104.50	133
0400	Combination clock & speaker		3.20	2.500		202	118		320	400
0600	Frequency generator		2	4		6,800	188		6,988	7,750
0800	Job time automatic stamp recorder, minimum		4	2		475	94		569	665
1000	Maximum		4	2		730	94		824	940
1200	Time stamp for correspondence, hand operated					365			365	400
1400	Fully automatic					530			530	585
1600	Master time clock system, clocks & bells, 20 room	4 Elec	.20	160		4,375	7,525		11,900	16,000
1800	50 room	"	.08	400		10,100	18,800		28,900	39,100
2000	Time clock, 100 cards in & out, 1 color	1 Elec	3.20	2.500		1,025	118		1,143	1,300
2200	2 colors		3.20	2.500		1,100	118		1,218	1,375
2400	With 3 circuit program device, minimum		2	4		400	188		588	720
2600	Maximum		2	4		585	188		773	920
2800	Metal rack for 25 cards		7	1.143		64.50	53.50		118	151
3000	Watchman's tour station		8	1		67.50	47		114.50	144
3200	Annunciator with zone indication		1	8		246	375		621	830
3400	Time clock with tape		1	8		655	375		1,030	1,275

Division Notes

		CREW	DAILY OUTPUT	LABOR-HOURS	UNIT	2009 BARE COSTS MAT.	LABOR	EQUIP.	TOTAL	TOTAL INCL O&P

Estimating Tips

- When estimating material costs for electronic safety and security systems, it is always prudent to obtain manufacturers' quotations for equipment prices and special installation requirements that affect the total cost.
- Fire alarm systems consist of control panels, annunciator panels, battery with rack, charger, and fire alarm actuating and indicating devices. Some fire alarm systems include speakers, telephone lines, door closer controls, and other components. Be careful not to overlook the costs related to installation for these items. Also be aware of costs for integrated automation instrumentation and terminal devices, control equipment, control wiring, and programming.
- Security equipment includes items such as CCTV, access control, and other detection and identification systems to perform alert and alarm functions. Be sure to consider the costs related to installation for this security equipment, such as for integrated automation instrumentation and terminal devices, control equipment, control wiring, and programming.

Reference Numbers

Reference numbers are shown in shaded boxes at the beginning of some major classifications. These numbers refer to related items in the Reference Section. The reference information may be an estimating procedure, an alternate pricing method, or technical information.

Note: Not all subdivisions listed here necessarily appear in this publication.

No part of this publication may be reproduced, stored in a retrieval system, or transmitted in any form or by any means without prior written permission of R.S. Means Company, Inc.

28 16 Intrusion Detection

28 16 16 – Intrusion Detection Systems Infrastructure

28 16 16.50 Intrusion Detection

		Crew	Daily Output	Labor-Hours	Unit	Material	2009 Bare Costs Labor	Equipment	Total	Total Incl O&P
0010	**INTRUSION DETECTION**, not including wires & conduits									
0100	Burglar alarm, battery operated, mechanical trigger	1 Elec	4	2	Ea.	272	94		366	440
0200	Electrical trigger		4	2		325	94		419	495
0400	For outside key control, add		8	1		77	47		124	155
0600	For remote signaling circuitry, add		8	1		122	47		169	205
0800	Card reader, flush type, standard		2.70	2.963		910	139		1,049	1,200
1000	Multi-code		2.70	2.963		1,175	139		1,314	1,500
1200	Door switches, hinge switch		5.30	1.509		57.50	71		128.50	169
1400	Magnetic switch		5.30	1.509		67.50	71		138.50	181
1600	Exit control locks, horn alarm		4	2		340	94		434	515
1800	Flashing light alarm		4	2		380	94		474	560
2000	Indicating panels, 1 channel		2.70	2.963		360	139		499	600
2200	10 channel	2 Elec	3.20	5		1,225	235		1,460	1,700
2400	20 channel		2	8		2,400	375		2,775	3,200
2600	40 channel		1.14	14.035		4,375	660		5,035	5,775
2800	Ultrasonic motion detector, 12 volt	1 Elec	2.30	3.478		225	163		388	490
3000	Infrared photoelectric detector		2.30	3.478		186	163		349	445
3200	Passive infrared detector		2.30	3.478		278	163		441	550
3400	Glass break alarm switch		8	1		46.50	47		93.50	121
3420	Switchmats, 30" x 5'		5.30	1.509		83	71		154	198
3440	30" x 25'		4	2		199	94		293	360
3460	Police connect panel		4	2		239	94		333	405
3480	Telephone dialer		5.30	1.509		375	71		446	520
3500	Alarm bell		4	2		76	94		170	224
3520	Siren		4	2		143	94		237	297
3540	Microwave detector, 10' to 200'		2	4		655	188		843	1,000
3560	10' to 350'		2	4		1,900	188		2,088	2,375

28 23 Video Surveillance

28 23 13 – Video Surveillance Control and Management Systems

28 23 13.10 Closed Circuit Television System

		Crew	Daily Output	Labor-Hours	Unit	Material	2009 Bare Costs Labor	Equipment	Total	Total Incl O&P
0010	**CLOSED CIRCUIT TELEVISION SYSTEM**									
2000	Surveillance, one station (camera & monitor)	2 Elec	2.60	6.154	Total	1,300	289		1,589	1,850
2200	For additional camera stations, add	1 Elec	2.70	2.963	Ea.	720	139		859	1,000
2400	Industrial quality, one station (camera & monitor)	2 Elec	2.60	6.154	Total	2,675	289		2,964	3,375
2600	For additional camera stations, add	1 Elec	2.70	2.963	Ea.	1,625	139		1,764	2,000
2610	For low light, add		2.70	2.963		1,325	139		1,464	1,650
2620	For very low light, add		2.70	2.963		9,675	139		9,814	10,900
2800	For weatherproof camera station, add		1.30	6.154		1,000	289		1,289	1,525
3000	For pan and tilt, add		1.30	6.154		2,600	289		2,889	3,300
3200	For zoom lens - remote control, add, minimum		2	4		2,400	188		2,588	2,925
3400	Maximum		2	4		8,775	188		8,963	9,950
3410	For automatic iris for low light, add		2	4		2,100	188		2,288	2,600
3600	Educational T.V. studio, basic 3 camera system, black & white,									
3800	electrical & electronic equip. only, minimum	4 Elec	.80	40	Total	12,500	1,875		14,375	16,500
4000	Maximum (full console)		.28	114		53,000	5,375		58,375	66,500
4100	As above, but color system, minimum		.28	114		70,500	5,375		75,875	85,500
4120	Maximum		.12	266		306,000	12,500		318,500	355,000
4200	For film chain, black & white, add	1 Elec	1	8	Ea.	14,300	375		14,675	16,300
4250	Color, add		.25	32		17,400	1,500		18,900	21,400
4400	For video tape recorders, add, minimum		1	8		3,000	375		3,375	3,850

28 23 Video Surveillance

28 23 13 – Video Surveillance Control and Management Systems

28 23 13.10 Closed Circuit Television System		Crew	Daily Output	Labor-Hours	Unit	Material	2009 Bare Costs Labor	Equipment	Total	Total Incl O&P
4600	Maximum	4 Elec	.40	80	Ea.	25,000	3,750		28,750	33,100

28 31 Fire Detection and Alarm

28 31 23 – Fire Detection and Alarm Annunciation Panels and Fire Stations

28 31 23.50 Alarm Panels and Devices

		Crew	Daily Output	Labor-Hours	Unit	Material	Labor	Equipment	Total	Incl O&P
0010	**ALARM PANELS AND DEVICES**, not including wires & conduits									
3594	Fire, alarm control panel									
3600	4 zone	2 Elec	2	8	Ea.	615	375		990	1,225
3800	8 zone		1	16		675	750		1,425	1,875
4000	12 zone		.67	23.988		2,300	1,125		3,425	4,200
4020	Alarm device	1 Elec	8	1		135	47		182	219
4050	Actuating device		8	1		325	47		372	425
4200	Battery and rack		4	2		805	94		899	1,025
4400	Automatic charger		8	1		495	47		542	615
4600	Signal bell		8	1		57.50	47		104.50	134
4800	Trouble buzzer or manual station		8	1		41	47		88	115
5600	Strobe and horn		5.30	1.509		105	71		176	222
5800	Fire alarm horn		6.70	1.194		40.50	56		96.50	128
6000	Door holder, electro-magnetic		4	2		86	94		180	235
6200	Combination holder and closer		3.20	2.500		475	118		593	700
6400	Code transmitter		4	2		765	94		859	980
6600	Drill switch		8	1		96	47		143	175
6800	Master box		2.70	2.963		3,425	139		3,564	3,975
7000	Break glass station		8	1		55.50	47		102.50	131
7800	Remote annunciator, 8 zone lamp		1.80	4.444		201	209		410	530
8000	12 zone lamp	2 Elec	2.60	6.154		345	289		634	810
8200	16 zone lamp	"	2.20	7.273		345	340		685	890

28 31 43 – Fire Detection Sensors

28 31 43.50 Fire and Heat Detectors

0010	**FIRE & HEAT DETECTORS**									
5000	Detector, rate of rise	1 Elec	8	1	Ea.	39	47		86	113
5100	Fixed temperature	"	8	1	"	32.50	47		79.50	106

28 31 46 – Smoke Detection Sensors

28 31 46.50 Smoke Detectors

0010	**SMOKE DETECTORS**									
5200	Smoke detector, ceiling type	1 Elec	6.20	1.290	Ea.	87.50	60.50		148	187
5400	Duct type	"	3.20	2.500	"	277	118		395	480

Division Notes

	CREW	DAILY OUTPUT	LABOR-HOURS	UNIT	MAT.	LABOR	EQUIP.	TOTAL	TOTAL INCL O&P

Estimating Tips

31 05 00 Common Work Results for Earthwork

- Estimating the actual cost of performing earthwork requires careful consideration of the variables involved. This includes items such as type of soil, whether water will be encountered, dewatering, whether banks need bracing, disposal of excavated earth, and length of haul to fill or spoil sites, etc. If the project has large quantities of cut or fill, consider raising or lowering the site to reduce costs, while paying close attention to the effect on site drainage and utilities.
- If the project has large quantities of fill, creating a borrow pit on the site can significantly lower the costs.
- It is very important to consider what time of year the project is scheduled for completion. Bad weather can create large cost overruns from dewatering, site repair, and lost productivity from cold weather.

Reference Numbers

Reference numbers are shown in shaded boxes at the beginning of some major classifications. These numbers refer to related items in the Reference Section. The reference information may be an estimating procedure, an alternate pricing method, or technical information.

Note: Not all subdivisions listed here necessarily appear in this publication.

No part of this publication may be reproduced, stored in a retrieval system, or transmitted in any form or by any means without prior written permission of R.S. Means Company, Inc.

31 46 Needle Beams

31 46 13 – Cantilever Needle Beams

31 46 13.10 Needle Beams		Crew	Daily Output	Labor-Hours	Unit	Material	2009 Bare Costs Labor	Equipment	Total	Total Incl O&P
0010	**NEEDLE BEAMS**									
0011	Incl. wood shoring 10' x 10' opening									
0400	Block, concrete, 8" thick	B-9	7.10	5.634	Ea.	45.50	180	27	252.50	360
0420	12" thick		6.70	5.970		52.50	191	28.50	272	385
0800	Brick, 4" thick with 8" backup block		5.70	7.018		52.50	225	33.50	311	445
1000	Brick, solid, 8" thick		6.20	6.452		45.50	206	31	282.50	405
1040	12" thick		4.90	8.163		52.50	261	39	352.50	505
1080	16" thick		4.50	8.889		67	284	42.50	393.50	560
2000	Add for additional floors of shoring	B-1	6	4		45.50	129		174.50	250

Estimating Tips

32 01 00 Operations and Maintenance of Exterior Improvements

- Recycling of asphalt pavement is becoming very popular and is an alternative to removal and replacement of asphalt pavement. It can be a good value engineering proposal if removed pavement can be recycled either at the site or another site that is reasonably close to the project site.

32 10 00 Bases, Ballasts, and Paving

- When estimating paving, keep in mind the project schedule. If an asphaltic paving project is in a colder climate and runs through to the spring, consider placing the base course in the autumn and then topping it in the spring, just prior to completion. This could save considerable costs in spring repair. Keep in mind that prices for asphalt and concrete are generally higher in the cold seasons.

32 90 00 Planting

- The timing of planting and guarantee specifications often dictate the costs for establishing tree and shrub growth and a stand of grass or ground cover. Establish the work performance schedule to coincide with the local planting season. Maintenance and growth guarantees can add from 20%–100% to the total landscaping cost. The cost to replace trees and shrubs can be as high as 5% of the total cost, depending on the planting zone, soil conditions, and time of year.

Reference Numbers

Reference numbers are shown in shaded boxes at the beginning of some major classifications. These numbers refer to related items in the Reference Section. The reference information may be an estimating procedure, an alternate pricing method, or technical information.

Note: Not all subdivisions listed here necessarily appear in this publication.

No part of this publication may be reproduced, stored in a retrieval system, or transmitted in any form or by any means without prior written permission of R.S. Means Company, Inc.

32 06 Schedules for Exterior Improvements

32 06 10 – Schedules for Bases, Ballasts, and Paving

32 06 10.10 Sidewalks, Driveways and Patios

		Crew	Daily Output	Labor-Hours	Unit	Material	2009 Bare Costs Labor	Equipment	Total	Total Incl O&P
0010	**SIDEWALKS, DRIVEWAYS AND PATIOS** No base									
0020	Asphaltic concrete, 2" thick	B-37	720	.067	S.Y.	6.05	2.21	.19	8.45	10.25
0100	2-1/2" thick	"	660	.073	"	7.65	2.41	.21	10.27	12.35
0300	Concrete, 3000 psi, CIP, 6 x 6 - W1.4 x W1.4 mesh,									
0310	broomed finish, no base, 4" thick	B-24	600	.040	S.F.	1.68	1.46		3.14	4.08
0350	5" thick		545	.044		2.24	1.61		3.85	4.92
0400	6" thick		510	.047		2.62	1.72		4.34	5.50
0450	For bank run gravel base, 4" thick, add	B-18	2500	.010		.64	.31	.02	.97	1.20
0520	8" thick, add	"	1600	.015		1.28	.48	.02	1.78	2.19
0550	Exposed aggregate finish, add to above, minimum	B-24	1875	.013		.15	.47		.62	.87
0600	Maximum	"	455	.053		.49	1.93		2.42	3.47
1000	Crushed stone, 1" thick, white marble	2 Clab	1700	.009		.22	.30		.52	.70
1050	Bluestone	"	1700	.009		.24	.30		.54	.73
1700	Redwood, prefabricated, 4' x 4' sections	2 Carp	316	.051		8.25	2.02		10.27	12.25
1750	Redwood planks, 1" thick, on sleepers	"	240	.067		5.80	2.66		8.46	10.50

32 06 10.20 Steps

		Crew	Daily Output	Labor-Hours	Unit	Material	Labor	Equipment	Total	Total Incl O&P
0010	**STEPS**									
0011	Incl. excav., borrow & concrete base as required									
0100	Brick steps	B-24	35	.686	LF Riser	11.40	25		36.40	50.50
0200	Railroad ties	2 Clab	25	.640		3.30	20		23.30	35
0300	Bluestone treads, 12" x 2" or 12" x 1-1/2"	B-24	30	.800		27	29.50		56.50	74
4000	Edging, redwood, 2" x 4"	2 Carp	330	.048	L.F.	2.48	1.94		4.42	5.70
4025	Steel edge strips, incl. stakes, 1/4" x 5"	B-1	390	.062		4.25	1.99		6.24	7.75
4050	Edging, landscape timber or railroad ties, 6" x 8"	2 Carp	170	.094		3.10	3.76		6.86	9.25

32 11 Base Courses

32 11 23 – Aggregate Base Courses

32 11 23.23 Base Course Drainage Layers

		Crew	Daily Output	Labor-Hours	Unit	Material	Labor	Equipment	Total	Total Incl O&P
0010	**BASE COURSE DRAINAGE LAYERS**									
0011	For roadways and large areas									
0050	Crushed 3/4" stone base, compacted, 3" deep	B-36C	5200	.008	S.Y.	4.23	.29	.63	5.15	5.80
0100	6" deep		5000	.008		8.45	.30	.66	9.41	10.50
0200	9" deep		4600	.009		12.70	.33	.71	13.74	15.30
0300	12" deep		4200	.010		16.95	.36	.78	18.09	20
0301	Crushed 1-1/2" stone base, compacted to 4" deep	B-36B	6000	.011		5.85	.39	.62	6.86	7.70
0302	6" deep		5400	.012		8.75	.44	.69	9.88	11
0303	8" deep		4500	.014		11.65	.52	.83	13	14.50
0304	12" deep		3800	.017		17.50	.62	.99	19.11	21.50
0350	Bank run gravel, spread and compacted									
0370	6" deep	B-32	6000	.005	S.Y.	5.35	.21	.31	5.87	6.50
0390	9" deep		4900	.007		8	.25	.38	8.63	9.60
0400	12" deep		4200	.008		10.65	.30	.44	11.39	12.65
6900	For small and irregular areas, add						50%	50%		
7000	Prepare and roll sub-base, small areas to 2500 S.Y.	B-32A	1500	.016	S.Y.		.61	.76	1.37	1.76

32 11 26 – Asphaltic Base Courses

32 11 26.19 Bituminous-Stabilized Base Courses

		Crew	Daily Output	Labor-Hours	Unit	Material	Labor	Equipment	Total	Total Incl O&P
0010	**BITUMINOUS-STABILIZED BASE COURSES**									
0020	And large paved areas									
8900	For small and irregular areas, add						50%	50%		

32 14 Unit Paving

32 14 13 – Precast Concrete Unit Paving

32 14 13.13 Interlocking Precast Concrete Unit Paving	Crew	Daily Output	Labor-Hours	Unit	Material	2009 Bare Costs Labor	Equipment	Total	Total Incl O&P
0010 **INTERLOCKING PRECAST CONCRETE UNIT PAVING**									
0020 "V" blocks for retaining soil	D-1	205	.078	S.F.	8.60	2.84		11.44	13.75

32 14 13.16 Precast Concrete Unit Paving Slabs

	Crew	Daily Output	Labor-Hours	Unit	Material	Labor	Equipment	Total	Total Incl O&P
0010 **PRECAST CONCRETE UNIT PAVING SLABS**									
0750 Exposed local aggregate, natural	2 Bric	250	.064	S.F.	6.45	2.59		9.04	11.05
0800 Colors		250	.064		7.05	2.59		9.64	11.70
0850 Exposed granite or limestone aggregate		250	.064		7.80	2.59		10.39	12.50
0900 Exposed white tumblestone aggregate	↓	250	.064	↓	4.61	2.59		7.20	9

32 14 16 – Brick Unit Paving

32 14 16.10 Brick Paving

	Crew	Daily Output	Labor-Hours	Unit	Material	Labor	Equipment	Total	Total Incl O&P
0010 **BRICK PAVING**									
0012 4" x 8" x 1-1/2", without joints (4.5 brick/S.F.)	D-1	110	.145	S.F.	2.78	5.30		8.08	11.05
0100 Grouted, 3/8" joint (3.9 brick/S.F.)		90	.178		3.38	6.45		9.83	13.50
0200 4" x 8" x 2-1/4", without joints (4.5 bricks/S.F.)		110	.145		3.74	5.30		9.04	12.10
0300 Grouted, 3/8" joint (3.9 brick/S.F.)	↓	90	.178		3.45	6.45		9.90	13.60
0500 Bedding, asphalt, 3/4" thick	B-25	5130	.017		.57	.59	.48	1.64	2.07
0540 Course washed sand bed, 1" thick	B-18	5000	.005		.28	.15	.01	.44	.55
0580 Mortar, 1" thick	D-1	300	.053		.64	1.94		2.58	3.64
0620 2" thick		200	.080		1.28	2.91		4.19	5.80
1500 Brick on 1" thick sand bed laid flat, 4.5 per S.F.		100	.160		2.99	5.80		8.79	12.10
2000 Brick pavers, laid on edge, 7.2 per S.F.		70	.229		2.74	8.30		11.04	15.60
2500 For 4" thick concrete bed and joints, add	↓	595	.027		1.19	.98		2.17	2.79
2800 For steam cleaning, add	A-1H	950	.008		.06	.27	.05	.38	.54

32 14 23 – Asphalt Unit Paving

32 14 23.10 Asphalt Blocks

	Crew	Daily Output	Labor-Hours	Unit	Material	Labor	Equipment	Total	Total Incl O&P
0010 **ASPHALT BLOCKS**									
0020 Rectangular, 6" x 12" x 1-1/4", w/bed & neopr. adhesive	D-1	135	.119	S.F.	7.65	4.31		11.96	15
0100 3" thick		130	.123		10.75	4.47		15.22	18.60
0300 Hexagonal tile, 8" wide, 1-1/4" thick		135	.119		7.65	4.31		11.96	15
0400 2" thick		130	.123		10.75	4.47		15.22	18.60
0500 Square, 8" x 8", 1-1/4" thick		135	.119		7.65	4.31		11.96	15
0600 2" thick		130	.123		10.75	4.47		15.22	18.60
0900 For exposed aggregate (ground finish) add					.53			.53	.58
0910 For colors, add	↓				.40			.40	.44

32 14 40 – Stone Paving

32 14 40.10 Stone Pavers

	Crew	Daily Output	Labor-Hours	Unit	Material	Labor	Equipment	Total	Total Incl O&P
0010 **STONE PAVERS**									
1100 Flagging, bluestone, irregular, 1" thick,	D-1	81	.198	S.F.	5.95	7.20		13.15	17.45
1150 Snapped random rectangular, 1" thick		92	.174		9	6.30		15.30	19.50
1200 1-1/2" thick		85	.188		10.80	6.85		17.65	22.50
1250 2" thick		83	.193		12.60	7		19.60	24.50
1300 Slate, natural cleft, irregular, 3/4" thick		92	.174		6.95	6.30		13.25	17.25
1350 Random rectangular, gauged, 1/2" thick		105	.152		15.05	5.55		20.60	25
1400 Random rectangular, butt joint, gauged, 1/4" thick	↓	150	.107		16.20	3.88		20.08	24
1450 For sand rubbed finish, add				↓	7.55			7.55	8.30
1500 For interior setting, add					25%				25%
1550 Granite blocks, 3-1/2" x 3-1/2" x 3-1/2"	D-1	92	.174	S.F.	10	6.30		16.30	20.50
1600 4" to 12" long, 3" to 5" wide, 3" to 5" thick		98	.163		8.35	5.95		14.30	18.15
1650 6" to 15" long, 3" to 6" wide, 3" to 5" thick	↓	105	.152	↓	4.45	5.55		10	13.30

32 18 Athletic and Recreational Surfacing

32 18 13 – Synthetic Grass Surfacing

32 18 13.10 Artificial Grass Surfacing

		Crew	Daily Output	Labor-Hours	Unit	Material	2009 Bare Costs Labor	Equipment	Total	Total Incl O&P
0010	**ARTIFICIAL GRASS SURFACING**									
0015	Not including asphalt base or drainage,									
0020	but including cushion pad, over 50,000 S.F.									
0200	1/2" pile and 5/16" cushion pad, standard	C-17	3200	.025	S.F.	10.30	1.03		11.33	12.90
0300	Deluxe		2560	.031		12.20	1.29		13.49	15.40
0500	1/2" pile and 5/8" cushion pad, standard		2844	.028		15	1.16		16.16	18.30
0600	Deluxe		2327	.034		16.40	1.42		17.82	20.50

32 18 23 – Athletic Surfacing

32 18 23.33 Running Track Surfacing

		Crew	Daily Output	Labor-Hours	Unit	Material	Labor	Equipment	Total	Total Incl O&P
0010	**RUNNING TRACK SURFACING**									
0020	Running track, asphalt, incl base, 3" thick	B-37	300	.160	S.Y.	20.50	5.30	.46	26.26	31
0100	Surface, latex rubber system, 3/8" thick, black	B-20	125	.192		9.30	6.80		16.10	20.50
0150	Colors		125	.192		16.50	6.80		23.30	28.50
0300	Urethane rubber system, 3/8" thick, black		120	.200		25	7.05		32.05	38
0400	Color coating		115	.209		30.50	7.40		37.90	45

32 18 23.53 Tennis Court Surfacing

		Crew	Daily Output	Labor-Hours	Unit	Material	Labor	Equipment	Total	Total Incl O&P
0010	**TENNIS COURT SURFACING**									
0020	Tennis court, asphalt, incl. base, 2-1/2" thick, one court	B-37	450	.107	S.Y.	28	3.54	.31	31.85	36.50
0800	Rubber-acrylic base resilient pavement	"	600	.080		52.50	2.65	.23	55.38	62.50
1000	Colored sealer, acrylic emulsion, 3 coats	2 Clab	800	.020		6.35	.63		6.98	7.95
1100	3 coat, 2 colors	"	900	.018		8.80	.56		9.36	10.50
1400	Posts for nets, 3-1/2" diameter with eye bolts	B-1	3.40	7.059	Pr.	210	228		438	585
1500	With pulley & reel		3.40	7.059	"	299	228		527	685
1700	Net, 42' long, nylon thread with binder		50	.480	Ea.	277	15.50		292.50	330
1800	All metal		6.50	3.692	"	490	119		609	725

32 31 Fences and Gates

32 31 13 – Chain Link Fences and Gates

32 31 13.35 Fence, Interior

		Crew	Daily Output	Labor-Hours	Unit	Material	Labor	Equipment	Total	Total Incl O&P
0010	**FENCE, INTERIOR**									
0011	Chain link, 6' high, post 10' O.C., 1-5/8" rail									
0050	6 ga. wire, galv. steel	2 Carp	230	.070	L.F.	17.90	2.78		20.68	24
0070	Aluminized steel		230	.070	"	21.50	2.78		24.28	28.50
1000	Add for corner post, 3" diam., galv. steel		37	.432	Ea.	89	17.30		106.30	125
1050	Aluminized steel		37	.432		114	17.30		131.30	153
1100	Add for braces, galv. steel		75	.213		27	8.50		35.50	43
1200	Aluminized steel		75	.213		31.50	8.50		40	47.50
2000	Gate for 6' high fence, 1-5/8" frame, 3' wide, galv. steel		16	1		151	40		191	228
2500	Aluminized steel		16	1		177	40		217	257
3000	5' high fence, 9 ga., 2" line post									
3100	10' O.C., 1-5/8" top rail									
3500	Galvanized steel	2 Carp	290	.055	L.F.	11.60	2.20		13.80	16.15
4000	Aluminized steel		290	.055	"	13.30	2.20		15.50	18.05
5000	Gate, 4' wide, 5' high, 2" frame, galv. steel		26	.615	Ea.	184	24.50		208.50	240
5500	Aluminized steel		26	.615	"	184	24.50		208.50	240
5600	Chain link, fabric, 2" mesh,									
5650	Galv. steel, no frame, on ceilings	2 Carp	1800	.009	S.F.	.79	.36		1.15	1.42
6000	Hardware cloth, 1/2" mesh no frame, on ceilings	"	2000	.008	"	.97	.32		1.29	1.57

32 31 Fences and Gates

32 31 26 – Wire Fences and Gates

32 31 26.10 Fences, Misc. Metal

		Crew	Daily Output	Labor-Hours	Unit	Material	2009 Bare Costs Labor	Equipment	Total	Total Incl O&P
0010	**FENCES, MISC. METAL**									
0012	Chicken wire, posts @ 4', 1" mesh, 4' high	B-80C	410	.059	L.F.	2.16	1.84	.49	4.49	5.75
0100	2" mesh, 6' high		350	.069		1.95	2.15	.58	4.68	6.10
0200	Galv. steel, 12 ga., 2" x 4" mesh, posts 5' O.C., 3' high		300	.080		3.12	2.51	.67	6.30	8.05
0300	5' high		300	.080		4.16	2.51	.67	7.34	9.20
0400	14 ga., 1" x 2" mesh, 3' high		300	.080		3.31	2.51	.67	6.49	8.25
0500	5' high		300	.080		4.58	2.51	.67	7.76	9.65

32 31 26.20 Wire Fencing, General

		Crew	Daily Output	Labor-Hours	Unit	Material	2009 Bare Costs Labor	Equipment	Total	Total Incl O&P
0010	**WIRE FENCING, GENERAL**									
0015	Barbed wire, galvanized, domestic steel, hi-tensile 15-1/2 ga.				M.L.F.	35			35	38.50
0020	Standard, 12-3/4 ga.					46.50			46.50	51.50
0210	Barbless wire, 2-strand galvanized, 12-1/2 ga.					46.50			46.50	51.50
0500	Helical razor ribbon, stainless steel, 18" dia x 18" spacing				C.L.F.	136			136	149
0600	Hardware cloth galv., 1/4" mesh, 23 ga., 2' wide				C.S.F.	61			61	67
0700	3' wide					59			59	65
0900	1/2" mesh, 19 ga., 2' wide					53.50			53.50	59
1000	4' wide					52.50			52.50	58
1200	Chain link fabric, steel, 2" mesh, 6 ga, galvanized					201			201	221
1300	9 ga, galvanized					99			99	109
1350	Vinyl coated					83			83	91.50
1360	Aluminized					129			129	142
1400	2-1/4" mesh, 11.5 ga, galvanized					67			67	73.50
1600	1-3/4" mesh (tennis courts), 11.5 ga (core), vinyl coated					95			95	104
1700	9 ga, galvanized					85.50			85.50	94
2100	Welded wire fabric, galvanized, 1" x 2", 14 ga.					57			57	62.50
2200	2" x 4", 12-1/2 ga.					19.45			19.45	21.50

32 31 29 – Wood Fences and Gates

32 31 29.20 Fence, Wood Rail

		Crew	Daily Output	Labor-Hours	Unit	Material	2009 Bare Costs Labor	Equipment	Total	Total Incl O&P
0010	**FENCE, WOOD RAIL**									
0012	Picket, No. 2 cedar, Gothic, 2 rail, 3' high	B-1	160	.150	L.F.	5.85	4.84		10.69	13.95
0050	Gate, 3'-6" wide	B-80C	9	2.667	Ea.	50.50	83.50	22.50	156.50	209
0400	3 rail, 4' high		150	.160	L.F.	6.75	5	1.35	13.10	16.65
0500	Gate, 3'-6" wide		9	2.667	Ea.	60.50	83.50	22.50	166.50	220
0600	Open rail, rustic, No. 1 cedar, 2 rail, 3' high		160	.150	L.F.	5.25	4.71	1.26	11.22	14.45
0650	Gate, 3' wide		9	2.667	Ea.	58.50	83.50	22.50	164.50	218
0700	3 rail, 4' high		150	.160	L.F.	6	5	1.35	12.35	15.85
0900	Gate, 3' wide		9	2.667	Ea.	75	83.50	22.50	181	236
1200	Stockade, No. 2 cedar, treated wood rails, 6' high		160	.150	L.F.	7.30	4.71	1.26	13.27	16.75
1250	Gate, 3' wide		9	2.667	Ea.	60	83.50	22.50	166	220
1300	No. 1 cedar, 3-1/4" cedar rails, 6' high		160	.150	L.F.	18.05	4.71	1.26	24.02	28.50
1500	Gate, 3' wide		9	2.667	Ea.	152	83.50	22.50	258	320
1520	Open rail, split, No. 1 cedar, 2 rail, 3' high		160	.150	L.F.	5.25	4.71	1.26	11.22	14.50
1540	3 rail, 4'-0" high		150	.160		6.75	5	1.35	13.10	16.70
3300	Board, shadow box, 1" x 6", treated pine, 6' high		160	.150		10.50	4.71	1.26	16.47	20
3400	No. 1 cedar, 6' high		150	.160		20.50	5	1.35	26.85	32
3900	Basket weave, No. 1 cedar, 6' high		160	.150		20.50	4.71	1.26	26.47	31
3950	Gate, 3'-6" wide	B-1	8	3	Ea.	140	97		237	305
4000	Treated pine, 6' high		150	.160	L.F.	13.45	5.15		18.60	23
4200	Gate, 3'-6" wide		9	2.667	Ea.	65.50	86		151.50	205

32 32 Retaining Walls

32 32 60 – Stone Retaining Walls

32 32 60.10 Retaining Walls, Stone

		Crew	Daily Output	Labor-Hours	Unit	Material	2009 Bare Costs Labor	2009 Bare Costs Equipment	Total	Total Incl O&P
0010	**RETAINING WALLS, STONE**									
0015	Including excavation, concrete footing and									
0020	stone 3' below grade. Price is exposed face area.									
0200	Decorative random stone, to 6' high, 1'-6" thick, dry set	D-1	35	.457	S.F.	47	16.60		63.60	77
0300	Mortar set		40	.400		47	14.55		61.55	74
0500	Cut stone, to 6' high, 1'-6" thick, dry set		35	.457		47	16.60		63.60	77
0600	Mortar set		40	.400		47	14.55		61.55	74

32 34 Fabricated Bridges

32 34 20 – Fabricated Pedestrian Bridges

32 34 20.10 Bridges, Pedestrian

		Crew	Daily Output	Labor-Hours	Unit	Material	Labor	Equipment	Total	Total Incl O&P
0010	**BRIDGES, PEDESTRIAN**									
0011	Spans over streams, roadways, etc.									
0020	including erection, not including foundations									
0050	Precast concrete, complete in place, 8' wide, 60' span	E-2	215	.260	S.F.	67.50	11.35	8.10	86.95	103
0100	100' span		185	.303		74	13.20	9.40	96.60	114
0300	Steel, trussed or arch spans, compl. in place, 8' wide, 40' span		320	.175		88	7.60	5.45	101.05	116
0400	50' span		395	.142		79	6.15	4.41	89.56	102
0500	60' span		465	.120		79	5.25	3.74	87.99	99.50
0600	80' span		570	.098		94	4.28	3.05	101.33	114
0700	100' span		465	.120		132	5.25	3.74	140.99	158
0800	120' span		365	.153		167	6.70	4.77	178.47	200
1100	10' wide, 80' span		640	.088		97.50	3.81	2.72	104.03	116
1200	120' span		415	.135		126	5.90	4.20	136.10	154
1600	Wood, laminated type, complete in place, 80' span	C-12	203	.236		58	9.30	3.79	71.09	82.50
1700	130' span	"	153	.314		60.50	12.35	5.05	77.90	91

32 91 Planting Preparation

32 91 13 – Soil Preparation

32 91 13.16 Mulching

		Crew	Daily Output	Labor-Hours	Unit	Material	Labor	Equipment	Total	Total Incl O&P
0010	**MULCHING**									
0100	Aged barks, 3" deep, hand spread	1 Clab	100	.080	S.Y.	2.88	2.53		5.41	7.10
0150	Skid steer loader	B-63	13.50	2.963	M.S.F.	320	98	11.05	429.05	515
0200	Hay, 1" deep, hand spread	1 Clab	475	.017	S.Y.	.47	.53		1	1.35
0250	Power mulcher, small	B-64	180	.089	M.S.F.	52	2.78	1.84	56.62	64
0350	Large	B-65	530	.030	"	52	.94	.88	53.82	60
0400	Humus peat, 1" deep, hand spread	1 Clab	700	.011	S.Y.	4.50	.36		4.86	5.50
0450	Push spreader	"	2500	.003	"	4.50	.10		4.60	5.10
0550	Tractor spreader	B-66	700	.011	M.S.F.	500	.45	.33	500.78	550
0600	Oat straw, 1" deep, hand spread	1 Clab	475	.017	S.Y.	.46	.53		.99	1.34
0650	Power mulcher, small	B-64	180	.089	M.S.F.	51	2.78	1.84	55.62	62.50
0700	Large	B-65	530	.030	"	51	.94	.88	52.82	58.50
0750	Add for asphaltic emulsion	B-45	1770	.009	Gal.	2.24	.33	.44	3.01	3.45
0800	Peat moss, 1" deep, hand spread	1 Clab	900	.009	S.Y.	1.85	.28		2.13	2.48
0850	Push spreader	"	2500	.003	"	1.85	.10		1.95	2.20
0950	Tractor spreader	B-66	700	.011	M.S.F.	206	.45	.33	206.78	227
1000	Polyethylene film, 6 mil.	2 Clab	2000	.008	S.Y.	.19	.25		.44	.60
1100	Redwood nuggets, 3" deep, hand spread	1 Clab	150	.053	"	4.65	1.69		6.34	7.70
1150	Skid steer loader	B-63	13.50	2.963	M.S.F.	515	98	11.05	624.05	735

32 91 Planting Preparation

32 91 13 – Soil Preparation

32 91 13.16 Mulching		Crew	Daily Output	Labor-Hours	Unit	Material	2009 Bare Costs Labor	Equipment	Total	Total Incl O&P
1200	Stone mulch, hand spread, ceramic chips, economy	1 Clab	125	.064	S.Y.	6.90	2.02		8.92	10.75
1250	Deluxe	"	95	.084	"	10.55	2.66		13.21	15.80
1300	Granite chips	B-1	10	2.400	C.Y.	34.50	77.50		112	158
1400	Marble chips		10	2.400		129	77.50		206.50	262
1600	Pea gravel		28	.857		68.50	27.50		96	118
1700	Quartz		10	2.400		166	77.50		243.50	300
1800	Tar paper, 15 Lb. felt	1 Clab	800	.010	S.Y.	.43	.32		.75	.96
1900	Wood chips, 2" deep, hand spread	"	220	.036	"	1.88	1.15		3.03	3.85
1950	Skid steer loader	B-63	20.30	1.970	M.S.F.	209	65	7.35	281.35	340

32 91 13.23 Structural Soil Mixing										
0010	**STRUCTURAL SOIL MIXING**									
0100	Rake topsoil, site material, harley rock rake, ideal	B-6	33	.727	M.S.F.		25	8.90	33.90	48
0200	Adverse	"	7	3.429			117	42	159	225
0300	Screened loam, york rake and finish, ideal	B-62	24	1			34	6.20	40.20	59
0400	Adverse	"	20	1.200			41	7.45	48.45	70.50

32 91 13.26 Planting Beds										
0010	**PLANTING BEDS**									
0100	Backfill planting pit, by hand, on site topsoil	2 Clab	18	.889	C.Y.		28		28	43.50
0200	Prepared planting mix, by hand	"	24	.667			21		21	32.50
0300	Skid steer loader, on site topsoil	B-62	340	.071			2.41	.44	2.85	4.17
0400	Prepared planting mix	"	410	.059			2	.36	2.36	3.46
1000	Excavate planting pit, by hand, sandy soil	2 Clab	16	1			31.50		31.50	49
1100	Heavy soil or clay	"	8	2			63		63	98
1200	1/2 C.Y. backhoe, sandy soil	B-11C	150	.107			3.89	1.96	5.85	8.10
1300	Heavy soil or clay	"	115	.139			5.10	2.55	7.65	10.55
2000	Mix planting soil, incl. loam, manure, peat, by hand	2 Clab	60	.267		39.50	8.45		47.95	56.50
2100	Skid steer loader	B-62	150	.160		39.50	5.45	1	45.95	53
3000	Pile sod, skid steer loader	"	2800	.009	S.Y.		.29	.05	.34	.51
3100	By hand	2 Clab	400	.040			1.26		1.26	1.96
4000	Remove sod, F.E. loader	B-10S	2000	.006			.23	.16	.39	.53
4100	Sod cutter	B-12K	3200	.005			.19	.35	.54	.67
4200	By hand	2 Clab	240	.067			2.11		2.11	3.27

32 93 Plants

32 93 10 – General Planting Costs

32 93 10.12 Travel										
0010	**TRAVEL** to all nursery items, for 10 to 20 miles, add									
0015	10 to 20 miles one way, add				All					5%
0100	30 to 50 miles one way, add				"					10%

32 93 13 – Ground Covers

32 93 13.20 Ground Cover and Vines										
0010	**GROUND COVER AND VINES** Planting only, no preparation									
0100	Ajuga, 1 year, bare root	B-1	9	2.667	C	70	86		156	210
0150	Potted, 2 year		6	4		255	129		384	480
0200	Berberis, potted, 2 year		6	4		270	129		399	495
0250	Cotoneaster, 15" to 18", shady areas, B & B		.60	40		765	1,300		2,065	2,850
0300	Boston ivy, on bank, 1 year, bare root		6	4		122	129		251	335
0350	Potted, 2 year		6	4		305	129		434	535
0400	English ivy, 1 year, bare root		9	2.667		71	86		157	211
0450	Potted, 2 year		6	4		118	129		247	330

32 93 Plants

32 93 13 – Ground Covers

32 93 13.20 Ground Cover and Vines

		Crew	Daily Output	Labor-Hours	Unit	Material	2009 Bare Costs Labor	Equipment	Total	Total Incl O&P
0500	Halls honeysuckle, 1 year, bare root	B-1	5	4.800	C	138	155		293	390
0550	Potted, 2 year		4	6		390	194		584	725
0600	Memorial rose, 9" to 12", 1 gallon container		3	8		95	258		353	505
0650	Potted, 2 gallon container		2	12		165	385		550	780
0700	Pachysandra, 1 year, bare root		10	2.400		14.20	77.50		91.70	136
0750	Potted, 2 year		6	4		100	129		229	310
0800	Vinca minor, 1 year, bare root		10	2.400		108	77.50		185.50	239
0850	Potted, 2 year		6	4		95	129		224	305
0900	Woodbine, on bank, 1/2 year, bare root		6	4		115	129		244	325
0950	Potted, 2 year		4	6		290	194		484	620
2000	Alternate method of figuring									
2100	Ajuga, field division, 4000/M.S.F.	B-1	.23	104	M.S.F.	2,225	3,375		5,600	7,675
2300	Boston ivy, 1 year, 60/M.S.F.		10	2.400		73	77.50		150.50	201
2400	English ivy, 1 yr., 500/M.S.F.		1.80	13.333		310	430		740	1,000
2500	Halls honeysuckle, 1 yr., 333/M.S.F.		1.50	16		175	515		690	995
2600	Memorial rose, 9"-12", 1 gal., 333/M.S.F.		.90	26.667		575	860		1,435	1,950
2700	Pachysandra, 1 yr., 4000/M.S.F.		.25	96		600	3,100		3,700	5,450
2800	Vinca minor, rooted cutting, 2000/M.S.F.		1	24		1,250	775		2,025	2,575
2900	Woodbine, 1 yr., 60/M.S.F.		10	2.400		73	77.50		150.50	201

32 93 33 – Shrubs

32 93 33.10 Shrubs and Trees

		Crew	Daily Output	Labor-Hours	Unit	Material	Labor	Equipment	Total	Total Incl O&P
0010	**SHRUBS AND TREES**									
0011	Evergreen, in prepared beds, B & B									
0100	Arborvitae pyramidal, 4'-5'	B-17	30	1.067	Ea.	47	36	20.50	103.50	130
0150	Globe, 12"-15"	B-1	96	.250		21	8.05		29.05	35.50
0200	Balsam, fraser, 6' - 7'	B-17	30	1.067		122	36	20.50	178.50	212
0300	Cedar, blue, 8'-10'		18	1.778		233	59.50	34.50	327	385
0350	Japanese, 4' - 5'		55	.582		99	19.50	11.30	129.80	151
0400	Cypress, hinoki, 15" - 18"	B-1	80	.300		61	9.70		70.70	82
0500	Hemlock, canadian, 2-1/2'-3'		36	.667		33.50	21.50		55	70.50
0550	Holly, Savannah, 8' - 10' H		9.68	2.479		475	80		555	650
0600	Juniper, andorra, 18"-24"		80	.300		16.75	9.70		26.45	33.50
0620	Wiltoni, 15"-18"		80	.300		18.10	9.70		27.80	35
0640	Skyrocket, 4-1/2'-5'	B-17	55	.582		55	19.50	11.30	85.80	103
0660	Blue pfitzer, 2'-2-1/2'	B-1	44	.545		25.50	17.60		43.10	55.50
0680	Ketleerie, 2-1/2'-3'		50	.480		33.50	15.50		49	61
0700	Pine, black, 2-1/2'-3'		50	.480		45.50	15.50		61	74
0720	Mugo, 18"-24"		60	.400		36	12.90		48.90	59.50
0740	White, 4'-5'	B-17	75	.427		56.50	14.30	8.25	79.05	93.50
0800	Spruce, blue, 18"-24"	B-1	60	.400		38	12.90		50.90	62
0820	Dwarf alberta, 18" - 24"	"	60	.400		25	12.90		37.90	47.50
0840	Norway, 4'-5'	B-17	75	.427		65	14.30	8.25	87.55	103
0900	Yew, denisforma, 12"-15"	B-1	60	.400		25.50	12.90		38.40	48.50
1000	Capitata, 18"-24"		30	.800		24	26		50	66
1100	Hicksi, 2'-2-1/2'		30	.800		32	26		58	75

32 93 33.20 Shrubs

		Crew	Daily Output	Labor-Hours	Unit	Material	Labor	Equipment	Total	Total Incl O&P
0010	**SHRUBS**									
0011	Broadleaf Evergreen, planted in prepared beds									
0100	Andromeda, 15"-18", container	B-1	96	.250	Ea.	26.50	8.05		34.55	41.50
0200	Azalea, 15" - 18", container		96	.250		32	8.05		40.05	47.50
0300	Barberry, 9"-12", container		130	.185		11.75	5.95		17.70	22
0400	Boxwood, 15"-18", B&B		96	.250		29	8.05		37.05	44

32 93 Plants

32 93 33 – Shrubs

32 93 33.20 Shrubs

		Crew	Daily Output	Labor-Hours	Unit	Material	2009 Bare Costs Labor	Equipment	Total	Total Incl O&P
0500	Euonymus, emerald gaiety, 12" to 15", container	B-1	115	.209	Ea.	21	6.75		27.75	33.50
0600	Holly, 15"-18", B & B		96	.250		18.60	8.05		26.65	33
0700	Leucothoe, 15" - 18", container		96	.250		15.50	8.05		23.55	29.50
0800	Mahonia, 18" - 24", container		80	.300		32	9.70		41.70	50
0900	Mount laurel, 18" - 24", B & B		80	.300		56	9.70		65.70	76.50
1000	Paxistema, 9 – 12" high		130	.185		18.25	5.95		24.20	29.50
1100	Rhododendron, 18"-24", container		48	.500		31.50	16.15		47.65	59.50
1200	Rosemary, 1 gal container		600	.040		62.50	1.29		63.79	71
2000	Deciduous, planted in prepared beds, amelanchier, 2'-3', B & B		57	.421		90	13.60		103.60	120
2100	Azalea, 15"-18", B & B		96	.250		24.50	8.05		32.55	39.50
2200	Barberry, 2' - 3', B & B		57	.421		23.50	13.60		37.10	47
2300	Bayberry, 2'-3', B & B		57	.421		28	13.60		41.60	51.50
2400	Boston ivy, 2 year, container		600	.040		17.65	1.29		18.94	21.50
2500	Corylus, 3' - 4', B & B	B-17	75	.427		40	14.30	8.25	62.55	75
2600	Cotoneaster, 15"-18", B & B	B-1	80	.300		16.50	9.70		26.20	33
2700	Deutzia, 12" - 15", B & B	"	96	.250		11.50	8.05		19.55	25
2800	Dogwood, 3'-4', B & B	B-17	40	.800		26	27	15.50	68.50	86.50
2900	Euonymus, alatus compacta, 15" to 18", container	B-1	80	.300		22	9.70		31.70	39
3000	Flowering almond, 2' - 3', container	"	36	.667		18	21.50		39.50	53.50
3100	Flowering currant, 3' - 4', container	B-17	75	.427		24	14.30	8.25	46.55	57.50
3200	Forsythia, 2'-3', container	B-1	60	.400		20	12.90		32.90	42
3300	Hibiscus, 3'-4', B & B	B-17	75	.427		15	14.30	8.25	37.55	47.50
3400	Honeysuckle, 3'-4', B & B	B-1	60	.400		23	12.90		35.90	45.50
3500	Hydrangea, 2'-3', B & B	"	57	.421		28.50	13.60		42.10	52.50
3600	Lilac, 3'-4', B & B	B-17	40	.800		27	27	15.50	69.50	87.50
3700	Mockorange, 3' - 4', B & B	B-1	36	.667		24	21.50		45.50	60
3800	Osier willow, 2' - 3', B & B		57	.421		20.50	13.60		34.10	43.50
3900	Privet, bare root, 18"-24"		80	.300		13.15	9.70		22.85	29.50
4000	Pyracantha, 2' - 3', container		80	.300		57	9.70		66.70	77.50
4100	Quince, 2'-3', B & B		57	.421		24	13.60		37.60	47.50
4200	Russian olive, 3'-4', B & B	B-17	75	.427		24	14.30	8.25	46.55	57
4300	Snowberry, 2' - 3', B & B	B-1	57	.421		12.70	13.60		26.30	35
4400	Spirea, 3'-4', B & B	"	70	.343		17.15	11.05		28.20	36
4500	Viburnum, 3'-4', B & B	B-17	40	.800		29.50	27	15.50	72	90.50
4600	Weigela, 3' - 4', B & B	B-1	70	.343		13.05	11.05		24.10	31.50

32 93 43 – Trees

32 93 43.20 Trees

			Crew	Daily Output	Labor-Hours	Unit	Material	2009 Bare Costs Labor	Equipment	Total	Total Incl O&P
0010	**TREES**										
0011	Deciduous, in prep. beds, balled & burlapped (B&B)										
0100	Ash, 2" caliper	G	B-17	8	4	Ea.	115	134	77.50	326.50	420
0200	Beech, 5'-6'	G		50	.640		223	21.50	12.40	256.90	292
0300	Birch, 6'-8', 3 stems	G		20	1.600		123	53.50	31	207.50	252
0400	Cherry, 6' - 8', 1" caliper	G		24	1.333		81.50	44.50	26	152	187
0500	Crabapple, 6'-8'	G		20	1.600		160	53.50	31	244.50	293
0600	Dogwood, 4'-5'	G		40	.800		68.50	27	15.50	111	134
0700	Eastern redbud 4'-5'	G		40	.800		133	27	15.50	175.50	205
0800	Elm, 8'-10'	G		20	1.600		119	53.50	31	203.50	248
0900	Ginkgo, 6'-7'	G		24	1.333		163	44.50	26	233.50	276
1000	Hawthorn, 8'-10', 1" caliper	G		20	1.600		136	53.50	31	220.50	267
1100	Honeylocust, 10'-12', 1-1/2" caliper	G		10	3.200		149	107	62	318	395
1200	Laburnum, 6' - 8', 1" caliper	G		24	1.333		55	44.50	26	125.50	158
1300	Larch, 8'	G		32	1		95	33.50	19.40	147.90	178

32 93 Plants

32 93 43 – Trees

32 93 43.20 Trees

		Crew	Daily Output	Labor-Hours	Unit	Material	2009 Bare Costs Labor	Equipment	Total	Total Incl O&P
1400	Linden, 8'-10', 1" caliper G	B-17	20	1.600	Ea.	107	53.50	31	191.50	235
1500	Magnolia, 4'-5' G		20	1.600		79.50	53.50	31	164	204
1600	Maple, red, 8'-10', 1-1/2" caliper G		10	3.200		174	107	62	343	425
1700	Mountain ash, 8'-10', 1" caliper G		16	2		181	67	39	287	345
1800	Oak, 2-1/2"-3" caliper G		6	5.333		258	179	103	540	670
1900	Pagoda, 6'-8' G		20	1.600		90	53.50	31	174.50	216
2000	Pear, 6'-8', 1" caliper G		20	1.600		133	53.50	31	217.50	263
2100	Planetree, 9'-11', 1-1/4" caliper G		10	3.200		105	107	62	274	350
2200	Plum, 6'-8', 1" caliper G		20	1.600		91	53.50	31	175.50	217
2300	Poplar, 9'-11', 1-1/4" caliper G		10	3.200		50	107	62	219	288
2400	Shadbush, 4'-5' G		60	.533		67.50	17.90	10.35	95.75	113
2500	Sumac, 2'-3' G		75	.427		23	14.30	8.25	45.55	56
2600	Tupelo, 5'-6' G		40	.800		50	27	15.50	92.50	113
2700	Tulip, 5'-6' G		40	.800		49	27	15.50	91.50	112
2800	Willow, 6'-8', 1" caliper G		20	1.600		64	53.50	31	148.50	187

32 94 Planting Accessories

32 94 13 – Landscape Edging

32 94 13.20 Edging

		Crew	Daily Output	Labor-Hours	Unit	Material	Labor	Equipment	Total	Total Incl O&P
0010	**EDGING**									
0050	Aluminum alloy, including stakes, 1/8" x 4", mill finish	B-1	390	.062	L.F.	3.10	1.99		5.09	6.50
0051	Black paint		390	.062		3.60	1.99		5.59	7.05
0052	Black anodized		390	.062		4.15	1.99		6.14	7.65
0060	3/16" x 4", mill finish		380	.063		4.46	2.04		6.50	8.05
0061	Black paint		380	.063		5.15	2.04		7.19	8.80
0062	Black anodized		380	.063		6	2.04		8.04	9.75
0070	1/8" x 5-1/2" mill finish		370	.065		4.50	2.09		6.59	8.20
0071	Black paint		370	.065		5.35	2.09		7.44	9.10
0072	Black anodized		370	.065		6.10	2.09		8.19	9.95
0080	3/16" x 5-1/2" mill finish		360	.067		6	2.15		8.15	9.95
0081	Black paint		360	.067		6.80	2.15		8.95	10.80
0082	Black anodized		360	.067		7.85	2.15		10	12
0100	Brick, set horizontally, 1-1/2 bricks per L.F.	D-1	370	.043		1.67	1.57		3.24	4.22
0150	Set vertically, 3 bricks per L.F.	"	135	.119		4.20	4.31		8.51	11.15
0200	Corrugated aluminum, roll, 4" wide	1 Carp	650	.012		.70	.49		1.19	1.53
0250	6" wide	"	550	.015		.88	.58		1.46	1.86
0350	Granite, 5" x 16", straight	B-29	300	.187		12.05	6.40	3.20	21.65	26.50
0400	Polyethylene grass barrier, 5" x 1/8"	D-1	400	.040		.80	1.45		2.25	3.08
0500	Precast scallops, green, 2" x 8" x 16"		400	.040		1.10	1.45		2.55	3.41
0550	2" x 8" x 16" other than green		400	.040		.88	1.45		2.33	3.17
0600	Railroad ties, 6" x 8"	2 Carp	170	.094		3.10	3.76		6.86	9.25
0650	7" x 9"		136	.118		3.44	4.70		8.14	11.10
0750	Redwood 2" x 4"		330	.048		2.27	1.94		4.21	5.50
0800	Steel edge strips, incl. stakes, 1/4" x 5"	B-1	390	.062		4.25	1.99		6.24	7.75
0850	3/16" x 4"	"	390	.062		3.36	1.99		5.35	6.75
0900	Hardwood, pressure treated, 4" x 6"	2 Carp	250	.064		2.50	2.56		5.06	6.70
0940	6" x 6"		200	.080		3.58	3.20		6.78	8.90
0980	6" x 8"		170	.094		4.08	3.76		7.84	10.35
1000	Pine, pressure treated, 1" x 4"		500	.032		.55	1.28		1.83	2.59
1040	2" x 4"		330	.048		.92	1.94		2.86	4.02
1080	4" x 6"		250	.064		3.08	2.56		5.64	7.35

32 94 Planting Accessories

32 94 13 – Landscape Edging

32 94 13.20 Edging		Crew	Daily Output	Labor-Hours	Unit	Material	2009 Bare Costs Labor	Equipment	Total	Total Incl O&P
1100	6" x 6"	2 Carp	200	.080	L.F.	4.62	3.20		7.82	10.05
1140	6" x 8"	↓	170	.094	↓	6.05	3.76		9.81	12.50

32 96 Transplanting

32 96 23 – Plant and Bulb Transplanting

32 96 23.23 Planting

		Crew	Daily Output	Labor-Hours	Unit	Material	Labor	Equipment	Total	Total Incl O&P
0010	**PLANTING**									
0012	Moving shrubs on site, 12" ball	B-62	28	.857	Ea.		29	5.35	34.35	51
0100	24" ball	"	22	1.091	"		37	6.80	43.80	64.50

32 96 23.43 Moving Trees

		Crew	Daily Output	Labor-Hours	Unit	Material	Labor	Equipment	Total	Total Incl O&P
0290	**MOVING TREES**, On site									
0300	Moving trees on site, 36" ball	B-6	3.75	6.400	Ea.		218	78.50	296.50	420
0400	60" ball	"	1	24	"		820	294	1,114	1,575

Division Notes

		CREW	DAILY OUTPUT	LABOR-HOURS	UNIT	2009 BARE COSTS				TOTAL INCL O&P
						MAT.	LABOR	EQUIP.	TOTAL	

Division 41 - Mat'l. Processing and Handling Equip.

Estimating Tips

Products such as conveyors, material handling cranes and hoists, as well as other items specified in this division, may require trained installers. The general contractor may not have any choice as to who will perform the installation or when it will be performed. Long lead times are often required for these products, making early decisions in purchasing and scheduling necessary.

The installation of this type of equipment may require the embedment of mounting hardware during construction of floors, structural walls, or interior walls/partitions. Electrical connections will require coordination with the electrical contractor.

Reference Numbers

Reference numbers are shown in shaded boxes at the beginning of some major classifications. These numbers refer to related items in the Reference Section. The reference information may be an estimating procedure, an alternate pricing method, or technical information.

Note: Not all subdivisions listed here necessarily appear in this publication.

41 21 Conveyors

41 21 23 – Piece Material Conveyors

41 21 23.16 Material Handling Conveyors

		Crew	Daily Output	Labor-Hours	Unit	Material	2009 Bare Costs Labor	2009 Bare Costs Equipment	Total	Total Incl O&P
0010	**MATERIAL HANDLING CONVEYORS**									
0020	Gravity fed, 2" rollers, 3" O.C.									
0050	10' sections with 2 supports, 600 lb. capacity, 18" wide				Ea.	450			450	495
0100	24" wide					515			515	570
0150	1400 lb. capacity, 18" wide					415			415	455
0200	24" wide					460			460	505
0350	Horizontal belt, center drive and takeup, 60 fpm									
0400	16" belt, 26.5' length	2 Mill	.50	32	Ea.	3,150	1,325		4,475	5,425
0450	24" belt, 41.5' length		.40	40		4,675	1,675		6,350	7,600
0500	61.5' length		.30	53.333		6,025	2,225		8,250	9,875
0600	Inclined belt, 10' rise with horizontal loader and									
0620	End idler assembly, 27.5' length, 18" belt	2 Mill	.30	53.333	Ea.	9,200	2,225		11,425	13,400
0700	24" belt	"	.15	106	"	10,300	4,425		14,725	17,900
3600	Monorail, overhead, manual, channel type									
3700	125 lb. per L.F.	1 Mill	26	.308	L.F.	10.25	12.80		23.05	30
3900	500 lb. per L.F.	"	21	.381	"	9.25	15.85		25.10	33.50
4000	Trolleys for above, 2 wheel, 125 lb. capacity				Ea.	77			77	84.50
4200	4 wheel, 250 lb. capacity					207			207	227
4300	8 wheel, 1,000 lb. capacity					410			410	450

41 22 Cranes and Hoists

41 22 23 – Hoists

41 22 23.10 Material Handling

		Crew	Daily Output	Labor-Hours	Unit	Material	2009 Bare Costs Labor	2009 Bare Costs Equipment	Total	Total Incl O&P
0010	**MATERIAL HANDLING**, cranes, hoists and lifts									
2100	Hoists, electric overhead, chain, hook hung, 15' lift, 1 ton cap.				Ea.	2,050			2,050	2,250
2200	3 ton capacity					2,575			2,575	2,825
2500	5 ton capacity					6,625			6,625	7,275
2600	For hand-pushed trolley, add					15%				
2700	For geared trolley, add					30%				
2800	For motor trolley, add					75%				
3000	For lifts over 15', 1 ton, add				L.F.	23.50			23.50	25.50
3100	5 ton, add				L.F.	52			52	57

Assemblies Section

Table of Contents

Table No.	Page
B SHELL	**433**
B1010 Floor Construction	
B1010 300 Balcony Floor Construction	434
B1010 700 Fireproofing	436
B2010 Exterior Walls	
B2010 100 Exterior Wall Construction	440
B2020 Exterior Windows	
B2020 100 Windows	443
B2030 Exterior Doors	
B2030 100 Glazed Doors & Entrances	446
B3020 Roof Openings	
B3020 100 Glazed Roof Openings	447
B3020 200 Roof Hatches	447
C INTERIORS	**449**
C1010 Partitions	
C1010 100 Fixed Partitions	450
C1010 200 Demountable Partitions	473
C1010 700 Interior Windows & Storefronts	474
C1020 Interior Doors	
C1020 100 Interior Doors	477
C1030 Fittings	
C1030 100 Fabricated Toilet Partitions	485
C2010 Stair Construction	
C2010 100 Regular Stairs	486
C3010 Wall Finishes	
C3010 200 Wall Finishes/Interior Walls	488

Table No.	Page
C3020 Floor Finishes	
C3020 400 Flooring	490
C3030 Ceiling Finishes	
C3030 100 Ceiling Finishes	492
C3030 200 Suspended Ceilings	493
D SERVICES	**495**
D1010 Elevators & Lifts	
D1010 100 Passenger Elevators	496
D1020 Escalators & Moving Walks	
D1020 100 Escalators	499
D1020 200 Moving Walks	499
D2010 Plumbing Fixtures	
D2010 100 Water Closets	500
D2010 200 Urinals	502
D2010 300 Lavatories	503
D2010 400 Sinks	504
D2010 800 Drinking Fountains/Coolers	507
D2010 900 Bidets/Other Fixtures	509
D3030 Cooling Generating Systems	
D3030 100 Chilled Water Systems	511
D3050 Terminal & Package Units	
D3050 100 Terminal Self-Contain. Units	512
D4010 Sprinklers	
D4010 300 Dry Sprinkler Systems	517
D4010 400 Wet Sprinkler Systems	520
D4020 Standpipes	
D4020 300 Standpipe Equipment	523
D4020 400 Fire Hose Equipment	527

Table No.	Page
D4090 Other Fire Protection Systems	
D4090 900 Misc. Other Fire Protection Systems	528
D5010 Electrical Service/Dist.	
D5010 100 High Tension Service & Distribution	529
D5020 Lighting & Branch Wiring	
D5020 100 Branch Wiring Devices	530
D5020 200 Lighting Equipment	533
D5030 Communic. & Security	
D5030 300 Telephone Systems	537
D5030 900 Other Communic. & Security Systems	538
E EQUIPMENT & FURNISHINGS	**541**
E2020 Moveable Furnishings	
E2020 400 Moveable Int. Landscaping	542

How to Use the Assemblies Cost Tables

The following is a detailed explanation of a sample Assemblies Cost Table. Most Assembly Tables are separated into three parts: 1) an illustration of the system to be estimated; 2) the components and related costs of a typical system; and 3) the costs for similar systems with dimensional and/or size variations. For costs of the components that comprise these systems, or assemblies, refer to the Unit Price Section. Next to each bold number below is the described item with the appropriate component of the sample entry following in parentheses. In most cases, if the work is to be subcontracted, the general contractor will need to add an additional markup (RSMeans suggests using 10%) to the "Total" figures.

System/Line Numbers (C1010 710 1000)

Each Assemblies Cost Line has been assigned a unique identification number based on the UNIFORMAT II classification system.

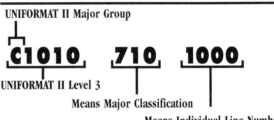

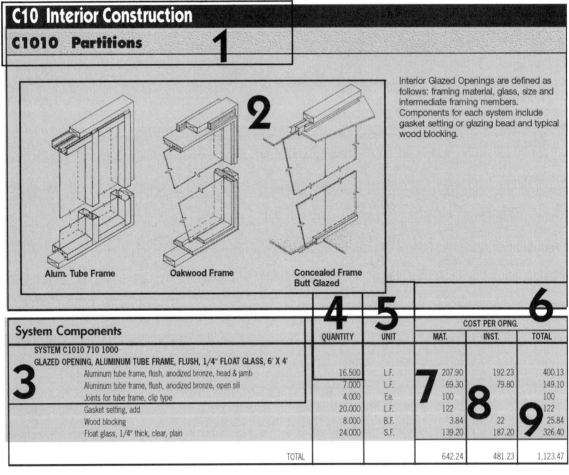

Interior Glazed Openings are defined as follows: framing material, glass, size and intermediate framing members. Components for each system include gasket setting or glazing bead and typical wood blocking.

System Components	QUANTITY	UNIT	COST PER OPNG.		
			MAT.	INST.	TOTAL
SYSTEM C1010 710 1000					
GLAZED OPENING, ALUMINUM TUBE FRAME, FLUSH, 1/4" FLOAT GLASS, 6' X 4'					
Aluminum tube frame, flush, anodized bronze, head & jamb	16.500	L.F.	207.90	192.23	400.13
Aluminum tube frame, flush, anodized bronze, open sill	7.000	L.F.	69.30	79.80	149.10
Joints for tube frame, clip type	4.000	Ea.	100		100
Gasket setting, add	20.000	L.F.	122		122
Wood blocking	8.000	B.F.	3.84	22	25.84
Float glass, 1/4" thick, clear, plain	24.000	S.F.	139.20	187.20	326.40
TOTAL			642.24	481.23	1,123.47

C1010 710		Interior Glazed Opening					
	FRAME	GLASS	OPENING-SIZE W X H	INTERMEDIATE MULLION	INTERMEDIATE HORIZONTAL	COST PER OPNG.		
						MAT.	INST.	TOTAL
1000	Aluminum flush	1/4" float	6'x4'	0	0	640	480	1,120
1040	Tube		12'x4'	3	0	1,575	1,025	2,600

430

2 Illustration

At the top of most assembly tables are an illustration, a brief description, and the design criteria used to develop the cost.

3 System Components

The components of a typical system are listed separately to show what has been included in the development of the total system price. The table below contains prices for other similar systems with dimensional and/or size variations.

4 Quantity

This is the number of line item units required for one system unit. For example, we assume that it will take 8 B.F. of wood blocking on a per-opening basis.

5 Unit of Measure for Each Item

The abbreviated designation indicates the unit of measure, as defined by industry standards, upon which the price of the component is based. For example, wood blocking is priced by the B.F. (board foot) while glass is priced by S.F. (square foot). For a complete listing of abbreviations, see the Reference Section.

6 Unit of Measure for Each System (Cost per Opening)

Costs shown in the three right-hand columns have been adjusted by the component quantity and unit of measure for the entire system. In this example, "COST PER OPENING." is the unit of measure for this system, or assembly.

7 Materials (642.24)

This column contains the Materials Cost of each component. These cost figures are bare costs plus 10% for profit.

8 Installation (481.23)

Installation includes labor and equipment plus the installing contractor's overhead and profit. Equipment costs are the bare rental costs plus 10% for profit. The labor overhead and profit are defined on the inside back cover of this book.

9 Total (1,123.47)

The figure in this column is the sum of the material and installation costs.

Material Cost	+	Installation Cost	=	Total
$642.24	+	$481.23	=	$1,123.47

No part of this publication may be reproduced, stored in a retrieval system, or transmitted in any form or by any means without prior written permission of R.S. Means Company, Inc.

B Shell

B10 Superstructure

B1010 Floor Construction

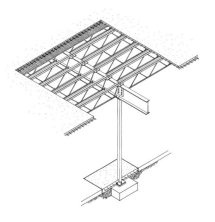

Mezzanine addition to existing building includes: Column footings; steel columns; structural steel; open web steel joists; uncoated 28 ga. steel slab forms; 2-1/2" concrete slab reinforced with welded wire fabric; steel trowel finish.

Design assumptions:

Structural steel is A36, high strength bolted. Slab form is 28 gauge, galvanized.
WWF 6 x 6 #10/#10
Conc. slab f'c = 3 ksi

System Components	QUANTITY	UNIT	COST EACH MAT.	COST EACH INST.	COST EACH TOTAL
SYSTEM B1010 310 0142					
MEZZANINE ADDITION TO EXISTING BUILDING; 100 PSF SUPERIMPOSED LOAD,					
SLAB FORM DECK, MTL. JOISTS, 2.5" CONC. SLAB, 3,000 PSI, 3,000 S.F.					
Saw cut existing slab, 6" thick	192.000	L.F.	622.08	1,520.64	2,142.72
Cut out existing slab, 4' x 4' openings	192.000	S.F.		2,409.60	2,409.60
Remove concrete debris	9.060	C.Y.		294.45	294.45
Excavate by hand	7.104	C.Y.		696.19	696.19
Backfill by hand	3.732	C.Y.		104.50	104.50
Compaction of backfill	3.732	C.Y.		71.09	71.09
Forms for footing, 3' x 3' x 1'	144.000	SFCA	570.24	2,606.40	3,176.64
Reinforcing bar for footing, 3#5 E.W. x 30" long	180.000	Lb.	160.20	100.80	261
Anchor bolts, 12", 2 per column	24.000	Ea.	61.44	170.40	231.84
Concrete, 3000 psi, footing	5.592	C.Y.	620.71		620.71
Place concrete in footing, direct chute	5.592	C.Y.		251.87	251.87
Premolded, bituminous fiber, 1/2" x 6"	192.000	L.F.	84.48	253.44	337.92
Reinforcing mesh for slab cutouts	1.920	C.S.F.	38.11	64.32	102.43
Concrete, 3500 psi, slab cutouts	2.784	C.Y.	317.38		317.38
Place concrete slab cutouts, direct chute	2.784	C.Y.		42.03	42.03
Machine trowel slab cutouts	192.000	S.F.		149.76	149.76
Grout for leveling plates	12.000	S.F.	94.20	153.60	247.80
Column, 4" x 4" x 1/4" x12'	12.000	Ea.	3,264	1,260	4,524
Structural steel W 21x50	80.000	L.F.	10,920	922.80	11,842.80
W 16x31	80.000	L.F.	6,780	811.20	7,591.20
Open web joists, H or K series	6.800	Ton	22,950	7,078.80	30,028.80
Slab form, steel 28 gauge, galvanized	30.000	C.S.F.	5,670	2,040	7,710
Concrete 3,000 psi, 2 1/2" slab, incl. premium delv. chg.	23.500	C.Y.	3,912.75		3,912.75
Welded wire fabric 6x6 #10/#10 (w1.4/w1.4)	30.000	C.S.F.	595.50	1,005	1,600.50
Place concrete	23.500	C.Y.		1,046.93	1,046.93
Monolithic steel trowel finish	30.000	C.S.F.		2,340	2,340
Curing with sprayed membrane curing compound	30.000	C.S.F.	172.50	247.50	420
TOTAL			56,833.59	25,641.32	82,474.91
COST PER S.F.			18.94	8.55	27.49

B10 Superstructure

B1010 Floor Construction

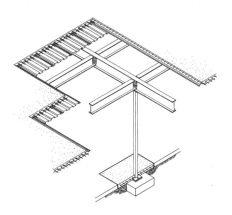

Mezzanine addition to existing building includes: Column footings; steel columns; structural steel; galvanized composite steel deck; 4" concrete slab reinforced with welded wire fabric; steel trowel finish.

Design assumptions:

Structural steel is A36, high strength bolted.
Composite deck is 22 gauge, galvanized.
WWF 6 x 6 #10/#10
Conc. slab f'c = 3 ksi

System Components	QUANTITY	UNIT	COST EACH MAT.	COST EACH INST.	COST EACH TOTAL
SYSTEM B1010 320 0152					
MEZZANINE ADDITION TO EXISTING BUILDING; 100 PSF SUPERIMPOSED LOAD					
COMPOSITE DECK, BEAMS, 4" CONC. SLAB, 3,000 PSI, 3,000 S.F.					
Saw cut existing slab, 6" thick	192.000	L.F.	622.08	1,520.64	2,142.72
Remove existing slab, 4' x 4' openings	192.000	S.F.		2,409.60	2,409.60
Remove concrete debris	9.060	C.Y.		294.45	294.45
Excavate by hand	7.104	C.Y.		696.19	696.19
Backfill by hand, no compaction, light soil	3.732	C.Y.		104.50	104.50
Compaction of backfill	3.732	C.Y.		71.09	71.09
Forms in place, footing, 3' x 3' x 1'	144.000	SFCA	570.24	2,606.40	3,176.64
Footings, #4 to #7 bars, 3#5 E.W. x 30" long	180.000	Lb.	160.20	100.80	261
Anchor bolts, 12" long, 2 per column	24.000	Ea.	61.44	170.40	231.84
Concrete, 3000 psi, footing	5.592	C.Y.	620.71		620.71
Place concrete in footings, direct chute	5.592	C.Y.		251.87	251.87
Premolded, bituminous fiber, 1/2" x 6"	192.000	L.F.	84.48	253.44	337.92
Structural steel W 12x19	225.000	L.F.	9,000	1,557	10,557
Reinforcing mesh for slab cutouts	1.920	C.S.F.	38.11	64.32	102.43
Concrete, 3500 psi, for slab cutouts	2.784	C.Y.	317.38		317.38
W 14x30	375.000	L.F.	30,656.25	3,802.51	34,458.76
Place concrete, slab cutouts, direct chute	2.784	C.Y.		42.03	42.03
W 16x31	80.000	L.F.	6,780	811.20	7,591.20
Machine trowel slab cutouts	192.000	S.F.		149.76	149.76
Grout for leveling plates	12.000	S.F.	94.20	153.60	247.80
Column, 4" x 4" x 1/4" x 12'	12.000	Ea.	3,264	1,260	4,524
W 18x50	80.000	L.F.	10,920	1,075.20	11,995.20
Metal deck, galvanized, 2" deep, 22 gauge.	30.000	C.S.F.	8,040	2,130	10,170
Concrete, 3,000 p.s.i., 4" slab, incl. premium delv. chg.	37.000	C.Y.	6,160.50		6,160.50
Welded wire fabric 6x6 #10/#10 (w1.4/w1.4)	30.000	C.S.F.	595.50	1,005	1,600.50
Place concrete	37.000	C.Y.		1,648.35	1,648.35
Monolithic steel trowel finish	30.000	C.S.F.		2,340	2,340
Curing with sprayed membrane curing compound	30.000	C.S.F.	172.50	247.50	420
TOTAL			78,157.59	24,765.85	102,923.44
COST PER S.F.			26.05	8.26	34.31

B10 Superstructure

B1010 Floor Construction

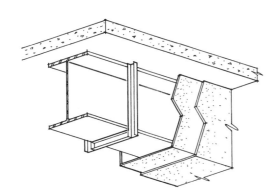

The table below lists fireproofing costs for steel beams by type, beam size, thickness and fire rating. Weights listed are for the fireproofing material only.

System Components	QUANTITY	UNIT	COST PER L.F.		
			MAT.	INST.	TOTAL
SYSTEM B1010 710 1300					
FIREPROOFING, 5/8" F.R. GYP. BOARD, 12"X 4" BEAM, 2" THICK, 2 HR. F.R.					
Corner bead for drywall, 1-1/4" x 1-1/4", galvanized	.020	C.L.F.	.40	2.84	3.24
L bead for drywall, galvanized	.020	C.L.F.	.43	3.30	3.73
Furring, beams & columns, 3/4" galv. channels, 24" O.C.	2.330	S.F.	.54	5.29	5.83
Drywall on beam, no finish, 2 layers at 5/8" thick	3.000	S.F.	2.67	9.90	12.57
Drywall, taping and finishing joints, add	3.000	S.F.	.15	1.50	1.65
TOTAL			4.19	22.83	27.02

B1010 710 Steel Beam Fireproofing

	ENCASEMENT SYSTEM	BEAM SIZE (IN.)	THICKNESS (IN.)	FIRE RATING (HRS.)	WEIGHT (P.L.F.)	COST PER L.F.		
						MAT.	INST.	TOTAL
1300	5/8" fire rated	12x4	2	2	15	4.19	23	27.19
1350	Gypsum board		2-5/8	3	24	5.85	28.50	34.35
1400		14x5	2	2	17	4.50	23.50	28
1450			2-5/8	3	27	6.45	30	36.45
1500		16x7	2	2	20	5.05	26.50	31.55
1550			2-5/8	3	31	6.50	26	32.50
1600		18x7-1/2	2	2	22	5.45	28.50	33.95
1650			2-5/8	3	34	7.75	36	43.75
1700	5/8" fire rated	24x9	2	2	27	6.75	35	41.75
1750	Gypsum board		2-5/8	3	42	9.55	44	53.55
1800		30x10-1/2	2	2	33	7.95	41	48.95
1850			2-5/8	3	51	11.25	51.50	62.75
1900	Gypsum	12x4	1-1/8	3	18	6	17.30	23.30
1950	Plaster on	14x5	1-1/8	3	21	6.80	19.50	26.30
2000	Metal lath	16x7	1-1/8	3	25	7.95	22.50	30.45
2050		18x7-1/2	1-1/8	3	32	9	25	34
2100		24x9	1-1/8	3	35	11.10	30.50	41.60
2150		30x10-1/2	1-1/8	3	44	13.35	36.50	49.85
2200	Perlite plaster	12x4	1-1/8	2	16	5.20	19.25	24.45
2250	On metal lath		1-1/4	3	20	5.50	20	25.50
2300			1-1/2	4	22	5.90	21.50	27.40
2350		14x5	1-1/8	2	18	6.75	23.50	30.25
2400			1-1/4	3	23	7.15	24.50	31.65
2450			1-1/2	4	25	7.90	26.50	34.40

B10 Superstructure

B1010 Floor Construction

B1010 710 Steel Beam Fireproofing

	ENCASEMENT SYSTEM	BEAM SIZE (IN.)	THICKNESS (IN.)	FIRE RATING (HRS.)	WEIGHT (P.L.F.)	COST PER L.F.		
						MAT.	INST.	TOTAL
2500	Perlite plaster	16x7	1-1/8	2	21	6.25	23.50	29.75
2550	On metal lath		1-1/4	3	26	6.55	24.50	31.05
2600			1-1/2	4	29	6.95	25.50	32.45
2650		18x7-1/2	1-1/8	2	26	7.75	27.50	35.25
2700			1-1/4	3	33	8.10	29	37.10
2750			1-1/2	4	36	8.90	31	39.90
2800		24x9	1-1/8	2	30	7.95	31	38.95
2850			1-1/4	3	38	8.30	32	40.30
2900			1-1/2	4	41	8.70	33	41.70
2950		30x10-1/2	1-1/8	2	37	10.30	39	49.30
3000			1-1/4	3	46	10.70	40	50.70
3050			1-1/2	4	51	11.50	41.50	53
3100	Sprayed Fiber	12x4	5/8	1	12	1.26	1.96	3.22
3150	(non asbestos)		1-1/8	2	21	2.26	3.51	5.77
3200			1-1/4	3	22	2.52	3.90	6.42
3250		14x5	5/8	1	14	1.26	1.96	3.22
3300			1-1/8	2	26	2.26	3.51	5.77
3350			1-1/4	3	29	2.52	3.90	6.42
3400		16x7	5/8	1	18	1.59	2.47	4.06
3450			1-1/8	2	32	2.86	4.44	7.30
3500			1-1/4	3	36	3.18	4.94	8.12
3550		18x7-1/2	5/8	1	20	1.76	2.73	4.49
3600			1-1/8	2	35	3.16	4.91	8.07
3650			1-1/4	3	39	3.49	5.40	8.89
3700		24x9	5/8	1	25	2.22	3.44	5.66
3750			1-1/8	2	45	4.02	6.20	10.22
3800			1-1/4	3	50	4.47	6.95	11.42
3850		30x10-1/2	5/8	1	31	2.69	4.18	6.87
3900			1-1/8	2	55	4.84	7.50	12.34
3950			1-1/4	3	61	5.40	8.35	13.75

B10 Superstructure
B1010 Floor Construction

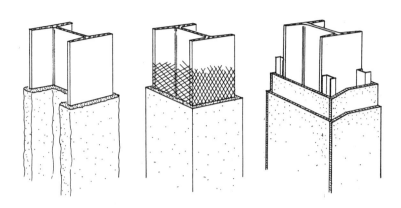

Listed below are costs per V.L.F. for fireproofing by material, column size, thickness and fire rating. Weights listed are for the fireproofing material only.

System Components	QUANTITY	UNIT	COST PER V.L.F.		
			MAT.	INST.	TOTAL
SYSTEM B1010 720 3450					
GYPSUM BOARD FIREPROOFING, 8″ STEEL COLUMN, 1/2″ THICK 2 HR. FIRERATING					
Furring channel, galvanized steel, 7/8″ deep	.040	C.L.F.	1.22	7.64	8.86
Corner bead, galvanized steel, 1″ x 1″	.040	C.L.F.	.61	4.96	5.57
Gypsum plaster board, 1/2″ thick	3.370	S.F.	1.18	4.95	6.13
Taping and finishing joints	3.370	S.F.	.17	1.69	1.86
TOTAL			3.18	19.24	22.42

B1010 720		Steel Column Fireproofing					
	ENCASEMENT SYSTEM	COLUMN SIZE (IN.)	THICKNESS (IN.)	FIRE RATING (HRS.)	WEIGHT (P.L.F.)	COST PER V.L.F.		
						MAT.	INST.	TOTAL
3450	Gypsum board	8	1/2	2	8	3.18	19.25	22.43
3500	1/2″ fire rated	10	1/2	2	11	3.35	20	23.35
3550	1 layer	14	1/2	2	18	3.44	20.50	23.94
3600	Gypsum board	8	1	3	14	4.43	24.50	28.93
3650	1/2″ fire rated	10	1	3	17	4.76	26	30.76
3700	2 layers	14	1	3	22	4.93	27	31.93
3750	Gypsum board	8	1-1/2	3	23	5.90	31	36.90
3800	1/2″ fire rated	10	1-1/2	3	27	6.65	34.50	41.15
3850	3 layers	14	1-1/2	3	35	7.40	37.50	44.90
3900	Sprayed fiber	8	1-1/2	2	6.3	3.93	6.10	10.03
3950	Direct application		2	3	8.3	5.40	8.40	13.80
4000			2-1/2	4	10.4	7	10.90	17.90
4050		10	1-1/2	2	7.9	4.75	7.40	12.15
4100			2	3	10.5	6.50	10.10	16.60
4150			2-1/2	4	13.1	8.40	13	21.40
4200		14	1-1/2	2	10.8	5.90	9.15	15.05
4250			2	3	14.5	8.05	12.50	20.55
4300			2-1/2	4	18	10.30	16	26.30
4350	3/4″ gypsum plaster	8	3/4	1	23	7.60	25	32.60
4400	On metal lath	10	3/4	1	28	8.60	28	36.60
4450		14	3/4	1	38	10.45	35	45.45
4500	Perlite plaster	8	1	2	18	8.15	27	35.15
4550	On metal lath		1-3/8	3	23	8.85	29.50	38.35
4600			1-3/4	4	35	10.40	36	46.40

B10 Superstructure

B1010 Floor Construction

B1010 720 Steel Column Fireproofing

	ENCASEMENT SYSTEM	COLUMN SIZE (IN.)	THICKNESS (IN.)	FIRE RATING (HRS.)	WEIGHT (P.L.F.)	COST PER V.L.F.		
						MAT.	INST.	TOTAL
4650	Perlite plaster	10	1	2	21	9.30	31.50	40.80
4700			1-3/8	3	27	10.45	36	46.45
4750			1-3/4	4	41	11.70	41	52.70
4800		14	1	2	29	11.60	39	50.60
4850			1-3/8	3	35	12.95	45	57.95
4900			1-3/4	4	53	13.45	46.50	59.95
4950	1/2 gypsum plaster	8	7/8	1	13	7.50	20.50	28
5000	On 3/8" gypsum lath	10	7/8	1	16	8.95	24	32.95
5050		14	7/8	1	21	10.95	29.50	40.45
5100	5/8" gypsum plaster	8	1	1-1/2	20	7.40	22.50	29.90
5150	On 3/8" gypsum lath	10	1	1-1/2	24	8.80	26.50	35.30
5200		14	1	1-1/2	33	10.75	32.50	43.25
5250	1"perlite plaster	8	1-3/8	2	23	8.10	23.50	31.60
5300	On 3/8" gypsum lath	10	1-3/8	2	28	9.30	27	36.30
5350		14	1-3/8	2	37	11.60	34	45.60
5400	1-3/8" perlite plaster	8	1-3/4	3	27	8.50	26	34.50
5450	On 3/8" gypsum lath	10	1-3/4	3	33	9.75	30.50	40.25
5500		14	1-3/4	3	43	12.15	38	50.15
5550	Concrete masonry	8	4-3/4	4	126	10.65	29	39.65
5600	Units 4" thick	10	4-3/4	4	166	13.35	36.50	49.85
5650	75% solid	14	4-3/4	4	262	16	43.50	59.50

B20 Exterior Enclosure

B2010 Exterior Walls

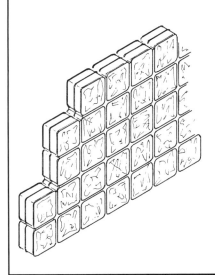

The table below lists costs per S.F. for glass block walls. Included in the costs are the following special accessories required for glass block walls.

Glass block accessories required for proper installation.

Wall ties: Galvanized double steel mesh full length of joint.

Fiberglass expansion joint at sides and top.

Silicone caulking: One gallon does 95 L.F.
Oakum: One lb. does 30 L.F.
Asphalt emulsion: One gallon does 600 L.F.

If block are not set in wall chase, use 2'-0" long wall anchors at 2'-0" O.C.

System Components	QUANTITY	UNIT	COST PER S.F.		
			MAT.	INST.	TOTAL
SYSTEM B2010 140 2300					
GLASS BLOCK, 4" THICK, 6" X 6" PLAIN, UNDER 1,000 S.F.					
Glass block, 4" thick, 6" x 6" plain, under 1000 S.F.	4.100	Ea.	22.50	19.60	42.10
Glass block, cleaning blocks after installation, both sides add	2.000	S.F.	.12	2.25	2.37
TOTAL			22.62	21.85	44.47

B2010 140	Glass Block	COST PER S.F.		
		MAT.	INST.	TOTAL
2300	Glass block 4" thick, 6"x6" plain, under 1,000 S.F.	22.50	22	44.50
2400	1,000 to 5,000 S.F.	22	18.95	40.95
2500	Over 5,000 S.F.	21.50	17.80	39.30
2600	Solar reflective, under 1,000 S.F.	31.50	29.50	61
2700	1,000 to 5,000 S.F.	31	25.50	56.50
2800	Over 5,000 S.F.	30	24	54
3500	8"x8" plain, under 1,000 S.F.	15.55	16.35	31.90
3600	1,000 to 5,000 S.F.	15.25	14.10	29.35
3700	Over 5,000 S.F.	14.75	12.75	27.50
3800	Solar reflective, under 1,000 S.F.	22	22	44
3900	1,000 to 5,000 S.F.	21.50	18.85	40.35
4000	Over 5,000 S.F.	20.50	16.95	37.45
5000	12"x12" plain, under 1,000 S.F.	17.70	15.15	32.85
5100	1,000 to 5,000 S.F.	17.35	12.75	30.10
5200	Over 5,000 S.F.	16.85	11.65	28.50
5300	Solar reflective, under 1,000 S.F.	25	20.50	45.50
5400	1,000 to 5,000 S.F.	24.50	16.95	41.45
5600	Over 5,000 S.F.	23.50	15.40	38.90
5800	3" thinline, 6"x6" plain, under 1,000 S.F.	16.65	22	38.65
5900	Over 5,000 S.F.	16.65	22	38.65
6000	Solar reflective, under 1,000 S.F.	23.50	29.50	53
6100	Over 5,000 S.F.	25	24	49
6200	8"x8" plain, under 1,000 S.F.	10.40	16.35	26.75
6300	Over 5,000 S.F.	10.05	12.75	22.80
6400	Solar reflective, under 1,000 S.F.	14.55	22	36.55
6500	Over 5,000 S.F.	14.05	16.95	31

B20 Exterior Enclosure

B2010 Exterior Walls

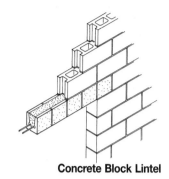

Concrete Block Lintel

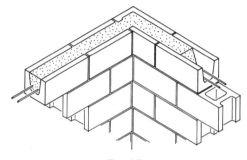

Bond Beam

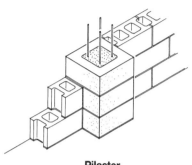

Pilaster

Concrete block specialties are divided into lintels, pilasters, and bond beams. Lintels are defined by span, thickness, height and wall loading. Span refers to the clear opening but the cost includes 8" of bearing at both ends.

Bond beams and pilasters are defined by height thickness and weight of the masonry unit itself. Components for bond beams also include grout and reinforcing. For pilasters, components include four #5 reinforcing bars and type N mortar.

System Components	QUANTITY	UNIT	COST PER LINTEL		
			MAT.	INST.	TOTAL
SYSTEM B2010 144 3100					
CONCRETE BLOCK LINTEL, 6"X8", LOAD 300 LB/L.F. 3'-4" SPAN					
Lintel blocks, 6" x 8" x 8"	7.000	Ea.	10.78	20.58	31.36
Joint reinforcing, #3 & #4 steel bars, horizontal	3.120	Lb.	2.65	3.40	6.05
Grouting, 8" deep, 6" thick, .15 C.F./L.F.	.700	C.F.	3.78	4.69	8.47
Temporary shoring, lintel forms	1.000	Set	.80	11.26	12.06
Temporary shoring, wood joists	1.000	Set		19.80	19.80
Temporary shoring, vertical members	1.000	Set		36	36
Mortar, masonry cement, 1:3 mix, type N	.270	C.F.	1.29	.09	1.38
TOTAL			19.30	95.82	115.12

B2010 144			Concrete Block Lintel				
	SPAN	THICKNESS (IN.)	HEIGHT (IN.)	WALL LOADING P.L.F.		COST PER LINTEL		
						MAT.	INST.	TOTAL
3100	3'-4"	6	8	300	R040513 -10	19.30	96	115.30
3150		6	16	1,000		30	103	133
3200		8	8	300	R040519 -50	21.50	103	124.50
3300		8	8	1,000		24	106	130
3400		8	16	1,000		34.50	112	146.50
3450	4'-0"	6	8	300	R042110 -50	22.50	106	128.50
3500		6	16	1,000		39	119	158
3600		8	8	300	R042210 -20	25	112	137
3700		8	16	1,000		39.50	121	160.50
3800	4'-8"	6	8	300		25	107	132
3900		6	16	1,000		39.50	118	157.50
4000		8	8	300		28.50	119	147.50
4100		8	16	1,000		45	130	175

B20 Exterior Enclosure

B2010 Exterior Walls

B2010 144 Concrete Block Lintel

	SPAN	THICKNESS (IN.)	HEIGHT (IN.)	WALL LOADING P.L.F.		COST PER LINTEL		
						MAT.	INST.	TOTAL
4200	5'-4"	6	8	300		31.50	120	151.50
4300		6	16	1,000		45.50	124	169.50
4400		8	8	300		35	131	166
4500		8	16	1,000		52	141	193
4600	6'-0"	6	8	300		39.50	145	184.50
4700		6	16	300		48	151	199
4800		6	16	1,000		52	157	209
4900		8	8	300		38.50	159	197.50
5000		8	16	1,000		58.50	172	230.50
5100	6'-8"	6	16	300		47	152	199
5200		6	16	1,000		51.50	158	209.50
5300		8	8	300		46.50	165	211.50
5400		8	16	1,000		64.50	182	246.50
5500	7'-4"	6	16	300		52	157	209
5600		6	16	1,000		68.50	169	237.50
5700		8	8	300		59.50	178	237.50
5800		8	16	300		78	190	268
5900		8	16	1,000		78	190	268
6000	8'-0"	6	16	300		56	164	220
6100		6	16	1,000		74	176	250
6200		8	16	300		86.50	199	285.50
6500		8	16	1,000		86.50	199	285.50

B2010 145 Concrete Block Specialties

	TYPE	HEIGHT (IN.)	THICKNESS (IN.)	WEIGHT (P.C.F.)		COST PER L.F.		
						MAT.	INST.	TOTAL
7100	Bond beam	8	8	125		5.65	6.60	12.25
7200				105		6.25	6.55	12.80
7300			12	125		6.95	8.30	15.25
7400				105		7.70	8.20	15.90
7500	Pilaster	8	16	125		13.05	22	35.05
7600			20	125		15.45	26	41.45

B20 Exterior Enclosure

B2020 Exterior Windows

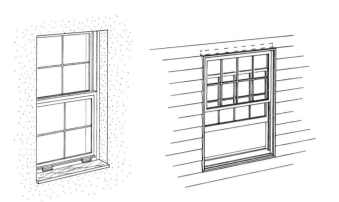

The table below lists window systems by material, type and size. Prices between sizes listed can be interpolated with reasonable accuracy. Prices include frame, hardware, and casing as illustrated in the component block below.

System Components	QUANTITY	UNIT	COST PER UNIT MAT.	COST PER UNIT INST.	COST PER UNIT TOTAL
SYSTEM B2020 102 3000					
WOOD, DOUBLE HUNG, STD. GLASS, 2'-8" X 4'-6"					
Framing rough opening, header w/jacks	14.680	B.F.	8.51	25.10	33.61
Casing, stock pine 11/16" x 2-1/2"	16.000	L.F.	23.36	32.96	56.32
Stool cap, pine, 11/16" x 3-1/2"	5.000	L.F.	12.50	12.40	24.90
Residential wood window, double hung, 2'-8" x 4'-6" standard glazed	1.000	Ea.	194	146	340
TOTAL			238.37	216.46	454.83

B2020 102 Wood Windows

	MATERIAL	TYPE	GLAZING	SIZE	DETAIL	MAT.	INST.	TOTAL
3000	Wood	double hung	std. glass	2'-8" x 4'-6"		238	216	454
3050				3'-0" x 5'-6"		315	248	563
3100			insul. glass	2'-8" x 4'-6"		255	216	471
3150				3'-0" x 5'-6"		340	248	588
3200		sliding	std. glass	3'-4" x 2'-7"		288	180	468
3250				4'-4" x 3'-3"		330	197	527
3300				5'-4" x 6'-0"		420	237	657
3350			insul. glass	3'-4" x 2'-7"		350	210	560
3400				4'-4" x 3'-3"		400	229	629
3450				5'-4" x 6'-0"		505	270	775
3500		awning	std. glass	2'-10" x 1'-9"		215	108	323
3600				4'-4" x 2'-8"		350	105	455
3700			insul. glass	2'-10" x 1'-9"		263	125	388
3800				4'-4" x 2'-8"		430	115	545
3900		casement	std. glass	1'-10" x 3'-2"	1 lite	320	142	462
3950				4'-2" x 4'-2"	2 lite	575	189	764
4000				5'-11" x 5'-2"	3 lite	870	244	1,114
4050				7'-11" x 6'-3"	4 lite	1,225	293	1,518
4100				9'-11" x 6'-3"	5 lite	1,600	340	1,940
4150			insul. glass	1'-10" x 3'-2"	1 lite	320	142	462
4200				4'-2" x 4'-2"	2 lite	590	189	779
4250				5'-11" x 5'-2"	3 lite	920	244	1,164
4300				7'-11" x 6'-3"	4 lite	1,300	293	1,593
4350				9'-11" x 6'-3"	5 lite	1,675	340	2,015

B20 Exterior Enclosure

B2020 Exterior Windows

B2020 102 Wood Windows

	MATERIAL	TYPE	GLAZING	SIZE	DETAIL	COST PER UNIT		
						MAT.	INST.	TOTAL
4400	Wood	picture	std. glass	4'-6" x 4'-6"		455	244	699
4450				5'-8" x 4'-6"		515	271	786
4500			insul. glass	4'-6" x 4'-6"		560	284	844
4550				5'-8" x 4'-6"		630	315	945
4600		fixed bay	std. glass	8' x 5'		1,600	445	2,045
4650				9'-9" x 5'-4"		1,125	630	1,755
4700			insul. glass	8' x 5'		2,150	445	2,595
4750				9'-9" x 5'-4"		1,225	630	1,855
4800		casement bay	std. glass	8' x 5'		1,600	510	2,110
4850			insul. glass	8' x 5'		1,800	510	2,310
4900		vert. bay	std. glass	8' x 5'		1,800	510	2,310
4950			insul. glass	8' x 5'		1,875	510	2,385

B2020 104 Steel Windows

	MATERIAL	TYPE	GLAZING	SIZE	DETAIL	COST PER UNIT		
						MAT.	INST.	TOTAL
5000	Steel	double hung	1/4" tempered	2'-8" x 4'-6"		835	170	1,005
5050				3'-4" x 5'-6"		1,275	259	1,534
5100			insul. glass	2'-8" x 4'-6"		855	194	1,049
5150				3'-4" x 5'-6"		1,300	296	1,596
5200		horz. pivoted	std. glass	2' x 2'		251	56.50	307.50
5250				3' x 3'		565	127	692
5300				4' x 4'		1,000	226	1,226
5350				6' x 4'		1,500	340	1,840
5400			insul. glass	2' x 2'		257	65	322
5450				3' x 3'		580	146	726
5500				4' x 4'		1,025	259	1,284
5550				6' x 4'		1,550	390	1,940
5600		picture window	std. glass	3' x 3'		355	127	482
5650				6' x 4'		940	340	1,280
5700			insul. glass	3' x 3'		370	146	516
5750				6' x 4'		980	390	1,370
5800		industrial security	std. glass	2'-9" x 4'-1"		755	159	914
5850				4'-1" x 5'-5"		1,475	315	1,790
5900			insul. glass	2'-9" x 4'-1"		775	182	957
5950				4'-1" x 5'-5"		1,525	360	1,885
6000		comm. projected	std. glass	3'-9" x 5'-5"		1,250	288	1,538
6050				6'-9" x 4'-1"		1,675	390	2,065
6100			insul. glass	3'-9" x 5'-5"		1,275	330	1,605
6150				6'-9" x 4'-1"		1,725	445	2,170
6200		casement	std. glass	4'-2" x 4'-2"	2 lite	985	246	1,231
6250			insul. glass	4'-2" x 4'-2"		1,000	281	1,281
6300			std. glass	5'-11" x 5'-2"	3 lite	1,850	435	2,285
6350			insul. glass	5'-11" x 5'-2"		1,900	495	2,395

B2020 106 Aluminum Windows

	MATERIAL	TYPE	GLAZING	SIZE	DETAIL	COST PER UNIT		
						MAT.	INST.	TOTAL
6400	Aluminum	awning	std. glass	3'-1" x 3'-2"		370	127	497
6450				4'-5" x 5'-3"		420	159	579
6500			insul. glass	3'-1" x 3'-2"		445	152	597
6550				4'-5" x 5'-3"		505	191	696

B20 Exterior Enclosure

B2020 Exterior Windows

B2020 106 Aluminum Windows

	MATERIAL	TYPE	GLAZING	SIZE	DETAIL	COST PER UNIT MAT.	COST PER UNIT INST.	COST PER UNIT TOTAL
6600	Aluminum	sliding	std. glass	3' x 2'		230	127	357
6650				5' x 3'		350	142	492
6700				8' x 4'		370	212	582
6750				9' x 5'		560	320	880
6800			insul. glass	3' x 2'		246	127	373
6850				5' x 3'		410	142	552
6900				8' x 4'		595	212	807
6950				9' x 5'		900	320	1,220
7000		single hung	std. glass	2' x 3'		218	127	345
7050				2'-8" x 6'-8"		385	159	544
7100				3'-4" x 5'-0"		315	142	457
7150			insul. glass	2' x 3'		264	127	391
7200				2'-8" x 6'-8"		495	159	654
7250				3'-4" x 5'		350	142	492
7300		double hung	std. glass	2' x 3'		289	85	374
7350				2'-8" x 6'-8"		860	252	1,112
7400				3'-4" x 5'		805	236	1,041
7450			insul. glass	2' x 3'		299	97	396
7500				2'-8" x 6'-8"		885	288	1,173
7550				3'-4" x 5'-0"		830	270	1,100
7600		casement	std. glass	3'-1" x 3'-2"		248	138	386
7650				4'-5" x 5'-3"		600	335	935
7700			insul. glass	3'-1" x 3'-2"		264	158	422
7750				4'-5" x 5'-3"		640	385	1,025
7800		hinged swing	std. glass	3' x 4'		530	170	700
7850				4' x 5'		885	283	1,168
7900			insul. glass	3' x 4'		550	194	744
7950				4' x 5'		915	325	1,240
8200		picture unit	std. glass	2'-0" x 3'-0"		159	85	244
8250				2'-8" x 6'-8"		470	252	722
8300				3'-4" x 5'-0"		440	236	676
8350			insul. glass	2'-0" x 3'-0"		169	97	266
8400				2'-8" x 6'-8"		500	288	788
8450				3'-4" x 5'-0"		470	270	740
8500		awning type	std. glass	3'-0" x 3'-0"	2 lite	445	91	536
8550				3'-0" x 4'-0"	3 lite	520	127	647
8600				3'-0" x 5'-4"	4 lite	625	127	752
8650				4'-0" x 5'-4"	4 lite	685	142	827
8700			insul. glass	3'-0" x 3'-0"	2 lite	475	91	566
8750				3'-0" x 4'-0"	3 lite	595	127	722
8800				3'-0" x 5'-4"	4 lite	735	127	862
8850				4'-0" x 5'-4"	4 lite	825	142	967

B20 Exterior Enclosure

B2030 Exterior Doors

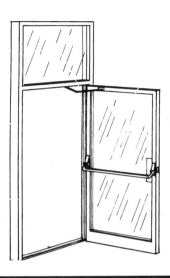

Costs are listed for exterior door systems by material, type and size. Prices between sizes listed can be interpolated with reasonable accuracy. Prices are per opening for a complete door system including frame.

B2030 110 — Glazed Doors, Steel or Aluminum

	MATERIAL	TYPE	DOORS	SPECIFICATION	OPENING	COST PER OPNG.		
						MAT.	INST.	TOTAL
6300	Alum. & glass	w/o transom	narrow stile	w/panic Hrdwre.	3'-0" x 7'-0"	1,500	925	2,425
6350				dbl. door, Hrdwre.	6'-0" x 7'-0"	2,600	1,525	4,125
6400			wide stile	hdwre.	3'-0" x 7'-0"	1,825	910	2,735
6450				dbl. door, Hdwre.	6'-0" x 7'-0"	3,550	1,825	5,375
6500			full vision	hdwre.	3'-0" x 7'-0"	2,225	1,450	3,675
6550				dbl. door, Hdwre.	6'-0" x 7'-0"	3,100	2,075	5,175
6900		w/transom	narrow stile	hdwre.	3'-0" x 10'-0"	2,125	1,050	3,175
6950				dbl. door, Hdwre.	6'-0" x 10'-0"	3,025	1,825	4,850
7000			wide stile	hdwre.	3'-0" x 10'-0"	2,450	1,275	3,725
7050				dbl. door, Hdwre.	6'-0" x 10'-0"	3,275	2,175	5,450
7100			full vision	hdwre.	3'-0" x 10'-0"	2,700	1,400	4,100
7150				dbl. door, Hdwre.	6'-0" x 10'-0"	3,550	2,400	5,950
7500		revolving	stock design	minimum	6'-10" x 7'-0"	20,600	3,400	24,000
7600				maximum	6'-10" x 7'-0"	43,000	5,675	48,675
7800		balanced	standard	economy	3'-0" x 7'-0"	6,375	1,425	7,800
7850				premium	3'-0" x 7'-0"	8,000	1,825	9,825
7900		mall front	sliding panels	alum. fin.	16'-0" x 9'-0"	3,325	720	4,045
7950					24'-0" x 9'-0"	4,825	1,325	6,150
8000				bronze fin.	16'-0" x 9'-0"	3,875	840	4,715
8050					24'-0" x 9'-0"	5,625	1,550	7,175
8100			fixed panels	alum. fin.	48'-0" x 9'-0"	9,000	1,025	10,025
8150				bronze fin.	48'-0" x 9'-0"	10,500	1,200	11,700
8200		sliding entrance	5' x 7' door	electric oper.	12'-0" x 7'-6"	8,475	1,325	9,800
8250		sliding patio	temp. glass	economy	6'-0" x 7'-0"	1,475	248	1,723

B30 Roofing

B3020 Roof Openings

Roof Hatch

Smoke Hatch

Skylight

B3020 110	Skylights	COST PER S.F.		
		MAT.	INST.	TOTAL
5100	Skylights, plastic domes, insul curbs, nom. size to 10 S.F., single glaze	25.50	12.15	37.65
5200	Double glazing	29	14.95	43.95
5300	10 S.F. to 20 S.F., single glazing	25	4.93	29.93
5400	Double glazing	24.50	6.20	30.70
5500	20 S.F. to 30 S.F., single glazing	21	4.19	25.19
5600	Double glazing	23	4.93	27.93
5700	30 S.F. to 65 S.F., single glazing	18.15	3.19	21.34
5800	Double glazing	23	4.19	27.19
6000	Sandwich panels fiberglass, 1-9/16" thick, 2 S.F. to 10 S.F.	17.60	9.75	27.35
6100	10 S.F. to 18 S.F.	15.75	7.35	23.10
6200	2-3/4" thick, 25 S.F. to 40 S.F.	25.50	6.60	32.10
6300	40 S.F. to 70 S.F.	21	5.90	26.90

B3020 210	Hatches	COST PER OPNG.		
		MAT.	INST.	TOTAL
0200	Roof hatches, with curb, and 1" fiberglass insulation, 2'-6"x3'-0", aluminum	1,075	195	1,270
0300	Galvanized steel 165 lbs.	655	195	850
0400	Primed steel 164 lbs.	550	195	745
0500	2'-6"x4'-6" aluminum curb and cover, 150 lbs.	935	216	1,151
0600	Galvanized steel 220 lbs.	660	216	876
0650	Primed steel 218 lbs.	910	216	1,126
0800	2'x6"x8'-0" aluminum curb and cover, 260 lbs.	1,825	295	2,120
0900	Galvanized steel, 360 lbs.	1,750	295	2,045
0950	Primed steel 358 lbs.	1,325	295	1,620
1200	For plexiglass panels, add to the above	440		440
2100	Smoke hatches, unlabeled not incl. hand winch operator, 2'-6"x3', galv	795	236	1,031
2200	Plain steel, 160 lbs.	665	236	901
2400	2'-6"x8'-0", galvanized steel, 360 lbs.	1,925	325	2,250
2500	Plain steel, 350 lbs.	1,450	325	1,775
3000	4'-0"x8'-0", double leaf low profile, aluminum cover, 359 lb.	2,850	243	3,093
3100	Galvanized steel 475 lbs.	2,425	243	2,668
3200	High profile, aluminum cover, galvanized curb, 361 lbs.	2,750	243	2,993

C Interiors

No part of this publication may be reproduced, stored in a retrieval system, or transmitted in any form or by any means without prior written permission of Reed Construction Data.

C10 Interior Construction

C1010 Partitions

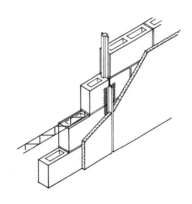

The Concrete Block Partition Systems are defined by weight and type of block, thickness, type of finish and number of sides finished. System components include joint reinforcing on alternate courses and vertical control joints.

System Components			COST PER S.F.		
	QUANTITY	UNIT	MAT.	INST.	TOTAL
SYSTEM C1010 102 1020					
CONC. BLOCK PARTITION, 8" X 16", 4" TK., 2 CT. GYP. PLASTER 2 SIDES					
Conc. block partition, 4" thick	1.000	S.F.	1.58	5.25	6.83
Control joint	.050	L.F.	.10	.13	.23
Horizontal joint reinforcing	.800	L.F.	.07	.06	.13
Gypsum plaster, 2 coat, on masonry	2.000	S.F.	.76	4.97	5.73
TOTAL			2.51	10.41	12.92

C1010 102 — Concrete Block Partitions - Regular Weight

	TYPE	THICKNESS (IN.)	TYPE FINISH	SIDES FINISHED		COST PER S.F.		
						MAT.	INST.	TOTAL
1000	Hollow	4	none	0	R015423 -10	1.75	5.45	7.20
1010			gyp. plaster 2 coat	1		2.13	7.90	10.03
1020				2	R040513 -10	2.51	10.40	12.91
1100			lime plaster - 2 coat	1		1.98	7.90	9.88
1150			lime portland - 2 coat	1	R040519 -50	2	7.90	9.90
1200			portland - 3 coat	1		2.02	8.30	10.32
1400			5/8" drywall	1	R042110 -50	2.38	7.15	9.53
1410				2		3.01	8.85	11.86
1500		6	none	0	R042210 -20	2.42	5.85	8.27
1510			gyp. plaster 2 coat	1		2.80	8.30	11.10
1520				2		3.18	10.80	13.98
1600			lime plaster - 2 coat	1		2.65	8.30	10.95
1650			lime portland - 2 coat	1		2.67	8.30	10.97
1700			portland - 3 coat	1		2.69	8.70	11.39
1900			5/8" drywall	1		3.05	7.55	10.60
1910				2		3.68	9.25	12.93
2000		8	none	0		2.55	6.20	8.75
2010			gyp. plaster 2 coat	1		2.93	8.70	11.63
2020			gyp. plaster 2 coat	2		3.31	11.20	14.51
2100			lime plaster - 2 coat	1		2.78	8.70	11.48
2150			lime portland - 2 coat	1		2.80	8.70	11.50
2200			portland - 3 coat	1		2.82	9.10	11.92
2400			5/8" drywall	1		3.18	7.90	11.08
2410				2		3.81	9.60	13.41

C10 Interior Construction

C1010 Partitions

C1010 102 Concrete Block Partitions - Regular Weight

	TYPE	THICKNESS (IN.)	TYPE FINISH	SIDES FINISHED		COST PER S.F.		
						MAT.	INST.	TOTAL
2500	Hollow	10	none	0		3.43	6.50	9.93
2510			gyp. plaster 2 coat	1		3.81	9	12.81
2520				2		4.19	11.45	15.64
2600			lime plaster - 2 coat	1		3.66	9	12.66
2650			lime portland - 2 coat	1		3.68	9	12.68
2700			portland - 3 coat	1		3.70	9.35	13.05
2900			5/8" drywall	1		4.06	8.20	12.26
2910				2		4.69	9.90	14.59
3000	Solid	2	none	0		1.61	5.40	7.01
3010			gyp. plaster	1		1.99	7.85	9.84
3020				2		2.37	10.35	12.72
3100			lime plaster - 2 coat	1		1.84	7.85	9.69
3150			lime portland - 2 coat	1		1.86	7.85	9.71
3200			portland - 3 coat	1		1.88	8.25	10.13
3400			5/8" drywall	1		2.24	7.10	9.34
3410				2		2.87	8.80	11.67
3500		4	none	0		2.15	5.65	7.80
3510			gyp. plaster	1		2.62	8.15	10.77
3520				2		2.91	10.60	13.51
3600			lime plaster - 2 coat	1		2.38	8.10	10.48
3650			lime portland - 2 coat	1		2.40	8.10	10.50
3700			portland - 3 coat	1		2.42	8.50	10.92
3900			5/8" drywall	1		2.78	7.35	10.13
3910				2		3.41	9.05	12.46
4000		6	none	0		2.40	6.05	8.45
4010			gyp. plaster	1		2.78	8.50	11.28
4020				2		3.16	11	14.16
4100			lime plaster - 2 coat	1		2.63	8.50	11.13
4150			lime portland - 2 coat	1		2.65	8.50	11.15
4200			portland - 3 coat	1		2.67	8.90	11.57
4400			5/8" drywall	1		3.03	7.75	10.78
4410				2		3.66	9.45	13.11

C1010 104 Concrete Block Partitions - Lightweight

	TYPE	THICKNESS (IN.)	TYPE FINISH	SIDES FINISHED		COST PER S.F.		
						MAT.	INST.	TOTAL
5000	Hollow	4	none	0		2.06	5.30	7.36
5010			gyp. plaster	1		2.44	7.75	10.19
5020				2		2.82	10.25	13.07
5100			lime plaster - 2 coat	1		2.29	7.75	10.04
5150			lime portland - 2 coat	1		2.31	7.75	10.06
5200			portland - 3 coat	1		2.33	8.15	10.48
5400			5/8" drywall	1		2.69	7	9.69
5410				2		3.32	8.70	12.02
5500		6	none	0		2.74	5.70	8.44
5510			gyp. plaster	1		3.12	8.15	11.27
5520			gyp. plaster	2		3.50	10.65	14.15
5600			lime plaster - 2 coat	1		2.97	8.15	11.12
5650			lime portland - 2 coat	1		2.99	8.15	11.14
5700			portland - 3 coat	1		3.01	8.55	11.56
5900			5/8" drywall	1		3.37	7.40	10.77
5910				2		4	9.10	13.10

C10 Interior Construction

C1010 Partitions

C1010 104 Concrete Block Partitions - Lightweight

	TYPE	THICKNESS (IN.)	TYPE FINISH	SIDES FINISHED		COST PER S.F.		
						MAT.	INST.	TOTAL
6000	Hollow	8	none	0		3.33	6.05	9.38
6010			gyp. plaster	1		3.71	8.55	12.26
6020				2		4.09	11.05	15.14
6100			lime plaster - 2 coat	1		3.56	8.55	12.11
6150			lime portland - 2 coat	1		3.58	8.55	12.13
6200			portland - 3 coat	1		3.60	8.95	12.55
6400			5/8" drywall	1		3.96	7.75	11.71
6410				2		4.59	9.45	14.04
6500		10	none	0		4.29	6.35	10.64
6700			portland - 3 coat	1		4.56	9.20	13.76
6900			5/8" drywall	1		4.92	8.05	12.97
6910				2		5.55	9.75	15.30
7000	Solid	4	none	0		2.70	5.55	8.25
7010			gyp. plaster	1		3.08	8	11.08
7020				2		3.46	10.50	13.96
7100			lime plaster - 2 coat	1		2.93	8	10.93
7150			lime portland - 2 coat	1		2.95	8	10.95
7200			portland - 3 coat	1		2.97	8.40	11.37
7400			5/8" drywall	1		3.33	7.25	10.58
7410				2		3.96	8.95	12.91
7500		6	none	0		3.10	6	9.10
7510			gyp. plaster	1		3.57	8.55	12.12
7520				2		3.86	10.95	14.81
7600			lime plaster - 2 coat	1		3.33	8.45	11.78
7650			lime portland - 2 coat	1		3.35	8.45	11.80
7700			portland - 3 coat	1		3.37	8.85	12.22
7900			5/8" drywall	1		3.73	7.70	11.43
7910				2		4.36	9.40	13.76
8000		8	none	0		4.11	6.40	10.51
8010			gyp. plaster	1		4.49	8.90	13.39
8020				2		4.87	11.40	16.27
8100			lime plaster - 2 coat	1		4.34	8.90	13.24
8150			lime portland - 2 coat	1		4.36	8.90	13.26
8200			portland - 3 coat	1		4.38	9.30	13.68
8400			5/8" drywall	1		4.74	8.10	12.84
8410				2		5.35	9.80	15.15

C10 Interior Construction

C1010 Partitions

Running Bond

Common Bond

Flemish Bond

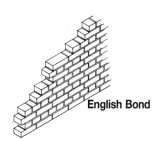
English Bond

Solid brick walls are defined in the following terms; type of brick, thickness, and bond. Sixteen different types of face bricks are presented, with single wythes shown in four different bonds.

System Components	QUANTITY	UNIT	COST PER S.F.		
			MAT.	INST.	TOTAL
SYSTEM C1010 108 1000					
BRICK WALL, COMMON, SINGLE WYTHE, 4″ THICK, RUNNING BOND					
Face brick wall, standard, running bond	1.000	S.F.	3.11	9.80	12.91
Wash brick	1.000	S.F.	.06	.88	.94
Control joint, backer rod	.100	L.F.		.11	.11
Control joint, sealant	.100	L.F.	.02	.27	.29
TOTAL			3.19	11.06	14.25

C1010 108			Solid Brick Walls - Single Wythe				
	TYPE	THICKNESS (IN.)	BOND			COST PER S.F.	
						MAT.	INST.	TOTAL

	TYPE	THICKNESS (IN.)	BOND			MAT.	INST.	TOTAL
1000	Common	4	running		R015423 -10	3.19	11.05	14.24
1010			common			4.05	12.80	16.85
1050			Flemish		R040513 -10	4.60	15.80	20.40
1100			English			5.30	16.80	22.10
1150	Standard	4	running		R040519 -50	6.25	11.50	17.75
1160			common			7.65	13.45	21.10
1200			Flemish		R042110 -20	8.60	16.30	24.90
1250			English			9.80	17.35	27.15
1300	Glazed	4	running		R042110 -50	14.30	12	26.30
1310			common			17.30	14.15	31.45
1350			Flemish			19.30	17.35	36.65
1400			English			21.50	18.60	40.10
1450	Engineer	4	running			4.15	9.90	14.05
1460			common			5.10	11.50	16.60
1500			Flemish			5.75	14.15	19.90
1550			English			6.60	14.90	21.50
1600	Economy	4	running			3.99	8.50	12.49
1610			common			4.87	9.90	14.77
1650			Flemish			5.50	12	17.50
1700			English			15.05	12.80	27.85

C10 Interior Construction

C1010 Partitions

C1010 108 — Solid Brick Walls - Single Wythe

	TYPE	THICKNESS (IN.)	BOND			COST PER S.F. MAT.	INST.	TOTAL
1750	Double	4	running			8.90	6.75	15.65
1760			common			10.75	7.80	18.55
1800			Flemish			12	9.45	21.45
1850			English			13.60	10.10	23.70
1900	Fire	4-1/2	running			10.10	9.90	20
1910			common			12.20	11.50	23.70
1950			Flemish			13.65	14.15	27.80
2000			English			15.50	14.90	30.40
2050	King	3-1/2	running			3.35	8.75	12.10
2060			common			4.10	10.25	14.35
2100			Flemish			5.20	12.50	17.70
2150			English			5.90	13.10	19
2200	Roman	4	running			6.15	10.25	16.40
2210			common			7.55	12	19.55
2250			Flemish			8.45	14.50	22.95
2300			English			9.70	15.80	25.50
2350	Norman	4	running			5.15	8.30	13.45
2360			common			6.30	9.60	15.90
2400			Flemish			15.60	11.75	27.35
2450			English			8.10	12.50	20.60
2500	Norwegian	4	running			4.03	7.25	11.28
2510			common			4.93	8.40	13.33
2550			Flemish			5.55	10.25	15.80
2600			English			6.35	10.85	17.20
2650	Utility	4	running			3.67	6.25	9.92
2660			common			4.48	7.20	11.68
2700			Flemish			5.05	8.75	13.80
2750			English			5.80	9.30	15.10
2800	Triple	4	running			2.68	5.55	8.23
2810			common			3.24	6.35	9.59
2850			Flemish			3.63	7.70	11.33
2900			English			4.13	8.10	12.23
2950	Scr	6	running			5.90	8.50	14.40
2960			common			7.25	9.90	17.15
3000			Flemish			8.20	12	20.20
3050			English			9.40	12.80	22.20
3100	Norwegian	6	running			4.90	7.45	12.35
3110			common			6	8.65	14.65
3150			Flemish			6.75	10.45	17.20
3200			English			7.70	11.05	18.75
3250	Jumbo	6	running			5.45	6.45	11.90
3260			common			6.65	7.45	14.10
3300	.		Flemish			7.50	9.05	16.55
3350			English			8.60	9.60	18.20

C10 Interior Construction

C1010 Partitions

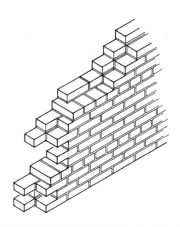

Solid brick walls are defined in the following terms; type of brick, thickness, and bond. Sixteen different types of face bricks are presented, with double wythes shown for common bond.

C1010 110 Solid Brick Walls - Double Wythe

	TYPE	THICKNESS (IN.)	COLLAR JOINT THICKNESS (IN.)			COST PER S.F.		
						MAT.	INST.	TOTAL
4100	Common	8	3/4	R015423 -10		7.05	19.40	26.45
4150	Standard	8	1/2			13.05	20	33.05
4200	Glazed	8	1/2	R040513 -10		29	21	50
4250	Engineer	8	1/2			8.85	17.60	26.45
4300	Economy	8	3/4	R040519 -50		8.50	14.95	23.45
4350	Double	8	3/4			11.10	11.85	22.95
4400	Fire	8	3/4	R042110 -20		20.50	17.60	38.10
4450	King	7	3/4			7.25	15.30	22.55
4500	Roman	8	1	R042110 -50		12.80	18.15	30.95
4550	Norman	8	3/4			10.85	14.60	25.45
4600	Norwegian	8	3/4			8.60	12.80	21.40
4650	Utility	8	3/4			7.90	10.90	18.80
4700	Triple	8	3/4			5.90	9.70	15.60
4750	Scr	12	3/4			12.30	15	27.30
4800	Norwegian	12	3/4			10.35	13.10	23.45
4850	Jumbo	12	3/4			11.40	11.30	22.70

C10 Interior Construction

C1010 Partitions

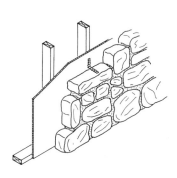

Systems are listed by costs per S.F. for stone veneer walls on various backup using different stone. Typical components for a system are shown in the component block below.

System Components	QUANTITY	UNIT	COST PER S.F.		
			MAT.	INST.	TOTAL
SYSTEM C1010 112 2000					
ASHLAR STONE VENEER 4", 2"X4" STUD 16" O.C. BACK UP, 8' HIGH					
Ashlar veneer, 4" thick, sawn face, split joints, low priced stone	1.000	S.F.	10.20	13.15	23.35
Partitions, 2" x 4" studs 8' high 16" O.C.	.920	B.F.	.36	1.24	1.60
Wall ties for stone veneer galv. corrg 7/8" x 7", 22 gauge	.700	Ea.	.08	.33	.41
Sheathing plywood on wall, CDX 1/2"	1.000	S.F.	.51	.88	1.39
TOTAL			11.15	15.60	26.75

C1010 112	Stone Veneer		COST PER S.F.		
			MAT.	INST.	TOTAL
2000	Ashlar stone veneer, 4" thick, wood stud backup, 8' high, 16" O.C.	R015423-10	11.15	15.60	26.75
2100	Metal stud backup, 8' high, 16" O.C.		12.45	15.85	28.30
2150	24" O.C.	R040513-10	12	15.45	27.45
2200	Conc. block backup, 4" thick		12.45	19.25	31.70
2300	6" thick		13.10	19.45	32.55
2350	8" thick		13.25	20	33.25
2400	10" thick		14.15	21	35.15
2500	12" thick		14.60	23	37.60
3100	wood stud backup, 10' high, 16" O.C.		28	21.50	49.50
3200	Metal stud backup, 10' high, 16" O.C.		29.50	21.50	51
3250	24" O.C.		29	21	50
3300	Conc. block backup, 10' high, 4" thick		29.50	25	54.50
3350	6" thick		30	25	55
3400	8" thick		30	25.50	55.50
3450	10" thick		31	27	58
3500	12" thick		31.50	28.50	60
4000	Indiana limestone 2" thick, sawn finish, wood stud backup, 10' high, 16" O.C.		39.50	11.80	51.30
4100	Metal stud backup, 10' high, 16" O.C.		41	12.05	53.05
4150	24" O.C.		40.50	11.65	52.15
4250	6" thick		41.50	15.65	57.15
4300	8" thick		41.50	16.20	57.70
4350	10" thick		42.50	17.40	59.90
4400	12" thick		43	19.10	62.10
4450	2" thick, smooth finish, wood stud backup, 8' high, 16" O.C.		39.50	11.80	51.30
4550	Metal stud backup, 8' high, 16" O.C.		41	11.90	52.90
4600	24" O.C.		40.50	11.45	51.95
4650	Conc. block backup, 4" thick		41	15.30	56.30
4700	6" thick		41.50	15.50	57
4750	8" thick		41.50	16.05	57.55
4800	10" thick		42.50	17.40	59.90

C10 Interior Construction

C1010 Partitions

C1010 112	Stone Veneer	COST PER S.F.		
		MAT.	INST.	TOTAL
4850	12" thick	43	19.10	62.10
5350	4" thick, smooth finish, wood stud backup, 8' high, 16" O.C.	46	11.80	57.80
5450	Metal stud backup, 8' high, 16" O.C.	47.50	12.05	59.55
5500	24" O.C.	47	11.65	58.65
5550	Conc. block backup, 4" thick	47.50	15.45	62.95
5600	6" thick	48	15.65	63.65
5650	8" thick	48	16.20	64.20
5700	10" thick	49	17.40	66.40
5750	12" thick	49.50	19.10	68.60
6000	Granite, grey or pink, 2" thick, wood stud backup, 8' high, 16" O.C.	23	23	46
6100	Metal studs, 8' high, 16" O.C.	24.50	23	47.50
6150	24" O.C.	24	23	47
6200	Conc. block backup, 4" thick	24.50	26.50	51
6250	6" thick	25	27	52
6300	8" thick	25	27.50	52.50
6350	10" thick	26	28.50	54.50
6400	12" thick	26.50	30.50	57
6900	4" thick, wood stud backup, 8' high, 16" O.C.	37.50	26.50	64
7000	Metal studs, 8' high, 16" O.C.	39	26.50	65.50
7050	24" O.C.	38.50	26.50	65
7100	Conc. block backup, 4" thick	39	30	69
7150	6" thick	39.50	30.50	70
7200	8" thick	39.50	31	70.50
7250	10" thick	40.50	32	72.50
7300	12" thick	41	34	75

C10 Interior Construction

C1010 Partitions

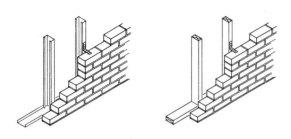

Brick veneer/ stud backup walls are defined in the following terms; type of brick and studs, stud spacing and bond. All systems include brick ties to the backup and control joints.

System Components	QUANTITY	UNIT	COST PER S.F.		
			MAT.	INST.	TOTAL
SYSTEM C1010 114 1100					
STANDARD BRICK VENEER, 2X4 WOOD STUDS @ 16"O.C., RUNNING BOND					
Standard brick wall, 4" thick, running bond	1.000	S.F.	6.15	10.25	16.40
Washing brick, smooth brick	1.000	S.F.	.06	.88	.94
Backer rod for control joints	.100	L.F.		.11	.11
Sealant for control joints	.100	L.F.	.02	.27	.29
Wall ties, corrugated, 7/8" x 7", 22 gauge	.300	Ea.	.03	.14	.17
Wood stud partition, backup, 2" x 4" @ 16" O.C.	1.000	S.F.	.34	.99	1.33
TOTAL			6.60	12.64	19.24

C1010 114				Brick Veneer/Wood Stud Backup				
	FACE BRICK	STUD BACKUP	STUD SPACING (IN.)	BOND		COST PER S.F.		
						MAT.	INST.	TOTAL
1100	Standard	2x4-wood	16	running	R015423 -10	6.60	12.65	19.25
1120				common		8	14.60	22.60
1140				Flemish	R040513 -10	8.95	17.45	26.40
1160				English		10.15	18.50	28.65
1400		2x6-wood	16	running	R040519 -50	6.85	12.75	19.60
1420				common		8.25	14.70	22.95
1440				Flemish	R042110 -20	9.20	17.55	26.75
1460				English		10.40	18.60	29
1500			24	running	R061636 -20	6.70	12.50	19.20
1520				common		8.10	14.45	22.55
1540				Flemish	R061110 -30	9.05	17.30	26.35
1560				English		10.25	18.35	28.60
1700	Glazed	2x4-wood	16	running		14.65	13.15	27.80
1720				common		17.65	15.30	32.95
1740				Flemish		19.65	18.50	38.15
1760				English		22	19.75	41.75
2000		2x6-wood	16	running		14.90	13.25	28.15
2020				common		17.90	15.40	33.30
2040				Flemish		19.90	18.60	38.50
2060				English		22	19.85	41.85
2100			24	running		14.75	13	27.75
2120				common		17.75	15.15	32.90
2140				Flemish		19.75	18.35	38.10
2160				English		22	19.60	41.60
2300	Engineer	2x4-wood	16	running		4.52	11.05	15.57
2320				common		5.45	12.65	18.10
2340				Flemish		6.10	15.30	21.40
2360				English		6.95	16.05	23

C10 Interior Construction

C1010 Partitions

C1010 114 — Brick Veneer/Wood Stud Backup

	FACE BRICK	STUD BACKUP	STUD SPACING (IN.)	BOND		COST PER S.F.		
						MAT.	INST.	TOTAL
2600	Engineer	2x6-wood	16	running		4.78	11.15	15.93
2620				common		5.70	12.75	18.45
2640				Flemish		6.35	15.40	21.75
2660				English		7.20	16.15	23.35
2700			24	running		4.64	10.90	15.54
2720				common		5.55	12.50	18.05
2740				Flemish		6.20	15.15	21.35
2760				English		7.05	15.90	22.95
2900	Roman	2x4-wood	16	running		6.50	11.40	17.90
2920				common		7.90	13.15	21.05
2940				Flemish		8.80	15.65	24.45
2960				English		10.05	16.95	27
3200		2x6-wood	16	running		6.75	11.50	18.25
3220				common		8.15	13.25	21.40
3240				Flemish		9.05	15.65	24.70
3260				English		10.30	17.05	27.35
3500	Norman	2x4-wood	16	running		5.50	9.45	14.95
3520				common		6.65	10.75	17.40
3540				Flemish		15.95	12.90	28.85
3560				English		8.45	13.65	22.10
3800		2x6-wood	16	running		5.75	9.55	15.30
3820				common		6.90	10.85	17.75
3840				Flemish		16.20	13	29.20
3860				English		8.70	13.75	22.45
4100	Norwegian	2x4-wood	16	running		4.40	8.40	12.80
4120				common		5.30	9.55	14.85
4140				Flemish		5.90	11.40	17.30
4160				English		6.70	12	18.70
4400		2x6-wood	16	running		4.66	8.50	13.16
4420				common		5.55	9.65	15.20
4440				Flemish		6.15	11.50	17.65
4460				English		6.95	12.10	19.05

C1010 115 — Brick Veneer/Metal Stud Backup

	FACE BRICK	STUD BACKUP	STUD SPACING (IN.)	BOND		COST PER S.F.		
						MAT.	INST.	TOTAL
5100	Standard	25ga.x6"NLB	24	running		6.80	12.35	19.15
5120				common		8.20	14.30	22.50
5140				Flemish		8.65	16.50	25.15
5160				English		10.35	18.20	28.55
5200		20ga.x3-5/8"NLB	16	running		6.95	12.95	19.90
5220				common		8.35	14.90	23.25
5240				Flemish		9.30	17.75	27.05
5260				English		10.50	18.80	29.30
5300			24	running		6.80	12.50	19.30
5320				common		8.20	14.45	22.65
5340				Flemish		9.15	17.30	26.45
5360				English		10.35	18.35	28.70
5400		16ga.x3-5/8"LB	16	running		7.90	13.15	21.05
5420				common		9.30	15.10	24.40
5440				Flemish		10.25	17.95	28.20
5460				English		11.45	19	30.45

C10 Interior Construction

C1010 Partitions

C1010 115 Brick Veneer/Metal Stud Backup

	FACE BRICK	STUD BACKUP	STUD SPACING (IN.)	BOND		COST PER S.F. MAT.	COST PER S.F. INST.	COST PER S.F. TOTAL
5500	Standard	16ga.x3-5/8"LB	24	running		7.45	12.75	20.20
5520				common		8.85	14.70	23.55
5540				Flemish		9.80	17.55	27.35
5560				English		11	18.60	29.60
5700	Glazed	25ga.x6"NLB	24	running		14.85	12.85	27.70
5720				common		17.85	15	32.85
5740				Flemish		19.85	18.20	38.05
5760				English		22	19.45	41.45
5800		20ga.x3-5/8"NLB	24	running		14.85	13	27.85
5820				common		17.85	15.15	33
5840				Flemish		19.85	18.35	38.20
5860				English		22	19.60	41.60
6000		16ga.x3-5/8"LB	16	running		15.95	13.65	29.60
6020				common		18.95	15.80	34.75
6040				Flemish		21	19	40
6060				English		23.50	20.50	44
6100			24	running		15.50	13.25	28.75
6120				common		18.50	15.40	33.90
6140				Flemish		20.50	18.60	39.10
6160				English		23	19.85	42.85
6300	Engineer	25ga.x6"NLB	24	running		4.73	10.75	15.48
6320				common		5.65	12.35	18
6340				Flemish		6.30	15	21.30
6360				English		7.15	15.75	22.90
6400		20ga.x3-5/8"NLB	16	running		4.88	11.35	16.23
6420				common		5.80	12.95	18.75
6440				Flemish		6.45	15.60	22.05
6460				English		7.30	16.35	23.65
6500			24	running		4.23	10.15	14.38
6520				common		5.65	12.50	18.15
6540				Flemish		6.30	15.15	21.45
6560				English		7.15	15.90	23.05
6600		16ga.x3-5/8"LB	16	running		5.85	11.55	17.40
6620				common		6.75	13.15	19.90
6640				Flemish		7.40	15.80	23.20
6660				English		8.25	16.55	24.80
6700			24	running		5.40	11.15	16.55
6720				common		6.30	12.75	19.05
6740				Flemish		6.95	15.40	22.35
6760				English		7.80	16.15	23.95
6900	Roman	25ga.x6"NLB	24	running		6.70	11.10	17.80
6920				common		8.10	12.85	20.95
6940				Flemish		9	15.35	24.35
6960				English		10.25	16.65	26.90
7000		20ga.x3-5/8"NLB	16	running		6.85	11.70	18.55
7020				common		8.25	13.45	21.70
7040				Flemish		9.15	15.95	25.10
7060				English		10.40	17.25	27.65
7100			24	running		6.70	11.25	17.95
7120				common		8.10	13	21.10
7140				Flemish		9	15.50	24.50
7160				English		10.25	16.80	27.05

C10 Interior Construction

C1010 Partitions

C1010 115 Brick Veneer/Metal Stud Backup

	FACE BRICK	STUD BACKUP	STUD SPACING (IN.)	BOND		COST PER S.F.		
						MAT.	INST.	TOTAL
7200	Roman	16ga.x3-5/8"LB	16	running		7.80	11.90	19.70
7220				common		9.20	13.65	22.85
7240				Flemish		10.10	16.15	26.25
7260				English		11.35	17.45	28.80
7300			24	running		7.35	11.50	18.85
7320				common		8.75	13.25	22
7340				Flemish		9.65	15.75	25.40
7360				English		10.90	17.05	27.95
7500	Norman	25ga.x6"NLB	24	running		5.70	9.15	14.85
7520				common		6.85	10.45	17.30
7540				Flemish		16.15	12.60	28.75
7560				English		8.65	13.35	22
7600		20ga.x3-5/8"NLB	24	running		5.70	9.30	15
7620				common		6.85	10.60	17.45
7640				Flemish		16.15	12.75	28.90
7660				English		8.65	13.50	22.15
7800		16ga.x3-5/8"LB	16	running		6.80	9.95	16.75
7820				common		7.95	11.25	19.20
7840				Flemish		17.25	13.40	30.65
7860				English		9.75	14.15	23.90
7900			24	running		6.35	9.55	15.90
7920				common		7.50	10.85	18.35
7940				Flemish		16.80	13	29.80
7960				English		9.30	13.75	23.05
8100	Norwegian	25ga.x6"NLB	24	running		4.61	8.10	12.71
8120				common		5.50	9.25	14.75
8140				Flemish		6.10	11.10	17.20
8160				English		6.90	11.70	18.60
8200		20ga.x3-5/8"NLB	16	running		4.76	8.70	13.46
8220				common		5.65	9.85	15.50
8240				Flemish		6.25	11.70	17.95
8260				English		7.05	12.30	19.35
8300			24	running		4.58	8.25	12.83
8320				common		5.50	9.40	14.90
8340				Flemish		6.10	11.25	17.35
8360				English		6.90	11.85	18.75
8400		16ga.x3-5/8"LB	16	running		5.70	8.90	14.60
8420				common		6.60	10.05	16.65
8440				Flemish		7.20	11.90	19.10
8460				English		8	12.50	20.50
8500			24	running		5.25	8.50	13.75
8520				common		6.15	9.65	15.80
8540				Flemish		6.75	11.50	18.25
8560				English		7.55	12.10	19.65

C10 Interior Construction

C1010 Partitions

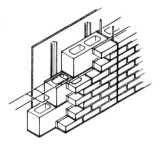
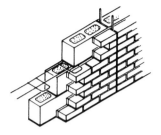

Brick face composite walls are defined in the following terms; type of face brick and backup masonry, thickness of backup masonry and finish. A special section is included on triple wythe construction at the end of the table. Seven types of face brick are shown with various thicknesses of backup.

System Components	QUANTITY	UNIT	COST PER S.F.		
			MAT.	INST.	TOTAL
SYSTEM C1010 116 1120					
COMPOSITE WALL, STD. BRICK FACE, 6" C.M.U. BACKUP, 5/8" DRYWALL FINISH					
Face brick veneer, standard, running bond	1.000	S.F.	6.15	10.25	16.40
Wash brick	1.000	S.F.	.06	.88	.94
Conc. block backup, 6" thick	1.000	S.F.	2.24	5.65	7.89
Ties	.300	Ea.	.08	.14	.22
Control joint	.050	L.F.	.07	.06	.13
Backer rod	.100	L.F.		.11	.11
Sealant	.100	L.F.	.02	.27	.29
Collar joint	1.000	S.F.	.39	.55	.94
Furring channels, 24" O.C.	1.000	S.F.	.23	1.20	1.43
Drywall, 5/8" thick, screwed to channels	1.000	S.F.	.40	.50	.90
TOTAL			9.64	19.61	29.25

C1010 116		Brick Face Composite Wall - Double Wythe					
	FACE BRICK	BACKUP MASONRY	BACKUP THICKNESS (IN.)	BACKUP FINISH		COST PER S.F.		
						MAT.	INST.	TOTAL
1000	Standard	common brick	4	none		9.90	22	31.90
1040		SCR brick	6	none		12.55	19.50	32.05
1080		conc. block	4	none		8.35	17.50	25.85
1120			6	5/8" drywall	R015423-10	9.65	19.60	29.25
1160				gyp. plaster		9.40	20	29.40
1200			8	5/8" drywall	R040513-10	9.75	19.95	29.70
1240				gyp. plaster		9.50	20.50	30
1280		L.W. block	4	none	R040519-50	8.65	17.35	26
1320			6	5/8" drywall		9.95	19.45	29.40
1360				gyp. plaster	R042110-50	9.70	20	29.70
1400			8	5/8" drywall		10.55	19.80	30.35
1440				gyp. plaster		10.30	20.50	30.80
1520		glazed block	4	none	R042110-10	16.10	18.80	34.90
1560			6	none		16.80	19.10	35.90
1640			8	none		16.90	19.50	36.40
1720		clay tile	4	none	R042110-20	14.40	16.80	31.20
1760			6	none		16.30	17.25	33.55
1800			8	none		18.95	17.90	36.85
1840		glazed tile	4	none	R042210-20	18.90	22.50	41.40
1880								

C10 Interior Construction

C1010 Partitions

C1010 116 Brick Face Composite Wall - Double Wythe

	FACE BRICK	BACKUP MASONRY	BACKUP THICKNESS (IN.)	BACKUP FINISH		COST PER S.F.		
						MAT.	INST.	TOTAL
2000	Glazed	common brick	4	none		17.95	22.50	40.45
2040		SCR brick	6	none		20.50	20	40.50
2080		conc. block	4	none		16.40	18	34.40
2120			6	5/8" drywall		17.70	20	37.70
2160				gyp. plaster		17.45	20.50	37.95
2200			8	5/8" drywall		17.80	20.50	38.30
2240				gyp. plaster		17.55	21	38.55
2280		L.W. block	4	none		16.70	17.85	34.55
2320			6	5/8" drywall		18	19.95	37.95
2360				gyp. plaster		17.75	20.50	38.25
2400			8	5/8" drywall		18.60	20.50	39.10
2440				gyp. plaster		18.35	21	39.35
2520		glazed block	4	none		24	19.30	43.30
2560			6	none		25	19.60	44.60
2640			8	none		25	20	45
2720		clay tile	4	none		22.50	17.30	39.80
2760			6	none		24.50	17.75	42.25
2800			8	none		27	18.40	45.40
2840		glazed tile	4	none		27	23	50
3000	Engineer	common brick	4	none		7.80	20.50	28.30
3040		SCR brick	6	none		10.50	17.90	28.40
3080		conc. block	4	none		6.25	15.90	22.15
3120			6	5/8" drywall		7.55	18	25.55
3160				gyp. plaster		7.30	18.80	26.10
3200			8	5/8" drywall		7.70	18.35	26.05
3240				gyp. plaster		7.45	19.15	26.60
3280		L.W. block	4	none		4.78	10.95	15.73
3320			6	5/8" drywall		7.90	17.85	25.75
3360				gyp. plaster		7.65	18.65	26.30
3400			8	5/8" drywall		8.45	18.20	26.65
3440				gyp. plaster		8.20	19	27.20
3520		glazed block	4	none		14.05	17.20	31.25
3560			6	none		14.75	17.50	32.25
3640			8	none		14.85	17.90	32.75
3720		clay tile	4	none		12.35	15.20	27.55
3760			6	none		14.25	15.65	29.90
3800			8	none		16.90	16.30	33.20
3840		glazed tile	4	none		16.85	21	37.85
4000	Roman	common brick	4	none		9.80	21	30.80
4040		SCR brick	6	none		12.45	18.25	30.70
4080		conc. block	4	none		8.25	16.25	24.50
4120			6	5/8" drywall		9.55	18.35	27.90
4160				gyp. plaster		9.30	19.15	28.45
4200			8	5/8" drywall		9.65	18.70	28.35
4240				gyp. plaster		9.40	19.50	28.90
4280		L.W. block	4	none		8.55	16.10	24.65
4320			6	5/8" drywall		9.85	18.20	28.05
4360				gyp. plaster		9.60	19	28.60
4400			8	5/8" drywall		10.45	18.55	29
4440				gyp. plaster		10.20	19.35	29.55

C10 Interior Construction

C1010 Partitions

C1010 116 Brick Face Composite Wall - Double Wythe

	FACE BRICK	BACKUP MASONRY	BACKUP THICKNESS (IN.)	BACKUP FINISH		COST PER S.F.		
						MAT.	INST.	TOTAL
4520	Roman	glazed block	4	none		16	17.55	33.55
4560			6	none		16.70	17.85	34.55
4640			8	none		16.80	18.25	35.05
4720		claytile	4	none		14.30	15.55	29.85
4760			6	none		16.20	16	32.20
4800			8	none		18.85	16.65	35.50
4840		glazed tile	4	none		18.80	21	39.80
5000	Norman	common brick	4	none		8.80	18.85	27.65
5040		SCR brick	6	none		11.45	16.30	27.75
5080		conc. block	4	none		7.25	14.30	21.55
5120			6	5/8" drywall		8.55	16.40	24.95
5160				gyp. plaster		8.30	17.20	25.50
5200			8	5/8" drywall		8.65	16.75	25.40
5240				gyp. plaster		8.40	17.55	25.95
5280		L.W. block	4	none		7.55	14.15	21.70
5320			6	5/8" drywall		8.90	16.45	25.35
5360				gyp. plaster		8.60	17.05	25.65
5400			8	5/8" drywall		9.45	16.60	26.05
5440				gyp. plaster		9.20	17.40	26.60
5520		glazed block	4	none		15	15.60	30.60
5560			6	none		15.70	15.90	31.60
5640			8	perlite		15.80	16.30	32.10
5720		clay tile	4	none		13.30	13.60	26.90
5760			6	none		15.20	14.05	29.25
5800			8	none		17.85	14.70	32.55
5840		glazed tile	4	none		17.80	19.30	37.10
6000	Norwegian	common brick	4	none		7.70	17.80	25.50
6040		SCR brick	6	none		10.35	15.25	25.60
6080		conc. block	4	none		6.15	13.25	19.40
6120			6	5/8" drywall		7.45	15.35	22.80
6160				gyp. plaster		7.20	16.15	23.35
6200			8	5/8" drywall		7.55	15.70	23.25
6240				gyp. plaster		7.30	16.50	23.80
6280		L.W. block	4	none		6.45	13.10	19.55
6320			6	5/8" drywall		7.75	15.20	22.95
6360				gyp. plaster		7.50	16	23.50
6400			8	5/8" drywall		8.35	15.55	23.90
6440				gyp. plaster		8.10	16.35	24.45
6520		glazed block	4	none		13.90	14.55	28.45
6560			6	none		14.60	14.85	29.45
6640			8	none		14.70	15.25	29.95
6720		clay tile	4	none		12.20	12.55	24.75
6760			6	none		14.10	13	27.10
6800			8	none		16.75	13.65	30.40
6840		glazed tile	4	none		16.70	18.25	34.95
7000	Utility	common brick	4	none		7.30	16.80	24.10
7040		SCR brick	6	none		10	14.25	24.25
7080		conc. block	4	none		5.80	12.25	18.05
7120			6	5/8" drywall		7.10	14.35	21.45
7160				gyp. plaster		6.85	15.15	22
7200			8	5/8" drywall		7.20	14.70	21.90
7240				gyp. plaster		6.95	15.50	22.45

C10 Interior Construction

C1010 Partitions

C1010 116 — Brick Face Composite Wall - Double Wythe

	FACE BRICK	BACKUP MASONRY	BACKUP THICKNESS (IN.)	BACKUP FINISH		COST PER S.F.		
						MAT.	INST.	TOTAL
7280	Utility	L.W. block	4	none		6.10	12.10	18.20
7320			6	5/8" drywall		7.40	14.20	21.60
7360				gyp. plaster		7.15	15	22.15
7400			8	5/8" drywall		8	14.55	22.55
7440				gyp. plaster		7.75	15.35	23.10
7520		glazed block	4	none		13.55	13.55	27.10
7560			6	none		14.25	13.85	28.10
7640			8	none		14.35	14.25	28.60
7720		clay tile	4	none		11.85	11.55	23.40
7760			6	none		13.75	12	25.75
7800			8	none		16.40	12.65	29.05
7840		glazed tile	4	none		16.35	17.25	33.60

C1010 117 — Brick Face Composite Wall - Triple Wythe

	FACE BRICK	BACKUP MASONRY	BACKUP THICKNESS (IN.)	BACKUP FINISH		COST PER S.F.		
						MAT.	INST.	TOTAL
8000	Standard	common brick	8	standard brick		16.50	32	48.50
8100		4" conc. brick	8	standard brick		15.95	33	48.95
8120		4" conc. brick	8	common brick		12.90	32.50	45.40
8200	Glazed	common brick	8	standard brick		24.50	32.50	57
8300		4" conc. brick	8	standard brick		24	33.50	57.50
8320		4" conc. brick	8	glazed brick		21	33	54
8400	Engineer	common brick	8	standard brick		14.45	30.50	44.95
8500		4" conc. brick	8	standard brick		13.85	31.50	45.35
8520		4" conc. brick	8	engineer brick		10.80	31	41.80
8600	Roman	common brick	8	standard brick		16.40	31	47.40
8700		4" conc. brick	8	standard brick		15.85	32	47.85
8720		4" conc. brick	8	Roman brick		12.80	31.50	44.30
8800	Norman	common, brick	8	standard brick		15.40	29	44.40
8900		4" conc. brick	8	standard brick		14.85	30	44.85
8920		4" conc. brick	8	Norman brick		11.80	29.50	41.30
9000	Norwegian	common brick	8	standard brick		14.30	28	42.30
9100		4" conc. brick	8	standard brick		13.75	29	42.75
9120		4" conc. brick	8	Norwegian brick		10.70	28.50	39.20
9200	Utility	common brick	8	standard brick		13.95	27	40.95
9300		4" conc. brick	8	standard brick		13.20	27.50	40.70
9320		4" conc. brick	8	utility brick		10.35	27.50	37.85

C10 Interior Construction

C1010 Partitions

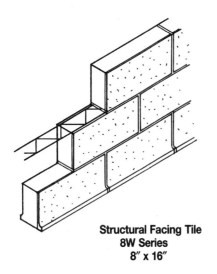

Structural Facing Tile
8W Series
8" x 16"

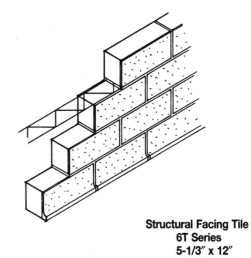

Structural Facing Tile
6T Series
5-1/3" x 12"

C1010 120	Tile Partitions		MAT.	INST.	TOTAL
1000	8W series 8"x16", 4" thick wall, reinf every 2 courses, glazed 1 side	R040513 -10	13.25	6.55	19.80
1100	Glazed 2 sides		14.80	6.95	21.75
1200	Glazed 2 sides, using 2 wythes of 2" thick tile		20.50	12.50	33
1300	6" thick wall, horizontal reinf every 2 courses, glazed 1 side		23.50	6.85	30.35
1400	Glazed 2 sides, each face different color, 2" and 4" tile		23.50	12.80	36.30
1500	8" thick wall, glazed 2 sides using 2 wythes of 4" thick tile		26.50	13.10	39.60
1600	10" thick wall, glazed 2 sides using 1 wythe of 4" tile & 1 wythe of 6" tile		37	13.40	50.40
1700	Glazed 2 sides cavity wall, using 2 wythes of 4" thick tile		26.50	13.10	39.60
1800	12" thick wall, glazed 2 sides using 2 wythes of 6" thick tile		47	13.70	60.70
1900	Glazed 2 sides cavity wall, using 2 wythes of 4" thick tile		26.50	13.10	39.60
2100	6T series 5-1/3"x12" tile, 4" thick, non load bearing glazed one side,		12.15	10.25	22.40
2200	Glazed two sides		15.60	11.55	27.15
2300	Glazed two sides, using two wythes of 2" thick tile		18.60	20	38.60
2400	6" thick, glazed one side		18.25	10.75	29
2500	Glazed two sides		21.50	12.20	33.70
2600	Glazed two sides using 2" thick tile and 4" thick tile		21.50	20.50	42
2700	8" thick, glazed one side		24	12.55	36.55
2800	Glazed two sides using two wythes of 4" thick tile		24.50	20.50	45
2900	Glazed two sides using 6" thick tile and 2" thick tile		27.50	21	48.50
3000	10" thick cavity wall, glazed two sides using two wythes of 4" tile		24.50	20.50	45
3100	12" thick, glazed two sides using 4" thick tile and 8" thick tile		36	23	59
3200	2" thick facing tile, glazed one side, on 6" concrete block		12.50	11.90	24.40
3300	On 8" concrete block		12.60	12.25	24.85
3400	On 10" concrete block		13.50	12.50	26

C10 Interior Construction

C1010 Partitions

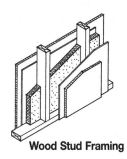
Wood Stud Framing

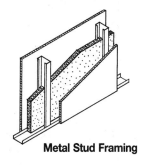
Metal Stud Framing

The Drywall Partitions/Stud Framing Systems are defined by type of drywall and number of layers, type and spacing of stud framing, and treatment on the opposite face. Components include taping and finishing.

Cost differences between regular and fire resistant drywall are negligible, and terminology is interchangeable. In some cases fiberglass insulation is included for additional sound deadening.

System Components	QUANTITY	UNIT	COST PER S.F. MAT.	INST.	TOTAL
SYSTEM C1010 124 1250					
DRYWALL PARTITION,5/8" F.R.1 SIDE,5/8" REG.1 SIDE,2"X4"STUDS,16" O.C.					
Gypsum plasterboard, nailed/screwed to studs, 5/8" F.R. fire resistant	1.000	S.F.	.40	.50	.90
Gypsum plasterboard, nailed/screwed to studs, 5/8" regular	1.000	S.F.	.37	.50	.87
Taping and finishing joints	2.000	S.F.	.10	1	1.10
Framing, 2 x 4 studs @ 16" O.C., 10' high	1.000	S.F.	.34	.99	1.33
TOTAL			1.21	2.99	4.20

C1010 124 — Drywall Partitions/Wood Stud Framing

	FACE LAYER	BASE LAYER	FRAMING	OPPOSITE FACE	INSULATION	MAT.	INST.	TOTAL
1200	5/8" FR drywall	none	2 x 4, @ 16" O.C.	same	0	1.24	2.99	4.23
1250		R061110 -30		5/8" reg. drywall	0	1.21	2.99	4.20
1300				nothing	0	.79	1.99	2.78
1400		1/4" SD gypsum	2 x 4 @ 16" O.C.	same	1-1/2" fiberglass	2.76	4.59	7.35
1450				5/8" FR drywall	1-1/2" fiberglass	2.35	4.04	6.39
1500				nothing	1-1/2" fiberglass	1.90	3.04	4.94
1600		resil. channels	2 x 4 @ 16", O.C.	same	1-1/2" fiberglass	2.29	5.80	8.09
1650				5/8" FR drywall	1-1/2" fiberglass	2.12	4.65	6.77
1700				nothing	1-1/2" fiberglass	1.67	3.65	5.32
1800		5/8" FR drywall	2 x 4 @ 24" O.C.	same	0	1.96	3.80	5.76
1850				5/8" FR drywall	0	1.56	3.30	4.86
1900				nothing	0	1.11	2.30	3.41
2050			on 6" plate	5/8" FR drywall		2.03	4	6.03
2100				nothing		1.54	2.92	4.46
2200		5/8" FR drywall	2 rows-2 x 4	same	2" fiberglass	3.13	5.50	8.63
2250			16"O.C.	5/8" FR drywall	2" fiberglass	2.73	4.98	7.71
2300				nothing	2" fiberglass	2.28	3.98	6.26
2400	5/8" WR drywall	none	2 x 4, @ 16" O.C.	same	0	1.48	2.99	4.47
2450				5/8" FR drywall	0	1.36	2.99	4.35
2500				nothing	0	.91	1.99	2.90
2600		5/8" FR drywall	2 x 4, @ 24" O.C.	same	0	2.20	3.80	6
2650				5/8" FR drywall	0	1.68	3.30	4.98
2700				nothing	0	1.23	2.30	3.53
2800	5/8 VF drywall	none	2 x 4, @ 16" O.C.	same	0	2.58	3.19	5.77
2850				5/8" FR drywall	0	1.91	3.09	5
2900				nothing	0	1.46	2.09	3.55

C10 Interior Construction

C1010 Partitions

C1010 124 — Drywall Partitions/Wood Stud Framing

	FACE LAYER	BASE LAYER	FRAMING	OPPOSITE FACE	INSULATION	COST PER S.F.		
						MAT.	INST.	TOTAL
3000	5/8 VF drywall	5/8" FR drywall	2 x 4, 24" O.C.	same	0	3.30	4	7.30
3050				5/8" FR drywall	0	2.23	3.40	5.63
3100				nothing	0	1.78	2.40	4.18
3200	1/2" reg drywall	3/8" reg drywall	2 x 4, @ 16" O.C.	same	0	1.80	3.99	5.79
3250				5/8" FR drywall	0	1.52	3.49	5.01
3310				nothing	0	1.07	2.49	3.56

C1010 126 — Drywall Partitions/Metal Stud Framing

	FACE LAYER	BASE LAYER	FRAMING	OPPOSITE FACE	INSULATION	COST PER S.F.		
						MAT.	INST.	TOTAL
5200	5/8" FR drywall	none	1-5/8" @ 24" O.C.	same	0	1.15	2.65	3.80
5250				5/8" reg. drywall	0	1.12	2.65	3.77
5300				nothing	0	.70	1.65	2.35
5400			3-5/8" @ 24" O.C.	same	0	1.26	2.67	3.93
5450				5/8" reg. drywall	0	1.23	2.67	3.90
5500				nothing	0	.81	1.67	2.48
5600		1/4" SD gypsum	1-5/8" @ 24" O.C.	same	0	1.97	3.75	5.72
5650				5/8" FR drywall	0	1.56	3.20	4.76
5700				nothing	0	1.11	2.20	3.31
5800			2-1/2" @ 24" O.C.	same	0	2.04	3.76	5.80
5850				5/8" FR drywall	0	1.63	3.21	4.84
5900				nothing	0	1.18	2.21	3.39
6000		5/8" FR drywall	2-1/2" @ 16" O.C.	same	0	2.35	4.27	6.62
6050				5/8" FR drywall	0	1.95	3.77	5.72
6100				nothing	0	1.50	2.77	4.27
6200			3-5/8" @ 24" O.C.	same	0	2.06	3.67	5.73
6250				5/8"FR drywall	3-1/2" fiberglass	2.29	3.54	5.83
6300				nothing	0	1.21	2.17	3.38
6400	5/8" WR drywall	none	1-5/8" @ 24" O.C.	same	0	1.39	2.65	4.04
6450				5/8" FR drywall	0	1.27	2.65	3.92
6500				nothing	0	.82	1.65	2.47
6600			3-5/8" @ 24" O.C.	same	0	1.50	2.67	4.17
6650				5/8" FR drywall	0	1.38	2.67	4.05
6700				nothing	0	.93	1.67	2.60
6800		5/8" FR drywall	2-1/2" @ 16" O.C.	same	0	2.59	4.27	6.86
6850				5/8" FR drywall	0	2.07	3.77	5.84
6900				nothing	0	1.62	2.77	4.39
7000			3-5/8" @ 24" O.C.	same	0	2.30	3.67	5.97
7050				5/8"FR drywall	3-1/2" fiberglass	2.41	3.54	5.95
7100				nothing	0	1.33	2.17	3.50
7200	5/8" VF drywall	none	1-5/8" @ 24" O.C.	same	0	2.49	2.85	5.34
7250				5/8" FR drywall	0	1.82	2.75	4.57
7300				nothing	0	1.37	1.75	3.12
7400			3-5/8" @ 24" O.C.	same	0	2.60	2.87	5.47
7450				5/8" FR drywall	0	1.93	2.77	4.70
7500				nothing	0	1.48	1.77	3.25
7600		5/8" FR drywall	2-1/2" @ 16" O.C.	same	0	3.69	4.47	8.16
7650				5/8" FR drywall	0	2.62	3.87	6.49
7700				nothing	0	2.17	2.87	5.04
7800			3-5/8" @ 24" O.C.	same	0	3.40	3.87	7.27
7850				5/8"FR drywall	3-1/2" fiberglass	2.96	3.64	6.60
7900				nothing	0	1.88	2.27	4.15

C10 Interior Construction

C1010 Partitions

C1010 128	Drywall Components	COST PER S.F.		
		MAT.	INST.	TOTAL
0140	Metal studs, 24" O.C. including track, load bearing, 18 gage, 2-1/2"	.87	.93	1.80
0160	3-5/8"	1.03	.95	1.98
0180	4"	.93	.96	1.89
0200	6"	1.38	.98	2.36
0220	16 gage, 2-1/2"	1	1.06	2.06
0240	3-5/8"	1.20	1.08	2.28
0260	4"	1.27	1.10	2.37
0280	6"	1.60	1.13	2.73
0300	Non load bearing, 25 gage, 1-5/8"	.25	.65	.90
0340	3-5/8"	.36	.67	1.03
0360	4"	.39	.67	1.06
0380	6"	.55	.68	1.23
0400	20 gage, 2-1/2"	.48	.83	1.31
0420	3-5/8"	.52	.84	1.36
0440	4"	.61	.84	1.45
0460	6"	.72	.85	1.57
0540	Wood studs including blocking, shoe and double top plate, 2"x4", 12"O.C.	.42	1.24	1.66
0560	16" O.C.	.34	.99	1.33
0580	24" O.C.	.26	.80	1.06
0600	2"x6", 12" O.C.	.75	1.42	2.17
0620	16" O.C.	.60	1.10	1.70
0640	24" O.C.	.46	.86	1.32
0642	Furring one side only, steel channels, 3/4", 12" O.C.	.38	1.79	2.17
0644	16" O.C.	.34	1.59	1.93
0646	24" O.C.	.23	1.20	1.43
0647	1-1/2" , 12" O.C.	.51	2	2.51
0648	16" O.C.	.46	1.75	2.21
0649	24" O.C.	.30	1.38	1.68
0650	Wood strips, 1" x 3", on wood, 12" O.C.	.38	.90	1.28
0651	16" O.C.	.29	.68	.97
0652	On masonry, 12" O.C.	.38	1	1.38
0653	16" O.C.	.29	.75	1.04
0654	On concrete, 12" O.C.	.38	1.91	2.29
0655	16" O.C.	.29	1.43	1.72
0665	Gypsum board, one face only, exterior sheathing, 1/2"	.67	.88	1.55
0680	Interior, fire resistant, 1/2"	.35	.50	.85
0700	5/8"	.40	.50	.90
0720	Sound deadening board 1/4"	.41	.55	.96
0740	Standard drywall 3/8"	.34	.50	.84
0760	1/2"	.34	.50	.84
0780	5/8"	.37	.50	.87
0800	Tongue & groove coreboard 1"	.97	2.06	3.03
0820	Water resistant, 1/2"	.53	.50	1.03
0840	5/8"	.52	.50	1.02
0860	Add for the following:, foil backing	.15		.15
0880	Fiberglass insulation, 3-1/2"	.63	.37	1
0900	6"	.85	.37	1.22
0920	Rigid insulation 1"	.54	.50	1.04
0940	Resilient furring @ 16" O.C.	.24	1.55	1.79
0960	Taping and finishing	.05	.50	.55
0980	Texture spray	.09	.53	.62
1000	Thin coat plaster	.09	.62	.71
1040	2"x4" staggered studs 2"x6" plates & blocking	.48	1.08	1.56

C10 Interior Construction

C1010 Partitions

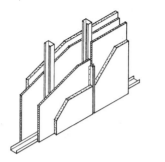
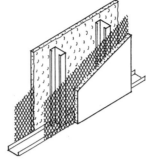

Plaster Partitions are defined as follows: type of plaster, type and spacing of framing, type of lath and treatment of opposite face.

Included in the system components are expansion joints. Metal studs are assumed to be non-loadbearing.

System Components	QUANTITY	UNIT	COST PER S.F.		
			MAT.	INST.	TOTAL
SYSTEM C1010 140 1000					
GYP PLASTER PART'N, 2 COATS, 2-1/2" MTL. STUD, 16"O.C., 3/8"GYP. LATH					
Gypsum plaster, 2 coats, on walls	2.000	S.F.	1.04	4.72	5.76
Gypsum lath, 3/8" thick, on metal studs	2.000	S.F.	1.34	1.24	2.58
Metal studs, 25 ga., 2-1/2" @ 16" O.C.	1.000	S.F.	.43	1.01	1.44
Expansion joint	.100	L.F.	.12	.16	.28
TOTAL			2.93	7.13	10.06

C1010 140   Plaster Partitions/Metal Stud Framing

	TYPE	FRAMING	LATH	OPPOSITE FACE		COST PER S.F.		
						MAT.	INST.	TOTAL
1000	2 coat gypsum	2-1/2" @ 16"O.C.	3/8" gypsum	same		2.93	7.15	10.08
1010				nothing		1.74	4.15	5.89
1100		3-1/4" @ 24"O.C.	1/2" gypsum	same		2.95	6.90	9.85
1110				nothing		1.72	3.86	5.58
1500	2 coat vermiculite	2-1/2" @ 16"O.C.	3/8" gypsum	same		2.88	7.75	10.63
1510				nothing		1.72	4.46	6.18
1600		3-1/4" @ 24"O.C.	1/2" gypsum	same		2.90	7.50	10.40
1610				nothing		1.70	4.17	5.87
2000	3 coat gypsum	2-1/2" @ 16"O.C.	3/8" gypsum	same		2.81	8.10	10.91
2010				nothing		1.68	4.63	6.31
2020			3.4 lb. diamond	same		2.59	8.10	10.69
2030				nothing		1.57	4.63	6.20
2040			2.75 lb. ribbed	same		2.35	8.10	10.45
2050				nothing		1.45	4.63	6.08
2100		3-1/4" @ 24"O.C.	1/2" gypsum	same		2.83	7.85	10.68
2110				nothing		1.66	4.34	6
2120			3.4 lb. ribbed	same		2.77	7.85	10.62
2130				nothing		1.62	4.34	5.96
2500	3 coat lime	2-1/2" @ 16"O.C.	3.4 lb. diamond	same		2.25	8.10	10.35
2510				nothing		1.40	4.63	6.03
2520			2.75 lb. ribbed	same		2.01	8.10	10.11
2530				nothing		1.28	4.63	5.91
2600		3-1/4" @ 24"O.C.	3.4 lb. ribbed	same		2.43	7.85	10.28
2610				nothing		1.45	4.34	5.79

C10 Interior Construction

C1010 Partitions

C1010 140 Plaster Partitions/Metal Stud Framing

	TYPE	FRAMING	LATH	OPPOSITE FACE		COST PER S.F.		
						MAT.	INST.	TOTAL
3000	3 coat Portland	2-1/2" @ 16"O.C.	3.4 lb. diamond	same		2.21	8.10	10.31
3010				nothing		1.38	4.63	6.01
3020			2.75 lb. ribbed	same		1.97	8.10	10.07
3030				nothing		1.26	4.63	5.89
3100		3-1/4" @ 24"O.C.	3.4 lb. ribbed	same		2.39	7.85	10.24
3110				nothing		1.43	4.34	5.77
3500	3 coat gypsum	2-1/2" @ 16"O.C.	3/8" gypsum	same		3.42	10.40	13.82
3510	W/med. Keenes			nothing		1.99	5.80	7.79
3520			3.4 lb. diamond	same		3.20	10.40	13.60
3530				nothing		1.88	5.80	7.68
3540			2.75 lb. ribbed	same		2.96	10.40	13.36
3550				nothing		1.76	5.80	7.56
3600		3-1/4" @ 24"O.C.	1/2" gypsum	same		3.44	10.15	13.59
3610				nothing		1.97	5.50	7.47
3620			3.4 lb. ribbed	same		3.38	10.15	13.53
3630				nothing		1.93	5.50	7.43
4000	3 coat gypsum	2-1/2" @ 16"O.C.	3/8" gypsum	same		3.43	11.55	14.98
4010	W/hard Keenes			nothing		1.99	6.35	8.34
4020			3.4 lb. diamond	same		3.21	11.55	14.76
4030				nothing		1.88	6.35	8.23
4040			2.75 lb. ribbed	same		2.97	11.55	14.52
4050				nothing		1.76	6.35	8.11
4100		3-1/4" @ 24"O.C.	1/2" gypsum	same		3.45	11.30	14.75
4110				nothing		1.97	6.05	8.02
4120			3.4 lb. ribbed	same		3.39	11.30	14.69
4130				nothing		1.93	6.05	7.98
4500	3 coat lime	2-1/2" @ 16"O.C.	3.4 lb. diamond	same		2.19	8.10	10.29
4510	Portland			nothing		1.37	4.63	6
4520			2.75 lb. ribbed	same		1.95	8.10	10.05
4530				nothing		1.25	4.63	5.88
4600		3-1/4" @ 24"O.C.	3.4 lb. ribbed	same		2.37	7.85	10.22
4610				nothing		1.42	4.34	5.76

C1010 142 Plaster Partitions/Wood Stud Framing

	TYPE	FRAMING	LATH	OPPOSITE FACE		COST PER S.F.		
						MAT.	INST.	TOTAL
5000	2 coat gypsum	2"x4" @ 16"O.C.	3/8" gypsum	same		2.84	6.95	9.79
5010				nothing		1.65	4.06	5.71
5100		2"x4" @ 24"O.C.	1/2" gypsum	same		2.85	6.85	9.70
5110				nothing		1.62	3.90	5.52
5500	2 coat vermiculite	2"x4" @ 16"O.C.	3/8" gypsum	same		2.79	7.60	10.39
5510				nothing		1.63	4.37	6
5600		2"x4" @ 24"O.C.	1/2" gypsum	same		2.80	7.50	10.30
5610				nothing		1.60	4.21	5.81

C10 Interior Construction

C1010 Partitions

C1010 142 Plaster Partitions/Wood Stud Framing

	TYPE	FRAMING	LATH	OPPOSITE FACE		COST PER S.F.		
						MAT.	INST.	TOTAL
6000	3 coat gypsum	2"x4" @ 16"O.C.	3/8" gypsum	same		2.72	7.95	10.67
6010				nothing		1.59	4.54	6.13
6020			3.4 lb. diamond	same		2.50	8	10.50
6030				nothing		1.48	4.57	6.05
6040			2.75 lb. ribbed	same		2.24	8.05	10.29
6050				nothing		1.35	4.59	5.94
6100		2"x4" @ 24"O.C.	1/2" gypsum	same		2.73	7.80	10.53
6110				nothing		1.56	4.38	5.94
6120			3.4 lb. ribbed	same		2.18	7.90	10.08
6130				nothing		1.28	4.42	5.70
6500	3 coat lime	2"x4" @ 16"O.C.	3.4 lb. diamond	same		2.16	8	10.16
6510				nothing		1.31	4.57	5.88
6520			2.75 lb. ribbed	same		1.90	8.05	9.95
6530				nothing		1.18	4.59	5.77
6600		2"x4" @ 24"O.C.	3.4 lb. ribbed	same		1.84	7.90	9.74
6610				nothing		1.11	4.42	5.53
7000	3 coat Portland	2"x4" @ 16"O.C.	3.4 lb. diamond	same		2.12	8	10.12
7010				nothing		1.29	4.57	5.86
7020			2.75 lb. ribbed	same		1.86	8.05	9.91
7030				nothing		1.16	4.59	5.75
7100		2"x4" @ 24"O.C.	3.4 lb. ribbed	same		2.29	7.95	10.24
7110				nothing		1.33	4.47	5.80
7500	3 coat gypsum	2"x4" @ 16"O.C.	3/8" gypsum	same		3.33	10.25	13.58
7510	W/med Keenes			nothing		1.90	5.70	7.60
7520			3.4 lb. diamond	same		3.11	10.30	13.41
7530				nothing		1.79	5.70	7.49
7540			2.75 lb. ribbed	same		2.85	10.35	13.20
7550				nothing		1.66	5.75	7.41
7600		2"x4" @ 24"O.C.	1/2" gypsum	same		3.34	10.10	13.44
7610				nothing		1.87	5.55	7.42
7620			3.4 lb. ribbed	same		3.28	10.25	13.53
7630				nothing		1.83	5.60	7.43
8000	3 coat gypsum	2"x4" @ 16"O.C.	3/8" gypsum	same		3.34	11.40	14.74
8010	W/hard Keenes			nothing		1.90	6.25	8.15
8020			3.4 lb. diamond	same		3.12	11.45	14.57
8030				nothing		1.79	6.30	8.09
8040			2.75 lb. ribbed	same		2.86	11.50	14.36
8050				nothing		1.66	6.30	7.96
8100		2"x4" @ 24"O.C.	1/2" gypsum	same		3.35	11.30	14.65
8110				nothing		1.87	6.10	7.97
8120			3.4 lb. ribbed	same		3.29	11.45	14.74
8130				nothing		1.83	6.20	8.03
8500	3 coat lime	2"x4" @ 16"O.C.	3.4 lb. diamond	same		2.10	8	10.10
8510	Portland			nothing		1.28	4.57	5.85
8520			2.75 lb. ribbed	same		1.84	8.05	9.89
8530				nothing		1.15	4.59	5.74
8600		2"x4" @ 24"O.C.	3.4 lb. ribbed	same		2.27	7.95	10.22
8610				nothing		1.32	4.47	5.79

C10 Interior Construction

C1010 Partitions

Folding Accordion   **Folding Leaf**   **Movable and Borrow Lites**

C1010 205	Partitions	MAT.	INST.	TOTAL
0360	Folding accordion, vinyl covered, acoustical, 3 lb. S.F., 17 ft max. hgt	28	9.90	37.90
0380	5 lb. per S.F. 27 ft max height	39	10.45	49.45
0400	5.5 lb. per S.F., 17 ft. max height	46	11	57
0420	Commercial, 1.75 lb per S.F., 8 ft. max height	24.50	4.41	28.91
0440	2.0 Lb per S.F., 17 ft. max height	25	6.60	31.60
0460	Industrial, 4.0 lb. per S.F. 27 ft max height	36	13.20	49.20
0480	Vinyl clad wood or steel, electric operation 6 psf	53	6.20	59.20
0500	Wood, non acoustic, birch or mahogany	28.50	3.30	31.80
0560	Folding leaf, aluminum framed acoustical 12 ft.high., 5.5 lb per S.F.,min.	39	16.50	55.50
0580	Maximum	47	33	80
0600	6.5 lb. per S.F., minimum	40.50	16.50	57
0620	Maximum	50.50	33	83.50
0640	Steel acoustical, 7.5 per S.F., vinyl faced, minimum	57.50	16.50	74
0660	Maximum	70	33	103
0680	Wood acoustic type, vinyl faced to 18' high 6 psf, minimum	54	16.50	70.50
0700	Average	64.50	22	86.50
0720	Maximum	83.50	33	116.50
0740	Formica or hardwood faced, minimum	55.50	16.50	72
0760	Maximum	59.50	33	92.50
0780	Wood, low acoustical type to 12 ft. high 4.5 psf	40.50	19.80	60.30
0840	Demountable, trackless wall, cork finish, semi acous, 1-5/8"th, min	42.50	3.05	45.55
0860	Maximum	40	5.20	45.20
0880	Acoustic, 2" thick, minimum	34	3.25	37.25
0900	Maximum	59	4.41	63.41
0920	In-plant modular office system, w/prehung steel door			
0940	3" thick honeycomb core panels			
0960	12' x 12', 2 wall	10.35	.62	10.97
0970	4 wall	9.65	.83	10.48
0980	16' x 16', 2 wall	10.10	.44	10.54
0990	4 wall	6.80	.44	7.24
1000	Gypsum, demountable, 3" to 3-3/4" thick x 9' high, vinyl clad	6.95	2.28	9.23
1020	Fabric clad	17.20	2.50	19.70
1040	1.75 system, vinyl clad hardboard, paper honeycomb core panel			
1060	1-3/4" to 2-1/2" thick x 9' high	11.65	2.28	13.93
1080	Unitized gypsum panel system, 2" to 2-1/2" thick x 9' high			
1100	Vinyl clad gypsum	15	2.28	17.28
1120	Fabric clad gypsum	24.50	2.50	27
1140	Movable steel walls, modular system			
1160	Unitized panels, 48" wide x 9' high			
1180	Baked enamel, pre-finished	16.85	1.83	18.68
1200	Fabric clad	24.50	1.96	26.46

C10 Interior Construction

C1010 Partitions

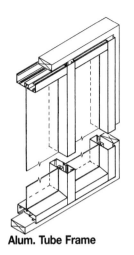
Alum. Tube Frame

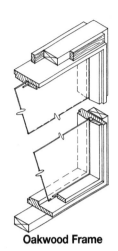
Oakwood Frame

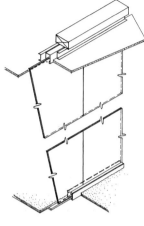
Concealed Frame
Butt Glazed

Interior Glazed Openings are defined as follows: framing material, glass, size and intermediate framing members. Components for each system include gasket setting or glazing bead and typical wood blocking.

System Components	QUANTITY	UNIT	COST PER OPNG.		
			MAT.	INST.	TOTAL
SYSTEM C1010 710 1000					
GLAZED OPENING, ALUMINUM TUBE FRAME, FLUSH, 1/4" FLOAT GLASS, 6' X 4'					
Aluminum tube frame, flush, anodized bronze, head & jamb	16.500	L.F.	207.90	192.23	400.13
Aluminum tube frame, flush, anodized bronze, open sill	7.000	L.F.	69.30	79.80	149.10
Joints for tube frame, clip type	4.000	Ea.	100		100
Gasket setting, add	20.000	L.F.	122		122
Wood blocking	8.000	B.F.	3.84	22	25.84
Float glass, 1/4" thick, clear, plain	24.000	S.F.	139.20	187.20	326.40
TOTAL			642.24	481.23	1,123.47

C1010 710		Interior Glazed Opening					
	FRAME	GLASS	OPENING-SIZE W X H	INTERMEDIATE MULLION	INTERMEDIATE HORIZONTAL	COST PER OPNG.		
						MAT.	INST.	TOTAL
1000	Aluminum flush	1/4" float	6'x4'	0	0	640	480	1,120
1040	Tube		12'x4'	3	0	1,575	1,025	2,600
1080	R088110 -10		4'x5'	0	0	585	415	1,000
1120			8'x5'	1	0	1,075	770	1,845
1160	R088110 -10		12'x5'	2	0	1,575	1,125	2,700
1240		3/8" float	9'x6'	2	0	1,750	1,300	3,050
1280			4'x8'-6"	0	1	1,125	830	1,955
1320			16'x10'	3	1	4,575	3,400	7,975
1400		1/4" tempered	6'x4'	0	0	700	480	1,180
1440			12'x4'	3	0	1,700	1,025	2,725
1480			4'x5'	0	0	630	415	1,045
1520			8'x5'	1	0	1,175	770	1,945
1560			12'x5'	2	0	1,725	1,125	2,850
1640		3/8" tempered	9'x6'	2	0	2,125	1,300	3,425
1680			4'x8'-6"	0	0	1,175	780	1,955
1720			16'x10'	3	0	4,975	3,200	8,175
1800		1/4" one way mirror	6'x4'	0	0	970	475	1,445
1840			12'x4'	3	0	2,225	1,025	3,250

C10 Interior Construction

C1010 Partitions

C1010 710 Interior Glazed Opening

	FRAME	GLASS	OPENING-SIZE W X H	INTERMEDIATE MULLION	INTERMEDIATE HORIZONTAL	COST PER OPNG. MAT.	COST PER OPNG. INST.	COST PER OPNG. TOTAL
1880	Aluminum flush	3/8" one way mirror	4'x5'	0	0	855	410	1,265
1920	Tube		8'x5'	1	0	1,625	755	2,380
1960			12'x5'	2	0	2,400	1,100	3,500
2040		3/8" laminated safety	9'x6'	2	0	2,425	1,275	3,700
2080			4'x8'-6"	0	0	1,375	765	2,140
2120			16'x10'	3	0	5,925	3,125	9,050
2200		1-3/16" bullet proof	6'x4'	0	0	2,900	1,725	4,625
2240			12'x4'	3	0	6,075	3,500	9,575
2280			4'x5'	0	0	2,475	1,450	3,925
2320			8'x5'	1	0	4,825	2,825	7,650
2360			12'x5'	2	0	7,175	4,200	11,375
2440		1" acoustical	9'x6'	2	0	3,250	1,150	4,400
2480			4'x8'-6"	0	0	1,925	695	2,620
2520			16'x10'	3	0	8,225	2,750	10,975
3000	Oakwood	1/4" float	6'x4'	0	0	288	360	648
3040			12'x4'	3	0	700	870	1,570
3080			4'x5'	0	0	247	293	540
3120			8'x5'	1	0	500	620	1,120
3160			12'x5'	2	0	755	945	1,700
3240		3/8" float	9'x6'	2	0	945	1,125	2,070
3280			4'x8'-6"	0	1	580	660	1,240
3320			16'x10'	3	1	2,700	3,250	5,950
3400		1/4" tempered	6'x4'	0	0	345	360	705
3440			12'x4'	3	0	815	870	1,685
3480			4'x5'	0	0	295	293	588
3520			8'x5'	1	0	595	620	1,215
3560			12'x5'	2	0	900	945	1,845
3640		3/8" tempered	9'x6'	2	0	1,300	1,125	2,425
3680			4'x8'-6"	0	0	740	595	1,335
3720			16'x10'	3	0	3,500	2,925	6,425
3800		1/4" one way mirror	6'x4'	0	0	615	350	965
3840			12'x4'	3	0	1,350	855	2,205
3880			4'x5'	0	0	520	286	806
3920			8'x5'	1	0	1,050	605	1,655
3960			12'x5'	2	0	1,575	920	2,495
4040		3/8" laminated safety	9'x6'	2	0	1,625	1,100	2,725
4080			4'x8'-6"	0	0	945	580	1,525
4120			16'x10'	3	0	4,450	2,850	7,300
4200		1-3/16" bullet proof	6'x4'	0	0	2,475	1,575	4,050
4240			12'x4'	3	0	5,075	3,300	8,375
4280			4'x5'	0	0	2,075	1,300	3,375
4360			12'x5'	2	0	6,225	3,975	10,200
4440		1" acoustical	9'x6'	2	0	2,325	955	3,280
4520			16'x10'	3	0	6,525	2,425	8,950

C10 Interior Construction

C1010 Partitions

C1010 710 Interior Glazed Opening

	FRAME	GLASS	OPENING-SIZE W X H	INTERMEDIATE MULLION	INTERMEDIATE HORIZONTAL	COST PER OPNG.		
						MAT.	INST.	TOTAL
5000	Concealed frame	1/2" float	6'x4'	1		680	870	1,550
5040	Butt glazed		9'x4'	2		980	1,250	2,230
5080			8'x5'	1		1,050	1,275	2,325
5120			16'x5'	3		2,025	2,400	4,425
5200		3/4" float	6'x8'	1		2,025	1,950	3,975
5240			9'x8'	2		2,975	2,800	5,775
5280			8'x10'	1		3,275	3,000	6,275
5320			16'x10'	3		6,375	5,650	12,025
5400		1/4" tempered	6'x4'	1		395	650	1,045
5440			9'x4'	2		555	915	1,470
5480			8'x5'	1		585	915	1,500
5520			16'x5'	3		1,075	1,675	2,750
5600		1/2" tempered	6'x8'	1		1,450	1,500	2,950
5640			9'x8'	2		2,100	2,100	4,200
5680			8'x10'	1		2,325	2,225	4,550
5720			16'x10'	3		4,450	4,125	8,575
5800		1/4" laminated safety	6'x4'	1		490	715	1,205
5840			9'x4'	2		700	1,000	1,700
5880			8'x5'	1		745	1,025	1,770
5920			16'x5'	3		1,400	1,875	3,275
6000		1/2" laminated safety	6'x8'	1		1,525	1,375	2,900
6040			9'x8'	2		2,225	1,925	4,150
6080			8'x10'	1		2,450	2,025	4,475
6120			16'x10'	3		4,700	3,700	8,400
6200		1-3/16" bullet proof	6'x4'	1		2,525	1,875	4,400
6240			9'x4'	2		3,750	2,750	6,500
6280			8'x5'	1		4,150	2,950	7,100
6320			16'x5'	3		8,175	5,725	13,900
6400		2" bullet proof	6'x8'	1		4,550	4,425	8,975
6440			9'x8'	2		6,750	6,500	13,250
6480			8'x10'	1		7,475	7,100	14,575
6520			16'x10'	3		14,800	13,900	28,700

C10 Interior Construction

C1020 Interior Doors

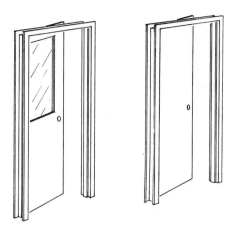

The Metal Door/Metal Frame Systems are defined as follows: door type, design and size, frame type and depth. Included in the components for each system is painting the door and frame. No hardware has been included in the systems.

System Components	QUANTITY	UNIT	COST EACH MAT.	COST EACH INST.	COST EACH TOTAL
SYSTEM C1020 114 1200					
STEEL DOOR, HOLLOW, 20 GA., HALF GLASS, 2'-8"X6'-8", D.W.FRAME, 4-7/8" DP					
Steel door, flush, hollow core, 1-3/4" thick, half glass, 20 ga., 2'-8"x6'-8"	1.000	Ea.	475	49.50	524.50
Steel frame, KD, 16 ga., drywall, 4-7/8" deep, 2'-8"x6'-8", single	1.000	Ea.	122	66	188
Float glass, 3/16" thick, clear, tempered	5.000	S.F.	33.25	36	69.25
Paint door and frame each side, primer	1.000	Ea.	5.16	84	89.16
Paint door and frame each side, 2 coats	1.000	Ea.	8.14	141	149.14
TOTAL			643.55	376.50	1,020.05

C1020 114 — Metal Door/Metal Frame

	TYPE	DESIGN	SIZE	FRAME	DEPTH	MAT.	INST.	TOTAL
1000	Flush-hollow	20 ga. full panel	2'-8" x 6'-8"	drywall K.D.	4-7/8"	470	345	815
1020				butt welded	8-3/4"	580	390	970
1160			6'-0" x 7'-0"	drywall K.D.	4-7/8"	920	560	1,480
1180				butt welded	8-3/4"	1,050	605	1,655
1200		20 ga. half glass	2'-8" x 6'-8"	drywall K.D.	4-7/8"	645	375	1,020
1220				butt welded	8-3/4"	755	420	1,175
1360			6'-0" x 7'-0"	drywall K.D.	4-7/8"	1,200	625	1,825
1380				butt welded	8-3/4"	1,325	670	1,995
1400		20 ga. - bottom louver	2'-8" x 6'-8"	drywall K.D.	4-7/8"	650	345	995
1420				butt welded	8-3/4"	760	390	1,150
1560			6'-0" x 7'-0"	drywall K.D.	4-7/8"	1,275	560	1,835
1580				butt welded	8-3/4"	1,400	605	2,005
1600		20 ga. - vision	2'-8" x 6'-8"	drywall K.D.	4-7/8"	580	355	935
1620				butt welded	8-3/4"	690	400	1,090
1760			6'-0" x 7'-0"	drywall K.D.	4-7/8"	1,150	575	1,725
1780				butt welded	8-3/4"	1,275	620	1,895
1800		18 ga. full panel	2'-8" x 6'-8"	drywall K.D.	4-7/8"	475	350	825
1820				butt welded	8-3/4"	585	395	980
1960			6'-0" x 7'-0"	drywall K.D.	4-7/8"	1,000	560	1,560
1980				butt welded	8-3/4"	1,150	605	1,755

C10 Interior Construction

C1020 Interior Doors

C1020 114 Metal Door/Metal Frame

	TYPE	DESIGN	SIZE	FRAME	DEPTH	COST EACH		
						MAT.	INST.	TOTAL
2000	Flush-hollow	18 ga. half glass	2'-8" x 6'-8"	drywall K.D.	4-7/8"	675	380	1,055
2020				butt welded	8-3/4"	785	425	1,210
2160			6'-0" x 7'-0"	drywall K.D.	4-7/8"	1,300	630	1,930
2180				butt welded	8-3/4"	1,425	675	2,100
2200		18 ga. bottom louver	2'-8" x 6'-8"	drywall K.D.	4-7/8"	655	350	1,005
2220				butt welded	8-3/4"	765	395	1,160
2360			6'-0" x 7'-0"	drywall K.D.	4-7/8"	1,375	560	1,935
2380				butt welded	8-3/4"	1,500	605	2,105
2400		18 ga. vision	2'-8" x 6'-8"	drywall K.D.	4-7/8"	585	355	940
2420				butt welded	8-3/4"	695	400	1,095
2560			6'-0" x 7'-0"	drywall K.D.	4-7/8"	1,225	575	1,800
2580				butt welded	8-3/4"	1,350	620	1,970
3000	Flush - composite	18 ga. full panel	3'-0" 6'-8"	drywall K.D.	4-7/8"	580	355	935
3020				butt welded	8-3/4"	690	400	1,090
3160			6'-0" x 7'-0"	drywall K.D.	4-7/8"	1,100	575	1,675
3180				butt welded	8-3/4"	1,250	620	1,870
3200		18 ga. half glass	3'-0" x 6'-8"	drywall K.D.	4-7/8"	760	390	1,150
3220				butt welded	8-3/4"	870	435	1,305
3360			6'-0" x 7'-0"	drywall K.D.	4-7/8"	1,550	640	2,190
3380				butt welded	8-3/4"	1,675	685	2,360
3400		18 ga. bottom louver	3'-0" x 6'-8"	drywall K.D.	4-7/8"	760	355	1,115
3420				butt welded	8-3/4"	870	400	1,270
3560			6'-0" x 7'-0"	drywall K.D.	4-7/8"	1,475	575	2,050
3580				butt welded	8-3/4"	1,600	620	2,220
3600		18 ga. vision	3'-0" x 6'-8"	drywall K.D.	4-7/8"	690	365	1,055
3620				butt welded	8-3/4"	800	410	1,210
3760			6'-0" x 7'-0"	drywall K.D.	4-7/8"	1,325	590	1,915
3780				butt welded	8-3/4"	1,450	635	2,085
4000	Embossed panel	24 ga., full panel	2'-8" x 6'-8"	drywall K.D.	4-7/8"	440	350	790
4020				butt welded	8-3/4"	520	400	920
4160			6'-0" x 7'-0	drywall K.D.	4-7/8"	840	575	1,415
4180				butt welded	8-3/4"	970	620	1,590
4200		24 ga., half glass	2'-8" x 6'-8"	drywall K.D.	4-7/8"	475	385	860
4220				butt welded	8-3/4"	660	445	1,105
4360			6'-0" x 7'-0"	drywall K.D.	4-7/8"	1,000	640	1,640
4380				butt welded	8-3/4"	1,150	685	1,835
4400	Raised plastic panel	24 ga. full panel	2'-8" x 6'-8"	drywall K.D.	4-7/8"	495	355	850
4420				butt welded	8-3/4"	550	400	950
4560			6'-0" x 7'-0"	drywall K.D.	4-7/8"	820	575	1,395
4580				butt welded	8-3/4"	950	620	1,570
5000		24 ga. half glass	2'-8" x 6'-8"	drywall K.D.	4-7/8"	515	385	900
5020				butt welded	8-3/4"	625	430	1,055
5160			6'-0" x 7'-0"	drywall K.D.	4-7/8"	1,025	640	1,665

C1020 116 Labeled Metal Door/Metal Frames

	TYPE	DESIGN	SIZE	FRAME	DEPTH	COST EACH		
						MAT.	INST.	TOTAL
6000	Hollow-1/2 hour	20 ga. full panel	2'-8" x 6'-8"	drywall K.D.	4-7/8"	580	345	925
6020				butt welded	8-3/4"	690	390	1,080
6160			6'-0" x 7'-0"	drywall K.D.	4-7/8"	1,050	560	1,610
6180				butt welded	8-3/4"	1,125	605	1,730
6200		20 ga. vision	2'-8" x 6'-8"	drywall K.D.	4-7/8"	710	355	1,065
6220				butt welded	8-3/4"	820	400	1,220

C10 Interior Construction

C1020 Interior Doors

C1020 116 Labeled Metal Door/Metal Frames

	TYPE	DESIGN	SIZE	FRAME	DEPTH	COST EACH		
						MAT.	INST.	TOTAL
6360	Hollow-1/2 hour	20 ga. vision	6'-0" x 7'-0"	drywall K.D.	4-7/8"	1,300	575	1,875
6380				butt welded	8-3/4"	1,400	620	2,020
6400		20 ga. bottom louver	2'-6" x 7'-0"	drywall K.D.	4-7/8"	760	345	1,105
6420				butt welded	8-3/4"	870	390	1,260
6560			6'-0" x 7'-0"	drywall K.D.	4-7/8"	1,400	560	1,960
6580				butt welded	8-3/4"	1,500	605	2,105
6600		18 ga. full panel	2'-8" x 7'-0"	drywall K.D.	4-7/8"	655	350	1,005
6620				butt welded	8-3/4"	765	395	1,160
6760			6'-0" x 7'-0"	drywall K.D.	4-7/8"	1,125	565	1,690
6780				butt welded	8-3/4"	1,225	610	1,835
6800		18 ga. vision	2'-8" x 7'-0"	drywall K.D.	4-7/8"	785	355	1,140
6820				butt welded	8-3/4"	895	400	1,295
6960			6'-0" x 7'-0"	drywall K.D.	4-7/8"	1,375	580	1,955
6980				butt welded	8-3/4"	1,475	625	2,100
7000		18 ga. bottom louver	2'-6" x 7'-0"	drywall K.D.	4-7/8"	835	350	1,185
7020				butt welded	8-3/4"	945	395	1,340
7160			6'-0" x 7'-0"	drywall K.D.	4-7/8"	1,475	565	2,040
7180				butt welded	8-3/4"	1,575	610	2,185
7200	Hollow-3 hour	18 ga. full panel	2'-8" x 7'-0"	butt K.D.	5-3/4"	655	350	1,005
7220				butt welded	8-3/4"	765	395	1,160
7360			6'-0" x 7'-0"	butt K.D.	5-3/4"	1,125	565	1,690
7380				butt welded	8-3/4"	1,225	610	1,835
8000	Composite-1-1/2 hour	20 ga. full panel	2'-8" x 6'-8"	drywall K.D.	4-3/8"	680	350	1,030
8020				butt welded	8-3/4"	790	395	1,185
8160			6'-0" x 7'-0"	drywall K.D.	4-7/8"	1,250	565	1,815
8180				butt welded	8-3/4"	1,325	610	1,935
8200		20 ga. vision	2'-8" x 6'-8"	drywall K.D.	4-7/8"	810	355	1,165
8220				butt welded	8-3/4"	920	400	1,320
8360			6'-0" x 7'-0"	drywall K.D.	4-7/8"	1,500	580	2,080
8400	Composite-3 hour	18 ga. full panel	2'-8" x 7'-0"	drywall K.D.	5-3/4"	770	355	1,125
8420				butt welded	8-3/4"	880	400	1,280
8560			6'-0" x 7'-0"	drywall K.D.	5-3/4"	1,350	575	1,925

C10 Interior Construction

C1020 Interior Doors

The Wood Door/Wood Frame Systems are defined as follows: door type, face material and size, frame type and depth. Included in the components for each system are door trim and finish. No hardware has been included in the system.

System Components	QUANTITY	UNIT	COST EACH		
			MAT.	INST.	TOTAL
SYSTEM C1020 120 1600					
WOOD DOOR, HOLLOW, LAUAN FACE, 2'-6"X6'-8", PINE FRAME, 3-5/8" DEEP					
Wood door, flush	1.000	Ea.	40	58.50	98.50
Interior frame, pine, 3/4" x 3-5/8" deep	16.000	L.F.	96.80	42.24	139.04
Door trim, pine, 2-1/2" wide	1.000	Ea.	25	84	109
Varnish 1 coat & sealer on wood trim, no sanding	55.000	S.F.	5.36	71.02	76.38
TOTAL			167.16	255.76	422.92

C1020 120 — Wood Door/Wood Frame

	TYPE	FACE	SIZE	FRAME	DEPTH	COST EACH		
						MAT.	INST.	TOTAL
1000	Hollow core/panel	moulded fiberboard	2'-6" x 6'-8"	pine	3-5/8"	201	410	611
1020					5-3/16"	192	465	657
1160			6'-0"x6'-8"	pine	3-5/8"	310	630	940
1180					5-3/16"	300	685	985
1600	Hollow core/flush	lauan	2'-8" x 6'-8"	pine	3-5/8"	167	256	423
1620					5-3/16"	157	274	431
1760			6'-0" x 6'-8"	pine	3-5/8"	271	400	671
1780					5-3/16"	258	425	683
1800		birch	2'-8" x 6'-8"	pine	3-5/8"	216	256	472
1820					5-3/16"	206	274	480
1960			6'-0" x 6'-8"	pine	3-5/8"	360	400	760
1980					5-3/16"	345	425	770
2000		oak	2'-8" x 6'-8"	oak	3-5/8"	137	218	355
2020					5-3/16"	221	277	498
2160			6'-0" x 6'-8"	oak	3-5/8"	350	405	755
2180					5-3/16"	365	430	795
2200		M.D. overlay	2'-6" x 6'-8"	pine	3-5/8"	238	410	648
2220		on hardboard			5-3/16"	230	465	695
2360			6'-0" x 6'-8"	pine	3-5/8"	420	630	1,050
2380					5-3/16"	405	685	1,090
2400		plastic laminate	2'-6" x 6'-8"	pine	3-5/8"	375	230	605
2420					5'3/16"	365	250	615
2560			6'-0" x 6'-8"	pine	3-5/8"	750	340	1,090
2580					5-3/16"	735	360	1,095

C10 Interior Construction

C1020 Interior Doors

C1020 120 Wood Door/Wood Frame

	TYPE	FACE	SIZE	FRAME	DEPTH	COST EACH		
						MAT.	INST.	TOTAL
3000	Particle core/flush	lauan	2'-8" x 6'-8"	pine	3-5/8"	218	263	481
3020					5-3/16"	207	281	488
3160			6'-0" x 7'-0"	pine	3-5/8"	365	440	805
3180					5-3/16"	350	460	810
3200		birch	2'-8" x 6'-8"	pine	3-5/8"	230	263	493
3220					5-3/16"	220	281	501
3360			6'-0" x 7'-0"	pine	3-5/8"	390	440	830
3380					5-3/16"	380	460	840
3400		oak	2'-8" x 6'-8"	oak	3-5/8"	221	266	487
3420					5-3/16"	236	284	520
3560			6'-0" x 7'-0"	oak	3-5/8"	390	440	830
3580					5-3/16"	410	465	875
3600		M.D. overlay	2'-6" x 6'-8"	pine	3-5/8"	249	420	669
3620		on hardboard			5-3/16"	241	475	716
3760			6'-0" x 7'-0"	pine	3-5/8"	415	680	1,095
3780					5-3/16"	405	735	1,140
3800		plastic laminate	2'-6" x 6'-8"	pine	3-5/8"	295	244	539
3820					5-3/16"	275	255	530
3960			6'-0" x 7'-0"	pine	3-5/8"	550	385	935
3980					5-3/16"	540	410	950
4400	Solid core/panel	solid pine	2'-6" x 6'-8"	pine	3-5/8"	315	420	735
4420					5-3/16"	305	475	780
4560			6'-0" x 6'-8"	pine	3-5/8"	580	665	1,245
4580					5-1/6"	565	720	1,285
5000	Solid core/flush	birch	2'-8" x 6'-8"	pine	3-5/8"	268	263	531
5020					5-3/16"	258	281	539
5160			6'-0" x 7'-0"	pine	3-5/8"	465	440	905
5180					5-3/16"	455	460	915
5200		oak	2'-8" x 6'-8"	oak	3-5/8"	259	266	525
5220					5-3/16"	274	284	558
5360			6'-0" x 7'-0"	oak	3-5/8"	470	440	910
5380					5-3/16"	485	465	950
5400		M.D. overlay	2'-8" x 6'-8"	pine	3-5/8"	293	420	713
5420		on hardboard			5-3/16"	284	475	759
5560			6'-0" x 7'-0"	pine	3-5/8"	505	680	1,185
5600		plastic laminate	2'-6" x 6'-8"	pine	3-5/8"	335	244	579
5620					5-3/16"	325	264	589
5760			6'-0" x 7'-0"	pine	3-5/8"	635	400	1,035

C10 Interior Construction

C1020 Interior Doors

The Wood Door/Metal Frame Systems are defined as follows; door type, face material and size, frame type and depth. Included in the components for each system are door and frame finishing. No hardware has been included in the systems.

System Components	QUANTITY	UNIT	COST EACH		
			MAT.	INST.	TOTAL
SYSTEM C1020 122 1600					
WOOD DOOR, HOLLOW, LAUAN FACE, 2'-6"X6'-8" FRAME, K.D., 4-7/8" DEEP					
Door, flush, hollow core, lauan face, 1-3/4" thick, 2'-6" x 6'-8"	1.000	Ea.	40	58.50	98.50
Frame, metal, drywall, 16 ga., 4-7/8" x 2'-6" x 6'-8"	1.000	Ea.	122	66	188
Varnish on wood door, clear	35.000	S.F.	2.80	37.10	39.90
Paint metal frame	16.000	L.F.	3.20	33.92	37.12
TOTAL			168	195.52	363.52

C1020 122 — Wood Door/Metal Frame

	TYPE	FACE	SIZE	FRAME	DEPTH	MAT.	INST.	TOTAL
1000	Hollow core/panel	moulded fiberboard	2'-6" x 6'-8"	drywall K.D.	4-7/8"	201	350	551
1020				butt welded	8-3/4"	315	395	710
1160			6'-0" x 7'-0"	drywall K.D.	4-7/8"	345	560	905
1180				butt welded	8-3/4"	475	605	1,080
1600	Hollow core/flush	lauan	2'-8" x 6'-8"	drywall K.D.	4-7/8"	168	196	364
1620				butt welded	8-3/4"	278	204	482
1760			6'-0" x 7'-0"	drywall K.D.	4-7/8"	300	330	630
1780				butt welded	8-3/4"	430	340	770
1800		birch	2'-8" x 6'-8"	drywall K.D.	4-7/8"	217	196	413
1820				butt welded	8-3/4"	325	204	529
1960			6'-0" x 7'-0"	drywall K.D.	4-7/8"	390	330	720
1980				butt welded	8-3/4"	515	340	855
2000		oak	2'-8" x 6'-8"	drywall K.D.	4-7/8"	227	196	423
2020				butt welded	8-3/4"	335	204	539
2160			6'-0" x 7'-0"	drywall K.D.	4-7/8"	405	330	735
2180				butt welded	8-3/4"	530	340	870
2200		M.D. overlay on hardboard	2'-6" x 6'-8"	drywall K.D.	4-7/8"	238	350	588
2220				butt welded	8-3/4"	350	395	745
2360			6'-0" x 7'-0"	drywall K.D.	4-7/8"	445	560	1,005
2380				butt welded	8-3/4"	575	605	1,180
2400		plastic laminate	2'-6" x 6'-8"	drywall K.D.	4-7/8"	370	162	532
2420				butt welded	8-3/4"	480	170	650
2560			6'-0" x 7'-0"	drywall K.D.	4-7/8"	775	257	1,032
2580				butt welded	8-3/4"	905	268	1,173

C10 Interior Construction

C1020 Interior Doors

C1020 122 Wood Door/Metal Frame

	TYPE	FACE	SIZE	FRAME	DEPTH	COST EACH		
						MAT.	INST.	TOTAL
3000	Particle core/flush	lauan	2'-8" x 6'-8"	drywall K.D.	4-7/8"	219	203	422
3020				butt welded	8-3/4"	330	212	542
3160			6'-0" x 7'-0"	drywall K.D.	4-7/8"	395	365	760
3180				butt welded	8-3/4"	525	380	905
3200		birch	2'-8" x 6'-8"	drywall K.D.	4-7/8"	231	203	434
3220				butt welded	8-3/4"	340	212	552
3360			6'-0" x 7'-0"	drywall K.D.	4-7/8"	420	365	785
3380				butt welded	8-3/4"	550	380	930
3400		oak	2'-8" x 6'-8"	drywall K.D.	4-7/8"	242	203	445
3420				butt welded	8-3/4"	350	212	562
3560			6'-0" x 7'-0"	drywall K.D.	4-7/8"	445	365	810
3580				butt welded	8-3/4"	575	380	955
3600		M.D. overlay	2'-6" x 6'-8"	drywall K.D.	4-7/8"	249	360	609
3620		on hardboard		butt welded	8-3/4"	360	405	765
3760			6'-0" x 7'-0"	drywall K.D.	4-7/8"	445	610	1,055
3780				butt welded	8-3/4"	575	655	1,230
3800		plastic laminate	2'-6" x 6'-8"	drywall K.D.	4-7/8"	292	176	468
3820				butt welded	8-3/4"	405	185	590
3960			6'-0" x 7'-0"	drywall K.D.	4-7/8"	575	305	880
3980				butt welded	8-3/4"	705	315	1,020
4400	Solid core/panel	solid pine	2'-6" x 6'-8"	drywall K.D.	4-7/8"	315	360	675
4420				butt welded	8-3/4"	425	405	830
4560			6'-0" x 7'-0"	drywall K.D.	4-7/8"	610	595	1,205
4580				butt welded	8-3/4"	730	460	1,190
5000	Solid core/flush	birch	2'-8" x 6'-8"	drywall K.D.	4-7/8"	269	203	472
5020				butt welded	8-3/4"	380	212	592
5160			6'-0" x 7'-0"	drywall K.D.	4-7/8"	495	365	860
5180				butt welded	8-3/4"	625	380	1,005
5200		oak	2'-8" x 6'-8"	drywall K.D.	4-7/8"	280	203	483
5220				butt welded	8-3/4"	390	212	602
5360			6'-0" x 7'-0"	drywall K.D.	4-7/8"	525	365	890
5380				butt welded	8-3/4"	650	380	1,030
5400		M.D. overlay	2'-6" x 6'-8"	drywall K.D.	4-7/8"	293	360	653
5420		on hardboard		butt welded	8-3/4"	405	405	810
5560			6'-0" x 7'-0"	drywall K.D.	4-7/8"	530	610	1,140
5580				butt welded	8-3/4"	660	655	1,315
5600		plastic laminate	2'-6" x 6'-8"	drywall K.D.	4-7/8"	335	176	511
5620				butt welded	8-3/4"	445	185	630
5760			6'-0" x 7'-0"	drywall K.D.	4-7/8"	655	305	960
5780				butt welded	8-3/4"	785	315	1,100

C1020 122 Labeled Wood Door/Metal Frame

	TYPE	FACE	SIZE	FRAME	DEPTH	COST EACH		
						MAT.	INST.	TOTAL
6000	1 hr/flush	birch	2'-8" x 6'-8"	drywall K.D.	4-7/8"	545	208	753
6020				butt welded	8-3/4"	650	217	867
6160			6'-0" x 7'-0"	drywall K.D.	4-7/8"	1,000	380	1,380
6180				butt welded	8-3/4"	1,100	390	1,490
6200		oak	2'-8" x 6'-8"	drywall K.D.	4-7/8"	540	208	748
6220				butt welded	8-3/4"	645	217	862
6360			6'-0" x 7'-0"	drywall K.D.	4-7/8"	980	380	1,360
6380				butt welded	8-3/4"	1,075	390	1,465

C10 Interior Construction

C1020 Interior Doors

C1020 122 Labeled Wood Door/Metal Frame

	TYPE	FACE	SIZE	FRAME	DEPTH	COST EACH		
						MAT.	INST.	TOTAL
6400	1 hr/flush	walnut	2'-8" x 6'-8"	drywall K.D.	4-7/8"	645	208	853
6420				butt welded	8-3/4"	750	217	967
6560			6'-0" x 7'-0"	drywall K.D.	4-7/8"	1,200	380	1,580
6580				butt welded	8-3/4"	1,300	390	1,690
6800		M.D. overlay	2'-6" x 6'-8"	drywall K.D.	4-7/8"	500	355	855
6820		on hardboard		butt welded	8-3/4"	610	400	1,010
6960			6'-0" x 7'-0"	drywall K.D.	4-7/8"	580	520	1,100
6980				butt welded	8-3/4"	675	565	1,240
7200		plastic laminate	2'-6" x 6'-8"	drywall K.D.	4-7/8"	595	176	771
7220				butt welded	8-3/4"	705	185	890
7360			6'-0" x 7'-0"	drywall K.D.	4-7/8"	1,100	305	1,405
7380				butt welded	8-3/4"	1,200	315	1,515
7600	1-1/2 hr/flush	birch	2'-8" x 6'-8"	drywall K.D.	4-7/8"	535	208	743
7620				butt welded	8-3/4"	640	217	857
7760			6'-0" x 7'-0"	drywall K.D.	4-7/8"	990	380	1,370
7780				butt welded	8-3/4"	1,075	390	1,465
7800		oak	2'-8" x 6'-8"	drywall K.D.	4-7/8"	505	208	713
7820				butt welded	8-3/4"	610	217	827
7960			6'-0" x 7'-0"	drywall K.D.	4-7/8"	910	380	1,290
7980				butt welded	8-3/4"	1,000	390	1,390
8000		walnut	2'-8" x 6'-8"	drywall K.D.	4-7/8"	620	208	828
8020				butt welded	8-3/4"	725	217	942
8160			6'-0" x 7'-0"	drywall K. D.	4 7/8"	1,150	380	1,530
8180				butt welded	8-3/4"	1,250	390	1,640
8400		M.D. overlay	2'-6" x 6'-8"	drywall K.D.	4-7/8"	525	157	682
8420		on hardboard		butt welded	8-3/4"	635	162	797
8560			6'-0" x 7'-0"	drywall K.D.	4-7/8"	950	276	1,226
8580				butt welded	8-3/4"	1,050	281	1,331
8800		plastic laminate	2'-6" x 6'-8"	drywall K.D.	4-7/8"	256	100	356
8820				butt welded	8-3/4"	365	108	473
8960			6'-0" x 7'-0"	drywall K.D.	4-7/8"	595	207	802
8980				butt welded	8-3/4"	460	136	596

C10 Interior Construction

C1030 Fittings

Toilet Units

Entrance Screens

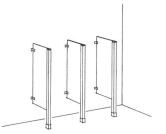
Urinal Screens

C1030 110	Toilet Partitions	MAT.	INST.	TOTAL
0380	Toilet partitions, cubicles, ceiling hung, marble	1,825	470	2,295
0400	Painted metal	515	248	763
0420	Plastic laminate	665	248	913
0460	Stainless steel	1,950	248	2,198
0480	Handicap addition	430		430
0520	Floor and ceiling anchored, marble	2,000	375	2,375
0540	Painted metal	535	198	733
0560	Plastic laminate	675	198	873
0600	Stainless steel	1,850	198	2,048
0620	Handicap addition	305		305
0660	Floor mounted marble	1,175	315	1,490
0680	Painted metal	595	142	737
0700	Plastic laminate	600	142	742
0740	Stainless steel	1,675	142	1,817
0760	Handicap addition	300		300
0780	Juvenile deduction	43.50		43.50
0820	Floor mounted with handrail marble	1,125	315	1,440
0840	Painted metal	485	165	650
0860	Plastic laminate	635	165	800
0900	Stainless steel	1,250	165	1,415
0920	Handicap addition	365		365
0960	Wall hung, painted metal	655	142	797
1020	Stainless steel	1,725	142	1,867
1040	Handicap addition	365		365
1080	Entrance screens, floor mounted, 54" high, marble	745	104	849
1100	Painted metal	256	66	322
1140	Stainless steel	935	66	1,001
1300	Urinal screens, floor mounted, 24" wide, laminated plastic	400	124	524
1320	Marble	630	147	777
1340	Painted metal	219	124	343
1380	Stainless steel	615	124	739
1428	Wall mounted wedge type, painted metal	141	99	240
1460	Stainless steel	595	99	694

C20 Stairs

C2010 Stair Construction

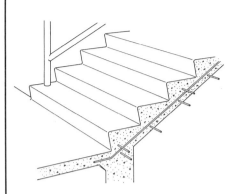

General Design: See reference section for code requirements. Maximum height between landings is 12'; usual stair angle is 20° to 50° with 30° to 35° best. Usual relation of riser to treads is:
- Riser + tread = 17.5.
- 2x (Riser) + tread = 25.
- Riser x tread = 70 or 75.

Maximum riser height is 7" for commercial, 8-1/4" for residential.
Usual riser height is 6-1/2" to 7-1/4".

Minimum tread width is 11" for commercial and 9" for residential.

For additional information please see reference section.

Cost Per Flight: Table below lists the cost per flight for 4'-0" wide stairs.
Side walls are not included.
Railings are included.

System Components	QUANTITY	UNIT	COST PER FLIGHT		
			MAT.	INST.	TOTAL
SYSTEM C2010 110 0560					
STAIRS, C.I.P. CONCRETE WITH LANDING, 12 RISERS					
Concrete in place, free standing stairs not incl. safety treads	48.000	L.F.	328.80	1,720.80	2,049.60
Concrete in place, free standing stair landing	32.000	S.F.	184	473.28	657.28
Stair tread C.I. abrasive 4" wide	48.000	L.F.	748.80	297.60	1,046.40
Industrial railing, welded, 2 rail 3'-6" high 1-1/2" pipe	18.000	L.F.	612	192.24	804.24
Wall railing with returns, steel pipe	17.000	L.F.	326.40	181.56	507.96
TOTAL			2,200	2,865.48	5,065.48

C2010 110	Stairs	COST PER FLIGHT		
		MAT.	INST.	TOTAL
0470	Stairs, C.I.P. concrete, w/o landing, 12 risers, w/o nosing	1,275	2,075	3,350
0480	With nosing	2,025	2,375	4,400
0550	W/landing, 12 risers, w/o nosing	1,450	2,575	4,025
0560	With nosing	2,200	2,875	5,075
0570	16 risers, w/o nosing	1,775	3,225	5,000
0580	With nosing	2,775	3,625	6,400
0590	20 risers, w/o nosing	2,100	3,875	5,975
0600	With nosing	3,350	4,375	7,725
0610	24 risers, w/o nosing	2,425	4,550	6,975
0620	With nosing	3,925	5,150	9,075
0630	Steel, grate type w/nosing & rails, 12 risers, w/o landing	6,425	1,075	7,500
0640	With landing	8,400	1,500	9,900
0660	16 risers, with landing	10,500	1,850	12,350
0680	20 risers, with landing	12,700	2,225	14,925
0700	24 risers, with landing	14,800	2,600	17,400
0710	Cement fill metal pan & picket rail, 12 risers, w/o landing	8,400	1,075	9,475
0720	With landing	11,000	1,650	12,650
0740	16 risers, with landing	13,800	2,000	15,800
0760	20 risers, with landing	16,600	2,350	18,950
0780	24 risers, with landing	19,400	2,725	22,125
0790	Cast iron tread & pipe rail, 12 risers, w/o landing	8,400	1,075	9,475
0800	With landing	11,000	1,650	12,650
0820	16 risers, with landing	13,800	2,000	15,800
0840	20 risers, with landing	16,600	2,350	18,950
0860	24 risers, with landing	19,400	2,725	22,125
0870	Pan tread & flat bar rail, pre-assembled, 12 risers, w/o landing	6,900	840	7,740
0880	With landing	12,500	1,300	13,800
0900	16 risers, with landing	13,800	1,475	15,275
0920	20 risers, with landing	16,100	1,725	17,825
0940	24 risers, with landing	18,400	2,025	20,425

C20 Stairs

C2010 Stair Construction

C2010 110	Stairs	COST PER FLIGHT		
		MAT.	INST.	TOTAL
0950	Spiral steel, industrial checkered plate 4'-6' dia., 12 risers	7,975	730	8,705
0960	16 risers	10,600	975	11,575
0970	20 risers	13,300	1,225	14,525
0980	24 risers	16,000	1,450	17,450
0990	Spiral steel, industrial checkered plate 6'-0" dia., 12 risers	9,725	820	10,545
1000	16 risers	13,000	1,075	14,075
1010	20 risers	16,200	1,375	17,575
1020	24 risers	19,400	1,650	21,050
1030	Aluminum, spiral, stock units, 5'-0" dia., 12 risers	9,000	730	9,730
1040	16 risers	12,000	975	12,975
1050	20 risers	15,000	1,225	16,225
1060	24 risers	18,000	1,450	19,450
1070	Custom 5'-0" dia., 12 risers	17,100	730	17,830
1080	16 risers	22,800	975	23,775
1090	20 risers	28,500	1,225	29,725
1100	24 risers	34,200	1,450	35,650
1120	Wood, prefab box type, oak treads, wood rails 3'-6" wide, 14 risers	2,100	415	2,515
1150	Prefab basement type, oak treads, wood rails 3'-0" wide, 14 risers	1,050	102	1,152

C30 Interior Finishes

C3010 Wall Finishes

C3010 230	Paint & Covering	COST PER S.F.		
		MAT.	INST.	TOTAL
0060	Painting, interior on plaster and drywall, brushwork, primer & 1 coat	.11	.65	.76
0080	Primer & 2 coats	.17	.86	1.03
0100	Primer & 3 coats	.24	1.06	1.30
0120	Walls & ceilings, roller work, primer & 1 coat	.11	.43	.54
0140	Primer & 2 coats	.17	.56	.73
0160	Woodwork incl. puttying, brushwork, primer & 1 coat	.11	.94	1.05
0180	Primer & 2 coats	.17	1.24	1.41
0200	Primer & 3 coats	.24	1.69	1.93
0260	Cabinets and casework, enamel, primer & 1 coat	.12	1.06	1.18
0280	Primer & 2 coats	.18	1.30	1.48
0300	Masonry or concrete, latex, brushwork, primer & 1 coat	.23	.88	1.11
0320	Primer & 2 coats	.30	1.26	1.56
0340	Addition for block filler	.15	1.08	1.23
0380	Fireproof paints, intumescent, 1/8" thick 3/4 hour	2.14	.86	3
0400	3/16" thick 1 hour	4.78	1.30	6.08
0420	7/16" thick 2 hour	6.15	3.01	9.16
0440	1-1/16" thick 3 hour	10.05	6.05	16.10
0480	Miscellaneous metal brushwork, exposed metal, primer & 1 coat	.10	1.06	1.16
0500	Gratings, primer & 1 coat	.26	1.32	1.58
0600	Pipes over 12" diameter	.49	4.22	4.71
0700	Structural steel, brushwork, light framing 300-500 S.F./Ton	.08	1.63	1.71
0720	Heavy framing 50-100 S.F./Ton	.08	.82	.90
0740	Spraywork, light framing 300-500 S.F./Ton	.09	.36	.45
0760	Heavy framing 50-100 S.F./Ton	.09	.40	.49
0800	Varnish, interior wood trim, no sanding sealer & 1 coat	.08	1.06	1.14
0820	Hardwood floor, no sanding 2 coats	.17	.22	.39
0840	Wall coatings, acrylic glazed coatings, minimum	.31	.80	1.11
0860	Maximum	.66	1.38	2.04
0880	Epoxy coatings, minimum	.41	.80	1.21
0900	Maximum	1.23	2.48	3.71
0940	Exposed epoxy aggregate, troweled on, 1/16" to 1/4" aggregate, minimum	.62	1.80	2.42
0960	Maximum	1.32	3.25	4.57
0980	1/2" to 5/8" aggregate, minimum	1.22	3.25	4.47
1000	Maximum	2.09	5.30	7.39
1020	1" aggregate, minimum	2.12	4.69	6.81
1040	Maximum	3.25	7.65	10.90
1060	Sprayed on, minimum	.56	1.43	1.99
1080	Maximum	1.06	2.91	3.97
1100	High build epoxy 50 mil, minimum	.68	1.08	1.76
1120	Maximum	1.17	4.44	5.61
1140	Laminated epoxy with fiberglass minimum	.74	1.43	2.17
1160	Maximum	1.31	2.91	4.22
1180	Sprayed perlite or vermiculite 1/16" thick, minimum	.26	.14	.40
1200	Maximum	.75	.66	1.41
1260	Wall coatings, vinyl plastic, minimum	.33	.57	.90
1280	Maximum	.83	1.76	2.59
1300	Urethane on smooth surface, 2 coats, minimum	.26	.37	.63
1320	Maximum	.58	.63	1.21
1340	3 coats, minimum	.35	.50	.85
1360	Maximum	.79	.90	1.69
1380	Ceramic-like glazed coating, cementitious, minimum	.48	.96	1.44
1400	Maximum	.81	1.22	2.03
1420	Resin base, minimum	.33	.66	.99
1440	Maximum	.54	1.28	1.82
1460	Wall coverings, aluminum foil	1.07	1.54	2.61
1480	Copper sheets, .025" thick, phenolic backing	7.40	1.77	9.17
1500	Vinyl backing	5.70	1.77	7.47
1520	Cork tiles, 12"x12", light or dark, 3/16" thick	4.68	1.77	6.45

C30 Interior Finishes

C3010 Wall Finishes

C3010 230	Paint & Covering	COST PER S.F.		
		MAT.	INST.	TOTAL
1540	5/16" thick	3.98	1.81	5.79
1560	Basketweave, 1/4" thick	6.15	1.77	7.92
1580	Natural, non-directional, 1/2" thick	7.35	1.77	9.12
1600	12"x36", granular, 3/16" thick	1.32	1.10	2.42
1620	1" thick	1.71	1.15	2.86
1640	12"x12", polyurethane coated, 3/16" thick	4.14	1.77	5.91
1660	5/16" thick	5.90	1.81	7.71
1661	Paneling, prefinished plywood, birch	1.49	2.36	3.85
1662	Mahogany, African	2.76	2.48	5.24
1664	Oak or cherry	3.54	2.48	6.02
1665	Rosewood	5.05	3.10	8.15
1666	Teak	3.54	2.48	6.02
1667	Chestnut	5.25	2.64	7.89
1668	Pecan	2.27	2.48	4.75
1669	Walnut	5.75	2.48	8.23
1670	Wood board, knotty pine, finished	1.89	4.10	5.99
1671	Rough sawn cedar	2.35	4.10	6.45
1672	Redwood	5.05	4.10	9.15
1673	Aromatic cedar	3.96	4.40	8.36
1680	Cork wallpaper, paper backed, natural	2.09	.89	2.98
1700	Color	2.92	.89	3.81
1720	Gypsum based, fabric backed, minimum	.88	.53	1.41
1740	Average	1.29	.59	1.88
1760	Maximum	1.43	.66	2.09
1780	Vinyl wall covering, fabric back, light weight	.72	.66	1.38
1800	Medium weight	.85	.89	1.74
1820	Heavy weight	1.75	.98	2.73
1840	Wall paper, double roll, solid pattern, avg. workmanship	.39	.66	1.05
1860	Basic pattern, avg. workmanship	.69	.79	1.48
1880	Basic pattern, quality workmanship	2.59	.98	3.57
1900	Grass cloths with lining paper, minimum	.81	1.06	1.87
1920	Maximum	2.61	1.21	3.82
1940	Ceramic tile, thin set, 4-1/4" x 4-1/4"	2.74	4.20	6.94
1960	12" x 12"	3.98	2.66	6.64

C3010 235	Paint Trim	COST PER L.F.		
		MAT.	INST.	TOTAL
2040	Painting, wood trim, to 6" wide, enamel, primer & 1 coat	.12	.53	.65
2060	Primer & 2 coats	.18	.67	.85
2080	Misc. metal brushwork, ladders	.52	5.30	5.82
2100	Pipes, to 4" dia.	.06	1.11	1.17
2120	6" to 8" dia.	.13	2.22	2.35
2140	10" to 12" dia.	.38	3.32	3.70
2160	Railings, 2" pipe	.18	2.64	2.82
2180	Handrail, single	.10	1.06	1.16
2185	Caulking & Sealants, Polyureathane, In place, 1 or 2 component, 1/2" X 1/4"	.39	3.39	3.78

C30 Interior Finishes

C3020 Floor Finishes

C3020 410	Tile & Covering	COST PER S.F.		
		MAT.	INST.	TOTAL
0060	Carpet tile, nylon, fusion bonded, 18" x 18" or 24" x 24", 24 oz.	3.55	.62	4.17
0080	35 oz.	4.16	.62	4.78
0100	42 oz.	5.55	.62	6.17
0140	Carpet, tufted, nylon, roll goods, 12' wide, 26 oz.	5.10	.71	5.81
0160	36 oz.	7.85	.71	8.56
0180	Woven, wool, 36 oz.	13.30	.71	14.01
0200	42 oz.	13.65	.71	14.36
0220	Padding, add to above, minimum	.56	.33	.89
0240	Maximum	.95	.33	1.28
0260	Composition flooring, acrylic, 1/4" thick	1.60	4.79	6.39
0280	3/8" thick	2.04	5.55	7.59
0300	Epoxy, minimum	2.74	3.68	6.42
0320	Maximum	3.32	5.10	8.42
0340	Epoxy terrazzo, minimum	5.80	5.40	11.20
0360	Maximum	8.50	7.20	15.70
0380	Mastic, hot laid, 1-1/2" thick, minimum	4.02	3.61	7.63
0400	Maximum	5.15	4.79	9.94
0420	Neoprene 1/4" thick, minimum	3.95	4.56	8.51
0440	Maximum	5.45	5.80	11.25
0460	Polyacrylate with ground granite 1/4", minimum	3.28	3.39	6.67
0480	Maximum	5.65	5.15	10.80
0500	Polyester with colored quart 2 chips 1/16", minimum	2.94	2.33	5.27
0520	Maximum	4.64	3.68	8.32
0540	Polyurethane with vinyl chips, minimum	7.70	2.33	10.03
0560	Maximum	11.10	2.90	14
0600	Concrete topping, granolithic concrete, 1/2" thick	.32	3.96	4.28
0620	1" thick	.63	4.07	4.70
0640	2" thick	1.26	4.68	5.94
0660	Heavy duty 3/4" thick, minimum	.43	6.15	6.58
0680	Maximum	.80	7.35	8.15
0700	For colors, add to above, minimum	.47	1.42	1.89
0720	Maximum	.78	1.56	2.34
0740	Exposed aggregate finish, minimum	.25	.72	.97
0760	Maximum	.73	.96	1.69
0780	Abrasives, .25 P.S.F. add to above, minimum	.48	.53	1.01
0800	Maximum	.67	.53	1.20
0820	Dust on coloring, add, minimum	.47	.34	.81
0840	Maximum	.78	.72	1.50
0860	Floor coloring using 0.6 psf powdered color, 1/2" integral, minimum	5.35	3.96	9.31
0880	Maximum	5.65	3.96	9.61
0900	Dustproofing, add, minimum	.19	.24	.43
0920	Maximum	.68	.34	1.02
0930	Paint	.30	1.26	1.56
0940	Hardeners, metallic add, minimum	.56	.53	1.09
0960	Maximum	1.67	.78	2.45
0980	Non-metallic, minimum	.25	.53	.78
1000	Maximum	.75	.78	1.53
1020	Integral topping and finish, 1:1:2 mix, 3/16" thick	.11	2.33	2.44
1040	1/2" thick	.28	2.46	2.74
1060	3/4" thick	.43	2.75	3.18
1080	1" thick	.57	3.11	3.68
1100	Terrazzo, minimum	3.16	14.40	17.56
1120	Maximum	5.90	18.05	23.95
1280	Resilient, asphalt tile, 1/8" thick on concrete, minimum	1.24	1.12	2.36
1300	Maximum	1.36	1.12	2.48
1320	On wood, add for felt underlay	.22		.22
1340	Cork tile, minimum	6.35	1.42	7.77
1360	Maximum	13.65	1.42	15.07

C30 Interior Finishes

C3020 Floor Finishes

C3020 410	Tile & Covering	COST PER S.F.		
		MAT.	INST.	TOTAL
1380	Polyethylene, in rolls, minimum	3.36	1.62	4.98
1400	Maximum	6.35	1.62	7.97
1420	Polyurethane, thermoset, minimum	5.05	4.46	9.51
1440	Maximum	6.05	8.95	15
1460	Rubber, sheet goods, minimum	6.90	3.72	10.62
1480	Maximum	11	4.96	15.96
1500	Tile, minimum	6.80	1.12	7.92
1520	Maximum	12.65	1.62	14.27
1540	Synthetic turf, minimum	4.95	2.13	7.08
1560	Maximum	12.40	2.35	14.75
1580	Vinyl, composition tile, minimum	.96	.89	1.85
1600	Maximum	2.35	.89	3.24
1620	Vinyl tile, minimum	3.25	.89	4.14
1640	Maximum	6.45	.89	7.34
1660	Sheet goods, minimum	3.85	1.79	5.64
1680	Maximum	6.55	2.23	8.78
1720	Tile, ceramic natural clay	5.15	4.36	9.51
1730	Marble, synthetic 12"x12"x5/8"	6.65	13.30	19.95
1740	Porcelain type, minimum	5.20	4.36	9.56
1760	Maximum	7.15	5.15	12.30
1800	Quarry tile, mud set, minimum	7.75	5.70	13.45
1820	Maximum	10	7.25	17.25
1840	Thin set, deduct		1.14	1.14
1860	Terrazzo precast, minimum	4.57	5.95	10.52
1880	Maximum	9.85	5.95	15.80
1900	Non-slip, minimum	19.45	14.70	34.15
1920	Maximum	21.50	19.60	41.10
2020	Wood block, end grain factory type, natural finish, 2" thick	5.10	3.96	9.06
2040	Fir, vertical grain, 1"x4", no finish, minimum	2.84	1.94	4.78
2060	Maximum	3.01	1.94	4.95
2080	Prefinished white oak, prime grade, 2-1/4" wide	7.05	2.92	9.97
2100	3-1/4" wide	9	2.68	11.68
2120	Maple strip, sanded and finished, minimum	4.51	4.25	8.76
2140	Maximum	5.45	4.25	9.70
2160	Oak strip, sanded and finished, minimum	4.36	4.25	8.61
2180	Maximum	5.25	4.25	9.50
2200	Parquetry, sanded and finished, minimum	5.70	4.43	10.13
2220	Maximum	6.30	6.30	12.60
2260	Add for sleepers on concrete, treated, 24" O.C., 1"x2"	2.50	2.55	5.05
2280	1"x3"	2.28	1.98	4.26
2300	2"x4"	.73	1	1.73
2320	2"x6"	.76	.76	1.52
2340	Underlayment, plywood, 3/8" thick	1.13	.66	1.79
2350	1/2" thick	1.10	.68	1.78
2360	5/8" thick	1.42	.71	2.13
2370	3/4" thick	1.46	.76	2.22
2380	Particle board, 3/8" thick	.45	.66	1.11
2390	1/2" thick	.50	.68	1.18
2400	5/8" thick	.59	.71	1.30
2410	3/4" thick	.72	.76	1.48
2420	Hardboard, 4' x 4', .215" thick	.55	.66	1.21

C30 Interior Finishes

C3030 Ceiling Finishes

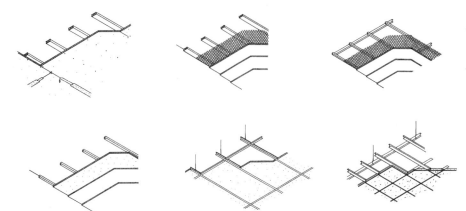

The Plaster and Acoustical Ceilings Systems are defined as follows: type, lath in the case of plaster, furring and support. Furring and support may consist of a grid and be listed either by its separate members or as a composite. Finish is included in the components of the plaster and drywall ceilings.

System Components	QUANTITY	UNIT	COST PER S.F. MAT.	COST PER S.F. INST.	COST PER S.F. TOTAL
SYSTEM C3030 105 2400					
GYPSUM PLASTER, 2 COATS, 3/8"GYP. LATH, WOOD FURRING, FRAMING					
Gypsum plaster 2 coats no lath, on ceilings	.110	S.Y.	.46	2.67	3.13
Gypsum lath plain/perforated nailed, 3/8" thick	.110	S.Y.	.67	.55	1.22
Add for ceiling installation	.110	S.Y.		.22	.22
Furring, 1" x 3" wood strips on wood joists	.750	L.F.	.29	1.07	1.36
Paint, primer and one coat	1.000	S.F.	.11	.43	.54
TOTAL			1.53	4.94	6.47

C3030 105 Plaster Ceilings

	TYPE	LATH	FURRING	SUPPORT		MAT.	INST.	TOTAL
2400	2 coat gypsum	3/8" gypsum	1"x3" wood, 16" O.C.	wood		1.53	4.94	6.47
2500	Painted			masonry		1.53	5.05	6.58
2600				concrete		1.53	5.60	7.13
2700	3 coat gypsum	3.4# metal	1"x3" wood, 16" O.C.	wood		1.62	5.35	6.97
2800	Painted			masonry		1.62	5.45	7.07
2900				concrete		1.62	6.05	7.67
3000	2 coat perlite	3/8" gypsum	1"x3" wood, 16" O.C.	wood		1.57	5.20	6.77
3100	Painted			masonry		1.57	5.30	6.87
3200				concrete		1.57	5.90	7.47
3300	3 coat perlite	3.4# metal	1"x3" wood, 16" O.C.	wood		1.46	5.30	6.76
3400	Painted			masonry		1.46	5.40	6.86
3500				concrete		1.46	6	7.46
3600	2 coat gypsum	3/8" gypsum	3/4" CRC, 12" O.C.	1-1/2" CRC, 48"O.C.		1.62	5.85	7.47
3700	Painted		3/4" CRC, 16" O.C.	1-1/2" CRC, 48"O.C.		1.58	5.30	6.88
3800			3/4" CRC, 24" O.C.	1 1/2" CRC, 48"O.C.		1.47	4.87	6.34
3900	2 coat perlite	3/8" gypsum	3/4" CRC, 12" O.C.	1-1/2" CRC, 48"O.C		1.66	6.35	8.01
4000	Painted		3/4" CRC, 16" O.C.	1-1/2" CRC, 48"O.C.		1.62	5.80	7.42
4100			3/4" CRC, 24" O.C.	1-1/2" CRC, 48"O.C.		1.51	5.35	6.86
4200	3 coat gypsum	3.4# metal	3/4" CRC, 12" O.C.	1-1/2" CRC, 36" O.C.		2.02	7.80	9.82
4300	Painted		3/4" CRC, 16" O.C.	1-1/2" CRC, 36" O.C.		1.98	7.25	9.23
4400			3/4" CRC, 24" O.C.	1-1/2" CRC, 36" O.C.		1.87	6.80	8.67

C30 Interior Finishes

C3030 Ceiling Finishes

C3030 105 — Plaster Ceilings

	TYPE	LATH	FURRING	SUPPORT		COST PER S.F.		
						MAT.	INST.	TOTAL
4500	3 coat perlite	3.4# metal	3/4" CRC, 12" O.C.	1-1/2" CRC, 36" O.C.		2.13	8.50	10.63
4600	Painted		3/4" CRC, 16" O.C.	1-1/2" CRC, 36" O.C.		2.09	7.95	10.04
4700			3/4" CRC, 24" O.C.	1-1/2" CRC, 36" O.C.		1.98	7.50	9.48

C3030 110 — Drywall Ceilings

	TYPE	FINISH	FURRING	SUPPORT		COST PER S.F.		
						MAT.	INST.	TOTAL
4800	1/2" F.R. drywall	painted and textured	1"x3" wood, 16" O.C.	wood		.89	3.06	3.95
4900				masonry		.89	3.15	4.04
5000				concrete		.89	3.76	4.65
5100	5/8" F.R. drywall	painted and textured	1"x3" wood, 16" O.C.	wood		.93	3.06	3.99
5200				masonry		.93	3.15	4.08
5300				concrete		.93	3.76	4.69
5400	1/2" F.R. drywall	painted and textured	7/8" resil. channels	24" O.C.		.75	2.96	3.71
5500			1"x2" wood	stud clips		.93	2.85	3.78
5600			1-5/8" metal studs	24" O.C.		.85	2.64	3.49
5700	5/8" F.R. drywall	painted and textured	1-5/8" metal studs	24" O.C.		.89	2.64	3.53
5702								

C3030 210 — Acoustical Ceilings

	TYPE	TILE	GRID	SUPPORT		COST PER S.F.		
						MAT.	INST.	TOTAL
5800	5/8" fiberglass board	24" x 48"	tee	suspended		1.72	1.48	3.20
5900		24" x 24"	tee	suspended		1.89	1.62	3.51
6000	3/4" fiberglass board	24" x 48"	tee	suspended		2.78	1.52	4.30
6100		24" x 24"	tee	suspended		2.95	1.66	4.61
6500	5/8" mineral fiber	12" x 12"	1"x3" wood, 12" O.C.	wood		2.42	3.07	5.49
6600				masonry		2.42	3.20	5.62
6700				concrete		2.42	4.01	6.43
6800	3/4" mineral fiber	12" x 12"	1"x3" wood, 12" O.C.	wood		2.44	3.07	5.51
6900				masonry		2.44	3.07	5.51
7000				concrete		2.44	3.07	5.51
7100	3/4" mineral fiber on	12" x 12"	25 ga. channels	runners		2.98	3.77	6.75
7102	5/8" F.R. drywall							
7200	5/8" plastic coated	12" x 12"		adhesive backed		2.45	1.65	4.10
7201	Mineral fiber							
7202								
7300	3/4" plastic coated	12" x 12"		adhesive backed		2.47	1.65	4.12
7301	Mineral fiber							
7302								
7400	3/4" mineral fiber	12" x 12"	conceal 2" bar &	suspended		2.66	3.72	6.38
7401			channels					
7402								

D Services

No part of this publication may be reproduced, stored in a retrieval system, or transmitted in any form or by any means without prior written permission of R.S. Means Company, Inc.

D10 Conveying

D1010 Elevators and Lifts

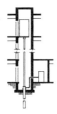

The hydraulic elevator obtains its motion from the movement of liquid under pressure in the piston connected to the car bottom. These pistons can provide travel to a maximum rise of 70′ and are sized for the intended load. As the rise reaches the upper limits the cost tends to exceed that of a geared electric unit.

System Components	QUANTITY	UNIT	COST EACH		
			MAT.	INST.	TOTAL
SYSTEM D1010 110 2000					
PASS. ELEV. HYDRAULIC 2500 LB. 5 FLOORS, 100 FPM					
Passenger elevator, hydraulic, 1500 lb capacity, 2 stop, 100 FPM	1.000	Ea.	43,800	13,500	57,300
Over 10′ travel height, passenger elevator, hydraulic, add	40.000	V.L.F.	39,200	7,440	46,640
Passenger elevator, hydraulic, 2500 lb. capacity over standard, add	1.000	Ea.	3,550		3,550
Over 2 stops, passenger elevator, hydraulic, add	3.000	Stop	8,475	15,000	23,475
Hall lantern	5.000	Ea.	2,975	845	3,820
Maintenance agreement for pass. elev. 9 months	1.000	Ea.	4,000		4,000
Position indicator at lobby	1.000	Ea.	685	42	727
TOTAL			102,685	36,827	139,512

D1010 110	Hydraulic		COST EACH		
			MAT.	INST.	TOTAL
1300	Pass. elev., 1500 lb., 2 Floors, 100 FPM	R142000 -10	49,000	13,800	62,800
1400	5 Floors, 100 FPM		99,000	36,800	135,800
1600	2000 lb., 2 Floors, 100 FPM	R142000 -20	51,000	13,800	64,800
1700	5 floors, 100 FPM		101,500	36,800	138,300
1900	2500 lb., 2 Floors, 100 FPM	R142000 -30	52,500	13,800	66,300
2000	5 floors, 100 FPM		102,500	36,800	139,300
2200	3000 lb., 2 Floors, 100 FPM	R142000 -40	53,500	13,800	67,300
2300	5 floors, 100 FPM		104,000	36,800	140,800
2500	3500 lb., 2 Floors, 100 FPM		56,000	13,800	69,800
2600	5 floors, 100 FPM		106,000	36,800	142,800
2800	4000 lb., 2 Floors, 100 FPM		57,500	13,800	71,300
2900	5 floors, 100 FPM		107,500	36,800	144,300
3100	4500 lb., 2 Floors, 100 FPM		59,000	13,800	72,800
3200	5 floors, 100 FPM		109,000	36,800	145,800
4000	Hospital elevators, 3500 lb., 2 Floors, 100 FPM		82,500	13,800	96,300
4100	5 Floors, 100 FPM		143,000	36,800	179,800
4300	4000 lb., 2 Floors, 100 FPM		82,500	13,800	96,300
4400	5 floors, 100 FPM		143,000	36,800	179,800
4600	4500 lb., 2 Floors, 100 FPM		90,000	13,800	103,800
4800	5 floors, 100 FPM		150,500	36,800	187,300
4900	5000 lb., 2 Floors, 100 FPM		92,500	13,800	106,300
5000	5 floors, 100 FPM		153,000	36,800	189,800
6700	Freight elevators (Class "B"), 3000 lb., 2 Floors, 50 FPM		132,000	17,600	149,600
6800	5 floors, 100 FPM		257,000	46,200	303,200
7000	4000 lb., 2 Floors, 50 FPM		141,000	17,600	158,600
7100	5 floors, 100 FPM		266,500	46,200	312,700
7500	10,000 lb., 2 Floors, 50 FPM		180,000	17,600	197,600
7600	5 floors, 100 FPM		305,000	46,200	351,200
8100	20,000 lb., 2 Floors, 50 FPM		248,000	17,600	265,600
8200	5 floors, 100 FPM		373,000	46,200	419,200

D10 Conveying

D1010 Elevators and Lifts

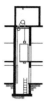

Geared traction elevators are the intermediate group both in cost and in operating areas. These are in buildings of four to fifteen floors and speed ranges from 200' to 500' per minute.

System Components	QUANTITY	UNIT	COST EACH		
			MAT.	INST.	TOTAL
SYSTEM D1010 140 1600					
PASSENGER 2500 LB., 5 FLOORS, 200 FPM					
Passenger elevator, geared, 2000 lb. capacity, 4 stop, 200 FPM	1.000	Ea.	100,500	27,000	127,500
Over 40' travel height, passenger elevator electric, add	10.000	V.L.F.	7,150	1,860	9,010
Passenger elevator, electric, 2500 lb. cap., add	1.000	Ea.	4,200		4,200
Over 4 stops, passenger elevator, electric, add	1.000	Stop	8,600	5,000	13,600
Hall lantern	5.000	Ea.	2,975	845	3,820
Maintenance agreement for passenger elevator, 9 months	1.000	Ea.	4,000		4,000
Position indicator at lobby	1.000	Ea.	685	42	727
TOTAL			128,110	34,747	162,857

D1010 140	Traction Geared Elevators		COST EACH		
			MAT.	INST.	TOTAL
1300	Passenger, 2000 Lb., 5 floors, 200 FPM	R142000 -10	124,000	34,700	158,700
1500	15 floors, 350 FPM		300,500	105,000	405,500
1600	2500 Lb., 5 floors, 200 FPM	R142000 -20	128,000	34,700	162,700
1800	15 floors, 350 FPM		305,000	105,000	410,000
1900	3000 Lb., 5 floors, 200 FPM	R142000 -30	129,000	34,700	163,700
2100	15 floors, 350 FPM		305,500	105,000	410,500
2200	3500 Lb., 5 floors, 200 FPM	R142000 -40	132,500	34,700	167,200
2400	15 floors, 350 FPM		309,000	105,000	414,000
2500	4000 Lb., 5 floors, 200 FPM		131,500	34,700	166,200
2700	15 floors, 350 FPM		308,500	105,000	413,500
2800	4500 Lb., 5 floors, 200 FPM		134,000	34,700	168,700
3000	15 floors, 350 FPM		310,500	105,000	415,500
3100	5000 Lb., 5 floors, 200 FPM		136,000	34,700	170,700
3300	15 floors, 350 FPM		312,500	105,000	417,500
4000	Hospital, 3500 Lb., 5 floors, 200 FPM		127,000	34,700	161,700
4200	15 floors, 350 FPM		301,500	105,000	406,500
4300	4000 Lb., 5 floors, 200 FPM		127,000	34,700	161,700
4500	15 floors, 350 FPM		301,500	105,000	406,500
4600	4500 Lb., 5 floors, 200 FPM		133,000	34,700	167,700
4800	15 floors, 350 FPM		307,500	105,000	412,500
4900	5000 Lb., 5 floors, 200 FPM		134,500	34,700	169,200
5100	15 floors, 350 FPM		309,000	105,000	414,000
6000	Freight, 4000 Lb., 5 floors, 50 FPM class 'B'		142,000	36,000	178,000
6200	15 floors, 200 FPM class 'B'		357,500	125,000	482,500
6300	8000 Lb., 5 floors, 50 FPM class 'B'		174,500	36,000	210,500
6500	15 floors, 200 FPM class 'B'		390,000	125,000	515,000
7000	10,000 Lb., 5 floors, 50 FPM class 'B'		192,500	36,000	228,500
7200	15 floors, 200 FPM class 'B'		556,000	125,000	681,000
8000	20,000 Lb., 5 floors, 50 FPM class 'B'		222,500	36,000	258,500
8200	15 floors, 200 FPM class 'B'		585,500	125,000	710,500

D10 Conveying

D1010 Elevators and Lifts

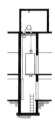

Gearless traction elevators are used in high rise situations where speeds to over 1000 FPM are required to move passengers efficiently. This type of installation is also the most costly.

System Components	QUANTITY	UNIT	COST EACH MAT.	COST EACH INST.	COST EACH TOTAL
SYSTEM D1010 150 1700					
PASSENGER, 2500 LB., 10 FLOORS, 200 FPM					
Passenger elevator, geared, 2000 lb capacity, 4 stop 100 FPM	1.000	Ea.	100,500	27,000	127,500
Over 10' travel height, passenger elevator electric, add	60.000	V.L.F.	42,900	11,160	54,060
Passenger elevator, electric, 2500 lb. cap., add	1.000	Ea.	4,200		4,200
Over 2 stops, passenger elevator, electric, add	6.000	Stop	51,600	30,000	81,600
Gearless variable voltage machinery, overhead	1.000	Ea.	116,000	19,300	135,300
Hall lantern	9.000	Ea.	5,950	1,690	7,640
Position indicator at lobby	1.000	Ea.	685	42	727
Fireman services control	1.000	Ea.	12,600	1,125	13,725
Emergency power manual switching, add	1.000	Ea.	550	169	719
Maintenance agreement for passenger elevator, 9 months	1.000	Ea.	4,000		4,000
TOTAL			338,985	90,486	429,471

D1010 150	Traction Gearless Elevators		MAT.	INST.	TOTAL
1700	Passenger, 2500 Lb., 10 floors, 200 FPM	R142000-10	339,000	90,500	429,500
1900	30 floors, 600 FPM		705,500	231,000	936,500
2000	3000 Lb., 10 floors, 200 FPM	R142000-20	340,000	90,500	430,500
2200	30 floors, 600 FPM		706,500	231,000	937,500
2300	3500 Lb., 10 floors, 200 FPM	R142000-30	343,000	90,500	433,500
2500	30 floors, 600 FPM		710,000	231,000	941,000
2700	50 floors, 800 FPM	R142000-40	1,048,500	371,500	1,420,000
2800	4000 lb., 10 floors, 200 FPM		342,500	90,500	433,000
3000	30 floors, 600 FPM		709,000	231,000	940,000
3200	50 floors, 800 FPM		1,048,000	371,500	1,419,500
3300	4500 lb., 10 floors, 200 FPM		344,500	90,500	435,000
3500	30 floors, 600 FPM		711,500	231,000	942,500
3700	50 floors, 800 FPM		1,050,000	371,500	1,421,500
3800	5000 lb., 10 floors, 200 FPM		346,500	90,500	437,000
4000	30 floors, 600 FPM		713,500	231,000	944,500
4200	50 floors, 800 FPM		1,052,000	371,500	1,423,500
6000	Hospital, 3500 Lb., 10 floors, 200 FPM		336,500	90,500	427,000
6200	30 floors, 600 FPM		699,000	231,000	930,000
6400	4000 Lb., 10 floors, 200 FPM		336,500	90,500	427,000
6600	30 floors, 600 FPM		699,000	231,000	930,000
6800	4500 Lb., 10 floors, 200 FPM		343,000	90,500	433,500
7000	30 floors, 600 FPM		705,000	231,000	936,000
7200	5000 Lb., 10 floors, 200 FPM		344,000	90,500	434,500
7400	30 floors, 600 FPM		706,000	231,000	937,000

D10 Conveying

D1020 Escalators and Moving Walks

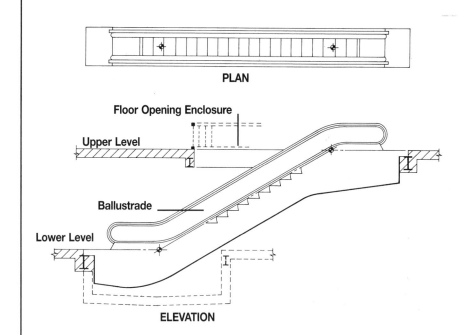

Moving stairs can be used for buildings where 600 or more people are to be carried to the second floor or beyond. Freight cannot be carried on escalators and at least one elevator must be available for this function.

Carrying capacity is 5000 to 8000 people per hour. Power requirement is 2 KW to 3 KW per hour and incline angle is 30°.

D1020 110				Moving Stairs				
	TYPE	HEIGHT	WIDTH	BALUSTRADE MATERIAL		COST EACH		
						MAT.	INST.	TOTAL
0100	Escalator	10ft	32"	glass	R143110 -10	99,500	37,500	137,000
0150				metal		105,500	37,500	143,000
0200			48"	glass	R143210 -20	108,000	37,500	145,500
0250				metal		114,500	37,500	152,000
0300		15ft	32"	glass		105,500	43,800	149,300
0350				metal		114,500	43,800	158,300
0400			48"	glass		112,000	43,800	155,800
0450				metal		114,500	43,800	158,300
0500		20ft	32"	glass		111,000	54,000	165,000
0550				metal		118,000	54,000	172,000
0600			48"	glass		121,000	54,000	175,000
0650				metal		126,500	54,000	180,500
0700		25ft	32"	glass		117,500	65,500	183,000
0750				metal		126,500	65,500	192,000
0800			48"	glass		136,500	65,500	202,000
0850				metal		126,500	65,500	192,000

D1020 210				Moving Walks				
	TYPE	STORY HEIGHT	DEGREE SLOPE	WIDTH	COST RANGE	COST PER L.F.		
						MAT.	INST.	TOTAL
1500	Ramp	10'-23'	12	3'-4"	minimum	2,525	740	3,265
1600					maximum	3,200	900	4,100
2000	Walk	0'	0	2'-0"	minimum	940	405	1,345
2500					maximum	1,300	595	1,895
3000		0	0	3'-4"	Minimun	2,125	595	2,720
3500					maximum	2,475	685	3,160

D20 Plumbing

D2010 Plumbing Fixtures

Systems are complete with trim seat and rough-in (supply, waste and vent) for connection to supply branches and waste mains.

One Piece Wall Hung

Supply

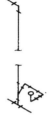
Waste/Vent

Floor Mount

System Components	QUANTITY	UNIT	COST EACH		
			MAT.	INST.	TOTAL
SYSTEM D2010 110 1880					
WATER CLOSET, VITREOUS CHINA, ELONGATED					
TANK TYPE, WALL HUNG, TWO PIECE		Ea.			
Water closet, tank type vit china wall hung 2 pc. w/seat supply & stop	1.000	Ea.	610	199	809
Pipe Steel galvanized, schedule 40, threaded, 2" diam.	4.000	L.F.	45.60	65.80	111.40
Pipe, CI soil, no hub, cplg 10' OC, hanger 5' OC, 4" diam.	2.000	L.F.	27.90	36.30	64.20
Pipe, coupling, standard coupling, CI soil, no hub, 4" diam.	2.000	Ea.	45	64	109
Copper tubing type L solder joint, hangar 10' O.C., 1/2" diam.	6.000	L.F.	25.26	43.50	68.76
Wrought copper 90° elbow for solder joints 1/2" diam.	2.000	Ea.	2.70	59	61.70
Wrought copper Tee for solder joints 1/2" diam.	1.000	Ea.	2.32	45	47.32
Supports/carrier, water closet, siphon jet, horiz, single, 4" waste	1.000	Ea.	550	110	660
TOTAL			1,308.78	622.60	1,931.38

D2010 110	Water Closet Systems		COST EACH		
			MAT.	INST.	TOTAL
1800	Water closet, vitreous china, elongated	R224000 -30			
1840	Tank type, wall hung				
1880	Close coupled two piece	R224000 -40	1,300	625	1,925
1920	Floor mount, one piece		915	660	1,575
1960	One piece low profile		775	660	1,435
2000	Two piece close coupled		535	660	1,195
2040	Bowl only with flush valve				
2080	Wall hung		1,175	705	1,880
2120	Floor mount		720	670	1,390
2160	Floor mount, ADA compliant with 18" high bowl		685	690	1,375

D20 Plumbing

D2010 Plumbing Fixtures

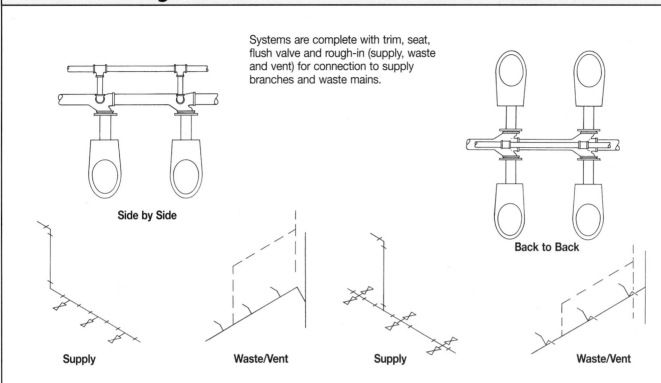

Systems are complete with trim, seat, flush valve and rough-in (supply, waste and vent) for connection to supply branches and waste mains.

Side by Side

Back to Back

Supply Waste/Vent Supply Waste/Vent

System Components	QUANTITY	UNIT	COST EACH		
			MAT.	INST.	TOTAL
SYSTEM D2010 120 1760					
WATER CLOSETS, BATTERY MOUNT, WALL HUNG, SIDE BY SIDE, FIRST CLOSET					
Water closet, bowl only w/flush valve, seat, wall hung	1.000	Ea.	315	182	497
Pipe, CI soil, no hub, cplg 10' OC, hanger 5' OC, 4" diam	3.000	L.F.	41.85	54.45	96.30
Coupling, standard, CI, soil, no hub, 4" diam	2.000	Ea.	45	64	109
Copper tubing type L, solder joints, hangers 10' OC, 1" diam	6.000	L.F.	59.70	51.60	111.30
Copper tubing, type DWV, solder joints, hangers 10'OC, 2" diam	6.000	L.F.	123	79.80	202.80
Wrought copper 90° elbow for solder joints 1" diam	1.000	Ea.	7.45	36.50	43.95
Wrought copper Tee for solder joints, 1" diam	1.000	Ea.	17.25	58.50	75.75
Support/carrier, siphon jet, horiz, adjustable single, 4" pipe	1.000	Ea.	550	110	660
Valve, gate, bronze, 125 lb, NRS, soldered 1" diam	1.000	Ea.	55.50	31	86.50
Wrought copper, DWV, 90° elbow, 2" diam	1.000	Ea.	25.50	58.50	84
TOTAL			1,240.25	726.35	1,966.60

D2010 120	Water Closets, Group		COST EACH		
			MAT.	INST.	TOTAL
1760	Water closets, battery mount, wall hung, side by side, first closet	R224000 -30	1,250	725	1,975
1800	Each additional water closet, add		1,150	685	1,835
3000	Back to back, first pair of closets	R224000 -40	2,075	965	3,040
3100	Each additional pair of closets, back to back		2,025	950	2,975

D20 Plumbing

D2010 Plumbing Fixtures

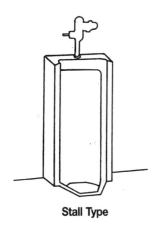
Stall Type

Systems are complete with trim, flush valve and rough-in (supply, waste and vent) for connection to supply branches and waste mains.

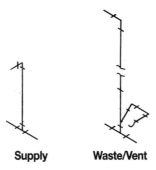
Supply   Waste/Vent

Wall Hung

System Components	QUANTITY	UNIT	COST EACH MAT.	COST EACH INST.	COST EACH TOTAL
SYSTEM D2010 210 2000					
URINAL, VITREOUS CHINA, WALL HUNG					
Urinal, wall hung, vitreous china, incl. hanger	1.000	Ea.	273	350	623
Pipe, steel, galvanized, schedule 40, threaded, 1-1/2" diam.	5.000	L.F.	42.75	65.75	108.50
Copper tubing type DWV, solder joint, hangers 10' OC, 2" diam.	3.000	L.F.	61.50	39.90	101.40
Combination Y & 1/8 bend for CI soil pipe, no hub, 3" diam.	1.000	Ea.	15.65		15.65
Pipe, CI, no hub, cplg. 10' OC, hanger 5' OC, 3" diam.	4.000	L.F.	43.20	65.80	109
Pipe coupling standard, CI soil, no hub, 3" diam.	3.000	Ea.	38	55	93
Copper tubing type L, solder joint, hanger 10' OC 3/4" diam.	5.000	L.F.	33	38.50	71.50
Wrought copper 90° elbow for solder joints 3/4" diam.	1.000	Ea.	3.04	31	34.04
Wrought copper Tee for solder joints, 3/4" diam.	1.000	Ea.	5.60	49	54.60
TOTAL			515.74	694.95	1,210.69

D2010 210	Urinal Systems		COST EACH MAT.	COST EACH INST.	COST EACH TOTAL
2000	Urinal, vitreous china, wall hung	R224000 -30	515	695	1,210
2040	Stall type		1,025	830	1,855
2060		R224000 -40			
2062					

D20 Plumbing

D2010 Plumbing Fixtures

Systems are complete with trim and rough-in (supply, waste and vent) to connect to supply branches and waste mains.

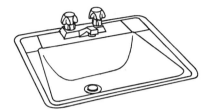
Vanity Top

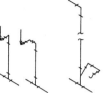
Supply Waste/Vent

Wall Hung

System Components	QUANTITY	UNIT	COST EACH		
			MAT.	INST.	TOTAL
SYSTEM D2010 310 1560					
LAVATORY W/TRIM, VANITY TOP, P.E. ON C.I., 20" X 18"					
Lavatory w/trim, PE on CI, white, vanity top, 20" x 18" oval	1.000	Ea.	254	165	419
Pipe, steel, galvanized, schedule 40, threaded, 1-1/4" diam.	4.000	L.F.	28.80	47.40	76.20
Copper tubing type DWV, solder joint, hanger 10' OC 1-1/4" diam.	4.000	L.F.	48	39	87
Wrought copper DWV, Tee, sanitary, 1-1/4" diam.	1.000	Ea.	26	65	91
P trap w/cleanout, 20 ga., 1-1/4" diam.	1.000	Ea.	79	32.50	111.50
Copper tubing type L, solder joint, hanger 10' OC 1/2" diam.	10.000	L.F.	42.10	72.50	114.60
Wrought copper 90° elbow for solder joints 1/2" diam.	2.000	Ea.	2.70	59	61.70
Wrought copper Tee for solder joints, 1/2" diam.	2.000	Ea.	4.64	90	94.64
Stop, chrome, angle supply, 1/2" diam.	2.000	Ea.	16.20	53	69.20
TOTAL			501.44	623.40	1,124.84

D2010 310	Lavatory Systems		COST EACH		
			MAT.	INST.	TOTAL
1560	Lavatory w/trim, vanity top, PE on CI 20" x 18", Vanity top by others.	R224000-30	500	625	1,125
1600	19" x 16" oval		420	625	1,045
1640	18" round	R224000-40	485	625	1,110
1680	Cultured marble, 19" x 17"		435	625	1,060
1720	25" x 19"		445	625	1,070
1760	Stainless, self-rimming, 25" x 22"		605	625	1,230
1800	17" x 22"		595	625	1,220
1840	Steel enameled, 20" x 17"		420	640	1,060
1880	19" round		420	640	1,060
1920	Vitreous china, 20" x 16"		530	655	1,185
1960	19" x 16"		535	655	1,190
2000	22" x 13"		540	655	1,195
2040	Wall hung, PE on CI, 18" x 15"		760	690	1,450
2080	19" x 17"		775	690	1,465
2120	20" x 18"		725	690	1,415
2160	Vitreous china, 18" x 15"		655	705	1,360
2200	19" x 17"		590	705	1,295
2240	24" x 20"		925	705	1,630

D20 Plumbing

D2010 Plumbing Fixtures

Systems are complete with trim and rough-in (supply, waste and vent) to connect to supply branches and waste mains.

Countertop Single Bowl

Supply

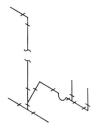
Waste/Vent

Countertop Double Bowl

System Components	QUANTITY	UNIT	COST EACH		
			MAT.	INST.	TOTAL
SYSTEM D2010 410 1720					
KITCHEN SINK W/TRIM, COUNTERTOP, P.E. ON C.I., 24″ X 21″, SINGLE BOWL					
Kitchen sink, countertop, PE on CI, 1 bowl, 24″ x 21″ OD	1.000	Ea.	299	188	487
Pipe, steel, galvanized, schedule 40, threaded, 1-1/4″ diam.	4.000	L.F.	28.80	47.40	76.20
Copper tubing, type DWV, solder, hangers 10′ OC 1-1/2″ diam.	6.000	L.F.	90.60	65.10	155.70
Wrought copper, DWV, Tee, sanitary, 1-1/2″ diam.	1.000	Ea.	32	73	105
P trap, standard, copper, 1-1/2″ diam.	1.000	Ea.	76.50	34.50	111
Copper tubing, type L, solder joints, hangers 10′ OC 1/2″ diam.	10.000	L.F.	42.10	72.50	114.60
Wrought copper 90° elbow for solder joints 1/2″ diam.	2.000	Ea.	2.70	59	61.70
Wrought copper Tee for solder joints, 1/2″ diam.	2.000	Ea.	4.64	90	94.64
Stop, angle supply, chrome, 1/2″ CTS	2.000	Ea.	16.20	53	69.20
TOTAL			592.54	682.50	1,275.04

D2010 410	Kitchen Sink Systems		COST EACH		
			MAT.	INST.	TOTAL
1720	Kitchen sink w/trim, countertop, PE on CI, 24″x21″, single bowl	R224000-30	595	685	1,280
1760	30″ x 21″ single bowl		600	685	1,285
1800	32″ x 21″ double bowl	R224000-40	735	735	1,470
1840	42″ x 21″ double bowl		890	745	1,635
1880	Stainless steel, 19″ x 18″ single bowl		855	685	1,540
1920	25″ x 22″ single bowl		920	685	1,605
1960	33″ x 22″ double bowl		1,225	735	1,960
2000	43″ x 22″ double bowl		1,400	745	2,145
2040	44″ x 22″ triple bowl		1,425	775	2,200
2080	44″ x 24″ corner double bowl		1,075	745	1,820
2120	Steel, enameled, 24″ x 21″ single bowl		455	685	1,140
2160	32″ x 21″ double bowl		505	735	1,240
2240	Raised deck, PE on CI, 32″ x 21″, dual level, double bowl		660	930	1,590
2280	42″ x 21″ dual level, triple bowl		1,250	1,025	2,275

D20 Plumbing

D2010 Plumbing Fixtures

Corrosion resistant laboratory sink systems are complete with trim and rough-in (supply, waste and vent) to connect to supply branches and waste mains.

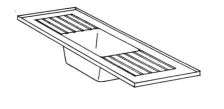
Laboratory Sink

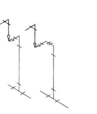
Supply

Waste/Vent

Polypropylene Cup Sink

System Components	QUANTITY	UNIT	COST EACH MAT.	COST EACH INST.	COST EACH TOTAL
SYSTEM D2010 430 1600					
LABORATORY SINK W/TRIM, POLYETHYLENE, SINGLE BOWL					
DOUBLE DRAINBOARD, 54" X 24" OD					
Sink w/trim, polyethylene, 1 bowl, 2 drainboards 54" x 24" OD	1.000	Ea.	1,150	350	1,500
Pipe, polypropylene, schedule 40, acid resistant 1-1/2" diam	10.000	L.F.	55	155	210
Tee, sanitary, polypropylene, acid resistant, 1-1/2" diam	1.000	Ea.	13.20	58.50	71.70
P trap, polypropylene, acid resistant, 1-1/2" diam	1.000	Ea.	43	34.50	77.50
Copper tubing type L, solder joint, hanger 10' OC 1/2" diam	10.000	L.F.	42.10	72.50	114.60
Wrought copper 90° elbow for solder joints 1/2" diam	2.000	Ea.	2.70	59	61.70
Wrought copper Tee for solder joints, 1/2" diam	2.000	Ea.	4.64	90	94.64
Stop, angle supply, chrome, 1/2" diam	2.000	Ea.	16.20	53	69.20
TOTAL			1,326.84	872.50	2,199.34

D2010 430	Laboratory Sink Systems	COST EACH MAT.	COST EACH INST.	COST EACH TOTAL
1580	Laboratory sink w/trim, polyethylene, single bowl,			
1600	Double drainboard, 54" x 24" O.D.	1,325	875	2,200
1640	Single drainboard, 47" x 24" O.D.	1,525	875	2,400
1680	70" x 24" O.D.	1,400	875	2,275
1760	Flanged, 14-1/2" x 14-1/2" O.D.	400	785	1,185
1800	18-1/2" x 18-1/2" O.D.	490	785	1,275
1840	23-1/2" x 20-1/2" O.D.	530	785	1,315
1920	Polypropylene, cup sink, oval, 7" x 4" O.D.	282	695	977
1960	10" x 4-1/2" O.D.	266	695	961

D20 Plumbing

D2010 Plumbing Fixtures

Wall Hung

Corrosion resistant laboratory sink systems are complete with trim and rough-in (supply, waste and vent) to connect to supply branches and waste mains.

Supply

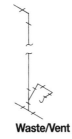
Waste/Vent

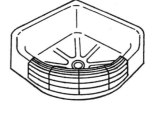
Corner, Floor

System Components	QUANTITY	UNIT	COST EACH		
			MAT.	INST.	TOTAL
SYSTEM D2010 440 4260					
SERVICE SINK, PE ON CI, CORNER FLOOR, 28"X28", W/RIM GUARD & TRIM					
Service sink, corner floor, PE on CI, 28" x 28", w/rim guard & trim	1.000	Ea.	690	239	929
Copper tubing type DWV, solder joint, hanger 10'OC 3" diam	6.000	L.F.	216	108.90	324.90
Copper tubing type DWV, solder joint, hanger 10'OC 2" diam	4.000	L.F.	82	53.20	135.20
Wrought copper DWV, Tee, sanitary, 3" diam.	1.000	Ea.	139	151	290
P trap with cleanout & slip joint, copper 3" diam	1.000	Ea.	284	53	337
Copper tubing, type L, solder joints, hangers 10' OC, 1/2" diam	10.000	L.F.	42.10	72.50	114.60
Wrought copper 90° elbow for solder joints 1/2" diam	2.000	Ea.	2.70	59	61.70
Wrought copper Tee for solder joints, 1/2" diam	2.000	Ea.	4.64	90	94.64
Stop, angle supply, chrome, 1/2" diam	2.000	Ea.	16.20	53	69.20
TOTAL			1,476.64	879.60	2,356.24

D2010 440	Service Sink Systems	COST EACH		
		MAT.	INST.	TOTAL
4260	Service sink w/trim, PE on CI, corner floor, 28" x 28", w/rim guard	1,475	880	2,355
4300	Wall hung w/rim guard, 22" x 18"	1,750	1,025	2,775
4340	24" x 20"	1,800	1,025	2,825
4380	Vitreous china, wall hung 22" x 20"	1,625	1,025	2,650

D20 Plumbing

D2010 Plumbing Fixtures

Wall Mounted, No Back

Systems are complete with trim and rough-in (supply, waste and vent) to connect to supply branches and waste mains.

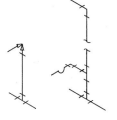

Supply   Waste/Vent

Wall Mounted, Low Back

System Components	QUANTITY	UNIT	COST EACH		
			MAT.	INST.	TOTAL
SYSTEM D2010 810 1800					
DRINKING FOUNTAIN, ONE BUBBLER, WALL MOUNTED					
NON RECESSED, BRONZE, NO BACK					
Drinking fountain, wall mount, bronze, 1 bubbler	1.000	Ea.	1,025	146	1,171
Copper tubing, type L, solder joint, hanger 10' OC 3/8" diam.	5.000	L.F.	17.90	34.75	52.65
Stop, supply, straight, chrome, 3/8" diam.	1.000	Ea.	6.75	24.50	31.25
Wrought copper 90° elbow for solder joints 3/8" diam.	1.000	Ea.	4.06	26.50	30.56
Wrought copper Tee for solder joints, 3/8" diam.	1.000	Ea.	6.85	42	48.85
Copper tubing, type DWV, solder joint, hanger 10' OC 1-1/4" diam.	4.000	L.F.	48	39	87
P trap, standard, copper drainage, 1-1/4" diam.	1.000	Ea.	79	32.50	111.50
Wrought copper, DWV, Tee, sanitary, 1-1/4" diam.	1.000	Ea.	26	65	91
TOTAL			1,213.56	410.25	1,623.81

D2010 810	Drinking Fountain Systems	COST EACH		
		MAT.	INST.	TOTAL
1740	Drinking fountain, one bubbler, wall mounted			
1760	Non recessed			
1800	Bronze, no back	1,225	410	1,635
1840	Cast iron, enameled, low back	830	410	1,240
1880	Fiberglass, 12" back	1,475	410	1,885
1920	Stainless steel, no back	1,525	410	1,935
1960	Semi-recessed, poly marble	1,050	410	1,460
2040	Stainless steel	1,175	410	1,585
2080	Vitreous china	860	410	1,270
2120	Full recessed, poly marble	1,175	410	1,585
2200	Stainless steel	1,150	410	1,560
2240	Floor mounted, pedestal type, aluminum	770	555	1,325
2320	Bronze	1,700	555	2,255
2360	Stainless steel	1,700	555	2,255

D20 Plumbing

D2010 Plumbing Fixtures

Systems are complete with trim and rough-in (supply, waste and vent) for connection to supply branches and waste mains.

Wall Hung

Supply

Waste/Vent

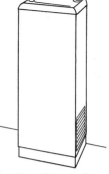
Floor Mounted

System Components	QUANTITY	UNIT	COST EACH		
			MAT.	INST.	TOTAL
SYSTEM D2010 820 1840					
WATER COOLER, ELECTRIC, SELF CONTAINED, WALL HUNG, 8.2 GPH					
Water cooler, wall mounted, 8.2 GPH	1.000	Ea.	710	263	973
Copper tubing type DWV, solder joint, hanger 10' OC 1-1/4" diam.	4.000	L.F.	48	39	87
Wrought copper DWV, Tee, sanitary 1-1/4" diam.	1.000	Ea.	26	65	91
P trap, copper drainage, 1-1/4" diam.	1.000	Ea.	79	32.50	111.50
Copper tubing type L, solder joint, hanger 10' OC 3/8" diam.	5.000	L.F.	17.90	34.75	52.65
Wrought copper 90° elbow for solder joints 3/8" diam.	1.000	Ea.	4.06	26.50	30.56
Wrought copper Tee for solder joints, 3/8" diam.	1.000	Ea.	6.85	42	48.85
Stop and waste, straightway, bronze, solder, 3/8" diam.	1.000	Ea.	6.75	24.50	31.25
TOTAL			898.56	527.25	1,425.81

D2010 820	Water Cooler Systems		COST EACH		
			MAT.	INST.	TOTAL
1840	Water cooler, electric, wall hung, 8.2 GPH	R224000 -30	900	525	1,425
1880	Dual height, 14.3 GPH		1,275	540	1,815
1920	Wheelchair type, 7.5 G.P.H.	R224000 -40	2,000	525	2,525
1960	Semi recessed, 8.1 G.P.H.		1,250	525	1,775
2000	Full recessed, 8 G.P.H.	R224000 -50	1,875	565	2,440
2040	Floor mounted, 14.3 G.P.H.		975	460	1,435
2080	Dual height, 14.3 G.P.H.		1,350	555	1,905
2120	Refrigerated compartment type, 1.5 G.P.H.		1,600	460	2,060

D20 Plumbing

D2010 Plumbing Fixtures

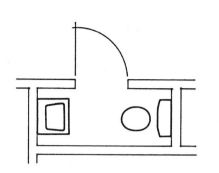

Two Fixture Bathroom Systems consisting of a lavatory, water closet, and rough-in service piping.
- Prices for plumbing and fixtures only.

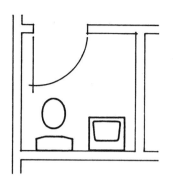

*Common wall is with an adjacent bathroom

System Components	QUANTITY	UNIT	COST EACH		
			MAT.	INST.	TOTAL
SYSTEM D2010 920 1180					
BATHROOM, LAVATORY & WATER CLOSET, 2 WALL PLUMBING, STAND ALONE					
Water closet, 2 Pc. close cpld. vit .china flr. mntd. w/seat, supply & stop	1.000	Ea.	203	199	402
Water closet, rough-in waste & vent	1.000	Set	300	345	645
Lavatory w/ftngs., wall hung, white, PE on CI, 20" x 18"	1.000	Ea.	325	132	457
Lavatory, rough-in waste & vent	1.000	Set	400	635	1,035
Copper tubing type L, solder joint, hanger 10' OC 1/2" diam.	10.000	L.F.	42.10	72.50	114.60
Pipe, steel, galvanized, schedule 40, threaded, 2" diam.	12.000	L.F.	136.80	197.40	334.20
Pipe, CI soil, no hub, coupling 10' OC, hanger 5' OC, 4" diam.	7.000	L.F.	88.55	134.05	222.60
TOTAL			1,495.45	1,714.95	3,210.40

D2010 920	Two Fixture Bathroom, Two Wall Plumbing		COST EACH		
			MAT.	INST.	TOTAL
1180	Bathroom, lavatory & water closet, 2 wall plumbing, stand alone		1,500	1,725	3,225
1201	Share common plumbing wall		1,350	1,475	2,825
1220					
1240					

D2010 922	Two Fixture Bathroom, One Wall Plumbing		COST EACH		
			MAT.	INST.	TOTAL
2220	Bathroom, lavatory & water closet, one wall plumbing, stand alone	R224000 -30	1,375	1,550	2,925
2240	Share common plumbing wall*		1,225	1,300	2,525
2260		R224000 -40			
2280					

D20 Plumbing

D2010 Plumbing Fixtures

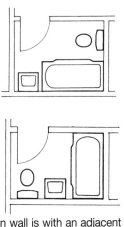
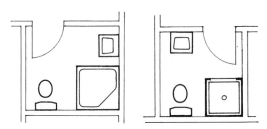
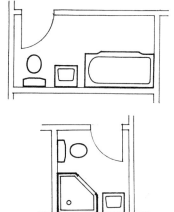

Three Fixture Bathroom Systems consisting of a lavatory, water closet, bathtub or shower and rough-in service piping.

- Prices for plumbing and fixtures only.

*Common wall is with an adjacent bathroom

System Components	QUANTITY	UNIT	COST EACH		
			MAT.	INST.	TOTAL
SYSTEM D2010 924 1170					
BATHROOM, LAVATORY, WATER CLOSET & BATHTUB					
ONE WALL PLUMBING, STAND ALONE					
Wtr closet, 2 pc close cpld vit china flr mntd w/seat supply & stop	1.000	Ea.	203	199	402
Water closet, rough-in waste & vent	1.000	Set	300	345	645
Lavatory w/ftngs, wall hung, white, PE on CI, 20" x 18"	1.000	Ea.	325	132	457
Lavatory, rough-in waste & vent	1.000	Set	400	635	1,035
Bathtub, white PE on CI, w/ftgs, mat bottom, recessed, 5' long	1.000	Ea.	890	239	1,129
Baths, rough-in waste and vent	1.000	Set	297	459	756
TOTAL			2,415	2,009	4,424

D2010 924	Three Fixture Bathroom, One Wall Plumbing		COST EACH		
			MAT.	INST.	TOTAL
1150	Bathroom, three fixture, one wall plumbing	R224000 -30			
1160	Lavatory, water closet & bathtub				
1170	Stand alone	R224000 -40	2,425	2,000	4,425
1180	Share common plumbing wall *		2,075	1,450	3,525

D2010 926	Three Fixture Bathroom, Two Wall Plumbing	COST EACH		
		MAT.	INST.	TOTAL
2130	Bathroom, three fixture, two wall plumbing			
2140	Lavatory, water closet & bathtub			
2160	Stand alone	2,425	2,025	4,450
2180	Long plumbing wall common *	2,175	1,625	3,800
3610	Lavatory, bathtub & water closet			
3620	Stand alone	2,650	2,325	4,975
3640	Long plumbing wall common *	2,475	2,100	4,575
4660	Water closet, corner bathtub & lavatory			
4680	Stand alone	3,525	2,050	5,575
4700	Long plumbing wall common *	3,200	1,550	4,750
6100	Water closet, stall shower & lavatory			
6120	Stand alone	2,275	2,300	4,575
6140	Long plumbing wall common *	2,125	2,125	4,250
7060	Lavatory, corner stall shower & water closet			
7080	Stand alone	3,075	2,050	5,125
7100	Short plumbing wall common *	2,525	1,375	3,900

D30 HVAC

D3030 Cooling Generating Systems

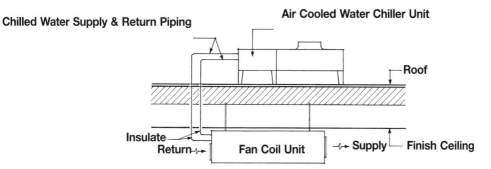

Design Assumptions: The chilled water, air cooled systems priced, utilize reciprocating hermetic compressors and propeller-type condenser fans. Piping with pumps and expansion tanks is included based on a two pipe system. No ducting is included and the fan-coil units are cooling only. Water treatment and balancing are not included. Chilled water piping is insulated. Area distribution is through the use of multiple fan coil units. Fewer but larger fan coil units with duct distribution would be approximately the same S.F. cost.

System Components	QUANTITY	UNIT	COST EACH		
			MAT.	INST.	TOTAL
SYSTEM D3030 110 1200					
PACKAGED CHILLER, AIR COOLED, WITH FAN COIL UNIT					
APARTMENT CORRIDORS, 3,000 S.F., 5.50 TON					
Fan coil air conditioning unit, cabinet mounted & filters chilled water	1.000	Ea.	3,345.23	489.41	3,834.64
Water chiller, air conditioning unit, reciprocating, air cooled,	1.000	Ea.	6,572.50	1,828.75	8,401.25
Chilled water unit coil connections	1.000	Ea.	980	1,325	2,305
Chilled water distribution piping	440.000	L.F.	10,340	18,040	28,380
TOTAL			21,237.73	21,683.16	42,920.89
COST PER S.F.			7.08	7.23	14.31

D3030 110	Chilled Water, Air Cooled Condenser Systems		COST PER S.F.		
			MAT.	INST.	TOTAL
1180	Packaged chiller, air cooled, with fan coil unit	R233400-10			
1200	Apartment corridors, 3,000 S.F., 5.50 ton		7.07	7.23	14.30
1360	40,000 S.F., 73.33 ton	R236000-20	4.16	3.20	7.36
1440	Banks and libraries, 3,000 S.F., 12.50 ton		10.60	8.50	19.10
1560	20,000 S.F., 83.33 ton		7.30	4.39	11.69
1680	Bars and taverns, 3,000 S.F., 33.25 ton		17.30	10.10	27.40
1760	10,000 S.F., 110.83 ton		12.75	2.50	15.25
1920	Bowling alleys, 3,000 S.F., 17.00 ton		13.15	9.55	22.70
2040	20,000 S.F., 113.33 ton		8.40	4.44	12.84
2160	Department stores, 3,000 S.F., 8.75 ton		9.55	8.05	17.60
2320	40,000 S.F., 116.66 ton		5.45	3.31	8.76
2400	Drug stores, 3,000 S.F., 20.00 ton		14.85	9.90	24.75
2520	20,000 S.F., 133.33 ton		10.40	4.68	15.08
2640	Factories, 2,000 S.F., 10.00 ton		9.30	8.15	17.45
2800	40,000 S.F., 133.33 ton		5.95	3.40	9.35
2880	Food supermarkets, 3,000 S.F., 8.50 ton		9.40	8	17.40
3040	40,000 S.F., 113.33 ton		5.20	3.32	8.52
3120	Medical centers, 3,000 S.F., 7.00 ton		8.50	7.80	16.30
3280	40,000 S.F., 93.33 ton		4.68	3.25	7.93
3360	Offices, 3,000 S.F., 9.50 ton		8.65	7.95	16.60
3520	40,000 S.F., 126.66 ton		5.85	3.43	9.28
3600	Restaurants, 3,000 S.F., 15.00 ton		11.85	8.85	20.70
3720	20,000 S.F., 100.00 ton		8.45	4.65	13.10
3840	Schools and colleges, 3,000 S.F., 11.50 ton		10.10	8.35	18.45
3960	20,000 S.F., 76.66 ton		7.10	4.43	11.53

D30 HVAC

D3050 Terminal & Package Units

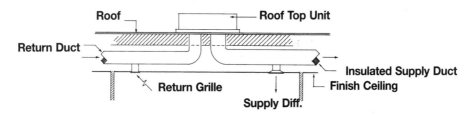

System Description: Rooftop single zone units are electric cooling and gas heat. Duct systems are low velocity, galvanized steel supply and return. Price variations between sizes are due to several factors. Jumps in the cost of the rooftop unit occur when the manufacturer shifts from the largest capacity unit on a small frame to the smallest capacity on the next larger frame, or changes from one compressor to two. As the unit capacity increases for larger areas the duct distribution grows in proportion. For most applications there is a tradeoff point where it is less expensive and more efficient to utilize smaller units with short simple distribution systems. Larger units also require larger initial supply and return ducts which can create a space problem. Supplemental heat may be desired in colder locations. The table below is based on one unit supplying the area listed. The 10,000 S.F. unit for bars and taverns is not listed because a nominal 110 ton unit would be required and this is above the normal single zone rooftop capacity.

System Components	QUANTITY	UNIT	COST EACH		
			MAT.	INST.	TOTAL
SYSTEM D3050 150 1280					
ROOFTOP, SINGLE ZONE, AIR CONDITIONER					
APARTMENT CORRIDORS, 500 S.F., .92 TON					
Rooftop air-conditioner, 1 zone, electric cool, standard controls, curb	1.000	Ea.	1,621.50	701.50	2,323
Ductwork package for rooftop single zone units	1.000	System	289.80	1,012	1,301.80
TOTAL			1,911.30	1,713.50	3,624.80
COST PER S.F.			3.82	3.43	7.25

*Size would suggest multiple units

D3050 150	Rooftop Single Zone Unit Systems		COST PER S.F.		
			MAT.	INST.	TOTAL
1260	Rooftop, single zone, air conditioner				
1280	Apartment corridors, 500 S.F., .92 ton	R236000-20	3.80	3.45	7.25
1480	10,000 S.F., 18.33 ton		2.61	2.33	4.94
1560	Banks or libraries, 500 S.F., 2.08 ton	R233400-10	8.65	7.75	16.40
1760	10,000 S.F., 41.67 ton		5.10	5.25	10.35
1840	Bars and taverns, 500 S.F. 5.54 ton		11.90	10.30	22.20
2000	5,000 S.F., 55.42 ton		11.30	7.90	19.20
2080	Bowling alleys, 500 S.F., 2.83 ton		7.60	8.70	16.30
2280	10,000 S.F., 56.67 ton		6.70	7.15	13.85
2360	Department stores, 500 S.F., 1.46 ton		6.05	5.45	11.50
2560	10,000 S.F., 29.17 ton		3.67	3.68	7.35
2640	Drug stores, 500 S.F., 3.33 ton		8.95	10.25	19.20
2840	10,000 S.F., 66.67 ton		7.85	8.40	16.25
2920	Factories, 500 S.F., 1.67 ton		6.90	6.20	13.10
3120	10,000 S.F., 33.33 ton		4.19	4.21	8.40
3200	Food supermarkets, 500 S.F., 1.42 ton		5.90	5.30	11.20
3400	10,000 S.F., 28.33 ton		3.56	3.58	7.14
3480	Medical centers, 500 S.F., 1.17 ton		4.85	4.35	9.20
3680	10,000 S.F., 23.33 ton		3.31	2.96	6.27
3760	Offices, 500 S.F., 1.58 ton		6.60	5.90	12.50
3960	10,000 S.F., 31.67 ton		3.99	3.99	7.98
4000	Restaurants, 500 S.F., 2.50 ton		10.40	9.30	19.70
4200	10,000 S.F., 50.00 ton		5.90	6.30	12.20
4240	Schools and colleges, 500 S.F., 1.92 ton		7.95	7.15	15.10
4440	10,000 S.F., 38.33 ton		4.68	4.84	9.52

D30 HVAC

D3050 Terminal & Package Units

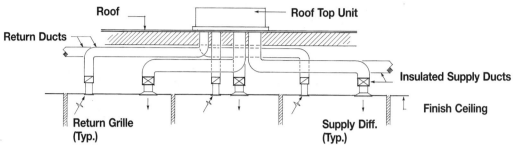

System Description: Rooftop units are multizone with up to 12 zones, and include electric cooling, gas heat, thermostats, filters, supply and return fans complete. Duct systems are low velocity, galvanized steel supply and return with insulated supplies.

Multizone units cost more per ton of cooling than single zone. However, they offer flexibility where load conditions are varied due to heat generating areas or exposure to radiational heating. For example, perimeter offices on the "sunny side" may require cooling at the same time "shady side" or central offices may require heating. It is possible to accomplish similar results using duct heaters in branches of the single zone unit. However, heater location could be a problem and total system operating energy efficiency could be lower.

System Components	QUANTITY	UNIT	COST EACH		
			MAT.	INST.	TOTAL
SYSTEM D3050 155 1280					
ROOFTOP, MULTIZONE, AIR CONDITIONER					
APARTMENT CORRIDORS, 3,000 S.F., 5.50 TON					
Rooftop multizone unit, standard controls, curb	1.000	Ea.	25,960	1,628	27,588
Ductwork package for rooftop multizone units	1.000	System	2,530	11,550	14,080
TOTAL			28,490	13,178	41,668
COST PER S.F.			9.50	4.39	13.89

Note A: Small single zone unit recommended
Note B: A combination of multizone units recommended

D3050 155	Rooftop Multizone Unit Systems	COST PER S.F.		
		MAT.	INST.	TOTAL
1240	Rooftop, multizone, air conditioner			
1260	Apartment corridors, 1,500 S.F., 2.75 ton. See Note A.			
1280	3,000 S.F., 5.50 ton	9.50	4.40	13.90
1440	25,000 S.F., 45.80 ton	6.35	4.25	10.60
1520	Banks or libraries, 1,500 S.F., 6.25 ton	21.50	10	31.50
1640	15,000 S.F., 62.50 ton	10.80	9.60	20.40
1720	25,000 S.F., 104.00 ton	9.75	9.55	19.30
1800	Bars and taverns, 1,500 S.F., 16.62 ton	46	14.35	60.35
1840	3,000 S.F., 33.24 ton	38.50	13.85	52.35
1880	10,000 S.F., 110.83 ton	23.50	13.80	37.30
2080	Bowling alleys, 1,500 S.F., 8.50 ton	29.50	13.60	43.10
2160	10,000 S.F., 56.70 ton	19.60	13.15	32.75
2240	20,000 S.F., 113.00 ton	13.25	12.95	26.20
2640	Drug stores, 1,500 S.F., 10.00 ton	34.50	15.95	50.45
2680	3,000 S.F., 20.00 ton	24	15.40	39.40
2760	15,000 S.F., 100.00 ton	15.60	15.30	30.90
3760	Offices, 1,500 S.F., 4.75 ton, See Note A			
3880	15,000 S.F., 47.50 ton	10.95	7.35	18.30
3960	25,000 S.F., 79.16 ton	8.20	7.30	15.50
4000	Restaurants, 1,500 S.F., 7.50 ton	26	12	38
4080	10,000 S.F., 50.00 ton	17.30	11.60	28.90
4160	20,000 S.F., 100.00 ton	11.70	11.50	23.20
4240	Schools and colleges, 1,500 S.F., 5.75 ton	19.85	9.20	29.05
4360	15,000 S.F., 57.50 ton	9.95	8.80	18.75
4441	25,000 S.F., 95.83 ton	9.40	8.80	18.20

D30 HVAC

D3050 Terminal & Package Units

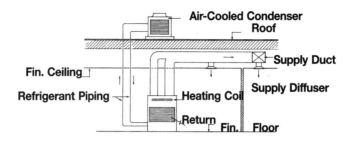

System Description: Self-contained air cooled units with remote air cooled condenser and interconnecting tubing. Systems for 1000 S.F. and up include duct and diffusers. Smaller units distribute air directly.

Returns are not ducted and supplies are not insulated.

Potential savings may be realized by using a single zone rooftop system or through-the-wall unit, especially in the smaller capacities, if the application permits.

Hot water or steam heating coils are included but piping to boiler and the boiler itself is not included.

Condenserless models are available for 15% less where remote refrigerant source is available.

System Components	QUANTITY	UNIT	COST EACH		
			MAT.	INST.	TOTAL
SYSTEM D3050 165 1320 SELF-CONTAINED, AIR COOLED UNIT APARTMENT CORRIDORS, 500 S.F., .92 TON					
Air cooled, package unit	1.000	Ea.	1,198.50	365.50	1,564
Ductwork package for water or air cooled packaged units	1.000	System	108.56	763.60	872.16
Refrigerant piping	1.000	System	506	492.20	998.20
Air cooled condenser, direct drive, propeller fan	1.000	Ea.	348.75	137.95	486.70
TOTAL			2,161.81	1,759.25	3,921.06
COST PER S.F.			4.32	3.52	7.84

D3050 165	Self-contained, Air Cooled Unit Systems		COST PER S.F.		
			MAT.	INST.	TOTAL
1300	Self-contained, air cooled unit	R236000			
1320	Apartment corridors, 500 S.F., .92 ton	-20	4.30	3.50	7.80
1480	10,000 S.F., 18.33 ton		3.15	2.84	5.99
1560	Banks or libraries, 500 S.F., 2.08 ton	R233400 -10	9.20	4.47	13.67
1720	10,000 S.F., 41.66 ton		8.30	6.35	14.65
1800	Bars and taverns, 500 S.F., 5.54 ton		19.30	9.75	29.05
1960	10,000 S.F., 110.00 ton		20.50	12.20	32.70
2040	Bowling alleys, 500 S.F., 2.83 ton		12.55	6.10	18.65
2200	10,000 S.F., 56.66 ton		10.70	8.60	19.30
2240	Department stores, 500 S.F., 1.46 ton		6.45	3.14	9.59
2400	10,000 S.F., 29.17 ton		5.10	4.43	9.53
2480	Drug stores, 500 S.F., 3.33 ton		14.75	7.20	21.95
2640	10,000 S.F., 66.66 ton		12.65	10.15	22.80
2720	Factories, 500 S.F., 1.66 ton		7.45	3.61	11.06
2880	10,000 S.F., 33.33 ton		5.85	5.05	10.90
3200	Medical centers, 500 S.F., 1.17 ton		5.20	2.52	7.72
3360	10,000 S.F., 23.33 ton		4.02	3.63	7.65
3440	Offices, 500 S.F., 1.58 ton		7	3.41	10.41
3600	10,000 S.F., 31.66 ton		5.60	4.81	10.41
3680	Restaurants, 500 S.F., 2.50 ton		11.05	5.40	16.45
3840	10,000 S.F., 50.00 ton		10.15	7.55	17.70
3920	Schools and colleges, 500 S.F., 1.92 ton		8.50	4.13	12.63
4080	10,000 S.F., 38.33 ton		7.65	5.85	13.50

D30 HVAC

D3050 Terminal & Package Units

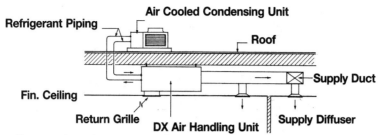

General: Split systems offer several important advantages which should be evaluated when a selection is to be made. They provide a greater degree of flexibility in component selection which permits an accurate match-up of the proper equipment size and type with the particular needs of the building. This allows for maximum use of modern energy saving concepts in heating and cooling. Outdoor installation of the air cooled condensing unit allows space savings in the building and also isolates the equipment operating sounds from building occupants.

Design Assumptions: The systems below are comprised of a direct expansion air handling unit and air cooled condensing unit with interconnecting copper tubing. Ducts and diffusers are also included for distribution of air. Systems are priced for cooling only. Heat can be added as desired either by putting hot water/steam coils into the air unit or into the duct supplying the particular area of need. Gas fired duct furnaces are also available. Refrigerant liquid line is insulated.

System Components	QUANTITY	UNIT	COST EACH		
			MAT.	INST.	TOTAL
SYSTEM D3050 170 1280					
SPLIT SYSTEM, AIR COOLED CONDENSING UNIT					
APARTMENT CORRIDORS, 1,000 S.F., 1.80 TON					
Fan coil AC unit, cabinet mntd. & filters direct expansion air cool	1.000	Ea.	411.75	129.93	541.68
Ductwork package, for split system, remote condensing unit	1.000	System	101.57	722.85	824.42
Refrigeration piping	1.000	System	406.26	814.35	1,220.61
Condensing unit, air cooled, incls. compressor & standard controls	1.000	Ea.	1,525	518.50	2,043.50
TOTAL			2,444.58	2,185.63	4,630.21
COST PER S.F.			2.44	2.19	4.63
*Cooling requirements would lead to choosing a water cooled unit					

D3050 170	Split Systems With Air Cooled Condensing Units	COST PER S.F.		
		MAT.	INST.	TOTAL
1260	Split system, air cooled condensing unit			
1280	Apartment corridors, 1,000 S.F., 1.83 ton	2.45	2.18	4.63
1440	20,000 S.F., 36.66 ton	2.58	3.08	5.66
1520	Banks and libraries, 1,000 S.F., 4.17 ton	4.02	5	9.02
1680	20,000 S.F., 83.32 ton	7.05	7.25	14.30
1760	Bars and taverns, 1,000 S.F., 11.08 ton	10.50	9.90	20.40
1880	10,000 S.F., 110.84 ton	16.05	11.55	27.60
2000	Bowling alleys, 1,000 S.F., 5.66 ton	5.70	9.40	15.10
2160	20,000 S.F., 113.32 ton	11.10	10.30	21.40
2320	Department stores, 1,000 S.F., 2.92 ton	2.86	3.46	6.32
2480	20,000 S.F., 58.33 ton	4.10	4.89	8.99
2560	Drug stores, 1,000 S.F., 6.66 ton	6.70	11.05	17.75
2720	20,000 S.F., 133.32 ton*			
2800	Factories, 1,000 S.F., 3.33 ton	3.25	3.94	7.19
2960	20,000 S.F., 66.66 ton	5.10	5.85	10.95
3040	Food supermarkets, 1,000 S.F., 2.83 ton	2.78	3.35	6.13
3200	20,000 S.F., 56.66 ton	3.98	4.76	8.74
3280	Medical centers, 1,000 S.F., 2.33 ton	2.46	2.72	5.18
3440	20,000 S.F., 46.66 ton	3.28	3.91	7.19
3520	Offices, 1,000 S.F., 3.17 ton	3.09	3.75	6.84
3680	20,000 S.F., 63.32 ton	4.85	5.55	10.40
3760	Restaurants, 1,000 S.F., 5.00 ton	5.05	8.30	13.35
3920	20,000 S.F., 100.00 ton	9.80	9.10	18.90
4000	Schools and colleges, 1,000 S.F., 3.83 ton	3.69	4.62	8.31
4160	20,000 S.F., 76.66 ton	5.90	6.70	12.60

D30 HVAC

D3050 Terminal & Package Units

Computer rooms impose special requirements on air conditioning systems. A prime requirement is reliability, due to the potential monetary loss that could be incurred by a system failure. A second basic requirement is the tolerance of control with which temperature and humidity are regulated, and dust eliminated. As the air conditioning system reliability is so vital, the additional cost of reserve capacity and redundant components is often justified.

System Descriptions: Computer areas may be environmentally controlled by one of three methods as follows:

1. Self-contained Units
These are units built to higher standards of performance and reliability. They usually contain alarms and controls to indicate component operation failure, filter change, etc. It should be remembered that these units in the room will occupy space that is relatively expensive to build and that all alterations and service of the equipment will also have to be accomplished within the computer area.

2. Decentralized Air Handling Units In operation these are similar to the self-contained units except that their cooling capability comes from remotely located refrigeration equipment as refrigerant or chilled water. As no compressors or refrigerating equipment are required in the air units, they are smaller and require less service than self-contained units. An added plus for this type of system occurs if some of the computer components themselves also require chilled water for cooling.

3. Central System Supply Cooling is obtained from a central source which, since it is not located within the computer room, may have excess capacity and permit greater flexibility without interfering with the computer components. System performance criteria must still be met.

Note: The costs shown below do not include an allowance for ductwork or piping.

D3050 185	Computer Room Cooling Units		MAT.	INST.	TOTAL
0560	Computer room unit, air cooled, includes remote condenser				
0580	3 ton	R236000 -20	17,200	2,125	19,325
0600	5 ton		18,400	2,375	20,775
0620	8 ton	R233400 -10	34,500	3,950	38,450
0640	10 ton		36,100	4,275	40,375
0660	15 ton		39,700	4,850	44,550
0680	20 ton		48,200	6,925	55,125
0700	23 ton		59,500	7,900	67,400
0800	Chilled water, for connection to existing chiller system				
0820	5 ton		13,200	1,450	14,650
0840	8 ton		13,300	2,125	15,425
0860	10 ton		13,400	2,175	15,575
0880	15 ton		14,200	2,225	16,425
0900	20 ton		15,000	2,325	17,325
0920	23 ton		16,000	2,625	18,625
1000	Glycol system, complete except for interconnecting tubing				
1020	3 ton		21,700	2,675	24,375
1040	5 ton		23,700	2,800	26,500
1060	8 ton		38,100	4,650	42,750
1080	10 ton		40,500	5,075	45,575
1100	15 ton		50,000	6,375	56,375
1120	20 ton		55,500	6,925	62,425
1140	23 ton		58,000	7,225	65,225
1240	Water cooled, not including condenser water supply or cooling tower				
1260	3 ton		17,500	1,725	19,225
1280	5 ton		19,100	1,975	21,075
1300	8 ton		31,300	3,225	34,525
1320	15 ton		38,000	3,950	41,950
1340	20 ton		40,900	4,375	45,275
1360	23 ton		42,800	4,750	47,550

D40 Fire Protection

D4010 Sprinklers

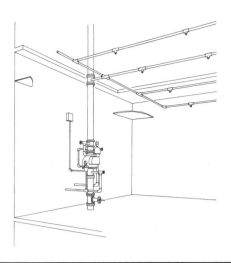

Dry Pipe System: A system employing automatic sprinklers attached to a piping system containing air under pressure, the release of which as from the opening of sprinklers permits the water pressure to open a valve known as a "dry pipe valve". The water then flows into the piping system and out the opened sprinklers.

All areas are assumed to be open.

System Components	QUANTITY	UNIT	COST EACH MAT.	COST EACH INST.	COST EACH TOTAL
SYSTEM D4010 310 0580					
DRY PIPE SPRINKLER, STEEL BLACK, SCH. 40 PIPE					
LIGHT HAZARD, ONE FLOOR, 2000 S.F.					
Valve, gate, iron body 125 lb., OS&Y, flanged, 4" pipe size	1.000	Ea.	420	262.50	682.50
Valve, swing check, bronze, 125 lb., regrinding disc, 2-1/2" pipe size	1.000	Ea.	420	52.50	472.50
Valve, angle, bronze, 150 lb., rising stem, threaded, 2" pipe size	1.000	Ea.	393.75	39.75	433.50
*Alarm valve, 2-1/2" pipe size	1.000	Ea.	1,012.50	258.75	1,271.25
Alarm, water motor, complete with gong	1.000	Ea.	251.25	108.75	360
Fire alarm horn, electric	1.000	Ea.	33.38	62.63	96.01
Valve swing check w/balldrip CI with brass trim, 4" pipe size	1.000	Ea.	210	258.75	468.75
Pipe, steel, black, schedule 40, 4" diam.	10.000	L.F.	149.25	226.20	375.45
Dry pipe valve, trim & gauges, 4" pipe size	1.000	Ea.	1,668.75	787.50	2,456.25
Pipe, steel, black, schedule 40, threaded, cplg & hngr 10'OC 2-1/2" diam.	20.000	L.F.	198.75	315	513.75
Pipe, steel, black, schedule 40, threaded, cplg & hngr 10'OC 2" diam.	12.500	L.F.	79.22	154.22	233.44
Pipe, steel, black, schedule 40, threaded, cplg & hngr 10'OC 1-1/4" diam.	37.500	L.F.	151.88	333.28	485.16
Pipe, steel, black, schedule 40, threaded, cplg & hngr 10'OC 1" diam.	112.000	L.F.	357.84	928.20	1,286.04
Pipe Tee, malleable iron black, 150 lb. threaded, 4" pipe size	2.000	Ea.	291	394.50	685.50
Pipe Tee, malleable iron black, 150 lb. threaded, 2-1/2" pipe size	2.000	Ea.	81.75	175.50	257.25
Pipe Tee, malleable iron black, 150 lb. threaded, 2" pipe size	1.000	Ea.	19.13	72	91.13
Pipe Tee, malleable iron black, 150 lb. threaded, 1-1/4" pipe size	5.000	Ea.	44.63	283.13	327.76
Pipe Tee, malleable iron black, 150 lb. threaded, 1" pipe size	4.000	Ea.	22.05	219	241.05
Pipe 90° elbow malleable iron black, 150 lb. threaded, 1" pipe size	6.000	Ea.	21.15	202.50	223.65
Sprinkler head dry 1/2" orifice 1" NPT, 3" to 4-3/4" length	12.000	Ea.	1,194	498	1,692
Air compressor, 200 Gal sprinkler system capacity, 1/3 HP	1.000	Ea.	573.75	333.75	907.50
*Standpipe connection, wall, flush, brs. w/plug & chain 2-1/2"x2-1/2"	1.000	Ea.	101.25	156	257.25
Valve gate bronze, 300 psi, NRS, class 150, threaded, 1" pipe size	1.000	Ea.	58.88	23.25	82.13
TOTAL			7,754.16	6,145.66	13,899.82
COST PER S.F.			3.88	3.07	6.95

D4010 310	Dry Pipe Sprinkler Systems		COST PER S.F. MAT.	COST PER S.F. INST.	COST PER S.F. TOTAL
0520	Dry pipe sprinkler systems, steel, black, sch. 40 pipe	R211313-10			
0530	Light hazard, one floor, 500 S.F.		7.25	5.25	12.50
0560	1000 S.F.	R211313-20	4.06	3.03	7.09
0580	2000 S.F.		3.88	3.06	6.94
0600	5000 S.F.		2.05	2.08	4.13
0620	10,000 S.F.		1.47	1.74	3.21
0640	50,000 S.F.		1.17	1.52	2.69
0660	Each additional floor, 500 S.F.		1.64	2.54	4.18

D40 Fire Protection

D4010 Sprinklers

D4010 310	Dry Pipe Sprinkler Systems	COST PER S.F.		
		MAT.	INST.	TOTAL
0680	1000 S.F.	1.35	2.08	3.43
0700	2000 S.F.	1.32	1.95	3.27
0720	5000 S.F.	1.10	1.65	2.75
0740	10,000 S.F.	1.01	1.52	2.53
0760	50,000 S.F.	.94	1.34	2.28
1000	Ordinary hazard, one floor, 500 S.F.	7.40	5.30	12.70
1020	1000 S.F.	4.18	3.07	7.25
1040	2000 S.F.	3.98	3.21	7.19
1060	5000 S.F.	2.34	2.22	4.56
1080	10,000 S.F.	1.93	2.29	4.22
1100	50,000 S.F.	1.70	2.15	3.85
1140	Each additional floor, 500 S.F.	1.79	2.60	4.39
1160	1000 S.F.	1.58	2.34	3.92
1180	2000 S.F.	1.56	2.13	3.69
1200	5000 S.F.	1.44	1.82	3.26
1220	10,000 S.F.	1.33	1.80	3.13
1240	50,000 S.F.	1.24	1.55	2.79
1500	Extra hazard, one floor, 500 S.F.	10.15	6.60	16.75
1520	1000 S.F.	6.20	4.73	10.93
1540	2000 S.F.	4.47	4.12	8.59
1560	5000 S.F.	2.77	3.10	5.87
1580	10,000 S.F.	2.83	2.95	5.78
1600	50,000 S.F.	2.99	2.83	5.82
1660	Each additional floor, 500 S.F.	2.37	3.23	5.60
1680	1000 S.F.	2.32	3.06	5.38
1700	2000 S.F.	2.17	3.07	5.24
1720	5000 S.F.	1.87	2.67	4.54
1740	10,000 S.F.	2.14	2.44	4.58
1760	50,000 S.F.	2.15	2.33	4.48
2020	Grooved steel, black, sch. 40 pipe, light hazard, one floor, 2000 S.F.	3.73	2.65	6.38
2060	10,000 S.F.	1.49	1.52	3.01
2100	Each additional floor, 2000 S.F.	1.31	1.57	2.88
2150	10,000 S.F.	1.03	1.30	2.33
2200	Ordinary hazard, one floor, 2000 S.F.	3.95	2.79	6.74
2250	10,000 S.F.	1.89	1.92	3.81
2300	Each additional floor, 2000 S.F.	1.53	1.71	3.24
2350	10,000 S.F.	1.44	1.71	3.15
2400	Extra hazard, one floor, 2000 S.F.	4.47	3.54	8.01
2450	10,000 S.F.	2.56	2.48	5.04
2500	Each additional floor, 2000 S.F.	2.18	2.54	4.72
2550	10,000 S.F.	1.98	2.18	4.16
3050	Grooved steel black sch. 10 pipe, light hazard, one floor, 2000 S.F.	3.66	2.62	6.28
3100	10,000 S.F.	1.43	1.49	2.92
3150	Each additional floor, 2000 S.F.	1.24	1.54	2.78
3200	10,000 S.F.	.97	1.27	2.24
3250	Ordinary hazard, one floor, 2000 S.F.	3.88	2.78	6.66
3300	10,000 S.F.	1.75	1.88	3.63
3350	Each additional floor, 2000 S.F.	1.46	1.70	3.16
3400	10,000 S.F.	1.30	1.67	2.97
3450	Extra hazard, one floor, 2000 S.F.	4.41	3.51	7.92
3500	10,000 S.F.	2.42	2.44	4.86
3550	Each additional floor, 2000 S.F.	2.12	2.51	4.63
3600	10,000 S.F.	1.90	2.15	4.05
4050	Copper tubing, type M, light hazard, one floor, 2000 S.F.	4.56	2.61	7.17
4100	10,000 S.F.	2.26	1.52	3.78
4150	Each additional floor, 2000 S.F.	2.16	1.56	3.72
4200	10,000 S.F.	1.80	1.31	3.11
4250	Ordinary hazard, one floor, 2000 S.F.	4.94	2.92	7.86

D40 Fire Protection

D4010 Sprinklers

D4010 310	Dry Pipe Sprinkler Systems	COST PER S.F.		
		MAT.	INST.	TOTAL
4300	10,000 S.F.	2.82	1.78	4.60
4350	Each additional floor, 2000 S.F.	2.96	1.82	4.78
4400	10,000 S.F.	2.30	1.54	3.84
4450	Extra hazard, one floor, 2000 S.F.	5.75	3.55	9.30
4500	10,000 S.F.	5.50	2.66	8.16
4550	Each additional floor, 2000 S.F.	3.47	2.55	6.02
4600	10,000 S.F.	3.96	2.34	6.30
5050	Copper tubing, type M, T-drill system, light hazard, one floor			
5060	2000 S.F.	4.58	2.45	7.03
5100	10,000 S.F.	2.14	1.27	3.41
5150	Each additional floor, 2000 S.F.	2.18	1.40	3.58
5200	10,000 S.F.	1.68	1.06	2.74
5250	Ordinary hazard, one floor, 2000 S.F.	4.75	2.51	7.26
5300	10,000 S.F.	2.73	1.60	4.33
5350	Each additional floor, 2000 S.F.	2.33	1.43	3.76
5400	10,000 S.F.	2.15	1.32	3.47
5450	Extra hazard, one floor, 2000 S.F.	5.20	2.94	8.14
5500	10,000 S.F.	4.55	1.95	6.50
5550	Each additional floor, 2000 S.F.	2.91	1.94	4.85
5600	10,000 S.F.	3.03	1.63	4.66

D40 Fire Protection

D4010 Sprinklers

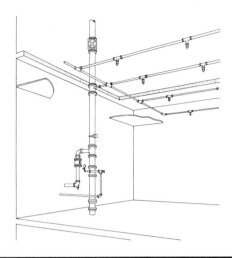

Wet Pipe System. A system employing automatic sprinklers attached to a piping system containing water and connected to a water supply so that water discharges immediately from sprinklers opened by heat from a fire.

All areas are assumed to be open.

System Components	QUANTITY	UNIT	COST EACH MAT.	COST EACH INST.	COST EACH TOTAL
SYSTEM D4010 410 0580					
WET PIPE SPRINKLER, STEEL, BLACK, SCH. 40 PIPE					
LIGHT HAZARD, ONE FLOOR, 2000 S.F.					
Valve, gate, iron body, 125 lb., OS&Y, flanged, 4" diam.	1.000	Ea.	420	262.50	682.50
Valve, swing check, bronze, 125 lb., regrinding disc, 2-1/2" pipe size	1.000	Ea.	420	52.50	472.50
Valve, angle, bronze, 150 lb., rising stem, threaded, 2" diam.	1.000	Ea.	393.75	39.75	433.50
*Alarm valve, 2-1/2" pipe size	1.000	Ea.	1,012.50	258.75	1,271.25
Alarm, water motor, complete with gong	1.000	Ea.	251.25	108.75	360
Valve, swing check, w/balldrip CI with brass trim 4" pipe size	1.000	Ea.	210	258.75	468.75
Pipe, steel, black, schedule 40, 4" diam.	10.000	L.F.	149.25	226.20	375.45
*Flow control valve, trim & gauges, 4" pipe size	1.000	Set	2,325	588.75	2,913.75
Fire alarm horn, electric	1.000	Ea.	33.38	62.63	96.01
Pipe, steel, black, schedule 40, threaded, cplg & hngr 10' OC, 2-1/2" diam.	20.000	L.F.	198.75	315	513.75
Pipe, steel, black, schedule 40, threaded, cplg & hngr 10' OC, 2" diam.	12.500	L.F.	79.22	154.22	233.44
Pipe, steel, black, schedule 40, threaded, cplg & hngr 10' OC, 1-1/4" diam.	37.500	L.F.	151.88	333.28	485.16
Pipe steel, black, schedule 40, threaded cplg & hngr 10' OC, 1" diam.	112.000	L.F.	357.84	928.20	1,286.04
Pipe Tee, malleable iron black, 150 lb. threaded, 4" pipe size	2.000	Ea.	291	394.50	685.50
Pipe Tee, malleable iron black, 150 lb. threaded, 2-1/2" pipe size	2.000	Ea.	81.75	175.50	257.25
Pipe Tee, malleable iron black, 150 lb. threaded, 2" pipe size	1.000	Ea.	19.13	72	91.13
Pipe Tee, malleable iron black, 150 lb. threaded, 1-1/4" pipe size	5.000	Ea.	44.63	283.13	327.76
Pipe Tee, malleable iron black, 150 lb. threaded, 1" pipe size	4.000	Ea.	22.05	219	241.05
Pipe 90° elbow, malleable iron black, 150 lb. threaded, 1" pipe size	6.000	Ea.	21.15	202.50	223.65
Sprinkler head, standard spray, brass 135°-286° F 1/2" NPT, 3/8" orifice	12.000	Ea.	151.20	432	583.20
Valve, gate, bronze, NRS, class 150, threaded, 1" pipe size	1.000	Ea.	58.88	23.25	82.13
*Standpipe connection, wall, single, flush w/plug & chain 2-1/2"x2-1/2"	1.000	Ea.	101.25	156	257.25
TOTAL			6,793.86	5,547.16	12,341.02
COST PER S.F.			3.40	2.77	6.17

*Not included in systems under 2000 S.F.

D4010 410	Wet Pipe Sprinkler Systems		COST PER S.F. MAT.	COST PER S.F. INST.	COST PER S.F. TOTAL
0520	Wet pipe sprinkler systems, steel, black, sch. 40 pipe	R211313-10			
0530	Light hazard, one floor, 500 S.F.		2.10	2.68	4.78
0560	1000 S.F.		3.27	2.75	6.02
0580	2000 S.F.	R211313-20	3.40	2.76	6.16
0600	5000 S.F.		1.65	1.94	3.59
0620	10,000 S.F.		1.10	1.67	2.77

D40 Fire Protection

D4010 Sprinklers

D4010 410	Wet Pipe Sprinkler Systems	COST PER S.F.		
		MAT.	INST.	TOTAL
0640	50,000 S.F.	.85	1.48	2.33
0660	Each additional floor, 500 S.F.	1.01	2.27	3.28
0680	1000 S.F.	.98	2.11	3.09
0700	2000 S.F.	.93	1.92	2.85
0720	5000 S.F.	.68	1.62	2.30
0740	10,000 S.F.	.64	1.50	2.14
0760	50,000 S.F.	.57	1.16	1.73
1000	Ordinary hazard, one floor, 500 S.F.	2.32	2.88	5.20
1020	1000 S.F.	3.24	2.71	5.95
1040	2000 S.F.	3.50	2.91	6.41
1060	5000 S.F.	1.84	2.08	3.92
1080	10,000 S.F.	1.41	2.21	3.62
1100	50,000 S.F.	1.17	2.07	3.24
1140	Each additional floor, 500 S.F.	1.27	2.57	3.84
1160	1000 S.F.	.95	2.09	3.04
1180	2000 S.F.	1.04	2.10	3.14
1200	5000 S.F.	1.05	1.98	3.03
1220	10,000 S.F.	.96	2.05	3.01
1240	50,000 S.F.	.87	1.81	2.68
1500	Extra hazard, one floor, 500 S.F.	6.15	4.44	10.59
1520	1000 S.F.	4.09	3.82	7.91
1540	2000 S.F.	3.73	3.91	7.64
1560	5000 S.F.	2.55	3.41	5.96
1580	10,000 S.F.	2.19	3.24	5.43
1600	50,000 S.F.	2.37	3.11	5.48
1660	Each additional floor, 500 S.F.	1.58	3.18	4.76
1680	1000 S.F.	1.53	3.01	4.54
1700	2000 S.F.	1.38	3.02	4.40
1720	5000 S.F.	1.19	2.67	3.86
1740	10,000 S.F.	1.37	2.44	3.81
1760	50,000 S.F.	1.36	2.32	3.68
2020	Grooved steel, black sch. 40 pipe, light hazard, one floor, 2000 S.F.	3.38	2.35	5.73
2060	10,000 S.F.	1.36	1.51	2.87
2100	Each additional floor, 2000 S.F.	.92	1.54	2.46
2150	10,000 S.F.	.66	1.28	1.94
2200	Ordinary hazard, one floor, 2000 S.F.	3.47	2.49	5.96
2250	10,000 S.F.	1.37	1.84	3.21
2300	Each additional floor, 2000 S.F.	1.01	1.68	2.69
2350	10,000 S.F.	.92	1.68	2.60
2400	Extra hazard, one floor, 2000 S.F.	3.72	3.22	6.94
2450	10,000 S.F.	1.77	2.38	4.15
2500	Each additional floor, 2000 S.F.	1.39	2.49	3.88
2550	10,000 S.F.	1.19	2.13	3.32
3050	Grooved steel black sch. 10 pipe, light hazard, one floor, 2000 S.F.	3.31	2.32	5.63
3100	10,000 S.F.	1.06	1.42	2.48
3150	Each additional floor, 2000 S.F.	.85	1.51	2.36
3200	10,000 S.F.	.60	1.25	1.85
3250	Ordinary hazard, one floor, 2000 S.F.	3.40	2.48	5.88
3300	10,000 S.F.	1.23	1.80	3.03
3350	Each additional floor, 2000 S.F.	.94	1.67	2.61
3400	10,000 S.F.	.78	1.64	2.42
3450	Extra hazard, one floor, 2000 S.F.	3.66	3.19	6.85
3500	10,000 S.F.	1.63	2.34	3.97
3550	Each additional floor, 2000 S.F.	1.33	2.46	3.79
3600	10,000 S.F.	1.11	2.10	3.21
4050	Copper tubing, type M, light hazard, one floor, 2000 S.F.	4.21	2.31	6.52
4100	10,000 S.F.	1.89	1.45	3.34
4150	Each additional floor, 2000 S.F.	1.77	1.53	3.30

D40 Fire Protection

D4010 Sprinklers

D4010 410	Wet Pipe Sprinkler Systems	COST PER S.F.		
		MAT.	INST.	TOTAL
4200	10,000 S.F.	1.43	1.29	2.72
4250	Ordinary hazard, one floor, 2000 S.F.	4.46	2.62	7.08
4300	10,000 S.F.	2.30	1.70	4
4350	Each additional floor, 2000 S.F.	2.09	1.73	3.82
4400	10,000 S.F.	1.78	1.51	3.29
4450	Extra hazard, one floor, 2000 S.F.	5	3.23	8.23
4500	10,000 S.F.	4.66	2.56	7.22
4550	Each additional floor, 2000 S.F.	2.68	2.50	5.18
4600	10,000 S.F.	3.17	2.29	5.46
5050	Copper tubing, type M, T-drill system, light hazard, one floor			
5060	2000 S.F.	4.23	2.15	6.38
5100	10,000 S.F.	1.77	1.20	2.97
5150	Each additional floor, 2000 S.F.	1.79	1.37	3.16
5200	10,000 S.F.	1.31	1.04	2.35
5250	Ordinary hazard, one floor, 2000 S.F.	4.27	2.21	6.48
5300	10,000 S.F.	2.21	1.52	3.73
5350	Each additional floor, 2000 S.F.	1.81	1.40	3.21
5400	10,000 S.F.	1.76	1.36	3.12
5450	Extra hazard, one floor, 2000 S.F.	4.45	2.62	7.07
5500	10,000 S.F.	3.73	1.85	5.58
5550	Each additional floor, 2000 S.F.	2.19	1.93	4.12
5600	10,000 S.F.	2.24	1.58	3.82

D40 Fire Protection

D4020 Standpipes

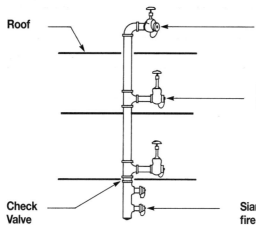

System Components	QUANTITY	UNIT	COST PER FLOOR		
			MAT.	INST.	TOTAL
SYSTEM D4020 310 0560					
WET STANDPIPE RISER, CLASS I, STEEL, BLACK, SCH. 40 PIPE, 10' HEIGHT					
4" DIAMETER PIPE, ONE FLOOR					
Pipe, steel, black, schedule 40, threaded, 4" diam.	20.000	L.F.	500	590	1,090
Pipe, Tee, malleable iron, black, 150 lb. threaded, 4" pipe size	2.000	Ea.	388	526	914
Pipe, 90° elbow, malleable iron, black, 150 lb. threaded 4" pipe size	1.000	Ea.	123	176	299
Pipe, nipple, steel, black, schedule 40, 2-1/2" pipe size x 3" long	2.000	Ea.	24.90	132	156.90
Fire valve, gate, 300 lb., brass w/handwheel, 2-1/2" pipe size	1.000	Ea.	171	82.50	253.50
Fire valve, pressure restricting, adj., rgh. brs., 2-1/2" pipe size	1.000	Ea.	304	165	469
Valve, swing check, w/ball drip, CI w/brs. ftngs., 4" pipe size	1.000	Ea.	280	345	625
Standpipe conn wall dble. flush brs. w/plugs & chains 2-1/2"x2-1/2"x4"	1.000	Ea.	470	208	678
Valve, swing check, bronze, 125 lb., regrinding disc, 2-1/2" pipe size	1.000	Ea.	560	70	630
Roof manifold, fire, w/valves & caps, horiz./vert. brs. 2-1/2"x2-1/2"x4"	1.000	Ea.	152	217	369
Fire, hydrolator, vent & drain, 2-1/2" pipe size	1.000	Ea.	64.50	48	112.50
Valve, gate, iron body 125 lb., OS&Y, threaded, 4" pipe size	1.000	Ea.	560	350	910
TOTAL			3,597.40	2,909.50	6,506.90

D4020 310	Wet Standpipe Risers, Class I		COST PER FLOOR		
			MAT.	INST.	TOTAL
0550	Wet standpipe risers, Class I, steel black sch. 40, 10' height				
0560	4" diameter pipe, one floor	R211313-10	3,600	2,900	6,500
0580	Additional floors		855	905	1,760
0600	6" diameter pipe, one floor	R211313-20	5,925	5,100	11,025
0620	Additional floors		1,450	1,425	2,875
0640	8" diameter pipe, one floor		8,850	6,175	15,025
0660	Additional floors		2,075	1,725	3,800
0680					

D4020 310	Wet Standpipe Risers, Class II	COST PER FLOOR		
		MAT.	INST.	TOTAL
1030	Wet standpipe risers, Class II, steel black sch. 40, 10' height			
1040	2" diameter pipe, one floor	1,425	1,050	2,475
1060	Additional floors	465	405	870
1080	2-1/2" diameter pipe, one floor	2,075	1,525	3,600
1100	Additional floors	535	470	1,005
1120				

D40 Fire Protection

D4020 Standpipes

D4020 310	Wet Standpipe Risers, Class III	COST PER FLOOR		
		MAT.	INST.	TOTAL
1530	Wet standpipe risers, Class III, steel black sch. 40, 10' height			
1540	4" diameter pipe, one floor	3,700	2,900	6,600
1560	Additional floors	725	755	1,480
1580	6" diameter pipe, one floor	6,025	5,100	11,125
1600	Additional floors	1,500	1,425	2,925
1620	8" diameter pipe, one floor	8,950	6,175	15,125
1640	Additional floors	2,125	1,725	3,850

D40 Fire Protection

D4020 Standpipes

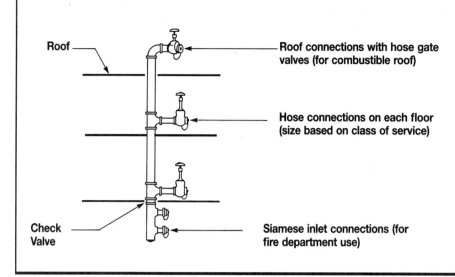

System Components	QUANTITY	UNIT	COST PER FLOOR		
			MAT.	INST.	TOTAL
SYSTEM D4020 330 0540					
DRY STANDPIPE RISER, CLASS I, PIPE, STEEL, BLACK, SCH 40, 10' HEIGHT					
4" DIAMETER PIPE, ONE FLOOR					
Pipe, steel, black, schedule 40, threaded, 4" diam.	20.000	L.F.	500	590	1,090
Pipe, Tee, malleable iron, black, 150 lb. threaded, 4" pipe size	2.000	Ea.	388	526	914
Pipe, 90° elbow, malleable iron, black, 150 lb. threaded 4" pipe size	1.000	Ea.	123	176	299
Pipe, nipple, steel, black, schedule 40, 2-1/2" pipe size x 3" long	2.000	Ea.	24.90	132	156.90
Fire valve gate NRS 300 lb., brass w/handwheel, 2-1/2" pipe size	1.000	Ea.	171	82.50	253.50
Fire valve, pressure restricting, adj., rgh. brs., 2-1/2" pipe size	1.000	Ea.	152	82.50	234.50
Standpipe conn wall dble. flush brs. w/plugs & chains 2-1/2"x2-1/2"x4"	1.000	Ea.	470	208	678
Valve swing check w/ball drip Cl w/brs. ftngs., 4"pipe size	1.000	Ea.	280	345	625
Roof manifold, fire, w/valves & caps, horiz./vert. brs. 2-1/2"x2-1/2"x4"	1.000	Ea.	152	217	369
TOTAL			2,260.90	2,359	4,619.90

D4020 330	Dry Standpipe Risers, Class I		COST PER FLOOR		
			MAT.	INST.	TOTAL
0530	Dry standpipe riser, Class I, steel black sch. 40, 10' height	R211313 -10			
0540	4" diameter pipe, one floor		2,250	2,350	4,600
0560	Additional floors		790	855	1,645
0580	6" diameter pipe, one floor	R211313 -20	4,325	4,050	8,375
0600	Additional floors		1,375	1,375	2,750
0620	8" diameter pipe, one floor		6,500	4,925	11,425
0640	Additional floors		2,000	1,675	3,675
0660					

D4020 330	Dry Standpipe Risers, Class II	COST PER FLOOR		
		MAT.	INST.	TOTAL
1030	Dry standpipe risers, Class II, steel black sch. 40, 10' height			
1040	2" diameter pipe, one floor	1,225	1,100	2,325
1060	Additional floor	400	355	755
1080	2-1/2" diameter pipe, one floor	1,650	1,275	2,925
1100	Additional floors	470	425	895
1120				

D40 Fire Protection

D4020 Standpipes

D4020 330	Dry Standpipe Risers, Class III	COST PER FLOOR		
		MAT.	INST.	TOTAL
1530	Dry standpipe risers, Class III, steel black sch. 40, 10' height			
1540	4" diameter pipe, one floor	2,300	2,325	4,625
1560	Additional floors	670	775	1,445
1580	6" diameter pipe, one floor	4,375	4,050	8,425
1600	Additional floors	1,425	1,375	2,800
1620	8" diameter pipe, one floor	6,550	4,925	11,475
1640	Additional floor	2,050	1,675	3,725

D40 Fire Protection

D4020 Standpipes

D4020 410	Fire Hose Equipment	COST EACH		
		MAT.	INST.	TOTAL
0100	Adapters, reducing, 1 piece, FxM, hexagon, cast brass, 2-1/2" x 1-1/2"	50		50
0200	Pin lug, 1-1/2" x 1"	43.50		43.50
0250	3" x 2-1/2"	113		113
0300	For polished chrome, add 75% mat.			
0400	Cabinets, D.S. glass in door, recessed, steel box, not equipped			
0500	Single extinguisher, steel door & frame	95.50	130	225.50
0550	Stainless steel door & frame	204	130	334
0600	Valve, 2-1/2" angle, steel door & frame	123	86.50	209.50
0650	Aluminum door & frame	149	86.50	235.50
0700	Stainless steel door & frame	200	86.50	286.50
0750	Hose rack assy, 2-1/2" x 1-1/2" valve & 100' hose, steel door & frame	229	173	402
0800	Aluminum door & frame	335	173	508
0850	Stainless steel door & frame	450	173	623
0900	Hose rack assy,& extinguisher, 2-1/2"x1-1/2" valve & hose, steel door & fram	228	208	436
0950	Aluminum	430	208	638
1000	Stainless steel	785	208	993
1550	Compressor, air, dry pipe system, automatic, 200 gal., 3/4 H.P.	765	445	1,210
1600	520 gal., 1 H.P.	805	445	1,250
1650	Alarm, electric pressure switch (circuit closer)	265	22	287
2500	Couplings, hose, rocker lug, cast brass, 1-1/2"	48		48
2550	2-1/2"	63		63
3000	Escutcheon plate, for angle valves, polished brass, 1-1/2"	12.15		12.15
3050	2-1/2"	29		29
3500	Fire pump, electric, w/controller, fittings, relief valve			
3550	4" pump, 30 H.P., 500 G.P.M.	21,800	3,250	25,050
3600	5" pump, 40 H.P., 1000 G.P.M.	31,800	3,675	35,475
3650	5" pump, 100 H.P., 1000 G.P.M.	35,700	4,075	39,775
3700	For jockey pump system, add	3,900	520	4,420
5000	Hose, per linear foot, synthetic jacket, lined,			
5100	300 lb. test, 1-1/2" diameter	2.63	.40	3.03
5150	2-1/2" diameter	4.38	.47	4.85
5200	500 lb. test, 1-1/2" diameter	2.72	.40	3.12
5250	2-1/2" diameter	4.72	.47	5.19
5500	Nozzle, plain stream, polished brass, 1-1/2" x 10"	43.50		43.50
5550	2-1/2" x 15" x 13/16" or 1-1/2"	78.50		78.50
5600	Heavy duty combination adjustable fog and straight stream w/handle 1-1/2"	405		405
5650	2-1/2" direct connection	505		505
6000	Rack, for 1-1/2" diameter hose 100 ft. long, steel	54	52	106
6050	Brass	66.50	52	118.50
6500	Reel, steel, for 50 ft. long 1-1/2" diameter hose	118	74.50	192.50
6550	For 75 ft. long 2-1/2" diameter hose	192	74.50	266.50
7050	Siamese, w/plugs & chains, polished brass, sidewalk, 4" x 2-1/2" x 2-1/2"	520	415	935
7100	6" x 2-1/2" x 2-1/2"	710	520	1,230
7200	Wall type, flush, 4" x 2-1/2" x 2-1/2"	470	208	678
7250	6" x 2-1/2" x 2-1/2"	655	226	881
7300	Projecting, 4" x 2-1/2" x 2-1/2"	430	208	638
7350	6" x 2-1/2" x 2-1/2"	725	226	951
7400	For chrome plate, add 15% mat.			
8000	Valves, angle, wheel handle, 300 Lb., rough brass, 1-1/2"	63.50	48	111.50
8050	2-1/2"	66	82.50	148.50
8100	Combination pressure restricting, 1-1/2"	77	48	125
8150	2-1/2"	153	82.50	235.50
8200	Pressure restricting, adjustable, satin brass, 1-1/2"	109	48	157
8250	2-1/2"	152	82.50	234.50
8300	Hydrolator, vent and drain, rough brass, 1-1/2"	64.50	48	112.50
8350	2-1/2"	64.50	48	112.50
8400	Cabinet assy, incls. adapter, rack, hose, and nozzle	675	315	990

D40 Fire Protection

D4090 Other Fire Protection Systems

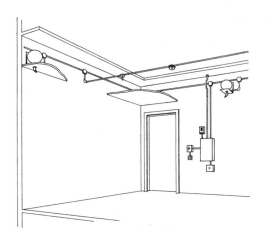

General: Automatic fire protection (suppression) systems other than water sprinklers may be desired for special environments, high risk areas, isolated locations or unusual hazards. Some typical applications would include:

Paint dip tanks
Securities vaults
Electronic data processing
Tape and data storage
Transformer rooms
Spray booths
Petroleum storage
High rack storage

Piping and wiring costs are dependent on the individual application and must be added to the component costs shown below.

All areas are assumed to be open.

D4090 910	Unit Components		MAT.	INST.	TOTAL
0020	Detectors with brackets	R211313 -10			
0040	Fixed temperature heat detector		36	70	106
0060	Rate of temperature rise detector		42.50	70	112.50
0080	Ion detector (smoke) detector		96	90.50	186.50
0200	Extinguisher agent				
0240	200 lb FM200, container		6,700	263	6,963
0280	75 lb carbon dioxide cylinder		1,100	176	1,276
0320	Dispersion nozzle				
0340	FM200 1-1/2" dispersion nozzle		58	42	100
0380	Carbon dioxide 3" x 5" dispersion nozzle		58	32.50	90.50
0420	Control station				
0440	Single zone control station with batteries		1,475	560	2,035
0470	Multizone (4) control station with batteries		2,825	1,125	3,950
0490					
0500	Electric mechanical release		144	286	430
0520					
0550	Manual pull station		52	97.50	149.50
0570					
0640	Battery standby power 10" x 10" x 17"		885	140	1,025
0700					
0740	Bell signalling device		63.50	70	133.50

D4090 920	FM200 Systems	MAT.	INST.	TOTAL
0820	Average FM200 system, minimum			1.52
0840	Maximum			3.03

528

D50 Electrical

D5010 Electrical Service/Distribution

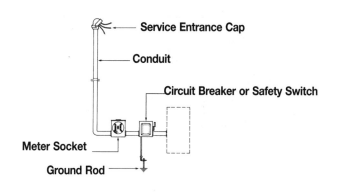

System Components	QUANTITY	UNIT	COST EACH		
			MAT.	INST.	TOTAL
SYSTEM D5010 120 0220					
SERVICE INSTALLATION, INCLUDES BREAKERS, METERING, 20' CONDUIT & WIRE					
3 PHASE, 4 WIRE, 60 A					
Circuit breaker, enclosed (NEMA 1), 600 volt, 3 pole, 60 A	1.000	Ea.	670	200	870
Meter socket, single position, 4 terminal, 100 A	1.000	Ea.	47.50	175	222.50
Rigid galvanized steel conduit, 3/4", including fittings	20.000	L.F.	71	140	211
Wire, 600V type XHHW, copper stranded #6	.900	C.L.F.	110.70	77.40	188.10
Service entrance cap 3/4" diameter	1.000	Ea.	11.55	43	54.55
Conduit LB fitting with cover, 3/4" diameter	1.000	Ea.	15.30	43	58.30
Ground rod, copper clad, 8' long, 3/4" diameter	1.000	Ea.	30.50	106	136.50
Ground rod clamp, bronze, 3/4" diameter	1.000	Ea.	7.35	17.50	24.85
Ground wire, bare armored, #6-1 conductor	.200	C.L.F.	28.80	62	90.80
TOTAL			992.70	863.90	1,856.60

D5010 120	Electric Service, 3 Phase - 4 Wire	COST EACH		
		MAT.	INST.	TOTAL
0200	Service installation, includes breakers, metering, 20' conduit & wire			
0220	3 phase, 4 wire, 120/208 volts, 60 A	995	865	1,860
0240	100 A	1,225	1,050	2,275
0280	200 A	1,875	1,600	3,475
0320	400 A	4,450	2,925	7,375
0360	600 A	8,400	3,975	12,375
0400	800 A	10,800	4,800	15,600
0440	1000 A	13,300	5,500	18,800
0480	1200 A	16,700	5,625	22,325
0520	1600 A	30,200	8,075	38,275
0560	2000 A	33,300	9,200	42,500
0570	Add 25% for 277/480 volt			
0580				
0610	1 phase, 3 wire, 120/240 volts, 100 A	525	940	1,465
0620	200 A	1,100	1,375	2,475

D50 Electrical

D5020 Lighting and Branch Wiring

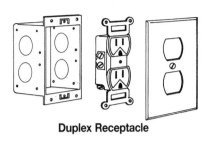

Duplex Receptacle

System Components	QUANTITY	UNIT	COST PER S.F.		
			MAT.	INST.	TOTAL
SYSTEM D5020 110 0200					
RECEPTACLES INCL. PLATE, BOX, CONDUIT, WIRE & TRANS. WHEN REQUIRED					
2.5 PER 1000 S.F., .3 WATTS PER S.F.					
Steel intermediate conduit, (IMC) 1/2" diam	167.000	L.F.	.37	.94	1.31
Wire 600V type THWN-THHN, copper solid #12	3.382	C.L.F.	.06	.17	.23
Wiring device, receptacle, duplex, 120V grounded, 15 amp	2.500	Ea.		.04	.04
Wall plate, 1 gang, brown plastic	2.500	Ea.		.02	.02
Steel outlet box 4" square	2.500	Ea.	.01	.07	.08
Steel outlet box 4" plaster rings	2.500	Ea.	.01	.02	.03
TOTAL			.45	1.26	1.71

D5020 110	Receptacle (by Wattage)	COST PER S.F.		
		MAT.	INST.	TOTAL
0190	Receptacles include plate, box, conduit, wire & transformer when required			
0200	2.5 per 1000 S.F., .3 watts per S.F.	.45	1.26	1.71
0240	With transformer	.52	1.32	1.84
0280	4 per 1000 S.F., .5 watts per S.F.	.50	1.47	1.97
0320	With transformer	.60	1.56	2.16
0360	5 per 1000 S.F., .6 watts per S.F.	.60	1.72	2.32
0400	With transformer	.73	1.84	2.57
0440	8 per 1000 S.F., .9 watts per S.F.	.62	1.90	2.52
0480	With transformer	.80	2.07	2.87
0520	10 per 1000 S.F., 1.2 watts per S.F.	.65	2.07	2.72
0560	With transformer	.94	2.35	3.29
0600	16.5 per 1000 S.F., 2.0 watts per S.F.	.76	2.59	3.35
0640	With transformer	1.26	3.07	4.33
0680	20 per 1000 S.F., 2.4 watts per S.F.	.80	2.82	3.62
0720	With transformer	1.39	3.38	4.77

D50 Electrical

D5020 Lighting and Branch Wiring

Underfloor Receptacle System

Description: Table D5020 115 includes installed costs of raceways and copper wire from panel to and including receptacle.

National Electrical Code prohibits use of undercarpet system in residential, school or hospital buildings. Can only be used with carpet squares.

Low density = (1) Outlet per 259 S.F. of floor area.

High density = (1) Outlet per 127 S.F. of floor area.

System Components	QUANTITY	UNIT	COST PER S.F.		
			MAT.	INST.	TOTAL
SYSTEM D5020 115 0200					
RECEPTACLE SYSTEMS, UNDERFLOOR DUCT, 5' ON CENTER, LOW DENSITY					
Underfloor duct 3-1/8" x 7/8" w/insert 24" on center	.190	L.F.	3.55	1.52	5.07
Vertical elbow for underfloor duct, 3-1/8", included					
Underfloor duct conduit adapter, 2" x 1-1/4", included					
Underfloor duct junction box, single duct, 3-1/8"	.003	Ea.	1.26	.42	1.68
Underfloor junction box carpet pan	.003	Ea.	1.05	.02	1.07
Underfloor duct outlet, high tension receptacle	.004	Ea.	.39	.28	.67
Wire 600V type THWN-THHN copper solid #12	.010	C.L.F.	.18	.51	.69
TOTAL			6.43	2.75	9.18

D5020 115	Receptacles, Floor	COST PER S.F.		
		MAT.	INST.	TOTAL
0200	Receptacle systems, underfloor duct, 5' on center, low density	6.45	2.75	9.20
0240	High density	7	3.54	10.54
0280	7' on center, low density	5.10	2.36	7.46
0320	High density	5.65	3.15	8.80
0400	Poke thru fittings, low density	1.15	1.34	2.49
0440	High density	2.26	2.65	4.91
0520	Telepoles, using Romex, low density	1.14	.82	1.96
0560	High density	2.27	1.66	3.93
0600	Using EMT, low density	1.21	1.08	2.29
0640	High density	2.40	2.19	4.59
0720	Conduit system with floor boxes, low density	1.09	.93	2.02
0760	High density	2.19	1.90	4.09
0840	Undercarpet power system, 3 conductor with 5 conductor feeder, low density	1.46	.33	1.79
0880	High density	2.81	.70	3.51

D50 Electrical

D5020 Lighting and Branch Wiring

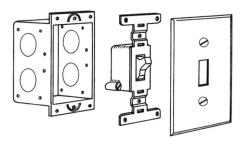

Description: Table D5020 130 includes the cost for switch, plate, box, conduit in slab or EMT exposed and copper wire. Add 20% for exposed conduit.

No power required for switches.

Federal energy guidelines recommend the maximum lighting area controlled per switch shall not exceed 1000 S.F. and that areas over 500 S.F. shall be so controlled that total illumination can be reduced by at least 50%.

System Components	QUANTITY	UNIT	COST PER S.F.		
			MAT.	INST.	TOTAL
SYSTEM D5020 130 0360					
WALL SWITCHES, 5.0 PER 1000 S.F.					
Steel, intermediate conduit (IMC), 1/2" diameter	88.000	L.F.	.20	.49	.69
Wire, 600V type THWN-THHN, copper solid #12	1.710	C.L.F.	.03	.09	.12
Toggle switch, single pole, 15 amp	5.000	Ea.	.03	.07	.10
Wall plate, 1 gang, brown plastic	5.000	Ea.		.04	.04
Steel outlet box 4" plaster rings	5.000	Ea.	.01	.14	.15
Plaster rings	5.000	Ea.	.02	.04	.06
TOTAL			.29	.87	1.16

D5020 130	Wall Switch by Sq. Ft.	COST PER S.F.		
		MAT.	INST.	TOTAL
0200	Wall switches, 1.0 per 1000 S.F.	.07	.20	.27
0240	1.2 per 1000 S.F.	.07	.22	.29
0280	2.0 per 1000 S.F.	.11	.32	.43
0320	2.5 per 1000 S.F.	.13	.41	.54
0360	5.0 per 1000 S.F.	.29	.87	1.16
0400	10.0 per 1000 S.F.	.58	1.76	2.34

D50 Electrical

D5020 Lighting and Branch Wiring

A. Strip Fixture

C. Recessed

B. Surface Mounted

D. Pendent Mounted

Design Assumptions:
1. A 100 footcandle average maintained level of illumination.
2. Ceiling heights range from 9' to 11'.
3. Average reflectance values are assumed for ceilings, walls and floors.
4. Cool white (CW) fluorescent lamps with 3150 lumens for 40 watt lamps and 6300 lumens for 8' slimline lamps.
5. Four 40 watt lamps per 4' fixture and two 8' lamps per 8' fixture.
6. Average fixture efficiency values and spacing to mounting height ratios.
7. Installation labor is average U.S. rate as of January 1.

System Components	QUANTITY	UNIT	COST PER S.F. MAT.	INST.	TOTAL
SYSTEM D5020 208 0520					
FLUORESCENT FIXTURES MOUNTED 9'-11" ABOVE FLOOR, 100 FC					
TYPE A, 8 FIXTURES PER 400 S.F.					
Steel intermediate conduit, (IMC) 1/2" diam.	.404	L.F.	.90	2.26	3.16
Wire, 600V, type THWN-THHN, copper, solid, #12	.008	C.L.F.	.14	.41	.55
Fluorescent strip fixture 8' long, surface mounted, two 75W SL	.020	Ea.	1.23	1.81	3.04
Steel outlet box 4" concrete	.020	Ea.	.33	.56	.89
Steel outlet box plate with stud, 4" concrete	.020	Ea.	.13	.14	.27
TOTAL			2.73	5.18	7.91

D5020 208	Fluorescent Fixtures (by Type)	MAT.	INST.	TOTAL
0520	Fluorescent fixtures, type A, 8 fixtures per 400 S.F.	2.73	5.20	7.93
0560	11 fixtures per 600 S.F.	2.60	5	7.60
0600	17 fixtures per 1000 S.F.	2.54	4.95	7.49
0640	23 fixtures per 1600 S.F.	2.33	4.68	7.01
0680	28 fixtures per 2000 S.F.	2.33	4.68	7.01
0720	41 fixtures per 3000 S.F.	2.29	4.67	6.96
0800	53 fixtures per 4000 S.F.	2.25	4.55	6.80
0840	64 fixtures per 5000 S.F.	2.25	4.55	6.80
0880	Type B, 11 fixtures per 400 S.F.	4.94	7.60	12.54
0920	15 fixtures per 600 S.F.	4.63	7.30	11.93
0960	24 fixtures per 1000 S.F.	4.53	7.25	11.78
1000	35 fixtures per 1600 S.F.	4.27	6.90	11.17
1040	42 fixtures per 2000 S.F.	4.19	6.95	11.14
1080	61 fixtures per 3000 S.F.	4.18	6.70	10.88
1160	80 fixtures per 4000 S.F.	4.09	6.85	10.94
1200	98 fixtures per 5000 S.F.	4.07	6.80	10.87
1240	Type C, 11 fixtures per 400 S.F.	4.05	8	12.05
1280	14 fixtures per 600 S.F.	3.67	7.45	11.12
1320	23 fixtures per 1000 S.F.	3.65	7.40	11.05
1360	34 fixtures per 1600 S.F.	3.56	7.30	10.86
1400	43 fixtures per 2000 S.F.	3.57	7.25	10.82
1440	63 fixtures per 3000 S.F.	3.49	7.15	10.64
1520	81 fixtures per 4000 S.F.	3.42	7.05	10.47
1560	101 fixtures per 5000 S.F.	3.42	7.05	10.47
1600	Type D, 8 fixtures per 400 S.F.	3.73	6.20	9.93
1640	12 fixtures per 600 S.F.	3.73	6.15	9.88
1680	19 fixtures per 1000 S.F.	3.60	6	9.60
1720	27 fixtures per 1600 S.F.	3.40	5.85	9.25
1760	34 fixtures per 2000 S.F.	3.38	5.80	9.18
1800	48 fixtures per 3000 S.F.	3.26	5.65	8.91
1880	64 fixtures per 4000 S.F.	3.26	5.65	8.91
1920	79 fixtures per 5000 S.F.	3.26	5.65	8.91

D50 Electrical

D5020 Lighting and Branch Wiring

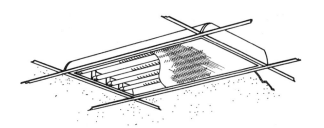

Type C. Recessed, mounted on grid ceiling suspension system, 2' x 4', four 40 watt lamps, acrylic prismatic diffusers.

5.3 watts per S.F. for 100 footcandles.
3 watts per S.F. for 57 footcandles.

System Components	QUANTITY	UNIT	COST PER S.F. MAT.	COST PER S.F. INST.	COST PER S.F. TOTAL
SYSTEM D5020 210 0200					
FLUORESCENT FIXTURES RECESS MOUNTED IN CEILING					
1 WATT PER S.F., 20 FC, 5 FIXTURES PER 1000 S.F.					
Steel intermediate conduit, (IMC) 1/2" diam.	.128	L.F.	.29	.72	1.01
Wire, 600 volt, type THW, copper, solid, #12	.003	C.L.F.	.05	.15	.20
Fluorescent fixture, recessed, 2'x 4', four 40W, w/ lens, for grid ceiling	.005	Ea.	.35	.60	.95
Steel outlet box 4" square	.005	Ea.	.08	.14	.22
Fixture whip, Greenfield w/#12 THHN wire	.005	Ea.	.03	.04	.07
TOTAL			.80	1.65	2.45

D5020 210	Fluorescent Fixtures (by Wattage)	MAT.	INST.	TOTAL
0190	Fluorescent fixtures recess mounted in ceiling			
0195	T-12, standard 40 watt lamps			
0200	1 watt per S.F., 20 FC, 5 fixtures @40 watts per 1000 S.F.	.80	1.65	2.45
0240	2 watt per S.F., 40 FC, 10 fixtures @40 watt per 1000 S.F.	1.60	3.23	4.83
0280	3 watt per S.F., 60 FC, 15 fixtures @40 watt per 1000 S.F	2.40	4.88	7.28
0320	4 watt per S.F., 80 FC, 20 fixtures @40 watt per 1000 S.F.	3.20	6.45	9.65
0400	5 watt per S.F., 100 FC, 25 fixtures @40 watt per 1000 S.F.	4	8.10	12.10
0450	T-8, energy saver 32 watt lamps			
0500	0.8 watt per S.F., 20 FC, 5 fixtures @32 watt per 1000 S.F.	.88	1.65	2.53
0520	1.6 watt per S.F., 40 FC, 10 fixtures @32 watt per 1000 S.F.	1.76	3.23	4.99
0540	2.4 watt per S.F., 60 FC, 15 fixtures @ 32 watt per 1000 S.F	2.64	4.88	7.52
0560	3.2 watt per S.F., 80 FC, 20 fixtures @32 watt per 1000 S.F.	3.52	6.45	9.97
0580	4 watt per S.F., 100 FC, 25 fixtures @32 watt per 1000 S.F.	4.40	8.10	12.50

D50 Electrical

D5020 Lighting and Branch Wiring

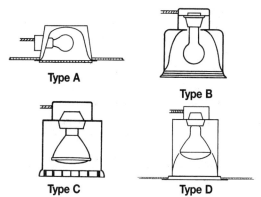

Type A. Recessed wide distribution reflector with flat glass lens 150 W.
 Maximum spacing = 1.2 x mounting height.
 13 watts per S.F. for 100 footcandles.

Type B. Recessed reflector down light with baffles 150 W.
 Maximum spacing = 0.8 x mounting height.
 18 watts per S.F. for 100 footcandles.

Type C. Recessed PAR–38 flood lamp with concentric louver 150 W.
 Maximum spacing = 0.5 x mounting height.
 19 watts per S.F. for 100 footcandles.

Type D. Recessed R–40 flood lamp with reflector skirt.
 Maximum spacing = 0.7 x mounting height.
 15 watts per S.F. for 100 footcandles.

System Components	QUANTITY	UNIT	COST PER S.F.		
			MAT.	INST.	TOTAL
SYSTEM D5020 214 0400					
INCANDESCENT FIXTURE RECESS MOUNTED, 100 FC					
TYPE A, 34 FIXTURES PER 400 S.F.					
Steel intermediate conduit, (IMC) 1/2" diam	1.060	L.F.	2.37	5.94	8.31
Wire, 600V, type THWN-THHN, copper, solid, #12	.033	C.L.F.	.58	1.68	2.26
Steel outlet box 4" square	.085	Ea.	1.39	2.38	3.77
Fixture whip, Greenfield w/#12 THHN wire	.085	Ea.	.56	.60	1.16
Incandescent fixture, recessed, w/lens, prewired, square trim, 200W	.085	Ea.	8.93	7.10	16.03
TOTAL			13.83	17.70	31.53

D5020 214	Incandescent Fixture (by Type)	COST PER S.F.		
		MAT.	INST.	TOTAL
0380	Incandescent fixture recess mounted, 100 FC			
0400	Type A, 34 fixtures per 400 S.F.	13.85	17.70	31.55
0440	49 fixtures per 600 S.F.	13.45	17.40	30.85
0480	63 fixtures per 800 S.F.	13.15	17.20	30.35
0520	90 fixtures per 1200 S.F.	12.70	16.85	29.55
0560	116 fixtures per 1600 S.F.	12.45	16.70	29.15
0600	143 fixtures per 2000 S.F.	12.35	16.65	29
0640	Type B, 47 fixtures per 400 S.F.	12.95	22.50	35.45
0680	66 fixtures per 600 S.F.	12.45	22	34.45
0720	88 fixtures per 800 S.F.	12.45	22	34.45
0760	127 fixtures per 1200 S.F.	12.20	21.50	33.70
0800	160 fixtures per 1600 S.F.	12.05	21.50	33.55
0840	206 fixtures per 2000 S.F.	12	21.50	33.50
0880	Type C, 51 fixtures per 400 S.F.	17.60	23.50	41.10
0920	74 fixtures per 600 S.F.	17.10	23	40.10
0960	97 fixtures per 800 S.F.	16.90	23	39.90
1000	142 fixtures per 1200 S.F.	16.60	22.50	39.10
1040	186 fixtures per 1600 S.F.	16.45	22.50	38.95
1080	230 fixtures per 2000 S.F.	16.35	22.50	38.85
1120	Type D, 39 fixtures per 400 S.F.	16.40	18.40	34.80
1160	57 fixtures per 600 S.F.	16	18.15	34.15
1200	75 fixtures per 800 S.F.	15.90	18.15	34.05
1240	109 fixtures per 1200 S.F.	15.50	17.95	33.45
1280	143 fixtures per 1600 S.F.	15.25	17.75	33
1320	176 fixtures per 2000 S.F.	15.15	17.70	32.85

D50 Electrical

D5020 Lighting and Branch Wiring

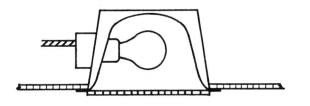

Type A. Recessed, wide distribution reflector with flat glass lens.

150 watt inside frost—2500 lumens per lamp.

PS-25 extended service lamp.

Maximum spacing = 1.2 x mounting height.

13 watts per S.F. for 100 footcandles.

System Components	QUANTITY	UNIT	COST PER S.F. MAT.	COST PER S.F. INST.	COST PER S.F. TOTAL
SYSTEM D5020 216 0200					
INCANDESCENT FIXTURE RECESS MOUNTED, TYPE A					
1 WATT PER S.F., 8 FC, 6 FIXT PER 1000 S.F.					
Steel intermediate conduit, (IMC) 1/2" diam	.091	L.F.	.20	.51	.71
Wire, 600V, type THWN-THHN, copper, solid, #12	.002	C.L.F.	.04	.10	.14
Incandescent fixture, recessed, w/lens, prewired, square trim, 200W	.006	Ea.	.63	.50	1.13
Steel outlet box 4" square	.006	Ea.	.10	.17	.27
Fixture whip, Greenfield w/#12 THHN wire	.006	Ea.	.04	.04	.08
TOTAL			1.01	1.32	2.33

D5020 216	Incandescent Fixture (by Wattage)	MAT.	INST.	TOTAL
0190	Incandescent fixture recess mounted, type A			
0200	1 watt per S.F., 8 FC, 6 fixtures per 1000 S.F.	1.01	1.32	2.33
0240	2 watt per S.F., 16 FC, 12 fixtures per 1000 S.F.	2.02	2.64	4.66
0280	3 watt per S.F., 24 FC, 18 fixtures, per 1000 S.F.	3	3.92	6.92
0320	4 watt per S.F., 32 FC, 24 fixtures per 1000 S.F.	4.01	5.25	9.26
0400	5 watt per S.F., 40 FC, 30 fixtures per 1000 S.F.	5	6.55	11.55

D50 Electrical

D5030 Communications and Security

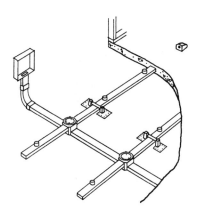

Description: System below includes telephone fitting installed. Does not include cable.

When poke thru fittings and telepoles are used for power, they can also be used for telephones at a negligible additional cost.

System Components	QUANTITY	UNIT	COST PER S.F. MAT.	COST PER S.F. INST.	COST PER S.F. TOTAL
SYSTEM D5030 310 0200					
TELEPHONE SYSTEMS, UNDERFLOOR DUCT, 5' ON CENTER, LOW DENSITY					
Underfloor duct 7-1/4" w/insert 2' O.C. 1-3/8" x 7-1/4" super duct	.190	L.F.	6.18	2.13	8.31
Vertical elbow for underfloor superduct, 7-1/4", included					
Underfloor duct conduit adapter, 2" x 1-1/4", included					
Underfloor duct junction box, single duct, 7-1/4" x 3 1/8"	.003	Ea.	1.47	.42	1.89
Underfloor junction box carpet pan	.003	Ea.	1.05	.02	1.07
Underfloor duct outlet, low tension	.004	Ea.	.39	.28	.67
TOTAL			9.09	2.85	11.94

D5030 310	Telephone Systems	COST PER S.F. MAT.	COST PER S.F. INST.	COST PER S.F. TOTAL
0200	Telephone systems, underfloor duct, 5' on center, low density	9.10	2.85	11.95
0240	5' on center, high density	9.50	3.13	12.63
0280	7' on center, low density	7.25	2.36	9.61
0320	7' on center, high density	7.65	2.64	10.29
0400	Poke thru fittings, low density	1.06	1.01	2.07
0440	High density	2.12	2.03	4.15
0520	Telepoles, low density	1.08	.60	1.68
0560	High density	2.16	1.21	3.37
0640	Conduit system with floor boxes, low density	1.23	.99	2.22
0680	High density	2.44	1.97	4.41
1020	Telephone wiring for offices & laboratories, 8 jacks/MSF	.56	2.42	2.98

D50 Electrical

D5030 Communications and Security

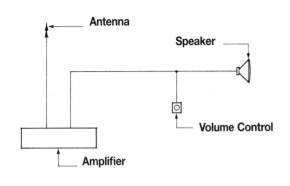

Sound System Includes AM-FM antenna, outlets, rigid conduit, and copper wire.
Fire Detection System Includes pull stations, signals, smoke and heat detectors, rigid conduit, and copper wire.
Intercom System Includes master and remote stations, rigid conduit, and copper wire.
Master Clock System Includes clocks, bells, rigid conduit, and copper wire.
Master TV Antenna Includes antenna, VHF-UHF reception and distribution, rigid conduit, and copper wire.

System Components	QUANTITY	UNIT	COST EACH MAT.	INST.	TOTAL
SYSTEM D5030 910 0220					
SOUND SYSTEM, INCLUDES OUTLETS, BOXES, CONDUIT & WIRE					
Steel intermediate conduit, (IMC) 1/2" diam	1200.000	L.F.	2,688	6,720	9,408
Wire sound shielded w/drain, #22-2 conductor	15.500	C.L.F.	465	1,085	1,550
Sound system speakers ceiling or wall	12.000	Ea.	1,452	840	2,292
Sound system volume control	12.000	Ea.	1,080	840	1,920
Sound system amplifier, 250 Watts	1.000	Ea.	1,800	560	2,360
Sound system antenna, AM FM	1.000	Ea.	225	140	365
Sound system monitor panel	1.000	Ea.	400	140	540
Sound system cabinet	1.000	Ea.	875	560	1,435
Steel outlet box 4" square	12.000	Ea.	31.44	336	367.44
Steel outlet box 4" plaster rings	12.000	Ea.	37.44	105	142.44
TOTAL			9,053.88	11,326	20,379.88

D5030 910	Communication & Alarm Systems		COST EACH MAT.	INST.	TOTAL
0200	Communication & alarm systems, includes outlets, boxes, conduit & wire				
0210	Sound system, 6 outlets		6,475	7,075	13,550
0220	12 outlets		9,050	11,300	20,350
0240	30 outlets		15,700	21,400	37,100
0280	100 outlets		47,800	71,500	119,300
0320	Fire detection systems, non-addressable, 12 detectors		3,125	5,875	9,000
0360	25 detectors		5,450	9,925	15,375
0400	50 detectors		10,700	19,700	30,400
0440	100 detectors		19,900	35,800	55,700
0450	Addressable type, 12 detectors		4,225	5,925	10,150
0452	25 detectors		7,675	10,000	17,675
0454	50 detectors		14,900	19,700	34,600
0456	100 detectors		28,600	36,100	64,700
0458	Fire alarm control panel, 8 zone		740	1,125	1,865
0459	12 zone		2,525	1,675	4,200
0460	Fire alarm command center, addressable without voice		2,600	980	3,580
0480	Intercom systems, 6 stations		3,625	4,850	8,475
0520	12 stations		7,375	9,650	17,025
0560	25 stations		12,700	18,600	31,300
0600	50 stations		24,500	34,900	59,400
0640	100 stations		48,300	68,000	116,300
0680	Master clock systems, 6 rooms		4,800	7,900	12,700
0720	12 rooms		7,125	13,500	20,925
0760	20 rooms		10,100	19,100	29,200
0800	30 rooms		17,100	35,100	52,200
0840	50 rooms		27,800	59,000	86,800

D50 Electrical

D5030 Communications and Security

D5030 910	Communication & Alarm Systems	COST EACH		
		MAT.	INST.	TOTAL
0880	100 rooms	54,000	117,500	171,500
0920	Master TV antenna systems, 6 outlets	3,250	5,000	8,250
0960	12 outlets	6,050	9,300	15,350
1000	30 outlets	12,400	21,600	34,000
1040	100 outlets	41,100	70,500	111,600

E Equipment & Furnishings

No part of this publication may be reproduced, stored in a retrieval system, or transmitted in any form or by any means without prior written permission of R.S. Means Company, Inc.

E20 Furnishings

E2020 Moveable Furnishings

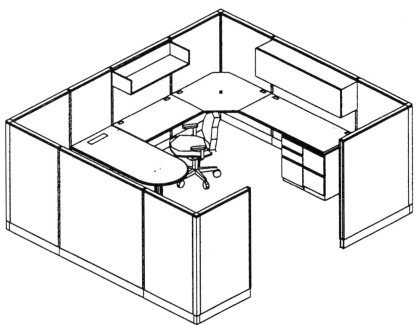

The work station shown here represents one of many types and sizes on the market. Costs can vary considerably due to the various parameters involved. For instance, wood systems can cost up to 40% more than metal or fabric systems. Generally, the cost per work station will decrease for "quantity" purchases.

Note: 4' raceway included in panels (not pre-wired)

System Components	QUANTITY	UNIT	COST EACH MAT.	COST EACH INST.	COST EACH TOTAL
SYSTEM E2020 410 2600					
WORK STATION - MANAGER					
Acoustical panel, aluminum frame, fabric covered, 30" wide, 64" high	1.000	Ea.	335		335
36" wide, 64" high	6.000	Ea.	2,160		2,160
42" wide, 64" high	2.000	Ea.	830		830
48" wide, 64" high	1.000	Ea.	440		440
60" wide, 64" high	2.000	Ea.	970		970
Straight connector kit, 64" high	7.000	Ea.	203		203
Ell connector kit, 64" high, 90" deep	4.000	Ea.	166		166
Panel end cover, 64" high	2.000	Ea.	58		58
Worksurface, 24" deep, 42" wide	1.000	Ea.	170		170
Worksurface, 24" deep, 60" wide	1.000	Ea.	228		228
Corner worksurface, 24" deep, 36" wide	1.000	Ea.	335		335
Peninsula worksurface, 36" wide, 66" long	1.000	Ea.	435		435
Peninsula support column, 29-1/2" high	1.000	Ea.	101		101
Cantilever bracket, 20" deep, 24" deep	3.000	Ea.	118.50		118.50
Pedestal spacer, 22" deep, 15" wide	2.000	Ea.	95		95
Pedestal, floorstanding, 2 box, 1 file	1.000	Ea.	330		330
Pedestal, floorstanding, 2 file, 22" deep	1.000	Ea.	310		310
Overhead storage cabinet with door, 60" wide	1.000	Ea.	415		415
Open bookshelf, 42" wide	1.000	Ea.	123		123
Task light, recessed, 42" - 48" wide	1.000	Ea.	139		139
Task light, recessed, 60" wide	1.000	Ea.	151		151
Electrical power harness, 36" wide	2.000	Ea.	189		189
Electrical power harness, 42" wide	1.000	Ea.	199		199
Electrical power harness, 60" wide	1.000	Ea.	99.50		99.50
Base power in-feed cable	1.000	Ea.	105		105
Duplex receptacle, circuit 1	2.000	Ea.	29		29
Duplex receptacle, circuit 2	2.000	Ea.	29		29
Installation	1.000	Ea.		495	495
TOTAL			8,763	495	9,258

E20 Furnishings

E2020 Moveable Furnishings

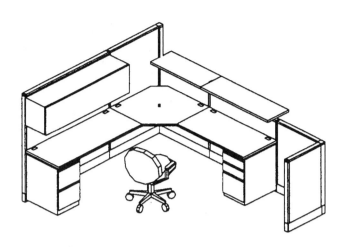

The work station shown here represents one of many types and sizes on the market. Costs can vary considerably due to the various parameters involved. For instance, wood systems can cost up to 40% more than metal or fabric systems. Generally, the cost per work station will decrease for "quantity" purchases.

System Components	QUANTITY	UNIT	COST EACH MAT.	COST EACH INST.	COST EACH TOTAL
SYSTEM E2020 420 2300					
WORK STATION - PROFESSIONAL/EXECUTIVE SECRETARY					
Acoustical panel, aluminum frame, fabric covered, 24" wide x 43" high	1.000	Ea.	253		253
30" wide x 43" high	1.000	Ea.	282		282
36" wide x 43" high	1.000	Ea.	310		310
48" wide x 43" high	1.000	Ea.	375		375
36" wide x 64" high	1.000	Ea.	360		360
60" wide x 64" high	1.000	Ea.	485		485
Straight connector kit, 43" high	2.000	Ea.	58		58
Straight connector kit, 64" high	1.000	Ea.	29		29
Ell connector kit, 43" high, 90" deep	1.000	Ea.	70.50		70.50
Ell connector kit, 64" high	1.000	Ea.	41.50		41.50
Panel end cover, 43" high	1.000	Ea.	29		29
Panel end cover, 64" high	1.000	Ea.	29		29
Finish end cover, variable height, 2-way	1.000	Ea.	43.50		43.50
Corner worksurface, 24" deep, 36" wide	1.000	Ea.	335		335
Worksurface, 24" deep, 48" wide	1.000	Ea.	182		182
Worksurface, 24" deep, 60" wide	1.000	Ea.	228		228
Cantilever bracket, 20" deep, 24" deep	2.000	Ea.	79		79
Countertop brackets, 1 pair	2.000	Ea.	32.90		32.90
Countertop, 15" deep, 36" wide	1.000	Ea.	133		133
Countertop, 15" deep, 48" wide	1.000	Ea.	149		149
Pedestal spacer, 22" deep, 15" wide	2.000	Ea.	95		95
Pedestal, floorstanding, 2box, 1 file, 22" deep	1.000	Ea.	330		330
Pedestal, floorstanding, 2 file, 22" deep	1.000	Ea.	310		310
Overhead storage cabinet with door, 60" wide	1.000	Ea.	415		415
Task light, recessed, 60" wide	1.000	Ea.	151		151
Electrical power harness, 36" wide	2.000	Ea.	189		189
Electrical power harness, 60" wide	1.000	Ea.	99.50		99.50
Base power in-feed cable	1.000	Ea.	105		105
Duplex receptacle, circuit 1	2.000	Ea.	29		29
Duplex receptacle, circuit 2	1.000	Ea.	14.50		14.50
Installation	1.000	Ea.		300	300
TOTAL			5,242.40	300	5,542.40

E20 Furnishings

E2020 Moveable Furnishings

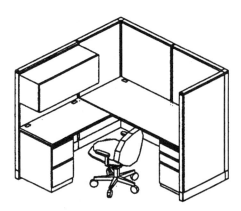

The work station shown here represents one of many types and sizes on the market. Costs can vary considerably due to the various parameters involved. For instance, wood systems can cost up to 40% more than metal or fabric systems. Generally, the cost per work station will decrease for "quantity" purchases.

System Components	QUANTITY	UNIT	COST EACH MAT.	COST EACH INST.	COST EACH TOTAL
SYSTEM E2020 430 1500					
WORK STATION - SECRETARY/CLERK					
Acoustical panel, aluminum frame, fabric covered, 30" wide x 64" high	2.000	Ea.	670		670
36" wide x 64" high	2.000	Ea.	720		720
42" wide x 64" high	1.000	Ea.	415		415
Straight connector kit, 64" high	2.000	Ea.	58		58
Ell connector kit, 64" high, 90" deep	2.000	Ea.	83		83
Panel end cover, 64" high	2.000	Ea.	58		58
Worksurface, 24" deep, 42" wide	1.000	Ea.	170		170
Worksurface, 72" wide, 30" deep	1.000	Ea.	290		290
Worksurface bracket kit, pair	2.000	Ea.	32.90		32.90
Flat bracket, 20" deep, 24" deep	1.000	Ea.	20.50		20.50
Pedestal spacer 22" deep, 15" wide	1.000	Ea.	47.50		47.50
Pedestal spacer, 28" deep, 15" wide	1.000	Ea.	57		57
Pedestal, 2 box, 1 file, 28" deep	1.000	Ea.	345		345
Pedestal, 2 file, 22" deep	1.000	Ea.	310		310
Overhead storage cabinet with door, 42" wide	1.000	Ea.	274		274
Task light, recessed, 42" - 48" wide	1.000	Ea.	139		139
Installation	1.000	Ea.		198	198
TOTAL			3,689.90	198	3,887.90

E20 Furnishings

E2020 Moveable Furnishings

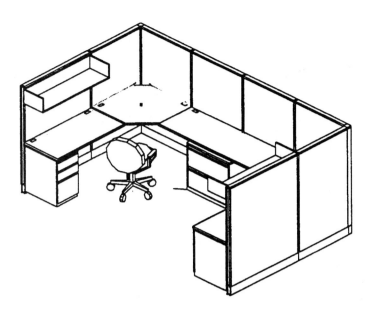

The work station shown here represents one of many types and sizes on the market. Costs can vary considerably due to the various parameters involved. For instance, wood systems can cost up to 40% more than metal or fabric systems. Generally, the cost per work station will decrease for "quantity" purchases.

System Components	QUANTITY	UNIT	COST EACH MAT.	COST EACH INST.	COST EACH TOTAL
SYSTEM E2020 440 1600					
WORK STATION - CLERICAL, TWO PERSON, UNIT					
Acoustical panel, aluminum frame, fabric covered, 36" wide, 64" high	6.000	Ea.	2,160		2,160
48" wide x 64" high	2.000	Ea.	880		880
Lateral file, 2 drawer, 30" wide	1.000	Ea.	435		435
Ell connector kit, 64" high, 90" deep	2.000	Ea.	83		83
Panel end cover, 64" high	2.000	Ea.	58		58
Corner worksurface, 36" wide x 24" deep	2.000	Ea.	670		670
Worksurface, 48" wide x 24" deep	2.000	Ea.	364		364
Worksurface, 72" wide x 24" deep	1.000	Ea.	257		257
Cantilever bracket, 20" deep, 24" deep	4.000	Ea.	158		158
Open bookshelf, 48" wide	2.000	Ea.	254		254
Pedestal spacer, 22" deep, 15" wide	2.000	Ea.	95		95
Pedestal, floorstanding, 2 box, 1 file, 22" deep	2.000	Ea.	660		660
Straight connector kit, 64" high	5.000	Ea.	145		145
Task light, recessed, 42" - 48" wide	2.000	Ea.	278		278
Electrical power harness, 36" wide	4.000	Ea.	378		378
Base power in-feed cable	1.000	Ea.	105		105
Duplex receptacle, circuit 1	2.000	Ea.	29		29
Duplex receptacle, circuit 2	2.000	Ea.	29		29
Installation	1.000	Ea.		395	395
TOTAL			7,038	395	7,433

E20 Furnishings

E2020 Moveable Furnishings

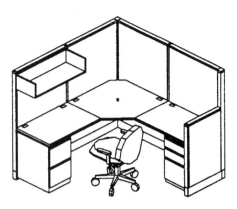

The work station shown here represents one of many types and sizes on the market. Costs can vary considerably due to the various parameters involved. For instance, wood systems can cost up to 40% more than metal or fabric systems. Generally, the cost per work station will decrease for "quantity" purchases.

System Components	QUANTITY	UNIT	COST EACH MAT.	COST EACH INST.	COST EACH TOTAL
SYSTEM E2020 450 1600					
WORK STATION - SECRETARY/WORD PROCESSING					
Acoustical panel, aluminum frame, fabric covered, 43" high, 24" wide	1.000	Ea.	253		253
Acoustical panel, 64" high, 36" wide	4.000	Ea.	1,440		1,440
Straight connector kit, 64" high	2.000	Ea.	58		58
Ell connector kit, 64" high, 90" deep	2.000	Ea.	83		83
Panel end cover, 43" high	1.000	Ea.	29		29
Panel end cover, 64" high	1.000	Ea.	29		29
Variable height finish end cover, 2-way	1.000	Ea.	43.50		43.50
Corner worksurface, 24" deep, 36" wide	1.000	Ea.	335		335
Worksurface, 24" deep, 36" wide	2.000	Ea.	276		276
Cantilever bracket, 20" deep, 24" deep	2.000	Ea.	79		79
Pedestal spacer, 22" deep, 15" wide	2.000	Ea.	95		95
Pedestal, floorstanding, 2 box, 1 file, 22" deep	1.000	Ea.	330		330
Pedestal, floorstanding, 2 file, 22" deep	1.000	Ea.	310		310
Open bookshelf, 36" wide	1.000	Ea.	117		117
Task light, recessed, 30" - 36" wide	1.000	Ea.	129		129
Electrical power harness, 36" wide	3.000	Ea.	283.50		283.50
Base power in-feed cable	1.000	Ea.	105		105
Duplex receptacle, circuit 1	2.000	Ea.	29		29
Duplex receptacle, circuit 2	1.000	Ea.	14.50		14.50
Installation	1.000	Ea.		248	248
TOTAL			4,038.50	248	4,286.50

Reference Section

All the reference information is in one section, making it easy to find what you need to know ... and easy to use the book on a daily basis. This section is visually identified by a vertical gray bar on the page edges.

In this Reference Section, we've included Equipment Rental Costs, a listing of rental and operating costs; Crew Listings, a full listing of all crews and equipment, and their costs; Historical Cost Indexes for cost comparisons over time; City Cost Indexes and Location Factors for adjusting costs to the region you are in; Reference Tables, where you will find explanations, estimating information and procedures, or technical data; Change Orders, information on pricing changes to contract documents; Square Foot Costs that allow you to make a rough estimate for the overall cost of a project; and an explanation of all the Abbreviations in the book.

Table of Contents

Construction Equipment Rental Costs	549
Crew Listings	561
Historical Cost Indexes	591
City Cost Indexes	592
Location Factors	612
Reference Tables	618
R01 General Requirements	618
R02 Existing Conditions	626
R03 Concrete	628
R04 Masonry	628
R05 Metals	630
R06 Wood, Plastics & Composites	631
R07 Thermal & Moisture Protection	631

Reference Tables (cont.)

R08 Openings	632
R09 Finishes	634
R14 Conveying Equipment	638
R21 Fire Suppression	640
R22 Plumbing	642
R23 Heating, Ventilating & Air Conditioning	645
Change Orders	648
Square Foot Costs	653
Abbreviations	663

Equipment Rental Costs

Estimating Tips

- This section contains the average costs to rent and operate hundreds of pieces of construction equipment. This is useful information when estimating the time and material requirements of any particular operation in order to establish a unit or total cost. Equipment costs include not only rental, but also operating costs for equipment under normal use.

Rental Costs

- Equipment rental rates are obtained from industry sources throughout North America–contractors, suppliers, dealers, manufacturers, and distributors.
- Rental rates vary throughout the country, with larger cities generally having lower rates. Lease plans for new equipment are available for periods in excess of six months, with a percentage of payments applying toward purchase.
- Monthly rental rates vary from 2% to 5% of the purchase price of the equipment depending on the anticipated life of the equipment and its wearing parts.
- Weekly rental rates are about 1/3 the monthly rates, and daily rental rates are about 1/3 the weekly rate.

Operating Costs

- The operating costs include parts and labor for routine servicing, such as repair and replacement of pumps, filters and worn lines. Normal operating expendables, such as fuel, lubricants, tires and electricity (where applicable), are also included.
- Extraordinary operating expendables with highly variable wear patterns, such as diamond bits and blades, are excluded. These costs can be found as material costs in the Unit Price section.
- The hourly operating costs listed do not include the operator's wages.

Equipment Cost/Day

- Any power equipment required by a crew is shown in the Crew Listings with a daily cost.
- The daily cost of equipment needed by a crew is based on dividing the weekly rental rate by 5 (number of working days in the week), and then adding the hourly operating cost times 8 (the number of hours in a day). This "Equipment Cost/Day" is shown in the far right column of the Equipment Rental pages.
- If equipment is needed for only one or two days, it is best to develop your own cost by including components for daily rent and hourly operating cost. This is important when the listed Crew for a task does not contain the equipment needed, such as a crane for lifting mechanical heating/cooling equipment up onto a roof.
- If the quantity of work is less than the crew's Daily Output shown for a Unit Price line item that includes a bare unit equipment cost, it is recommended to estimate one day's rental cost and operating cost for equipment shown in the Crew Listing for that line item.

Mobilization/Demobilization

- The cost to move construction equipment from an equipment yard or rental company to the jobsite and back again is not included in equipment rental costs listed in the Reference section, nor in the bare equipment cost of any Unit Price line item, nor in any equipment costs shown in the Crew listings.
- Mobilization (to the site) and demobilization (from the site) costs can be found in the Unit Price section.
- If a piece of equipment is already at the jobsite, it is not appropriate to utilize mobil./demob. costs again in an estimate.

No part of this publication may be reproduced, stored in a retrieval system, or transmitted in any form or by any means without prior written permission of R.S. Means Company, Inc.

01 54 | Construction Aids

01 54 33 | Equipment Rental

			UNIT	HOURLY OPER. COST	RENT PER DAY	RENT PER WEEK	RENT PER MONTH	EQUIPMENT COST/DAY	
10	0010	**CONCRETE EQUIPMENT RENTAL** without operators R015433-10							10
	0200	Bucket, concrete lightweight, 1/2 C.Y.	Ea.	.65	20	60	180	17.20	
	0300	1 C.Y.		.70	24.50	73	219	20.20	
	0400	1-1/2 C.Y.		.90	33	99	297	27	
	0500	2 C.Y.		1.00	40	120	360	32	
	0580	8 C.Y.		5.50	262	785	2,350	201	
	0600	Cart, concrete, self propelled, operator walking, 10 C.F.		2.65	56.50	170	510	55.20	
	0700	Operator riding, 18 C.F.		4.30	86.50	260	780	86.40	
	0800	Conveyer for concrete, portable, gas, 16" wide, 26' long		10.00	123	370	1,100	154	
	0900	46' long		10.40	150	450	1,350	173.20	
	1000	56' long		10.50	158	475	1,425	179	
	1100	Core drill, electric, 2-1/2 H.P., 1" to 8" bit diameter		1.72	65.50	196	590	52.95	
	1150	11 HP, 8" to 18" cores		5.15	115	345	1,025	110.20	
	1200	Finisher, concrete floor, gas, riding trowel, 96" wide		9.95	145	435	1,300	166.60	
	1300	Gas, walk-behind, 3 blade, 36" trowel		1.65	19.65	59	177	25	
	1400	4 blade, 48" trowel		3.20	28	84	252	42.40	
	1500	Float, hand-operated (Bull float) 48" wide		.08	13.65	41	123	8.85	
	1570	Curb builder, 14 H.P., gas, single screw		11.85	243	730	2,200	240.80	
	1590	Double screw		12.55	285	855	2,575	271.40	
	1600	Grinder, concrete and terrazzo, electric, floor		2.45	103	310	930	81.60	
	1700	Wall grinder		1.23	51.50	155	465	40.85	
	1800	Mixer, powered, mortar and concrete, gas, 6 C.F., 18 H.P.		6.95	118	355	1,075	126.60	
	1900	10 C.F., 25 H.P.		8.65	143	430	1,300	155.20	
	2000	16 C.F.		8.95	167	500	1,500	171.60	
	2100	Concrete, stationary, tilt drum, 2 C.Y.		6.10	232	695	2,075	187.80	
	2120	Pump, concrete, truck mounted 4" line 80' boom		22.85	925	2,775	8,325	737.80	
	2140	5" line, 110' boom		29.75	1,225	3,695	11,100	977	
	2160	Mud jack, 50 C.F. per hr.		6.10	130	390	1,175	126.80	
	2180	225 C.F. per hr.		7.95	147	440	1,325	151.60	
	2190	Shotcrete pump rig, 12 CY/hr		12.60	228	685	2,050	237.80	
	2600	Saw, concrete, manual, gas, 18 H.P.		5.15	36.50	110	330	63.20	
	2650	Self-propelled, gas, 30 H.P.		10.25	98.50	295	885	141	
	2700	Vibrators, concrete, electric, 60 cycle, 2 H.P.		.46	8.35	25	75	8.70	
	2800	3 H.P.		.64	10.35	31	93	11.30	
	2900	Gas engine, 5 H.P.		1.55	14.35	43	129	21	
	3000	8 H.P.		2.10	15.35	46	138	26	
	3050	Vibrating screed, gas engine, 8 H.P.		3.32	73	219	655	70.35	
	3120	Concrete transit mixer, 6 x 4, 250 H.P., 8 C.Y., rear discharge		47.15	575	1,720	5,150	721.20	
	3200	Front discharge		55.50	710	2,125	6,375	869	
	3300	6 x 6, 285 H.P., 12 C.Y., rear discharge		54.60	665	2,000	6,000	836.80	
	3400	Front discharge		57.05	715	2,150	6,450	886.40	
20	0010	**EARTHWORK EQUIPMENT RENTAL** without operators R015433-10							20
	0040	Aggregate spreader, push type 8' to 12' wide	Ea.	2.45	25.50	77	231	35	
	0045	Tailgate type, 8' wide		2.35	33.50	100	300	38.80	
	0055	Earth auger, truck-mounted, for fence & sign posts, utility poles		13.50	505	1,520	4,550	412	
	0060	For borings and monitoring wells		38.20	670	2,005	6,025	706.60	
	0070	Portable, trailer mounted		2.65	26	78	234	36.80	
	0075	Truck-mounted, for caissons, water wells		85.35	3,075	9,250	27,800	2,533	
	0080	Horizontal boring machine, 12" to 36" diameter, 45 H.P.		20.30	202	605	1,825	283.40	
	0090	12" to 48" diameter, 65 H.P.		28.65	360	1,075	3,225	444.20	
	0095	Auger, for fence posts, gas engine, hand held		.50	5.65	17	51	7.40	
	0100	Excavator, diesel hydraulic, crawler mounted, 1/2 C.Y. cap.		20.95	380	1,135	3,400	394.60	
	0120	5/8 C.Y. capacity		25.40	525	1,570	4,700	517.20	
	0140	3/4 C.Y. capacity		29.70	600	1,805	5,425	598.60	
	0150	1 C.Y. capacity		36.90	625	1,870	5,600	669.20	
	0200	1-1/2 C.Y. capacity		45.10	840	2,525	7,575	865.80	
	0300	2 C.Y. capacity		61.75	1,150	3,440	10,300	1,182	

01 54 | Construction Aids

01 54 33 | Equipment Rental

		UNIT	HOURLY OPER. COST	RENT PER DAY	RENT PER WEEK	RENT PER MONTH	EQUIPMENT COST/DAY	
0320	2-1/2 C.Y. capacity	Ea.	83.30	1,600	4,765	14,300	1,619	20
0325	3-1/2 C.Y. capacity		111.30	2,125	6,360	19,100	2,162	
0330	4-1/2 C.Y. capacity		134.65	2,600	7,790	23,400	2,635	
0335	6 C.Y. capacity		168.45	2,875	8,660	26,000	3,080	
0340	7 C.Y. capacity		173.20	3,125	9,350	28,100	3,256	
0342	Excavator attachments, bucket thumbs		2.70	215	645	1,925	150.60	
0345	Grapples		2.45	188	565	1,700	132.60	
0347	Hydraulic hammer for boom mounting, 5000 ft-lb		12.10	405	1,220	3,650	340.80	
0349	11,000 ft-lb		20.40	730	2,190	6,575	601.20	
0350	Gradall type, truck mounted, 3 ton @ 15' radius, 5/8 C.Y.		49.60	935	2,810	8,425	958.80	
0370	1 C.Y. capacity		55.10	1,150	3,445	10,300	1,130	
0400	Backhoe-loader, 40 to 45 H.P., 5/8 C.Y. capacity		11.90	227	680	2,050	231.20	
0450	45 H.P. to 60 H.P., 3/4 C.Y. capacity		15.85	278	835	2,500	293.80	
0460	80 H.P., 1-1/4 C.Y. capacity		19.00	320	965	2,900	345	
0470	112 H.P., 1-1/2 C.Y. capacity		25.45	465	1,400	4,200	483.60	
0482	Backhoe-loader attachment, compactor, 20,000 lb		4.70	127	380	1,150	113.60	
0485	Hydraulic hammer, 750 ft-lbs		2.80	90	270	810	76.40	
0486	Hydraulic hammer, 1200 ft-lbs		5.25	185	555	1,675	153	
0500	Brush chipper, gas engine, 6" cutter head, 35 H.P.		8.95	108	325	975	136.60	
0550	12" cutter head, 130 H.P.		14.95	177	530	1,600	225.60	
0600	15" cutter head, 165 H.P.		22.45	217	650	1,950	309.60	
0750	Bucket, clamshell, general purpose, 3/8 C.Y.		1.10	36.50	110	330	30.80	
0800	1/2 C.Y.		1.20	45	135	405	36.60	
0850	3/4 C.Y.		1.35	55	165	495	43.80	
0900	1 C.Y.		1.40	58.50	175	525	46.20	
0950	1-1/2 C.Y.		2.25	80	240	720	66	
1000	2 C.Y.		2.40	88.50	265	795	72.20	
1010	Bucket, dragline, medium duty, 1/2 C.Y.		.65	23.50	70	210	19.20	
1020	3/4 C.Y.		.65	24.50	74	222	20	
1030	1 C.Y.		.70	26.50	80	240	21.60	
1040	1-1/2 C.Y.		1.10	40	120	360	32.80	
1050	2 C.Y.		1.15	45	135	405	36.20	
1070	3 C.Y.		1.65	55	165	495	46.20	
1200	Compactor, manually guided 2-drum vibratory smooth roller, 7.5 H.P.		5.65	167	500	1,500	145.20	
1250	Rammer/tamper, gas, 8"		2.30	41.50	125	375	43.40	
1260	15"		2.55	48.50	145	435	49.40	
1300	Vibratory plate, gas, 18" plate, 3000 lb. blow		2.25	23	69	207	31.80	
1350	21" plate, 5000 lb. blow		2.80	28	84	252	39.20	
1370	Curb builder/extruder, 14 H.P., gas, single screw		11.85	243	730	2,200	240.80	
1390	Double screw		12.55	285	855	2,575	271.40	
1500	Disc harrow attachment, for tractor		.39	65.50	197	590	42.50	
1810	Feller buncher, shearing & accumulating trees, 100 H.P.		28.55	555	1,665	5,000	561.40	
1860	Grader, self-propelled, 25,000 lb.		24.85	450	1,345	4,025	467.80	
1910	30,000 lb.		29.65	520	1,565	4,700	550.20	
1920	40,000 lb.		45.50	995	2,985	8,950	961	
1930	55,000 lb.		59.10	1,400	4,215	12,600	1,316	
1950	Hammer, pavement breaker, self-propelled, diesel, 1000 to 1250 lb		23.55	340	1,025	3,075	393.40	
2000	1300 to 1500 lb.		35.33	685	2,050	6,150	692.65	
2050	Pile driving hammer, steam or air, 4150 ft.-lb. @ 225 BPM		9.35	470	1,405	4,225	355.80	
2100	8750 ft.-lb. @ 145 BPM		11.35	650	1,955	5,875	481.80	
2150	15,000 ft.-lb. @ 60 BPM		12.95	790	2,375	7,125	578.60	
2200	24,450 ft.-lb. @ 111 BPM		13.95	885	2,655	7,975	642.60	
2250	Leads, 60' high for pile driving hammers up to 20,000 ft.-lb.		2.85	78	234	700	69.60	
2300	90' high for hammers over 20,000 ft.-lb.		4.35	140	420	1,250	118.80	
2350	Diesel type hammer, 22,400 ft.-lb.		25.75	640	1,920	5,750	590	
2400	41,300 ft.-lb.		32.70	695	2,080	6,250	677.60	
2450	141,000 ft.-lb.		49.25	1,200	3,570	10,700	1,108	
2500	Vib. elec. hammer/extractor, 200 KW diesel generator, 34 H.P.		41.05	665	2,000	6,000	728.40	

01 54 | Construction Aids

01 54 33 | Equipment Rental

		UNIT	HOURLY OPER. COST	RENT PER DAY	RENT PER WEEK	RENT PER MONTH	EQUIPMENT COST/DAY
2550	80 H.P.	Ea.	73.75	980	2,940	8,825	1,178
2600	150 H.P.		139.15	1,900	5,720	17,200	2,257
2700	Extractor, steam or air, 700 ft.-lb.		17.15	510	1,530	4,600	443.20
2750	1000 ft.-lb.		19.55	625	1,875	5,625	531.40
2800	Log chipper, up to 22" diam, 600 H.P.		40.05	445	1,340	4,025	588.40
2850	Logger, for skidding & stacking logs, 150 H.P.		44.95	895	2,690	8,075	897.60
2860	Mulcher, diesel powered, trailer mounted		17.60	218	655	1,975	271.80
2900	Rake, spring tooth, with tractor		11.34	278	835	2,500	257.70
3000	Roller, vibratory, tandem, smooth drum, 20 H.P.		7.35	132	395	1,175	137.80
3050	35 H.P.		9.50	262	785	2,350	233
3100	Towed type vibratory compactor, smooth drum, 50 H.P.		22.75	380	1,145	3,425	411
3150	Sheepsfoot, 50 H.P.		25.05	440	1,325	3,975	465.40
3170	Landfill compactor, 220 HP		69.10	1,450	4,315	12,900	1,416
3200	Pneumatic tire roller, 80 H.P.		13.10	340	1,015	3,050	307.80
3250	120 H.P.		19.90	600	1,805	5,425	520.20
3300	Sheepsfoot vibratory roller, 240 H.P.		56.65	1,150	3,445	10,300	1,142
3320	340 H.P.		75.60	1,425	4,240	12,700	1,453
3350	Smooth drum vibratory roller, 75 H.P.		19.70	520	1,560	4,675	469.60
3400	125 H.P.		25.70	650	1,945	5,825	594.60
3410	Rotary mower, brush, 60", with tractor		15.30	252	755	2,275	273.40
3420	Rototiller, walk-behind, gas, 5 H.P.		2.43	62.50	188	565	57.05
3422	8 H.P.		3.89	100	300	900	91.10
3440	Scrapers, towed type, 7 CY capacity		4.90	105	315	945	102.20
3450	10 C.Y. capacity		6.20	173	520	1,550	153.60
3500	15 C.Y. capacity		6.85	190	570	1,700	168.80
3525	Self-propelled, single engine, 14 C.Y. capacity		83.90	1,400	4,190	12,600	1,509
3550	Dual engine, 21 C.Y. capacity		119.25	1,625	4,850	14,600	1,924
3600	31 C.Y. capacity		165.60	2,600	7,765	23,300	2,878
3640	44 C.Y. capacity		195.70	3,150	9,430	28,300	3,452
3650	Elevating type, single engine, 11 CY capacity		56.70	1,075	3,205	9,625	1,095
3700	22 C.Y. capacity		105.60	2,125	6,410	19,200	2,127
3710	Screening plant 110 H.P. w/ 5' x 10' screen		30.75	410	1,230	3,700	492
3720	5' x 16' screen		32.90	515	1,540	4,625	571.20
3850	Shovel, crawler-mounted, front-loading, 7 C.Y. capacity		171.90	2,850	8,575	25,700	3,090
3855	12 C.Y. capacity		244.40	3,650	10,965	32,900	4,148
3860	Shovel/backhoe bucket, 1/2 C.Y.		2.10	60	180	540	52.80
3870	3/4 C.Y.		2.15	68.50	205	615	58.20
3880	1 C.Y.		2.25	76.50	230	690	64
3890	1-1/2 C.Y.		2.40	91.50	275	825	74.20
3910	3 C.Y.		2.70	127	380	1,150	97.60
3950	Stump chipper, 18" deep, 30 H.P.		8.99	215	646	1,950	201.10
4110	Tractor, crawler, with bulldozer, torque converter, diesel 80 H.P.		21.90	375	1,120	3,350	399.20
4150	105 H.P.		31.70	580	1,735	5,200	600.60
4200	140 H.P.		36.50	650	1,950	5,850	682
4260	200 H.P.		55.85	1,050	3,175	9,525	1,082
4310	300 H.P.		71.45	1,425	4,255	12,800	1,423
4360	410 H.P.		95.30	1,800	5,435	16,300	1,849
4370	500 H.P.		127.20	2,400	7,230	21,700	2,464
4380	700 H.P.		204.15	4,325	12,995	39,000	4,232
4400	Loader, crawler, torque conv., diesel, 1-1/2 C.Y., 80 H.P.		20.00	345	1,035	3,100	367
4450	1-1/2 to 1-3/4 C.Y., 95 H.P.		23.75	455	1,365	4,100	463
4510	1-3/4 to 2-1/4 C.Y., 130 H.P.		35.05	810	2,435	7,300	767.40
4530	2-1/2 to 3-1/4 C.Y., 190 H.P.		47.75	1,025	3,040	9,125	990
4560	3-1/2 to 5 C.Y., 275 H.P.		68.30	1,500	4,520	13,600	1,450
4610	Front end loader, 4WD, articulated frame, 1 to 1-1/4 C.Y., 70 H.P.		14.25	208	625	1,875	239
4620	1-1/2 to 1-3/4 C.Y., 95 H.P.		18.90	295	885	2,650	328.20
4650	1-3/4 to 2 C.Y., 130 H.P.		21.60	345	1,035	3,100	379.80
4710	2-1/2 to 3-1/2 C.Y., 145 H.P.		22.85	355	1,065	3,200	395.80

01 54 | Construction Aids

01 54 33 | Equipment Rental

			UNIT	HOURLY OPER. COST	RENT PER DAY	RENT PER WEEK	RENT PER MONTH	EQUIPMENT COST/DAY
20	4730	3 to 4-1/2 C.Y., 185 H.P.	Ea.	29.75	505	1,515	4,550	541
	4760	5-1/4 to 5-3/4 C.Y., 270 H.P.		48.05	785	2,350	7,050	854.40
	4810	7 to 9 C.Y., 475 H.P.		81.85	1,400	4,210	12,600	1,497
	4870	9 - 11 C.Y., 620 H.P.		112.20	2,200	6,635	19,900	2,225
	4880	Skid steer loader, wheeled, 10 C.F., 30 H.P. gas		8.15	140	420	1,250	149.20
	4890	1 C.Y., 78 H.P., diesel		14.95	233	700	2,100	259.60
	4892	Skid-steer attachment, auger		.54	89.50	268	805	57.90
	4893	Backhoe		.67	112	336	1,000	72.55
	4894	Broom		.70	117	352	1,050	76
	4895	Forks		.24	39.50	118	355	25.50
	4896	Grapple		.57	95.50	287	860	61.95
	4897	Concrete hammer		1.06	176	528	1,575	114.10
	4898	Tree spade		.80	133	398	1,200	86
	4899	Trencher		.77	128	385	1,150	83.15
	4900	Trencher, chain, boom type, gas, operator walking, 12 H.P.		3.95	46.50	140	420	59.60
	4910	Operator riding, 40 H.P.		15.50	295	885	2,650	301
	5000	Wheel type, diesel, 4' deep, 12" wide		65.30	810	2,435	7,300	1,009
	5100	6' deep, 20" wide		77.15	1,900	5,680	17,000	1,753
	5150	Chain type, diesel, 5' deep, 8" wide		28.90	580	1,745	5,225	580.20
	5200	Diesel, 8' deep, 16" wide		74.70	1,975	5,895	17,700	1,777
	5210	Tree spade, self-propelled		16.78	267	800	2,400	294.25
	5250	Truck, dump, 2-axle, 12 ton, 8 CY payload, 220 H.P.		24.55	217	650	1,950	326.40
	5300	Three axle dump, 16 ton, 12 CY payload, 400 H.P.		43.25	310	935	2,800	533
	5350	Dump trailer only, rear dump, 16-1/2 C.Y.		4.70	130	390	1,175	115.60
	5400	20 C.Y.		5.15	148	445	1,325	130.20
	5450	Flatbed, single axle, 1-1/2 ton rating		19.30	66.50	200	600	194.40
	5500	3 ton rating		23.20	95	285	855	242.60
	5550	Off highway rear dump, 25 ton capacity		59.95	1,325	4,000	12,000	1,280
	5600	35 ton capacity		55.80	1,175	3,490	10,500	1,144
	5610	50 ton capacity		76.40	1,625	4,855	14,600	1,582
	5620	65 ton capacity		78.80	1,600	4,820	14,500	1,594
	5630	100 ton capacity		101.35	2,050	6,180	18,500	2,047
	6000	Vibratory plow, 25 H.P., walking		6.65	61.50	185	555	90.20
40	0010	**GENERAL EQUIPMENT RENTAL** without operators						
	0150	Aerial lift, scissor type, to 15' high, 1000 lb. cap., electric (R015433-10)	Ea.	2.70	56.50	170	510	55.60
	0160	To 25' high, 2000 lb. capacity		3.15	80	240	720	73.20
	0170	Telescoping boom to 40' high, 500 lb. capacity, gas		17.10	320	960	2,875	328.80
	0180	To 45' high, 500 lb. capacity		18.10	365	1,100	3,300	364.80
	0190	To 60' high, 600 lb. capacity		20.45	505	1,510	4,525	465.60
	0195	Air compressor, portable, 6.5 CFM, electric		.47	12.65	38	114	11.35
	0196	Gasoline		.96	19	57	171	19.10
	0200	Towed type, gas engine, 60 C.F.M.		11.70	48.50	145	435	122.60
	0300	160 C.F.M.		13.60	50	150	450	138.80
	0400	Diesel engine, rotary screw, 250 C.F.M.		12.80	100	300	900	162.40
	0500	365 C.F.M.		17.15	123	370	1,100	211.20
	0550	450 C.F.M.		21.75	153	460	1,375	266
	0600	600 C.F.M.		38.05	217	650	1,950	434.40
	0700	750 C.F.M.		38.20	223	670	2,000	439.60
	0800	For silenced models, small sizes, add to rent		3%	5%	5%	5%	
	0900	Large sizes, add to rent		5%	7%	7%	7%	
	0930	Air tools, breaker, pavement, 60 lb.	Ea.	.40	8.35	25	75	8.20
	0940	80 lb.		.40	8.35	25	75	8.20
	0950	Drills, hand (jackhammer) 65 lb.		.50	16	48	144	13.60
	0960	Track or wagon, swing boom, 4" drifter		49.20	760	2,285	6,850	850.60
	0970	5" drifter		60.00	825	2,470	7,400	974
	0975	Track mounted quarry drill, 6" diameter drill		82.25	1,250	3,780	11,300	1,414
	0980	Dust control per drill		.99	19	57	171	19.30

01 54 | Construction Aids

01 54 33 | Equipment Rental

		Description	UNIT	HOURLY OPER. COST	RENT PER DAY	RENT PER WEEK	RENT PER MONTH	EQUIPMENT COST/DAY	
40	0990	Hammer, chipping, 12 lb.	Ea.	.45	22.50	67	201	17	40
	1000	Hose, air with couplings, 50' long, 3/4" diameter		.03	4.33	13	39	2.85	
	1100	1" diameter		.04	6.35	19	57	4.10	
	1200	1-1/2" diameter		.06	9.65	29	87	6.30	
	1300	2" diameter		.13	21.50	65	195	14.05	
	1400	2-1/2" diameter		.12	19.65	59	177	12.75	
	1410	3" diameter		.15	25	75	225	16.20	
	1450	Drill, steel, 7/8" x 2'		.05	8.35	25	75	5.40	
	1460	7/8" x 6'		.06	9.35	28	84	6.10	
	1520	Moil points		.02	3.67	11	33	2.35	
	1525	Pneumatic nailer w/accessories		.45	30	90	270	21.60	
	1530	Sheeting driver for 60 lb. breaker		.04	6	18	54	3.90	
	1540	For 90 lb. breaker		.12	8	24	72	5.75	
	1550	Spade, 25 lb.		.40	6	18	54	6.80	
	1560	Tamper, single, 35 lb.		.50	33.50	100	300	24	
	1570	Triple, 140 lb.		.75	50	150	450	36	
	1580	Wrenches, impact, air powered, up to 3/4" bolt		.30	8	24	72	7.20	
	1590	Up to 1-1/4" bolt		.45	21	63	189	16.20	
	1600	Barricades, barrels, reflectorized, 1 to 50 barrels		.03	4.60	13.80	41.50	3	
	1610	100 to 200 barrels		.02	3.53	10.60	32	2.30	
	1620	Barrels with flashers, 1 to 50 barrels		.03	5.25	15.80	47.50	3.40	
	1630	100 to 200 barrels		.03	4.20	12.60	38	2.75	
	1640	Barrels with steady burn type C lights		.04	7	21	63	4.50	
	1650	Illuminated board, trailer mounted, with generator		.70	120	360	1,075	77.60	
	1670	Portable barricade, stock, with flashers, 1 to 6 units		.03	5.25	15.80	47.50	3.40	
	1680	25 to 50 units		.03	4.90	14.70	44	3.20	
	1690	Butt fusion machine, electric		20.85	55	165	495	199.80	
	1695	Electro fusion machine		15.35	40	120	360	146.80	
	1700	Carts, brick, hand powered, 1000 lb. capacity		.38	63	189	565	40.85	
	1800	Gas engine, 1500 lb., 7-1/2' lift		4.73	119	357	1,075	109.25	
	1822	Dehumidifier, medium, 6 lb/hr, 150 CFM		.85	51.50	155	465	37.80	
	1824	Large, 18 lb/hr, 600 CFM		1.64	100	300	900	73.10	
	1830	Distributor, asphalt, trailer mtd, 2000 gal., 38 H.P. diesel		9.15	340	1,020	3,050	277.20	
	1840	3000 gal., 38 H.P. diesel		10.40	370	1,105	3,325	304.20	
	1850	Drill, rotary hammer, electric, 1-1/2" diameter		.70	22	66	198	18.80	
	1860	Carbide bit for above		.05	9	27	81	5.80	
	1865	Rotary, crawler, 250 H.P.		122.45	2,075	6,240	18,700	2,228	
	1870	Emulsion sprayer, 65 gal., 5 H.P. gas engine		3.20	99	297	890	85	
	1880	200 gal., 5 H.P. engine		6.85	165	495	1,475	153.80	
	1930	Floodlight, mercury vapor, or quartz, on tripod, 1000 watt		.34	13	39	117	10.50	
	1940	2000 watt		.59	24.50	73	219	19.30	
	1950	Floodlights, trailer mounted with generator, 1 - 300 watt light		3.00	70	210	630	66	
	1960	2 - 1000 watt lights		4.15	107	320	960	97.20	
	2000	4 - 300 watt lights		3.65	86.50	260	780	81.20	
	2005	Foam spray rig, incl. box trailer, compressor, generator, proportioner		24.90	480	1,445	4,325	488.20	
	2020	Forklift, straight mast, 12' lift, 5000 lb., 2 wheel drive, gas		22.00	198	595	1,775	295	
	2040	21' lift, 5000 lb., 4 wheel drive, diesel		16.65	247	740	2,225	281.20	
	2050	For rough terrain, 42' lift, 35' reach, 9000 lb., 110 HP		22.50	445	1,330	4,000	446	
	2060	For plant, 4 T. capacity, 80 H.P., 2 wheel drive, gas		13.55	105	315	945	171.40	
	2080	10 T. capacity, 120 H.P., 2 wheel drive, diesel		18.65	183	550	1,650	259.20	
	2100	Generator, electric, gas engine, 1.5 KW to 3 KW		3.10	10.35	31	93	31	
	2200	5 KW		3.95	14.35	43	129	40.20	
	2300	10 KW		7.45	33.50	100	300	79.60	
	2400	25 KW		9.10	65	195	585	111.80	
	2500	Diesel engine, 20 KW		9.25	73.50	220	660	118	
	2600	50 KW		16.60	98.50	295	885	191.80	
	2700	100 KW		31.90	120	360	1,075	327.20	
	2800	250 KW		63.05	232	695	2,075	643.40	

01 54 | Construction Aids

01 54 33 | Equipment Rental

			UNIT	HOURLY OPER. COST	RENT PER DAY	RENT PER WEEK	RENT PER MONTH	EQUIPMENT COST/DAY	
40	2850	Hammer, hydraulic, for mounting on boom, to 500 ft.-lb.	Ea.	2.15	68.50	205	615	58.20	40
	2860	1000 ft.-lb.		3.65	112	335	1,000	96.20	
	2900	Heaters, space, oil or electric, 50 MBH		2.50	6.65	20	60	24	
	3000	100 MBH		4.52	9	27	81	41.55	
	3100	300 MBH		14.52	36.50	110	330	138.15	
	3150	500 MBH		29.33	45	135	405	261.65	
	3200	Hose, water, suction with coupling, 20' long, 2" diameter		.02	3	9	27	1.95	
	3210	3" diameter		.03	4.67	14	42	3.05	
	3220	4" diameter		.03	5.35	16	48	3.45	
	3230	6" diameter		.11	18.35	55	165	11.90	
	3240	8" diameter		.20	33.50	100	300	21.60	
	3250	Discharge hose with coupling, 50' long, 2" diameter		.01	1.67	5	15	1.10	
	3260	3" diameter		.02	2.67	8	24	1.75	
	3270	4" diameter		.02	3.67	11	33	2.35	
	3280	6" diameter		.06	9.65	29	87	6.30	
	3290	8" diameter		.20	33.50	100	300	21.60	
	3295	Insulation blower		.22	6	18	54	5.35	
	3300	Ladders, extension type, 16' to 36' long		.14	23	69	207	14.90	
	3400	40' to 60' long		.20	33.50	100	300	21.60	
	3405	Lance for cutting concrete		2.97	104	313	940	86.35	
	3407	Lawn mower, rotary, 22", 5HP		2.48	65	195	585	58.85	
	3408	48" self propelled		4.48	128	385	1,150	112.85	
	3410	Level, laser type, for pipe and sewer leveling		1.46	97.50	292	875	70.10	
	3430	Electronic		.75	50	150	450	36	
	3440	Laser type, rotating beam for grade control		1.17	77.50	233	700	55.95	
	3460	Builders level with tripod and rod		.08	14	42	126	9.05	
	3500	Light towers, towable, with diesel generator, 2000 watt		3.65	86.50	260	780	81.20	
	3600	4000 watt		4.15	107	320	960	97.20	
	3700	Mixer, powered, plaster and mortar, 6 C.F., 7 H.P.		2.10	19.65	59	177	28.60	
	3800	10 C.F., 9 H.P.		2.25	31.50	95	285	37	
	3850	Nailer, pneumatic		.45	30	90	270	21.60	
	3900	Paint sprayers complete, 8 CFM		.67	44.50	134	400	32.15	
	4000	17 CFM		1.17	77.50	233	700	55.95	
	4020	Pavers, bituminous, rubber tires, 8' wide, 50 H.P., diesel		35.70	1,050	3,155	9,475	916.60	
	4030	10' wide, 150 H.P.		75.65	1,700	5,095	15,300	1,624	
	4050	Crawler, 8' wide, 100 H.P., diesel		75.65	1,775	5,355	16,100	1,676	
	4060	10' wide, 150 H.P.		84.80	2,175	6,505	19,500	1,979	
	4070	Concrete paver, 12' to 24' wide, 250 H.P.		81.00	1,575	4,755	14,300	1,599	
	4080	Placer-spreader-trimmer, 24' wide, 300 H.P.		116.90	2,600	7,825	23,500	2,500	
	4100	Pump, centrifugal gas pump, 1-1/2" diam., 65 GPM		3.35	45	135	405	53.80	
	4200	2" diameter, 130 GPM		4.55	50	150	450	66.40	
	4300	3" diameter, 250 GPM		4.80	51.50	155	465	69.40	
	4400	6" diameter, 1500 GPM		25.55	175	525	1,575	309.40	
	4500	Submersible electric pump, 1-1/4" diameter, 55 GPM		.39	15.65	47	141	12.50	
	4600	1-1/2" diameter, 83 GPM		.42	18	54	162	14.15	
	4700	2" diameter, 120 GPM		1.20	22.50	67	201	23	
	4800	3" diameter, 300 GPM		2.05	41.50	125	375	41.40	
	4900	4" diameter, 560 GPM		8.90	163	490	1,475	169.20	
	5000	6" diameter, 1590 GPM		13.05	222	665	2,000	237.40	
	5100	Diaphragm pump, gas, single, 1-1/2" diameter		1.24	45.50	136	410	37.10	
	5200	2" diameter		3.65	56.50	170	510	63.20	
	5300	3" diameter		3.65	56.50	170	510	63.20	
	5400	Double, 4" diameter		4.90	78.50	235	705	86.20	
	5450	Pressure Washer 5 Ga.Pm,3000 PSI		4.00	35	105	315	53	
	5500	Trash pump, self-priming, gas, 2" diameter		3.85	20	60	180	42.80	
	5600	Diesel, 4" diameter		10.25	58.50	175	525	117	
	5650	Diesel, 6" diameter		34.00	130	390	1,175	350	
	5655	Grout Pump		13.80	93.50	280	840	166.40	

01 54 | Construction Aids

01 54 33 | Equipment Rental

			UNIT	HOURLY OPER. COST	RENT PER DAY	RENT PER WEEK	RENT PER MONTH	EQUIPMENT COST/DAY	
40	5700	Salamanders, L.P. gas fired, 100,000 BTU	Ea.	4.56	15.35	46	138	45.70	40
	5705	50,000 BTU		3.40	8.65	26	78	32.40	
	5720	Sandblaster, portable, open top, 3 C.F. capacity		.55	27	81	243	20.60	
	5730	6 C.F. capacity		.85	40	120	360	30.80	
	5740	Accessories for above		.12	19.65	59	177	12.75	
	5750	Sander, floor		.81	18.65	56	168	17.70	
	5760	Edger		.75	25.50	77	231	21.40	
	5800	Saw, chain, gas engine, 18" long		1.90	17.65	53	159	25.80	
	5900	Hydraulic powered, 36" long		.60	55	165	495	37.80	
	5950	60" long		.65	56.50	170	510	39.20	
	6000	Masonry, table mounted, 14" diameter, 5 H.P.		1.25	53	159	475	41.80	
	6050	Portable cut-off, 8 H.P.		2.05	28	84	252	33.20	
	6100	Circular, hand held, electric, 7-1/4" diameter		.19	4	12	36	3.90	
	6200	12" diameter		.26	7.35	22	66	6.50	
	6250	Wall saw, w/hydraulic power, 10 H.P.		7.50	60	180	540	96	
	6275	Shot blaster, walk behind, 20" wide		4.60	300	905	2,725	217.80	
	6280	Sidewalk broom, walk-behind		2.31	57	171	515	52.70	
	6300	Steam cleaner, 100 gallons per hour		2.90	46.50	140	420	51.20	
	6310	200 gallons per hour		4.15	55	165	495	66.20	
	6340	Tar Kettle/Pot, 400 gallon		6.15	80	240	720	97.20	
	6350	Torch, cutting, acetylene-oxygen, 150' hose		.50	21.50	65	195	17	
	6360	Hourly operating cost includes tips and gas		9.45				75.60	
	6410	Toilet, portable chemical		.11	19	57	171	12.30	
	6420	Recycle flush type		.14	23	69	207	14.90	
	6430	Toilet, fresh water flush, garden hose,		.16	26.50	79	237	17.10	
	6440	Hoisted, non-flush, for high rise		.14	22.50	68	204	14.70	
	6450	Toilet, trailers, minimum		.24	39.50	118	355	25.50	
	6460	Maximum		.72	119	358	1,075	77.35	
	6465	Tractor, farm with attachment		14.40	262	785	2,350	272.20	
	6500	Trailers, platform, flush deck, 2 axle, 25 ton capacity		4.75	107	320	960	102	
	6600	40 ton capacity		6.20	150	450	1,350	139.60	
	6700	3 axle, 50 ton capacity		6.70	165	495	1,475	152.60	
	6800	75 ton capacity		8.35	218	655	1,975	197.80	
	6810	Trailer mounted cable reel for H.V. line work		4.84	231	692	2,075	177.10	
	6820	Trailer mounted cable tensioning rig		9.59	455	1,370	4,100	350.70	
	6830	Cable pulling rig		69.62	2,575	7,710	23,100	2,099	
	6900	Water tank, engine driven discharge, 5000 gallons		6.25	143	430	1,300	136	
	6925	10,000 gallons		8.50	202	605	1,825	189	
	6950	Water truck, off highway, 6000 gallons		66.95	775	2,320	6,950	999.60	
	7010	Tram car for H.V. line work, powered, 2 conductor		6.48	125	375	1,125	126.85	
	7020	Transit (builder's level) with tripod		.08	14	42	126	9.05	
	7030	Trench box, 3000 lbs. 6'x8'		.56	93	279	835	60.30	
	7040	7200 lbs. 6'x20'		1.05	175	525	1,575	113.40	
	7050	8000 lbs., 8' x 16'		.95	158	475	1,425	102.60	
	7060	9500 lbs., 8'x20'		1.16	194	581	1,750	125.50	
	7065	11,000 lbs., 8'x24'		1.27	212	637	1,900	137.55	
	7070	12,000 lbs., 10' x 20'		1.71	285	855	2,575	184.70	
	7100	Truck, pickup, 3/4 ton, 2 wheel drive		10.35	56.50	170	510	116.80	
	7200	4 wheel drive		10.65	71.50	215	645	128.20	
	7250	Crew carrier, 9 passenger		14.70	86.50	260	780	169.60	
	7290	Flat bed truck, 20,000 G.V.W.		15.55	122	365	1,100	197.40	
	7300	Tractor, 4 x 2, 220 H.P.		21.80	190	570	1,700	288.40	
	7410	330 H.P.		32.15	262	785	2,350	414.20	
	7500	6 x 4, 380 H.P.		36.85	305	920	2,750	478.80	
	7600	450 H.P.		44.90	370	1,110	3,325	581.20	
	7620	Vacuum truck, hazardous material, 2500 gallon		10.80	310	925	2,775	271.40	
	7625	5,000 gallon		16.90	435	1,300	3,900	395.20	
	7640	Tractor, with A frame, boom and winch, 225 H.P.		24.35	267	800	2,400	354.80	

01 54 | Construction Aids

01 54 33 | Equipment Rental

			UNIT	HOURLY OPER. COST	RENT PER DAY	RENT PER WEEK	RENT PER MONTH	EQUIPMENT COST/DAY	
40	7650	Vacuum, H.E.P.A., 16 gal., wet/dry	Ea.	.82	18	54	162	17.35	40
	7655	55 gal, wet/dry		.83	27	81	243	22.85	
	7660	Water tank, portable		.17	28.50	85.50	257	18.45	
	7690	Sewer/catch basin vacuum, 14 CY, 1500 Gallon		21.45	650	1,950	5,850	561.60	
	7700	Welder, electric, 200 amp		3.63	17.65	53	159	39.65	
	7800	300 amp		5.36	21.50	64	192	55.70	
	7900	Gas engine, 200 amp		12.85	25.50	77	231	118.20	
	8000	300 amp		14.70	27.50	83	249	134.20	
	8100	Wheelbarrow, any size		.07	11.65	35	105	7.55	
	8200	Wrecking ball, 4000 lb.		2.10	73.50	220	660	60.80	
50	0010	**HIGHWAY EQUIPMENT RENTAL** without operators							50
	0050	Asphalt batch plant, portable drum mixer, 100 ton/hr.	Ea.	63.70	1,450	4,330	13,000	1,376	
	0060	200 ton/hr.		71.10	1,525	4,565	13,700	1,482	
	0070	300 ton/hr.		82.55	1,800	5,395	16,200	1,739	
	0100	Backhoe attachment, long stick, up to 185 HP, 10.5' long		.32	21.50	64	192	15.35	
	0140	Up to 250 HP, 12' long		.35	23	69	207	16.60	
	0180	Over 250 HP, 15' long		.47	31	93	279	22.35	
	0200	Special dipper arm, up to 100 HP, 32' long		.96	63.50	191	575	45.90	
	0240	Over 100 HP, 33' long		1.19	79.50	238	715	57.10	
	0280	Catch basin/sewer cleaning truck, 3 ton, 9 CY, 1000 Gal		34.00	405	1,210	3,625	514	
	0300	Concrete batch plant, portable, electric, 200 CY/Hr		17.20	510	1,530	4,600	443.60	
	0520	Grader/dozer attachment, ripper/scarifier, rear mounted, up to 135 HP		3.10	65	195	585	63.80	
	0540	Up to 180 HP		3.70	83.50	250	750	79.60	
	0580	Up to 250 HP		4.05	95	285	855	89.40	
	0700	Pvmt. removal bucket, for hyd. excavator, up to 90 HP		1.65	48.50	145	435	42.20	
	0740	Up to 200 HP		1.85	70	210	630	56.80	
	0780	Over 200 HP		1.95	83.50	250	750	65.60	
	0900	Aggregate spreader, self-propelled, 187 HP		54.90	695	2,080	6,250	855.20	
	1000	Chemical spreader, 3 C.Y.		2.75	43.50	130	390	48	
	1900	Hammermill, traveling, 250 HP		92.70	1,850	5,540	16,600	1,850	
	2000	Horizontal borer, 3" diam, 13 HP gas driven		5.45	56.50	170	510	77.60	
	2150	Horizontal directional drill, 20,000 lb. thrust, 78 H.P. diesel		26.15	700	2,095	6,275	628.20	
	2160	30,000 lb. thrust, 115 H.P. diesel		32.45	1,050	3,185	9,550	896.60	
	2170	50,000 lb. thrust, 170 H.P. diesel		45.85	1,350	4,080	12,200	1,183	
	2190	Mud trailer for HDD, 1500 gallon, 175 H.P., gas		24.40	147	440	1,325	283.20	
	2200	Hydromulchers, gas power, 3000 gal., for truck mounting		17.00	258	775	2,325	291	
	2400	Joint & crack cleaner, walk behind, 25 HP		3.10	50	150	450	54.80	
	2500	Filler, trailer mounted, 400 gal., 20 HP		7.90	218	655	1,975	194.20	
	3000	Paint striper, self propelled, double line, 30 HP		6.25	162	485	1,450	147	
	3200	Post drivers, 6" I-Beam frame, for truck mounting		13.40	435	1,305	3,925	368.20	
	3400	Road sweeper, self propelled, 8' wide, 90 HP		31.95	575	1,730	5,200	601.60	
	3450	Road sweeper, vacuum assisted, 4 CY, 220 Gal		56.60	630	1,895	5,675	831.80	
	4000	Road mixer, self-propelled, 130 HP		39.70	770	2,315	6,950	780.60	
	4100	310 HP		71.00	2,225	6,640	19,900	1,896	
	4220	Cold mix paver, incl pug mill and bitumen tank, 165 HP		86.60	2,175	6,515	19,500	1,996	
	4250	Paver, asphalt, wheel or crawler, 130 H.P., diesel		83.50	2,125	6,340	19,000	1,936	
	4300	Paver, road widener, gas 1' to 6', 67 HP		40.10	825	2,480	7,450	816.80	
	4400	Diesel, 2' to 14', 88 HP		52.30	1,050	3,175	9,525	1,053	
	4600	Slipform pavers, curb and gutter, 2 track, 75 HP		36.80	765	2,300	6,900	754.40	
	4700	4 track, 165 HP		45.85	815	2,445	7,325	855.80	
	4800	Median barrier, 215 HP		46.50	845	2,535	7,600	879	
	4901	Trailer, low bed, 75 ton capacity		9.00	218	655	1,975	203	
	5000	Road planer, walk behind, 10" cutting width, 10 HP		2.90	30.50	91	273	41.40	
	5100	Self propelled, 12" cutting width, 64 HP		8.25	127	380	1,150	142	
	5120	Traffic line remover, metal ball blaster, truck mounted, 115 HP		46.45	760	2,285	6,850	828.60	
	5140	Grinder, truck mounted, 115 HP		49.55	805	2,420	7,250	880.40	
	5160	Walk-behind, 11 HP		3.25	46.50	140	420	54	
	5200	Pavement profiler, 4' to 6' wide, 450 HP		210.55	3,400	10,220	30,700	3,728	

01 54 | Construction Aids

01 54 33 | Equipment Rental

			UNIT	HOURLY OPER. COST	RENT PER DAY	RENT PER WEEK	RENT PER MONTH	EQUIPMENT COST/DAY	
50	5300	8' to 10' wide, 750 HP	Ea.	333.90	4,725	14,160	42,500	5,503	50
	5400	Roadway plate, steel, 1"x8'x20'		.07	11.35	34	102	7.35	
	5600	Stabilizer, self-propelled, 150 HP		39.45	610	1,830	5,500	681.60	
	5700	310 HP		66.45	1,350	4,060	12,200	1,344	
	5800	Striper, thermal, truck mounted 120 gal. paint, 150 H.P.		49.90	505	1,520	4,550	703.20	
	6000	Tar kettle, 330 gal., trailer mounted		5.77	61.50	185	555	83.15	
	7000	Tunnel locomotive, diesel, 8 to 12 ton		26.70	585	1,750	5,250	563.60	
	7005	Electric, 10 ton		23.25	665	2,000	6,000	586	
	7010	Muck cars, 1/2 C.Y. capacity		1.75	23	69	207	27.80	
	7020	1 C.Y. capacity		1.95	31.50	94	282	34.40	
	7030	2 C.Y. capacity		2.10	36.50	110	330	38.80	
	7040	Side dump, 2 C.Y. capacity		2.30	45	135	405	45.40	
	7050	3 C.Y. capacity		3.10	51.50	155	465	55.80	
	7060	5 C.Y. capacity		4.40	65	195	585	74.20	
	7100	Ventilating blower for tunnel, 7-1/2 H.P.		1.35	51.50	155	465	41.80	
	7110	10 H.P.		1.57	53.50	160	480	44.55	
	7120	20 H.P.		2.58	69.50	208	625	62.25	
	7140	40 H.P.		4.56	98.50	295	885	95.50	
	7160	60 H.P.		6.90	152	455	1,375	146.20	
	7175	75 H.P.		8.81	202	607	1,825	191.90	
	7180	200 H.P.		19.95	305	910	2,725	341.60	
	7800	Windrow loader, elevating		47.65	1,350	4,080	12,200	1,197	
60	0010	**LIFTING AND HOISTING EQUIPMENT RENTAL** without operators	R015433 -10						60
	0120	Aerial lift truck, 2 person, to 80'	Ea.	25.55	725	2,180	6,550	640.40	
	0140	Boom work platform, 40' snorkel	R015433 -15	15.40	260	780	2,350	279.20	
	0150	Crane, flatbed mntd, 3 ton cap.		13.50	195	585	1,750	225	
	0200	Crane, climbing, 106' jib, 6000 lb. capacity, 410 FPM	R312316 -45	37.40	1,475	4,400	13,200	1,179	
	0300	101' jib, 10,250 lb. capacity, 270 FPM		43.30	1,850	5,580	16,700	1,462	
	0500	Tower, static, 130' high, 106' jib, 6200 lb. capacity at 400 FPM		40.85	1,700	5,090	15,300	1,345	
	0600	Crawler mounted, lattice boom, 1/2 C.Y., 15 tons at 12' radius		30.66	615	1,850	5,550	615.30	
	0700	3/4 C.Y., 20 tons at 12' radius		40.88	770	2,310	6,925	789.05	
	0800	1 C.Y., 25 tons at 12' radius		54.50	1,025	3,080	9,250	1,052	
	0900	1-1/2 C.Y., 40 tons at 12' radius		54.50	1,025	3,095	9,275	1,055	
	1000	2 C.Y., 50 tons at 12' radius		54.20	1,125	3,380	10,100	1,110	
	1100	3 C.Y., 75 tons at 12' radius		72.95	1,525	4,560	13,700	1,496	
	1200	100 ton capacity, 60' boom		71.55	1,725	5,195	15,600	1,611	
	1300	165 ton capacity, 60' boom		89.60	1,925	5,750	17,300	1,867	
	1400	200 ton capacity, 70' boom		111.30	2,350	7,085	21,300	2,307	
	1500	350 ton capacity, 80' boom		158.35	3,775	11,335	34,000	3,534	
	1600	Truck mounted, lattice boom, 6 x 4, 20 tons at 10' radius		44.77	1,200	3,610	10,800	1,080	
	1700	25 tons at 10' radius		48.31	1,300	3,930	11,800	1,172	
	1800	30 tons at 10' radius		52.66	1,400	4,180	12,500	1,257	
	1900	40 tons at 12' radius		56.59	1,450	4,370	13,100	1,327	
	2000	60 tons at 15' radius		64.84	1,550	4,620	13,900	1,443	
	2050	82 tons at 15' radius		73.58	1,650	4,940	14,800	1,577	
	2100	90 tons at 15' radius		83.16	1,800	5,380	16,100	1,741	
	2200	115 tons at 15' radius		94.14	2,000	6,020	18,100	1,957	
	2300	150 tons at 18' radius		77.90	2,100	6,335	19,000	1,890	
	2350	165 tons at 18' radius		112.04	2,250	6,720	20,200	2,240	
	2400	Truck mounted, hydraulic, 12 ton capacity		49.10	625	1,880	5,650	768.80	
	2500	25 ton capacity		49.30	660	1,975	5,925	789.40	
	2550	33 ton capacity		50.35	695	2,080	6,250	818.80	
	2560	40 ton capacity		48.50	670	2,015	6,050	791	
	2600	55 ton capacity		72.00	915	2,745	8,225	1,125	
	2700	80 ton capacity		85.50	1,025	3,060	9,175	1,296	
	2720	100 ton capacity		104.90	1,550	4,685	14,100	1,776	
	2740	120 ton capacity		83.95	2,000	5,975	17,900	1,867	
	2760	150 ton capacity		89.85	2,175	6,490	19,500	2,017	

01 54 | Construction Aids

01 54 33 | Equipment Rental

			UNIT	HOURLY OPER. COST	RENT PER DAY	RENT PER WEEK	RENT PER MONTH	EQUIPMENT COST/DAY	
60	2800	Self-propelled, 4 x 4, with telescoping boom, 5 ton	Ea.	14.65	235	705	2,125	258.20	60
	2900	12-1/2 ton capacity		34.60	540	1,625	4,875	601.80	
	3000	15 ton capacity		32.30	575	1,725	5,175	603.40	
	3050	20 ton capacity		37.95	660	1,985	5,950	700.60	
	3100	25 ton capacity		38.85	675	2,030	6,100	716.80	
	3150	40 ton capacity		50.15	895	2,680	8,050	937.20	
	3200	Derricks, guy, 20 ton capacity, 60' boom, 75' mast		33.55	360	1,078	3,225	484	
	3300	100' boom, 115' mast		51.95	615	1,850	5,550	785.60	
	3400	Stiffleg, 20 ton capacity, 70' boom, 37' mast		35.80	465	1,400	4,200	566.40	
	3500	100' boom, 47' mast		54.68	745	2,240	6,725	885.45	
	3550	Helicopter, small, lift to 1250 lbs. maximum, w/pilot		105.67	2,900	8,710	26,100	2,587	
	3600	Hoists, chain type, overhead, manual, 3/4 ton		.10	1	3	9	1.40	
	3900	10 ton		.70	9.65	29	87	11.40	
	4000	Hoist and tower, 5000 lb. cap., portable electric, 40' high		4.65	207	621	1,875	161.40	
	4100	For each added 10' section, add		.10	16.35	49	147	10.60	
	4200	Hoist and single tubular tower, 5000 lb. electric, 100' high		6.26	289	867	2,600	223.50	
	4300	For each added 6'-6" section, add		.17	27.50	83	249	17.95	
	4400	Hoist and double tubular tower, 5000 lb., 100' high		6.70	320	955	2,875	244.60	
	4500	For each added 6'-6" section, add		.19	31	93	279	20.10	
	4550	Hoist and tower, mast type, 6000 lb., 100' high		7.26	330	990	2,975	256.10	
	4570	For each added 10' section, add		.12	19.65	59	177	12.75	
	4600	Hoist and tower, personnel, electric, 2000 lb., 100' @ 125 FPM		14.74	880	2,640	7,925	645.90	
	4700	3000 lb., 100' @ 200 FPM		16.88	995	2,990	8,975	733.05	
	4800	3000 lb., 150' @ 300 FPM		18.63	1,125	3,340	10,000	817.05	
	4900	4000 lb., 100' @ 300 FPM		19.36	1,125	3,410	10,200	836.90	
	5000	6000 lb., 100' @ 275 FPM		20.98	1,200	3,580	10,700	883.85	
	5100	For added heights up to 500', add	L.F.	.01	1.67	5	15	1.10	
	5200	Jacks, hydraulic, 20 ton	Ea.	.05	1.67	5	15	1.40	
	5500	100 ton		.35	10.65	32	96	9.20	
	6100	Jacks, hydraulic, climbing w/ 50' jackrods, control console, 30 ton cap		1.79	119	357	1,075	85.70	
	6150	For each added 10' jackrod section, add		.05	3.33	10	30	2.40	
	6300	50 ton capacity		2.87	191	574	1,725	137.75	
	6350	For each added 10' jackrod section, add		.06	4	12	36	2.90	
	6500	125 ton capacity		7.55	505	1,510	4,525	362.40	
	6550	For each added 10' jackrod section, add		.52	34.50	103	310	24.75	
	6600	Cable jack, 10 ton capacity with 200' cable		1.50	99.50	299	895	71.80	
	6650	For each added 50' of cable, add		.17	11.35	34	102	8.15	
70	0010	**WELLPOINT EQUIPMENT RENTAL** without operators	R015433 -10						70
	0020	Based on 2 months rental							
	0100	Combination jetting & wellpoint pump, 60 H.P. diesel	Ea.	21.79	295	884	2,650	351.10	
	0200	High pressure gas jet pump, 200 H.P., 300 psi	"	57.29	252	756	2,275	609.50	
	0300	Discharge pipe, 8" diameter	L.F.	.01	.48	1.44	4.32	.35	
	0350	12" diameter		.01	.71	2.12	6.35	.50	
	0400	Header pipe, flows up to 150 G.P.M., 4" diameter		.01	.43	1.29	3.87	.35	
	0500	400 G.P.M., 6" diameter		.01	.51	1.54	4.62	.40	
	0600	800 G.P.M., 8" diameter		.01	.71	2.12	6.35	.50	
	0700	1500 G.P.M., 10" diameter		.01	.74	2.22	6.65	.50	
	0800	2500 G.P.M., 12" diameter		.02	1.40	4.21	12.65	1	
	0900	4500 G.P.M., 16" diameter		.03	1.80	5.39	16.15	1.30	
	0950	For quick coupling aluminum and plastic pipe, add		.03	1.86	5.58	16.75	1.35	
	1100	Wellpoint, 25' long, with fittings & riser pipe, 1-1/2" or 2" diameter	Ea.	.06	3.71	11.13	33.50	2.70	
	1200	Wellpoint pump, diesel powered, 4" diameter, 20 H.P.		8.77	170	510	1,525	172.15	
	1300	6" diameter, 30 H.P.		12.22	211	632	1,900	224.15	
	1400	8" suction, 40 H.P.		16.47	289	867	2,600	305.15	
	1500	10" suction, 75 H.P.		26.59	340	1,013	3,050	415.30	
	1600	12" suction, 100 H.P.		37.34	540	1,620	4,850	622.70	
	1700	12" suction, 175 H.P.		57.89	590	1,770	5,300	817.10	

01 54 | Construction Aids

01 54 33 | Equipment Rental

			UNIT	HOURLY OPER. COST	RENT PER DAY	RENT PER WEEK	RENT PER MONTH	EQUIPMENT COST/DAY	
80	0010	**MARINE EQUIPMENT RENTAL** without operators	R015433 -10						80
	0200	Barge, 400 Ton, 30' wide x 90' long		Ea.	14.80	1,050	3,140	9,425	746.40
	0240	800 Ton, 45' wide x 90' long			17.95	1,275	3,800	11,400	903.60
	2000	Tugboat, diesel, 100 HP			27.75	195	585	1,750	339
	2040	250 HP			56.05	360	1,075	3,225	663.40
	2080	380 HP			115.60	1,075	3,195	9,575	1,564

Crews

Crew No.	Bare Costs		Incl. Subs O & P		Cost Per Labor-Hour	
Crew A-1	Hr.	Daily	Hr.	Daily	Bare Costs	Incl. O&P
1 Building Laborer	$31.60	$252.80	$49.00	$392.00	$31.60	$49.00
1 Concrete saw, gas manual		63.20		69.52	7.90	8.69
8 L.H., Daily Totals		$316.00		$461.52	$39.50	$57.69
Crew A-1A	Hr.	Daily	Hr.	Daily	Bare Costs	Incl. O&P
1 Skilled Worker	$40.85	$326.80	$63.25	$506.00	$40.85	$63.25
1 Shot Blaster, 20"		217.80		239.58	27.23	29.95
8 L.H., Daily Totals		$544.60		$745.58	$68.08	$93.20
Crew A-1B	Hr.	Daily	Hr.	Daily	Bare Costs	Incl. O&P
1 Building Laborer	$31.60	$252.80	$49.00	$392.00	$31.60	$49.00
1 Concrete Saw		141.00		155.10	17.63	19.39
8 L.H., Daily Totals		$393.80		$547.10	$49.23	$68.39
Crew A-1C	Hr.	Daily	Hr.	Daily	Bare Costs	Incl. O&P
1 Building Laborer	$31.60	$252.80	$49.00	$392.00	$31.60	$49.00
1 Chain saw, gas, 18"		25.80		28.38	3.23	3.55
8 L.H., Daily Totals		$278.60		$420.38	$34.83	$52.55
Crew A-1D	Hr.	Daily	Hr.	Daily	Bare Costs	Incl. O&P
1 Building Laborer	$31.60	$252.80	$49.00	$392.00	$31.60	$49.00
1 Vibrating plate, gas, 18"		31.80		34.98	3.98	4.37
8 L.H., Daily Totals		$284.60		$426.98	$35.58	$53.37
Crew A-1E	Hr.	Daily	Hr.	Daily	Bare Costs	Incl. O&P
1 Building Laborer	$31.60	$252.80	$49.00	$392.00	$31.60	$49.00
1 Vibratory Plate, Gas, 21"		39.20		43.12	4.90	5.39
8 L.H., Daily Totals		$292.00		$435.12	$36.50	$54.39
Crew A-1F	Hr.	Daily	Hr.	Daily	Bare Costs	Incl. O&P
1 Building Laborer	$31.60	$252.80	$49.00	$392.00	$31.60	$49.00
1 Rammer/tamper, gas, 8"		43.40		47.74	5.42	5.97
8 L.H., Daily Totals		$296.20		$439.74	$37.02	$54.97
Crew A-1G	Hr.	Daily	Hr.	Daily	Bare Costs	Incl. O&P
1 Building Laborer	$31.60	$252.80	$49.00	$392.00	$31.60	$49.00
1 Rammer/tamper, gas, 15"		49.40		54.34	6.17	6.79
8 L.H., Daily Totals		$302.20		$446.34	$37.77	$55.79
Crew A-1H	Hr.	Daily	Hr.	Daily	Bare Costs	Incl. O&P
1 Building Laborer	$31.60	$252.80	$49.00	$392.00	$31.60	$49.00
1 Exterior Steam Cleaner		51.20		56.32	6.40	7.04
8 L.H., Daily Totals		$304.00		$448.32	$38.00	$56.04
Crew A-1J	Hr.	Daily	Hr.	Daily	Bare Costs	Incl. O&P
1 Building Laborer	$31.60	$252.80	$49.00	$392.00	$31.60	$49.00
1 Cultivator, Walk-Behind, 5 H.P.		57.05		62.76	7.13	7.84
8 L.H., Daily Totals		$309.85		$454.76	$38.73	$56.84
Crew A-1K	Hr.	Daily	Hr.	Daily	Bare Costs	Incl. O&P
1 Building Laborer	$31.60	$252.80	$49.00	$392.00	$31.60	$49.00
1 Cultivator, Walk-Behind, 8 H.P.		91.10		100.21	11.39	12.53
8 L.H., Daily Totals		$343.90		$492.21	$42.99	$61.53
Crew A-1M	Hr.	Daily	Hr.	Daily	Bare Costs	Incl. O&P
1 Building Laborer	$31.60	$252.80	$49.00	$392.00	$31.60	$49.00
1 Snow Blower, Walk-Behind		52.70		57.97	6.59	7.25
8 L.H., Daily Totals		$305.50		$449.97	$38.19	$56.25

Crew No.	Bare Costs		Incl. Subs O & P		Cost Per Labor-Hour	
Crew A-2	Hr.	Daily	Hr.	Daily	Bare Costs	Incl. O&P
2 Laborers	$31.60	$505.60	$49.00	$784.00	$31.38	$48.55
1 Truck Driver (light)	30.95	247.60	47.65	381.20		
1 Flatbed Truck, Gas, 1.5 Ton		194.40		213.84	8.10	8.91
24 L.H., Daily Totals		$947.60		$1379.04	$39.48	$57.46
Crew A-2A	Hr.	Daily	Hr.	Daily	Bare Costs	Incl. O&P
2 Laborers	$31.60	$505.60	$49.00	$784.00	$31.38	$48.55
1 Truck Driver (light)	30.95	247.60	47.65	381.20		
1 Flatbed Truck, Gas, 1.5 Ton		194.40		213.84		
1 Concrete Saw		141.00		155.10	13.98	15.37
24 L.H., Daily Totals		$1088.60		$1534.14	$45.36	$63.92
Crew A-2B	Hr.	Daily	Hr.	Daily	Bare Costs	Incl. O&P
1 Truck Driver (light)	$30.95	$247.60	$47.65	$381.20	$30.95	$47.65
1 Flatbed Truck, Gas, 1.5 Ton		194.40		213.84	24.30	26.73
8 L.H., Daily Totals		$442.00		$595.04	$55.25	$74.38
Crew A-3A	Hr.	Daily	Hr.	Daily	Bare Costs	Incl. O&P
1 Truck Driver (light)	$30.95	$247.60	$47.65	$381.20	$30.95	$47.65
1 Pickup truck, 4 x 4, 3/4 ton		128.20		141.02	16.02	17.63
8 L.H., Daily Totals		$375.80		$522.22	$46.98	$65.28
Crew A-3B	Hr.	Daily	Hr.	Daily	Bare Costs	Incl. O&P
1 Equip. Oper. (medium)	$41.35	$330.80	$62.15	$497.20	$36.65	$55.65
1 Truck Driver (heavy)	31.95	255.60	49.15	393.20		
1 Dump Truck, 12 C.Y., 400 H.P.		533.00		586.30		
1 F.E. Loader, W.M.,2.5 C.Y.		395.80		435.38	58.05	63.85
16 L.H., Daily Totals		$1515.20		$1912.08	$94.70	$119.51
Crew A-3C	Hr.	Daily	Hr.	Daily	Bare Costs	Incl. O&P
1 Equip. Oper. (light)	$39.05	$312.40	$58.70	$469.60	$39.05	$58.70
1 Loader, Skid Steer, 78 H.P.		259.60		285.56	32.45	35.70
8 L.H., Daily Totals		$572.00		$755.16	$71.50	$94.39
Crew A-3D	Hr.	Daily	Hr.	Daily	Bare Costs	Incl. O&P
1 Truck Driver, Light	$30.95	$247.60	$47.65	$381.20	$30.95	$47.65
1 Pickup truck, 4 x 4, 3/4 ton		128.20		141.02		
1 Flatbed Trailer, 25 Ton		102.00		112.20	28.77	31.65
8 L.H., Daily Totals		$477.80		$634.42	$59.73	$79.30
Crew A-3E	Hr.	Daily	Hr.	Daily	Bare Costs	Incl. O&P
1 Equip. Oper. (crane)	$42.55	$340.40	$63.95	$511.60	$37.25	$56.55
1 Truck Driver (heavy)	31.95	255.60	49.15	393.20		
1 Pickup truck, 4 x 4, 3/4 ton		128.20		141.02	8.01	8.81
16 L.H., Daily Totals		$724.20		$1045.82	$45.26	$65.36
Crew A-3F	Hr.	Daily	Hr.	Daily	Bare Costs	Incl. O&P
1 Equip. Oper. (crane)	$42.55	$340.40	$63.95	$511.60	$37.25	$56.55
1 Truck Driver (heavy)	31.95	255.60	49.15	393.20		
1 Pickup truck, 4 x 4, 3/4 ton		128.20		141.02		
1 Truck Tractor, 6x4, 380 H.P.		478.80		526.68		
1 Lowbed Trailer, 75 Ton		203.00		223.30	50.63	55.69
16 L.H., Daily Totals		$1406.00		$1795.80	$87.88	$112.24

Crews

Crew No.	Bare Costs		Incl. Subs O & P		Cost Per Labor-Hour	
Crew A-3G	Hr.	Daily	Hr.	Daily	Bare Costs	Incl. O&P
1 Equip. Oper. (crane)	$42.55	$340.40	$63.95	$511.60	$37.25	$56.55
1 Truck Driver (heavy)	31.95	255.60	49.15	393.20		
1 Pickup truck, 4 x 4, 3/4 ton		128.20		141.02		
1 Truck Tractor, 6x4, 450 H.P.		581.20		639.32		
1 Lowbed Trailer, 75 Ton		203.00		223.30	57.02	62.73
16 L.H., Daily Totals		$1508.40		$1908.44	$94.28	$119.28
Crew A-3H	Hr.	Daily	Hr.	Daily	Bare Costs	Incl. O&P
1 Equip. Oper. (crane)	$42.55	$340.40	$63.95	$511.60	$42.55	$63.95
1 Hyd. crane, 12 Ton (daily)		1018.00		1119.80	127.25	139.97
8 L.H., Daily Totals		$1358.40		$1631.40	$169.80	$203.93
Crew A-3I	Hr.	Daily	Hr.	Daily	Bare Costs	Incl. O&P
1 Equip. Oper. (crane)	$42.55	$340.40	$63.95	$511.60	$42.55	$63.95
1 Hyd. crane, 25 Ton (daily)		1054.00		1159.40	131.75	144.93
8 L.H., Daily Totals		$1394.40		$1671.00	$174.30	$208.88
Crew A-3J	Hr.	Daily	Hr.	Daily	Bare Costs	Incl. O&P
1 Equip. Oper. (crane)	$42.55	$340.40	$63.95	$511.60	$42.55	$63.95
1 Hyd. crane, 40 Ton (daily)		1058.00		1163.80	132.25	145.47
8 L.H., Daily Totals		$1398.40		$1675.40	$174.80	$209.43
Crew A-3K	Hr.	Daily	Hr.	Daily	Bare Costs	Incl. O&P
1 Equip. Oper. (crane)	$42.55	$340.40	$63.95	$511.60	$39.67	$59.63
1 Equip. Oper. Oiler	36.80	294.40	55.30	442.40		
1 Hyd. crane, 55 Ton (daily)		1491.00		1640.10		
1 P/U Truck, 3/4 Ton (daily)		137.80		151.58	101.80	111.98
16 L.H., Daily Totals		$2263.60		$2745.68	$141.47	$171.60
Crew A-3L	Hr.	Daily	Hr.	Daily	Bare Costs	Incl. O&P
1 Equip. Oper. (crane)	$42.55	$340.40	$63.95	$511.60	$39.67	$59.63
1 Equip. Oper. Oiler	36.80	294.40	55.30	442.40		
1 Hyd. crane, 80 Ton (daily)		1704.00		1874.40		
1 P/U Truck, 3/4 Ton (daily)		137.80		151.58	115.11	126.62
16 L.H., Daily Totals		$2476.60		$2979.98	$154.79	$186.25
Crew A-3M	Hr.	Daily	Hr.	Daily	Bare Costs	Incl. O&P
1 Equip. Oper. (crane)	$42.55	$340.40	$63.95	$511.60	$39.67	$59.63
1 Equip. Oper. Oiler	36.80	294.40	55.30	442.40		
1 Hyd. crane, 100 Ton (daily)		2399.00		2638.90		
1 P/U Truck, 3/4 Ton (daily)		137.80		151.58	158.55	174.41
16 L.H., Daily Totals		$3171.60		$3744.48	$198.22	$234.03
Crew A-3N	Hr.	Daily	Hr.	Daily	Bare Costs	Incl. O&P
1 Equip. Oper. (crane)	$42.55	$340.40	$63.95	$511.60	$42.55	$63.95
1 Tower crane (monthly)		1022.00		1124.20	127.75	140.53
8 L.H., Daily Totals		$1362.40		$1635.80	$170.30	$204.47
Crew A-3P	Hr.	Daily	Hr.	Daily	Bare Costs	Incl. O&P
1 Equip. Oper., Light	$39.05	$312.40	$58.70	$469.60	$39.05	$58.70
1 A.T. Forklift, 42' lift		446.00		490.60	55.75	61.33
8 L.H., Daily Totals		$758.40		$960.20	$94.80	$120.03
Crew A-4	Hr.	Daily	Hr.	Daily	Bare Costs	Incl. O&P
2 Carpenters	$39.95	$639.20	$61.95	$991.20	$38.37	$58.88
1 Painter, Ordinary	35.20	281.60	52.75	422.00		
24 L.H., Daily Totals		$920.80		$1413.20	$38.37	$58.88

Crew No.	Bare Costs		Incl. Subs O & P		Cost Per Labor-Hour	
Crew A-5	Hr.	Daily	Hr.	Daily	Bare Costs	Incl. O&P
2 Laborers	$31.60	$505.60	$49.00	$784.00	$31.53	$48.85
.25 Truck Driver (light)	30.95	61.90	47.65	95.30		
.25 Flatbed Truck, Gas, 1.5 Ton		48.60		53.46	2.70	2.97
18 L.H., Daily Totals		$616.10		$932.76	$34.23	$51.82
Crew A-6	Hr.	Daily	Hr.	Daily	Bare Costs	Incl. O&P
1 Instrument Man	$40.85	$326.80	$63.25	$506.00	$39.73	$60.80
1 Rodman/Chainman	38.60	308.80	58.35	466.80		
1 Laser Transit/Level		70.10		77.11	4.38	4.82
16 L.H., Daily Totals		$705.70		$1049.91	$44.11	$65.62
Crew A-7	Hr.	Daily	Hr.	Daily	Bare Costs	Incl. O&P
1 Chief Of Party	$50.20	$401.60	$77.15	$617.20	$43.22	$66.25
1 Instrument Man	40.85	326.80	63.25	506.00		
1 Rodman/Chainman	38.60	308.80	58.35	466.80		
1 Laser Transit/Level		70.10		77.11	2.92	3.21
24 L.H., Daily Totals		$1107.30		$1667.11	$46.14	$69.46
Crew A-8	Hr.	Daily	Hr.	Daily	Bare Costs	Incl. O&P
1 Chief of Party	$50.20	$401.60	$77.15	$617.20	$42.06	$64.28
1 Instrument Man	40.85	326.80	63.25	506.00		
2 Rodmen/Chainmen	38.60	617.60	58.35	933.60		
1 Laser Transit/Level		70.10		77.11	2.19	2.41
32 L.H., Daily Totals		$1416.10		$2133.91	$44.25	$66.68
Crew A-9	Hr.	Daily	Hr.	Daily	Bare Costs	Incl. O&P
1 Asbestos Foreman	$44.60	$356.80	$69.90	$559.20	$44.16	$69.20
7 Asbestos Workers	44.10	2469.60	69.10	3869.60		
64 L.H., Daily Totals		$2826.40		$4428.80	$44.16	$69.20
Crew A-10A	Hr.	Daily	Hr.	Daily	Bare Costs	Incl. O&P
1 Asbestos Foreman	$44.60	$356.80	$69.90	$559.20	$44.27	$69.37
2 Asbestos Workers	44.10	705.60	69.10	1105.60		
24 L.H., Daily Totals		$1062.40		$1664.80	$44.27	$69.37
Crew A-10B	Hr.	Daily	Hr.	Daily	Bare Costs	Incl. O&P
1 Asbestos Foreman	$44.60	$356.80	$69.90	$559.20	$44.23	$69.30
3 Asbestos Workers	44.10	1058.40	69.10	1658.40		
32 L.H., Daily Totals		$1415.20		$2217.60	$44.23	$69.30
Crew A-10C	Hr.	Daily	Hr.	Daily	Bare Costs	Incl. O&P
3 Asbestos Workers	$44.10	$1058.40	$69.10	$1658.40	$44.10	$69.10
1 Flatbed Truck, Gas, 1.5 Ton		194.40		213.84	8.10	8.91
24 L.H., Daily Totals		$1252.80		$1872.24	$52.20	$78.01
Crew A-10D	Hr.	Daily	Hr.	Daily	Bare Costs	Incl. O&P
2 Asbestos Workers	$44.10	$705.60	$69.10	$1105.60	$41.89	$64.36
1 Equip. Oper. (crane)	42.55	340.40	63.95	511.60		
1 Equip. Oper. Oiler	36.80	294.40	55.30	442.40		
1 Hydraulic Crane, 33 Ton		818.80		900.68	25.59	28.15
32 L.H., Daily Totals		$2159.20		$2960.28	$67.47	$92.51
Crew A-11	Hr.	Daily	Hr.	Daily	Bare Costs	Incl. O&P
1 Asbestos Foreman	$44.60	$356.80	$69.90	$559.20	$44.16	$69.20
7 Asbestos Workers	44.10	2469.60	69.10	3869.60		
2 Chip. Hammers, 12 Lb., Elec.		34.00		37.40	.53	.58
64 L.H., Daily Totals		$2860.40		$4466.20	$44.69	$69.78

Crews

Crew No.	Bare Costs		Incl. Subs O & P		Cost Per Labor-Hour	
Crew A-12	Hr.	Daily	Hr.	Daily	Bare Costs	Incl. O&P
1 Asbestos Foreman	$44.60	$356.80	$69.90	$559.20	$44.16	$69.20
7 Asbestos Workers	44.10	2469.60	69.10	3869.60		
1 Trk-mtd vac, 14 CY, 1500 Gal.		561.60		617.76		
1 Flatbed Truck, 20,000 GVW		197.40		217.14	11.86	13.05
64 L.H., Daily Totals		$3585.40		$5263.70	$56.02	$82.25
Crew A-13	Hr.	Daily	Hr.	Daily	Bare Costs	Incl. O&P
1 Equip. Oper. (light)	$39.05	$312.40	$58.70	$469.60	$39.05	$58.70
1 Trk-mtd vac, 14 CY, 1500 Gal.		561.60		617.76		
1 Flatbed Truck, 20,000 GVW		197.40		217.14	94.88	104.36
8 L.H., Daily Totals		$1071.40		$1304.50	$133.93	$163.06
Crew B-1	Hr.	Daily	Hr.	Daily	Bare Costs	Incl. O&P
1 Labor Foreman (outside)	$33.60	$268.80	$52.10	$416.80	$32.27	$50.03
2 Laborers	31.60	505.60	49.00	784.00		
24 L.H., Daily Totals		$774.40		$1200.80	$32.27	$50.03
Crew B-1A	Hr.	Daily	Hr.	Daily	Bare Costs	Incl. O&P
1 Laborer Foreman	$33.60	$268.80	$52.10	$416.80	$32.27	$50.03
2 Laborers	31.60	505.60	49.00	784.00		
2 Cutting Torches		34.00		37.40		
2 Gases		151.20		166.32	7.72	8.49
24 L.H., Daily Totals		$959.60		$1404.52	$39.98	$58.52
Crew B-1B	Hr.	Daily	Hr.	Daily	Bare Costs	Incl. O&P
1 Laborer Foreman	$33.60	$268.80	$52.10	$416.80	$34.84	$53.51
2 Laborers	31.60	505.60	49.00	784.00		
1 Equip. Oper. (crane)	42.55	340.40	63.95	511.60		
2 Cutting Torches		34.00		37.40		
2 Gases		151.20		166.32		
1 Hyd. Crane, 12 Ton		768.80		845.68	29.81	32.79
32 L.H., Daily Totals		$2068.80		$2761.80	$64.65	$86.31
Crew B-2	Hr.	Daily	Hr.	Daily	Bare Costs	Incl. O&P
1 Labor Foreman (outside)	$33.60	$268.80	$52.10	$416.80	$32.00	$49.62
4 Laborers	31.60	1011.20	49.00	1568.00		
40 L.H., Daily Totals		$1280.00		$1984.80	$32.00	$49.62
Crew B-3	Hr.	Daily	Hr.	Daily	Bare Costs	Incl. O&P
1 Labor Foreman (outside)	$33.60	$268.80	$52.10	$416.80	$33.67	$51.76
2 Laborers	31.60	505.60	49.00	784.00		
1 Equip. Oper. (med.)	41.35	330.80	62.15	497.20		
2 Truck Drivers (heavy)	31.95	511.20	49.15	786.40		
1 Crawler Loader, 3 C.Y.		990.00		1089.00		
2 Dump Trucks 12 C.Y., 400 H.P.		1066.00		1172.60	42.83	47.12
48 L.H., Daily Totals		$3672.40		$4746.00	$76.51	$98.88
Crew B-3A	Hr.	Daily	Hr.	Daily	Bare Costs	Incl. O&P
4 Laborers	$31.60	$1011.20	$49.00	$1568.00	$33.55	$51.63
1 Equip. Oper. (med.)	41.35	330.80	62.15	497.20		
1 Hyd. Excavator, 1.5 C.Y.		865.80		952.38	21.65	23.81
40 L.H., Daily Totals		$2207.80		$3017.58	$55.20	$75.44
Crew B-3B	Hr.	Daily	Hr.	Daily	Bare Costs	Incl. O&P
2 Laborers	$31.60	$505.60	$49.00	$784.00	$34.13	$52.33
1 Equip. Oper. (med.)	41.35	330.80	62.15	497.20		
1 Truck Driver (heavy)	31.95	255.60	49.15	393.20		
1 Backhoe Loader, 80 H.P.		345.00		379.50		
1 Dump Truck, 12 C.Y., 400 H.P.		533.00		586.30	27.44	30.18
32 L.H., Daily Totals		$1970.00		$2640.20	$61.56	$82.51

Crew No.	Bare Costs		Incl. Subs O & P		Cost Per Labor-Hour	
Crew B-3C	Hr.	Daily	Hr.	Daily	Bare Costs	Incl. O&P
3 Laborers	$31.60	$758.40	$49.00	$1176.00	$34.04	$52.29
1 Equip. Oper. (med.)	41.35	330.80	62.15	497.20		
1 Crawler Loader, 4 C.Y.		1450.00		1595.00	45.31	49.84
32 L.H., Daily Totals		$2539.20		$3268.20	$79.35	$102.13
Crew B-4	Hr.	Daily	Hr.	Daily	Bare Costs	Incl. O&P
1 Labor Foreman (outside)	$33.60	$268.80	$52.10	$416.80	$31.99	$49.54
4 Laborers	31.60	1011.20	49.00	1568.00		
1 Truck Driver (heavy)	31.95	255.60	49.15	393.20		
1 Truck Tractor, 220 H.P.		288.40		317.24		
1 Flatbed Trailer, 40 Ton		139.60		153.56	8.92	9.81
48 L.H., Daily Totals		$1963.60		$2848.80	$40.91	$59.35
Crew B-5	Hr.	Daily	Hr.	Daily	Bare Costs	Incl. O&P
1 Labor Foreman (outside)	$33.60	$268.80	$52.10	$416.80	$34.67	$53.20
4 Laborers	31.60	1011.20	49.00	1568.00		
2 Equip. Oper. (med.)	41.35	661.60	62.15	994.40		
1 Air Compressor, 250 cfm		162.40		178.64		
2 Breakers, Pavement, 60 lb.		16.40		18.04		
2 -50' Air Hoses, 1.5"		12.60		13.86		
1 Crawler Loader, 3 C.Y.		990.00		1089.00	21.10	23.21
56 L.H., Daily Totals		$3123.00		$4278.74	$55.77	$76.41
Crew B-5A	Hr.	Daily	Hr.	Daily	Bare Costs	Incl. O&P
1 Foreman	$33.60	$268.80	$52.10	$416.80	$34.07	$52.28
6 Laborers	31.60	1516.80	49.00	2352.00		
2 Equip. Oper. (med.)	41.35	661.60	62.15	994.40		
1 Equip. Oper. (light)	39.05	312.40	58.70	469.60		
2 Truck Drivers (heavy)	31.95	511.20	49.15	786.40		
1 Air Compressor, 365 cfm		211.20		232.32		
2 Breakers, Pavement, 60 lb.		16.40		18.04		
8 -50' Air Hoses, 1"		32.80		36.08		
2 Dump Trucks, 8 C.Y., 220 H.P.		652.80		718.08	9.51	10.46
96 L.H., Daily Totals		$4184.00		$6023.72	$43.58	$62.75
Crew B-5B	Hr.	Daily	Hr.	Daily	Bare Costs	Incl. O&P
1 Powderman	$40.85	$326.80	$63.25	$506.00	$36.57	$55.83
2 Equip. Oper. (med.)	41.35	661.60	62.15	994.40		
3 Truck Drivers (heavy)	31.95	766.80	49.15	1179.60		
1 F.E. Loader, W.M.,2.5 C.Y.		395.80		435.38		
3 Dump Trucks, 12 C.Y., 400 H.P.		1599.00		1758.90		
1 Air Compressor, 365 cfm		211.20		232.32	45.96	50.55
48 L.H., Daily Totals		$3961.20		$5106.60	$82.53	$106.39
Crew B-5C	Hr.	Daily	Hr.	Daily	Bare Costs	Incl. O&P
3 Laborers	$31.60	$758.40	$49.00	$1176.00	$34.92	$53.34
1 Equip. Oper. (medium)	41.35	330.80	62.15	497.20		
2 Truck Drivers (heavy)	31.95	511.20	49.15	786.40		
1 Equip. Oper. (crane)	42.55	340.40	63.95	511.60		
1 Equip. Oper. Oiler	36.80	294.40	55.30	442.40		
2 Dump Trucks, 12 C.Y., 400 H.P.		1066.00		1172.60		
1 Crawler Loader, 4 C.Y.		1450.00		1595.00		
1 S.P. Crane, 4x4, 25 Ton		716.80		788.48	50.51	55.56
64 L.H., Daily Totals		$5468.00		$6969.68	$85.44	$108.90
Crew B-6	Hr.	Daily	Hr.	Daily	Bare Costs	Incl. O&P
2 Laborers	$31.60	$505.60	$49.00	$784.00	$34.08	$52.23
1 Equip. Oper. (light)	39.05	312.40	58.70	469.60		
1 Backhoe Loader, 48 H.P.		293.80		323.18	12.24	13.47
24 L.H., Daily Totals		$1111.80		$1576.78	$46.33	$65.70

563

Crews

Crew No.	Bare Costs		Incl. Subs O & P		Cost Per Labor-Hour	
Crew B-6A	**Hr.**	**Daily**	**Hr.**	**Daily**	**Bare Costs**	**Incl. O&P**
.5 Labor Foreman (outside)	$33.60	$134.40	$52.10	$208.40	$35.90	$54.88
1 Laborer	31.60	252.80	49.00	392.00		
1 Equip. Oper. (med.)	41.35	330.80	62.15	497.20		
1 Vacuum Trk.,5000 Gal.		395.20		434.72	19.76	21.74
20 L.H., Daily Totals		$1113.20		$1532.32	$55.66	$76.62
Crew B-6B	**Hr.**	**Daily**	**Hr.**	**Daily**	**Bare Costs**	**Incl. O&P**
2 Labor Foremen (out)	$33.60	$537.60	$52.10	$833.60	$32.27	$50.03
4 Laborers	31.60	1011.20	49.00	1568.00		
1 S.P. Crane, 4x4, 5 Ton		258.20		284.02		
1 Flatbed Truck, Gas, 1.5 Ton		194.40		213.84		
1 Butt Fusion Machine		199.80		219.78	13.59	14.95
48 L.H., Daily Totals		$2201.20		$3119.24	$45.86	$64.98
Crew B-7	**Hr.**	**Daily**	**Hr.**	**Daily**	**Bare Costs**	**Incl. O&P**
1 Labor Foreman (outside)	$33.60	$268.80	$52.10	$416.80	$33.56	$51.71
4 Laborers	31.60	1011.20	49.00	1568.00		
1 Equip. Oper. (med.)	41.35	330.80	62.15	497.20		
1 Brush Chipper, 12", 130 H.P.		225.60		248.16		
1 Crawler Loader, 3 C.Y.		990.00		1089.00		
2 Chain Saws, Gas, 36" Long		75.60		83.16	26.90	29.59
48 L.H., Daily Totals		$2902.00		$3902.32	$60.46	$81.30
Crew B-7A	**Hr.**	**Daily**	**Hr.**	**Daily**	**Bare Costs**	**Incl. O&P**
2 Laborers	$31.60	$505.60	$49.00	$784.00	$34.08	$52.23
1 Equip. Oper. (light)	39.05	312.40	58.70	469.60		
1 Rake w/Tractor		257.70		283.47		
2 Chain Saws, gas, 18"		51.60		56.76	12.89	14.18
24 L.H., Daily Totals		$1127.30		$1593.83	$46.97	$66.41
Crew B-8	**Hr.**	**Daily**	**Hr.**	**Daily**	**Bare Costs**	**Incl. O&P**
1 Labor Foreman (outside)	$33.60	$268.80	$52.10	$416.80	$35.02	$53.70
2 Laborers	31.60	505.60	49.00	784.00		
2 Equip. Oper. (med.)	41.35	661.60	62.15	994.40		
1 Equip. Oper. Oiler	36.80	294.40	55.30	442.40		
2 Truck Drivers (heavy)	31.95	511.20	49.15	786.40		
1 Hyd. Crane, 25 Ton		789.40		868.34		
1 Crawler Loader, 3 C.Y.		990.00		1089.00		
2 Dump Trucks, 12 C.Y., 400 H.P.		1066.00		1172.60	44.46	48.91
64 L.H., Daily Totals		$5087.00		$6553.94	$79.48	$102.41
Crew B-9	**Hr.**	**Daily**	**Hr.**	**Daily**	**Bare Costs**	**Incl. O&P**
1 Labor Foreman (outside)	$33.60	$268.80	$52.10	$416.80	$32.00	$49.62
4 Laborers	31.60	1011.20	49.00	1568.00		
1 Air Compressor, 250 cfm		162.40		178.64		
2 Breakers, Pavement, 60 lb.		16.40		18.04		
2 -50' Air Hoses, 1.5"		12.60		13.86	4.79	5.26
40 L.H., Daily Totals		$1471.40		$2195.34	$36.78	$54.88
Crew B-9A	**Hr.**	**Daily**	**Hr.**	**Daily**	**Bare Costs**	**Incl. O&P**
2 Laborers	$31.60	$505.60	$49.00	$784.00	$31.72	$49.05
1 Truck Driver (heavy)	31.95	255.60	49.15	393.20		
1 Water Tanker, 5000 Gal.		136.00		149.60		
1 Truck Tractor, 220 H.P.		288.40		317.24		
2 -50' Discharge Hoses, 3"		3.50		3.85	17.83	19.61
24 L.H., Daily Totals		$1189.10		$1647.89	$49.55	$68.66

Crew No.	Bare Costs		Incl. Subs O & P		Cost Per Labor-Hour	
Crew B-9B	**Hr.**	**Daily**	**Hr.**	**Daily**	**Bare Costs**	**Incl. O&P**
2 Laborers	$31.60	$505.60	$49.00	$784.00	$31.72	$49.05
1 Truck Driver (heavy)	31.95	255.60	49.15	393.20		
2 -50' Discharge Hoses, 3"		3.50		3.85		
1 Water Tanker, 5000 Gal.		136.00		149.60		
1 Truck Tractor, 220 H.P.		288.40		317.24		
1 Pressure Washer		53.00		58.30	20.04	22.04
24 L.H., Daily Totals		$1242.10		$1706.19	$51.75	$71.09
Crew B-9D	**Hr.**	**Daily**	**Hr.**	**Daily**	**Bare Costs**	**Incl. O&P**
1 Labor Foreman (Outside)	$33.60	$268.80	$52.10	$416.80	$32.00	$49.62
4 Common Laborers	31.60	1011.20	49.00	1568.00		
1 Air Compressor, 250 cfm		162.40		178.64		
2 -50' Air Hoses, 1.5"		12.60		13.86		
2 Air Powered Tampers		48.00		52.80	5.58	6.13
40 L.H., Daily Totals		$1503.00		$2230.10	$37.58	$55.75
Crew B-10	**Hr.**	**Daily**	**Hr.**	**Daily**	**Bare Costs**	**Incl. O&P**
1 Equip. Oper. (med.)	$41.35	$330.80	$62.15	$497.20	$38.10	$57.77
.5 Laborer	31.60	126.40	49.00	196.00		
12 L.H., Daily Totals		$457.20		$693.20	$38.10	$57.77
Crew B-10A	**Hr.**	**Daily**	**Hr.**	**Daily**	**Bare Costs**	**Incl. O&P**
1 Equip. Oper. (med.)	$41.35	$330.80	$62.15	$497.20	$38.10	$57.77
.5 Laborer	31.60	126.40	49.00	196.00		
1 Roller, 2-Drum, W.B., 7.5 H.P.		145.20		159.72	12.10	13.31
12 L.H., Daily Totals		$602.40		$852.92	$50.20	$71.08
Crew B-10B	**Hr.**	**Daily**	**Hr.**	**Daily**	**Bare Costs**	**Incl. O&P**
1 Equip. Oper. (med.)	$41.35	$330.80	$62.15	$497.20	$38.10	$57.77
.5 Laborer	31.60	126.40	49.00	196.00		
1 Dozer, 200 H.P.		1082.00		1190.20	90.17	99.18
12 L.H., Daily Totals		$1539.20		$1883.40	$128.27	$156.95
Crew B-10C	**Hr.**	**Daily**	**Hr.**	**Daily**	**Bare Costs**	**Incl. O&P**
1 Equip. Oper. (med.)	$41.35	$330.80	$62.15	$497.20	$38.10	$57.77
.5 Laborer	31.60	126.40	49.00	196.00		
1 Dozer, 200 H.P.		1082.00		1190.20		
1 Vibratory Roller, Towed, 23 Ton		411.00		452.10	124.42	136.86
12 L.H., Daily Totals		$1950.20		$2335.50	$162.52	$194.63
Crew B-10D	**Hr.**	**Daily**	**Hr.**	**Daily**	**Bare Costs**	**Incl. O&P**
1 Equip. Oper. (med.)	$41.35	$330.80	$62.15	$497.20	$38.10	$57.77
.5 Laborer	31.60	126.40	49.00	196.00		
1 Dozer, 200 H.P.		1082.00		1190.20		
1 Sheepsft. Roller, Towed		465.40		511.94	128.95	141.85
12 L.H., Daily Totals		$2004.60		$2395.34	$167.05	$199.61
Crew B-10E	**Hr.**	**Daily**	**Hr.**	**Daily**	**Bare Costs**	**Incl. O&P**
1 Equip. Oper. (med.)	$41.35	$330.80	$62.15	$497.20	$38.10	$57.77
.5 Laborer	31.60	126.40	49.00	196.00		
1 Tandem Roller, 5 Ton		137.80		151.58	11.48	12.63
12 L.H., Daily Totals		$595.00		$844.78	$49.58	$70.40
Crew B-10F	**Hr.**	**Daily**	**Hr.**	**Daily**	**Bare Costs**	**Incl. O&P**
1 Equip. Oper. (med.)	$41.35	$330.80	$62.15	$497.20	$38.10	$57.77
.5 Laborer	31.60	126.40	49.00	196.00		
1 Tandem Roller, 10 Ton		233.00		256.30	19.42	21.36
12 L.H., Daily Totals		$690.20		$949.50	$57.52	$79.13

Crews

Crew No.	Bare Costs		Incl. Subs O & P		Cost Per Labor-Hour	
Crew B-10G	Hr.	Daily	Hr.	Daily	Bare Costs	Incl. O&P
1 Equip. Oper. (med.)	$41.35	$330.80	$62.15	$497.20	$38.10	$57.77
.5 Laborer	31.60	126.40	49.00	196.00		
1 Sheepsft. Roll., 240 H.P.		1142.00		1256.20	95.17	104.68
12 L.H., Daily Totals		$1599.20		$1949.40	$133.27	$162.45
Crew B-10H	Hr.	Daily	Hr.	Daily	Bare Costs	Incl. O&P
1 Equip. Oper. (med.)	$41.35	$330.80	$62.15	$497.20	$38.10	$57.77
.5 Laborer	31.60	126.40	49.00	196.00		
1 Diaphragm Water Pump, 2"		63.20		69.52		
1 -20' Suction Hose, 2"		1.95		2.15		
2 -50' Discharge Hoses, 2"		2.20		2.42	5.61	6.17
12 L.H., Daily Totals		$524.55		$767.28	$43.71	$63.94
Crew B-10I	Hr.	Daily	Hr.	Daily	Bare Costs	Incl. O&P
1 Equip. Oper. (med.)	$41.35	$330.80	$62.15	$497.20	$38.10	$57.77
.5 Laborer	31.60	126.40	49.00	196.00		
1 Diaphragm Water Pump, 4"		86.20		94.82		
1 -20' Suction Hose, 4"		3.45		3.79		
2 -50' Discharge Hoses, 4"		4.70		5.17	7.86	8.65
12 L.H., Daily Totals		$551.55		$796.99	$45.96	$66.42
Crew B-10J	Hr.	Daily	Hr.	Daily	Bare Costs	Incl. O&P
1 Equip. Oper. (med.)	$41.35	$330.80	$62.15	$497.20	$38.10	$57.77
.5 Laborer	31.60	126.40	49.00	196.00		
1 Centrifugal Water Pump, 3"		69.40		76.34		
1 -20' Suction Hose, 3"		3.05		3.36		
2 -50' Discharge Hoses, 3"		3.50		3.85	6.33	6.96
12 L.H., Daily Totals		$533.15		$776.75	$44.43	$64.73
Crew B-10K	Hr.	Daily	Hr.	Daily	Bare Costs	Incl. O&P
1 Equip. Oper. (med.)	$41.35	$330.80	$62.15	$497.20	$38.10	$57.77
.5 Laborer	31.60	126.40	49.00	196.00		
1 Centr. Water Pump, 6"		309.40		340.34		
1 -20' Suction Hose, 6"		11.90		13.09		
2 -50' Discharge Hoses, 6"		12.60		13.86	27.82	30.61
12 L.H., Daily Totals		$791.10		$1060.49	$65.92	$88.37
Crew B-10L	Hr.	Daily	Hr.	Daily	Bare Costs	Incl. O&P
1 Equip. Oper. (med.)	$41.35	$330.80	$62.15	$497.20	$38.10	$57.77
.5 Laborer	31.60	126.40	49.00	196.00		
1 Dozer, 80 H.P.		399.20		439.12	33.27	36.59
12 L.H., Daily Totals		$856.40		$1132.32	$71.37	$94.36
Crew B-10M	Hr.	Daily	Hr.	Daily	Bare Costs	Incl. O&P
1 Equip. Oper. (med.)	$41.35	$330.80	$62.15	$497.20	$38.10	$57.77
.5 Laborer	31.60	126.40	49.00	196.00		
1 Dozer, 300 H.P.		1423.00		1565.30	118.58	130.44
12 L.H., Daily Totals		$1880.20		$2258.50	$156.68	$188.21
Crew B-10N	Hr.	Daily	Hr.	Daily	Bare Costs	Incl. O&P
1 Equip. Oper. (med.)	$41.35	$330.80	$62.15	$497.20	$38.10	$57.77
.5 Laborer	31.60	126.40	49.00	196.00		
1 F.E. Loader, T.M., 1.5 C.Y		367.00		403.70	30.58	33.64
12 L.H., Daily Totals		$824.20		$1096.90	$68.68	$91.41
Crew B-10O	Hr.	Daily	Hr.	Daily	Bare Costs	Incl. O&P
1 Equip. Oper. (med.)	$41.35	$330.80	$62.15	$497.20	$38.10	$57.77
.5 Laborer	31.60	126.40	49.00	196.00		
1 F.E. Loader, T.M., 2.25 C.Y.		767.40		844.14	63.95	70.34
12 L.H., Daily Totals		$1224.60		$1537.34	$102.05	$128.11

Crew No.	Bare Costs		Incl. Subs O & P		Cost Per Labor-Hour	
Crew B-10P	Hr.	Daily	Hr.	Daily	Bare Costs	Incl. O&P
1 Equip. Oper. (med.)	$41.35	$330.80	$62.15	$497.20	$38.10	$57.77
.5 Laborer	31.60	126.40	49.00	196.00		
1 Crawler Loader, 3 C.Y.		990.00		1089.00	82.50	90.75
12 L.H., Daily Totals		$1447.20		$1782.20	$120.60	$148.52
Crew B-10Q	Hr.	Daily	Hr.	Daily	Bare Costs	Incl. O&P
1 Equip. Oper. (med.)	$41.35	$330.80	$62.15	$497.20	$38.10	$57.77
.5 Laborer	31.60	126.40	49.00	196.00		
1 Crawler Loader, 4 C.Y.		1450.00		1595.00	120.83	132.92
12 L.H., Daily Totals		$1907.20		$2288.20	$158.93	$190.68
Crew B-10R	Hr.	Daily	Hr.	Daily	Bare Costs	Incl. O&P
1 Equip. Oper. (med.)	$41.35	$330.80	$62.15	$497.20	$38.10	$57.77
.5 Laborer	31.60	126.40	49.00	196.00		
1 F.E. Loader, W.M., 1 C.Y.		239.00		262.90	19.92	21.91
12 L.H., Daily Totals		$696.20		$956.10	$58.02	$79.67
Crew B-10S	Hr.	Daily	Hr.	Daily	Bare Costs	Incl. O&P
1 Equip. Oper. (med.)	$41.35	$330.80	$62.15	$497.20	$38.10	$57.77
.5 Laborer	31.60	126.40	49.00	196.00		
1 F.E. Loader, W.M., 1.5 C.Y.		328.20		361.02	27.35	30.09
12 L.H., Daily Totals		$785.40		$1054.22	$65.45	$87.85
Crew B-10T	Hr.	Daily	Hr.	Daily	Bare Costs	Incl. O&P
1 Equip. Oper. (med.)	$41.35	$330.80	$62.15	$497.20	$38.10	$57.77
.5 Laborer	31.60	126.40	49.00	196.00		
1 F.E. Loader, W.M., 2.5 C.Y.		395.80		435.38	32.98	36.28
12 L.H., Daily Totals		$853.00		$1128.58	$71.08	$94.05
Crew B-10U	Hr.	Daily	Hr.	Daily	Bare Costs	Incl. O&P
1 Equip. Oper. (med.)	$41.35	$330.80	$62.15	$497.20	$38.10	$57.77
.5 Laborer	31.60	126.40	49.00	196.00		
1 F.E. Loader, W.M., 5.5 C.Y.		854.40		939.84	71.20	78.32
12 L.H., Daily Totals		$1311.60		$1633.04	$109.30	$136.09
Crew B-10V	Hr.	Daily	Hr.	Daily	Bare Costs	Incl. O&P
1 Equip. Oper. (med.)	$41.35	$330.80	$62.15	$497.20	$38.10	$57.77
.5 Laborer	31.60	126.40	49.00	196.00		
1 Dozer, 700 H.P.		4232.00		4655.20	352.67	387.93
12 L.H., Daily Totals		$4689.20		$5348.40	$390.77	$445.70
Crew B-10W	Hr.	Daily	Hr.	Daily	Bare Costs	Incl. O&P
1 Equip. Oper. (med.)	$41.35	$330.80	$62.15	$497.20	$38.10	$57.77
.5 Laborer	31.60	126.40	49.00	196.00		
1 Dozer, 105 H.P.		600.60		660.66	50.05	55.06
12 L.H., Daily Totals		$1057.80		$1353.86	$88.15	$112.82
Crew B-10X	Hr.	Daily	Hr.	Daily	Bare Costs	Incl. O&P
1 Equip. Oper. (med.)	$41.35	$330.80	$62.15	$497.20	$38.10	$57.77
.5 Laborer	31.60	126.40	49.00	196.00		
1 Dozer, 410 H.P.		1849.00		2033.90	154.08	169.49
12 L.H., Daily Totals		$2306.20		$2727.10	$192.18	$227.26
Crew B-10Y	Hr.	Daily	Hr.	Daily	Bare Costs	Incl. O&P
1 Equip. Oper. (med.)	$41.35	$330.80	$62.15	$497.20	$38.10	$57.77
.5 Laborer	31.60	126.40	49.00	196.00		
1 Vibr. Roller, Towed, 12 Ton		469.60		516.56	39.13	43.05
12 L.H., Daily Totals		$926.80		$1209.76	$77.23	$100.81

Crews

Crew No.	Bare Costs		Incl. Subs O & P		Cost Per Labor-Hour	
Crew B-11A	Hr.	Daily	Hr.	Daily	Bare Costs	Incl. O&P
1 Equipment Oper. (med.)	$41.35	$330.80	$62.15	$497.20	$36.48	$55.58
1 Laborer	31.60	252.80	49.00	392.00		
1 Dozer, 200 H.P.		1082.00		1190.20	67.63	74.39
16 L.H., Daily Totals		$1665.60		$2079.40	$104.10	$129.96
Crew B-11B	Hr.	Daily	Hr.	Daily	Bare Costs	Incl. O&P
1 Equipment Oper. (light)	$39.05	$312.40	$58.70	$469.60	$35.33	$53.85
1 Laborer	31.60	252.80	49.00	392.00		
1 Air Powered Tamper		24.00		26.40		
1 Air Compressor, 365 cfm		211.20		232.32		
2 -50' Air Hoses, 1.5"		12.60		13.86	15.49	17.04
16 L.H., Daily Totals		$813.00		$1134.18	$50.81	$70.89
Crew B-11C	Hr.	Daily	Hr.	Daily	Bare Costs	Incl. O&P
1 Equipment Oper. (med.)	$41.35	$330.80	$62.15	$497.20	$36.48	$55.58
1 Laborer	31.60	252.80	49.00	392.00		
1 Backhoe Loader, 48 H.P.		293.80		323.18	18.36	20.20
16 L.H., Daily Totals		$877.40		$1212.38	$54.84	$75.77
Crew B-11J	Hr.	Daily	Hr.	Daily	Bare Costs	Incl. O&P
1 Equipment Oper. (med.)	$41.35	$330.80	$62.15	$497.20	$36.48	$55.58
1 Laborer	31.60	252.80	49.00	392.00		
1 Grader, 30,000 Lbs.		550.20		605.22		
1 Ripper, beam & 1 shank		79.60		87.56	39.36	43.30
16 L.H., Daily Totals		$1213.40		$1581.98	$75.84	$98.87
Crew B-11K	Hr.	Daily	Hr.	Daily	Bare Costs	Incl. O&P
1 Equipment Oper. (med.)	$41.35	$330.80	$62.15	$497.20	$36.48	$55.58
1 Laborer	31.60	252.80	49.00	392.00		
1 Trencher, Chain Type, 8' D		1777.00		1954.70	111.06	122.17
16 L.H., Daily Totals		$2360.60		$2843.90	$147.54	$177.74
Crew B-11L	Hr.	Daily	Hr.	Daily	Bare Costs	Incl. O&P
1 Equipment Oper. (med.)	$41.35	$330.80	$62.15	$497.20	$36.48	$55.58
1 Laborer	31.60	252.80	49.00	392.00		
1 Grader, 30,000 Lbs.		550.20		605.22	34.39	37.83
16 L.H., Daily Totals		$1133.80		$1494.42	$70.86	$93.40
Crew B-11M	Hr.	Daily	Hr.	Daily	Bare Costs	Incl. O&P
1 Equipment Oper. (med.)	$41.35	$330.80	$62.15	$497.20	$36.48	$55.58
1 Laborer	31.60	252.80	49.00	392.00		
1 Backhoe Loader, 80 H.P.		345.00		379.50	21.56	23.72
16 L.H., Daily Totals		$928.60		$1268.70	$58.04	$79.29
Crew B-11N	Hr.	Daily	Hr.	Daily	Bare Costs	Incl. O&P
1 Labor Foreman	$33.60	$268.80	$52.10	$416.80	$34.22	$52.37
2 Equipment Operators (med.)	41.35	661.60	62.15	994.40		
6 Truck Drivers (hvy.)	31.95	1533.60	49.15	2359.20		
1 F.E. Loader, W.M., 5.5 C.Y.		854.40		939.84		
1 Dozer, 410 H.P.		1849.00		2033.90		
6 Dump Trucks, Off Hwy., 50 Ton		9492.00		10441.20	169.38	186.32
72 L.H., Daily Totals		$14659.40		$17185.34	$203.60	$238.69
Crew B-11Q	Hr.	Daily	Hr.	Daily	Bare Costs	Incl. O&P
1 Equipment Operator (med.)	$41.35	$330.80	$62.15	$497.20	$38.10	$57.77
.5 Laborer	31.60	126.40	49.00	196.00		
1 Dozer, 140 H.P.		682.00		750.20	56.83	62.52
12 L.H., Daily Totals		$1139.20		$1443.40	$94.93	$120.28

Crew No.	Bare Costs		Incl. Subs O & P		Cost Per Labor-Hour	
Crew B-11R	Hr.	Daily	Hr.	Daily	Bare Costs	Incl. O&P
1 Equipment Operator (med.)	$41.35	$330.80	$62.15	$497.20	$38.10	$57.77
.5 Laborer	31.60	126.40	49.00	196.00		
1 Dozer, 200 H.P.		1082.00		1190.20	90.17	99.18
12 L.H., Daily Totals		$1539.20		$1883.40	$128.27	$156.95
Crew B-11S	Hr.	Daily	Hr.	Daily	Bare Costs	Incl. O&P
1 Equipment Operator (med.)	$41.35	$330.80	$62.15	$497.20	$38.10	$57.77
.5 Laborer	31.60	126.40	49.00	196.00		
1 Dozer, 300 H.P.		1423.00		1565.30		
1 Ripper, beam & 1 shank		79.60		87.56	125.22	137.74
12 L.H., Daily Totals		$1959.80		$2346.06	$163.32	$195.51
Crew B-11T	Hr.	Daily	Hr.	Daily	Bare Costs	Incl. O&P
1 Equipment Operator (med.)	$41.35	$330.80	$62.15	$497.20	$38.10	$57.77
.5 Laborer	31.60	126.40	49.00	196.00		
1 Dozer, 410 H.P.		1849.00		2033.90		
1 Ripper, beam & 2 shanks		89.40		98.34	161.53	177.69
12 L.H., Daily Totals		$2395.60		$2825.44	$199.63	$235.45
Crew B-11U	Hr.	Daily	Hr.	Daily	Bare Costs	Incl. O&P
1 Equipment Operator (med.)	$41.35	$330.80	$62.15	$497.20	$38.10	$57.77
.5 Laborer	31.60	126.40	49.00	196.00		
1 Dozer, 520 H.P.		2464.00		2710.40	205.33	225.87
12 L.H., Daily Totals		$2921.20		$3403.60	$243.43	$283.63
Crew B-11V	Hr.	Daily	Hr.	Daily	Bare Costs	Incl. O&P
3 Laborers	$31.60	$758.40	$49.00	$1176.00	$31.60	$49.00
1 Roller, 2-Drum, W.B., 7.5 H.P.		145.20		159.72	6.05	6.66
24 L.H., Daily Totals		$903.60		$1335.72	$37.65	$55.66
Crew B-11W	Hr.	Daily	Hr.	Daily	Bare Costs	Incl. O&P
1 Equipment Operator (med.)	$41.35	$330.80	$62.15	$497.20	$32.70	$50.22
1 Common Laborer	31.60	252.80	49.00	392.00		
10 Truck Drivers (hvy.)	31.95	2556.00	49.15	3932.00		
1 Dozer, 200 H.P.		1082.00		1190.20		
1 Vibratory Roller, Towed, 23 Ton		411.00		452.10		
10 Dump Trucks, 8 C.Y., 220 H.P.		3264.00		3590.40	49.55	54.51
96 L.H., Daily Totals		$7896.60		$10053.90	$82.26	$104.73
Crew B-11Y	Hr.	Daily	Hr.	Daily	Bare Costs	Incl. O&P
1 Labor Foreman (Outside)	$33.60	$268.80	$52.10	$416.80	$35.07	$53.73
5 Common Laborers	31.60	1264.00	49.00	1960.00		
3 Equipment Operators (med.)	41.35	992.40	62.15	1491.60		
1 Dozer, 80 H.P.		399.20		439.12		
2 Rollers, 2-Drum, W.B., 7.5 H.P.		290.40		319.44		
4 Vibratory Plates, Gas, 21"		156.80		172.48	11.76	12.93
72 L.H., Daily Totals		$3371.60		$4799.44	$46.83	$66.66
Crew B-12A	Hr.	Daily	Hr.	Daily	Bare Costs	Incl. O&P
1 Equip. Oper. (crane)	$42.55	$340.40	$63.95	$511.60	$37.08	$56.48
1 Laborer	31.60	252.80	49.00	392.00		
1 Hyd. Excavator, 1 C.Y.		669.20		736.12	41.83	46.01
16 L.H., Daily Totals		$1262.40		$1639.72	$78.90	$102.48
Crew B-12B	Hr.	Daily	Hr.	Daily	Bare Costs	Incl. O&P
1 Equip. Oper. (crane)	$42.55	$340.40	$63.95	$511.60	$37.08	$56.48
1 Laborer	31.60	252.80	49.00	392.00		
1 Hyd. Excavator, 1.5 C.Y.		865.80		952.38	54.11	59.52
16 L.H., Daily Totals		$1459.00		$1855.98	$91.19	$116.00

Reed Construction Data
The leader in construction information and BIM solutions

AIA StrategicPartner

Reed Construction Data, Inc. is a leading provider of construction information and building information modeling (BIM) solutions. The company's portfolio of information products and services is designed specifically to help construction industry professionals advance their business with timely and accurate project, product, and cost data. Reed Construction Data is a division of Reed Business Information, a member of the Reed Elsevier PLC group of companies.

Cost Information
RSMeans, the undisputed market leader in construction costs, provides current cost and estimating information through its innovative MeansCostworks.com® web-based solution. In addition, RSMeans publishes annual cost books, estimating software, and a rich library of reference books. RSMeans also conducts a series of professional seminars and provides construction cost consulting for owners, manufacturers, designers, and contractors to sharpen personal skills and maximize the effective use of cost estimating and management tools.

Project Data
Reed Construction Data assembles one of the largest databases of public and private project data for use by contractors, distributors, and building product manufacturers in the U.S. and Canadian markets. In addition, Reed Construction Data is the North American construction community's premier resource for project leads and bid documents. Reed Bulletin and Reed CONNECT™ provide project leads and project data through all stages of construction for many of the country's largest public sector, commercial, industrial, and multi-family residential projects.

Research and Analytics
Reed Construction Data's forecasting tools cover most aspects of the construction business in the U.S. and Canada. With a vast network of resources, Reed Construction Data is uniquely qualified to give you the information you need to keep your business profitable.

SmartBIM Solutions
Reed Construction Data has emerged as the leader in the field of building information modeling (BIM) with products and services that have helped to advance the evolution of BIM. Through an in-depth SmartBIM Object Creation Program, BPMs can rely on Reed to create high-quality, real world objects embedded with superior cost data (RSMeans). In addition, Reed has made it easy for architects to manage these manufacturer-specific objects as well as generic objects in Revit with the SmartBIM Library.

SmartBuilding Index
The leading industry source for product research, product documentation, BIM objects, design ideas, and source locations. Search, select, and specify available building products with our online directories of manufacturer profiles, MANU-SPEC, SPEC-DATA, guide specs, manufacturer catalogs, building codes, historical project data, and BIM objects.

SmartBuilding Studio
A vibrant online resource library which brings together a comprehensive catalog of commercial interior finishes and products under a single standard for high-definition imagery, product data, and searchable attributes.

Associated Construction Publications (ACP)
Reed Construction Data's regional construction magazines cover the nation through a network of 14 regional magazines. Serving the construction market for more than 100 years, our magazines are a trusted source of news and information in the local and national construction communities.

For more information, please visit our website at www.reedconstructiondata.com

The leader in construction information and BIM solutions.

Crews

Crew No.	Bare Costs		Incl. Subs O & P		Cost Per Labor-Hour	
Crew B-12C	Hr.	Daily	Hr.	Daily	Bare Costs	Incl. O&P
1 Equip. Oper. (crane)	$42.55	$340.40	$63.95	$511.60	$37.08	$56.48
1 Laborer	31.60	252.80	49.00	392.00		
1 Hyd. Excavator, 2 C.Y.		1182.00		1300.20	73.88	81.26
16 L.H., Daily Totals		$1775.20		$2203.80	$110.95	$137.74
Crew B-12D	Hr.	Daily	Hr.	Daily	Bare Costs	Incl. O&P
1 Equip. Oper. (crane)	$42.55	$340.40	$63.95	$511.60	$37.08	$56.48
1 Laborer	31.60	252.80	49.00	392.00		
1 Hyd. Excavator, 3.5 C.Y.		2162.00		2378.20	135.13	148.64
16 L.H., Daily Totals		$2755.20		$3281.80	$172.20	$205.11
Crew B-12E	Hr.	Daily	Hr.	Daily	Bare Costs	Incl. O&P
1 Equip. Oper. (crane)	$42.55	$340.40	$63.95	$511.60	$37.08	$56.48
1 Laborer	31.60	252.80	49.00	392.00		
1 Hyd. Excavator, .5 C.Y.		394.60		434.06	24.66	27.13
16 L.H., Daily Totals		$987.80		$1337.66	$61.74	$83.60
Crew B-12F	Hr.	Daily	Hr.	Daily	Bare Costs	Incl. O&P
1 Equip. Oper. (crane)	$42.55	$340.40	$63.95	$511.60	$37.08	$56.48
1 Laborer	31.60	252.80	49.00	392.00		
1 Hyd. Excavator, .75 C.Y.		598.60		658.46	37.41	41.15
16 L.H., Daily Totals		$1191.80		$1562.06	$74.49	$97.63
Crew B-12G	Hr.	Daily	Hr.	Daily	Bare Costs	Incl. O&P
1 Equip. Oper. (crane)	$42.55	$340.40	$63.95	$511.60	$37.08	$56.48
1 Laborer	31.60	252.80	49.00	392.00		
1 Crawler Crane, 15 Ton		615.30		676.83		
1 Clamshell Bucket, .5 C.Y.		36.60		40.26	40.74	44.82
16 L.H., Daily Totals		$1245.10		$1620.69	$77.82	$101.29
Crew B-12H	Hr.	Daily	Hr.	Daily	Bare Costs	Incl. O&P
1 Equip. Oper. (crane)	$42.55	$340.40	$63.95	$511.60	$37.08	$56.48
1 Laborer	31.60	252.80	49.00	392.00		
1 Crawler Crane, 25 Ton		1052.00		1157.20		
1 Clamshell Bucket, 1 C.Y.		46.20		50.82	68.64	75.50
16 L.H., Daily Totals		$1691.40		$2111.62	$105.71	$131.98
Crew B-12I	Hr.	Daily	Hr.	Daily	Bare Costs	Incl. O&P
1 Equip. Oper. (crane)	$42.55	$340.40	$63.95	$511.60	$37.08	$56.48
1 Laborer	31.60	252.80	49.00	392.00		
1 Crawler Crane, 20 Ton		789.05		867.96		
1 Dragline Bucket, .75 C.Y.		20.00		22.00	50.57	55.62
16 L.H., Daily Totals		$1402.25		$1793.56	$87.64	$112.10
Crew B-12J	Hr.	Daily	Hr.	Daily	Bare Costs	Incl. O&P
1 Equip. Oper. (crane)	$42.55	$340.40	$63.95	$511.60	$37.08	$56.48
1 Laborer	31.60	252.80	49.00	392.00		
1 Gradall, 5/8 C.Y.		958.80		1054.68	59.92	65.92
16 L.H., Daily Totals		$1552.00		$1958.28	$97.00	$122.39
Crew B-12K	Hr.	Daily	Hr.	Daily	Bare Costs	Incl. O&P
1 Equip. Oper. (crane)	$42.55	$340.40	$63.95	$511.60	$37.08	$56.48
1 Laborer	31.60	252.80	49.00	392.00		
1 Gradall, 3 Ton, 1 C.Y.		1130.00		1243.00	70.63	77.69
16 L.H., Daily Totals		$1723.20		$2146.60	$107.70	$134.16
Crew B-12L	Hr.	Daily	Hr.	Daily	Bare Costs	Incl. O&P
1 Equip. Oper. (crane)	$42.55	$340.40	$63.95	$511.60	$37.08	$56.48
1 Laborer	31.60	252.80	49.00	392.00		
1 Crawler Crane, 15 Ton		615.30		676.83		
1 F.E. Attachment, .5 C.Y.		52.80		58.08	41.76	45.93
16 L.H., Daily Totals		$1261.30		$1638.51	$78.83	$102.41
Crew B-12M	Hr.	Daily	Hr.	Daily	Bare Costs	Incl. O&P
1 Equip. Oper. (crane)	$42.55	$340.40	$63.95	$511.60	$37.08	$56.48
1 Laborer	31.60	252.80	49.00	392.00		
1 Crawler Crane, 20 Ton		789.05		867.96		
1 F.E. Attachment, .75 C.Y.		58.20		64.02	52.95	58.25
16 L.H., Daily Totals		$1440.45		$1835.58	$90.03	$114.72
Crew B-12N	Hr.	Daily	Hr.	Daily	Bare Costs	Incl. O&P
1 Equip. Oper. (crane)	$42.55	$340.40	$63.95	$511.60	$37.08	$56.48
1 Laborer	31.60	252.80	49.00	392.00		
1 Crawler Crane, 25 Ton		1052.00		1157.20		
1 F.E. Attachment, 1 C.Y.		64.00		70.40	69.75	76.72
16 L.H., Daily Totals		$1709.20		$2131.20	$106.83	$133.20
Crew B-12O	Hr.	Daily	Hr.	Daily	Bare Costs	Incl. O&P
1 Equip. Oper. (crane)	$42.55	$340.40	$63.95	$511.60	$37.08	$56.48
1 Laborer	31.60	252.80	49.00	392.00		
1 Crawler Crane, 40 Ton		1055.00		1160.50		
1 F.E. Attachment, 1.5 C.Y.		74.20		81.62	70.58	77.63
16 L.H., Daily Totals		$1722.40		$2145.72	$107.65	$134.11
Crew B-12P	Hr.	Daily	Hr.	Daily	Bare Costs	Incl. O&P
1 Equip. Oper. (crane)	$42.55	$340.40	$63.95	$511.60	$37.08	$56.48
1 Laborer	31.60	252.80	49.00	392.00		
1 Crawler Crane, 40 Ton		1055.00		1160.50		
1 Dragline Bucket, 1.5 C.Y.		32.80		36.08	67.99	74.79
16 L.H., Daily Totals		$1681.00		$2100.18	$105.06	$131.26
Crew B-12Q	Hr.	Daily	Hr.	Daily	Bare Costs	Incl. O&P
1 Equip. Oper. (crane)	$42.55	$340.40	$63.95	$511.60	$37.08	$56.48
1 Laborer	31.60	252.80	49.00	392.00		
1 Hyd. Excavator, 5/8 C.Y.		517.20		568.92	32.33	35.56
16 L.H., Daily Totals		$1110.40		$1472.52	$69.40	$92.03
Crew B-12S	Hr.	Daily	Hr.	Daily	Bare Costs	Incl. O&P
1 Equip. Oper. (crane)	$42.55	$340.40	$63.95	$511.60	$37.08	$56.48
1 Laborer	31.60	252.80	49.00	392.00		
1 Hyd. Excavator, 2.5 C.Y.		1619.00		1780.90	101.19	111.31
16 L.H., Daily Totals		$2212.20		$2684.50	$138.26	$167.78
Crew B-12T	Hr.	Daily	Hr.	Daily	Bare Costs	Incl. O&P
1 Equip. Oper. (crane)	$42.55	$340.40	$63.95	$511.60	$37.08	$56.48
1 Laborer	31.60	252.80	49.00	392.00		
1 Crawler Crane, 75 Ton		1496.00		1645.60		
1 F.E. Attachment, 3 C.Y.		97.60		107.36	99.60	109.56
16 L.H., Daily Totals		$2186.80		$2656.56	$136.68	$166.04
Crew B-12V	Hr.	Daily	Hr.	Daily	Bare Costs	Incl. O&P
1 Equip. Oper. (crane)	$42.55	$340.40	$63.95	$511.60	$37.08	$56.48
1 Laborer	31.60	252.80	49.00	392.00		
1 Crawler Crane, 75 Ton		1496.00		1645.60		
1 Dragline Bucket, 3 C.Y.		46.20		50.82	96.39	106.03
16 L.H., Daily Totals		$2135.40		$2600.02	$133.46	$162.50

Crews

Crew No.	Bare Costs		Incl. Subs O & P		Cost Per Labor-Hour	
Crew B-13	Hr.	Daily	Hr.	Daily	Bare Costs	Incl. O&P
1 Labor Foreman (outside)	$33.60	$268.80	$52.10	$416.80	$34.19	$52.48
4 Laborers	31.60	1011.20	49.00	1568.00		
1 Equip. Oper. (crane)	42.55	340.40	63.95	511.60		
1 Equip. Oper. Oiler	36.80	294.40	55.30	442.40		
1 Hyd. Crane, 25 Ton		789.40		868.34	14.10	15.51
56 L.H., Daily Totals		$2704.20		$3807.14	$48.29	$67.98
Crew B-13A	Hr.	Daily	Hr.	Daily	Bare Costs	Incl. O&P
1 Foreman	$33.60	$268.80	$52.10	$416.80	$34.77	$53.24
2 Laborers	31.60	505.60	49.00	784.00		
2 Equipment Operators (med.)	41.35	661.60	62.15	994.40		
2 Truck Drivers (heavy)	31.95	511.20	49.15	786.40		
1 Crawler Crane, 75 Ton		1496.00		1645.60		
1 Crawler Loader, 4 C.Y.		1450.00		1595.00		
2 Dump Trucks, 8 C.Y., 220 H.P.		652.80		718.08	64.26	70.69
56 L.H., Daily Totals		$5546.00		$6940.28	$99.04	$123.93
Crew B-13B	Hr.	Daily	Hr.	Daily	Bare Costs	Incl. O&P
1 Labor Foreman (outside)	$33.60	$268.80	$52.10	$416.80	$34.19	$52.48
4 Laborers	31.60	1011.20	49.00	1568.00		
1 Equip. Oper. (crane)	42.55	340.40	63.95	511.60		
1 Equip. Oper. Oiler	36.80	294.40	55.30	442.40		
1 Hyd. Crane, 55 Ton		1125.00		1237.50	20.09	22.10
56 L.H., Daily Totals		$3039.80		$4176.30	$54.28	$74.58
Crew B-13C	Hr.	Daily	Hr.	Daily	Bare Costs	Incl. O&P
1 Labor Foreman (outside)	$33.60	$268.80	$52.10	$416.80	$34.19	$52.48
4 Laborers	31.60	1011.20	49.00	1568.00		
1 Equip. Oper. (crane)	42.55	340.40	63.95	511.60		
1 Equip. Oper. Oiler	36.80	294.40	55.30	442.40		
1 Crawler Crane, 100 Ton		1611.00		1772.10	28.77	31.64
56 L.H., Daily Totals		$3525.80		$4710.90	$62.96	$84.12
Crew B-13D	Hr.	Daily	Hr.	Daily	Bare Costs	Incl. O&P
1 Laborer	$31.60	$252.80	$49.00	$392.00	$37.08	$56.48
1 Equip. Oper. (crane)	42.55	340.40	63.95	511.60		
1 Hyd. Excavator, 1 C.Y.		669.20		736.12		
1 Trench Box		113.40		124.74	48.91	53.80
16 L.H., Daily Totals		$1375.80		$1764.46	$85.99	$110.28
Crew B-13E	Hr.	Daily	Hr.	Daily	Bare Costs	Incl. O&P
1 Laborer	$31.60	$252.80	$49.00	$392.00	$37.08	$56.48
1 Equip. Oper. (crane)	42.55	340.40	63.95	511.60		
1 Hyd. Excavator, 1.5 C.Y.		865.80		952.38		
1 Trench Box		113.40		124.74	61.20	67.32
16 L.H., Daily Totals		$1572.40		$1980.72	$98.28	$123.80
Crew B-13F	Hr.	Daily	Hr.	Daily	Bare Costs	Incl. O&P
1 Laborer	$31.60	$252.80	$49.00	$392.00	$37.08	$56.48
1 Equip. Oper. (crane)	42.55	340.40	63.95	511.60		
1 Hyd. Excavator, 3.5 C.Y.		2162.00		2378.20		
1 Trench Box		113.40		124.74	142.21	156.43
16 L.H., Daily Totals		$2868.60		$3406.54	$179.29	$212.91
Crew B-13G	Hr.	Daily	Hr.	Daily	Bare Costs	Incl. O&P
1 Laborer	$31.60	$252.80	$49.00	$392.00	$37.08	$56.48
1 Equip. Oper. (crane)	42.55	340.40	63.95	511.60		
1 Hyd. Excavator, .75 C.Y.		598.60		658.46		
1 Trench Box		113.40		124.74	44.50	48.95
16 L.H., Daily Totals		$1305.20		$1686.80	$81.58	$105.43

Crew No.	Bare Costs		Incl. Subs O & P		Cost Per Labor-Hour	
Crew B-13H	Hr.	Daily	Hr.	Daily	Bare Costs	Incl. O&P
1 Laborer	$31.60	$252.80	$49.00	$392.00	$37.08	$56.48
1 Equip. Oper. (crane)	42.55	340.40	63.95	511.60		
1 Gradall, 5/8 C.Y.		958.80		1054.68		
1 Trench Box		113.40		124.74	67.01	73.71
16 L.H., Daily Totals		$1665.40		$2083.02	$104.09	$130.19
Crew B-13I	Hr.	Daily	Hr.	Daily	Bare Costs	Incl. O&P
1 Laborer	$31.60	$252.80	$49.00	$392.00	$37.08	$56.48
1 Equip. Oper. (crane)	42.55	340.40	63.95	511.60		
1 Gradall, 3 Ton, 1 C.Y.		1130.00		1243.00		
1 Trench Box		113.40		124.74	77.71	85.48
16 L.H., Daily Totals		$1836.60		$2271.34	$114.79	$141.96
Crew B-13J	Hr.	Daily	Hr.	Daily	Bare Costs	Incl. O&P
1 Laborer	$31.60	$252.80	$49.00	$392.00	$37.08	$56.48
1 Equip. Oper. (crane)	42.55	340.40	63.95	511.60		
1 Hyd. Excavator, 2.5 C.Y.		1619.00		1780.90		
1 Trench Box		113.40		124.74	108.28	119.10
16 L.H., Daily Totals		$2325.60		$2809.24	$145.35	$175.58
Crew B-14	Hr.	Daily	Hr.	Daily	Bare Costs	Incl. O&P
1 Labor Foreman (outside)	$33.60	$268.80	$52.10	$416.80	$33.17	$51.13
4 Laborers	31.60	1011.20	49.00	1568.00		
1 Equip. Oper. (light)	39.05	312.40	58.70	469.60		
1 Backhoe Loader, 48 H.P.		293.80		323.18	6.12	6.73
48 L.H., Daily Totals		$1886.20		$2777.58	$39.30	$57.87
Crew B-14A	Hr.	Daily	Hr.	Daily	Bare Costs	Incl. O&P
1 Equip. Oper. (crane)	$42.55	$340.40	$63.95	$511.60	$38.90	$58.97
.5 Laborer	31.60	126.40	49.00	196.00		
1 Hyd. Excavator, 4.5 C.Y.		2635.00		2898.50	219.58	241.54
12 L.H., Daily Totals		$3101.80		$3606.10	$258.48	$300.51
Crew B-14B	Hr.	Daily	Hr.	Daily	Bare Costs	Incl. O&P
1 Equip. Oper. (crane)	$42.55	$340.40	$63.95	$511.60	$38.90	$58.97
.5 Laborer	31.60	126.40	49.00	196.00		
1 Hyd. Excavator, 6 C.Y.		3080.00		3388.00	256.67	282.33
12 L.H., Daily Totals		$3546.80		$4095.60	$295.57	$341.30
Crew B-14C	Hr.	Daily	Hr.	Daily	Bare Costs	Incl. O&P
1 Equip. Oper. (crane)	$42.55	$340.40	$63.95	$511.60	$38.90	$58.97
.5 Laborer	31.60	126.40	49.00	196.00		
1 Hyd. Excavator, 7 C.Y.		3256.00		3581.60	271.33	298.47
12 L.H., Daily Totals		$3722.80		$4289.20	$310.23	$357.43
Crew B-14F	Hr.	Daily	Hr.	Daily	Bare Costs	Incl. O&P
1 Equip. Oper. (crane)	$42.55	$340.40	$63.95	$511.60	$38.90	$58.97
.5 Laborer	31.60	126.40	49.00	196.00		
1 Hyd. Shovel, 7 C.Y.		3090.00		3399.00	257.50	283.25
12 L.H., Daily Totals		$3556.80		$4106.60	$296.40	$342.22
Crew B-14G	Hr.	Daily	Hr.	Daily	Bare Costs	Incl. O&P
1 Equip. Oper. (crane)	$42.55	$340.40	$63.95	$511.60	$38.90	$58.97
.5 Laborer	31.60	126.40	49.00	196.00		
1 Hyd. Shovel, 12 C.Y.		4148.00		4562.80	345.67	380.23
12 L.H., Daily Totals		$4614.80		$5270.40	$384.57	$439.20

Crews

Crew No.	Bare Costs		Incl. Subs O & P		Cost Per Labor-Hour	
Crew B-14J	Hr.	Daily	Hr.	Daily	Bare Costs	Incl. O&P
1 Equip. Oper. (med.)	$41.35	$330.80	$62.15	$497.20	$38.10	$57.77
.5 Laborer	31.60	126.40	49.00	196.00		
1 F.E. Loader, 8 C.Y.		1497.00		1646.70	124.75	137.22
12 L.H., Daily Totals		$1954.20		$2339.90	$162.85	$194.99
Crew B-14K	Hr.	Daily	Hr.	Daily	Bare Costs	Incl. O&P
1 Equip. Oper. (med.)	$41.35	$330.80	$62.15	$497.20	$38.10	$57.77
.5 Laborer	31.60	126.40	49.00	196.00		
1 F.E. Loader, 10 C.Y.		2225.00		2447.50	185.42	203.96
12 L.H., Daily Totals		$2682.20		$3140.70	$223.52	$261.73
Crew B-15	Hr.	Daily	Hr.	Daily	Bare Costs	Incl. O&P
1 Equipment Oper. (med)	$41.35	$330.80	$62.15	$497.20	$34.59	$52.84
.5 Laborer	31.60	126.40	49.00	196.00		
2 Truck Drivers (heavy)	31.95	511.20	49.15	786.40		
2 Dump Trucks, 12 C.Y., 400 H.P.		1066.00		1172.60		
1 Dozer, 200 H.P.		1082.00		1190.20	76.71	84.39
28 L.H., Daily Totals		$3116.40		$3842.40	$111.30	$137.23
Crew B-16	Hr.	Daily	Hr.	Daily	Bare Costs	Incl. O&P
1 Labor Foreman (outside)	$33.60	$268.80	$52.10	$416.80	$32.19	$49.81
2 Laborers	31.60	505.60	49.00	784.00		
1 Truck Driver (heavy)	31.95	255.60	49.15	393.20		
1 Dump Truck, 12 C.Y., 400 H.P.		533.00		586.30	16.66	18.32
32 L.H., Daily Totals		$1563.00		$2180.30	$48.84	$68.13
Crew B-17	Hr.	Daily	Hr.	Daily	Bare Costs	Incl. O&P
2 Laborers	$31.60	$505.60	$49.00	$784.00	$33.55	$51.46
1 Equip. Oper. (light)	39.05	312.40	58.70	469.60		
1 Truck Driver (heavy)	31.95	255.60	49.15	393.20		
1 Backhoe Loader, 48 H.P.		293.80		323.18		
1 Dump Truck, 8 C.Y., 220 H.P.		326.40		359.04	19.38	21.32
32 L.H., Daily Totals		$1693.80		$2329.02	$52.93	$72.78
Crew B-17A	Hr.	Daily	Hr.	Daily	Bare Costs	Incl. O&P
2 Laborer Foremen	$33.60	$537.60	$52.10	$833.60	$34.05	$52.78
6 Laborers	31.60	1516.80	49.00	2352.00		
1 Skilled Worker Foreman	42.85	342.80	66.35	530.80		
1 Skilled Worker	40.85	326.80	63.25	506.00		
80 L.H., Daily Totals		$2724.00		$4222.40	$34.05	$52.78
Crew B-18	Hr.	Daily	Hr.	Daily	Bare Costs	Incl. O&P
1 Labor Foreman (outside)	$33.60	$268.80	$52.10	$416.80	$32.27	$50.03
2 Laborers	31.60	505.60	49.00	784.00		
1 Vibratory Plate, Gas, 21"		39.20		43.12	1.63	1.80
24 L.H., Daily Totals		$813.60		$1243.92	$33.90	$51.83
Crew B-19	Hr.	Daily	Hr.	Daily	Bare Costs	Incl. O&P
1 Pile Driver Foreman	$40.50	$324.00	$65.75	$526.00	$39.55	$62.37
4 Pile Drivers	38.50	1232.00	62.50	2000.00		
2 Equip. Oper. (crane)	42.55	680.80	63.95	1023.20		
1 Equip. Oper. Oiler	36.80	294.40	55.30	442.40		
1 Crawler Crane, 40 Ton		1055.00		1160.50		
1 Lead, 90' high		118.80		130.68		
1 Hammer, Diesel, 22k Ft-Lb		590.00		649.00	27.56	30.32
64 L.H., Daily Totals		$4295.00		$5931.78	$67.11	$92.68

Crew No.	Bare Costs		Incl. Subs O & P		Cost Per Labor-Hour	
Crew B-19A	Hr.	Daily	Hr.	Daily	Bare Costs	Incl. O&P
1 Pile Driver Foreman	$40.50	$324.00	$65.75	$526.00	$39.55	$62.37
4 Pile Drivers	38.50	1232.00	62.50	2000.00		
2 Equip. Oper. (crane)	42.55	680.80	63.95	1023.20		
1 Equip. Oper. Oiler	36.80	294.40	55.30	442.40		
1 Crawler Crane, 75 Ton		1496.00		1645.60		
1 Lead, 90' high		118.80		130.68		
1 Hammer, Diesel, 41k Ft-Lb		677.60		745.36	35.82	39.40
64 L.H., Daily Totals		$4823.60		$6513.24	$75.37	$101.77
Crew B-20	Hr.	Daily	Hr.	Daily	Bare Costs	Incl. O&P
1 Labor Foreman (out)	$33.60	$268.80	$52.10	$416.80	$35.35	$54.78
1 Skilled Worker	40.85	326.80	63.25	506.00		
1 Laborer	31.60	252.80	49.00	392.00		
24 L.H., Daily Totals		$848.40		$1314.80	$35.35	$54.78
Crew B-20A	Hr.	Daily	Hr.	Daily	Bare Costs	Incl. O&P
1 Labor Foreman	$33.60	$268.80	$52.10	$416.80	$38.24	$58.20
1 Laborer	31.60	252.80	49.00	392.00		
1 Plumber	48.75	390.00	73.15	585.20		
1 Plumber Apprentice	39.00	312.00	58.55	468.40		
32 L.H., Daily Totals		$1223.60		$1862.40	$38.24	$58.20
Crew B-21	Hr.	Daily	Hr.	Daily	Bare Costs	Incl. O&P
1 Labor Foreman (out)	$33.60	$268.80	$52.10	$416.80	$36.38	$56.09
1 Skilled Worker	40.85	326.80	63.25	506.00		
1 Laborer	31.60	252.80	49.00	392.00		
.5 Equip. Oper. (crane)	42.55	170.20	63.95	255.80		
.5 S.P. Crane, 4x4, 5 Ton		129.10		142.01	4.61	5.07
28 L.H., Daily Totals		$1147.70		$1712.61	$40.99	$61.16
Crew B-21A	Hr.	Daily	Hr.	Daily	Bare Costs	Incl. O&P
1 Labor Foreman	$33.60	$268.80	$52.10	$416.80	$39.10	$59.35
1 Laborer	31.60	252.80	49.00	392.00		
1 Plumber	48.75	390.00	73.15	585.20		
1 Plumber Apprentice	39.00	312.00	58.55	468.40		
1 Equip. Oper. (crane)	42.55	340.40	63.95	511.60		
1 S.P. Crane, 4x4, 12 Ton		601.80		661.98	15.05	16.55
40 L.H., Daily Totals		$2165.80		$3035.98	$54.15	$75.90
Crew B-21B	Hr.	Daily	Hr.	Daily	Bare Costs	Incl. O&P
1 Laborer Foreman	$33.60	$268.80	$52.10	$416.80	$34.19	$52.61
3 Laborers	31.60	758.40	49.00	1176.00		
1 Equip. Oper. (crane)	42.55	340.40	63.95	511.60		
1 Hyd. Crane, 12 Ton		768.80		845.68	19.22	21.14
40 L.H., Daily Totals		$2136.40		$2950.08	$53.41	$73.75
Crew B-21C	Hr.	Daily	Hr.	Daily	Bare Costs	Incl. O&P
1 Laborer Foreman	$33.60	$268.80	$52.10	$416.80	$34.19	$52.48
4 Laborers	31.60	1011.20	49.00	1568.00		
1 Equip. Oper. (crane)	42.55	340.40	63.95	511.60		
1 Equip. Oper. Oiler	36.80	294.40	55.30	442.40		
2 Cutting Torches		34.00		37.40		
2 Gases		151.20		166.32		
1 Lattice Boom Crane, 90 Ton		1741.00		1915.10	34.40	37.84
56 L.H., Daily Totals		$3841.00		$5057.62	$68.59	$90.31

Crews

Crew No.	Bare Costs		Incl. Subs O & P		Cost Per Labor-Hour	
Crew B-22	Hr.	Daily	Hr.	Daily	Bare Costs	Incl. O&P
1 Labor Foreman (out)	$33.60	$268.80	$52.10	$416.80	$36.79	$56.62
1 Skilled Worker	40.85	326.80	63.25	506.00		
1 Laborer	31.60	252.80	49.00	392.00		
.75 Equip. Oper. (crane)	42.55	255.30	63.95	383.70		
.75 S.P. Crane, 4x4, 5 Ton		193.65		213.01	6.46	7.10
30 L.H., Daily Totals		$1297.35		$1911.52	$43.24	$63.72

Crew No.	Bare Costs		Incl. Subs O & P		Cost Per Labor-Hour	
Crew B-22A	Hr.	Daily	Hr.	Daily	Bare Costs	Incl. O&P
1 Labor Foreman (out)	$33.60	$268.80	$52.10	$416.80	$35.70	$55.01
1 Skilled Worker	40.85	326.80	63.25	506.00		
2 Laborers	31.60	505.60	49.00	784.00		
.75 Equipment Oper. (crane)	42.55	255.30	63.95	383.70		
.75 S.P. Crane, 4x4, 5 Ton		193.65		213.01		
1 Generator, 5 kW		40.20		44.22		
1 Butt Fusion Machine		199.80		219.78	11.41	12.55
38 L.H., Daily Totals		$1790.15		$2567.51	$47.11	$67.57

Crew B-22B	Hr.	Daily	Hr.	Daily	Bare Costs	Incl. O&P
1 Skilled Worker	$40.85	$326.80	$63.25	$506.00	$36.23	$56.13
1 Laborer	31.60	252.80	49.00	392.00		
1 Electro Fusion Machine		146.80		161.48	9.18	10.09
16 L.H., Daily Totals		$726.40		$1059.48	$45.40	$66.22

Crew B-23	Hr.	Daily	Hr.	Daily	Bare Costs	Incl. O&P
1 Labor Foreman (outside)	$33.60	$268.80	$52.10	$416.80	$32.00	$49.62
4 Laborers	31.60	1011.20	49.00	1568.00		
1 Drill Rig, Truck-Mounted		2533.00		2786.30		
1 Flatbed Truck, Gas, 3 Ton		242.60		266.86	69.39	76.33
40 L.H., Daily Totals		$4055.60		$5037.96	$101.39	$125.95

Crew B-23A	Hr.	Daily	Hr.	Daily	Bare Costs	Incl. O&P
1 Labor Foreman (outside)	$33.60	$268.80	$52.10	$416.80	$35.52	$54.42
1 Laborer	31.60	252.80	49.00	392.00		
1 Equip. Operator (medium)	41.35	330.80	62.15	497.20		
1 Drill Rig, Truck-Mounted		2533.00		2786.30		
1 Pickup Truck, 3/4 Ton		116.80		128.48	110.41	121.45
24 L.H., Daily Totals		$3502.20		$4220.78	$145.93	$175.87

Crew B-23B	Hr.	Daily	Hr.	Daily	Bare Costs	Incl. O&P
1 Labor Foreman (outside)	$33.60	$268.80	$52.10	$416.80	$35.52	$54.42
1 Laborer	31.60	252.80	49.00	392.00		
1 Equip. Operator (medium)	41.35	330.80	62.15	497.20		
1 Drill Rig, Truck-Mounted		2533.00		2786.30		
1 Pickup Truck, 3/4 Ton		116.80		128.48		
1 Centr. Water Pump, 6"		309.40		340.34	123.30	135.63
24 L.H., Daily Totals		$3811.60		$4561.12	$158.82	$190.05

Crew B-24	Hr.	Daily	Hr.	Daily	Bare Costs	Incl. O&P
1 Cement Finisher	$38.30	$306.40	$56.05	$448.40	$36.62	$55.67
1 Laborer	31.60	252.80	49.00	392.00		
1 Carpenter	39.95	319.60	61.95	495.60		
24 L.H., Daily Totals		$878.80		$1336.00	$36.62	$55.67

Crew B-25	Hr.	Daily	Hr.	Daily	Bare Costs	Incl. O&P
1 Labor Foreman	$33.60	$268.80	$52.10	$416.80	$34.44	$52.87
7 Laborers	31.60	1769.60	49.00	2744.00		
3 Equip. Oper. (med.)	41.35	992.40	62.15	1491.60		
1 Asphalt Paver, 130 H.P.		1936.00		2129.60		
1 Tandem Roller, 10 Ton		233.00		256.30		
1 Roller, Pneum. Whl, 12 Ton		307.80		338.58	28.15	30.96
88 L.H., Daily Totals		$5507.60		$7376.88	$62.59	$83.83

Crew No.	Bare Costs		Incl. Subs O & P		Cost Per Labor-Hour	
Crew B-25B	Hr.	Daily	Hr.	Daily	Bare Costs	Incl. O&P
1 Labor Foreman	$33.60	$268.80	$52.10	$416.80	$35.02	$53.64
7 Laborers	31.60	1769.60	49.00	2744.00		
4 Equip. Oper. (medium)	41.35	1323.20	62.15	1988.80		
1 Asphalt Paver, 130 H.P.		1936.00		2129.60		
2 Tandem Rollers, 10 Ton		466.00		512.60		
1 Roller, Pneum. Whl, 12 Ton		307.80		338.58	28.23	31.05
96 L.H., Daily Totals		$6071.40		$8130.38	$63.24	$84.69

Crew B-25C	Hr.	Daily	Hr.	Daily	Bare Costs	Incl. O&P
1 Labor Foreman	$33.60	$268.80	$52.10	$416.80	$35.18	$53.90
3 Laborers	31.60	758.40	49.00	1176.00		
2 Equip. Oper. (medium)	41.35	661.60	62.15	994.40		
1 Asphalt Paver, 130 H.P.		1936.00		2129.60		
1 Tandem Roller, 10 Ton		233.00		256.30	45.19	49.71
48 L.H., Daily Totals		$3857.80		$4973.10	$80.37	$103.61

Crew B-26	Hr.	Daily	Hr.	Daily	Bare Costs	Incl. O&P
1 Labor Foreman (outside)	$33.60	$268.80	$52.10	$416.80	$35.34	$54.48
6 Laborers	31.60	1516.80	49.00	2352.00		
2 Equip. Oper. (med.)	41.35	661.60	62.15	994.40		
1 Rodman (reinf.)	44.55	356.40	72.85	582.80		
1 Cement Finisher	38.30	306.40	56.05	448.40		
1 Grader, 30,000 Lbs.		550.20		605.22		
1 Paving Mach. & Equip.		2500.00		2750.00	34.66	38.13
88 L.H., Daily Totals		$6160.20		$8149.62	$70.00	$92.61

Crew B-26A	Hr.	Daily	Hr.	Daily	Bare Costs	Incl. O&P
1 Labor Foreman (outside)	$33.60	$268.80	$52.10	$416.80	$35.34	$54.48
6 Laborers	31.60	1516.80	49.00	2352.00		
2 Equip. Oper. (med.)	41.35	661.60	62.15	994.40		
1 Rodman (reinf.)	44.55	356.40	72.85	582.80		
1 Cement Finisher	38.30	306.40	56.05	448.40		
1 Grader, 30,000 Lbs.		550.20		605.22		
1 Paving Mach. & Equip.		2500.00		2750.00		
1 Concrete Saw		141.00		155.10	36.26	39.89
88 L.H., Daily Totals		$6301.20		$8304.72	$71.60	$94.37

Crew B-26B	Hr.	Daily	Hr.	Daily	Bare Costs	Incl. O&P
1 Labor Foreman (outside)	$33.60	$268.80	$52.10	$416.80	$35.84	$55.12
6 Laborers	31.60	1516.80	49.00	2352.00		
3 Equip. Oper. (med.)	41.35	992.40	62.15	1491.60		
1 Rodman (reinf.)	44.55	356.40	72.85	582.80		
1 Cement Finisher	38.30	306.40	56.05	448.40		
1 Grader, 30,000 Lbs.		550.20		605.22		
1 Paving Mach. & Equip.		2500.00		2750.00		
1 Concrete Pump, 110' Boom		977.00		1074.70	41.95	46.15
96 L.H., Daily Totals		$7468.00		$9721.52	$77.79	$101.27

Crew B-27	Hr.	Daily	Hr.	Daily	Bare Costs	Incl. O&P
1 Labor Foreman (outside)	$33.60	$268.80	$52.10	$416.80	$32.10	$49.77
3 Laborers	31.60	758.40	49.00	1176.00		
1 Berm Machine		271.40		298.54	8.48	9.33
32 L.H., Daily Totals		$1298.60		$1891.34	$40.58	$59.10

Crew B-28	Hr.	Daily	Hr.	Daily	Bare Costs	Incl. O&P
2 Carpenters	$39.95	$639.20	$61.95	$991.20	$37.17	$57.63
1 Laborer	31.60	252.80	49.00	392.00		
24 L.H., Daily Totals		$892.00		$1383.20	$37.17	$57.63

Crews

Crew No.	Bare Costs		Incl. Subs O & P		Cost Per Labor-Hour	
Crew B-29	Hr.	Daily	Hr.	Daily	Bare Costs	Incl. O&P
1 Labor Foreman (outside)	$33.60	$268.80	$52.10	$416.80	$34.19	$52.48
4 Laborers	31.60	1011.20	49.00	1568.00		
1 Equip. Oper. (crane)	42.55	340.40	63.95	511.60		
1 Equip. Oper. Oiler	36.80	294.40	55.30	442.40		
1 Gradall, 5/8 C.Y.		958.80		1054.68	17.12	18.83
56 L.H., Daily Totals		$2873.60		$3993.48	$51.31	$71.31

Crew B-30	Hr.	Daily	Hr.	Daily	Bare Costs	Incl. O&P
1 Equip. Oper. (med.)	$41.35	$330.80	$62.15	$497.20	$35.08	$53.48
2 Truck Drivers (heavy)	31.95	511.20	49.15	786.40		
1 Hyd. Excavator, 1.5 C.Y.		865.80		952.38		
2 Dump Trucks, 12 C.Y., 400 H.P.		1066.00		1172.60	80.49	88.54
24 L.H., Daily Totals		$2773.80		$3408.58	$115.58	$142.02

Crew B-31	Hr.	Daily	Hr.	Daily	Bare Costs	Incl. O&P
1 Labor Foreman (outside)	$33.60	$268.80	$52.10	$416.80	$33.67	$52.21
3 Laborers	31.60	758.40	49.00	1176.00		
1 Carpenter	39.95	319.60	61.95	495.60		
1 Air Compressor, 250 cfm		162.40		178.64		
1 Sheeting Driver		5.75		6.33		
2 -50' Air Hoses, 1.5"		12.60		13.86	4.52	4.97
40 L.H., Daily Totals		$1527.55		$2287.22	$38.19	$57.18

Crew B-32	Hr.	Daily	Hr.	Daily	Bare Costs	Incl. O&P
1 Laborer	$31.60	$252.80	$49.00	$392.00	$38.91	$58.86
3 Equip. Oper. (med.)	41.35	992.40	62.15	1491.60		
1 Grader, 30,000 Lbs.		550.20		605.22		
1 Tandem Roller, 10 Ton		233.00		256.30		
1 Dozer, 200 H.P.		1082.00		1190.20	58.29	64.12
32 L.H., Daily Totals		$3110.40		$3935.32	$97.20	$122.98

Crew B-32A	Hr.	Daily	Hr.	Daily	Bare Costs	Incl. O&P
1 Laborer	$31.60	$252.80	$49.00	$392.00	$38.10	$57.77
2 Equip. Oper. (medium)	41.35	661.60	62.15	994.40		
1 Grader, 30,000 Lbs.		550.20		605.22		
1 Roller, Vibratory, 25 Ton		594.60		654.06	47.70	52.47
24 L.H., Daily Totals		$2059.20		$2645.68	$85.80	$110.24

Crew B-32B	Hr.	Daily	Hr.	Daily	Bare Costs	Incl. O&P
1 Laborer	$31.60	$252.80	$49.00	$392.00	$38.10	$57.77
2 Equip. Oper. (medium)	41.35	661.60	62.15	994.40		
1 Dozer, 200 H.P.		1082.00		1190.20		
1 Roller, Vibratory, 25 Ton		594.60		654.06	69.86	76.84
24 L.H., Daily Totals		$2591.00		$3230.66	$107.96	$134.61

Crew B-32C	Hr.	Daily	Hr.	Daily	Bare Costs	Incl. O&P
1 Labor Foreman	$33.60	$268.80	$52.10	$416.80	$36.81	$56.09
2 Laborers	31.60	505.60	49.00	784.00		
3 Equip. Oper. (medium)	41.35	992.40	62.15	1491.60		
1 Grader, 30,000 Lbs.		550.20		605.22		
1 Tandem Roller, 10 Ton		233.00		256.30		
1 Dozer, 200 H.P.		1082.00		1190.20	38.86	42.74
48 L.H., Daily Totals		$3632.00		$4744.12	$75.67	$98.84

Crew B-33A	Hr.	Daily	Hr.	Daily	Bare Costs	Incl. O&P
1 Equip. Oper. (med.)	$41.35	$330.80	$62.15	$497.20	$38.56	$58.39
.5 Laborer	31.60	126.40	49.00	196.00		
.25 Equip. Oper. (med.)	41.35	82.70	62.15	124.30		
1 Scraper, Towed, 7 C.Y.		102.20		112.42		
1.25 Dozers, 300 H.P.		1778.75		1956.63	134.35	147.79
14 L.H., Daily Totals		$2420.85		$2886.55	$172.92	$206.18

Crew B-33B	Hr.	Daily	Hr.	Daily	Bare Costs	Incl. O&P
1 Equip. Oper. (med.)	$41.35	$330.80	$62.15	$497.20	$38.56	$58.39
.5 Laborer	31.60	126.40	49.00	196.00		
.25 Equip. Oper. (med.)	41.35	82.70	62.15	124.30		
1 Scraper, Towed, 10 C.Y.		153.60		168.96		
1.25 Dozers, 300 H.P.		1778.75		1956.63	138.03	151.83
14 L.H., Daily Totals		$2472.25		$2943.09	$176.59	$210.22

Crew B-33C	Hr.	Daily	Hr.	Daily	Bare Costs	Incl. O&P
1 Equip. Oper. (med.)	$41.35	$330.80	$62.15	$497.20	$38.56	$58.39
.5 Laborer	31.60	126.40	49.00	196.00		
.25 Equip. Oper. (med.)	41.35	82.70	62.15	124.30		
1 Scraper, Towed, 15 C.Y.		168.80		185.68		
1.25 Dozers, 300 H.P.		1778.75		1956.63	139.11	153.02
14 L.H., Daily Totals		$2487.45		$2959.80	$177.68	$211.41

Crew B-33D	Hr.	Daily	Hr.	Daily	Bare Costs	Incl. O&P
1 Equip. Oper. (med.)	$41.35	$330.80	$62.15	$497.20	$38.56	$58.39
.5 Laborer	31.60	126.40	49.00	196.00		
.25 Equip. Oper. (med.)	41.35	82.70	62.15	124.30		
1 S.P. Scraper, 14 C.Y.		1509.00		1659.90		
.25 Dozer, 300 H.P.		355.75		391.32	133.20	146.52
14 L.H., Daily Totals		$2404.65		$2868.72	$171.76	$204.91

Crew B-33E	Hr.	Daily	Hr.	Daily	Bare Costs	Incl. O&P
1 Equip. Oper. (med.)	$41.35	$330.80	$62.15	$497.20	$38.56	$58.39
.5 Laborer	31.60	126.40	49.00	196.00		
.25 Equip. Oper. (med.)	41.35	82.70	62.15	124.30		
1 S.P. Scraper, 21 C.Y.		1924.00		2116.40		
.25 Dozer, 300 H.P.		355.75		391.32	162.84	179.12
14 L.H., Daily Totals		$2819.65		$3325.22	$201.40	$237.52

Crew B-33F	Hr.	Daily	Hr.	Daily	Bare Costs	Incl. O&P
1 Equip. Oper. (med.)	$41.35	$330.80	$62.15	$497.20	$38.56	$58.39
.5 Laborer	31.60	126.40	49.00	196.00		
.25 Equip. Oper. (med.)	41.35	82.70	62.15	124.30		
1 Elev. Scraper, 11 C.Y.		1095.00		1204.50		
.25 Dozer, 300 H.P.		355.75		391.32	103.63	113.99
14 L.H., Daily Totals		$1990.65		$2413.32	$142.19	$172.38

Crew B-33G	Hr.	Daily	Hr.	Daily	Bare Costs	Incl. O&P
1 Equip. Oper. (med.)	$41.35	$330.80	$62.15	$497.20	$38.56	$58.39
.5 Laborer	31.60	126.40	49.00	196.00		
.25 Equip. Oper. (med.)	41.35	82.70	62.15	124.30		
1 Elev. Scraper, 22 C.Y.		2127.00		2339.70		
.25 Dozer, 300 H.P.		355.75		391.32	177.34	195.07
14 L.H., Daily Totals		$3022.65		$3548.53	$215.90	$253.47

Crew B-33H	Hr.	Daily	Hr.	Daily	Bare Costs	Incl. O&P
.5 Laborer	$31.60	$126.40	$49.00	$196.00	$38.56	$58.39
1 Equipment Operator (med.)	41.35	330.80	62.15	497.20		
.25 Equipment Operator (med.)	41.35	82.70	62.15	124.30		
1 S.P. Scraper, 44 C.Y.		3452.00		3797.20		
.25 Dozer, 410 H.P.		462.25		508.48	279.59	307.55
14 L.H., Daily Totals		$4454.15		$5123.18	$318.15	$365.94

Crew B-33J	Hr.	Daily	Hr.	Daily	Bare Costs	Incl. O&P
1 Equipment Operator (med.)	$41.35	$330.80	$62.15	$497.20	$41.35	$62.15
1 S.P. Scraper, 14 C.Y.		1509.00		1659.90	188.63	207.49
8 L.H., Daily Totals		$1839.80		$2157.10	$229.97	$269.64

Crews

Crew No.	Bare Costs		Incl. Subs O & P		Cost Per Labor-Hour	
Crew B-33K	Hr.	Daily	Hr.	Daily	Bare Costs	Incl. O&P
1 Equipment Operator (med.)	$41.35	$330.80	$62.15	$497.20	$38.56	$58.39
.25 Equipment Operator (med.)	41.35	82.70	62.15	124.30		
.5 Laborer	31.60	126.40	49.00	196.00		
1 S.P. Scraper, 31 C.Y.		2878.00		3165.80		
.25 Dozer, 410 H.P.		462.25		508.48	238.59	262.45
14 L.H., Daily Totals		$3880.15		$4491.77	$277.15	$320.84
Crew B-34A	Hr.	Daily	Hr.	Daily	Bare Costs	Incl. O&P
1 Truck Driver (heavy)	$31.95	$255.60	$49.15	$393.20	$31.95	$49.15
1 Dump Truck, 8 C.Y., 220 H.P.		326.40		359.04	40.80	44.88
8 L.H., Daily Totals		$582.00		$752.24	$72.75	$94.03
Crew B-34B	Hr.	Daily	Hr.	Daily	Bare Costs	Incl. O&P
1 Truck Driver (heavy)	$31.95	$255.60	$49.15	$393.20	$31.95	$49.15
1 Dump Truck, 12 C.Y., 400 H.P.		533.00		586.30	66.63	73.29
8 L.H., Daily Totals		$788.60		$979.50	$98.58	$122.44
Crew B-34C	Hr.	Daily	Hr.	Daily	Bare Costs	Incl. O&P
1 Truck Driver (heavy)	$31.95	$255.60	$49.15	$393.20	$31.95	$49.15
1 Truck Tractor, 6x4, 380 H.P.		478.80		526.68		
1 Dump Trailer, 16.5 C.Y.		115.60		127.16	74.30	81.73
8 L.H., Daily Totals		$850.00		$1047.04	$106.25	$130.88
Crew B-34D	Hr.	Daily	Hr.	Daily	Bare Costs	Incl. O&P
1 Truck Driver (heavy)	$31.95	$255.60	$49.15	$393.20	$31.95	$49.15
1 Truck Tractor, 6x4, 380 H.P.		478.80		526.68		
1 Dump Trailer, 20 C.Y.		130.20		143.22	76.13	83.74
8 L.H., Daily Totals		$864.60		$1063.10	$108.08	$132.89
Crew B-34E	Hr.	Daily	Hr.	Daily	Bare Costs	Incl. O&P
1 Truck Driver (heavy)	$31.95	$255.60	$49.15	$393.20	$31.95	$49.15
1 Dump Truck, Off Hwy., 25 Ton		1280.00		1408.00	160.00	176.00
8 L.H., Daily Totals		$1535.60		$1801.20	$191.95	$225.15
Crew B-34F	Hr.	Daily	Hr.	Daily	Bare Costs	Incl. O&P
1 Truck Driver (heavy)	$31.95	$255.60	$49.15	$393.20	$31.95	$49.15
1 Dump Truck, Off Hwy., 35 Ton		1144.00		1258.40	143.00	157.30
8 L.H., Daily Totals		$1399.60		$1651.60	$174.95	$206.45
Crew B-34G	Hr.	Daily	Hr.	Daily	Bare Costs	Incl. O&P
1 Truck Driver (heavy)	$31.95	$255.60	$49.15	$393.20	$31.95	$49.15
1 Dump Truck, Off Hwy., 50 Ton		1582.00		1740.20	197.75	217.53
8 L.H., Daily Totals		$1837.60		$2133.40	$229.70	$266.68
Crew B-34H	Hr.	Daily	Hr.	Daily	Bare Costs	Incl. O&P
1 Truck Driver (heavy)	$31.95	$255.60	$49.15	$393.20	$31.95	$49.15
1 Dump Truck, Off Hwy., 65 Ton		1594.00		1753.40	199.25	219.18
8 L.H., Daily Totals		$1849.60		$2146.60	$231.20	$268.32
Crew B-34J	Hr.	Daily	Hr.	Daily	Bare Costs	Incl. O&P
1 Truck Driver (heavy)	$31.95	$255.60	$49.15	$393.20	$31.95	$49.15
1 Dump Truck, Off Hwy., 100 Ton		2047.00		2251.70	255.88	281.46
8 L.H., Daily Totals		$2302.60		$2644.90	$287.82	$330.61
Crew B-34K	Hr.	Daily	Hr.	Daily	Bare Costs	Incl. O&P
1 Truck Driver (heavy)	$31.95	$255.60	$49.15	$393.20	$31.95	$49.15
1 Truck Tractor, 6x4, 450 H.P.		581.20		639.32		
1 Lowbed Trailer, 75 Ton		203.00		223.30	98.03	107.83
8 L.H., Daily Totals		$1039.80		$1255.82	$129.97	$156.98

Crew No.	Bare Costs		Incl. Subs O & P		Cost Per Labor-Hour	
Crew B-34L	Hr.	Daily	Hr.	Daily	Bare Costs	Incl. O&P
1 Equip. Oper. (light)	$39.05	$312.40	$58.70	$469.60	$39.05	$58.70
1 Flatbed Truck, Gas, 1.5 Ton		194.40		213.84	24.30	26.73
8 L.H., Daily Totals		$506.80		$683.44	$63.35	$85.43
Crew B-34N	Hr.	Daily	Hr.	Daily	Bare Costs	Incl. O&P
1 Truck Driver (heavy)	$31.95	$255.60	$49.15	$393.20	$31.95	$49.15
1 Dump Truck, 8 C.Y., 220 H.P.		326.40		359.04		
1 Flatbed Trailer, 40 Ton		139.60		153.56	58.25	64.08
8 L.H., Daily Totals		$721.60		$905.80	$90.20	$113.22
Crew B-34P	Hr.	Daily	Hr.	Daily	Bare Costs	Incl. O&P
1 Pipe Fitter	$49.35	$394.80	$74.05	$592.40	$40.55	$61.28
1 Truck Driver (light)	30.95	247.60	47.65	381.20		
1 Equip. Oper. (medium)	41.35	330.80	62.15	497.20		
1 Flatbed Truck, Gas, 3 Ton		242.60		266.86		
1 Backhoe Loader, 48 H.P.		293.80		323.18	22.35	24.59
24 L.H., Daily Totals		$1509.60		$2060.84	$62.90	$85.87
Crew B-34Q	Hr.	Daily	Hr.	Daily	Bare Costs	Incl. O&P
1 Pipe Fitter	$49.35	$394.80	$74.05	$592.40	$40.95	$61.88
1 Truck Driver (light)	30.95	247.60	47.65	381.20		
1 Eqip. Oper. (crane)	42.55	340.40	63.95	511.60		
1 Flatbed Trailer, 25 Ton		102.00		112.20		
1 Dump Truck, 8 C.Y., 220 H.P.		326.40		359.04		
1 Hyd. Crane, 25 Ton		789.40		868.34	50.74	55.82
24 L.H., Daily Totals		$2200.60		$2824.78	$91.69	$117.70
Crew B-34R	Hr.	Daily	Hr.	Daily	Bare Costs	Incl. O&P
1 Pipe Fitter	$49.35	$394.80	$74.05	$592.40	$40.95	$61.88
1 Truck Driver (light)	30.95	247.60	47.65	381.20		
1 Eqip. Oper. (crane)	42.55	340.40	63.95	511.60		
1 Flatbed Trailer, 25 Ton		102.00		112.20		
1 Dump Truck, 8 C.Y., 220 H.P.		326.40		359.04		
1 Hyd. Crane, 25 Ton		789.40		868.34		
1 Hyd. Excavator, 1 C.Y.		669.20		736.12	78.63	86.49
24 L.H., Daily Totals		$2869.80		$3560.90	$119.58	$148.37
Crew B-34S	Hr.	Daily	Hr.	Daily	Bare Costs	Incl. O&P
2 Pipe Fitters	$49.35	$789.60	$74.05	$1184.80	$43.30	$65.30
1 Truck Driver (heavy)	31.95	255.60	49.15	393.20		
1 Eqip. Oper. (crane)	42.55	340.40	63.95	511.60		
1 Flatbed Trailer, 40 Ton		139.60		153.56		
1 Truck Tractor, 6x4, 380 H.P.		478.80		526.68		
1 Hyd. Crane, 80 Ton		1296.00		1425.60		
1 Hyd. Excavator, 2 C.Y.		1182.00		1300.20	96.76	106.44
32 L.H., Daily Totals		$4482.00		$5495.64	$140.06	$171.74
Crew B-34T	Hr.	Daily	Hr.	Daily	Bare Costs	Incl. O&P
2 Pipe Fitters	$49.35	$789.60	$74.05	$1184.80	$43.30	$65.30
1 Truck Driver (heavy)	31.95	255.60	49.15	393.20		
1 Eqip. Oper. (crane)	42.55	340.40	63.95	511.60		
1 Flatbed Trailer, 40 Ton		139.60		153.56		
1 Truck Tractor, 6x4, 380 H.P.		478.80		526.68		
1 Hyd. Crane, 80 Ton		1296.00		1425.60	59.83	65.81
32 L.H., Daily Totals		$3300.00		$4195.44	$103.13	$131.11

Crews

Crew No.		Bare Costs		Incl. Subs O & P		Cost Per Labor-Hour	
Crew B-35	Hr.	Daily	Hr.	Daily	Bare Costs	Incl. O&P
1 Laborer Foreman (out)	$33.60	$268.80	$52.10	$416.80	$39.02	$59.46
1 Skilled Worker	40.85	326.80	63.25	506.00		
1 Welder (plumber)	48.75	390.00	73.15	585.20		
1 Laborer	31.60	252.80	49.00	392.00		
1 Equip. Oper. (crane)	42.55	340.40	63.95	511.60		
1 Equip. Oper. Oiler	36.80	294.40	55.30	442.40		
1 Welder, electric, 300 amp		55.70		61.27		
1 Hyd. Excavator, .75 C.Y.		598.60		658.46	13.63	14.99
48 L.H., Daily Totals		$2527.50		$3573.73	$52.66	$74.45
Crew B-35A	Hr.	Daily	Hr.	Daily	Bare Costs	Incl. O&P
1 Laborer Foreman (out)	$33.60	$268.80	$52.10	$416.80	$37.96	$57.96
2 Laborers	31.60	505.60	49.00	784.00		
1 Skilled Worker	40.85	326.80	63.25	506.00		
1 Welder (plumber)	48.75	390.00	73.15	585.20		
1 Equip. Oper. (crane)	42.55	340.40	63.95	511.60		
1 Equip. Oper. Oiler	36.80	294.40	55.30	442.40		
1 Welder, gas engine, 300 amp		134.20		147.62		
1 Crawler Crane, 75 Ton		1496.00		1645.60	29.11	32.02
56 L.H., Daily Totals		$3756.20		$5039.22	$67.08	$89.99
Crew B-36	Hr.	Daily	Hr.	Daily	Bare Costs	Incl. O&P
1 Labor Foreman (outside)	$33.60	$268.80	$52.10	$416.80	$35.90	$54.88
2 Laborers	31.60	505.60	49.00	784.00		
2 Equip. Oper. (med.)	41.35	661.60	62.15	994.40		
1 Dozer, 200 H.P.		1082.00		1190.20		
1 Aggregate Spreader		35.00		38.50		
1 Tandem Roller, 10 Ton		233.00		256.30	33.75	37.13
40 L.H., Daily Totals		$2786.00		$3680.20	$69.65	$92.00
Crew B-36A	Hr.	Daily	Hr.	Daily	Bare Costs	Incl. O&P
1 Labor Foreman (outside)	$33.60	$268.80	$52.10	$416.80	$37.46	$56.96
2 Laborers	31.60	505.60	49.00	784.00		
4 Equip. Oper. (med.)	41.35	1323.20	62.15	1988.80		
1 Dozer, 200 H.P.		1082.00		1190.20		
1 Aggregate Spreader		35.00		38.50		
1 Tandem Roller, 10 Ton		233.00		256.30		
1 Roller, Pneum. Whl, 12 Ton		307.80		338.58	29.60	32.56
56 L.H., Daily Totals		$3755.40		$5013.18	$67.06	$89.52
Crew B-36B	Hr.	Daily	Hr.	Daily	Bare Costs	Incl. O&P
1 Labor Foreman (outside)	$33.60	$268.80	$52.10	$416.80	$36.77	$55.98
2 Laborers	31.60	505.60	49.00	784.00		
4 Equip. Oper. (medium)	41.35	1323.20	62.15	1988.80		
1 Truck Driver, Heavy	31.95	255.60	49.15	393.20		
1 Grader, 30,000 Lbs.		550.20		605.22		
1 F.E. Loader, crl, 1.5 C.Y.		463.00		509.30		
1 Dozer, 300 H.P.		1423.00		1565.30		
1 Roller, Vibratory, 25 Ton		594.60		654.06		
1 Truck Tractor, 6x4, 450 H.P.		581.20		639.32		
1 Water Tanker, 5000 Gal.		136.00		149.60	58.56	64.42
64 L.H., Daily Totals		$6101.20		$7705.60	$95.33	$120.40

Crew No.		Bare Costs		Incl. Subs O & P		Cost Per Labor-Hour	
Crew B-36C	Hr.	Daily	Hr.	Daily	Bare Costs	Incl. O&P
1 Labor Foreman (outside)	$33.60	$268.80	$52.10	$416.80	$37.92	$57.54
3 Equip. Oper. (medium)	41.35	992.40	62.15	1491.60		
1 Truck Driver, Heavy	31.95	255.60	49.15	393.20		
1 Grader, 30,000 Lbs.		550.20		605.22		
1 Dozer, 300 H.P.		1423.00		1565.30		
1 Roller, Vibratory, 25 Ton		594.60		654.06		
1 Truck Tractor, 6x4, 450 H.P.		581.20		639.32		
1 Water Tanker, 5000 Gal.		136.00		149.60	82.13	90.34
40 L.H., Daily Totals		$4801.80		$5915.10	$120.05	$147.88
Crew B-37	Hr.	Daily	Hr.	Daily	Bare Costs	Incl. O&P
1 Labor Foreman (outside)	$33.60	$268.80	$52.10	$416.80	$33.17	$51.13
4 Laborers	31.60	1011.20	49.00	1568.00		
1 Equip. Oper. (light)	39.05	312.40	58.70	469.60		
1 Tandem Roller, 5 Ton		137.80		151.58	2.87	3.16
48 L.H., Daily Totals		$1730.20		$2605.98	$36.05	$54.29
Crew B-38	Hr.	Daily	Hr.	Daily	Bare Costs	Incl. O&P
1 Labor Foreman (outside)	$33.60	$268.80	$52.10	$416.80	$35.44	$54.19
2 Laborers	31.60	505.60	49.00	784.00		
1 Equip. Oper. (light)	39.05	312.40	58.70	469.60		
1 Equip. Oper. (medium)	41.35	330.80	62.15	497.20		
1 Backhoe Loader, 48 H.P.		293.80		323.18		
1 Hyd.Hammer, (1200 lb.)		153.00		168.30		
1 F.E. Loader, W.M., 4 C.Y.		541.00		595.10		
1 Pavt. Rem. Bucket		56.80		62.48	26.11	28.73
40 L.H., Daily Totals		$2462.20		$3316.66	$61.56	$82.92
Crew B-39	Hr.	Daily	Hr.	Daily	Bare Costs	Incl. O&P
1 Labor Foreman (outside)	$33.60	$268.80	$52.10	$416.80	$33.17	$51.13
4 Laborers	31.60	1011.20	49.00	1568.00		
1 Equip. Oper. (light)	39.05	312.40	58.70	469.60		
1 Air Compressor, 250 cfm		162.40		178.64		
2 Breakers, Pavement, 60 lb.		16.40		18.04		
2 -50' Air Hoses, 1.5"		12.60		13.86	3.99	4.39
48 L.H., Daily Totals		$1783.80		$2664.94	$37.16	$55.52
Crew B-40	Hr.	Daily	Hr.	Daily	Bare Costs	Incl. O&P
1 Pile Driver Foreman (out)	$40.50	$324.00	$65.75	$526.00	$39.55	$62.37
4 Pile Drivers	38.50	1232.00	62.50	2000.00		
2 Equip. Oper. (crane)	42.55	680.80	63.95	1023.20		
1 Equip. Oper. Oiler	36.80	294.40	55.30	442.40		
1 Crawler Crane, 40 Ton		1055.00		1160.50		
1 Vibratory Hammer & Gen.		2257.00		2482.70	51.75	56.92
64 L.H., Daily Totals		$5843.20		$7634.80	$91.30	$119.29
Crew B-40B	Hr.	Daily	Hr.	Daily	Bare Costs	Incl. O&P
1 Laborer Foreman	$33.60	$268.80	$52.10	$416.80	$34.63	$53.06
3 Laborers	31.60	758.40	49.00	1176.00		
1 Equip. Oper. (crane)	42.55	340.40	63.95	511.60		
1 Equip. Oper. Oiler	36.80	294.40	55.30	442.40		
1 Lattice Boom Crane, 40 Ton		1327.00		1459.70	27.65	30.41
48 L.H., Daily Totals		$2989.00		$4006.50	$62.27	$83.47
Crew B-41	Hr.	Daily	Hr.	Daily	Bare Costs	Incl. O&P
1 Labor Foreman (outside)	$33.60	$268.80	$52.10	$416.80	$32.70	$50.53
4 Laborers	31.60	1011.20	49.00	1568.00		
.25 Equip. Oper. (crane)	42.55	85.10	63.95	127.90		
.25 Equip. Oper. Oiler	36.80	73.60	55.30	110.60		
.25 Crawler Crane, 40 Ton		263.75		290.13	5.99	6.59
44 L.H., Daily Totals		$1702.45		$2513.43	$38.69	$57.12

Crews

Crew No.	Bare Costs		Incl. Subs O & P		Cost Per Labor-Hour	
Crew B-42	Hr.	Daily	Hr.	Daily	Bare Costs	Incl. O&P
1 Labor Foreman (outside)	$33.60	$268.80	$52.10	$416.80	$35.51	$55.88
4 Laborers	31.60	1011.20	49.00	1568.00		
1 Equip. Oper. (crane)	42.55	340.40	63.95	511.60		
1 Equip. Oper. Oiler	36.80	294.40	55.30	442.40		
1 Welder	44.70	357.60	79.65	637.20		
1 Hyd. Crane, 25 Ton		789.40		868.34		
1 Welder, gas engine, 300 amp		134.20		147.62		
1 Horz. Boring Csg. Mch.		444.20		488.62	21.37	23.51
64 L.H., Daily Totals		$3640.20		$5080.58	$56.88	$79.38
Crew B-43	Hr.	Daily	Hr.	Daily	Bare Costs	Incl. O&P
1 Labor Foreman (outside)	$33.60	$268.80	$52.10	$416.80	$34.63	$53.06
3 Laborers	31.60	758.40	49.00	1176.00		
1 Equip. Oper. (crane)	42.55	340.40	63.95	511.60		
1 Equip. Oper. Oiler	36.80	294.40	55.30	442.40		
1 Drill Rig, Truck-Mounted		2533.00		2786.30	52.77	58.05
48 L.H., Daily Totals		$4195.00		$5333.10	$87.40	$111.11
Crew B-44	Hr.	Daily	Hr.	Daily	Bare Costs	Incl. O&P
1 Pile Driver Foreman	$40.50	$324.00	$65.75	$526.00	$38.90	$61.58
4 Pile Drivers	38.50	1232.00	62.50	2000.00		
2 Equip. Oper. (crane)	42.55	680.80	63.95	1023.20		
1 Laborer	31.60	252.80	49.00	392.00		
1 Crawler Crane, 40 Ton		1055.00		1160.50		
1 Lead, 60' high		69.60		76.56		
1 Hammer, diesel, 15K Ft.-Lbs.		578.60		636.46	26.61	29.27
64 L.H., Daily Totals		$4192.80		$5814.72	$65.51	$90.86
Crew B-45	Hr.	Daily	Hr.	Daily	Bare Costs	Incl. O&P
1 Equip. Oper. (med.)	$41.35	$330.80	$62.15	$497.20	$36.65	$55.65
1 Truck Driver (heavy)	31.95	255.60	49.15	393.20		
1 Dist. Tanker, 3000 Gallon		304.20		334.62		
1 Truck Tractor, 6x4, 380 H.P.		478.80		526.68	48.94	53.83
16 L.H., Daily Totals		$1369.40		$1751.70	$85.59	$109.48
Crew B-46	Hr.	Daily	Hr.	Daily	Bare Costs	Incl. O&P
1 Pile Driver Foreman	$40.50	$324.00	$65.75	$526.00	$35.38	$56.29
2 Pile Drivers	38.50	616.00	62.50	1000.00		
3 Laborers	31.60	758.40	49.00	1176.00		
1 Chain Saw, Gas, 36" Long		37.80		41.58	.79	.87
48 L.H., Daily Totals		$1736.20		$2743.58	$36.17	$57.16
Crew B-47	Hr.	Daily	Hr.	Daily	Bare Costs	Incl. O&P
1 Blast Foreman	$33.60	$268.80	$52.10	$416.80	$34.75	$53.27
1 Driller	31.60	252.80	49.00	392.00		
1 Equip. Oper. (light)	39.05	312.40	58.70	469.60		
1 Air Track Drill, 4"		850.60		935.66		
1 Air Compressor, 600 cfm		434.40		477.84		
2 -50' Air Hoses, 3"		32.40		35.64	54.89	60.38
24 L.H., Daily Totals		$2151.40		$2727.54	$89.64	$113.65
Crew B-47A	Hr.	Daily	Hr.	Daily	Bare Costs	Incl. O&P
1 Drilling Foreman	$33.60	$268.80	$52.10	$416.80	$37.65	$57.12
1 Equip. Oper. (heavy)	42.55	340.40	63.95	511.60		
1 Oiler	36.80	294.40	55.30	442.40		
1 Air Track Drill, 5"		974.00		1071.40	40.58	44.64
24 L.H., Daily Totals		$1877.60		$2442.20	$78.23	$101.76

Crew No.	Bare Costs		Incl. Subs O & P		Cost Per Labor-Hour	
Crew B-47C	Hr.	Daily	Hr.	Daily	Bare Costs	Incl. O&P
1 Laborer	$31.60	$252.80	$49.00	$392.00	$35.33	$53.85
1 Equip. Oper. (light)	39.05	312.40	58.70	469.60		
1 Air Compressor, 750 cfm		439.60		483.56		
2 -50' Air Hoses, 3"		32.40		35.64		
1 Air Track Drill, 4"		850.60		935.66	82.66	90.93
16 L.H., Daily Totals		$1887.80		$2316.46	$117.99	$144.78
Crew B-47E	Hr.	Daily	Hr.	Daily	Bare Costs	Incl. O&P
1 Laborer Foreman	$33.60	$268.80	$52.10	$416.80	$32.10	$49.77
3 Laborers	31.60	758.40	49.00	1176.00		
1 Flatbed Truck, Gas, 3 Ton		242.60		266.86	7.58	8.34
32 L.H., Daily Totals		$1269.80		$1859.66	$39.68	$58.11
Crew B-47G	Hr.	Daily	Hr.	Daily	Bare Costs	Incl. O&P
1 Laborer Foreman	$33.60	$268.80	$52.10	$416.80	$33.96	$52.20
2 Laborers	31.60	505.60	49.00	784.00		
1 Equip. Oper. (light)	39.05	312.40	58.70	469.60		
1 Air Track Drill, 4"		850.60		935.66		
1 Air Compressor, 600 cfm		434.40		477.84		
2 -50' Air Hoses, 3"		32.40		35.64		
1 Grout Pump		166.40		183.04	46.37	51.01
32 L.H., Daily Totals		$2570.60		$3302.58	$80.33	$103.21
Crew B-48	Hr.	Daily	Hr.	Daily	Bare Costs	Incl. O&P
1 Labor Foreman (outside)	$33.60	$268.80	$52.10	$416.80	$35.26	$53.86
3 Laborers	31.60	758.40	49.00	1176.00		
1 Equip. Oper. (crane)	42.55	340.40	63.95	511.60		
1 Equip. Oper. Oiler	36.80	294.40	55.30	442.40		
1 Equip. Oper. (light)	39.05	312.40	58.70	469.60		
1 Centr. Water Pump, 6"		309.40		340.34		
1 -20' Suction Hose, 6"		11.90		13.09		
1 -50' Discharge Hose, 6"		6.30		6.93		
1 Drill Rig, Truck-Mounted		2533.00		2786.30	51.08	56.19
56 L.H., Daily Totals		$4835.00		$6163.06	$86.34	$110.05
Crew B-49	Hr.	Daily	Hr.	Daily	Bare Costs	Incl. O&P
1 Labor Foreman (outside)	$33.60	$268.80	$52.10	$416.80	$36.65	$56.48
3 Laborers	31.60	758.40	49.00	1176.00		
2 Equip. Oper. (crane)	42.55	680.80	63.95	1023.20		
2 Equip. Oper. Oilers	36.80	588.80	55.30	884.80		
1 Equip. Oper. (light)	39.05	312.40	58.70	469.60		
2 Pile Drivers	38.50	616.00	62.50	1000.00		
1 Hyd. Crane, 25 Ton		789.40		868.34		
1 Centr. Water Pump, 6"		309.40		340.34		
1 -20' Suction Hose, 6"		11.90		13.09		
1 -50' Discharge Hose, 6"		6.30		6.93		
1 Drill Rig, Truck-Mounted		2533.00		2786.30	41.48	45.63
88 L.H., Daily Totals		$6875.20		$8985.40	$78.13	$102.11
Crew B-50	Hr.	Daily	Hr.	Daily	Bare Costs	Incl. O&P
2 Pile Driver Foremen	$40.50	$648.00	$65.75	$1052.00	$37.76	$59.76
6 Pile Drivers	38.50	1848.00	62.50	3000.00		
2 Equip. Oper. (crane)	42.55	680.80	63.95	1023.20		
1 Equip. Oper. Oiler	36.80	294.40	55.30	442.40		
3 Laborers	31.60	758.40	49.00	1176.00		
1 Crawler Crane, 40 Ton		1055.00		1160.50		
1 Lead, 60' high		69.60		76.56		
1 Hammer, diesel, 15K Ft.-Lbs.		578.60		636.46		
1 Air Compressor, 600 cfm		434.40		477.84		
2 -50' Air Hoses, 3"		32.40		35.64		
1 Chain Saw, Gas, 36" Long		37.80		41.58	19.71	21.68
112 L.H., Daily Totals		$6437.40		$9122.18	$57.48	$81.45

Crews

Crew No.	Bare Costs		Incl. Subs O & P		Cost Per Labor-Hour	
Crew B-51	Hr.	Daily	Hr.	Daily	Bare Costs	Incl. O&P
1 Labor Foreman (outside)	$33.60	$268.80	$52.10	$416.80	$31.82	$49.29
4 Laborers	31.60	1011.20	49.00	1568.00		
1 Truck Driver (light)	30.95	247.60	47.65	381.20		
1 Flatbed Truck, Gas, 1.5 Ton		194.40		213.84	4.05	4.46
48 L.H., Daily Totals		$1722.00		$2579.84	$35.88	$53.75
Crew B-52	Hr.	Daily	Hr.	Daily	Bare Costs	Incl. O&P
1 Carpenter Foreman	$41.95	$335.60	$65.05	$520.40	$36.85	$56.79
1 Carpenter	39.95	319.60	61.95	495.60		
3 Laborers	31.60	758.40	49.00	1176.00		
1 Cement Finisher	38.30	306.40	56.05	448.40		
.5 Rodman (reinf.)	44.55	178.20	72.85	291.40		
.5 Equip. Oper. (med.)	41.35	165.40	62.15	248.60		
.5 Crawler Loader, 3 C.Y.		495.00		544.50	8.84	9.72
56 L.H., Daily Totals		$2558.60		$3724.90	$45.69	$66.52
Crew B-53	Hr.	Daily	Hr.	Daily	Bare Costs	Incl. O&P
1 Equip. Oper. (light)	$39.05	$312.40	$58.70	$469.60	$39.05	$58.70
1 Trencher, Chain, 12 H.P.		59.60		65.56	7.45	8.20
8 L.H., Daily Totals		$372.00		$535.16	$46.50	$66.89
Crew B-54	Hr.	Daily	Hr.	Daily	Bare Costs	Incl. O&P
1 Equip. Oper. (light)	$39.05	$312.40	$58.70	$469.60	$39.05	$58.70
1 Trencher, Chain, 40 H.P.		301.00		331.10	37.63	41.39
8 L.H., Daily Totals		$613.40		$800.70	$76.67	$100.09
Crew B-54A	Hr.	Daily	Hr.	Daily	Bare Costs	Incl. O&P
.17 Labor Foreman (outside)	$33.60	$45.70	$52.10	$70.86	$40.22	$60.69
1 Equipment Operator (med.)	41.35	330.80	62.15	497.20		
1 Wheel Trencher, 67 H.P.		1009.00		1109.90	107.80	118.58
9.36 L.H., Daily Totals		$1385.50		$1677.96	$148.02	$179.27
Crew B-54B	Hr.	Daily	Hr.	Daily	Bare Costs	Incl. O&P
.25 Labor Foreman (outside)	$33.60	$67.20	$52.10	$104.20	$39.80	$60.14
1 Equipment Operator (med.)	41.35	330.80	62.15	497.20		
1 Wheel Trencher, 150 H.P.		1753.00		1928.30	175.30	192.83
10 L.H., Daily Totals		$2151.00		$2529.70	$215.10	$252.97
Crew B-54C	Hr.	Daily	Hr.	Daily	Bare Costs	Incl. O&P
1 Laborer	$31.60	$252.80	$49.00	$392.00	$36.48	$55.58
1 Equipment Operator (med.)	41.35	330.80	62.15	497.20		
1 Wheel Trencher, 67 H.P.		1009.00		1109.90	63.06	69.37
16 L.H., Daily Totals		$1592.60		$1999.10	$99.54	$124.94
Crew B-55	Hr.	Daily	Hr.	Daily	Bare Costs	Incl. O&P
2 Laborers	$31.60	$505.60	$49.00	$784.00	$31.38	$48.55
1 Truck Driver (light)	30.95	247.60	47.65	381.20		
1 Truck-mounted earth auger		706.60		777.26		
1 Flatbed Truck, Gas, 3 Ton		242.60		266.86	39.55	43.51
24 L.H., Daily Totals		$1702.40		$2209.32	$70.93	$92.06
Crew B-56	Hr.	Daily	Hr.	Daily	Bare Costs	Incl. O&P
1 Laborer	$31.60	$252.80	$49.00	$392.00	$35.33	$53.85
1 Equip. Oper. (light)	39.05	312.40	58.70	469.60		
1 Air Track Drill, 4"		850.60		935.66		
1 Air Compressor, 600 cfm		434.40		477.84		
1 -50' Air Hose, 3"		16.20		17.82	81.33	89.46
16 L.H., Daily Totals		$1866.40		$2292.92	$116.65	$143.31

Crew No.	Bare Costs		Incl. Subs O & P		Cost Per Labor-Hour	
Crew B-57	Hr.	Daily	Hr.	Daily	Bare Costs	Incl. O&P
1 Labor Foreman (outside)	$33.60	$268.80	$52.10	$416.80	$35.87	$54.67
2 Laborers	31.60	505.60	49.00	784.00		
1 Equip. Oper. (crane)	42.55	340.40	63.95	511.60		
1 Equip. Oper. (light)	39.05	312.40	58.70	469.60		
1 Equip. Oper. Oiler	36.80	294.40	55.30	442.40		
1 Crawler Crane, 25 Ton		1052.00		1157.20		
1 Clamshell Bucket, 1 C.Y.		46.20		50.82		
1 Centr. Water Pump, 6"		309.40		340.34		
1 -20' Suction Hose, 6"		11.90		13.09		
20 -50' Discharge Hoses, 6"		126.00		138.60	32.20	35.42
48 L.H., Daily Totals		$3267.10		$4324.45	$68.06	$90.09
Crew B-58	Hr.	Daily	Hr.	Daily	Bare Costs	Incl. O&P
2 Laborers	$31.60	$505.60	$49.00	$784.00	$34.08	$52.23
1 Equip. Oper. (light)	39.05	312.40	58.70	469.60		
1 Backhoe Loader, 48 H.P.		293.80		323.18		
1 Small Helicopter, w/pilot		2587.00		2845.70	120.03	132.04
24 L.H., Daily Totals		$3698.80		$4422.48	$154.12	$184.27
Crew B-59	Hr.	Daily	Hr.	Daily	Bare Costs	Incl. O&P
1 Truck Driver (heavy)	$31.95	$255.60	$49.15	$393.20	$31.95	$49.15
1 Truck Tractor, 220 H.P.		288.40		317.24		
1 Water Tanker, 5000 Gal.		136.00		149.60	53.05	58.35
8 L.H., Daily Totals		$680.00		$860.04	$85.00	$107.51
Crew B-59A	Hr.	Daily	Hr.	Daily	Bare Costs	Incl. O&P
2 Laborers	$31.60	$505.60	$49.00	$784.00	$31.72	$49.05
1 Truck Driver (heavy)	31.95	255.60	49.15	393.20		
1 Water Tanker, 5000 Gal.		136.00		149.60		
1 Truck Tractor, 220 H.P.		288.40		317.24	17.68	19.45
24 L.H., Daily Totals		$1185.60		$1644.04	$49.40	$68.50
Crew B-60	Hr.	Daily	Hr.	Daily	Bare Costs	Incl. O&P
1 Labor Foreman (outside)	$33.60	$268.80	$52.10	$416.80	$36.32	$55.25
2 Laborers	31.60	505.60	49.00	784.00		
1 Equip. Oper. (crane)	42.55	340.40	63.95	511.60		
2 Equip. Oper. (light)	39.05	624.80	58.70	939.20		
1 Equip. Oper. Oiler	36.80	294.40	55.30	442.40		
1 Crawler Crane, 40 Ton		1055.00		1160.50		
1 Lead, 60' high		69.60		76.56		
1 Hammer, diesel, 15K Ft.-Lbs.		578.60		636.46		
1 Backhoe Loader, 48 H.P.		293.80		323.18	35.66	39.23
56 L.H., Daily Totals		$4031.00		$5290.70	$71.98	$94.48
Crew B-61	Hr.	Daily	Hr.	Daily	Bare Costs	Incl. O&P
1 Labor Foreman (outside)	$33.60	$268.80	$52.10	$416.80	$33.49	$51.56
3 Laborers	31.60	758.40	49.00	1176.00		
1 Equip. Oper. (light)	39.05	312.40	58.70	469.60		
1 Cement Mixer, 2 C.Y.		187.80		206.58		
1 Air Compressor, 160 cfm		138.80		152.68	8.16	8.98
40 L.H., Daily Totals		$1666.20		$2421.66	$41.66	$60.54
Crew B-62	Hr.	Daily	Hr.	Daily	Bare Costs	Incl. O&P
2 Laborers	$31.60	$505.60	$49.00	$784.00	$34.08	$52.23
1 Equip. Oper. (light)	39.05	312.40	58.70	469.60		
1 Loader, Skid Steer, 30 H.P., gas		149.20		164.12	6.22	6.84
24 L.H., Daily Totals		$967.20		$1417.72	$40.30	$59.07

Crews

Crew No.		Bare Costs		Incl. Subs O & P		Cost Per Labor-Hour	
Crew B-63	Hr.	Daily	Hr.	Daily	Bare Costs	Incl. O&P
4 Laborers	$31.60	$1011.20	$49.00	$1568.00	$33.09	$50.94
1 Equip. Oper. (light)	39.05	312.40	58.70	469.60		
1 Loader, Skid Steer, 30 H.P., gas		149.20		164.12	3.73	4.10
40 L.H., Daily Totals		$1472.80		$2201.72	$36.82	$55.04

Crew B-64	Hr.	Daily	Hr.	Daily	Bare Costs	Incl. O&P
1 Laborer	$31.60	$252.80	$49.00	$392.00	$31.27	$48.33
1 Truck Driver (light)	30.95	247.60	47.65	381.20		
1 Power Mulcher (small)		136.60		150.26		
1 Flatbed Truck, Gas, 1.5 Ton		194.40		213.84	20.69	22.76
16 L.H., Daily Totals		$831.40		$1137.30	$51.96	$71.08

Crew B-65	Hr.	Daily	Hr.	Daily	Bare Costs	Incl. O&P
1 Laborer	$31.60	$252.80	$49.00	$392.00	$31.27	$48.33
1 Truck Driver (light)	30.95	247.60	47.65	381.20		
1 Power Mulcher (large)		271.80		298.98		
1 Flatbed Truck, Gas, 1.5 Ton		194.40		213.84	29.14	32.05
16 L.H., Daily Totals		$966.60		$1286.02	$60.41	$80.38

Crew B-66	Hr.	Daily	Hr.	Daily	Bare Costs	Incl. O&P
1 Equip. Oper. (light)	$39.05	$312.40	$58.70	$469.60	$39.05	$58.70
1 Loader-Backhoe		231.20		254.32	28.90	31.79
8 L.H., Daily Totals		$543.60		$723.92	$67.95	$90.49

Crew B-67	Hr.	Daily	Hr.	Daily	Bare Costs	Incl. O&P
1 Millwright	$41.60	$332.80	$61.05	$488.40	$40.33	$59.88
1 Equip. Oper. (light)	39.05	312.40	58.70	469.60		
1 Forklift, R/T, 4,000 Lb.		281.20		309.32	17.57	19.33
16 L.H., Daily Totals		$926.40		$1267.32	$57.90	$79.21

Crew B-68	Hr.	Daily	Hr.	Daily	Bare Costs	Incl. O&P
2 Millwrights	$41.60	$665.60	$61.05	$976.80	$40.75	$60.27
1 Equip. Oper. (light)	39.05	312.40	58.70	469.60		
1 Forklift, R/T, 4,000 Lb.		281.20		309.32	11.72	12.89
24 L.H., Daily Totals		$1259.20		$1755.72	$52.47	$73.16

Crew B-69	Hr.	Daily	Hr.	Daily	Bare Costs	Incl. O&P
1 Labor Foreman (outside)	$33.60	$268.80	$52.10	$416.80	$34.63	$53.06
3 Laborers	31.60	758.40	49.00	1176.00		
1 Equip Oper. (crane)	42.55	340.40	63.95	511.60		
1 Equip Oper. Oiler	36.80	294.40	55.30	442.40		
1 Hyd. Crane, 80 Ton		1296.00		1425.60	27.00	29.70
48 L.H., Daily Totals		$2958.00		$3972.40	$61.63	$82.76

Crew B-69A	Hr.	Daily	Hr.	Daily	Bare Costs	Incl. O&P
1 Labor Foreman	$33.60	$268.80	$52.10	$416.80	$34.67	$52.88
3 Laborers	31.60	758.40	49.00	1176.00		
1 Equip. Oper. (medium)	41.35	330.80	62.15	497.20		
1 Concrete Finisher	38.30	306.40	56.05	448.40		
1 Curb/Gutter Paver, 2-Track		754.40		829.84	15.72	17.29
48 L.H., Daily Totals		$2418.80		$3368.24	$50.39	$70.17

Crew B-69B	Hr.	Daily	Hr.	Daily	Bare Costs	Incl. O&P
1 Labor Foreman	$33.60	$268.80	$52.10	$416.80	$34.67	$52.88
3 Laborers	31.60	758.40	49.00	1176.00		
1 Equip. Oper. (medium)	41.35	330.80	62.15	497.20		
1 Cement Finisher	38.30	306.40	56.05	448.40		
1 Curb/Gutter Paver, 4-Track		855.80		941.38	17.83	19.61
48 L.H., Daily Totals		$2520.20		$3479.78	$52.50	$72.50

Crew No.		Bare Costs		Incl. Subs O & P		Cost Per Labor-Hour	
Crew B-70	Hr.	Daily	Hr.	Daily	Bare Costs	Incl. O&P
1 Labor Foreman (outside)	$33.60	$268.80	$52.10	$416.80	$36.06	$55.08
3 Laborers	31.60	758.40	49.00	1176.00		
3 Equip. Oper. (med.)	41.35	992.40	62.15	1491.60		
1 Grader, 30,000 Lbs.		550.20		605.22		
1 Ripper, beam & 1 shank		79.60		87.56		
1 Road Sweeper, S.P., 8' wide		601.60		661.76		
1 F.E. Loader, W.M., 1.5 C.Y.		328.20		361.02	27.85	30.64
56 L.H., Daily Totals		$3579.20		$4799.96	$63.91	$85.71

Crew B-70A	Hr.	Daily	Hr.	Daily	Bare Costs	Incl. O&P
1 Laborer	$31.60	$252.80	$49.00	$392.00	$39.40	$59.52
4 Equip. Oper. (med.)	41.35	1323.20	62.15	1988.80		
1 Grader, 40,000 Lbs.		961.00		1057.10		
1 F.E. Loader, W.M.,2.5 C.Y.		395.80		435.38		
1 Dozer, 80 H.P.		399.20		439.12		
1 Roller, Pneum. Whl, 12 Ton		307.80		338.58	51.59	56.75
40 L.H., Daily Totals		$3639.80		$4650.98	$91.00	$116.27

Crew B-71	Hr.	Daily	Hr.	Daily	Bare Costs	Incl. O&P
1 Labor Foreman (outside)	$33.60	$268.80	$52.10	$416.80	$36.06	$55.08
3 Laborers	31.60	758.40	49.00	1176.00		
3 Equip. Oper. (med.)	41.35	992.40	62.15	1491.60		
1 Pvmt. Profiler, 750 H.P.		5503.00		6053.30		
1 Road Sweeper, S.P., 8' wide		601.60		661.76		
1 F.E. Loader, W.M., 1.5 C.Y.		328.20		361.02	114.87	126.36
56 L.H., Daily Totals		$8452.40		$10160.48	$150.94	$181.44

Crew B-72	Hr.	Daily	Hr.	Daily	Bare Costs	Incl. O&P
1 Labor Foreman (outside)	$33.60	$268.80	$52.10	$416.80	$36.73	$55.96
3 Laborers	31.60	758.40	49.00	1176.00		
4 Equip. Oper. (med.)	41.35	1323.20	62.15	1988.80		
1 Pvmt. Profiler, 750 H.P.		5503.00		6053.30		
1 Hammermill, 250 H.P.		1850.00		2035.00		
1 Windrow Loader		1197.00		1316.70		
1 Mix Paver 165 H.P.		1996.00		2195.60		
1 Roller, Pneum. Whl, 12 Ton		307.80		338.58	169.59	186.55
64 L.H., Daily Totals		$13204.20		$15520.78	$206.32	$242.51

Crew B-73	Hr.	Daily	Hr.	Daily	Bare Costs	Incl. O&P
1 Labor Foreman (outside)	$33.60	$268.80	$52.10	$416.80	$37.94	$57.61
2 Laborers	31.60	505.60	49.00	784.00		
5 Equip. Oper. (med.)	41.35	1654.00	62.15	2486.00		
1 Road Mixer, 310 H.P.		1896.00		2085.60		
1 Tandem Roller, 10 Ton		233.00		256.30		
1 Hammermill, 250 H.P.		1850.00		2035.00		
1 Grader, 30,000 Lbs.		550.20		605.22		
.5 F.E. Loader, W.M., 1.5 C.Y.		164.10		180.51		
.5 Truck Tractor, 220 H.P.		144.20		158.62		
.5 Water Tanker, 5000 Gal.		68.00		74.80	76.65	84.31
64 L.H., Daily Totals		$7333.90		$9082.85	$114.59	$141.92

Crews

Crew No.	Bare Costs		Incl. Subs O & P		Cost Per Labor-Hour	
Crew B-74	Hr.	Daily	Hr.	Daily	Bare Costs	Incl. O&P
1 Labor Foreman (outside)	$33.60	$268.80	$52.10	$416.80	$36.81	$56.00
1 Laborer	31.60	252.80	49.00	392.00		
4 Equip. Oper. (med.)	41.35	1323.20	62.15	1988.80		
2 Truck Drivers (heavy)	31.95	511.20	49.15	786.40		
1 Grader, 30,000 Lbs.		550.20		605.22		
1 Ripper, beam & 1 shank		79.60		87.56		
2 Stabilizers, 310 H.P.		2688.00		2956.80		
1 Flatbed Truck, Gas, 3 Ton		242.60		266.86		
1 Chem. Spreader, Towed		48.00		52.80		
1 Roller, Vibratory, 25 Ton		594.60		654.06		
1 Water Tanker, 5000 Gal.		136.00		149.60		
1 Truck Tractor, 220 H.P.		288.40		317.24	72.30	79.53
64 L.H., Daily Totals		$6983.40		$8674.14	$109.12	$135.53
Crew B-75	Hr.	Daily	Hr.	Daily	Bare Costs	Incl. O&P
1 Labor Foreman (outside)	$33.60	$268.80	$52.10	$416.80	$37.51	$56.98
1 Laborer	31.60	252.80	49.00	392.00		
4 Equip. Oper. (med.)	41.35	1323.20	62.15	1988.80		
1 Truck Driver (heavy)	31.95	255.60	49.15	393.20		
1 Grader, 30,000 Lbs.		550.20		605.22		
1 Ripper, beam & 1 shank		79.60		87.56		
2 Stabilizers, 310 H.P.		2688.00		2956.80		
1 Dist. Tanker, 3000 Gallon		304.20		334.62		
1 Truck Tractor, 6x4, 380 H.P.		478.80		526.68		
1 Roller, Vibratory, 25 Ton		594.60		654.06	83.85	92.23
56 L.H., Daily Totals		$6795.80		$8355.74	$121.35	$149.21
Crew B-76	Hr.	Daily	Hr.	Daily	Bare Costs	Incl. O&P
1 Dock Builder Foreman	$40.50	$324.00	$65.75	$526.00	$39.43	$62.38
5 Dock Builders	38.50	1540.00	62.50	2500.00		
2 Equip. Oper. (crane)	42.55	680.80	63.95	1023.20		
1 Equip. Oper. Oiler	36.80	294.40	55.30	442.40		
1 Crawler Crane, 50 Ton		1110.00		1221.00		
1 Barge, 400 Ton		746.40		821.04		
1 Hammer, diesel, 15K Ft.-Lbs.		578.60		636.46		
1 Lead, 60' high		69.60		76.56		
1 Air Compressor, 600 cfm		434.40		477.84		
2 -50' Air Hoses, 3"		32.40		35.64	41.27	45.40
72 L.H., Daily Totals		$5810.60		$7760.14	$80.70	$107.78
Crew B-76A	Hr.	Daily	Hr.	Daily	Bare Costs	Incl. O&P
1 Laborer Foreman	$33.60	$268.80	$52.10	$416.80	$33.87	$52.04
5 Laborers	31.60	1264.00	49.00	1960.00		
1 Equip. Oper. (crane)	42.55	340.40	63.95	511.60		
1 Equip. Oper. Oiler	36.80	294.40	55.30	442.40		
1 Crawler Crane, 50 Ton		1110.00		1221.00		
1 Barge, 400 Ton		746.40		821.04	29.01	31.91
64 L.H., Daily Totals		$4024.00		$5372.84	$62.88	$83.95
Crew B-77	Hr.	Daily	Hr.	Daily	Bare Costs	Incl. O&P
1 Labor Foreman	$33.60	$268.80	$52.10	$416.80	$31.87	$49.35
3 Laborers	31.60	758.40	49.00	1176.00		
1 Truck Driver (light)	30.95	247.60	47.65	381.20		
1 Crack Cleaner, 25 H.P.		54.80		60.28		
1 Crack Filler, Trailer Mtd.		194.20		213.62		
1 Flatbed Truck, Gas, 3 Ton		242.60		266.86	12.29	13.52
40 L.H., Daily Totals		$1766.40		$2514.76	$44.16	$62.87

Crew No.	Bare Costs		Incl. Subs O & P		Cost Per Labor-Hour	
Crew B-78	Hr.	Daily	Hr.	Daily	Bare Costs	Incl. O&P
1 Labor Foreman	$33.60	$268.80	$52.10	$416.80	$31.82	$49.29
4 Laborers	31.60	1011.20	49.00	1568.00		
1 Truck Driver (light)	30.95	247.60	47.65	381.20		
1 Paint Striper, S.P.		147.00		161.70		
1 Flatbed Truck, Gas, 3 Ton		242.60		266.86		
1 Pickup Truck, 3/4 Ton		116.80		128.48	10.55	11.61
48 L.H., Daily Totals		$2034.00		$2923.04	$42.38	$60.90
Crew B-78A	Hr.	Daily	Hr.	Daily	Bare Costs	Incl. O&P
1 Equip. Oper. (light)	$39.05	$312.40	$58.70	$469.60	$39.05	$58.70
1 Line Rem., (metal balls) 115 H.P.		828.60		911.46	103.58	113.93
8 L.H., Daily Totals		$1141.00		$1381.06	$142.63	$172.63
Crew B-78B	Hr.	Daily	Hr.	Daily	Bare Costs	Incl. O&P
2 Laborers	$31.60	$505.60	$49.00	$784.00	$32.43	$50.08
.25 Equip. Oper. (light)	39.05	78.10	58.70	117.40		
1 Pickup Truck, 3/4 Ton		116.80		128.48		
1 Line Rem., 11 H.P., walk behind		54.00		59.40		
.25 Road Sweeper, S.P., 8' wide		150.40		165.44	17.84	19.63
18 L.H., Daily Totals		$904.90		$1254.72	$50.27	$69.71
Crew B-79	Hr.	Daily	Hr.	Daily	Bare Costs	Incl. O&P
1 Labor Foreman	$33.60	$268.80	$52.10	$416.80	$31.87	$49.35
3 Laborers	31.60	758.40	49.00	1176.00		
1 Truck Driver (light)	30.95	247.60	47.65	381.20		
1 Thermo. Striper, T.M.		703.20		773.52		
1 Flatbed Truck, Gas, 3 Ton		242.60		266.86		
2 Pickup Trucks, 3/4 Ton		233.60		256.96	29.48	32.43
40 L.H., Daily Totals		$2454.20		$3271.34	$61.35	$81.78
Crew B-79A	Hr.	Daily	Hr.	Daily	Bare Costs	Incl. O&P
1.5 Equip. Oper. (light)	$39.05	$468.60	$58.70	$704.40	$39.05	$58.70
.5 Line Remov. (grinder) 115 H.P.		440.20		484.22		
1 Line Rem. (metal balls) 115 H.P.		828.60		911.46	105.73	116.31
12 L.H., Daily Totals		$1737.40		$2100.08	$144.78	$175.01
Crew B-80	Hr.	Daily	Hr.	Daily	Bare Costs	Incl. O&P
1 Labor Foreman	$33.60	$268.80	$52.10	$416.80	$33.80	$51.86
1 Laborer	31.60	252.80	49.00	392.00		
1 Truck Driver (light)	30.95	247.60	47.65	381.20		
1 Equip. Oper. (light)	39.05	312.40	58.70	469.60		
1 Flatbed Truck, Gas, 3 Ton		242.60		266.86		
1 Earth Auger, Truck-Mtd.		412.00		453.20	20.46	22.50
32 L.H., Daily Totals		$1736.20		$2379.66	$54.26	$74.36
Crew B-80A	Hr.	Daily	Hr.	Daily	Bare Costs	Incl. O&P
3 Laborers	$31.60	$758.40	$49.00	$1176.00	$31.60	$49.00
1 Flatbed Truck, Gas, 3 Ton		242.60		266.86	10.11	11.12
24 L.H., Daily Totals		$1001.00		$1442.86	$41.71	$60.12
Crew B-80B	Hr.	Daily	Hr.	Daily	Bare Costs	Incl. O&P
3 Laborers	$31.60	$758.40	$49.00	$1176.00	$33.46	$51.42
1 Equip. Oper. (light)	39.05	312.40	58.70	469.60		
1 Crane, Flatbed Mounted, 3 Ton		225.00		247.50	7.03	7.73
32 L.H., Daily Totals		$1295.80		$1893.10	$40.49	$59.16

Crews

Crew No.	Bare Costs		Incl. Subs O & P		Cost Per Labor-Hour	
Crew B-80C	Hr.	Daily	Hr.	Daily	Bare Costs	Incl. O&P
2 Laborers	$31.60	$505.60	$49.00	$784.00	$31.38	$48.55
1 Truck Driver (light)	30.95	247.60	47.65	381.20		
1 Flatbed Truck, Gas, 1.5 Ton		194.40		213.84		
1 Manual fence post auger, gas		7.40		8.14	8.41	9.25
24 L.H., Daily Totals		$955.00		$1387.18	$39.79	$57.80
Crew B-81	Hr.	Daily	Hr.	Daily	Bare Costs	Incl. O&P
1 Laborer	$31.60	$252.80	$49.00	$392.00	$34.97	$53.43
1 Equip. Oper. (med.)	41.35	330.80	62.15	497.20		
1 Truck Driver (heavy)	31.95	255.60	49.15	393.20		
1 Hydromulcher, T.M.		291.00		320.10		
1 Truck Tractor, 220 H.P.		288.40		317.24	24.14	26.56
24 L.H., Daily Totals		$1418.60		$1919.74	$59.11	$79.99
Crew B-82	Hr.	Daily	Hr.	Daily	Bare Costs	Incl. O&P
1 Laborer	$31.60	$252.80	$49.00	$392.00	$35.33	$53.85
1 Equip. Oper. (light)	39.05	312.40	58.70	469.60		
1 Horiz. Borer, 6 H.P.		77.60		85.36	4.85	5.34
16 L.H., Daily Totals		$642.80		$946.96	$40.17	$59.19
Crew B-82A	Hr.	Daily	Hr.	Daily	Bare Costs	Incl. O&P
1 Laborer	$31.60	$252.80	$49.00	$392.00	$35.33	$53.85
1 Equip. Oper. (light)	39.05	312.40	58.70	469.60		
1 Flatbed Truck, Gas, 3 Ton		242.60		266.86		
1 Flatbed Trailer, 25 Ton		102.00		112.20		
1 Horiz. Dir. Drill, 20k lb. thrust		628.20		691.02	60.80	66.88
16 L.H., Daily Totals		$1538.00		$1931.68	$96.13	$120.73
Crew B-82B	Hr.	Daily	Hr.	Daily	Bare Costs	Incl. O&P
2 Laborers	$31.60	$505.60	$49.00	$784.00	$34.08	$52.23
1 Equip. Oper. (light)	39.05	312.40	58.70	469.60		
1 Flatbed Truck, Gas, 3 Ton		242.60		266.86		
1 Flatbed Trailer, 25 Ton		102.00		112.20		
1 Horiz. Dir. Drill, 30k lb. thrust		896.60		986.26	51.72	56.89
24 L.H., Daily Totals		$2059.20		$2618.92	$85.80	$109.12
Crew B-82C	Hr.	Daily	Hr.	Daily	Bare Costs	Incl. O&P
2 Laborers	$31.60	$505.60	$49.00	$784.00	$34.08	$52.23
1 Equip. Oper. (light)	39.05	312.40	58.70	469.60		
1 Flatbed Truck, Gas, 3 Ton		242.60		266.86		
1 Flatbed Trailer, 25 Ton		102.00		112.20		
1 Horiz. Dir. Drill, 50k lb. thrust		1183.00		1301.30	63.65	70.02
24 L.H., Daily Totals		$2345.60		$2933.96	$97.73	$122.25
Crew B-82D	Hr.	Daily	Hr.	Daily	Bare Costs	Incl. O&P
1 Equip. Oper. (light)	$39.05	$312.40	$58.70	$469.60	$39.05	$58.70
1 Mud Trailer for HDD, 1500 gallon		283.20		311.52	35.40	38.94
8 L.H., Daily Totals		$595.60		$781.12	$74.45	$97.64
Crew B-83	Hr.	Daily	Hr.	Daily	Bare Costs	Incl. O&P
1 Tugboat Captain	$41.35	$330.80	$62.15	$497.20	$36.48	$55.58
1 Tugboat Hand	31.60	252.80	49.00	392.00		
1 Tugboat, 250 H.P.		663.40		729.74	41.46	45.61
16 L.H., Daily Totals		$1247.00		$1618.94	$77.94	$101.18
Crew B-84	Hr.	Daily	Hr.	Daily	Bare Costs	Incl. O&P
1 Equip. Oper. (med.)	$41.35	$330.80	$62.15	$497.20	$41.35	$62.15
1 Rotary Mower/Tractor		273.40		300.74	34.17	37.59
8 L.H., Daily Totals		$604.20		$797.94	$75.53	$99.74

Crew No.	Bare Costs		Incl. Subs O & P		Cost Per Labor-Hour	
Crew B-85	Hr.	Daily	Hr.	Daily	Bare Costs	Incl. O&P
3 Laborers	$31.60	$758.40	$49.00	$1176.00	$33.62	$51.66
1 Equip. Oper. (med.)	41.35	330.80	62.15	497.20		
1 Truck Driver (heavy)	31.95	255.60	49.15	393.20		
1 Aerial Lift Truck, 80'		640.40		704.44		
1 Brush Chipper, 12", 130 H.P.		225.60		248.16		
1 Pruning Saw, Rotary		6.50		7.15	21.81	23.99
40 L.H., Daily Totals		$2217.30		$3026.15	$55.43	$75.65
Crew B-86	Hr.	Daily	Hr.	Daily	Bare Costs	Incl. O&P
1 Equip. Oper. (med.)	$41.35	$330.80	$62.15	$497.20	$41.35	$62.15
1 Stump Chipper, S.P.		201.10		221.21	25.14	27.65
8 L.H., Daily Totals		$531.90		$718.41	$66.49	$89.80
Crew B-86A	Hr.	Daily	Hr.	Daily	Bare Costs	Incl. O&P
1 Equip. Oper. (medium)	$41.35	$330.80	$62.15	$497.20	$41.35	$62.15
1 Grader, 30,000 Lbs.		550.20		605.22	68.78	75.65
8 L.H., Daily Totals		$881.00		$1102.42	$110.13	$137.80
Crew B-86B	Hr.	Daily	Hr.	Daily	Bare Costs	Incl. O&P
1 Equip. Oper. (medium)	$41.35	$330.80	$62.15	$497.20	$41.35	$62.15
1 Dozer, 200 H.P.		1082.00		1190.20	135.25	148.78
8 L.H., Daily Totals		$1412.80		$1687.40	$176.60	$210.93
Crew B-87	Hr.	Daily	Hr.	Daily	Bare Costs	Incl. O&P
1 Laborer	$31.60	$252.80	$49.00	$392.00	$39.40	$59.52
4 Equip. Oper. (med.)	41.35	1323.20	62.15	1988.80		
2 Feller Bunchers, 100 H.P.		1122.80		1235.08		
1 Log Chipper, 22" Tree		588.40		647.24		
1 Dozer, 105 H.P.		600.60		660.66		
1 Chain Saw, Gas, 36" Long		37.80		41.58	58.74	64.61
40 L.H., Daily Totals		$3925.60		$4965.36	$98.14	$124.13
Crew B-88	Hr.	Daily	Hr.	Daily	Bare Costs	Incl. O&P
1 Laborer	$31.60	$252.80	$49.00	$392.00	$39.96	$60.27
6 Equip. Oper. (med.)	41.35	1984.80	62.15	2983.20		
2 Feller Bunchers, 100 H.P.		1122.80		1235.08		
1 Log Chipper, 22" Tree		588.40		647.24		
2 Log Skidders, 50 H.P.		1795.20		1974.72		
1 Dozer, 105 H.P.		600.60		660.66		
1 Chain Saw, Gas, 36" Long		37.80		41.58	74.01	81.42
56 L.H., Daily Totals		$6382.40		$7934.48	$113.97	$141.69
Crew B-89	Hr.	Daily	Hr.	Daily	Bare Costs	Incl. O&P
1 Equip. Oper. (light)	$39.05	$312.40	$58.70	$469.60	$35.00	$53.17
1 Truck Driver (light)	30.95	247.60	47.65	381.20		
1 Flatbed Truck, Gas, 3 Ton		242.60		266.86		
1 Concrete Saw		141.00		155.10		
1 Water Tank, 65 Gal.		18.45		20.30	25.13	27.64
16 L.H., Daily Totals		$962.05		$1293.06	$60.13	$80.82
Crew B-89A	Hr.	Daily	Hr.	Daily	Bare Costs	Incl. O&P
1 Skilled Worker	$40.85	$326.80	$63.25	$506.00	$36.23	$56.13
1 Laborer	31.60	252.80	49.00	392.00		
1 Core Drill (large)		110.20		121.22	6.89	7.58
16 L.H., Daily Totals		$689.80		$1019.22	$43.11	$63.70

Crews

Crew No.	Bare Costs		Incl. Subs O & P		Cost Per Labor-Hour	
Crew B-89B	Hr.	Daily	Hr.	Daily	Bare Costs	Incl. O&P
1 Equip. Oper. (light)	$39.05	$312.40	$58.70	$469.60	$35.00	$53.17
1 Truck Driver, Light	30.95	247.60	47.65	381.20		
1 Wall Saw, Hydraulic, 10 H.P.		96.00		105.60		
1 Generator, Diesel, 100 kW		327.20		359.92		
1 Water Tank, 65 Gal.		18.45		20.30		
1 Flatbed Truck, Gas, 3 Ton		242.60		266.86	42.77	47.04
16 L.H., Daily Totals		$1244.25		$1603.47	$77.77	$100.22

Crew No.	Bare Costs		Incl. Subs O & P		Cost Per Labor-Hour	
Crew B-90	Hr.	Daily	Hr.	Daily	Bare Costs	Incl. O&P
1 Labor Foreman (outside)	$33.60	$268.80	$52.10	$416.80	$33.80	$51.85
3 Laborers	31.60	758.40	49.00	1176.00		
2 Equip. Oper. (light)	39.05	624.80	58.70	939.20		
2 Truck Drivers (heavy)	31.95	511.20	49.15	786.40		
1 Road Mixer, 310 H.P.		1896.00		2085.60		
1 Dist. Truck, 2000 Gal.		277.20		304.92	33.96	37.35
64 L.H., Daily Totals		$4336.40		$5708.92	$67.76	$89.20

Crew B-90A	Hr.	Daily	Hr.	Daily	Bare Costs	Incl. O&P
1 Labor Foreman	$33.60	$268.80	$52.10	$416.80	$37.46	$56.96
2 Laborers	31.60	505.60	49.00	784.00		
4 Equip. Oper. (medium)	41.35	1323.20	62.15	1988.80		
2 Graders, 30,000 Lbs.		1100.40		1210.44		
1 Tandem Roller, 10 Ton		233.00		256.30		
1 Roller, Pneum. Whl, 12 Ton		307.80		338.58	29.31	32.24
56 L.H., Daily Totals		$3738.80		$4994.92	$66.76	$89.19

Crew B-90B	Hr.	Daily	Hr.	Daily	Bare Costs	Incl. O&P
1 Labor Foreman	$33.60	$268.80	$52.10	$416.80	$36.81	$56.09
2 Laborers	31.60	505.60	49.00	784.00		
3 Equip. Oper. (medium)	41.35	992.40	62.15	1491.60		
1 Tandem Roller, 10 Ton		233.00		256.30		
1 Roller, Pneum. Whl, 12 Ton		307.80		338.58		
1 Road Mixer, 310 H.P.		1896.00		2085.60	50.77	55.84
48 L.H., Daily Totals		$4203.60		$5372.88	$87.58	$111.94

Crew B-91	Hr.	Daily	Hr.	Daily	Bare Costs	Incl. O&P
1 Labor Foreman (outside)	$33.60	$268.80	$52.10	$416.80	$36.77	$55.98
2 Laborers	31.60	505.60	49.00	784.00		
4 Equip. Oper. (med.)	41.35	1323.20	62.15	1988.80		
1 Truck Driver (heavy)	31.95	255.60	49.15	393.20		
1 Dist. Tanker, 3000 Gallon		304.20		334.62		
1 Truck Tractor, 6x4, 380 H.P.		478.80		526.68		
1 Aggreg. Spreader, S.P.		855.20		940.72		
1 Roller, Pneum. Whl, 12 Ton		307.80		338.58		
1 Tandem Roller, 10 Ton		233.00		256.30	34.05	37.45
64 L.H., Daily Totals		$4532.20		$5979.70	$70.82	$93.43

Crew B-92	Hr.	Daily	Hr.	Daily	Bare Costs	Incl. O&P
1 Labor Foreman (outside)	$33.60	$268.80	$52.10	$416.80	$32.10	$49.77
3 Laborers	31.60	758.40	49.00	1176.00		
1 Crack Cleaner, 25 H.P.		54.80		60.28		
1 Air Compressor, 60 cfm		122.60		134.86		
1 Tar Kettle, T.M.		83.15		91.47		
1 Flatbed Truck, Gas, 3 Ton		242.60		266.86	15.72	17.30
32 L.H., Daily Totals		$1530.35		$2146.26	$47.82	$67.07

Crew B-93	Hr.	Daily	Hr.	Daily	Bare Costs	Incl. O&P
1 Equip. Oper. (med.)	$41.35	$330.80	$62.15	$497.20	$41.35	$62.15
1 Feller Buncher, 100 H.P.		561.40		617.54	70.17	77.19
8 L.H., Daily Totals		$892.20		$1114.74	$111.53	$139.34

Crew No.	Bare Costs		Incl. Subs O & P		Cost Per Labor-Hour	
Crew B-94A	Hr.	Daily	Hr.	Daily	Bare Costs	Incl. O&P
1 Laborer	$31.60	$252.80	$49.00	$392.00	$31.60	$49.00
1 Diaphragm Water Pump, 2"		63.20		69.52		
1 -20' Suction Hose, 2"		1.95		2.15		
2 -50' Discharge Hoses, 2"		2.20		2.42	8.42	9.26
8 L.H., Daily Totals		$320.15		$466.08	$40.02	$58.26

Crew B-94B	Hr.	Daily	Hr.	Daily	Bare Costs	Incl. O&P
1 Laborer	$31.60	$252.80	$49.00	$392.00	$31.60	$49.00
1 Diaphragm Water Pump, 4"		86.20		94.82		
1 -20' Suction Hose, 4"		3.45		3.79		
2 -50' Discharge Hoses, 4"		4.70		5.17	11.79	12.97
8 L.H., Daily Totals		$347.15		$495.79	$43.39	$61.97

Crew B-94C	Hr.	Daily	Hr.	Daily	Bare Costs	Incl. O&P
1 Laborer	$31.60	$252.80	$49.00	$392.00	$31.60	$49.00
1 Centrifugal Water Pump, 3"		69.40		76.34		
1 -20' Suction Hose, 3"		3.05		3.36		
2 -50' Discharge Hoses, 3"		3.50		3.85	9.49	10.44
8 L.H., Daily Totals		$328.75		$475.55	$41.09	$59.44

Crew B-94D	Hr.	Daily	Hr.	Daily	Bare Costs	Incl. O&P
1 Laborer	$31.60	$252.80	$49.00	$392.00	$31.60	$49.00
1 Centr. Water Pump, 6"		309.40		340.34		
1 -20' Suction Hose, 6"		11.90		13.09		
2 -50' Discharge Hoses, 6"		12.60		13.86	41.74	45.91
8 L.H., Daily Totals		$586.70		$759.29	$73.34	$94.91

Crew C-1	Hr.	Daily	Hr.	Daily	Bare Costs	Incl. O&P
3 Carpenters	$39.95	$958.80	$61.95	$1486.80	$37.86	$58.71
1 Laborer	31.60	252.80	49.00	392.00		
32 L.H., Daily Totals		$1211.60		$1878.80	$37.86	$58.71

Crew C-2	Hr.	Daily	Hr.	Daily	Bare Costs	Incl. O&P
1 Carpenter Foreman (out)	$41.95	$335.60	$65.05	$520.40	$38.89	$60.31
4 Carpenters	39.95	1278.40	61.95	1982.40		
1 Laborer	31.60	252.80	49.00	392.00		
48 L.H., Daily Totals		$1866.80		$2894.80	$38.89	$60.31

Crew C-2A	Hr.	Daily	Hr.	Daily	Bare Costs	Incl. O&P
1 Carpenter Foreman (out)	$41.95	$335.60	$65.05	$520.40	$38.62	$59.33
3 Carpenters	39.95	958.80	61.95	1486.80		
1 Cement Finisher	38.30	306.40	56.05	448.40		
1 Laborer	31.60	252.80	49.00	392.00		
48 L.H., Daily Totals		$1853.60		$2847.60	$38.62	$59.33

Crew C-3	Hr.	Daily	Hr.	Daily	Bare Costs	Incl. O&P
1 Rodman Foreman	$46.55	$372.40	$76.10	$608.80	$40.88	$65.53
4 Rodmen (reinf.)	44.55	1425.60	72.85	2331.20		
1 Equip. Oper. (light)	39.05	312.40	58.70	469.60		
2 Laborers	31.60	505.60	49.00	784.00		
3 Stressing Equipment		27.60		30.36		
.5 Grouting Equipment		75.80		83.38	1.62	1.78
64 L.H., Daily Totals		$2719.40		$4307.34	$42.49	$67.30

Crew C-4	Hr.	Daily	Hr.	Daily	Bare Costs	Incl. O&P
1 Rodman Foreman	$46.55	$372.40	$76.10	$608.80	$45.05	$73.66
3 Rodmen (reinf.)	44.55	1069.20	72.85	1748.40		
3 Stressing Equipment		27.60		30.36	.86	.95
32 L.H., Daily Totals		$1469.20		$2387.56	$45.91	$74.61

Crews

Crew No.	Bare Costs		Incl. Subs O & P		Cost Per Labor-Hour	
Crew C-5	**Hr.**	**Daily**	**Hr.**	**Daily**	**Bare Costs**	**Incl. O&P**
1 Rodman Foreman	$46.55	$372.40	$76.10	$608.80	$43.44	$69.54
4 Rodmen (reinf.)	44.55	1425.60	72.85	2331.20		
1 Equip. Oper. (crane)	42.55	340.40	63.95	511.60		
1 Equip. Oper. Oiler	36.80	294.40	55.30	442.40		
1 Hyd. Crane, 25 Ton		789.40		868.34	14.10	15.51
56 L.H., Daily Totals		$3222.20		$4762.34	$57.54	$85.04
Crew C-6	**Hr.**	**Daily**	**Hr.**	**Daily**	**Bare Costs**	**Incl. O&P**
1 Labor Foreman (outside)	$33.60	$268.80	$52.10	$416.80	$33.05	$50.69
4 Laborers	31.60	1011.20	49.00	1568.00		
1 Cement Finisher	38.30	306.40	56.05	448.40		
2 Gas Engine Vibrators		52.00		57.20	1.08	1.19
48 L.H., Daily Totals		$1638.40		$2490.40	$34.13	$51.88
Crew C-7	**Hr.**	**Daily**	**Hr.**	**Daily**	**Bare Costs**	**Incl. O&P**
1 Labor Foreman (outside)	$33.60	$268.80	$52.10	$416.80	$34.23	$52.29
5 Laborers	31.60	1264.00	49.00	1960.00		
1 Cement Finisher	38.30	306.40	56.05	448.40		
1 Equip. Oper. (med.)	41.35	330.80	62.15	497.20		
1 Equip. Oper. (oiler)	36.80	294.40	55.30	442.40		
2 Gas Engine Vibrators		52.00		57.20		
1 Concrete Bucket, 1 C.Y.		20.20		22.22		
1 Hyd. Crane, 55 Ton		1125.00		1237.50	16.63	18.29
72 L.H., Daily Totals		$3661.60		$5081.72	$50.86	$70.58
Crew C-7A	**Hr.**	**Daily**	**Hr.**	**Daily**	**Bare Costs**	**Incl. O&P**
1 Labor Foreman (outside)	$33.60	$268.80	$52.10	$416.80	$31.94	$49.42
5 Laborers	31.60	1264.00	49.00	1960.00		
2 Truck Drivers (Heavy)	31.95	511.20	49.15	786.40		
2 Conc. Transit Mixers		1772.80		1950.08	27.70	30.47
64 L.H., Daily Totals		$3816.80		$5113.28	$59.64	$79.89
Crew C-7B	**Hr.**	**Daily**	**Hr.**	**Daily**	**Bare Costs**	**Incl. O&P**
1 Labor Foreman (outside)	$33.60	$268.80	$52.10	$416.80	$33.87	$52.04
5 Laborers	31.60	1264.00	49.00	1960.00		
1 Equipment Oper. (crane)	42.55	340.40	63.95	511.60		
1 Equipment Oiler	36.80	294.40	55.30	442.40		
1 Conc. Bucket, 2 C.Y.		32.00		35.20		
1 Lattice Boom Crane, 165 Ton		2240.00		2464.00	35.50	39.05
64 L.H., Daily Totals		$4439.60		$5830.00	$69.37	$91.09
Crew C-7C	**Hr.**	**Daily**	**Hr.**	**Daily**	**Bare Costs**	**Incl. O&P**
1 Labor Foreman (outside)	$33.60	$268.80	$52.10	$416.80	$34.29	$52.67
5 Laborers	31.60	1264.00	49.00	1960.00		
2 Equipment Operators (medium)	41.35	661.60	62.15	994.40		
2 F.E. Loaders, W.M., 4 C.Y.		1082.00		1190.20	16.91	18.60
64 L.H., Daily Totals		$3276.40		$4561.40	$51.19	$71.27
Crew C-7D	**Hr.**	**Daily**	**Hr.**	**Daily**	**Bare Costs**	**Incl. O&P**
1 Labor Foreman (outside)	$33.60	$268.80	$52.10	$416.80	$33.28	$51.32
5 Laborers	31.60	1264.00	49.00	1960.00		
1 Equip. Oper. (med.)	41.35	330.80	62.15	497.20		
1 Concrete Conveyer		179.00		196.90	3.20	3.52
56 L.H., Daily Totals		$2042.60		$3070.90	$36.48	$54.84

Crew No.	Bare Costs		Incl. Subs O & P		Cost Per Labor-Hour	
Crew C-8	**Hr.**	**Daily**	**Hr.**	**Daily**	**Bare Costs**	**Incl. O&P**
1 Labor Foreman (outside)	$33.60	$268.80	$52.10	$416.80	$35.19	$53.34
3 Laborers	31.60	758.40	49.00	1176.00		
2 Cement Finishers	38.30	612.80	56.05	896.80		
1 Equip. Oper. (med.)	41.35	330.80	62.15	497.20		
1 Concrete Pump (small)		737.80		811.58	13.18	14.49
56 L.H., Daily Totals		$2708.60		$3798.38	$48.37	$67.83
Crew C-8A	**Hr.**	**Daily**	**Hr.**	**Daily**	**Bare Costs**	**Incl. O&P**
1 Labor Foreman (outside)	$33.60	$268.80	$52.10	$416.80	$34.17	$51.87
3 Laborers	31.60	758.40	49.00	1176.00		
2 Cement Finishers	38.30	612.80	56.05	896.80		
48 L.H., Daily Totals		$1640.00		$2489.60	$34.17	$51.87
Crew C-8B	**Hr.**	**Daily**	**Hr.**	**Daily**	**Bare Costs**	**Incl. O&P**
1 Labor Foreman (outside)	$33.60	$268.80	$52.10	$416.80	$33.95	$52.25
3 Laborers	31.60	758.40	49.00	1176.00		
1 Equip. Oper. (med.)	41.35	330.80	62.15	497.20		
1 Vibrating Power Screed		70.35		77.39		
1 Roller, Vibratory, 25 Ton		594.60		654.06		
1 Dozer, 200 H.P.		1082.00		1190.20	43.67	48.04
40 L.H., Daily Totals		$3104.95		$4011.65	$77.62	$100.29
Crew C-8C	**Hr.**	**Daily**	**Hr.**	**Daily**	**Bare Costs**	**Incl. O&P**
1 Labor Foreman (outside)	$33.60	$268.80	$52.10	$416.80	$34.67	$52.88
3 Laborers	31.60	758.40	49.00	1176.00		
1 Cement Finisher	38.30	306.40	56.05	448.40		
1 Equip. Oper. (med.)	41.35	330.80	62.15	497.20		
1 Shotcrete Rig, 12 C.Y./Hr		237.80		261.58	4.95	5.45
48 L.H., Daily Totals		$1902.20		$2799.98	$39.63	$58.33
Crew C-8D	**Hr.**	**Daily**	**Hr.**	**Daily**	**Bare Costs**	**Incl. O&P**
1 Labor Foreman (outside)	$33.60	$268.80	$52.10	$416.80	$35.64	$53.96
1 Laborer	31.60	252.80	49.00	392.00		
1 Cement Finisher	38.30	306.40	56.05	448.40		
1 Equipment Oper. (light)	39.05	312.40	58.70	469.60		
1 Air Compressor, 250 cfm		162.40		178.64		
2 -50' Air Hoses, 1"		8.20		9.02	5.33	5.86
32 L.H., Daily Totals		$1311.00		$1914.46	$40.97	$59.83
Crew C-8E	**Hr.**	**Daily**	**Hr.**	**Daily**	**Bare Costs**	**Incl. O&P**
1 Labor Foreman (outside)	$33.60	$268.80	$52.10	$416.80	$35.64	$53.96
1 Laborer	31.60	252.80	49.00	392.00		
1 Cement Finisher	38.30	306.40	56.05	448.40		
1 Equipment Oper. (light)	39.05	312.40	58.70	469.60		
1 Air Compressor, 250 cfm		162.40		178.64		
2 -50' Air Hoses, 1"		8.20		9.02		
1 Concrete Pump (small)		737.80		811.58	28.39	31.23
32 L.H., Daily Totals		$2048.80		$2726.04	$64.03	$85.19
Crew C-10	**Hr.**	**Daily**	**Hr.**	**Daily**	**Bare Costs**	**Incl. O&P**
1 Laborer	$31.60	$252.80	$49.00	$392.00	$36.07	$53.70
2 Cement Finishers	38.30	612.80	56.05	896.80		
24 L.H., Daily Totals		$865.60		$1288.80	$36.07	$53.70
Crew C-10B	**Hr.**	**Daily**	**Hr.**	**Daily**	**Bare Costs**	**Incl. O&P**
3 Laborers	$31.60	$758.40	$49.00	$1176.00	$34.28	$51.82
2 Cement Finishers	38.30	612.80	56.05	896.80		
1 Concrete Mixer, 10 C.F.		155.20		170.72		
2 Trowels, 48" Walk-Behind		84.80		93.28	6.00	6.60
40 L.H., Daily Totals		$1611.20		$2336.80	$40.28	$58.42

Crews

Crew No.	Bare Costs		Incl. Subs O & P		Cost Per Labor-Hour	
Crew C-10C	Hr.	Daily	Hr.	Daily	Bare Costs	Incl. O&P
1 Laborer	$31.60	$252.80	$49.00	$392.00	$36.07	$53.70
2 Cement Finishers	38.30	612.80	56.05	896.80		
1 Trowel, 48" Walk-Behind		42.40		46.64	1.77	1.94
24 L.H., Daily Totals		$908.00		$1335.44	$37.83	$55.64
Crew C-10D	Hr.	Daily	Hr.	Daily	Bare Costs	Incl. O&P
1 Laborer	$31.60	$252.80	$49.00	$392.00	$36.07	$53.70
2 Cement Finishers	38.30	612.80	56.05	896.80		
1 Vibrating Power Screed		70.35		77.39		
1 Trowel, 48" Walk-Behind		42.40		46.64	4.70	5.17
24 L.H., Daily Totals		$978.35		$1412.83	$40.76	$58.87
Crew C-10E	Hr.	Daily	Hr.	Daily	Bare Costs	Incl. O&P
1 Laborer	$31.60	$252.80	$49.00	$392.00	$36.07	$53.70
2 Cement Finishers	38.30	612.80	56.05	896.80		
1 Vibrating Power Screed		70.35		77.39		
1 Cement Trowel, 96" Ride-On		166.60		183.26	9.87	10.86
24 L.H., Daily Totals		$1102.55		$1549.44	$45.94	$64.56
Crew C-11	Hr.	Daily	Hr.	Daily	Bare Costs	Incl. O&P
1 Struc. Steel Foreman	$46.70	$373.60	$83.20	$665.60	$43.81	$75.59
6 Struc. Steel Workers	44.70	2145.60	79.65	3823.20		
1 Equip. Oper. (crane)	42.55	340.40	63.95	511.60		
1 Equip. Oper. Oiler	36.80	294.40	55.30	442.40		
1 Lattice Boom Crane, 150 Ton		1890.00		2079.00	26.25	28.88
72 L.H., Daily Totals		$5044.00		$7521.80	$70.06	$104.47
Crew C-12	Hr.	Daily	Hr.	Daily	Bare Costs	Incl. O&P
1 Carpenter Foreman (out)	$41.95	$335.60	$65.05	$520.40	$39.33	$60.64
3 Carpenters	39.95	958.80	61.95	1486.80		
1 Laborer	31.60	252.80	49.00	392.00		
1 Equip. Oper. (crane)	42.55	340.40	63.95	511.60		
1 Hyd. Crane, 12 Ton		768.80		845.68	16.02	17.62
48 L.H., Daily Totals		$2656.40		$3756.48	$55.34	$78.26
Crew C-13	Hr.	Daily	Hr.	Daily	Bare Costs	Incl. O&P
1 Struc. Steel Worker	$44.70	$357.60	$79.65	$637.20	$43.12	$73.75
1 Welder	44.70	357.60	79.65	637.20		
1 Carpenter	39.95	319.60	61.95	495.60		
1 Welder, gas engine, 300 amp		134.20		147.62	5.59	6.15
24 L.H., Daily Totals		$1169.00		$1917.62	$48.71	$79.90
Crew C-14	Hr.	Daily	Hr.	Daily	Bare Costs	Incl. O&P
1 Carpenter Foreman (out)	$41.95	$335.60	$65.05	$520.40	$39.01	$60.75
5 Carpenters	39.95	1598.00	61.95	2478.00		
4 Laborers	31.60	1011.20	49.00	1568.00		
4 Rodmen (reinf.)	44.55	1425.60	72.85	2331.20		
2 Cement Finishers	38.30	612.80	56.05	896.80		
1 Equip. Oper. (crane)	42.55	340.40	63.95	511.60		
1 Equip. Oper. Oiler	36.80	294.40	55.30	442.40		
1 Hyd. Crane, 80 Ton		1296.00		1425.60	9.00	9.90
144 L.H., Daily Totals		$6914.00		$10174.00	$48.01	$70.65

Crew No.	Bare Costs		Incl. Subs O & P		Cost Per Labor-Hour	
Crew C-14A	Hr.	Daily	Hr.	Daily	Bare Costs	Incl. O&P
1 Carpenter Foreman (out)	$41.95	$335.60	$65.05	$520.40	$40.09	$62.55
16 Carpenters	39.95	5113.60	61.95	7929.60		
4 Rodmen (reinf.)	44.55	1425.60	72.85	2331.20		
2 Laborers	31.60	505.60	49.00	784.00		
1 Cement Finisher	38.30	306.40	56.05	448.40		
1 Equip. Oper. (med.)	41.35	330.80	62.15	497.20		
1 Gas Engine Vibrator		26.00		28.60		
1 Concrete Pump (small)		737.80		811.58	3.82	4.20
200 L.H., Daily Totals		$8781.40		$13350.98	$43.91	$66.75
Crew C-14B	Hr.	Daily	Hr.	Daily	Bare Costs	Incl. O&P
1 Carpenter Foreman (out)	$41.95	$335.60	$65.05	$520.40	$40.02	$62.30
16 Carpenters	39.95	5113.60	61.95	7929.60		
4 Rodmen (reinf.)	44.55	1425.60	72.85	2331.20		
2 Laborers	31.60	505.60	49.00	784.00		
2 Cement Finishers	38.30	612.80	56.05	896.80		
1 Equip. Oper. (med.)	41.35	330.80	62.15	497.20		
1 Gas Engine Vibrator		26.00		28.60		
1 Concrete Pump (small)		737.80		811.58	3.67	4.04
208 L.H., Daily Totals		$9087.80		$13799.38	$43.69	$66.34
Crew C-14C	Hr.	Daily	Hr.	Daily	Bare Costs	Incl. O&P
1 Carpenter Foreman (out)	$41.95	$335.60	$65.05	$520.40	$38.25	$59.61
6 Carpenters	39.95	1917.60	61.95	2973.60		
2 Rodmen (reinf.)	44.55	712.80	72.85	1165.60		
4 Laborers	31.60	1011.20	49.00	1568.00		
1 Cement Finisher	38.30	306.40	56.05	448.40		
1 Gas Engine Vibrator		26.00		28.60	.23	.26
112 L.H., Daily Totals		$4309.60		$6704.60	$38.48	$59.86
Crew C-14D	Hr.	Daily	Hr.	Daily	Bare Costs	Incl. O&P
1 Carpenter Foreman (out)	$41.95	$335.60	$65.05	$520.40	$39.72	$61.68
18 Carpenters	39.95	5752.80	61.95	8920.80		
2 Rodmen (reinf.)	44.55	712.80	72.85	1165.60		
2 Laborers	31.60	505.60	49.00	784.00		
1 Cement Finisher	38.30	306.40	56.05	448.40		
1 Equip. Oper. (med.)	41.35	330.80	62.15	497.20		
1 Gas Engine Vibrator		26.00		28.60		
1 Concrete Pump (small)		737.80		811.58	3.82	4.20
200 L.H., Daily Totals		$8707.80		$13176.58	$43.54	$65.88
Crew C-14E	Hr.	Daily	Hr.	Daily	Bare Costs	Incl. O&P
1 Carpenter Foreman (out)	$41.95	$335.60	$65.05	$520.40	$39.38	$62.13
2 Carpenters	39.95	639.20	61.95	991.20		
4 Rodmen (reinf.)	44.55	1425.60	72.85	2331.20		
3 Laborers	31.60	758.40	49.00	1176.00		
1 Cement Finisher	38.30	306.40	56.05	448.40		
1 Gas Engine Vibrator		26.00		28.60	.30	.33
88 L.H., Daily Totals		$3491.20		$5495.80	$39.67	$62.45
Crew C-14F	Hr.	Daily	Hr.	Daily	Bare Costs	Incl. O&P
1 Laborer Foreman (out)	$33.60	$268.80	$52.10	$416.80	$36.29	$54.04
2 Laborers	31.60	505.60	49.00	784.00		
6 Cement Finishers	38.30	1838.40	56.05	2690.40		
1 Gas Engine Vibrator		26.00		28.60	.36	.40
72 L.H., Daily Totals		$2638.80		$3919.80	$36.65	$54.44

Crews

Crew No.	Bare Costs		Incl. Subs O & P		Cost Per Labor-Hour	
Crew C-14G	Hr.	Daily	Hr.	Daily	Bare Costs	Incl. O&P
1 Laborer Foreman (out)	$33.60	$268.80	$52.10	$416.80	$35.71	$53.47
2 Laborers	31.60	505.60	49.00	784.00		
4 Cement Finishers	38.30	1225.60	56.05	1793.60		
1 Gas Engine Vibrator		26.00		28.60	.46	.51
56 L.H., Daily Totals		$2026.00		$3023.00	$36.18	$53.98

Crew C-14H	Hr.	Daily	Hr.	Daily	Bare Costs	Incl. O&P
1 Carpenter Foreman (out)	$41.95	$335.60	$65.05	$520.40	$39.38	$61.14
2 Carpenters	39.95	639.20	61.95	991.20		
1 Rodman (reinf.)	44.55	356.40	72.85	582.80		
1 Laborer	31.60	252.80	49.00	392.00		
1 Cement Finisher	38.30	306.40	56.05	448.40		
1 Gas Engine Vibrator		26.00		28.60	.54	.60
48 L.H., Daily Totals		$1916.40		$2963.40	$39.92	$61.74

Crew C-14L	Hr.	Daily	Hr.	Daily	Bare Costs	Incl. O&P
1 Carpenter Foreman (out)	$41.95	$335.60	$65.05	$520.40	$37.20	$57.40
6 Carpenters	39.95	1917.60	61.95	2973.60		
4 Laborers	31.60	1011.20	49.00	1568.00		
1 Cement Finisher	38.30	306.40	56.05	448.40		
1 Gas Engine Vibrator		26.00		28.60	.27	.30
96 L.H., Daily Totals		$3596.80		$5539.00	$37.47	$57.70

Crew C-15	Hr.	Daily	Hr.	Daily	Bare Costs	Incl. O&P
1 Carpenter Foreman (out)	$41.95	$335.60	$65.05	$520.40	$37.53	$57.88
2 Carpenters	39.95	639.20	61.95	991.20		
3 Laborers	31.60	758.40	49.00	1176.00		
2 Cement Finishers	38.30	612.80	56.05	896.80		
1 Rodman (reinf.)	44.55	356.40	72.85	582.80		
72 L.H., Daily Totals		$2702.40		$4167.20	$37.53	$57.88

Crew C-16	Hr.	Daily	Hr.	Daily	Bare Costs	Incl. O&P
1 Labor Foreman (outside)	$33.60	$268.80	$52.10	$416.80	$37.27	$57.67
3 Laborers	31.60	758.40	49.00	1176.00		
2 Cement Finishers	38.30	612.80	56.05	896.80		
1 Equip. Oper. (med.)	41.35	330.80	62.15	497.20		
2 Rodmen (reinf.)	44.55	712.80	72.85	1165.60		
1 Concrete Pump (small)		737.80		811.58	10.25	11.27
72 L.H., Daily Totals		$3421.40		$4963.98	$47.52	$68.94

Crew C-17	Hr.	Daily	Hr.	Daily	Bare Costs	Incl. O&P
2 Skilled Worker Foremen	$42.85	$685.60	$66.35	$1061.60	$41.25	$63.87
8 Skilled Workers	40.85	2614.40	63.25	4048.00		
80 L.H., Daily Totals		$3300.00		$5109.60	$41.25	$63.87

Crew C-17A	Hr.	Daily	Hr.	Daily	Bare Costs	Incl. O&P
2 Skilled Worker Foremen	$42.85	$685.60	$66.35	$1061.60	$41.27	$63.87
8 Skilled Workers	40.85	2614.40	63.25	4048.00		
.125 Equip. Oper. (crane)	42.55	42.55	63.95	63.95		
.125 Hyd. Crane, 80 Ton		162.00		178.20	2.00	2.20
81 L.H., Daily Totals		$3504.55		$5351.75	$43.27	$66.07

Crew C-17B	Hr.	Daily	Hr.	Daily	Bare Costs	Incl. O&P
2 Skilled Worker Foremen	$42.85	$685.60	$66.35	$1061.60	$41.28	$63.87
8 Skilled Workers	40.85	2614.40	63.25	4048.00		
.25 Equip. Oper. (crane)	42.55	85.10	63.95	127.90		
.25 Hyd. Crane, 80 Ton		324.00		356.40		
.25 Trowel, 48" Walk-Behind		10.60		11.66	4.08	4.49
82 L.H., Daily Totals		$3719.70		$5605.56	$45.36	$68.36

Crew No.	Bare Costs		Incl. Subs O & P		Cost Per Labor-Hour	
Crew C-17C	Hr.	Daily	Hr.	Daily	Bare Costs	Incl. O&P
2 Skilled Worker Foremen	$42.85	$685.60	$66.35	$1061.60	$41.30	$63.87
8 Skilled Workers	40.85	2614.40	63.25	4048.00		
.375 Equip. Oper. (crane)	42.55	127.65	63.95	191.85		
.375 Hyd. Crane, 80 Ton		486.00		534.60	5.86	6.44
83 L.H., Daily Totals		$3913.65		$5836.05	$47.15	$70.31

Crew C-17D	Hr.	Daily	Hr.	Daily	Bare Costs	Incl. O&P
2 Skilled Worker Foremen	$42.85	$685.60	$66.35	$1061.60	$41.31	$63.87
8 Skilled Workers	40.85	2614.40	63.25	4048.00		
.5 Equip. Oper. (crane)	42.55	170.20	63.95	255.80		
.5 Hyd. Crane, 80 Ton		648.00		712.80	7.71	8.49
84 L.H., Daily Totals		$4118.20		$6078.20	$49.03	$72.36

Crew C-17E	Hr.	Daily	Hr.	Daily	Bare Costs	Incl. O&P
2 Skilled Worker Foremen	$42.85	$685.60	$66.35	$1061.60	$41.25	$63.87
8 Skilled Workers	40.85	2614.40	63.25	4048.00		
1 Hyd. Jack with Rods		85.70		94.27	1.07	1.18
80 L.H., Daily Totals		$3385.70		$5203.87	$42.32	$65.05

Crew C-18	Hr.	Daily	Hr.	Daily	Bare Costs	Incl. O&P
.125 Labor Foreman (out)	$33.60	$33.60	$52.10	$52.10	$31.82	$49.34
1 Laborer	31.60	252.80	49.00	392.00		
1 Concrete Cart, 10 C.F.		55.20		60.72	6.13	6.75
9 L.H., Daily Totals		$341.60		$504.82	$37.96	$56.09

Crew C-19	Hr.	Daily	Hr.	Daily	Bare Costs	Incl. O&P
.125 Labor Foreman (out)	$33.60	$33.60	$52.10	$52.10	$31.82	$49.34
1 Laborer	31.60	252.80	49.00	392.00		
1 Concrete Cart, 18 C.F.		86.40		95.04	9.60	10.56
9 L.H., Daily Totals		$372.80		$539.14	$41.42	$59.90

Crew C-20	Hr.	Daily	Hr.	Daily	Bare Costs	Incl. O&P
1 Labor Foreman (outside)	$33.60	$268.80	$52.10	$416.80	$33.91	$51.91
5 Laborers	31.60	1264.00	49.00	1960.00		
1 Cement Finisher	38.30	306.40	56.05	448.40		
1 Equip. Oper. (med.)	41.35	330.80	62.15	497.20		
2 Gas Engine Vibrators		52.00		57.20		
1 Concrete Pump (small)		737.80		811.58	12.34	13.57
64 L.H., Daily Totals		$2959.80		$4191.18	$46.25	$65.49

Crew C-21	Hr.	Daily	Hr.	Daily	Bare Costs	Incl. O&P
1 Labor Foreman (outside)	$33.60	$268.80	$52.10	$416.80	$33.91	$51.91
5 Laborers	31.60	1264.00	49.00	1960.00		
1 Cement Finisher	38.30	306.40	56.05	448.40		
1 Equip. Oper. (med.)	41.35	330.80	62.15	497.20		
2 Gas Engine Vibrators		52.00		57.20		
1 Concrete Conveyer		179.00		196.90	3.61	3.97
64 L.H., Daily Totals		$2401.00		$3576.50	$37.52	$55.88

Crew C-22	Hr.	Daily	Hr.	Daily	Bare Costs	Incl. O&P
1 Rodman Foreman	$46.55	$372.40	$76.10	$608.80	$44.70	$72.84
4 Rodmen (reinf.)	44.55	1425.60	72.85	2331.20		
.125 Equip. Oper. (crane)	42.55	42.55	63.95	63.95		
.125 Equip. Oper. Oiler	36.80	36.80	55.30	55.30		
.125 Hyd. Crane, 25 Ton		98.67		108.54	2.35	2.58
42 L.H., Daily Totals		$1976.03		$3167.79	$47.05	$75.42

Crews

Crew No.		Bare Costs		Incl. Subs O & P		Cost Per Labor-Hour	
Crew C-23	Hr.	Daily	Hr.	Daily	Bare Costs	Incl. O&P	
2 Skilled Worker Foremen	$42.85	$685.60	$66.35	$1061.60	$41.02	$63.15	
6 Skilled Workers	40.85	1960.80	63.25	3036.00			
1 Equip. Oper. (crane)	42.55	340.40	63.95	511.60			
1 Equip. Oper. Oiler	36.80	294.40	55.30	442.40			
1 Lattice Boom Crane, 90 Ton		1741.00		1915.10	21.76	23.94	
80 L.H., Daily Totals		$5022.20		$6966.70	$62.78	$87.08	

Crew C-23A	Hr.	Daily	Hr.	Daily	Bare Costs	Incl. O&P
1 Labor Foreman (outside)	$33.60	$268.80	$52.10	$416.80	$35.23	$53.87
2 Laborers	31.60	505.60	49.00	784.00		
1 Equip. Oper. (crane)	42.55	340.40	63.95	511.60		
1 Equip. Oper. Oiler	36.80	294.40	55.30	442.40		
1 Crawler Crane, 100 Ton		1611.00		1772.10		
3 Conc. Buckets, 8 C.Y.		603.00		663.30	55.35	60.88
40 L.H., Daily Totals		$3623.20		$4590.20	$90.58	$114.76

Crew C-24	Hr.	Daily	Hr.	Daily	Bare Costs	Incl. O&P
2 Skilled Worker Foremen	$42.85	$685.60	$66.35	$1061.60	$41.02	$63.15
6 Skilled Workers	40.85	1960.80	63.25	3036.00		
1 Equip. Oper. (crane)	42.55	340.40	63.95	511.60		
1 Equip. Oper. Oiler	36.80	294.40	55.30	442.40		
1 Lattice Boom Crane, 150 Ton		1890.00		2079.00	23.63	25.99
80 L.H., Daily Totals		$5171.20		$7130.60	$64.64	$89.13

Crew C-25	Hr.	Daily	Hr.	Daily	Bare Costs	Incl. O&P
2 Rodmen (reinf.)	$44.55	$712.80	$72.85	$1165.60	$34.95	$57.77
2 Rodmen Helpers	25.35	405.60	42.70	683.20		
32 L.H., Daily Totals		$1118.40		$1848.80	$34.95	$57.77

Crew C-27	Hr.	Daily	Hr.	Daily	Bare Costs	Incl. O&P
2 Cement Finishers	$38.30	$612.80	$56.05	$896.80	$38.30	$56.05
1 Concrete Saw		141.00		155.10	8.81	9.69
16 L.H., Daily Totals		$753.80		$1051.90	$47.11	$65.74

Crew C-28	Hr.	Daily	Hr.	Daily	Bare Costs	Incl. O&P
1 Cement Finisher	$38.30	$306.40	$56.05	$448.40	$38.30	$56.05
1 Portable Air Compressor, Gas		19.10		21.01	2.39	2.63
8 L.H., Daily Totals		$325.50		$469.41	$40.69	$58.68

Crew D-1	Hr.	Daily	Hr.	Daily	Bare Costs	Incl. O&P
1 Bricklayer	$40.50	$324.00	$61.45	$491.60	$36.33	$55.10
1 Bricklayer Helper	32.15	257.20	48.75	390.00		
16 L.H., Daily Totals		$581.20		$881.60	$36.33	$55.10

Crew D-2	Hr.	Daily	Hr.	Daily	Bare Costs	Incl. O&P
3 Bricklayers	$40.50	$972.00	$61.45	$1474.80	$37.41	$56.88
2 Bricklayer Helpers	32.15	514.40	48.75	780.00		
.5 Carpenter	39.95	159.80	61.95	247.80		
44 L.H., Daily Totals		$1646.20		$2502.60	$37.41	$56.88

Crew D-3	Hr.	Daily	Hr.	Daily	Bare Costs	Incl. O&P
3 Bricklayers	$40.50	$972.00	$61.45	$1474.80	$37.29	$56.64
2 Bricklayer Helpers	32.15	514.40	48.75	780.00		
.25 Carpenter	39.95	79.90	61.95	123.90		
42 L.H., Daily Totals		$1566.30		$2378.70	$37.29	$56.64

Crew D-4	Hr.	Daily	Hr.	Daily	Bare Costs	Incl. O&P
1 Bricklayer	$40.50	$324.00	$61.45	$491.60	$35.96	$54.41
2 Bricklayer Helpers	32.15	514.40	48.75	780.00		
1 Equip. Oper. (light)	39.05	312.40	58.70	469.60		
1 Grout Pump, 50 C.F./Hr.		126.80		139.48	3.96	4.36
32 L.H., Daily Totals		$1277.60		$1880.68	$39.92	$58.77

Crew D-5	Hr.	Daily	Hr.	Daily	Bare Costs	Incl. O&P
1 Bricklayer	$40.50	$324.00	$61.45	$491.60	$40.50	$61.45
8 L.H., Daily Totals		$324.00		$491.60	$40.50	$61.45

Crew D-6	Hr.	Daily	Hr.	Daily	Bare Costs	Incl. O&P
3 Bricklayers	$40.50	$972.00	$61.45	$1474.80	$36.47	$55.37
3 Bricklayer Helpers	32.15	771.60	48.75	1170.00		
.25 Carpenter	39.95	79.90	61.95	123.90		
50 L.H., Daily Totals		$1823.50		$2768.70	$36.47	$55.37

Crew D-7	Hr.	Daily	Hr.	Daily	Bare Costs	Incl. O&P
1 Tile Layer	$38.10	$304.80	$55.80	$446.40	$34.08	$49.90
1 Tile Layer Helper	30.05	240.40	44.00	352.00		
16 L.H., Daily Totals		$545.20		$798.40	$34.08	$49.90

Crew D-8	Hr.	Daily	Hr.	Daily	Bare Costs	Incl. O&P
3 Bricklayers	$40.50	$972.00	$61.45	$1474.80	$37.16	$56.37
2 Bricklayer Helpers	32.15	514.40	48.75	780.00		
40 L.H., Daily Totals		$1486.40		$2254.80	$37.16	$56.37

Crew D-9	Hr.	Daily	Hr.	Daily	Bare Costs	Incl. O&P
3 Bricklayers	$40.50	$972.00	$61.45	$1474.80	$36.33	$55.10
3 Bricklayer Helpers	32.15	771.60	48.75	1170.00		
48 L.H., Daily Totals		$1743.60		$2644.80	$36.33	$55.10

Crew D-10	Hr.	Daily	Hr.	Daily	Bare Costs	Incl. O&P
1 Bricklayer Foreman	$42.50	$340.00	$64.45	$515.60	$39.42	$59.65
1 Bricklayer	40.50	324.00	61.45	491.60		
1 Bricklayer Helper	32.15	257.20	48.75	390.00		
1 Equip. Oper. (crane)	42.55	340.40	63.95	511.60		
1 S.P. Crane, 4x4, 12 Ton		601.80		661.98	18.81	20.69
32 L.H., Daily Totals		$1863.40		$2570.78	$58.23	$80.34

Crew D-11	Hr.	Daily	Hr.	Daily	Bare Costs	Incl. O&P
1 Bricklayer Foreman	$42.50	$340.00	$64.45	$515.60	$38.38	$58.22
1 Bricklayer	40.50	324.00	61.45	491.60		
1 Bricklayer Helper	32.15	257.20	48.75	390.00		
24 L.H., Daily Totals		$921.20		$1397.20	$38.38	$58.22

Crew D-12	Hr.	Daily	Hr.	Daily	Bare Costs	Incl. O&P
1 Bricklayer Foreman	$42.50	$340.00	$64.45	$515.60	$36.83	$55.85
1 Bricklayer	40.50	324.00	61.45	491.60		
2 Bricklayer Helpers	32.15	514.40	48.75	780.00		
32 L.H., Daily Totals		$1178.40		$1787.20	$36.83	$55.85

Crew D-13	Hr.	Daily	Hr.	Daily	Bare Costs	Incl. O&P
1 Bricklayer Foreman	$42.50	$340.00	$64.45	$515.60	$38.30	$58.22
1 Bricklayer	40.50	324.00	61.45	491.60		
2 Bricklayer Helpers	32.15	514.40	48.75	780.00		
1 Carpenter	39.95	319.60	61.95	495.60		
1 Equip. Oper. (crane)	42.55	340.40	63.95	511.60		
1 S.P. Crane, 4x4, 12 Ton		601.80		661.98	12.54	13.79
48 L.H., Daily Totals		$2440.20		$3456.38	$50.84	$72.01

Crews

Crew No.		Bare Costs		Incl. Subs O & P		Cost Per Labor-Hour	
Crew E-1	Hr.	Daily	Hr.	Daily	Bare Costs	Incl. O&P
1 Welder Foreman	$46.70	$373.60	$83.20	$665.60	$43.48	$73.85
1 Welder	44.70	357.60	79.65	637.20		
1 Equip. Oper. (light)	39.05	312.40	58.70	469.60		
1 Welder, gas engine, 300 amp		134.20		147.62	5.59	6.15
24 L.H., Daily Totals		$1177.80		$1920.02	$49.08	$80.00
Crew E-2	Hr.	Daily	Hr.	Daily	Bare Costs	Incl. O&P
1 Struc. Steel Foreman	$46.70	$373.60	$83.20	$665.60	$43.55	$74.44
4 Struc. Steel Workers	44.70	1430.40	79.65	2548.80		
1 Equip. Oper. (crane)	42.55	340.40	63.95	511.60		
1 Equip. Oper. Oiler	36.80	294.40	55.30	442.40		
1 Lattice Boom Crane, 90 Ton		1741.00		1915.10	31.09	34.20
56 L.H., Daily Totals		$4179.80		$6083.50	$74.64	$108.63
Crew E-3	Hr.	Daily	Hr.	Daily	Bare Costs	Incl. O&P
1 Struc. Steel Foreman	$46.70	$373.60	$83.20	$665.60	$45.37	$80.83
1 Struc. Steel Worker	44.70	357.60	79.65	637.20		
1 Welder	44.70	357.60	79.65	637.20		
1 Welder, gas engine, 300 amp		134.20		147.62	5.59	6.15
24 L.H., Daily Totals		$1223.00		$2087.62	$50.96	$86.98
Crew E-4	Hr.	Daily	Hr.	Daily	Bare Costs	Incl. O&P
1 Struc. Steel Foreman	$46.70	$373.60	$83.20	$665.60	$45.20	$80.54
3 Struc. Steel Workers	44.70	1072.80	79.65	1911.60		
1 Welder, gas engine, 300 amp		134.20		147.62	4.19	4.61
32 L.H., Daily Totals		$1580.60		$2724.82	$49.39	$85.15
Crew E-5	Hr.	Daily	Hr.	Daily	Bare Costs	Incl. O&P
2 Struc. Steel Foremen	$46.70	$747.20	$83.20	$1331.20	$44.09	$76.36
5 Struc. Steel Workers	44.70	1788.00	79.65	3186.00		
1 Equip. Oper. (crane)	42.55	340.40	63.95	511.60		
1 Welder	44.70	357.60	79.65	637.20		
1 Equip. Oper. Oiler	36.80	294.40	55.30	442.40		
1 Lattice Boom Crane, 90 Ton		1741.00		1915.10		
1 Welder, gas engine, 300 amp		134.20		147.62	23.44	25.78
80 L.H., Daily Totals		$5402.80		$8171.12	$67.53	$102.14
Crew E-6	Hr.	Daily	Hr.	Daily	Bare Costs	Incl. O&P
3 Struc. Steel Foremen	$46.70	$1120.80	$83.20	$1996.80	$44.09	$76.50
9 Struc. Steel Workers	44.70	3218.40	79.65	5734.80		
1 Equip. Oper. (crane)	42.55	340.40	63.95	511.60		
1 Welder	44.70	357.60	79.65	637.20		
1 Equip. Oper. Oiler	36.80	294.40	55.30	442.40		
1 Equip. Oper. (light)	39.05	312.40	58.70	469.60		
1 Lattice Boom Crane, 90 Ton		1741.00		1915.10		
1 Welder, gas engine, 300 amp		134.20		147.62		
1 Air Compressor, 160 cfm		138.80		152.68		
2 Impact Wrenches		32.40		35.64	15.99	17.59
128 L.H., Daily Totals		$7690.40		$12043.44	$60.08	$94.09
Crew E-7	Hr.	Daily	Hr.	Daily	Bare Costs	Incl. O&P
1 Struc. Steel Foreman	$46.70	$373.60	$83.20	$665.60	$44.09	$76.36
4 Struc. Steel Workers	44.70	1430.40	79.65	2548.80		
1 Equip. Oper. (crane)	42.55	340.40	63.95	511.60		
1 Equip. Oper. Oiler	36.80	294.40	55.30	442.40		
1 Welder Foreman	46.70	373.60	83.20	665.60		
2 Welders	44.70	715.20	79.65	1274.40		
1 Lattice Boom Crane, 90 Ton		1741.00		1915.10		
2 Welders, gas engine, 300 amp		268.40		295.24	25.12	27.63
80 L.H., Daily Totals		$5537.00		$8318.74	$69.21	$103.98
Crew E-8	Hr.	Daily	Hr.	Daily	Bare Costs	Incl. O&P
1 Struc. Steel Foreman	$46.70	$373.60	$83.20	$665.60	$43.80	$75.50
4 Struc. Steel Workers	44.70	1430.40	79.65	2548.80		
1 Welder Foreman	46.70	373.60	83.20	665.60		
4 Welders	44.70	1430.40	79.65	2548.80		
1 Equip. Oper. (crane)	42.55	340.40	63.95	511.60		
1 Equip. Oper. Oiler	36.80	294.40	55.30	442.40		
1 Equip. Oper. (light)	39.05	312.40	58.70	469.60		
1 Lattice Boom Crane, 90 Ton		1741.00		1915.10		
4 Welders, gas engine, 300 amp		536.80		590.48	21.90	24.09
104 L.H., Daily Totals		$6833.00		$10357.98	$65.70	$99.60
Crew E-9	Hr.	Daily	Hr.	Daily	Bare Costs	Incl. O&P
2 Struc. Steel Foremen	$46.70	$747.20	$83.20	$1331.20	$44.09	$76.50
5 Struc. Steel Workers	44.70	1788.00	79.65	3186.00		
1 Welder Foreman	46.70	373.60	83.20	665.60		
5 Welders	44.70	1788.00	79.65	3186.00		
1 Equip. Oper. (crane)	42.55	340.40	63.95	511.60		
1 Equip. Oper. Oiler	36.80	294.40	55.30	442.40		
1 Equip. Oper. (light)	39.05	312.40	58.70	469.60		
1 Lattice Boom Crane, 90 Ton		1741.00		1915.10		
5 Welders, gas engine, 300 amp		671.00		738.10	18.84	20.73
128 L.H., Daily Totals		$8056.00		$12445.60	$62.94	$97.23
Crew E-10	Hr.	Daily	Hr.	Daily	Bare Costs	Incl. O&P
1 Welder Foreman	$46.70	$373.60	$83.20	$665.60	$45.70	$81.42
1 Welder	44.70	357.60	79.65	637.20		
1 Welder, gas engine, 300 amp		134.20		147.62		
1 Flatbed Truck, Gas, 3 Ton		242.60		266.86	23.55	25.91
16 L.H., Daily Totals		$1108.00		$1717.28	$69.25	$107.33
Crew E-11	Hr.	Daily	Hr.	Daily	Bare Costs	Incl. O&P
2 Painters, Struc. Steel	$36.00	$576.00	$65.35	$1045.60	$35.66	$59.60
1 Building Laborer	31.60	252.80	49.00	392.00		
1 Equip. Oper. (light)	39.05	312.40	58.70	469.60		
1 Air Compressor, 250 cfm		162.40		178.64		
1 Sandblaster, portable, 3 C.F.		20.60		22.66		
1 Set Sand Blasting Accessories		12.75		14.03	6.12	6.73
32 L.H., Daily Totals		$1336.95		$2122.53	$41.78	$66.33
Crew E-12	Hr.	Daily	Hr.	Daily	Bare Costs	Incl. O&P
1 Welder Foreman	$46.70	$373.60	$83.20	$665.60	$42.88	$70.95
1 Equip. Oper. (light)	39.05	312.40	58.70	469.60		
1 Welder, gas engine, 300 amp		134.20		147.62	8.39	9.23
16 L.H., Daily Totals		$820.20		$1282.82	$51.26	$80.18
Crew E-13	Hr.	Daily	Hr.	Daily	Bare Costs	Incl. O&P
1 Welder Foreman	$46.70	$373.60	$83.20	$665.60	$44.15	$75.03
.5 Equip. Oper. (light)	39.05	156.20	58.70	234.80		
1 Welder, gas engine, 300 amp		134.20		147.62	11.18	12.30
12 L.H., Daily Totals		$664.00		$1048.02	$55.33	$87.33
Crew E-14	Hr.	Daily	Hr.	Daily	Bare Costs	Incl. O&P
1 Welder Foreman	$46.70	$373.60	$83.20	$665.60	$46.70	$83.20
1 Welder, gas engine, 300 amp		134.20		147.62	16.77	18.45
8 L.H., Daily Totals		$507.80		$813.22	$63.48	$101.65
Crew E-16	Hr.	Daily	Hr.	Daily	Bare Costs	Incl. O&P
1 Welder Foreman	$46.70	$373.60	$83.20	$665.60	$45.70	$81.42
1 Welder	44.70	357.60	79.65	637.20		
1 Welder, gas engine, 300 amp		134.20		147.62	8.39	9.23
16 L.H., Daily Totals		$865.40		$1450.42	$54.09	$90.65

Crews

Crew No.		Bare Costs		Incl. Subs O & P		Cost Per Labor-Hour	
Crew E-17	Hr.	Daily	Hr.	Daily	Bare Costs	Incl. O&P
1 Structural Steel Foreman	$46.70	$373.60	$83.20	$665.60	$45.70	$81.42
1 Structural Steel Worker	44.70	357.60	79.65	637.20		
16 L.H., Daily Totals		$731.20		$1302.80	$45.70	$81.42
Crew E-18	Hr.	Daily	Hr.	Daily	Bare Costs	Incl. O&P
1 Structural Steel Foreman	$46.70	$373.60	$83.20	$665.60	$44.43	$76.86
3 Structural Steel Workers	44.70	1072.80	79.65	1911.60		
1 Equipment Operator (med.)	41.35	330.80	62.15	497.20		
1 Lattice Boom Crane, 20 Ton		1080.00		1188.00	27.00	29.70
40 L.H., Daily Totals		$2857.20		$4262.40	$71.43	$106.56
Crew E-19	Hr.	Daily	Hr.	Daily	Bare Costs	Incl. O&P
1 Structural Steel Worker	$44.70	$357.60	$79.65	$637.20	$43.48	$73.85
1 Structural Steel Foreman	46.70	373.60	83.20	665.60		
1 Equip. Oper. (light)	39.05	312.40	58.70	469.60		
1 Lattice Boom Crane, 20 Ton		1080.00		1188.00	45.00	49.50
24 L.H., Daily Totals		$2123.60		$2960.40	$88.48	$123.35
Crew E-20	Hr.	Daily	Hr.	Daily	Bare Costs	Incl. O&P
1 Structural Steel Foreman	$46.70	$373.60	$83.20	$665.60	$43.69	$75.09
5 Structural Steel Workers	44.70	1788.00	79.65	3186.00		
1 Equip. Oper. (crane)	42.55	340.40	63.95	511.60		
1 Oiler	36.80	294.40	55.30	442.40		
1 Lattice Boom Crane, 40 Ton		1327.00		1459.70	20.73	22.81
64 L.H., Daily Totals		$4123.40		$6265.30	$64.43	$97.90
Crew E-22	Hr.	Daily	Hr.	Daily	Bare Costs	Incl. O&P
1 Skilled Worker Foreman	$42.85	$342.80	$66.35	$530.80	$41.52	$64.28
2 Skilled Workers	40.85	653.60	63.25	1012.00		
24 L.H., Daily Totals		$996.40		$1542.80	$41.52	$64.28
Crew E-24	Hr.	Daily	Hr.	Daily	Bare Costs	Incl. O&P
3 Structural Steel Workers	$44.70	$1072.80	$79.65	$1911.60	$43.86	$75.28
1 Equipment Operator (medium)	41.35	330.80	62.15	497.20		
1 Hyd. Crane, 25 Ton		789.40		868.34	24.67	27.14
32 L.H., Daily Totals		$2193.00		$3277.14	$68.53	$102.41
Crew E-25	Hr.	Daily	Hr.	Daily	Bare Costs	Incl. O&P
1 Welder Foreman	$46.70	$373.60	$83.20	$665.60	$46.70	$83.20
1 Cutting Torch		17.00		18.70		
1 Gases		75.60		83.16	11.57	12.73
8 L.H., Daily Totals		$466.20		$767.46	$58.27	$95.93
Crew F-3	Hr.	Daily	Hr.	Daily	Bare Costs	Incl. O&P
4 Carpenters	$39.95	$1278.40	$61.95	$1982.40	$40.47	$62.35
1 Equip. Oper. (crane)	42.55	340.40	63.95	511.60		
1 Hyd. Crane, 12 Ton		768.80		845.68	19.22	21.14
40 L.H., Daily Totals		$2387.60		$3339.68	$59.69	$83.49
Crew F-4	Hr.	Daily	Hr.	Daily	Bare Costs	Incl. O&P
4 Carpenters	$39.95	$1278.40	$61.95	$1982.40	$39.86	$61.17
1 Equip. Oper. (crane)	42.55	340.40	63.95	511.60		
1 Equip. Oper. Oiler	36.80	294.40	55.30	442.40		
1 Hyd. Crane, 55 Ton		1125.00		1237.50	23.44	25.78
48 L.H., Daily Totals		$3038.20		$4173.90	$63.30	$86.96
Crew F-5	Hr.	Daily	Hr.	Daily	Bare Costs	Incl. O&P
1 Carpenter Foreman	$41.95	$335.60	$65.05	$520.40	$40.45	$62.73
3 Carpenters	39.95	958.80	61.95	1486.80		
32 L.H., Daily Totals		$1294.40		$2007.20	$40.45	$62.73

Crew No.		Bare Costs		Incl. Subs O & P		Cost Per Labor-Hour	
Crew F-6	Hr.	Daily	Hr.	Daily	Bare Costs	Incl. O&P
2 Carpenters	$39.95	$639.20	$61.95	$991.20	$37.13	$57.17
2 Building Laborers	31.60	505.60	49.00	784.00		
1 Equip. Oper. (crane)	42.55	340.40	63.95	511.60		
1 Hyd. Crane, 12 Ton		768.80		845.68	19.22	21.14
40 L.H., Daily Totals		$2254.00		$3132.48	$56.35	$78.31
Crew F-7	Hr.	Daily	Hr.	Daily	Bare Costs	Incl. O&P
2 Carpenters	$39.95	$639.20	$61.95	$991.20	$35.77	$55.48
2 Building Laborers	31.60	505.60	49.00	784.00		
32 L.H., Daily Totals		$1144.80		$1775.20	$35.77	$55.48
Crew G-1	Hr.	Daily	Hr.	Daily	Bare Costs	Incl. O&P
1 Roofer Foreman	$36.25	$290.00	$61.10	$488.80	$31.99	$53.90
4 Roofers, Composition	34.25	1096.00	57.70	1846.40		
2 Roofer Helpers	25.35	405.60	42.70	683.20		
1 Application Equipment		173.20		190.52		
1 Tar Kettle/Pot		97.20		106.92		
1 Crew Truck		169.60		186.56	7.86	8.64
56 L.H., Daily Totals		$2231.60		$3502.40	$39.85	$62.54
Crew G-2	Hr.	Daily	Hr.	Daily	Bare Costs	Incl. O&P
1 Plasterer	$36.15	$289.20	$54.55	$436.40	$33.35	$50.77
1 Plasterer Helper	32.30	258.40	48.75	390.00		
1 Building Laborer	31.60	252.80	49.00	392.00		
1 Grout Pump, 50 C.F./Hr.		126.80		139.48	5.28	5.81
24 L.H., Daily Totals		$927.20		$1357.88	$38.63	$56.58
Crew G-2A	Hr.	Daily	Hr.	Daily	Bare Costs	Incl. O&P
1 Roofer, composition	$34.25	$274.00	$57.70	$461.60	$30.40	$49.80
1 Roofer Helper	25.35	202.80	42.70	341.60		
1 Building Laborer	31.60	252.80	49.00	392.00		
1 Foam spray rig, trailer-mtd.		488.20		537.02		
1 Pickup Truck, 3/4 Ton		116.80		128.48	25.21	27.73
24 L.H., Daily Totals		$1334.60		$1860.70	$55.61	$77.53
Crew G-3	Hr.	Daily	Hr.	Daily	Bare Costs	Incl. O&P
2 Sheet Metal Workers	$47.20	$755.20	$72.65	$1162.40	$39.40	$60.83
2 Building Laborers	31.60	505.60	49.00	784.00		
32 L.H., Daily Totals		$1260.80		$1946.40	$39.40	$60.83
Crew G-4	Hr.	Daily	Hr.	Daily	Bare Costs	Incl. O&P
1 Labor Foreman (outside)	$33.60	$268.80	$52.10	$416.80	$32.27	$50.03
2 Building Laborers	31.60	505.60	49.00	784.00		
1 Flatbed Truck, Gas, 1.5 Ton		194.40		213.84		
1 Air Compressor, 160 cfm		138.80		152.68	13.88	15.27
24 L.H., Daily Totals		$1107.60		$1567.32	$46.15	$65.31
Crew G-5	Hr.	Daily	Hr.	Daily	Bare Costs	Incl. O&P
1 Roofer Foreman	$36.25	$290.00	$61.10	$488.80	$31.09	$52.38
2 Roofers, Composition	34.25	548.00	57.70	923.20		
2 Roofer Helpers	25.35	405.60	42.70	683.20		
1 Application Equipment		173.20		190.52	4.33	4.76
40 L.H., Daily Totals		$1416.80		$2285.72	$35.42	$57.14
Crew G-6A	Hr.	Daily	Hr.	Daily	Bare Costs	Incl. O&P
2 Roofers Composition	$34.25	$548.00	$57.70	$923.20	$34.25	$57.70
1 Small Compressor, Electric		11.35		12.48		
2 Pneumatic Nailers		43.20		47.52	3.41	3.75
16 L.H., Daily Totals		$602.55		$983.21	$37.66	$61.45

Crews

Crew No.	Bare Costs		Incl. Subs O & P		Cost Per Labor-Hour	
Crew G-7	Hr.	Daily	Hr.	Daily	Bare Costs	Incl. O&P
1 Carpenter	$39.95	$319.60	$61.95	$495.60	$39.95	$61.95
1 Small Compressor, Electric		11.35		12.48		
1 Pneumatic Nailer		21.60		23.76	4.12	4.53
8 L.H., Daily Totals		$352.55		$531.85	$44.07	$66.48
Crew H-1	Hr.	Daily	Hr.	Daily	Bare Costs	Incl. O&P
2 Glaziers	$38.60	$617.60	$58.35	$933.60	$41.65	$69.00
2 Struc. Steel Workers	44.70	715.20	79.65	1274.40		
32 L.H., Daily Totals		$1332.80		$2208.00	$41.65	$69.00
Crew H-2	Hr.	Daily	Hr.	Daily	Bare Costs	Incl. O&P
2 Glaziers	$38.60	$617.60	$58.35	$933.60	$36.27	$55.23
1 Building Laborer	31.60	252.80	49.00	392.00		
24 L.H., Daily Totals		$870.40		$1325.60	$36.27	$55.23
Crew H-3	Hr.	Daily	Hr.	Daily	Bare Costs	Incl. O&P
1 Glazier	$38.60	$308.80	$58.35	$466.80	$34.45	$52.58
1 Helper	30.30	242.40	46.80	374.40		
16 L.H., Daily Totals		$551.20		$841.20	$34.45	$52.58
Crew J-1	Hr.	Daily	Hr.	Daily	Bare Costs	Incl. O&P
3 Plasterers	$36.15	$867.60	$54.55	$1309.20	$34.61	$52.23
2 Plasterer Helpers	32.30	516.80	48.75	780.00		
1 Mixing Machine, 6 C.F.		126.60		139.26	3.17	3.48
40 L.H., Daily Totals		$1511.00		$2228.46	$37.77	$55.71
Crew J-2	Hr.	Daily	Hr.	Daily	Bare Costs	Incl. O&P
3 Plasterers	$36.15	$867.60	$54.55	$1309.20	$34.77	$52.29
2 Plasterer Helpers	32.30	516.80	48.75	780.00		
1 Lather	35.55	284.40	52.60	420.80		
1 Mixing Machine, 6 C.F.		126.60		139.26	2.64	2.90
48 L.H., Daily Totals		$1795.40		$2649.26	$37.40	$55.19
Crew J-3	Hr.	Daily	Hr.	Daily	Bare Costs	Incl. O&P
1 Terrazzo Worker	$37.70	$301.60	$55.20	$441.60	$34.27	$50.17
1 Terrazzo Helper	30.85	246.80	45.15	361.20		
1 Terrazzo Grinder, Electric		81.60		89.76		
1 Terrazzo Mixer		171.60		188.76	15.82	17.41
16 L.H., Daily Totals		$801.60		$1081.32	$50.10	$67.58
Crew K-1	Hr.	Daily	Hr.	Daily	Bare Costs	Incl. O&P
1 Carpenter	$39.95	$319.60	$61.95	$495.60	$35.45	$54.80
1 Truck Driver (light)	30.95	247.60	47.65	381.20		
1 Flatbed Truck, Gas, 3 Ton		242.60		266.86	15.16	16.68
16 L.H., Daily Totals		$809.80		$1143.66	$50.61	$71.48
Crew K-2	Hr.	Daily	Hr.	Daily	Bare Costs	Incl. O&P
1 Struc. Steel Foreman	$46.70	$373.60	$83.20	$665.60	$40.78	$70.17
1 Struc. Steel Worker	44.70	357.60	79.65	637.20		
1 Truck Driver (light)	30.95	247.60	47.65	381.20		
1 Flatbed Truck, Gas, 3 Ton		242.60		266.86	10.11	11.12
24 L.H., Daily Totals		$1221.40		$1950.86	$50.89	$81.29
Crew L-1	Hr.	Daily	Hr.	Daily	Bare Costs	Incl. O&P
1 Electrician	$47.00	$376.00	$69.95	$559.60	$47.88	$71.55
1 Plumber	48.75	390.00	73.15	585.20		
16 L.H., Daily Totals		$766.00		$1144.80	$47.88	$71.55

Crew No.	Bare Costs		Incl. Subs O & P		Cost Per Labor-Hour	
Crew L-2	Hr.	Daily	Hr.	Daily	Bare Costs	Incl. O&P
1 Carpenter	$39.95	$319.60	$61.95	$495.60	$35.13	$54.38
1 Carpenter Helper	30.30	242.40	46.80	374.40		
16 L.H., Daily Totals		$562.00		$870.00	$35.13	$54.38
Crew L-3	Hr.	Daily	Hr.	Daily	Bare Costs	Incl. O&P
1 Carpenter	$39.95	$319.60	$61.95	$495.60	$43.52	$66.63
.5 Electrician	47.00	188.00	69.95	279.80		
.5 Sheet Metal Worker	47.20	188.80	72.65	290.60		
16 L.H., Daily Totals		$696.40		$1066.00	$43.52	$66.63
Crew L-3A	Hr.	Daily	Hr.	Daily	Bare Costs	Incl. O&P
1 Carpenter Foreman (outside)	$41.95	$335.60	$65.05	$520.40	$43.70	$67.58
.5 Sheet Metal Worker	47.20	188.80	72.65	290.60		
12 L.H., Daily Totals		$524.40		$811.00	$43.70	$67.58
Crew L-4	Hr.	Daily	Hr.	Daily	Bare Costs	Incl. O&P
2 Skilled Workers	$40.85	$653.60	$63.25	$1012.00	$37.33	$57.77
1 Helper	30.30	242.40	46.80	374.40		
24 L.H., Daily Totals		$896.00		$1386.40	$37.33	$57.77
Crew L-5	Hr.	Daily	Hr.	Daily	Bare Costs	Incl. O&P
1 Struc. Steel Foreman	$46.70	$373.60	$83.20	$665.60	$44.68	$77.91
5 Struc. Steel Workers	44.70	1788.00	79.65	3186.00		
1 Equip. Oper. (crane)	42.55	340.40	63.95	511.60		
1 Hyd. Crane, 25 Ton		789.40		868.34	14.10	15.51
56 L.H., Daily Totals		$3291.40		$5231.54	$58.77	$93.42
Crew L-5A	Hr.	Daily	Hr.	Daily	Bare Costs	Incl. O&P
1 Structural Steel Foreman	$46.70	$373.60	$83.20	$665.60	$44.66	$76.61
2 Structural Steel Workers	44.70	715.20	79.65	1274.40		
1 Equip. Oper. (crane)	42.55	340.40	63.95	511.60		
1 S.P. Crane, 4x4, 25 Ton		716.80		788.48	22.40	24.64
32 L.H., Daily Totals		$2146.00		$3240.08	$67.06	$101.25
Crew L-5B	Hr.	Daily	Hr.	Daily	Bare Costs	Incl. O&P
1 Structural Steel Foreman	$46.70	$373.60	$83.20	$665.60	$45.35	$72.19
2 Structural Steel Workers	44.70	715.20	79.65	1274.40		
2 Electricians	47.00	752.00	69.95	1119.20		
2 Steamfitters/Pipefitters	49.35	789.60	74.05	1184.80		
1 Equip. Oper. (crane)	42.55	340.40	63.95	511.60		
1 Equip. Oper. Oiler	36.80	294.40	55.30	442.40		
1 Hyd. Crane, 80 Ton		1296.00		1425.60	18.00	19.80
72 L.H., Daily Totals		$4561.20		$6623.60	$63.35	$91.99
Crew L-6	Hr.	Daily	Hr.	Daily	Bare Costs	Incl. O&P
1 Plumber	$48.75	$390.00	$73.15	$585.20	$48.17	$72.08
.5 Electrician	47.00	188.00	69.95	279.80		
12 L.H., Daily Totals		$578.00		$865.00	$48.17	$72.08
Crew L-7	Hr.	Daily	Hr.	Daily	Bare Costs	Incl. O&P
2 Carpenters	$39.95	$639.20	$61.95	$991.20	$38.57	$59.39
1 Building Laborer	31.60	252.80	49.00	392.00		
.5 Electrician	47.00	188.00	69.95	279.80		
28 L.H., Daily Totals		$1080.00		$1663.00	$38.57	$59.39
Crew L-8	Hr.	Daily	Hr.	Daily	Bare Costs	Incl. O&P
2 Carpenters	$39.95	$639.20	$61.95	$991.20	$41.71	$64.19
.5 Plumber	48.75	195.00	73.15	292.60		
20 L.H., Daily Totals		$834.20		$1283.80	$41.71	$64.19

Crews

Crew No.	Bare Costs		Incl. Subs O & P		Cost Per Labor-Hour	
Crew L-9	**Hr.**	**Daily**	**Hr.**	**Daily**	**Bare Costs**	**Incl. O&P**
1 Labor Foreman (inside)	$32.10	$256.80	$49.80	$398.40	$36.33	$58.32
2 Building Laborers	31.60	505.60	49.00	784.00		
1 Struc. Steel Worker	44.70	357.60	79.65	637.20		
.5 Electrician	47.00	188.00	69.95	279.80		
36 L.H., Daily Totals		$1308.00		$2099.40	$36.33	$58.32
Crew L-10	**Hr.**	**Daily**	**Hr.**	**Daily**	**Bare Costs**	**Incl. O&P**
1 Structural Steel Foreman	$46.70	$373.60	$83.20	$665.60	$44.65	$75.60
1 Structural Steel Worker	44.70	357.60	79.65	637.20		
1 Equip. Oper. (crane)	42.55	340.40	63.95	511.60		
1 Hyd. Crane, 12 Ton		768.80		845.68	32.03	35.24
24 L.H., Daily Totals		$1840.40		$2660.08	$76.68	$110.84
Crew L-11	**Hr.**	**Daily**	**Hr.**	**Daily**	**Bare Costs**	**Incl. O&P**
2 Wreckers	$31.60	$505.60	$54.15	$866.40	$36.20	$57.74
1 Equip. Oper. (crane)	42.55	340.40	63.95	511.60		
1 Equip. Oper. (light)	39.05	312.40	58.70	469.60		
1 Hyd. Excavator, 2.5 C.Y.		1619.00		1780.90		
1 Loader, Skid Steer, 78 H.P.		259.60		285.56	58.71	64.58
32 L.H., Daily Totals		$3037.00		$3914.06	$94.91	$122.31
Crew M-1	**Hr.**	**Daily**	**Hr.**	**Daily**	**Bare Costs**	**Incl. O&P**
3 Elevator Constructors	$56.60	$1358.40	$84.30	$2023.20	$53.77	$80.09
1 Elevator Apprentice	45.30	362.40	67.45	539.60		
5 Hand Tools		57.00		62.70	1.78	1.96
32 L.H., Daily Totals		$1777.80		$2625.50	$55.56	$82.05
Crew M-3	**Hr.**	**Daily**	**Hr.**	**Daily**	**Bare Costs**	**Incl. O&P**
1 Electrician Foreman (out)	$49.00	$392.00	$72.90	$583.20	$45.37	$68.04
1 Common Laborer	31.60	252.80	49.00	392.00		
.25 Equipment Operator, Medium	41.35	82.70	62.15	124.30		
1 Elevator Constructor	56.60	452.80	84.30	674.40		
1 Elevator Apprentice	45.30	362.40	67.45	539.60		
.25 S.P. Crane, 4x4, 20 Ton		175.15		192.66	5.15	5.67
34 L.H., Daily Totals		$1717.85		$2506.17	$50.52	$73.71
Crew M-4	**Hr.**	**Daily**	**Hr.**	**Daily**	**Bare Costs**	**Incl. O&P**
1 Electrician Foreman (out)	$49.00	$392.00	$72.90	$583.20	$44.96	$67.44
1 Common Laborer	31.60	252.80	49.00	392.00		
.25 Equipment Operator, Crane	42.55	85.10	63.95	127.90		
.25 Equipment Operator, Oiler	36.80	73.60	55.30	110.60		
1 Elevator Constructor	56.60	452.80	84.30	674.40		
1 Elevator Apprentice	45.30	362.40	67.45	539.60		
.25 S.P. Crane, 4x4, 40 Ton		234.30		257.73	6.51	7.16
36 L.H., Daily Totals		$1853.00		$2685.43	$51.47	$74.60
Crew Q-1	**Hr.**	**Daily**	**Hr.**	**Daily**	**Bare Costs**	**Incl. O&P**
1 Plumber	$48.75	$390.00	$73.15	$585.20	$43.88	$65.85
1 Plumber Apprentice	39.00	312.00	58.55	468.40		
16 L.H., Daily Totals		$702.00		$1053.60	$43.88	$65.85
Crew Q-1A	**Hr.**	**Daily**	**Hr.**	**Daily**	**Bare Costs**	**Incl. O&P**
.25 Plumber Foreman (out)	$50.75	$101.50	$76.20	$152.40	$49.15	$73.76
1 Plumber	48.75	390.00	73.15	585.20		
10 L.H., Daily Totals		$491.50		$737.60	$49.15	$73.76

Crew No.	Bare Costs		Incl. Subs O & P		Cost Per Labor-Hour	
Crew Q-1C	**Hr.**	**Daily**	**Hr.**	**Daily**	**Bare Costs**	**Incl. O&P**
1 Plumber	$48.75	$390.00	$73.15	$585.20	$43.03	$64.62
1 Plumber Apprentice	39.00	312.00	58.55	468.40		
1 Equip. Oper. (medium)	41.35	330.80	62.15	497.20		
1 Trencher, Chain Type, 8' D		1777.00		1954.70	74.04	81.45
24 L.H., Daily Totals		$2809.80		$3505.50	$117.08	$146.06
Crew Q-2	**Hr.**	**Daily**	**Hr.**	**Daily**	**Bare Costs**	**Incl. O&P**
2 Plumbers	$48.75	$780.00	$73.15	$1170.40	$45.50	$68.28
1 Plumber Apprentice	39.00	312.00	58.55	468.40		
24 L.H., Daily Totals		$1092.00		$1638.80	$45.50	$68.28
Crew Q-3	**Hr.**	**Daily**	**Hr.**	**Daily**	**Bare Costs**	**Incl. O&P**
1 Plumber Foreman (inside)	$49.25	$394.00	$73.90	$591.20	$46.44	$69.69
2 Plumbers	48.75	780.00	73.15	1170.40		
1 Plumber Apprentice	39.00	312.00	58.55	468.40		
32 L.H., Daily Totals		$1486.00		$2230.00	$46.44	$69.69
Crew Q-4	**Hr.**	**Daily**	**Hr.**	**Daily**	**Bare Costs**	**Incl. O&P**
1 Plumber Foreman (inside)	$49.25	$394.00	$73.90	$591.20	$46.44	$69.69
1 Plumber	48.75	390.00	73.15	585.20		
1 Welder (plumber)	48.75	390.00	73.15	585.20		
1 Plumber Apprentice	39.00	312.00	58.55	468.40		
1 Welder, electric, 300 amp		55.70		61.27	1.74	1.91
32 L.H., Daily Totals		$1541.70		$2291.27	$48.18	$71.60
Crew Q-5	**Hr.**	**Daily**	**Hr.**	**Daily**	**Bare Costs**	**Incl. O&P**
1 Steamfitter	$49.35	$394.80	$74.05	$592.40	$44.42	$66.67
1 Steamfitter Apprentice	39.50	316.00	59.30	474.40		
16 L.H., Daily Totals		$710.80		$1066.80	$44.42	$66.67
Crew Q-6	**Hr.**	**Daily**	**Hr.**	**Daily**	**Bare Costs**	**Incl. O&P**
2 Steamfitters	$49.35	$789.60	$74.05	$1184.80	$46.07	$69.13
1 Steamfitter Apprentice	39.50	316.00	59.30	474.40		
24 L.H., Daily Totals		$1105.60		$1659.20	$46.07	$69.13
Crew Q-7	**Hr.**	**Daily**	**Hr.**	**Daily**	**Bare Costs**	**Incl. O&P**
1 Steamfitter Foreman (inside)	$49.85	$398.80	$74.80	$598.40	$47.01	$70.55
2 Steamfitters	49.35	789.60	74.05	1184.80		
1 Steamfitter Apprentice	39.50	316.00	59.30	474.40		
32 L.H., Daily Totals		$1504.40		$2257.60	$47.01	$70.55
Crew Q-8	**Hr.**	**Daily**	**Hr.**	**Daily**	**Bare Costs**	**Incl. O&P**
1 Steamfitter Foreman (inside)	$49.85	$398.80	$74.80	$598.40	$47.01	$70.55
1 Steamfitter	49.35	394.80	74.05	592.40		
1 Welder (steamfitter)	49.35	394.80	74.05	592.40		
1 Steamfitter Apprentice	39.50	316.00	59.30	474.40		
1 Welder, electric, 300 amp		55.70		61.27	1.74	1.91
32 L.H., Daily Totals		$1560.10		$2318.87	$48.75	$72.46
Crew Q-9	**Hr.**	**Daily**	**Hr.**	**Daily**	**Bare Costs**	**Incl. O&P**
1 Sheet Metal Worker	$47.20	$377.60	$72.65	$581.20	$42.48	$65.38
1 Sheet Metal Apprentice	37.75	302.00	58.10	464.80		
16 L.H., Daily Totals		$679.60		$1046.00	$42.48	$65.38
Crew Q-10	**Hr.**	**Daily**	**Hr.**	**Daily**	**Bare Costs**	**Incl. O&P**
2 Sheet Metal Workers	$47.20	$755.20	$72.65	$1162.40	$44.05	$67.80
1 Sheet Metal Apprentice	37.75	302.00	58.10	464.80		
24 L.H., Daily Totals		$1057.20		$1627.20	$44.05	$67.80

Crews

Crew No.	Bare Costs Hr.	Bare Costs Daily	Incl. Subs O & P Hr.	Incl. Subs O & P Daily	Cost Per Labor-Hour Bare Costs	Cost Per Labor-Hour Incl. O&P
Crew Q-11					Bare Costs	Incl. O&P
1 Sheet Metal Foreman (inside)	$47.70	$381.60	$73.40	$587.20	$44.96	$69.20
2 Sheet Metal Workers	47.20	755.20	72.65	1162.40		
1 Sheet Metal Apprentice	37.75	302.00	58.10	464.80		
32 L.H., Daily Totals		$1438.80		$2214.40	$44.96	$69.20
Crew Q-12	Hr.	Daily	Hr.	Daily	Bare Costs	Incl. O&P
1 Sprinkler Installer	$48.15	$385.20	$72.25	$578.00	$43.33	$65.03
1 Sprinkler Apprentice	38.50	308.00	57.80	462.40		
16 L.H., Daily Totals		$693.20		$1040.40	$43.33	$65.03
Crew Q-13	Hr.	Daily	Hr.	Daily	Bare Costs	Incl. O&P
1 Sprinkler Foreman (inside)	$48.65	$389.20	$73.00	$584.00	$45.86	$68.83
2 Sprinkler Installers	48.15	770.40	72.25	1156.00		
1 Sprinkler Apprentice	38.50	308.00	57.80	462.40		
32 L.H., Daily Totals		$1467.60		$2202.40	$45.86	$68.83
Crew Q-14	Hr.	Daily	Hr.	Daily	Bare Costs	Incl. O&P
1 Asbestos Worker	$44.10	$352.80	$69.10	$552.80	$39.70	$62.20
1 Asbestos Apprentice	35.30	282.40	55.30	442.40		
16 L.H., Daily Totals		$635.20		$995.20	$39.70	$62.20
Crew Q-15	Hr.	Daily	Hr.	Daily	Bare Costs	Incl. O&P
1 Plumber	$48.75	$390.00	$73.15	$585.20	$43.88	$65.85
1 Plumber Apprentice	39.00	312.00	58.55	468.40		
1 Welder, electric, 300 amp		55.70		61.27	3.48	3.83
16 L.H., Daily Totals		$757.70		$1114.87	$47.36	$69.68
Crew Q-16	Hr.	Daily	Hr.	Daily	Bare Costs	Incl. O&P
2 Plumbers	$48.75	$780.00	$73.15	$1170.40	$45.50	$68.28
1 Plumber Apprentice	39.00	312.00	58.55	468.40		
1 Welder, electric, 300 amp		55.70		61.27	2.32	2.55
24 L.H., Daily Totals		$1147.70		$1700.07	$47.82	$70.84
Crew Q-17	Hr.	Daily	Hr.	Daily	Bare Costs	Incl. O&P
1 Steamfitter	$49.35	$394.80	$74.05	$592.40	$44.42	$66.67
1 Steamfitter Apprentice	39.50	316.00	59.30	474.40		
1 Welder, electric, 300 amp		55.70		61.27	3.48	3.83
16 L.H., Daily Totals		$766.50		$1128.07	$47.91	$70.50
Crew Q-17A	Hr.	Daily	Hr.	Daily	Bare Costs	Incl. O&P
1 Steamfitter	$49.35	$394.80	$74.05	$592.40	$43.80	$65.77
1 Steamfitter Apprentice	39.50	316.00	59.30	474.40		
1 Equip. Oper. (crane)	42.55	340.40	63.95	511.60		
1 Hyd. Crane, 12 Ton		768.80		845.68		
1 Welder, electric, 300 amp		55.70		61.27	34.35	37.79
24 L.H., Daily Totals		$1875.70		$2485.35	$78.15	$103.56
Crew Q-18	Hr.	Daily	Hr.	Daily	Bare Costs	Incl. O&P
2 Steamfitters	$49.35	$789.60	$74.05	$1184.80	$46.07	$69.13
1 Steamfitter Apprentice	39.50	316.00	59.30	474.40		
1 Welder, electric, 300 amp		55.70		61.27	2.32	2.55
24 L.H., Daily Totals		$1161.30		$1720.47	$48.39	$71.69
Crew Q-19	Hr.	Daily	Hr.	Daily	Bare Costs	Incl. O&P
1 Steamfitter	$49.35	$394.80	$74.05	$592.40	$45.28	$67.77
1 Steamfitter Apprentice	39.50	316.00	59.30	474.40		
1 Electrician	47.00	376.00	69.95	559.60		
24 L.H., Daily Totals		$1086.80		$1626.40	$45.28	$67.77
Crew Q-20	Hr.	Daily	Hr.	Daily	Bare Costs	Incl. O&P
1 Sheet Metal Worker	$47.20	$377.60	$72.65	$581.20	$43.38	$66.29
1 Sheet Metal Apprentice	37.75	302.00	58.10	464.80		
.5 Electrician	47.00	188.00	69.95	279.80		
20 L.H., Daily Totals		$867.60		$1325.80	$43.38	$66.29
Crew Q-21	Hr.	Daily	Hr.	Daily	Bare Costs	Incl. O&P
2 Steamfitters	$49.35	$789.60	$74.05	$1184.80	$46.30	$69.34
1 Steamfitter Apprentice	39.50	316.00	59.30	474.40		
1 Electrician	47.00	376.00	69.95	559.60		
32 L.H., Daily Totals		$1481.60		$2218.80	$46.30	$69.34
Crew Q-22	Hr.	Daily	Hr.	Daily	Bare Costs	Incl. O&P
1 Plumber	$48.75	$390.00	$73.15	$585.20	$43.88	$65.85
1 Plumber Apprentice	39.00	312.00	58.55	468.40		
1 Hyd. Crane, 12 Ton		768.80		845.68	48.05	52.85
16 L.H., Daily Totals		$1470.80		$1899.28	$91.92	$118.71
Crew Q-22A	Hr.	Daily	Hr.	Daily	Bare Costs	Incl. O&P
1 Plumber	$48.75	$390.00	$73.15	$585.20	$40.48	$61.16
1 Plumber Apprentice	39.00	312.00	58.55	468.40		
1 Laborer	31.60	252.80	49.00	392.00		
1 Equip. Oper. (crane)	42.55	340.40	63.95	511.60		
1 Hyd. Crane, 12 Ton		768.80		845.68	24.02	26.43
32 L.H., Daily Totals		$2064.00		$2802.88	$64.50	$87.59
Crew Q-23	Hr.	Daily	Hr.	Daily	Bare Costs	Incl. O&P
1 Plumber Foreman	$50.75	$406.00	$76.20	$609.60	$46.95	$70.50
1 Plumber	48.75	390.00	73.15	585.20		
1 Equip. Oper. (medium)	41.35	330.80	62.15	497.20		
1 Lattice Boom Crane, 20 Ton		1080.00		1188.00	45.00	49.50
24 L.H., Daily Totals		$2206.80		$2880.00	$91.95	$120.00
Crew R-1	Hr.	Daily	Hr.	Daily	Bare Costs	Incl. O&P
1 Electrician Foreman	$47.50	$380.00	$70.70	$565.60	$41.52	$62.36
3 Electricians	47.00	1128.00	69.95	1678.80		
2 Helpers	30.30	484.80	46.80	748.80		
48 L.H., Daily Totals		$1992.80		$2993.20	$41.52	$62.36
Crew R-1A	Hr.	Daily	Hr.	Daily	Bare Costs	Incl. O&P
1 Electrician	$47.00	$376.00	$69.95	$559.60	$38.65	$58.38
1 Helper	30.30	242.40	46.80	374.40		
16 L.H., Daily Totals		$618.40		$934.00	$38.65	$58.38
Crew R-2	Hr.	Daily	Hr.	Daily	Bare Costs	Incl. O&P
1 Electrician Foreman	$47.50	$380.00	$70.70	$565.60	$41.66	$62.59
3 Electricians	47.00	1128.00	69.95	1678.80		
2 Helpers	30.30	484.80	46.80	748.80		
1 Equip. Oper. (crane)	42.55	340.40	63.95	511.60		
1 S.P. Crane, 4x4, 5 Ton		258.20		284.02	4.61	5.07
56 L.H., Daily Totals		$2591.40		$3788.82	$46.27	$67.66
Crew R-3	Hr.	Daily	Hr.	Daily	Bare Costs	Incl. O&P
1 Electrician Foreman	$47.50	$380.00	$70.70	$565.60	$46.31	$69.05
1 Electrician	47.00	376.00	69.95	559.60		
.5 Equip. Oper. (crane)	42.55	170.20	63.95	255.80		
.5 S.P. Crane, 4x4, 5 Ton		129.10		142.01	6.46	7.10
20 L.H., Daily Totals		$1055.30		$1523.01	$52.77	$76.15

Crews

Crew No.	Bare Costs		Incl. Subs O & P		Cost Per Labor-Hour	
Crew R-4	Hr.	Daily	Hr.	Daily	Bare Costs	Incl. O&P
1 Struc. Steel Foreman	$46.70	$373.60	$83.20	$665.60	$45.56	$78.42
3 Struc. Steel Workers	44.70	1072.80	79.65	1911.60		
1 Electrician	47.00	376.00	69.95	559.60		
1 Welder, gas engine, 300 amp		134.20		147.62	3.36	3.69
40 L.H., Daily Totals		$1956.60		$3284.42	$48.91	$82.11

Crew R-5	Hr.	Daily	Hr.	Daily	Bare Costs	Incl. O&P
1 Electrician Foreman	$47.50	$380.00	$70.70	$565.60	$40.97	$61.60
4 Electrician Linemen	47.00	1504.00	69.95	2238.40		
2 Electrician Operators	47.00	752.00	69.95	1119.20		
4 Electrician Groundmen	30.30	969.60	46.80	1497.60		
1 Crew Truck		169.60		186.56		
1 Flatbed Truck, 20,000 GVW		197.40		217.14		
1 Pickup Truck, 3/4 Ton		116.80		128.48		
.2 Hyd. Crane, 55 Ton		225.00		247.50		
.2 Hyd. Crane, 12 Ton		153.76		169.14		
.2 Earth Auger, Truck-Mtd.		82.40		90.64		
1 Tractor w/Winch		354.80		390.28	14.77	16.25
88 L.H., Daily Totals		$4905.36		$6850.54	$55.74	$77.85

Crew R-6	Hr.	Daily	Hr.	Daily	Bare Costs	Incl. O&P
1 Electrician Foreman	$47.50	$380.00	$70.70	$565.60	$40.97	$61.60
4 Electrician Linemen	47.00	1504.00	69.95	2238.40		
2 Electrician Operators	47.00	752.00	69.95	1119.20		
4 Electrician Groundmen	30.30	969.60	46.80	1497.60		
1 Crew Truck		169.60		186.56		
1 Flatbed Truck, 20,000 GVW		197.40		217.14		
1 Pickup Truck, 3/4 Ton		116.80		128.48		
.2 Hyd. Crane, 55 Ton		225.00		247.50		
.2 Hyd. Crane, 12 Ton		153.76		169.14		
.2 Earth Auger, Truck-Mtd.		82.40		90.64		
1 Tractor w/Winch		354.80		390.28		
3 Cable Trailers		531.30		584.43		
.5 Tensioning Rig		175.35		192.88		
.5 Cable Pulling Rig		1049.50		1154.45	34.73	38.20
88 L.H., Daily Totals		$6661.51		$8782.30	$75.70	$99.80

Crew R-7	Hr.	Daily	Hr.	Daily	Bare Costs	Incl. O&P
1 Electrician Foreman	$47.50	$380.00	$70.70	$565.60	$33.17	$50.78
5 Electrician Groundmen	30.30	1212.00	46.80	1872.00		
1 Crew Truck		169.60		186.56	3.53	3.89
48 L.H., Daily Totals		$1761.60		$2624.16	$36.70	$54.67

Crew R-8	Hr.	Daily	Hr.	Daily	Bare Costs	Incl. O&P
1 Electrician Foreman	$47.50	$380.00	$70.70	$565.60	$41.52	$62.36
3 Electrician Linemen	47.00	1128.00	69.95	1678.80		
2 Electrician Groundmen	30.30	484.80	46.80	748.80		
1 Pickup Truck, 3/4 Ton		116.80		128.48		
1 Crew Truck		169.60		186.56	5.97	6.56
48 L.H., Daily Totals		$2279.20		$3308.24	$47.48	$68.92

Crew R-9	Hr.	Daily	Hr.	Daily	Bare Costs	Incl. O&P
1 Electrician Foreman	$47.50	$380.00	$70.70	$565.60	$38.71	$58.47
1 Electrician Lineman	47.00	376.00	69.95	559.60		
2 Electrician Operators	47.00	752.00	69.95	1119.20		
4 Electrician Groundmen	30.30	969.60	46.80	1497.60		
1 Pickup Truck, 3/4 Ton		116.80		128.48		
1 Crew Truck		169.60		186.56	4.47	4.92
64 L.H., Daily Totals		$2764.00		$4057.04	$43.19	$63.39

Crew R-10	Hr.	Daily	Hr.	Daily	Bare Costs	Incl. O&P
1 Electrician Foreman	$47.50	$380.00	$70.70	$565.60	$44.30	$66.22
4 Electrician Linemen	47.00	1504.00	69.95	2238.40		
1 Electrician Groundman	30.30	242.40	46.80	374.40		
1 Crew Truck		169.60		186.56		
3 Tram Cars		380.55		418.61	11.46	12.61
48 L.H., Daily Totals		$2676.55		$3783.57	$55.76	$78.82

Crew R-11	Hr.	Daily	Hr.	Daily	Bare Costs	Incl. O&P
1 Electrician Foreman	$47.50	$380.00	$70.70	$565.60	$44.24	$66.21
4 Electricians	47.00	1504.00	69.95	2238.40		
1 Equip. Oper. (crane)	42.55	340.40	63.95	511.60		
1 Common Laborer	31.60	252.80	49.00	392.00		
1 Crew Truck		169.60		186.56		
1 Hyd. Crane, 12 Ton		768.80		845.68	16.76	18.43
56 L.H., Daily Totals		$3415.60		$4739.84	$60.99	$84.64

Crew R-12	Hr.	Daily	Hr.	Daily	Bare Costs	Incl. O&P
1 Carpenter Foreman	$40.45	$323.60	$62.75	$502.00	$37.52	$58.94
4 Carpenters	39.95	1278.40	61.95	1982.40		
4 Common Laborers	31.60	1011.20	49.00	1568.00		
1 Equip. Oper. (med.)	41.35	330.80	62.15	497.20		
1 Steel Worker	44.70	357.60	79.65	637.20		
1 Dozer, 200 H.P.		1082.00		1190.20		
1 Pickup Truck, 3/4 Ton		116.80		128.48	13.62	14.98
88 L.H., Daily Totals		$4500.40		$6505.48	$51.14	$73.93

Crew R-13	Hr.	Daily	Hr.	Daily	Bare Costs	Incl. O&P
1 Electrician Foreman	$47.50	$380.00	$70.70	$565.60	$44.94	$67.02
3 Electricians	47.00	1128.00	69.95	1678.80		
.25 Equip. Oper. (crane)	42.55	85.10	63.95	127.90		
1 Equipment Oiler	36.80	294.40	55.30	442.40		
.25 Hydraulic Crane, 33 Ton		204.70		225.17	4.87	5.36
42 L.H., Daily Totals		$2092.20		$3039.87	$49.81	$72.38

Crew R-15	Hr.	Daily	Hr.	Daily	Bare Costs	Incl. O&P
1 Electrician Foreman	$47.50	$380.00	$70.70	$565.60	$45.76	$68.20
4 Electricians	47.00	1504.00	69.95	2238.40		
1 Equipment Oper. (light)	39.05	312.40	58.70	469.60		
1 Aerial Lift Truck		328.80		361.68	6.85	7.54
48 L.H., Daily Totals		$2525.20		$3635.28	$52.61	$75.73

Crew R-18	Hr.	Daily	Hr.	Daily	Bare Costs	Incl. O&P
.25 Electrician Foreman	$47.50	$95.00	$70.70	$141.40	$36.76	$55.76
1 Electrician	47.00	376.00	69.95	559.60		
2 Helpers	30.30	484.80	46.80	748.80		
26 L.H., Daily Totals		$955.80		$1449.80	$36.76	$55.76

Crew R-19	Hr.	Daily	Hr.	Daily	Bare Costs	Incl. O&P
.5 Electrician Foreman	$47.50	$190.00	$70.70	$282.80	$47.10	$70.10
2 Electricians	47.00	752.00	69.95	1119.20		
20 L.H., Daily Totals		$942.00		$1402.00	$47.10	$70.10

Crew R-21	Hr.	Daily	Hr.	Daily	Bare Costs	Incl. O&P
1 Electrician Foreman	$47.50	$380.00	$70.70	$565.60	$46.98	$69.94
3 Electricians	47.00	1128.00	69.95	1678.80		
.1 Equip. Oper. (med.)	41.35	33.08	62.15	49.72		
.1 S.P. Crane, 4x4, 25 Ton		71.68		78.85	2.19	2.40
32.8 L.H., Daily Totals		$1612.76		$2372.97	$49.17	$72.35

Crews

Crew No.	Bare Costs		Incl. Subs O & P		Cost Per Labor-Hour	
Crew R-22	Hr.	Daily	Hr.	Daily	Bare Costs	Incl. O&P
.66 Electrician Foreman	$47.50	$250.80	$70.70	$373.30	$39.90	$60.12
2 Helpers	30.30	484.80	46.80	748.80		
2 Electricians	47.00	752.00	69.95	1119.20		
37.28 L.H., Daily Totals		$1487.60		$2241.30	$39.90	$60.12
Crew R-30	Hr.	Daily	Hr.	Daily	Bare Costs	Incl. O&P
.25 Electrician Foreman (out)	$49.00	$98.00	$72.90	$145.80	$37.68	$57.28
1 Electrician	47.00	376.00	69.95	559.60		
2 Laborers, (Semi-Skilled)	31.60	505.60	49.00	784.00		
26 L.H., Daily Totals		$979.60		$1489.40	$37.68	$57.28
Crew R-31	Hr.	Daily	Hr.	Daily	Bare Costs	Incl. O&P
1 Electrician	$47.00	$376.00	$69.95	$559.60	$47.00	$69.95
1 Core Drill, Electric, 2.5 H.P.		52.95		58.24	6.62	7.28
8 L.H., Daily Totals		$428.95		$617.85	$53.62	$77.23
Crew W-41E	Hr.	Daily	Hr.	Daily	Bare Costs	Incl. O&P
.5 Plumber Foreman (out)	$50.75	$203.00	$76.20	$304.80	$42.29	$64.10
1 Plumber	48.75	390.00	73.15	585.20		
1 Laborer	31.60	252.80	49.00	392.00		
20 L.H., Daily Totals		$845.80		$1282.00	$42.29	$64.10

Historical Cost Indexes

The table below lists both the RSMeans Historical Cost Index based on Jan. 1, 1993 = 100 as well as the computed value of an index based on Jan. 1, 2009 costs. Since the Jan. 1, 2009 figure is estimated, space is left to write in the actual index figures as they become available through either the quarterly "RSMeans Construction Cost Indexes" or as printed in the "Engineering News-Record." To compute the actual index based on Jan. 1, 2009 = 100, divide the Historical Cost Index for a particular year by the actual Jan. 1, 2009 Construction Cost Index. Space has been left to advance the index figures as the year progresses.

Year	Historical Cost Index Jan. 1, 1993 = 100		Current Index Based on Jan. 1, 2009 = 100		Year	Historical Cost Index Jan. 1, 1993 = 100	Current Index Based on Jan. 1, 2009 = 100		Year	Historical Cost Index Jan. 1, 1993 = 100	Current Index Based on Jan. 1, 2009 = 100	
	Est.	Actual	Est.	Actual		Actual	Est.	Actual		Actual	Est.	Actual
Oct 2009					July 1994	104.4	60.3		July 1976	46.9	27.1	
July 2009					1993	101.7	54.7		1975	44.8	24.1	
April 2009					1992	99.4	53.5		1974	41.4	22.3	
Jan 2009	185.9		100.0	100.0	1991	96.8	52.1		1973	37.7	20.3	
July 2008		180.4	97.0		1990	94.3	50.7		1972	34.8	18.7	
2007		169.4	91.1		1989	92.1	49.6		1971	32.1	17.3	
2006		162.0	87.1		1988	89.9	48.3		1970	28.7	15.4	
2005		151.6	81.5		1987	87.7	47.2		1969	26.9	14.5	
2004		143.7	77.3		1986	84.2	45.3		1968	24.9	13.4	
2003		132.0	71.0		1985	82.6	44.4		1967	23.5	12.6	
2002		128.7	69.2		1984	82.0	44.1		1966	22.7	12.2	
2001		125.1	67.3		1983	80.2	43.1		1965	21.7	11.7	
2000		120.9	65.0		1982	76.1	41.0		1964	21.2	11.4	
1999		117.6	63.3		1981	70.0	37.6		1963	20.7	11.1	
1998		115.1	61.9		1980	62.9	33.8		1962	20.2	10.9	
1997		112.8	60.7		1979	57.8	31.1		1961	19.8	10.7	
1996		110.2	59.3		1978	53.5	28.8		1960	19.7	10.6	
1995		107.6	57.9		1977	49.5	26.6		1959	19.3	10.4	

Adjustments to Costs

The Historical Cost Index can be used to convert National Average building costs at a particular time to the approximate building costs for some other time.

Example:

Estimate and compare construction costs for different years in the same city.

To estimate the National Average construction cost of a building in 1970, knowing that it cost $900,000 in 2009:

INDEX in 1970 = 28.7

INDEX in 2009 = 185.9

Note: The City Cost Indexes for Canada can be used to convert U.S. National averages to local costs in Canadian dollars.

Time Adjustment using the Historical Cost Indexes:

$$\frac{\text{Index for Year A}}{\text{Index for Year B}} \times \text{Cost in Year B} = \text{Cost in Year A}$$

$$\frac{\text{INDEX 1970}}{\text{INDEX 2009}} \times \text{Cost 2009} = \text{Cost 1970}$$

$$\frac{28.7}{185.9} \times \$900{,}000 = .154 \times \$900{,}000 = \$138{,}600$$

The construction cost of the building in 1970 is $138,600.

City Cost Indexes

How to Use the City Cost Indexes

What you should know before you begin

RSMeans City Cost Indexes (CCI) are an extremely useful tool to use when you want to compare costs from city to city and region to region.

This publication contains average construction cost indexes for 316 U.S. and Canadian cities covering over 930 three-digit zip code locations.

Keep in mind that a City Cost Index number is a percentage ratio of a specific city's cost to the national average cost of the same item at a stated time period.

In other words, these index figures represent relative construction factors (or, if you prefer, multipliers) for Material and Installation costs, as well as the weighted average for Total In Place costs for each CSI MasterFormat division. Installation costs include both labor and equipment rental costs. When estimating equipment rental rates only, for a specific location, use 01543 CONTRACTOR EQUIPMENT index.

The 30 City Average Index is the average of 30 major U.S. cities and serves as a National Average.

Index figures for both material and installation are based on the 30 major city average of 100 and represent the cost relationship as of July 1, 2008. The index for each division is computed from representative material and labor quantities for that division. The weighted average for each city is a weighted total of the components listed above it, but does not include relative productivity between trades or cities.

As changes occur in local material prices, labor rates, and equipment rental rates, (including fuel costs) the impact of these changes should be accurately measured by the change in the City Cost Index for each particular city (as compared to the 30 City Average).

Therefore, if you know (or have estimated) building costs in one city today, you can easily convert those costs to expected building costs in another city.

In addition, by using the Historical Cost Index, you can easily convert National Average building costs at a particular time to the approximate building costs for some other time. The City Cost Indexes can then be applied to calculate the costs for a particular city.

Quick Calculations

Location Adjustment Using the City Cost Indexes:

$$\frac{\text{Index for City A}}{\text{Index for City B}} \times \text{Cost in City B} = \text{Cost in City A}$$

Time Adjustment for the National Average Using the Historical Cost Index:

$$\frac{\text{Index for Year A}}{\text{Index for Year B}} \times \text{Cost in Year B} = \text{Cost in Year A}$$

Adjustment from the National Average:

$$\frac{\text{Index for City A}}{100} \times \text{National Average Cost} = \text{Cost in City A}$$

Since each of the other RSMeans publications contains many different items, any *one* item multiplied by the particular city index may give incorrect results. However, the larger the number of items compiled, the closer the results should be to actual costs for that particular city.

The City Cost Indexes for Canadian cities are calculated using Canadian material and equipment prices and labor rates, in Canadian dollars. Therefore, indexes for Canadian cities can be used to convert U.S. National Average prices to local costs in Canadian dollars.

How to use this section

1. Compare costs from city to city.

In using the RSMeans Indexes, remember that an index number is not a fixed number but a ratio: It's a percentage ratio of a building component's cost at any stated time to the National Average cost of that same component at the same time period. Put in the form of an equation:

$$\frac{\text{Specific City Cost}}{\text{National Average Cost}} \times 100 = \text{City Index Number}$$

Therefore, when making cost comparisons between cities, do not subtract one city's index number from the index number of another city and read the result as a percentage difference. Instead, divide one city's index number by that of the other city. The resulting number may then be used as a multiplier to calculate cost differences from city to city.

The formula used to find cost differences between cities for the purpose of comparison is as follows:

$$\frac{\text{City A Index}}{\text{City B Index}} \times \text{City B Cost (Known)} = \text{City A Cost (Unknown)}$$

In addition, you can use RSMeans CCI to calculate and compare costs division by division between cities using the same basic formula. (Just be sure that you're comparing similar divisions.)

2. Compare a specific city's construction costs with the National Average.

When you're studying construction location feasibility, it's advisable to compare a prospective project's cost index with an index of the National Average cost.

For example, divide the weighted average index of construction costs of a specific city by that of the 30 City Average, which = 100.

$$\frac{\text{City Index}}{100} = \% \text{ of National Average}$$

As a result, you get a ratio that indicates the relative cost of construction in that city in comparison with the National Average.

3. Convert U.S. National Average to actual costs in Canadian City.

$$\frac{\text{Index for Canadian City}}{100} \times \text{National Average Cost} = \text{Cost in Canadian City in \$ CAN}$$

4. Adjust construction cost data based on a National Average.
When you use a source of construction cost data which is based on a National Average (such as RSMeans cost data publications), it is necessary to adjust those costs to a specific location.

$$\frac{\text{City Index}}{100} \times \text{``Book'' Cost Based on National Average Costs} = \text{City Cost (Unknown)}$$

5. When applying the City Cost Indexes to demolition projects, use the appropriate division installation index. For example, for removal of existing doors and windows, use Division 8 (Openings) index.

What you might like to know about how we developed the Indexes

The information presented in the CCI is organized according to the Construction Specifications Institute (CSI) MasterFormat 2004.

To create a reliable index, RSMeans researched the building type most often constructed in the United States and Canada. Because it was concluded that no one type of building completely represented the building construction industry, nine different types of buildings were combined to create a composite model.

The exact material, labor and equipment quantities are based on detailed analysis of these nine building types, then each quantity is weighted in proportion to expected usage. These various material items, labor hours, and equipment rental rates are thus combined to form a composite building representing as closely as possible the actual usage of materials, labor and equipment used in the North American Building Construction Industry.

The following structures were chosen to make up that composite model:

1. Factory, 1 story
2. Office, 2-4 story
3. Store, Retail
4. Town Hall, 2-3 story
5. High School, 2-3 story
6. Hospital, 4-8 story
7. Garage, Parking
8. Apartment, 1-3 story
9. Hotel/Motel, 2-3 story

For the purposes of ensuring the timeliness of the data, the components of the index for the composite model have been streamlined. They currently consist of:

- specific quantities of 66 commonly used construction materials;
- specific labor-hours for 21 building construction trades; and
- specific days of equipment rental for 6 types of construction equipment (normally used to install the 66 material items by the 21 trades.) Fuel costs and routine maintenance costs are included in the equipment costs.

A sophisticated computer program handles the updating of all costs for each city on a quarterly basis. Material and equipment price quotations are gathered quarterly from 316 cities in the United States and Canada. These prices and the latest negotiated labor wage rates for 21 different building trades are used to compile the quarterly update of the City Cost Index.

The 30 major U.S. cities used to calculate the National Average are:

Atlanta, GA
Baltimore, MD
Boston, MA
Buffalo, NY
Chicago, IL
Cincinnati, OH
Cleveland, OH
Columbus, OH
Dallas, TX
Denver, CO
Detroit, MI
Houston, TX
Indianapolis, IN
Kansas City, MO
Los Angeles, CA
Memphis, TN
Milwaukee, WI
Minneapolis, MN
Nashville, TN
New Orleans, LA
New York, NY
Philadelphia, PA
Phoenix, AZ
Pittsburgh, PA
St. Louis, MO
San Antonio, TX
San Diego, CA
San Francisco, CA
Seattle, WA
Washington, DC

What the CCI does not indicate

The weighted average for each city is a total of the divisional components weighted to reflect typical usage, but it does not include the productivity variations between trades or cities.

In addition, the CCI does not take into consideration factors such as the following:

- managerial efficiency
- competitive conditions
- automation
- restrictive union practices
- unique local requirements
- regional variations due to specific building codes

City Cost Indexes

		UNITED STATES ANYTOWN			ALABAMA BIRMINGHAM			ALABAMA HUNTSVILLE			ALABAMA MOBILE			ALABAMA MONTGOMERY			ALABAMA TUSCALOOSA		
DIVISION		MAT.	INST.	TOTAL	MAT.	INST.	TOTAL	MAT.	INST.	TOTAL	MAT.	INST.	TOTAL	MAT.	INST.	TOTAL	MAT.	INST.	TOTAL
015433	CONTRACTOR EQUIPMENT		100.0	100.0		100.5	100.5		100.4	100.4		98.9	98.9		98.9	98.9		100.4	100.4
0241, 31 - 34	SITE & INFRASTRUCTURE, DEMOLITION	100.0	100.0	100.0	86.5	91.9	90.2	82.2	91.4	88.5	95.7	87.7	90.2	93.7	87.8	89.6	82.7	90.8	88.2
0310	Concrete Forming & Accessories	100.0	100.0	100.0	91.0	72.9	75.3	94.9	67.1	70.9	95.0	59.2	64.1	92.7	49.9	55.7	94.8	53.8	59.4
0320	Concrete Reinforcing	100.0	100.0	100.0	89.7	80.3	85.8	89.7	76.3	84.2	92.5	52.8	76.3	92.5	78.9	87.0	89.7	79.3	85.4
0330	Cast-in-Place Concrete	100.0	100.0	100.0	99.2	71.4	88.6	89.7	68.5	81.6	94.7	64.3	83.1	97.0	52.0	79.8	93.4	54.5	78.6
03	CONCRETE	100.0	100.0	100.0	92.8	74.8	84.6	88.6	70.6	80.4	91.5	61.3	77.8	92.5	57.8	76.7	90.4	60.4	76.8
04	MASONRY	100.0	100.0	100.0	87.3	81.1	83.6	86.8	66.0	74.3	87.5	60.7	71.4	86.0	41.3	59.2	85.9	52.9	66.1
05	METALS	100.0	100.0	100.0	99.9	91.5	97.4	101.3	90.4	98.1	100.1	80.0	94.2	99.9	88.7	96.6	100.5	89.2	97.2
06	WOOD, PLASTICS & COMPOSITES	100.0	100.0	100.0	91.7	70.1	79.9	94.6	67.4	79.7	94.8	59.1	75.2	91.3	49.6	68.4	94.6	53.7	72.1
07	THERMAL & MOISTURE PROTECTION	100.0	100.0	100.0	94.8	85.8	91.3	92.4	78.2	86.9	92.4	77.4	86.6	91.8	68.2	82.6	92.7	72.1	84.7
08	OPENINGS	100.0	100.0	100.0	96.8	73.7	91.1	97.1	64.9	89.1	97.1	56.5	87.0	97.1	57.2	87.2	97.1	63.8	88.8
0920	Plaster & Gypsum Board	100.0	100.0	100.0	103.6	69.6	81.5	98.3	66.7	77.8	100.7	58.1	73.0	103.6	48.3	67.7	100.7	50.8	68.3
0950, 0980	Ceilings & Acoustic Treatment	100.0	100.0	100.0	97.1	69.6	80.0	100.4	66.7	79.6	100.4	58.1	74.2	97.1	48.3	66.9	100.4	50.8	69.7
0960	Flooring	100.0	100.0	100.0	103.5	67.6	93.8	103.5	60.6	91.9	113.4	62.8	99.7	111.6	33.2	90.4	103.5	42.1	86.9
0970, 0990	Wall Finishes & Painting/Coating	100.0	100.0	100.0	100.1	63.9	78.4	100.1	60.5	76.4	104.2	62.4	79.1	100.1	61.3	76.8	100.1	52.8	71.7
09	FINISHES	100.0	100.0	100.0	97.7	70.6	83.4	97.4	64.8	80.2	102.5	59.9	79.9	101.0	46.9	72.4	97.7	50.8	72.9
COVERS	DIVS. 10 - 14, 25, 28, 43, 44	100.0	100.0	100.0	100.0	87.4	97.5	100.0	83.7	96.8	100.0	81.4	96.3	100.0	78.7	95.8	100.0	80.9	96.2
21, 22, 23	FIRE SUPPRESSION, PLUMBING & HVAC	100.0	100.0	100.0	99.9	67.3	86.9	99.8	62.1	84.8	99.8	63.7	85.4	99.9	40.6	76.3	99.9	40.4	76.2
26, 27, 3370	ELECTRICAL, COMMUNICATIONS & UTIL.	100.0	100.0	100.0	108.6	60.1	85.0	100.5	67.3	84.3	100.5	58.4	80.0	103.4	71.0	87.6	100.0	60.1	80.5
MF2004	WEIGHTED AVERAGE	100.0	100.0	100.0	98.3	74.9	88.5	97.0	71.1	86.1	97.9	65.9	84.5	98.0	59.2	81.7	97.0	60.5	81.7

		ALASKA ANCHORAGE			ALASKA FAIRBANKS			ALASKA JUNEAU			ARIZONA FLAGSTAFF			ARIZONA MESA/TEMPE			ARIZONA PHOENIX		
DIVISION		MAT.	INST.	TOTAL	MAT.	INST.	TOTAL	MAT.	INST.	TOTAL	MAT.	INST.	TOTAL	MAT.	INST.	TOTAL	MAT.	INST.	TOTAL
015433	CONTRACTOR EQUIPMENT		113.0	113.0		113.0	113.0		113.0	113.0		96.2	96.2		99.2	99.2		99.8	99.8
0241, 31 - 34	SITE & INFRASTRUCTURE, DEMOLITION	140.1	126.1	130.5	126.3	126.2	126.3	139.2	126.1	130.2	81.1	102.1	95.4	86.2	104.8	98.9	86.7	106.0	99.9
0310	Concrete Forming & Accessories	131.5	114.1	116.5	137.2	116.5	119.3	133.2	114.0	116.6	103.2	65.4	70.6	103.8	60.0	66.0	104.8	69.2	74.0
0320	Concrete Reinforcing	136.8	103.8	123.3	139.4	103.8	124.9	104.9	103.7	104.4	99.9	79.4	91.5	98.8	79.4	90.9	97.4	79.7	90.2
0330	Cast-in-Place Concrete	168.8	111.3	146.8	148.0	112.6	134.5	169.7	111.2	147.3	96.0	77.4	88.9	104.8	65.6	89.8	104.9	77.9	94.6
03	CONCRETE	138.1	110.7	125.6	125.3	112.1	119.3	132.2	110.6	122.4	121.0	72.3	98.9	108.5	65.9	89.1	108.1	74.3	92.7
04	MASONRY	207.3	116.1	152.7	216.8	117.7	157.5	198.8	116.1	149.3	97.8	61.8	76.3	110.7	47.6	72.9	97.8	64.9	78.1
05	METALS	124.0	100.9	117.3	124.1	101.1	117.4	124.1	100.8	117.3	95.3	74.9	89.3	93.5	75.4	88.2	95.1	76.6	89.7
06	WOOD, PLASTICS & COMPOSITES	114.3	113.6	113.9	118.5	115.7	117.0	114.3	113.6	113.9	105.5	65.2	83.4	96.2	65.3	79.3	97.1	70.0	82.2
07	THERMAL & MOISTURE PROTECTION	181.2	111.1	153.9	178.9	113.9	153.6	179.5	111.1	152.9	98.9	67.0	86.5	107.4	58.4	88.3	107.3	68.1	92.1
08	OPENINGS	125.0	109.9	121.3	122.2	111.0	119.4	122.2	109.8	119.1	100.8	68.9	92.9	96.3	66.7	89.0	97.3	71.5	90.9
0920	Plaster & Gypsum Board	139.3	113.8	122.7	156.6	116.0	130.2	139.3	113.8	122.7	85.7	64.0	71.6	84.9	64.0	71.3	87.3	68.8	75.2
0950, 0980	Ceilings & Acoustic Treatment	125.5	113.8	118.3	131.4	113.8	121.8	125.5	113.8	118.3	112.0	64.0	82.3	101.2	64.0	78.2	108.7	68.8	84.0
0960	Flooring	158.9	120.7	148.6	158.9	120.7	148.6	158.9	120.7	148.6	98.3	44.5	83.7	102.9	55.5	90.1	103.2	58.0	91.0
0970, 0990	Wall Finishes & Painting/Coating	155.7	112.9	130.0	155.7	121.0	134.8	155.7	112.9	130.0	90.4	55.1	69.2	94.0	55.1	70.7	94.0	57.2	71.9
09	FINISHES	148.5	115.6	131.1	150.3	118.2	133.3	147.3	115.6	130.5	101.0	59.8	79.2	100.2	58.1	78.0	102.4	65.3	82.8
COVERS	DIVS. 10 - 14, 25, 28, 41, 43, 44	100.0	110.2	102.0	100.0	111.1	102.2	100.0	110.2	102.0	100.0	81.5	96.3	100.0	76.7	95.4	100.0	82.4	96.5
21, 22, 23	FIRE SUPPRESSION, PLUMBING & HVAC	100.5	103.0	101.5	100.4	109.8	104.2	100.5	94.4	98.1	100.2	76.8	90.9	100.1	66.6	86.8	100.1	76.9	90.9
26, 27, 3370	ELECTRICAL, COMMUNICATIONS & UTIL.	140.0	109.5	125.1	148.2	109.5	129.4	141.5	109.5	125.9	101.6	61.3	81.9	92.5	61.3	77.3	100.3	64.5	82.9
MF2004	WEIGHTED AVERAGE	128.3	110.3	120.8	127.6	112.6	121.3	127.0	108.5	119.2	101.4	71.6	88.9	99.2	66.7	85.6	99.9	74.0	89.0

		ARIZONA PRESCOTT			ARIZONA TUCSON			ARKANSAS FORT SMITH			ARKANSAS JONESBORO			ARKANSAS LITTLE ROCK			ARKANSAS PINE BLUFF		
DIVISION		MAT.	INST.	TOTAL	MAT.	INST.	TOTAL	MAT.	INST.	TOTAL	MAT.	INST.	TOTAL	MAT.	INST.	TOTAL	MAT.	INST.	TOTAL
015433	CONTRACTOR EQUIPMENT		96.2	96.2		99.2	99.2		92.0	92.0		105.6	105.6		92.0	92.0		92.0	92.0
0241, 31 - 34	SITE & INFRASTRUCTURE, DEMOLITION	69.8	100.8	91.0	82.0	105.6	98.1	78.5	90.7	86.9	101.1	96.1	97.7	87.0	90.8	89.6	80.6	90.7	87.5
0310	Concrete Forming & Accessories	99.4	49.6	56.4	104.1	68.8	73.6	99.9	40.9	48.9	88.5	55.8	60.3	92.2	68.5	71.7	81.9	68.2	70.1
0320	Concrete Reinforcing	99.9	74.3	89.4	81.4	79.4	80.6	99.6	69.8	87.4	95.3	71.7	85.7	99.7	66.5	86.2	99.7	66.4	86.1
0330	Cast-in-Place Concrete	95.9	59.1	81.8	107.9	77.8	96.4	89.9	68.1	81.5	85.6	62.4	76.7	90.3	68.3	81.9	82.4	68.2	77.0
03	CONCRETE	106.7	57.9	84.5	106.3	74.0	91.6	89.0	56.8	74.4	85.8	62.5	75.2	88.8	68.4	79.5	86.9	68.2	78.4
04	MASONRY	98.1	50.9	69.8	95.8	61.8	75.4	97.4	57.0	73.2	93.6	48.6	66.6	94.5	57.0	72.0	118.5	57.0	81.6
05	METALS	95.3	69.2	87.7	94.3	75.5	88.8	102.8	72.3	93.9	96.8	83.8	93.0	98.7	71.7	90.7	101.3	71.4	92.5
06	WOOD, PLASTICS & COMPOSITES	101.1	47.9	71.9	96.4	70.0	81.9	103.9	36.4	66.9	90.3	56.5	71.8	98.0	73.2	84.4	83.1	73.2	77.6
07	THERMAL & MOISTURE PROTECTION	97.6	55.0	81.0	108.7	64.4	91.5	99.2	52.7	81.1	102.5	56.3	84.5	98.5	56.6	82.2	98.0	56.6	81.9
08	OPENINGS	100.8	53.1	89.0	92.6	71.5	87.4	97.2	44.6	84.2	99.1	60.4	89.5	97.2	65.1	89.3	92.5	65.1	85.7
0920	Plaster & Gypsum Board	83.1	46.0	59.0	90.4	68.8	76.3	86.9	34.7	53.0	98.2	55.1	70.2	83.1	72.7	76.4	80.6	72.7	75.5
0950, 0980	Ceilings & Acoustic Treatment	110.3	46.0	70.6	102.1	68.8	81.5	92.8	34.7	56.9	91.6	55.1	69.1	90.3	72.7	79.4	88.6	72.7	78.8
0960	Flooring	96.7	44.3	82.5	93.7	44.5	80.4	112.0	68.2	100.1	74.9	59.9	70.8	113.1	68.2	101.0	101.8	68.2	92.7
0970, 0990	Wall Finishes & Painting/Coating	90.4	42.4	61.6	94.0	55.1	70.7	98.5	61.9	76.5	87.4	55.6	68.3	98.5	63.5	77.5	98.5	63.5	77.5
09	FINISHES	98.6	47.0	71.3	98.3	62.6	79.4	95.6	47.1	70.0	88.2	56.3	71.4	95.0	68.9	81.2	90.9	68.9	79.3
COVERS	DIVS. 10 - 14, 25, 28, 41, 43, 44	100.0	77.8	95.6	100.0	82.4	96.5	100.0	69.7	94.0	100.0	56.6	91.4	100.0	74.0	94.9	100.0	74.0	94.9
21, 22, 23	FIRE SUPPRESSION, PLUMBING & HVAC	100.2	71.2	88.7	100.1	68.8	87.7	100.1	48.1	79.5	100.3	48.1	79.6	100.1	66.4	86.7	100.1	50.8	80.5
26, 27, 3370	ELECTRICAL, COMMUNICATIONS & UTIL.	101.3	61.2	81.8	94.8	58.8	77.3	95.8	75.3	85.8	103.8	52.2	78.7	100.5	76.6	88.9	95.1	76.6	86.1
MF2004	WEIGHTED AVERAGE	99.2	63.7	84.3	98.0	70.6	86.5	97.4	60.1	81.8	96.9	60.2	81.5	97.1	69.9	85.7	96.9	66.6	84.2

City Cost Indexes

	DIVISION	ARKANSAS TEXARKANA			CALIFORNIA ANAHEIM			BAKERSFIELD			FRESNO			LOS ANGELES			OAKLAND		
		MAT.	INST.	TOTAL	MAT.	INST.	TOTAL	MAT.	INST.	TOTAL	MAT.	INST.	TOTAL	MAT.	INST.	TOTAL	MAT.	INST.	TOTAL
015433	CONTRACTOR EQUIPMENT		90.7	90.7		102.2	102.2		100.7	100.7		100.7	100.7		100.3	100.3		104.6	104.6
0241, 31 - 34	SITE & INFRASTRUCTURE, DEMOLITION	95.9	87.8	90.4	101.8	109.4	107.0	108.0	107.8	107.9	107.6	107.2	107.4	98.5	110.1	106.4	141.8	105.6	117.1
0310	Concrete Forming & Accessories	88.6	46.1	51.9	106.2	126.7	123.9	102.1	126.1	122.9	102.5	126.4	123.2	104.2	126.7	123.6	108.4	142.5	137.8
0320	Concrete Reinforcing	99.2	69.7	87.2	93.2	118.2	103.4	105.0	118.3	110.4	88.9	118.5	101.0	104.4	118.6	110.2	98.2	119.2	106.8
0330	Cast-in-Place Concrete	89.9	49.0	74.2	101.2	123.5	109.7	108.3	122.4	113.7	106.7	117.0	110.6	108.8	122.0	113.9	132.9	120.7	128.2
03	CONCRETE	86.1	52.5	70.8	104.4	122.9	112.8	107.6	122.3	114.3	105.9	120.5	112.5	108.7	122.4	115.0	122.1	129.1	125.3
04	MASONRY	97.4	33.2	59.0	88.8	115.3	104.7	109.8	117.4	114.3	112.8	118.2	116.1	95.5	121.8	111.2	156.6	127.6	139.2
05	METALS	93.7	69.9	86.8	106.4	103.8	105.6	101.3	103.1	101.8	106.5	102.7	105.4	104.0	103.2	103.8	98.6	106.9	101.2
06	WOOD, PLASTICS & COMPOSITES	91.9	49.2	68.4	94.5	126.5	112.1	87.6	126.3	109.0	99.6	127.0	114.6	86.8	126.3	108.5	106.4	146.2	128.2
07	THERMAL & MOISTURE PROTECTION	98.9	45.9	78.3	102.5	116.7	108.1	100.5	109.8	104.1	96.4	112.5	102.7	104.2	118.8	109.9	107.3	125.6	114.4
08	OPENINGS	97.7	52.0	86.4	104.5	121.5	108.7	102.0	119.4	106.3	103.8	118.1	107.3	98.3	121.4	104.0	104.2	133.2	111.4
0920	Plaster & Gypsum Board	84.8	47.9	60.8	105.6	127.3	119.7	100.8	127.3	118.0	101.3	127.8	118.5	104.4	127.3	119.3	108.3	147.3	133.6
0950, 0980	Ceilings & Acoustic Treatment	94.5	47.9	65.7	114.8	127.3	122.6	116.4	127.3	123.2	119.5	127.8	124.6	126.1	127.3	126.9	120.5	147.3	137.0
0960	Flooring	104.4	57.2	91.6	114.2	113.6	114.1	114.2	87.9	107.1	121.1	133.7	124.5	102.3	113.6	105.3	117.6	116.2	117.2
0970, 0990	Wall Finishes & Painting/Coating	98.5	32.6	59.0	104.7	105.8	105.3	106.2	101.6	103.5	124.4	100.4	110.0	97.5	111.7	106.0	108.1	144.5	129.9
09	FINISHES	94.2	46.6	69.0	109.0	122.4	116.1	110.5	117.3	114.1	114.9	126.0	120.7	109.6	122.9	116.6	115.6	139.4	128.2
COVERS	DIVS. 10 - 14, 25, 28, 41, 43, 44	100.0	41.1	88.3	100.0	114.8	102.9	100.0	115.2	103.0	100.0	126.3	105.2	100.0	114.3	102.8	100.0	130.1	106.0
21, 22, 23	FIRE SUPPRESSION, PLUMBING & HVAC	100.1	35.6	74.5	100.1	115.5	106.3	100.1	99.9	100.0	100.3	111.6	104.8	100.3	115.5	106.3	100.3	143.0	117.3
26, 27, 3370	ELECTRICAL, COMMUNICATIONS & UTIL.	97.0	42.0	70.2	91.5	105.2	98.2	104.2	97.6	101.0	90.7	94.5	92.5	100.4	114.5	107.3	104.9	130.0	117.1
MF2004	WEIGHTED AVERAGE	95.9	48.8	76.1	101.5	114.7	107.0	103.3	109.4	105.8	103.2	112.6	107.2	102.2	116.7	108.3	108.8	129.5	117.5

	DIVISION	CALIFORNIA OXNARD			REDDING			RIVERSIDE			SACRAMENTO			SAN DIEGO			SAN FRANCISCO		
		MAT.	INST.	TOTAL	MAT.	INST.	TOTAL	MAT.	INST.	TOTAL	MAT.	INST.	TOTAL	MAT.	INST.	TOTAL	MAT.	INST.	TOTAL
015433	CONTRACTOR EQUIPMENT		100.0	100.0		100.6	100.6		101.4	101.4		102.9	102.9		99.6	99.6		108.1	108.1
0241, 31 - 34	SITE & INFRASTRUCTURE, DEMOLITION	108.0	106.3	106.9	112.2	106.2	108.1	100.1	107.8	105.4	111.5	110.7	110.9	105.6	104.1	104.6	144.3	111.2	121.7
0310	Concrete Forming & Accessories	105.7	126.6	123.8	105.4	126.0	123.2	106.4	126.5	123.7	106.9	127.0	124.2	108.1	111.0	110.6	108.2	144.2	139.3
0320	Concrete Reinforcing	105.0	117.9	110.3	101.9	118.0	108.5	104.1	118.0	109.8	90.8	118.6	102.1	99.3	118.2	107.0	111.8	119.9	115.1
0330	Cast-in-Place Concrete	107.5	122.7	113.3	121.8	116.7	119.8	105.0	123.4	112.0	112.3	118.0	114.5	107.0	105.8	106.6	136.3	122.7	131.1
03	CONCRETE	107.5	122.6	114.3	116.4	120.1	118.1	106.2	122.8	113.7	110.9	121.1	115.6	108.5	109.9	109.1	126.4	130.8	128.4
04	MASONRY	114.7	109.9	111.8	117.9	110.1	113.3	87.8	110.1	101.6	125.0	118.3	121.0	98.7	111.5	106.4	157.2	136.0	144.5
05	METALS	101.1	103.1	101.7	105.9	101.6	104.7	106.6	103.3	105.6	94.5	103.4	97.1	103.5	103.1	103.4	104.9	110.5	106.5
06	WOOD, PLASTICS & COMPOSITES	94.3	126.6	112.0	94.5	127.0	112.3	94.5	126.5	112.1	100.3	127.2	115.1	95.3	107.6	102.0	106.4	146.4	128.3
07	THERMAL & MOISTURE PROTECTION	105.8	112.5	108.4	106.2	108.9	107.3	102.8	111.8	106.3	112.8	114.5	113.4	108.4	105.6	107.3	109.9	138.3	120.9
08	OPENINGS	100.9	121.6	106.0	103.5	118.9	107.3	103.0	121.5	107.6	117.4	120.2	118.1	99.9	109.8	102.3	108.6	135.8	115.3
0920	Plaster & Gypsum Board	105.6	127.3	119.7	104.0	127.8	119.4	105.0	127.3	119.5	103.4	127.8	119.2	117.3	107.6	111.0	110.7	147.3	134.5
0950, 0980	Ceilings & Acoustic Treatment	121.4	127.3	125.1	126.4	127.8	127.2	118.9	127.3	124.1	118.6	127.8	124.3	107.2	107.6	107.4	127.8	147.3	139.8
0960	Flooring	112.7	113.6	113.0	113.5	109.6	112.4	118.4	113.6	117.1	117.0	114.3	116.3	105.4	107.6	106.0	117.6	126.2	119.9
0970, 0990	Wall Finishes & Painting/Coating	105.5	96.5	100.1	105.5	104.7	105.0	102.7	105.8	104.6	105.2	114.1	110.7	103.8	111.7	108.5	108.1	156.9	137.4
09	FINISHES	111.8	121.4	116.9	113.4	122.1	118.0	110.7	122.4	116.9	112.2	124.3	118.6	107.6	110.1	108.9	117.8	142.9	131.1
COVERS	DIVS. 10 - 14, 25, 28, 41, 43, 44	100.0	115.0	103.0	100.0	126.3	105.2	100.0	114.7	102.9	100.0	126.9	105.3	100.0	111.7	102.3	100.0	130.7	106.1
21, 22, 23	FIRE SUPPRESSION, PLUMBING & HVAC	100.2	115.6	106.3	100.2	103.5	101.5	100.1	115.5	106.2	100.1	112.9	105.2	100.3	113.8	105.6	100.3	164.9	126.0
26, 27, 3370	ELECTRICAL, COMMUNICATIONS & UTIL.	96.4	99.5	97.9	99.4	98.1	98.8	91.6	96.3	93.9	99.6	103.7	101.6	97.7	104.1	100.8	104.9	162.2	132.8
MF2004	WEIGHTED AVERAGE	102.9	112.8	107.0	105.8	109.8	107.5	101.6	112.7	106.3	104.9	114.6	108.9	102.4	108.8	105.1	111.2	141.3	123.8

	DIVISION	CALIFORNIA SAN JOSE			SANTA BARBARA			STOCKTON			VALLEJO			COLORADO COLORADO SPRINGS			DENVER		
		MAT.	INST.	TOTAL	MAT.	INST.	TOTAL	MAT.	INST.	TOTAL	MAT.	INST.	TOTAL	MAT.	INST.	TOTAL	MAT.	INST.	TOTAL
015433	CONTRACTOR EQUIPMENT		102.4	102.4		100.7	100.7		100.6	100.6		103.0	103.0		94.5	94.5		98.3	98.3
0241, 31 - 34	SITE & INFRASTRUCTURE, DEMOLITION	144.1	103.4	116.3	107.9	107.8	107.9	105.7	107.1	106.7	110.6	110.1	110.2	89.9	94.0	92.7	89.0	102.2	98.1
0310	Concrete Forming & Accessories	107.9	143.1	138.3	106.5	126.4	123.7	106.0	126.3	123.6	107.3	141.4	136.8	93.6	80.7	82.5	99.9	82.1	84.5
0320	Concrete Reinforcing	92.9	119.3	103.7	105.0	118.2	110.4	105.6	118.1	110.7	97.8	118.8	106.4	106.9	79.6	95.8	106.9	79.7	95.8
0330	Cast-in-Place Concrete	127.8	120.7	125.1	107.1	122.6	113.0	107.0	117.2	110.9	122.3	119.1	121.1	101.8	82.9	94.5	96.3	84.2	91.7
03	CONCRETE	114.6	129.5	121.4	107.3	122.4	114.2	107.4	120.5	113.4	117.0	127.9	122.0	108.0	81.5	96.0	102.5	82.6	93.5
04	MASONRY	148.2	128.9	136.6	111.1	114.7	113.2	114.3	110.2	111.9	88.9	121.4	108.3	101.4	75.9	86.1	103.3	79.0	88.7
05	METALS	100.5	109.6	103.2	101.5	103.2	102.0	102.9	102.9	102.9	97.7	104.4	99.6	101.7	86.0	97.1	104.7	86.1	98.9
06	WOOD, PLASTICS & COMPOSITES	101.9	145.9	126.1	94.3	126.6	112.0	96.1	127.0	113.1	98.0	145.9	124.3	94.4	84.4	88.9	102.1	84.2	92.3
07	THERMAL & MOISTURE PROTECTION	102.1	135.8	115.2	101.9	109.6	104.9	105.8	110.1	107.4	110.0	122.1	114.7	100.0	80.2	92.3	98.9	76.6	90.2
08	OPENINGS	93.9	135.5	104.2	102.0	121.6	106.8	101.2	120.1	105.9	117.0	135.5	121.6	99.5	85.2	96.0	99.4	85.1	95.9
0920	Plaster & Gypsum Board	100.9	147.3	131.0	105.6	127.3	119.7	106.0	127.8	120.1	108.2	147.3	133.6	82.4	83.9	83.4	95.8	83.9	88.1
0950, 0980	Ceilings & Acoustic Treatment	110.5	147.3	133.3	121.4	127.3	125.1	121.4	127.8	125.3	128.0	147.3	139.9	107.2	83.9	92.8	106.3	83.9	92.4
0960	Flooring	106.2	126.2	111.6	114.2	87.9	107.1	114.2	105.4	111.8	125.9	126.2	126.0	104.0	76.5	96.6	109.2	88.8	103.7
0970, 0990	Wall Finishes & Painting/Coating	106.2	146.2	130.2	105.5	96.5	100.1	105.5	105.1	105.3	105.1	128.2	119.0	105.7	46.6	70.2	106.0	76.5	88.3
09	FINISHES	108.6	141.4	126.0	112.5	117.0	114.8	112.3	121.4	117.2	115.6	139.1	128.0	100.0	76.7	87.7	103.1	83.0	92.4
COVERS	DIVS. 10 - 14, 25, 28, 41, 43, 44	100.0	129.5	105.8	100.0	115.0	103.0	100.0	126.3	105.2	100.0	128.9	105.7	100.0	87.8	97.6	100.0	88.6	97.7
21, 22, 23	FIRE SUPPRESSION, PLUMBING & HVAC	100.2	145.6	118.3	100.2	115.5	106.3	100.2	103.7	101.6	100.2	117.5	107.1	100.1	75.5	90.2	100.0	86.7	94.7
26, 27, 3370	ELECTRICAL, COMMUNICATIONS & UTIL.	103.1	141.0	121.6	88.4	100.3	94.2	99.1	111.7	105.2	95.6	116.5	105.8	104.9	83.3	94.3	106.7	85.4	96.3
MF2004	WEIGHTED AVERAGE	106.0	132.5	117.1	101.9	112.9	106.5	103.5	112.0	107.0	104.1	121.6	111.4	101.5	81.0	92.9	101.8	85.5	95.0

City Cost Indexes

DIVISION		COLORADO													CONNECTICUT					
		FORT COLLINS			GRAND JUNCTION			GREELEY			PUEBLO			BRIDGEPORT			BRISTOL		
		MAT.	INST.	TOTAL	MAT.	INST.	TOTAL	MAT.	INST.	TOTAL	MAT.	INST.	TOTAL	MAT.	INST.	TOTAL	MAT.	INST.	TOTAL
015433	CONTRACTOR EQUIPMENT		96.0	96.0		98.9	98.9		96.0	96.0		95.8	95.8		100.4	100.4		100.4	100.4
0241, 31 - 34	SITE & INFRASTRUCTURE, DEMOLITION	99.8	96.8	97.8	119.3	99.5	105.7	87.3	95.5	92.9	111.1	93.0	98.7	98.6	104.2	102.5	97.8	104.2	102.2
0310	Concrete Forming & Accessories	100.5	74.6	78.2	110.1	75.0	79.8	98.5	47.2	54.2	107.5	81.0	84.6	96.4	121.9	118.4	96.4	121.6	118.2
0320	Concrete Reinforcing	107.9	77.7	95.6	107.5	77.6	95.3	107.8	77.2	95.3	103.6	79.6	93.8	100.4	127.1	111.3	100.4	127.1	111.3
0330	Cast-in-Place Concrete	111.9	78.1	98.9	118.5	79.0	103.4	93.5	56.1	79.2	106.2	84.0	97.7	100.0	120.2	107.7	93.7	120.1	103.8
03	CONCRETE	114.9	76.8	97.6	119.9	77.2	100.5	100.2	57.0	80.6	109.6	82.1	97.1	107.4	122.0	114.0	104.5	121.8	112.4
04	MASONRY	112.1	52.7	76.5	139.9	54.8	88.9	106.6	35.0	63.7	103.5	74.2	86.0	106.6	132.0	121.8	99.2	132.0	118.8
05	METALS	99.8	81.5	94.5	98.1	80.9	93.1	99.8	80.1	94.0	99.7	86.8	96.0	94.9	123.3	103.2	94.9	123.1	103.1
06	WOOD, PLASTICS & COMPOSITES	101.9	78.6	89.1	103.9	78.7	90.1	99.4	46.9	70.6	101.2	84.7	92.2	101.4	119.6	111.4	101.4	119.6	111.4
07	THERMAL & MOISTURE PROTECTION	99.7	64.4	86.0	103.6	61.1	87.1	99.1	54.1	81.6	103.2	79.6	94.0	94.5	127.7	107.4	94.6	124.1	106.1
08	OPENINGS	95.3	81.8	92.0	99.3	81.9	95.0	95.3	64.6	87.7	93.8	85.4	91.7	104.9	129.4	110.9	104.9	127.4	110.4
0920	Plaster & Gypsum Board	94.3	78.0	83.7	105.1	78.0	87.5	93.1	45.2	62.0	80.6	83.9	82.7	104.8	119.5	114.4	104.8	119.5	114.4
0950, 0980	Ceilings & Acoustic Treatment	100.5	78.0	86.6	106.2	78.0	88.8	100.5	45.2	66.3	117.8	83.9	96.8	99.1	119.5	111.8	99.1	119.5	111.8
0960	Flooring	109.0	61.4	96.2	119.4	61.4	103.7	107.8	61.4	95.2	115.8	88.8	108.5	96.8	125.6	104.6	96.8	111.9	100.9
0970, 0990	Wall Finishes & Painting/Coating	106.0	46.2	70.1	108.9	76.5	89.5	106.0	28.8	59.7	108.9	43.2	69.5	91.7	116.7	106.7	91.7	116.7	106.7
09	FINISHES	102.0	69.4	84.8	110.7	73.4	91.0	100.8	46.5	72.1	108.6	78.8	92.9	100.5	121.9	111.8	100.5	119.5	110.5
COVERS	DIVS. 10 - 14, 25, 28, 41, 43, 44	100.0	85.9	97.2	100.0	87.0	97.4	100.0	78.8	95.8	100.0	88.6	97.8	100.0	111.7	102.3	100.0	111.7	102.3
21, 22, 23	FIRE SUPPRESSION, PLUMBING & HVAC	100.0	79.1	91.7	99.9	65.2	86.1	100.0	74.1	89.7	99.9	68.0	87.2	100.2	114.4	105.9	100.2	114.4	105.8
26, 27, 3370	ELECTRICAL, COMMUNICATIONS & UTIL.	101.2	85.3	93.5	94.7	61.7	78.7	101.2	85.3	93.5	95.7	77.6	86.9	97.0	109.9	103.3	97.0	109.6	103.2
MF2004	WEIGHTED AVERAGE	102.2	77.3	91.7	104.9	72.0	91.1	99.7	67.0	86.0	101.3	79.0	91.9	100.3	118.6	108.0	99.6	118.0	107.3

DIVISION		CONNECTICUT																	
		HARTFORD			NEW BRITAIN			NEW HAVEN			NORWALK			STAMFORD			WATERBURY		
		MAT.	INST.	TOTAL	MAT.	INST.	TOTAL	MAT.	INST.	TOTAL	MAT.	INST.	TOTAL	MAT.	INST.	TOTAL	MAT.	INST.	TOTAL
015433	CONTRACTOR EQUIPMENT		100.4	100.4		100.4	100.4		100.6	100.6		100.4	100.4		100.4	100.4		100.4	100.4
0241, 31 - 34	SITE & INFRASTRUCTURE, DEMOLITION	95.1	104.2	101.3	98.0	104.2	102.2	97.9	104.6	102.5	98.4	104.2	102.4	99.0	104.3	102.6	98.2	104.2	102.3
0310	Concrete Forming & Accessories	94.5	121.6	117.9	96.6	121.6	118.2	96.1	121.6	118.1	96.4	122.1	118.6	96.4	122.4	118.8	96.4	121.8	118.3
0320	Concrete Reinforcing	100.4	127.1	111.3	100.4	127.1	111.3	100.4	127.1	111.2	100.4	127.2	111.3	100.4	127.3	111.3	100.4	127.1	111.3
0330	Cast-in-Place Concrete	96.1	120.1	105.3	95.2	120.1	104.7	96.9	111.0	102.3	98.4	128.8	110.0	100.0	128.9	111.1	100.0	120.1	107.7
03	CONCRETE	105.5	121.8	112.9	105.2	121.8	112.7	120.8	118.7	119.8	106.7	125.1	115.0	107.4	125.3	115.5	107.4	121.9	114.0
04	MASONRY	99.4	132.0	118.9	101.2	132.0	119.6	99.4	132.0	118.9	98.8	133.7	119.7	99.7	133.7	120.0	99.7	132.0	119.0
05	METALS	99.5	123.1	106.4	91.6	123.1	100.8	91.8	123.0	100.9	94.9	123.6	103.3	94.9	123.9	103.4	94.9	123.2	103.2
06	WOOD, PLASTICS & COMPOSITES	99.9	119.6	110.7	101.4	119.6	111.4	101.4	119.6	111.4	101.4	119.6	111.4	101.4	119.6	111.4	101.4	119.6	111.4
07	THERMAL & MOISTURE PROTECTION	95.1	124.1	106.4	94.6	124.5	106.2	94.7	123.0	105.7	94.7	129.5	108.2	94.6	129.5	108.2	94.6	124.5	106.2
08	OPENINGS	105.6	127.4	111.0	104.9	127.4	110.4	104.9	129.4	110.9	104.9	129.4	110.9	104.9	129.4	110.9	104.9	129.4	110.9
0920	Plaster & Gypsum Board	106.1	119.5	114.8	104.8	119.5	114.4	104.8	119.5	114.4	104.8	119.5	114.4	104.8	119.5	114.4	104.8	119.5	114.4
0950, 0980	Ceilings & Acoustic Treatment	99.1	119.5	111.8	99.1	119.5	111.8	99.1	119.5	111.8	99.1	119.5	111.8	99.1	119.5	111.8	99.1	119.5	111.8
0960	Flooring	96.8	111.9	100.9	96.8	111.9	100.9	96.8	125.6	104.6	96.8	120.6	103.2	96.8	120.6	103.2	96.8	125.6	104.6
0970, 0990	Wall Finishes & Painting/Coating	91.7	116.7	106.7	91.7	116.7	106.7	91.7	116.7	106.7	91.7	116.7	106.7	91.7	116.7	106.7	91.7	116.7	106.7
09	FINISHES	100.2	119.5	110.4	100.5	119.5	110.5	100.5	121.9	111.8	100.5	121.0	111.3	100.6	121.0	111.4	100.4	121.9	111.7
COVERS	DIVS. 10 - 14, 25, 28, 41, 43, 44	100.0	111.7	102.3	100.0	111.7	102.3	100.0	111.7	102.3	100.0	111.7	102.3	100.0	111.7	102.3	100.0	111.7	102.3
21, 22, 23	FIRE SUPPRESSION, PLUMBING & HVAC	100.1	114.4	105.8	100.2	114.4	105.8	100.2	114.4	105.8	100.2	114.5	105.9	100.2	114.5	105.9	100.2	114.4	105.9
26, 27, 3370	ELECTRICAL, COMMUNICATIONS & UTIL.	96.9	110.5	103.5	97.1	109.6	103.2	97.0	109.6	103.1	97.0	108.3	102.5	97.0	154.5	125.0	96.6	109.9	103.1
MF2004	WEIGHTED AVERAGE	100.5	118.1	107.9	99.2	118.0	107.1	101.0	118.0	108.2	99.9	119.0	107.9	100.0	125.4	110.7	99.9	118.5	107.7

DIVISION		D.C.			DELAWARE			FLORIDA											
		WASHINGTON			WILMINGTON			DAYTONA BEACH			FORT LAUDERDALE			JACKSONVILLE			MELBOURNE		
		MAT.	INST.	TOTAL	MAT.	INST.	TOTAL	MAT.	INST.	TOTAL	MAT.	INST.	TOTAL	MAT.	INST.	TOTAL	MAT.	INST.	TOTAL
015433	CONTRACTOR EQUIPMENT		103.6	103.6		114.7	114.7		98.9	98.9		93.2	93.2		98.9	98.9		98.9	98.9
0241, 31 - 34	SITE & INFRASTRUCTURE, DEMOLITION	110.3	91.8	97.7	91.3	109.0	103.4	116.1	89.5	97.9	101.4	78.7	85.9	116.2	88.7	97.4	124.6	88.7	100.0
0310	Concrete Forming & Accessories	101.3	81.9	84.5	99.5	102.8	102.4	95.8	77.7	80.2	94.0	68.1	71.6	95.5	59.1	64.1	92.0	79.7	81.4
0320	Concrete Reinforcing	113.5	91.1	104.3	99.5	100.1	99.7	92.5	82.2	88.3	92.5	80.4	87.6	92.5	51.9	75.9	93.5	82.2	88.9
0330	Cast-in-Place Concrete	127.6	88.5	112.7	98.3	89.1	94.8	91.9	77.2	86.3	96.4	74.4	88.0	92.8	66.0	82.5	110.8	79.5	98.8
03	CONCRETE	121.1	87.3	105.8	102.8	98.5	100.8	90.4	79.4	85.4	92.4	74.0	84.0	90.8	61.8	77.6	101.5	81.1	92.2
04	MASONRY	101.7	81.9	89.8	107.6	93.8	99.3	86.3	72.3	77.9	86.6	68.2	75.6	86.2	58.9	69.8	83.7	76.3	79.2
05	METALS	98.0	109.3	101.3	99.6	114.7	104.0	101.7	94.5	99.6	101.6	94.4	99.5	100.2	81.2	94.7	110.7	94.9	106.1
06	WOOD, PLASTICS & COMPOSITES	100.9	81.2	90.1	99.8	104.2	102.2	95.7	81.0	87.6	91.6	65.6	77.4	95.7	57.9	74.9	91.4	81.0	85.7
07	THERMAL & MOISTURE PROTECTION	105.3	84.7	97.3	99.8	102.9	101.0	95.1	79.8	89.1	95.1	82.3	90.1	95.3	64.7	83.4	95.5	85.9	91.7
08	OPENINGS	107.0	92.7	103.4	95.2	108.6	98.5	99.7	74.9	93.6	97.1	66.1	89.4	99.7	52.1	88.0	98.9	80.9	94.5
0920	Plaster & Gypsum Board	111.0	80.6	91.2	111.0	104.3	106.6	100.7	80.8	87.8	100.3	64.9	77.3	100.7	56.9	72.2	97.0	80.8	86.5
0950, 0980	Ceilings & Acoustic Treatment	107.4	80.6	90.8	94.7	104.3	100.6	100.4	80.8	88.3	100.4	64.9	78.4	100.4	56.9	73.5	96.2	80.8	86.7
0960	Flooring	127.7	100.2	120.3	82.1	98.2	86.5	117.2	79.9	107.1	117.2	65.4	103.2	117.2	55.9	100.6	114.0	79.9	104.7
0970, 0990	Wall Finishes & Painting/Coating	125.2	87.4	102.5	95.8	98.0	97.1	115.3	76.7	92.1	110.4	54.2	76.7	115.3	49.6	75.9	115.3	105.7	109.5
09	FINISHES	111.0	85.3	97.5	100.5	101.4	101.0	106.3	78.4	91.6	103.9	65.1	83.4	106.3	57.2	80.4	104.5	82.7	93.0
COVERS	DIVS. 10 - 14, 25, 28, 41, 43, 44	100.0	98.9	99.8	100.0	105.5	101.1	100.0	83.5	96.7	100.0	86.3	97.3	100.0	79.0	95.8	100.0	85.0	97.0
21, 22, 23	FIRE SUPPRESSION, PLUMBING & HVAC	100.1	93.8	97.6	100.2	112.4	105.1	99.9	69.7	87.8	99.9	67.2	86.9	99.9	48.1	79.3	99.9	71.9	88.7
26, 27, 3370	ELECTRICAL, COMMUNICATIONS & UTIL.	103.4	101.5	102.5	105.3	108.1	106.6	99.2	62.9	81.6	99.2	75.5	87.7	98.9	68.2	84.0	100.5	72.9	87.1
MF2004	WEIGHTED AVERAGE	104.6	92.8	99.7	100.6	105.9	102.8	99.1	76.5	89.6	98.5	73.6	88.0	98.8	63.1	83.8	102.0	80.0	92.7

City Cost Indexes

DIVISION		FLORIDA															
		MIAMI			ORLANDO			PANAMA CITY			PENSACOLA			ST. PETERSBURG			TALLAHASSEE		
		MAT.	INST.	TOTAL	MAT.	INST.	TOTAL	MAT.	INST.	TOTAL	MAT.	INST.	TOTAL	MAT.	INST.	TOTAL	MAT.	INST.	TOTAL
015433	CONTRACTOR EQUIPMENT		93.2	93.2		98.9	98.9		98.9	98.9		98.9	98.9		98.9	98.9		98.9	98.9
0241, 31 - 34	SITE & INFRASTRUCTURE, DEMOLITION	104.8	79.0	87.2	118.7	88.3	98.0	131.2	86.4	100.6	130.8	87.9	101.5	117.0	87.8	97.0	114.3	87.6	96.0
0310	Concrete Forming & Accessories	98.8	68.8	72.9	95.5	79.4	81.6	94.8	43.6	50.6	92.8	55.9	60.9	93.6	49.9	55.9	97.3	45.9	52.9
0320	Concrete Reinforcing	92.5	80.6	87.6	92.5	79.0	87.0	96.6	48.8	77.1	99.0	48.7	78.5	95.8	55.4	79.3	92.5	51.6	75.8
0330	Cast-in-Place Concrete	104.6	75.8	93.6	121.1	76.9	104.2	97.5	46.3	77.9	120.5	62.6	98.4	104.0	61.8	87.8	102.3	54.3	83.9
03	CONCRETE	96.5	74.8	86.6	105.3	79.5	93.5	98.7	47.1	75.2	109.1	58.6	86.1	96.6	56.9	78.5	95.3	51.9	75.6
04	MASONRY	85.9	72.0	77.6	90.3	72.3	79.5	90.5	44.5	62.9	109.4	53.0	75.7	130.7	48.6	81.5	89.0	47.1	63.9
05	METALS	107.0	94.5	103.4	107.8	92.8	103.4	101.5	69.5	92.1	102.6	80.0	96.0	105.0	81.7	98.2	94.7	80.7	90.6
06	WOOD, PLASTICS & COMPOSITES	97.6	65.6	80.1	95.4	83.9	89.1	94.6	43.7	66.7	93.0	57.6	73.5	93.1	49.4	69.1	93.5	44.4	66.6
07	THERMAL & MOISTURE PROTECTION	103.2	79.5	94.0	96.2	80.1	89.9	95.5	47.4	76.8	95.5	59.2	81.3	95.0	55.6	79.7	100.5	74.5	90.4
08	OPENINGS	97.1	66.4	89.5	99.7	73.0	93.1	97.4	36.6	82.3	97.3	53.8	86.6	98.3	50.1	86.4	100.2	45.6	86.7
0920	Plaster & Gypsum Board	96.3	64.9	75.9	109.7	83.8	92.9	98.8	42.2	62.0	100.0	56.5	71.8	98.0	48.0	65.6	114.3	42.9	67.9
0950, 0980	Ceilings & Acoustic Treatment	97.1	64.9	77.2	100.4	83.8	90.2	94.6	42.2	62.2	96.2	56.5	71.7	94.6	48.0	65.8	100.4	42.9	64.8
0960	Flooring	124.6	68.9	109.4	117.4	79.9	107.2	116.6	46.9	97.8	110.9	59.4	97.0	115.6	60.3	100.7	117.2	50.3	99.1
0970, 0990	Wall Finishes & Painting/Coating	110.4	54.2	76.7	115.3	58.7	81.4	115.3	41.0	70.7	115.3	49.0	75.5	115.3	43.3	72.1	115.3	43.8	72.4
09	FINISHES	105.0	66.2	84.5	107.6	78.1	92.0	105.9	44.0	73.2	104.4	55.7	78.7	104.1	50.7	75.9	107.8	45.5	74.9
COVERS	DIVS. 10 - 14, 25, 28, 41, 43, 44	100.0	86.9	97.4	100.0	83.7	96.8	100.0	46.9	89.5	100.0	48.0	89.7	100.0	56.8	91.4	100.0	71.7	94.4
21, 22, 23	FIRE SUPPRESSION, PLUMBING & HVAC	99.9	72.1	88.8	99.8	64.3	85.7	99.9	37.8	75.2	99.9	57.1	82.8	99.9	55.6	82.3	99.8	42.2	76.9
26, 27, 3370	ELECTRICAL, COMMUNICATIONS & UTIL.	107.1	76.3	92.1	107.0	40.4	74.6	97.9	41.8	70.6	101.9	58.2	80.6	99.5	47.0	73.9	106.8	36.7	72.6
MF2004	WEIGHTED AVERAGE	101.2	75.4	90.3	103.1	72.0	90.0	100.3	48.6	78.6	102.9	61.3	85.4	102.3	58.1	83.7	99.6	53.0	80.1

DIVISION		FLORIDA			GEORGIA														
		TAMPA			ALBANY			ATLANTA			AUGUSTA			COLUMBUS			MACON		
		MAT.	INST.	TOTAL	MAT.	INST.	TOTAL	MAT.	INST.	TOTAL	MAT.	INST.	TOTAL	MAT.	INST.	TOTAL	MAT.	INST.	TOTAL
015433	CONTRACTOR EQUIPMENT		98.9	98.9		94.3	94.3		93.4	93.4		92.7	92.7		94.3	94.3		102.3	102.3
0241, 31 - 34	SITE & INFRASTRUCTURE, DEMOLITION	117.3	88.5	97.6	101.7	81.2	87.6	104.1	94.1	97.3	100.4	90.9	93.9	101.6	81.9	88.1	102.1	94.0	96.5
0310	Concrete Forming & Accessories	96.7	82.8	84.7	94.9	49.5	55.7	97.6	76.1	79.0	95.1	55.1	60.6	94.8	62.4	66.8	94.2	59.4	64.1
0320	Concrete Reinforcing	92.5	86.2	89.9	92.1	89.8	91.2	98.4	91.7	95.6	99.3	69.3	87.0	92.5	90.9	91.9	93.7	90.2	92.2
0330	Cast-in-Place Concrete	101.6	67.4	88.5	98.1	51.4	80.3	112.0	74.5	97.7	105.7	53.6	85.8	97.8	61.1	83.7	96.4	53.5	80.0
03	CONCRETE	94.9	79.0	87.7	93.1	59.2	77.7	101.6	78.6	91.1	97.3	57.9	79.4	93.0	68.6	81.9	92.6	64.5	79.8
04	MASONRY	87.5	73.0	78.8	89.4	45.4	63.1	87.7	73.7	79.3	87.9	45.1	62.3	88.4	64.0	73.8	101.5	43.2	66.6
05	METALS	103.9	95.5	101.5	101.0	89.0	97.5	93.6	83.5	90.7	92.4	71.5	86.3	100.5	92.8	98.2	96.1	92.0	94.9
06	WOOD, PLASTICS & COMPOSITES	97.0	86.4	91.2	94.6	44.5	67.1	97.3	77.4	86.4	94.5	55.6	73.1	94.6	61.0	76.2	103.4	59.9	79.5
07	THERMAL & MOISTURE PROTECTION	95.3	90.3	93.3	92.3	60.5	80.0	94.3	78.2	88.0	93.9	56.1	79.2	92.5	66.4	82.3	91.1	62.6	81.4
08	OPENINGS	99.7	77.8	94.3	97.1	51.0	85.7	99.1	75.4	93.2	93.3	54.8	83.8	97.1	62.1	88.4	95.2	60.5	86.6
0920	Plaster & Gypsum Board	100.7	86.4	91.4	98.0	43.0	62.3	118.8	77.0	91.7	117.7	54.3	76.6	100.5	60.1	74.3	108.5	58.9	76.3
0950, 0980	Ceilings & Acoustic Treatment	100.4	86.4	91.8	99.6	43.0	64.6	109.8	77.0	89.5	110.7	54.3	75.8	99.6	60.1	75.2	95.7	58.9	73.0
0960	Flooring	117.2	60.3	101.8	117.2	34.6	94.9	89.4	71.8	84.6	88.3	44.1	76.3	117.2	56.5	100.8	91.2	40.7	77.5
0970, 0990	Wall Finishes & Painting/Coating	115.3	61.6	83.1	110.4	46.5	72.1	102.3	86.2	92.9	102.8	42.0	66.3	110.4	54.9	77.1	112.7	51.8	76.1
09	FINISHES	106.4	76.7	90.7	103.3	44.8	72.4	100.7	76.2	87.7	100.2	51.2	74.3	103.6	59.7	80.4	92.4	54.1	72.2
COVERS	DIVS. 10 - 14, 25, 28, 41, 43, 44	100.0	84.7	97.0	100.0	81.4	96.3	100.0	85.7	97.2	100.0	59.5	92.0	100.0	83.3	96.7	100.0	81.1	96.3
21, 22, 23	FIRE SUPPRESSION, PLUMBING & HVAC	99.9	89.3	95.7	99.8	65.6	86.2	100.0	78.0	91.2	100.0	73.6	89.5	99.9	55.9	82.4	99.9	65.7	86.3
26, 27, 3370	ELECTRICAL, COMMUNICATIONS & UTIL.	99.1	47.0	73.8	94.5	58.3	76.9	98.7	76.7	88.0	99.6	68.7	84.5	94.6	61.4	78.4	93.1	63.8	78.8
MF2004	WEIGHTED AVERAGE	100.1	78.7	91.1	97.9	62.0	82.9	98.2	79.2	90.2	96.9	64.3	83.2	97.8	66.8	84.8	96.5	66.6	83.9

DIVISION		GEORGIA			HAWAII						IDAHO					
		SAVANNAH			VALDOSTA			HONOLULU			STATES & POSS., GUAM			BOISE			LEWISTON		
		MAT.	INST.	TOTAL	MAT.	INST.	TOTAL	MAT.	INST.	TOTAL	MAT.	INST.	TOTAL	MAT.	INST.	TOTAL	MAT.	INST.	TOTAL
015433	CONTRACTOR EQUIPMENT		95.0	95.0		94.3	94.3		101.7	101.7		166.8	166.8		101.1	101.1		97.4	97.4
0241, 31 - 34	SITE & INFRASTRUCTURE, DEMOLITION	104.1	81.4	88.6	112.2	81.1	90.9	137.0	107.0	116.5	175.5	104.4	126.9	77.2	101.1	93.6	84.6	100.5	95.4
0310	Concrete Forming & Accessories	95.0	54.7	60.2	84.2	47.4	52.4	116.5	134.3	131.9	110.8	65.5	71.9	101.1	75.5	79.0	119.6	63.6	71.2
0320	Concrete Reinforcing	93.4	69.5	83.6	94.4	48.1	75.5	105.0	112.4	108.1	185.6	35.3	124.3	102.2	72.7	90.2	109.4	95.6	103.8
0330	Cast-in-Place Concrete	114.0	52.5	90.5	96.0	54.8	80.3	174.6	120.7	154.0	177.7	117.2	154.6	96.8	81.3	90.9	108.8	90.1	101.7
03	CONCRETE	100.7	58.2	81.4	96.9	51.9	76.4	139.4	124.4	132.6	156.7	78.9	121.3	105.8	77.2	92.8	116.7	79.3	99.7
04	MASONRY	88.4	55.8	68.9	94.2	50.6	68.1	126.5	125.4	125.9	193.1	50.3	107.6	129.1	64.4	90.4	131.3	85.4	103.8
05	METALS	96.5	82.9	92.5	100.1	75.6	92.9	137.8	105.6	128.4	147.9	85.3	129.6	100.9	76.3	93.7	94.3	87.3	92.2
06	WOOD, PLASTICS & COMPOSITES	109.0	52.4	77.9	81.9	43.4	60.8	109.8	139.0	125.8	109.7	67.2	86.4	93.5	76.2	84.0	103.4	55.7	77.2
07	THERMAL & MOISTURE PROTECTION	95.1	57.8	80.6	92.7	63.3	81.2	115.3	122.7	118.2	123.5	73.1	103.9	95.4	71.0	85.9	146.2	78.1	119.7
08	OPENINGS	98.2	50.6	86.4	92.6	40.8	79.8	106.1	129.4	111.9	105.6	56.9	93.6	92.6	71.5	87.4	113.1	62.9	100.7
0920	Plaster & Gypsum Board	97.1	51.2	67.3	92.8	41.9	59.7	136.6	140.3	138.9	209.0	53.9	108.2	81.2	75.2	77.3	136.2	54.1	82.8
0950, 0980	Ceilings & Acoustic Treatment	100.4	51.2	70.0	94.6	41.9	62.0	119.7	140.2	132.3	244.1	53.9	126.5	112.2	75.2	89.3	143.8	54.1	88.3
0960	Flooring	118.0	52.0	100.1	109.3	43.4	91.5	164.8	125.5	154.2	164.5	52.0	134.1	100.3	71.3	92.5	142.5	89.7	128.2
0970, 0990	Wall Finishes & Painting/Coating	112.3	52.6	76.5	110.4	37.0	66.4	104.0	139.2	125.1	105.5	41.3	67.0	98.1	45.4	66.5	119.6	67.5	88.3
09	FINISHES	103.9	53.6	77.3	99.9	44.9	70.8	133.9	134.7	134.3	200.4	61.4	126.9	100.8	72.2	85.7	168.1	67.5	114.9
COVERS	DIVS. 10 - 14, 25, 28, 41, 43, 44	100.0	61.0	92.3	100.0	59.8	92.0	100.0	116.4	103.2	100.0	91.3	98.3	100.0	71.5	94.4	100.0	74.7	95.0
21, 22, 23	FIRE SUPPRESSION, PLUMBING & HVAC	99.9	59.6	83.9	99.9	40.8	76.4	100.2	107.3	103.0	102.8	44.9	79.8	99.9	69.2	87.7	101.0	83.4	94.0
26, 27, 3370	ELECTRICAL, COMMUNICATIONS & UTIL.	97.4	60.5	79.4	92.7	36.0	65.1	111.7	121.3	116.4	159.7	48.1	105.4	97.4	77.1	87.5	84.8	78.0	81.5
MF2004	WEIGHTED AVERAGE	98.8	62.0	83.3	97.8	51.0	78.2	118.7	118.8	118.7	138.1	64.3	107.1	100.6	74.8	89.8	109.0	80.4	97.0

City Cost Indexes

		IDAHO						ILLINOIS									
	DIVISION	POCATELLO			TWIN FALLS			CHICAGO			DECATUR			EAST ST. LOUIS			JOLIET		
		MAT.	INST.	TOTAL	MAT.	INST.	TOTAL	MAT.	INST.	TOTAL	MAT.	INST.	TOTAL	MAT.	INST.	TOTAL	MAT.	INST.	TOTAL
015433	CONTRACTOR EQUIPMENT		101.1	101.1		101.1	101.1		94.1	94.1		101.3	101.3		108.1	108.1		92.0	92.0
0241, 31 - 34	SITE & INFRASTRUCTURE, DEMOLITION	78.5	101.1	94.0	84.9	99.1	94.6	98.1	95.3	96.2	88.3	95.1	93.0	104.6	98.0	100.1	97.9	93.6	94.9
0310	Concrete Forming & Accessories	101.2	75.2	78.8	102.0	32.3	41.8	100.5	153.2	146.1	98.1	108.8	107.4	93.2	110.3	107.9	102.5	134.2	129.9
0320	Concrete Reinforcing	102.6	73.0	90.5	104.4	71.2	90.8	106.3	157.6	127.2	99.0	96.9	98.1	101.0	105.9	103.0	106.3	146.9	122.9
0330	Cast-in-Place Concrete	99.6	81.1	92.5	102.1	43.4	79.7	105.9	145.2	120.9	99.4	108.5	102.9	93.0	107.7	98.6	105.8	138.0	118.1
03	CONCRETE	105.2	77.0	92.4	112.9	44.8	81.9	104.5	149.8	125.1	97.3	106.6	101.5	89.1	109.2	98.2	104.6	136.9	119.3
04	MASONRY	126.4	70.6	93.0	129.2	33.8	72.1	94.4	152.2	129.0	72.1	109.4	94.5	75.6	113.5	98.3	97.2	139.5	122.5
05	METALS	108.9	76.4	99.4	108.9	73.0	98.4	90.1	130.7	102.0	95.0	104.6	97.8	92.7	117.4	99.9	88.2	124.1	98.7
06	WOOD, PLASTICS & COMPOSITES	93.5	76.2	84.0	94.1	31.3	59.6	105.2	152.3	131.1	102.9	108.2	105.8	100.2	109.1	105.1	107.0	131.5	120.5
07	THERMAL & MOISTURE PROTECTION	95.5	66.3	84.1	96.2	41.7	75.0	100.7	142.7	117.1	99.2	102.3	100.4	93.6	108.4	99.3	100.6	134.6	113.8
08	OPENINGS	93.3	65.7	86.5	96.6	38.8	82.3	101.6	155.3	114.8	96.9	105.3	99.0	86.6	113.9	93.3	99.4	141.2	109.8
0920	Plaster & Gypsum Board	78.7	75.2	76.4	78.4	28.6	46.1	92.0	154.1	132.3	98.5	108.4	104.9	95.1	109.4	104.4	89.4	132.6	117.5
0950, 0980	Ceilings & Acoustic Treatment	117.8	75.2	91.5	111.2	28.6	60.1	101.5	154.1	134.0	93.4	108.4	102.7	87.6	109.4	101.1	101.5	132.6	120.8
0960	Flooring	103.6	71.3	94.8	104.9	48.3	89.6	92.0	149.5	107.5	100.2	111.2	103.2	109.6	118.3	111.9	91.6	138.4	104.2
0970, 0990	Wall Finishes & Painting/Coating	98.0	47.2	67.5	98.1	27.8	55.9	86.5	148.0	123.4	93.0	106.6	101.2	100.6	103.3	102.2	84.6	135.3	115.0
09	FINISHES	102.8	72.4	86.7	102.2	33.8	66.0	93.8	152.7	124.9	97.1	109.9	103.8	96.2	110.9	104.0	93.2	135.2	115.4
COVERS	DIVS. 10 - 14, 25, 28, 41, 43, 44	100.0	71.5	94.4	100.0	41.3	88.4	100.0	126.8	105.3	100.0	95.4	99.1	100.0	96.9	99.4	100.0	121.3	104.2
21, 22, 23	FIRE SUPPRESSION, PLUMBING & HVAC	99.9	69.2	87.7	99.9	57.6	83.1	99.9	133.8	113.4	99.9	101.2	100.4	99.9	96.5	98.5	100.0	125.8	110.3
26, 27, 3370	ELECTRICAL, COMMUNICATIONS & UTIL.	93.8	72.8	83.6	87.9	63.6	76.1	97.5	135.8	116.1	96.5	91.2	93.9	93.5	100.0	96.7	96.3	118.0	106.9
MF2004	WEIGHTED AVERAGE	101.6	74.5	90.2	102.5	54.0	82.1	97.9	138.3	114.9	96.2	102.5	98.8	93.9	105.7	98.8	97.4	126.9	109.8

		ILLINOIS									INDIANA								
	DIVISION	PEORIA			ROCKFORD			SPRINGFIELD			ANDERSON			BLOOMINGTON			EVANSVILLE		
		MAT.	INST.	TOTAL	MAT.	INST.	TOTAL	MAT.	INST.	TOTAL	MAT.	INST.	TOTAL	MAT.	INST.	TOTAL	MAT.	INST.	TOTAL
015433	CONTRACTOR EQUIPMENT		100.5	100.5		100.5	100.5		101.3	101.3		96.9	96.9		87.5	87.5		113.5	113.5
0241, 31 - 34	SITE & INFRASTRUCTURE, DEMOLITION	97.2	94.2	95.2	96.3	95.6	95.8	95.5	95.2	95.3	83.4	94.7	91.1	74.8	94.8	88.5	79.4	121.3	108.0
0310	Concrete Forming & Accessories	96.5	109.8	108.0	100.6	119.9	117.3	98.7	108.7	107.4	94.9	78.7	80.9	99.9	76.1	79.4	93.9	81.8	83.4
0320	Concrete Reinforcing	99.0	107.5	102.4	93.5	127.0	107.2	99.0	100.1	99.4	95.1	76.5	87.5	86.1	76.5	82.2	94.4	78.7	88.0
0330	Cast-in-Place Concrete	97.1	103.1	99.4	99.4	116.6	106.0	88.4	103.8	94.3	95.0	79.6	89.1	95.9	76.5	88.5	91.7	89.6	90.9
03	CONCRETE	96.1	107.2	101.1	96.3	120.0	107.1	92.1	105.6	98.2	90.2	79.3	85.2	97.9	76.3	88.1	98.6	84.2	92.1
04	MASONRY	112.5	109.2	110.5	87.6	110.3	101.2	76.8	110.3	96.8	85.8	80.4	82.6	86.5	74.2	79.1	82.8	84.6	83.9
05	METALS	95.0	109.6	99.2	95.0	119.6	102.2	97.4	106.0	99.9	100.2	87.8	96.6	94.3	76.1	89.0	87.0	86.1	86.8
06	WOOD, PLASTICS & COMPOSITES	102.9	106.1	104.7	102.9	117.7	111.1	104.0	107.4	105.8	106.6	78.4	91.1	115.7	75.4	93.6	95.2	80.5	87.1
07	THERMAL & MOISTURE PROTECTION	99.3	107.5	102.5	102.0	116.7	107.7	99.0	105.8	101.6	96.2	78.8	89.4	92.8	79.0	87.4	96.8	85.3	92.3
08	OPENINGS	97.1	107.5	99.6	97.1	119.8	102.7	98.1	105.7	100.0	95.6	78.1	91.3	103.1	76.5	96.5	95.5	79.1	91.5
0920	Plaster & Gypsum Board	98.5	106.2	103.5	98.5	118.3	111.4	94.9	107.6	103.1	95.7	78.1	84.3	88.8	75.2	80.0	84.7	79.1	81.1
0950, 0980	Ceilings & Acoustic Treatment	93.4	106.2	101.4	93.4	118.3	108.8	91.8	107.6	101.6	91.1	78.1	83.1	79.9	75.2	77.0	85.3	79.1	81.5
0960	Flooring	100.2	110.2	102.9	100.2	113.6	103.8	100.2	102.7	100.9	92.9	87.0	91.3	104.8	84.1	99.2	98.3	82.6	94.0
0970, 0990	Wall Finishes & Painting/Coating	93.0	112.8	104.9	93.0	122.6	110.8	93.0	106.2	100.9	95.1	69.4	79.7	91.6	85.5	87.9	97.7	82.0	88.3
09	FINISHES	97.2	110.3	104.1	97.2	119.2	108.8	96.0	107.8	102.3	91.4	79.3	85.0	93.7	78.6	85.7	92.4	81.9	86.9
COVERS	DIVS. 10 - 14, 25, 28, 41, 43, 44	100.0	102.0	100.4	100.0	102.4	100.5	100.0	95.6	99.1	100.0	89.9	98.0	100.0	80.3	96.1	100.0	85.6	97.1
21, 22, 23	FIRE SUPPRESSION, PLUMBING & HVAC	99.9	105.9	102.3	100.0	111.9	104.7	99.8	102.1	100.7	99.9	74.0	89.6	99.7	78.3	91.2	99.9	80.6	92.2
26, 27, 3370	ELECTRICAL, COMMUNICATIONS & UTIL.	95.6	95.1	95.4	96.0	122.8	109.0	101.8	96.9	99.4	83.2	93.0	88.0	98.6	78.8	89.0	93.8	87.3	90.6
MF2004	WEIGHTED AVERAGE	98.2	104.9	101.0	97.1	115.0	104.7	97.0	103.4	99.7	94.8	82.5	89.6	96.9	78.8	89.3	94.4	86.5	91.1

		INDIANA																	
	DIVISION	FORT WAYNE			GARY			INDIANAPOLIS			MUNCIE			SOUTH BEND			TERRE HAUTE		
		MAT.	INST.	TOTAL	MAT.	INST.	TOTAL	MAT.	INST.	TOTAL	MAT.	INST.	TOTAL	MAT.	INST.	TOTAL	MAT.	INST.	TOTAL
015433	CONTRACTOR EQUIPMENT		96.9	96.9		96.9	96.9		92.0	92.0		96.5	96.5		105.8	105.8		113.5	113.5
0241, 31 - 34	SITE & INFRASTRUCTURE, DEMOLITION	84.5	94.5	91.3	83.9	97.6	93.2	83.0	98.1	93.2	74.9	95.1	88.7	86.0	95.0	92.1	81.1	121.4	108.7
0310	Concrete Forming & Accessories	93.6	74.1	76.7	95.0	106.7	105.1	95.5	83.7	85.3	92.2	78.4	80.3	96.5	77.7	80.3	95.0	80.8	82.7
0320	Concrete Reinforcing	95.1	75.8	87.2	95.1	91.8	93.8	94.5	80.9	88.9	95.3	76.5	87.6	95.1	72.9	86.1	94.4	83.8	90.0
0330	Cast-in-Place Concrete	100.9	78.3	92.3	99.2	107.2	102.3	91.4	85.9	89.3	100.8	80.8	93.2	96.2	82.3	90.9	88.8	87.6	88.3
03	CONCRETE	92.9	76.6	85.5	92.2	104.0	97.6	91.7	83.8	88.1	97.5	79.5	89.3	87.4	79.9	84.0	101.0	84.0	93.3
04	MASONRY	89.4	77.8	82.5	87.0	104.4	97.4	93.5	83.6	87.5	88.5	80.4	83.6	86.0	79.0	81.8	89.4	81.2	84.5
05	METALS	100.2	86.9	96.3	100.2	98.8	99.8	102.2	80.7	95.9	95.9	87.5	93.5	100.2	98.0	99.5	87.7	88.6	88.0
06	WOOD, PLASTICS & COMPOSITES	106.3	73.1	88.1	104.5	107.0	105.9	103.5	83.3	92.4	109.8	78.2	92.5	103.7	77.1	89.1	97.3	80.2	87.9
07	THERMAL & MOISTURE PROTECTION	96.0	81.7	90.4	95.2	104.4	98.8	94.4	83.9	90.3	95.2	78.9	88.9	91.2	83.8	88.3	96.8	80.9	90.7
08	OPENINGS	95.6	72.7	90.0	95.6	104.7	97.9	103.6	83.4	98.6	97.9	78.0	93.0	88.9	76.2	85.8	96.1	80.9	92.3
0920	Plaster & Gypsum Board	94.9	72.6	80.4	90.5	107.8	101.7	94.0	83.0	86.8	84.5	78.1	80.4	88.6	76.7	80.9	84.7	78.8	80.9
0950, 0980	Ceilings & Acoustic Treatment	91.1	72.6	79.7	91.1	107.8	101.4	95.3	83.0	87.7	80.8	78.1	79.1	87.0	76.7	80.6	85.3	78.8	81.3
0960	Flooring	92.9	76.2	88.4	92.9	112.5	98.2	92.7	89.6	91.9	97.7	87.0	94.8	91.3	83.4	89.2	98.3	87.3	95.3
0970, 0990	Wall Finishes & Painting/Coating	95.1	76.1	83.7	95.1	104.9	100.9	95.1	90.8	92.5	91.6	69.4	78.3	94.6	81.4	86.7	97.7	84.1	89.6
09	FINISHES	91.2	74.6	82.5	90.7	108.0	99.8	92.5	85.6	88.9	90.9	79.1	84.7	89.0	79.1	83.8	92.4	82.3	87.1
COVERS	DIVS. 10 - 14, 25, 28, 41, 43, 44	100.0	79.4	95.9	100.0	86.2	97.3	100.0	91.7	98.4	100.0	89.3	97.9	100.0	80.7	96.1	100.0	92.8	98.6
21, 22, 23	FIRE SUPPRESSION, PLUMBING & HVAC	99.9	73.7	89.5	99.9	100.5	100.2	99.9	85.7	94.2	99.7	74.0	89.4	99.8	80.4	92.1	99.9	82.7	93.1
26, 27, 3370	ELECTRICAL, COMMUNICATIONS & UTIL.	84.0	77.4	80.8	96.0	97.4	96.7	98.3	93.0	95.7	89.1	79.7	84.5	103.7	82.7	93.5	92.1	83.7	88.0
MF2004	WEIGHTED AVERAGE	95.3	78.4	88.2	96.3	101.5	98.5	98.0	86.7	93.2	95.5	80.6	89.3	95.6	83.1	90.4	95.0	86.6	91.5

City Cost Indexes

	DIVISION	IOWA																	
		CEDAR RAPIDS			COUNCIL BLUFFS			DAVENPORT			DES MOINES			DUBUQUE			SIOUX CITY		
		MAT.	INST.	TOTAL	MAT.	INST.	TOTAL	MAT.	INST.	TOTAL	MAT.	INST.	TOTAL	MAT.	INST.	TOTAL	MAT.	INST.	TOTAL
015433	CONTRACTOR EQUIPMENT		96.8	96.8		96.1	96.1		99.9	99.9		101.2	101.2		95.9	95.9		99.9	99.9
0241, 31 - 34	SITE & INFRASTRUCTURE, DEMOLITION	94.5	96.5	95.9	101.6	92.4	95.3	92.6	99.8	97.5	85.1	99.7	95.1	92.5	93.6	93.3	103.8	95.9	98.4
0310	Concrete Forming & Accessories	100.0	77.3	80.4	82.1	54.4	58.2	99.7	87.9	89.5	100.7	72.6	76.4	83.5	68.8	70.8	100.0	62.1	67.3
0320	Concrete Reinforcing	102.3	80.6	93.4	104.3	75.4	92.5	102.3	92.5	98.3	102.3	77.1	92.0	100.9	80.4	92.6	102.3	60.8	85.4
0330	Cast-in-Place Concrete	105.3	79.0	95.3	109.5	77.0	97.1	101.4	91.7	97.7	97.2	92.5	95.4	103.1	89.1	97.8	104.4	54.0	85.2
03	CONCRETE	100.6	79.3	90.9	103.0	67.6	86.9	98.8	90.7	95.1	95.8	81.3	89.2	97.6	79.0	89.1	100.0	60.3	81.9
04	MASONRY	102.5	75.8	86.5	102.9	76.2	86.9	98.6	84.9	90.4	94.1	69.4	79.3	103.4	67.3	81.8	96.2	53.0	70.3
05	METALS	87.7	91.0	88.7	93.1	88.5	91.8	87.7	99.9	91.3	87.9	90.6	88.7	86.5	90.6	87.7	87.7	81.0	85.8
06	WOOD, PLASTICS & COMPOSITES	108.4	76.4	90.8	87.1	47.8	65.5	108.4	86.6	96.4	108.2	72.1	88.4	88.9	68.4	77.7	108.4	62.4	83.1
07	THERMAL & MOISTURE PROTECTION	101.7	77.6	92.3	101.1	64.4	86.9	101.2	86.7	95.6	102.4	74.4	91.6	101.4	68.4	88.6	101.2	52.2	82.2
08	OPENINGS	99.0	79.8	94.2	98.0	63.1	89.4	99.0	89.5	96.6	99.0	77.0	93.6	98.0	77.4	92.9	99.0	59.9	89.3
0920	Plaster & Gypsum Board	98.5	75.8	83.8	89.9	46.2	61.5	98.5	86.2	90.5	88.9	71.2	77.4	90.3	67.6	75.6	98.5	61.1	74.2
0950, 0980	Ceilings & Acoustic Treatment	113.2	75.8	90.1	109.1	46.2	70.2	113.2	86.2	96.5	109.1	71.2	85.6	109.1	67.6	83.4	113.2	61.1	81.0
0960	Flooring	132.2	50.0	110.0	105.8	40.8	88.2	116.8	93.1	110.4	116.5	37.3	95.1	121.3	34.8	97.9	117.3	52.7	99.9
0970, 0990	Wall Finishes & Painting/Coating	105.6	73.3	86.2	98.4	70.2	81.5	102.0	88.5	93.9	102.0	76.4	86.6	104.6	73.3	85.8	103.0	61.0	77.8
09	FINISHES	114.3	71.0	91.4	103.7	51.5	76.1	109.5	88.8	98.5	106.0	65.6	84.6	108.7	62.0	84.0	111.2	60.2	84.2
COVERS	DIVS. 10 - 14, 25, 28, 41, 43, 44	100.0	84.7	97.0	100.0	83.5	96.7	100.0	88.0	97.6	100.0	83.3	96.7	100.0	81.5	96.3	100.0	80.5	96.1
21, 22, 23	FIRE SUPPRESSION, PLUMBING & HVAC	100.2	81.3	92.7	100.2	77.8	91.3	100.2	88.0	95.4	100.1	70.5	88.4	100.2	73.0	89.4	100.2	72.4	89.1
26, 27, 3370	ELECTRICAL, COMMUNICATIONS & UTIL.	96.5	78.6	87.7	101.6	82.4	92.3	93.9	87.3	90.7	109.1	72.8	91.4	100.2	75.0	87.9	96.5	71.5	84.3
MF2004	WEIGHTED AVERAGE	98.8	80.8	91.2	99.6	74.6	89.1	97.7	90.2	94.5	98.2	76.7	89.2	97.9	75.8	88.6	98.4	68.8	86.0

	DIVISION	IOWA			KANSAS														
		WATERLOO			DODGE CITY			KANSAS CITY			SALINA			TOPEKA			WICHITA		
		MAT.	INST.	TOTAL	MAT.	INST.	TOTAL	MAT.	INST.	TOTAL	MAT.	INST.	TOTAL	MAT.	INST.	TOTAL	MAT.	INST.	TOTAL
015433	CONTRACTOR EQUIPMENT		99.9	99.9		101.9	101.9		96.4	96.4		101.9	101.9		100.5	100.5		101.9	101.9
0241, 31 - 34	SITE & INFRASTRUCTURE, DEMOLITION	93.1	95.0	94.4	111.7	90.9	97.5	91.9	85.4	87.5	101.4	91.3	94.5	95.7	89.1	91.2	94.2	91.1	92.0
0310	Concrete Forming & Accessories	100.4	43.7	51.5	95.3	61.1	65.8	100.0	99.0	99.1	92.8	49.4	55.2	99.0	46.2	53.4	98.7	51.0	57.4
0320	Concrete Reinforcing	102.3	80.2	93.3	107.2	54.8	85.8	101.8	88.5	96.4	106.5	70.6	91.9	99.0	95.2	97.4	99.0	78.0	90.4
0330	Cast-in-Place Concrete	105.3	45.9	82.6	115.4	53.9	91.9	90.5	96.2	92.7	100.1	46.8	79.7	91.5	51.2	76.1	87.3	55.0	75.0
03	CONCRETE	100.6	53.0	79.0	113.8	58.8	88.8	95.7	96.3	96.0	101.7	54.2	80.1	93.6	58.8	77.8	91.6	58.9	76.8
04	MASONRY	97.6	55.6	72.5	104.7	51.3	72.8	105.5	94.8	99.1	119.4	45.0	74.9	99.6	58.9	75.2	93.5	58.3	72.4
05	METALS	87.7	89.0	88.1	92.4	79.3	88.6	97.2	96.9	97.1	92.3	85.6	90.3	97.4	96.0	97.0	97.4	87.2	94.4
06	WOOD, PLASTICS & COMPOSITES	109.0	43.4	73.0	96.3	67.8	80.6	102.4	100.4	101.3	92.8	52.0	70.4	98.2	43.7	68.3	99.8	49.8	72.4
07	THERMAL & MOISTURE PROTECTION	101.0	48.4	80.5	99.9	57.5	83.4	98.5	97.3	98.0	99.5	53.1	81.4	100.4	64.6	86.5	99.0	59.4	83.6
08	OPENINGS	94.7	56.2	85.2	96.8	57.9	87.2	95.8	92.5	95.0	96.7	51.9	85.6	97.1	59.6	87.8	97.1	57.1	87.2
0920	Plaster & Gypsum Board	98.5	41.4	61.4	95.6	66.5	76.7	92.1	100.3	97.4	94.9	50.2	65.8	93.8	41.5	59.8	90.9	47.9	62.9
0950, 0980	Ceilings & Acoustic Treatment	113.2	41.4	68.8	86.8	66.5	74.2	87.6	100.3	95.5	86.8	50.2	64.1	87.6	41.5	59.1	85.9	47.9	62.4
0960	Flooring	118.7	52.9	100.9	99.5	43.9	84.5	89.2	97.8	91.5	97.6	32.1	79.9	101.1	40.8	84.8	100.0	70.6	92.0
0970, 0990	Wall Finishes & Painting/Coating	103.0	32.0	60.4	93.0	44.7	64.0	100.6	77.8	86.9	93.0	44.7	64.0	93.0	60.4	73.4	93.0	55.1	70.3
09	FINISHES	110.1	43.3	74.8	96.1	57.0	75.4	94.0	95.9	95.0	94.7	45.1	68.5	95.3	44.8	68.6	94.2	54.6	73.3
COVERS	DIVS. 10 - 14, 25, 28, 41, 43, 44	100.0	74.6	95.0	100.0	46.0	89.3	100.0	74.5	94.9	100.0	56.1	91.3	100.0	59.3	91.9	100.0	58.5	91.8
21, 22, 23	FIRE SUPPRESSION, PLUMBING & HVAC	100.2	39.9	76.2	99.9	59.4	83.8	99.9	90.2	96.0	99.9	59.3	83.8	99.9	63.9	85.6	99.8	57.5	83.0
26, 27, 3370	ELECTRICAL, COMMUNICATIONS & UTIL.	96.5	64.9	81.1	96.7	71.8	84.5	101.8	98.7	100.3	96.4	71.8	84.4	108.4	72.2	90.7	103.2	71.8	87.9
MF2004	WEIGHTED AVERAGE	97.8	58.4	81.2	99.9	64.0	84.8	98.3	93.7	96.4	98.7	61.6	83.1	98.9	66.1	85.1	97.6	64.9	83.9

	DIVISION	KENTUCKY												LOUISIANA					
		BOWLING GREEN			LEXINGTON			LOUISVILLE			OWENSBORO			ALEXANDRIA			BATON ROUGE		
		MAT.	INST.	TOTAL	MAT.	INST.	TOTAL	MAT.	INST.	TOTAL	MAT.	INST.	TOTAL	MAT.	INST.	TOTAL	MAT.	INST.	TOTAL
015433	CONTRACTOR EQUIPMENT		94.2	94.2		101.0	101.0		94.2	94.2		113.5	113.5		90.7	90.7		89.6	89.6
0241, 31 - 34	SITE & INFRASTRUCTURE, DEMOLITION	67.9	96.2	87.2	75.3	99.3	91.7	65.8	95.9	86.4	79.4	121.8	108.4	103.3	88.8	93.1	117.8	86.5	96.4
0310	Concrete Forming & Accessories	85.2	83.5	83.7	97.8	75.8	78.8	93.5	82.0	83.6	90.6	75.0	77.1	84.2	39.7	45.8	100.0	63.7	68.7
0320	Concrete Reinforcing	85.5	88.2	86.6	94.4	93.4	94.0	94.4	93.8	94.1	85.6	91.5	88.0	101.1	63.1	85.6	100.9	63.3	85.6
0330	Cast-in-Place Concrete	83.7	106.4	92.4	90.4	84.0	87.9	94.3	78.8	88.4	86.3	103.0	92.7	93.9	42.6	74.3	91.5	62.6	80.4
03	CONCRETE	90.5	92.4	91.4	93.1	82.3	88.2	94.7	83.3	89.5	99.8	88.2	93.9	90.9	46.6	70.7	93.7	63.9	80.1
04	MASONRY	89.1	76.0	81.3	86.6	68.3	75.6	87.8	80.4	83.4	86.1	74.3	79.0	115.2	49.6	75.9	106.9	54.3	75.4
05	METALS	92.0	87.2	90.6	93.4	90.4	92.5	100.8	89.6	97.5	83.9	90.0	85.7	92.4	75.4	87.4	105.1	73.7	95.9
06	WOOD, PLASTICS & COMPOSITES	91.0	82.5	86.4	101.2	75.1	86.9	98.6	82.5	89.8	91.4	71.3	80.4	86.6	38.8	60.3	104.1	69.2	84.9
07	THERMAL & MOISTURE PROTECTION	85.3	86.6	85.8	97.1	89.0	94.0	91.1	80.8	87.1	96.8	78.5	89.7	99.3	49.9	80.1	101.3	59.7	85.1
08	OPENINGS	95.1	80.4	91.5	96.1	80.0	92.1	96.1	85.4	93.4	93.3	78.1	89.6	99.3	45.9	86.1	102.6	62.3	92.6
0920	Plaster & Gypsum Board	80.3	82.2	81.5	87.3	73.5	78.4	83.8	82.2	82.8	80.3	69.6	73.4	81.9	37.1	52.8	99.4	68.4	79.3
0950, 0980	Ceilings & Acoustic Treatment	85.3	82.2	83.4	89.5	73.5	79.6	86.1	82.2	83.7	77.0	69.6	72.4	92.0	37.1	58.0	98.6	68.4	80.0
0960	Flooring	94.1	85.0	91.7	99.0	62.2	89.1	97.5	71.4	90.5	96.6	85.0	93.5	102.0	64.0	91.7	107.2	68.7	96.8
0970, 0990	Wall Finishes & Painting/Coating	97.7	72.4	82.5	97.7	72.2	82.4	97.7	79.8	87.0	97.7	92.7	94.7	98.5	36.6	61.4	106.6	42.0	67.8
09	FINISHES	90.7	82.5	86.3	94.1	72.1	82.5	91.9	79.6	85.4	89.4	78.0	83.4	93.0	43.1	66.6	102.0	63.1	81.4
COVERS	DIVS. 10 - 14, 25, 28, 41, 43, 44	100.0	69.7	94.0	100.0	93.4	98.7	100.0	92.9	98.6	100.0	89.3	97.9	100.0	59.3	91.9	100.0	77.9	95.6
21, 22, 23	FIRE SUPPRESSION, PLUMBING & HVAC	99.9	85.3	94.1	99.9	68.9	87.6	99.9	83.4	93.4	99.9	73.0	89.2	100.1	55.4	82.3	99.9	54.4	81.8
26, 27, 3370	ELECTRICAL, COMMUNICATIONS & UTIL.	92.8	79.4	86.2	93.2	76.6	85.1	104.3	79.4	92.2	92.2	80.4	86.5	93.4	55.5	75.0	107.6	62.9	85.8
MF2004	WEIGHTED AVERAGE	93.6	84.6	89.8	95.1	78.5	88.1	97.1	83.9	91.6	93.4	83.2	89.1	97.0	56.0	79.8	102.2	63.8	86.1

City Cost Indexes

	DIVISION	LOUISIANA												MAINE					
		LAKE CHARLES			MONROE			NEW ORLEANS			SHREVEPORT			AUGUSTA			BANGOR		
		MAT.	INST.	TOTAL	MAT.	INST.	TOTAL	MAT.	INST.	TOTAL	MAT.	INST.	TOTAL	MAT.	INST.	TOTAL	MAT.	INST.	TOTAL
015433	CONTRACTOR EQUIPMENT		89.6	89.6		90.7	90.7		91.0	91.0		90.7	90.7		100.4	100.4		100.4	100.4
0241, 31 - 34	SITE & INFRASTRUCTURE, DEMOLITION	119.6	86.2	96.8	103.3	87.9	92.8	122.9	89.8	100.2	105.0	87.8	93.2	81.6	100.5	94.5	80.8	98.1	92.6
0310	Concrete Forming & Accessories	103.4	54.1	60.8	83.6	40.4	46.3	104.8	66.4	71.7	99.2	42.6	50.3	95.0	88.6	89.5	90.4	83.8	84.7
0320	Concrete Reinforcing	100.9	60.4	84.4	100.0	63.1	84.9	100.9	64.9	86.2	99.6	63.2	84.8	82.0	99.0	88.9	81.7	99.2	88.8
0330	Cast-in-Place Concrete	93.4	62.6	81.6	93.9	60.7	81.2	92.5	71.9	84.6	93.3	48.9	76.3	83.5	58.8	74.0	77.5	59.5	70.6
03	CONCRETE	94.8	59.1	78.6	90.6	53.0	73.5	94.4	68.7	82.7	90.6	49.9	72.1	100.6	79.8	91.1	97.0	77.8	88.3
04	MASONRY	105.0	51.8	73.1	109.9	50.2	74.2	106.3	59.7	78.4	101.0	51.9	71.6	97.6	45.5	66.4	111.9	52.0	76.0
05	METALS	99.3	72.2	91.4	92.4	74.3	87.1	113.3	76.7	102.6	88.6	74.2	84.4	82.7	83.3	82.9	82.4	79.5	81.6
06	WOOD, PLASTICS & COMPOSITES	108.8	56.3	80.0	85.9	39.1	60.2	105.0	69.2	85.3	103.1	42.5	69.8	99.9	97.7	98.7	94.8	90.4	92.4
07	THERMAL & MOISTURE PROTECTION	102.9	57.1	85.1	99.3	53.5	81.5	103.7	62.8	87.8	98.6	52.4	80.7	94.4	54.1	78.7	94.8	54.4	79.1
08	OPENINGS	102.6	55.8	91.0	99.3	50.2	87.2	103.4	68.1	94.7	97.2	48.0	85.0	101.7	77.4	95.7	101.6	74.9	95.0
0920	Plaster & Gypsum Board	105.2	55.1	72.6	81.5	37.5	52.9	104.0	68.4	80.9	86.3	40.9	56.8	106.9	96.8	100.4	100.4	89.3	93.2
0950, 0980	Ceilings & Acoustic Treatment	97.8	55.1	71.4	92.0	37.5	58.3	96.9	68.4	79.3	90.3	40.9	59.8	90.0	96.8	94.2	90.8	89.3	89.9
0960	Flooring	107.4	64.0	95.6	101.5	64.0	91.4	107.7	68.7	97.2	111.4	68.7	99.9	96.8	37.7	80.8	94.6	47.0	81.7
0970, 0990	Wall Finishes & Painting/Coating	106.6	38.7	65.8	98.5	43.7	65.6	108.1	63.4	81.3	98.5	36.6	61.4	91.7	31.7	55.7	91.7	29.2	54.2
09	FINISHES	102.5	54.2	77.0	92.8	44.2	67.1	102.9	66.8	83.8	96.3	46.2	69.8	97.3	74.7	85.3	96.0	72.5	83.6
COVERS	DIVS. 10 - 14, 25, 28, 41, 43, 44	100.0	76.4	95.3	100.0	54.5	91.0	100.0	79.8	96.0	100.0	69.0	93.9	100.0	58.2	91.7	100.0	70.4	94.1
21, 22, 23	FIRE SUPPRESSION, PLUMBING & HVAC	100.0	55.9	82.5	100.1	49.1	79.8	100.0	66.1	86.5	100.0	55.3	82.3	100.0	69.1	87.7	100.2	67.9	87.3
26, 27, 3370	ELECTRICAL, COMMUNICATIONS & UTIL.	95.3	61.1	78.7	95.2	56.7	76.5	96.3	68.6	82.8	102.6	68.7	86.1	98.3	80.1	89.5	95.3	80.0	87.9
MF2004	WEIGHTED AVERAGE	100.1	61.1	83.7	96.9	56.0	79.7	102.9	69.6	88.9	96.8	59.2	81.0	96.1	74.3	86.9	95.8	73.7	86.5

	DIVISION	MAINE						MARYLAND						MASSACHUSETTS					
		LEWISTON			PORTLAND			BALTIMORE			HAGERSTOWN			BOSTON			BROCKTON		
		MAT.	INST.	TOTAL	MAT.	INST.	TOTAL	MAT.	INST.	TOTAL	MAT.	INST.	TOTAL	MAT.	INST.	TOTAL	MAT.	INST.	TOTAL
015433	CONTRACTOR EQUIPMENT		100.4	100.4		100.4	100.4		104.8	104.8		101.9	101.9		105.7	105.7		102.3	102.3
0241, 31 - 34	SITE & INFRASTRUCTURE, DEMOLITION	78.2	98.1	91.8	79.6	98.1	92.2	96.0	97.2	96.8	85.7	94.6	91.8	85.3	106.8	100.0	82.9	103.8	97.2
0310	Concrete Forming & Accessories	95.3	83.8	85.3	94.4	83.8	85.2	101.7	77.9	81.1	91.6	79.4	81.1	99.3	137.7	132.5	99.5	122.8	119.6
0320	Concrete Reinforcing	100.4	99.2	99.9	100.4	99.2	99.9	100.9	86.7	95.1	89.2	73.9	83.0	99.3	146.1	118.4	100.4	145.7	118.9
0330	Cast-in-Place Concrete	79.1	59.5	71.6	100.6	59.5	84.9	95.2	79.8	89.3	81.0	60.1	73.0	100.0	144.7	117.1	95.2	142.8	113.4
03	CONCRETE	97.6	77.8	88.6	107.6	77.8	94.0	101.3	81.3	92.2	87.0	72.8	80.5	110.3	140.6	124.1	107.9	133.2	119.4
04	MASONRY	96.5	52.0	69.8	96.8	52.0	70.0	96.9	71.5	81.7	100.6	73.3	84.3	114.9	155.6	139.3	111.4	146.4	132.4
05	METALS	85.3	79.5	83.6	86.5	79.5	84.5	97.3	97.9	97.5	93.8	91.7	93.2	96.8	126.2	105.4	93.8	123.2	102.4
06	WOOD, PLASTICS & COMPOSITES	100.6	90.4	95.0	99.8	90.4	94.6	97.9	79.9	88.0	86.7	79.8	82.9	104.1	136.6	121.9	102.9	120.1	112.3
07	THERMAL & MOISTURE PROTECTION	94.6	54.4	79.0	97.2	54.4	80.5	97.6	80.2	90.8	96.9	68.5	85.8	94.8	148.7	115.8	94.9	142.3	113.3
08	OPENINGS	104.9	74.9	97.5	104.9	74.9	97.5	93.7	85.8	91.8	90.9	78.9	88.0	109.0	141.0	110.8	98.9	129.0	106.3
0920	Plaster & Gypsum Board	105.5	89.3	95.0	111.1	89.3	96.9	104.6	79.4	88.2	99.8	79.4	86.5	108.0	137.1	127.0	103.1	120.0	114.1
0950, 0980	Ceilings & Acoustic Treatment	99.1	89.3	93.0	96.6	89.3	92.1	95.4	79.4	85.5	96.3	79.4	85.9	98.2	137.1	122.3	100.1	120.0	112.4
0960	Flooring	97.5	47.0	83.9	96.8	47.0	83.3	89.9	78.2	86.7	85.4	72.9	82.0	97.3	167.4	116.2	97.8	167.4	116.6
0970, 0990	Wall Finishes & Painting/Coating	91.7	29.2	54.2	91.7	29.2	54.2	94.6	81.0	86.4	94.6	35.7	59.3	94.4	158.2	132.7	93.9	130.3	115.7
09	FINISHES	99.1	72.5	85.1	99.0	72.5	85.0	96.0	77.8	86.4	93.4	73.7	83.0	99.1	145.7	123.7	99.0	131.6	116.3
COVERS	DIVS. 10 - 14, 25, 28, 41, 43, 44	100.0	70.4	94.1	100.0	70.4	94.1	100.0	86.0	97.2	100.0	88.7	97.8	100.0	121.9	104.3	100.0	118.1	103.6
21, 22, 23	FIRE SUPPRESSION, PLUMBING & HVAC	100.2	67.9	87.3	100.0	67.9	87.2	99.9	88.5	95.4	99.9	87.9	95.1	100.2	131.5	112.6	100.2	108.5	103.5
26, 27, 3370	ELECTRICAL, COMMUNICATIONS & UTIL.	97.2	80.0	88.8	103.7	80.0	92.2	98.2	91.7	95.0	95.4	81.4	88.6	96.3	137.1	116.2	95.4	99.5	97.4
MF2004	WEIGHTED AVERAGE	96.4	73.7	86.9	98.5	73.7	88.1	98.2	86.0	93.1	94.9	81.6	89.3	100.5	135.9	115.4	99.2	120.7	108.2

	DIVISION	MASSACHUSETTS																	
		FALL RIVER			HYANNIS			LAWRENCE			LOWELL			NEW BEDFORD			PITTSFIELD		
		MAT.	INST.	TOTAL	MAT.	INST.	TOTAL	MAT.	INST.	TOTAL	MAT.	INST.	TOTAL	MAT.	INST.	TOTAL	MAT.	INST.	TOTAL
015433	CONTRACTOR EQUIPMENT		103.2	103.2		102.3	102.3		102.3	102.3		100.4	100.4		103.2	103.2		100.4	100.4
0241, 31 - 34	SITE & INFRASTRUCTURE, DEMOLITION	82.0	103.7	96.8	79.8	103.8	96.2	83.6	103.8	97.4	82.5	103.8	97.0	80.5	103.7	96.4	83.6	101.8	96.0
0310	Concrete Forming & Accessories	99.5	122.5	119.4	91.6	122.2	118.0	99.3	122.8	119.6	96.5	122.9	119.3	99.5	122.5	119.4	96.5	103.9	102.9
0320	Concrete Reinforcing	100.4	122.5	109.4	80.5	122.4	97.6	99.5	131.4	112.5	100.4	131.6	113.1	100.4	122.5	109.4	83.3	111.1	94.7
0330	Cast-in-Place Concrete	92.1	144.9	112.3	86.9	144.3	108.8	96.0	140.7	113.1	87.3	140.8	107.7	81.1	144.9	105.5	95.2	117.3	103.7
03	CONCRETE	106.4	129.5	116.9	97.5	129.2	111.9	108.0	129.9	118.0	99.1	129.8	113.1	101.3	129.5	114.1	99.4	109.4	103.9
04	MASONRY	111.2	148.3	133.4	110.2	148.4	133.0	110.8	143.8	130.6	97.1	146.5	126.7	110.3	148.3	133.0	97.7	114.9	108.0
05	METALS	93.8	114.5	99.9	90.5	114.1	97.4	91.4	117.7	99.1	91.4	115.9	98.5	93.8	114.5	99.9	91.2	102.0	94.4
06	WOOD, PLASTICS & COMPOSITES	102.9	120.3	112.5	93.7	120.0	108.2	102.9	120.1	112.3	102.3	120.1	112.1	102.9	120.3	112.5	102.3	104.7	103.6
07	THERMAL & MOISTURE PROTECTION	94.9	137.4	111.4	94.5	138.2	111.5	94.9	141.3	112.9	94.7	141.9	113.0	94.9	137.4	111.4	94.7	111.7	101.3
08	OPENINGS	98.9	117.9	103.6	95.0	117.7	100.6	98.9	122.4	104.7	104.9	122.4	109.2	98.9	117.9	103.6	104.9	106.7	105.3
0920	Plaster & Gypsum Board	103.1	120.0	114.1	96.1	120.0	111.6	105.5	120.0	115.0	105.5	120.0	115.0	103.1	120.0	114.1	105.5	104.1	104.6
0950, 0980	Ceilings & Acoustic Treatment	100.1	120.0	112.4	91.7	120.0	109.2	99.1	120.0	112.1	99.1	120.0	112.1	100.1	120.0	112.4	99.1	104.1	102.2
0960	Flooring	97.6	167.4	116.5	94.1	167.4	113.9	96.8	167.4	115.9	96.8	167.4	115.9	97.6	167.4	116.5	97.0	131.4	106.3
0970, 0990	Wall Finishes & Painting/Coating	93.9	130.3	115.7	93.9	127.1	113.8	91.8	130.3	114.9	91.7	127.1	112.9	93.9	130.3	115.7	91.7	125.5	111.9
09	FINISHES	99.0	131.8	116.4	94.8	131.3	114.1	98.7	131.6	116.1	98.7	131.3	115.9	98.9	131.8	116.3	98.7	109.4	104.4
COVERS	DIVS. 10 - 14, 25, 28, 41, 43, 44	100.0	118.8	103.7	100.0	118.1	103.6	100.0	118.2	103.6	100.0	118.2	103.6	100.0	118.8	103.7	100.0	94.7	98.9
21, 22, 23	FIRE SUPPRESSION, PLUMBING & HVAC	100.2	108.3	103.4	100.2	108.3	103.4	100.0	112.9	105.3	100.2	122.9	109.2	100.2	108.3	103.4	100.2	87.5	95.1
26, 27, 3370	ELECTRICAL, COMMUNICATIONS & UTIL.	95.2	99.4	97.3	93.5	99.4	96.4	96.7	125.4	110.7	97.1	125.4	110.9	96.3	99.4	97.8	97.1	90.8	94.1
MF2004	WEIGHTED AVERAGE	99.0	118.9	107.3	96.2	118.7	105.7	98.9	123.6	109.3	97.7	125.7	109.5	98.4	118.9	107.0	97.8	101.0	99.1

City Cost Indexes

		MASSACHUSETTS							MICHIGAN								
		SPRINGFIELD			WORCESTER			ANN ARBOR			DEARBORN			DETROIT			FLINT		
DIVISION		MAT.	INST.	TOTAL	MAT.	INST.	TOTAL	MAT.	INST.	TOTAL	MAT.	INST.	TOTAL	MAT.	INST.	TOTAL	MAT.	INST.	TOTAL
015433	CONTRACTOR EQUIPMENT		100.4	100.4		100.4	100.4		111.3	111.3		111.3	111.3		98.5	98.5		111.3	111.3
0241, 31 - 34	SITE & INFRASTRUCTURE, DEMOLITION	83.0	102.1	96.0	82.9	103.7	97.1	77.1	96.5	90.4	76.9	96.7	90.4	90.9	98.3	95.9	67.3	95.7	86.7
0310	Concrete Forming & Accessories	96.7	105.4	104.2	97.0	126.0	122.0	98.2	110.1	108.5	98.1	119.5	116.6	99.1	119.6	116.8	101.2	95.0	95.8
0320	Concrete Reinforcing	100.4	109.0	103.9	100.4	141.6	117.2	96.1	128.4	109.3	96.1	128.7	109.4	95.4	128.7	109.0	96.1	127.9	109.1
0330	Cast-in-Place Concrete	90.8	119.2	101.7	90.3	142.6	110.3	85.0	114.5	96.3	83.1	118.0	96.4	91.3	118.0	101.5	85.6	100.4	91.3
03	CONCRETE	100.8	110.3	105.1	100.5	133.6	115.6	89.6	115.3	101.3	88.7	120.7	103.3	92.5	119.5	104.7	90.1	103.6	96.2
04	MASONRY	97.4	118.2	109.9	96.9	146.5	126.6	92.7	113.2	105.0	92.5	119.5	108.7	91.8	119.5	108.4	92.7	101.3	97.8
05	METALS	93.8	101.3	96.0	93.8	118.6	101.1	86.1	121.4	96.4	86.1	122.4	96.7	89.9	104.6	94.2	86.1	119.0	95.7
06	WOOD, PLASTICS & COMPOSITES	102.3	104.7	103.6	102.8	125.4	115.2	96.7	109.5	103.7	96.7	119.9	109.4	97.2	119.9	109.7	99.9	93.6	96.4
07	THERMAL & MOISTURE PROTECTION	94.7	113.1	101.8	94.7	134.8	110.3	102.5	111.9	106.2	101.1	121.3	109.0	99.3	121.3	107.8	100.2	96.9	99.0
08	OPENINGS	104.9	106.1	105.2	104.9	130.8	111.3	94.5	109.6	98.2	94.5	115.9	99.6	96.2	117.4	101.4	94.5	97.0	95.1
0920	Plaster & Gypsum Board	105.5	104.1	104.6	105.5	125.5	118.5	85.4	108.7	100.5	85.4	119.6	107.6	85.4	119.6	107.6	86.9	92.2	90.3
0950, 0980	Ceilings & Acoustic Treatment	99.1	104.1	102.2	99.1	125.5	115.4	88.7	108.7	101.1	88.7	119.6	107.8	89.7	119.6	108.2	88.7	92.2	90.9
0960	Flooring	96.7	131.4	106.1	96.8	163.3	114.8	92.7	120.3	100.2	92.4	126.2	101.5	92.4	126.2	101.6	92.5	82.5	89.8
0970, 0990	Wall Finishes & Painting/Coating	93.3	104.5	100.0	91.7	130.3	114.9	92.9	102.4	98.6	92.9	108.3	102.1	94.4	108.3	102.7	92.9	88.8	90.4
09	FINISHES	98.8	110.3	104.8	98.7	133.9	117.3	90.8	111.2	101.6	90.7	120.1	106.2	92.0	120.1	106.8	90.2	91.4	90.8
COVERS	DIVS. 10 - 14, 25, 28, 41, 43, 44	100.0	95.9	99.2	100.0	101.5	100.3	100.0	106.7	101.3	100.0	109.0	101.8	100.0	109.0	101.8	100.0	94.3	98.9
21, 22, 23	FIRE SUPPRESSION, PLUMBING & HVAC	100.2	101.9	100.9	100.2	105.4	102.3	99.9	99.5	99.7	99.9	110.7	104.2	99.9	112.5	105.0	99.9	101.4	100.5
26, 27, 3370	ELECTRICAL, COMMUNICATIONS & UTIL.	97.2	90.8	94.1	97.2	95.7	96.5	95.2	76.9	86.3	95.2	116.0	105.3	96.5	115.9	106.0	95.2	99.3	97.2
MF2004	WEIGHTED AVERAGE	98.4	104.5	101.0	98.3	118.9	107.0	93.7	104.5	98.2	93.5	115.5	102.7	95.3	114.2	103.2	93.4	100.8	96.5

		MICHIGAN															MINNESOTA		
		GRAND RAPIDS			KALAMAZOO			LANSING			MUSKEGON			SAGINAW			DULUTH		
DIVISION		MAT.	INST.	TOTAL	MAT.	INST.	TOTAL	MAT.	INST.	TOTAL	MAT.	INST.	TOTAL	MAT.	INST.	TOTAL	MAT.	INST.	TOTAL
015433	CONTRACTOR EQUIPMENT		101.4	101.4		101.4	101.4		111.3	111.3		101.4	101.4		111.3	111.3		101.1	101.1
0241, 31 - 34	SITE & INFRASTRUCTURE, DEMOLITION	81.3	81.7	81.6	80.0	82.4	81.7	85.4	95.2	92.1	77.9	82.3	80.9	70.4	94.8	87.1	93.3	102.7	99.7
0310	Concrete Forming & Accessories	98.1	72.0	75.5	97.7	89.5	90.7	101.8	95.1	96.0	98.3	81.1	83.4	98.3	86.8	88.4	97.1	114.3	112.0
0320	Concrete Reinforcing	94.4	83.7	90.0	94.4	85.7	90.8	96.1	127.6	108.9	95.0	85.0	90.9	96.1	127.4	108.9	102.0	101.7	101.9
0330	Cast-in-Place Concrete	91.4	90.4	91.0	92.1	104.6	96.9	87.4	99.4	92.0	90.0	99.0	93.4	84.0	93.6	87.7	107.7	104.9	106.6
03	CONCRETE	93.6	80.4	87.6	96.0	93.5	94.9	90.9	103.3	96.5	91.6	87.7	89.8	89.2	97.6	93.0	100.8	109.3	104.7
04	MASONRY	88.8	47.0	63.8	91.9	86.7	88.8	86.5	96.5	92.5	90.5	73.9	80.6	94.0	89.5	91.3	102.1	116.9	110.9
05	METALS	93.8	80.7	90.0	92.3	85.1	90.2	84.7	118.2	94.5	90.1	83.7	88.2	86.2	117.5	95.3	89.3	119.0	98.0
06	WOOD, PLASTICS & COMPOSITES	94.9	73.7	83.2	97.3	89.9	93.2	99.3	94.1	96.4	93.7	79.6	86.0	92.6	86.2	89.1	111.8	114.3	113.2
07	THERMAL & MOISTURE PROTECTION	94.4	56.5	79.7	92.7	87.5	90.7	101.4	93.8	98.4	91.9	76.2	85.8	101.0	88.2	96.0	102.3	115.3	107.3
08	OPENINGS	94.1	66.1	87.2	90.7	83.4	88.9	94.5	97.3	95.2	89.9	76.2	86.5	92.4	93.0	92.6	94.7	116.1	100.0
0920	Plaster & Gypsum Board	84.3	68.4	74.0	85.8	85.2	85.4	83.8	92.8	89.6	72.6	74.6	73.9	85.4	84.6	84.9	94.5	115.3	108.0
0950, 0980	Ceilings & Acoustic Treatment	87.8	68.4	75.8	89.5	85.2	86.8	87.1	92.8	90.6	91.1	74.6	80.9	88.7	84.6	86.2	96.3	115.3	108.1
0960	Flooring	98.1	38.0	81.8	97.9	74.2	91.5	102.6	87.9	98.6	97.1	70.0	89.8	92.7	79.9	89.3	107.5	123.6	111.9
0970, 0990	Wall Finishes & Painting/Coating	97.7	37.1	61.4	97.7	75.3	84.3	105.2	91.7	97.1	97.0	53.8	71.1	92.9	83.8	87.4	91.3	100.9	97.1
09	FINISHES	93.1	61.8	76.5	93.4	85.2	89.1	94.8	93.3	94.0	91.2	75.6	83.0	90.5	84.8	87.5	101.7	115.2	108.9
COVERS	DIVS. 10 - 14, 25, 28, 41, 43, 44	100.0	93.8	98.8	100.0	98.9	99.8	100.0	94.5	98.9	100.0	97.5	99.5	100.0	91.6	98.3	100.0	95.1	99.0
21, 22, 23	FIRE SUPPRESSION, PLUMBING & HVAC	99.9	56.3	82.6	99.9	82.3	92.9	99.8	99.8	99.8	99.8	84.2	93.6	99.9	84.2	93.6	99.8	102.4	100.9
26, 27, 3370	ELECTRICAL, COMMUNICATIONS & UTIL.	103.2	53.6	79.1	92.6	81.8	87.3	101.2	77.2	89.5	93.2	79.1	86.3	92.9	72.8	83.1	102.5	96.9	99.7
MF2004	WEIGHTED AVERAGE	96.1	65.2	83.1	94.8	85.6	91.0	94.4	96.9	95.5	93.5	81.4	88.4	93.0	89.9	91.7	98.2	108.2	102.4

		MINNESOTA												MISSISSIPPI					
		MINNEAPOLIS			ROCHESTER			SAINT PAUL			ST. CLOUD			BILOXI			GREENVILLE		
DIVISION		MAT.	INST.	TOTAL	MAT.	INST.	TOTAL	MAT.	INST.	TOTAL	MAT.	INST.	TOTAL	MAT.	INST.	TOTAL	MAT.	INST.	TOTAL
015433	CONTRACTOR EQUIPMENT		104.1	104.1		101.1	101.1		101.1	101.1		100.8	100.8		99.5	99.5		99.5	99.5
0241, 31 - 34	SITE & INFRASTRUCTURE, DEMOLITION	91.9	107.7	102.7	90.9	101.7	98.3	95.7	103.5	101.0	83.7	105.3	98.5	103.7	89.6	94.1	107.6	89.7	95.4
0310	Concrete Forming & Accessories	99.4	135.9	131.0	99.8	104.9	104.2	93.5	125.6	121.2	85.1	123.3	118.1	93.6	61.8	66.1	81.0	74.8	75.6
0320	Concrete Reinforcing	102.2	122.7	110.6	102.0	122.2	110.3	98.4	122.6	108.3	103.3	122.2	111.0	92.5	63.2	80.6	100.3	70.6	88.1
0330	Cast-in-Place Concrete	111.9	121.2	115.4	109.7	101.3	106.5	113.9	120.4	116.4	95.6	118.8	104.5	106.8	60.3	89.0	107.2	63.7	90.6
03	CONCRETE	104.4	128.4	115.3	101.6	107.8	104.6	106.4	123.6	114.2	91.9	121.9	105.5	97.1	63.2	81.7	100.6	71.4	87.3
04	MASONRY	100.9	131.8	119.4	100.1	115.7	109.5	109.8	131.8	123.0	105.0	124.3	116.5	88.7	50.1	65.6	129.1	56.0	85.3
05	METALS	91.8	131.8	103.5	89.2	129.7	101.0	88.8	131.3	101.2	90.7	129.1	101.9	97.6	87.9	94.7	95.7	91.2	94.4
06	WOOD, PLASTICS & COMPOSITES	115.2	135.9	126.6	115.2	102.8	108.4	107.8	122.3	115.7	94.4	120.8	108.9	94.9	65.7	78.8	78.9	79.7	79.4
07	THERMAL & MOISTURE PROTECTION	102.5	133.8	114.7	102.6	102.3	102.5	103.6	131.8	114.6	101.1	113.7	106.0	92.6	62.2	80.7	92.7	68.8	83.4
08	OPENINGS	98.0	142.0	108.9	94.7	124.0	102.0	92.3	134.5	102.8	91.2	133.8	101.8	97.1	66.2	89.4	96.4	71.8	90.3
0920	Plaster & Gypsum Board	98.9	137.5	124.0	98.5	103.4	101.7	91.8	123.6	112.4	83.9	122.2	108.8	104.0	64.9	78.6	92.8	79.5	84.2
0950, 0980	Ceilings & Acoustic Treatment	99.7	137.5	123.1	98.0	103.4	101.4	98.0	123.6	111.9	69.9	122.2	102.2	100.4	64.9	78.5	94.6	79.5	85.2
0960	Flooring	105.3	126.6	111.0	107.7	92.5	103.6	101.2	126.6	108.0	105.0	126.6	110.9	117.2	59.2	101.5	108.0	59.2	94.8
0970, 0990	Wall Finishes & Painting/Coating	98.9	128.0	116.4	94.2	101.5	98.6	98.9	119.8	111.4	103.2	128.0	118.1	110.4	49.9	74.1	110.4	75.7	89.6
09	FINISHES	102.7	134.3	119.4	102.6	102.8	102.7	99.4	125.3	113.0	93.2	124.8	109.9	104.3	60.3	81.1	99.1	72.9	85.3
COVERS	DIVS. 10 - 14, 25, 28, 41, 43, 44	100.0	106.0	101.2	100.0	96.9	99.4	100.0	104.0	100.8	100.0	98.8	99.8	100.0	69.6	94.0	100.0	73.5	94.8
21, 22, 23	FIRE SUPPRESSION, PLUMBING & HVAC	99.8	117.1	106.7	99.8	98.0	99.1	99.8	112.5	104.8	99.5	115.8	106.0	99.9	50.0	80.0	99.9	57.4	83.0
26, 27, 3370	ELECTRICAL, COMMUNICATIONS & UTIL.	101.9	115.3	108.5	100.6	85.7	93.4	100.1	102.4	101.2	100.8	102.4	101.6	99.9	55.5	78.3	99.2	73.2	86.5
MF2004	WEIGHTED AVERAGE	99.3	124.2	109.8	98.1	104.7	100.9	98.6	118.7	107.0	95.8	117.5	104.9	98.5	62.6	83.4	100.0	70.9	87.8

City Cost Indexes

		MISSISSIPPI						MISSOURI											
DIVISION		JACKSON			MERIDIAN			CAPE GIRARDEAU			COLUMBIA			JOPLIN			KANSAS CITY		
		MAT.	INST.	TOTAL	MAT.	INST.	TOTAL	MAT.	INST.	TOTAL	MAT.	INST.	TOTAL	MAT.	INST.	TOTAL	MAT.	INST.	TOTAL
015433	CONTRACTOR EQUIPMENT		99.5	99.5		99.5	99.5		107.1	107.1		108.1	108.1		105.9	105.9		100.9	100.9
0241, 31 - 34	SITE & INFRASTRUCTURE, DEMOLITION	98.9	89.7	92.6	98.7	90.0	92.7	93.4	93.9	93.7	102.8	95.7	97.9	104.2	98.8	100.5	96.7	93.5	94.5
0310	Concrete Forming & Accessories	93.0	67.4	70.9	79.8	74.1	74.9	86.6	74.2	75.9	88.2	66.7	69.6	104.3	69.2	74.0	103.3	107.7	107.1
0320	Concrete Reinforcing	92.5	61.7	79.9	99.0	61.7	83.8	105.1	83.1	96.1	102.9	113.6	107.3	100.8	70.9	88.6	95.8	114.7	103.5
0330	Cast-in-Place Concrete	99.2	63.7	85.6	101.1	66.3	87.8	94.2	83.8	90.2	89.0	85.3	87.6	100.3	69.9	88.7	93.2	107.9	98.8
03	CONCRETE	93.5	66.5	81.3	93.6	70.4	83.1	93.6	80.9	87.8	86.6	83.8	85.3	97.9	71.0	85.8	93.9	109.3	100.9
04	MASONRY	89.8	56.0	69.6	88.4	60.6	71.8	119.3	77.3	94.1	131.0	84.4	103.1	93.1	58.8	72.6	98.8	107.6	104.1
05	METALS	97.5	87.6	94.6	95.6	87.8	93.3	94.2	104.9	97.4	90.0	118.4	98.3	94.9	87.6	92.8	102.4	113.4	105.6
06	WOOD, PLASTICS & COMPOSITES	93.7	69.8	80.6	77.8	76.3	77.0	88.5	72.1	79.5	94.8	59.6	75.5	104.2	68.2	84.4	103.5	107.6	105.8
07	THERMAL & MOISTURE PROTECTION	94.4	67.7	84.0	92.3	70.4	83.8	102.0	78.1	92.7	93.7	81.5	88.9	96.6	64.8	84.2	96.3	106.2	100.1
08	OPENINGS	97.4	63.9	89.2	96.4	73.9	90.8	98.8	72.5	92.3	91.4	86.9	90.3	91.5	70.4	86.3	97.5	110.3	100.7
0920	Plaster & Gypsum Board	96.8	69.2	78.9	92.8	75.9	81.8	95.2	71.0	79.5	93.6	58.1	70.5	97.0	66.8	77.4	93.7	107.7	102.8
0950, 0980	Ceilings & Acoustic Treatment	97.1	69.2	79.8	94.6	75.9	83.0	86.1	71.0	76.8	87.6	58.1	69.3	85.0	61.9	70.7	90.8	107.7	101.3
0960	Flooring	117.6	59.2	101.8	106.9	59.2	94.0	86.9	67.2	81.5	106.5	88.5	101.7	122.7	45.2	101.7	98.9	99.9	99.2
0970, 0990	Wall Finishes & Painting/Coating	110.4	75.7	89.6	110.4	87.5	96.7	98.3	77.8	86.0	100.6	75.5	85.5	94.6	40.3	62.0	99.5	120.1	111.9
09	FINISHES	102.7	67.1	83.9	98.1	73.4	85.0	90.8	72.3	81.1	94.9	68.1	80.7	103.1	61.8	81.3	98.3	107.5	103.2
COVERS	DIVS. 10 - 14, 25, 28, 41, 43, 44	100.0	72.3	94.5	100.0	74.7	95.0	100.0	60.5	92.2	100.0	94.1	98.8	100.0	69.5	94.0	100.0	99.4	99.9
21, 22, 23	FIRE SUPPRESSION, PLUMBING & HVAC	99.9	63.4	85.4	99.9	65.9	86.4	99.9	89.9	95.9	99.9	91.3	96.5	100.1	52.6	81.2	100.0	104.1	101.6
26, 27, 3370	ELECTRICAL, COMMUNICATIONS & UTIL.	104.6	73.2	89.3	98.9	58.9	79.4	100.9	108.3	104.5	96.0	85.3	90.8	90.9	68.8	80.1	101.2	105.2	103.1
MF2004	WEIGHTED AVERAGE	98.5	69.9	86.5	96.8	70.9	85.9	98.2	87.4	93.7	96.3	88.1	92.9	97.1	68.0	84.9	99.2	106.1	102.1

		MISSOURI						MONTANA								
DIVISION		SPRINGFIELD			ST. JOSEPH			ST. LOUIS			BILLINGS			BUTTE			GREAT FALLS		
		MAT.	INST.	TOTAL	MAT.	INST.	TOTAL	MAT.	INST.	TOTAL	MAT.	INST.	TOTAL	MAT.	INST.	TOTAL	MAT.	INST.	TOTAL
015433	CONTRACTOR EQUIPMENT		101.2	101.2		99.5	99.5		108.0	108.0		99.1	99.1		98.7	98.7		98.7	98.7
0241, 31 - 34	SITE & INFRASTRUCTURE, DEMOLITION	95.9	91.6	92.9	98.2	88.6	91.7	93.2	97.5	96.1	91.0	97.7	95.6	99.9	96.7	97.7	103.6	97.5	99.4
0310	Concrete Forming & Accessories	101.2	71.7	75.7	103.1	79.4	82.6	98.8	105.3	104.4	96.9	65.3	69.6	85.4	69.7	71.8	96.8	66.7	70.8
0320	Concrete Reinforcing	99.0	113.3	104.8	94.6	104.7	98.7	96.1	109.4	101.5	102.3	70.6	89.4	111.0	70.7	94.6	102.3	70.5	89.3
0330	Cast-in-Place Concrete	96.6	62.9	83.8	93.2	103.3	97.0	94.2	112.3	101.1	108.7	69.2	93.6	119.7	70.9	101.0	126.5	68.4	104.3
03	CONCRETE	96.1	77.4	87.6	93.6	93.4	93.5	92.5	109.3	100.2	102.3	68.8	87.1	106.2	71.3	90.4	110.3	69.1	91.6
04	MASONRY	88.7	78.1	82.4	98.4	90.3	93.6	99.5	110.3	106.0	120.5	71.1	90.9	116.3	74.4	91.2	120.3	73.2	92.1
05	METALS	93.8	105.3	97.2	98.6	107.2	101.1	98.9	119.5	104.9	106.1	86.5	100.3	98.9	86.8	95.4	102.8	86.5	98.1
06	WOOD, PLASTICS & COMPOSITES	103.2	72.6	86.4	104.2	74.6	88.0	102.4	102.5	102.5	102.6	65.0	82.0	90.4	70.0	79.2	103.8	65.0	82.5
07	THERMAL & MOISTURE PROTECTION	98.2	73.3	88.5	96.6	92.3	94.9	101.9	107.8	104.2	101.3	69.3	88.9	101.2	73.8	90.5	101.7	65.8	87.8
08	OPENINGS	97.1	81.0	93.1	96.2	89.9	94.7	97.7	109.7	100.7	97.7	62.6	89.0	95.6	65.8	88.2	99.0	63.1	90.1
0920	Plaster & Gypsum Board	96.9	71.5	80.4	98.4	73.5	82.2	103.2	102.6	102.8	93.6	64.1	74.4	91.0	69.3	76.9	98.5	64.1	76.1
0950, 0980	Ceilings & Acoustic Treatment	87.6	71.5	77.7	90.0	73.5	79.8	91.1	102.6	98.2	106.7	64.1	80.3	111.6	69.3	85.4	113.2	64.1	82.8
0960	Flooring	109.2	45.2	91.9	102.0	97.0	100.7	94.0	102.4	96.3	110.1	61.4	96.9	106.9	83.2	100.5	115.5	83.2	106.7
0970, 0990	Wall Finishes & Painting/Coating	95.0	57.7	72.6	95.1	85.4	89.3	98.3	109.2	104.8	100.5	53.8	72.5	98.4	58.9	74.7	98.4	55.9	72.9
09	FINISHES	98.4	64.9	80.7	99.4	81.8	90.1	95.3	104.3	100.1	105.0	62.9	82.8	104.5	71.1	86.9	108.8	68.2	87.3
COVERS	DIVS. 10 - 14, 25, 28, 41, 43, 44	100.0	89.1	97.8	100.0	92.8	98.6	100.0	103.9	100.8	100.0	73.6	94.8	100.0	74.7	95.0	100.0	74.9	95.0
21, 22, 23	FIRE SUPPRESSION, PLUMBING & HVAC	99.9	68.3	87.4	100.1	90.1	96.1	99.9	108.7	103.4	100.1	76.1	90.6	100.2	69.9	88.2	100.2	74.3	89.9
26, 27, 3370	ELECTRICAL, COMMUNICATIONS & UTIL.	99.7	73.1	86.7	99.0	84.0	91.7	103.6	111.7	107.6	95.1	75.5	85.6	102.4	72.6	87.9	95.2	72.5	84.1
MF2004	WEIGHTED AVERAGE	97.3	77.6	89.1	98.3	90.2	94.9	98.6	108.8	102.9	101.9	74.6	90.4	101.6	75.0	90.4	103.1	74.8	91.2

		MONTANA			NEBRASKA												
DIVISION		HELENA			MISSOULA			GRAND ISLAND			LINCOLN			NORTH PLATTE			OMAHA		
		MAT.	INST.	TOTAL	MAT.	INST.	TOTAL	MAT.	INST.	TOTAL	MAT.	INST.	TOTAL	MAT.	INST.	TOTAL	MAT.	INST.	TOTAL
015433	CONTRACTOR EQUIPMENT		98.7	98.7		98.7	98.7		100.5	100.5		100.5	100.5		100.5	100.5		91.9	91.9
0241, 31 - 34	SITE & INFRASTRUCTURE, DEMOLITION	100.7	96.6	97.9	81.3	96.6	91.8	99.7	90.7	93.6	89.3	90.7	90.3	100.6	90.2	93.5	80.1	90.7	87.3
0310	Concrete Forming & Accessories	97.1	66.1	70.3	88.2	61.0	64.7	97.9	82.8	84.8	102.5	65.5	70.5	97.8	81.6	83.8	94.4	72.4	75.4
0320	Concrete Reinforcing	105.8	70.5	91.4	112.9	80.2	99.6	107.8	73.6	93.8	99.0	73.0	88.4	109.2	74.3	95.0	103.6	73.2	91.2
0330	Cast-in-Place Concrete	106.7	67.0	91.5	88.1	66.0	79.7	116.1	74.6	100.2	92.6	77.3	86.7	116.1	71.2	98.9	99.3	79.2	91.6
03	CONCRETE	101.8	68.4	86.6	80.0	67.5	77.6	108.2	78.8	94.8	94.3	72.1	84.2	108.5	77.2	94.3	97.1	75.3	87.2
04	MASONRY	113.0	71.6	88.2	139.5	69.1	97.3	107.5	84.8	93.9	96.6	74.6	83.4	92.2	97.1	95.2	101.8	80.2	88.9
05	METALS	102.0	86.4	97.5	95.9	90.8	94.4	90.1	87.7	89.4	95.0	86.8	92.6	90.3	88.4	89.8	94.3	78.7	89.8
06	WOOD, PLASTICS & COMPOSITES	103.9	65.0	82.6	94.0	60.6	75.6	99.8	84.6	91.4	102.7	61.5	80.1	99.6	84.6	91.4	93.7	71.3	81.4
07	THERMAL & MOISTURE PROTECTION	101.4	73.7	90.6	100.5	70.8	89.0	99.0	80.5	91.8	98.5	75.3	89.5	99.0	87.9	94.6	94.4	82.9	89.9
08	OPENINGS	98.5	61.8	89.4	95.6	63.1	87.6	90.7	77.9	87.5	96.2	61.0	87.5	90.0	77.2	86.8	99.3	71.2	92.4
0920	Plaster & Gypsum Board	91.2	64.1	73.6	91.7	59.4	70.8	94.9	83.9	87.8	91.4	60.0	71.0	95.1	83.9	87.8	93.7	70.7	78.8
0950, 0980	Ceilings & Acoustic Treatment	108.2	64.1	80.9	111.6	59.4	79.3	86.8	83.9	85.0	89.3	60.0	71.2	87.6	83.9	85.3	102.1	70.7	82.7
0960	Flooring	115.5	66.3	102.2	109.0	83.2	102.0	98.5	120.2	104.3	100.2	93.7	98.4	98.4	112.5	102.2	128.1	54.5	108.2
0970, 0990	Wall Finishes & Painting/Coating	98.4	49.7	69.2	98.4	54.7	72.2	93.0	77.7	83.8	93.0	89.5	90.9	93.0	73.3	81.2	158.4	71.7	106.4
09	FINISHES	106.4	64.2	84.1	104.2	63.8	82.8	95.2	88.6	91.7	95.5	72.6	83.4	95.4	86.1	90.5	116.0	60.3	90.8
COVERS	DIVS. 10 - 14, 25, 28, 41, 43, 44	100.0	86.3	97.3	100.0	83.7	96.8	100.0	88.2	97.7	100.0	85.5	97.1	100.0	82.2	96.5	100.0	85.3	97.1
21, 22, 23	FIRE SUPPRESSION, PLUMBING & HVAC	100.1	70.3	88.3	100.2	66.6	86.8	99.9	80.1	92.1	99.8	80.1	92.0	99.9	78.2	91.3	99.7	80.0	91.8
26, 27, 3370	ELECTRICAL, COMMUNICATIONS & UTIL.	101.0	72.5	87.1	100.3	76.7	88.8	89.9	74.0	82.1	103.8	74.0	89.3	91.8	89.9	90.9	103.6	83.2	93.7
MF2004	WEIGHTED AVERAGE	101.8	73.5	89.9	99.0	73.2	88.2	97.3	82.4	91.0	97.6	77.1	89.0	96.8	85.0	91.8	99.5	78.8	90.8

602

City Cost Indexes

DIVISION		NEVADA									NEW HAMPSHIRE								
		CARSON CITY			LAS VEGAS			RENO			MANCHESTER			NASHUA			PORTSMOUTH		
		MAT.	INST.	TOTAL	MAT.	INST.	TOTAL	MAT.	INST.	TOTAL	MAT.	INST.	TOTAL	MAT.	INST.	TOTAL	MAT.	INST.	TOTAL
015433	CONTRACTOR EQUIPMENT		101.1	101.1		101.1	101.1		101.1	101.1		100.4	100.4		100.4	100.4		100.4	100.4
0241, 31 - 34	SITE & INFRASTRUCTURE, DEMOLITION	68.7	103.0	92.2	63.5	106.4	92.9	61.8	103.0	90.0	82.1	98.4	93.2	84.7	98.4	94.1	79.1	99.7	93.2
0310	Concrete Forming & Accessories	101.7	92.8	94.0	103.7	110.5	109.6	98.7	92.9	93.7	94.5	81.3	83.1	96.7	81.3	83.4	86.0	83.7	84.0
0320	Concrete Reinforcing	107.5	113.2	109.8	99.6	125.4	110.2	102.1	124.4	111.2	100.4	93.2	97.4	100.4	93.2	97.4	80.3	93.3	85.6
0330	Cast-in-Place Concrete	109.5	91.2	102.5	106.6	115.0	109.8	116.4	91.2	106.8	101.0	117.1	107.2	88.2	117.1	99.3	83.6	120.5	97.9
03	CONCRETE	110.8	96.1	104.1	108.0	114.4	110.9	112.8	98.2	106.1	107.8	96.0	102.4	101.9	96.0	99.2	92.7	98.5	95.4
04	MASONRY	123.0	78.0	96.0	117.2	103.7	109.1	123.8	78.0	96.4	98.0	87.9	91.9	97.8	87.9	91.8	93.3	93.1	93.2
05	METALS	94.9	101.3	96.7	101.1	110.0	103.7	95.3	105.6	98.3	93.8	90.6	92.9	93.8	90.6	92.8	90.4	93.6	91.4
06	WOOD, PLASTICS & COMPOSITES	93.3	93.6	93.5	93.3	108.8	101.8	89.4	93.6	91.7	98.9	80.4	88.7	102.3	80.4	90.3	89.6	80.4	84.6
07	THERMAL & MOISTURE PROTECTION	99.4	87.1	94.6	115.4	102.4	110.4	100.8	87.1	95.5	94.2	92.3	93.4	94.8	92.3	93.8	94.5	113.5	101.9
08	OPENINGS	93.3	106.4	96.5	94.8	117.7	100.5	93.6	106.6	96.8	104.9	77.3	98.1	104.9	77.3	98.1	105.5	72.2	97.3
0920	Plaster & Gypsum Board	78.5	93.2	88.1	84.5	109.0	100.4	78.7	93.2	88.2	108.7	78.9	89.4	105.5	78.9	88.3	97.0	78.9	85.3
0950, 0980	Ceilings & Acoustic Treatment	107.8	93.2	98.8	121.2	109.0	113.6	102.2	93.2	102.6	96.6	78.9	85.7	99.1	78.9	86.6	90.8	78.9	83.5
0960	Flooring	104.9	57.3	92.0	98.8	98.1	98.6	104.5	57.3	91.8	97.9	100.4	98.5	96.8	100.4	97.8	91.8	100.4	94.1
0970, 0990	Wall Finishes & Painting/Coating	99.2	87.7	92.3	100.4	118.4	111.2	98.1	87.7	91.9	91.7	96.7	94.7	91.7	96.7	94.7	91.7	96.7	94.7
09	FINISHES	101.0	85.1	92.6	102.4	108.7	105.7	102.4	85.1	93.2	98.7	86.0	92.0	99.1	86.0	92.2	94.1	87.2	90.5
COVERS	DIVS. 10 - 14, 25, 28, 41, 43, 44	100.0	109.7	101.9	100.0	104.5	100.9	100.0	109.7	101.9	100.0	69.9	94.0	100.0	69.9	94.0	100.0	71.7	94.4
21, 22, 23	FIRE SUPPRESSION, PLUMBING & HVAC	99.9	87.3	94.9	100.0	110.2	104.1	99.9	87.5	95.0	100.0	85.0	94.1	100.2	85.0	94.2	100.2	87.7	95.2
26, 27, 3370	ELECTRICAL, COMMUNICATIONS & UTIL.	98.8	99.4	99.1	99.0	128.0	113.1	94.5	99.4	96.9	96.7	83.4	90.2	97.0	83.4	90.4	95.2	83.4	89.5
MF2004	WEIGHTED AVERAGE	99.9	93.1	97.1	101.1	112.0	105.7	99.8	93.9	97.3	99.1	87.7	94.3	98.6	87.7	94.0	95.9	90.1	93.5

DIVISION		NEW JERSEY																	
		CAMDEN			ELIZABETH			JERSEY CITY			NEWARK			PATERSON			TRENTON		
		MAT.	INST.	TOTAL	MAT.	INST.	TOTAL	MAT.	INST.	TOTAL	MAT.	INST.	TOTAL	MAT.	INST.	TOTAL	MAT.	INST.	TOTAL
015433	CONTRACTOR EQUIPMENT		98.5	98.5		100.4	100.4		98.5	98.5		100.4	100.4		100.4	100.4		98.1	98.1
0241, 31 - 34	SITE & INFRASTRUCTURE, DEMOLITION	87.9	103.7	98.7	100.8	104.6	103.4	87.9	104.4	99.2	103.4	104.6	104.2	99.1	104.5	102.8	87.6	104.3	99.0
0310	Concrete Forming & Accessories	96.7	120.5	117.2	104.7	121.1	118.9	96.7	121.2	117.9	93.3	121.2	117.4	95.5	121.1	117.6	93.9	120.8	117.1
0320	Concrete Reinforcing	100.4	114.2	106.0	77.3	119.5	94.5	100.4	119.5	108.2	100.4	119.5	108.2	100.4	119.5	108.2	100.4	113.2	105.6
0330	Cast-in-Place Concrete	76.7	123.8	94.7	84.5	125.1	100.0	76.7	125.2	95.2	98.1	125.1	108.4	97.8	125.1	108.2	87.3	125.0	101.7
03	CONCRETE	96.5	119.4	106.9	97.8	121.4	108.5	96.5	121.2	107.8	106.3	121.2	113.2	106.3	121.3	113.2	101.4	119.8	109.7
04	MASONRY	91.2	120.3	108.6	110.7	120.3	116.4	88.6	120.3	107.6	99.9	120.3	112.1	95.4	120.3	110.3	87.2	120.3	107.0
05	METALS	93.6	102.4	96.2	90.2	108.8	95.6	93.7	106.5	97.4	93.7	108.9	98.1	88.9	108.8	94.7	88.9	102.3	92.8
06	WOOD, PLASTICS & COMPOSITES	102.3	120.8	112.5	116.2	120.8	118.8	102.3	120.8	112.5	100.9	120.8	111.8	104.1	120.8	113.3	99.3	120.8	111.1
07	THERMAL & MOISTURE PROTECTION	94.4	120.6	104.6	95.1	125.8	107.0	94.4	125.8	106.6	94.6	125.8	106.7	95.0	120.7	105.0	90.6	120.7	102.3
08	OPENINGS	104.9	117.4	108.0	105.9	119.9	109.4	104.9	119.9	108.6	110.9	119.9	113.1	110.9	119.9	113.1	105.4	117.1	108.3
0920	Plaster & Gypsum Board	105.5	120.8	115.5	109.3	120.8	116.8	105.5	120.8	115.5	108.7	120.8	116.6	105.5	120.8	115.5	108.7	120.8	116.6
0950, 0980	Ceilings & Acoustic Treatment	99.1	120.8	112.5	90.8	120.8	109.4	99.1	120.8	112.5	96.6	120.8	111.6	99.1	120.8	112.5	96.6	120.8	111.6
0960	Flooring	96.8	136.7	107.6	101.1	151.8	114.8	96.8	151.8	111.7	97.0	151.8	111.8	96.8	151.8	111.7	97.2	151.8	111.9
0970, 0990	Wall Finishes & Painting/Coating	91.7	124.2	111.2	91.6	126.9	112.8	91.7	126.9	112.8	91.6	126.9	112.8	91.6	126.9	112.8	91.7	124.2	111.2
09	FINISHES	99.5	124.6	112.8	100.1	127.5	114.6	99.5	127.5	114.3	99.6	127.5	114.4	99.6	127.5	114.4	99.3	127.2	114.0
COVERS	DIVS. 10 - 14, 25, 28, 41, 43, 44	100.0	110.2	102.0	100.0	118.0	103.6	100.0	118.0	103.6	100.0	118.0	103.6	100.0	118.0	103.6	100.0	110.1	102.0
21, 22, 23	FIRE SUPPRESSION, PLUMBING & HVAC	100.2	113.9	105.6	100.2	120.6	108.3	100.2	120.6	108.3	100.2	122.8	109.2	100.2	120.6	108.3	100.3	121.8	108.8
26, 27, 3370	ELECTRICAL, COMMUNICATIONS & UTIL.	97.5	134.0	115.3	95.0	137.5	115.6	98.8	141.1	119.4	99.0	141.1	119.5	98.8	137.5	117.6	101.4	139.2	119.8
MF2004	WEIGHTED AVERAGE	97.8	117.8	106.2	98.6	121.5	108.2	97.8	122.2	108.0	100.5	122.0	109.5	99.3	121.8	108.8	97.6	120.6	107.3

DIVISION		NEW MEXICO															NEW YORK		
		ALBUQUERQUE			FARMINGTON			LAS CRUCES			ROSWELL			SANTA FE			ALBANY		
		MAT.	INST.	TOTAL	MAT.	INST.	TOTAL	MAT.	INST.	TOTAL	MAT.	INST.	TOTAL	MAT.	INST.	TOTAL	MAT.	INST.	TOTAL
015433	CONTRACTOR EQUIPMENT		112.1	112.1		112.1	112.1		90.9	90.9		112.1	112.1		112.1	112.1		112.9	112.9
0241, 31 - 34	SITE & INFRASTRUCTURE, DEMOLITION	78.2	108.0	98.6	84.5	108.0	100.6	89.2	91.1	90.5	89.0	108.0	102.0	84.0	108.0	100.4	73.4	105.3	95.2
0310	Concrete Forming & Accessories	99.8	69.1	73.3	99.8	69.1	73.3	97.0	67.4	71.4	99.8	68.8	73.1	98.9	69.1	73.2	96.1	87.4	88.6
0320	Concrete Reinforcing	102.8	70.9	89.8	112.3	70.9	95.4	106.4	50.6	83.7	111.4	50.9	86.7	110.2	70.9	94.2	99.1	94.3	97.1
0330	Cast-in-Place Concrete	108.9	77.6	96.9	109.9	77.6	97.6	95.6	69.2	85.5	101.3	77.5	92.2	125.5	77.6	107.2	83.4	101.9	90.5
03	CONCRETE	109.5	73.5	93.1	113.8	73.5	95.5	91.2	65.7	79.6	114.0	69.5	93.8	118.7	73.5	98.1	99.0	94.8	97.1
04	MASONRY	113.9	65.3	84.8	120.3	65.3	87.3	107.2	62.7	80.6	121.7	65.3	87.9	116.6	65.3	85.9	91.8	99.1	96.2
05	METALS	101.4	89.8	98.0	99.1	89.8	96.4	99.0	73.7	91.6	99.1	80.8	93.7	99.1	89.8	96.4	94.7	105.7	97.9
06	WOOD, PLASTICS & COMPOSITES	93.4	69.5	80.3	93.5	69.5	80.3	84.5	68.3	75.6	93.5	69.5	80.3	92.8	69.5	80.0	95.1	84.5	89.3
07	THERMAL & MOISTURE PROTECTION	100.3	74.8	90.4	100.6	74.8	90.6	86.5	69.1	79.7	100.9	74.8	90.7	100.2	74.8	90.3	89.5	91.8	90.4
08	OPENINGS	93.4	71.8	88.1	96.2	71.8	90.1	86.0	64.9	80.8	92.2	65.6	85.6	92.4	71.8	87.3	96.0	82.8	92.7
0920	Plaster & Gypsum Board	83.2	68.0	73.3	76.1	68.0	70.8	78.3	68.0	71.6	76.1	68.0	70.8	84.5	68.0	73.8	107.9	83.7	92.2
0950, 0980	Ceilings & Acoustic Treatment	112.4	68.0	84.9	107.8	68.0	83.2	101.3	68.0	80.7	107.8	68.0	83.2	107.8	68.0	83.2	92.4	83.7	87.0
0960	Flooring	102.1	69.0	93.2	104.5	69.0	94.9	138.2	65.2	118.4	104.5	69.0	94.9	104.5	69.0	94.9	83.8	86.7	84.6
0970, 0990	Wall Finishes & Painting/Coating	102.6	54.9	74.0	98.1	54.9	72.2	90.8	54.9	69.3	98.1	54.9	72.2	98.1	54.9	72.2	82.4	79.3	80.5
09	FINISHES	101.7	67.5	83.6	100.4	67.5	83.0	115.2	65.8	89.0	100.8	67.5	83.2	101.6	67.5	83.6	94.3	85.8	89.8
COVERS	DIVS. 10 - 14, 25, 28, 41, 43, 44	100.0	75.1	95.1	100.0	75.1	95.1	100.0	71.8	94.4	100.0	75.1	95.1	100.0	75.1	95.1	100.0	93.2	98.7
21, 22, 23	FIRE SUPPRESSION, PLUMBING & HVAC	100.0	72.3	89.0	99.9	72.3	88.9	100.2	71.7	88.9	99.9	72.0	88.8	99.9	72.3	88.9	100.0	93.0	97.3
26, 27, 3370	ELECTRICAL, COMMUNICATIONS & UTIL.	88.6	73.8	81.4	86.8	73.8	80.5	88.0	57.1	72.9	87.5	73.8	80.9	101.6	73.8	88.1	106.9	93.9	100.6
MF2004	WEIGHTED AVERAGE	99.8	76.0	89.8	100.3	76.0	90.1	97.0	68.5	85.0	100.3	74.3	89.3	102.0	76.0	91.1	97.4	94.7	96.3

City Cost Indexes

		NEW YORK																	
	DIVISION	BINGHAMTON			BUFFALO			HICKSVILLE			NEW YORK			RIVERHEAD			ROCHESTER		
		MAT.	INST.	TOTAL	MAT.	INST.	TOTAL	MAT.	INST.	TOTAL	MAT.	INST.	TOTAL	MAT.	INST.	TOTAL	MAT.	INST.	TOTAL
015433	CONTRACTOR EQUIPMENT		117.7	117.7		98.1	98.1		109.6	109.6		110.2	110.2		109.6	109.6		114.6	114.6
0241, 31 - 34	SITE & INFRASTRUCTURE, DEMOLITION	95.6	96.0	95.9	95.5	99.3	98.1	114.6	121.4	119.3	130.7	121.2	124.2	115.6	121.4	119.6	72.0	105.3	94.8
0310	Concrete Forming & Accessories	101.0	79.0	82.0	97.1	113.7	111.4	90.2	144.0	136.7	107.0	184.8	174.2	94.9	144.0	137.3	98.0	92.4	93.2
0320	Concrete Reinforcing	98.1	94.4	96.6	98.4	102.3	100.0	99.4	184.3	134.0	106.7	190.7	140.9	101.2	184.3	135.2	99.2	89.2	95.1
0330	Cast-in-Place Concrete	102.9	90.2	98.1	110.7	116.6	112.9	96.3	156.1	119.1	108.8	168.8	131.7	97.9	156.1	120.1	99.5	99.1	99.4
03	CONCRETE	99.7	87.7	94.2	107.4	111.8	109.4	103.5	154.3	126.6	112.0	177.7	141.9	104.3	154.3	127.1	111.1	96.7	103.9
04	MASONRY	107.0	81.0	91.5	109.3	118.4	114.8	110.9	159.6	140.0	107.8	171.5	145.9	116.4	159.6	142.3	105.3	95.8	99.6
05	METALS	91.3	114.7	98.1	94.1	95.5	94.5	101.1	138.4	112.0	108.4	140.9	117.9	101.5	138.4	112.3	90.8	105.9	95.2
06	WOOD, PLASTICS & COMPOSITES	105.3	79.7	91.3	98.4	113.4	106.6	87.5	140.7	116.7	110.5	188.2	153.2	92.9	140.7	119.2	95.4	92.8	94.0
07	THERMAL & MOISTURE PROTECTION	103.1	80.5	94.3	99.1	109.0	102.9	106.4	150.4	123.5	109.3	167.8	132.1	107.0	150.4	123.9	94.9	95.6	95.2
08	OPENINGS	91.0	78.8	88.0	95.2	102.2	96.9	88.8	150.6	104.1	98.0	177.2	117.5	88.8	150.6	104.1	98.8	87.9	96.1
0920	Plaster & Gypsum Board	119.9	78.5	93.0	99.0	113.6	108.5	107.3	142.2	129.9	128.3	191.1	169.1	109.4	142.2	130.7	98.2	92.5	94.5
0950, 0980	Ceilings & Acoustic Treatment	96.6	78.5	85.4	98.5	113.6	107.8	79.1	142.2	118.1	109.1	191.1	159.8	81.6	142.2	119.1	97.8	92.5	94.5
0960	Flooring	94.4	81.9	91.0	89.1	117.2	96.7	94.2	160.0	112.0	93.3	169.6	113.9	95.5	160.0	112.9	79.3	101.1	85.2
0970, 0990	Wall Finishes & Painting/Coating	86.7	85.2	85.8	86.1	112.6	102.0	111.0	154.5	137.1	90.3	156.7	130.2	111.0	154.5	137.1	84.7	96.0	91.5
09	FINISHES	97.3	79.8	88.1	92.5	115.0	104.4	104.7	147.1	127.1	109.2	180.1	146.7	105.9	147.1	127.7	92.1	94.6	93.5
COVERS	DIVS. 10 - 14, 25, 28, 41, 43, 44	100.0	93.3	98.7	100.0	107.5	101.5	100.0	128.9	105.7	100.0	139.6	107.8	100.0	128.9	105.7	100.0	88.4	97.7
21, 22, 23	FIRE SUPPRESSION, PLUMBING & HVAC	100.4	79.0	91.9	99.9	97.1	98.8	99.8	146.9	118.5	100.2	164.6	125.8	99.9	146.9	118.6	99.9	89.6	95.8
26, 27, 3370	ELECTRICAL, COMMUNICATIONS & UTIL.	102.2	86.8	94.7	99.8	98.5	99.1	102.2	149.3	125.1	109.4	173.1	140.4	103.8	149.3	126.0	101.0	86.2	93.8
MF2004	WEIGHTED AVERAGE	98.1	86.9	93.4	99.0	104.8	101.5	101.2	146.4	120.1	106.1	164.6	130.7	102.0	146.4	120.6	98.4	94.1	96.6

		NEW YORK																	
	DIVISION	SCHENECTADY			SYRACUSE			UTICA			WATERTOWN			WHITE PLAINS			YONKERS		
		MAT.	INST.	TOTAL	MAT.	INST.	TOTAL	MAT.	INST.	TOTAL	MAT.	INST.	TOTAL	MAT.	INST.	TOTAL	MAT.	INST.	TOTAL
015433	CONTRACTOR EQUIPMENT		112.9	112.9		112.9	112.9		112.9	112.9		112.9	112.9		109.7	109.7		109.7	109.7
0241, 31 - 34	SITE & INFRASTRUCTURE, DEMOLITION	73.4	105.4	95.3	94.4	104.9	101.6	71.6	102.8	92.9	80.1	105.4	97.4	120.5	114.7	116.5	130.0	114.5	119.4
0310	Concrete Forming & Accessories	100.7	88.1	89.8	99.9	88.9	90.4	101.2	85.6	87.8	86.2	92.4	91.6	106.4	128.8	125.7	106.7	128.8	125.8
0320	Concrete Reinforcing	97.7	94.3	96.3	99.1	94.7	97.3	99.1	91.4	96.0	99.7	94.4	97.6	99.8	183.6	134.0	103.8	183.6	136.3
0330	Cast-in-Place Concrete	94.5	102.9	97.7	95.5	96.9	96.0	87.4	90.1	88.4	101.8	101.0	101.5	95.9	133.8	110.4	107.3	133.8	117.4
03	CONCRETE	104.1	95.3	100.1	103.0	93.7	98.8	101.1	89.3	95.7	113.8	96.6	106.0	100.8	139.6	118.4	110.8	139.6	123.9
04	MASONRY	91.2	100.7	96.9	98.0	96.0	96.8	89.8	85.7	87.4	91.0	98.5	95.5	101.8	133.1	120.6	107.3	133.1	122.7
05	METALS	94.8	105.7	98.0	94.9	104.8	97.6	92.9	103.5	96.0	92.9	104.8	96.4	95.9	132.6	106.6	104.7	132.6	112.8
06	WOOD, PLASTICS & COMPOSITES	101.7	84.5	92.3	101.1	87.2	93.7	101.7	87.7	94.0	84.0	91.2	87.9	111.6	129.2	121.3	111.4	129.2	121.2
07	THERMAL & MOISTURE PROTECTION	90.1	92.6	91.1	100.2	94.7	98.0	89.9	90.2	90.0	90.2	96.4	92.6	109.7	141.0	121.9	110.0	141.0	122.1
08	OPENINGS	96.0	82.8	92.7	94.0	82.1	91.0	96.0	83.9	93.0	96.0	88.3	94.1	92.5	142.9	105.0	96.0	143.5	107.7
0920	Plaster & Gypsum Board	115.0	83.7	94.7	115.0	86.4	96.4	115.0	87.0	96.8	107.9	90.6	96.7	121.9	129.9	127.1	127.9	129.9	129.2
0950, 0980	Ceilings & Acoustic Treatment	96.6	83.7	88.6	96.6	86.4	90.3	96.6	87.0	90.7	96.6	90.6	92.9	83.3	129.9	112.1	107.5	129.9	121.3
0960	Flooring	83.4	86.7	84.3	85.1	88.4	86.0	83.4	93.3	86.1	76.1	93.3	80.8	89.7	164.6	110.0	89.3	164.6	109.6
0970, 0990	Wall Finishes & Painting/Coating	82.4	79.3	80.5	88.3	91.0	89.9	82.4	80.1	81.0	82.4	89.0	86.4	88.2	154.5	128.0	88.2	154.5	128.0
09	FINISHES	95.9	86.2	90.8	97.5	88.7	92.9	96.0	86.6	91.0	93.5	92.3	92.8	100.3	137.7	120.1	107.4	137.7	123.4
COVERS	DIVS. 10 - 14, 25, 28, 41, 43, 44	100.0	93.9	98.8	100.0	97.5	99.5	100.0	90.0	98.0	100.0	94.9	99.0	100.0	122.9	104.5	100.0	122.9	104.5
21, 22, 23	FIRE SUPPRESSION, PLUMBING & HVAC	100.2	93.9	97.7	100.2	86.9	94.9	100.2	81.7	92.9	100.2	76.0	90.6	100.4	127.3	111.1	100.0	127.3	111.1
26, 27, 3370	ELECTRICAL, COMMUNICATIONS & UTIL.	102.2	93.9	98.2	102.2	91.0	96.8	99.7	90.4	95.2	102.2	90.9	96.7	99.0	137.5	117.8	106.8	137.5	121.7
MF2004	WEIGHTED AVERAGE	97.7	95.3	96.7	98.7	93.1	96.4	96.7	89.5	93.6	98.4	92.2	95.8	99.8	132.8	113.6	104.7	132.8	116.5

		NORTH CAROLINA																	
	DIVISION	ASHEVILLE			CHARLOTTE			DURHAM			FAYETTEVILLE			GREENSBORO			RALEIGH		
		MAT.	INST.	TOTAL	MAT.	INST.	TOTAL	MAT.	INST.	TOTAL	MAT.	INST.	TOTAL	MAT.	INST.	TOTAL	MAT.	INST.	TOTAL
015433	CONTRACTOR EQUIPMENT		98.0	98.0		98.0	98.0		102.1	102.1		102.1	102.1		102.1	102.1		102.1	102.1
0241, 31 - 34	SITE & INFRASTRUCTURE, DEMOLITION	106.5	78.0	87.0	110.4	78.0	88.3	106.6	85.3	92.0	105.4	85.5	91.8	106.5	85.5	92.1	110.6	85.5	93.5
0310	Concrete Forming & Accessories	96.6	41.4	48.9	101.2	43.4	51.3	99.0	41.9	49.7	96.1	42.7	49.9	98.8	42.6	50.2	99.3	43.5	51.1
0320	Concrete Reinforcing	101.1	45.2	78.3	101.6	42.2	77.4	102.7	52.8	82.3	104.9	53.0	83.7	101.6	52.7	81.6	101.6	53.0	81.7
0330	Cast-in-Place Concrete	103.2	49.5	82.7	116.0	51.6	91.4	104.9	49.7	83.8	108.2	52.9	87.1	104.1	51.1	83.8	118.9	56.5	95.1
03	CONCRETE	106.2	47.1	79.3	111.9	48.1	82.9	106.8	48.8	80.4	108.3	50.2	81.9	106.2	49.5	80.4	113.1	51.8	85.2
04	MASONRY	85.3	41.9	59.3	92.1	45.9	64.4	90.9	37.7	59.0	89.0	42.9	61.4	89.0	38.4	58.7	91.3	43.9	62.9
05	METALS	88.3	77.6	85.2	92.5	76.7	87.9	102.1	80.7	95.8	106.4	80.7	98.9	95.7	80.3	91.2	93.3	80.7	89.6
06	WOOD, PLASTICS & COMPOSITES	96.4	40.8	65.9	101.7	43.0	69.5	98.9	42.3	67.8	95.2	42.3	66.1	98.6	42.2	67.7	98.6	43.0	68.1
07	THERMAL & MOISTURE PROTECTION	102.3	45.2	80.1	99.8	46.8	79.2	103.1	47.3	81.4	102.2	45.8	80.2	102.8	44.8	80.3	102.0	46.1	80.2
08	OPENINGS	94.2	40.5	80.9	98.3	42.1	84.4	98.3	44.8	85.1	94.3	44.8	82.0	98.3	44.7	85.0	95.1	44.9	82.6
0920	Plaster & Gypsum Board	106.6	38.4	62.3	108.1	40.7	64.3	110.6	40.0	64.8	110.0	40.0	64.5	112.3	39.9	65.3	108.1	40.7	64.3
0950, 0980	Ceilings & Acoustic Treatment	92.6	38.4	59.1	95.1	40.7	61.5	97.6	40.0	62.0	93.4	40.0	60.4	97.6	39.9	61.9	95.1	40.7	61.5
0960	Flooring	107.3	46.7	90.9	110.9	46.9	93.6	111.5	46.7	94.0	107.4	46.7	91.0	111.5	43.2	93.0	111.5	46.7	94.0
0970, 0990	Wall Finishes & Painting/Coating	114.4	36.8	67.8	114.4	42.0	71.0	114.4	38.6	68.9	114.4	36.8	67.8	114.4	35.8	67.2	114.4	40.6	70.1
09	FINISHES	101.8	41.6	70.0	103.5	43.7	71.9	104.8	42.1	71.6	102.5	42.5	70.7	105.1	41.7	71.6	104.0	43.5	72.0
COVERS	DIVS. 10 - 14, 25, 28, 41, 43, 44	100.0	75.0	95.1	100.0	75.6	95.2	100.0	70.2	94.1	100.0	71.1	94.3	100.0	75.3	95.1	100.0	71.5	94.4
21, 22, 23	FIRE SUPPRESSION, PLUMBING & HVAC	100.2	41.5	76.9	99.9	41.7	76.8	100.3	40.2	76.4	100.1	41.4	76.8	100.2	41.5	76.8	100.0	39.6	76.0
26, 27, 3370	ELECTRICAL, COMMUNICATIONS & UTIL.	97.7	40.2	69.7	100.3	52.3	76.9	98.6	50.9	75.4	97.2	46.2	72.4	97.5	40.4	69.7	104.9	40.1	73.3
MF2004	WEIGHTED AVERAGE	97.6	49.6	77.5	100.2	52.2	80.0	101.1	51.7	80.4	101.1	52.1	80.5	99.8	50.7	79.2	100.7	51.3	80.0

City Cost Indexes

DIVISION		NORTH CAROLINA						NORTH DAKOTA										
		WILMINGTON			WINSTON-SALEM			BISMARCK			FARGO			GRAND FORKS			MINOT		
		MAT.	INST.	TOTAL	MAT.	INST.	TOTAL	MAT.	INST.	TOTAL	MAT.	INST.	TOTAL	MAT.	INST.	TOTAL	MAT.	INST.	TOTAL
015433	CONTRACTOR EQUIPMENT		98.0	98.0		102.1	102.1		98.7	98.7		98.7	98.7		98.7	98.7		98.7	98.7
0241, 31 - 34	SITE & INFRASTRUCTURE, DEMOLITION	107.6	78.0	87.3	106.8	85.4	92.2	96.1	97.3	96.9	95.6	97.3	96.8	104.5	94.4	97.6	101.8	97.3	98.7
0310	Concrete Forming & Accessories	98.0	41.9	49.5	100.0	41.7	49.6	94.7	45.5	52.2	95.6	45.8	52.6	91.7	39.5	46.6	87.6	60.6	64.3
0320	Concrete Reinforcing	101.9	51.1	81.2	101.6	41.9	77.2	111.0	69.7	94.1	102.3	68.9	88.7	109.3	70.1	93.3	113.0	70.1	95.5
0330	Cast-in-Place Concrete	102.8	50.8	82.9	106.7	48.5	84.4	101.0	53.0	82.7	97.2	54.8	81.0	104.9	50.4	84.1	104.9	51.9	84.6
03	CONCRETE	106.2	48.8	80.1	107.5	46.2	79.6	100.1	54.1	79.2	105.6	54.8	82.5	106.5	50.4	81.0	105.5	60.4	85.0
04	MASONRY	75.1	39.7	53.9	89.2	39.5	59.5	100.2	58.0	74.9	101.6	43.4	66.8	102.7	65.2	80.2	103.0	65.2	80.4
05	METALS	88.0	79.8	85.6	93.3	76.2	88.3	92.7	80.8	89.2	94.9	79.7	90.4	92.7	77.0	88.1	92.9	81.6	89.6
06	WOOD, PLASTICS & COMPOSITES	98.1	41.6	67.1	98.6	41.5	67.3	87.0	40.7	61.6	86.9	41.3	61.8	83.7	37.5	58.3	79.2	60.8	69.1
07	THERMAL & MOISTURE PROTECTION	102.3	45.2	80.1	102.8	44.4	80.1	101.6	50.6	81.8	100.2	48.3	80.0	102.5	52.8	83.2	102.3	54.4	83.7
08	OPENINGS	94.3	43.3	81.7	98.3	41.4	84.2	99.1	46.4	86.1	99.1	46.8	86.1	99.1	40.9	84.7	99.3	57.4	88.9
0920	Plaster & Gypsum Board	107.5	39.3	63.2	112.3	39.2	64.8	99.2	38.8	60.0	101.6	39.4	61.2	106.4	35.5	60.4	104.9	59.7	75.5
0950, 0980	Ceilings & Acoustic Treatment	93.4	39.3	59.9	97.6	39.2	61.5	139.6	38.8	77.3	139.6	39.4	77.7	142.1	35.5	76.2	142.1	59.7	91.1
0960	Flooring	108.2	46.7	91.6	111.5	46.5	94.0	115.9	67.4	102.8	115.6	40.2	95.2	114.2	40.2	94.2	111.5	80.8	103.2
0970, 0990	Wall Finishes & Painting/Coating	114.4	35.7	67.2	114.4	37.5	68.3	98.4	33.8	59.7	98.4	69.7	81.2	98.4	25.8	54.9	98.4	28.3	56.4
09	FINISHES	102.4	41.8	70.4	105.1	42.0	71.7	115.4	46.9	79.2	115.7	45.8	78.7	117.1	37.4	75.0	115.9	60.8	86.8
COVERS	DIVS. 10 - 14, 25, 28, 41, 43, 44	100.0	70.8	94.2	100.0	74.9	95.0	100.0	80.0	96.0	100.0	80.1	96.1	100.0	40.9	88.3	100.0	82.2	96.5
21, 22, 23	FIRE SUPPRESSION, PLUMBING & HVAC	100.2	41.2	76.7	100.2	40.1	76.3	100.3	60.9	84.6	100.3	64.6	86.1	100.4	39.8	76.3	100.4	60.4	84.5
26, 27, 3370	ELECTRICAL, COMMUNICATIONS & UTIL.	98.0	38.8	69.2	97.5	44.0	71.5	103.0	66.2	85.1	104.9	66.3	86.1	97.4	62.5	80.4	100.6	67.4	84.4
MF2004	WEIGHTED AVERAGE	97.2	49.6	77.2	99.5	50.0	78.7	100.1	62.9	84.5	101.4	62.0	84.9	100.8	55.1	81.6	100.9	67.3	86.8

DIVISION		OHIO																
		AKRON			CANTON			CINCINNATI			CLEVELAND			COLUMBUS			DAYTON		
		MAT.	INST.	TOTAL	MAT.	INST.	TOTAL	MAT.	INST.	TOTAL	MAT.	INST.	TOTAL	MAT.	INST.	TOTAL	MAT.	INST.	TOTAL
015433	CONTRACTOR EQUIPMENT		95.4	95.4		95.4	95.4		100.2	100.2		96.1	96.1		95.3	95.3		95.1	95.1
0241, 31 - 34	SITE & INFRASTRUCTURE, DEMOLITION	89.0	102.8	98.4	89.1	102.8	98.5	73.9	105.2	95.3	89.0	104.5	99.6	88.5	100.1	96.4	73.0	104.9	94.8
0310	Concrete Forming & Accessories	101.0	96.3	96.9	101.0	85.4	87.5	96.2	85.4	86.8	101.1	103.3	103.0	98.3	83.6	85.6	96.1	80.6	82.7
0320	Concrete Reinforcing	98.0	93.3	96.1	98.0	77.1	89.5	89.7	84.8	87.7	98.6	93.8	96.6	92.8	86.2	90.1	89.7	80.5	85.9
0330	Cast-in-Place Concrete	93.2	100.7	96.1	94.1	98.3	95.7	81.6	87.3	83.8	91.4	109.9	98.5	89.5	93.0	90.8	75.9	90.4	81.5
03	CONCRETE	95.0	96.4	95.6	95.4	87.8	92.0	87.8	86.1	87.0	94.3	102.9	98.2	92.6	87.1	90.1	85.2	83.8	84.6
04	MASONRY	90.0	94.7	92.8	90.7	87.0	88.4	75.7	92.4	85.7	94.4	107.2	102.1	93.1	93.2	93.1	75.2	88.1	82.9
05	METALS	92.1	82.3	89.2	92.1	75.5	87.2	97.5	88.1	94.8	93.5	85.7	91.3	91.9	83.4	89.4	96.8	79.1	91.6
06	WOOD, PLASTICS & COMPOSITES	96.9	96.8	96.9	97.2	85.5	90.8	93.8	82.8	87.8	96.0	100.8	98.6	98.1	81.2	88.8	95.0	78.1	85.7
07	THERMAL & MOISTURE PROTECTION	104.4	97.5	101.7	105.3	93.7	100.8	88.8	93.5	96.8	103.2	110.2	106.0	100.2	93.8	97.7	104.3	89.3	98.5
08	OPENINGS	107.6	96.0	104.7	101.7	75.2	95.1	97.8	82.9	94.1	97.8	98.2	97.9	102.0	81.4	95.6	98.0	78.3	93.1
0920	Plaster & Gypsum Board	90.7	96.4	94.4	91.5	84.6	87.0	92.6	82.5	86.0	89.9	100.5	96.8	88.0	80.5	83.1	92.8	77.6	82.9
0950, 0980	Ceilings & Acoustic Treatment	92.3	96.4	94.8	92.3	84.6	87.5	99.2	82.5	88.9	90.6	100.5	96.8	91.1	80.5	84.5	100.2	77.6	86.2
0960	Flooring	102.6	88.2	98.7	102.8	80.9	96.9	106.6	93.6	103.1	102.4	103.3	102.6	93.6	91.5	93.0	109.3	86.2	103.0
0970, 0990	Wall Finishes & Painting/Coating	105.7	108.0	107.1	105.7	81.4	91.1	106.9	88.9	96.1	105.7	112.3	109.7	97.3	94.2	95.4	106.9	85.5	94.1
09	FINISHES	99.9	96.3	98.0	100.1	84.2	91.7	99.4	86.8	92.7	99.5	104.2	102.0	93.6	85.6	89.4	100.4	81.6	90.5
COVERS	DIVS. 10 - 14, 25, 28, 41, 43, 44	100.0	88.8	97.8	100.0	68.1	93.7	100.0	91.9	98.4	100.0	103.5	100.7	100.0	93.2	98.6	100.0	90.5	98.1
21, 22, 23	FIRE SUPPRESSION, PLUMBING & HVAC	99.9	95.5	98.2	99.9	81.4	92.6	100.0	87.0	94.9	99.9	105.5	102.1	99.7	89.8	95.8	101.0	85.2	94.7
26, 27, 3370	ELECTRICAL, COMMUNICATIONS & UTIL.	97.5	87.7	92.7	96.8	88.2	92.6	94.7	80.0	87.5	97.3	104.2	100.7	98.1	85.1	91.8	93.1	83.3	88.3
MF2004	WEIGHTED AVERAGE	97.8	93.7	96.1	97.3	85.0	92.1	95.3	88.1	92.3	97.2	102.6	99.5	96.3	88.6	93.0	95.2	85.5	91.1

DIVISION		OHIO												OKLAHOMA					
		LORAIN			SPRINGFIELD			TOLEDO			YOUNGSTOWN			ENID			LAWTON		
		MAT.	INST.	TOTAL	MAT.	INST.	TOTAL	MAT.	INST.	TOTAL	MAT.	INST.	TOTAL	MAT.	INST.	TOTAL	MAT.	INST.	TOTAL
015433	CONTRACTOR EQUIPMENT		95.4	95.4		95.1	95.1		97.4	97.4		95.4	95.4		81.5	81.5		82.1	82.1
0241, 31 - 34	SITE & INFRASTRUCTURE, DEMOLITION	88.4	102.3	97.9	73.3	103.2	93.8	87.7	99.5	95.8	88.9	103.4	98.8	109.5	91.1	96.9	104.6	92.2	96.1
0310	Concrete Forming & Accessories	101.1	95.1	95.9	96.1	82.2	84.1	98.3	102.8	102.2	101.1	89.9	91.4	96.3	36.5	44.6	99.8	49.3	56.1
0320	Concrete Reinforcing	98.0	93.7	96.2	89.7	85.8	88.1	92.8	93.1	93.0	98.0	88.2	94.0	99.3	77.2	90.3	99.6	77.2	90.4
0330	Cast-in-Place Concrete	88.7	105.6	95.2	78.0	89.9	82.5	89.5	106.0	95.8	92.3	102.3	96.1	92.6	49.2	76.0	89.6	49.2	74.2
03	CONCRETE	92.9	97.7	95.1	86.1	85.3	85.8	92.6	101.5	96.7	94.6	93.2	94.0	92.1	49.2	72.6	88.9	54.8	73.4
04	MASONRY	86.7	100.9	95.2	75.4	87.3	82.5	102.1	100.3	101.0	90.2	93.6	92.3	103.5	55.6	74.8	97.4	55.6	72.3
05	METALS	92.6	83.9	90.1	96.8	80.8	92.1	91.7	89.1	91.0	92.1	80.6	88.8	95.9	70.0	88.3	99.5	70.1	90.9
06	WOOD, PLASTICS & COMPOSITES	96.9	93.9	95.2	96.3	80.9	87.9	98.1	103.6	101.1	96.9	88.6	92.4	100.9	33.3	63.8	103.9	50.5	74.6
07	THERMAL & MOISTURE PROTECTION	105.2	104.2	104.8	104.2	89.3	98.4	101.8	107.2	103.9	105.5	94.8	101.3	99.5	57.9	83.3	99.3	59.7	83.9
08	OPENINGS	101.7	94.4	99.9	95.9	81.1	92.2	98.1	98.5	98.2	101.7	89.7	98.7	95.5	46.5	83.4	97.2	55.8	87.0
0920	Plaster & Gypsum Board	90.7	93.3	92.4	92.6	80.5	84.7	88.0	103.7	98.2	90.7	87.9	88.9	84.5	31.6	50.2	86.9	49.4	62.6
0950, 0980	Ceilings & Acoustic Treatment	92.3	93.3	92.9	100.2	80.5	88.0	91.1	103.7	98.9	92.3	87.9	89.5	83.6	31.6	51.5	92.8	49.4	66.0
0960	Flooring	102.8	108.5	104.3	109.3	86.2	103.0	92.8	98.8	94.4	102.8	93.1	100.2	109.3	46.1	92.2	112.0	46.1	94.2
0970, 0990	Wall Finishes & Painting/Coating	105.7	112.3	109.7	106.9	85.5	94.1	97.3	104.5	101.6	105.7	93.8	98.5	98.5	58.9	74.7	98.5	58.9	74.7
09	FINISHES	100.0	99.4	99.7	100.4	83.0	91.2	93.4	102.6	98.2	100.1	90.7	95.1	94.4	38.6	64.9	97.2	48.7	71.5
COVERS	DIVS. 10 - 14, 25, 28, 41, 43, 44	100.0	100.2	100.0	100.0	90.5	98.1	100.0	91.3	98.3	100.0	87.7	97.6	100.0	65.0	93.1	100.0	67.0	93.5
21, 22, 23	FIRE SUPPRESSION, PLUMBING & HVAC	99.9	84.1	93.6	101.0	84.8	94.6	99.7	98.4	99.2	99.9	86.1	94.5	100.1	63.6	85.6	100.1	63.6	85.6
26, 27, 3370	ELECTRICAL, COMMUNICATIONS & UTIL.	97.0	91.6	94.4	93.1	82.2	87.8	98.0	105.0	101.4	97.0	86.0	91.6	95.5	66.6	81.4	97.1	66.6	82.3
MF2004	WEIGHTED AVERAGE	96.9	93.6	95.5	95.1	85.7	91.2	96.5	99.7	97.9	97.2	89.7	94.1	97.4	59.7	81.6	97.8	62.5	83.0

City Cost Indexes

DIVISION		OKLAHOMA									OREGON								
		MUSKOGEE			OKLAHOMA CITY			TULSA			EUGENE			MEDFORD			PORTLAND		
		MAT.	INST.	TOTAL	MAT.	INST.	TOTAL	MAT.	INST.	TOTAL	MAT.	INST.	TOTAL	MAT.	INST.	TOTAL	MAT.	INST.	TOTAL
015433	CONTRACTOR EQUIPMENT		90.7	90.7		82.3	82.3		90.7	90.7		100.7	100.7		100.7	100.7		100.7	100.7
0241, 31 - 34	SITE & INFRASTRUCTURE, DEMOLITION	95.0	87.8	90.1	104.0	92.7	96.3	101.6	89.3	93.2	106.9	104.9	105.5	116.1	104.9	108.4	109.7	104.9	106.5
0310	Concrete Forming & Accessories	101.0	32.8	42.0	98.9	42.7	50.4	100.8	42.8	50.7	105.7	100.0	100.8	101.4	100.0	100.2	107.1	100.4	101.3
0320	Concrete Reinforcing	99.3	34.3	72.8	99.6	77.2	90.4	99.6	77.1	90.4	102.3	95.5	99.5	99.8	95.5	98.0	103.1	95.8	100.1
0330	Cast-in-Place Concrete	83.3	38.8	66.3	92.3	52.0	76.9	90.8	48.1	74.5	104.9	102.8	104.1	108.3	102.8	106.2	107.8	103.0	106.0
03	CONCRETE	84.8	36.7	62.9	90.1	52.4	73.2	89.5	52.4	72.6	105.7	100.0	103.1	111.8	100.0	106.4	107.3	100.2	104.1
04	MASONRY	113.5	48.9	74.8	97.1	57.5	73.4	96.8	57.5	73.3	114.5	98.1	104.7	111.6	98.1	103.5	115.5	103.2	108.1
05	METALS	95.8	60.3	85.4	101.5	70.0	92.3	98.9	81.5	93.8	91.3	95.9	92.6	90.9	95.9	92.3	92.3	96.6	93.6
06	WOOD, PLASTICS & COMPOSITES	106.0	33.0	65.9	103.1	40.7	68.8	104.5	41.9	70.2	94.2	100.0	97.3	88.4	100.0	94.7	95.5	100.0	97.9
07	THERMAL & MOISTURE PROTECTION	99.1	43.7	77.6	96.8	59.5	82.3	99.1	56.8	82.7	105.9	91.5	100.0	106.0	93.5	101.1	105.1	99.9	103.1
08	OPENINGS	95.5	32.0	79.8	97.2	50.5	85.7	97.2	50.8	85.7	99.2	101.3	99.7	101.8	101.3	101.7	96.8	101.3	97.9
0920	Plaster & Gypsum Board	88.4	31.1	51.2	83.1	39.3	54.6	86.9	40.4	56.7	102.3	99.7	100.7	100.9	99.7	100.1	107.8	99.7	102.6
0950, 0980	Ceilings & Acoustic Treatment	92.8	31.1	54.7	90.3	39.3	58.7	92.8	40.4	60.4	103.3	99.7	101.1	118.9	99.7	107.0	107.0	99.7	102.5
0960	Flooring	113.2	37.2	92.6	112.0	46.1	94.2	111.8	48.0	94.6	112.4	99.0	108.8	107.0	99.0	107.0	109.9	99.0	106.9
0970, 0990	Wall Finishes & Painting/Coating	98.5	31.6	58.3	98.5	58.9	74.7	98.5	46.6	67.3	110.1	67.7	84.7	110.1	62.1	81.3	109.6	72.7	87.5
09	FINISHES	97.0	33.7	63.5	95.9	43.4	68.2	96.6	43.1	68.3	107.4	96.5	101.6	110.8	95.9	102.9	108.3	97.1	102.3
COVERS	DIVS. 10 - 14, 25, 28, 41, 43, 44	100.0	64.2	92.9	100.0	66.6	93.4	100.0	66.9	93.5	100.0	101.3	100.3	100.0	101.3	100.3	100.0	101.4	100.3
21, 22, 23	FIRE SUPPRESSION, PLUMBING & HVAC	100.1	25.8	70.6	100.0	64.6	86.0	100.1	53.1	81.4	100.2	99.5	99.9	100.2	99.5	99.9	100.1	104.3	101.8
26, 27, 3370	ELECTRICAL, COMMUNICATIONS & UTIL.	95.2	30.7	63.8	103.0	66.6	85.3	97.1	40.1	69.4	99.3	95.1	97.2	102.8	85.0	94.1	99.6	99.8	99.7
MF2004	WEIGHTED AVERAGE	96.9	41.6	73.7	98.7	61.7	83.1	97.6	56.3	80.3	100.7	98.4	99.7	102.3	97.0	100.1	101.1	101.0	101.0

DIVISION		OREGON			PENNSYLVANIA														
		SALEM			ALLENTOWN			ALTOONA			ERIE			HARRISBURG			PHILADELPHIA		
		MAT.	INST.	TOTAL	MAT.	INST.	TOTAL	MAT.	INST.	TOTAL	MAT.	INST.	TOTAL	MAT.	INST.	TOTAL	MAT.	INST.	TOTAL
015433	CONTRACTOR EQUIPMENT		100.7	100.7		112.9	112.9		112.9	112.9		112.9	112.9		112.1	112.1		96.8	96.8
0241, 31 - 34	SITE & INFRASTRUCTURE, DEMOLITION	103.5	104.9	104.5	93.2	104.4	100.9	98.1	104.1	102.2	94.4	104.8	101.5	83.6	102.8	96.7	103.0	99.5	99.3
0310	Concrete Forming & Accessories	104.5	100.2	100.7	99.3	112.3	110.5	84.5	80.5	81.0	99.8	88.7	90.0	92.5	84.3	85.4	100.8	135.8	131.0
0320	Concrete Reinforcing	103.2	95.8	100.2	99.1	103.5	101.0	96.1	92.7	94.7	98.1	93.8	96.4	99.1	100.9	99.8	101.3	140.6	117.3
0330	Cast-in-Place Concrete	94.7	102.9	97.9	86.5	100.6	91.9	96.4	84.4	91.8	94.8	86.3	91.6	96.1	93.8	95.2	109.1	127.5	116.1
03	CONCRETE	101.1	100.1	100.7	97.8	107.3	102.1	93.1	85.7	89.7	92.1	90.2	91.2	100.1	92.2	96.5	108.2	133.1	119.6
04	MASONRY	119.7	103.2	109.8	94.9	99.5	97.7	97.0	74.6	83.6	87.2	90.9	89.4	95.0	87.8	90.7	96.6	134.8	119.5
05	METALS	91.7	96.4	93.1	95.2	119.3	102.2	89.1	112.0	95.8	89.3	112.6	96.1	97.0	117.6	103.0	102.3	127.7	109.7
06	WOOD, PLASTICS & COMPOSITES	91.6	100.0	96.2	101.1	116.0	109.3	80.0	82.7	81.5	97.3	87.8	92.1	94.9	83.0	88.4	100.0	137.2	120.5
07	THERMAL & MOISTURE PROTECTION	102.8	95.6	100.0	100.2	114.4	105.7	99.4	88.0	95.0	99.6	94.5	97.6	105.4	105.4	105.4	99.5	133.9	112.9
08	OPENINGS	98.7	101.3	99.3	94.0	110.9	98.2	88.4	88.3	88.3	88.5	89.5	88.7	94.0	90.8	93.2	97.7	141.5	108.5
0920	Plaster & Gypsum Board	95.4	99.7	98.2	112.6	116.3	115.0	106.3	81.8	90.4	112.6	87.1	96.0	109.7	82.1	91.8	103.8	138.5	126.3
0950, 0980	Ceilings & Acoustic Treatment	104.7	99.7	101.6	87.4	116.3	105.3	91.6	81.8	85.5	87.4	87.1	87.2	89.9	82.1	85.1	96.4	138.5	122.4
0960	Flooring	112.6	99.0	108.9	85.1	94.7	87.7	79.2	51.1	71.6	86.5	76.1	83.7	85.7	90.0	86.8	84.3	136.6	98.5
0970, 0990	Wall Finishes & Painting/Coating	110.1	67.7	84.7	88.3	81.1	84.0	84.2	105.7	97.1	94.5	89.4	91.4	88.3	87.7	88.0	97.3	144.5	125.6
09	FINISHES	106.5	96.5	101.2	95.0	105.5	100.6	93.7	76.8	84.8	96.4	86.0	90.9	94.1	85.2	89.4	100.4	137.3	119.9
COVERS	DIVS. 10 - 14, 25, 28, 41, 43, 44	100.0	101.3	100.3	100.0	103.4	100.7	100.0	94.0	98.8	100.0	98.7	99.8	100.0	88.9	97.8	100.0	120.3	104.0
21, 22, 23	FIRE SUPPRESSION, PLUMBING & HVAC	100.0	99.6	99.9	100.2	109.1	103.8	99.8	84.8	93.8	99.8	86.4	94.5	100.1	91.6	96.7	100.0	127.6	111.0
26, 27, 3370	ELECTRICAL, COMMUNICATIONS & UTIL.	107.7	95.1	101.6	101.3	96.3	98.9	90.6	98.0	94.2	92.4	85.6	89.1	100.0	84.4	92.4	98.7	137.5	117.6
MF2004	WEIGHTED AVERAGE	101.0	99.1	100.2	97.7	106.5	101.4	94.3	89.4	92.2	94.2	92.0	93.3	97.9	93.1	95.9	101.0	130.0	113.2

DIVISION		PENNSYLVANIA												PUERTO RICO			RHODE ISLAND		
		PITTSBURGH			READING			SCRANTON			YORK			SAN JUAN			PROVIDENCE		
		MAT.	INST.	TOTAL	MAT.	INST.	TOTAL	MAT.	INST.	TOTAL	MAT.	INST.	TOTAL	MAT.	INST.	TOTAL	MAT.	INST.	TOTAL
015433	CONTRACTOR EQUIPMENT		111.7	111.7		114.5	114.5		112.9	112.9		112.1	112.1		85.9	85.9		102.0	102.0
0241, 31 - 34	SITE & INFRASTRUCTURE, DEMOLITION	108.6	105.1	106.2	100.1	107.9	105.4	93.7	104.3	100.9	82.4	102.8	96.4	118.7	82.9	94.2	81.6	103.3	96.4
0310	Concrete Forming & Accessories	99.7	93.0	93.9	100.2	84.7	86.8	99.4	82.7	85.0	82.3	84.5	84.2	90.3	18.4	28.2	98.4	112.2	110.3
0320	Concrete Reinforcing	96.1	104.8	99.7	99.5	97.5	98.7	99.1	104.7	101.4	98.1	100.9	99.3	180.4	10.8	111.2	100.4	113.9	105.9
0330	Cast-in-Place Concrete	96.2	91.5	94.4	79.7	96.6	86.1	90.4	88.9	89.8	85.2	94.1	88.6	100.0	31.3	73.7	94.2	117.2	103.0
03	CONCRETE	97.7	95.9	96.9	94.1	92.8	93.5	99.6	90.5	95.5	98.1	92.5	95.5	114.5	22.7	72.8	107.3	113.8	110.3
04	MASONRY	86.6	93.2	90.5	97.8	91.1	93.8	95.2	94.8	95.0	95.1	88.4	91.1	88.1	15.1	44.4	107.4	123.6	117.1
05	METALS	93.9	118.6	101.1	98.8	116.2	103.9	97.1	119.1	103.5	94.7	117.6	101.4	107.3	37.5	86.9	93.8	108.6	98.1
06	WOOD, PLASTICS & COMPOSITES	96.0	93.2	94.5	100.2	82.0	90.2	101.1	78.9	88.9	86.7	83.0	84.7	95.1	18.6	53.1	102.0	110.1	106.4
07	THERMAL & MOISTURE PROTECTION	102.6	96.6	100.3	99.1	107.7	102.5	100.1	98.2	99.4	99.3	105.7	101.8	131.3	21.7	88.6	93.0	114.7	101.4
08	OPENINGS	91.8	102.6	94.5	95.2	90.6	94.1	94.0	89.9	93.0	90.8	90.8	90.8	152.2	15.3	118.4	99.5	112.2	102.6
0920	Plaster & Gypsum Board	99.1	92.8	95.0	107.4	81.2	90.4	115.0	77.8	90.9	108.9	82.1	91.5	119.4	15.5	51.9	103.3	109.7	107.5
0950, 0980	Ceilings & Acoustic Treatment	85.5	92.8	90.0	86.3	81.2	83.2	96.6	77.8	85.0	89.1	82.1	84.8	219.3	15.5	93.2	95.9	109.7	104.4
0960	Flooring	97.3	99.5	97.9	82.5	91.3	84.9	85.1	107.4	91.1	81.1	90.0	83.5	199.5	14.7	149.6	97.2	126.8	105.2
0970, 0990	Wall Finishes & Painting/Coating	97.0	105.7	102.2	95.8	99.0	97.7	88.3	98.0	94.1	88.3	87.7	88.0	198.5	13.7	87.6	93.9	120.7	110.0
09	FINISHES	96.3	94.8	95.5	98.0	86.1	91.7	97.5	88.0	92.5	92.5	85.3	88.7	198.7	17.4	102.9	98.0	115.9	107.5
COVERS	DIVS. 10 - 14, 25, 28, 41, 43, 44	100.0	98.5	99.7	100.0	96.7	99.4	100.0	97.5	99.5	100.0	89.2	97.8	100.0	19.5	84.1	100.0	98.2	99.6
21, 22, 23	FIRE SUPPRESSION, PLUMBING & HVAC	99.9	93.8	97.5	100.2	105.6	102.3	100.2	88.8	95.7	100.2	91.9	96.9	104.1	13.7	68.1	100.0	107.2	102.9
26, 27, 3370	ELECTRICAL, COMMUNICATIONS & UTIL.	96.7	98.0	97.3	100.9	88.0	94.6	101.3	95.8	98.7	94.9	84.4	89.8	122.6	13.8	69.6	95.9	94.5	95.2
MF2004	WEIGHTED AVERAGE	96.9	98.7	97.6	98.5	97.6	98.1	98.5	95.2	97.1	96.1	93.3	94.9	120.1	23.9	79.7	98.8	109.1	103.2

City Cost Indexes

DIVISION		SOUTH CAROLINA																	SOUTH DAKOTA		
		CHARLESTON			COLUMBIA			FLORENCE			GREENVILLE			SPARTANBURG			ABERDEEN		
		MAT.	INST.	TOTAL	MAT.	INST.	TOTAL	MAT.	INST.	TOTAL	MAT.	INST.	TOTAL	MAT.	INST.	TOTAL	MAT.	INST.	TOTAL
015433	CONTRACTOR EQUIPMENT		101.7	101.7		101.7	101.7		101.7	101.7		101.7	101.7		101.7	101.7		98.7	98.7
0241, 31 - 34	SITE & INFRASTRUCTURE, DEMOLITION	101.6	85.5	90.5	102.8	85.4	90.9	113.3	85.4	94.2	107.6	85.0	92.2	107.4	85.0	92.1	90.7	95.3	93.8
0310	Concrete Forming & Accessories	98.5	45.4	52.6	100.5	44.3	52.0	86.2	44.5	50.2	98.2	44.3	51.7	101.5	44.3	52.1	94.9	43.0	50.0
0320	Concrete Reinforcing	101.6	64.9	86.6	101.6	64.7	86.5	101.2	64.8	86.4	101.1	51.7	81.0	101.1	51.7	81.0	109.5	44.4	82.9
0330	Cast-in-Place Concrete	95.5	54.1	79.7	102.3	54.7	84.1	81.4	53.9	70.9	81.4	53.8	70.8	81.4	53.8	70.8	100.0	49.9	80.8
03	CONCRETE	102.2	53.8	80.2	105.4	53.6	81.8	102.5	53.4	80.1	100.8	50.9	78.1	101.0	50.9	78.2	99.9	47.2	75.9
04	MASONRY	96.8	42.6	64.4	91.5	40.3	60.9	79.6	42.6	57.4	77.6	42.6	56.6	79.6	42.6	57.4	104.3	60.4	78.0
05	METALS	88.9	82.3	87.0	88.9	81.5	86.7	87.9	81.8	86.1	87.9	77.3	84.8	87.9	77.3	84.8	99.1	70.6	90.8
06	WOOD, PLASTICS & COMPOSITES	98.6	44.3	68.8	100.9	43.8	69.5	83.8	43.8	61.8	98.4	43.8	68.4	102.5	43.8	70.2	101.5	42.2	68.9
07	THERMAL & MOISTURE PROTECTION	102.5	46.4	80.6	101.6	45.0	79.6	102.7	46.3	80.8	102.7	46.3	80.8	102.7	46.3	80.8	101.1	53.0	82.4
08	OPENINGS	98.3	48.3	85.9	98.3	48.1	85.9	94.3	48.1	82.8	94.2	45.0	82.0	94.2	45.0	82.0	94.8	43.0	82.0
0920	Plaster & Gypsum Board	114.8	42.1	67.5	108.1	41.5	64.9	102.7	41.5	63.0	108.1	41.5	64.8	109.9	41.5	65.5	100.3	40.4	61.4
0950, 0980	Ceilings & Acoustic Treatment	97.6	42.1	63.2	95.1	41.5	61.9	93.4	41.5	61.3	92.6	41.5	61.0	92.6	41.5	61.0	112.0	40.4	67.7
0960	Flooring	111.5	65.3	99.0	107.9	48.9	92.0	101.4	48.9	87.2	108.7	64.1	96.6	110.4	64.1	97.9	116.2	56.4	100.1
0970, 0990	Wall Finishes & Painting/Coating	114.4	45.0	72.8	111.9	45.0	71.7	114.4	45.0	72.8	114.4	45.0	72.8	114.4	45.0	72.8	98.4	42.7	65.0
09	FINISHES	105.8	48.6	75.6	102.8	45.0	72.3	101.3	45.5	71.8	103.6	48.1	74.3	104.4	48.1	74.6	108.5	45.0	75.0
COVERS	DIVS. 10 - 14, 25, 28, 41, 43, 44	100.0	68.9	93.8	100.0	68.6	93.8	100.0	68.6	93.8	100.0	68.6	93.8	100.0	68.6	93.8	100.0	51.0	90.3
21, 22, 23	FIRE SUPPRESSION, PLUMBING & HVAC	100.2	44.8	78.1	99.9	39.6	76.0	100.2	39.7	76.1	100.2	38.0	75.5	100.2	38.0	75.5	100.1	47.9	79.4
26, 27, 3370	ELECTRICAL, COMMUNICATIONS & UTIL.	97.2	103.6	100.3	101.3	53.0	77.8	95.2	50.4	73.4	97.3	54.5	76.4	97.3	54.5	76.4	97.9	59.7	79.3
MF2004	WEIGHTED AVERAGE	98.4	62.2	83.2	98.6	53.3	79.6	96.6	53.3	78.4	96.7	52.9	78.3	96.9	52.9	78.4	99.9	56.3	81.6

DIVISION		SOUTH DAKOTA						TENNESSEE								
		PIERRE			RAPID CITY			SIOUX FALLS			CHATTANOOGA			JACKSON			JOHNSON CITY		
		MAT.	INST.	TOTAL	MAT.	INST.	TOTAL	MAT.	INST.	TOTAL	MAT.	INST.	TOTAL	MAT.	INST.	TOTAL	MAT.	INST.	TOTAL
015433	CONTRACTOR EQUIPMENT		98.7	98.7		98.7	98.7		99.4	99.4		103.7	103.7		103.8	103.8		98.8	98.8
0241, 31 - 34	SITE & INFRASTRUCTURE, DEMOLITION	89.8	95.3	93.6	88.9	95.4	93.3	90.5	96.6	94.7	102.0	95.5	97.6	100.7	93.9	96.1	111.5	86.6	94.5
0310	Concrete Forming & Accessories	93.5	43.9	50.7	101.2	41.7	49.8	94.3	46.7	53.2	94.6	42.0	49.2	88.1	34.0	41.4	82.1	42.7	48.1
0320	Concrete Reinforcing	109.0	55.9	87.3	102.3	55.7	83.3	102.3	55.9	83.4	84.7	59.9	74.5	87.3	37.7	67.2	85.3	58.6	74.4
0330	Cast-in-Place Concrete	92.0	48.0	75.2	96.4	48.6	78.1	91.9	49.0	75.5	97.3	46.6	78.0	102.7	40.9	79.1	78.3	50.5	67.7
03	CONCRETE	95.4	49.1	74.3	96.5	48.3	74.6	93.0	50.7	73.8	90.6	49.1	71.7	93.9	39.5	69.2	95.8	50.5	75.2
04	MASONRY	99.8	60.0	76.0	101.1	55.9	74.0	97.2	58.1	73.8	97.7	38.4	62.2	109.4	29.9	61.8	108.7	38.3	66.5
05	METALS	99.1	75.7	92.3	100.8	75.7	93.4	100.9	76.2	93.7	103.4	83.8	97.7	98.1	72.8	90.7	100.4	83.4	95.4
06	WOOD, PLASTICS & COMPOSITES	99.3	43.0	68.4	105.2	39.1	68.9	100.2	46.0	70.5	99.3	42.3	68.0	86.0	33.9	57.4	74.0	43.3	57.1
07	THERMAL & MOISTURE PROTECTION	100.9	51.3	81.6	101.5	52.9	82.6	102.6	54.7	84.0	95.4	54.6	79.6	92.0	36.3	70.3	90.2	54.7	76.4
08	OPENINGS	98.0	47.0	85.4	99.0	44.9	85.6	99.2	48.6	86.7	99.4	48.4	86.8	100.5	35.4	84.4	95.3	49.3	83.9
0920	Plaster & Gypsum Board	94.3	41.3	59.9	100.2	37.2	59.3	99.8	44.4	63.8	82.6	40.8	55.4	90.6	32.0	52.5	96.2	41.8	60.8
0950, 0980	Ceilings & Acoustic Treatment	107.0	41.3	66.4	116.2	37.2	67.3	109.5	44.4	69.2	97.1	40.8	62.3	95.1	32.0	56.1	93.5	41.8	61.5
0960	Flooring	115.5	40.2	95.1	115.5	71.0	103.5	115.5	76.7	105.0	99.5	44.3	84.6	91.7	21.5	72.7	93.8	45.6	80.7
0970, 0990	Wall Finishes & Painting/Coating	98.4	51.2	70.1	98.4	51.2	70.1	98.4	51.2	70.1	108.8	41.3	68.3	100.4	27.9	56.9	106.1	48.4	71.5
09	FINISHES	106.2	43.3	73.0	109.2	47.4	76.5	107.7	52.4	78.5	93.8	41.9	66.4	94.0	31.1	60.7	95.6	43.3	68.0
COVERS	DIVS. 10 - 14, 25, 28, 41, 43, 44	100.0	75.8	95.2	100.0	75.7	95.2	100.0	76.2	95.3	100.0	46.6	89.4	100.0	50.4	90.2	100.0	53.0	90.7
21, 22, 23	FIRE SUPPRESSION, PLUMBING & HVAC	100.0	47.5	79.2	100.1	48.0	79.4	100.0	46.2	78.6	100.0	39.7	76.0	99.9	48.8	79.6	99.7	52.2	80.8
26, 27, 3370	ELECTRICAL, COMMUNICATIONS & UTIL.	100.9	51.8	77.0	94.6	51.8	73.8	102.9	69.1	86.5	106.5	69.3	88.4	104.1	45.8	75.7	95.0	53.9	75.0
MF2004	WEIGHTED AVERAGE	99.5	56.5	81.4	99.7	56.5	81.5	99.9	60.1	83.2	99.4	55.0	80.8	99.0	47.9	77.5	98.4	55.4	80.3

DIVISION		TENNESSEE									TEXAS								
		KNOXVILLE			MEMPHIS			NASHVILLE			ABILENE			AMARILLO			AUSTIN		
		MAT.	INST.	TOTAL	MAT.	INST.	TOTAL	MAT.	INST.	TOTAL	MAT.	INST.	TOTAL	MAT.	INST.	TOTAL	MAT.	INST.	TOTAL
015433	CONTRACTOR EQUIPMENT		98.8	98.8		101.9	101.9		104.9	104.9		90.7	90.7		90.7	90.7		90.0	90.0
0241, 31 - 34	SITE & INFRASTRUCTURE, DEMOLITION	88.0	86.6	87.1	90.4	91.5	91.1	96.9	98.6	98.1	102.0	88.4	92.7	103.2	89.3	93.7	91.2	88.8	89.5
0310	Concrete Forming & Accessories	93.7	42.7	49.6	94.8	60.5	65.2	94.0	64.0	68.1	97.9	39.2	47.2	98.2	49.2	55.9	91.9	54.1	59.3
0320	Concrete Reinforcing	84.7	58.6	74.1	91.0	65.6	80.6	91.0	64.5	80.2	96.6	53.3	78.9	96.6	48.0	76.8	97.4	46.2	76.5
0330	Cast-in-Place Concrete	91.4	46.8	74.4	86.6	66.5	78.9	90.5	67.4	81.7	94.2	42.3	74.4	93.3	47.5	75.8	91.5	48.0	74.9
03	CONCRETE	88.1	49.2	70.4	87.1	65.3	77.2	89.3	66.9	79.1	90.3	44.4	69.4	93.0	49.6	71.6	84.5	51.4	69.5
04	MASONRY	75.1	38.3	53.0	87.8	68.0	76.0	83.4	62.6	71.0	100.4	50.3	70.4	100.1	45.8	67.6	94.0	48.8	66.9
05	METALS	104.0	83.6	98.1	105.1	91.0	101.1	105.1	89.4	100.5	102.8	70.5	93.4	102.8	68.4	92.8	98.1	65.3	88.5
06	WOOD, PLASTICS & COMPOSITES	87.6	43.3	63.3	91.7	61.1	74.9	97.1	64.9	79.4	100.6	38.3	66.4	99.0	51.7	73.0	96.5	58.1	75.4
07	THERMAL & MOISTURE PROTECTION	88.4	54.2	75.1	90.1	67.5	81.3	93.4	64.6	82.2	99.2	46.2	78.6	99.4	45.1	78.3	87.0	51.3	73.1
08	OPENINGS	92.2	49.3	81.6	98.4	63.3	89.7	97.6	65.6	89.7	93.2	42.3	80.6	93.2	46.2	81.6	101.2	54.7	89.7
0920	Plaster & Gypsum Board	102.7	41.8	63.1	91.4	60.1	71.1	98.7	64.0	76.2	86.9	36.6	54.3	82.7	50.5	61.8	89.9	57.1	68.6
0950, 0980	Ceilings & Acoustic Treatment	94.3	41.8	61.8	98.7	60.1	74.8	96.1	64.0	76.3	92.8	36.6	58.1	90.3	50.5	65.7	86.3	57.1	68.3
0960	Flooring	99.2	45.6	84.7	94.2	40.5	79.6	101.6	71.6	93.5	112.0	64.7	99.2	111.8	57.3	97.1	94.6	43.9	80.9
0970, 0990	Wall Finishes & Painting/Coating	106.1	48.4	71.5	101.3	57.0	74.7	108.4	66.4	83.2	97.1	49.2	68.4	97.1	34.1	59.3	92.7	38.5	60.0
09	FINISHES	90.7	43.3	65.6	90.8	55.8	72.3	99.3	65.7	81.5	96.6	44.5	69.0	95.7	49.1	71.1	89.5	50.6	68.9
COVERS	DIVS. 10 - 14, 25, 28, 41, 43, 44	100.0	53.0	90.7	100.0	79.0	95.9	100.0	79.9	96.0	100.0	73.3	94.7	100.0	66.5	93.4	100.0	67.0	93.5
21, 22, 23	FIRE SUPPRESSION, PLUMBING & HVAC	99.7	52.2	80.8	99.9	66.6	86.6	99.9	78.9	91.5	100.1	39.9	76.2	100.0	48.3	79.5	100.0	55.5	82.3
26, 27, 3370	ELECTRICAL, COMMUNICATIONS & UTIL.	100.8	58.3	80.1	103.2	68.7	86.4	103.0	65.4	84.7	97.3	43.8	71.2	101.9	68.6	85.7	104.2	67.6	86.4
MF2004	WEIGHTED AVERAGE	95.8	55.8	79.0	97.5	70.1	86.0	98.5	73.6	88.0	98.2	50.8	78.3	98.6	56.8	81.0	96.5	59.1	80.8

607

City Cost Indexes

		TEXAS																	
		BEAUMONT			CORPUS CHRISTI			DALLAS			EL PASO			FORT WORTH			HOUSTON		
DIVISION		MAT.	INST.	TOTAL	MAT.	INST.	TOTAL	MAT.	INST.	TOTAL	MAT.	INST.	TOTAL	MAT.	INST.	TOTAL	MAT.	INST.	TOTAL
015433	CONTRACTOR EQUIPMENT		92.0	92.0		96.8	96.8		99.4	99.4		90.7	90.7		90.7	90.7		100.1	100.1
0241, 31 - 34	SITE & INFRASTRUCTURE, DEMOLITION	98.8	89.0	92.1	122.7	82.5	95.2	119.7	88.1	98.1	100.2	87.7	91.7	104.7	89.0	93.9	123.0	85.7	97.5
0310	Concrete Forming & Accessories	104.7	48.2	55.9	98.0	35.7	44.1	94.8	59.3	64.2	97.3	43.2	50.6	93.7	58.5	63.3	94.1	65.9	69.8
0320	Concrete Reinforcing	96.4	39.4	73.2	96.7	44.8	75.5	96.7	52.9	78.8	96.6	43.5	75.0	96.6	52.7	78.7	98.0	58.2	81.8
0330	Cast-in-Place Concrete	94.3	52.5	78.3	98.9	45.0	78.3	92.3	55.4	78.2	86.4	41.0	69.1	98.2	51.5	80.3	97.2	68.9	86.4
03	CONCRETE	92.7	49.3	72.9	90.0	42.8	68.5	88.7	58.4	75.0	86.7	43.7	67.1	92.0	55.9	75.6	92.2	67.2	80.8
04	MASONRY	101.3	56.1	74.2	83.8	48.5	62.7	102.2	58.6	76.1	98.0	46.2	67.0	94.7	58.5	73.0	96.1	60.8	75.0
05	METALS	102.3	65.8	91.6	97.6	76.1	91.3	102.5	82.2	96.6	102.6	63.5	91.1	99.5	70.4	91.0	106.9	87.5	101.2
06	WOOD, PLASTICS & COMPOSITES	111.4	48.3	76.8	115.1	34.9	71.0	98.7	60.2	77.6	100.2	45.8	70.3	101.1	60.1	78.6	96.8	66.7	80.3
07	THERMAL & MOISTURE PROTECTION	104.5	55.7	85.5	95.3	44.3	75.5	93.2	63.4	81.6	95.7	52.4	78.8	100.4	52.1	81.6	96.8	67.4	85.4
08	OPENINGS	94.9	44.3	82.4	108.6	36.1	90.7	103.4	54.0	91.2	93.2	41.3	80.3	87.1	53.9	78.9	104.5	63.6	94.4
0920	Plaster & Gypsum Board	96.3	47.1	64.3	93.9	32.9	54.3	90.5	59.2	70.1	85.2	44.4	58.7	86.2	59.2	68.6	94.8	65.9	76.0
0950, 0980	Ceilings & Acoustic Treatment	104.4	47.1	69.0	87.9	32.9	53.9	94.2	59.2	72.5	90.3	44.4	61.9	92.8	59.2	72.0	105.8	65.9	81.1
0960	Flooring	111.1	70.9	100.2	107.1	43.4	89.9	100.6	53.2	87.8	112.0	62.4	98.6	145.6	42.6	117.7	98.5	58.7	87.7
0970, 0990	Wall Finishes & Painting/Coating	93.9	46.0	65.1	105.5	53.5	74.3	103.5	53.1	73.2	97.1	32.7	58.5	98.5	48.6	68.6	101.4	61.2	77.3
09	FINISHES	95.0	51.8	72.2	96.3	38.2	65.6	98.4	57.5	76.7	95.7	46.0	69.4	107.1	54.7	79.4	99.9	64.0	80.9
COVERS	DIVS. 10 - 14, 25, 28, 41, 43, 44	100.0	80.4	96.1	100.0	77.8	95.6	100.0	79.2	95.9	100.0	64.9	93.0	100.0	78.9	95.8	100.0	86.0	97.2
21, 22, 23	FIRE SUPPRESSION, PLUMBING & HVAC	100.0	59.1	83.7	99.9	40.6	76.3	99.9	63.8	85.5	100.0	34.9	74.1	100.0	53.6	81.6	99.9	71.9	88.8
26, 27, 3370	ELECTRICAL, COMMUNICATIONS & UTIL.	95.0	65.0	80.4	97.6	50.1	74.5	97.3	69.2	83.6	98.6	51.9	75.9	98.1	60.3	79.7	97.2	68.5	83.2
MF2004	WEIGHTED AVERAGE	98.4	60.1	82.3	98.4	50.4	78.3	99.4	66.2	85.4	97.5	49.8	77.5	98.0	60.7	82.4	100.7	71.2	88.3

		TEXAS																	
		LAREDO			LUBBOCK			ODESSA			SAN ANTONIO			WACO			WICHITA FALLS		
DIVISION		MAT.	INST.	TOTAL	MAT.	INST.	TOTAL	MAT.	INST.	TOTAL	MAT.	INST.	TOTAL	MAT.	INST.	TOTAL	MAT.	INST.	TOTAL
015433	CONTRACTOR EQUIPMENT		90.0	90.0		98.4	98.4		90.7	90.7		91.8	91.8		90.7	90.7		90.7	90.7
0241, 31 - 34	SITE & INFRASTRUCTURE, DEMOLITION	90.5	88.6	89.2	130.8	84.3	99.0	102.2	88.8	93.0	90.2	92.3	91.6	100.8	88.9	92.7	101.6	88.5	92.6
0310	Concrete Forming & Accessories	92.6	35.9	43.6	96.9	38.6	46.5	97.9	36.7	45.0	92.7	55.6	60.6	98.5	38.8	46.9	98.5	39.6	47.6
0320	Concrete Reinforcing	97.4	44.9	75.9	97.7	53.1	79.5	96.6	53.0	78.8	104.4	49.3	81.9	96.6	46.1	76.0	96.6	47.3	76.5
0330	Cast-in-Place Concrete	76.2	59.8	69.9	94.3	47.7	76.5	94.2	44.1	75.0	74.7	67.5	71.9	85.0	52.9	72.7	90.8	47.3	74.2
03	CONCRETE	81.2	47.3	65.8	89.5	46.7	70.0	90.3	43.8	69.2	82.0	59.5	71.7	86.1	46.5	68.1	88.8	45.3	69.0
04	MASONRY	91.7	49.6	66.5	99.7	44.4	66.6	100.4	44.7	67.0	91.5	60.6	73.0	97.0	55.3	72.0	97.5	56.3	72.8
05	METALS	99.1	64.5	89.0	106.5	82.3	99.4	102.1	69.5	92.6	99.7	69.4	90.9	102.7	67.4	92.4	102.6	71.3	93.5
06	WOOD, PLASTICS & COMPOSITES	95.8	34.7	62.3	100.4	38.4	66.4	100.6	36.5	65.4	95.8	54.0	72.9	106.9	34.0	66.9	106.9	38.3	69.2
07	THERMAL & MOISTURE PROTECTION	91.4	49.4	75.1	89.9	46.8	73.1	99.2	42.5	77.2	91.4	64.9	81.1	99.7	47.2	79.3	99.7	51.0	80.8
08	OPENINGS	101.9	36.6	85.7	104.2	40.7	88.5	93.2	39.3	79.9	103.9	53.6	91.4	87.1	35.0	74.2	87.1	42.2	76.0
0920	Plaster & Gypsum Board	90.9	32.9	53.3	87.4	36.6	54.4	86.9	34.7	53.0	90.9	52.9	66.2	86.9	32.1	51.3	86.9	36.6	54.2
0950, 0980	Ceilings & Acoustic Treatment	87.9	32.9	53.9	94.5	36.6	58.7	92.8	34.7	56.9	87.9	52.9	66.3	92.8	32.1	55.3	92.8	36.6	58.0
0960	Flooring	91.6	43.4	78.6	104.0	36.1	85.6	112.0	35.7	91.4	91.6	66.5	84.8	145.6	33.1	115.2	146.7	75.0	127.3
0970, 0990	Wall Finishes & Painting/Coating	92.7	50.0	67.1	108.4	30.6	61.7	97.1	30.6	57.2	92.7	50.0	67.1	98.5	31.6	58.3	102.1	50.0	70.8
09	FINISHES	89.2	37.7	62.0	98.8	36.3	65.7	96.6	35.1	64.1	89.2	55.9	71.6	106.9	35.5	69.1	107.5	46.6	75.3
COVERS	DIVS. 10 - 14, 25, 28, 41, 43, 44	100.0	71.7	94.4	100.0	73.1	94.7	100.0	64.1	92.9	100.0	78.0	95.7	100.0	75.8	95.2	100.0	64.9	93.0
21, 22, 23	FIRE SUPPRESSION, PLUMBING & HVAC	99.9	36.6	74.7	99.6	45.9	78.2	100.1	33.4	73.6	99.9	67.0	86.8	100.1	54.1	81.9	100.1	45.6	78.5
26, 27, 3370	ELECTRICAL, COMMUNICATIONS & UTIL.	99.3	64.9	82.6	95.8	41.2	69.2	97.4	38.3	68.7	99.8	64.9	82.8	97.4	71.9	85.0	99.7	60.7	80.7
MF2004	WEIGHTED AVERAGE	95.8	51.7	77.3	100.1	51.1	79.5	98.1	46.3	76.3	96.2	65.4	83.3	97.8	56.8	80.6	98.5	55.3	80.3

		UTAH												VERMONT					
		LOGAN			OGDEN			PROVO			SALT LAKE CITY			BURLINGTON			RUTLAND		
DIVISION		MAT.	INST.	TOTAL	MAT.	INST.	TOTAL	MAT.	INST.	TOTAL	MAT.	INST.	TOTAL	MAT.	INST.	TOTAL	MAT.	INST.	TOTAL
015433	CONTRACTOR EQUIPMENT		101.4	101.4		101.4	101.4		100.7	100.7		101.4	101.4		100.4	100.4		100.4	100.4
0241, 31 - 34	SITE & INFRASTRUCTURE, DEMOLITION	88.6	101.7	97.5	77.5	101.7	94.0	85.2	100.4	95.6	77.3	101.7	93.9	76.5	96.8	90.3	76.6	96.8	90.4
0310	Concrete Forming & Accessories	105.2	55.4	62.2	105.3	55.4	62.2	106.5	55.5	62.5	104.7	55.5	62.2	93.5	55.8	60.9	97.0	56.3	61.8
0320	Concrete Reinforcing	102.6	74.0	90.9	102.3	74.0	90.7	110.9	74.0	95.9	104.6	74.0	92.1	100.4	81.1	92.5	100.4	81.2	92.5
0330	Cast-in-Place Concrete	92.3	71.4	84.3	93.7	71.4	85.2	92.4	71.4	84.4	102.4	71.4	90.6	94.7	107.0	99.1	88.1	107.2	95.4
03	CONCRETE	112.9	65.1	91.2	102.6	65.1	85.6	113.1	65.2	91.3	122.1	65.2	96.2	104.5	79.2	91.6	101.9	79.2	91.6
04	MASONRY	113.2	57.9	80.1	106.9	57.9	77.6	118.6	57.9	82.3	120.8	57.9	83.2	107.1	53.8	75.2	85.9	53.8	66.7
05	METALS	100.8	75.6	93.5	101.3	75.6	93.8	99.0	75.7	92.2	106.0	75.7	97.1	95.4	79.7	90.8	93.8	80.3	89.8
06	WOOD, PLASTICS & COMPOSITES	85.2	52.3	67.1	85.2	52.3	67.1	86.6	52.3	67.7	86.9	52.3	67.9	97.3	55.9	74.6	102.8	55.9	77.1
07	THERMAL & MOISTURE PROTECTION	99.5	63.9	85.7	98.3	63.9	84.9	101.3	63.9	86.8	101.8	63.9	87.1	94.1	54.6	78.7	94.1	63.0	82.0
08	OPENINGS	88.2	54.8	79.9	88.2	54.8	79.9	92.5	54.8	83.1	90.1	54.8	81.3	104.9	55.5	92.6	104.9	55.5	92.6
0920	Plaster & Gypsum Board	76.3	50.4	59.5	76.3	50.4	59.5	76.7	50.4	59.6	78.7	50.4	60.3	105.1	53.6	71.6	104.0	53.6	71.2
0950, 0980	Ceilings & Acoustic Treatment	108.7	50.4	72.6	108.7	50.4	72.6	108.7	50.4	72.6	103.1	50.4	70.5	92.5	53.6	68.5	93.3	53.6	68.7
0960	Flooring	104.5	53.5	90.7	102.2	53.5	89.0	105.3	53.5	91.3	103.3	53.5	89.9	97.0	61.8	87.5	96.8	61.8	87.4
0970, 0990	Wall Finishes & Painting/Coating	98.1	43.1	65.1	98.1	43.1	65.1	98.1	60.7	75.7	100.7	60.7	76.7	91.7	35.7	58.1	91.7	35.7	58.1
09	FINISHES	101.5	52.9	75.8	99.7	52.9	75.0	102.2	54.9	77.2	99.7	54.9	76.0	96.4	54.0	74.0	96.3	54.0	74.0
COVERS	DIVS. 10 - 14, 25, 28, 41, 43, 44	100.0	64.0	92.9	100.0	64.0	92.9	100.0	64.0	92.9	100.0	64.0	92.9	100.0	81.9	96.4	100.0	81.9	96.4
21, 22, 23	FIRE SUPPRESSION, PLUMBING & HVAC	99.9	69.5	87.8	99.9	69.5	87.8	99.9	69.5	87.8	100.0	69.5	87.9	100.0	64.9	86.1	100.2	65.0	86.2
26, 27, 3370	ELECTRICAL, COMMUNICATIONS & UTIL.	94.5	66.8	81.0	94.9	66.8	81.2	95.1	69.7	82.8	98.2	69.7	84.3	99.2	62.8	81.5	96.9	62.8	80.3
MF2004	WEIGHTED AVERAGE	100.3	67.3	86.4	98.4	67.3	85.3	100.8	67.8	87.0	102.9	67.9	88.2	99.3	67.7	86.1	97.5	68.1	85.2

City Cost Indexes

		\multicolumn{18}{c}{VIRGINIA}
	DIVISION	ALEXANDRIA			ARLINGTON			NEWPORT NEWS			NORFOLK			PORTSMOUTH			RICHMOND		
		MAT.	INST.	TOTAL	MAT.	INST.	TOTAL	MAT.	INST.	TOTAL	MAT.	INST.	TOTAL	MAT.	INST.	TOTAL	MAT.	INST.	TOTAL
015433	CONTRACTOR EQUIPMENT		102.9	102.9		102.1	102.1		105.8	105.8		106.2	106.2		105.8	105.8		105.8	105.8
0241, 31 - 34	SITE & INFRASTRUCTURE, DEMOLITION	116.5	89.1	97.8	128.2	86.7	99.8	110.7	87.7	95.0	111.7	88.4	95.8	109.2	87.0	94.0	111.5	88.1	95.5
0310	Concrete Forming & Accessories	94.1	76.1	78.6	93.4	74.1	76.7	98.7	72.5	76.1	100.4	72.6	76.4	88.3	59.5	63.4	98.2	65.5	70.0
0320	Concrete Reinforcing	90.2	82.2	86.9	101.8	78.7	92.4	101.6	71.3	89.2	101.6	71.4	89.3	101.2	71.3	89.0	101.6	71.8	89.4
0330	Cast-in-Place Concrete	106.8	82.9	97.7	103.9	80.3	94.9	104.6	65.6	89.7	105.6	65.6	96.1	103.6	65.0	88.8	106.2	60.2	88.6
03	CONCRETE	108.2	80.8	95.7	113.7	77.9	97.4	106.4	71.2	90.4	111.4	71.2	93.1	105.3	65.1	87.0	107.1	66.3	88.5
04	MASONRY	91.6	72.3	80.1	104.3	69.2	83.3	97.7	58.4	74.2	103.1	58.4	76.3	102.5	58.0	75.9	96.0	60.9	75.0
05	METALS	97.2	96.7	97.0	95.9	86.6	93.2	97.2	91.7	95.6	96.2	91.7	94.9	96.2	89.8	94.3	99.3	92.4	97.3
06	WOOD, PLASTICS & COMPOSITES	97.7	75.7	85.6	94.4	75.7	84.2	98.6	77.4	87.0	100.5	77.4	87.8	86.4	60.1	72.0	97.4	68.4	81.5
07	THERMAL & MOISTURE PROTECTION	101.5	80.6	93.4	103.2	72.5	91.3	102.5	61.3	86.4	99.9	61.3	84.8	102.5	58.7	85.5	101.5	61.8	86.0
08	OPENINGS	98.3	78.3	93.3	96.2	72.3	90.3	98.3	69.5	91.2	98.3	69.5	91.2	98.4	60.1	88.9	98.3	65.7	90.2
0920	Plaster & Gypsum Board	112.3	74.7	87.9	109.0	74.7	86.7	112.3	75.6	88.5	108.8	75.6	87.2	105.3	57.7	74.4	117.0	66.4	84.1
0950, 0980	Ceilings & Acoustic Treatment	97.6	74.7	83.4	93.4	74.7	81.8	97.6	75.6	84.0	97.6	75.6	84.0	97.6	57.7	72.9	100.9	66.4	79.5
0960	Flooring	111.5	88.9	105.4	109.5	61.7	96.6	111.5	56.7	96.7	110.9	56.7	96.3	111.5	71.7	94.2	110.9	79.8	102.5
0970, 0990	Wall Finishes & Painting/Coating	127.6	83.3	101.0	127.6	83.3	101.0	114.4	48.3	74.8	114.4	64.6	84.5	114.4	64.6	84.5	114.4	67.6	86.3
09	FINISHES	105.9	78.9	91.6	105.1	73.4	88.4	105.1	67.6	85.3	104.4	69.4	85.9	101.5	61.9	80.6	106.2	68.7	86.4
COVERS	DIVS. 10 - 14, 25, 28, 41, 43, 44	100.0	89.2	97.9	100.0	80.8	96.2	100.0	80.8	96.2	100.0	80.7	96.2	100.0	71.9	94.4	100.0	79.0	95.8
21, 22, 23	FIRE SUPPRESSION, PLUMBING & HVAC	100.2	88.5	95.5	100.2	82.4	93.1	100.2	66.9	86.9	100.0	65.9	86.4	100.2	65.9	86.5	100.0	61.7	84.8
26, 27, 3370	ELECTRICAL, COMMUNICATIONS & UTIL.	97.6	98.2	97.9	95.1	93.4	94.3	97.2	63.8	80.9	105.5	60.4	83.5	95.4	60.4	78.4	104.2	69.8	87.5
MF2004	WEIGHTED AVERAGE	100.7	86.0	94.5	101.5	80.8	92.8	100.5	70.9	88.1	101.9	70.5	88.7	99.8	67.4	86.2	101.7	70.1	88.4

		VIRGINIA			\multicolumn{15}{c}{WASHINGTON}
	DIVISION	ROANOKE			EVERETT			RICHLAND			SEATTLE			SPOKANE			TACOMA		
		MAT.	INST.	TOTAL	MAT.	INST.	TOTAL	MAT.	INST.	TOTAL	MAT.	INST.	TOTAL	MAT.	INST.	TOTAL	MAT.	INST.	TOTAL
015433	CONTRACTOR EQUIPMENT		102.1	102.1		101.3	101.3		95.9	95.9		101.8	101.8		95.9	95.9		101.3	101.3
0241, 31 - 34	SITE & INFRASTRUCTURE, DEMOLITION	108.2	86.2	93.2	98.0	110.8	106.7	104.1	96.0	98.5	102.3	110.1	107.6	103.4	96.0	98.3	101.4	111.1	108.0
0310	Concrete Forming & Accessories	98.5	69.0	73.0	111.8	100.6	102.1	117.8	80.6	85.7	103.4	102.7	102.8	123.3	80.6	86.4	103.4	102.5	102.6
0320	Concrete Reinforcing	101.6	65.7	86.9	113.0	93.3	105.0	99.5	86.1	94.0	111.7	93.3	104.2	100.2	86.2	94.5	111.7	93.3	104.2
0330	Cast-in-Place Concrete	117.6	65.2	97.6	101.0	100.9	101.0	111.2	86.8	101.9	106.1	108.7	107.1	115.6	86.8	104.6	103.9	108.6	105.7
03	CONCRETE	112.5	68.5	92.4	98.0	98.8	98.3	103.1	83.8	94.3	100.7	102.5	101.5	105.6	83.8	95.7	99.7	102.3	100.9
04	MASONRY	98.2	61.2	76.1	138.6	96.3	113.3	115.6	81.3	95.1	133.2	100.8	113.8	116.3	83.7	96.8	133.0	100.8	113.7
05	METALS	97.0	88.0	94.3	109.5	88.6	103.4	91.2	83.8	89.0	111.2	90.3	105.1	93.7	83.9	90.8	111.2	88.8	104.7
06	WOOD, PLASTICS & COMPOSITES	98.6	73.3	84.7	103.9	102.1	102.9	96.8	79.9	87.5	94.8	102.1	98.8	105.7	79.9	91.5	93.7	102.1	98.3
07	THERMAL & MOISTURE PROTECTION	102.4	60.0	85.9	105.1	93.7	100.6	151.3	80.0	123.6	105.0	97.3	102.0	148.0	80.9	121.9	104.7	96.4	101.5
08	OPENINGS	98.3	65.2	90.1	103.7	97.9	102.3	111.6	74.8	102.5	106.0	97.9	104.0	112.2	74.8	103.0	104.4	97.9	102.8
0920	Plaster & Gypsum Board	112.3	72.1	86.2	112.7	102.1	105.8	140.1	78.9	100.4	107.3	102.1	103.9	137.2	78.9	99.4	109.7	102.1	104.8
0950, 0980	Ceilings & Acoustic Treatment	97.6	72.1	81.8	105.1	102.1	103.3	111.4	78.9	91.3	108.2	102.1	104.4	106.7	78.9	89.5	108.4	102.1	104.5
0960	Flooring	111.5	43.9	93.2	122.4	102.7	117.1	107.6	38.8	89.0	114.7	102.9	111.5	108.0	70.2	97.8	115.3	102.9	111.9
0970, 0990	Wall Finishes & Painting/Coating	114.4	47.8	74.4	109.3	81.6	92.7	109.0	67.3	84.0	109.3	89.6	97.5	108.9	67.3	83.9	109.3	89.6	97.5
09	FINISHES	105.0	62.9	82.7	110.2	99.4	104.5	122.6	70.8	95.2	108.1	101.4	104.5	121.3	77.2	98.0	108.5	101.4	104.8
COVERS	DIVS. 10 - 14, 25, 28, 41, 43, 44	100.0	71.9	94.4	100.0	98.2	99.6	100.0	76.1	95.3	100.0	101.4	100.3	100.0	76.0	95.3	100.0	101.4	100.3
21, 22, 23	FIRE SUPPRESSION, PLUMBING & HVAC	100.2	53.2	81.5	100.2	96.4	98.7	100.6	96.4	98.9	100.2	107.0	102.9	100.5	82.1	93.2	100.2	98.8	99.7
26, 27, 3370	ELECTRICAL, COMMUNICATIONS & UTIL.	97.2	44.4	71.5	103.3	90.3	97.0	97.0	90.8	93.9	103.1	99.4	101.3	95.3	76.3	86.1	103.1	97.1	100.2
MF2004	WEIGHTED AVERAGE	101.2	63.7	85.4	105.0	96.7	101.5	104.1	85.7	96.4	105.4	101.7	103.9	104.5	81.8	95.0	105.1	99.6	102.8

		\multicolumn{6}{c}{WASHINGTON}	\multicolumn{12}{c}{WEST VIRGINIA}
	DIVISION	VANCOUVER			YAKIMA			CHARLESTON			HUNTINGTON			PARKERSBURG			WHEELING		
		MAT.	INST.	TOTAL	MAT.	INST.	TOTAL	MAT.	INST.	TOTAL	MAT.	INST.	TOTAL	MAT.	INST.	TOTAL	MAT.	INST.	TOTAL
015433	CONTRACTOR EQUIPMENT		98.2	98.2		101.3	101.3		102.1	102.1		102.1	102.1		102.1	102.1		102.1	102.1
0241, 31 - 34	SITE & INFRASTRUCTURE, DEMOLITION	113.5	100.9	104.9	104.4	109.4	107.8	103.9	89.4	94.0	107.3	90.3	95.7	112.4	89.4	96.7	113.1	88.9	96.6
0310	Concrete Forming & Accessories	104.4	91.6	93.4	103.7	94.9	96.1	104.3	84.5	87.2	99.7	89.4	90.8	90.5	83.3	84.3	92.2	83.6	84.8
0320	Concrete Reinforcing	112.6	92.9	104.6	112.2	85.7	101.4	101.6	81.1	93.2	101.6	86.3	95.4	100.2	88.1	95.2	99.5	84.0	93.1
0330	Cast-in-Place Concrete	116.1	98.0	109.2	111.1	82.4	100.1	99.4	99.8	99.5	110.0	104.1	108.0	103.1	95.7	100.3	103.1	96.4	100.6
03	CONCRETE	109.1	94.0	102.2	104.2	88.5	97.1	104.3	90.0	97.8	109.2	94.6	102.6	109.2	89.4	100.2	109.2	89.0	100.0
04	MASONRY	133.4	93.0	109.2	125.2	63.4	88.2	94.7	89.4	91.5	97.6	86.6	91.0	81.7	83.9	83.1	105.3	81.5	91.0
05	METALS	108.4	89.5	102.9	109.4	84.2	102.0	97.2	96.7	97.1	97.3	98.2	97.5	95.8	98.7	96.6	95.9	97.4	96.4
06	WOOD, PLASTICS & COMPOSITES	86.4	91.6	89.3	94.0	102.1	98.4	104.8	82.8	92.7	98.6	87.8	92.7	88.3	81.2	84.4	90.1	82.8	86.1
07	THERMAL & MOISTURE PROTECTION	105.3	86.7	98.1	104.8	76.6	93.9	99.7	87.7	95.0	102.7	88.3	97.1	102.5	84.7	95.6	102.9	83.8	95.5
08	OPENINGS	102.2	91.4	99.6	103.8	84.5	99.1	99.5	77.5	94.1	98.3	81.5	94.1	98.9	78.3	93.8	99.8	82.1	95.4
0920	Plaster & Gypsum Board	107.8	91.5	97.2	109.1	102.1	104.6	107.9	82.0	91.1	111.0	87.1	95.5	105.2	80.4	89.1	105.6	82.0	90.3
0950, 0980	Ceilings & Acoustic Treatment	103.5	91.5	96.1	103.4	102.1	102.6	94.3	82.0	86.7	92.6	87.1	89.2	91.8	80.4	84.7	91.8	82.0	85.7
0960	Flooring	121.2	71.4	107.7	116.5	58.3	100.8	111.5	105.0	109.7	111.3	93.6	106.5	105.7	95.6	103.0	106.8	94.5	103.5
0970, 0990	Wall Finishes & Painting/Coating	115.7	64.1	84.7	109.3	67.3	84.1	114.4	87.1	98.0	114.4	87.1	98.0	114.4	87.2	98.1	114.4	91.3	100.5
09	FINISHES	107.0	84.7	95.2	107.8	86.0	96.3	103.5	88.5	95.6	103.5	89.9	96.3	101.5	85.7	93.1	101.9	86.2	93.6
COVERS	DIVS. 10 - 14, 25, 28, 41, 43, 44	100.0	72.3	94.5	100.0	95.7	99.2	100.0	83.9	96.8	100.0	85.7	97.2	100.0	83.8	96.8	100.0	93.5	98.7
21, 22, 23	FIRE SUPPRESSION, PLUMBING & HVAC	100.3	100.9	100.6	100.2	93.0	97.4	100.0	83.0	93.3	100.2	84.9	94.1	100.1	87.3	95.0	100.2	90.6	96.4
26, 27, 3370	ELECTRICAL, COMMUNICATIONS & UTIL.	109.7	97.6	103.8	106.1	90.8	98.6	102.5	88.0	95.4	97.2	95.8	96.5	97.7	94.3	96.0	94.8	92.8	93.8
MF2004	WEIGHTED AVERAGE	106.5	93.8	101.1	105.3	88.0	98.0	100.4	87.8	95.1	100.6	90.3	96.3	99.6	88.7	95.0	100.6	89.2	95.8

City Cost Indexes

	DIVISION	WISCONSIN																	
		EAU CLAIRE			GREEN BAY			KENOSHA			LA CROSSE			MADISON			MILWAUKEE		
		MAT.	INST.	TOTAL	MAT.	INST.	TOTAL	MAT.	INST.	TOTAL	MAT.	INST.	TOTAL	MAT.	INST.	TOTAL	MAT.	INST.	TOTAL
015433	CONTRACTOR EQUIPMENT		101.0	101.0		98.7	98.7		98.3	98.3		101.0	101.0		100.4	100.4		89.4	89.4
0241, 31 - 34	SITE & INFRASTRUCTURE, DEMOLITION	90.8	104.0	99.8	94.1	99.5	97.8	97.0	103.1	101.1	84.4	104.1	97.9	91.0	106.5	101.6	91.9	97.4	95.7
0310	Concrete Forming & Accessories	97.0	94.4	94.7	105.0	93.6	95.2	104.1	104.4	104.4	84.4	94.9	93.5	95.9	95.7	95.7	99.6	113.8	111.8
0320	Concrete Reinforcing	100.8	100.0	100.5	98.9	85.6	93.5	100.4	102.3	101.1	100.5	88.2	95.5	100.6	82.8	95.7	100.6	102.7	101.5
0330	Cast-in-Place Concrete	97.8	94.4	96.5	101.2	96.2	99.3	108.2	100.4	105.2	87.9	94.5	90.4	97.3	98.2	97.6	97.1	108.5	101.5
03	CONCRETE	97.7	95.8	96.8	98.3	93.4	96.1	102.5	102.6	102.5	89.7	93.8	91.6	96.9	95.4	96.2	97.2	109.0	102.6
04	MASONRY	89.5	95.1	92.8	119.3	94.2	104.3	100.4	107.7	104.8	88.7	98.0	94.2	101.2	104.0	102.9	103.2	117.5	111.8
05	METALS	91.9	101.6	94.8	94.1	95.7	94.5	99.3	102.6	100.3	91.8	96.5	93.2	101.1	95.7	99.5	101.8	96.2	100.2
06	WOOD, PLASTICS & COMPOSITES	113.5	94.8	103.3	115.8	94.8	104.3	112.2	103.9	107.6	98.3	94.8	96.4	107.0	94.7	100.3	112.3	113.6	113.0
07	THERMAL & MOISTURE PROTECTION	100.1	85.2	94.3	102.0	83.5	94.8	99.8	97.3	98.8	99.6	85.2	94.0	95.5	92.9	94.5	97.8	112.1	103.4
08	OPENINGS	100.7	93.5	98.9	98.8	91.8	97.1	98.4	105.9	100.3	100.7	82.9	96.3	103.3	95.2	101.3	105.6	111.3	107.0
0920	Plaster & Gypsum Board	100.4	95.0	96.9	92.7	95.0	94.2	85.0	104.4	97.6	94.8	95.0	94.9	92.1	95.0	94.0	96.5	114.3	108.1
0950, 0980	Ceilings & Acoustic Treatment	99.9	95.0	96.9	92.5	95.0	94.0	82.6	104.4	96.1	98.3	95.0	96.2	85.3	95.0	91.3	86.8	114.3	103.8
0960	Flooring	97.0	106.7	99.6	116.7	106.7	114.0	122.8	112.2	119.9	89.5	108.7	94.7	98.8	105.0	100.5	107.4	120.5	110.9
0970, 0990	Wall Finishes & Painting/Coating	89.2	91.8	90.8	101.0	74.5	85.1	102.8	101.8	102.2	89.2	71.5	78.6	93.6	93.5	93.5	95.6	113.1	106.1
09	FINISHES	99.4	96.8	98.0	103.8	94.6	99.0	101.3	106.3	103.9	95.6	95.6	95.6	94.8	97.6	96.3	98.6	115.5	107.5
COVERS	DIVS. 10 - 14, 25, 28, 41, 43, 44	100.0	87.3	97.5	100.0	86.6	97.3	100.0	102.3	100.5	100.0	88.3	97.7	100.0	87.9	97.6	100.0	105.1	101.0
21, 22, 23	FIRE SUPPRESSION, PLUMBING & HVAC	100.1	80.1	92.2	100.4	83.8	93.8	100.1	88.5	95.5	100.1	81.3	92.7	99.7	89.3	95.6	99.9	101.4	100.5
26, 27, 3370	ELECTRICAL, COMMUNICATIONS & UTIL.	101.9	87.7	94.9	96.6	83.4	90.2	98.6	92.9	95.8	102.1	87.7	95.1	101.3	88.9	95.3	99.6	104.3	101.9
MF2004	WEIGHTED AVERAGE	97.9	92.0	95.4	99.7	90.4	95.8	100.0	99.3	99.7	96.3	91.3	94.2	99.5	95.0	97.6	100.3	106.4	102.9

	DIVISION	WISCONSIN			WYOMING								CANADA					
		RACINE			CASPER			CHEYENNE			ROCK SPRINGS			CALGARY, ALBERTA			EDMONTON, ALBERTA		
		MAT.	INST.	TOTAL	MAT.	INST.	TOTAL	MAT.	INST.	TOTAL	MAT.	INST.	TOTAL	MAT.	INST.	TOTAL	MAT.	INST.	TOTAL
015433	CONTRACTOR EQUIPMENT		100.4	100.4		101.1	101.1		101.1	101.1		101.1	101.1		106.4	106.4		106.4	106.4
0241, 31 - 34	SITE & INFRASTRUCTURE, DEMOLITION	91.0	107.3	102.1	92.6	100.1	97.7	85.8	100.2	95.6	82.2	98.9	93.6	124.3	104.8	111.0	154.6	104.8	120.6
0310	Concrete Forming & Accessories	97.4	104.6	103.6	103.9	45.0	53.0	105.5	61.3	67.3	101.8	40.4	48.7	127.0	90.2	95.2	128.4	90.2	95.4
0320	Concrete Reinforcing	100.6	102.3	101.3	109.3	46.0	83.5	103.0	46.9	80.1	111.4	46.7	85.0	147.9	63.2	113.4	147.9	63.2	113.4
0330	Cast-in-Place Concrete	97.5	100.2	98.5	102.7	76.1	92.5	98.0	76.3	89.7	99.0	56.8	82.9	186.7	101.0	153.9	208.9	101.0	167.6
03	CONCRETE	97.1	102.6	99.6	108.1	57.1	84.9	105.3	64.5	86.7	106.7	48.4	80.2	157.0	89.4	126.3	167.5	89.4	131.9
04	MASONRY	103.3	107.7	105.9	107.3	39.3	66.6	104.0	54.2	74.2	162.7	48.6	94.4	213.5	91.4	140.4	202.0	91.4	135.8
05	METALS	100.2	102.8	100.9	98.3	64.7	88.5	99.5	66.3	89.8	95.5	64.1	86.3	151.1	88.1	132.7	151.6	88.0	133.0
06	WOOD, PLASTICS & COMPOSITES	110.3	103.9	106.8	97.7	42.4	67.3	99.3	63.7	79.8	96.1	39.1	64.8	114.3	89.5	100.7	110.9	89.5	99.2
07	THERMAL & MOISTURE PROTECTION	99.5	98.6	99.1	97.6	51.9	79.8	99.4	58.1	83.3	100.7	49.9	80.9	123.0	87.9	109.3	123.7	87.9	109.8
08	OPENINGS	103.3	105.9	103.9	92.7	41.9	80.2	94.8	53.5	84.6	99.1	40.1	84.5	92.0	79.9	89.0	92.0	79.9	89.0
0920	Plaster & Gypsum Board	95.6	104.4	101.3	86.7	40.2	56.5	82.1	62.2	69.2	84.1	36.7	53.3	162.0	88.6	114.3	158.2	88.6	113.0
0950, 0980	Ceilings & Acoustic Treatment	83.4	104.4	96.4	112.1	40.2	67.6	106.0	62.2	78.9	108.8	36.7	64.2	144.8	88.6	110.0	148.1	88.6	111.3
0960	Flooring	104.5	112.2	106.6	103.5	37.0	85.5	108.0	59.4	94.9	107.0	52.3	92.2	134.4	88.8	122.1	135.6	88.8	122.9
0970, 0990	Wall Finishes & Painting/Coating	93.5	101.8	98.5	98.0	49.5	68.9	101.4	49.5	70.2	97.3	31.7	57.9	109.6	100.7	104.3	109.7	92.2	99.2
09	FINISHES	96.7	106.3	101.8	104.7	42.9	72.1	104.8	60.0	81.1	102.3	40.4	69.6	132.7	91.6	111.0	135.2	90.7	111.6
COVERS	DIVS. 10 - 14, 25, 28, 41, 43, 44	100.0	102.3	100.5	100.0	79.4	95.9	100.0	82.0	96.4	100.0	65.3	93.1	140.0	102.3	132.5	140.0	102.3	132.5
21, 22, 23	FIRE SUPPRESSION, PLUMBING & HVAC	99.9	93.7	97.5	100.0	60.7	84.4	99.9	59.9	84.0	99.9	57.3	83.0	100.1	87.6	95.1	100.1	87.6	95.2
26, 27, 3370	ELECTRICAL, COMMUNICATIONS & UTIL.	98.0	96.2	97.1	101.2	59.3	80.8	97.6	72.7	85.5	93.3	60.4	77.3	125.2	84.7	105.5	118.5	84.7	102.1
MF2004	WEIGHTED AVERAGE	99.5	101.3	100.2	100.6	58.4	82.9	100.0	65.9	85.7	102.0	56.5	82.9	130.7	89.9	113.6	131.8	89.8	114.1

	DIVISION	CANADA																	
		HALIFAX, NOVA SCOTIA			HAMILTON, ONTARIO			KITCHENER, ONTARIO			LAVAL, QUEBEC			LONDON, ONTARIO			MONTREAL, QUEBEC		
		MAT.	INST.	TOTAL	MAT.	INST.	TOTAL	MAT.	INST.	TOTAL	MAT.	INST.	TOTAL	MAT.	INST.	TOTAL	MAT.	INST.	TOTAL
015433	CONTRACTOR EQUIPMENT		102.9	102.9		107.3	107.3		103.8	103.8		104.0	104.0		104.1	104.1		105.6	105.6
0241, 31 - 34	SITE & INFRASTRUCTURE, DEMOLITION	99.3	99.2	99.2	114.9	111.1	112.3	101.3	104.7	103.6	94.4	101.1	99.0	114.5	104.9	108.0	101.1	101.0	101.0
0310	Concrete Forming & Accessories	96.4	76.4	79.1	129.6	95.4	100.1	117.4	87.7	91.8	129.2	86.1	91.9	129.7	89.0	94.5	135.9	93.7	99.5
0320	Concrete Reinforcing	154.5	61.2	116.4	161.2	90.3	132.3	110.2	90.2	102.0	156.6	79.7	125.2	130.4	85.3	112.0	146.2	93.3	124.9
0330	Cast-in-Place Concrete	171.3	75.7	134.7	151.9	101.4	132.6	140.3	82.6	118.6	134.5	95.5	119.6	149.3	99.4	130.2	174.7	101.9	146.9
03	CONCRETE	150.7	74.0	115.8	143.6	96.6	122.2	121.6	86.7	105.7	134.5	88.5	113.6	136.2	92.8	116.5	151.6	96.7	126.6
04	MASONRY	173.5	82.9	119.2	181.0	101.5	133.4	161.9	97.8	123.5	160.9	85.4	115.7	182.4	99.1	132.5	167.3	93.7	123.2
05	METALS	120.4	81.4	109.0	127.9	94.2	118.1	116.1	93.8	109.6	105.5	89.3	100.8	125.3	92.1	115.6	123.5	94.8	115.1
06	WOOD, PLASTICS & COMPOSITES	86.1	75.5	80.3	115.5	94.6	104.0	109.3	86.1	96.6	127.9	85.7	104.8	115.5	87.0	99.9	127.9	93.9	109.3
07	THERMAL & MOISTURE PROTECTION	106.0	76.2	94.4	120.1	96.5	110.9	110.0	93.2	103.5	105.1	88.3	98.5	120.9	93.2	110.1	117.1	96.9	109.3
08	OPENINGS	82.8	68.9	79.4	92.0	91.4	91.9	83.3	85.2	83.7	92.0	74.7	87.7	93.2	85.4	91.2	92.0	82.1	89.6
0920	Plaster & Gypsum Board	163.1	74.4	105.4	195.8	94.3	129.9	152.4	85.5	109.0	145.7	84.9	106.2	199.5	86.4	126.1	152.5	93.2	114.0
0950, 0980	Ceilings & Acoustic Treatment	105.4	74.4	86.2	120.6	94.3	104.3	105.4	85.5	93.1	95.4	84.9	88.9	138.5	86.4	106.3	102.0	93.2	96.5
0960	Flooring	110.1	64.5	97.8	134.4	94.8	123.7	127.8	94.8	118.9	132.0	95.4	122.1	134.6	94.8	123.9	133.9	95.4	123.5
0970, 0990	Wall Finishes & Painting/Coating	109.6	74.4	88.5	109.6	100.5	104.2	109.6	89.7	97.7	109.6	88.6	97.0	109.6	97.0	102.0	109.6	104.2	106.4
09	FINISHES	114.2	74.2	93.1	130.8	95.8	112.3	118.0	89.0	102.6	116.0	88.2	101.3	135.6	90.6	111.8	119.2	95.8	106.9
COVERS	DIVS. 10 - 14, 25, 28, 41, 43, 44	140.0	73.8	126.9	140.0	106.6	133.4	140.0	104.6	133.0	140.0	91.6	130.4	140.0	105.2	133.1	140.0	94.1	130.9
21, 22, 23	FIRE SUPPRESSION, PLUMBING & HVAC	99.2	76.1	90.0	100.0	91.9	96.8	99.2	88.9	95.1	99.2	92.7	96.6	100.0	89.0	95.6	100.1	94.4	97.8
26, 27, 3370	ELECTRICAL, COMMUNICATIONS & UTIL.	128.4	74.6	102.2	128.8	96.1	112.9	124.1	93.4	109.1	123.5	74.4	99.6	123.3	93.4	108.8	129.8	86.1	108.5
MF2004	WEIGHTED AVERAGE	119.0	78.0	101.8	123.4	96.9	112.3	114.5	92.2	105.1	114.5	87.6	103.2	122.0	93.3	110.0	121.7	93.8	110.0

610

City Cost Indexes

		CANADA																	
DIVISION		OSHAWA, ONTARIO			OTTAWA, ONTARIO			QUEBEC, QUEBEC			REGINA, SASKATCHEWAN			SASKATOON, SASKATCHEWAN			ST CATHARINES, ONTARIO		
		MAT.	INST.	TOTAL	MAT.	INST.	TOTAL	MAT.	INST.	TOTAL	MAT.	INST.	TOTAL	MAT.	INST.	TOTAL	MAT.	INST.	TOTAL
015433	CONTRACTOR EQUIPMENT		103.8	103.8		103.8	103.8		106.0	106.0		101.2	101.2		101.2	101.2		102.2	102.2
0241, 31 - 34	SITE & INFRASTRUCTURE, DEMOLITION	114.1	104.0	107.2	111.7	104.6	106.9	99.7	101.1	100.7	113.0	96.8	101.9	107.4	97.0	100.3	102.1	101.9	102.0
0310	Concrete Forming & Accessories	123.3	87.7	92.6	130.2	91.7	97.0	137.1	94.0	99.8	106.0	59.5	65.8	106.0	59.4	65.7	115.3	92.6	95.7
0320	Concrete Reinforcing	174.5	85.3	138.1	160.1	88.8	131.0	140.1	94.0	121.3	125.7	62.6	100.0	118.8	62.6	95.9	111.1	90.2	102.6
0330	Cast-in-Place Concrete	162.9	87.0	133.9	153.0	98.5	132.2	148.3	102.3	130.7	156.3	70.3	123.4	142.0	70.3	114.6	134.4	101.2	121.7
03	CONCRETE	151.0	87.4	122.1	143.9	93.7	121.1	138.1	96.9	119.4	131.1	64.8	100.9	123.1	64.7	96.5	118.6	95.3	108.0
04	MASONRY	165.4	93.8	122.5	179.7	97.2	130.3	174.3	93.7	126.0	166.4	63.6	104.8	165.9	63.6	104.6	161.4	101.5	125.5
05	METALS	107.1	92.7	102.9	126.2	93.7	116.7	125.2	95.1	116.4	105.5	76.4	97.0	105.5	76.3	97.0	106.4	93.7	102.7
06	WOOD, PLASTICS & COMPOSITES	116.4	85.9	99.6	115.2	90.9	101.9	128.6	94.0	109.6	95.8	57.8	74.9	94.2	57.8	74.2	106.9	91.5	98.4
07	THERMAL & MOISTURE PROTECTION	110.9	86.4	101.4	123.4	92.7	111.5	118.6	97.1	110.2	106.0	63.5	89.5	104.9	62.5	88.4	110.0	93.4	103.5
08	OPENINGS	90.8	85.8	89.6	92.0	88.4	91.1	92.0	89.6	91.4	87.0	55.0	79.2	86.0	55.6	78.5	82.7	88.5	84.2
0920	Plaster & Gypsum Board	155.3	85.3	109.8	234.7	90.5	141.1	191.0	93.2	127.4	165.1	56.2	94.3	145.5	56.2	87.5	137.2	91.1	107.2
0950, 0980	Ceilings & Acoustic Treatment	99.6	85.3	90.7	131.4	90.5	106.1	93.6	93.2	93.3	119.5	56.2	80.3	119.5	56.2	80.3	99.6	91.1	94.3
0960	Flooring	132.0	97.2	122.6	134.4	93.4	123.4	134.4	95.4	123.9	119.8	60.7	103.8	119.8	60.7	103.8	126.0	94.8	117.6
0970, 0990	Wall Finishes & Painting/Coating	109.6	104.9	106.8	109.6	91.4	98.7	110.2	104.2	106.6	109.6	64.5	82.6	109.6	55.0	76.9	109.6	100.5	104.2
09	FINISHES	119.0	91.1	104.2	138.2	92.0	113.8	123.1	95.9	108.7	122.0	59.8	89.1	119.1	58.7	87.2	114.0	94.0	103.3
COVERS	DIVS. 10 - 14, 25, 28, 41, 43, 44	140.0	104.7	133.0	140.0	103.1	132.7	140.0	94.3	130.9	140.0	69.7	126.1	140.0	69.7	126.1	140.0	81.2	128.4
21, 22, 23	FIRE SUPPRESSION, PLUMBING & HVAC	99.2	103.2	100.8	100.2	89.4	95.9	99.9	94.4	97.7	99.4	77.6	90.8	99.3	77.6	90.7	99.2	90.6	95.8
26, 27, 3370	ELECTRICAL, COMMUNICATIONS & UTIL.	125.3	92.7	109.4	119.7	94.4	107.4	121.7	86.1	104.4	126.7	64.4	96.4	127.0	64.4	96.5	126.3	94.5	110.8
MF2004	WEIGHTED AVERAGE	117.9	94.7	108.2	122.7	93.9	110.6	120.1	94.2	109.2	115.1	69.9	96.1	113.6	69.8	95.2	112.3	94.2	104.7

		CANADA																	
DIVISION		ST JOHNS, NEWFOUNDLAND			THUNDER BAY, ONTARIO			TORONTO, ONTARIO			VANCOUVER, BRITISH COLUMBIA			WINDSOR, ONTARIO			WINNIPEG, MANITOBA		
		MAT.	INST.	TOTAL	MAT.	INST.	TOTAL	MAT.	INST.	TOTAL	MAT.	INST.	TOTAL	MAT.	INST.	TOTAL	MAT.	INST.	TOTAL
015433	CONTRACTOR EQUIPMENT		104.8	104.8		102.2	102.2		104.1	104.1		112.7	112.7		102.2	102.2		106.4	106.4
0241, 31 - 34	SITE & INFRASTRUCTURE, DEMOLITION	115.7	99.5	104.7	107.6	101.8	103.7	139.0	105.5	116.1	121.8	107.4	111.9	97.2	101.5	100.1	119.3	100.9	106.7
0310	Concrete Forming & Accessories	103.8	64.3	69.7	123.3	93.0	97.1	131.1	100.7	104.8	124.8	78.0	84.4	123.3	90.0	94.5	130.3	65.5	74.3
0320	Concrete Reinforcing	163.8	57.7	120.5	99.3	89.6	95.4	157.0	91.0	130.1	156.7	72.3	122.3	108.9	88.9	100.7	147.9	56.0	110.4
0330	Cast-in-Place Concrete	172.4	77.6	136.2	148.0	99.8	129.6	147.4	109.5	132.9	163.0	92.2	135.9	137.7	100.8	123.6	183.5	72.3	141.0
03	CONCRETE	164.6	68.7	121.0	126.3	94.8	112.0	140.8	101.8	123.1	153.2	82.7	121.1	120.2	93.7	108.2	155.5	67.1	115.3
04	MASONRY	164.6	65.1	105.1	162.1	101.3	125.7	193.0	106.7	141.4	177.4	81.1	119.8	161.5	99.9	124.6	180.0	62.1	109.4
05	METALS	108.2	79.1	99.7	106.2	92.7	102.3	127.7	94.9	118.2	165.9	89.2	143.5	106.3	93.4	102.5	159.9	78.6	136.1
06	WOOD, PLASTICS & COMPOSITES	95.2	64.3	78.3	116.4	91.6	102.8	116.4	99.4	107.1	114.3	75.1	92.8	116.4	88.3	101.0	114.2	66.0	87.7
07	THERMAL & MOISTURE PROTECTION	109.3	63.4	91.5	110.2	94.8	104.2	121.0	101.2	113.3	132.4	82.1	112.8	110.0	93.6	103.6	114.1	67.3	95.9
08	OPENINGS	98.2	60.2	88.8	81.7	87.9	83.3	90.8	96.0	92.1	93.8	75.8	89.3	81.5	86.6	82.7	92.0	60.7	84.3
0920	Plaster & Gypsum Board	168.5	62.7	99.8	166.5	91.2	117.6	177.3	99.3	126.6	154.9	73.5	102.0	156.7	87.8	111.9	157.0	64.1	96.7
0950, 0980	Ceilings & Acoustic Treatment	104.6	62.7	78.7	95.4	91.2	92.8	132.6	99.3	112.0	141.0	73.5	99.3	95.4	87.8	90.7	121.9	64.1	86.2
0960	Flooring	114.8	54.9	98.6	132.0	54.6	111.1	135.6	100.5	126.1	134.8	92.0	123.2	132.0	95.5	122.1	133.3	67.3	115.5
0970, 0990	Wall Finishes & Painting/Coating	109.6	64.5	82.6	109.6	93.2	99.8	111.5	104.9	107.5	109.6	88.7	97.1	109.6	92.8	99.5	109.7	48.1	72.7
09	FINISHES	117.4	62.5	88.3	119.2	86.7	102.0	133.2	101.2	116.3	130.7	81.2	104.5	117.5	91.3	103.6	125.4	64.2	93.1
COVERS	DIVS. 10 - 14, 25, 28, 41, 43, 44	140.0	71.2	126.4	140.0	81.6	128.4	140.0	108.5	133.8	140.0	99.9	132.1	140.0	80.6	128.2	140.0	71.7	126.5
21, 22, 23	FIRE SUPPRESSION, PLUMBING & HVAC	99.6	64.2	85.5	99.2	90.7	95.8	99.9	95.9	98.3	99.9	73.3	89.3	99.2	90.2	95.6	99.9	65.6	86.3
26, 27, 3370	ELECTRICAL, COMMUNICATIONS & UTIL.	122.9	65.8	95.1	124.1	92.7	108.8	122.0	96.8	109.8	124.7	75.0	100.5	130.2	94.4	112.8	127.9	67.0	98.3
MF2004	WEIGHTED AVERAGE	119.9	69.2	98.6	113.5	92.9	104.9	123.6	99.8	113.6	131.2	82.0	110.5	113.0	93.2	104.7	129.6	69.6	104.4

Location Factors

Costs shown in RSMeans cost data publications are based on national averages for materials and installation. To adjust these costs to a specific location, simply multiply the base cost by the factor and divide by 100 for that city. The data is arranged alphabetically by state and postal zip code numbers. For a city not listed, use the factor for a nearby city with similar economic characteristics.

STATE/ZIP	CITY	MAT.	INST.	TOTAL
ALABAMA				
350-352	Birmingham	98.3	74.9	88.5
354	Tuscaloosa	97.0	60.5	81.7
355	Jasper	97.2	54.0	79.0
356	Decatur	97.0	59.9	81.4
357-358	Huntsville	97.0	71.1	86.1
359	Gadsden	96.9	59.4	81.2
360-361	Montgomery	98.0	59.2	81.7
362	Anniston	96.3	52.9	78.1
363	Dothan	96.9	52.5	78.3
364	Evergreen	96.4	57.3	80.0
365-366	Mobile	97.9	65.9	84.5
367	Selma	96.6	54.4	78.9
368	Phenix City	97.3	59.4	81.4
369	Butler	96.8	55.7	79.5
ALASKA				
995-996	Anchorage	128.3	110.3	120.8
997	Fairbanks	127.6	112.6	121.3
998	Juneau	127.0	108.5	119.2
999	Ketchikan	139.6	108.1	126.4
ARIZONA				
850,853	Phoenix	99.9	74.0	89.0
852	Mesa/Tempe	99.2	66.7	85.6
855	Globe	99.5	63.1	84.2
856-857	Tucson	98.0	70.6	86.5
859	Show Low	99.5	64.2	84.7
860	Flagstaff	101.4	71.6	88.9
863	Prescott	99.2	63.7	84.3
864	Kingman	97.5	69.6	85.8
865	Chambers	97.5	64.4	83.6
ARKANSAS				
716	Pine Bluff	96.9	66.6	84.2
717	Camden	94.6	44.6	73.6
718	Texarkana	95.9	48.8	76.1
719	Hot Springs	93.8	47.9	74.5
720-722	Little Rock	97.1	69.9	85.7
723	West Memphis	96.0	59.6	80.8
724	Jonesboro	96.9	60.2	81.5
725	Batesville	94.7	54.3	77.7
726	Harrison	95.9	55.0	78.8
727	Fayetteville	93.4	54.7	77.1
728	Russellville	94.5	56.7	78.6
729	Fort Smith	97.4	60.1	81.8
CALIFORNIA				
900-902	Los Angeles	102.2	116.7	108.3
903-905	Inglewood	97.6	107.9	101.9
906-908	Long Beach	99.3	108.0	102.9
910-912	Pasadena	98.0	107.9	102.2
913-916	Van Nuys	101.2	107.9	104.0
917-918	Alhambra	100.2	107.9	103.4
919-921	San Diego	102.4	108.8	105.1
922	Palm Springs	99.2	106.8	102.4
923-924	San Bernardino	96.9	107.2	101.2
925	Riverside	101.6	112.7	106.3
926-927	Santa Ana	98.9	107.5	102.5
928	Anaheim	101.5	114.7	107.0
930	Oxnard	102.9	112.8	107.0
931	Santa Barbara	101.9	112.9	106.5
932-933	Bakersfield	103.3	109.4	105.8
934	San Luis Obispo	102.5	105.3	103.7
935	Mojave	99.6	104.3	101.6
936-938	Fresno	103.2	112.6	107.2
939	Salinas	103.3	116.4	108.8
940-941	San Francisco	111.2	141.3	123.8
942,956-958	Sacramento	104.9	114.6	108.9
943	Palo Alto	103.5	123.7	112.0
944	San Mateo	106.4	129.3	116.0
945	Vallejo	104.1	121.6	111.4
946	Oakland	108.8	129.5	117.5
947	Berkeley	108.3	123.8	114.8
948	Richmond	107.4	123.9	114.3
949	San Rafael	108.7	124.5	115.3
950	Santa Cruz	108.4	116.6	111.8

STATE/ZIP	CITY	MAT.	INST.	TOTAL
CALIFORNIA (CONT'D)				
951	San Jose	106.0	132.5	117.1
952	Stockton	103.5	112.0	107.0
953	Modesto	103.4	111.8	106.9
954	Santa Rosa	103.3	126.1	112.9
955	Eureka	104.7	109.6	106.7
959	Marysville	104.0	111.2	107.0
960	Redding	105.8	109.8	107.5
961	Susanville	104.6	110.0	106.8
COLORADO				
800-802	Denver	101.8	85.5	95.0
803	Boulder	98.6	82.0	91.6
804	Golden	100.7	80.8	92.4
805	Fort Collins	102.2	77.3	91.7
806	Greeley	99.7	67.0	86.0
807	Fort Morgan	99.2	80.6	91.4
808-809	Colorado Springs	101.5	81.0	92.9
810	Pueblo	101.3	79.0	91.9
811	Alamosa	102.4	75.3	91.0
812	Salida	102.2	76.2	91.3
813	Durango	102.8	76.0	91.6
814	Montrose	101.4	74.4	90.1
815	Grand Junction	104.9	72.0	91.1
816	Glenwood Springs	102.4	77.7	92.1
CONNECTICUT				
060	New Britain	99.2	118.0	107.1
061	Hartford	100.5	118.1	107.9
062	Willimantic	99.8	117.5	107.2
063	New London	95.8	117.1	104.8
064	Meriden	97.9	118.5	106.6
065	New Haven	101.0	118.0	108.2
066	Bridgeport	100.3	118.6	108.0
067	Waterbury	99.9	118.5	107.7
068	Norwalk	99.9	119.0	107.9
069	Stamford	100.0	125.4	110.7
D.C.				
200-205	Washington	104.6	92.8	99.7
DELAWARE				
197	Newark	100.5	105.8	102.7
198	Wilmington	100.6	105.9	102.8
199	Dover	101.2	105.8	103.1
FLORIDA				
320,322	Jacksonville	98.8	63.1	83.8
321	Daytona Beach	99.1	76.5	89.6
323	Tallahassee	99.6	53.0	80.1
324	Panama City	100.3	48.6	78.6
325	Pensacola	102.9	61.3	85.4
326,344	Gainesville	100.5	66.0	86.0
327-328,347	Orlando	103.1	72.0	90.0
329	Melbourne	102.0	80.0	92.7
330-332,340	Miami	101.2	75.4	90.3
333	Fort Lauderdale	98.5	73.6	88.0
334,349	West Palm Beach	97.2	69.8	85.7
335-336,346	Tampa	100.1	78.7	91.1
337	St. Petersburg	102.3	58.1	83.7
338	Lakeland	99.2	77.8	90.2
339,341	Fort Myers	98.6	70.1	86.6
342	Sarasota	100.4	71.1	88.1
GEORGIA				
300-303,399	Atlanta	98.2	79.2	90.2
304	Statesboro	97.9	48.1	77.0
305	Gainesville	96.5	63.2	82.5
306	Athens	95.9	66.9	83.7
307	Dalton	98.0	52.3	78.8
308-309	Augusta	96.9	64.3	83.2
310-312	Macon	96.5	66.6	83.9
313-314	Savannah	98.8	62.0	83.3
315	Waycross	97.7	57.2	80.7
316	Valdosta	97.8	51.0	78.2
317,398	Albany	97.9	62.0	82.9
318-319	Columbus	97.8	66.8	84.8

Location Factors

STATE/ZIP	CITY	MAT.	INST.	TOTAL
HAWAII				
967	Hilo	114.2	118.7	116.1
968	Honolulu	118.7	118.7	118.7
STATES & POSS.				
969	Guam	138.1	64.3	107.1
IDAHO				
832	Pocatello	101.6	74.5	90.2
833	Twin Falls	102.5	54.0	82.1
834	Idaho Falls	99.9	59.8	83.0
835	Lewiston	109.0	80.4	97.0
836-837	Boise	100.6	74.8	89.8
838	Coeur d'Alene	108.3	77.4	95.3
ILLINOIS				
600-603	North Suburban	97.4	123.7	108.5
604	Joliet	97.4	126.9	109.8
605	South Suburban	97.4	124.5	108.8
606-608	Chicago	97.9	138.3	114.9
609	Kankakee	93.6	107.0	99.2
610-611	Rockford	97.1	115.0	104.7
612	Rock Island	94.7	99.5	96.7
613	La Salle	95.9	106.3	100.3
614	Galesburg	95.7	99.0	97.1
615-616	Peoria	98.2	104.9	101.0
617	Bloomington	95.1	106.4	99.8
618-619	Champaign	98.4	105.0	101.2
620-622	East St. Louis	93.9	105.7	98.8
623	Quincy	94.9	94.8	94.9
624	Effingham	94.3	95.9	95.0
625	Decatur	96.2	102.5	98.8
626-627	Springfield	97.0	103.4	99.7
628	Centralia	92.4	100.5	95.8
629	Carbondale	92.1	94.1	92.9
INDIANA				
460	Anderson	94.8	82.5	89.6
461-462	Indianapolis	98.0	86.7	93.2
463-464	Gary	96.3	101.5	98.5
465-466	South Bend	95.6	83.1	90.4
467-468	Fort Wayne	95.3	78.4	88.2
469	Kokomo	92.5	80.8	87.6
470	Lawrenceburg	91.1	76.9	85.1
471	New Albany	92.4	74.4	84.9
472	Columbus	94.7	79.5	88.3
473	Muncie	95.5	80.6	89.3
474	Bloomington	96.9	78.8	89.3
475	Washington	92.7	82.4	88.4
476-477	Evansville	94.4	86.5	91.1
478	Terre Haute	95.0	86.6	91.5
479	Lafayette	94.3	82.1	89.2
IOWA				
500-503,509	Des Moines	98.2	76.7	89.2
504	Mason City	95.7	61.4	81.3
505	Fort Dodge	96.0	57.1	79.7
506-507	Waterloo	97.8	58.4	81.2
508	Creston	96.2	61.2	81.5
510-511	Sioux City	98.4	68.8	86.0
512	Sibley	96.9	47.6	76.2
513	Spencer	98.5	46.3	76.6
514	Carroll	95.6	51.4	77.0
515	Council Bluffs	99.6	74.6	89.1
516	Shenandoah	96.3	49.3	76.6
520	Dubuque	97.9	75.8	88.6
521	Decorah	96.5	48.8	76.5
522-524	Cedar Rapids	98.8	80.8	91.2
525	Ottumwa	96.6	69.3	85.1
526	Burlington	95.8	70.7	85.2
527-528	Davenport	97.7	90.2	94.5
KANSAS				
660-662	Kansas City	98.3	93.7	96.4
664-666	Topeka	98.9	66.1	85.1
667	Fort Scott	96.6	71.4	86.0
668	Emporia	96.7	59.7	81.2
669	Belleville	98.4	61.0	82.7
670-672	Wichita	97.6	64.9	83.9
673	Independence	98.1	63.3	83.5
674	Salina	98.7	61.6	83.1
675	Hutchinson	93.8	61.2	80.1
676	Hays	97.6	62.5	82.9
677	Colby	98.4	62.3	83.3

STATE/ZIP	CITY	MAT.	INST.	TOTAL
KANSAS (CONT'D)				
678	Dodge City	99.9	64.0	84.8
679	Liberal	97.5	62.2	82.7
KENTUCKY				
400-402	Louisville	97.1	83.9	91.6
403-405	Lexington	95.1	78.5	88.1
406	Frankfort	96.3	78.9	89.0
407-409	Corbin	92.3	65.8	81.2
410	Covington	92.8	99.3	95.5
411-412	Ashland	91.7	96.7	93.8
413-414	Campton	93.1	66.2	81.8
415-416	Pikeville	94.0	81.7	88.8
417-418	Hazard	92.4	58.7	78.3
420	Paducah	90.9	87.2	89.4
421-422	Bowling Green	93.6	84.6	89.8
423	Owensboro	93.4	83.2	89.1
424	Henderson	90.7	86.8	89.0
425-426	Somerset	90.6	70.3	82.1
427	Elizabethtown	90.1	83.6	87.3
LOUISIANA				
700-701	New Orleans	102.9	69.6	88.9
703	Thibodaux	100.4	63.7	85.0
704	Hammond	97.6	58.4	81.2
705	Lafayette	100.0	58.5	82.6
706	Lake Charles	100.1	61.1	83.7
707-708	Baton Rouge	102.2	63.8	86.1
710-711	Shreveport	96.8	59.2	81.0
712	Monroe	96.9	56.0	79.7
713-714	Alexandria	97.0	56.0	79.8
MAINE				
039	Kittery	93.0	74.4	85.2
040-041	Portland	98.5	73.7	88.1
042	Lewiston	96.4	73.7	86.9
043	Augusta	96.1	74.3	86.9
044	Bangor	95.8	73.7	86.5
045	Bath	94.2	74.4	85.9
046	Machias	93.7	73.9	85.4
047	Houlton	93.9	74.4	85.7
048	Rockland	92.9	74.5	85.2
049	Waterville	94.2	74.4	85.9
MARYLAND				
206	Waldorf	100.5	69.5	87.5
207-208	College Park	100.4	79.4	91.6
209	Silver Spring	99.7	75.5	89.5
210-212	Baltimore	98.2	86.0	93.1
214	Annapolis	98.7	80.2	90.9
215	Cumberland	94.0	80.4	88.3
216	Easton	95.5	42.8	73.4
217	Hagerstown	94.9	81.6	89.3
218	Salisbury	95.8	50.3	76.7
219	Elkton	93.1	62.9	80.4
MASSACHUSETTS				
010-011	Springfield	98.4	104.5	101.0
012	Pittsfield	97.8	101.0	99.1
013	Greenfield	95.7	102.3	98.4
014	Fitchburg	94.4	117.6	104.1
015-016	Worcester	98.3	118.9	107.0
017	Framingham	93.9	126.1	107.4
018	Lowell	97.7	125.7	109.5
019	Lawrence	98.9	123.6	109.3
020-022, 024	Boston	100.5	135.9	115.4
023	Brockton	99.2	120.7	108.2
025	Buzzards Bay	93.3	118.2	103.8
026	Hyannis	96.2	118.7	105.7
027	New Bedford	98.4	118.9	107.0
MICHIGAN				
480,483	Royal Oak	91.4	103.9	96.6
481	Ann Arbor	93.7	104.5	98.2
482	Detroit	95.3	114.2	103.2
484-485	Flint	93.4	100.8	96.5
486	Saginaw	93.0	89.9	91.7
487	Bay City	93.1	90.0	91.8
488-489	Lansing	94.4	96.9	95.5
490	Battle Creek	94.5	88.2	91.9
491	Kalamazoo	94.8	85.6	91.0
492	Jackson	92.7	91.6	92.3
493,495	Grand Rapids	96.1	65.2	83.1
494	Muskegon	93.5	81.4	88.4

Location Factors

STATE/ZIP	CITY	MAT.	INST.	TOTAL		STATE/ZIP	CITY	MAT.	INST.	TOTAL
MICHIGAN (CONT'D)						**NEW HAMPSHIRE (CONT'D)**				
496	Traverse City	92.3	69.1	82.5		032-033	Concord	97.0	83.8	91.5
497	Gaylord	93.3	72.1	84.4		034	Keene	95.1	53.4	77.6
498-499	Iron Mountain	95.3	83.2	90.2		035	Littleton	95.0	60.6	80.5
MINNESOTA						036	Charleston	94.6	50.8	76.2
550-551	Saint Paul	98.6	118.7	107.0		037	Claremont	93.7	50.8	75.7
553-555	Minneapolis	99.3	124.2	109.8		038	Portsmouth	95.9	90.1	93.5
556-558	Duluth	98.2	108.2	102.4		**NEW JERSEY**				
MINNESOTA (CONT'd)						070-071	Newark	100.5	122.0	109.5
559	Rochester	98.1	104.7	100.9		072	Elizabeth	98.6	121.5	108.2
560	Mankato	96.0	103.2	99.0		073	Jersey City	97.8	122.2	108.0
561	Windom	94.8	77.6	87.6		074-075	Paterson	99.3	121.8	108.8
562	Willmar	94.3	82.8	89.5		076	Hackensack	97.3	122.3	107.8
563	St. Cloud	95.8	117.5	104.9		077	Long Branch	97.1	121.2	107.2
564	Brainerd	95.9	99.5	97.4		078	Dover	97.6	122.0	107.8
565	Detroit Lakes	97.7	94.6	96.4		079	Summit	97.6	121.5	107.6
566	Bemidji	97.1	95.9	96.6		080,083	Vineland	95.7	117.6	104.9
567	Thief River Falls	96.7	91.7	94.6		081	Camden	97.8	117.8	106.2
MISSISSIPPI						082,084	Atlantic City	96.3	117.6	105.2
386	Clarksdale	96.6	60.4	81.4		085-086	Trenton	97.6	120.6	107.3
387	Greenville	100.0	70.9	87.8		087	Point Pleasant	97.7	120.8	107.4
388	Tupelo	97.9	63.1	83.2		088-089	New Brunswick	98.1	121.4	107.9
389	Greenwood	97.8	59.9	81.9		**NEW MEXICO**				
390-392	Jackson	98.5	69.9	86.5		870-872	Albuquerque	99.8	76.0	89.8
393	Meridian	96.8	70.9	85.9		873	Gallup	99.9	76.0	89.9
394	Laurel	97.8	63.7	83.5		874	Farmington	100.3	76.0	90.1
395	Biloxi	98.5	62.6	83.4		875	Santa Fe	102.0	76.0	91.1
396	Mccomb	96.3	59.3	80.8		877	Las Vegas	98.4	75.8	88.9
397	Columbus	97.7	61.0	82.3		878	Socorro	97.9	76.0	88.7
MISSOURI						879	Truth/Consequences	98.1	70.9	86.7
630-631	St. Louis	98.6	108.8	102.9		880	Las Cruces	97.0	68.5	85.0
633	Bowling Green	97.1	90.3	94.2		881	Clovis	98.5	74.2	88.3
634	Hannibal	95.9	80.1	89.3		882	Roswell	100.3	74.3	89.3
635	Kirksville	98.7	73.3	88.0		883	Carrizozo	100.7	76.0	90.3
636	Flat River	98.1	91.1	95.1		884	Tucumcari	99.3	74.2	88.8
637	Cape Girardeau	98.2	87.4	93.7		**NEW YORK**				
638	Sikeston	96.1	77.2	88.2		100-102	New York	106.1	164.6	130.7
639	Poplar Bluff	95.6	77.7	88.1		103	Staten Island	101.9	162.0	127.1
640-641	Kansas City	99.2	106.1	102.1		104	Bronx	100.0	162.0	126.0
644-645	St. Joseph	98.3	90.2	94.9		105	Mount Vernon	99.8	132.8	113.6
646	Chillicothe	95.1	68.5	83.9		106	White Plains	99.8	132.8	113.6
647	Harrisonville	94.7	98.4	96.2		107	Yonkers	104.7	132.8	116.5
648	Joplin	97.1	68.0	84.9		108	New Rochelle	100.1	132.8	113.8
650-651	Jefferson City	95.4	86.0	91.5		109	Suffern	99.9	121.8	109.1
652	Columbia	96.3	88.1	92.9		110	Queens	101.3	161.9	126.7
653	Sedalia	95.4	81.7	89.6		111	Long Island City	102.9	161.9	127.6
654-655	Rolla	94.1	73.0	85.2		112	Brooklyn	103.1	161.9	127.8
656-658	Springfield	97.3	77.6	89.1		113	Flushing	103.0	161.9	127.7
MONTANA						114	Jamaica	101.2	161.9	126.7
590-591	Billings	101.9	74.6	90.4		115,117,118	Hicksville	101.2	146.4	120.1
592	Wolf Point	101.1	71.5	88.7		116	Far Rockaway	103.1	161.9	127.8
593	Miles City	98.9	72.8	88.0		119	Riverhead	102.0	146.4	120.6
594	Great Falls	103.1	74.8	91.2		120-122	Albany	97.4	94.7	96.3
595	Havre	100.0	72.8	88.6		123	Schenectady	97.7	95.3	96.7
596	Helena	101.8	73.5	89.9		124	Kingston	100.8	112.5	105.7
597	Butte	101.6	75.0	90.4		125-126	Poughkeepsie	100.0	127.3	111.5
598	Missoula	99.0	73.2	88.2		127	Monticello	99.3	115.7	106.2
599	Kalispell	97.9	72.0	87.0		128	Glens Falls	92.2	91.0	91.7
NEBRASKA						129	Plattsburgh	96.7	85.9	92.2
680-681	Omaha	99.5	78.8	90.8		130-132	Syracuse	98.7	93.1	96.4
683-685	Lincoln	97.6	77.1	89.0		133-135	Utica	96.7	89.5	93.6
686	Columbus	95.5	76.4	87.5		136	Watertown	98.4	92.2	95.8
687	Norfolk	97.2	80.4	90.1		137-139	Binghamton	98.1	86.9	93.4
688	Grand Island	97.3	82.4	91.0		140-142	Buffalo	99.0	104.8	101.5
689	Hastings	96.5	84.8	91.6		143	Niagara Falls	96.7	101.5	98.7
690	Mccook	96.4	75.1	87.5		144-146	Rochester	98.4	94.1	96.6
691	North Platte	96.8	85.0	91.8		147	Jamestown	95.6	82.5	90.1
692	Valentine	98.7	73.6	88.2		148-149	Elmira	95.5	84.4	90.8
693	Alliance	98.6	71.6	87.3		**NORTH CAROLINA**				
NEVADA						270,272-274	Greensboro	99.8	50.7	79.2
889-891	Las Vegas	101.1	112.0	105.7		271	Winston-Salem	99.5	50.0	78.7
893	Ely	99.5	72.0	88.0		275-276	Raleigh	100.7	51.3	80.0
894-895	Reno	99.8	93.9	97.3		277	Durham	101.1	51.7	80.4
897	Carson City	99.9	93.1	97.1		278	Rocky Mount	96.9	41.6	73.7
898	Elko	98.2	77.6	89.6		279	Elizabeth City	97.7	43.0	74.7
NEW HAMPSHIRE						280	Gastonia	98.5	49.6	78.0
030	Nashua	98.6	87.7	94.0		281-282	Charlotte	100.2	52.2	80.0
031	Manchester	99.1	87.7	94.3		283	Fayetteville	101.1	52.1	80.5
						284	Wilmington	97.2	49.6	77.2
						285	Kinston	95.2	43.3	73.4

Location Factors

STATE/ZIP	CITY	MAT.	INST.	TOTAL
NORTH CAROLINA (CONT'D)				
286	Hickory	95.5	45.9	74.7
287-288	Asheville	97.6	49.6	77.5
289	Murphy	96.3	35.7	70.9
NORTH DAKOTA				
580-581	Fargo	101.4	62.0	84.9
582	Grand Forks	100.8	55.1	81.6
583	Devils Lake	100.0	57.3	82.0
584	Jamestown	100.0	49.1	78.7
585	Bismarck	100.1	62.9	84.5
586	Dickinson	100.9	60.1	83.7
587	Minot	100.9	67.3	86.8
588	Williston	99.4	60.1	82.9
OHIO				
430-432	Columbus	96.3	88.6	93.0
433	Marion	92.6	83.5	88.8
434-436	Toledo	96.5	99.7	97.9
437-438	Zanesville	93.1	82.5	88.7
439	Steubenville	94.1	91.6	93.0
440	Lorain	96.9	93.6	95.5
441	Cleveland	97.2	102.6	99.5
442-443	Akron	97.8	93.7	96.1
444-445	Youngstown	97.2	89.7	94.1
446-447	Canton	97.3	85.0	92.1
448-449	Mansfield	94.4	88.3	91.9
450	Hamilton	95.0	85.8	91.1
451-452	Cincinnati	95.3	88.1	92.3
453-454	Dayton	95.2	85.5	91.1
455	Springfield	95.1	85.7	91.2
456	Chillicothe	93.8	91.4	92.8
457	Athens	96.7	75.9	88.0
458	Lima	97.0	85.3	92.1
OKLAHOMA				
730-731	Oklahoma City	98.7	61.7	83.1
734	Ardmore	95.2	61.1	80.9
735	Lawton	97.8	62.5	83.0
736	Clinton	96.6	59.7	81.1
737	Enid	97.4	59.7	81.6
738	Woodward	95.4	59.8	80.4
739	Guymon	96.4	32.0	69.4
740-741	Tulsa	97.6	56.3	80.3
743	Miami	94.1	64.5	81.7
744	Muskogee	96.9	41.6	73.7
745	Mcalester	93.8	52.4	76.5
746	Ponca City	94.4	60.4	80.1
747	Durant	94.4	60.4	80.2
748	Shawnee	95.9	57.8	79.9
749	Poteau	93.5	62.8	80.6
OREGON				
970-972	Portland	101.1	101.0	101.0
973	Salem	101.0	99.1	100.2
974	Eugene	100.7	98.4	99.7
975	Medford	102.3	97.0	100.1
976	Klamath Falls	102.2	96.7	99.9
977	Bend	101.0	98.5	100.0
978	Pendleton	96.1	99.0	97.3
979	Vale	93.8	90.3	92.3
PENNSYLVANIA				
150-152	Pittsburgh	96.9	98.7	97.6
153	Washington	93.8	98.4	95.7
154	Uniontown	94.1	95.5	94.7
155	Bedford	95.1	89.3	92.6
156	Greensburg	95.0	96.8	95.8
157	Indiana	93.9	95.6	94.6
158	Dubois	95.5	93.2	94.5
159	Johnstown	95.1	93.2	94.3
160	Butler	91.9	96.3	93.8
161	New Castle	91.9	94.7	93.1
162	Kittanning	92.4	98.6	95.0
163	Oil City	91.9	92.9	92.3
164-165	Erie	94.2	92.0	93.3
166	Altoona	94.3	89.4	92.2
167	Bradford	95.4	90.7	93.4
168	State College	94.9	90.1	92.9
169	Wellsboro	96.0	90.7	93.8
170-171	Harrisburg	97.9	93.1	95.9
172	Chambersburg	95.5	88.9	92.7
173-174	York	96.1	93.3	94.9
175-176	Lancaster	94.2	87.9	91.6

STATE/ZIP	CITY	MAT.	INST.	TOTAL
PENNSYLVANIA (CONT'D)				
177	Williamsport	92.9	81.1	87.9
178	Sunbury	95.0	92.2	93.8
179	Pottsville	94.1	92.3	93.4
180	Lehigh Valley	95.5	111.4	102.2
181	Allentown	97.7	106.5	101.4
182	Hazleton	94.9	92.4	93.9
183	Stroudsburg	94.8	98.8	96.5
184-185	Scranton	98.5	95.2	97.1
186-187	Wilkes-Barre	94.7	92.7	93.8
188	Montrose	94.4	93.7	94.1
189	Doylestown	94.4	118.6	104.5
190-191	Philadelphia	101.0	130.0	113.2
193	Westchester	97.2	119.9	106.8
194	Norristown	96.2	127.4	109.3
195-196	Reading	98.5	97.6	98.1
PUERTO RICO				
009	San Juan	120.1	23.9	79.7
RHODE ISLAND				
028	Newport	97.7	109.1	102.5
029	Providence	98.8	109.1	103.2
SOUTH CAROLINA				
290-292	Columbia	98.6	53.3	79.6
293	Spartanburg	96.9	52.9	78.4
294	Charleston	98.4	62.2	83.2
295	Florence	96.6	53.3	78.4
296	Greenville	96.7	52.9	78.3
297	Rock Hill	96.1	50.9	77.1
298	Aiken	97.1	70.0	85.7
299	Beaufort	97.9	46.2	76.2
SOUTH DAKOTA				
570-571	Sioux Falls	99.9	60.1	83.2
572	Watertown	98.1	55.3	80.1
573	Mitchell	97.1	55.2	79.5
574	Aberdeen	99.9	56.3	81.6
575	Pierre	99.5	56.5	81.4
576	Mobridge	97.7	55.0	79.8
577	Rapid City	99.7	56.5	81.5
TENNESSEE				
370-372	Nashville	98.5	73.6	88.0
373-374	Chattanooga	99.4	55.0	80.8
375,380-381	Memphis	97.5	70.1	86.0
376	Johnson City	98.4	55.4	80.3
377-379	Knoxville	95.8	55.8	79.0
382	McKenzie	97.1	55.2	79.5
383	Jackson	99.0	47.9	77.5
384	Columbia	95.6	57.0	79.4
385	Cookeville	97.0	58.1	80.6
TEXAS				
750	McKinney	98.7	51.5	78.8
751	Waxahackie	98.6	54.2	80.0
752-753	Dallas	99.4	66.2	85.4
754	Greenville	98.7	38.4	73.4
755	Texarkana	97.6	50.8	77.9
756	Longview	98.0	39.9	73.6
757	Tyler	98.8	54.2	80.1
758	Palestine	95.0	40.3	72.0
759	Lufkin	95.7	43.9	74.0
760-761	Fort Worth	98.0	60.7	82.4
762	Denton	97.6	48.1	76.8
763	Wichita Falls	98.5	55.3	80.3
764	Eastland	96.9	39.4	72.8
765	Temple	95.6	48.7	75.9
766-767	Waco	97.8	56.8	80.6
768	Brownwood	98.0	37.2	72.5
769	San Angelo	97.6	45.3	75.7
770-772	Houston	100.7	71.2	88.3
773	Huntsville	98.8	37.9	73.2
774	Wharton	100.2	42.2	75.8
775	Galveston	98.2	69.1	86.0
776-777	Beaumont	98.4	60.1	82.3
778	Bryan	95.4	62.6	81.6
779	Victoria	100.1	44.3	76.7
780	Laredo	95.8	51.7	77.3
781-782	San Antonio	96.2	65.4	83.3
783-784	Corpus Christi	98.4	50.4	78.3
785	McAllen	98.0	45.1	75.8
786-787	Austin	96.5	59.1	80.8

Location Factors

STATE/ZIP	CITY	MAT.	INST.	TOTAL
TEXAS (CONT'D)				
788	Del Rio	97.6	31.9	70.1
789	Giddings	94.9	39.5	71.7
790-791	Amarillo	98.6	56.8	81.0
792	Childress	97.6	49.6	77.4
793-794	Lubbock	100.1	51.1	79.5
795-796	Abilene	98.2	50.8	78.3
797	Midland	100.0	47.4	77.9
798-799,885	El Paso	97.5	49.8	77.5
UTAH				
840-841	Salt Lake City	102.9	67.9	88.2
842,844	Ogden	98.4	67.3	85.3
843	Logan	100.3	67.3	86.4
845	Price	100.7	46.5	78.0
846-847	Provo	100.8	67.8	87.0
VERMONT				
050	White River Jct.	96.3	56.6	79.7
051	Bellows Falls	94.8	64.0	81.9
052	Bennington	95.2	66.8	83.2
053	Brattleboro	95.6	69.0	84.4
054	Burlington	99.3	67.7	86.1
056	Montpelier	95.9	68.1	84.2
057	Rutland	97.5	68.1	85.2
058	St. Johnsbury	96.3	57.1	79.9
059	Guildhall	95.1	56.9	79.0
VIRGINIA				
220-221	Fairfax	100.2	82.2	92.6
222	Arlington	101.5	80.8	92.8
223	Alexandria	100.7	86.0	94.5
224-225	Fredericksburg	98.8	73.8	88.3
226	Winchester	99.4	66.5	85.6
227	Culpeper	99.3	72.0	87.9
228	Harrisonburg	99.6	68.2	86.4
229	Charlottesville	100.0	66.6	85.9
230-232	Richmond	101.7	70.1	88.4
233-235	Norfolk	101.9	70.5	88.7
236	Newport News	100.5	70.9	88.1
237	Portsmouth	99.8	67.4	86.2
238	Petersburg	99.9	70.1	87.4
239	Farmville	99.0	56.9	81.3
240-241	Roanoke	101.2	63.7	85.4
242	Bristol	98.7	57.4	81.4
243	Pulaski	98.4	54.7	80.1
244	Staunton	99.3	62.8	84.0
245	Lynchburg	99.4	68.1	86.2
246	Grundy	98.8	54.1	80.0
WASHINGTON				
980-981,987	Seattle	105.4	101.7	103.9
982	Everett	105.0	96.7	101.5
983-984	Tacoma	105.1	99.6	102.8
985	Olympia	103.1	99.5	101.6
986	Vancouver	106.5	93.8	101.1
988	Wenatchee	104.8	81.4	95.0
989	Yakima	105.3	88.0	98.0
990-992	Spokane	104.5	81.8	95.0
993	Richland	104.1	85.7	96.4
994	Clarkston	102.9	82.2	94.2
WEST VIRGINIA				
247-248	Bluefield	97.5	77.3	89.0
249	Lewisburg	99.3	82.0	92.0
250-253	Charleston	100.4	87.8	95.1
254	Martinsburg	98.9	76.5	89.5
255-257	Huntington	100.6	90.3	96.3
258-259	Beckley	97.3	86.0	92.6
260	Wheeling	100.6	89.2	95.8
261	Parkersburg	99.6	88.7	95.0
262	Buckhannon	98.9	88.9	94.7
263-264	Clarksburg	99.3	87.8	94.5
265	Morgantown	99.4	88.5	94.8
266	Gassaway	98.7	88.8	94.5
267	Romney	98.8	82.3	91.9
268	Petersburg	98.6	84.6	92.7
WISCONSIN				
530,532	Milwaukee	100.3	106.4	102.9
531	Kenosha	100.0	99.3	99.7
534	Racine	99.5	101.3	100.2
535	Beloit	99.3	94.6	97.3
537	Madison	99.5	95.0	97.6

STATE/ZIP	CITY	MAT.	INST.	TOTAL
WISCONSIN (CONT'D)				
538	Lancaster	96.9	90.3	94.1
539	Portage	95.6	94.3	95.0
540	New Richmond	95.5	93.6	94.7
541-543	Green Bay	99.7	90.4	95.8
544	Wausau	94.9	88.5	92.2
545	Rhinelander	98.0	88.7	94.1
546	La Crosse	96.3	91.3	94.2
547	Eau Claire	97.9	92.0	95.4
548	Superior	95.4	95.7	95.5
549	Oshkosh	95.5	89.9	93.2
WYOMING				
820	Cheyenne	100.0	65.9	85.7
821	Yellowstone Nat'l Park	97.5	58.6	81.2
822	Wheatland	98.6	58.6	81.8
823	Rawlins	100.2	58.4	82.6
824	Worland	98.1	56.5	80.6
825	Riverton	99.2	56.5	81.3
826	Casper	100.6	58.4	82.9
827	Newcastle	97.9	58.4	81.3
828	Sheridan	100.7	60.5	83.8
829-831	Rock Springs	102.0	56.5	82.9
CANADIAN FACTORS (reflect Canadian currency)				
ALBERTA				
	Calgary	130.7	89.9	113.6
	Edmonton	131.8	89.8	114.1
	Fort McMurray	127.7	92.6	112.9
	Lethbridge	121.9	92.0	109.3
	Lloydminster	116.4	88.8	104.8
	Medicine Hat	116.4	88.1	104.5
	Red Deer	116.9	88.1	104.8
BRITISH COLUMBIA				
	Kamloops	117.2	90.8	106.1
	Prince George	118.3	90.8	106.7
	Vancouver	131.2	82.0	110.5
	Victoria	118.3	78.8	101.7
MANITOBA				
	Brandon	116.4	76.1	99.5
	Portage la Prairie	116.5	74.7	98.9
	Winnipeg	129.6	69.6	104.4
NEW BRUNSWICK				
	Bathurst	114.7	67.8	95.0
	Dalhousie	114.6	67.8	95.0
	Fredericton	117.1	72.4	98.3
	Moncton	115.0	68.7	95.5
	Newcastle	114.7	67.8	95.0
	St. John	117.3	72.4	98.4
NEWFOUNDLAND				
	Corner Brook	119.7	68.2	98.0
	St. Johns	119.9	69.2	98.6
NORTHWEST TERRITORIES				
	Yellowknife	121.4	85.7	106.4
NOVA SCOTIA				
	Bridgewater	116.1	75.0	98.9
	Dartmouth	117.5	75.0	99.7
	Halifax	119.0	78.0	101.8
	New Glasgow	115.5	75.0	98.5
	Sydney	113.1	75.0	97.1
	Truro	115.5	75.0	98.5
	Yarmouth	115.4	75.0	98.5
ONTARIO				
	Barrie	119.2	93.0	108.2
	Brantford	118.4	96.9	109.4
	Cornwall	118.5	93.3	107.9
	Hamilton	123.4	96.9	112.3
	Kingston	119.4	93.4	108.5
	Kitchener	114.5	92.2	105.1
	London	122.0	93.3	110.0
	North Bay	118.4	91.2	107.0
	Oshawa	117.9	94.7	108.2
	Ottawa	122.7	93.9	110.6
	Owen Sound	119.4	91.2	107.6
	Peterborough	118.4	93.2	107.8
	Sarnia	117.9	97.5	109.3

Location Factors

STATE/ZIP	CITY	MAT.	INST.	TOTAL
ONTARIO (CONT'D)				
	Sault Ste Marie	112.9	92.6	104.4
	St. Catharines	112.3	94.2	104.7
	Sudbury	112.7	92.9	104.4
	Thunder Bay	113.5	92.9	104.9
	Timmins	118.6	91.2	107.1
	Toronto	123.6	99.8	113.6
	Windsor	113.0	93.2	104.7
PRINCE EDWARD ISLAND				
	Charlottetown	117.8	63.3	94.9
	Summerside	117.2	63.3	94.6
QUEBEC				
	Cap-de-la-Madeleine	115.3	87.8	103.8
	Charlesbourg	115.3	87.8	103.8
	Chicoutimi	114.0	93.5	105.4
	Gatineau	114.9	87.6	103.4
	Granby	115.1	87.6	103.5
	Hull	115.0	87.6	103.5
	Joliette	115.5	87.8	103.9
	Laval	114.5	87.6	103.2
	Montreal	121.7	93.8	110.0
	Quebec	120.1	94.2	109.2
	Rimouski	114.4	93.5	105.6
	Rouyn-Noranda	114.8	87.6	103.4
	Saint Hyacinthe	114.2	87.6	103.0
	Sherbrooke	115.2	87.6	103.6
	Sorel	115.4	87.8	103.8
	St. Jerome	114.8	87.6	103.4
	Trois Rivieres	115.4	87.8	103.8
SASKATCHEWAN				
	Moose Jaw	113.5	69.9	95.2
	Prince Albert	112.5	68.1	93.9
	Regina	115.1	69.9	96.1
	Saskatoon	113.6	69.8	95.2
YUKON				
	Whitehorse	113.2	68.9	94.6

General Requirements   R0111 Summary of Work

R011105-05   Tips for Accurate Estimating

1. Use pre-printed or columnar forms for orderly sequence of dimensions and locations and for recording telephone quotations.
2. Use only the front side of each paper or form except for certain pre-printed summary forms.
3. Be consistent in listing dimensions: For example, length x width x height. This helps in rechecking to ensure that, the total length of partitions is appropriate for the building area.
4. Use printed (rather than measured) dimensions where given.
5. Add up multiple printed dimensions for a single entry where possible.
6. Measure all other dimensions carefully.
7. Use each set of dimensions to calculate multiple related quantities.
8. Convert foot and inch measurements to decimal feet when listing. Memorize decimal equivalents to .01 parts of a foot (1/8" equals approximately .01').
9. Do not "round off" quantities until the final summary.
10. Mark drawings with different colors as items are taken off.
11. Keep similar items together, different items separate.
12. Identify location and drawing numbers to aid in future checking for completeness.
13. Measure or list everything on the drawings or mentioned in the specifications.
14. It may be necessary to list items not called for to make the job complete.
15. Be alert for: Notes on plans such as N.T.S. (not to scale); changes in scale throughout the drawings; reduced size drawings; discrepancies between the specifications and the drawings.
16. Develop a consistent pattern of performing an estimate. For example:
    a. Start the quantity takeoff at the lower floor and move to the next higher floor.
    b. Proceed from the main section of the building to the wings.
    c. Proceed from south to north or vice versa, clockwise or counterclockwise.
    d. Take off floor plan quantities first, elevations next, then detail drawings.
17. List all gross dimensions that can be either used again for different quantities, or used as a rough check of other quantities for verification (exterior perimeter, gross floor area, individual floor areas, etc.).
18. Utilize design symmetry or repetition (repetitive floors, repetitive wings, symmetrical design around a center line, similar room layouts, etc.). Note: Extreme caution is needed here so as not to omit or duplicate an area.
19. Do not convert units until the final total is obtained. For instance, when estimating concrete work, keep all units to the nearest cubic foot, then summarize and convert to cubic yards.
20. When figuring alternatives, it is best to total all items involved in the basic system, then total all items involved in the alternates. Therefore you work with positive numbers in all cases. When adds and deducts are used, it is often confusing whether to add or subtract a portion of an item; especially on a complicated or involved alternate.

General Requirements   R0121 Allowances

R012153-10   Repair and Remodeling

Cost figures are based on new construction utilizing the most cost-effective combination of labor, equipment and material with the work scheduled in proper sequence to allow the various trades to accomplish their work in an efficient manner.

The costs for repair and remodeling work must be modified due to the following factors that may be present in any given repair and remodeling project.

1. Equipment usage curtailment due to the physical limitations of the project, with only hand-operated equipment being used.
2. Increased requirement for shoring and bracing to hold up the building while structural changes are being made and to allow for temporary storage of construction materials on above-grade floors.
3. Material handling becomes more costly due to having to move within the confines of an enclosed building. For multi-story construction, low capacity elevators and stairwells may be the only access to the upper floors.
4. Large amount of cutting and patching and attempting to match the existing construction is required. It is often more economical to remove entire walls rather than create many new door and window openings. This sort of trade-off has to be carefully analyzed.
5. Cost of protection of completed work is increased since the usual sequence of construction usually cannot be accomplished.
6. Economies of scale usually associated with new construction may not be present. If small quantities of components must be custom fabricated due to job requirements, unit costs will naturally increase. Also, if only small work areas are available at a given time, job scheduling between trades becomes difficult and subcontractor quotations may reflect the excessive start-up and shut-down phases of the job.
7. Work may have to be done on other than normal shifts and may have to be done around an existing production facility which has to stay in production during the course of the repair and remodeling.
8. Dust and noise protection of adjoining non-construction areas can involve substantial special protection and alter usual construction methods.
9. Job may be delayed due to unexpected conditions discovered during demolition or removal. These delays ultimately increase construction costs.
10. Piping and ductwork runs may not be as simple as for new construction. Wiring may have to be snaked through walls and floors.
11. Matching "existing construction" may be impossible because materials may no longer be manufactured. Substitutions may be expensive.
12. Weather protection of existing structure requires additional temporary structures to protect building at openings.
13. On small projects, because of local conditions, it may be necessary to pay a tradesman for a minimum of four hours for a task that is completed in one hour.

All of the above areas can contribute to increased costs for a repair and remodeling project. Each of the above factors should be considered in the planning, bidding and construction stage in order to minimize the increased costs associated with repair and remodeling jobs.

General Requirements — R0129 Payment Procedures

R012909-80 Sales Tax by State

State sales tax on materials is tabulated below (5 states have no sales tax). Many states allow local jurisdictions, such as a county or city, to levy additional sales tax.

Some projects may be sales tax exempt, particularly those constructed with public funds.

State	Tax (%)	State	Tax (%)	State	Tax (%)	State	Tax (%)
Alabama	4	Illinois	6.25	Montana	0	Rhode Island	7
Alaska	0	Indiana	6	Nebraska	5.5	South Carolina	6
Arizona	5.6	Iowa	5	Nevada	6.5	South Dakota	4
Arkansas	6	Kansas	5.3	New Hampshire	0	Tennessee	7
California	7.25	Kentucky	6	New Jersey	7	Texas	6.25
Colorado	2.9	Louisiana	4	New Mexico	5	Utah	4.65
Connecticut	6	Maine	5	New York	4	Vermont	6
Delaware	0	Maryland	6	North Carolina	4.25	Virginia	5
District of Columbia	5.75	Massachusetts	5	North Dakota	5	Washington	6.5
Florida	6	Michigan	6	Ohio	5.5	West Virginia	6
Georgia	4	Minnesota	6.5	Oklahoma	4.5	Wisconsin	5
Hawaii	4	Mississippi	7	Oregon	0	Wyoming	4
Idaho	6	Missouri	4.225	Pennsylvania	6	Average	4.91%

Sales Tax by Province (Canada)

GST - a value-added tax, which the government imposes on most goods and services provided in or imported into Canada. PST - a retail sales tax, which five of the provinces impose on the price of most goods and some services. QST - a value-added tax, similar to the federal GST, which Quebec imposes. HST - Three provinces have combined their retail sales tax with the federal GST into one harmonized tax.

Province	PST (%)	QST (%)	GST (%)	HST (%)
Alberta	0	0	6	0
British Columbia	7	0	6	0
Manitoba	7	0	6	0
New Brunswick	0	0	0	13
Newfoundland	0	0	0	13
Northwest Territories	0	0	6	0
Nova Scotia	0	0	0	13
Ontario	8	0	6	0
Prince Edward Island	10	0	6	0
Quebec	0	7.5	6	0
Saskatchewan	5	0	6	0
Yukon	0	0	6	0

R012909-80 Unemployment Taxes and Social Security Taxes

State Unemployment Tax rates vary not only from state to state, but also with the experience rating of the contractor. The Federal Unemployment Tax rate is 6.2% of the first $7,000 of wages. This is reduced by a credit of up to 5.4% for timely payment to the state. The minimum Federal Unemployment Tax is 0.8% after all credits.

Social Security (FICA) for 2009 is estimated at time of publication to be 7.65% of wages up to $102,000.

General Requirements — R0129 Payment Procedures

R012909-90 Overtime

One way to improve the completion date of a project or eliminate negative float from a schedule is to compress activity duration times. This can be achieved by increasing the crew size or working overtime with the proposed crew.

To determine the costs of working overtime to compress activity duration times, consider the following examples. Below is an overtime efficiency and cost chart based on a five, six, or seven day week with an eight through twelve hour day. Payroll percentage increases for time and one half and double time are shown for the various working days.

Days per Week	Hours per Day	Production Efficiency				Average 4 Weeks	Payroll Cost Factors	
		1st Week	2nd Week	3rd Week	4th Week		@ 1-1/2 Times	@ 2 Times
5	8	100%	100%	100%	100%	100 %	100 %	100 %
	9	100	100	95	90	96.25	105.6	111.1
	10	100	95	90	85	91.25	110.0	120.0
	11	95	90	75	65	81.25	113.6	127.3
	12	90	85	70	60	76.25	116.7	133.3
6	8	100	100	95	90	96.25	108.3	116.7
	9	100	95	90	85	92.50	113.0	125.9
	10	95	90	85	80	87.50	116.7	133.3
	11	95	85	70	65	78.75	119.7	139.4
	12	90	80	65	60	73.75	122.2	144.4
7	8	100	95	85	75	88.75	114.3	128.6
	9	95	90	80	70	83.75	118.3	136.5
	10	90	85	75	65	78.75	121.4	142.9
	11	85	80	65	60	72.50	124.0	148.1
	12	85	75	60	55	68.75	126.2	152.4

General Requirements — R0131 Project Management & Coordination

R013113-40 Builder's Risk Insurance

Builder's Risk Insurance is insurance on a building during construction. Premiums are paid by the owner or the contractor. Blasting, collapse and underground insurance would raise total insurance costs above those listed. Floater policy for materials delivered to the job runs $.75 to $1.25 per $100 value. Contractor equipment insurance runs $.50 to $1.50 per $100 value. Insurance for miscellaneous tools to $1,500 value runs from $3.00 to $7.50 per $100 value.

Tabulated below are New England Builder's Risk insurance rates in dollars per $100 value for $1,000 deductible. For $25,000 deductible, rates can be reduced 13% to 34%. On contracts over $1,000,000, rates may be lower than those tabulated. Policies are written annually for the total completed value in place. For "all risk" insurance (excluding flood, earthquake and certain other perils) add $.025 to total rates below.

Coverage	Frame Construction (Class 1)		Brick Construction (Class 4)		Fire Resistive (Class 6)	
	Range	Average	Range	Average	Range	Average
Fire Insurance	$.350 to $.850	$.600	$.158 to $.189	$.174	$.052 to $.080	$.070
Extended Coverage	.115 to .200	.158	.080 to .105	.101	.081 to .105	.100
Vandalism	.012 to .016	.014	.008 to .011	.011	.008 to .011	.010
Total Annual Rate	$.477 to $1.066	$.772	$.246 to $.305	$.286	$.141 to $.196	$.180

General Requirements — R0131 Project Management & Coordination

R013113-50 General Contractor's Overhead

There are two distinct types of overhead on a construction project: Project Overhead and Main Office Overhead. Project Overhead includes those costs at a construction site not directly associated with the installation of construction materials. Examples of Project Overhead costs include the following:

1. Superintendent
2. Construction office and storage trailers
3. Temporary sanitary facilities
4. Temporary utilities
5. Security fencing
6. Photographs
7. Clean up
8. Performance and payment bonds

The above Project Overhead items are also referred to as General Requirements and therefore are estimated in Division 1. Division 1 is the first division listed in the CSI MasterFormat but it is usually the last division estimated. The sum of the costs in Divisions 1 through 49 is referred to as the sum of the direct costs.

All construction projects also include indirect costs. The primary components of indirect costs are the contractor's Main Office Overhead and profit. The amount of the Main Office Overhead expense varies depending on the following:

1. Owner's compensation
2. Project managers and estimator's wages
3. Clerical support wages
4. Office rent and utilities
5. Corporate legal and accounting costs
6. Advertising
7. Automobile expenses
8. Association dues
9. Travel and entertainment expenses

These costs are usually calculated as a percentage of annual sales volume. This percentage can range from 35% for a small contractor doing less than $500,000 to 5% for a large contractor with sales in excess of $100 million.

General Requirements — R0131 Project Management & Coordination

R013113-60 Workers' Compensation Insurance Rates by Trade

The table below tabulates the national averages for Workers' Compensation insurance rates by trade and type of building. The average "Insurance Rate" is multiplied by the "% of Building Cost" for each trade. This produces the "Workers' Compensation Cost" by % of total labor cost, to be added for each trade by building type to determine the weighted average Workers' Compensation rate for the building types analyzed.

Trade	Insurance Rate (% Labor Cost) Range	Insurance Rate (% Labor Cost) Average	% of Building Cost Office Bldgs.	% of Building Cost Schools & Apts.	% of Building Cost Mfg.	Workers' Compensation Office Bldgs.	Workers' Compensation Schools & Apts.	Workers' Compensation Mfg.
Excavation, Grading, etc.	4.2% to 17.5%	10.0%	4.8%	4.9%	4.5%	0.48%	0.49%	0.45%
Piles & Foundations	5.9 to 40.3	20.1	7.1	5.2	8.7	1.43	1.05	1.75
Concrete	5.1 to 26.8	14.6	5.0	14.8	3.7	0.73	2.16	0.54
Masonry	4.8 to 43.3	14.4	6.9	7.5	1.9	0.99	1.08	0.27
Structural Steel	5.9 to 104.1	37.9	10.7	3.9	17.6	4.06	1.48	6.67
Miscellaneous & Ornamental Metals	3.4 to 22.8	10.7	2.8	4.0	3.6	0.30	0.43	0.39
Carpentry & Millwork	5.9 to 53.2	17.8	3.7	4.0	0.5	0.66	0.71	0.09
Metal or Composition Siding	5.9 to 38	16.6	2.3	0.3	4.3	0.38	0.05	0.71
Roofing	5.9 to 77.1	31.2	2.3	2.6	3.1	0.72	0.81	0.97
Doors & Hardware	4.9 to 32	11.6	0.9	1.4	0.4	0.10	0.16	0.05
Sash & Glazing	5.9 to 32.4	13.9	3.5	4.0	1.0	0.49	0.56	0.14
Lath & Plaster	3.3 to 43.9	13.6	3.3	6.9	0.8	0.45	0.94	0.11
Tile, Marble & Floors	3.1 to 17.9	9.1	2.6	3.0	0.5	0.24	0.27	0.05
Acoustical Ceilings	2.6 to 45.6	10.6	2.4	0.2	0.3	0.25	0.02	0.03
Painting	4.7 to 29.6	12.5	1.5	1.6	1.6	0.19	0.20	0.20
Interior Partitions	5.9 to 53.2	17.8	3.9	4.3	4.4	0.69	0.77	0.78
Miscellaneous Items	2.1 to 168.2	16.0	5.2	3.7	9.7	0.83	0.59	1.55
Elevators	2.8 to 11.8	6.6	2.1	1.1	2.2	0.14	0.07	0.15
Sprinklers	2.5 to 14.3	7.8	0.5	—	2.0	0.04	—	0.16
Plumbing	2.9 to 12.4	7.8	4.9	7.2	5.2	0.38	0.56	0.41
Heat., Vent., Air Conditioning	4.3 to 23	11.6	13.5	11.0	12.9	1.57	1.28	1.50
Electrical	2.8 to 11.5	6.5	10.1	8.4	11.1	0.66	0.55	0.72
Total	2.1% to 168.2%	—	100.0%	100.0%	100.0%	15.78%	14.23%	17.69%
Overall Weighted Average		15.90%						

Workers' Compensation Insurance Rates by States

The table below lists the weighted average Workers' Compensation base rate for each state with a factor comparing this with the national average of 15.5%.

State	Weighted Average	Factor	State	Weighted Average	Factor	State	Weighted Average	Factor
Alabama	24.2%	156	Kentucky	18.4%	119	North Dakota	13.7%	88
Alaska	21.7	140	Louisiana	28.3	183	Ohio	14.8	95
Arizona	9.5	61	Maine	15.4	99	Oklahoma	14.6	94
Arkansas	13.2	85	Maryland	16.7	108	Oregon	13.5	87
California	19.6	126	Massachusetts	12.5	81	Pennsylvania	13.8	89
Colorado	13.2	85	Michigan	17.3	112	Rhode Island	21.2	137
Connecticut	21.0	135	Minnesota	25.7	166	South Carolina	18.3	118
Delaware	15.1	97	Mississippi	19.1	123	South Dakota	19.1	123
District of Columbia	13.7	88	Missouri	17.2	111	Tennessee	15.1	97
Florida	13.8	89	Montana	16.6	107	Texas	12.2	79
Georgia	26.2	169	Nebraska	23.0	148	Utah	12.3	79
Hawaii	17.0	110	Nevada	11.6	75	Vermont	18.0	116
Idaho	10.8	70	New Hampshire	20.0	129	Virginia	11.7	75
Illinois	21.6	139	New Jersey	13.4	86	Washington	9.8	63
Indiana	5.9	38	New Mexico	18.5	119	West Virginia	9.2	59
Iowa	11.5	74	New York	12.6	81	Wisconsin	15.0	97
Kansas	8.9	57	North Carolina	18.9	122	Wyoming	6.5	42
Weighted Average for U.S. is			15.9% of payroll = 100%					

Rates in the following table are the base or manual costs per $100 of payroll for Workers' Compensation in each state. Rates are usually applied to straight time wages only and not to premium time wages and bonuses.

The weighted average skilled worker rate for 35 trades is 15.5%. For bidding purposes, apply the full value of Workers' Compensation directly to total labor costs, or if labor is 38%, materials 42% and overhead and profit 20% of total cost, carry 38/80 x 15.5% =7.4% of cost (before overhead and profit) into overhead. Rates vary not only from state to state but also with the experience rating of the contractor.

Rates are the most current available at the time of publication.

General Requirements — R0131 Project Management & Coordination

R013113-060 Workers' Compensation Insurance Rates by Trade and State (cont.)

State	Carpentry — 3 stories or less 5651	Carpentry — interior cab. work 5437	Carpentry — general 5403	Concrete Work — NOC 5213	Concrete Work — flat (flr., sdwk.) 5221	Electrical Wiring — inside 5190	Excavation — earth NOC 6217	Excavation — rock 6217	Glaziers 5462	Insulation Work 5479	Lathing 5443	Masonry 5022	Painting & Decorating 5474	Pile Driving 6003	Plastering 5480	Plumbing 5183	Roofing 5551	Sheet Metal Work (HVAC) 5538	Steel Erection — door & sash 5102	Steel Erection — inter., ornam. 5102	Steel Erection — structure 5040	Steel Erection — NOC 5057	Tile Work — (interior ceramic) 5348	Waterproofing 9014	Wrecking 5701
AL	37.98	22.82	32.46	16.20	11.27	10.68	11.16	11.16	21.26	16.28	11.93	27.00	25.81	32.65	18.85	12.27	58.14	22.96	11.01	11.01	55.68	25.37	13.31	6.56	55.68
AK	14.73	12.27	17.44	11.07	9.71	8.72	12.14	12.14	32.36	27.78	9.74	43.27	21.95	32.40	43.85	9.97	35.12	8.61	9.39	9.39	43.01	33.00	6.76	6.61	43.01
AZ	11.60	6.45	15.40	8.56	4.45	4.59	4.90	4.90	6.93	12.48	6.61	6.51	6.33	11.45	5.78	5.24	13.38	7.38	10.66	10.66	21.10	15.34	3.17	2.59	21.10
AR	14.82	7.45	16.26	12.17	6.48	5.02	7.76	7.76	9.82	16.49	5.93	9.93	11.50	16.48	16.10	5.15	23.09	13.09	6.77	6.77	32.38	25.50	6.08	3.74	32.38
CA	31.95	31.95	31.95	11.89	11.89	7.98	13.79	13.79	15.74	12.50	11.24	16.21	17.08	15.27	20.53	12.03	44.43	15.60	12.46	12.46	17.17	20.88	8.54	17.08	20.88
CO	13.57	8.92	12.33	11.57	7.87	5.28	10.34	10.34	9.75	16.84	6.02	14.01	9.02	14.31	9.58	7.43	26.09	11.61	9.31	9.31	32.62	17.58	7.06	5.03	17.58
CT	15.67	14.66	28.76	24.79	10.09	8.01	12.91	12.91	22.98	15.94	21.94	22.24	17.61	22.88	14.48	11.16	41.95	15.61	18.57	18.57	42.72	23.45	11.58	5.41	42.72
DE	15.17	15.17	11.46	11.51	9.49	5.89	9.32	9.32	13.11	11.46	13.11	12.77	26.77	18.75	13.11	7.87	26.71	9.78	12.28	12.28	26.77	12.28	9.29	12.77	26.77
DC	11.01	12.78	10.19	9.66	9.14	6.01	12.54	12.54	17.02	8.18	9.40	11.97	7.51	12.11	12.48	10.38	19.21	8.93	10.61	10.61	37.21	16.49	17.86	4.00	37.21
FL	13.05	10.47	14.39	15.31	6.97	6.64	8.30	8.30	9.91	9.24	5.98	11.32	9.73	38.24	18.70	6.75	22.19	12.68	8.85	8.85	27.89	14.56	6.55	5.17	27.89
GA	35.68	20.54	25.67	16.85	13.26	10.82	17.38	17.38	18.77	23.64	15.09	21.06	27.77	27.36	22.83	11.4	54.71	20.76	19.93	19.93	64.78	42.21	12.42	8.50	64.78
HI	17.38	11.62	28.20	13.74	12.27	7.10	7.73	7.73	20.55	21.19	10.94	17.87	11.33	20.15	15.61	6.06	33.24	8.02	11.11	11.11	31.71	21.70	9.78	11.54	31.71
ID	9.16	6.04	10.35	11.25	6.13	3.70	5.72	5.72	8.70	7.67	8.89	7.59	8.45	12.13	13.37	5.36	30.27	8.89	5.72	5.72	29.16	10.28	9.79	4.75	29.16
IL	24.93	14.46	20.78	26.59	12.10	9.62	10.06	10.06	19.18	17.48	15.59	19.27	11.62	34.85	15.88	11.07	31.06	14.96	16.05	16.05	72.39	25.69	14.02	4.84	72.39
IN	9.65	4.92	7.29	5.10	3.22	2.84	4.87	4.87	6.28	7.44	2.62	4.84	4.72	6.44	3.72	2.91	10.88	4.45	3.43	3.43	12.12	7.04	3.09	2.47	12.12
IA	10.43	10.02	10.34	12.12	7.00	4.96	6.27	6.27	9.42	7.06	5.58	9.61	6.65	9.78	7.33	6.46	22.58	7.52	6.15	6.15	30.21	35.21	7.42	3.87	35.21
KS	10.05	7.85	9.89	7.19	6.42	3.37	4.41	4.41	9.22	7.45	4.68	6.91	7.57	11.31	6.25	4.85	15.06	7.14	4.82	4.82	26.19	13.21	6.35	3.46	13.21
KY	18.72	14.07	25.30	13.39	8.00	5.43	9.15	9.15	17.37	17.59	11.24	9.12	13.18	27.72	14.41	7.06	38.25	19.58	12.32	12.32	54.03	22.56	13.67	5.74	56.16
LA	22.02	24.93	53.17	26.43	15.61	9.88	17.49	17.49	20.14	21.43	24.51	28.33	29.58	31.25	22.37	8.64	77.12	21.20	18.98	18.98	51.76	24.21	13.77	13.78	66.41
ME	13.79	10.40	28.55	20.94	8.74	6.31	9.05	9.05	18.14	11.82	7.90	14.76	12.73	16.70	11.79	10.56	22.87	11.61	9.73	9.73	31.67	23.11	7.00	7.12	31.67
MD	15.05	11.06	14.11	15.67	6.67	9.24	9.20	9.20	20.03	15.02	8.83	11.43	7.48	20.09	15.02	6.11	34.74	13.54	9.51	9.51	59.03	28.60	7.60	5.36	28.60
MA	6.80	5.60	11.46	19.51	6.57	3.20	4.19	4.19	8.96	10.06	6.35	10.81	4.79	14.68	5.08	3.98	32.80	5.15	7.61	7.61	45.71	36.69	6.21	2.09	25.88
MI	18.40	11.39	20.64	19.45	8.95	4.72	10.20	10.20	12.41	14.75	13.59	15.54	13.64	40.27	15.18	6.84	30.67	10.00	10.16	10.16	40.27	22.21	11.11	4.87	40.27
MN	19.53	19.80	41.48	14.65	15.73	7.53	14.95	14.95	15.68	11.90	22.05	19.18	16.95	24.63	22.05	10.88	70.18	14.23	10.58	10.58	104.13	33.30	14.55	6.68	36.03
MS	21.07	11.11	22.29	15.49	9.26	6.28	11.91	11.91	13.07	14.85	7.91	14.09	13.14	35.34	31.72	9.80	44.11	16.91	14.23	14.23	36.11	25.03	9.84	4.18	36.11
MO	29.99	11.16	13.98	18.91	10.39	7.64	9.17	9.17	11.13	15.17	7.93	15.01	12.53	19.12	17.02	10.05	29.98	13.19	11.31	11.31	34.56	35.84	11.18	6.51	34.56
MT	17.12	11.15	22.57	12.37	12.79	6.39	15.54	15.54	10.67	32.03	11.06	13.75	9.61	25.18	10.02	8.75	37.61	11.53	8.70	8.70	27.45	15.73	7.53	7.46	15.73
NE	24.65	17.47	21.80	25.07	15.13	10.38	16.90	16.90	23.70	31.50	12.20	22.22	20.07	22.50	21.67	12.35	35.22	19.42	16.05	16.05	45.40	35.27	11.10	6.38	47.05
NV	15.25	7.14	11.58	10.76	8.23	7.17	9.77	9.77	10.19	8.12	5.27	7.50	7.97	13.33	7.30	5.98	14.08	18.10	8.67	8.67	25.10	21.54	5.39	5.55	21.54
NH	25.92	13.36	18.90	26.79	14.17	6.94	14.52	14.52	13.07	21.36	9.63	24.34	7.77	18.99	11.71	10.13	49.23	13.59	15.27	15.27	52.24	18.83	11.37	6.09	52.24
NJ	14.34	8.94	14.34	16.01	9.99	4.76	8.35	8.35	8.55	12.65	11.64	13.09	11.92	16.03	11.64	6.17	37.77	6.49	9.46	9.46	23.72	13.59	6.99	5.72	22.31
NM	18.91	7.15	20.11	17.36	10.05	7.51	11.20	11.20	21.26	13.72	8.39	15.34	11.64	21.86	13.44	8.13	36.37	13.47	13.69	13.69	55.78	38.81	6.59	6.29	55.78
NY	12.72	6.45	12.67	16.02	11.64	6.03	8.29	8.29	10.97	7.98	12.10	16.02	9.89	12.77	9.03	6.88	27.98	10.46	8.84	8.84	23.45	12.72	6.66	6.04	8.73
NC	18.76	14.56	15.36	18.15	6.88	11.50	11.70	11.70	15.21	16.00	15.28	12.51	12.90	17.50	15.68	9.16	29.70	15.90	10.44	10.44	79.66	26.21	9.28	5.55	79.66
ND	11.17	11.17	11.17	6.42	6.42	3.63	6.64	6.64	11.17	11.17	7.51	7.71	6.61	22.78	7.51	5.22	23.53	5.22	22.78	22.78	22.78	22.78	11.17	23.53	10.98
OH	11.04	8.52	11.68	12.33	9.99	6.14	8.70	8.70	7.86	16.94	45.64	12.71	12.37	19.06	3.28	6.99	31.63	9.71	7.59	7.59	29.23	16.59	9.60	6.42	16.59
OK	13.42	9.17	11.76	11.76	6.34	5.73	10.83	10.83	16.07	19.47	8.45	10.02	8.47	20.25	12.19	7.09	21.13	9.05	13.76	13.76	39.72	23.97	6.50	5.71	39.72
OR	15.43	8.10	15.69	14.22	8.97	5.14	9.33	9.33	15.22	11.80	7.60	14.07	10.63	16.17	12.73	5.50	26.30	9.41	7.93	7.93	26.94	18.78	11.04	4.54	26.94
PA	13.18	13.18	11.38	14.21	10.46	5.95	8.27	8.27	11.47	11.38	11.47	12.22	13.31	15.78	11.47	7.33	27.42	7.84	14.30	14.30	21.31	14.30	8.02	12.22	21.37
RI	19.53	11.65	18.07	18.23	16.24	4.43	10.38	10.38	12.85	22.78	11.97	25.11	24.13	37.66	17.25	8.52	33.92	10.29	14.07	14.07	59.49	37.50	14.36	7.70	59.49
SC	22.54	15.41	20.91	14.80	7.92	10.92	13.16	13.16	16.12	14.69	11.32	12.88	16.52	19.48	16.00	11.47	45.24	12.95	12.99	12.99	32.57	26.39	9.62	6.68	47.05
SD	27.77	8.13	29.58	25.56	9.55	5.52	11.01	11.01	12.16	12.05	7.70	11.31	9.53	21.32	12.44	11.75	20.73	12.48	8.98	8.98	80.86	40.98	8.74	4.98	80.86
TN	15.68	13.11	16.06	13.89	7.97	7.35	13.68	13.68	13.09	11.80	7.30	15.60	12.63	24.89	16.02	7.66	22.25	12.77	7.17	7.17	29.18	23.47	10.15	4.65	29.18
TX	11.25	8.89	11.25	9.00	7.26	6.15	8.54	8.54	9.65	12.87	6.36	10.91	8.98	22.43	8.98	6.75	19.59	12.84	9.09	9.09	29.68	13.37	7.07	6.69	9.96
UT	15.74	8.88	12.97	9.81	8.47	4.57	11.59	11.59	14.70	12.49	6.35	11.25	8.37	12.47	7.26	6.18	24.99	12.64	7.16	7.16	25.67	14.67	6.80	4.27	18.37
VT	15.70	10.75	17.37	17.27	7.21	6.98	12.59	12.59	19.75	17.78	8.33	15.82	11.62	16.74	10.78	10.44	36.43	11.59	12.51	12.51	44.95	42.80	11.24	7.28	44.95
VA	11.05	8.02	9.16	11.05	5.31	5.52	7.57	7.57	10.25	9.90	14.34	9.62	8.99	10.75	10.22	5.83	22.52	7.65	8.37	8.37	33.52	19.66	4.74	3.04	33.52
WA	8.48	8.48	8.48	7.32	7.32	2.81	6.69	6.69	12.27	9.77	8.43	9.09	10.93	16.12	9.95	4.17	16.36	4.27	7.60	7.60	7.60	7.60	8.45	16.36	7.60
WV	10.65	7.30	10.23	8.24	4.41	4.70	6.22	6.22	7.89	7.25	6.11	7.98	7.90	11.90	7.93	4.90	18.05	5.74	5.58	5.58	24.17	13.20	5.65	2.66	17.34
WI	9.99	11.82	14.45	11.15	9.69	6.02	8.19	8.19	10.44	13.58	6.25	15.31	13.03	16.47	9.96	6.24	37.36	7.21	9.36	9.36	28.23	44.93	13.80	4.86	28.23
WY	5.87	5.87	5.87	5.87	5.87	5.87	5.87	5.87	5.87	5.87	5.87	5.87	5.87	5.87	5.87	5.87	5.87	5.87	5.87	5.87	5.87	5.87	5.87	5.87	5.87
AVG.	16.63	11.62	17.80	14.58	9.14	6.46	10.01	10.01	13.89	14.44	10.63	14.37	12.49	20.09	13.60	7.84	31.18	11.57	10.74	10.74	37.94	23.15	9.13	6.69	34.13

General Requirements — R0131 Project Management & Coordination

R013113-060 Workers' Compensation (cont.) (Canada in Canadian dollars)

Province		Alberta	British Columbia	Manitoba	Ontario	New Brunswick	Newfndld. & Labrador	Northwest Territories	Nova Scotia	Prince Edward Island	Quebec	Saskat-chewan	Yukon
Carpentry—3 stories or less	Rate	7.32	4.74	4.36	4.35	3.79	10.08	3.84	7.82	5.96	15.05	6.23	8.89
	Code	42143	721028	40102	723	4226	4226	4-41	4226	401	80110	B1317	202
Carpentry—interior cab. work	Rate	2.18	4.90	4.36	4.35	4.40	4.69	3.84	4.52	3.71	15.05	3.51	8.89
	Code	42133	721021	40102	723	4279	4270	4-41	4274	402	80110	B11-27	202
CARPENTRY—general	Rate	7.32	4.74	4.36	4.35	3.79	4.69	3.84	7.82	5.96	15.05	6.23	8.89
	Code	42143	721028	40102	723	4226	4299	4-41	4226	401	80110	B1317	202
CONCRETE WORK—NOC	Rate	4.41	4.81	7.46	16.02	3.79	10.08	3.84	4.52	5.96	17.99	6.24	4.67
	Code	42104	721010	40110	748	4224	4224	4-41	4224	401	80100	B13-14	203
CONCRETE WORK—flat (flr. sidewalk)	Rate	4.41	4.81	7.46	16.02	3.79	10.08	3.84	4.52	5.96	17.99	6.24	4.67
	Code	42104	721010	40110	748	4224	4224	4-41	4224	401	80100	B13-14	203
ELECTRICAL Wiring—inside	Rate	2.13	1.77	2.51	3.79	2.17	3.01	3.52	2.31	3.71	6.12	3.51	4.67
	Code	42124	721019	40203	704	4261	4261	4-46	4261	402	80170	B11-05	206
EXCAVATION—earth NOC	Rate	2.51	3.56	3.76	4.55	2.62	4.37	3.64	3.34	3.82	7.13	3.93	4.67
	Code	40604	721031	40706	711	4214	4214	4-43	4214	404	80030	R11-06	207
EXCAVATION—rock	Rate	2.51	3.56	3.76	4.55	2.62	4.37	3.64	3.34	3.82	7.13	3.93	4.67
	Code	40604	721031	40706	711	4214	4214	4-43	4214	404	80030	R11-06	207
GLAZIERS	Rate	3.16	3.62	4.36	8.90	4.55	6.81	3.84	5.21	3.71	17.08	6.23	4.67
	Code	42121	715020	40109	751	4233	4233	4-41	4233	402	80150	B13-04	212
INSULATION WORK	Rate	2.57	8.42	4.36	8.90	4.55	6.81	3.84	5.21	5.96	15.05	5.98	8.89
	Code	42184	721029	40102	751	4234	4234	4-41	4234	401	80110	B12-07	202
LATHING	Rate	5.59	6.78	4.36	4.35	4.40	4.69	3.84	4.52	3.71	15.05	6.23	8.89
	Code	42135	721033	40102	723	4273	4279	4-41	4271	402	80110	B13-16	202
MASONRY	Rate	4.41	6.78	4.36	11.15	4.55	6.81	3.84	5.21	5.96	17.99	6.23	8.89
	Code	42102	721037	40102	741	4231	4231	4-41	4231	401	80100	B13-18	202
PAINTING & DECORATING	Rate	4.35	4.94	3.18	6.75	4.40	4.69	3.84	4.52	3.71	15.05	5.98	8.89
	Code	42111	721041	40105	719	4275	4275	4-41	4275	402	80110	B12-01	202
PILE DRIVING	Rate	4.41	5.18	3.76	6.34	3.79	9.78	3.64	4.64	5.96	7.13	6.23	8.89
	Code	42159	722004	40706	732	4221	4221	4-43	4221	401	80030	B13-10	202
PLASTERING	Rate	5.59	6.78	5.08	6.75	4.40	4.69	3.84	4.52	3.71	15.05	5.98	8.89
	Code	42135	721042	40108	719	4271	4271	4-41	4271	402	80110	B12-21	202
PLUMBING	Rate	2.13	3.11	3.05	4.02	2.32	3.37	3.52	2.43	3.71	6.93	3.51	4.67
	Code	42122	721043	40204	707	4241	4241	4-46	4241	402	80160	B11-01	214
ROOFING	Rate	8.12	8.68	7.02	12.98	4.55	10.08	3.84	10.57	5.96	23.01	6.23	8.89
	Code	42118	721036	40403	728	4236	4236	4-41	4236	401	80130	B13-20	202
SHEET METAL WORK (HVAC)	Rate	2.13	3.11	7.02	4.02	2.32	3.37	3.52	2.43	3.71	6.93	3.51	4.67
	Code	42117	721043	40402	707	4244	4244	4-46	4244	402	80160	B11-07	208
STEEL ERECTION—door & sash	Rate	2.57	14.07	11.32	16.02	3.79	10.08	3.84	5.21	5.96	25.32	6.23	8.89
	Code	42106	722005	40502	748	4227	4227	4-41	4227	401	80080	B13-22	202
STEEL ERECTION—inter., ornam.	Rate	2.57	14.07	11.32	16.02	3.79	10.08	3.84	5.21	5.96	25.32	6.23	8.89
	Code	42106	722005	40502	748	4227	4227	4-41	4227	401	80080	B13-22	202
STEEL ERECTION—structure	Rate	2.57	14.07	11.32	16.02	3.79	10.08	3.84	5.21	5.96	25.32	6.23	8.89
	Code	42106	722005	40502	748	4227	4227	4-41	4227	401	80080	B13-22	202
STEEL ERECTION—NOC	Rate	2.57	14.07	11.32	16.02	3.79	10.08	3.84	5.21	5.96	25.32	6.23	8.89
	Code	42106	722005	40502	748	4227	4227	4-41	4227	401	80080	B13-22	202
TILE WORK—inter. (ceramic)	Rate	3.63	5.62	1.97	6.75	4.40	10.08	3.84	4.52	3.71	15.05	6.23	8.89
	Code	42113	721054	40103	719	4276	4276	4-41	4276	402	80110	B13-01	202
WATERPROOFING	Rate	4.35	4.93	4.36	4.35	4.55	4.69	3.84	5.21	3.71	23.01	5.98	8.89
	Code	42139	721016	40102	723	4239	4299	4-41	4239	402	80130	B12-17	202
WRECKING	Rate	2.51	4.59	6.59	16.02	2.62	4.39	3.64	3.34	5.96	15.05	6.23	8.89
	Code	40604	721005	40106	748	4211	4211	4-43	4211	401	80110	B13-09	202

General Requirements R0154 Construction Aids

R015423-10 Steel Tubular Scaffolding

On new construction, tubular scaffolding is efficient up to 60' high or five stories. Above this it is usually better to use a hung scaffolding if construction permits. Swing scaffolding operations may interfere with tenants. In this case, the tubular is more practical at all heights.

In repairing or cleaning the front of an existing building the cost of tubular scaffolding per S.F. of building front increases as the height increases above the first tier. The first tier cost is relatively high due to leveling and alignment.

The minimum efficient crew for erecting and dismantling is three workers. They can set up and remove 18 frame sections per day up to 5 stories high. For 6 to 12 stories high, a crew of four is most efficient. Use two or more on top and two on the bottom for handing up or hoisting. They can also set up and remove 18 frame sections per day. At 7' horizontal spacing, this will run about 800 S.F. per day of erecting and dismantling. Time for placing and removing planks must be added to the above. A crew of three can place and remove 72 planks per day up to 5 stories. For over 5 stories, a crew of four can place and remove 80 planks per day.

The table below shows the number of pieces required to erect tubular steel scaffolding for 1000 S.F. of building frontage. This area is made up of a scaffolding system that is 12 frames (11 bays) long by 2 frames high.

For jobs under twenty-five frames, add 50% to rental cost. Rental rates will be lower for jobs over three months duration. Large quantities for long periods can reduce rental rates by 20%.

Description of Component	Number of Pieces for 1000 S.F. of Building Front	Unit
5' Wide Standard Frame, 6'-4" High	24	Ea.
Leveling Jack & Plate	24	
Cross Brace	44	
Side Arm Bracket, 21"	12	
Guardrail Post	12	
Guardrail, 7' section	22	
Stairway Section	2	
Stairway Starter Bar	1	
Stairway Inside Handrail	2	
Stairway Outside Handrail	2	
Walk-Thru Frame Guardrail	2	

Scaffolding is often used as falsework over 15' high during construction of cast-in-place concrete beams and slabs. Two foot wide scaffolding is generally used for heavy beam construction. The span between frames depends upon the load to be carried with a maximum span of 5'.

Heavy duty shoring frames with a capacity of 10,000#/leg can be spaced up to 10' O.C. depending upon form support design and loading.

Scaffolding used as horizontal shoring requires less than half the material required with conventional shoring.

On new construction, erection is done by carpenters.

Rolling towers supporting horizontal shores can reduce labor and speed the job. For maintenance work, catwalks with spans up to 70' can be supported by the rolling towers.

Existing Conditions — R0241 Demolition

R024119-10 Demolition Defined

Whole Building Demolition - Demolition of the whole building with no concern for any particular building element, component, or material type being demolished. This type of demolition is accomplished with large pieces of construction equipment that break up the structure, load it into trucks and haul it to a disposal site, but disposal or dump fees are not included. Demolition of below-grade foundation elements, such as footings, foundation walls, grade beams, slabs on grade, etc., is not included. Certain mechanical equipment containing flammable liquids or ozone-depleting refrigerants, electric lighting elements, communication equipment components, and other building elements may contain hazardous waste, and must be removed, either selectively or carefully, as hazardous waste before the building can be demolished.

Foundation Demolition - Demolition of below-grade foundation footings, foundation walls, grade beams, and slabs on grade. This type of demolition is accomplished by hand or pneumatic hand tools, and does not include saw cutting, or handling, loading, hauling, or disposal of the debris.

Gutting - Removal of building interior finishes and electrical/mechanical systems down to the load-bearing and sub-floor elements of the rough building frame, with no concern for any particular building element, component, or material type being demolished. This type of demolition is accomplished by hand or pneumatic hand tools, and includes loading into trucks, but not hauling, disposal or dump fees, scaffolding, or shoring. Certain mechanical equipment containing flammable liquids or ozone-depleting refrigerants, electric lighting elements, communication equipment components, and other building elements may contain hazardous waste, and must be removed, either selectively or carefully, as hazardous waste, before the building is gutted.

Selective Demolition - Demolition of a selected building element, component, or finish, with some concern for surrounding or adjacent elements, components, or finishes (see the first Subdivision (s) at the beginning of appropriate Divisions). This type of demolition is accomplished by hand or pneumatic hand tools, and does not include handling, loading, storing, hauling, or disposal of the debris, scaffolding, or shoring. "Gutting" methods may be used in order to save time, but damage that is caused to surrounding or adjacent elements, components, or finishes may have to be repaired at a later time.

Careful Removal - Removal of a piece of service equipment, building element or component, or material type, with great concern for both the removed item and surrounding or adjacent elements, components or finishes. The purpose of careful removal may be to protect the removed item for later re-use, preserve a higher salvage value of the removed item, or replace an item while taking care to protect surrounding or adjacent elements, components, connections, or finishes from cosmetic and/or structural damage. An approximation of the time required to perform this type of removal is 1/3 to 1/2 the time it would take to install a new item of like kind (see Reference Number R220105-10). This type of removal is accomplished by hand or pneumatic hand tools, and does not include loading, hauling, or storing the removed item, scaffolding, shoring, or lifting equipment.

Cutout Demolition - Demolition of a small quantity of floor, wall, roof, or other assembly, with concern for the appearance and structural integrity of the surrounding materials. This type of demolition is accomplished by hand or pneumatic hand tools, and does not include saw cutting, handling, loading, hauling, or disposal of debris, scaffolding, or shoring.

Rubbish Handling - Work activities that involve handling, loading or hauling of debris. Generally, the cost of rubbish handling must be added to the cost of all types of demolition, with the exception of whole building demolition.

Minor Site Demolition - Demolition of site elements outside the footprint of a building. This type of demolition is accomplished by hand or pneumatic hand tools, or with larger pieces of construction equipment, and may include loading a removed item onto a truck (check the Crew for equipment used). It does not include saw cutting, hauling or disposal of debris, and, sometimes, handling or loading.

R024119-20 Dumpsters

Dumpster rental costs on construction sites are presented in two ways.

The cost per week rental includes the delivery of the dumpster; its pulling or emptying once per week, and its final removal. The assumption is made that the dumpster contractor could choose to empty a dumpster by simply bringing in an empty unit and removing the full one. These costs also include the disposal of the materials in the dumpster.

The Alternate Pricing can be used when actual planned conditions are not approximated by the weekly numbers. For example, these lines can be used when a dumpster is needed for 4 weeks and will need to be emptied 2 or 3 times per week. Conversely the Alternate Pricing lines can be used when a dumpster will be rented for several weeks or months but needs to be emptied only a few times over this period.

Existing Conditions — R0282 Asbestos Remediation

R028213-20 Asbestos Removal Process

Asbestos removal is accomplished by a specialty contractor who understands the federal and state regulations regarding the handling and disposal of the material. The process of asbestos removal is divided into many individual steps. An accurate estimate can be calculated only after all the steps have been priced.

The steps are generally as follows:
1. Obtain an asbestos abatement plan from an industrial hygienist.
2. Monitor the air quality in and around the removal area and along the path of travel between the removal area and transport area. This establishes the background contamination.
3. Construct a two part decontamination chamber at entrance to removal area.
4. Install a HEPA filter to create a negative pressure in the removal area.
5. Install wall, floor and ceiling protection as required by the plan, usually 2 layers of fireproof 6 mil polyethylene.
6. Industrial hygienist visually inspects work area to verify compliance with plan.
7. Provide temporary supports for conduit and piping affected by the removal process.
8. Proceed with asbestos removal and bagging process. Monitor air quality as described in Step #2. Discontinue operations when contaminate levels exceed applicable standards.
9. Document the legal disposal of materials in accordance with EPA standards.
10. Thoroughly clean removal area including all ledges, crevices and surfaces.
11. Post abatement inspection by industrial hygienist to verify plan compliance.
12. Provide a certificate from a licensed industrial hygienist attesting that contaminate levels are within acceptable standards before returning area to regular use.

Existing Conditions — R0283 Lead Remediation

R028319-60 Lead Paint Remediation Methods

Lead paint remediation can be accomplished by the following methods.
1. Abrasive blast
2. Chemical stripping
3. Power tool cleaning with vacuum collection system
4. Encapsulation
5. Remove and replace
6. Enclosure

Each of these methods has strengths and weakness depending on the specific circumstances of the project. The following is an overview of each method.

1. **Abrasive blasting** is usually accomplished with sand or recyclable metallic blast. Before work can begin, the area must be contained to ensure the blast material with lead does not escape to the atmosphere. The use of vacuum blast greatly reduces the containment requirements. Lead abatement equipment that may be associated with this work includes a negative air machine. In addition, it is necessary to have an industrial hygienist monitor the project on a continual basis. When the work is complete, the spent blast sand with lead must be disposed of as a hazardous material. If metallic shot was used, the lead is separated from the shot and disposed of as hazardous material. Worker protection includes disposable clothing and respiratory protection.

2. **Chemical stripping** requires strong chemicals be applied to the surface to remove the lead paint. Before the work can begin, the area under/adjacent to the work area must be covered to catch the chemical and removed lead. After the chemical is applied to the painted surface it is usually covered with paper. The chemical is left in place for the specified period, then the paper with lead paint is pulled or scraped off. The process may require several chemical applications. The paper with chemicals and lead paint adhered to it, plus the containment and loose scrapings collected by a HEPA (High Efficiency Particulate Air Filter) vac, must be disposed of as a hazardous material. The chemical stripping process usually requires a neutralizing agent and several wash downs after the paint is removed. Worker protection includes a neoprene or other compatible protective clothing and respiratory protection with face shield. An industrial hygienist is required intermittently during the process.

3. **Power tool cleaning** is accomplished using shrouded needle blasting guns. The shrouding with different end configurations is held up against the surface to be cleaned. The area is blasted with hardened needles and the shroud captures the lead with a HEPA vac and deposits it in a holding tank. An industrial hygienist monitors the project, protective clothing and a respirator is required until air samples prove otherwise. When the work is complete the lead must be disposed of as a hazardous material.

4. **Encapsulation** is a method that leaves the well bonded lead paint in place after the peeling paint has been removed. Before the work can begin, the area under/adjacent to the work must be covered to catch the scrapings. The scraped surface is then washed with a detergent and rinsed. The prepared surface is covered with approximately 10 mils of paint. A reinforcing fabric can also be embedded in the paint covering. The scraped paint and containment must be disposed of as a hazardous material. Workers must wear protective clothing and respirators.

5. **Remove and replace** is an effective way to remove lead paint from windows, gypsum walls and concrete masonry surfaces. The painted materials are removed and new materials are installed. Workers should wear a respirator and tyvek suit. The demolished materials must be disposed of as hazardous waste if it fails the TCLP (Toxicity Characteristic Leachate Process) test.

6. **Enclosure** is the process that permanently seals lead painted materials in place. This process has many applications such as covering lead painted drywall with new drywall, covering exterior construction with tyvek paper then residing, or covering lead painted structural members with aluminum or plastic. The seams on all enclosing materials must be securely sealed. An industrial hygienist monitors the project, and protective clothing and a respirator is required until air samples prove otherwise.

All the processes require clearance monitoring and wipe testing as required by the hygienist.

Concrete — R0331 Structural Concrete

R033105-20 Materials for One C.Y. of Concrete

This is an approximate method of figuring quantities of cement, sand and coarse aggregate for a field mix with waste allowance included.

With crushed gravel as coarse aggregate, to determine barrels of cement required, divide 10 by total mix; that is, for 1:2:4 mix, 10 divided by 7 = 1-3/7 barrels.

If the coarse aggregate is crushed stone, use 10-1/2 instead of 10 as given for gravel.

To determine tons of sand required, multiply barrels of cement by parts of sand and then by 0.2; that is, for the 1:2:4 mix, as above, 1-3/7 x 2 x .2 = .57 tons.

Tons of crushed gravel are in the same ratio to tons of sand as parts in the mix, or 4/2 x .57 = 1.14 tons.

1 bag cement = 94#	1 C.Y. sand or crushed gravel = 2700#	1 C.Y. crushed stone = 2575#
4 bags = 1 barrel	1 ton sand or crushed gravel = 20 C.F.	1 ton crushed stone = 21 C.F.

Average carload of cement is 692 bags; of sand or gravel is 56 tons.

Do not stack stored cement over 10 bags high.

Masonry — R0405 Common Work Results for Masonry

R040513-10 Cement Mortar (material only)

Type N - 1:1:6 mix by volume. Use everywhere above grade except as noted below. - 1:3 mix using conventional masonry cement which saves handling two separate bagged materials.

Type M - 1:1/4:3 mix by volume, or 1 part cement, 1/4 (10% by wt.) lime, 3 parts sand. Use for heavy loads and where earthquakes or hurricanes may occur. Also for reinforced brick, sewers, manholes and everywhere below grade.

Mix Proportions by Volume and Compressive Strength of Mortar

Where Used	Mortar Type	Allowable Proportions by Volume				Compressive Strength @ 28 days
		Portland Cement	Masonry Cement	Hydrated Lime	Masonry Sand	
Plain Masonry	M	1	1	—	6	2500 psi
		1	—	1/4	3	
	S	1/2	1	—	4	1800 psi
		1	—	1/4 to 1/2	4	
	N	—	1	—	3	750 psi
		1	—	1/2 to 1-1/4	6	
	O	—	1	—	3	350 psi
		1	—	1-1/4 to 2-1/2	9	
	K	1	—	2-1/2 to 4	12	75 psi
Reinforced Masonry	PM	1	1	—	6	2500 psi
	PL	1	—	1/4 to 1/2	4	2500 psi

Note: The total aggregate should be between 2.25 to 3 times the sum of the cement and lime used.

The labor cost to mix the mortar is included in the productivity and labor cost of unit price lines in unit cost sections for brickwork, blockwork and stonework.

The material cost of mixed mortar is included in the material cost of those same unit price lines and includes the cost of renting and operating a 10 C.F. mixer at the rate of 200 C.F. per day.

There are two types of mortar color used. One type is the inert additive type with about 100 lbs. per M brick as the typical quantity required. These colors are also available in smaller-batch-sized bags (1 lb. to 15 lb.) which can be placed directly into the mixer without measuring. The other type is premixed and replaces the masonry cement. Dark green color has the highest cost.

R040519-50 Masonry Reinforcing

Horizontal joint reinforcing helps prevent wall cracks where wall movement may occur and in many locations is required by code. Horizontal joint reinforcing is generally not considered to be structural reinforcing and an unreinforced wall may still contain joint reinforcing.

Reinforcing strips come in 10' and 12' lengths and in truss and ladder shapes, with and without drips. Field labor runs between 2.7 to 5.3 hours per 1000 L.F. for wall thicknesses up to 12".

The wire meets ASTM A82 for cold drawn steel wire and the typical size is 9 ga. sides and ties with 3/16" diameter also available. Typical finish is mill galvanized with zinc coating at .10 oz. per S.F. Class I (.40 oz. per S.F.) and Class III (.80 oz per S.F.) are also available, as is hot dipped galvanizing at 1.50 oz. per S.F.

Masonry — R0421 Clay Unit Masonry

R042110-10 Economy in Bricklaying

Have adequate supervision. Be sure bricklayers are always supplied with materials so there is no waiting. Place best bricklayers at corners and openings.

Use only screened sand for mortar. Otherwise, labor time will be wasted picking out pebbles. Use seamless metal tubs for mortar as they do not leak or catch the trowel. Locate stack and mortar for easy wheeling.

Have brick delivered for stacking. This makes for faster handling, reduces chipping and breakage, and requires less storage space. Many dealers will deliver select common in 2' x 3' x 4' pallets or face brick packaged. This affords quick handling with a crane or forklift and easy tonging in units of ten, which reduces waste.

Use wider bricks for one wythe wall construction. Keep scaffolding away from wall to allow mortar to fall clear and not stain wall.

On large jobs develop specialized crews for each type of masonry unit.

Consider designing for prefabricated panel construction on high rise projects.

Avoid excessive corners or openings. Each opening adds about 50% to labor cost for area of opening.

Bolting stone panels and using window frames as stops reduces labor costs and speeds up erection.

R042110-50 Brick, Block & Mortar Quantities

Type Brick	Nominal Size (incl. mortar) L x H x W	Modular Coursing	Number of Brick per S.F.	C.F. of Mortar per M Bricks, Waste Included 3/8" Joint	C.F. of Mortar per M Bricks, Waste Included 1/2" Joint	Bond Type	Description	Factor
Standard	8 x 2-2/3 x 4	3C=8"	6.75	10.3	12.9	Common	full header every fifth course	+20%
Economy	8 x 4 x 4	1C=4"	4.50	11.4	14.6		full header every sixth course	+16.7%
Engineer	8 x 3-1/5 x 4	5C=16"	5.63	10.6	13.6	English	full header every second course	+50%
Fire	9 x 2-1/2 x 4-1/2	2C=5"	6.40	550 # Fireclay	—	Flemish	alternate headers every course	+33.3%
Jumbo	12 x 4 x 6 or 8	1C=4"	3.00	23.8	30.8		every sixth course	+5.6%
Norman	12 x 2-2/3 x 4	3C=8"	4.50	14.0	17.9	Header = W x H exposed		+100%
Norwegian	12 x 3-1/5 x 4	5C=16"	3.75	14.6	18.6	Rowlock = H x W exposed		+100%
Roman	12 x 2 x 4	2C=4"	6.00	13.4	17.0	Rowlock stretcher = L x W exposed		+33.3%
SCR	12 x 2-2/3 x 6	3C=8"	4.50	21.8	28.0	Soldier = H x L exposed		—
Utility	12 x 4 x 4	1C=4"	3.00	15.4	19.6	Sailor = W x L exposed		-33.3%

Concrete Blocks Nominal Size	Approximate Weight per S.F. Standard	Approximate Weight per S.F. Lightweight	Blocks per 100 S.F.	Mortar per M block, waste included Partitions	Mortar per M block, waste included Back up
2" x 8" x 16"	20 PSF	15 PSF	113	27 C.F.	36 C.F.
4"	30	20		41	51
6"	42	30		56	66
8"	55	38		72	82
10"	70	47		87	97
12"	85	55		102	112

Masonry — R0422 Concrete Unit Masonry

R042210-20 Concrete Block

The material cost of special block such as corner, jamb and head block can be figured at the same price as ordinary block of same size. Labor on specials is about the same as equal-sized regular block.

Bond beam and 16" high lintel blocks are more expensive than regular units of equal size. Lintel blocks are 8" long and either 8" or 16" high.

Use of motorized mortar spreader box will speed construction of continuous walls.

Hollow non-load-bearing units are made according to ASTM C129 and hollow load-bearing units according to ASTM C90.

Metals — R0505 Common Work Results for Metals

R050521-20 Welded Structural Steel

Usual weight reductions with welded design run 10% to 20% compared with bolted or riveted connections. This amounts to about the same total cost compared with bolted structures since field welding is more expensive than bolts. For normal spans of 18' to 24' figure 6 to 7 connections per ton.

Trusses — For welded trusses add 4% to weight of main members for connections. Up to 15% less steel can be expected in a welded truss compared to one that is shop bolted. Cost of erection is the same whether shop bolted or welded.

General — Typical electrodes for structural steel welding are E6010, E6011, E60T and E70T. Typical buildings vary between 2# to 8# of weld rod per ton of steel. Buildings utilizing continuous design require about three times as much welding as conventional welded structures. In estimating field erection by welding, it is best to use the average linear feet of weld per ton to arrive at the welding cost per ton. The type, size and position of the weld will have a direct bearing on the cost per linear foot. A typical field welder will deposit 1.8# to 2# of weld rod per hour manually. Using semiautomatic methods can increase production by as much as 50% to 75%.

Metals — R0531 Steel Decking

R053100-10 Decking Descriptions

General - All Deck Products

Steel deck is made by cold forming structural grade sheet steel into a repeating pattern of parallel ribs. The strength and stiffness of the panels are the result of the ribs and the material properties of the steel. Deck lengths can be varied to suit job conditions, but because of shipping considerations, are usually less than 40 feet. Standard deck width varies with the product used but full sheets are usually 12", 18", 24", 30", or 36". Deck is typically furnished in a standard width with the ends cut square. Any cutting for width, such as at openings or for angular fit, is done at the job site.

Deck is typically attached to the building frame with arc puddle welds, self-drilling screws, or powder or pneumatically driven pins. Sheet to sheet fastening is done with screws, button punching (crimping), or welds.

Composite Floor Deck

After installation and adequate fastening, floor deck serves several purposes. It (a) acts as a working platform, (b) stabilizes the frame, (c) serves as a concrete form for the slab, and (d) reinforces the slab to carry the design loads applied during the life of the building. Composite decks are distinguished by the presence of shear connector devices as part of the deck. These devices are designed to mechanically lock the concrete and deck together so that the concrete and the deck work together to carry subsequent floor loads. These shear connector devices can be rolled-in embossments, lugs, holes, or wires welded to the panels. The deck profile can also be used to interlock concrete and steel.

Composite deck finishes are either galvanized (zinc coated) or phosphatized/painted. Galvanized deck has a zinc coating on both the top and bottom surfaces. The phosphatized/painted deck has a bare (phosphatized) top surface that will come into contact with the concrete. This bare top surface can be expected to develop rust before the concrete is placed. The bottom side of the deck has a primer coat of paint.

Composite floor deck is normally installed so the panel ends do not overlap on the supporting beams. Shear lugs or panel profile shape often prevent a tight metal to metal fit if the panel ends overlap; the air gap caused by overlapping will prevent proper fusion with the structural steel supports when the panel end laps are shear stud welded.

Adequate end bearing of the deck must be obtained as shown on the drawings. If bearing is actually less in the field than shown on the drawings, further investigation is required.

Roof Deck

Roof deck is not designed to act compositely with other materials. Roof deck acts alone in transferring horizontal and vertical loads into the building frame. Roof deck rib openings are usually narrower than floor deck rib openings. This provides adequate support of rigid thermal insulation board.

Roof deck is typically installed to endlap approximately 2" over supports. However, it can be butted (or lapped more than 2") to solve field fit problems. Since designers frequently use the installed deck system as part of the horizontal bracing system (the deck as a diaphragm), any fastening substitution or change should be approved by the designer. Continuous perimeter support of the deck is necessary to limit edge deflection in the finished roof and may be required for diaphragm shear transfer.

Standard roof deck finishes are galvanized or primer painted. The standard factory applied paint for roof deck is a primer paint and is not intended to weather for extended periods of time. Field painting or touching up of abrasions and deterioration of the primer coat or other protective finishes is the responsibility of the contractor.

Cellular Deck

Cellular deck is made by attaching a bottom steel sheet to a roof deck or composite floor deck panel. Cellular deck can be used in the same manner as floor deck. Electrical, telephone, and data wires are easily run through the chase created between the deck panel and the bottom sheet.

When used as part of the electrical distribution system, the cellular deck must be installed so that the ribs line up and create a smooth cell transition at abutting ends. The joint that occurs at butting cell ends must be taped or otherwise sealed to prevent wet concrete from seeping into the cell. Cell interiors must be free of welding burrs, or other sharp intrusions, to prevent damage to wires.

When used as a roof deck, the bottom flat plate is usually left exposed to view. Care must be maintained during erection to keep good alignment and prevent damage.

Cellular deck is sometimes used with the flat plate on the top side to provide a flat working surface. Installation of the deck for this purpose requires special methods for attachment to the frame because the flat plate, now on the top, can prevent direct access to the deck material that is bearing on the structural steel. It may be advisable to treat the flat top surface to prevent slipping.

Cellular deck is always furnished galvanized or painted over galvanized.

Form Deck

Form deck can be any floor or roof deck product used as a concrete form. Connections to the frame are by the same methods used to anchor floor and roof deck. Welding washers are recommended when welding deck that is less than 20 gauge thickness.

Form deck is furnished galvanized, prime painted, or uncoated. Galvanized deck must be used for those roof deck systems where form deck is used to carry a lightweight insulating concrete fill.

Wood, Plastics & Comp. — R0611 Wood Framing

R061110-30 Lumber Product Material Prices

The price of forest products fluctuates widely from location to location and from season to season depending upon economic conditions. The bare material prices in the unit cost sections of the book show the National Average material prices in effect Jan. 1 of this book year. It must be noted that lumber prices in general may change significantly during the year.

Availability of certain items depends upon geographic location and must be checked prior to firm-price bidding.

Wood, Plastics & Comp. — R0616 Sheathing

R061636-20 Plywood

There are two types of plywood used in construction: interior, which is moisture-resistant but not waterproofed, and exterior, which is waterproofed.

The grade of the exterior surface of the plywood sheets is designated by the first letter: A, for smooth surface with patches allowed; B, for solid surface with patches and plugs allowed; C, which may be surface plugged or may have knot holes up to 1″ wide; and D, which is used only for interior type plywood and may have knot holes up to 2-1/2″ wide. "Structural Grade" is specifically designed for engineered applications such as box beams. All CC & DD grades have roof and floor spans marked on them.

Underlayment-grade plywood runs from 1/4″ to 1-1/4″ thick. Thicknesses 5/8″ and over have optional tongue and groove joints which eliminate the need for blocking the edges. Underlayment 19/32″ and over may be referred to as Sturd-i-Floor.

The price of plywood can fluctuate widely due to geographic and economic conditions.

Typical uses for various plywood grades are as follows:

AA-AD Interior — cupboards, shelving, paneling, furniture
BB Plyform — concrete form plywood
CDX — wall and roof sheathing
Structural — box beams, girders, stressed skin panels
AA-AC Exterior — fences, signs, siding, soffits, etc.
Underlayment — base for resilient floor coverings
Overlaid HDO — high density for concrete forms & highway signs
Overlaid MDO — medium density for painting, siding, soffits & signs
303 Siding — exterior siding, textured, striated, embossed, etc.

Thermal & Moist. Protec. — R0784 Firestopping

R078413-30 Firestopping

Firestopping is the sealing of structural, mechanical, electrical and other penetrations through fire-rated assemblies. The basic components of firestop systems are safing insulation and firestop sealant on both sides of wall penetrations and the top side of floor penetrations.

Pipe penetrations are assumed to be through concrete, grout, or joint compound and can be sleeved or unsleeved. Costs for the penetrations and sleeves are not included. An annular space of 1″ is assumed. Escutcheons are not included.

Metallic pipe is assumed to be copper, aluminum, cast iron or similar metallic material. Insulated metallic pipe is assumed to be covered with a thermal insulating jacket of varying thickness and materials.

Non-metallic pipe is assumed to be PVC, CPVC, FR Polypropylene or similar plastic piping material. Intumescent firestop sealant or wrap strips are included. Collars on both sides of wall penetrations and a sheet metal plate on the underside of floor penetrations are included.

Ductwork is assumed to be sheet metal, stainless steel or similar metallic material. Duct penetrations are assumed to be through concrete, grout or joint compound. Costs for penetrations and sleeves are not included. An annular space of 1/2″ is assumed.

Multi-trade openings include costs for sheet metal forms, firestop mortar, wrap strips, collars and sealants as necessary.

Structural penetrations joints are assumed to be 1/2″ or less. CMU walls are assumed to be within 1-1/2″ of metal deck. Drywall walls are assumed to be tight to the underside of metal decking.

Metal panel, glass or curtain wall systems include a spandrel area of 5′ filled with mineral wool foil-faced insulation. Fasteners and stiffeners are included.

Openings — R0813 Metal Doors

R081313-20 Steel Door Selection Guide

Standard steel doors are classified into four levels, as recommended by the Steel Door Institute in the chart below. Each of the four levels offers a range of construction models and designs, to meet architectural requirements for preference and appearance, including full flush, seamless, and stile & rail. Recommended minimum gauge requirements are also included.

For complete standard steel door construction specifications and available sizes, refer to the Steel Door Institute Technical Data Series, ANSI A250.8-98 (SDI-100), and ANSI A250.4-94 Test Procedure and Acceptance Criteria for Physical Endurance of Steel Door and Hardware Reinforcements.

Level	Model		Construction	For Full Flush or Seamless		
				Min. Gauge	Thickness (in)	Thickness (mm)
I	Standard Duty	1	Full Flush	20	0.032	0.8
		2	Seamless			
II	Heavy Duty	1	Full Flush	18	0.042	1.0
		2	Seamless			
III	Extra Heavy Duty	1	Full Flush	16	0.053	1.3
		2	Seamless			
		3	*Stile & Rail			
IV	Maximum Duty	1	Full Flush	14	0.067	1.6
		2	Seamless			

*Stiles & rails are 16 gauge; flush panels, when specified, are 18 gauge

Openings — R0871 Door Hardware

R087120-10 Hinges

All closer equipped doors should have ball bearing hinges. Lead lined or extremely heavy doors require special strength hinges. Usually 1-1/2 pair of hinges are used per door up to 7'-6" high openings. Table below shows typical hinge requirements.

Use Frequency	Type Hinge Required	Type of Opening	Type of Structure
High	Heavy weight ball bearing	Entrances	Banks, Office buildings, Schools, Stores & Theaters
		Toilet Rooms	Office buildings and Schools
Average	Standard weight ball bearing	Entrances	Dwellings
		Corridors	Office buildings and Schools
		Toilet Rooms	Stores
Low	Plain bearing	Interior	Dwellings

Door Thickness	Weight of Doors in Pounds per Square Foot				
	White Pine	Oak	Hollow Core	Solid Core	Hollow Metal
1-3/8"	3psf	6psf	1-1/2psf	3-1/2 — 4psf	6-1/2psf
1-3/4"	3-1/2	7	2-1/2	4-1/2 — 5-1/4	6-1/2
2-1/4"	4-1/2	9	—	5-1/2 — 6-3/4	6-1/2

Openings — R0881 Glass Glazing

R088110-10 Glazing Productivity

Some glass sizes are estimated by the "united inch" (height + width). The table below shows the number of lights glazed in an eight-hour period by the crew size indicated, for glass up to 1/4" thick. Square or nearly square lights are more economical on a S.F. basis. Long slender lights will have a high S.F. installation cost. For insulated glass reduce production by 33%. For 1/2" float glass reduce production by 50%. Production time for glazing with two glaziers per day averages: 1/4" float glass 120 S.F.; 1/2" float glass 55 S.F.; 1/2" insulated glass 95 S.F.; 3/4" insulated glass 75 S.F.

Glazing Method	United Inches per Light							
	40"	60"	80"	100"	135"	165"	200"	240"
Number of Men in Crew	1	1	1	1	2	3	3	4
Industrial sash, putty	60	45	24	15	18	—	—	—
With stops, putty bed	50	36	21	12	16	8	4	3
Wood stops, rubber	40	27	15	9	11	6	3	2
Metal stops, rubber	30	24	14	9	9	6	3	2
Structural glass	10	7	4	3	—	—	—	—
Corrugated glass	12	9	7	4	4	4	3	—
Storefronts	16	15	13	11	7	6	4	4
Skylights, putty glass	60	36	21	12	16	—	—	—
Thiokol set	15	15	11	9	9	6	3	2
Vinyl set, snap on	18	18	13	12	12	7	5	4
Maximum area per light	2.8 S.F.	6.3 S.F.	11.1 S.F.	17.4 S.F.	31.6 S.F.	47 S.F.	69 S.F.	100 S.F.

Finishes — R0920 Plaster & Gypsum Board

R092000-50 Lath, Plaster and Gypsum Board

Gypsum board lath is available in 3/8" thick x 16" wide x 4' long sheets as a base material for multi-layer plaster applications. It is also available as a base for either multi-layer or veneer plaster applications in 1/2" and 5/8" thick–4' wide x 8', 10' or 12' long sheets. Fasteners are screws or blued ring shank nails for wood framing and screws for metal framing.

Metal lath is available in diamond mesh pattern with flat or self-furring profiles. Paper backing is available for applications where excessive plaster waste needs to be avoided. A slotted mesh ribbed lath should be used in areas where the span between structural supports is greater than normal. Most metal lath comes in 27" x 96" sheets. Diamond mesh weighs 1.75, 2.5 or 3.4 pounds per square yard, slotted mesh lath weighs 2.75 or 3.4 pounds per square yard. Metal lath can be nailed, screwed or tied in place.

Many **accessories** are available. Corner beads, flat reinforcing strips, casing beads, control and expansion joints, furring brackets and channels are some examples. Note that accessories are not included in plaster or stucco line items.

Plaster is defined as a material or combination of materials that when mixed with a suitable amount of water, forms a plastic mass or paste. When applied to a surface, the paste adheres to it and subsequently hardens, preserving in a rigid state the form or texture imposed during the period of elasticity.

Gypsum plaster is made from ground calcined gypsum. It is mixed with aggregates and water for use as a base coat plaster.

Vermiculite plaster is a fire-retardant plaster covering used on steel beams, concrete slabs and other heavy construction materials. Vermiculite is a group name for certain clay minerals, hydrous silicates or aluminum, magnesium and iron that have been expanded by heat.

Perlite plaster is a plaster using perlite as an aggregate instead of sand. Perlite is a volcanic glass that has been expanded by heat.

Gauging plaster is a mix of gypsum plaster and lime putty that when applied produces a quick drying finish coat.

Veneer plaster is a one or two component gypsum plaster used as a thin finish coat over special gypsum board.

Keenes cement is a white cementitious material manufactured from gypsum that has been burned at a high temperature and ground to a fine powder. Alum is added to accelerate the set. The resulting plaster is hard and strong and accepts and maintains a high polish, hence it is used as a finishing plaster.

Stucco is a Portland cement based plaster used primarily as an exterior finish.

Plaster is used on both interior and exterior surfaces. Generally it is applied in multiple-coat systems. A three-coat system uses the terms scratch, brown and finish to identify each coat. A two-coat system uses base and finish to describe each coat. Each type of plaster and application system has attributes that are chosen by the designer to best fit the intended use.

Gypsum Plaster	2 Coat, 5/8" Thick		3 Coat, 3/4" Thick		
Quantities for 100 S.Y.	Base	Finish	Scratch	Brown	Finish
	1:3 Mix	2:1 Mix	1:2 Mix	1:3 Mix	2:1 Mix
Gypsum plaster	1,300 lb.		1,350 lb.	650 lb.	
Sand	1.75 C.Y.		1.85 C.Y.	1.35 C.Y.	
Finish hydrated lime		340 lb.			340 lb.
Gauging plaster		170 lb.			170 lb.

Vermiculite or Perlite Plaster	2 Coat, 5/8" Thick		3 Coat, 3/4" Thick		
Quantities for 100 S.Y.	Base	Finish	Scratch	Brown	Finish
Gypsum plaster	1,250 lb.		1,450 lb.	800 lb.	
Vermiculite or perlite	7.8 bags		8.0 bags	3.3 bags	
Finish hydrated lime		340 lb.			340 lb.
Gauging plaster		170 lb.			170 lb.

Stucco–Three-Coat System	On Wood	On
Quantities for 100 S.Y.	Frame	Masonry
Portland cement	29 bags	21 bags
Sand	2.6 C.Y.	2.0 C.Y.
Hydrated lime	180 lb.	120 lb.

Finishes

R0929 Gypsum Board

R092910-10 Levels of Gypsum Drywall Finish

In the past, contract documents often used phrases such as "industry standard" and "workmanlike finish" to specify the expected quality of gypsum board wall and ceiling installations. The vagueness of these descriptions led to unacceptable work and disputes.

In order to resolve this problem, four major trade associations concerned with the manufacture, erection, finish and decoration of gypsum board wall and ceiling systems have developed an industry-wide *Recommended Levels of Gypsum Board Finish*.

The finish of gypsum board walls and ceilings for specific final decoration is dependent on a number of factors. A primary consideration is the location of the surface and the degree of decorative treatment desired. Painted and unpainted surfaces in warehouses and other areas where appearance is normally not critical may simply require the taping of wallboard joints and 'spotting' of fastener heads. Blemish-free, smooth, monolithic surfaces often intended for painted and decorated walls and ceilings in habitated structures, ranging from single-family dwellings through monumental buildings, require additional finishing prior to the application of the final decoration.

Other factors to be considered in determining the level of finish of the gypsum board surface are (1) the type of angle of surface illumination (both natural and artificial lighting), and (2) the paint and method of application or the type and finish of wallcovering specified as the final decoration. Critical lighting conditions, gloss paints, and thin wallcoverings require a higher level of gypsum board finish than do heavily textured surfaces which are subsequently painted or surfaces which are to be decorated with heavy grade wallcoverings.

The following descriptions were developed jointly by the Association of the Wall and Ceiling Industries-International (AWCI), Ceiling & Interior Systems Construction Association (CISCA), Gypsum Association (GA), and Painting and Decorating Contractors of America (PDCA) as a guide.

Level 0: No taping, finishing, or accessories required. This level of finish may be useful in temporary construction or whenever the final decoration has not been determined.

Level 1: All joints and interior angles shall have tape set in joint compound. Surface shall be free of excess joint compound. Tool marks and ridges are acceptable. Frequently specified in plenum areas above ceilings, in attics, in areas where the assembly would generally be concealed or in building service corridors, and other areas not normally open to public view.

Level 2: All joints and interior angles shall have tape embedded in joint compound and wiped with a joint knife leaving a thin coating of joint compound over all joints and interior angles. Fastener heads and accessories shall be covered with a coat of joint compound. Surface shall be free of excess joint compound. Tool marks and ridges are acceptable. Joint compound applied over the body of the tape at the time of tape embedment shall be considered a separate coat of joint compound and shall satisfy the conditions of this level. Specified where water-resistant gypsum backing board is used as a substrate for tile; may be specified in garages, warehouse storage, or other similar areas where surface appearance is not of primary concern.

Level 3: All joints and interior angles shall have tape embedded in joint compound and one additional coat of joint compound applied over all joints and interior angles. Fastener heads and accessories shall be covered with two separate coats of joint compound. All joint compound shall be smooth and free of tool marks and ridges. Typically specified in appearance areas which are to receive heavy- or medium-texture (spray or hand applied) finishes before final painting, or where heavy-grade wallcoverings are to be applied as the final decoration. This level of finish is not recommended where smooth painted surfaces or light to medium wallcoverings are specified.

Level 4: All joints and interior angles shall have tape embedded in joint compound and two separate coats of joint compound applied over all flat joints and one separate coat of joint compound applied over interior angles. Fastener heads and accessories shall be covered with three separate coats of joint compound. All joint compound shall be smooth and free of tool marks and ridges. This level should be specified where flat paints, light textures, or wallcoverings are to be applied. In critical lighting areas, flat paints applied over light textures tend to reduce joint photographing. Gloss, semi-gloss, and enamel paints are not recommended over this level of finish. The weight, texture, and sheen level of wallcoverings applied over this level of finish should be carefully evaluated. Joints and fasteners must be adequately concealed if the wallcovering material is lightweight, contains limited pattern, has a gloss finish, or any combination of these finishes is present. Unbacked vinyl wallcoverings are not recommended over this level of finish.

Level 5: All joints and interior angles shall have tape embedded in joint compound and two separate coats of joint compound applied over all flat joints and one separate coat of joint compound applied over interior angles. Fastener heads and accessories shall be covered with three separate coats of joint compound. A thin skim coat of joint compound or a material manufactured especially for this purpose, shall be applied to the entire surface. The surface shall be smooth and free of tool marks and ridges. This level of finish is highly recommended where gloss, semi-gloss, enamel, or nontextured flat paints are specified or where severe lighting conditions occur. This highest quality finish is the most effective method to provide a uniform surface and minimize the possibility of joint photographing and of fasteners showing through the final decoration.

Finishes — R0966 Terrazzo Flooring

R096613-10 Terrazzo Floor

The table below lists quantities required for 100 S.F. of 5/8" terrazzo topping, either bonded or not bonded.

Description	Bonded to Concrete 1-1/8" Bed, 1:4 Mix	Not Bonded 2-1/8" Bed and 1/4" Sand
Portland cement, 94 lb. Bag	6 bags	8 bags
Sand	10 C.F.	20 C.F.
Divider strips, 4' squares	50 L.F.	50 L.F.
Terrazzo fill, 50 lb. Bag	12 bags	12 bags
15 Lb. tarred felt		1 C.S.F.
Mesh 2 x 2 #14 galvanized		1 C.S.F.
Crew J-3	0.77 days	0.87 days

2' x 2' panels require 1.00 L.F. divider strip per S.F.
3' x 3' panels require 0.67 L.F. divider strip per S.F.
4' x 4' panels require 0.50 L.F. divider strip per S.F.
5' x 5' panels require 0.40 L.F. divider strip per S.F.
6' x 6' panels require 0.33 L.F. divider strip per S.F.

Finishes — R0972 Wall Coverings

R097223-10 Wall Covering

The table below lists the quantities required for 100 S.F. of wall covering.

Description	Medium-Priced Paper	Expensive Paper
Paper	1.6 dbl. rolls	1.6 dbl. rolls
Wall sizing	0.25 gallon	0.25 gallon
Vinyl wall paste	0.6 gallon	0.6 gallon
Apply sizing	0.3 hour	0.3 hour
Apply paper	1.2 hours	1.5 hours

Most wallpapers now come in double rolls only.
To remove old paper, allow 1.3 hours per 100 S.F.

Finishes — R0991 Painting

R099100-10 Painting Estimating Techniques

Proper estimating methodology is needed to obtain an accurate painting estimate. There is no known reliable shortcut or square foot method. The following steps should be followed:

- List all surfaces to be painted, with an accurate quantity (area) of each. Items having similar surface condition, finish, application method and accessibility may be grouped together.
- List all the tasks required for each surface to be painted, including surface preparation, masking, and protection of adjacent surfaces. Surface preparation may include minor repairs, washing, sanding and puttying.
- Select the proper Means line for each task. Review and consider all adjustments to labor and materials for type of paint and location of work. Apply the height adjustment carefully. For instance, when applying the adjustment for work over 8' high to a wall that is 12' high, apply the adjustment only to the area between 8' and 12' high, and not to the entire wall.

When applying more than one percent (%) adjustment, apply each to the base cost of the data, rather than applying one percentage adjustment on top of the other.

When estimating the cost of painting walls and ceilings remember to add the brushwork for all cut-ins at inside corners and around windows and doors as a LF measure. One linear foot of cut-in with brush equals one square foot of painting.

All items for spray painting include the labor for roll-back.

Deduct for openings greater than 100 SF, or openings that extend from floor to ceiling and are greater than 5' wide. Do not deduct small openings.

The cost of brushes, rollers, ladders and spray equipment are considered to be part of a painting contractor's overhead, and should not be added to the estimate. The cost of rented equipment such as scaffolding and swing staging should be added to the estimate.

R099100-20 Painting

Item	Coat	One Gallon Covers			In 8 Hours a Laborer Covers			Labor-Hours per 100 S.F.		
		Brush	Roller	Spray	Brush	Roller	Spray	Brush	Roller	Spray
Paint wood siding	prime	250 S.F.	225 S.F.	290 S.F.	1150 S.F.	1300 S.F.	2275 S.F.	.695	.615	.351
	others	270	250	290	1300	1625	2600	.615	.492	.307
Paint exterior trim	prime	400	—	—	650	—	—	1.230	—	—
	1st	475	—	—	800	—	—	1.000	—	—
	2nd	520	—	—	975	—	—	.820	—	—
Paint shingle siding	prime	270	255	300	650	975	1950	1.230	.820	.410
	others	360	340	380	800	1150	2275	1.000	.695	.351
Stain shingle siding	1st	180	170	200	750	1125	2250	1.068	.711	.355
	2nd	270	250	290	900	1325	2600	.888	.603	.307
Paint brick masonry	prime	180	135	160	750	800	1800	1.066	1.000	.444
	1st	270	225	290	815	975	2275	.981	.820	.351
	2nd	340	305	360	815	1150	2925	.981	.695	.273
Paint interior plaster or drywall	prime	400	380	495	1150	2000	3250	.695	.400	.246
	others	450	425	495	1300	2300	4000	.615	.347	.200
Paint interior doors and windows	prime	400	—	—	650	—	—	1.230	—	—
	1st	425	—	—	800	—	—	1.000	—	—
	2nd	450	—	—	975	—	—	.820	—	—

Conveying Equipment — R1420 Elevators

R142000-10 Freight Elevators

Capacities run from 2,000 lbs. to over 100,000 lbs. with 3,000 lbs. to 10,000 lbs. most common. Travel speeds are generally lower and control less intricate than on passenger elevators. Costs in the Unit Price section are for hydraulic and geared elevators.

R142000-20 Elevator Selective Costs   See R142000-40 for cost development.

	Passenger		Freight		Hospital	
A. Base Unit	Hydraulic	Electric	Hydraulic	Electric	Hydraulic	Electric
Capacity	1,500 lb.	2,000 lb.	2,000 lb.	4,000 lb.	4,000 lb.	4,000 lb.
Speed	100 F.P.M.	200 F.P.M.	100 F.P.M.	200 F.P.M.	100 F.P.M.	200 F.P.M.
#Stops/Travel Ft.	2/12	4/40	2/20	4/40	2/20	4/40
Push Button Oper.	Yes	Yes	Yes	Yes	Yes	Yes
Telephone Box & Wire	"	"	"	"	"	"
Emergency Lighting	"	"	No	No	"	"
Cab	Plastic Lam. Walls	Plastic Lam. Walls	Painted Steel	Painted Steel	Plastic Lam. Walls	Plastic Lam. Walls
Cove Lighting	Yes	Yes	No	No	Yes	Yes
Floor	V.C.T.	V.C.T.	Wood w/Safety Treads	Wood w/Safety Treads	V.C.T.	V.C.T.
Doors, & Speedside Slide	Yes	Yes	Yes	Yes	Yes	Yes
Gates, Manual	No	No	No	No	No	No
Signals, Lighted Buttons	Car and Hall	Car and Hall	Car and Hall	Car and Hall	Car and Hall	Car and Hall
O.H. Geared Machine	N.A.	Yes	N.A.	Yes	N.A.	Yes
Variable Voltage Contr.	"	"	N.A.	"	"	"
Emergency Alarm	Yes	"	Yes	"	Yes	"
Class "A" Loading	N.A.	N.A.	"	"	N.A.	N.A.

R142000-30 Passenger Elevators

Electric elevators are used generally but hydraulic elevators can be used for lifts up to 70' and where large capacities are required. Hydraulic speeds are limited to 200 F.P.M. but cars are self leveling at the stops. On low rises, hydraulic installation runs about 15% less than standard electric types but on higher rises this installation cost advantage is reduced. Maintenance of hydraulic elevators is about the same as electric type but underground portion is not included in the maintenance contract.

In electric elevators there are several control systems available, the choice of which will be based upon elevator use, size, speed and cost criteria. The two types of drives are geared for low speeds and gearless for 450 F.P.M. and over.

The tables on the preceding pages illustrate typical installed costs of the various types of elevators available.

R142000-40 Elevator Cost Development

To price a new car or truck from the factory, you must start with the manufacturer's basic model, then add or exchange optional equipment and features. The same is true for pricing elevators.

Requirement: One-passenger elevator, five-story hydraulic, 2,500 lb. capacity, 12' floor to floor, speed 150 F.P.M., emergency power switching and maintenance contract.

Example:

Description	Adjustment
A. Base Elevator: Hydraulic Passenger, 1500 lb. Capacity, 100 fpm, 2 Stops, Standard Finish	1 Ea.
B. Capacity Adjustment (2,500 lb.)	1 Ea.
C. Excess Travel Adjustment: 48' Total Travel (4 x 12') minus 12' Base Unit Travel =	36 V.L.F.
D. Stops Adjustment: 5 Total Stops minus 2 Stops (Base Unit) =	3 Stops
E. Speed Adjustment (150 F.P.M.)	1 Ea.
F. Options:	
1. Intercom Service	1 Ea.
2. Emergency Power Switching, Automatic	1 Ea.
3. Stainless Steel Entrance Doors	5 Ea.
4. Maintenance Contract (12 Months)	1 Ea.
5. Position Indicator for main floor level (none indicated in Base Unit)	1 Ea.

Conveying Equipment — R1431 Escalators

R143110-10 Escalators

Moving stairs can be used for buildings where 600 or more people are to be carried to the second floor or beyond. Freight cannot be carried on escalators and at least one elevator must be available for this function.

Carrying capacity is 5,000 to 8,000 people per hour. Power requirement is 2 to 3 KW per hour and incline angle is 30°.

Conveying Equipment — R1432 Moving Walks

R143210-20 Moving Ramps and Walks

These are a specialized form of conveyor 3' to 6' wide with capacities of 3,600 to 18,000 persons per hour. Maximum speed is 140 F.P.M. and normal incline is 0° to 15°.

Local codes will determine the maximum angle. Outdoor units would require additional weather protection.

Fire Suppression   R2113 Fire-Suppression Sprinkler Systems

R211313-10   Sprinkler Systems (Automatic)

Sprinkler systems may be classified by type as follows:

1. **Wet Pipe System.** A system employing automatic sprinklers attached to a piping system containing water and connected to a water supply so that water discharges immediately from sprinklers opened by a fire.
2. **Dry Pipe System.** A system employing automatic sprinklers attached to a piping system containing air under pressure, the release of which as from the opening of sprinklers permits the water pressure to open a valve known as a "dry pipe valve". The water then flows into the piping system and out the opened sprinklers.
3. **Pre-Action System.** A system employing automatic sprinklers attached to a piping system containing air that may or may not be under pressure, with a supplemental heat responsive system of generally more sensitive characteristics than the automatic sprinklers themselves, installed in the same areas as the sprinklers; actuation of the heat responsive system, as from a fire, opens a valve which permits water to flow into the sprinkler piping system and to be discharged from any sprinklers which may be open.
4. **Deluge System.** A system employing open sprinklers attached to a piping system connected to a water supply through a valve which is opened by the operation of a heat responsive system installed in the same areas as the sprinklers. When this valve opens, water flows into the piping system and discharges from all sprinklers attached thereto.
5. **Combined Dry Pipe and Pre-Action Sprinkler System.** A system employing automatic sprinklers attached to a piping system containing air under pressure with a supplemental heat responsive system of generally more sensitive characteristics than the automatic sprinklers themselves, installed in the same areas as the sprinklers; operation of the heat responsive system, as from a fire, actuates tripping devices which open dry pipe valves simultaneously and without loss of air pressure in the system. Operation of the heat responsive system also opens approved air exhaust valves at the end of the feed main which facilitates the filling of the system with water which usually precedes the opening of sprinklers. The heat responsive system also serves as an automatic fire alarm system.
6. **Limited Water Supply System.** A system employing automatic sprinklers and conforming to these standards but supplied by a pressure tank of limited capacity.
7. **Chemical Systems.** Systems using halon, carbon dioxide, dry chemical or high expansion foam as selected for special requirements. Agent may extinguish flames by chemically inhibiting flame propagation, suffocate flames by excluding oxygen, interrupting chemical action of oxygen uniting with fuel or sealing and cooling the combustion center.
8. **Firecycle System.** Firecycle is a fixed fire protection sprinkler system utilizing water as its extinguishing agent. It is a time delayed, recycling, preaction type which automatically shuts the water off when heat is reduced below the detector operating temperature and turns the water back on when that temperature is exceeded. The system senses a fire condition through a closed circuit electrical detector system which controls water flow to the fire automatically. Batteries supply up to 90 hour emergency power supply for system operation. The piping system is dry (until water is required) and is monitored with pressurized air. Should any leak in the system piping occur, an alarm will sound, but water will not enter the system until heat is sensed by a firecycle detector.

Area coverage sprinkler systems may be laid out and fed from the supply in any one of several patterns as shown below. It is desirable, if possible, to utilize a central feed and achieve a shorter flow path from the riser to the furthest sprinkler. This permits use of the smallest sizes of pipe possible with resulting savings.

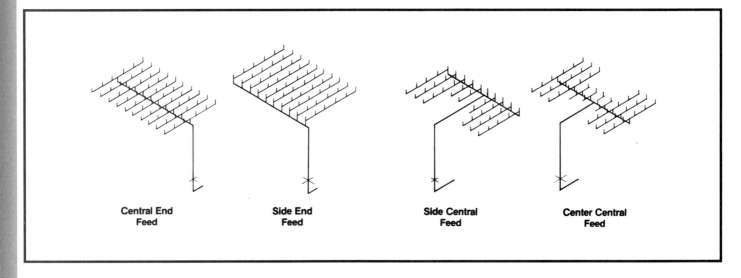

Central End Feed   Side End Feed   Side Central Feed   Center Central Feed

Fire Suppression — R2113 Fire-Suppression Sprinkler Systems

R211313-20 System Classification

System Classification
Rules for installation of sprinkler systems vary depending on the classification of occupancy falling into one of three categories as follows:

Light Hazard Occupancy
The protection area allotted per sprinkler should not exceed 225 S.F., with the maximum distance between lines and sprinklers on lines being 15'. The sprinklers do not need to be staggered. Branch lines should not exceed eight sprinklers on either side of a cross main. Each large area requiring more than 100 sprinklers and without a sub-dividing partition should be supplied by feed mains or risers sized for ordinary hazard occupancy.
Maximum system area = 52,000 S.F.

Included in this group are:
- Churches
- Clubs
- Educational
- Hospitals
- Institutional
- Libraries (except large stack rooms)
- Museums
- Nursing Homes
- Offices
- Residential
- Restaurants
- Theaters and Auditoriums (except stages and prosceniums)
- Unused Attics

Ordinary Hazard Occupancy
The protection area allotted per sprinkler shall not exceed 130 S.F. of noncombustible ceiling and 130 S.F. of combustible ceiling. The maximum allowable distance between sprinkler lines and sprinklers on line is 15'. Sprinklers shall be staggered if the distance between heads exceeds 12'. Branch lines should not exceed eight sprinklers on either side of a cross main.
Maximum system area = 52,000 S.F.

Included in this group are:

Group 1
- Automotive Parking and Showrooms
- Bakeries
- Beverage manufacturing
- Canneries
- Dairy Products Manufacturing/Processing
- Electronic Plans
- Glass and Glass Products Manufacturing
- Laundries
- Restaurant Service Areas

Group 2
- Cereal Mills
- Chemical Plants—Ordinary
- Confectionery Products
- Distilleries
- Dry Cleaners
- Feed Mills
- Horse Stables
- Leather Goods Manufacturing
- Libraries—Large Stack Room Areas
- Machine Shops
- Metal Working
- Mercantile
- Paper and Pulp Mills
- Paper Process Plants
- Piers and Wharves
- Post Offices
- Printing and Publishing
- Repair Garages
- Stages
- Textile Manufacturing
- Tire Manufacturing
- Tobacco Products Manufacturing
- Wood Machining
- Wood Product Assembly

Extra Hazard Occupancy
The protection area allotted per sprinkler shall not exceed 100 S.F. of noncombustible ceiling and 100 S.F. of combustible ceiling. The maximum allowable distance between lines and between sprinklers on lines is 12'. Sprinklers on alternate lines shall be staggered if the distance between sprinklers on lines exceeds 8'. Branch lines should not exceed six sprinklers on either side of a cross main.
Maximum system area:
  Design by pipe schedule = 25,000 S.F.
  Design by hydraulic calculation = 40,000 S.F.

Included in this group are:

Group 1
- Aircraft hangars
- Combustible Hydraulic Fluid Use Area
- Die Casting
- Metal Extruding
- Plywood/Particle Board Manufacturing
- Printing (inks with flash points < 100 degrees F)
- Rubber Reclaiming, Compounding, Drying, Milling, Vulcanizing
- Saw Mills
- Textile Picking, Opening, Blending, Garnetting, Carding, Combing of Cotton, Synthetics, Wood Shoddy, or Burlap
- Upholstering with Plastic Foams

Group 2
- Asphalt Saturating
- Flammable Liquids Spraying
- Flow Coating
- Manufactured/Modular Home Building Assemblies (where finished enclosure is present and has combustible interiors)
- Open Oil Quenching
- Plastics Processing
- Solvent Cleaning
- Varnish and Paint Dipping

Plumbing — R2201 Operation & Maintenance of Plumbing

R220105-10 Demolition (Selective vs. Removal for Replacement)

Demolition can be divided into two basic categories.

One type of demolition involves the removal of material with no concern for its replacement. The labor-hours to estimate this work are found under "Selective Demolition" in the Fire Protection, Plumbing and HVAC Divisions. It is selective in that individual items or all the material installed as a system or trade grouping such as plumbing or heating systems are removed. This may be accomplished by the easiest way possible, such as sawing, torch cutting, or sledge hammer as well as simple unbolting.

The second type of demolition is the removal of some item for repair or replacement. This removal may involve careful draining, opening of unions, disconnecting and tagging of electrical connections, capping of pipes/ducts to prevent entry of debris or leakage of the material contained as well as transport of the item away from its in-place location to a truck/dumpster. An approximation of the time required to accomplish this type of demolition is to use half of the time indicated as necessary to install a new unit. For example; installation of a new pump might be listed as requiring 6 labor-hours so if we had to estimate the removal of the old pump we would allow an additional 3 hours for a total of 9 hours. That is, the complete replacement of a defective pump with a new pump would be estimated to take 9 labor-hours.

Plumbing

R2240 Plumbing Fixtures

R224000-30 Minimum Plumbing Fixture Requirements

Minimum Plumbing Fixture Requirements

Type of Building or Occupancy (2)	Water Closets (14) (Fixtures per Person) Male	Water Closets (14) (Fixtures per Person) Female	Urinals (5,10) (Fixtures per Person) Male	Urinals (5,10) (Fixtures per Person) Female	Lavatories (Fixtures per Person) Male	Lavatories (Fixtures per Person) Female	Bathtubs or Showers (Fixtures per Person)	Drinking Fountains (Fixtures per Person) (3, 13)
Assembly Places- Theatres, Auditoriums, Convention Halls, etc.-for permanent employee use	1: 1 - 15 2: 16 - 35 3: 36 - 55 Over 55, add 1 fixture for each additional 40 persons	1: 1 - 15 2: 16 - 35 3: 36 - 55	0: 1 - 9 1: 10 - 50 Add one fixture for each additional 50 males		1 per 40	1 per 40		
Assembly Places- Theatres, Auditoriums, Convention Halls, etc. - for public use	1: 1 - 100 2: 101 - 200 3: 201 - 400 Over 400, add 1 fixture for each additional 500 males and 1 for each additional 125 females	3: 1 - 50 4: 51 - 100 8: 101 - 200 11: 201 - 400	1: 1 - 100 2: 101 - 200 3: 201 - 400 4: 401 - 600 Over 600, add 1 fixture for each additional 300 males		1: 1 - 200 2: 201 - 400 3: 401 - 750 Over 750, add 1 fixture for each additional 500 persons	1: 1 - 200 2: 201 - 400 3: 401 - 750		1: 1 - 150 2: 151 - 400 3: 401 - 750 Over 750, add one fixture for each additional 500 persons
Dormitories (9) School or Labor	1 per 10 Add 1 fixture for each additional 25 males (over 10) and 1 for each additional 20 females (over 8)	1 per 8	1 per 25 Over 150, add 1 fixture for each additional 50 males		1 per 12 Over 12 add 1 fixture for each additional 20 males and 1 for each 15 additional females	1 per 12	1 per 8 For females add 1 bathtub per 30. Over 150, add 1 per 20	1 per 150 (12)
Dormitories- for Staff Use	1: 1 - 15 2: 16 - 35 3: 36 - 55 Over 55, add 1 fixture for each additional 40 persons	1: 1 - 15 3: 16 - 35 4: 36 - 55	1 per 50		1 per 40	1 per 40	1 per 8	
Dwellings: Single Dwelling Multiple Dwelling or Apartment House	1 per dwelling 1 per dwelling or apartment unit				1 per dwelling 1 per dwelling or apartment unit		1 per dwelling 1 per dwelling or apartment unit	
Hospital Waiting rooms	1 per room				1 per room			1 per 150 (12)
Hospitals- for employee use	1: 1 - 15 2: 16 - 35 3: 36 - 55 Over 55, add 1 fixture for each additional 40 persons	1: 1 - 15 3: 16 - 35 4: 36 - 55	0: 1 - 9 1: 10 - 50 Add 1 fixture for each additional 50 males		1 per 40	1 per 40		
Hospitals: Individual Room Ward Room	1 per room 1 per 8 patients				1 per room 1 per 10 patients		1 per room 1 per 20 patients	1 per 150 (12)
Industrial (6) Warehouses Workshops, Foundries and similar establishments- for employee use	1: 1 -10 2: 11 - 25 3: 26 - 50 4: 51 - 75 5: 76 - 100 Over 100, add 1 fixture for each additional 30 persons	1: 1 -10 2: 11 - 25 3: 26 - 50 4: 51 - 75 5: 76 - 100			Up to 100, per 10 persons Over 100, 1 per 15 persons (7, 8)		1 shower for each 15 persons exposed to excessive heat or to skin contamination with poisonous, infectious or irritating material	1 per 150 (12)
Institutional - Other than Hospitals or Penal Institutions (on each occupied floor)	1 per 25	1 per 20	0: 1 - 9 1: 10 - 50 Add 1 fixture for each additional 50 males		1 per 10	1 per 10	1 per 8	1 per 150 (12)
Institutional - Other than Hospitals or Penal Institutions (on each occupied floor)- for employee use	1: 1 - 15 2: 16 - 35 3: 36 - 55 Over 55, add 1 fixture for each additional 40 persons	1: 1 - 15 3: 16 - 35 4: 36 - 55	0: 1 - 9 1: 10 - 50 Add 1 fixture for each additional 50 males		1 per 40	1 per 40	1 per 8	1 per 150 (12)
Office or Public Buildings	1: 1 - 100 2: 101 - 200 3: 201 - 400 Over 400, add 1 fixture for each additional 500 males and 1 for each additional 150 females	3: 1 - 50 4: 51 - 100 8: 101 - 200 11: 201 - 400	1: 1 - 100 2: 101 - 200 3: 201 - 400 4: 401 - 600 Over 600, add 1 fixture for each additional 300 males		1: 1 - 200 2: 201 - 400 3: 401 - 750 Over 750, add 1 fixture for each additional 500 persons	1: 1 - 200 2: 201 - 400 3: 401 - 750		1 per 150 (12)
Office or Public Buildings - for employee use	1: 1 - 15 2: 16 - 35 3: 36 - 55 Over 55, add 1 fixture for each additional 40 persons	1: 1 - 15 3: 16 - 35 4: 36 - 55	0: 1 - 9 1: 10 - 50 Add 1 fixture for each additional 50 males		1 per 40	1 per 40		

Plumbing — R2240 Plumbing Fixtures

R224000-30 Minimum Plumbing Fixture Requirements (cont.)

Minimum Plumbing Fixture Requirements

Type of Building or Occupancy	Water Closets (14) (Fixtures per Person)		Urinals (5, 10) (Fixtures per Person)		Lavatories (Fixtures per Person)		Bathtubs or Showers (Fixtures per Person)	Drinking Fountains (Fixtures per Person) (3, 13)
	Male	Female	Male	Female	Male	Female		
Penal Institutions - for employee use	1: 1 - 15 2: 16 - 35 3: 36 - 55 Over 55, add 1 fixture for each additional 40 persons	1: 1 - 15 3: 16 - 35 4: 36 - 55	0: 1 - 9 1: 10 - 50 Add 1 fixture for each additional 50 males		1 per 40	1 per 40		1 per 150 (12)
Penal Institutions - for prison use Cell	1 per cell				1 per cell			1 per cellblock floor
Exercise room	1 per exercise room		1 per exercise room		1 per exercise room			1 per exercise room
Restaurants, Pubs and Lounges (11)	1: 1 - 50 2: 51 - 150 3: 151 - 300 Over 300, add 1 fixture for each additional 200 persons	1: 1 - 50 2: 51 - 150 4: 151 - 300	1: 1 - 150 Over 150, add 1 fixture for each additional 150 males		1: 1 - 150 2: 151 - 200 3: 201 - 400 Over 400, add 1 fixture for each additional 400 persons	1: 1 - 150 2: 151 - 200 3: 201 - 400		
Schools - for staff use All Schools	1: 1 - 15 2: 16 - 35 3: 36 - 55 Over 55, add 1 fixture for each additional 40 persons	1: 1 - 15 3: 16 - 35 4: 36 - 55	1 per 50		1 per 40	1 per 40		
Schools - for student use: Nursery	1: 1 - 20 2: 21 - 50 Over 50, add 1 fixture for each additional 50 persons	1: 1 - 20 2: 21 - 50			1: 1 - 25 2: 26 - 50 Over 50, add 1 fixture for each additional 50 persons	1: 1 - 25 2: 26 - 50		1 per 150 (12)
Elementary	1 per 30	1 per 25	1 per 75		1 per 35	1 per 35		1 per 150 (12)
Secondary	1 per 40	1 per 30	1 per 35		1 per 40	1 per 40		1 per 150 (12)
Others (Colleges, Universities, Adult Centers, etc.)	1 per 40	1 per 30	1 per 35		1 per 40	1 per 40		1 per 150 (12)
Worship Places Educational and Activities Unit	1 per 150	1 per 75	1 per 150		1 per 2 water closets			1 per 150 (12)
Worship Places Principal Assembly Place	1 per 150	1 per 75	1 per 150		1 per 2 water closets			1 per 150 (12)

Notes:
1. The figures shown are based upon one (1) fixture being the minimum required for the number of persons indicated or any fraction thereof.
2. Building categories not shown on this table shall be considered separately by the Administrative Authority.
3. Drinking fountains shall not be installed in toilet rooms.
4. Laundry trays. One (1) laundry tray or one (1) automatic washer standpipe for each dwelling unit or one (1) laundry trays or one (1) automatic washer standpipes, or combination thereof, for each twelve (12) apartments. Kitchen sinks, one (1) for each dwelling or apartment unit.
5. For each urinal added in excess of the minimum required, one water closet may be deducted. The number of water closets shall not be reduced to less than two-thirds (2/3) of the minimum requirement.
6. As required by ANSI Z4.1-1968, Sanitation in Places of Employment.
7. Where there is exposure to skin contamination with poisonous, infectious, or irritating materials, provide one (1) lavatory for each five (5) persons.
8. Twenty-four (24) lineal inches of wash sink or eighteen (18) inches of a circular basin, when provided with water outlets for such space shall be considered equivalent to one (1) lavatory.
9. Laundry trays, one (1) for each fifty (50) persons. Service sinks, one (1) for each hundred (100) persons.
10. General. In applying this schedule of facilities, consideration shall be given to the accessibility of the fixtures. Conformity purely on a numerical basis may not result in an installation suited to the need of the individual establishment. For example, schools should be provided with toilet facilities on each floor having classrooms.
    a. Surrounding materials, wall and floor space to a point two (2) feet in front of urinal lip and four (4) feet above the floor, and at least two (2) feet to each side of the urinal shall be lined with non-absorbent materials.
    b. Trough urinals shall be prohibited.
11. A restaurant is defined as a business which sells food to be consumed on the premises.
    a. The number of occupants for a drive-in restaurant shall be considered as equal to the number of parking stalls.
    b. Employee toilet facilities shall not to be included in the above restaurant requirements. Hand washing facilities shall be available in the kitchen for employees.
12. Where food is consumed indoors, water stations may be substituted for drinking fountains. Offices, or public buildings for use by more than six (6) persons shall have one (1) drinking fountain for the first one hundred fifty (150) persons and one additional fountain for each three hundred (300) persons thereafter.
13. There shall be a minimum of one (1) drinking fountain per occupied floor in schools, theaters, auditoriums, dormitories, offices of public building.
14. The total number of water closets for females shall be at least equal to the total number of water closets and urinals required for males.

Plumbing — R2240 Plumbing Fixtures

R224000-40 Plumbing Fixture Installation Time

Item	Rough-In	Set	Total Hours	Item	Rough-In	Set	Total Hours
Bathtub	5	5	10	Shower head only	2	1	3
Bathtub and shower, cast iron	6	6	12	Shower drain	3	1	4
Fire hose reel and cabinet	4	2	6	Shower stall, slate		15	15
Floor drain to 4 inch diameter	3	1	4	Slop sink	5	3	8
Grease trap, single, cast iron	5	3	8	Test 6 fixtures			14
Kitchen gas range		4	4	Urinal, wall	6	2	8
Kitchen sink, single	4	4	8	Urinal, pedestal or floor	6	4	10
Kitchen sink, double	6	6	12	Water closet and tank	4	3	7
Laundry tubs	4	2	6	Water closet and tank, wall hung	5	3	8
Lavatory wall hung	5	3	8	Water heater, 45 gals. gas, automatic	5	2	7
Lavatory pedestal	5	3	8	Water heaters, 65 gals. gas, automatic	5	2	7
Shower and stall	6	4	10	Water heaters, electric, plumbing only	4	2	6

Fixture prices in front of book are based on the cost per fixture set in place. The rough-in cost, which must be added for each fixture, includes carrier, if required, some supply, waste and vent pipe connecting fittings and stops. The lengths of rough-in pipe are nominal runs which would connect to the larger runs and stacks. The supply runs and DWV runs and stacks must be accounted for in separate entries. In the eastern half of the United States it is common for the plumber to carry these to a point 5′ outside the building.

R224000-50 Water Cooler Application

Type of Service	Requirement
Office, School or Hospital	12 persons per gallon per hour
Office, Lobby or Department Store	4 or 5 gallons per hour per fountain
Light manufacturing	7 persons per gallon per hour
Heavy manufacturing	5 persons per gallon per hour
Hot heavy manufacturing	4 persons per gallon per hour
Hotel	.08 gallons per hour per room
Theatre	1 gallon per hour per 100 seats

Heating, Ventilating & A.C. — R2334 HVAC Fans

R233400-10 Recommended Ventilation Air Changes

Table below lists range of time in minutes per change for various types of facilities.

Facility	Min/Change	Facility	Min/Change	Facility	Min/Change
Assembly Halls	2-10	Dance Halls	2-10	Laundries	1-3
Auditoriums	2-10	Dining Rooms	3-10	Markets	2-10
Bakeries	2-3	Dry Cleaners	1-5	Offices	2-10
Banks	3-10	Factories	2-5	Pool Rooms	2-5
Bars	2-5	Garages	2-10	Recreation Rooms	2-10
Beauty Parlors	2-5	Generator Rooms	2-5	Sales Rooms	2-10
Boiler Rooms	1-5	Gymnasiums	2-10	Theaters	2-8
Bowling Alleys	2-10	Kitchens-Hospitals	2-5	Toilets	2-5
Churches	5-10	Kitchens-Restaurant	1-3	Transformer Rooms	1-5

CFM air required for changes = Volume of room in cubic feet ÷ Minutes per change.

Heating, Ventilating & A.C. R2350 Central Heating Equipment

R235000-30 The Basics of a Heating System

The function of a heating system is to achieve and maintain a desired temperature in a room or building by replacing the amount of heat being dissipated. There are four kinds of heating systems: hot-water, steam, warm-air and electric resistance. Each has certain essential and similar elements with the exception of electric resistance heating.

The basic elements of a heating system are:

A. A **combustion chamber** in which fuel is burned and heat transferred to a conveying medium.

B. The **"fluid"** used for conveying the heat (water, steam or air).

C. **Conductors** or pipes for transporting the fluid to specific desired locations.

D. A means of disseminating the heat, sometimes called **terminal units**.

A. The **combustion chamber** in a furnace heats air which is then distributed. This is called a warm-air system.

The combustion chamber in a boiler heats water which is either distributed as hot water or steam and this is termed a hydronic system.

The maximum allowable working pressures are limited by ASME "Code for Heating Boilers" to 15 PSI for steam and 160 PSI for hot water heating boilers, with a maximum temperature limitation of 250° F. Hot water boilers are generally rated for a working pressure of 30 PSI. High pressure boilers are governed by the ASME "Code for Power Boilers" which is used almost universally for boilers operating over 15 PSIG. High pressure boilers used for a combination of heating/process loads are usually designed for 150 PSIG.

Boiler ratings are usually indicated as either Gross or Net Output. The Gross Load is equal to the Net Load plus a piping and pickup allowance. When this allowance cannot be determined, divide the gross output rating by 1.25 for a value equal to or greater than the net heat loss requirement of the building.

B. Of the three **fluids** used, steam carries the greatest amount of heat per unit volume. This is due to the fact that it gives up its latent heat of vaporization at a temperature considerably above room temperature. Another advantage is that the pressure to produce a positive circulation is readily available. Piping conducts the steam to terminal units and returns condensate to the boiler.

The **steam system** is well adapted to large buildings because of its positive circulation, its comparatively economical installation and its ability to deliver large quantities of heat. Nearly all large office buildings, stores, hotels, and industrial buildings are so heated, in addition to many residences.

Hot water, when used as the heat carrying fluid, gives up a portion of its sensible heat and then returns to the boiler or heating apparatus for reheating. As the heat conveyed by each pound of water is about one-fiftieth of the heat conveyed by a pound of steam, it is necessary to circulate about fifty times as much water as steam by weight (although only one-thirtieth as much by volume). The hot water system is usually, although not necessarily, designed to operate at temperatures below that of the ordinary steam system and so the amount of heat transfer surface must be correspondingly greater. A temperature of 190° F to 200° F is normally the maximum. Circulation in small buildings may depend on the difference in density between hot water and the cool water returning to the boiler; circulating pumps are normally used to maintain a desired rate of flow. Pumps permit a greater degree of flexibility and better control.

In **warm-air** furnace systems, cool air is taken from one or more points in the building, passed over the combustion chamber and flue gas passages and then distributed through a duct system. A disadvantage of this system is that the ducts take up much more building volume than steam or hot water pipes. Advantages of this system are the relative ease with which humidification can be accomplished by the evaporation of water as the air circulates through the heater, and the lack of need for expensive disseminating units as the warm air simply becomes part of the interior atmosphere of the building.

C. Conductors (pipes and ducts) have been lightly treated in the discussion of conveying fluids. For more detailed information such as sizing and distribution methods, the reader is referred to technical publications such as the American Society of Heating, Refrigerating and Air-Conditioning Engineers "Handbook of Fundamentals."

D. Terminal units come in an almost infinite variety of sizes and styles, but the basic principles of operation are very limited. As previously mentioned, warm-air systems require only a simple register or diffuser to mix heated air with that present in the room. Special application items such as radiant coils and infrared heaters are available to meet particular conditions but are not usually considered for general heating needs. Most heating is accomplished by having air flow over coils or pipes containing the heat transporting medium (steam, hot-water, electricity). These units, while varied, may be separated into two general types, (1) radiator/convectors and (2) unit heaters.

Radiator/convectors may be cast, fin-tube or pipe assemblies. They may be direct, indirect, exposed, concealed or mounted within a cabinet enclosure, upright or baseboard style. These units are often collectively referred to as "radiatiors" or "radiation" although none gives off heat either entirely by radiation or by convection but rather a combination of both. The air flows over the units as a gravity "current." It is necessary to have one or more heat-emitting units in each room. The most efficient placement is low along an outside wall or under a window to counteract the cold coming into the room and achieve an even distribution.

In contrast to radiator/convectors which operate most effectively against the walls of smaller rooms, **unit heaters** utilize a fan to move air over heating coils and are very effective in locations of relatively large volume. Unit heaters, while usually suspended overhead, may be floor mounted. They also may take in fresh outside air for ventilation. The heat distributed by unit heaters may be from a remote source and conveyed by a fluid or it may be from the combustion of fuel in each individual heater. In the latter case the only piping required would be for fuel, however, a vent for the products of combustion would be necessary.

The following list gives may of the advantages of unit heaters for applications other than office or residential:

a. Large capacity so smaller number of units are required, **b.** Piping system simplified, **c.** Space saved where they are located overhead out of the way, **d.** Rapid heating directed where needed with effective wide distribution, **e.** Difference between floor and ceiling temperature reduced, **f.** Circulation of air obtained, and ventilation with introduction of fresh air possible, **g.** Heat output flexible and easily controlled.

Heating, Ventilating & A.C. — R2360 Central Cooling Equipment

R236000-20 Air Conditioning Requirements

BTU's per hour per S.F. of floor area and S.F. per ton of air conditioning.

Type of Building	BTU/Hr per S.F.	S.F. per Ton	Type of Building	BTU/Hr per S.F.	S.F. per Ton	Type of Building	BTU per S.F.	S.F. per Ton
Apartments, Individual	26	450	Dormitory, Rooms	40	300	Libraries	50	240
Corridors	22	550	Corridors	30	400	Low Rise Office, Exterior	38	320
Auditoriums & Theaters	40	300/18*	Dress Shops	43	280	Interior	33	360
Banks	50	240	Drug Stores	80	150	Medical Centers	28	425
Barber Shops	48	250	Factories	40	300	Motels	28	425
Bars & Taverns	133	90	High Rise Office—Ext. Rms.	46	263	Office (small suite)	43	280
Beauty Parlors	66	180	Interior Rooms	37	325	Post Office, Individual Office	42	285
Bowling Alleys	68	175	Hospitals, Core	43	280	Central Area	46	260
Churches	36	330/20*	Perimeter	46	260	Residences	20	600
Cocktail Lounges	68	175	Hotel, Guest Rooms	44	275	Restaurants	60	200
Computer Rooms	141	85	Corridors	30	400	Schools & Colleges	46	260
Dental Offices	52	230	Public Spaces	55	220	Shoe Stores	55	220
Dept. Stores, Basement	34	350	Industrial Plants, Offices	38	320	Shop'g. Ctrs., Supermarkets	34	350
Main Floor	40	300	General Offices	34	350	Retail Stores	48	250
Upper Floor	30	400	Plant Areas	40	300	Specialty	60	200

*Persons per ton
12,000 BTU = 1 ton of air conditioning

Change Orders

Change Order Considerations
A Change Order is a written document, usually prepared by the design professional, and signed by the owner, the architect/engineer and the contractor. A change order states the agreement of the parties to: an addition, deletion, or revision in the work; an adjustment in the contract sum, if any; or an adjustment in the contract time, if any. Change orders, or "extras" in the construction process occur after execution of the construction contract and impact architects/engineers, contractors and owners.

Change orders that are properly recognized and managed can ensure orderly, professional and profitable progress for all who are involved in the project. There are many causes for change orders and change order requests. In all cases, change orders or change order requests should be addressed promptly and in a precise and prescribed manner. The following paragraphs include information regarding change order pricing and procedures.

The Causes of Change Orders
Reasons for issuing change orders include:
- Unforeseen field conditions that require a change in the work
- Correction of design discrepancies, errors or omissions in the contract documents
- Owner-requested changes, either by design criteria, scope of work, or project objectives
- Completion date changes for reasons unrelated to the construction process
- Changes in building code interpretations, or other public authority requirements that require a change in the work
- Changes in availability of existing or new materials and products

Procedures
Properly written contract documents must include the correct change order procedures for all parties—owners, design professionals and contractors—to follow in order to avoid costly delays and litigation.

Being "in the right" is not always a sufficient or acceptable defense. The contract provisions requiring notification and documentation must be adhered to within a defined or reasonable time frame.

The appropriate method of handling change orders is by a written proposal and acceptance by all parties involved. Prior to starting work on a project, all parties should identify their authorized agents who may sign and accept change orders, as well as any limits placed on their authority.

Time may be a critical factor when the need for a change arises. For such cases, the contractor might be directed to proceed on a "time and materials" basis, rather than wait for all paperwork to be processed—a delay that could impede progress. In this situation, the contractor must still follow the prescribed change order procedures, including but not limited to, notification and documentation.

All forms used for change orders should be dated and signed by the proper authority. Lack of documentation can be very costly, especially if legal judgments are to be made and if certain field personnel are no longer available. For time and material change orders, the contractor should keep accurate daily records of all labor and material allocated to the change. Forms that can be used to document change order work are available in *Means Forms for Building Construction Professionals*.

Owners or awarding authorities who do considerable and continual building construction (such as the federal government) realize the inevitability of change orders for numerous reasons, both predictable and unpredictable. As a result, the federal government, the American Institute of Architects (AIA), the Engineers Joint Contract Documents Committee (EJCDC) and other contractor, legal and technical organizations have developed standards and procedures to be followed by all parties to achieve contract continuance and timely completion, while being financially fair to all concerned.

In addition to the change order standards put forth by industry associations, there are also many books available on the subject.

Pricing Change Orders
When pricing change orders, regardless of their cause, the most significant factor is when the change occurs. The need for a change may be perceived in the field or requested by the architect/engineer *before* any of the actual installation has begun, or may evolve or appear *during* construction when the item of work in question is partially installed. In the latter cases, the original sequence of construction is disrupted, along with all contiguous and supporting systems. Change orders cause the greatest impact when they occur *after* the installation has been completed and must be uncovered, or even replaced. Post-completion changes may be caused by necessary design changes, product failure, or changes in the owner's requirements that are not discovered until the building or the systems begin to function.

Specified procedures of notification and record keeping must be adhered to and enforced regardless of the stage of construction: *before*, *during*, or *after* installation. Some bidding documents anticipate change orders by requiring that unit prices including overhead and profit percentages—for additional as well as deductible changes—be listed. Generally these unit prices do not fully take into account the ripple effect, or impact on other trades, and should be used for general guidance only.

When pricing change orders, it is important to classify the time frame in which the change occurs. There are two basic time frames for change orders: *pre-installation change orders*, which occur before the start of construction, and *post-installation change orders*, which involve reworking after the original installation. Change orders that occur between these stages may be priced according to the extent of work completed using a combination of techniques developed for pricing *pre-* and *post-installation* changes.

The following factors are the basis for a check list to use when preparing a change order estimate.

Factors To Consider When Pricing Change Orders

As an estimator begins to prepare a change order, the following questions should be reviewed to determine their impact on the final price.

General

- Is the change order work *pre-installation* or *post-installation*?

  Change order work costs vary according to how much of the installation has been completed. Once workers have the project scoped in their mind, even though they have not started, it can be difficult to refocus. Consequently they may spend more than the normal amount of time understanding the change. Also, modifications to work in place such as trimming or refitting usually take more time than was initially estimated. The greater the amount of work in place, the more reluctant workers are to change it. Psychologically they may resent the change and as a result the rework takes longer than normal. Post-installation change order estimates must include demolition of existing work as required to accomplish the change. If the work is performed at a later time, additional obstacles such as building finishes may be present which must be protected. Regardless of whether the change occurs pre-installation or post-installation, attempt to isolate the identifiable factors and price them separately. For example, add shipping costs that may be required pre-installation or any demolition required post-installation. Then analyze the potential impact on productivity of psychological and/or learning curve factors and adjust the output rates accordingly. One approach is to break down the typical workday into segments and quantify the impact on each segment. The following chart may be useful as a guide:

	Activities (Productivity) Expressed as Percentages of a Workday		
Task	Means Mechanical Cost Data (for New Construction)	Pre-Installation Change Orders	Post-Installation Change Orders
1. Study plans	3%	6%	6%
2. Material procurement	3%	3%	3%
3. Receiving and storing	3%	3%	3%
4. Mobilization	5%	5%	5%
5. Site movement	5%	5%	8%
6. Layout and marking	8%	10%	12%
7. Actual installation	64%	59%	54%
8. Clean-up	3%	3%	3%
9. Breaks—non-productive	6%	6%	6%
Total	100%	100%	100%

Change Order Installation Efficiency

The labor-hours expressed (for new construction) are based on average installation time, using an efficiency level of approximately 60-65%. For change order situations, adjustments to this efficiency level should reflect the daily labor-hour allocation for that particular occurrence.

If any of the specific percentages expressed in the above chart do not apply to a particular project situation, then those percentage points should be reallocated to the appropriate task(s). Example: Using data for new construction, assume there is no new material being utilized. The percentages for Tasks 2 and 3 would therefore be reallocated to other tasks. If the time required for Tasks 2 and 3 can now be applied to installation, we can add the time allocated for *Material Procurement* and *Receiving and Storing* to the *Actual Installation* time for new construction, thereby increasing the Actual Installation percentage.

This chart shows that, due to reduced productivity, labor costs will be higher than those for new construction by 5% to 15% for pre-installation change orders and by 15% to 25% for post-installation change orders. Each job and change order is unique and must be examined individually. Many factors, covered elsewhere in this section, can each have a significant impact on productivity and change order costs. All such factors should be considered in every case.

- Will the change substantially delay the original completion date?

  A significant change in the project may cause the original completion date to be extended. The extended schedule may subject the contractor to new wage rates dictated by relevant labor contracts. Project supervision and other project overhead must also be extended beyond the original completion date. The schedule extension may also put installation into a new weather season. For example, underground piping scheduled for October installation was delayed until January. As a result, frost penetrated the trench area, thereby changing the degree of difficulty of the task. Changes and delays may have a ripple effect throughout the project. This effect must be analyzed and negotiated with the owner.

- What is the net effect of a deduct change order?

  In most cases, change orders resulting in a deduction or credit reflect only bare costs. The contractor may retain the overhead and profit based on the original bid.

Materials

- Will you have to pay more or less for the new material, required by the change order, than you paid for the original purchase?

  The same material prices or discounts will usually apply to materials purchased for change orders as new construction. In some instances, however, the contractor may forfeit the advantages of competitive pricing for change orders. Consider the following example:

  A contractor purchased over $20,000 worth of fan coil units for an installation, and obtained the maximum discount. Some time later it was determined the project required an additional matching unit. The contractor has to purchase this unit from the original supplier to ensure a match. The supplier at this time may not discount the unit because of the small quantity, and the fact that he is no longer in a competitive situation. The impact of quantity on purchase can add between 0% and 25% to material prices and/or subcontractor quotes.

- If materials have been ordered or delivered to the job site, will they be subject to a cancellation charge or restocking fee?

  Check with the supplier to determine if ordered materials are subject to a cancellation charge. Delivered materials not used as result of a change order may be subject to a restocking fee if returned to the supplier. Common restocking charges run between 20% and 40%. Also, delivery charges to return the goods to the supplier must be added.

Labor

- How efficient is the existing crew at the actual installation?

  Is the same crew that performed the initial work going to do the change order? Possibly the change consists of the installation of a unit identical to one already installed; therefore the change should take less time. Be sure to consider this potential productivity increase and modify the productivity rates accordingly.

- If the crew size is increased, what impact will that have on supervision requirements?

  Under most bargaining agreements or management practices, there is a point at which a working foreman is replaced by a nonworking foreman. This replacement increases project overhead by adding a nonproductive worker. If additional workers are added to accelerate the project or to perform changes while maintaining the schedule, be sure to add additional supervision time if warranted. Calculate the hours involved and the additional cost directly if possible.

- What are the other impacts of increased crew size?

  The larger the crew, the greater the potential for productivity to decrease. Some of the factors that cause this productivity loss are: overcrowding (producing restrictive conditions in the working space), and possibly a shortage of any special tools and equipment required. Such factors affect not only the crew working on the elements directly involved in the change order, but other crews whose movement may also be hampered.

  As the crew increases, check its basic composition for changes by the addition or deletion of apprentices or nonworking foreman and quantify the potential effects of equipment shortages or other logistical factors.

- As new crews, unfamiliar with the project, are brought onto the site, how long will it take them to become oriented to the project requirements?

  The orientation time for a new crew to become 100% effective varies with the site and type of project. Orientation is easiest at a new construction site, and most difficult at existing, very restrictive renovation sites. The type of work also affects orientation time. When all elements of the work are exposed, such as concrete or masonry work, orientation is decreased. When the work is concealed or less visible, such as existing electrical systems, orientation takes longer. Usually orientation can be accomplished in one day or less. Costs for added orientation should be itemized and added to the total estimated cost.

- How much actual production can be gained by working overtime?

  Short term overtime can be used effectively to accomplish more work in a day. However, as overtime is scheduled to run beyond several weeks, studies have shown marked decreases in output. The following chart shows the effect of long term overtime on worker efficiency. If the anticipated change requires extended overtime to keep the job on schedule, these factors can be used as a guide to predict the impact on time and cost. Add project overhead, particularly supervision, that may also be incurred.

Days per Week	Hours per Day	Production Efficiency					Payroll Cost Factors	
		1 Week	2 Weeks	3 Weeks	4 Weeks	Average 4 Weeks	@ 1-1/2 Times	@ 2 Times
5	8	100%	100%	100%	100%	100%	100%	100%
	9	100	100	95	90	96.25	105.6	111.1
	10	100	95	90	85	91.25	110.0	120.0
	11	95	90	75	65	81.25	113.6	127.3
	12	90	85	70	60	76.25	116.7	133.3
6	8	100	100	95	90	96.25	108.3	116.7
	9	100	95	90	85	92.50	113.0	125.9
	10	95	90	85	80	87.50	116.7	133.3
	11	95	85	70	65	78.75	119.7	139.4
	12	90	80	65	60	73.75	122.2	144.4
7	8	100	95	85	75	88.75	114.3	128.6
	9	95	90	80	70	83.75	118.3	136.5
	10	90	85	75	65	78.75	121.4	142.9
	11	85	80	65	60	72.50	124.0	148.1
	12	85	75	60	55	68.75	126.2	152.4

Effects of Overtime

Caution: Under many labor agreements, Sundays and holidays are paid at a higher premium than the normal overtime rate.

The use of long-term overtime is counterproductive on almost any construction job; that is, the longer the period of overtime, the lower the actual production rate. Numerous studies have been conducted, and while they have resulted in slightly different numbers, all reach the same conclusion. The figure above tabulates the effects of overtime work on efficiency.

As illustrated, there can be a difference between the *actual* payroll cost per hour and the *effective* cost per hour for overtime work. This is due to the reduced production efficiency with the increase in weekly hours beyond 40. This difference between actual and effective cost results from overtime work over a prolonged period. Short-term overtime work does not result in as great a reduction in efficiency, and in such cases, effective cost may not vary significantly from the actual payroll cost. As the total hours per week are increased on a regular basis, more time is lost because of fatigue, lowered morale, and an increased accident rate.

As an example, assume a project where workers are working 6 days a week, 10 hours per day. From the figure above (based on productivity studies), the average effective productive hours over a four-week period are:

$$0.875 \times 60 = 52.5$$

Depending upon the locale and day of week, overtime hours may be paid at time and a half or double time. For time and a half, the overall (average) *actual* payroll cost (including regular and overtime hours) is determined as follows:

$$\frac{40 \text{ reg. hrs.} + (20 \text{ overtime hrs.} \times 1.5)}{60 \text{ hrs.}} = 1.167$$

Based on 60 hours, the payroll cost per hour will be 116.7% of the normal rate at 40 hours per week. However, because the effective production (efficiency) for 60 hours is reduced to the equivalent of 52.5 hours, the effective cost of overtime is calculated as follows:

For time and a half:

$$\frac{40 \text{ reg. hrs.} + (20 \text{ overtime hrs.} \times 1.5)}{52.5 \text{ hrs.}} = 1.33$$

Installed cost will be 133% of the normal rate (for labor).

Thus, when figuring overtime, the actual cost per unit of work will be higher than the apparent overtime payroll dollar increase, due to the reduced productivity of the longer workweek. These efficiency calculations are true only for those cost factors determined by hours worked. Costs that are applied weekly or monthly, such as equipment rentals, will not be similarly affected.

Equipment

- What equipment is required to complete the change order?

Change orders may require extending the rental period of equipment already on the job site, or the addition of special equipment brought in to accomplish the change work. In either case, the additional rental charges and operator labor charges must be added.

Summary

The preceding considerations and others you deem appropriate should be analyzed and applied to a change order estimate. The impact of each should be quantified and listed on the estimate to form an audit trail.

Change orders that are properly identified, documented, and managed help to ensure the orderly, professional and profitable progress of the work. They also minimize potential claims or disputes at the end of the project.

Estimating Tips

- The cost figures in this Square Foot Cost section were derived from approximately 11,200 projects contained in the RSMeans database of completed construction projects. They include the contractor's overhead and profit, but do not generally include architectural fees or land costs. The figures have been adjusted to January of the current year. New projects are added to our files each year, and outdated projects are discarded. For this reason, certain costs may not show a uniform annual progression. In no case are all subdivisions of a project listed.

- These projects were located throughout the U.S. and reflect a tremendous variation in square foot (S.F.) and cubic foot (C.F.) costs. This is due to differences, not only in labor and material costs, but also in individual owners' requirements. For instance, a bank in a large city would have different features than one in a rural area. This is true of all the different types of buildings analyzed. Therefore, caution should be exercised when using these Square Foot costs. For example, for court houses, costs in the database are local court house costs and will not apply to the larger, more elaborate federal court houses. As a general rule, the projects in the 1/4 column do not include any site work or equipment, while the projects in the 3/4 column may include both equipment and site work. The median figures do not generally include site work.

- None of the figures "go with" any others. All individual cost items were computed and tabulated separately. Thus, the sum of the median figures for Plumbing, HVAC and Electrical will not normally total up to the total Mechanical and Electrical costs arrived at by separate analysis and tabulation of the projects.

- Each building was analyzed as to total and component costs and percentages. The figures were arranged in ascending order with the results tabulated as shown. The 1/4 column shows that 25% of the projects had lower costs and 75% had higher. The 3/4 column shows that 75% of the projects had lower costs and 25% had higher. The median column shows that 50% of the projects had lower costs and 50% had higher.

- There are two times when square foot costs are useful. The first is in the conceptual stage when no details are available. Then square foot costs make a useful starting point. The second is after the bids are in and the costs can be worked back into their appropriate units for information purposes. As soon as details become available in the project design, the square foot approach should be discontinued and the project priced as to its particular components. When more precision is required, or for estimating the replacement cost of specific buildings, the current edition of *RSMeans Square Foot Costs* should be used.

- In using the figures in this section, it is recommended that the median column be used for preliminary figures if no additional information is available. The median figures, when multiplied by the total city construction cost index figures (see City Cost Indexes) and then multiplied by the project size modifier at the end of this section, should present a fairly accurate base figure, which would then have to be adjusted in view of the estimator's experience, local economic conditions, code requirements, and the owner's particular requirements. There is no need to factor the percentage figures, as these should remain constant from city to city. All tabulations mentioning air conditioning had at least partial air conditioning.

- The editors of this book would greatly appreciate receiving cost figures on one or more of your recent projects, which would then be included in the averages for next year. All cost figures received will be kept confidential, except that they will be averaged with other similar projects to arrive at Square Foot cost figures for next year's book. See the last page of the book for details and the discount available for submitting one or more of your projects.

50 17 | Square Foot Costs

50 17 00 | S.F. Costs

			UNIT	UNIT COSTS 1/4	MEDIAN	3/4	% OF TOTAL 1/4	MEDIAN	3/4
01	0010	**APARTMENTS Low Rise (1 to 3 story)**	S.F.	67	84.50	112			
	0020	Total project cost	C.F.	6	8	9.85			
	0100	Site work	S.F.	5.75	7.85	13.75	6.05%	10.55%	14.05%
	0500	Masonry		1.32	3.07	5.30	1.54%	3.67%	6.35%
	1500	Finishes		7.10	9.75	12.10	9.05%	10.75%	12.85%
	1800	Equipment		2.19	3.32	4.94	2.73%	4.03%	5.95%
	2720	Plumbing		5.20	6.70	8.55	6.65%	8.95%	10.05%
	2770	Heating, ventilating, air conditioning		3.33	4.10	6.05	4.20%	5.60%	7.60%
	2900	Electrical		3.88	5.15	6.90	5.20%	6.65%	8.40%
	3100	Total: Mechanical & Electrical		13.45	17.10	21.50	15.90%	18.05%	23%
	9000	Per apartment unit, total cost	Apt.	62,500	95,500	140,500			
	9500	Total: Mechanical & Electrical	"	11,800	18,600	24,300			
02	0010	**APARTMENTS Mid Rise (4 to 7 story)**	S.F.	89	107	133			
	0020	Total project costs	C.F.	6.95	9.60	13.10			
	0100	Site work	S.F.	3.56	7.05	12.65	5.25%	6.70%	9.15%
	0500	Masonry		5.90	8.15	11.15	5.10%	7.25%	10.50%
	1500	Finishes		11.20	14.60	18.40	10.55%	13.45%	17.70%
	1800	Equipment		2.58	3.90	5.05	2.54%	3.48%	4.31%
	2500	Conveying equipment		1.91	2.44	2.96	1.94%	2.27%	2.69%
	2720	Plumbing		5.20	8.35	8.85	5.70%	7.20%	8.95%
	2900	Electrical		5.85	7.95	9.65	6.65%	7.20%	8.95%
	3100	Total: Mechanical & Electrical		18.75	23.50	28.50	18.50%	21%	23%
	9000	Per apartment unit, total cost	Apt.	100,500	118,500	196,500			
	9500	Total: Mechanical & Electrical	"	19,000	22,000	27,800			
03	0010	**APARTMENTS High Rise (8 to 24 story)**	S.F.	101	116	139			
	0020	Total project costs	C.F.	9.80	11.40	14.55			
	0100	Site work	S.F.	3.66	5.90	8.30	2.58%	4.84%	6.15%
	0500	Masonry		5.85	10.60	13.20	4.74%	9.65%	11.05%
	1500	Finishes		11.20	14	16.50	9.75%	11.80%	13.70%
	1800	Equipment		3.24	3.99	5.30	2.78%	3.49%	4.35%
	2500	Conveying equipment		2.29	3.48	4.73	2.23%	2.78%	3.37%
	2720	Plumbing		7.40	8.75	12.25	6.80%	7.20%	10.45%
	2900	Electrical		6.95	8.75	11.85	6.45%	7.65%	8.80%
	3100	Total: Mechanical & Electrical		21	26.50	32	17.95%	22.50%	24.50%
	9000	Per apartment unit, total cost	Apt.	105,000	115,500	160,000			
	9500	Total: Mechanical & Electrical	"	22,700	25,900	27,400			
04	0010	**AUDITORIUMS**	S.F.	104	141	203			
	0020	Total project costs	C.F.	6.55	9.15	13.10			
	2720	Plumbing	S.F.	6.35	9.20	11.15	5.85%	7.20%	8.70%
	2900	Electrical		8.10	11.70	15.75	6.80%	8.95%	11.30%
	3100	Total: Mechanical & Electrical		23	46	56	24.50%	30.50%	31.50%
05	0010	**AUTOMOTIVE SALES**	S.F.	77.50	105	130			
	0020	Total project costs	C.F.	5.10	6.15	7.95			
	2720	Plumbing	S.F.	3.54	6.15	6.70	2.89%	6.05%	6.50%
	2770	Heating, ventilating, air conditioning		5.45	8.35	9	4.61%	10%	10.35%
	2900	Electrical		6.25	9.80	13.30	7.25%	9.80%	12.15%
	3100	Total: Mechanical & Electrical		19.60	28	33.50	19.15%	20.50%	22%
06	0010	**BANKS**	S.F.	151	189	240			
	0020	Total project costs	C.F.	10.85	14.75	19.50			
	0100	Site work	S.F.	17.35	26.50	38	7.75%	12.45%	17%
	0500	Masonry		7.90	15.15	20.50	3.30%	6.95%	10.35%
	1500	Finishes		13.45	19.50	24	5.80%	8.45%	11.25%
	1800	Equipment		5.70	12.60	26	1.34%	5.95%	10.65%
	2720	Plumbing		4.76	6.80	9.95	2.82%	3.90%	4.93%
	2770	Heating, ventilating, air conditioning		9.05	12.10	16.10	4.86%	7.15%	8.50%
	2900	Electrical		14.35	19.20	25	8.20%	10.20%	12.15%
	3100	Total: Mechanical & Electrical		33.50	45.50	53	16.25%	19.40%	23%
	3500	See also Divisions 11020 & 11030 (MF2004 11 16 00 & 11 17 00)							

50 17 | Square Foot Costs

50 17 00 | S.F. Costs

			UNIT	UNIT COSTS 1/4	UNIT COSTS MEDIAN	UNIT COSTS 3/4	% OF TOTAL 1/4	% OF TOTAL MEDIAN	% OF TOTAL 3/4	
13	0010	**CHURCHES**	S.F.	102	130	168				13
	0020	Total project costs	C.F.	6.35	8	10.55				
	1800	Equipment	S.F.	1.22	2.93	6.25	.95%	2.11%	4.50%	
	2720	Plumbing		3.99	5.55	8.20	3.51%	4.96%	6.25%	
	2770	Heating, ventilating, air conditioning		9.30	12.15	17.20	7.50%	10%	12%	
	2900	Electrical		8.60	11.80	15.90	7.35%	8.80%	11%	
	3100	Total: Mechanical & Electrical		26.50	34.50	46	18.30%	22%	24.50%	
	3500	See also Division 11040 (MF2004 11 91 00)								
15	0010	**CLUBS, COUNTRY**	S.F.	110	132	167				15
	0020	Total project costs	C.F.	8.85	10.80	14.90				
	2720	Plumbing	S.F.	6.65	9.85	22.50	5.60%	7.90%	10%	
	2900	Electrical		8.65	11.85	15.45	7%	8.95%	11%	
	3100	Total: Mechanical & Electrical		46	57.50	60.50	19%	26.50%	29.50%	
17	0010	**CLUBS, SOCIAL Fraternal**	S.F.	87.50	126	169				17
	0020	Total project costs	C.F.	5.50	8.30	9.90				
	2720	Plumbing	S.F.	5.50	6.85	10.40	5.60%	6.90%	8.55%	
	2770	Heating, ventilating, air conditioning		7.95	9.60	12.35	8.20%	9.25%	14.40%	
	2900	Electrical		6.60	10.85	12.40	6.50%	9.50%	10.55%	
	3100	Total: Mechanical & Electrical		19.45	37	47	21%	23%	23.50%	
18	0010	**CLUBS, Y.M.C.A.**	S.F.	110	143	179				18
	0020	Total project costs	C.F.	5.05	8.50	12.65				
	2720	Plumbing	S.F.	6.95	13.85	15.55	5.65%	7.60%	10.85%	
	2900	Electrical		8.80	11	15.15	6.05%	7.60%	9.25%	
	3100	Total: Mechanical & Electrical		33.50	37.50	51.50	18.40%	21.50%	28.50%	
19	0010	**COLLEGES Classrooms & Administration**	S.F.	111	152	201				19
	0020	Total project costs	C.F.	8.10	11.80	18.20				
	0500	Masonry	S.F.	8.15	15.35	18.85	5.65%	8.25%	10.50%	
	2720	Plumbing		5.65	11.60	20.50	5.10%	6.60%	8.95%	
	2900	Electrical		9.25	14.05	19.20	7.70%	9.85%	12%	
	3100	Total: Mechanical & Electrical		37	51.50	61.50	24%	28%	31.50%	
21	0010	**COLLEGES Science, Engineering, Laboratories**	S.F.	207	242	280				21
	0020	Total project costs	C.F.	11.85	17.30	19.65				
	1800	Equipment	S.F.	11.50	26	28.50	2%	6.45%	12.65%	
	2900	Electrical		17.05	23.50	37	7.10%	9.40%	12.10%	
	3100	Total: Mechanical & Electrical		63.50	75	116	28.50%	31.50%	41%	
	3500	See also Division 11600 (MF2004 11 53 00)								
23	0010	**COLLEGES Student Unions**	S.F.	132	179	216				23
	0020	Total project costs	C.F.	7.35	9.65	11.90				
	3100	Total: Mechanical & Electrical	S.F.	49.50	53.50	63.50	23.50%	26%	29%	
25	0010	**COMMUNITY CENTERS**	S.F.	109	134	181				25
	0020	Total project costs	C.F.	7.10	10.15	13.15				
	1800	Equipment	S.F.	2.64	4.46	7.10	1.48%	3.01%	5.45%	
	2720	Plumbing		5.15	9	12.30	4.85%	7%	8.95%	
	2770	Heating, ventilating, air conditioning		8.25	12.10	17.20	6.80%	10.35%	12.90%	
	2900	Electrical		8.90	11.90	17.05	7.30%	9%	10.45%	
	3100	Total: Mechanical & Electrical		31	38	54	20%	25%	31%	
28	0010	**COURT HOUSES**	S.F.	157	180	233				28
	0020	Total project costs	C.F.	12.05	14.40	18.15				
	2720	Plumbing	S.F.	7.45	10.45	15	5.95%	7.45%	8.20%	
	2900	Electrical		16.70	18.50	27	8.90%	10.45%	11.55%	
	3100	Total: Mechanical & Electrical		42.50	58	64	22.50%	27.50%	30.50%	
30	0010	**DEPARTMENT STORES**	S.F.	58	78.50	99				30
	0020	Total project costs	C.F.	3.11	4.03	5.50				
	2720	Plumbing	S.F.	1.81	2.28	3.47	1.82%	4.21%	5.90%	
	2770	Heating, ventilating, air conditioning		5.30	8.15	12.25	8.20%	9.10%	14.80%	

50 17 | Square Foot Costs

50 17 00 | S.F. Costs

				UNIT COSTS			% OF TOTAL		
			UNIT	1/4	MEDIAN	3/4	1/4	MEDIAN	3/4
30	2900	Electrical	S.F.	6.65	9.15	10.80	9.05%	12.15%	14.95%
	3100	Total: Mechanical & Electrical	↓	11.75	15	26.50	13.20%	21.50%	50%
31	0010	**DORMITORIES Low Rise (1 to 3 story)**	S.F.	109	143	185			
	0020	Total project costs	C.F.	6.20	10.05	15.05			
	2720	Plumbing	S.F.	6.60	8.85	11.20	8.05%	9%	9.65%
	2770	Heating, ventilating, air conditioning		7	8.40	11.15	4.61%	8.05%	10%
	2900	Electrical		7.15	10.95	14.85	6.25%	8.65%	9.50%
	3100	Total: Mechanical & Electrical	↓	36.50	40.50	63.50	21.50%	24%	27%
	9000	Per bed, total cost	Bed	46,700	52,000	111,000			
32	0010	**DORMITORIES Mid Rise (4 to 8 story)**	S.F.	135	176	218			
	0020	Total project costs	C.F.	14.90	16.35	19.60			
	2900	Electrical	S.F.	14.35	16.30	22	8.20%	10.20%	11.95%
	3100	Total: Mechanical & Electrical	"	40	41.50	81.50	25.50%	34.50%	37.50%
	9000	Per bed, total cost	Bed	19,200	43,900	111,000			
34	0010	**FACTORIES**	S.F.	51	76.50	117			
	0020	Total project costs	C.F.	3.28	4.89	8.10			
	0100	Site work	S.F.	5.85	10.65	16.85	6.95%	11.45%	17.95%
	2720	Plumbing		2.76	5.15	8.50	3.73%	6.05%	8.10%
	2770	Heating, ventilating, air conditioning		5.35	7.70	10.40	5.25%	8.45%	11.35%
	2900	Electrical		6.35	10.05	15.35	8.10%	10.50%	14.20%
	3100	Total: Mechanical & Electrical	↓	18.15	24.50	37	21%	28.50%	35.50%
36	0010	**FIRE STATIONS**	S.F.	102	139	182			
	0020	Total project costs	C.F.	5.95	8.15	10.85			
	0500	Masonry	S.F.	15	27	34.50	8.60%	11.65%	16.45%
	1140	Roofing		3.31	8.95	10.20	1.90%	4.94%	5.05%
	1580	Painting		2.57	3.84	3.93	1.37%	1.57%	2.07%
	1800	Equipment		1.31	2.57	4.47	.74%	1.98%	3.54%
	2720	Plumbing		5.70	9.15	13.25	5.85%	7.35%	9.45%
	2770	Heating, ventilating, air conditioning		5.60	9.10	14.10	5.15%	7.40%	9.40%
	2900	Electrical		7.20	12.55	16.95	6.85%	8.65%	10.60%
	3100	Total: Mechanical & Electrical	↓	35.50	45.50	52.50	18.75%	23%	27%
37	0010	**FRATERNITY HOUSES & Sorority Houses**	S.F.	102	131	179			
	0020	Total project costs	C.F.	10.10	10.55	12.70			
	2720	Plumbing	S.F.	7.65	8.80	16.10	6.80%	8%	10.85%
	2900	Electrical		6.70	14.45	17.70	6.60%	9.90%	10.65%
	3100	Total: Mechanical & Electrical	↓	17.90	26	31		15.10%	15.90%
38	0010	**FUNERAL HOMES**	S.F.	107	146	265			
	0020	Total project costs	C.F.	10.95	12.15	23.50			
	2900	Electrical	S.F.	4.73	8.65	9.50	3.58%	4.44%	5.95%
	3100	Total: Mechanical & Electrical	↓	16.75	24.50	34	12.90%	12.90%	12.90%
39	0010	**GARAGES, COMMERCIAL (Service)**	S.F.	60.50	93.50	129			
	0020	Total project costs	C.F.	3.99	5.90	8.55			
	1800	Equipment	S.F.	3.42	7.70	12	2.21%	4.62%	6.80%
	2720	Plumbing		4.20	6.45	11.80	5.45%	7.85%	10.65%
	2730	Heating & ventilating		5.50	7.30	9.90	5.25%	6.85%	8.20%
	2900	Electrical		5.75	8.80	12.65	7.15%	9.25%	10.85%
	3100	Total: Mechanical & Electrical	↓	12.70	24.50	36	12.35%	17.40%	26%
40	0010	**GARAGES, MUNICIPAL (Repair)**	S.F.	89	119	167			
	0020	Total project costs	C.F.	5.55	7.05	12.15			
	0500	Masonry	S.F.	8.35	16.35	25.50	5.60%	9.15%	12.50%
	2720	Plumbing		4	7.65	14.45	3.59%	6.70%	7.95%
	2730	Heating & ventilating		6.85	9.90	19.10	6.15%	7.45%	13.50%
	2900	Electrical		6.60	10.35	14.90	6.65%	8.15%	11.15%
	3100	Total: Mechanical & Electrical	↓	30.50	42	61	21.50%	25.50%	28.50%

50 17 | Square Foot Costs

50 17 00 | S.F. Costs

			UNIT	UNIT COSTS 1/4	UNIT COSTS MEDIAN	UNIT COSTS 3/4	% OF TOTAL 1/4	% OF TOTAL MEDIAN	% OF TOTAL 3/4	
41	0010	**GARAGES, PARKING**	S.F.	34.50	50.50	87				41
	0020	Total project costs	C.F.	3.26	4.42	6.45				
	2720	Plumbing	S.F.	.98	1.52	2.35	1.72%	2.70%	3.85%	
	2900	Electrical		1.90	2.33	3.67	4.33%	5.20%	6.30%	
	3100	Total: Mechanical & Electrical	▼	3.89	5.40	6.75	7%	8.90%	11.05%	
	3200									
	9000	Per car, total cost	Car	14,600	18,400	23,400				
43	0010	**GYMNASIUMS**	S.F.	96.50	129	174				43
	0020	Total project costs	C.F.	4.81	6.50	8				
	1800	Equipment	S.F.	2.29	4.30	8.25	1.81%	3.30%	6.70%	
	2720	Plumbing		6.10	7.25	9.35	4.65%	6.40%	7.75%	
	2770	Heating, ventilating, air conditioning		6.55	10	20	5.15%	9.05%	11.10%	
	2900	Electrical		7.40	10.10	13.30	6.60%	8.30%	10.30%	
	3100	Total: Mechanical & Electrical	▼	27	36.50	43.50	19.75%	23.50%	29%	
	3500	See also Division 11480 (MF2004 11 67 00)								
46	0010	**HOSPITALS**	S.F.	185	228	315				46
	0020	Total project costs	C.F.	14.05	17.45	25				
	1800	Equipment	S.F.	4.70	9.05	15.60	.96%	2.63%	5%	
	2720	Plumbing		15.95	22.50	28.50	7.60%	9.10%	10.85%	
	2770	Heating, ventilating, air conditioning		23.50	30	40.50	7.80%	12.95%	16.65%	
	2900	Electrical		20	26.50	39	9.85%	11.55%	13.90%	
	3100	Total: Mechanical & Electrical	▼	57	81.50	123	27%	33.50%	36.50%	
	9000	Per bed or person, total cost	Bed	214,500	295,500	340,500				
	9900	See also Division 11700 (MF2004 11 71 00)								
48	0010	**HOUSING For the Elderly**	S.F.	91	115	142				48
	0020	Total project costs	C.F.	6.50	9	11.50				
	0100	Site work	S.F.	6.35	9.85	14.45	5.05%	7.90%	12.10%	
	0500	Masonry		2.77	10.35	15.15	1.30%	6.05%	11%	
	1800	Equipment		2.20	3.03	4.82	1.88%	3.23%	4.43%	
	2510	Conveying systems		2.21	2.98	4.03	1.78%	2.20%	2.81%	
	2720	Plumbing		6.75	8.60	10.85	8.15%	9.55%	10.50%	
	2730	Heating, ventilating, air conditioning		3.47	4.91	7.35	3.30%	5.60%	7.25%	
	2900	Electrical		6.80	9.20	11.80	7.30%	8.50%	10.25%	
	3100	Total: Mechanical & Electrical	▼	23.50	28	37	18.10%	22.50%	29%	
	9000	Per rental unit, total cost	Unit	84,500	99,000	110,500				
	9500	Total: Mechanical & Electrical	"	18,800	21,700	25,300				
50	0010	**HOUSING Public (Low Rise)**	S.F.	76.50	106	138				50
	0020	Total project costs	C.F.	6.80	8.50	10.55				
	0100	Site work	S.F.	9.75	14.05	22.50	8.35%	11.75%	16.50%	
	1800	Equipment		2.08	3.39	5.15	2.26%	3.03%	4.24%	
	2720	Plumbing		5.50	7.30	9.25	7.15%	9.05%	11.60%	
	2730	Heating, ventilating, air conditioning		2.77	5.40	5.90	4.26%	6.05%	6.45%	
	2900	Electrical		4.63	6.90	9.55	5.10%	6.55%	8.25%	
	3100	Total: Mechanical & Electrical	▼	22	28.50	31.50	14.50%	17.55%	26.50%	
	9000	Per apartment, total cost	Apt.	84,000	95,500	120,000				
	9500	Total: Mechanical & Electrical	"	17,900	22,100	24,500				
51	0010	**ICE SKATING RINKS**	S.F.	65.50	153	168				51
	0020	Total project costs	C.F.	4.81	4.92	5.65				
	2720	Plumbing	S.F.	2.44	4.58	4.69	3.12%	3.23%	5.65%	
	2900	Electrical		7	10.75	11.35	6.30%	10.15%	15.05%	
	3100	Total: Mechanical & Electrical	▼	11.65	16.50	20.50	18.95%	18.95%	18.95%	
52	0010	**JAILS**	S.F.	199	257	330				52
	0020	Total project costs	C.F.	17.95	25	29.50				
	1800	Equipment	S.F.	7.75	23	39	2.80%	5.55%	10.35%	
	2720	Plumbing		19.15	25.50	34	7%	8.90%	13.35%	
	2770	Heating, ventilating, air conditioning		17.95	24	46.50	7.50%	9.45%	17.75%	
	2900	Electrical	▼	20.50	27.50	35	8.20%	11.55%	14.95%	

50 17 | Square Foot Costs

50 17 00 | S.F. Costs

			UNIT	UNIT COSTS 1/4	UNIT COSTS MEDIAN	UNIT COSTS 3/4	% OF TOTAL 1/4	% OF TOTAL MEDIAN	% OF TOTAL 3/4	
52	3100	Total: Mechanical & Electrical	S.F.	54	99	117	28%	30%	34%	52
53	0010	**LIBRARIES**	S.F.	126	161	207				53
	0020	Total project costs	C.F.	8.60	10.80	13.75				
	0500	Masonry	S.F.	9.85	17.40	29	5.80%	7.60%	11.80%	
	1800	Equipment		1.72	4.63	7	.37%	1.50%	4.16%	
	2720	Plumbing		4.65	6.75	9.15	3.38%	4.60%	5.70%	
	2770	Heating, ventilating, air conditioning		10.30	17.40	22.50	7.80%	10.95%	12.80%	
	2900	Electrical		12.85	16.65	21	8.30%	10.30%	11.95%	
	3100	Total: Mechanical & Electrical		38	48	58.50	19.65%	23%	26.50%	
54	0010	**LIVING, ASSISTED**	S.F.	116	137	162				54
	0020	Total project costs	C.F.	9.80	11.45	13				
	0500	Masonry	S.F.	3.42	4.08	4.79	2.37%	3.16%	3.86%	
	1800	Equipment		2.65	3.08	3.94	2.12%	2.45%	2.66%	
	2720	Plumbing		9.75	13.05	13.55	6.05%	8.15%	10.60%	
	2770	Heating, ventilating, air conditioning		11.60	12.10	13.25	7.95%	9.35%	9.70%	
	2900	Electrical		11.40	12.60	14.55	9%	10%	10.70%	
	3100	Total: Mechanical & Electrical		32	37.50	43	26%	29%	31.50%	
55	0010	**MEDICAL CLINICS**	S.F.	118	146	185				55
	0020	Total project costs	C.F.	8.70	11.25	14.95				
	1800	Equipment	S.F.	3.20	6.70	10.45	1.05%	2.94%	6.35%	
	2720	Plumbing		7.85	11.10	14.80	6.15%	8.40%	10.10%	
	2770	Heating, ventilating, air conditioning		9.40	12.30	18.10	6.65%	8.85%	11.35%	
	2900	Electrical		10.20	14.45	18.90	8.10%	10%	12.25%	
	3100	Total: Mechanical & Electrical		32.50	44	60.50	22.50%	27%	33.50%	
	3500	See also Division 11700 (MF2004 11 71 00)								
57	0010	**MEDICAL OFFICES**	S.F.	112	138	169				57
	0020	Total project costs	C.F.	8.30	11.25	15.25				
	1800	Equipment	S.F.	3.68	7.25	10.35	.70%	5.10%	7.05%	
	2720	Plumbing		6.15	9.50	12.80	5.60%	6.80%	8.50%	
	2770	Heating, ventilating, air conditioning		7.45	10.75	14.20	6.10%	8%	9.70%	
	2900	Electrical		9.10	13.05	18.10	7.65%	9.80%	11.70%	
	3100	Total: Mechanical & Electrical		26	35.50	51.50	19.35%	23%	29%	
59	0010	**MOTELS**	S.F.	70.50	102	133				59
	0020	Total project costs	C.F.	6.25	8.40	13.70				
	2720	Plumbing	S.F.	7.15	9.05	10.80	9.45%	10.60%	12.55%	
	2770	Heating, ventilating, air conditioning		4.34	6.50	11.60	5.60%	5.60%	10%	
	2900	Electrical		6.65	8.40	10.45	7.45%	9.05%	10.45%	
	3100	Total: Mechanical & Electrical		22.50	28.50	48.50	18.50%	24%	25.50%	
	5000									
	9000	Per rental unit, total cost	Unit	35,800	68,000	73,500				
	9500	Total: Mechanical & Electrical	"	6,975	10,600	12,300				
60	0010	**NURSING HOMES**	S.F.	110	142	174				60
	0020	Total project costs	C.F.	8.65	10.80	14.80				
	1800	Equipment	S.F.	3.47	4.60	7.70	2.02%	3.62%	4.99%	
	2720	Plumbing		9.45	14.30	17.25	8.75%	10.10%	12.70%	
	2770	Heating, ventilating, air conditioning		9.95	15.10	20	9.70%	11.45%	11.80%	
	2900	Electrical		10.90	13.60	18.55	9.40%	10.55%	12.45%	
	3100	Total: Mechanical & Electrical		26	36.50	61	26%	29.50%	30.50%	
	9000	Per bed or person, total cost	Bed	48,900	61,000	79,000				
61	0010	**OFFICES Low Rise (1 to 4 story)**	S.F.	93	120	155				61
	0020	Total project costs	C.F.	6.65	9.15	12.05				
	0100	Site work	S.F.	7.20	12.60	18.85	5.90%	9.70%	13.60%	
	0500	Masonry		3.46	7.25	13.15	2.62%	5.50%	8.45%	
	1800	Equipment		.99	1.93	5.25	.69%	1.50%	3.63%	
	2720	Plumbing		3.32	5.15	7.50	3.66%	4.50%	6.10%	
	2770	Heating, ventilating, air conditioning		7.35	10.25	15	7.20%	10.30%	11.70%	
	2900	Electrical		7.55	10.75	15.20	7.45%	9.60%	11.35%	

50 17 | Square Foot Costs

50 17 00 | S.F. Costs

			UNIT	UNIT COSTS 1/4	UNIT COSTS MEDIAN	UNIT COSTS 3/4	% OF TOTAL 1/4	% OF TOTAL MEDIAN	% OF TOTAL 3/4	
61	3100	Total: Mechanical & Electrical	S.F.	20.50	27.50	41	18%	22%	27%	61
62	0010	**OFFICES Mid Rise (5 to 10 story)**	S.F.	98.50	119	157				62
	0020	Total project costs	C.F.	6.95	8.90	12.60				
	2720	Plumbing	S.F.	2.97	4.61	6.60	2.83%	3.74%	4.50%	
	2770	Heating, ventilating, air conditioning		7.45	10.65	17.05	7.65%	9.40%	11%	
	2900	Electrical		7.30	9.35	12.95	6.35%	7.80%	10%	
	3100	Total: Mechanical & Electrical		18.90	24.50	46.50	19.15%	21.50%	27.50%	
63	0010	**OFFICES High Rise (11 to 20 story)**	S.F.	121	152	187				63
	0020	Total project costs	C.F.	8.45	10.55	15.15				
	2900	Electrical	S.F.	7.35	8.95	13.30	5.80%	7.85%	10.50%	
	3100	Total: Mechanical & Electrical		23.50	31.50	53.50	16.90%	23.50%	34%	
64	0010	**POLICE STATIONS**	S.F.	145	190	240				64
	0020	Total project costs	C.F.	11.55	14.15	19.35				
	0500	Masonry	S.F.	13.60	24	30	6.70%	9.10%	11.35%	
	1800	Equipment		2.30	10.10	16.05	.98%	3.35%	6.70%	
	2720	Plumbing		8.10	16.15	20	5.65%	6.90%	10.75%	
	2770	Heating, ventilating, air conditioning		12.65	16.80	23	5.85%	10.55%	11.70%	
	2900	Electrical		15.85	22.50	30	9.80%	11.85%	14.80%	
	3100	Total: Mechanical & Electrical		52	62.50	84.50	25%	31.50%	32.50%	
65	0010	**POST OFFICES**	S.F.	114	141	180				65
	0020	Total project costs	C.F.	6.90	8.70	9.90				
	2720	Plumbing	S.F.	5.15	6.40	8.05	4.24%	5.30%	5.60%	
	2770	Heating, ventilating, air conditioning		8.05	9.95	11.05	6.65%	7.15%	9.35%	
	2900	Electrical		9.45	13.30	15.75	7.25%	9%	11%	
	3100	Total: Mechanical & Electrical		27.50	35.50	40.50	16.25%	18.80%	22%	
66	0010	**POWER PLANTS**	S.F.	795	1,000	1,925				66
	0020	Total project costs	C.F.	22	47.50	102				
	2900	Electrical	S.F.	56	119	177	9.30%	12.75%	21.50%	
	8100	Total: Mechanical & Electrical		140	455	1,025	32.50%	32.50%	52.50%	
67	0010	**RELIGIOUS EDUCATION**	S.F.	92	120	148				67
	0020	Total project costs	C.F.	5.10	7.35	9.15				
	2720	Plumbing	S.F.	3.85	5.45	7.70	4.40%	5.30%	7.10%	
	2770	Heating, ventilating, air conditioning		9.75	11	15.55	10.05%	11.45%	12.35%	
	2900	Electrical		7.30	10.30	13.65	7.60%	9.05%	10.35%	
	3100	Total: Mechanical & Electrical		29.50	39	47.50	22%	23%	27%	
69	0010	**RESEARCH Laboratories & Facilities**	S.F.	138	198	288				69
	0020	Total project costs	C.F.	10.40	20	24				
	1800	Equipment	S.F.	6.10	12.05	29.50	.94%	4.58%	8.80%	
	2720	Plumbing		13.85	17.70	28.50	6.15%	8.30%	10.80%	
	2770	Heating, ventilating, air conditioning		12.40	41.50	49.50	7.25%	16.50%	17.50%	
	2900	Electrical		16.15	26.50	43.50	9.55%	11.50%	15.40%	
	3100	Total: Mechanical & Electrical		49	92	132	29.50%	37%	42%	
70	0010	**RESTAURANTS**	S.F.	134	172	224				70
	0020	Total project costs	C.F.	11.25	14.75	19.35				
	1800	Equipment	S.F.	8.65	21	32	6.10%	13%	15.65%	
	2720	Plumbing		10.60	12.85	16.85	6.10%	8.15%	9%	
	2770	Heating, ventilating, air conditioning		13.45	18.60	24	9.20%	12%	12.40%	
	2900	Electrical		14.15	17.45	22.50	8.35%	10.55%	11.55%	
	3100	Total: Mechanical & Electrical		44	47	60.50	21%	24.50%	29.50%	
	9000	Per seat unit, total cost	Seat	4,900	6,550	7,750				
	9500	Total: Mechanical & Electrical	"	1,225	1,625	1,925				
72	0010	**RETAIL STORES**	S.F.	62.50	84	111				72
	0020	Total project costs	C.F.	4.23	6.05	8.40				
	2720	Plumbing	S.F.	2.26	3.78	6.45	3.26%	4.60%	6.80%	
	2770	Heating, ventilating, air conditioning		4.89	6.70	10.05	6.75%	8.75%	10.15%	
	2900	Electrical		5.65	7.70	11.10	7.25%	9.90%	11.65%	
	3100	Total: Mechanical & Electrical		14.95	19.20	26.50	17.05%	21%	23.50%	

50 17 | Square Foot Costs

50 17 00 | S.F. Costs

			UNIT	UNIT COSTS 1/4	UNIT COSTS MEDIAN	UNIT COSTS 3/4	% OF TOTAL 1/4	% OF TOTAL MEDIAN	% OF TOTAL 3/4	
74	0010	**SCHOOLS Elementary**	S.F.	101	125	155				74
	0020	Total project costs	C.F.	6.65	8.55	11.05				
	0500	Masonry	S.F.	9.10	15.70	23.50	5.80%	11%	14.95%	
	1800	Equipment		2.78	4.70	8.75	1.89%	3.32%	4.71%	
	2720	Plumbing		5.85	8.30	11.05	5.70%	7.15%	9.35%	
	2730	Heating, ventilating, air conditioning		8.80	14	19.55	8.15%	10.80%	14.90%	
	2900	Electrical		9.60	12.70	15.95	8.40%	10.05%	11.70%	
	3100	Total: Mechanical & Electrical		34	43	52.50	25%	27.50%	30%	
	9000	Per pupil, total cost	Ea.	11,700	17,400	50,000				
	9500	Total: Mechanical & Electrical	"	3,300	4,175	14,500				
76	0010	**SCHOOLS Junior High & Middle**	S.F.	103	129	157				76
	0020	Total project costs	C.F.	6.65	8.65	9.70				
	0500	Masonry	S.F.	13.10	16.90	19.75	8%	11.40%	14.30%	
	1800	Equipment		3.36	5.40	8.15	1.81%	3.26%	4.86%	
	2720	Plumbing		6.10	7.55	9.35	5.30%	6.80%	7.25%	
	2770	Heating, ventilating, air conditioning		12.20	14.85	26	8.90%	11.55%	14.20%	
	2900	Electrical		10.30	12.40	15.95	7.90%	9.35%	10.60%	
	3100	Total: Mechanical & Electrical		33.50	43	53	23.50%	27%	29.50%	
	9000	Per pupil, total cost	Ea.	13,300	17,500	23,500				
78	0010	**SCHOOLS Senior High**	S.F.	108	133	167				78
	0020	Total project costs	C.F.	6.60	9.70	15.65				
	1800	Equipment	S.F.	2.87	6.75	9.50	1.88%	2.98%	4.80%	
	2720	Plumbing		6.15	9.20	16.80	5.60%	6.90%	8.30%	
	2770	Heating, ventilating, air conditioning		12.50	14.35	27.50	8.95%	11.60%	15%	
	2900	Electrical		10.95	14.25	21	8.65%	10.15%	11.95%	
	3100	Total: Mechanical & Electrical		36.50	42.50	71.50	23.50%	26.50%	28.50%	
	9000	Per pupil, total cost	Ea.	10,300	21,000	26,200				
80	0010	**SCHOOLS Vocational**	S.F.	88.50	128	159				80
	0020	Total project costs	C.F.	5.50	7.90	10.90				
	0500	Masonry	S.F.	5.20	12.80	19.60	3.53%	6.70%	10.95%	
	1800	Equipment		2.77	6.90	9.60	1.24%	3.10%	4.26%	
	2720	Plumbing		5.65	8.45	12.40	5.40%	6.90%	8.55%	
	2770	Heating, ventilating, air conditioning		7.90	14.75	24.50	8.60%	11.90%	14.65%	
	2900	Electrical		9.20	12.05	16.60	8.45%	10.95%	13.20%	
	3100	Total: Mechanical & Electrical		32	35.50	61	23.50%	27.50%	31%	
	9000	Per pupil, total cost	Ea.	12,300	33,000	49,200				
83	0010	**SPORTS ARENAS**	S.F.	77.50	103	159				83
	0020	Total project costs	C.F.	4.20	7.50	9.70				
	2720	Plumbing	S.F.	4.49	6.80	14.35	4.35%	6.35%	9.40%	
	2770	Heating, ventilating, air conditioning		9.65	11.40	15.85	8.80%	10.20%	13.55%	
	2900	Electrical		8.05	10.95	14.15	8.60%	9.90%	12.25%	
	3100	Total: Mechanical & Electrical		20	35	47	21.50%	25%	27.50%	
85	0010	**SUPERMARKETS**	S.F.	71.50	83	97				85
	0020	Total project costs	C.F.	3.98	4.81	7.30				
	2720	Plumbing	S.F.	3.99	5.05	5.85	5.40%	6%	7.45%	
	2770	Heating, ventilating, air conditioning		5.85	7.80	9.50	8.60%	8.65%	9.60%	
	2900	Electrical		8.95	10.30	12.15	10.40%	12.45%	13.60%	
	3100	Total: Mechanical & Electrical		23	25	32.50	20.50%	26.50%	31%	
86	0010	**SWIMMING POOLS**	S.F.	116	194	415				86
	0020	Total project costs	C.F.	9.25	11.55	12.60				
	2720	Plumbing	S.F.	10.70	12.20	16.60	4.80%	9.70%	20.50%	
	2900	Electrical		8.70	14.10	20.50	5.75%	6.95%	7.60%	
	3100	Total: Mechanical & Electrical		21	55	73.50	11.15%	14.10%	23.50%	
87	0010	**TELEPHONE EXCHANGES**	S.F.	154	225	285				87
	0020	Total project costs	C.F.	9.55	15.35	21				
	2720	Plumbing	S.F.	6.50	10	14.65	4.52%	5.80%	6.90%	
	2770	Heating, ventilating, air conditioning		15.05	30.50	37.50	11.80%	16.05%	18.40%	

660

50 17 | Square Foot Costs

50 17 00 | S.F. Costs

			UNIT	UNIT COSTS			% OF TOTAL		
				1/4	MEDIAN	3/4	1/4	MEDIAN	3/4
87	2900	Electrical	S.F.	15.65	25	44	10.90%	14%	17.85%
	3100	Total: Mechanical & Electrical	↓	46	87.50	124	29.50%	33.50%	44.50%
91	0010	**THEATERS**	S.F.	96.50	120	183			
	0020	Total project costs	C.F.	4.46	6.60	9.70			
	2720	Plumbing	S.F.	3.22	3.49	14.25	2.92%	4.70%	6.80%
	2770	Heating, ventilating, air conditioning		9.40	11.35	14.05	8%	12.25%	13.40%
	2900	Electrical		8.45	11.40	23	8.05%	9.95%	12.25%
	3100	Total: Mechanical & Electrical	↓	21.50	32.50	66.50	23%	26.50%	27.50%
94	0010	**TOWN HALLS City Halls & Municipal Buildings**	S.F.	108	137	179			
	0020	Total project costs	C.F.	9.85	11.80	16.60			
	2720	Plumbing	S.F.	4.50	8.40	15.50	4.31%	5.95%	7.95%
	2770	Heating, ventilating, air conditioning		8.15	16.15	23.50	7.05%	9.05%	13.45%
	2900	Electrical		10.25	14.60	19.95	8.05%	9.45%	11.65%
	3100	Total: Mechanical & Electrical	↓	35.50	45	69	22%	26.50%	31%
97	0010	**WAREHOUSES & Storage Buildings**	S.F.	40.50	60	86			
	0020	Total project costs	C.F.	2.11	3.30	5.45			
	0100	Site work	S.F.	4.15	8.25	12.45	6.05%	12.95%	19.85%
	0500	Masonry		2.42	5.70	12.35	3.73%	7.40%	12.30%
	1800	Equipment		.65	1.39	7.80	.91%	1.82%	5.55%
	2720	Plumbing		1.34	2.41	4.50	2.90%	4.80%	6.55%
	2730	Heating, ventilating, air conditioning		1.53	4.32	5.80	2.41%	5%	8.90%
	2900	Electrical		2.38	4.48	7.40	5.15%	7.20%	10.10%
	3100	Total: Mechanical & Electrical	↓	6.65	10.20	20	12.75%	18.90%	26%
99	0010	**WAREHOUSE & OFFICES Combination**	S.F.	49.50	66	90.50			
	0020	Total project costs	C.F.	2.53	3.67	5.45			
	1800	Equipment	S.F.	.86	1.66	2.47	.52%	1.20%	2.40%
	2720	Plumbing		1.91	3.39	4.96	3.74%	4.76%	6.30%
	2770	Heating, ventilating, air conditioning		3.02	4.72	6.60	5%	5.65%	10.05%
	2900	Electrical		3.31	4.93	7.75	5.75%	8%	10%
	3100	Total: Mechanical & Electrical	↓	9.25	14.25	22.50	14.40%	19.95%	24.50%

Square Foot Project Size Modifier

One factor that affects the S.F. cost of a particular building is the size. In general, for buildings built to the same specifications in the same locality, the larger building will have the lower S.F. cost. This is due mainly to the decreasing contribution of the exterior walls plus the economy of scale usually achievable in larger buildings. The Area Conversion Scale shown below will give a factor to convert costs for the typical size building to an adjusted cost for the particular project.

The Square Foot Base Size lists the median costs, most typical project size in our accumulated data, and the range in size of the projects.

The Size Factor for your project is determined by dividing your project area in S.F. by the typical project size for the particular Building Type. With this factor, enter the Area Conversion Scale at the appropriate Size Factor and determine the appropriate cost multiplier for your building size.

Example: Determine the cost per S.F. for a 100,000 S.F. Mid-rise apartment building.

$$\frac{\text{Proposed building area} = 100,000 \text{ S.F.}}{\text{Typical size from below} = 50,000 \text{ S.F.}} = 2.00$$

Enter Area Conversion scale at 2.0, intersect curve, read horizontally the appropriate cost multiplier of .94. Size adjusted cost becomes .94 x $107.00 = $101.00 based on national average costs.

Note: For Size Factors less than .50, the Cost Multiplier is 1.1
For Size Factors greater than 3.5, the Cost Multiplier is .90

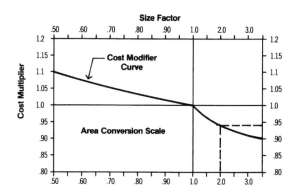

Square Foot Base Size

Building Type	Median Cost per S.F.	Typical Size Gross S.F.	Typical Range Gross S.F.	Building Type	Median Cost per S.F.	Typical Size Gross S.F.	Typical Range Gross S.F.
Apartments, Low Rise	$ 84.50	21,000	9,700 - 37,200	Jails	$ 257.00	40,000	5,500 - 145,000
Apartments, Mid Rise	107.00	50,000	32,000 - 100,000	Libraries	161.00	12,000	7,000 - 31,000
Apartments, High Rise	116.00	145,000	95,000 - 600,000	Living, Assisted	137.00	32,300	23,500 - 50,300
Auditoriums	141.00	25,000	7,600 - 39,000	Medical Clinics	146.00	7,200	4,200 - 15,700
Auto Sales	105.00	20,000	10,800 - 28,600	Medical Offices	138.00	6,000	4,000 - 15,000
Banks	189.00	4,200	2,500 - 7,500	Motels	102.00	40,000	15,800 - 120,000
Churches	130.00	17,000	2,000 - 42,000	Nursing Homes	142.00	23,000	15,000 - 37,000
Clubs, Country	132.00	6,500	4,500 - 15,000	Offices, Low Rise	120.00	20,000	5,000 - 80,000
Clubs, Social	126.00	10,000	6,000 - 13,500	Offices, Mid Rise	119.00	120,000	20,000 - 300,000
Clubs, YMCA	143.00	28,300	12,800 - 39,400	Offices, High Rise	152.00	260,000	120,000 - 800,000
Colleges (Class)	152.00	50,000	15,000 - 150,000	Police Stations	190.00	10,500	4,000 - 19,000
Colleges (Science Lab)	242.00	45,600	16,600 - 80,000	Post Offices	141.00	12,400	6,800 - 30,000
College (Student Union)	179.00	33,400	16,000 - 85,000	Power Plants	1000.00	7,500	1,000 - 20,000
Community Center	134.00	9,400	5,300 - 16,700	Religious Education	120.00	9,000	6,000 - 12,000
Court Houses	180.00	32,400	17,800 - 106,000	Research	198.00	19,000	6,300 - 45,000
Dept. Stores	78.50	90,000	44,000 - 122,000	Restaurants	172.00	4,400	2,800 - 6,000
Dormitories, Low Rise	143.00	25,000	10,000 - 95,000	Retail Stores	84.00	7,200	4,000 - 17,600
Dormitories, Mid Rise	176.00	85,000	20,000 - 200,000	Schools, Elementary	125.00	41,000	24,500 - 55,000
Factories	76.50	26,400	12,900 - 50,000	Schools, Jr. High	129.00	92,000	52,000 - 119,000
Fire Stations	139.00	5,800	4,000 - 8,700	Schools, Sr. High	133.00	101,000	50,500 - 175,000
Fraternity Houses	131.00	12,500	8,200 - 14,800	Schools, Vocational	128.00	37,000	20,500 - 82,000
Funeral Homes	146.00	10,000	4,000 - 20,000	Sports Arenas	103.00	15,000	5,000 - 40,000
Garages, Commercial	93.50	9,300	5,000 - 13,600	Supermarkets	83.00	44,000	12,000 - 60,000
Garages, Municipal	119.00	8,300	4,500 - 12,600	Swimming Pools	194.00	20,000	10,000 - 32,000
Garages, Parking	50.50	163,000	76,400 - 225,300	Telephone Exchange	225.00	4,500	1,200 - 10,600
Gymnasiums	129.00	19,200	11,600 - 41,000	Theaters	120.00	10,500	8,800 - 17,500
Hospitals	228.00	55,000	27,200 - 125,000	Town Halls	137.00	10,800	4,800 - 23,400
House (Elderly)	115.00	37,000	21,000 - 66,000	Warehouses	60.00	25,000	8,000 - 72,000
Housing (Public)	106.00	36,000	14,400 - 74,400	Warehouse & Office	66.00	25,000	8,000 - 72,000
Ice Rinks	153.00	29,000	27,200 - 33,600				

Abbreviations

A	Area Square Feet; Ampere	BTUH	BTU per Hour	Cwt.	100 Pounds
ABS	Acrylonitrile Butadiene Stryrene; Asbestos Bonded Steel	B.U.R.	Built-up Roofing	C.W.X.	Cool White Deluxe
		BX	Interlocked Armored Cable	C.Y.	Cubic Yard (27 cubic feet)
A.C.	Alternating Current; Air-Conditioning; Asbestos Cement; Plywood Grade A & C	°C	degree centegrade	C.Y./Hr.	Cubic Yard per Hour
		c	Conductivity, Copper Sweat	Cyl.	Cylinder
		C	Hundred; Centigrade	d	Penny (nail size)
		C/C	Center to Center, Cedar on Cedar	D	Deep; Depth; Discharge
A.C.I.	American Concrete Institute	C-C	Center to Center	Dis., Disch.	Discharge
AD	Plywood, Grade A & D	Cab.	Cabinet	Db.	Decibel
Addit.	Additional	Cair.	Air Tool Laborer	Dbl.	Double
Adj.	Adjustable	Calc	Calculated	DC	Direct Current
af	Audio-frequency	Cap.	Capacity	DDC	Direct Digital Control
A.G.A.	American Gas Association	Carp.	Carpenter	Demob.	Demobilization
Agg.	Aggregate	C.B.	Circuit Breaker	d.f.u.	Drainage Fixture Units
A.H.	Ampere Hours	C.C.A.	Chromate Copper Arsenate	D.H.	Double Hung
A hr.	Ampere-hour	C.C.F.	Hundred Cubic Feet	DHW	Domestic Hot Water
A.H.U.	Air Handling Unit	cd	Candela	DI	Ductile Iron
A.I.A.	American Institute of Architects	cd/sf	Candela per Square Foot	Diag.	Diagonal
AIC	Ampere Interrupting Capacity	CD	Grade of Plywood Face & Back	Diam., Dia	Diameter
Allow.	Allowance	CDX	Plywood, Grade C & D, exterior glue	Distrib.	Distribution
alt.	Altitude			Div.	Division
Alum.	Aluminum	Cefi.	Cement Finisher	Dk.	Deck
a.m.	Ante Meridiem	Cem.	Cement	D.L.	Dead Load; Diesel
Amp.	Ampere	CF	Hundred Feet	DLH	Deep Long Span Bar Joist
Anod.	Anodized	C.F.	Cubic Feet	Do.	Ditto
Approx.	Approximate	CFM	Cubic Feet per Minute	Dp.	Depth
Apt.	Apartment	c.g.	Center of Gravity	D.P.S.T.	Double Pole, Single Throw
Asb.	Asbestos	CHW	Chilled Water; Commercial Hot Water	Dr.	Drive
A.S.B.C.	American Standard Building Code			Drink.	Drinking
Asbe.	Asbestos Worker	C.I.	Cast Iron	D.S.	Double Strength
ASCE.	American Society of Civil Engineers	C.I.P.	Cast in Place	D.S.A.	Double Strength A Grade
A.S.H.R.A.E.	American Society of Heating, Refrig. & AC Engineers	Circ.	Circuit	D.S.B.	Double Strength B Grade
		C.L.	Carload Lot	Dty.	Duty
A.S.M.E.	American Society of Mechanical Engineers	Clab.	Common Laborer	DWV	Drain Waste Vent
		Clam	Common maintenance laborer	DX	Deluxe White, Direct Expansion
A.S.T.M.	American Society for Testing and Materials	C.L.F.	Hundred Linear Feet	dyn	Dyne
		CLF	Current Limiting Fuse	e	Eccentricity
Attchmt.	Attachment	CLP	Cross Linked Polyethylene	E	Equipment Only; East
Avg., Ave.	Average	cm	Centimeter	Ea.	Each
A.W.G.	American Wire Gauge	CMP	Corr. Metal Pipe	E.B.	Encased Burial
AWWA	American Water Works Assoc.	C.M.U.	Concrete Masonry Unit	Econ.	Economy
Bbl.	Barrel	CN	Change Notice	E.C.Y	Embankment Cubic Yards
B&B	Grade B and Better; Balled & Burlapped	Col.	Column	EDP	Electronic Data Processing
		CO_2	Carbon Dioxide	EIFS	Exterior Insulation Finish System
B.&S.	Bell and Spigot	Comb.	Combination	E.D.R.	Equiv. Direct Radiation
B.&W.	Black and White	Compr.	Compressor	Eq.	Equation
b.c.c.	Body-centered Cubic	Conc.	Concrete	EL	elevation
B.C.Y.	Bank Cubic Yards	Cont.	Continuous; Continued, Container	Elec.	Electrician; Electrical
BE	Bevel End	Corr.	Corrugated	Elev.	Elevator; Elevating
B.F.	Board Feet	Cos	Cosine	EMT	Electrical Metallic Conduit; Thin Wall Conduit
Bg. cem.	Bag of Cement	Cot	Cotangent		
BHP	Boiler Horsepower; Brake Horsepower	Cov.	Cover	Eng.	Engine, Engineered
		C/P	Cedar on Paneling	EPDM	Ethylene Propylene Diene Monomer
B.I.	Black Iron	CPA	Control Point Adjustment		
Bit., Bitum.	Bituminous	Cplg.	Coupling	EPS	Expanded Polystyrene
Bit., Conc.	Bituminous Concrete	C.P.M.	Critical Path Method	Eqhv.	Equip. Oper., Heavy
Bk.	Backed	CPVC	Chlorinated Polyvinyl Chloride	Eqlt.	Equip. Oper., Light
Bkrs.	Breakers	C.Pr.	Hundred Pair	Eqmd.	Equip. Oper., Medium
Bldg.	Building	CRC	Cold Rolled Channel	Eqmm.	Equip. Oper., Master Mechanic
Blk.	Block	Creos.	Creosote	Eqol.	Equip. Oper., Oilers
Bm.	Beam	Crpt.	Carpet & Linoleum Layer	Equip.	Equipment
Boil.	Boilermaker	CRT	Cathode-ray Tube	ERW	Electric Resistance Welded
B.P.M.	Blows per Minute	CS	Carbon Steel, Constant Shear Bar Joist	E.S.	Energy Saver
BR	Bedroom			Est.	Estimated
Brg.	Bearing	Csc	Cosecant	esu	Electrostatic Units
Brhe.	Bricklayer Helper	C.S.F.	Hundred Square Feet	E.W.	Each Way
Bric.	Bricklayer	CSI	Construction Specifications Institute	EWT	Entering Water Temperature
Brk.	Brick			Excav.	Excavation
Brng.	Bearing	C.T.	Current Transformer	Exp.	Expansion, Exposure
Brs.	Brass	CTS	Copper Tube Size	Ext.	Exterior
Brz.	Bronze	Cu	Copper, Cubic	Extru.	Extrusion
Bsn.	Basin	Cu. Ft.	Cubic Foot	f.	Fiber stress
Btr.	Better	cw	Continuous Wave	F	Fahrenheit; Female; Fill
Btu	British Thermal Unit	C.W.	Cool White; Cold Water	Fab.	Fabricated

Abbreviations

FBGS	Fiberglass	H.P.	Horsepower; High Pressure	LE	Lead Equivalent
F.C.	Footcandles	H.P.F.	High Power Factor	LED	Light Emitting Diode
f.c.c.	Face-centered Cubic	Hr.	Hour	L.F.	Linear Foot
f'c.	Compressive Stress in Concrete; Extreme Compressive Stress	Hrs./Day	Hours per Day	L.F. Nose	Linear Foot of Stair Nosing
		HSC	High Short Circuit	L.F. Rsr	Linear Foot of Stair Riser
F.E.	Front End	Ht.	Height	Lg.	Long; Length; Large
FEP	Fluorinated Ethylene Propylene (Teflon)	Htg.	Heating	L & H	Light and Heat
		Htrs.	Heaters	LH	Long Span Bar Joist
F.G.	Flat Grain	HVAC	Heating, Ventilation & Air-Conditioning	L.H.	Labor Hours
F.H.A.	Federal Housing Administration			L.L.	Live Load
Fig.	Figure	Hvy.	Heavy	L.L.D.	Lamp Lumen Depreciation
Fin.	Finished	HW	Hot Water	lm	Lumen
Fixt.	Fixture	Hyd.;Hydr.	Hydraulic	lm/sf	Lumen per Square Foot
Fl. Oz.	Fluid Ounces	Hz.	Hertz (cycles)	lm/W	Lumen per Watt
Flr.	Floor	I.	Moment of Inertia	L.O.A.	Length Over All
F.M.	Frequency Modulation; Factory Mutual	IBC	International Building Code	log	Logarithm
		I.C.	Interrupting Capacity	L-O-L	Lateralolet
Fmg.	Framing	ID	Inside Diameter	long.	longitude
Fdn.	Foundation	I.D.	Inside Dimension; Identification	L.P.	Liquefied Petroleum; Low Pressure
Fori.	Foreman, Inside	I.F.	Inside Frosted	L.P.F.	Low Power Factor
Foro.	Foreman, Outside	I.M.C.	Intermediate Metal Conduit	LR	Long Radius
Fount.	Fountain	In.	Inch	L.S.	Lump Sum
fpm	Feet per Minute	Incan.	Incandescent	Lt.	Light
FPT	Female Pipe Thread	Incl.	Included; Including	Lt. Ga.	Light Gauge
Fr.	Frame	Int.	Interior	L.T.L.	Less than Truckload Lot
F.R.	Fire Rating	Inst.	Installation	Lt. Wt.	Lightweight
FRK	Foil Reinforced Kraft	Insul.	Insulation/Insulated	L.V.	Low Voltage
FRP	Fiberglass Reinforced Plastic	I.P.	Iron Pipe	M	Thousand; Material; Male; Light Wall Copper Tubing
FS	Forged Steel	I.P.S.	Iron Pipe Size		
FSC	Cast Body; Cast Switch Box	I.P.T.	Iron Pipe Threaded	M^2CA	Meters Squared Contact Area
Ft.	Foot; Feet	I.W.	Indirect Waste	m/hr.; M.H.	Man-hour
Ftng.	Fitting	J	Joule	mA	Milliampere
Ftg.	Footing	J.I.C.	Joint Industrial Council	Mach.	Machine
Ft lb.	Foot Pound	K	Thousand; Thousand Pounds; Heavy Wall Copper Tubing, Kelvin	Mag. Str.	Magnetic Starter
Furn.	Furniture			Maint.	Maintenance
FVNR	Full Voltage Non-Reversing	K.A.H.	Thousand Amp. Hours	Marb.	Marble Setter
FXM	Female by Male	KCMIL	Thousand Circular Mils	Mat; Mat'l.	Material
Fy.	Minimum Yield Stress of Steel	KD	Knock Down	Max.	Maximum
g	Gram	K.D.A.T.	Kiln Dried After Treatment	MBF	Thousand Board Feet
G	Gauss	kg	Kilogram	MBH	Thousand BTU's per hr.
Ga.	Gauge	kG	Kilogauss	MC	Metal Clad Cable
Gal., gal.	Gallon	kgf	Kilogram Force	M.C.F.	Thousand Cubic Feet
gpm, GPM	Gallon per Minute	kHz	Kilohertz	M.C.F.M.	Thousand Cubic Feet per Minute
Galv.	Galvanized	Kip.	1000 Pounds	M.C.M.	Thousand Circular Mils
Gen.	General	KJ	Kiljoule	M.C.P.	Motor Circuit Protector
G.F.I.	Ground Fault Interrupter	K.L.	Effective Length Factor	MD	Medium Duty
Glaz.	Glazier	K.L.F.	Kips per Linear Foot	M.D.O.	Medium Density Overlaid
GPD	Gallons per Day	Km	Kilometer	Med.	Medium
GPH	Gallons per Hour	K.S.F.	Kips per Square Foot	MF	Thousand Feet
GPM	Gallons per Minute	K.S.I.	Kips per Square Inch	M.F.B.M.	Thousand Feet Board Measure
GR	Grade	kV	Kilovolt	Mfg.	Manufacturing
Gran.	Granular	kVA	Kilovolt Ampere	Mfrs.	Manufacturers
Grnd.	Ground	K.V.A.R.	Kilovar (Reactance)	mg	Milligram
H	High Henry	KW	Kilowatt	MGD	Million Gallons per Day
H.C.	High Capacity	KWh	Kilowatt-hour	MGPH	Thousand Gallons per Hour
H.D.	Heavy Duty; High Density	L	Labor Only; Length; Long; Medium Wall Copper Tubing	MH, M.H.	Manhole; Metal Halide; Man-Hour
H.D.O.	High Density Overlaid			MHz	Megahertz
H.D.P.E.	high density polyethelene	Lab.	Labor	Mi.	Mile
Hdr.	Header	lat	Latitude	MI	Malleable Iron; Mineral Insulated
Hdwe.	Hardware	Lath.	Lather	mm	Millimeter
Help.	Helper Average	Lav.	Lavatory	Mill.	Millwright
HEPA	High Efficiency Particulate Air Filter	lb.; #	Pound	Min., min.	Minimum, minute
		L.B.	Load Bearing; L Conduit Body	Misc.	Miscellaneous
Hg	Mercury	L. & E.	Labor & Equipment	ml	Milliliter, Mainline
HIC	High Interrupting Capacity	lb./hr.	Pounds per Hour	M.L.F.	Thousand Linear Feet
HM	Hollow Metal	lb./L.F.	Pounds per Linear Foot	Mo.	Month
HMWPE	high molecular weight polyethylene	lbf/sq.in.	Pound-force per Square Inch	Mobil.	Mobilization
		L.C.L.	Less than Carload Lot	Mog.	Mogul Base
H.O.	High Output	L.C.Y.	Loose Cubic Yard	MPH	Miles per Hour
Horiz.	Horizontal	Ld.	Load	MPT	Male Pipe Thread

Abbreviations

MRT	Mile Round Trip	Pl.	Plate	S.F.C.A.	Square Foot Contact Area
ms	Millisecond	Plah.	Plasterer Helper	S.F. Flr.	Square Foot of Floor
M.S.F.	Thousand Square Feet	Plas.	Plasterer	S.F.G.	Square Foot of Ground
Mstz.	Mosaic & Terrazzo Worker	Pluh.	Plumbers Helper	S.F. Hor.	Square Foot Horizontal
M.S.Y.	Thousand Square Yards	Plum.	Plumber	S.F.R.	Square Feet of Radiation
Mtd., mtd.	Mounted	Ply.	Plywood	S.F. Shlf.	Square Foot of Shelf
Mthe.	Mosaic & Terrazzo Helper	p.m.	Post Meridiem	S4S	Surface 4 Sides
Mtng.	Mounting	Pntd.	Painted	Shee.	Sheet Metal Worker
Mult.	Multi; Multiply	Pord.	Painter, Ordinary	Sin.	Sine
M.V.A.	Million Volt Amperes	pp	Pages	Skwk.	Skilled Worker
M.V.A.R.	Million Volt Amperes Reactance	PP, PPL	Polypropylene	SL	Saran Lined
MV	Megavolt	P.P.M.	Parts per Million	S.L.	Slimline
MW	Megawatt	Pr.	Pair	Sldr.	Solder
MXM	Male by Male	P.E.S.B.	Pre-engineered Steel Building	SLH	Super Long Span Bar Joist
MYD	Thousand Yards	Prefab.	Prefabricated	S.N.	Solid Neutral
N	Natural; North	Prefin.	Prefinished	S-O-L	Socketolet
nA	Nanoampere	Prop.	Propelled	sp	Standpipe
NA	Not Available; Not Applicable	PSF, psf	Pounds per Square Foot	S.P.	Static Pressure; Single Pole; Self-Propelled
N.B.C.	National Building Code	PSI, psi	Pounds per Square Inch		
NC	Normally Closed	PSIG	Pounds per Square Inch Gauge	Spri.	Sprinkler Installer
N.E.M.A.	National Electrical Manufacturers Assoc.	PSP	Plastic Sewer Pipe	spwg	Static Pressure Water Gauge
		Pspr.	Painter, Spray	S.P.D.T.	Single Pole, Double Throw
NEHB	Bolted Circuit Breaker to 600V.	Psst.	Painter, Structural Steel	SPF	Spruce Pine Fir
N.L.B.	Non-Load-Bearing	P.T.	Potential Transformer	S.P.S.T.	Single Pole, Single Throw
NM	Non-Metallic Cable	P. & T.	Pressure & Temperature	SPT	Standard Pipe Thread
nm	Nanometer	Ptd.	Painted	Sq.	Square; 100 Square Feet
No.	Number	Ptns.	Partitions	Sq. Hd.	Square Head
NO	Normally Open	Pu	Ultimate Load	Sq. In.	Square Inch
N.O.C.	Not Otherwise Classified	PVC	Polyvinyl Chloride	S.S.	Single Strength; Stainless Steel
Nose.	Nosing	Pvmt.	Pavement	S.S.B.	Single Strength B Grade
N.P.T.	National Pipe Thread	Pwr.	Power	sst, ss	Stainless Steel
NQOD	Combination Plug-on/Bolt on Circuit Breaker to 240V.	Q	Quantity Heat Flow	Sswk.	Structural Steel Worker
		Qt.	Quart	Sswl.	Structural Steel Welder
N.R.C.	Noise Reduction Coefficient/ Nuclear Regulator Commission	Quan., Qty.	Quantity	St.;Stl.	Steel
		Q.C.	Quick Coupling	S.T.C.	Sound Transmission Coefficient
N.R.S.	Non Rising Stem	r	Radius of Gyration	Std.	Standard
ns	Nanosecond	R	Resistance	Stg.	Staging
nW	Nanowatt	R.C.P.	Reinforced Concrete Pipe	STK	Select Tight Knot
OB	Opposing Blade	Rect.	Rectangle	STP	Standard Temperature & Pressure
OC	On Center	Reg.	Regular	Stpi.	Steamfitter, Pipefitter
OD	Outside Diameter	Reinf.	Reinforced	Str.	Strength; Starter; Straight
O.D.	Outside Dimension	Req'd.	Required	Strd.	Stranded
ODS	Overhead Distribution System	Res.	Resistant	Struct.	Structural
O.G.	Ogee	Resi.	Residential	Sty.	Story
O.H.	Overhead	Rgh.	Rough	Subj.	Subject
O&P	Overhead and Profit	RGS	Rigid Galvanized Steel	Subs.	Subcontractors
Oper.	Operator	R.H.W.	Rubber, Heat & Water Resistant; Residential Hot Water	Surf.	Surface
Opng.	Opening			Sw.	Switch
Orna.	Ornamental	rms	Root Mean Square	Swbd.	Switchboard
OSB	Oriented Strand Board	Rnd.	Round	S.Y.	Square Yard
O.S.&Y.	Outside Screw and Yoke	Rodm.	Rodman	Syn.	Synthetic
Ovhd.	Overhead	Rofc.	Roofer, Composition	S.Y.P.	Southern Yellow Pine
OWG	Oil, Water or Gas	Rofp.	Roofer, Precast	Sys.	System
Oz.	Ounce	Rohe.	Roofer Helpers (Composition)	t.	Thickness
P.	Pole; Applied Load; Projection	Rots.	Roofer, Tile & Slate	T	Temperature; Ton
p.	Page	R.O.W.	Right of Way	Tan	Tangent
Pape.	Paperhanger	RPM	Revolutions per Minute	T.C.	Terra Cotta
P.A.P.R.	Powered Air Purifying Respirator	R.S.	Rapid Start	T & C	Threaded and Coupled
PAR	Parabolic Reflector	Rsr	Riser	T.D.	Temperature Difference
Pc., Pcs.	Piece, Pieces	RT	Round Trip	Tdd	Telecommunications Device for the Deaf
P.C.	Portland Cement; Power Connector	S.	Suction; Single Entrance; South		
P.C.F.	Pounds per Cubic Foot	SC	Screw Cover	T.E.M.	Transmission Electron Microscopy
P.C.M.	Phase Contrast Microscopy	SCFM	Standard Cubic Feet per Minute	TFE	Tetrafluoroethylene (Teflon)
P.E.	Professional Engineer; Porcelain Enamel; Polyethylene; Plain End	Scaf.	Scaffold	T. & G.	Tongue & Groove; Tar & Gravel
		Sch., Sched.	Schedule		
		S.C.R.	Modular Brick	Th., Thk.	Thick
Perf.	Perforated	S.D.	Sound Deadening	Thn.	Thin
PEX	Cross linked polyethylene	S.D.R.	Standard Dimension Ratio	Thrded	Threaded
Ph.	Phase	S.E.	Surfaced Edge	Tilf.	Tile Layer, Floor
P.I.	Pressure Injected	Sel.	Select	Tilh.	Tile Layer, Helper
Pile.	Pile Driver	S.E.R., S.E.U.	Service Entrance Cable	THHN	Nylon Jacketed Wire
Pkg.	Package	S.F.	Square Foot	THW.	Insulated Strand Wire

Abbreviations

THWN	Nylon Jacketed Wire	USP	United States Primed	Wrck.	Wrecker
T.L.	Truckload	UTP	Unshielded Twisted Pair	W.S.P.	Water, Steam, Petroleum
T.M.	Track Mounted	V	Volt	WT., Wt.	Weight
Tot.	Total	V.A.	Volt Amperes	WWF	Welded Wire Fabric
T-O-L	Threadolet	V.C.T.	Vinyl Composition Tile	XFER	Transfer
T.S.	Trigger Start	VAV	Variable Air Volume	XFMR	Transformer
Tr.	Trade	VC	Veneer Core	XHD	Extra Heavy Duty
Transf.	Transformer	Vent.	Ventilation	XHHW, XLPE	Cross-Linked Polyethylene Wire Insulation
Trhv.	Truck Driver, Heavy	Vert.	Vertical		
Trlr	Trailer	V.F.	Vinyl Faced	XLP	Cross-linked Polyethylene
Trlt.	Truck Driver, Light	V.G.	Vertical Grain	Y	Wye
TTY	Teletypewriter	V.H.F.	Very High Frequency	yd	Yard
TV	Television	VHO	Very High Output	yr	Year
T.W.	Thermoplastic Water Resistant Wire	Vib.	Vibrating	Δ	Delta
		V.L.F.	Vertical Linear Foot	%	Percent
UCI	Uniform Construction Index	Vol.	Volume	~	Approximately
UF	Underground Feeder	VRP	Vinyl Reinforced Polyester	Ø	Phase; diameter
UGND	Underground Feeder	W	Wire; Watt; Wide; West	@	At
U.H.F.	Ultra High Frequency	w/	With	#	Pound; Number
U.I.	United Inch	W.C.	Water Column; Water Closet	<	Less Than
U.L.	Underwriters Laboratory	W.F.	Wide Flange	>	Greater Than
Uld.	unloading	W.G.	Water Gauge	Z	zone
Unfin.	Unfinished	Wldg.	Welding		
URD	Underground Residential Distribution	W. Mile	Wire Mile		
		W-O-L	Weldolet		
US	United States	W.R.	Water Resistant		

Index

A

Abatement asbestos 22
ABC extinguisher 237
Abrasive aluminum oxide 30
   floor 33
   floor tile 186
   silicon carbide 30, 33
   terrazzo 194
   tile 185
   tread 187
A/C packaged terminal 374
Accelerator sprinkler system 333
Accent light 400
   light connector 400
   light tranformer 400
Access control 165
   control card type 165
   door and panels 140
   door fire rated 140
   door floor 141
   door metal 140
   door roof 121
   door stainless steel 140
   floor 198
   road and parking area 15
Accessories bathroom .. 185, 233, 234
   door 165
   drywall 184
   plaster 179
   toilet 233
Accessory fireplace 235
   furniture 283
   masonry 41, 43
Accordion door 138
Acid proof floor 189, 195
Acoustic ceiling board 187
Acoustical and plaster ceiling .. 492, 493
   batt 200
   block 47
   ceiling 187, 188, 493
   door 144, 231
   enclosure 314
   folding partition 231
   glass 170
   panel 201, 229
   partition 229, 231
   room 312
   sealant 124, 184
   treatment 200
   underlayment 201
   wall tile 188
   wallboard 184
   window wall 147
Acrylic ceiling 187
   floor 195
   wall coating 217
   wallcovering 199
   wood block 190
Adapter fire hose swivel 329
Adhesive cement 193
   wallpaper 199
Adjustable astragal 164
Admixture cement 33
Admixtures concrete 30
Adobe brick 49
Aerator 355
Aggregate exposed 33
   masonry 42
Air balancing 358
   cleaner electronic 370
   compressor control system 360
   compressor dental 262
   compressor sprinkler system .. 333
   conditioner cooling & heating . 374
conditioner direct expansion .. 376
conditioner fan coil 376
conditioner gas heat 373
conditioner packaged term. ... 374
conditioner receptacle 388
conditioner removal 358
conditioner rooftop 373
conditioner thru-wall 374
conditioner wiring 390
conditioning 647
conditioning computer 374
conditioning fan 363
conditioning ventilating .. 360, 364, 368
cooled condensing unit 373
curtain 366
curtain door 366
filter 369, 370
filter industrial 370
filter roll type 369
filters washable 369
filtration 22
handling fan 363
handling troffer 398
handling unit 376
register 368
return grille 367
sampling 22
supply pneumatic control 360
supply register 368
tube system 326
vent automatic 361
wall 232
Air conditioner air cooled .. 514-516
   chilled water 511, 516
   direct expansion 512, 514, 515
   fan coil 511
   self-contained 514
   single zone 512
   split system 515
   water cooled 516
Air conditioning computer 516
   drug store 513
Aircraft cable 63
Airfoil fan centrifugal 364
Airless sprayer 22
Air-source heat pumps 375
Alarm exit control 410
   fire 332, 410
   panel & devices 411
   residential 389
   sprinkler 411
   standpipe 332, 411
   valve sprinkler 328
   water motor 332
Alteration fee 8
Alternating tread ladder 77
Aluminum and glass door 446
   astragal 163
   bench 307
   blind 268
   cast tread 79
   ceiling tile 188
   column 80, 108
   column base 80
   column cap 80
   cross 264
   curtain wall 148
   curtain wall glazed 148
   diffuser perforated 367
   directory board 223
   door 129, 133, 144, 145
   door frame 148
   ductwork 362
   edging 424
   entrance 148
foil 115, 117, 201
folding 473
grating tread 79
grille 367
handrail 198
nail 88
register 368
rivet 59
salvage 338
sash 149
shade 273
siding paint 205
sign 223
storefront 148
tile 118, 187
transom 148
tube frame 146
weatherstrip 163
weld rod 59
window demolition 128
window wall 475
windows 149, 150, 444, 445
Anchor bolt 43
   brick 43
   buck 43
   expansion 58
   framing 90
   hollow wall 58
   machinery 59
   masonry 43
   partition 43
   rafter 90
   rigid 43
   steel 43
   stone 43
   toggle bolt 58
Anechoic chamber 312
Angle corner 232
Apartment call system 406
Appliance 248
   plumbing 341, 342, 354
   residential 246, 248, 389
Approach ramp 198
Apron wood 101
Architectural fee 8
Area clean-up 24
Armored cable 383
Artificial grass surfacing 418
   turf 418
Artwork 268
Asbestos abatement 22
   demolition 24
   disposal 25
   removal 23
   removal process 626
Ash conveyor 263
   receiver 234, 283
   urn 283
Ashlar stone 52
   veneer 50
Asphalt block 417
   block floor 417
   cutting 21
   pavement 418
   primer 193
   sheathing 100
   shingle 117
   sidewalk 416
Asphaltic emulsion 420
Aspirator dental 262
Astragal adjustable 164
   aluminum 163
   magnetic 164
   molding 102
   one piece 163
   overlapping 164
rubber 163, 164
split 164
steel 163
Athletic equipment 259
   paving 418
Atomizer water 22
Attic stair 250
Audio masking 315
Audiometric room 312
Auditorium chair 301
Automatic fire suppression 334
   flush 350
   opener 156
   operator 156
   teller 245
   timed thermostat 361
   wall switches 403
   washing machine 249
Autopsy equipment 261, 263
Awning window 150, 151
Axe fire 330
Axial flow fan 363

B

Backer rod 124
Backerboard 181
Backfill planting pit 421
Backsplash 279
   countertop 279
Backstop basketball 260
   electric 260
Bag disposable 22
   glove 22
Baked enamel door 131
   enamel frame 130
Balanced door 147, 446
Balancing air 358
   water 358
Balcony fire escape 77
Baler 264
Balustrade painting 213
Band joist framing 71
   molding 101
   riser 258
Bank air conditioning 515
   counter 245
   equipment 245
   window 245
Banquet booth 295
Baptistry 264
Bar bell 259
   front 314
   grab 233
   joist painting 65
   panic 158
   parallel 261
   restaurant 290
   touch 158
   towel 234
   tub 233
   Zee 185
Barber equipment 246
Bare floor conduit 400
Bark mulch 420
   mulch redwood 420
Barrel bolts 156
Barricade 16
   tape 16
Barrier security 244
   separation 22, 27
   X-ray 316
Barriers and enclosures 16
Base cabinet 274, 275
   carpet 197

667

Index

ceramic tile	185
column	90
course	416
course drainage layers	416
cove	192
gravel	416
masonry	48
molding	101
quarry tile	186
resilient	192
road	416
sink	274
stone	52, 416
terrazzo	194
vanity	277
wood	101
Baseball scoreboard	260
Baseboard demolition	88
heat	376
heat electric	377
heater electric	377
Baseplate scaffold	14
shoring	14
Basic finish materials	172
Basketball backstop	260
scoreboard	260
Bath communal	310
hospital	351
paraffin	261
perineal	351
spa	310
steam	313
whirlpool	261, 310
Bathroom	345
accessories	185, 233, 234
exhaust fan	365
faucet	346
fixture	343-346
heater & fan	366
soaking	345
system	510
Bathtub bar	233
enclosure	227
pier	351
removal	338
Bathtubs residential	345
Batt acoustical	200
insulation	113, 114
thermal	200
Battery inverter	394
Bead blast demo	173
board insulation	311
casing	184
corner	179, 184
parting	103
Beam & girder framing	91
bondcrete	180
ceiling	108
concrete placement	32
drywall	182
fireproofing	122
hanger	90
mantel	108
plaster	179, 180
steel fireproofing	436
wood	91, 92
Beams load-bear. stud wall box	67
Bed comforter	283
molding	101
Bedding brick	417
Bedpan cleanser	351
Bedspread	283
Bell bar	259
system	393
Belt material handling	428
Bench aluminum	307
fiberglass	307
folding	302
park	307
planter	306
players	307
wood	307
work	258
Benches	307
Bevel glass	166
siding	119
Bicycle trainer	259
Bidet	343
rough-in	343
Bi-fold door	137
Bin parts	239
Bi-passing closet door	137
Birch door	137-139
molding	102
paneling	104
wood frame	109
Bit drill	36
Bituminous block	417
Blackboard	229
Blanket curing	34
insulation	113, 359
sound attenuation	200
Blast curtains	272
demo bead	173
floor shot	173
Blasting water	40
Bleacher telescoping	302
Blind venetian	268
window	268
Blinds vertical	269
Block acoustical	47
asphalt	417
bituminous	417
Block, brick and mortar	629
Block cap	48
column	46
concrete	46, 47, 629
corner	48
filler	214
floor	191
glass	48, 49
glazed	48
lightweight	47
lintel	46
lintel concrete	441
partition	47, 48, 452
reflective	49
sill	48
slotted	47
solar screen	48
wall glass	440
Blocking	91
carpentry	91
wood	94
Bloodbank refrigeration	257
Blower	364
pneumatic tube	326
Blown in cellulose	115
in fiberglass	115
in insulation	115
Blueboard	181
partition	175
Bluestone	50
sidewalk	416
sill	53
step	416
Board & batten siding	119
batten fence	419
bulletin	222
ceiling	187
control	222, 259
directory	223, 224
gypsum	182, 184
insulation	112
paneling	105
partition gypsum	175
partition NLB gypsum	174
sheathing	100
valance	275
verge	102
Boiler control system	360
demolition	358
insulation	359
removal	358
Bollard light	402
Bolt anchor	43
flush	156
steel	90
Bolts barrel	156
remove	57
Bond performance	11
Bondcrete	180
Bookcase	240, 275
Bookshelf	298
Booster fan	364
Booth banquet	295
fixed	295
freestanding	295
mounted	296
telephone	224
Border light	402
Bottle storage	251
Bow window	153
Bowl toilet	343
Box beams load-bear. stud wall	67
connector EMT	384, 385
locker	237
low voltage relay	392
mail	238
safety deposit	245
spring	289
stair	105
storage	13
termination	406
Boxed beam headers	71
Brace cross	61
Bracing	91
let-in	91
load-bearing stud wall	66
metal joist	70
roof rafter	73
shoring	14
Bracket scaffold	14
Brass hinge	161, 162
salvage	338
screw	89
Break glass station	411
ties concrete	34
Breaker circuit	394
Breeching insulation	359
Brick	45
adobe	49
anchor	43
bedding	417
Brick, block and mortar	629
Brick chimney simulated	234
demolition	41, 172
economy	44
edging	424
engineer	45
face	44
floor	417
molding	102
paving	417
saw	40
sidewalk	417
sill	53
simulated	54
stair	50
step	416
structural	44
veneer	44
veneer demolition	42
veneer wall	458-460
wall	455
wall panel	49
wall solid	453
Bricklaying	629
Bridges pedestrian	420
Bridging	91
joist	64
load-bearing stud wall	66
metal joist	70
roof rafter	73
steel	91
wood	91
Broiler	253
Bronze cast tread	79
cross	264
hinge	161
letter	223
plaque	224
push-pull plate	160
Broom cabinet	275
finish concrete	33
Brown coat	195
Brownstone	52
Bubbler	353
drinking	353, 354
Buck anchor	43
rough	93
Buggy concrete	32
Builder's risk insurance	620
Building air conditioning	511
air conditioning	511-516
demolition	20
directory	223, 406
greenhouse	314
hardware	157
insulation	114
model	8
paper	117
permit	12
prefabricated	314
sprinkler	333
temporary	13
wall exterior	440
Built-in shower	346
Bulb incandescent	404
Bullet resistant window	245
Bulletin board	222
board cork	222
Bulletproof glass	170
Bumper dock	244
door	160
wall	160, 232
Burglar alarm indicating panel	410
Burlap curing	34
rubbing concrete	34
Bush hammer	34
hammer finish concrete	34
Butyl caulking	124
BX cable	383

C

Cabinet base	274, 275
broom	275
casework	275, 276
convector heater	378
corner base	274
corner wall	275
demolition	88

Index

door	273, 276, 277	
fire equipment		236
hardboard		274
hardware		277, 278
heater electric		378
hinge		278
hose rack		236
hotel		234
key		165, 278
kitchen		274, 279
laboratory		278
medicine		234
metal		279
oven		275
portable		251
school		278
shower		226
stain		209
storage		278
strip		406
valve		236
varnish		209
wall		274, 279
wardrobe		240
Cable aircraft		63
armored		383
BX		383
electric		383
flat		382
MC		383
railing		81
safety railing		63
undercarpet system		382
Cafe door		138
Calcium silicate insulation		359
silicate pipe covering		340
Call system apartment		406
system emergency		406
systems nurse		407
Camera TV closed circuit		410
Canopy		240
framing		96
Cant roof		95
Cantilever needle beams		414
Cap block		48
post		90
Carbon dioxide exting.		237, 334
Carborundum rub finish		34
Card catalog file		298
Carousel compactor		264
Carpet		196, 197
base		197
computer room		198
felt pad		197
floor		197
lighting		400
nylon		197
pad		196
padding		196
removal		172
stair		197
tile		197
tile removal		172
urethane pad		197
wool		197
Carrel		297
Carrier ceiling		177
channel		189
Cart concrete		32
mounted extinguisher		237
Carving stone		52
Car wash water heater		342
Case display		278
exhibit		278
refrigerated		250
work		275

Casement window		149-151, 153
window vinyl		155
Casework		279
cabinet		275, 276
custom		274
demolition		88
ground		98
metal		165, 278
painting		209
stain		209
varnish		209
Cash register		245
Casing bead		184
wood		101
Cast in place concrete		31
in place terrazzo		194
iron bench		307
iron damper		235
iron stair		486, 487
iron tread		79
trim lock		159
Caster scaffold		14
Cast iron weld rod		59
Catch door		277
Catwalk scaffold		15
Caulking		124
masonry		125
polyurethane		124
sealant		124
Cavity wall insulation		114
Cedar closet		104
fence		419
paneling		105
post		108
roof plank		99
shingle		117
siding		119
Ceiling		187
acoustical		187, 188
beam		108
board		187
board acoustic		187
board fiberglass		187
bondcrete		180
carrier		177
demolition		172
diffuser		367
drywall		182
eggcrate		187
fan		364
framing		92
furring		98, 176, 177
hatch		141
heater		250
insulation		115
lath		178
luminous		188, 403
molding		101
painting		213-215
plaster		179
plaster and acoustical		493
polystyrene		311
register		368
stair		250
support		60
suspended		177, 182, 188
suspension system		189
tile		188
woven wire		228
Cell prison		314
Cellar wine		251
Cellular glass insulation		338
Cellulose blown in		115
insulation		114
Cement adhesive		193
admixture		33

color		42
fill stair		486
Keene		180
masonry		42
masonry unit		46, 47
mortar		186, 628
parging		112
plaster		311
terrazzo portland		194
underlayment self-leveling		35
vermiculite		122
Cementitious waterproofing		112
Central vacuum		247
Centrifugal airfoil fan		364
fan		364
Ceramic coating		217
mulch		421
tile		185, 186
tile demolition		173
tile floor		185
Chain core rope		244
hoist		428
hoist door		144
link fabric		418
link fence		418, 419
link fence paint		204
steel		79
Chair		296
barber		246
dental		262
folding		287
hydraulic		262
life guard		310
lifts		325
locker bench		238
molding		102
movie		256, 301, 302
portable		258
rail demolition		88
seating		301
stack		287
Chalkboard		220
electric		221
freestanding		222
liquid chalk		220
metal		229
sliding		221
swing leaf		221
wall hung		220
Chamber anechoic		312
decontamination		22, 27
echo		314
Chandelier		399
Chan-l-wire system		392
Channel carrier		189
frame		130
furring		176, 184
siding		119
steel		179
Charge powder		59
Charges disposal		25
Charging desk		298
Check floor		160
valve		328
Checkout counter		245
scanner		245
supermarket		245
Chemical cleaning masonry		40
dry extinguisher		237
media filter		370
Chest		297
Chime door		393
Chimney accessories		234
all fuel vent		370
demolition		41
metal		234

screen		235
simulated brick		234
vent fitting		370
Chips wood		421
Chlorination system		355
Church equipment		264
pew		302
Chute mail		239
package		326
refuse		326
rubbish		21
Chutes		326
linen		326
Cinema equipment		256
Circline fixture		400
Circuit breaker		394
Clapboard painting		205
Classroom seating		301
Clean control joint		33
room		311
Cleaning & disposal equipment		255
masonry		40
metal		57
up		17
Cleanout door		235
Clean-up area		24
Clevis hooks		79
Clinic service sink		351
Clip plywood		90
wire rope		61
Clock		283
& bell system		407
systems		407
timer		389
Closed circuit television system		410
circuit T.V.		245
Closer concealed		157
electronic		157
floor		157
holder		157
Closers door		157
Closet cedar		104
door		137, 138
pole		102
rod		240
water		343, 348
Cloth hardware		418, 419
Clothes dryer commercial		247
CMU		46
Coat brown		195
glaze		214
hook		233
rack		240, 275
scratch		195
Coating ceramic		217
epoxy		33, 217
wall		217
Coffee urn		252
Coiling door and grilles		141
Coin laundry water heater		342
operated gate		244
CO_2 extinguisher		237
Cold storage door		142, 143
Collection box lobby		238
Collector solar energy system		372
Colonial door		137, 139
wood frame		109
Color floor		33
wheel		402
Column		108
aluminum		80
base		90
base aluminum		80
block		46
bondcrete		180
cap aluminum		80

669

Index

concrete 31
concrete placement 32
demolition 41
drywall 182
fireproof 122
lath 178
plaster 179, 180
removal 41
stone 52
wood 92, 108
Columns framing 92
ornamental 82
structural 60
Combination device 386
storm door 137, 139
Comforter bed 283
Commercial air conditioner .. 511
air conditioning 512-516
door 131, 136, 141
electric service 529
folding partition 231
gas water heater 342
lavatory 349
refrigeration 311
water heater 341
water heater electric 341
Common brick 44
nail 88, 89
Communal bath 310
Communicating lockset ... 158, 159
Communication system 406
Compactor 263
residential 249
waste 264
Compartments shower 226
toilet 224
Compensation workers' 10
Component control 361
Composite brick wall 462-464
door 131
insulation 116
wall 462
Composition flooring ... 190, 195
flooring removal 172
Compound dustproofing 30
Compressor tank mounted 360
Computer air conditioning ... 374
floor 198
room carpet 198
Concealed closer 157
Concrete 628
admixtures 30
block 46, 47, 629
block back-up 46
block demolition 20
block insulation 114
block painting 214
block partition 450, 452
block planter 417
break ties 34
broom finish 33
buggy 32
burlap rubbing 34
bush hammer finish 34
cart 32
cast in place 31
column 31
core drilling 35
curing 34
curing compound 34
cutout 20
demolition 20, 30
finishing 33
fireproofing column 438
float finish 33, 34
floor patching 30

furring 98
granolithic finish 33
hand trowel finish 33
hardener 33
impact drilling 36
integral colors 30
integral topping 33
integral waterproofing 30
machine trowel finish 33
materials 628
membrane curing 34
pan fill stair 31
placement beam 32
placement column 32
placement wall 32
placing 32
planter 306
pumped 32
sandblast finish 34
scarify 172
sidewalk 416
slab cutting 35
stair 32, 487
stair finish 33
stair tread insert 31
topping 33
unit paving slabs precast ... 417
wall 31
wall patching 30
walls cutting 35
water curing 34
wheeling 32
Condensing unit air cooled .. 373
Conductive floor 194-196
terrazzo 195
Conductor 381
Conduit bare floor 400
electrical 384
Confessional 264
Connection standpipe 328
Connector accent light 400
joist 90
timber 90
Conserving water closet 348
Construction aids 13
cost index 9
management fee 8
photography 12
temporary 16
time 8
Contingencies 8
Continuous hinge 163
Contractor overhead 17
Control board 222, 259
component 361
joint 33
joint clean 33
joint sawn 33
joint sealant 34
package firecycle 332
pressure sprinkler 333
radiator supply 361
system air compressor 360
system boiler 360
system cooling 360
system electronic 360
system split system 360
system ventilator 360
systems pneumatic 360
valve heating 361
wire low-voltage 393
Controller time system 407
Convection oven 253
Convector cabinet heater 378
heater electric 377
Conveyor 428

ash 263
door 366
material handling 428
system 325
Cooking equipment 253
range 248
Cooler 311
beverage 251, 255
door 366
evaporator 374
water 354
Cooling computer 516
control system 360
Coping 51
removal 41
Copper cable 383
flashing 120, 121
roof 120
salvage 338
wall covering 198
wire 384
Core drilling concrete 35
Coreboard 183
Cork floor 193
insulation 311
tile 193
wall tile 198
Corner base cabinet 274
bead 179, 184
block 48
guard 232
protection 232
wall cabinet 275
Cornice molding 101
painting 213
stone 52
Correspondence lift 325
Corridor air conditioning ... 515
Corrosion resistant sink 349
Cost control 12
mark-up 11
Cot prison 296
Counter bank 245
checkout 245
door 141
hospital 279
top 279, 280
top demolition 88
window 238
Countertop backsplash 279
sink 345
Coupling EMT 384
fire hose 329
Course drainage layers base . 416
Court handball 313
racquetball 313
squash 313
Cove base 192
base ceramic tile 185
base terrazzo 194
molding 101
molding scotia 101
Cover ground 421, 422
pool 310
stair tread 16
Covering and paint wall .. 488, 489
and tile floor 490, 491
calcium silicate pipe 340
wall 636
Crane 428
Crematory 263
Critical path schedule 12
Cross brace 61
wall 264
Crowbar fire 330
Crown molding 101, 102

Crushed stone sidewalk 416
Cubicle 227
curtain 227
detention 155
shower 346
toilet 226
Cup sink 349, 505
Cupric oxychloride floor 195
Curb roof 95
terrazzo 194
Curing blanket 34
compound concrete 34
concrete 34
concrete membrane 34
concrete water 34
paper 117
Curtain air 366
cubicle 227
divider 260
gymnasium divider 260
rod 233
sound 315
track 259
valance 269-271
wall 148
wall glass 148
Curtains blast 272
Custom casework 274
Cutout demolition 20
slab 20
Cutting asphalt 21
concrete slabs 35
concrete walls 35
masonry 21
saw 35
wood 21
Cylinder lockset 158
recore 158, 159

D

Dairy case 250
Damper fireplace 235
Darkroom 312
door 143
equipment 248
revolving 312
sink 248
Day gate 143
Deadbolt 158, 160
Deadlocking latch 159
Deciduous shrub 423
tree 424
Deck edge form 66
roof 99, 100
slab form 66
wood 99
Decking 630
form 66
Decontamination chamber .. 22, 27
enclosure 22, 24, 27
equipment 22
Decorative hinge 162
Decorator device 386
switch 386
Deep therapy room 315
Deep longspan joist 64
Dehumidifier 250
Dehydrator package sprinkler 333
Delicatessen case 250
Delivery charge 15
Deluge sprinkler monit. panel 333
valve assembly sprinkler 333
Demo specialties 220
thermal & moisture protect. . 112

Index

Demolition 42, 173, 626
asbestos 24
baseboard 88
boiler 358
brick 41, 172
cabinet 88
casework 88
ceiling 172
ceramic tile 173
chimney 41
column 41
concrete 20, 30
concrete block 20
door 128
drywall 172
ductwork 358
electrical 380, 381
enclosure 24
fireplace 42
flooring 172
framing 85
glass 128
granite 42
gutting 21
HVAC 358
joist 85
lath 172
masonry 20, 41, 173
metal 57
metal stud 173
millwork 88
mold contaminated area 27
paneling 88
partition 173
plaster 172, 173
plenum 173
plumbing 338
plywood 172, 173
post 86
rafter 86
railing 88
roofing 24, 112
saw cutting 21
selective 84, 128, 242
steel 57
steel window 128
stucco 173
terra cotta 173
terrazzo 173
tile 172
torch cutting 21
trim 88
wall 173
walls and partitions 173
window 128
wood 172
Demountable partition 228
Dental equipment 262
Depository night 245
Desk 296, 297
accessory 284
Detection intrusion 410
system 410
Detector infrared 410
motion 410
smoke 411
temperature rise 411
ultrasonic motion 410
Detectors fire & heat 411
Detention cubicle 155
doors and frames 144
equipment 245
Developing tank 248
Device combination 386
decorator 386
exit 158
GFI 388
panic 157, 158
receptacle 387
residential 385
wiring 392, 393
Diamond lath 178
Diaper station 233
Diffuser ceiling 367
opposed blade damper 367
perforated aluminum 367
rectangular 367
steel 367
T bar mount 367
Dimmer switch 386, 393
Direct expansion A/C 376
Directory 223
board 223, 224
building 223, 406
shelf 224
Disappearing stairs 250
Dishwasher 249, 255
Dispenser hot water 341
napkin 233
soap 233
toilet tissue 233, 234
towel 233, 234
Dispensing equipment food 254
Dispersion nozzle 334
Display case 278
Disposable bag 22
Disposal asbestos 25
charges 25
garbage 249
waste 263
Distiller water 261
Divider curtain 260
office 230
strip terrazzo 194
Diving board 310
Dock bumper 244
Dome roof 149
Door accessories 165
accordion 138
acoustical 144, 231
air curtain 366
aluminum 129, 133
and grilles coiling 141
and panels access 140
and window matls 128
automatic entrance 147
balanced 147
bell residential 388
bell system 393
bi-fold 137
bi-passing closet 137
birch 138
bullet resistant 245
bumper 160
cabinet 273, 276, 277
cafe 138
catch 277
chain hoist 144
chime 393
cleanout 235
closers 157
closet 138
cold storage 142
combination storm 137
commercial 131, 136
composite 131
conveyor 366
cooler 366
counter 141
darkroom 143
demolition 128
double 132
dutch 137
dutch oven 235
entrance 129, 137, 146
fiberglass 139
fire 132, 135, 142, 411
fire rated access 140
flexible 145
floor 141
flush 134, 135, 137
flush wood 134
folding 231
folding accordion 138
frame 109, 129, 130
frame interior 109
frame lead lined 315
freezer 366
french 139
garage 144
glass 141, 147
grille 367
handle 277
hardware 157
industrial 142
jamb molding 102
kick plate 163
knob 159
labeled 132, 135
metal 129, 144
metal access 140
mirror 168
molding 102
moulded 136
opener 128, 142, 156
operator 128, 156
overhead 144, 145
overhead commercial 144
panel 136
paneled 132
partition 229
passage 138, 139
prefinished 134, 135
pre-hung 139
prison 144
pull 160
refrigerator 142, 143
release 406
removal 128
residential 109, 132, 137
revolving 143, 147
revolving entrance 147
rolling 142
roof 121
rough buck 94
sauna 313
sectional 144
security vault 144
shower 227
side light 139
sidelight 132, 133
sill 103, 109, 160
sliding 133, 141
special 140, 144
stain 209
stainless steel (access) 140
steel 131, 132, 144, 632
stop 146, 160
swing 139
switch burglar alarm 410
telescoping 145
threshold 109
traffic 145
varnish 209
vault 143
vertical lift 144, 145
weatherstrip 165
weatherstrip garage 165
wire partition 228
wood 133, 134, 136, 139
Doors & windows exterior paint 206
& windows interior paint 210, 211
and frames detention 144
Dormitory furniture 297
Double acting door 145
brick 45
hung window 150, 152, 154
weight hinge 162
Drapery 269-271
hardware 271, 272
installation 271
rings 271
Drawer track 278
type cabinet 279
wood 277
Dressing unit 297
Drill bit 36
console dental 262
shop 258
wood 90
Drilling plaster 175
steel 57
Drinking bubbler 353, 354
fountain 353, 354
fountain deck rough-in 354
fountain floor rough-in 354
fountain handicap 353
fountain wall rough-in 353, 354
Drive-up window 245
Driveway 416
Drop pan ceiling 188
Drug store air conditioning 511, 515
Dry chemical extinguisher 237
fall painting 215
pipe sprinkler head 333
wall leaded 315
wall partition 468
Drycleaner 246
Dryer commercial clothes 247
darkroom 248
hand 233
receptacle 388
residential 249
vent 249
Drywall 182
accessories 184
ceiling 182, 493
column 182
cutout 20
demolition 172, 173
finish 635
fireproofing beam 436
frame 129, 130
gypsum 182, 184
high abuse 184
nail 88
painting 213, 215
partition 175, 467, 468
partition NLB 174
removal 173
screw 185
Duck tarpaulin 16
Duct flexible insulated 363
flexible noninsulated 363
grille 198
heater electric 376
humidifier 378
insulation 359
liner 363
liner non-fiberous 363
Ductless split system 375
Ducts flexible 363
Ductwork 362
aluminum 362

Index

demolition 358
fabric coated flexible 363
fabricated 362
fiberglass 362
galvanized 362
laboratory 257
metal 362
rectangular 362
rigid 362
Dumbbell 259
Dumbwaiter 320
electric 320
Dump charges 21
Dumpsters 21, 626
Duplex receptacle 393
Dust collector shop 258
Dustproofing 33
compound 30
Dutch door 137
oven door 235
DX coil evaporator 374

E

Ecclesiastical equipment 264
Echo chamber 314
Economizer shower head water 346
Economy brick 44
Edging 424
aluminum 424
precast 424
wood 424
Educational TV studio 410
Eggcrate ceiling 187
framed grille 368
grille lay in 368
T-bar grille 368
Electric backstop 260
baseboard heater 377
bed 299
cabinet heater 378
cable 383
chalkboard 221
commercial water heater 341
convector heater 377
duct heater 376
dumbwaiter 320
fixture 396, 397, 400, 402, 404
fixture hanger 395
heater 250
heating 377
hinge 162
hoist 428
lamp 403
log 235
metallic tubing 384
panelboard 391
service 394
service installation 529
stair 250
switch 392, 394
unit heater 377
water heater 341
Electrical conduit 384
demolition 380, 381
fee 8
laboratory 258
undercarpet 381
wire 384
Electronic air cleaner 370
closer 157
control system 360
Elevator 320, 321, 638
cab 322
controls 323
freight 496-498
geared traction 497
gearless traction 498
hospital 496-498
hydraulic 496
options 323
passenger 496-498
shaft wall 174
Embossed print door 136
Emergency call system 406
door pivot 162
door stop 162
lighting 400, 401
Employer liability 10
EMT 384
box connector 384, 385
coupling 384
fitting 384
Emulsion adhesive 193
asphaltic 420
Encapsulation 25
pipe 25
Enclosure acoustical 314
bathtub 227
decontamination 22, 24, 27
demolition 24
shower 227
telephone 224
Energy circulator air solar 372
saving lighting devices 403
system controller solar 372
system heat exchanger solar 372
Engineer brick 45
Engineered stone countertops 281
Engineering fee 8
Entrance aluminum 148
and storefront 147
door 129, 137, 139, 146
frame 109
hinge 161
lock 160
screen 225
sliding 147
strip 145
Epoxy coating 33, 217
floor 195, 196
grating 110
grout 185, 186
terrazzo 195, 196
wall coating 217
Equipment 245
bank 245
barber 246
cinema 256
cleaning & disposal 255
darkroom 248
dental 262
detention 245
ecclesiastical 264
fire 329, 527
gymnasium 260
health club 259
hospital 261
insulation 359
laboratory 257, 278
laundry 246, 247
loading dock 244
medical 261, 263
pad 31
parking 80, 296, 302, 314
safety 352
security and vault 244
shop 258
stage 258
swimming pool 310
theater and stage 258
waste handling 263
Escalator 324, 499, 639
Escape fire 77
Escutcheon plate 331, 332
sprinkler 332
Estimate electrical heating 377
Estimating 618
Evaporator cooler 374
DX coil 374
Evergreen shrub 422
tree 422
Excavation planting pit 421
Exercise equipment 259
ladder 259
rope 259
weight 259
Exhaust hood 249, 257
vent 368
Exhauster roof 364
Exhibit case 278
Exit control alarm 410
device 158
light 401
lighting 401
trim 158
Expansion anchor 58
joint 179
Expense office 11
Explosionproof fixture 401
fixture incandescent 401
lighting 401
Exposed aggregate 33, 416
aggregate coating 217
Exterior door frame 109
molding 101
plaster 180
pre-hung door 139
residential door 137
signage 223
tile 186
wood frame 109
Extinguisher ABC 237
carbon dioxide 237
chemical dry 237
co_2 237
fire 236, 334
installation 237
pressurized fire 237
standard 237
Extra work 11
Extractor industrial 246
Eye wash fountain 352
wash portable 352
Eye/face wash safety equipment 352

F

Fabric 269
chain link 418
flashing 121
stile 164
upholstery 268
wire 419
Fabricated ductwork 362
Face brick 44
wash fountain 352
Faceted glass 166
Facing panel 53
stone 52
tile structural 45
Factor 8
Factory air conditioning 511
Fall painting dry 215
Fan 363
air conditioning 363
air handling 363
axial flow 363
bathroom exhaust 365
booster 364
ceiling 364
centrifugal 364
coil air conditioning 376
in-line 364
kitchen exhaust 366
low sound 363
paddle 390
propeller exhaust 365
residential 365, 389
roof 364
utility set 364
vaneaxial 363
ventilation 389
ventilator 363
wall 365
wall exhaust 365
wiring 389
Fascia board 94
board demolition 85
wood 102
Fast food equipment 252
Fastener timber 88
wood 89
Faucet & fitting 346, 347, 350, 351
bathroom 346
gooseneck 351
laundry 347
lavatory 347
medical 351
Fee architectural 8
engineering 8
Felt carpet pad 196
Fence and gate 418
board batten 419
cedar 419
chain link 418, 419
helical topping 419
interior 418
picket paint 204
treated pine 419
wire 419
wood rail 419
Fences 204
misc. metal 419
Fencing wire 419
Fiberboard insulation 115
Fiberglass bench 307
blown in 115
ceiling board 187
cross 264
door 139, 144
ductwork 362
insulation 112, 113, 115
panel 16, 201
pipe covering 340
planter 306
reinforced ceiling 311
reinforced plastic panel 199
shade 273
shower stall 226
trash receptacle 306
tread 79
wall lamination 217
wool 114
Field office expense 13
personnel 10
Fieldstone 50
File system 285
Filler block 214
strip 164
Fillet welding 58
Film equipment 248, 256

Index

polyethylene 420
security 169
Filter air 370
   chemical media 370
   mechanical media 369
   swimming pool 355
Filtration air 22
Finish floor 191, 216
   hardware 157
   Keene cement 180
   lime 42
   nail 88
   refrigeration 311
Finishes wall 198
Finishing concrete 33
Fir column 108
   floor 190
   molding 102
   roof plank 99
Fire & heat detectors 411
   alarm 332, 410, 411
   axe 330
   call pullbox 411
   crowbar 330
   damage repair 203
   detection system 538
   door 132, 135, 142, 411, 483
   door frame 130
   equipment 329
   equipment cabinet 236
   escape 77
   escape balcony 77
   escape disappearing 77
   escape stairs 77
   extinguisher 236, 334
   extinguisher portable 236
   extinguishing . . . 517, 519-523, 525, 526-528
   extinguishing system 237, 329, 330, 333
   horn 411
   hose 329
   hose adapter 329
   hose coupling 329
   hose gate valve 331
   hose nipple 330
   hose nozzle 330
   hose storage cabinet 236
   hose valve 331
   hose wye 332
   hydrant 328
   protection 236
   protection kitchen 369
   protection sys. classification . . . 641
   pump 335
   rated closer 157
   rated tile 45
   resistant ceiling 493
   resistant drywall 182, 183
   resistant wall 174
   signal bell 411
   sprinkler head 333
   suppression automatic 334
   valve 331
   valve hydrant 331
Firecycle system control pkg. . . . 332
Fireplace accessory 235
   box 54
   damper 235
   demolition 42
   form 235
   free standing 235
   mantel 108
   masonry 54
   prefabricated 234, 235
Fireproofing 122

plaster 122
plastic 122
spray 122
steel beam 436
steel column 438, 439
Firestop wood 93
Firestopping 122, 631
Fitting EMT 384
   poke-thru 385
   vent chimney 370
Fixed booth 295
Fixture bathroom 343-346
   electric 396, 397
   explosionproof 401
   fluorescent 397, 400
   fluorescent installed 533
   hanger 395
   incandescent 398
   incandescent vaportight 399
   institution 352
   interior light 395-398, 401
   landscape 402
   metal halide 398
   metal halide explosionproof . . 401
   mirror light 399
   plumbing . 341, 343, 344, 349, 353, 354
   prison 352
   removal 338
   residential 388, 400
   sodium high pressure 401
   type/SF incandescent 535
   vandalproof 399
   vaporproof 396
   watt/SF incandescent 536
Flagging 189, 417
   slate 417
Flashing 120, 121
   copper 120, 121
   fabric 121
   lead coated 121
   paperbacked 121
   stainless 120
   vent chimney 371
Flexible door 145
   ducts 363
   ductwork fabric coated 363
   insulated duct 363
   noninsulated duct 363
   sign 224
   sprinkler connector 332
Float finish concrete 33, 34
   glass 166
Floating floor 191
   pin 161
Floor 189
   abrasive 33
   access 198
   acid proof 189
   asphalt block 417
   brick 189, 417
   carpet 197, 490, 491
   check 160
   cleaning 17
   closer 157
   color 33
   composition 490, 491
   concrete topping 491
   conductive 194-196
   cork 193
   door 141
   epoxy 195, 196
   finish 216
   flagging 417
   framing removal 21
   grating 110

hardener 30
hatch 141
marble 52
mastic 195
mat 284
nail 88
neoprene 195
oak 191
paint 212
parquet 190
pedestal 198
plywood 99
polyacrylate 196
polyester 196
polyethylene 197
polyurethane 198
portable 258
quarry tile 186
refrigeration 311
resilient 490, 491
rubber 193
sealer 30
slate 52
stain 212
subfloor 99
terrazzo 194, 636
tile and covering 490
tile terrazzo 194
topping 33
underlayment 99, 490, 491
varnish 212
vinyl 193
wood 190, 490, 491
wood athletic 191
wood composition 190
Flooring 189
   composition 195
   demolition 172
Flue screen 235
Fluid heat transfer 372
Fluorescent fixture 397, 400, 533
   fixture system 533
   light by wattage 534
Fluoroscopy room 315
Flush automatic 350
   bolt 156
   door 134, 135, 137
   tube framing 146
   valve 350
   wood door 134
FM200 fire extinguisher 334
   fire suppression 528
Foam glass insulation 113
   insulation 114
   pipe covering 340
   urethane 125
Foam-water system components . 334
Foil aluminum 115, 117, 201
   barrier 201
Folder laundry 246
Folding accordion door 138
   accordion partition 231
   bench 302
   chair 287
   door 231
   door shower 227
   gate 228
   partition 231
   table 290
Food dispensing equipment . . . 254
   mixer 252
   service equipment 253
   warmer 254
Football scoreboard 260
Footing spread 31
Form deck edge 66

decking 66
fireplace 235
Foundation mat 31
Fountain drinking 353, 354
   eye wash 352
   face wash 352
   lighting 355
   pump 355
   wash 350
Frame baked enamel 130
   door 109, 129, 130
   drywall 129, 130
   entrance 109
   fire door 130
   labeled 130
   metal 129
   metal butt 130
   scaffold 14
   steel 129
   welded 130
   window 152
   wood 109, 275
Framing anchor 90
   band joist 71
   beam & girder 91
   canopy 96
   columns 92
   demolition 85
   joist 93
   lightweight 60
   lightweight angle 60
   lightweight channel 61
   load-bearing stud partition . . . 67
   load-bearing stud wall 67
   metal joist 71
   NLB partition 176, 177
   removal 84
   roof metal rafter 74
   roof metal truss 75
   roof rafters 94
   roof soffits 75
   sill 95
   sleepers 96
   slotted channel 61
   treated lumber 96
   tube 146
   wall 96
   window wall 146
   wood 91, 92
Freestanding booth 295
   chalkboard 222
   office furniture 300
Freezer 250, 251, 311
   door 366
Freight elevators 638
French door 139
Frequency filter radio 317
Front vault 143
FRP panel 199
Fryer 253
Full vision door 129
   vision glass 167
Fume hood 257
Furnace wall 371
Furnishings site 306
Furniture 284
   accessory 283
   dormitory 297
   freestanding office 300
   hospital 299
   hotel 289
   library 298
   office 285-288
   panel hung office 299
   restaurant 290-292
   school 296

Index

systems ... 301
Furring and lathing ... 175
   ceiling ... 98, 176, 177
   channel ... 176, 184
   metal ... 175
   steel ... 176
   wall ... 98, 176
Fusible link closer ... 157

G

Galvanized ductwork ... 362
Garage door ... 144
   door weatherstrip ... 165
Garbage disposal ... 249, 255
Gas fired space heater ... 371
   heat air conditioner ... 373
   incinerator ... 263
   log ... 235
   water heater ... 342
   water heater commercial ... 342
Gasket neoprene ... 124
   toilet ... 348
Gate day ... 143
   folding ... 228
   security ... 228
Gauging plaster ... 180
General contractor's overhead ... 621
Generator steam ... 261
GFI receptacle ... 388
Girder joist ... 65
   wood ... 92
Glass ... 166, 168, 169
   acoustical ... 170
   and glazing ... 170
   bead ... 146
   bevel ... 166
   block ... 48, 49
   block skylight ... 156
   break alarm ... 410
   bulletin board ... 222
   bulletproof ... 170
   curtain wall ... 148
   demolition ... 128
   door ... 129, 141, 147
   door astragal ... 164
   door shower ... 227
   faceted ... 166
   filler ... 354
   float ... 166
   full vision ... 167
   heat reflective ... 167
   insulating ... 167
   insulation cellular ... 338
   laminated ... 170
   lead ... 315
   lined water heater ... 249
   low emissivity ... 166
   mirror ... 168, 233
   mosaic ... 187
   obscure ... 167
   patterned ... 167
   reflective ... 168
   safety ... 167
   sandblast ... 166
   shower stall ... 226
   spandrel ... 168
   tempered ... 166
   tile ... 168
   tinted ... 166
   window ... 168
   window wall ... 147
   wire ... 168
Glassware washer ... 257
Glaze coat ... 214

Glazed aluminum curtain wall ... 148
   block ... 48
   brick ... 45
   ceramic tile ... 185
   wall coating ... 217
Glazing ... 166
   plastic ... 169
   polycarbonate ... 169
   productivity ... 633
Glove bag ... 23
   box ... 257
Golf tee surface ... 198
Gooseneck faucet ... 351
Grab bar ... 233
Granite ... 51
   building ... 51
   chips ... 421
   conductive floor ... 196
   demolition ... 42
   edging ... 424
   paver ... 51
   paving block ... 417
   sidewalk ... 417
Granolithic finish concrete ... 33
Grass cloth wallpaper ... 199
Grating floor ... 110
   stair ... 77
Gravel base ... 416
   pea ... 421
Greenhouse ... 314
   cooling ... 314
Grid spike ... 90
Griddle ... 253
Grille air return ... 367
   aluminum ... 367
   decorative wood ... 107
   door ... 367
   duct ... 198
   eggcrate framed ... 368
   lay-in T-bar ... 368
   painting ... 212
   plastic ... 368
   roll up ... 141
   top coiling ... 141
   window ... 154
Grinder shop ... 258
Ground ... 98
   cover ... 421, 422
   fault indicating receptacle ... 393
Group shower ... 350
   wash fountain ... 350
Grout epoxy ... 185, 186
   tile ... 185
   topping ... 35
Guard corner ... 232
   lamp ... 404
   service ... 16
   snow ... 122
   sprinkler head ... 333
   wall ... 232
   window ... 80
Guardrail scaffold ... 14
Guards wall & corner ... 232
Gutter swimming pool ... 310
Gutting ... 21
   demolition, ... 21
Gym floor underlayment ... 191
   mat ... 260
Gymnasium divider curtain ... 260
   equipment ... 259, 260
   floor ... 191, 198
   squash court ... 313
Gypsum block demolition ... 20
   board ... 182, 184
   board accessories ... 184
   board ceiling ... 492, 493

board leaded ... 315
board partition ... 175
board partition NLB ... 174
board removal ... 173
board system ... 174
drywall ... 182, 184
fabric wallcovering ... 199
lath ... 178
lath nail ... 89
partition ... 228
partition NLB ... 175
plaster ... 179
plaster ceiling ... 492
plaster partition ... 470
restoration ... 172
shaft wall ... 174
sheathing ... 100
weatherproof ... 100

H

Half mortise hinge ... 161
   round molding ... 102
Hammer bush ... 34
Hand carved door ... 133
   dryer ... 233
   split shake ... 118
   trowel finish concrete ... 33
Handball court ... 313
Handicap drinking fountain ... 353
   lavatory ... 351
   lever ... 160
   opener ... 156
   ramp ... 32
   shower ... 346, 351
   tub shower ... 346
   water cooler ... 354
Handle door ... 277
Handling material ... 428
   unit air ... 376
   waste ... 263
Handrail ... 233
   aluminum ... 198
   wood ... 102, 107
Handwasher-dryer module ... 349
Hanger beam ... 90
   fixture ... 395
   joist ... 90
Hanging lintel ... 61
   wire ... 189
Hardboard cabinet ... 274
   paneling ... 103
   tempered ... 103
   underlayment ... 99
Hardener concrete ... 33
   floor ... 30
Hardware ... 133, 156
   cabinet ... 277, 278
   cloth ... 418, 419
   door ... 157
   drapery ... 271, 272
   finish ... 157
   panic ... 158
   rough ... 91
   window ... 156, 157
Hardwood carrel ... 297
   floor stage ... 258
   grille ... 107
Hat and coat strip ... 233
   rack ... 240
Hatch ceiling ... 141
   floor ... 141
   roof ... 121, 447
   smoke ... 121
Hay ... 420

Hazardous waste handling ... 264
Head sprinkler ... 332
Header wood ... 96
Headers boxed beam ... 71
   load-bearing stud wall ... 67
Health club equipment ... 259
Hearth ... 54
Heat baseboard ... 376
   electric baseboard ... 377
   greenhouse ... 314
   pump ... 375
   pump residential ... 390
   pumps air-source ... 375
   pumps water-source ... 375
   recovery ... 263
   reflective glass ... 167
   temporary ... 12
   therapy ... 261
   transfer fluid ... 372
   transfer package ... 343
Heated doorway curtain ... 366
Heater & fan bathroom ... 366
   cabinet convector ... 378
   electric ... 250
   floor mounted space ... 371
   sauna ... 313
   unit ... 371
   water ... 249, 341, 342
Heating ... 340, 371, 372, 646
   control valve ... 361
   electric ... 377
   estimate electrical ... 377
   hydronic ... 377
   insulation ... 359
Heavy construction ... 420
   duty shoring ... 14
Hemlock column ... 108
High abuse drywall ... 184
   build coating ... 217
   intensity discharge lamp ... 403
   intensity discharge lighting ... 395, 396, 401
   output lamp ... 395
   pressure fixture sodium ... 395, 396, 398
   pressure sodium lighting ... 395, 398
   rib lath ... 178
   rise glazing ... 167
Highway paver ... 417
Hinge brass ... 161, 162
   bronze ... 161
   cabinet ... 278
   continuous ... 163
   decorative ... 162
   electric ... 162
   entrance ... 161
   half mortise ... 161
   hospital ... 163
   kalamein door ... 162
   mortise ... 160, 161
   paumelle ... 162
   prison ... 162
   residence ... 162
   residential ... 161
   school ... 162, 163
   security ... 162
   special ... 162
   steel ... 160-162
   surface ... 161, 162
Hinges ... 632
Hip rafter ... 95
Hockey scoreboard ... 260
Hoist ... 428
   chain ... 428
   electric ... 428
   overhead ... 428

Index

Holder closer 157
Holdown 90
Hollow concrete block partition . 450
    core door 134, 138
    metal 129
    metal door 131, 477-479
    metal stud partition 178
    wall anchor 58
Hood exhaust 257
    fire protection 369
    fume 257
    range 249
Hook coat 233
    robe 233
Hooks clevis 79
Hook-up electric service 529
Hopper refuse 326
Horn fire 411
Hose adapter fire 329
    bibb sillcock 348
    equipment 329, 330
    fire 329
    nozzle 330
    rack 331
    rack cabinet 236
    reel 330
    valve cabinet 236
Hospital bath 351
    cabinet 279
    door hardware 157
    equipment 261
    furniture 299
    hinge 163
    kitchen equipment 254
    partition 227
    sink 351
    tip pin 161
    water closet 351
    whirlpool 351
Hot tub 310
    water dispenser 341
    water heating 376
    water-steam exchange 343
Hotel cabinet 234
    furniture 289
    lockset 158, 159
House fan whole 365
    telephone 406
Housewrap 117
Hubbard tank 261
Humidifier 250, 378
    duct 378
    room 378
Humus peat 420
Hutch 297
HVAC demolition 358
Hydrant fire 328
    screw type valve 331
    tool 330
    wall 328
Hydrated lime 42
Hydraulic chair 262
Hydronic heating 377

I

Ice machine 255
Icemaker 248, 255
Impact drilling concrete 36
In insulation blown 115
Incandescent bulb 404
    explosionproof fixture 401
    exterior lamp 404
    fixture 398
    fixture system 535

interior lamp 404
Incinerator gas 263
    waste 263
Inclined ramp 325
Incubator 257
Index construction cost 9
Indicating panel burglar alarm .. 410
Industrial address system 406
    air filter 370
    door 141, 142
    folding partition 231
    lighting 397
    railing 79
    safety fixture 352
    window 150
Infrared broiler 253
    detector 410
In-line fan 364
Installation drapery 271
    extinguisher 237
Institution fixture 352
Insulated glass spandrel 168
    protectors ADA 338
Insulating glass 167
Insulation 112, 115, 340, 359
    batt 114
    blanket 113, 359
    board 112
    boiler 359
    breeching 359
    building 114
    calcium silicate 359
    cavity wall 114
    cellulose 114
    composite 116
    duct 359
    equipment 359
    fiberglass 112, 113, 115
    foam 114
    foam glass 113
    masonry 114
    mineral fiber 114
    pipe 338, 340
    polystyrene 113, 114
    refrigeration 311
    removal 23
    roof 115
    roof deck 115
    spray 115
    vapor barrier 117
    vapor proof 338, 339
    vermiculite 114
    wall 113
    water heater 338
Insurance 10, 620
    builder risk 10
    public liability 10
Intake exhaust vent 369
Integral colors concrete 30
    topping concrete 33
    waterproofing concrete 30
Intercom 406
    system 538
Interior door frame 109
    finish 488, 489
    glazing 474
    HID fixture 395
    light fixture 395-398, 401
    lighting 396
    plants 302, 303
    pre-hung door 139
    residential door 137
    window wall 475, 476
    wood frame 109
Interlocking precast conc. unit .. 417
Interval timer 386

Intrusion detection 410
    system 410
Inverted bucket steam trap 361
Inverter battery 394
Iron jamb hinge 162
Ironer laundry 246
Ironing center 234
Ironspot brick 189
I.V. track system 227

J

Jack rafter 95
Jail equipment 80
Jeweler safe 244
Job condition 9
Joint control 33
    expansion 179
    sealer 124
Jointer shop 258
Joist bridging 64
    connector 90
    deep longspan 64
    demolition 85
    framing 93
    girder 65
    hanger 90
    longspan 64
    metal framing 71
    open web 64
    removal 85
    sister 93
    truss 65
    web stiffeners 72
    wood 93
Jumbo brick 44, 45

K

Kalamein door 136
    door hinge 162
Keene cement 180
Kettle 253
Key cabinet 165, 278
    keeper 238
Keyless lock 159
Kick plate 163
    plate door 163
Kiln vocational 258
King brick 45
Kitchen cabinet 274, 279
    equipment 250, 252, 253
    equipment fee 8
    exhaust fan 366
    sink 345
    sink faucet 347
    unit 247
K-lath 178
Knee action mixing valve 351
Knob door 159
Kraft paper 117

L

Labeled door 132, 135
    frame 130
Labor index 9
Laboratory cabinet 278
    equipment 257, 278
    sink 257, 349
    table 257
Ladder alternating tread 77
    exercise 259

rolling 15
ship 77
swimming pool 310
towel 234
vertical metal 77
Lag screw 58
Laminated epoxy & fiberglass .. 217
    glass 170
    lead 315
Lamp 282
    guard 404
    high intensity discharge ... 403
    high output 395
    incandescent exterior 404
    incandescent interior 404
    mercury vapor 403
    metal halide 403
    post 80
    quartz line 404
    shade 282
    slimline 403
    sodium high pressure 404
    sodium low pressure 404
Lampholder 393
Lamphouse 256
Landing metal pan 76
    stair 195
Landscape fixture 402
    light 402
    surface 197
Latch deadlocking 159
    set 158, 159
Latex caulking 124
    underlayment 192
Lath demolition 172
    gypsum 178
    metal 174, 178
Lath, plaster and gypsum board . 634
Lath rib 178
Lathe shop 258
Lattice molding 102
Lauan door 138
Laundry equipment 246, 247
    faucet 347
    sink 346
    tray 346
    water heater 342
    water heater coin 342
Lava stone 51
Lavatory commercial 349
    faucet 347
    handicap 351
    removal 338
    residential 344, 349
    sink 344
    vanity top 344
    wall hung 345
Lavatories 503
Lawn bed preparation 421
Lazy susan 274, 275
Lead barrier 201
    coated flashing 121
    flashing 120
    glass 315
    gypsum board 315
    lined darkroom 312
    lined door frame 315
    paint encapsulation 25
    paint remediation methods .. 627
    paint removal 26
    plastic 315, 316
    roof 120
    sheets 315
    shielding 315
Leader line wye 332
Lean-to type greenhouse 314

Index

Lectern	264
Lecture hall seating	301
Lens movie	256
Let-in bracing	91
Letter sign	223
slot	238
Leveling jack shoring	14
Lever handicap	160
Lexan	169
Liability employer	10
insurance	621
Library air conditioning	514
equipment	297, 298
furniture	298
shelf	298
Life guard chair	310
Lift	325
correspondence	325
parcel	325
wheelchair	325
Light accent	400
bollard	402
border	402
dental	262
exit	401
fixture interior	395-398, 401
fixture troffer	398
landscape	402
nurse call	407
safety	400
stand	22
strobe	402
support	60
switch	532
temporary	12
track	402
underwater	310
Lighting	395-397, 399, 404
carpet	400
darkroom	248
emergency	400, 401
exit	401
explosionproof	401
fountain	355
high intensity discharge	395, 396, 401
high pressure sodium	398
incandescent	398
industrial	397
interior	396
metal halide	396, 398
outlet	388
picture	399
residential	388
stage	402
stair	400
strip	397
surgical	261
theatrical	402
track	402
Lightning suppressor	385
Lightweight angle framing	60
block	47
channel framing	61
framing	60
natural stone	51
Lime finish	42
hydrated	42
Limestone	51
Linen chutes	326
collector	326
wallcovering	199
Liner duct	363
non-fiberous duct	363
Lint collector	246
Lintel	51
block	46
concrete block	441, 442
hanging	61
precast concrete	34
steel	61
Load center residential	385
Load-bearing stud part. framing	67
stud wall bracing	66
stud wall bridging	66
stud wall framing	67
stud wall headers	67
Loader vacuum	22
Loading dock equipment	244
Lobby collection box	238
Lock electric release	159, 160
entrance	160
keyless	159
tile	284
time	143
tubular	159
Locker metal	237
plastic	238
steel	237
wall mount	238
wire mesh	237
Locking receptacle	388
Lockset communicating	158, 159
cylinder	158
hotel	158, 159
mortise	159
Log electric	235
gas	235
Longspan joist	64
Louver redwood	170
ventilation	170
wood	107
Louvered door	138, 140
Low sound fan	363
Low voltage silicon rectifier	392
switching	392
switchplate	392
transformer	392
wiring	393
Lumber core paneling	104
product prices	631
Luminous ceiling	188, 403
panel	187

M

Machine screw	59
trowel finish concrete	33
Machinery anchor	59
Magazine shelving	298
Magnesium oxychloride	122
Magnetic astragal	163, 164
Mahogany door	133
Mail box	238
box call system	406
chute	239
slot	238
Main office expense	11
Mall front	148
Management fee construction	8
Manifold roof fire valve	328
Mantel beam	108
fireplace	108
Map rail	222
Maple floor	191
wall	313
Marble	52
chips	421
countertop	280
floor	52
screen	226
shower stall	226
sill	53
sink	503
soffit	52
stair	52
synthetic	190
tile	190
Mark-up cost	11
Masking	41
Mason scaffold	13
Masonry accessory	41, 43
aggregate	42
anchor	43
base	48
caulking	125
cement	42
cleaning	40
color	42
cutting	21
demolition	20, 41, 173
fireplace	54
furring	98
insulation	114
nail	89
painting	214
panel	49
partition	450
patching	40
pointing	40
reinforcing	628
removal	41
sill	53
step	416
toothing	20, 40
wall	420
wall tie	43
Massage table	259
Master clock system	407, 538
T.V. antenna system	539
Mastic floor	195
Mat floor	284
foundation	31
gym	260
wall	260
Material handling	428
handling belt	428
handling system	428
index	9
Materials concrete	628
Mattress	289
MC cable	383
Meat case	250
Mechanical	642
equipment demolition	358
fee	8
media filter	369
Medical center air conditioning	514
equipment	261, 263
exam equipment	261
faucet	351
X-ray	263
Medicine cabinet	234
Mercury vapor lamp	403
Mesh partition	228
security	179
stucco	178
wire	419
Metal bookshelf	298
butt frame	130
cabinet	278, 279
casework	165, 278
chalkboard	221, 229
chimney	234
cleaning	57
demolition	57
door	129, 144, 477-479
door residential	132
ductwork	362
facing panel	119
frame	129
framing parapet	74
furring	175
halide explosionproof fixture	401
halide fixture	398
halide lamp	403
halide lighting	396, 398
hollow	129
in field painting	217
joist bracing	70
joist bridging	70
joist framing	71
lath	174, 178
locker	237
molding	179
pan ceiling	188
pan landing	76
pan stair	76
pipe removal	338
plate stair	77
plumbing fixture	353
pressure washing	57
rafter framing	74
roof parapet framing	74
sandblasting	57
sash	150
screen	151
sheet	120
shelf	239
sign	223
steam cleaning	57
stud	174
stud demolition	173
stud NLB	176, 177
support assemblies	175
threshold	160
tile	187
toilet partition	225, 226
trash receptacle	283
truss framing	75
water blasting	57
window	149-151
wire brushing	57
Metallic foil	115
hardener	33
Meter Venturi flow	362
Mezzanine	434, 435
Microphone	406
Microwave detector	410
oven	248
Millwork	101, 275
demolition	88
Mineral fiber ceiling	187, 188
fiber insulation	114
wool blown in	115
Mirror	168, 233
ceiling board	187
door	168
glass	168
light fixture	399
plexiglass	169
wall	168, 169
Miscellaneous painting	204, 209
Mix planting pit	421
Mixer food	252
Mixing valve	347
valve shower	346
Mobile shelving	298
X-ray	263
Mobilization or demob.	15
Model building	8
Modification to cost	9
Modular office	473

Index

office system 228, 229
Module handwasher-dryer 349
   tub/shower 345
Mold abatement work area 27
   contaminated area demolition .. 27
Molding 101
   base 101
   brick 102
   ceiling 101
   chair 102
   cornice 101
   exterior 101
   hardboard 103
   metal 179
   pine 101
   trim 102
   window and door 102
   wood transition 191
Money safe 244
Monitor support 60
Monolithic terrazzo 194
Monorail 428
Mop holder strip 233
   sink 349
Mortar 42
Mortar, brick and block 629
Mortar cement 186, 628
   masonry cement 42
   pigments 42
   Portland cement 42
   restoration 42
   sand 43
   thinset 186
Mortise hinge 160, 161
   lockset 159
Mortuary equipment 263
   refrigeration 263
Mosaic glass 187
Moss peat 420
Motion detector 410
Motor support 60
Moulded door 136
Mounted booth 296
Mounting board plywood 100
Movable louver blind 272
   office partition 228
Movie equipment 256
   lens 256
   projector 256
   screen 256
Moving ramp 325
   ramps and walks 639
   shrub 425
   stairs & walks 324
   tree 425
   walk 325, 499
Mulch bark 420
   ceramic 421
   stone 421
Mulching 420
Mullion vertical 146
Multizone air cond. rooftop .. 373
Muntin window 154
Mushroom ventilator stationary .. 368
Music room 312
Mylar tarpaulin 16

N

Nail 88, 89
   lead head 315
Nailer steel 93
   wood 93
Napkin dispenser 233
Needle beams cantilever 414

Neoprene floor 195
   gasket 124
Net safety 13
   tennis court 418
Newspaper rack 298
Night depository 245
Nightstand 297
Nipple fire hose 330
Non removable pin 161
Non template hinge 162
Norwegian brick 44, 45
Nosing rubber 192
   safety 192
   stair 192, 194
Nozzle dispersion 334
   fire hose 330
   fog 330
   playpipe 330
Nurse call light 407
   call systems 407
   speaker station 407
   station cabinet 279
Nursing home bed 299
Nuts remove 57
Nylon carpet 104, 197, 200

O

Oak door frame 109
   floor 191
   molding 102
   paneling 104
   threshold 109
Obscure glass 167
Occupancy sensors 403
Office divider 230
   expense 11
   floor 198
   furniture 285-288
   panel 229, 230
   partition 228
   partition movable 228
   safe 244
   system modular 229
   trailer 13
Oil fired space heater 371
   water heater 342
Olive knuckle hinge 162
Omitted work 11
One piece astragal 163
Onyx 194
Opener automatic 156
   door 128, 142, 156
   handicap 156
Opening roof frame 61
Open web joist 64
Operable partition 232
Operating room equipment 262
Operator automatic 156
Options elevator 323
Ornamental aluminum rail 81
   columns 82
   glass rail 81
   railing 81
   wrought iron rail 81
OSHA testing 24
Outlet box plastic 385
   box steel 385
   lighting 388
Oven 253
   cabinet 275
   convection 253
   microwave 248
Overbed table 299
Overhaul 21

Overhead & profit 11
   commercial door 144
   contractor 17, 621
   distribution system 391
   door 144, 145
   hoist 428
   support 129
Overlapping astragal 164
Overlay face door 134-136
Overpass 420
Overtime 9, 11, 620
Oxygen lance cutting 22

P

Package chute 326
   receiver 245
Packaged terminal air cond. .. 374
Packaging waste 25
Pad equipment 31
Padding carpet 196
Paddle blade air circulator .. 365
   fan 390
Paint aluminum siding 205
   and covering wall .. 488, 489
   cabinets 488, 489
   casework 488, 489
   ceiling 488, 489
   chain link fence 204
   doors & windows exterior . 206
   doors & windows int. 210, 211
   encapsulation lead 25
   fence picket 204
   floor 212
   floor concrete 212
   floor wood 212
   miscellaneous metal . 488, 489
   removal 26
   siding 205
   structural steel ... 488, 489
   trim exterior 207
   walls masonry exterior .. 208
   woodwork 488, 489
Painting 637
   balustrade 213
   bar joist 65
   casework 209
   ceiling 213-215
   clapboard 205
   concrete block 214
   cornice 213
   decking 205
   drywall 213, 215
   grille 212
   masonry 214
   metal in field 217
   miscellaneous 204, 209
   pipe 212, 213
   plaster 213, 215
   railing 204
   shutters 204
   siding 205
   stair stringers 205
   steel 217
   steel siding 205
   stucco 205
   trellis/lattice 205
   trim 213
   truss 213
   wall 205, 214, 215
   window 209, 210
Paints & coatings 201
Pallet rack 239
Pan tread stair 486, 487
Panel acoustical 201, 229

brick wall 49
   door 136, 137
   facing 53
   fiberglass 16
   fire 411
   FRP 199
   hung office furniture ... 299
   luminous 187
   masonry 49
   metal facing 119
   office 229, 230
   portable 229
   sound 314
   spandrel 168
   vision 155
   wall 54
   woven wire 228
Panelboard 391
   electric 391
   w/circuit breaker 391
Paneled door 132
   pine door 138
Paneling 103
   board 105
   cutout 20
   demolition 88
   hardboard 103
   plywood 104
   wood 103
Panelized shingle 118
Panic bar 158
   device 157, 158
   device door hardware 157
   trim 158
Paper building 117
   sheathing 117
Paper-backed flashing 121
Paperhanging 199
Paperholder 234
Paraffin bath 261
Parallel bar 259, 261
Parapet metal framing 74
Parcel lift 325
Parging cement 112
Park bench 307
Parking equipment . 80, 296, 302, 314
Parquet floor 190
Particle board underlayment .. 99
   core door 134, 135
Parting bead 103
Partition 228, 229
   accordion 473
   acoustical 229, 231, 473
   anchor 43
   bathroom 485
   block 47, 48
   blueboard 175
   bulletproof 245
   commercial 473
   demolition 173
   demountable 228
   door 229
   drywall 175, 469
   folding 231, 473
   folding leaf 231
   framing load bearing stud 67
   framing NLB 176, 177
   gypsum 228
   gypsum board 467, 469
   handicap 485
   hospital 227
   industrial 473
   marble 485
   mesh 228
   metal 485
   movable 473

677

Index

movable office 228	Pin powder 59	window 155	panel 229
NLB drywall 174	Pine door 137	Plate escutcheon 331, 332	partition 232
NLB gypsum 175	door frame 109	glass greenhouse 314	post 244
office 228	fireplace mantel 108	push-pull 160	stage 258
operable 232	floor 191	shear 90	Portland cement terrazzo 194
plaster 174	molding 101	wall switch 393	Post 244
plastic laminate 485	shelving 240	Platform telescoping 258	cap 90
porcelain enamel 485	siding 119	Plating zinc 89	cedar 108
portable 232	stair tread 105	Players bench 307	demolition 86
refrigeration 311	Pipe & fittings 347, 348	Playground surface 193	lamp 80
shower 52, 226	covering 338	Playpipe fire hose 330	portable 244
steel 229, 231	covering calcium silicate .. 340	Plenum demolition 173	tennis court 418
steel folding 473	covering fiberglass 340	Plexiglass 169	wood 92
stud 178	encapsulation 25	mirror 169	Postal specialty 238
support 60	insulation 338, 340	Plug wall 43	Potters wheel 258
thin plaster 175	insulation removal 23	Plugmold raceway 391	Poured insulation 114
toilet 52, 224-226, 485	painting 212, 213	Plumbing appliance ... 341, 342, 354	Powder actuated tool 59
urinal 485	rail aluminum 78	demolition 338	charge 59
vinyl folding 473	rail galvanized 78	fixture ... 341, 343, 344, 349, 353,	pin 59
wall NLB 174, 175	rail stainless 78	354, 504, 507, 645	Power conditioner transformer .. 394
wire 228	rail steel 78	fixture metal 353	supply uninterruptible ... 394
wire mesh 228	rail wall 78	fixture requirements 643, 644	system undercarpet 382
wood folding 473	railing 78	fixtures removal 338	temporary 12
wood frame 94	removal metal 338	laboratory 258	wiring 394
woven wire 228	Pivot reinforced hinge 162	Plywood 631	Precast bridge 420
Parts bin 239	Pivoted window 150	clip 90	concrete lintel 34
Passage door 138, 139	Placing concrete 32	demolition 172, 173	concrete unit paving slabs 417
Passenger elevators 638	Plain tube framing 146	floor 99	concrete window sill 35
Patching concrete floor 30	Planer shop 258	mounting board 100	edging 424
concrete wall 30	Plank roof 99	paneling 104	receptor 227
masonry 40	scaffolding 14	sheathing roof & walls 100	terrazzo 194
Patient nurse call 407	sheathing 100	shelving 240	Prefabricated building 314
Patio door 141	Plant bed preparation 421	siding 119	fireplace 234
Patterned glass 167	Planter 303-306	soffit 103	stair 105
Paumelle hinge 162	bench 306	subfloor 99	Prefinished door 134, 135
Pavement 417	concrete 306	underlayment 99	floor 191
asphalt 418	fiberglass 306	Pneumatic control systems 360	shelving 240
sealer 418	Planting 425	stud driver 59	Pre-hung door 139
Paver floor 189	Plants and planters 302	tube 245	Preparation lawn bed 421
highway 417	interior 302, 303	tube system 326	plant bed 421
Paving athletic 418	transplanting 425	Pocket door frame 109	surface 201, 203
block granite 417	Plaque 224	Pointing masonry 40	Pressure restricting valve ... 331
brick 417	Plaster 174	Poke-thru fitting 385	sodium lighting high 395
Pea gravel 421	accessories 179	Pole closet 102	wash 202
Peat humus 420	and acoustical ceiling 492, 493	portable decorative 244	washing metal 57
moss 420	beam 180	Police connect panel 410	Pressurized fire extinguisher 237
Pedestal floor 198	ceiling 179, 492, 493	Polyacrylate floor 196	Prices lumber products 631
type seating 301	cement 311	Polycarbonate glazing 169	Primer asphalt 193
Pedestrian bridges 420	column 180	Polyester floor 196	steel 217
traffic control 244	cutout 20	Polyethylene film 420	Prison door 144
Pegboard 103	demolition 172, 173	floor 197	equipment 144
Perforated ceiling 188	drilling 175	pool cover 310	fixture 352
Performance bond 11	gauging 180	sink 349	hinge 162
Perineal bath 351	ground 98	tarpaulin 16	toilet 245
Perlite insulation 113	gypsum 179	Polypropylene shower 346	Prisons work in 9
plaster 179, 493	painting 213, 215	sink 349	Process air handling fan 363
plaster ceiling 492, 493	partition 174, 470, 471	Polystyrene ceiling 188, 311	Produce case 250
sprayed 217	partition thin 175	ceiling panel 187	Productivity glazing 633
Permit building 12	perlite 179	insulation 113, 114, 311	Progress schedule 12
Personal respirator 22	thincoat 181	Polysulfide caulking 124	Project overhead 11
Personnel field 10	vermiculite 179	Polyurethane caulking 124	Projected window 150, 151
protection 13	wall 179	floor 196, 198	Projection screen 221, 256
Pew church 302	Plasterboard 182, 492, 493	varnish 217	Projector movie 256
Phone booth 224	Plastic faced hardboard 103	Polyvinyl soffit 103	Propeller exhaust fan 365
Photography 12	fireproofing 122	tarpaulin 16	Protection corner 232
construction 12	glazing 169	Pool accessory 310	fire 236
time lapse 12	grating 110	cover 310	radiation 317
Physician's scale 261	grille 368	cover polyethylene 310	stile 163, 164
Picket railing 78	laminate door 134	filtration swimming 355	winter 16
Picture lighting 399	lead 315, 316	Porcelain enamel sink 503	worker 22
window 149, 150, 152	locker 238	tile 185	Protectors ADA insulated 338
Pier bathtub 351	outlet box 385	Porch molding 102	PT/AC units 374
Pigments mortar 42	sign 224	Portable cabinet 251	Public address system 406
Pilaster toilet partition 226	sink 349	eye wash 352	Pull door 277
wood column 108	skylight 155	fire extinguisher 236	Pulpit church 264
Pile sod 421	trash receptacle 283	floor 192	Pump fire 335

678

Index

fountain 355
   heat 375
   sump 249
Pumped concrete 32
Push button lock 158
Push-pull plate 160
Putting surface 198
Puttying 201
PVC blind 269
   sheet 198
   siding 120

Q

Quarry tile 186
Quarter round molding 102
Quartz 421
   line lamp 404
Quoins 51, 52

R

Raceway 382, 384, 385, 391
   plugmold 391
   surface 391
   wiremold 391
Rack coat 240, 275
   hat 240
   hose 331
   pallet 239
Racquetball court 313
Radiation protection 317
Radiator supply control 361
   thermostat control sys. 360
Radio frequency filter 317
   frequency shielding 316
Radiology equipment 263
Rafter 95
   anchor 90
   demolition 86
   framing metal 74
   metal bracing 73
   metal bridging 73
Rail aluminum pipe 78
   crash 232
   galvanized pipe 78
   map 222
   ornamental aluminum 81
   ornamental glass 81
   ornamental wrought iron 81
   stainless pipe 78
   steel pipe 78
   trolley 232
   wall pipe 78
Railing cable 81
   church 264
   demolition 88
   industrial 79
   ornamental 81
   picket 78
   pipe 78
   wood 102, 105, 107
Railroad tie 416, 424
   tie step 416
Raised floor 198
Rake topsoil 421
Ramp approach 198
   handicap 32
   moving 325
   temporary 15
Ranch plank floor 191
Range cooking 248
   hood 249
   receptacle 388, 393

restaurant 253
Razor wire 419
Reading table 298
Ready mix concrete 618
Receiver ash 283
   trash 283
Receptacle air conditioner 388
   device 387
   dryer 388
   duplex 393
   GFI 388
   ground fault indicating 393
   locking 388
   range 388, 393
   telephone 388
   television 388
   trash 306
   waste 234
   weatherproof 388
Receptor precast 227
   shower 226, 350
   terrazzo 227
Recessed mat 284
Recorder videotape 410
Recore cylinder 158, 159
Rectangular diffuser 367
   ductwork 362
Rectifier low voltage silicon . . . 392
Redwood bark mulch 420
   louver 170
   paneling 105
   siding 119
   trim 102
   tub 310
   wine cellar 251
Refinish floor 191
Reflective block 49
   glass 168
   insulation 115
Refrigerant removal 25
Refrigerated case 250
   wine cellar 251
Refrigeration 311
   bloodbank 257
   commercial 250, 251
   floor 311
   insulation 311
   mortuary 263
   panel fiberglass 311
   partition 311
   residential 248
   walk-in 311
Refrigerator compartment cooler . 354
   door 142, 143
Refuse chute 326
   hopper 326
Register air supply 368
   cash 245
   return 368
   steel 368
   wall 368
Reinforced plastic panel fbrgls. . . 199
Relay box low voltage 392
   frame low voltage 393
Release door 406
   emergency sprinkler 332
Relief vent ventilator 368
Remote power pack 403
Removal air conditioner 358
   asbestos 23
   bathtub 338
   boiler 358
   fixture 338
   insulation 23
   lavatory 338
   paint 26

pipe insulation 23
plumbing fixtures 338
refrigerant 25
shingle 112
sink 338
sod 421
urinal 338
VAT 24
water closet 338
water fountain 338
water heater 338
window 128
Remove bolts 57
   nuts 57
Rendering 8
Repair fire damage 203
Repellent water 214
Residence hinge 162
Residential alarm 389
   appliance 246, 248, 389
   applications 385
   bathtubs 345
   device 385
   door 109, 132, 137
   door bell 388
   dryer 249
   elevator 321
   fan 365, 389
   fixture 388, 400
   folding partition 231
   greenhouse 314
   gutting 21
   heat pump 390
   hinge 161
   lavatory 344, 349
   lighting 388
   load center 385
   lock 158
   refrigeration 248
   roof jack 366
   service 385
   sinks 345
   smoke detector 389
   stair 105
   switch 385
   transition 366
   wall cap 366
   washer 249
   water heater 390
   wiring 385, 389
Resilient base 192
   floor 193
   pavement 418
Respirator 22
   personal 22
Resquared shingle 118
Restaurant air conditioning 513
   furniture 290-292
   range 253
Restoration gypsum 172
   mortar 42
   window 26
Retaining walls stone 420
Retarder vapor 117
Return register 368
Revolving darkroom 312
   door 143, 147
   entrance door 147
Rewind table 256
Rib lath 178
Ridge board 95
Rigid anchor 43
Ring split 90
   toothed 90
Rings drapery 271
Riser rubber 192

stair 195
standpipe 523, 526
terrazzo 195
Rivet 59
   aluminum 59
Road base 416
Robe hook 233
Rod backer 124
   closet 240
   curtain 233
   tie 61
   weld 59
Roll type air filter 369
   up grille 141
Rolling door 142, 148
   ladder 15
   roof 149
   service door 142
   tower scaffold 15
Roman brick 45
Roof cant 95
   copper 120
   deck 99
   deck insulation 115
   dome 149
   exhauster 364
   fan 364
   fire valve manifold 328
   flashing vent chimney 371
   frame opening 61
   framing removal 21
   hatch 121
   insulation 115
   jack residential 366
   lead 120
   metal rafter framing 74
   metal truss framing 75
   nail 89
   rafter 95
   rafter bracing 73
   rafter bridging 73
   rafters framing 94
   rolling 149
   sheathing 100
   sheet metal 120
   skylight 155
   soffits framing 75
   tile 118
   truss 101
Roofing demolition 24, 112
Rooftop air conditioner 373
   air conditioner 512, 513
   multizone 513
   multizone air cond. 373
Room acoustical 312
   audiometric 312
   clean 311
   equipment operating 262
   humidifier 378
Rope decorative 244
   exercise 259
   steel wire 63
Rosewood door 133
Rosin paper 117
Rough buck 93
   hardware 91
   stone wall 50
Rough-in bidet 343
   drinking fountain deck 354
   drinking fountain floor 354
   drinking fountain wall . . . 353, 354
   sink corrosion resistant 349
   sink countertop 345
   sink cup 349
   sink raised deck 345
   sink service floor 349

Index

sink service wall	350
tub	346
Round diffuser	367
rail fence	419
table	290
Rowing machine	259
Rub finish carborundum	34
Rubber astragal	163, 164
base	192
floor	193
floor tile	193
nosing	192
pavement	418
pipe insulation	340
riser	192
sheet	193
stair	192
threshold	160
tile	194
Rubbing wall	34
Rubbish chute	21
handling	21
Running track	191, 192
track surfacing	418

S

Safe office	244
Safety deposit box	245
equipment	352
equipment laboratory	257
fixture industrial	352
glass	167
light	400
net	13
nosing	192
railing cable	63
shower	352
switch	394
Sales tax	10, 619
Sampling air	22
Sand	42, 43
screened	43
Sandblast finish concrete	34
glass	166
masonry	41
Sandblasting metal	57
Sanding	201
floor	191
Sandstone	52
flagging	417
Sandwich panel skylight	155
Sanitary base cove	185
Sash aluminum	149
metal	150
security	150
steel	150
wood	152
Sauna	313
door	313
Saw brick	40
cutting	35
shop	258
table	258
Sawing	40
Sawn control joint	33
Scaffold baseplate	14
bracket	14
caster	14
catwalk	15
frame	14
guardrail	14
mason	13
rolling tower	15
specialties	15

stairway	14
wood plank	14
Scaffolding	625
plank	14
tubular	13
Scale physician's	261
Scanner checkout	245
Scarify concrete	172
Schedule	12
board	223
critical path	12
progress	12
School air conditioning	513
cabinet	278
door hardware	157
equipment	260
furniture	296
hinge	162, 163
T.V.	410
Scissor gate	228
Scoreboard baseball	260
SCR brick	45
Scrape after damage	203
Scratch coat	195
Screen chimney	235
entrance	225
fence	419
metal	151
molding	102
projection	221, 256
security	80, 151
sight	229
squirrel and bird	235
urinal	226
window	151, 154
wood	154
Screened loam	421
sand	43
Screw brass	89
drywall	185
lag	58
machine	59
sheet metal	89
steel	89
wood	89
Scrub station	262
Seal security	163
Sealant	117
acoustical	184
caulking	124
control joint	34
tape	124
Sealer floor	30
joint	124
pavement	418
Seamless floor	195
Seat toilet	343
water closet	343, 344
Seating	288
church	264
movie	256
Sectional door	144
Security and vault equipment	244
barrier	244
film	169
gate	228
hinge	162
mesh	179
sash	150
screen	80, 151
seal	163
shower	353
turnstile	244
vault door	144
Selective demolition	20, 41, 84, 128, 242, 243, 268

Separation barrier	22, 27
Separator soil	424
Service door	142
electric	394
entrance electric	529
residential	385
sink	349
sink clinic	351
sink faucet	350
Shade	269-271, 273
lamp	282
Shaft wall	174
Shake wood	118
Shear plate	90
wall	100
Sheathing	99, 100
asphalt	100
gypsum	100
paper	117
roof	100
roof & walls plywood	100
wall	100
Sheet floor	196
metal	120
metal screw	89
Sheets lead	315
Shelf bathroom	233
bin	239
directory	224
library	298
metal	239
Shellac door	209
Shelving	240
mobile	298
pine	240
plywood	240
prefinished	240
refrigeration	312
steel	239
storage	239
wood	240
Shield undercarpet	383
Shielding lead	315
radio frequency	316
Shift work	9
Shingle	117
asphalt	117
panelized	118
removal	112
stain	205
strip	117
wood	117
Ship ladder	77
Shipping door	366
Shock absorbing door	145
Shop drill	258
equipment	258
Shoring baseplate	14
bracing	14
heavy duty	14
leveling jack	14
Shot blast floor	173
Shower arm	346
built-in	346
by-pass valve	347
compartments	226
cubicle	346
door	227
enclosure	227
glass door	227
group	350
handicap	346, 351
partition	52, 226
polypropylene	346
receptor	226, 350
safety	352

security	353
spray	347
stall	346
surround	227
Shower-head water economizer	346
Shower/tub control set	347
Shower/tub valve spout set	347
Shrub broadleaf evergreen	422
deciduous	423
evergreen	422
moving	425
Shrubs and trees	422
Shutter	272
Siamese	328
Side light door	139
Sidelight	109
door	132, 133
Sidewalk	189, 417
asphalt	416
brick	417
concrete	416
Sidewalks driveways and patios	416
Siding bevel	119
cedar	119
nail	89
paint	205
painting	205
plywood	119
redwood	119
removal	112
stain	205
wood	119
Sign	223, 224
aluminum	223
flexible	224
letter	223
Signage exterior	223
Signal bell fire	411
Silicon carbide abrasive	30, 33
Silicone caulking	124
Sill	51, 95
block	48
door	103, 109, 160
framing	95
masonry	53
precast concrete window	35
quarry tile	187
stone	50, 53
Sillcock hose bibb	348
Simulated brick	54
stone	54
Single hung window	149, 150
zone rooftop unit	373
Sink barber	246
base	274
corrosion resistant rough-in	349
corrosion resistant	349
countertop	345
countertop rough-in	345
cup	349
cup rough-in	349
darkroom	248
faucet kitchen	347
hospital	351
kitchen	345
laboratory	257, 349
laundry	346
lavatory	344
mop	349
plastic	349
polyethylene	349
polypropylene	349
raised deck rough-in	345
removal	338
service	349
service floor rough-in	349

Index

service wall rough-in 350
slop 346, 349
waste treatment 341
Sinks residential 345
Siren 410
Sister joist 93
Site furnishings 306
  improvement 307
Skylight 155
  roof 155
Skyroof 149
Slab cutout 20
  edge form 66
  on grade 32
Slate 52, 53
  flagging 417
  sidewalk 417
  sill 53
  stair 53
  tile 190
Slatwall 104, 200
Sleeper 96
Sleepers framing 96
Slide swimming pool 310
Sliding chalkboard 221
  door 133, 141
  door shower 227
  entrance 147
  glass door 141
  mirror 234
  panel door 148
  patio door 446
  window 150, 153
Slop sink 346, 349
Slot letter 238
Slotted block 47
  channel framing 61
Small tools 15
Smoke detector 411
  hatch 121, 447
  vent 121
Snow guard 122
Soaking bathroom 345
  tub 345
Soap dispenser 233
  holder 234
  tank 233
Socket wire rope 61
Sodium high pressure fixture .. 395,
              396, 398, 401
  high pressure lamp 404
  low pressure lamp 404
Soffit 96
  drywall 182
  marble 52
  plaster 179
  plywood 103
  stucco 180
  wood 103
Soil separator 424
Solar energy 372
  energy circulator air 372
  energy system collector .. 372
  energy system controller . 372
  energy system heat exchanger 372
  film glass 169
  screen block 48
Solid brick wall 454
  concrete block partition .. 451
  core door 135
  surface countertops 280
  wood door 136
Sound attenuation 200
  curtain 315
  movie 256
  panel 314

proof enclosure 312
system 406, 538
system speaker 406
Spa bath 310
Space heater floor mounted ... 371
Spandrel glass 168
Speaker movie 256
  sound system 406
  station nurse 407
Special construction 315
  door 140, 144
  hinge 162
  systems 407, 410
Specialties 233
  demo 220
  piping HVAC 361
  scaffold 15
  telephone 224
Spike grid 90
Spiral stair 105
Splicer movie 256
Split astragal 164
  rail fence 419
  ring 90
  system control system .. 360
  system ductless 375
Spotlight 402
Spray fireproofing 122
  insulation 115
  shower 347
  substrate 24
Sprayer airless 22
Spread footing 31
Spring bolt astragal 163
  bronze weatherstrip 164
  hinge 163
Sprinkler alarm 411
  cabinet 332
  connector flexible 332
  control pressure 333
  dehydrator package 333
  dry pipe 517
  dry pipe system 518
  escutcheon 332
  head 332, 333
  head guard 333
  head wrench 333
  line tester 328
  monitoring panel deluge . 333
  release emergency 332
  system 640
  system accelerator 333
  systems 332
  valve 328, 333
  wet pipe system 520, 521, 522
Squash court 313
Stack chair 287
Stage equipment 258
  lighting 402
  portable 258
Stain cabinet 209
  casework 209
  door 209
  floor 212
  shingle 205
  siding 119, 205
  truss 213
Stainless duct 362
  flashing 120
  screen 225
  sign 223
  steel corner guard 232
  steel cot 296
  steel cross 265
  steel hinge 161, 162
  steel shelf 234

steel sink 503, 504
steel storefront 148
weld rod 60
Stair basement 105
  brick 50
  carpet 197
  ceiling 250
  climber 325
  concrete 32
  concrete pan fill 31
  electric 250
  finish concrete 33
  grating 77
  landing 195
  lighting 400
  marble 52
  metal pan 76
  metal plate 77
  nosing 192, 194
  prefabricated 105
  removal 87
  residential 105
  riser 195
  riser vinyl 192
  rubber 192
  slate 53
  spiral 105
  stringer 94, 195
  temporary protection 16
  terrazzo 194
  tread 50, 79, 192
  tread insert concrete 31
  tread terrazzo 195
  tread tile 186
  wood 105-107
Stairlift wheelchair 325
Stairs 487
  & walks moving 324
  disappearing 250
  fire escape 77
Stairway door hardware 157
  scaffold 14
Stairwork and handrails ... 107
Stall shower 346
  toilet 225, 226
  type urinal 348
  urinal 348
Stamp time 407
Standard extinguisher 237
Standpipe alarm 332, 411
  connection 328
  equipment 527
Station diaper 233
  hospital 261
  transfer 264
Stationary ventilator 368
  ventilator mushroom ... 368
Steam bath 313
  clean masonry 41
  cleaning metal 57
  humidifier 378
  jacketed kettle 253
  pressure valve 361
  trap 361
Steamer 253
Steel anchor 43
  astragal 163
  beam fireproofing 436
  blind 268
  bolt 90
  bridge 420
  bridging 91
  chain 79
  channel 179
  column fireproofing ... 438
  corner guard 232

demolition 57
diffuser 367
door 131, 132, 139, 141, 144,
              477-479, 632
drilling 57
edging 416, 424
frame 129
furring 176
grate stair 487
hinge 160-162
lath 178
lintel 61
locker 237
nailer 93
painting 217
partition 229, 231
primer 217
register 368
sash 150
screw 89
shelving 239
sink 503, 504
stud NLB 176, 177
weld rod 59
window 150, 151, 444
window demolition 128
wire rope 63
Steeple tip pin 161
Step bluestone 416
  brick 416
  masonry 416
  railroad tie 416
  stone 51
Sterilizer barber 246
  dental 262
  medical 261
Stiffeners joist web 72
Stile fabric 164
  protection 163, 164
Stone anchor 43
  ashlar 52
  base 52, 416
  cast 54
  column 52
  countertops engineered . 281
  faced wall 456
  floor 52
  mulch 421
  paver 51
  pavers 417
  retaining walls 420
  sill 50, 53
  simulated 54
  step 51
  stool 53
  tread 50
  veneer wall 456
  wall 50, 420
Stool cap 103
  stone 53
  window 52, 53
Stop door 102
  valve 348
  water supply 348
Storage bottle 251
  box 13
  cabinet 278
  door cold 143
  shelving 239
Storefront 147
  aluminum 148
Stove 236, 253
  wood burning 236
Strainer Y type iron body ... 361
Strap tie 90
Straw 420

681

Index

Stringer stair 94, 195
   stair terrazzo 195
Strip cabinet 406
   entrance 145
   filler 164
   floor 190
   footing 31
   lighting 397
   shingle 117
Strobe light 402
Structural brick 44
   columns 60
   face tile 45
   facing tile 45
   tile 45
   welding 58
Stucco 180
   demolition 173
   mesh 178
   painting 205
Stud back up wall 458
   demolition 87
   driver pneumatic 59
   metal 174
   NLB metal 176, 177
   NLB steel 176, 177
   NLB wall 176, 177
   partition 94, 178
   partition framing load bearing . . 67
   wall 94, 175, 467, 468, 470-472
   wall blocking load-bearing 66
   wall box beams load-bearing . . . 67
   wall bracing load-bearing 66
   wall framing load bearing 67
   wall headers load-bearing 67
   wall wood 174
   wood 174
Subcontractor O & P 11
Subfloor 99
   adhesive 99
   plywood 99
   wood 99
Sump pump 249
Supermarket checkout 245
   scanner 245
Support ceiling 60
   light 60
   monitor 60
   motor 60
   partition 60
   X-ray 60
Suppressor lightning 385
Surface countertops solid 280
   hinge 161, 162
   landscape 197
   playground 193
   preparation 201, 203
   raceway 391
Surfactant 24
Surgery equipment 262
   table 261
Surgical lighting 261
Surround shower 227
   tub 227
Surveillance system TV 410
Suspended ceiling 177, 182, 188
Suspension mounted heater 371
   system 492, 493
   system ceiling 189, 493
Swimming pool 310
   pool equipment 310
   pool filter 355
   pool filtration 355
   pool hydraulic lift 310
   pool ladder 310
   pool ramp 310

Swing clear hinge 163
   door 139
Switch box plastic 385
   decorator 386
   dimmer 386, 393
   electric 392, 394
   general duty 394
   plate wall 393
   residential 385
   safety 394
   time 394
   toggle 393
Switching low-voltage 392
Switchplate low-voltage 392
Swivel adapter fire hose 329
Synthetic marble 190
   turf 193
System ceiling suspension 189
   chan-l-wire 392
   communication and alarm 538
   components foam-water 334
   control 360
   drinking fountain 507
   dry pipe sprinkler 518, 519
   electric service 529
   fire detection 538
   fire extinguishing . . 237, 329, 330, 333
   fire sprinkler 517, 519-522
   fluorescent fixture 533
   fluorescent watt/SF 534
   fm200 fire suppression 528
   incandescent fixture 536
   incandescent fixture/SF 535
   intercom 538
   master clock 538
   overhead distribution 391
   packaged water-heater 342
   receptacle 531
   receptacle watt/SF 530
   riser dry standpipe 526
   sprinkler 333, 334
   standpipe riser dry 525, 526
   standpipe riser wet 523
   telephone underfloor 537
   tube 326
   T.V. 406
   T.V. surveillance 410
   UHF 406
   undercarpet data processing . . 383
   undercarpet power 382
   undercarpet telephone 382
   underfloor duct 531
   VHF 406
   wall exterior glass block 440
   wall switch 532
   wet pipe sprinkler 521, 522
Systems clock 407
   furniture 301
   sprinkler 332

T

Table base 292
   folding 290
   laboratory 257
   massage 259
   overbed 299
   physical therapy 261
   reading 298
   rewind 256
   round 290
   saw 258
   surgery 261
   top 293-295

Tank darkroom 248
   developing 248
   hubbard 261
   soap 233
   water storage solar 372
Tape barricade 16
   sealant 124
Tar paper 421
Tarpaulin 16
   duck 16
   mylar 16
   polyethylene 16
   polyvinyl 16
Tavern air-conditioning 511, 513
Tax 10
   sales 10
   social security 10
   unemployment 10
T-bar grille eggcrate 368
   grille lay-in 368
   mount diffuser 367
Teak floor 190
   molding 102
   paneling 104
Telephone booth 224
   enclosure 224
   house 406
   receptacle 388
   specialties 224
   undercarpet 382
Telescoping bleacher 302
   door 145
   platform 258
Television equipment 406
   receptacle 388
Teller automatic 245
   window 245
Temperature rise detector 411
Tempered glass 166
   glass greenhouse 314
   hardboard 103
Temporary building 13
   construction 12, 16
   facility 16
   heat 12
   light 12
   utilities 12
Tennis court fence 419
   court net 418
   court post 418
   court surface 198
   court surfacing 418
Terminal A/C packaged 374
   air conditioner packaged 374
Termination box 406
Terra cotta 46
   cotta demolition 20, 173
Terrazzo 194
   abrasive 194
   base 194
   conductive 195
   demolition 173
   epoxy 195, 196
   floor 194, 636
   monolithic 194
   precast 194
   receptor 227
   stair 194
   venetian 194
   wainscot 194, 195
Tester sprinkler line 328
Testing OSHA 24
Theater and stage equipment . . 258
Theatrical lighting 402
Thermal & moist. protect. demo . 112
   batt 200

Thermometer 257
Thermostat 361
   automatic timed 361
   control sys. radiator 360
   wire 390
Thimble wire rope 61
Thin plaster partition 175
Thincoat plaster 181
Thinset ceramic tile 185
   mortar 186
Threshold 160
   door 109
   stone 52
   wood 103
Thru-wall air conditioner 374
Tie rafter 95
   railroad 416, 424
   rod 61
   strap 90
   wall 43
Tier locker 237
Tile 185, 193
   abrasive 185
   aluminum 118
   and covering floor 490, 491
   carpet 197
   ceiling 188, 492
   ceramic 185, 186
   cork 193
   cork wall 198
   demolition 172
   exterior 186
   fire rated 45
   glass 168, 187
   grout 185
   marble 190
   metal 187
   porcelain 185
   quarry 186
   roof 118
   rubber 194
   slate 190
   stainless steel 187
   stair tread 186
   structural 45
   terra cotta 46
   vinyl 193
   wall 185
   window sill 187
Timber connector 90
   fastener 88
Time lapse photography 12
   lock 143
   stamp 407
   switch 394
   system controller 407
Timer clock 389
   interval 386
Tin clad door 136
Tinted glass 166
Toaster 254
Toggle switch 393
Toggle-bolt anchor 58
Toilet accessories 233
   accessory 233
   bowl 343
   compartments 224
   door hardware 157
   gasket 348
   partition 52, 224-226
   partition removal 174
   prison 245
   seat 343
   stall 225, 226
   tissue dispenser 233, 234
Tool hydrant 330

Index

powder actuated 59
Tools small 15
Toothed ring 90
Toothing masonry 20, 40
Top coiling grille 141
   counter 279, 280
   demolition counter 88
   vanity 281
Topping concrete 33
   epoxy 196
   floor 33
   grout 35
Torch cutting demolition 21
Touch bar 158
Tour station watchman 407
Towel bar 234
   dispenser 233, 234
Track curtain 259
   drawer 278
   light 402
   lighting 402
   running 191, 192
   surface 198
   traverse 271
Traffic control pedestrian .. 244
   door 145
Trailer office 13
Trainer bicycle 259
Tranformer accent light ... 400
Transfer station 264
Transformer low-voltage ... 392
   UPS 394
Transition molding wood .. 191
   residential 366
Transom aluminum 148
   lite frame 130
   windows 154
Trap inverted bucket steam .. 361
   rock surface 33
   steam 361
Trash compactor 255
   receiver 283
   receptacle 306
   receptacle fiberglass ... 306
   receptacle wood 306
Traverse 271, 272
   track 271
Travertine 52
Tray laundry 346
Tread abrasive 187
   aluminum cast 79
   aluminum grating 79
   bronze cast 79
   cast iron 79
   fiberglass 79
   rubber 192
   stair 50, 79
   stone 52, 53
   vinyl 192
Treadmill 259, 260
Treated lumber framing 96
   pine fence 419
Treatment acoustical 200
Trees deciduous 423, 424
   evergreen 422
   moving 425
Trim demolition 88
   exit 158
   exterior 102
   painting 213
   panic 158
   redwood 102
   tile 185
   wood 102
Triple brick 45
   weight hinge 162

Troffer air handling 398
   light fixture 398
Trolley rail 232
Truss framing metal 75
   joist 65
   painting 213
   plate 90
   roof 101
   stain 213
   varnish 213
Tub bar 233
   hot 310
   redwood 310
   rough-in 346
   shower handicap 346
   soaking 345
   surround 227
Tube framing 146
   pneumatic 245
   system 326
   system air 326
   system pneumatic 326
Tubing electric metallic ... 384
Tub-shower module 345
Tubular lock 159
   scaffolding 13
   steel door 139
Tumbler holder 234
Turf artificial 418
   synthetic 193
Turnbuckle wire rope 62
Turned column 108
Turnstile 244
   security 244
TV closed circuit camera .. 410
   system 406
T.V. systems 406
Two piece astragal 164

U

UHF system 406
Ultrasonic cleaner 262
   motion detector 410
Undercarpet electrical 381
   power system 382
   shield 383
   telephone system 382
Underfloor duct telephone system
   537
Underlayment 99
   acoustical 201
   gym floor 191
   hardboard 99
   latex 192
   self-leveling cement ... 35
Underwater light 310
Unemployment tax 10
Uninterruptable power supply .. 394
Uninterruptible power supply .. 394
Unit heater 371
   heater electric 377
   kitchen 247
Upholstry fabric 268
Urethane foam 125, 200
   insulation 311
   wall coating 217
Urinal 348
   removal 338
   screen 226
   stall 348
   stall type 348
   system 502
   wall hung 348
Urn ash 283

Utensil washer medical ... 261
Utilities temporary 12
Utility brick 45
   set fan 364
   temporary 12

V

Vacuum central 247
   cleaning 247
   loader 22
Valance board 275
   curtain 269-271
Valley rafter 95
Valve assembly sprinkler deluge . 333
   cabinet 236
   cap fire 331
   check 328
   fire 331
   fire hose 331
   flush 350
   heating control 361
   hot water radiator 361
   hydrant fire 331
   hydrant screw type ... 331
   mixing 347
   shower by-pass 347
   shower mixing 346
   spout set shower-tub .. 347
   sprinkler 328, 333
   sprinkler alarm 328
   steam pressure 361
   stop 348
   water supply 348
Vandalproof fixture 399
Vane-axial fan 363
Vanity base 277
   top 281
   top lavatory 344, 503
Vapor barrier 117
   barrier sheathing 100
   proof insulation .. 338, 339
   retarder 117
Vaportight fixture incandescent . 399
Varnish cabinet 209
   casework 209
   door 209
   floor 212, 216
   polyurethane 217
   truss 213
Vat removal 24
Vault door 143
   front 143
Vaulting side horse 260
Vct removal 172
Veneer ashlar 50
   brick 44
   core paneling 104
   granite 51
   removal 42
   wall stone 456, 457
Venetian blind 268
   terrazzo 194, 195
Vent automatic air 361
   chimney all fuel 370
   chimney fitting 370
   chimney flashing 371
   dryer 249
   exhaust 368
   intake-exhaust 369
   smoke 121
   wave guide 317
Ventilating air conditioning . 360, 362, 364, 368
Ventilation 645

fan 389
   louver 170
Ventilator control system .. 360
   fan 363
   mushroom stationary .. 368
   relief vent 368
   stationary mushroom .. 368
Venturi flow meter 362
Verge board 102
Vermiculite cement 122
   insulation 114
   plaster 179
Vertical blinds 269
   lift door 144, 145
   metal ladder 77
VHF system 406
Videotape recorder 410
Vinyl casement window .. 155
   composition floor 193
   faced wallboard 182
   floor 193
   lead barrier 201
   stair riser 192
   stair tread 192
   tile 193
   tread 192
   wall coating 217
   wallpaper 199
Vision panel 155
Vitreous china lavatory .. 503
Vitreous-china service sink . 506
Vocational kiln 258
   shop equipment 258

W

Wainscot ceramic tile 185
   molding 102
   quarry tile 187
   terrazzo 194, 195
Walk 416
   moving 325
Walk-in refrigeration 311
Walkway luminaire 402
Wall & corner guards 232
   blocking load-bearing stud .. 66
   box beams load-bearing stud .. 67
   bracing load-bearing stud .. 66
   bumper 160, 232
   cabinet 274, 278, 279
   canopy 240
   cap residential 366
   ceramic tile 186
   coating 217
   concrete 31
   concrete placement ... 32
   covering 636
   cross 264
   curtain 148
   cutout 20
   demolition 173
   drywall 182
   exhaust fan 365
   fan 365
   finishes 198
   framing 96
   framing load bearing stud .. 67
   framing removal 21
   furnace 371
   furring 98, 176
   glass block 440
   guard 232
   headers load-bearing stud .. 67
   heater 250
   hung lavatory 345

683

Index

hung urinal 348
hydrant 328
insulation 112, 113, 115
lath 178
masonry 420
mat 260
mirror 168, 169
NLB partition 174, 175
painting 205, 214, 215
panel 54
panel brick 49
panel woven wire 228
paneling 104
plaster 179
plug 43
register 368
rubbing 34
shaft 174
shear 100
sheathing 100
stone 50
stone faced metal stud ... 456, 457
stone faced wood stud 456
stucco 180
stud 94, 174, 175
stud NLB 176, 177
switch plate 393
tie 43
tie masonry 43
tile 185
tile cork 198
urn ash receiver 234
window 147, 148
wood stud 174
Wallboard acoustical 184
Wallcovering 198
 acrylic 199
 gypsum fabric 199
Wallguard 232
Wallpaper 199, 636
 grass cloth 199
 vinyl 199
 wall 488, 489
Walls and partitions demolition .. 173
Walnut door frame 109
 floor 190
Wardrobe 240, 297
 cabinet 240
 wood 276
Wash fountain 350
 fountain group 350
 safety equipment eye/face ... 352
Washable air filters 369
Washer 90
 commercial 247
 darkroom 248
 residential 249
Washing machine automatic 249
Waste compactor 264
 disposal 263
 handling 263
 handling equipment 263
 incinerator 263
 packaging 25
 receptacle 234
 treatment sink 341
Watchman service 16
 tour station 407
Water atomizer 22
 balancing 358
 blasting 40
 blasting metal 57
 bubbler 507
 closet 343, 348
 closet hospital 351
 closet removal 338

closet seat 343, 344
cooler 354, 508, 645
cooler handicap 354
dispenser hot 341
distiller 261
fountain removal 338
heater 249, 342, 390
heater commercial 341
heater electric 341
heater gas 342
heater insulation 338
heater laundry 342
heater oil 342
heater removal 338
heater wrap kit 338
heating hot 376
motor alarm 332
pump 249
pump fire 335
repellent 214
storage solar tank 372
supply stop 348
supply valve 348
Water-closet 501
 battery mount 501
 conserving 348
 system 500
Water-heater 342
 car-wash 342
 system packaged 342
Waterproofing 112
 cementitious 112
Water-source heat pumps 375
Wave guide vent 317
Weatherproof receptacle 388
Weatherstrip 142, 164
 aluminum 163
 door 165
Weatherstripping 164
Weight exercise 259
 lifting multi station 260
Weld rod 59
 rod aluminum 59
 rod cast-iron 59
 rod stainless 60
 rod steel 59
Welded frame 130
 structural steel 630
Welder arc 258
Welding fillet 58
 structural 58
Wheel color 402
 potters 258
Wheelchair lift 325
 stairlift 325
Whirlpool bath 261, 310
 hospital 351
Whole house fan 365
Wide throw hinge 163
Window 149
 aluminum 149, 150
 and door molding 102
 awning 151
 bank 245
 blind 268
 bullet resistant 245
 casement 150, 151, 153
 casing 101
 counter 238
 demolition 128
 double hung 150, 152, 154
 drive-up 245
 frame 152
 glass 168
 grille 154
 guard 80

hardware 156, 157
industrial 150
metal 149, 150
muntin 154
painting 209, 210
picture 150, 152
pivoted 150
plastic 155
precast concrete sill 35
projected 150
removal 128
restoration 26
screen 151, 154
sill 53
sill tile 187
sliding 153
steel 150, 151
stool 52, 53
teller 245
trim set 103
wall 147
wall framing 146
wood 151-153
Windows transom 154
Wine cellar 251
Winter protection 16
Wire brushing metal 57
 copper 384
 electrical 384
 fence 419
 fencing 419
 glass 168
 hanging 189
 low-voltage control 393
 mesh 180, 419
 mesh locker 237
 mesh partition 228
 partition 228
 razor 419
 rope clip 61
 rope socket 61
 rope thimble 61
 rope turnbuckle 62
 thermostat 390
 THWN-THHN 384
Wiremold raceway 391
Wiring air conditioner 390
 device 392, 393
 fan 389
 low-voltage 393
 power 394
 residential 385, 389
Wood base 101
 beam 91, 92
 bench 307
 blind 272
 block floor 190
 block floor demolition .. 173
 blocking 94
 bridge 420
 bridging 91
 cabinet 278
 canopy 96
 casing 101
 chips 421
 column 92, 108
 cutting 21
 deck 99
 demolition 172
 door .. 133, 134, 136, 137, 139, 480
                            , 481-484
 drawer 277
 edging 424
 fascia 102
 fastener 89
 fiber insulation 114

fiber sheathing 100
fiber underlayment 99
floor 190
floor demolition 173
folding partition 231
frame 109, 275
framing 91, 92
furring 98
girder 92
handrail 102
joist 93
louver 107
nailer 93
panel door 136
paneling 103
parquet 190
partition 94, 472
plank scaffold 14
planter 306
rail fence 419
railing 107
roof deck 99
sash 152
screen 154
screw 89
shade 273
shake 118
sheathing 100
shelving 240
shingle 117
sidewalk 416
siding 119
siding demolition 112
sill 95
soffit 103
stair 105-107
storm door 139
stud 174
subfloor 99
threshold 103
trash receptacle 306
trim 102
veneer wallpaper 199
wardrobe 276
window 151-153, 443, 444
window demolition 128
Wool carpet 197
 fiberglass 114
Work extra 11
 in prisons 9
 station 542-546
Worker protection 22
Workers' compensation 10
Workers' compensation ... 622-624
Woven wire partition 228
Wrench sprinkler head 333
Wrestling mat 260
Wye fire hose 332

X

X-ray barrier 316
 dental 262
 medical 263
 mobile 263
 protection 315
 support 60

Y

Y type iron body strainer 361
Yellow pine floor 191

Index

Z

Z bar suspension 189
Zee bar 185
Zinc divider strip 194
    plating 89
    terrazzo strip 194
    weatherstrip 164

Notes

Notes

Notes

Notes

Division Notes

		CREW	DAILY OUTPUT	LABOR-HOURS	UNIT	\multicolumn{4}{c	}{2000 BARE COSTS}	TOTAL INCL O&P
						MAT.	LABOR	EQUIP.	TOTAL	

Division Notes

		CREW	DAILY OUTPUT	LABOR-HOURS	UNIT	MAT.	LABOR	EQUIP.	TOTAL	TOTAL INCL O&P
							2000 BARE COSTS			

Division Notes

	CREW	DAILY OUTPUT	LABOR-HOURS	UNIT	2000 BARE COSTS				TOTAL INCL O&P
					MAT.	LABOR	EQUIP.	TOTAL	

Division Notes

		CREW	DAILY OUTPUT	LABOR-HOURS	UNIT	2000 BARE COSTS				TOTAL INCL O&P
						MAT.	LABOR	EQUIP.	TOTAL	

Reed Construction Data/RSMeans— a tradition of excellence in construction cost information and services since 1942

Table of Contents
Annual Cost Guides
Reference Books
Seminars
Electronic Data
New Titles
Order Form

For more information visit the RSMeans website at www.rsmeans.com

Book Selection Guide

The following table provides definitive information on the content of each cost data publication. The number of lines of data provided in each unit price or assemblies division, as well as the number of reference tables and crews, is listed for each book. The presence of other elements such as equipment rental costs, historical cost indexes, city cost indexes, square foot models, or cross-referenced indexes is also indicated. You can use the table to help select the RSMeans book that has the quantity and type of information you most need in your work.

Unit Cost Divisions	Building Construction Costs	Mechanical	Electrical	Repair & Remodel	Square Foot	Site Work Landsc.	Assemblies	Interior	Concrete Masonry	Open Shop	Heavy Construction	Light Commercial	Facilities Construction	Plumbing	Residential
1	553	372	385	474		489		301	437	552	483	220	967	386	157
2	555	239	60	544		782		333	189	553	680	409	1022	262	217
3	1471	193	179	830		1321		266	1803	1467	1459	304	1386	162	309
4	878	24	0	677		683		557	1082	854	576	469	1125	0	385
5	1847	197	152	951		796		1023	751	1816	1045	809	1837	275	724
6	1870	78	80	1458		461		1191	310	1849	395	1562	1484	22	2076
7	1395	164	71	1418		506		513	439	1393	0	1091	1451	175	852
8	1941	61	10	1969		298		1677	673	1906	0	1389	2134	0	1319
9	1800	72	26	1606		292		1899	392	1761	0	1513	2035	54	1357
10	978	17	10	582		226		800	156	978	29	465	1072	233	212
11	998	209	160	476		126		848	28	983	0	223	1012	170	110
12	547	0	2	303		248		1691	12	536	19	349	1736	23	317
13	730	119	111	247		341		255	69	702	253	80	740	74	79
14	281	36	0	230		28		260	0	280	0	12	300	21	6
21	78	0	16	31		0		252	0	78	0	60	370	375	0
22	1090	6955	154	1049		1409		596	20	1069	1634	751	6824	8656	572
23	1189	7103	609	776		159		682	38	1103	110	649	5143	1860	425
26	1246	493	9825	991		743		1053	55	1234	527	1060	9761	438	557
27	75	0	200	34		9		67	0	75	35	48	198	0	4
28	87	59	94	64		0		69	0	72	0	42	109	45	27
31	1904	1162	1031	1225		2731		7	1635	1848	2804	1027	1964	1087	1033
32	737	46	0	603		3849		360	271	710	1366	353	1370	158	412
33	514	1020	429	199		1957		0	225	512	1883	114	1547	1189	112
34	99	0	20	4		134		0	31	55	113	0	119	0	0
35	18	0	0	0		166		0	0	18	281	0	83	0	0
41	52	0	0	24		7		22	0	52	30	0	59	14	0
44	98	90	0	0		32		0	0	23	32	0	98	105	0
Totals	23031	18709	13264	16765		17793		14722	8616	22479	13754	12999	45946	15784	11262

Assembly Divisions	Building Construction Costs	Mechanical	Electrical	Repair & Remodel	Square Foot	Site Work Landscape	Assemblies	Interior	Concrete Masonry	Open Shop	Heavy Construction	Light Commercial	Facilities Construction	Plumbing	Asm Div	Residential
A		15	0	182	150	530	598	0	536		570	154	24	0	1	374
B		0	0	808	2471	0	5577	329	1914		368	2014	143	0	2	217
C		0	0	630	864	0	1229	1567	146		0	765	242	0	3	588
D		1066	810	694	1779	0	2408	765	0		0	1270	1011	1068	4	861
E		0	0	85	260	0	297	5	0		0	257	5	0	5	392
F		0	0	0	123	0	124	0	0		0	123	3	0	6	358
G		528	294	311	202	2206	668	0	535		1194	200	293	685	7	301
															8	760
															9	80
															10	0
															11	0
															12	0
Totals		1609	1104	2710	5849	2736	10901	2666	3131		2132	4783	1721	1753		3931

Reference Section	Building Construction Costs	Mechanical	Electrical	Repair & Remodel	Square Foot	Site Work Landscape	Assemblies	Interior	Concrete Masonry	Open Shop	Heavy Construction	Light Commercial	Facilities Construction	Plumbing	Residential
Reference Tables	yes	yes	yes	yes	no	yes	yes	yes	yes	yes	yes	yes	yes	yes	yes
Models					105							46			32
Crews	482	482	482	462		482		482	482	460	482	460	462	482	460
Equipment Rental Costs	yes	yes	yes	yes		yes		yes	yes	yes	yes	yes	yes	yes	yes
Historical Cost Indexes	yes	yes	yes	yes	yes	yes	yes	yes	yes	yes	yes	yes	yes	yes	no
City Cost Indexes	yes	yes	yes	yes	yes	yes	yes	yes	yes	yes	yes	yes	yes	yes	yes

Annual Cost Guides

For more information visit the RSMeans website at www.rsmeans.com

RSMeans Building Construction Cost Data 2009

Offers you unchallenged unit price reliability in an easy-to-use arrangement. Whether used for complete, finished estimates or for periodic checks, it supplies more cost facts better and faster than any comparable source. Over 20,000 unit prices for 2009. The City Cost Indexes and Location Factors cover over 930 areas, for indexing to any project location in North America. Order and get *RSMeans Quarterly Update Service* FREE. You'll have year-long access to the RSMeans Estimating **HOTLINE** FREE with your subscription. Expert assistance when using RSMeans data is just a phone call away.

$154.95 per copy
Available Oct. 2008
Catalog no. 60019

Unit prices organized according to MasterFormat 2004!

RSMeans Mechanical Cost Data 2009
- **HVAC**
- **Controls**

Total unit and systems price guidance for mechanical construction... materials, parts, fittings, and complete labor cost information. Includes prices for piping, heating, air conditioning, ventilation, and all related construction.

Plus new 2009 unit costs for:

- Over 2,500 installed HVAC/controls assemblies components
- "On-site" Location Factors for over 930 cities and towns in the U.S. and Canada
- Crews, labor, and equipment

$154.95 per copy
Available Oct. 2008
Catalog no. 60029

RSMeans Plumbing Cost Data 2009

Comprehensive unit prices and assemblies for plumbing, irrigation systems, commercial and residential fire protection, point-of-use water heaters, and the latest approved materials. This publication and its companion, *RSMeans Mechanical Cost Data*, provide full-range cost estimating coverage for all the mechanical trades.

Now contains more lines of no-hub CI soil pipe fittings, more flange-type escutcheons, fiberglass pipe insulation in a full range of sizes for 2-1/2" and 3" wall thicknesses, 220 lines of grease duct, and much more.

$154.95 per copy
Available Oct. 2008
Catalog no. 60219

Unit prices organized according to MasterFormat 2004!

RSMeans Facilities Construction Cost Data 2009

For the maintenance and construction of commercial, industrial, municipal, and institutional properties. Costs are shown for new and remodeling construction and are broken down into materials, labor, equipment, and overhead and profit. Special emphasis is given to sections on mechanical, electrical, furnishings, site work, building maintenance, finish work, and demolition.

More than 43,000 unit costs, plus assemblies costs and a comprehensive Reference Section are included.

$368.95 per copy
Available Nov. 2008
Catalog no. 60209

For more information visit the RSMeans website at www.rsmeans.com

Annual Cost Guides

RSMeans Square Foot Costs 2009

It's Accurate and Easy To Use!

- **Updated price information** based on nationwide figures from suppliers, estimators, labor experts, and contractors
- "How-to-Use" sections, with **clear examples** of commercial, residential, industrial, and institutional structures
- Realistic graphics, offering true-to-life illustrations of building projects
- Extensive information on using square foot cost data, including sample estimates and alternate pricing methods

$168.95 per copy
Available Oct. 2008
Catalog no. 60059

RSMeans Repair & Remodeling Cost Data 2009

Commercial/Residential

Use this valuable tool to estimate commercial and residential renovation and remodeling.

Includes: New costs for hundreds of unique methods, materials, and conditions that only come up in repair and remodeling, PLUS:

- Unit costs for over 15,000 construction components
- Installed costs for over 90 assemblies
- Over 930 "on-site" localization factors for the U.S. and Canada.

Unit prices organized according to MasterFormat 2004!

$132.95 per copy
Available Nov. 2008
Catalog no. 60049

RSMeans Electrical Cost Data 2009

Pricing information for every part of electrical cost planning. More than 13,000 unit and systems costs with design tables; clear specifications and drawings; engineering guides; illustrated estimating procedures; complete labor-hour and materials costs for better scheduling and procurement; and the latest electrical products and construction methods.

- A variety of special electrical systems including cathodic protection
- Costs for maintenance, demolition, HVAC/mechanical, specialties, equipment, and more

$154.95 per copy
Available Oct. 2008
Catalog no. 60039

Unit prices organized according to MasterFormat 2004!

RSMeans Electrical Change Order Cost Data 2009

RSMeans Electrical Change Order Cost Data provides you with electrical unit prices exclusively for pricing change orders—based on the recent, direct experience of contractors and suppliers. Analyze and check your own change order estimates against the experience others have had doing the same work. It also covers productivity analysis and change order cost justifications. With useful information for calculating the effects of change orders and dealing with their administration.

$154.95 per copy
Available Dec. 2008
Catalog no. 60239

Annual Cost Guides

For more information visit the RSMeans website at www.rsmeans.com

RSMeans Assemblies Cost Data 2009

RSMeans Assemblies Cost Data takes the guesswork out of preliminary or conceptual estimates. Now you don't have to try to calculate the assembled cost by working up individual component costs. We've done all the work for you.

Presents detailed illustrations, descriptions, specifications, and costs for every conceivable building assembly—240 types in all—arranged in the easy-to-use UNIFORMAT II system. Each illustrated "assembled" cost includes a complete grouping of materials and associated installation costs, including the installing contractor's overhead and profit.

$245.95 per copy
Available Oct. 2008
Catalog no. 60069

RSMeans Open Shop Building Construction Cost Data 2009

The latest costs for accurate budgeting and estimating of new commercial and residential construction... renovation work... change orders... cost engineering.

RSMeans Open Shop "BCCD" will assist you to:
- Develop benchmark prices for change orders
- Plug gaps in preliminary estimates and budgets
- Estimate complex projects
- Substantiate invoices on contracts
- Price ADA-related renovations

Unit prices organized according to MasterFormat 2004!

$154.95 per copy
Available Dec. 2008
Catalog no. 60159

RSMeans Residential Cost Data 2009

Contains square foot costs for 30 basic home models with the look of today, plus hundreds of custom additions and modifications you can quote right off the page. With costs for the 100 residential systems you're most likely to use in the year ahead. Complete with blank estimating forms, sample estimates, and step-by-step instructions.

Now contains line items for cultured stone and brick, PVC trim lumber, and TPO roofing.

$132.95 per copy
Available Oct. 2008
Catalog no. 60179

Unit prices organized according to MasterFormat 2004!

RSMeans Site Work & Landscape Cost Data 2009

Includes unit and assemblies costs for earthwork, sewerage, piped utilities, site improvements, drainage, paving, trees & shrubs, street openings/repairs, underground tanks, and more. Contains 57 tables of assemblies costs for accurate conceptual estimates.

Includes:
- Estimating for infrastructure improvements
- Environmentally-oriented construction
- ADA-mandated handicapped access
- Hazardous waste line items

$154.95 per copy
Available Nov. 2008
Catalog no. 60289

RSMeans Facilities Maintenance & Repair Cost Data 2009

RSMeans Facilities Maintenance & Repair Cost Data gives you a complete system to manage and plan your facility repair and maintenance costs and budget efficiently. Guidelines for auditing a facility and developing an annual maintenance plan. Budgeting is included, along with reference tables on cost and management, and information on frequency and productivity of maintenance operations.

The only nationally recognized source of maintenance and repair costs. Developed in cooperation with the Civil Engineering Research Laboratory (CERL) of the Army Corps of Engineers.

$336.95 per copy
Available Dec. 2008
Catalog no. 60309

RSMeans Light Commercial Cost Data 2009

Specifically addresses the light commercial market, which is a specialized niche in the construction industry. Aids you, the owner/designer/contractor, in preparing all types of estimates—from budgets to detailed bids. Includes new advances in methods and materials.

Assemblies Section allows you to evaluate alternatives in the early stages of design/planning.

Over 11,000 unit costs ensure that you have the prices you need... when you need them.

Unit prices organized according to MasterFormat 2004!

$132.95 per copy
Available Nov. 2008
Catalog no. 60189

For more information
visit the RSMeans website
at www.rsmeans.com

Annual Cost Guides

RSMeans Concrete & Masonry Cost Data 2009

Provides you with cost facts for virtually all concrete/masonry estimating needs, from complicated formwork to various sizes and face finishes of brick and block—all in great detail. The comprehensive Unit Price Section contains more than 7,500 selected entries. Also contains an Assemblies [Cost] Section, and a detailed Reference Section that supplements the cost data.

Unit prices organized according to MasterFormat 2004!

$139.95 per copy
Available Dec. 2008
Catalog no. 60119

RSMeans Labor Rates for the Construction Industry 2009

Complete information for estimating labor costs, making comparisons, and negotiating wage rates by trade for over 300 U.S. and Canadian cities. With 46 construction trades listed by local union number in each city, and historical wage rates included for comparison. Each city chart lists the county and is alphabetically arranged with handy visual flip tabs for quick reference.

$336.95 per copy
Available Dec. 2008
Catalog no. 60129

RSMeans Construction Cost Indexes 2009

What materials and labor costs will change unexpectedly this year? By how much?

- Breakdowns for 316 major cities
- National averages for 30 key cities
- Expanded five major city indexes
- Historical construction cost indexes

$335.00 per year (subscription)
Catalog no. 50149

$83.25 individual quarters
Catalog no. 60149 A,B,C,D

RSMeans Interior Cost Data 2009

Provides you with prices and guidance needed to make accurate interior work estimates. Contains costs on materials, equipment, hardware, custom installations, furnishings, and labor costs... for new and remodel commercial and industrial interior construction, including updated information on office furnishings, and reference information.

Unit prices organized according to MasterFormat 2004!

$154.95 per copy
Available Nov. 2008
Catalog no. 60099

RSMeans Heavy Construction Cost Data 2009

A comprehensive guide to heavy construction costs. Includes costs for highly specialized projects such as tunnels, dams, highways, airports, and waterways. Information on labor rates, equipment, and materials costs is included. Features unit price costs, systems costs, and numerous reference tables for costs and design.

$154.95 per copy
Available Dec. 2008
Catalog no. 60169

Unit prices organized according to MasterFormat 2004!

Reference Books

For more information visit the RSMeans website at www.rsmeans.com

Value Engineering: Practical Applications
For Design, Construction, Maintenance & Operations
by Alphonse Dell'Isola, PE

A tool for immediate application—for engineers, architects, facility managers, owners, and contractors. Includes making the case for VE—the management briefing; integrating VE into planning, budgeting, and design; conducting life cycle costing; using VE methodology in design review and consultant selection; case studies; VE workbook; and a life cycle costing program on disk.

$79.95 per copy
Over 450 pages, illustrated, softcover
Catalog no. 67319A

Facilities Operations & Engineering Reference
by the Association for Facilities Engineering and RSMeans

An all-in-one technical reference for planning and managing facility projects and solving day-to-day operations problems. Selected as the official certified plant engineer reference, this handbook covers financial analysis, maintenance, HVAC and energy efficiency, and more.

$54.98 per copy
Over 700 pages, illustrated, hardcover
Catalog no. 67318

The Building Professional's Guide to Contract Documents
3rd Edition
by Waller S. Poage, AIA, CSI, CVS

A comprehensive reference for owners, design professionals, contractors, and students

- Structure your documents for maximum efficiency.
- Effectively communicate construction requirements.
- Understand the roles and responsibilities of construction professionals.
- Improve methods of project delivery.

$64.95 per copy, 400 pages
Diagrams and construction forms, hardcover
Catalog no. 67261A

Complete Book of Framing
by Scot Simpson

This straightforward, easy-to-learn method will help framers, carpenters, and handy homeowners build their skills in rough carpentry and framing. Shows how to frame all the parts of a house: floors, walls, roofs, door & window openings, and stairs—with hundreds of color photographs and drawings that show every detail.

The book gives beginners all the basics they need to go from zero framing knowledge to a journeyman level.

$29.95 per copy
352 pages, softcover
Catalog no. 67353

Cost Planning & Estimating for Facilities Maintenance

In this unique book, a team of facilities management authorities shares their expertise on:

- Evaluating and budgeting maintenance operations
- Maintaining and repairing key building components
- Applying *RSMeans Facilities Maintenance & Repair Cost Data* to your estimating

Covers special maintenance requirements of the ten major building types.

$89.95 per copy
Over 475 pages, hardcover
Catalog no. 67314

Life Cycle Costing for Facilities
by Alphonse Dell'Isola and Dr. Steven Kirk

Guidance for achieving higher quality design and construction projects at lower costs! Cost-cutting efforts often sacrifice quality to yield the cheapest product. Life cycle costing enables building designers and owners to achieve both. The authors of this book show how LCC can work for a variety of projects — from roads to HVAC upgrades to different types of buildings.

$99.95 per copy
450 pages, hardcover
Catalog no. 67341

Planning & Managing Interior Projects 2nd Edition
by Carol E. Farren, CFM

Expert guidance on managing renovation & relocation projects

This book guides you through every step in relocating to a new space or renovating an old one. From initial meeting through design and construction, to post-project administration, it helps you get the most for your company or client. Includes sample forms, spec lists, agreements, drawings, and much more!

$34.98 per copy
200 pages, softcover
Catalog no. 67245A

Construction Business Management
by Nick Ganaway

Only 43% of construction firms stay in business after four years. Make sure your company thrives with valuable guidance from a pro with 25 years of success as a commercial contractor. Find out what it takes to build all aspects of a business that is profitable, enjoyable, and enduring. With a bonus chapter on retail construction.

$49.95 per copy
200 pages, softcover
Catalog no. 67352

For more information visit the RSMeans website at www.rsmeans.com

Reference Books

Interior Home Improvement Costs 9th Edition

Updated estimates for the most popular remodeling and repair projects—from small, do-it-yourself jobs to major renovations and new construction. Includes: kitchens & baths; new living space from your attic, basement, or garage; new floors, paint, and wallpaper; tearing out or building new walls; closets, stairs, and fireplaces; new energy-saving improvements, home theaters, and more!

$24.95 per copy
250 pages, illustrated, softcover
Catalog no. 67308E

Exterior Home Improvement Costs 9th Edition

Updated estimates for the most popular remodeling and repair projects—from small, do-it-yourself jobs, to major renovations and new construction. Includes: curb appeal projects—landscaping, patios, porches, driveways, and walkways; new windows and doors; decks, greenhouses, and sunrooms; room additions and garages; roofing, siding, and painting; "green" improvements to save energy & water.

$24.95 per copy
Over 275 pages, illustrated, softcover
Catalog no. 67309E

Builder's Essentials: Plan Reading & Material Takeoff
For Residential and Light Commercial Construction

by Wayne J. DelPico

A valuable tool for understanding plans and specs, and accurately calculating material quantities. Step-by-step instructions and takeoff procedures based on a full set of working drawings.

$35.95 per copy
Over 420 pages, softcover
Catalog no. 67307

Means Unit Price Estimating Methods
New 4th Edition

This new edition includes up-to-date cost data and estimating examples, updated to reflect changes to the CSI numbering system and new features of RSMeans cost data. It describes the most productive, universally accepted ways to estimate, and uses checklists and forms to illustrate shortcuts and timesavers. A model estimate demonstrates procedures. A new chapter explores computer estimating alternatives.

$59.95 per copy
Over 350 pages, illustrated, softcover
Catalog no. 67303B

Total Productive Facilities Management

by Richard W. Sievert, Jr.

Today, facilities are viewed as strategic resources. . . elevating the facility manager to the role of asset manager supporting the organization's overall business goals. Now, Richard Sievert Jr., in this well-articulated guidebook, sets forth a new operational standard for the facility manager's emerging role. . . a comprehensive program for managing facilities as a true profit center.

$29.98 per copy
275 pages, softcover
Catalog no. 67321

Green Building: Project Planning & Cost Estimating 2nd Edition

This new edition has been completely updated with the latest in green building technologies, design concepts, standards, and costs. Now includes a 2009 Green Building *CostWorks* CD with more than 300 green building assemblies and over 5,000 unit price line items for sustainable building. The new edition is also full-color with all new case studies—plus a new chapter on deconstruction, a key aspect of green building.

$129.95 per copy
350 pages, softcover
Catalog no. 67338A

Concrete Repair and Maintenance Illustrated
by Peter Emmons

Hundreds of illustrations show users how to analyze, repair, clean, and maintain concrete structures for optimal performance and cost effectiveness. From parking garages to roads and bridges to structural concrete, this comprehensive book describes the causes, effects, and remedies for concrete wear and failure. Invaluable for planning jobs, selecting materials, and training employees, this book is a must-have for concrete specialists, general contractors, facility managers, civil and structural engineers, and architects.

$69.95 per copy
300 pages, illustrated, softcover
Catalog no. 67146

Reference Books

For more information visit the RSMeans website at www.rsmeans.com

Means Illustrated Construction Dictionary Condensed, 2nd Edition

The best portable dictionary for office or field use—an essential tool for contractors, architects, insurance and real estate personnel, facility managers, homeowners, and anyone who needs quick, clear definitions for construction terms. The second edition has been further enhanced with updates and hundreds of new terms and illustrations . . . in keeping with the most recent developments in the construction industry.

Now with a quick-reference Spanish section. Includes tools and equipment, materials, tasks, and more!

$59.95 per copy
Over 500 pages, softcover
Catalog no. 67282A

Means Repair & Remodeling Estimating 4th Edition
By Edward B. Wetherill and RSMeans

This important reference focuses on the unique problems of estimating renovations of existing structures, and helps you determine the true costs of remodeling through careful evaluation of architectural details and a site visit.

New section on disaster restoration costs.

$69.95 per copy
Over 450 pages, illustrated, hardcover
Catalog no. 67265B

Facilities Planning & Relocation
by David D. Owen

A complete system for planning space needs and managing relocations. Includes step-by-step manual, over 50 forms, and extensive reference section on materials and furnishings.

New lower price and user-friendly format.

$49.98 per copy
Over 450 pages, softcover
Catalog no. 67301

Means Square Foot & UNIFORMAT Assemblies Estimating Methods 3rd Edition

Develop realistic square foot and assemblies costs for budgeting and construction funding. The new edition features updated guidance on square foot and assemblies estimating using UNIFORMAT II. An essential reference for anyone who performs conceptual estimates.

$69.95 per copy
Over 300 pages, illustrated, softcover
Catalog no. 67145B

Means Electrical Estimating Methods 3rd Edition

Expanded edition includes sample estimates and cost information in keeping with the latest version of the CSI MasterFormat and UNIFORMAT II. Complete coverage of fiber optic and uninterruptible power supply electrical systems, broken down by components, and explained in detail. Includes a new chapter on computerized estimating methods. A practical companion to *RSMeans Electrical Cost Data*.

$64.95 per copy
Over 325 pages, hardcover
Catalog no. 67230B

Means Mechanical Estimating Methods 4th Edition

Completely updated, this guide assists you in making a review of plans, specs, and bid packages, with suggestions for takeoff procedures, listings, substitutions, and pre-bid scheduling for all components of HVAC. Includes suggestions for budgeting labor and equipment usage. Compares materials and construction methods to allow you to select the best option.

$64.95 per copy
Over 350 pages, illustrated, softcover
Catalog no. 67294B

Means ADA Compliance Pricing Guide 2nd Edition
by Adaptive Environments and RSMeans

Completely updated and revised to the new 2004 *Americans with Disabilities Act Accessibility Guidelines*, this book features more than 70 of the most commonly needed modifications for ADA compliance. Projects range from installing ramps and walkways, widening doorways and entryways, and installing and refitting elevators, to relocating light switches and signage.

$79.95 per copy
Over 350 pages, illustrated, softcover
Catalog no. 67310A

Project Scheduling & Management for Construction 3rd Edition
by David R. Pierce, Jr.

A comprehensive yet easy-to-follow guide to construction project scheduling and control—from vital project management principles through the latest scheduling, tracking, and controlling techniques. The author is a leading authority on scheduling, with years of field and teaching experience at leading academic institutions. Spend a few hours with this book and come away with a solid understanding of this essential management topic.

$64.95 per copy
Over 300 pages, illustrated, hardcover
Catalog no. 67247B

For more information
visit the RSMeans website
at www.rsmeans.com

Reference Books

The Practice of Cost Segregation Analysis
by Bruce A. Desrosiers and Wayne J. DelPico

This expert guide walks you through the practice of cost segregation analysis, which enables property owners to defer taxes and benefit from "accelerated cost recovery" through depreciation deductions on assets that are properly identified and classified.

With a glossary of terms, sample cost segregation estimates for various building types, key information resources, and updates via a dedicated website, this book is a critical resource for anyone involved in cost segregation analysis.

$99.95 per copy
Over 225 pages
Catalog no. 67345

Preventive Maintenance for Multi-Family Housing
by John C. Maciha

Prepared by one of the nation's leading experts on multi-family housing.

This complete PM system for apartment and condominium communities features expert guidance, checklists for buildings and grounds maintenance tasks and their frequencies, a reusable wall chart to track maintenance, and a dedicated website featuring customizable electronic forms. A must-have for anyone involved with multi-family housing maintenance and upkeep.

$89.95 per copy
225 pages
Catalog no. 67346

How to Estimate with Means Data & CostWorks
New 3rd Edition
by RSMeans and Saleh A. Mubarak, Ph.D.

New 3rd Edition—fully updated with new chapters, plus new CD with updated *CostWorks* cost data and MasterFormat organization. Includes all major construction items—with more than 300 exercises and two sets of plans that show how to estimate for a broad range of construction items and systems—including general conditions and equipment costs.

$59.95 per copy
272 pages, softcover
Includes CostWorks CD
Catalog no. 67324B

Job Order Contracting
Expediting Construction Project Delivery
by Allen Henderson

Expert guidance to help you implement JOC—fast becoming the preferred project delivery method for repair and renovation, minor new construction, and maintenance projects in the public sector and in many states and municipalities. The author, a leading JOC expert and practitioner, shows how to:

• Establish a JOC program
• Evaluate proposals and award contracts
• Handle general requirements and estimating
• Partner for maximum benefits

$89.95 per copy
192 pages, illustrated, hardcover
Catalog no. 67348

Builder's Essentials: Estimating Building Costs
For the Residential & Light Commercial Contractor
by Wayne J. DelPico

Step-by-step estimating methods for residential and light commercial contractors. Includes a detailed look at every construction specialty—explaining all the components, takeoff units, and labor needed for well-organized, complete estimates. Covers correctly interpreting plans and specifications, and developing accurate and complete labor and material costs.

$29.95 per copy
Over 400 pages, illustrated, softcover
Catalog no. 67343

Building & Renovating Schools

This all-inclusive guide covers every step of the school construction process—from initial planning, needs assessment, and design, right through moving into the new facility. A must-have resource for anyone concerned with new school construction or renovation. With square foot cost models for elementary, middle, and high school facilities, and real-life case studies of recently completed school projects.

The contributors to this book—architects, construction project managers, contractors, and estimators who specialize in school construction—provide start-to-finish, expert guidance on the process.

$69.98 per copy
Over 425 pages, hardcover
Catalog no. 67342

Reference Books

For more information visit the RSMeans website at www.rsmeans.com

The Homeowner's Guide to Mold By Michael Pugliese

Expert guidance to protect your health and your home.

Mold, whether caused by leaks, humidity or flooding, is a real health and financial issue—for homeowners and contractors. This full-color book explains:
- Construction and maintenance practices to prevent mold
- How to inspect for and remove mold
- Mold remediation procedures and costs
- What to do after a flood
- How to deal with insurance companies

$21.95 per copy
144 pages, softcover
Catalog no. 67344

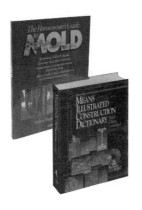

Means Illustrated Construction Dictionary
Unabridged 3rd Edition, with CD-ROM

Long regarded as the industry's finest, *Means Illustrated Construction Dictionary* is now even better. With the addition of over 1,000 new terms and hundreds of new illustrations, it is the clear choice for the most comprehensive and current information. The companion CD-ROM that comes with this new edition adds many extra features: larger graphics, expanded definitions, and links to both CSI MasterFormat numbers and product information.

$99.95 per copy
Over 790 pages, illustrated, hardcover
Catalog no. 67292A

Means Landscape Estimating Methods
New 5th Edition

Answers questions about preparing competitive landscape construction estimates, with up-to-date cost estimates and the new MasterFormat classification system. Expanded and revised to address the latest materials and methods, including new coverage on approaches to green building. Includes:
- Step-by-step explanation of the estimating process
- Sample forms and worksheets that save time and prevent errors

$64.95 per copy
Over 350 pages, softcover
Catalog no. 67295C

Means Plumbing Estimating Methods 3rd Edition
by Joseph Galeno and Sheldon Greene

Updated and revised! This practical guide walks you through a plumbing estimate, from basic materials and installation methods through change order analysis. *Plumbing Estimating Methods* covers residential, commercial, industrial, and medical systems, and features sample takeoff and estimate forms and detailed illustrations of systems and components.

$29.98 per copy
330+ pages, softcover
Catalog no. 67283B

Understanding & Negotiating Construction Contracts
by Kit Werremeyer

Take advantage of the author's 30 years' experience in small-to-large (including international) construction projects. Learn how to identify, understand, and evaluate high risk terms and conditions typically found in all construction contracts—then negotiate to lower or eliminate the risk, improve terms of payment, and reduce exposure to claims and disputes. The author avoids "legalese" and gives real-life examples from actual projects.

$69.95 per copy
300 pages, softcover
Catalog no. 67350

Means Estimating Handbook
2nd Edition

Updated Second Edition answers virtually any estimating technical question—all organized by CSI MasterFormat. This comprehensive reference covers the full spectrum of technical data required to estimate construction costs. The book includes information on sizing, productivity, equipment requirements, code-mandated specifications, design standards, and engineering factors.

$99.95 per copy
Over 900 pages, hardcover
Catalog no. 67276A

Means Spanish/English Construction Dictionary 2nd Edition
by RSMeans and the International Code Council

This expanded edition features thousands of the most common words and useful phrases in the construction industry with easy-to-follow pronunciations in both Spanish and English. Over 800 new terms, phrases, and illustrations have been added. It also features a new stand-alone "Safety & Emergencies" section, with colored pages for quick access.

Unique to this dictionary are the systems illustrations showing the relationship of components in the most common building systems for all major trades.

$23.95 per copy
Over 400 pages
Catalog no. 67327A

How Your House Works
by Charlie Wing

A must-have reference for every homeowner, handyman, and contractor—for repair, remodeling, and new construction. This book uncovers the mysteries behind just about every major appliance and building element in your house—from electrical, heating and AC, to plumbing, framing, foundations and appliances. Clear, "exploded" drawings show exactly how things should be put together and how they function—what to check if they don't work, and what you can do that might save you having to call in a professional.

$21.95 per copy
160 pages, softcover
Catalog no. 67351

For more information visit the RSMeans website at www.rsmeans.com

Professional Development

Means CostWorks® Training

This one-day course helps users become more familiar with the functionality of *Means CostWorks* program. Each menu, icon, screen, and function found in the program is explained in depth. Time is devoted to hands-on estimating exercises.

Some of what you'll learn:
- Search the database utilizing all navigation methods
- Export RSMeans Data to your preferred spreadsheet format
- View crews, assembly components, and much more!
- Automatically regionalize the database

You are required to bring your own laptop computer for this course.

When you register for this course you will receive an outline for your laptop requirements.

Unit Price Estimating

This interactive *two-day* seminar teaches attendees how to interpret project information and process it into final, detailed estimates with the greatest accuracy level.

The most important credential an estimator can take to the job is the ability to visualize construction, and estimate accurately.

Some of what you'll learn:
- Interpreting the design in terms of cost
- The most detailed, time-tested methodology for accurate pricing
- Key cost drivers—material, labor, equipment, staging, and subcontracts
- Understanding direct and indirect costs for accurate job cost accounting and change order management

Who should attend: Corporate and government estimators and purchasers, architects, engineers... and others needing to produce accurate project estimates.

Square Foot & Assemblies Estimating

This *two-day* course teaches attendees how to quickly deliver accurate square foot estimates using limited budget and design information.

Some of what you'll learn:
- How square foot costing gets the estimate done faster
- Taking advantage of a "systems" or "assemblies" format
- The RSMeans "building assemblies/square foot cost approach"
- How to create a reliable preliminary systems estimate using bare-bones design information

Who should attend: Facilities managers, facilities engineers, estimators, planners, developers, construction finance professionals... and others needing to make quick, accurate construction cost estimates at commercial, government, educational, and medical facilities.

Repair & Remodeling Estimating

This *two-day* seminar emphasizes all the underlying considerations unique to repair/remodeling estimating and presents the correct methods for generating accurate, reliable R&R project costs using the unit price and assemblies methods.

Some of what you'll learn:
- Estimating considerations—like labor-hours, building code compliance, working within existing structures, purchasing materials in smaller quantities, unforeseen deficiencies
- Identifying problems and providing solutions to estimating building alterations
- Rules for factoring in minimum labor costs, accurate productivity estimates, and allowances for project contingencies
- R&R estimating examples calculated using unit price and assemblies data

Who should attend: Facilities managers, plant engineers, architects, contractors, estimators, builders... and others who are concerned with the proper preparation and/or evaluation of repair and remodeling estimates.

Mechanical & Electrical Estimating

This *two-day* course teaches attendees how to prepare more accurate and complete mechanical/electrical estimates, avoiding the pitfalls of omission and double-counting, while understanding the composition and rationale within the RSMeans Mechanical/Electrical database.

Some of what you'll learn:
- The unique way mechanical and electrical systems are interrelated
- M&E estimates–conceptual, planning, budgeting, and bidding stages
- Order of magnitude, square foot, assemblies, and unit price estimating
- Comparative cost analysis of equipment and design alternatives

Who should attend: Architects, engineers, facilities managers, mechanical and electrical contractors... and others needing a highly reliable method for developing, understanding, and evaluating mechanical and electrical contracts.

Green Building Planning & Construction

In this *two-day* course, learn about tailoring a building and its placement on the site to the local climate, site conditions, culture, and community. Includes information to reduce resource consumption and augment resource supply.

Some of what you'll learn:
- Green technologies, materials, systems, and standards
- Energy efficiencies with energy modeling tools
- Cost vs. value of green products over their life cycle
- Low-cost strategies and economic incentives and funding
- Health, comfort, and productivity goals and techniques

Who should attend: Contractors, project managers, building owners, building officials, healthcare and insurance professionals.

Facilities Maintenance & Repair Estimating

This *two-day* course teaches attendees how to plan, budget, and estimate the cost of ongoing and preventive maintenance and repair for existing buildings and grounds.

Some of what you'll learn:
- The most financially favorable maintenance, repair, and replacement scheduling and estimating
- Auditing and value engineering facilities
- Preventive planning and facilities upgrading
- Determining both in-house and contract-out service costs
- Annual, asset-protecting M&R plan

Who should attend: Facility managers, maintenance supervisors, buildings and grounds superintendents, plant managers, planners, estimators... and others involved in facilities planning and budgeting.

Practical Project Management for Construction Professionals

In this *two-day* course, acquire the essential knowledge and develop the skills to effectively and efficiently execute the day-to-day responsibilities of the construction project manager.

Covers:
- General conditions of the construction contract
- Contract modifications: change orders and construction change directives
- Negotiations with subcontractors and vendors
- Effective writing: notification and communications
- Dispute resolution: claims and liens

Who should attend: Architects, engineers, owner's representatives, project managers.

Facilities Repair & Remodeling Estimating

In this *two-day* course, professionals working in facilities management can get help with their daily challenges to establish budgets for all phase of a project

Some of what you'll learn:
- Determine the full scope of a project
- Identify the scope of risks & opportunities
- Creative solutions to estimating issues
- Organizing estimates for presentation & discussion
- Special techniques for repair/remodel and maintenance projects
- Negotiating project change orders

Who should attend: Facility managers, engineers, contractors, facility trades-people, planners & project managers

Professional Development

For more information visit the RSMeans website at www.rsmeans.com

Scheduling and Project Management

This *two-day* course teaches attendees the most current and proven scheduling and management techniques needed to bring projects in on time and on budget.

Some of what you'll learn:
- Crucial phases of planning and scheduling
- How to establish project priorities and develop realistic schedules and management techniques
- Critical Path and Precedence Methods
- Special emphasis on cost control

Who should attend: Construction project managers, supervisors, engineers, estimators, contractors... and others who want to improve their project planning, scheduling, and management skills.

Understanding & Negotiating Construction Contracts

In this *two-day* course, learn how to protect the assets of your company by justifying or eliminating commercial risk through negotiation of a contract's terms and conditions.

Some of what you'll learn:
- Myths and paradigms of contracts
- Scope of work, terms of payment, and scheduling
- Dispute resolution
- Why clients love CLAIMS!
- Negotiating issues and much more

Who should attend: Contractors & subcontractors, material suppliers, project managers, risk & insurance managers, procurement managers, owners and facility managers, corporate executives.

Site Work & Heavy Construction Estimating

This *two-day* course teaches attendees how to estimate earthwork, site utilities, foundations, and site improvements, using the assemblies and the unit price methods.

Some of what you'll learn:
- Basic site work and heavy construction estimating skills
- Estimating foundations, utilities, earthwork, and site improvements
- Correct equipment usage, quality control, and site investigation for estimating purposes

Who should attend: Project managers, design engineers, estimators, and contractors doing site work and heavy construction.

Assessing Scope of Work for Facility Construction Estimating

This *two-day* course is a practical training program that addresses the vital importance of understanding the SCOPE of projects in order to produce accurate cost estimates in a facility repair and remodeling environment.

Some of what you'll learn:
- Discussions of site visits, plans/specs, record drawings of facilities, and site-specific lists
- Review of CSI divisions, including means, methods, materials, and the challenges of scoping each topic
- Exercises in SCOPE identification and SCOPE writing for accurate estimating of projects
- Hands-on exercises that require SCOPE, take-off, and pricing

Who should attend: Corporate and government estimators, planners, facility managers, and others who need to produce accurate project estimates.

2009 RSMeans Seminar Schedule

Note: call for exact dates and details.

Location	Dates
Las Vegas, NV	March
Washington, DC	April
Denver, CO	May
San Francisco, CA	June
Washington, DC	September
Dallas, TX	September
Las Vegas, NV	October
Atlantic City, NJ	October
Orlando, FL	November
San Diego, CA	December

1-800-334-3509, ext. 5115

For more information visit the RSMeans website at www.rsmeans.com

Professional Development

Registration Information

Register early... Save up to $100! Register 30 days before the start date of a seminar and save $100 off your total fee. *Note: This discount can be applied only once per order. It cannot be applied to team discount registrations or any other special offer.*

How to register Register by phone today! The RSMeans toll-free number for making reservations is **1-800-334-3509, ext. 5115.**

Individual seminar registration fee - $935. *Means CostWorks®* **training registration fee - $375.** To register by mail, complete the registration form and return, with your full fee, to: RSMeans Seminars, 63 Smiths Lane, Kingston, MA 02364.

Government pricing All federal government employees save off the regular seminar price. Other promotional discounts cannot be combined with the government discount.

Team discount program for two to four seminar registrations. Call for pricing: 1-800-334-3509, ext. 5115

Multiple course discounts When signing up for two or more courses, call for pricing.

Refund policy Cancellations will be accepted up to ten business days prior to the seminar start. There are no refunds for cancellations received later than ten working days prior to the first day of the seminar. A $150 processing fee will be applied for all cancellations. Written notice of cancellation is required. Substitutions can be made at any time before the session starts. **No-shows are subject to the full seminar fee.**

AACE approved courses Many seminars described and offered here have been approved for 14 hours (1.4 recertification credits) of credit by the AACE International Certification Board toward meeting the continuing education requirements for recertification as a Certified Cost Engineer/Certified Cost Consultant.

AIA Continuing Education We are registered with the AIA Continuing Education System (AIA/CES) and are committed to developing quality learning activities in accordance with the CES criteria. Many seminars meet the AIA/CES criteria for Quality Level 2. AIA members may receive (14) learning units (LUs) for each two-day RSMeans course.

NASBA CPE sponsor credits We are part of the National Registry of CPE sponsors. Attendees may be eligible for (16) CPE credits.

Daily course schedule The first day of each seminar session begins at 8:30 a.m. and ends at 4:30 p.m. The second day begins at 8:00 a.m. and ends at 4:00 p.m. Participants are urged to bring a hand-held calculator, since many actual problems will be worked out in each session.

Continental breakfast Your registration includes the cost of a continental breakfast, and a morning and afternoon refreshment break. These informal segments allow you to discuss topics of mutual interest with other seminar attendees. (You are free to make your own lunch and dinner arrangements.)

Hotel/transportation arrangements RSMeans arranges to hold a block of rooms at most host hotels. To take advantage of special group rates when making your reservation, be sure to mention that you are attending the RSMeans seminar. You are, of course, free to stay at the lodging place of your choice. **(Hotel reservations and transportation arrangements should be made directly by seminar attendees.)**

Important Class sizes are limited, so please register as soon as possible.

Note: Pricing subject to change.

Registration Form

ADDS-1000

Call 1-800-334-3509, ext. 5115 to register or FAX this form 1-800-632-6732. Visit our Web site: www.rsmeans.com

Please register the following people for the RSMeans construction seminars as shown here. We understand that we must make our own hotel reservations if overnight stays are necessary.

☐ Full payment of $_____ enclosed.

☐ Bill me

Please print name of registrant(s)
(To appear on certificate of completion)

_____

_____

_____

_____

P.O. #: _____
GOVERNMENT AGENCIES MUST SUPPLY PURCHASE ORDER NUMBER OR TRAINING FORM.

Firm name _____

Address _____

City/State/Zip _____

Telephone no. _____ Fax no. _____

E-mail address _____

Charge registration(s) to: ☐ MasterCard ☐ VISA ☐ American Express

Account no. _____ Exp. date _____

Cardholder's signature _____

Seminar name _____

Seminar City _____

Please mail check to: RSMeans Seminars, 63 Smiths Lane, P.O. Box 800, Kingston, MA 02364 USA

ELECTRONIC PRODUCTS

Harness the full power of RSMeans online with
MeansCostWorks.com

MeansCostWorks.com
Manage your estimates in a secure environment... available 24/7 from a web-enabled computer.

To Learn MORE call 1.800.334.3509 or go to: www.MeansCostWorks.com
View the online demo...Get a FREE TRIAL Subscription!

MeansCostWorks.com from RSMeans is web-based software for estimating construction costs on commercial and residential construction projects. It provides fast and accurate construction estimating using the latest construction cost data available from RSMeans. *It's an invaluable tool for construction, engineering and design professionals.*

Introducing MeansCostsWorks.com
Now you can access and use the world's largest construction cost database for estimating interior projects... verifying bids, managing change orders, budgeting, designing, and more.

MeansCostWorks.com is current, comprehensive, and very easy to use. It combines exhaustive cost data for virtually every construction component and task with state-of-the-art estimating and robust customization.

- Estimate faster and more accurately than you ever thought possible.
- Collaborate seamlessly with authorized team members.
- Easily print, e-mail, or export your work product to your Microsoft EXCEL program.

MeansData™
Software Integration
CONSTRUCTION COSTS FOR SOFTWARE APPLICATIONS
Your construction estimating software is only as good as your cost data.

Have you ever wondered why so many projects are over budget, over scope, and take longer to design and build than expected?

DProfiler with RSMeans is the solution that reduces your costs, the risks to the owner, and gives you a competitive advantage.

Use DProfiler with RSMeans cost data at the start of a project to establish the project scope and budget.

www.beck-technology.com

 **RSMeans**

For more information visit the RSMeans website at www.rsmeans.com

New Titles

The Gypsum Construction Handbook by USG

An invaluable reference of construction procedures for gypsum drywall, cement board, veneer plaster, and conventional plaster. This book includes the newest product developments, installation methods, fire- and sound-rated construction information, illustrated framing-to-finish application instructions, estimating and planning information, and more. Great for architects and engineers; contractors, builders, and dealers; apprentices and training programs; building inspectors and code officials; and anyone interested in gypsum construction. Features information on tools and safety practices, a glossary of construction terms, and a list of important agencies and associations. Also available in Spanish.

$29.95
Over 500 pages, illustrated, softcover
Catalog no. 67357

BIM Handbook
A Guide to Building Information Modeling for Owners, Managers, Designers, Engineers and Contractors
by Chuck Eastman, Paul Teicholz, Rafael Sacks, Kathleen Liston

Building Information Modeling (BIM) is a new approach to design, construction, and facility management where a digital depiction of the building process is used to facilitate the exchange of information in digital format.

The BIM Handbook provides an in-depth understanding of BIM technologies, the business and organizational issues associated with its implementation, and the profound advantages that effective use of BIM can provide to all members of a project team. Also includes a rich set of informative BIM case studies.

$85.00
504 pages
Catalog no. 67356

Universal Design Ideas for Style, Comfort & Safety
by RSMeans and Lexicon Consulting, Inc.

Incorporating universal design when building or remodeling helps people of any age and physical ability more fully and safely enjoy their living spaces. This book shows how universal design can be artfully blended into the most attractive homes. It discusses specialized products like adjustable countertops and chair lifts, as well as simple ways to enhance a home's safety and comfort. With color photos and expert guidance, every area of the home is covered. Includes budget estimates that give an idea how much projects will cost.

$21.95
160 pages, illustrated, softcover
Catalog no. 67354

Green Home Improvement
by Daniel D. Chiras, PhD

With energy costs rising and environmental awareness increasing, many people are looking to make their homes greener. This book makes it easy, with 65 projects and actual costs and projected savings that will help homeowners prioritize their green improvements.

Projects range from simple water savers that cost only a few dollars, to bigger-ticket items such as efficient appliances and HVAC systems, green roofing, flooring, landscaping, and much more. With color photos and cost estimates, each project compares options and describes the work involved, the benefits, and the savings.

$24.95
320 pages, illustrated, softcover
Catalog no. 67355

Residential & Light Commercial Construction Standards, 3rd Edition, Updated
by RSMeans and contributing authors

This book provides authoritative requirements and recommendations compiled from leading professional associations, industry publications, and building code organizations. This all-in-one reference helps establish a standard for workmanship, quickly resolve disputes, and avoid defect claims. Updated third edition includes new coverage of green building, seismic, hurricane, and mold-resistant construction, plus updated national building code requirements.

Covers all areas of building construction, including paving; concrete; framing; finish carpentry; insulation; roofing and siding; doors and windows; drywall; ceilings and floors; wall coverings; plumbing; HVAC; electrical; and more!

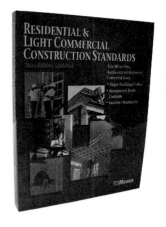

$59.95
Over 550 pages, illustrated, softcover
Catalog no. 67322B

2009 Order Form

ORDER TOLL FREE 1-800-334-3509
OR FAX 1-800-632-6732

Qty.	Book no.	COST ESTIMATING BOOKS	Unit Price	Total
	60069	Assemblies Cost Data 2009	$245.95	
	60019	Building Construction Cost Data 2009	154.95	
	60119	Concrete & Masonry Cost Data 2009	136.95	
	50149	Construction Cost Indexes 2009 (subscription)	335.00	
	60149A	Construction Cost Index–January 2009	83.75	
	60149B	Construction Cost Index–April 2009	83.75	
	60149C	Construction Cost Index–July 2009	83.75	
	60149D	Construction Cost Index–October 2009	83.75	
	60349	Contr. Pricing Guide: Resid. R & R Costs 2009	39.95	
	60339	Contr. Pricing Guide: Resid. Detailed 2009	39.95	
	60329	Contr. Pricing Guide: Resid. Sq. Ft. 2009	39.95	
	60239	Electrical Change Order Cost Data 2009	154.95	
	60039	Electrical Cost Data 2009	154.95	
	60209	Facilities Construction Cost Data 2009	368.95	
	60309	Facilities Maintenance & Repair Cost Data 2009	336.95	
	60169	Heavy Construction Cost Data 2009	154.95	
	60099	Interior Cost Data 2009	154.95	
	60129	Labor Rates for the Const. Industry 2009	336.95	
	60189	Light Commercial Cost Data 2009	132.95	
	60029	Mechanical Cost Data 2009	154.95	
	60159	Open Shop Building Const. Cost Data 2009	154.95	
	60219	Plumbing Cost Data 2009	154.95	
	60049	Repair & Remodeling Cost Data 2009	132.95	
	60179	Residential Cost Data 2009	132.95	
	60289	Site Work & Landscape Cost Data 2009	154.95	
	60059	Square Foot Costs 2009	168.95	
	62018	Yardsticks for Costing (2008)	149.95	
	62019	Yardsticks for Costing (2009)	154.95	
		REFERENCE BOOKS		
	67310A	ADA Compliance Pricing Guide, 2nd Ed.	79.95	
	67330	Bldrs Essentials: Adv. Framing Methods	12.98	
	67329	Bldrs Essentials: Best Bus. Practices for Bldrs	29.95	
	67298A	Bldrs Essentials: Framing/Carpentry 2nd Ed.	15.98	
	67298AS	Bldrs Essentials: Framing/Carpentry Spanish	15.98	
	67307	Bldrs Essentials: Plan Reading & Takeoff	35.95	
	67342	Building & Renovating Schools	69.98	
	67339	Building Security: Strategies & Costs	44.98	
	67353	Complete Book of Framing	29.95	
	67146	Concrete Repair & Maintenance Illustrated	69.95	
	67352	Construction Business Management	49.95	
	67314	Cost Planning & Est. for Facil. Maint.	89.95	
	67328A	Designing & Building with the IBC, 2nd Ed.	99.95	
	67230B	Electrical Estimating Methods, 3rd Ed.	64.95	
	64777A	Environmental Remediation Est. Methods, 2nd Ed.	49.98	
	67343	Estimating Bldg. Costs for Resi. & Lt. Comm.	29.95	
	67276A	Estimating Handbook, 2nd Ed.	$99.95	
	67318	Facilities Operations & Engineering Reference	54.98	
	67301	Facilities Planning & Relocation	49.98	
	67338A	Green Building: Proj. Planning & Cost Est., 2nd Ed.	129.95	
	67355	Green Home Improvement	24.95	

Qty.	Book no.	REFERENCE BOOKS (Cont.)	Unit Price	Total
	67357	The Gypsum Construction Handbook	29.95	
	67357S	The Gypsum Construction Handbook (Spanish)	29.95	
	67323	Historic Preservation: Proj. Planning & Est.	49.98	
	67349	Home Addition & Renovation Project Costs	19.98	
	67308E	Home Improvement Costs–Int. Projects, 9th Ed.	24.95	
	67309E	Home Improvement Costs–Ext. Projects, 9th Ed.	24.95	
	67344	Homeowner's Guide to Mold	21.95	
	67324B	How to Est.w/Means Data & CostWorks, 3rd Ed.	59.95	
	67351	How Your House Works	21.95	
	67282A	Illustrated Const. Dictionary, Condensed, 2nd Ed.	59.95	
	67292A	Illustrated Const. Dictionary, w/CD-ROM, 3rd Ed.	99.95	
	67348	Job Order Contracting	89.95	
	67347	Kitchen & Bath Project Costs	19.98	
	67295C	Landscape Estimating Methods, 5th Ed.	64.95	
	67341	Life Cycle Costing for Facilities	99.95	
	67294B	Mechanical Estimating Methods, 4th Ed.	64.95	
	67245A	Planning & Managing Interior Projects, 2nd Ed.	34.98	
	67283B	Plumbing Estimating Methods, 3rd Ed.	29.98	
	67345	Practice of Cost Segregation Analysis	99.95	
	67337	Preventive Maint. for Higher Education Facilities	149.95	
	67346	Preventive Maint. for Multi-Family Housing	89.95	
	67326	Preventive Maint. Guidelines for School Facil.	149.95	
	67247B	Project Scheduling & Management for Constr. 3rd Ed.	64.95	
	67265B	Repair & Remodeling Estimating Methods, 4th Ed.	69.95	
	67322B	Resi. & Light Commercial Const. Stds., 3rd Ed.	59.95	
	67327A	Spanish/English Construction Dictionary, 2nd Ed.	23.95	
	67145B	Sq. Ft. & Assem. Estimating Methods, 3rd Ed.	69.95	
	67321	Total Productive Facilities Management	29.98	
	67350	Understanding and Negotiating Const. Contracts	69.95	
	67303B	Unit Price Estimating Methods, 4th Ed.	59.95	
	67354	Universal Design	21.95	
	67319A	Value Engineering: Practical Applications	79.95	

MA residents add 5% state sales tax	
Shipping & Handling**	
Total (U.S. Funds)*	

Prices are subject to change and are for U.S. delivery only. *Canadian customers may call for current prices. **Shipping & handling charges: Add 7% of total order for check and credit card payments. Add 9% of total order for invoiced orders.

Send order to: **ADDV-1000**

Name (please print) _____

Company _____

☐ **Company**
☐ **Home Address** _____

_____

City/State/Zip _____

Phone # _____ P.O. # _____

(Must accompany all orders being billed)

Mail to: RSMeans, P.O. Box 800, Kingston, MA 02364

RSMeans Project Cost Report

By filling out this report, your project data will contribute to the database that supports the RSMeans Project Cost Square Foot Data. When you fill out this form, RSMeans will provide a $30 discount off one of the RSMeans products advertised in the preceding pages. Please complete the form including all items where you have cost data, and all the items marked (✔).

$30.00 Discount per product for each report you submit.

Project Description (No remodeling projects, please.)

✔ Building Use (Office, School...) _____

✔ Address (City, State) _____

_____

✔ Building Area (SF) _____

✔ Frame (Wood, Steel...) _____

✔ Exterior Wall (Brick, Tilt-up...) _____

✔ Basement: (check one) ☐ Full ☐ Partial ☐ None

✔ Number Stories _____

✔ Floor-to-Floor Height _____

✔ Volume (C.F.) _____

   % Air Conditioned _____ Tons _____

Total Project Cost $ _____

Owner _____

Architect _____

General Contractor _____

✔ Bid Date _____

✔ Typical Bay Size _____

✔ Capacity _____

✔ Labor Force: _____ % Union _____ % Non-Union

✔ Project Description (Circle one number in each line)

   1. Economy   2. Average   3. Custom   4. Luxury
   1. Square   2. Rectangular   3. Irregular   4. Very Irregular

Comments _____

_____

A	✔ General Conditions	$	
B	✔ Site Work	$	
C	✔ Concrete	$	
D	✔ Masonry	$	
E	✔ Metals	$	
F	✔ Wood & Plastics	$	
G	✔ Thermal & Moisture Protection	$	
GR	Roofing & Flashing	$	
H	✔ Doors and Windows	$	
J	✔ Finishes	$	
JP	Painting & Wall Covering	$	

K	✔ Specialties	$	
L	✔ Equipment	$	
M	✔ Furnishings	$	
N	✔ Special Construction	$	
P	✔ Conveying Systems	$	
Q	✔ Mechanical	$	
QP	Plumbing	$	
QB	HVAC	$	
R	✔ Electrical	$	
S	✔ Mech./Elec. Combined	$	

Please specify the RSMeans product you wish to receive. Complete the address information.

Product Name _____

Product Number _____

Your Name _____

Title _____

Company _____

   ☐ Company
   ☐ Home   Street Address _____

City, State, Zip _____

Return by mail or Fax 888-492-6770.

Method of Payment:

Credit Card # _____

Expiration Date _____

Check _____

Purchase Order _____

R.S. Means Company, Inc.
Square Foot Costs Department
P.O. Box 800
Kingston, MA 02364-9988